Encyclopedia of Earth Sciences

Editorial Board

ENCYCLOPEDIA OF EARTH SCIENCES

E. Julius Dasch

Editor in Chief

Volume 2

MACMILLAN REFERENCE USA
Simon & Schuster Macmillan
NEW YORK

Simon & Schuster and Prentice Hall International
LONDON MEXICO CITY NEW DELHI SINGAPORE SYDNEY TORONTO

Simon & Schuster Macmillan
1633 Broadway
New York, NY 10019

Library of Congress Catalog Card Number: 96-11302

PRINTED IN THE UNITED STATES OF AMERICA

Printing Number

1 2 3 4 5 6 7 8 9 10

LIBRARY OF CONGRESS CATALOGING-IN-PUBLICATION DATA

Encyclopedia of Earth Sciences / E. Julius Dasch, editor in chief

p. cm.

Includes bibliographical references and index.

ISBN 0-02-883000-8 (set). — ISBN 0-02-897112-4 (v. 1). — ISBN 0-02-897114-0 (v. 2)

1. Earth Sciences — Encyclopedias. I. Dasch, E. Julius.

QE5.E5137 1996 96-11302

550′ .3—dc20 CIP

This paper meets the requirements of ANSI-NISO Z39.48-1992 (Permanence of Paper)

Table 1. Elements Arranged by Atomic Number

Name	*Symbol*	*Atomic Number*	*Units*[1]	*CI Chondrites*[2]	*Bulk Earth*[3]	*Core*[4]	*Primitive Mantle*[5]	*Oceanic Crust*[6]	*Continental Crust*[7]			*Oceans*[8]	*Atmosphere*[9]
									Bulk	*Lower*	*Upper*		
Hydrogen	H	1	mg/g	20.2	0.078	‡						108	0.21
Helium	He	2	ng/g	10	1.3							0.005	723
Lithium	Li	3	μg/g	1.50	2.7		1.6		13	11	20	0.17	
Beryllium	Be	4	ng/g	24.9	56		68		1,500	1,000	3,000	0.0005	
Boron	B	5	μg/g	0.87	0.47		0.30		10	8.3	15	4.6	
Carbon	C	6	mg/g	34.5	0.35	‡	0.12					0.025	0.131
Nitrogen	N	7	μg/g	3,180	9.1		2					17	7.538×10^{5}
Oxygen	O	8	mg/g	464	285.0	‡						857	233.0
Fluorine	F	9	μg/g	60.7	53		25					1	
Neon	Ne	10	pg/g	201	25							100	1.27×10^{7}
Sodium	Na	11	mg/g	5.00	1.58		2.67	19.9	23.0	20.8	28.9	10.8	
Magnesium	Mg	12	mg/g	98.9	132.1		228	45.7	32.0	38.0	13.3	1.28	
Aluminum	Al	13	mg/g	8.68	17.7		23.5	15.3	84.1	85.2	80.4	0.0008	
Silicon	Si	14	mg/g	106.4	143.4		210	236	268	254	308	5.6	
Phosphorus	P	15	μg/g	1,220	2,150	5,900	90				700	120	
Sulfur	S	16	mg/g	62.5	18.4	‡	0.25					0.905	
Chlorine	Cl	17	μg/g	704	25		17					19,400	
Argon	Ar	18	ng/g	397	70.4							630	1.29×10^{7}
Potassium	K	19	μg/g	558	170		240	884	9,100	2,800	28,000	400	
Calcium	Ca	20	mg/g	9.28	19.3		25.3	80.8	52.9	60.7	30.0	0.412	
Scandium	Sc	21	μg/g	5.82	12.1		16.2	41.4	30	36	11	0.00004	
Titanium	Ti	22	mg/g	0.436	1.03		1.205	9.68	5.4	6.0	3.0	2×10^{-8}	
Vanadium	V	23	μg/g	56.5	103		82		230	285	60	0.002	
Chromium	Cr	24	mg/g	2.66	4.78		2.625		0.185	0.235	0.035	2×10^{-7}	
Manganese	Mn	25	mg/g	1.99	0.59		1.045		1.4	1.7	0.6	4×10^{-6}	
Iron	Fe	26	mg/g	190.4	358.7	891	62.6	81.0	70.7	82.4	35.0	0.00006	
Cobalt	Co	27	μg/g	502	940	2,450	105	47.1	29	35	10	0.0019	
Nickel	Ni	28	mg/g	11.0	20.4	53.7	1.96	0.150	0.105	0.135	0.02	0.00047	
Copper	Cu	29	μg/g	126	57	100	30	74.4	75	90	25	0.25	
Zinc	Zn	30	μg/g	312	93	160	55		80	83	71	0.39	
Gallium	Ga	31	μg/g	10.0	5.5	7.8	4.0		18	18	17	0.0002	
Germanium	Ge	32	μg/g	32.7	13.8	37	1.1		1.6	1.6	1.6	0.00007	
Arsenic	As	33	μg/g	1.86	3.6	10	0.05		1.0	0.8	1.5	0.003	
Selenium	Se	34	μg/g	18.6	6.1	17	0.075		50	50	50	0.005	
Bromine	Br	35	ng/g	3,570	134		50					67,400	
Krypton	Kr	36	pg/g	57	4.3							300	3.23×10^{6}
Rubidium	Rb	37	μg/g	2.30	0.58		0.600	1.26	32	5.3	112	0.12	

Table 1. Continued

Name	Symbol	Atomic Number	Units[1]	CI Chondrites[2]	Bulk Earth[3]	Core[4]	Primitive Mantle[5]	Oceanic Crust[6]	Continental Crust[7]			Oceans[8]	Atmosphere[9]
									Bulk	Lower	Upper		
Strontium	Sr	38	μg/g	7.80	18.2		19.9	113	260	230	350	8	
Yttrium	Y	39	μg/g	1.56	3.29		4.30	35.8	20	19	22	0.0003	
Zirconium	Zr	40	μg/g	3.94	7.42		10.5	104	100	70	190		
Niobium	Nb	41	μg/g	0.246	0.501		0.658	3.51	11	6	25	0.00002	
Molybdenum	Mo	42	μg/g	0.928	2.96	8.2	0.050		1.0	0.8	1.5	0.008	
Technetium	Tc	43	*										
Ruthenium	Ru	44	μg/g	0.712	1.48	4.2	0.0050						
Rhodium	Rh	45	μg/g	0.134	0.32	0.90	0.0009						
Palladium	Pd	46	ng/g	560	1,000	2,800	3.9		1	1	0.5		
Silver	Ag	47	ng/g	199	80	210	8		80	90	50	0.14	
Cadmium	Cd	48	ng/g	686	21		40		98	98	98	0.11	
Indium	In	49	ng/g	80	2.7		11		50	50	50		
Tin	Sn	50	μg/g	1.72	0.71	1.8	0.13	1.38	2.5	1.5	5.5	0.0003	
Antimony	Sb	51	ng/g	142	64	170	5.5		200	200	200	0.5	
Tellurium	Te	52	ng/g	2,320	940	2,600	12						
Iodine	I	53	ng/g	433	17		10					60	
Xenon	Xe	54	pg/g	190	10							100	394
Cesium	Cs	55	ng/g	187	59		21	14.1	1,000	100	3,700	0.5	
Barium	Ba	56	μg/g	2.34	5.1		6.60	13.9	250	150	550	0.04	
Lanthanum	La	57	μg/g	0.2347	0.48		0.648	3.90	16	11	30	0.0003	
Cerium	Ce	58	μg/g	0.6032	1.28		1.675	12.0	33	23	64	0.0004	
Praseodymium	Pr	59	μg/g	0.0891	0.162		0.254	2.07	3.9	2.8	7.1		
Neodymium	Nd	60	μg/g	0.4524	0.87		1.25	11.2	16	12.7	26		
Promethium	Pm	61	*										
Samarium	Sm	62	μg/g	0.1471	0.26		0.406	3.75	3.5	3.17	4.5		
Europium	Eu	63	μg/g	0.056	0.100		0.154	1.34	1.1	1.17	0.88		
Gadolinium	Gd	64	μg/g	0.1966	0.37		0.544	5.08	3.3	3.13	3.8		
Terbium	Tb	65	μg/g	0.0363	0.067		0.099	0.885	0.60	0.59	0.64		
Dysprosium	Dy	66	μg/g	0.2427	0.45		0.674	6.30	3.7	3.6	3.5		
Holmium	Ho	67	μg/g	0.0556	0.101		0.149	1.34	0.78	0.77	0.80		
Erbium	Er	68	μg/g	0.1589	0.29		0.438	4.14	2.2	2.2	2.3		
Thulium	Tm	69	μg/g	0.0242	0.044		0.068	0.621	0.32	0.32	0.33		
Ytterbium	Yb	70	μg/g	0.1625	0.29		0.441	3.90	2.2	2.2	2.2		
Lutetium	Lu	71	μg/g	0.0243	0.049		0.0675	0.589	0.30	0.29	0.32		
Hafnium	Hf	72	μg/g	0.104	0.29		0.283	2.97	3.0	2.1	5.8		
Tantalum	Ta	73	ng/g	14.2	29		37	192	1,000	600	2,200		
Tungsten	W	74	ng/g	92.6	250	650	29		1,000	700	2,000	0.12	
Rhenium	Re	75	ng/g	36.5	76	210	0.28		0.4	0.4	0.4		
Osmium	Os	76	ng/g	486	1,110	3,100	3.4		0.05	0.05	0.05		

Table 1. Continued

Name	Symbol	Atomic Number	Units[1]	Cl Chondrites[2]	Bulk Earth[3]	Core[4]	Primitive Mantle[5]	Oceanic Crust[6]	Continental Crust[7]			Oceans[8]	Atmosphere[9]
									Bulk	*Lower*	*Upper*		
Iridium	Ir	77	ng/g	481	1,060	3,000	3.2		0.1	0.13	0.02		
Platinum	Pt	78	ng/g	990	2,100	5,900	7.1						
Gold	Au	79	ng/g	140	290	820	1.0		3.0	3.4	1.8	0.2	
Mercury	Hg	80	ng/g	258	9.9	8.7	10					0.2	
Thallium	Tl	81	ng/g	142	4.9		3.5		360	230	750		
Lead	Pb	82	ng/g	2,470	141		150	489	8,000	4,000	20,000	1	
Bismuth	Bi	83	ng/g	114	3.7		2.5		60	38	127	0.02	
Polonium	Po	84	†										
Astatine	At	85	†										
Radon	Rn	86	†										
Francium	Fr	87	†										
Radium	Ra	88	†										
Actinium	Ac	89	†										
Thorium	Th	90	ng/g	29.4	65		79.5	187	3,500	1,060	10,700	0.05	
Protactinium	Pa	91	†										
Uranium	U	92	ng/g	8.1	18		20.3	71.1	910	280	2,800	3	

[1] The units are SI units rather than the more familiar parts per million (ppm), parts per billion (ppb), etc. These tables use SI units in part because they are the standard, but more important because "billion" does not have a single definition; it means 10^9 in American usage, but 10^{12} in British usage. *See* "Measurements and Their Conversion in the Earth Sciences" in the frontmatter.

[2] Cl Chondrites: primitive meteorite composition, modified from E. Anders and N. Grevese, *Geochimica et Cosmochimica Acta* 53 (1989): 197–214. *See also* entry METEORITES.

[3] Bulk Earth: modified from E. Anders, *Philosophical Transactions of the Royal Society of London* A 285 (1977):23–40. *See also* the entry EARTH, COMPOSITION OF.

[4] Core: estimated by mass balance from the bulk Earth and primitive mantle compositions.

[5] Primitive Mantle: from W. F. McDonough and S.-S. Sun, *Chemical Geology* 120 (1995):223–253.

[6] Oceanic Crust: average normal mid-ocean ridge basalt of A. W. Hofmann, *Earth and Planetary Science Letters* 90 (1988):297–314. *See also* the entry OCEANIC CRUST, STRUCTURE OF.

[7] Continental Crust—Bulk, Lower and Upper: from S. R. Taylor and S. M. McLennan, *Reviews of Geophysics* 33 (1995):241–265.

[8] Oceans: modified from the entry OCEANOGRAPHY, CHEMICAL, and D. A. Ross, *Introduction to Oceanography*, 4th ed. (1988).

[9] Atmosphere: modified from the entry EARTH'S ATMOSPHERE, CHEMICAL COMPOSITION OF, and M. Ozima and F. A. Podosek, *Nobel Gas Geochemistry* (1983).

* No long-lived radioactive ($>10^8$ years) or stable nuclei of these elements exist.

† No long-lived radioactive ($>10^8$ years) or stable nuclei of these elements exist. Their abundances in the earth are steady-state concentrations in the ^{232}Th, ^{235}U, and ^{238}U decay schemes.

‡ Earth's core contains about 10% of a light element, of which these are the most likely candidates. See the entry CORE, COMPOSITION OF.

Table 2. Elements Arranged Alphabetically by Name

Name	Symbol	Atomic Number	Units[1]	CI Chondrites[2]	Bulk Earth[3]	Core[4]	Primitive Mantle[5]	Oceanic Crust[6]	Continental Crust[7] Bulk	Continental Crust[7] Lower	Continental Crust[7] Upper	Oceans[8]	Atmosphere[9]
Actinium	Ac	89	†										
Aluminum	Al	13	mg/g	8.68	17.7		23.5	15.3	84.1	85.2	80.4	0.0008	
Antimony	Sb	51	ng/g	142	64	170	5.5		200	200	200	0.5	
Argon	Ar	18	ng/g	397	70.4							630	1.29×10^7
Arsenic	As	33	µg/g	1.86	3.6	10	0.05		1.0	0.8	1.5	0.0003	
Astatine	At	85	†										
Barium	Ba	56	µg/g	2.34	5.1		6.60	13.9	250	150	550	0.04	
Beryllium	Be	4	ng/g	24.9	56		68		1,500	1,000	3,000	0.0005	
Bismuth	Bi	83	ng/g	114	3.7		2.5		60	38	127	0.02	
Boron	B	5	µg/g	0.87	0.47		0.30		10	8.3	15	4.6	
Bromine	Br	35	ng/g	3,570	134		50					67,400	
Cadmium	Cd	48	ng/g	686	21		40		98	98	98	0.11	
Calcium	Ca	20	mg/g	9.28	19.3		25.3	80.8	52.9	60.7	30.0	0.412	
Carbon	C	6	mg/g	34.5	0.35	‡	0.12					0.025	0.131
Cerium	Ce	58	µg/g	0.6032	1.28		1.675	12.0	33	23	64	0.0004	
Cesium	Cs	55	ng/g	187	59		21	14.1	1,000	100	3,700	0.5	
Chlorine	Cl	17	µg/g	704	25		17					19,400	
Chromium	Cr	24	mg/g	2.66	4.78		2.625		0.185	0.235	0.035	2×10^{-7}	
Cobalt	Co	27	µg/g	502	940	2,450	105	47.1	29	35	10	0.0019	
Copper	Cu	29	µg/g	126	57	100	30	74.4	75	90	25	0.25	
Dysprosium	Dy	66	µg/g	0.2427	0.45		0.674	6.30	3.7	3.6	3.5		
Erbium	Er	68	µg/g	0.1589	0.29		0.438	4.14	2.2	2.2	2.3		
Europium	Eu	63	µg/g	0.056	0.100		0.154	1.34	1.1	1.17	0.88		
Fluorine	F	9	µg/g	60.7	53		25					1	
Francium	Fr	87	†										
Gadolinium	Gd	64	µg/g	0.1966	0.37		0.544	5.08	3.3	3.13	3.8		
Gallium	Ga	31	µg/g	10.0	5.5	7.8	4.0		18	18	17	0.0002	
Germanium	Ge	32	µg/g	32.7	13.8	37	1.1		1.6	1.6	1.6	0.00007	
Gold	Au	79	ng/g	140	290	820	1.0		3.0	3.4	1.8	0.2	
Hafnium	Hf	72	µg/g	0.104	0.29		0.283	2.97	3.0	2.1	5.8		
Helium	He	2	ng/g	10	1.3							0.005	723
Holmium	Ho	67	µg/g	0.0556	0.101		0.149	1.34	0.78	0.77	0.80		
Hydrogen	H	1	mg/g	20.2	0.078	‡						108	0.21
Indium	In	49	ng/g	80	2.7		11		50	50	50		
Iodine	I	53	ng/g	443	17		10					60	
Iridium	Ir	77	ng/g	481	1,060	3,000	3.2		0.1	0.13	0.02		
Iron	Fe	26	mg/g	190.4	358.7	891	62.6	81.0	70.7	82.4	35.0	0.00006	

Table 2. Continued

Name	Symbol	Atomic Number	Units[1]	Cl Chondrites[2]	Bulk Earth[3]	Core[4]	Primitive Mantle[5]	Oceanic Crust[6]	Continental Crust[7]			Oceans[8]	Atmosphere[9]
									Bulk	Lower	Upper		
Krypton	Kr	36	pg/g	57	4.3							300	3.23×10^8
Lanthanum	La	57	μg/g	0.2347	0.48		0.648	3.90	16	11	30	0.0003	
Lead	Pb	82	ng/g	2,470	141		150	489	8,000	4,000	20,000	1	
Lithium	Li	3	μg/g	1.50	2.7		1.6		13	11	20	0.17	
Lutetium	Lu	71	μg/g	0.0243	0.049		0.0675	0.589	0.30	0.29	0.32		
Magnesium	Mg	12	mg/g	98.8	132.1		228	45.7	32.0	38.0	13.3	1.28	
Manganese	Mn	25	mg/g	1.99	0.59		1.045		1.4	1.7	0.6	4×10^{-6}	
Mercury	Hg	80	ng/g	258	9.9	8.7	10					0.2	
Molybdenum	Mo	42	μg/g	0.928	2.96	8.2	0.050		1.0	0.8	1.5	0.008	
Neodymium	Nd	60	μg/g	0.4524	0.87		1.25	11.2	16	12.7	26		
Neon	Ne	10	pg/g	201	25							100	1.27×10^7
Nickel	Ni	28	mg/g	11.0	20.4	53.7	1.96	0.150	0.105	0.135	0.02	0.00047	
Niobium	Nb	41	μg/g	0.246	0.501		0.658	3.51	11	6	25	0.00002	
Nitrogen	N	7	μg/g	3,180	9.1		2					17	7.538×10^5
Osmium	Os	76	ng/g	486	1,100	3,100	3.4		0.05	0.05	0.05		
Oxygen	O	8	mg/g	464	285.0	‡						857	233.0
Palladium	Pd	46	ng/g	560	1,000	2,800	3.9		1	1	0.5		
Phosphorus	P	15	μg/g	1,220	2,150	5,900	90				700	120	
Platinum	Pt	78	ng/g	900	2,100	5,900	7.1						
Polonium	Po	84	†										
Potassium	K	19	μg/g	558	170		240	884	9,100	2,800	28,000	400	
Praseodymium	Pr	59	μg/g	0.0891	0.162		0.254	2.07	3.9	2.8	7.1		
Promethium	Pm	61	*										
Protactinium	Pa	91	†										
Radium	Ra	88	†										
Radon	Rn	86	†										
Rhenium	Re	75	ng/g	36.5	76	210	0.28		0.4	0.4	0.4		
Rhodium	Rh	45	μg/g	0.134	0.32	0.90	0.0009						
Rubidium	Rb	37	μg/g	2.30	0.58		0.600	1.26	32	5.3	112	0.12	
Ruthenium	Ru	44	μg/g	0.712	1.48	4.2	0.0050						
Samarium	Sm	62	μg/g	0.1471	0.26		0.406	3.75	3.5	3.17	4.5		
Scandium	Sc	21	μg/g	5.82	12.1		16.2	41.4	30	36	11	0.00004	
Selenium	Se	34	μg/g	18.6	6.1	17	0.075		50	50	50	0.005	
Silicon	Si	14	mg/g	106.4	143.4		210	236	268	254	308	5.6	
Silver	Ag	47	ng/g	199	80	210	8		80	90	50	0.14	
Sodium	Na	11	mg/g	5.00	1.58		2.67	19.9	23.0	20.8	28.9	10.8	

Table 2. Continued

Name	Symbol	Atomic Number	Units[1]	Cl Chondrites[2]	Bulk Earth[3]	Core[4]	Primitive Mantle[5]	Oceanic Crust[6]	Continental Crust[7]			Oceans[8]	Atmosphere[9]
									Bulk	Lower	Upper		
Strontium	Sr	38	μg/g	7.80	18.2		19.9	113	260	230	350	8	
Sulfur	S	16	mg/g	62.5	18.4	‡	0.25					0.905	
Tantalum	Ta	73	ng/g	14.2	29		37	192	1,000	600	2,200		
Technetium	Tc	43	*										
Tellurium	Te	52	ng/g	2,320	940	2,600	12						
Terbium	Tb	65	μg/g	0.0363	0.067		0.099	0.885	0.60	0.59	0.64		
Thallium	Tl	81	ng/g	142	4.9		3.5		360	230	750		
Thorium	Th	90	ng/g	29.4	65		79.5	187	3,500	1,060	10,700	0.05	
Thulium	Tm	69	μg/g	0.0242	0.044		0.068	0.621	0.32	0.32	0.33		
Tin	Sn	50	μg/g	1.72	0.71	1.8	0.13	1.38	2.5	1.5	5.5	0.0003	
Titanium	Ti	22	mg/g	0.436	1.03		1.205	9.68	5.4	6.0	3.0	2×10^{-8}	
Tungsten	W	74	ng/g	92.6	250	650	29		1,000	700	2,000	0.12	
Uranium	U	92	ng/g	8.1	18		20.3	71.1	910	280	2,800	3	
Vanadium	V	23	μg/g	56.5	103		82		230	285	60	0.002	
Xenon	Xe	54	pg/g	190	10							100	394
Ytterbium	Yb	70	μg/g	0.1625	0.29		0.441	3.90	2.2	2.2	2.2		
Yttrium	Y	39	μg/g	1.56	3.29		4.30	35.8	20	19	22	0.0003	
Zinc	Zn	30	μg/g	312	93	160	55		80	83	71	0.39	
Zirconium	Zr	40	μg/g	3.94	7.42		10.5	104	100	70	190		

[1] The units are SI units rather than the more familiar parts per million (ppm), parts per billion (ppb), etc. These tables use SI units in part because they are the standard, but more important because "billion" does not have a single definition; it means 10^9 in American usage, but 10^{12} in British usage. *See* "Measurements and Their Conversion in the Earth Sciences" in the frontmatter.

[2] Cl Chondrites: primitive meteorite composition, modified from E. Anders and N. Grevese, *Geochimica et Cosmochimica Acta* 53 (1989): 197–214. *See also* entry METEORITES.

[3] Bulk Earth: modified from E. Anders, *Philosophical Transactions of the Royal Society of London* A 285 (1977):23–40. *See also* the entry EARTH, COMPOSITION OF.

[4] Core: estimated by mass balance from the bulk Earth and primitive mantle compositions.

[5] Primitive Mantle: from W. F. McDonough and S.-S. Sun, *Chemical Geology* 120 (1995):223–253.

[6] Oceanic Crust: average normal mid-ocean ridge basalt of A. W. Hofmann, *Earth and Planetary Science Letters* 90 (1988):297–314. *See also* the entry OCEANIC CRUST, STRUCTURE OF.

[7] Continental Crust—Bulk, Lower and Upper: from S. R. Taylor and S. M. McLennan, *Reviews of Geophysics* 33 (1995):241–265.

[8] Oceans: modified from the entry OCEANOGRAPHY, CHEMICAL, and D. A. Ross, *Introduction to Oceanography*, 4th ed. (1988).

[9] Atmosphere: modified from the entry EARTH'S ATMOSPHERE, CHEMICAL COMPOSITION OF, and M. Ozima and F. A. Podosek, *Nobel Gas Geochemistry* (1983).

* No long-lived radioactive ($>10^8$ years) or stable nuclei of these elements exist.

† No long-lived radioactive ($>10^8$ years) or stable nuclei of these elements exist. Their abundances in the earth are steady-state concentrations in the ^{232}Th, ^{235}U, and ^{238}U decay schemes.

‡ Earth's core contains about 10% of a light element, of which these are the most likely candidates. See the entry CORE, COMPOSITION OF.

ACRONYMS AND STANDARD ABBREVIATIONS

A/E architectural/engineering
AABW Antarctic Bottom Water
AAG Association of American Geographers
AAIW Antarctic Intermediate Water
ABET Accreditation Board for Engineers and Technologists
ACCP Atlantic Climate Change Program
ADEOS Advanced Earth Observing System
AEG Association of Engineering Geologists
AGI American Geological Institute
API American Petroleum Institute
ASCE American Society of Civil Engineers
ATLAS Atmospheric Laboratory for Applications and Science
ATOC Acoustic Thermometry for Ocean Climate
AU astronomical units
b.y. billion years
B.A. bachelor of arts
B.C.E. before common era
B.S. bachelor of sciences
Bbo barrels of oil
BBO billion barrels of oil
BMP best management practices
BOD biological oxygen demand
BP before present
C.E. common era
ca. circa
CalTech California Institute of Technology
CD-ROM compact disc read-only memory
CMB core-mantle boundary
CME coronal mass ejection
COCORP Consortium for Continental Reflection Profiling
CofM center of mass
ct carat
CZCS Coastal Zone Color Scanner
D/H deuterium-to-hydrogen ratio
DOD Department of Defense
DOE Department of Energy
DSDP Deep Sea Drilling Project
DTA differential thermal analysis
ED electrodialysis
EDR electrodialysis-reversal
EDX (or EDS) energy dispersive X-ray spectroscopy
EELS electron energy loss spectroscopy
EGs engineering geologists
EOS Earth Observing System
EOSDIS EOS Data and Information System
EPA Environmental Protection Agency
ERBE Earth Radiation Budget Experiment
ESA European Space Agency
FAMOUS French-American Mid-Ocean Undersea Study
Ga billion years
GAC granular activated charcoal
GCM General Circulation Model
GEs geological engineers
GIS geographic information system
GISP Greenland Ice Sheet Project
GPS Global Positioning System
GSA Geological Society of America
GWP Global Warming Potential
ICB inner core boundary
ICBM intercontinental ballistic missile
ICES International Council for the Exploration of the Sea
IGY International Geophysical Year
IPCC Intergovernmental Panel on Climate Change
JOI Joint Oceanographic Institutions
JOIDES Joint Oceanographic Institutions for Deep Earth Sampling

Ka thousand years
kya thousands of years ago
LAGEOS Laser Geodynamics Satellite
LANDSAT Land Remote-Sensing Satellite
LDG Libyan Desert Glass
LIL large-ion-lithophile
LITE Lidar In-Space Technology Experiment
LVZ lower velocity zone
M.A. master of arts
Ma million years
MCL maximum containment level
MED multiple effect distillation
MESUR Mars Environmental Survey Mission
MHD magnetohydrodynamics
MM Modified Mercalli
MORB mid-ocean ridge basalt
MSF multistage flash
MSW municipal solid waste
MTPE Mission to Planet Earth
mya millions of years ago
NADW North Atlantic Deep Water
NAPL non-aqueous phase liquid
NAS National Academy of Sciences
NASA National Aeronautics and Space Administration
NEOs near-Earth objects
NMR nuclear magnetic resonance
NOAA National Oceanic and Atmospheric Administration
NPC National Petroleum Council
NPDES National Pollutant Discharge Elimination System
NSCAT NASA Scatterometer
NSF National Science Foundation
NTIS National Technical Information Service
NWS National Weather Service
ODP Ocean Drilling Project
PCE pyrometric cone equivalent
PDR precision depth recorder
PGE platinum-group elements
PGM platinum-group minerals
PICs products of incomplete combustion
PIXE proton-induced X-ray emission
ppb parts per billion
ppm parts per million
ppt parts per thousand
PREM Preliminary Reference Earth Model
PSU practical salinity unit
PV photovoltaic
RDF refuse-derived fuel
REE rare-earth elements
RO reverse osmosis
ROV remotely operated vehicle
RPM regulatory program manager
SAR Synthetic Aperture Radar
SDI Strategic Defense Initiative
SeaWiFS Sea-Viewing Wide Field Sensor
SEM scanning electron microscope
SETI Search for Extraterrestrial Intelligence
SIMS secondary ion mass spectrometry
SLR satellite laser ranging
SNC shergottite, nakhlite, and chassignite
SOC synthetic organic chemical
SPS solar power satellite
SST sea surface topography
STEM scanning transmission electron microscope
SYNROC synthetic rock
TDS total dissolved solids
TEM transmission electron microscope
TGA thermal gravimetric analysis
TMDL total maximum daily load
TOF time of flight
TOGA Tropical Ocean–Global Atmosphere
TOMS Total Ozone Mapping Spectrometer
TOPEX Ocean Topography Experiment
TRMM Tropical Rainfall Measuring Mission
TVA Tennessee Valley Authority
UARS Upper Atmosphere Research Satellite
UNEP United Nations Environment Program
UNESCO United Nations Educational, Scientific, and Cultural Organization
USBR U.S. Bureau of Reclamation
USGCRP U.S. Global Change Research Program
USGS U.S. Geological Survey
VC vapor compression
VLBI Very Long Baseline Interferometry
VOC volatile organic chemical
WOCE World Ocean Circulation Experiment
WWSSN World Wid Standarized Seismograph Network
WWW World Wide Web

M

MAGMA

Magma is hot, mobile natural fluid, formed inside Earth, that cools and solidifies to solid rock (igneous rock). "Hot" means hundreds of degrees Celsius (°C) above the earth's surface temperatures. Lava, magma that erupts onto the earth's surface, solidifies to form volcanic rocks like basalt. Most magmas solidify inside the earth and form plutonic rocks like granite. Almost all magmas contain solids (rock fragments and growing crystals) and gases (dissolved and as bubbles). Also, almost all magmas are silicates, composed principally of silicon and oxygen (like Earth itself).

The most abundant kind of magma is basaltic, like the lavas that erupt in Hawaii and Iceland. Basaltic magmas have relatively less silicon, relatively more magnesium, iron, and calcium, and little sodium or potassium. Basaltic magmas are relatively hot (1,050–1,100°C) and fluid (like molasses). Because basalt magma is so fluid, it can flow easily through cracks in the earth's crust, and erupt as lava at the earth's surface. Also because basalt magma is so fluid, basalt lava flows tend to be thin and long, extending to tens or hundred of kilometers. When cooled slowly, basaltic magmas solidify to plutonic rocks like gabbro. Typical minerals in basaltic rocks are plagioclase, pyroxenes, and olivine. The ocean floors are mostly basalt, and most ocean islands and island arcs are the tops of huge basalt volcanoes. On land, basalt is most common where the continents have been broken, "rifted," as in East Africa, western North America, and the Rhine area of Europe.

Granitic magmas (rich in silicon, aluminum, sodium, and potassium) are much less abundant than basaltic magmas. Granitic magmas are relatively cool (550–650°C) and extremely viscous. Because they are so viscous, granitic magmas cannot flow easily through cracks in the earth's crust and rarely erupt as lavas; most often, they stay in the crust and cool slowly to granite rock. When granitic magmas do erupt as lavas, they tend to form short thick flows of rhyolite rock or obsidian (almost all glass). However, most granitic magmas contain dissolved water and carbon dioxide, and can "bubble over" to form violent volcanic eruptions of bubbly lava fragments (pumice), gas, and shards of volcanic glass. Eruptions like these can produce very fluid lavas of volcanic glass fragments suspended in hot turbulent gas. These pyroclastic flows can move very rapidly and are very destructive. Typical minerals in granitic rocks are quartz, albite, microcline (commonly pink), and micas. Granitic magmas form almost exclusively on the continents.

Magmas originate when rock melts inside the earth. Earth's interior is almost all solid rock. Magma only forms where temperature is unusually high, pressure is unusually low, or where volatiles (like water) are unusually abundant. Most basaltic magmas form in the earth's mantle (deeper than 30 km), and may percolate directly to the earth's surface to erupt as basalt lava. Or, basaltic magma may stop in the earth's crust and may melt it to produce granitic magmas.

In the solar system, available information indicates that almost all magmas are or were basaltic. On the Moon, basalt lava flows make up the dark areas (the maria), and the bright areas (highlands) are made of plagioclase-rich gabbro (anorthosite). The surface of Mercury has smooth plains similar to the Moon's maria. Basalt lava flows cover most of the surface of Venus, as shown by chemical analyses of the surface by the *Venera* and *VEGA* spacecraft and by *Magellan* spacecraft radar imagery. From the sampling of meteorites, magmas on asteroids were all basaltic. Finally, Mars has huge basalt volcanoes and extensive basalt lava flows, as shown on images from the *Viking* orbiter spacecraft. Martian soil is derived from basalt lava, and the Martian meteorites are all basaltic.

Bibliography

CHESTERMAN, C. W. *The Audubon Society Field Guide to North American Rocks and Minerals.* New York, 1978.

HESS, P. C. *Origins of Igneous Rocks.* Cambridge, MA, 1989.

PHILPOTTS, A. R. *Principles of Igneous and Metamorphic Petrology.* NJ, 1990.

ALLAN TREIMAN

MAGMATIC DIFFERENTIATION

See Igneous Processes

MAMMALS

The mammals, formally named the class Mammalia by Carolus Linnaeus in 1758, are the group of vertebrate animals that in the present day are warm-blooded, bear fur, and suckle their young with milk after birth. The over four thousand species of living mammals can be divided into three major groups. The Monotremata, specifically the duckbill platypus and echidnas of the Australian, New Guinea, and Tasmanian regions, are often considered the most "primitive" (i.e., retaining many ancestral mammalian characteristics). Monotremes lay eggs, yet nurse their young and bear fur. Marsupialia, a large and diverse group of living mammals whose members are concentrated in the Australian and South American regions, includes the familiar kangaroos and opossums. Marsupials give birth to extremely immature live young that subsequently continue to develop in a pouch on the mother's body, where they suckle and grow. The Eutheria, or placental mammals, includes the vast majority of living mammals. The placental mammals have well-developed placentas (the organ that unites the fetus to its mother's uterus before birth), which allow them to give birth to relatively mature live young, lack pouches as seen in marsupials, and differ from monotremes and marsupials in details of the skeleton and teeth. Living placentals are found around the globe and occupy many niches on land (from moles to elephants), in the air (bats), and under the water (such as whales, dolphins, and seals). Modern humans, *Homo sapiens,* are a species of placental mammal.

Earliest Mammals

Numerous fossil remains of mammals are known, many belonging to extinct species, families, and orders. Indeed, some fossil mammals cannot be assigned to any of the three infraclasses of living mammals. Mammals are usually considered most closely related to now extinct Mesozoic "reptiles" generally known as the cynodonts (also known as the mammal-like reptiles). In particular, the tritylodonts, forms of herbivorous cynodonts, appear to be very closely related to the earliest true mammals. The oldest known reputed mammal is *Adelobasileus cromptoni,* based on a single incomplete skull found in Texas, dating back to the Late Triassic (about 225 million years ago [mya]). The earliest mammals are distinguished on the basis of details of the anatomy of the skull (particularly the ear region and jaw), dentition, and postcranial skeleton. In the case of *Adelobasileus cromptoni* the teeth and postcranial skeleton are not known, and the possibility remains that it may actually represent a type of "cynodont" previously known only from teeth, jaws, or limb-bone remains. Of slightly later age than *Adelobasileus,* a number of very early mammal remains are known from the end of the Triassic in western Europe. By the early Jurassic mammal remains are known from Europe, North America, Africa, and Asia.

The earliest mammals were tiny creatures (some with a head and body length of only about ten cm) that may have generally resembled living shrews in appearance. Many early mammals may have fed on insects, eggs, and other small prey. They probably lived in thick vegetation or close to the ground among rocks and were characterized by nocturnal behavior patterns. Their small size and habits may have helped them avoid becoming the major prey of carnivorous dinosaurs (*see* DINOSAURS).

Through the course of the late Triassic, Jurassic, and Cretaceous the mammals increased in numbers and diversity. In Mesozoic rocks the remains of a number of types of mammals (such as triconodonts, docodonts, amphilestids, and other groups) are found that apparently went extinct without leaving any descendants. Toward the end of the Mesozoic, probably in the latest Jurassic or earliest Cretaceous (about 140 to 150 mya) the ancestors of the monotremes and the multituberculates (discussed further below) first arose. Perhaps slightly later in the Cretaceous the common ancestor of marsupials and placentals evolved.

Multituberculates, Monotremes, Marsupials

The Cenozoic is often referred to as the "Age of Mammals." The end of the Mesozoic (about 65 mya) was marked by the extinction of the dinosaurs, and mammals quickly came to dominate terrestrial faunas. An extremely important group of mammals during the late Mesozoic and early Cenozoic was the order Multituberculata (subclass Allotheria). The earliest possible multituberculate remains are known from the late Triassic, and undisputed true members of the group are known from the late Jurassic. They survived for over a 100 Ma, making them longer lived than any other major group of mammals, and only went extinct in the middle Cenozoic (about 35 mya). Multituberculates may have superficially looked like rodents, and they ranged from mouse- to woodchuck-size. Various species appear to have been herbivorous and/or omnivorous. After reaching their greatest diversity during the Paleocene (the first 10 Ma of the Cenozoic), the multituberculates underwent a decline and finally disappeared completely. They may have suffered from competition with newly evolving herbivorous mammals such as rodents and early primates.

The monotremes have never been diverse or dominant. Only a handful of species are extant, and their fossil record consists of only a smattering of remains, including a possible lower jaw found in early Cretaceous rocks of Australia.

Marsupials and placentals (collectively known as therians) had a common ancestor during the early Cretaceous; by the late Cretaceous the two groups had diverged geographically and evolutionarily. Marsupials are well known from the Cretaceous of North America, may have originated in North, Central, or South America, and appear to have migrated north to Europe (the northern Atlantic Ocean had not yet opened) and south to South America and the Australian region via Antarctica (at this time these three continents were in contact with one another). Marsupials had their greatest radiations in South America and the Australian region, though they are known from North America and Europe. A fossil marsupial, related to South American forms, has been found on Antarctica; fossil remains of the group are known from North Africa, and a single fossil marsupial tooth has been reported from central Asia.

After the late Cretaceous, with the separation of the two American continents, marsupials formed an important component of the South American fauna. Likewise, the marsupials became isolated in the Australian region and underwent a diverse evolutionary radiation unimpeded by competition from placental mammals. In both areas diverse marsupial forms evolved, including herbivorous, carnivorous, and omnivorous forms. The South American *Thylacosmilus* is a marsupial version of a placental "saber-toothed tiger," and in the Australian region marsupial forms as diverse as the Tasmanian wolf, Tasmanian devil, wombats, kangaroos, bandicoots, and koalas are found. A continuing problem for the study of Australian marsupials is the lack of fossil evidence before about 30 mya.

Placental Mammals

Since the beginning of the Cenozoic, placental (eutherian) mammals have dominated the terrestrial vertebrate faunas of all the continents except Antarctica and Australia (where marsupials have prevailed). Remains that may represent the earliest placentals have been found in early Cretaceous rocks of Texas and Mongolia, and by the late Cretaceous well-identified and undisputed placental remains are present in fossil deposits. It has been suggested that the late Cretaceous *Kennalestes* of central Asia is very similar in form to the probable common ancestor of all later placentals. *Kennalestes* was a small animal (the skull is only three centimeters long) that probably resembled the modern tree shrew in life. The earliest placentals are usually considered to have been generally insectivorous in diet and habits.

During the early part of the Cenozoic (about 65 to 40 mya) the placental mammals underwent a spectacular radiation. Many groups and families evolved, including ancestors of most of the living orders of mammals as well as many archaic and now extinct forms. Extinct groups of placental mammals include the Pantodonta, Dinocerata (uintatheres), Taeniodonta, Tillodonta, Embrithopoda (arsinoitheres), and many others. Of special note is the fact that a number of independent, and now extinct, placental groups (such as the notoungulates, xenungulates, toxodonts, and litopterns) evolved in isolated South America. These South American ungulates appear to mimic horses, rhinos, camels, hippos, elephants, and other forms that evolved in other parts of the world.

Today about eighteen orders of living placental mammals are recognized, most of which can trace their ancestral lineage back to the early Cenozoic. The placental mammals are an extremely diverse and successful group. Placental mammals inhabit the land, have invaded the air, and flourish in the oceans. Mammals, in the form of humans, are the first earthly organisms to have visited an extraterrestrial world (the Moon). The major types of living placentals are: Edentata (also known as Xenarthra, sloths, armadillos, and relatives), Pholidota (pangolins or scaly anteaters), Insectivora (various types of hedgehogs, shrews, and moles), Scandentia (tree shrews), Dermoptera (flying lemurs), Chiroptera (bats), Primates (lemurs, monkeys, apes, humans), Carnivora (dog family, cat family, bears, ferrets, minks, wolverines, badgers, skunks, otters, and so on; the Pinnipedia—seals, sea lions, walruses, and relatives—are also often included within the Carnivora), Cetacea (dolphins, porpoises, and whales), Sirenia (sea cows, dugongs, manatees), Proboscidea (elephants), Perissodactyla (tapirs, horses, rhinoceroses, and relatives), Artiodactyla (pigs, hippopotamuses, goats, sheep, camels, antelopes, deer, cows, and relatives), Rodentia (mice, rats, squirrels, woodchucks, prairie dogs, beavers, capybaras, and so on), Lagomorpha (rabbits, hares, pikas), Hyracoidea (hyraxes, dassies, conies), Tubulidentata (aardvarks), and Macroscelidea (elephant shrews—sometimes included within the Insectivora). A major focus of much current research is just exactly how the members of these groups are related evolutionarily to one another.

Bibliography

Carroll, R. L. *Vertebrate Paleontology and Evolution.* New York, 1988.

Honacki, J. H.; K. E. Kinman; and J. W. Koeppl, eds. *Mammal Species of the World.* Lawrence, KS, 1982.

Lillegraven, J. A.; Z. Kielan-Jaworowska; and W. A. Clemens, eds. *Mesozoic Mammals: The First Two-Thirds of Mammalian History.* Berkeley, CA, 1979.

Lucas, S. G., and Z. Luo. "*Adelobasileus* from the Upper Triassic of West Texas: The Oldest Mammal." *Journal of Vertebrate Paleontology* 13 (1993):309–334.

Romer, A. S. *Vertebrate Paleontology,* 3rd ed. Chicago, 1966.

Savage, D. E., and D. E. Russell. *Mammalian Paleofaunas of the World.* Reading, MA, 1983.

Schoch, R. M., ed. *Vertebrate Paleontology.* New York, 1984.

Simpson, G. G. *The Principles of Classification and a Classification of Mammals.* New York, 1945.

Walker, E. P., ed. *Mammals of the World,* 3rd ed. Baltimore, 1975.

Woodburne, M. O., ed. *Cenozoic Mammals of North America: Geochronology and Biostratigraphy.* Berkeley, CA, 1987.

ROBERT M. SCHOCH

MANGANESE DEPOSITS

Manganese is used extensively in the manufacture of steel, where about six kilos of manganese are used to make one ton of steel. It is also used in various alloys and in a variety of processes that utilize its unique chemical properties.

The formation of manganese deposits requires that two general conditions be met: manganese must be concentrated at least 200 times its average crustal abundance of about 0.15 percent, and iron cannot be concentrated to any appreciable degree from its average crustal abundance of about 5 percent (Force and Cannon, 1988). Because manganese forms extremely insoluble compounds (MnO_2) in the presence of oxygen, concentration of manganese in seawater can occur only in oxygen-deficient (anoxic) waters. This condition may develop in deeper portions of ocean basins where Mn^{+2} ions remain in solution. The separation of iron and manganese may be accomplished in chemically reducing (anoxic) environments where black shales accumulate. Iron may be largely removed from the seawater as pyrite (FeS_2), which accumulates in the black shale. Manganese sulfide (MnS) is relatively soluble and does not precipitate with iron sulfides. Manganese may be precipitated where the iron-depleted, manganese-rich water rises into oxygen-rich near-surface waters along coastlines.

The manganese is usually precipitated as one or more manganese oxides: pyrolusite (MnO_2), manganite [MnO(OH)], and a complex mixture of manganese oxides (called "psilomelane" or romanechite [$BaMn^{2+}Mn^{4+}{}_8O_{16}(OH)_4$] sometimes appropriately called "wad").

The shallow-water origin of these deposits is indicated by the typical presence of concentrically banded oolites and pisolites in the ores, such as those in the large deposits of Groote Eylandt, in northern Australia. Manganese deposits at several localities north of the Black Sea—at Nikapol, Ukraine; Chiatura, Georgia; and at other nearby deposits—form some of the largest concentrations of manganese in the world. In addition to oolitic and pisolitic oxide deposits, manganese carbonates, rhodochrosite ($MnCO_3$), and kutnohorite [$CaMn(CO_3)_2$], are present. The manganese carbonates may be primary deposits formed in deeper (reducing) water environments, or may result from earlier limestones being replaced by Mn^{2+} in solution in the seawaters.

Deposits in the Kalahari manganese field in South Africa are probably the largest manganese reserve on land in the world (Miyano and Beukes, 1987). They are Early Proterozoic (2100–2500 Ma) and are therefore much older than the Cretaceous to Oligocene (30–65 Ma) of the Australian and Russian deposits. The Kalahari deposits differ from the younger manganese deposits in that they occur as manganese-rich layers interbedded with sedimentary iron-formation of the Hotazel formation (Miyano and Beukes, 1987).

The deposits are believed to have originated as fine muds of manganese silicate (braunite [Mn^{2+} (Mn^{3+}, $Fe^{3+})_6$ SiO_{12}] and kutnohorite [$CaMn(CO_3)_2$]), which constitute most of the ore reserves. The fine grain size and thinly laminated nature of the deposits suggest that they formed in relatively deep water. This primary rock has been enriched by rising hydrothermal fluids that converted the original fine-grained rock into a coarser-grained mixture of braunite-II [$Ca(Mn^{3+}, Fe^{3+})_7SiO_{12}$], hausmannite ($Mn^{2+}Mn^{3+}{}_2O_4$), and bixbyite [$(Mn^{3+}, Fe^{3+})_2O_3$]. The manganese and associated iron in these deposits may be of volcanic origin.

Huge reserves of manganese have been discovered on the ocean floor, where large areas are covered by nodules composed of a mixture of manganese and iron oxides and other metalliferous sediments. The nodules are concentrically layered disk-shaped structures that range in size from less than one centimeter to more than a meter, but most are in the 2 to 5 cm range. The nodules contain 20 to 30 percent manganese oxides, with variable amounts of iron and 2 to 3 percent combined copper, nickel, and cobalt. The nodules presumably form by precipitation in concentric layers around sand grains, sharks' teeth, or some other nucleus. The manganese, and other elements in the nodules, is generally believed to be related to submarine volcanism along the mid-ocean ridges, for they are most abundant near active ridges (Bonatti and Nayudu, 1965; Bostrom and Peterson, 1966). Estimates of recoverable nodules on the seafloor range from 50 billion to more than a trillion tons, therefore, these deposits represent a very large reserve of metals.

Recent studies (e.g., Robbins et al., 1992) indicate that deposition of manganese in modern environments is accomplished by biological oxidation of manganese in solution by the so-called iron-bacteria. Additional studies by Schmidt and Robbins (1992) suggest that microbial precipitation of manganese may also have been important in the formation of manganese deposits in certain ancient sedimentary sequences.

Bibliography

BONATTI, E., and Y. R. NAYUDU. "The Origin of Manganese Nodules on the Ocean Floor." *American Journal of Science* 263 (1965):17–39.

BOSTROM, K., and M. N. A. PETERSON. "Precipitates from Hydrothermal Exhalations of the East Pacific Rise." *Economic Geology* 61 (1966):1258–1265.

FORCE, E. R., and W. F. CANNON. "Depositional Model for Shallow-marine Manganese Deposits Around Black Shale Basins." *Economic Geology* 83 (1988):93–117.

MIYANO, T., and N. J. BEUKES. "Physicochemical Environments for the Formation of Quartz-free Manganese Oxide Ores from the Early Proterozoic Hotazel Formation, Kalahari Manganese Field, South Africa." *Economic Geology* 82 (1987):706–718.

ROBBINS, E. I.; J. P. D'AGOSTINO; J. OSTWALD; D. S. FANNING; V. CARTER; and R. L. VAN HOVEN. "Manganese Nodules and Microbial Oxidation of Manganese in the Huntley Meadows Wetland, Virginia." In *Biomineralization Processes: Iron, Manganese*, eds. H. C. W. Skinner and R. W. Fitzpatrick. Cremlingen, 1992.

SCHMIDT, R. G., and E. I. ROBBINS. "New Evidence of an Organic Contribution to Manganese Precipitation in Iron Formation and Review of Sedimentary Conditions in the Cuyuna North Range, Minnesota." In *Biomineralization Processes: Iron, Manganese*, eds. H. C. W. Skinner and R. W. Fitzpatrick. Cremlingen, 1992.

GENE L. LA BERGE

MANTLE, STRUCTURE OF

The mantle is the region of Earth that lies above the molten iron outer core and beneath the crust. This shell of rock has an average thickness of 2,867 kilometers (km) and an average density of 4.5 million grams per cubic meter (Mg/m^3). One Mg/m^3 is equivalent to 1,000 kilograms per cubic meter (1,000 kg/m^3) or, in more familiar units, 1 gram per cubic centimeter (1 g/cm^3). The mantle contains about two-thirds of the mass of the planet and takes up about five-sixths of its volume. It is made up primarily of compounds of magnesium (Mg), iron (Fe), silicon (Si), and oxygen (O); hence, it is denser than the overlying crust, which is depleted in magnesium and iron and enriched in silicon, aluminum (Al), and other lighter elements. On human timescales, the mantle behaves as an elastic body. Seismic waves generated by earthquakes travel through the mantle. By analyzing these waves, seismologists can learn quite a bit about the structure of the mantle. On timescales of thousands to millions of years, mantle rocks respond to stresses by ductile flow. Density variations arising primarily from temperature and compositional differences drive convective motions within the mantle, accompanied by plate motions, earthquakes, and volcanic activity. The dynamics of the mantle depend on the interaction between its density structure, which generates stresses, and its mechanical structure, which responds to these stresses.

This article describes our estimates of the average mantle structure as a function of radius. (Lateral variations in mantle structure are discussed in the entry SEISMIC TOMOGRAPHY.) Since most of what is known about mantle structure is derived from seismology, we begin with the seismic structure—seismic velocities and density as a function of radius. Next the petrologic structure inferred from the seismic structure is discussed, followed by an estimate of thermal structure. Discussion of the strength (rheology) of the mantle, which is determined by the interaction of the petrologic and thermal structures, follows. Variations in these properties as a function of radius are given in Table 1, plotted in Figure 1, and discussed in the following sections.

Seismic Velocity Structure

There are several different kinds of seismic waves that sample the mantle. The fastest traveling waves are called P-waves because they are the first, or *P*rimary, arrivals on a seismogram. P-waves are compressional waves; like sound waves, they can travel through either solids or liquids. S-waves, which are generally the *S*econdary arrivals on seismograms, are shear waves; they are unable to propagate through the liquid core. Earthquakes also set off vibrations of the entire planet, called free oscillations, that are analogous to the ringing of a bell, as well as surface waves, which are confined to the near-surface regions. By analyzing the arrival times of P-waves, S-waves, and surface waves, and by determining the frequencies of Earth's free oscillations, seismologists can deter-

Table 1. Mantle Structure

Region	*Main Phases*	*Radius (km)*	*Density (Mg/m^3)*	*V_P (km/s)*	*V_S (km/s)*	*Pressure (GPa)*	*Temper-ature (K)*	*Strength (MPa)*	
Lower mantle (D″)	Perovskite Periclase Metal	3480	5.57	13.72	7.26	137	4400	1	
		3600	5.51	13.69	7.27	130	3689	3	
		3700	5.46	13.60	7.23	124	3096	10	
Lower mantle	Perovskite Periclase	3800	5.41	13.48	7.19	118	2750	20	
		4000	5.31	13.25	7.10	107	2666	20	Meso-
		4400	5.11	12.78	6.92	86	2499	20	sphere
		4800	4.90	12.29	6.73	66	2331	20	
		5200	4.68	11.73	6.50	47	2164	20	
		5600	4.44	11.07	6.24	28	1996	20	
		5711	4.38	10.75	5.95	24	2000	20	
Transition zone	Spinel Garnet Pyroxene	5711	3.99	10.27	5.57	24	2000	2	
		5800	3.94	10.01	5.43	20	1933	2	
		5900	3.81	9.50	5.14	16	1811	2	
		5971	3.72	9.13	4.93	13	1700	2	Astheno-sphere
Upper mantle	Olivine Pyroxene Garnet	5971	3.54	8.91	4.77	13	1700	0.1	
		6100	3.46	8.66	4.68	9	1608	0.1	
		6200	3.40	7.97	4.46	5	1575	0.1	
		6275	3.35	8.04	4.48	3	1550	0.1	
		6300	3.35	8.06	4.49	2	1525	10	Litho-
		6325	3.41	8.09	4.50	1	1067	100	sphere
		6340	3.44	8.10	4.50	1	793	250	
		6347	3.46	8.11	4.50	1	672	180	

mine P-wave velocity, S-wave velocity, and density as a function of radius. Once density is known, the pressure can also be calculated.

The variations in density, P-wave velocity (V_P) and S-wave velocity (V_S), are plotted as a function of radius for the outer half of the earth in Figure 1. These variations define several subregions within the mantle. The boundary between the mantle and the underlying fluid core is at a radius of 3,480 km, where there are abrupt changes in density and seismic velocities. Changes in properties with radius are smooth throughout the region extending from the core-mantle boundary to a radius of 5,711 km (depth of 660 km), where there are sharp changes, or discontinuities, in density and seismic velocities. This region of smooth variations is called the lower mantle. The variations result primarily from the effects of the increase in pressure with depth, from a value of 24 GPa (240,000 atmospheres) at the top of the lower mantle to 137 GPa (1.37 million atmospheres) at the core-mantle boundary. The average density just above the core-mantle boundary (CMB) is 5.57 Mg/m^3, decreasing with decreasing pressure to 4.38 Mg/m^3 at the top of the lower mantle.

In the part of the lower mantle extending several hundred kilometers just above the core-mantle boundary, there are large variations in properties that cannot be explained just by the effects of pressure. This region was termed the "D″ region" by the seismologists who first noticed its effects on the arrival times of seismic waves. The D″ region shows large variations in structure laterally, as well as radially. There may be chemical reactions occurring between the metallic core and the silicate mantle that result in the D″ layer having a composition intermediate between that of the mantle and the core.

The region above the lower mantle, between a radius of 5,711 km and 5,971 km (depth of 400 km), is the transition zone. There is an exceptionally large change in density with radius in this region resulting primarily from pressure-induced phase transitions, discussed below. Strictly speak-

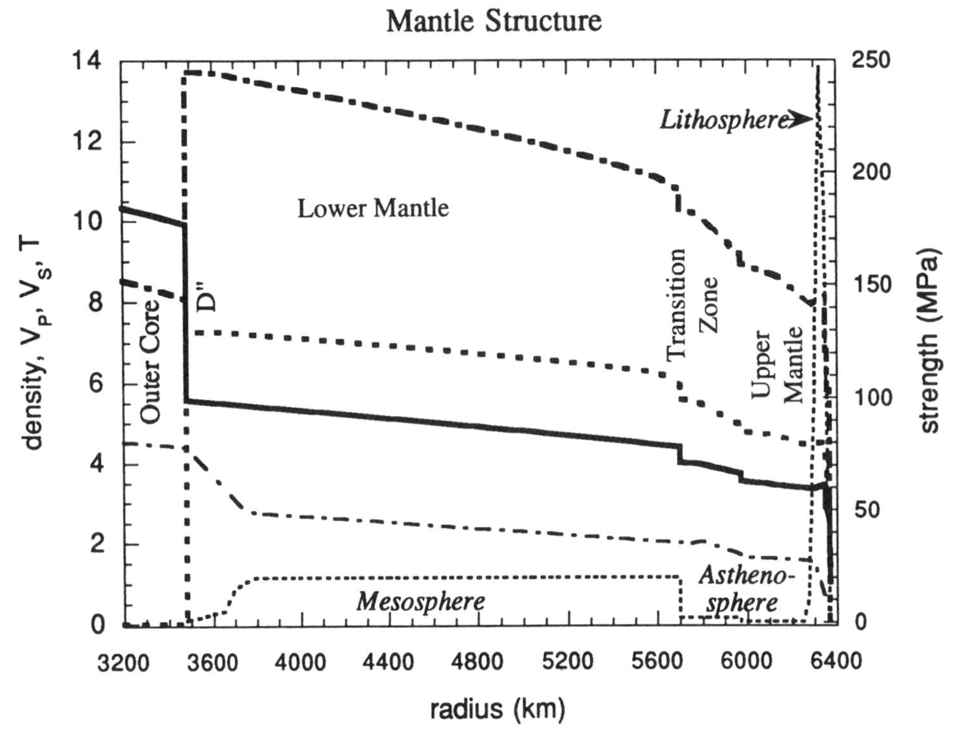

Figure 1. The variation of density (Mg/m^3; heavy solid line), P-wave velocity, V_P (km/s; heavy dot-dashed line), S-wave velocity, V_S (km/s; heavy dotted line), temperature (1,000 K; light dot-dashed line), and strength (MPa; light dotted line) as a function of radius. The petrologic regions are indicated in regular type; the "dynamic" regions are indicated in italics.

ing, the upper mantle is the region that lies above the transition zone and below the crust, although "upper mantle" is sometimes used to include the transition zone as well. The boundary between the crust and the upper mantle, named the Mohorovičić discontinuity after the seismologist who discovered it, lies at an average depth of 24 km (radius = 6,347 km), although the depth to the "Moho" varies from place to place by more than 70 km, with the thickest crust under the high Tibetan plateau and the thinnest crust beneath the ocean basins. The density in the upper mantle just above 400-km depth is 3.54 Mg/m^3. Density decreases with radius, reaching a value of 3.38 Mg/m^3 at a depth of about 100 km before increasing, due to a decrease in temperature to a value of 3.42 Mg/m^3 at the top of the upper mantle.

Petrologic Structure

The mantle is denser than the core upon which it floats because the mantle has a chemical composition that is more enriched in oxygen, magnesium, and silicon than the dominantly metallic and heavier core. In addition to the kind of atoms that make up a material—its chemical composition—the way these atoms are arranged—its petrology, or the phases it contains—is important in determining a material's density. A familiar example of how different phases of a material can have different densities is H_2O, which has a density of 1.0 Mg/m^3 in its liquid phase (water), a density of 0.9 Mg/m^3 in its solid phase (ice), and a density of 0.0007 Mg/m^3 in its gaseous phase (water vapor)—less than the average density of the atmosphere. The ocean is more dense than the atmosphere not because the ocean is made up of an intrinsically denser compound, but because this compound is arranged in an intrinsically denser phase (liquid) than are the atoms in the gaseous atmosphere.

The pressure a material is subjected to affects its density and seismic velocities in two ways. First, increasing pressure increases the density of material by squeezing the atoms closer together, at the same time increasing its seismic velocities by increasing its stiffness. In addition, changes in pressure can cause changes in phase, where the atoms in a compound are rearranged. For example, the materials making up the lower part of the mantle have their atoms arranged in more closely packed geometric structures than those in the upper mantle, where the minerals are characterized by a more loosely packed arrangement of atoms. The density of the mantle increases by a factor of 5/3 from the uppermost to the lowermost mantle. Of this increase, a factor of 4/3 is the result of the ordinary effects of compression and a factor of 5/4 is the result of rearrangement of atoms into different phases.

The effects of increases in temperature on density and seismic velocities are opposite to the effects of increases in pressure. Increasing temperature leads to thermal expansion, resulting in a decrease in density, and to a decrease in stiffness. In addition, changes in temperature can lead to changes in phase, with higher temperatures generally favoring less densely packed atomic arrangements. Because both pressure and temperature increase with depth in the earth, their effects offset each other to some extent. Within thermal boundary layers, regions with large temperature gradients, the effects of temperature variations predominate; elsewhere, effects of pressure variations are generally more important.

The lower mantle is made up primarily of minerals consisting of $MgSiO_3$ in the perovskite structure and MgO in a cubic structure. The variations of density and seismic velocities with radius through the lower mantle are quite smooth because they result primarily from the large but smooth pressure variations with radius—squeezing a substance increases both its density and its seismic velocities. The predicted variations in density and seismic velocities are very close to those predicted for the effects of pressure and temperature acting on a material of uniform chemical composition. There are, however, some subtle differences between observations and predictions that some geophysicists attribute to a slight change in chemical composition with depth. Others attribute these subtle differences to errors in the extrapolations of either material properties or temperatures.

The density and seismic velocity changes that occur abruptly at a depth of 660 km are primarily the result of minerals changing structure from the perovskite structure characteristic of the lower mantle (density of 4.38 Mg/m^3) to the spinel and garnet structures (density of 4.08 Mg/m^3) that characterize the transition zone. The density and seismic velocities within the transition zone decrease relatively rapidly with increasing radius as the result of additional distributed phase changes, with minerals in the garnet structure changing to the pyroxene structure, reaching a density of 3.77 Mg/m^3 at the top of the transition zone. At a depth of 400 km (radius of 5,971), there is another important phase change, with the material in the upper mantle above 400-km depth predominantly in the olivine structure. Some garnet remains, with pyroxenes and other important minerals.

Geophysicists can carry out well-calibrated experiments at the pressures characteristic of the transition zone and can demonstrate that these phase changes occur at the appropriate pressures. But in the mantle, the discontinuities are sharper than would be predicted by extrapolating laboratory measurements. The observed sharpness of these discontinuities has led some to propose that they represent discontinuities in composition, as well as phase, leading to the hypothesis that the lower mantle is compositionally distinct from the upper mantle. Others prefer the explanation that these phase changes are sharp because they are not at thermodynamic equilibrium.

Thermal Structure

The temperature at the top of the core must be hot enough to melt an iron-rich metallic alloy at high pressure. This temperature is estimated to be 4,400 ± 600 degrees Kelvin (K). Just above the core-mantle boundary, in the D″ region, the increase in seismic velocity with depth is more gradual than in the remainder of the lower mantle. This is the result that would be expected if there were a rapid increase in temperature with depth that partially offsets the effects of increasing pressure. The D″ layer represents a thermal boundary layer between the hotter core and the cooler mantle. D″ may also be enriched in iron relative to the rest of the lower mantle, forming a compositionally distinct layer. Such a distinct composition would allow it to remain dense despite its higher temperatures—there may be a temperature drop of as much as 1,400 K across D″ (see Figure 1). Such a thermal boundary layer is a likely spot for mantle plumes to originate. The heat transported by conduction out of the core and across this region is primarily the heat released by cooling of the outer core and freezing of the inner core.

The decrease in temperature with radius from the top of D″ to the bottom of the transition zone is primarily the result of the decrease in pressure across this region. (Squeezing a material increases its temperature and lowering the pressure decreases the temperature.) There are lateral variations in temperature of hundreds of degrees Kelvin associated with mantle convection, in addition to these radial variations.

The temperature throughout the transition zone is constrained by laboratory experiments that determine the temperatures required in order that the phase transitions occur at the depths and pressures determined from seismology. The temperature decreases with radius from 2,000 ± 200 K at the base of the transition zone to 1,700 ± 150 K at the top of this layer. There is a further gradual decrease to 1,550 ± 100 K at a depth of about 100 km resulting from a decrease in pressure over this interval. There is a much more rapid decrease in temperature with radius through the outer 100 km of the planet through the thermal boundary layer that is associated with the tectonic plates. At depths shallower than about 100 km the pressure is sufficiently low and the temperature sufficiently high beneath mid-oceanic ridges and island arcs that partial melting of the mantle can occur. Lateral variations in temperature approach 1,000 K in the upper 100 km.

Strength and Dynamic Structure of the Mantle

The material making up most of the mantle, like the children's toy Silly Putty, deforms over time when subjected to the stresses that result from gravity acting on variations in density. The way in which a material deforms in response to applied stress is called its rheology. The rheology, or strength, of Earth material depends on composition, phase, pressure, and temperature. Convection in the earth's mantle drives plate motions. The details of the interaction of the stresses generated by density differences and the resistance to deformation determine the style of tectonic activity.

The strength of the solid Earth varies dramatically as a function of position. The rheology of the rocks that make up the mantle (as well as the overlying crust) depends on temperature, pressure, composition, and the level of the stress applied. Near the surface, where temperature and pressure are both relatively low, when the stresses are sufficiently high, rocks fail by brittle failure. This is the process that generates earthquakes. Near the surface, at temperatures below about 750 K, the strength of the rocks is determined mainly by friction resisting motion on faults, with negligible dependence on composition. In the absence of fluids, the strength is approximately a third of the weight of the overlying rocks. But many faults fail at lower stresses, perhaps because they are lubricated by high pore fluid pressures. There also seems to be a relationship between the distance that a fault has slipped over geologic time and its strength, with major plate boundary faults, which have accommodated substantial displacements, weaker than other faults.

At higher temperatures, rocks deform by ductile flow, again like Silly Putty, and can be characterized by an effective viscosity. For ductile materials, the "strength" depends on the rate at which they deform. The strengths tabulated and plotted here are for a rate of deformation of 1 percent per million years—a deformation rate characteristic for the convecting mantle. Mantle rocks become increasingly ductile with increasing temperature, with their strength decreasing by about a factor of ten for each increase in temperature by 100 K. Because temperature increases rapidly with depth near the surface of the earth, there is an accompanying rapid decrease in strength with depth. The cold, strong rocks near the surface make up the lithosphere. This is a layer that forms the tectonic plates. The weaker, more ductile layer beneath is the asthenosphere. Because of the great contrast in strength, the lithospheric plates move over the asthenosphere without deforming, except at weak plate boundaries. The thickness of the lithosphere is on average about 75 km, but its thickness varies greatly, from only about 10 km beneath mid-oceanic ridges to hundreds of kilometers beneath old continents. In most places the entire crust and part of the upper mantle are included within this layer, which is defined according to its strength, not according to its composition.

Beneath about 75–100 km depth, the temperature increases more slowly with depth, and the effect of increasing pressure or rock strength becomes important. The phase changes that occur across the transition zone also affect the strength. The net result is that the strength of mantle rock increases with depth, as shown in the figure. The stronger region beneath the asthenosphere is called the mesosphere. The D″ region is probably weaker than the overlying mantle because of the rapid increase in temperature with depth across this layer.

Most boundaries in the "dynamic" structure are determined primarily by the thermal structure, rather than the petrological structure. The mechanically strong lithosphere includes both crust and upper mantle. The D″ layer is weak primarily because it is a hot thermal boundary layer. But the boundary between the asthenosphere and the mesosphere is likely a result of the difference in strength of minerals making up the transition zone and the rest of the mantle.

Although the qualitative variation of strength with depth is reasonably well known, the quantitative values of the strength distribution are uncertain by a factor of three or more. The main reason for this uncertainty is that it is impossible to carry out experiments at strain rates as low as those associated with mantle convection. Laboratory experiments are typically carried out at rates of deformation of 1 percent per month or more—ten million times faster than applies in the mantle. The variation of strength as a function of depth in the mantle is constrained by the rebound of the earth's surface that is still going on more than ten thousand years after the glaciers melted. But this postglacial rebound is sensitive only to averages in strength over great depths. Laboratory experiments provide constraints on where changes in strength should occur, while geophysical observations provide constraints on the average strength.

Bibliography

ANDERSON, D. L. *Theory of the Earth.* Boston, 1989.

BENZ, H. M., and J. E. VIDALE. "Sharpness of Upper-Mantle Discontinuities Determined from High-frequency Reflections." *Nature* 365 (1993):147–150.

JEANLOZ, R. "The Mantle in Sharper Focus." *Nature* 365 (1993):110–111.

JEANLOZ, R., and S. MORRIS. "Temperature Distribution in the Crust and Mantle." *Annual Reviews of Earth and Planetary Sciences* 14 (1986):377–415.

STACEY, F. D. *Physics of the Earth,* 3d ed. Brisbane, Australia, 1992.

BRADFORD H. HAGER

MANTLE CONVECTION AND PLUMES

Mantle convection is the mechanism whereby solid material in the iron-magnesium-silicate outer shell of the earth is caused to circulate or connect, owing to the destabilizing temperature difference that exists between the planet's surface and the boundary between the solid mantle and the liquid outer core. Although the mantle is nominally "solid" in the sense that it is capable of transmitting elastic shear waves (*see* EARTHQUAKES AND SEISMICITY; MANTLE, STRUCTURE OF) on sufficiently long timescales, it behaves as a very viscous fluid and is therefore capable of thermal convection. This process is extremely well understood on the basis of intensive laboratory and theoretical analyses that have been performed since the beginning of the twentieth century. In a very viscous fluid for which the ratio of the momentum diffusivity to the thermal diffusivity is extremely large (this ratio is usually referred to as the Prandtl number), the circulation induced either in the plane layer geometry that would be employed in a laboratory experiment or the spherical shell geometry of a planetary mantle is controlled entirely by the Rayleigh number. This latter number is a nondimensional number that measures the ratio of the destabilizing influence of the vertical temperature gradient to the stabilizing influence of the dissipative processes (thermal and momentum diffusion). The Rayleigh number (Ra) may be written in the form of the following equation:

$$R_a^B = \frac{g\alpha\Delta T d^3}{\kappa\nu} \quad (1)$$

in which g is the gravitational acceleration, α is the coefficient of thermal expansion, ΔT is the temperature difference across the layer (high temperature on the bottom, low temperature on the top), d is the layer thickness, κ is the thermal diffusivity ($\kappa = k/\rho C_p$, where κ is the thermal conductivity, ρ the mean density, and C_p the heat capacity as constant pressure), and ν is the so-called kinematic viscosity ($\nu = \mu/\rho$, where μ is the molecular viscosity and ρ the mean density).

When a fluid (a substance with a low resistance to flow) is heated entirely from below and cooled from above by the imposition of constant temperatures on both bounding surfaces, the configuration is referred to as the Bénard (or heated from below) configuration, to distinguish the resulting convection process from that which occurs when heat is applied to the fluid differently. The most important alternative heating mechanism, especially from the perspective of understanding the mantle convection phenomenon, is associated with heating entirely from within the volume of the fluid itself. Because the earth's mantle contains significant concentrations of the long-lived radioactive isotopes U, Th, and K (uranium, thorium, and potassium) and because the decay of these isotopes leads to heating of the material in which the decay takes place (*see* HEAT BUDGET OF THE EARTH), so-called heated-from-within flows are germane to understanding the mantle convection phenomenon. For purely heated-from-within convection in which the upper surface of the layer is held at constant temperature while the bottom boundary is adiabatic, the form of the Rayleigh number that determines circulation properties takes the following form:

$$R_a^Q = \frac{g\alpha Q d^5}{\kappa^2 \nu} \qquad (2)$$

in which Q is the rate of heat addition per unit volume of the fluid and the remaining parameters are the same as those that appeared previously in (1).

For either heating configuration, and in either plane layer or spherical geometry, the fluid remains at rest until a certain value of the Rayleigh number is exceeded, a value that is referred to as the "critical" Rayleigh number. For the plane layer and for heating from below and within these critical Rayleigh numbers are, respectively, 658 and 1,708. For Rayleigh numbers slightly above these limits, the motion of the fluid is slow, but as the Rayleigh number increases, say, by increasing the rate of heating, the circulation becomes increasingly vigorous. For the mantle of the earth, the effective Rayleigh number is expected to be close to the value of 10^7, which is approximately 10,000 times higher than the critical value that must be exceeded before thermal convection begins. This number may be evaluated by using the definition (1) for R_a^B and assuming $g \simeq 10\ \text{m s}^{-2}$, $\alpha \simeq 3 \times 10^{-5}\ \text{K}^{-1}$, $\Delta T \simeq 3{,}000$ K, $d \simeq 3{,}000$ km, $\kappa \simeq 2 \times 10^{-6}\ \text{m}^2\ \text{s}^{-1}$, and $\nu \simeq 10^{18}\ \text{m}^2\ \text{s}^{-1}$. Of these parameters the one for which the value is most often debated is the kinematic viscosity ν. The process by which mantle rocks flow is called creep. Creep is basically the continuous, slow deformation of the crystal lattice, and it occurs via a variety of mechanisms. Since the creep of a solid is a thermally activated process that is strongly dependent upon the details of the creep mechanism (which may be either linear or nonlinear), it is difficult to measure ν directly with confidence in the laboratory. The effective value of the viscosity of the mantle that is required to evaluate the Rayleigh number is therefore determined by appeal to geophysical data, principally those related to the large-scale recovery of the Canadian and Fennoscandian shield regions to the gravitational imbalance created by the disintegration of the huge ice sheets that covered these regions at last glacial maximum approximately 21,000 years ago (*see* ISOSTASY; EARTH AS A DYNAMIC SYSTEM). Since the postglacial "rebound" of the crust that follows deglaciation occurs at a rate that is determined by the viscosity of the mantle, observations of the rate and geometric form of the rebound may be employed to infer mantle viscosity. The value so derived yields the estimated Rayleigh number of the mantle, which shows that convection must be occurring.

In both laboratory experiments and numerical simulations it is observed that the nature of a thermally forced convective circulation changes markedly as the Rayleigh number increases above the critical value. For heated-from-below convection, boundary layer arguments may be invoked to show that typical flow velocities increase with two-thirds power of the Rayleigh number for Rayleigh numbers sufficiently large that well-defined horizontal and vertical boundary layers exist in the temperature field. The vertical boundary layers are usually referred to as thermal "plumes," and the pattern of convection for heated-from-below flow is such that individual convection cells are bounded by hot upwelling and cold downwelling plumes. These plumes play a vital role in the convection process because it is precisely through the buoyancy of anomalously hot material and the negative bouyancy of anomalously cold material that gravitational potential energy is converted into kinetic energy against the action of the dissipative processes. As the Rayleigh number increases beyond approximately ten times the critical value, the horizontal dimensions of the plumes decrease. For plane-layer heated from below flow, this rate of decrease is as the one-third power of the Rayleigh number. For heated-from-within convection with an adiabatic lower boundary no hot upwelling plumes exist, only cold downwellings. In the earth the convective circulation is probably of mixed type and the relative proportion of the heating from below and within remains an issue of active debate.

Depending upon the mode of heating and the mechanical boundary conditions subject to which the convective circulation evolves, the cell-bounding hot upwelling and cold downwelling plumes tend to become strongly time-dependent above a Rayleigh number that is approximately 100 times the critical value for the onset of motion. Typically this time-dependent regime occurs at a lower Rayleigh number in three-dimensional convection than it does in numerical simulations, in which the evolution of the flow is restricted to two space dimensions. Since the Rayleigh number of the mantle is approximately 10,000 times the critical value, it is to be expected on these grounds that the flow may be highly time-dependent.

In the above discussion, the earth's mantle is assumed to be a simple fluid with constant viscosity. In the actual earth, there are of course circumstances that are expected to significantly modify the nature of the convection process from this simple model. Foremost among these circumstances may be the fact that mantle viscosity must be strongly temperature-dependent because of the thermally activated nature of all solid-state creep mechanisms, which cause the viscosity to decrease at higher temperature. Hot upwelling plumes will therefore have anomalously low viscosity and, quite probably, enhanced ascent velocities relative to ambient mantle. Furthermore, and perhaps more important, the viscosity of the material that constitutes the cold horizontal lithospheric boundary layer will be extremely high. The plate-like nature of the horizontal velocity field involved in surface "plate-tectonics" exists as a consequence of the extremely high viscosity of the near-surface material (*see* PLATE TECTONICS). The issue of the influence of the surface plates upon the mantle convection process is one of the most outstanding and important in this area of geophysical science. Understanding of the mantle convection process, including the role of the plumes, will be incomplete until such time as a detailed understanding of the role of the surface plates is understood.

Of equal importance in distinguishing mantle convection from convection in a simple fluid in the laboratory, or in a computer simulation, concerns the fact that the mantle contains a number of phase transitions in which both the "packing" of the atomic constituents of the individual molecules and the dominant molecules themselves change as depth increases due to increasing pressure. The planetary mantle contains two principal transitions of this kind, one exothermic (heat-producing) and involving a transition of the mineral olivine to a spinel structure that occurs near 400-km depth, and a second that is endothermic (heat-absorbing) and involves a disproportionation reaction in which the spinel structure phase is further transformed into a mixture of perovskite and mangesiowüstite. The latter transition occurs near a depth of 660 km, a region known as the transition zone (*see* MANTLE STRUCTURE OF). Recent numerical simulations of the mantle convection process in the presence of these transitions demonstrate that the endothermic transition is by far the most important and that its influence may be sufficiently strong to cause the circulation to assume a predominantly layered style in which the regions above and below the 660-km horizon convect in separate cellular patterns with only weak mixing between them. The simulations show that the layered state is episodically disrupted, however, by intense "avalanches" of material that descend from the base of the "transition zone" into the lower mantle.

The idea of episodic layering that has been developed on the basis of computer simulations of convective mixing and the connected idea that layering may be strongly enhanced as a consequence of crustal differentiation early in Earth's history has reopened an old question: What is the source depth of the intra-plate thermal plumes that appear to be responsible for the creation of chains of volcanic islands such as the Hawaiian-Emperor chain in the Pacific? If the convective circulation is layered, then maximum depth of the plume source would be 660 km, whereas if the circulation is whole-mantle in style, then the maximum depth of the plume source would be the depth to the core-mantle boundary. The issue of "layered" or "whole-mantle" convection thus has critical implications for the understanding of thermal plumes and the surface "hot-spots" to which the plumes give rise when they impact the lithosphere. Surface plates and thermal plumes might usefully be viewed as defining distinct spatial scales of the mantle convective circulation.

Bibliography

ANDERSON, D. L. *Theory of the Earth.* Boston, 1989.
FAWLER, C. M. R. *The Solid Earth: An Introduction to Global Geophysics.* Cambridge, Eng., 1992.
PELTIER, W. R., ed. *Mantle Convection.* New York, 1989.

W. RICHARD PELTIER

MAPPING, GEOLOGIC

Perhaps the most important activity in the science of geology is the making of geologic maps. Geologic maps use a combination of colors, lines, and symbols to depict the distribution of the various geologic materials that make up the landscape. They contain descriptive information about the type of bedrock (sedimentary, igneous, and metamorphic rock), the character and origin of surficial deposits (water-lain alluvium, windblown sand dunes, glacial till, etc.), and the geologic structure of the rocks (faults and folds). A geologic map usually includes descriptive text that contains an interpretation of how the geologic materials are related in space and time. Two concepts that are widely used to determine the relative ages of geologic features are the principles of superposition and crosscutting relationships. The principle of superposition applies to materials originally deposited in horizontal layers, and states that younger deposits overlie older deposits. The arrangement of older to younger deposits in a chronological sequence is referred to as stratigraphy. The principal of crosscutting relationships states that a crosscutting feature, such as an igneous intrusion or a fault, is younger than the feature it cuts. Field observations of these two principles are essential to a synthesis of geologic history, which is one of the goals of geologic mapping (*see also* GEOLOGIC TIME).

Geologic maps (Figure 1) provide a basic framework of earth science data that is used for a variety of applications, including: (1) petroleum and mineral exploration; (2) location of construction materials; (3) identification of soils; (4) location of groundwater; (5) selection of transportation routes and land fills; and (6) identification of geologic hazards. Geologic mapping is conducted principally by the U.S. Geological Survey (USGS), state geological surveys, and universities, often at a scale of 1:24,000 using a topographic map of a 7.5-minute quadrangle as a base. These geologic quadrangle maps are often used in compilation of regional or state geologic maps at smaller scales such as 1:100,000 or 1:250,000, which, in turn, are compiled for the 1:2,500,000-scale geologic map of the United States.

Systematic geologic mapping of a 7.5-minute quadrangle (145 km^2) typically takes several months. The field geologist covers the area extensively by vehicle and on foot in search of exposures of bedrock and unconsolidated surficial deposits that overlie bedrock. Although many geologic maps display both bedrock and surficial units, field geologists tend to specialize in either bedrock or surficial geologic mapping, as these deal with two greatly different types of materials.

In bedrock geologic mapping, the fundamental source of information is exposed bedrock. Exposures of rock come in two main forms: (1) outcrops, which are naturally occurring; and (2) artificially created exposures such as roadcuts and quarries. Outcrops occur in a wide variety of settings. In the humid, low-lying areas of the southeastern United States, bedrock is found mainly in streambeds, whereas in the glaciated northeastern United States bedrock is commonly found on ridge crests. Outcrops are most abundant in the arid western United States and locally dominate the landscape, as anyone who has been to the Grand Canyon or Yosemite National Park can attest. Roadcuts along highways have become a vital additional source of data to twentieth-century geologists, particularly in the eastern United States where outcrops are less abundant. A roadcut can provide valuable stratigraphic and structural data where bedrock is not exposed.

In examining bedrock exposures during the creation of a geologic map, field geologists employ a variety of tools that include: (1) a USGS topographic map of a 7.5-minute quadrangle and/or steroscopic aerial photographs that are used for orientation and location of outcrops; (2) a notebook and pen for recording observations, measurements, and sketches; (3) a specially designed field compass for geologic mapping; (4) a hand lens for rock and mineral identification; and (5) a rock hammer for removal of samples or for exposing fresher rock. Other equipment may include a rucksack for carrying rock samples, food, and water; a medicine dropper and a bottle filled with dilute hydrochloric acid for determining the presence of calcium carbonate in rock; a small magnet for determining the presence of magnetite in rock; safety goggles; a measuring tape; and a camera. Field geologists usually wear rugged outdoor clothing, including sturdy hiking boots. A typical geologic mapping routine progresses as follows:

1. Note the location of the outcrop on the topographic map. Number the location on the map and in the notebook, creating what is called a field station.
2. Wearing safety goggles, hit the outcrop with the rock hammer to expose unweathered rock. Examine the fresh surface with a hand lens to identify the component mineral crystals. Enter in the notebook a description of the rock, including its color, grain size, mineralogy, and structure, and possibly a sketch of the outcrop.
3. Use the compass to measure the strike (direction) and dip (inclination) of bedding, foliation, and any other structures, and record these data in the notebook.
4. Break off a piece of the outcrop as a hand specimen, if desirable, and label the rock sample using the field station number.

Surficial deposits overlie bedrock and are geologically young, generally flat-lying, relatively thin, and poorly exposed. The means of examining and mapping these deposits are quite different from the means for mapping bedrock. Because they are unconsolidated, a shovel is more useful than a rock hammer in exposing fresh material. Surficial deposits can be examined in natural exposures such as stream incisions and landslides. Man-made exposures such as quarries, gravel pits, and excavations are also very helpful. These use of an auger or continuous-core drill may be necessary for adequate mapping of flat-lying surficial deposits in lowland areas. Deposits such as alluvium, colluvium, stream terraces, and glacial till are generally not well exposed, but they produce distinctive landforms that can be traced on topographic maps and aerial photographs.

Field geologists gradually construct a geologic map by accumulating data across the map area on traverses. The resulting map (Figure 1a) shows symbols representing structural data measured at each field station and contacts (thin black lines) between bedrock or surficial formations that have different ages and (or) characteristics. The accuracy of formation contacts on geologic maps is a function of the quantity and distribution of the field data. In areas of good exposure, contacts are likely to be more accurately located than in areas of poor exposure. For the latter, remote-sensing techniques such as airborne magnetic, gravity, or radioactivity surveys become useful tools to identify contacts.

Field observations are augmented by laboratory studies, which could include examination and identification of fossils, microscopic analysis of minerals in thin sections of rock samples (petrography), X-ray analysis of clay minerals, analysis of grain size of unconsolidated material, and geochemical analyses. Radioactive isotopes of elements such as uranium and potassium and their daughter products in rock samples can be measured to determine ages of igneous crystallization and metamorphic cooling (*see* GEOLOGIC TIME, MEASUREMENT OF).

One of the most important components on most published geologic maps is the cross section (Figure 1b), which is a diagram of a vertical slice through the earth showing subsurface geology. Cross sections add an important third dimension to the visual representation of the geology of a map area and better illustrate certain geologic structures such as folds and faults. A typical map may have two or more cross sections that extend hundreds to thousands of feet below the earth's surface. The locations of these diagrams are marked on the map by the use of section lines identified by letter symbols such as A-A′ and C-C′ (Figure 1a). Cross sections must match the surface geology mapped along the line of section, particularly the location and attitude of formation contacts. Below the earth's surface, cross sections become more interpretive and may portray structures such as folds or buried faults. Cross sections are important contributions to the understanding of the stratigraphy, structure, and geologic history of an area.

In preparing the geologic map for publication, geologists write a description of map units (Figure 1c) based on the field observations of each formation as well as petrography and laboratory analyses. Each description is keyed to the corresponding map unit by a box containing a letter symbol and a color that is assigned to the unit on the map (Figure 1c). An explanation of map symbols (Figure 1d) describes the structure symbols plotted at each field station. A correlation of map units (Figure 1e) shows the inferred stratigraphic or structural rela-

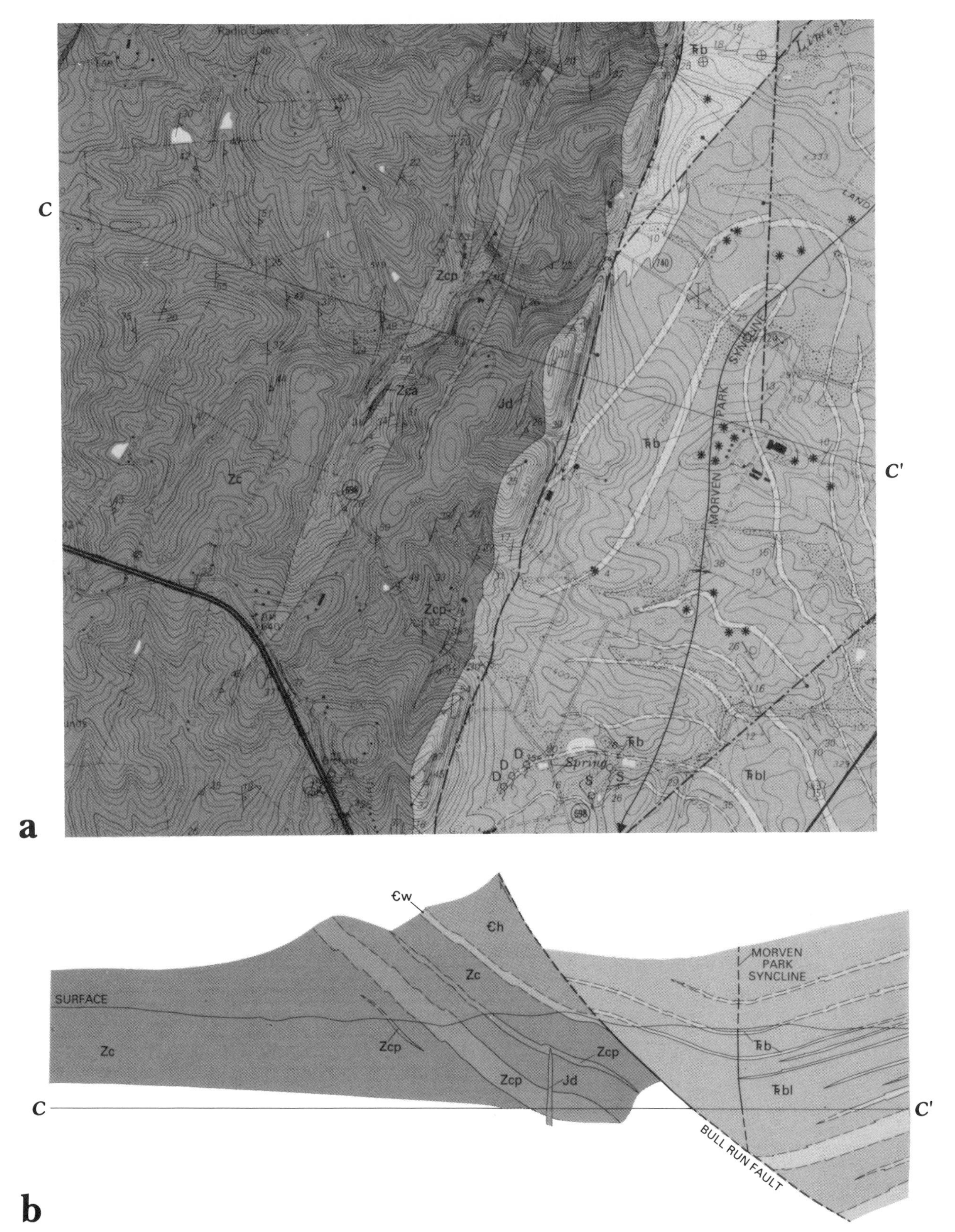

Figure 1. Elements of a geologic map. a. Portion of a geologic map showing location of cross section C-C′. b. Cross section C-C′. c. Description of map units. d. Explanation of map symbols. e. Correlation of map units.

DESCRIPTION OF MAP UNITS

MESOZOIC SEDIMENTARY ROCKS OF THE CULPEPER BASIN

TRb — **Balls Bluff Siltstone (Upper Triassic)**—Red to reddish-brown, massive to well-bedded, feldspathic,.calcareous sandy siltstone and lesser shale, with fine-grained sandstone at base. Gradational with underlying Poolesville Member of the Manassas Sandstone (TRmp) and intertongues laterally with Leesburg Member (TRbl); as much as 1,200 m thick

TRbl — Leesburg Member—Gray-weathering, crudely bedded carbonate conglomerate with conspicuous subangular to subrounded boulders, cobbles, and pebbles of gray and pink lower Paleozoic limestone and dolostone in reddish-brown pebbly sandstone and sandy siltstone matrix. Has minor partings and intercalations of calcareous sandstone and siltstone. Forms abundant hummocky outcrops. Interfingers with sandy Balls Bluff Siltstone to south and east; as much as 1,200 m thick

c

EXPLANATION OF MAP SYMBOLS

Contact—Dashed where approximate; dotted where under water, projected across metadiabase, or inferred in area of no exposure

Fault—Dashed where approximate; dotted where under water; unmarked where movement sense unknown

Strike-slip fault—Movement sense shown by arrows

Reverse fault—Sawteeth on upthrown side

Normal fault—Bar and ball on downthrown side

PLANAR FEATURES

(Where two or more planar symbols are combined, their intersection marks the point of observation)

Strike and dip of bedding—Ball indicates facing direction known from sedimentary structures

32 — Inclined

23 — Overturned

Horizontal

d

CORRELATION OF MAP UNITS

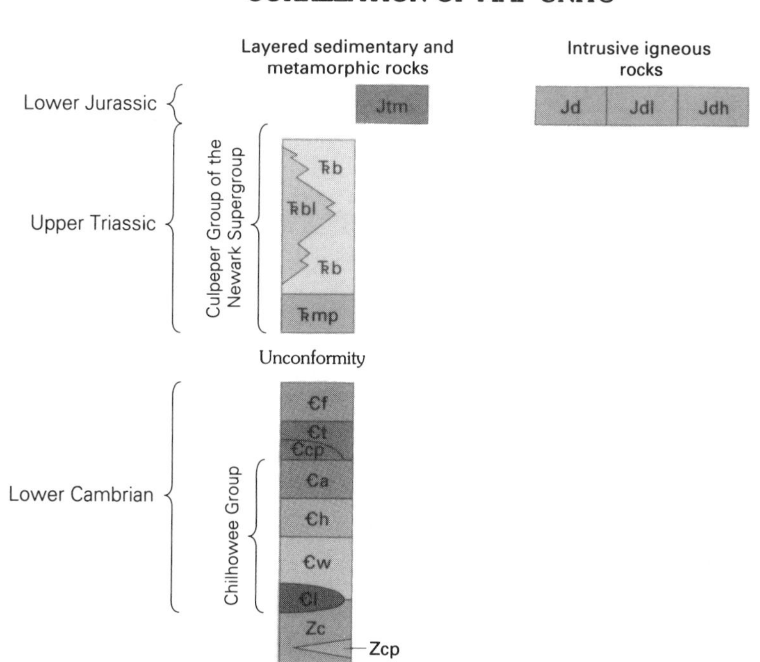

e

tionships among the different formations, determined by using the principles of superposition and cross-cutting relationships, and shows their positions relative to the geologic timescale.

Comprehensive geologic maps have an accompanying text that describes in more detail the various aspects of the geology shown on the map and cross sections, presents the results of structural or petrological analyses, and provides a synthesis of the geologic history of the area based on all of the available data. Information contained in geologic maps of small areas becomes part of the geologic synthesis of larger regions and allows scientists to achieve a greater understanding of the history of the earth.

Bibliography

Bernknopf, R. L., et al. *Societal Value of Geologic Maps.* U.S. Geological Survey Circular 1111. Denver, 1993.

Compton, R. R. *Manual of Field Geology.* New York, 1962.

———. *Geology in the Field.* New York, 1985.

Cvancara, A. M. *A Field Manual for the Amateur Geologist.* New York, 1985.

Thompson, M. M. *Maps for America.* Washington, DC, 1979.

WILLIAM C. BURTON
SCOTT SOUTHWORTH

MAPPING, PLANETARY

Maps are fundamental tools in any exploration, whether of Earth or of other planets. The first humans undoubtedly scratched maplike sketches in the dirt to show fellow hunters where to find game. Today, airline pilots and tourists alike decide their routes by using maps, and the success of their respective journeys depends on the quality of the maps they use. Similarly, scientists and engineers must use maps to plan their expeditions and to explain the data they collect.

There are many kinds of maps. Weather maps, for example, are especially important for some research, but they are valid only so long as the weather does not change. Jupiter, Saturn, Uranus, and Neptune are made of gases, whose patterns change constantly. This discussion is confined to the mapping of planets and moons with more stable surfaces.

The first step in making a planetary map is to derive a control net by computing the positions of selected features on the planet. One can do this by sketching a globe of latitudes and longitudes on overlapping global pictures. Each sketch is then refined until each surface feature appears at the same latitude and longitude on every sketch. In practice, this "sketching" is done mathematically by computers. Trajectory information from spacecraft navigation engineers and observations by Earth-based astronomers are also included in the computations.

Few photographs can properly be called maps. A photograph of the Moon taken through a telescope is a map of sorts because one can measure distances and directions with some precision. The Moon, however, always presents very nearly the same face to Earth. A collection of pictures taken from a spacecraft as it sped past a planet is a different dataset entirely! No two of these pictures have the same scale, and one cannot simply use a ruler to figure out the sizes and distributions of features as one could with an accurate map. Nevertheless, pictures taken from spacecraft are the fundamental data from which planetary maps are made. Each of the pictures must be transformed in shape from the spacecraft's perspective to the vertically overhead view of a map. Once this is done, the pictures can be assembled like a jigsaw puzzle into a mosaic that is consistent with the computed positions of features in the control network. An artist-cartographer may make a shaded-relief painting of the surface of the planet for some special maps.

Ideal maps show feature elevations. Radar scientists have compiled a comprehensive grid of topographic elevations on Venus by making altitude sounds with radar. This dataset resulted in the most uniform and complete topographic map ever made of any planet in the solar system (including Earth). Radar mapping has not yet been possible on other planets, however, and the stereoscopic (three-dimensional) pictures normally used for topographic mapping are difficult to take with planetary mapping cameras. The few topographic maps that have been made, however, provide intriguing information. The Olympus Mons volcano on Mars, for example, is nearly 26 km high (about three times as high as Mount Everest). The Vallis Marineris canyon on Mars is about 4 km deep, about two and a half times the depth of the Grand Canyon in Arizona. There are cliffs on Miranda, a satellite of Uranus, that are nearly 20 km high. Miranda is so tiny (484 km in diameter) and its gravity so weak that a careless astronaut who tripped at the top of one of the cliffs would fall for about eleven minutes. Although the fall would be slow, the impact at the bottom would be only a little less than that of a skydiver with a failed parachute on Earth.

Maps of spherical or spheroidal planets are familiar and well understood, but most asteroids and small satellites are not at all spherical. Mapping an asteroid would be like mapping a potato, if only someone would invent a "potatographic map projection." Scientists who study these bodies therefore rely on individual pictures, or, for analytical work, on digital maps. A digital map is made by combining pictures of the bodies with models of their shapes. With this kind of map, a computer can display a view from any point in space, whether a spacecraft had taken a picture there or not.

Geologic maps of planets show the area of coverage and the age relationships of various layers or "units" of materials. Each unit is color coded for clarity, making geologic maps some of the most colorful products of planetary exploration. Various lines and symbols are used to show the geologic structures that have shaped the surface. Extensive technical explanations of the probable nature and origin of the materials are given in the margin of the map.

Geologists make these maps to explain the processes that shape the surfaces of planets (*see* MAPPING, GEOLOGIC). Careful examination of images can reveal that a layer of "something" (figuring out what may come later) lies on top of a layer of "something else." The top layer must therefore have been deposited more recently than the one beneath it. The layer with the most meteorite impact craters is likely to have been exposed for the longest time. Piece by piece, scientist-detectives use clues like these to learn the history of the processes that shaped other worlds.

It is not enough to know a sequence of events, however. A map is required to show where each of these things happened, as well as when. The nature of a blanket of material on the flank of the giant lunar crater Copernicus has a different meaning than that of soil on a lunar plain far from any large crater.

We usually think of maps as large sheets of paper covered with lines and symbols. While there will always be a need for a basic set of these kinds of maps, cartographic datasets have become too large to manage on paper. Many terrestrial and planetary maps are therefore being made in digital form, for use with computer workstations. Maps of this kind are now being published on CD-ROM disks as well as on conventional paper formats.

Bibliography

GREELEY, R., and BATSON, R. M. *Planetary Mapping.* New York, 1990.

RAYMOND M. BATSON

MARS

Mars, named for the Roman god of war, has captured human imagination throughout the centuries and continues to play a dominant role in space exploration plans. The planet's reddish color, caused by abundant iron oxide ("rust") on its surface, has long drawn people's attention and has earned it its nickname of the "Red Planet."

Orbital and physical characteristics of Mars are listed in Table 1. Telescopic observations reveal

Table 1. Orbital and Physical Characteristics of Mars

Orbital Properties	
Average distance from the Sun	228×10^6 km
Maximum distance from the Sun	249×10^6 km
Minimum distance from the Sun	206×10^6 km
Minimum distance from Earth	55×10^6 km
Orbital eccentricity	0.0934
Orbital inclination	1.8504°
Orbital period	687 days
Physical Properties	
Rotation period	24h 37m 22.7s
Mass	6.42×10^{23} kg
Equatorial diameter	6,780 km
Mean density	3.93 g/cm^3
Obliquity of rotation axis	25.19°
Surface gravity	3.7 m/s^2

the existence of permanent bright and dark regions (albedo features), bright polar caps, and a thin atmosphere composed primarily of carbon dioxide in which clouds of water vapor and dust occasionally are observed. The planet is surrounded by two small, irregularly-shaped moons called Phobos and Deimos (*see* SATELLITES, SMALL).

The Martian Atmosphere

Mars is surrounded by a thin atmosphere composed of carbon dioxide (95%), nitrogen (2.7%), and argon (1.6%), with traces of oxygen (O_2), water, neon (Ne), and ozone (O_3). The actual amounts of some species vary due to the seasonal cycles of carbon dioxide (CO_2), water (H_2O), and dust as the polar caps vary in size. The atmospheric circulation is driven primarily by the seasonal transfer of CO_2 and H_2O between the poles, atmosphere, and surface. The major effects of weather systems on Mars are changes in temperature and pressure. The atmospheric pressure on the Martian surface is only 6 millibars (about one-hundredth that of Earth at sea level), precluding the existence of liquid water on the surface under present conditions.

Winds associated with solar heating of the surface and atmosphere can be strong enough to entrain and carry dust. Most of these dust storms cover only localized regions of the planet, but occasionally some can grow into global-wide storms. The global-wide dust storms often occur during perihelion, when the southern hemisphere is experiencing summer. Global dust storms can occur for several months and, upon dissipation, drop dust over the entire planet, dramatically reducing the contrast between regions of varying albedo.

Life on Mars?

Earth-based telescopic observations revealed that Mars has several Earth-like properties, and many people over the years have believed in the possibility of martian life. The belief in martian life reached its peak around the turn of the twentieth century with the controversy over the existence of canals, thin dark lines seen on the martian surface by some astronomers and interpreted to be waterways constructed by intelligent martian life. The canal controversy was not resolved until spacecraft images of the planet became available in the 1960s and 1970s. Spacecraft imagery revealed no evidence of past or present martian life. The canals are nothing more than optical illusions and the seasonal albedo changes previously attributed to vegetation were found instead to result from the shifting of dust by winds.

The search for life on Mars continued, however, because the discovery by *Mariner 9* of water-formed channels on the planet's surface led to renewed hope of microbial life forms. In 1976, the United States landed two robotic laboratories, the *Viking 1* and *Viking 2* Landers, in the northern hemisphere of Mars to search for life. The *Viking*

Table 2. Successful Spacecraft Missions to Mars

Spacecraft	*Country*	*Launch*	*Comments*
Mariner 4	USA	11/28/64	Flyby 7/14/65
Mariner 6	USA	02/24/69	Flyby 7/31/69
Mariner 7	USA	03/27/69	Flyby 8/05/69
Mars 2	USSR	05/19/71	Orbit insertion 11/27/71 Lander crashed
Mars 3	USSR	05/28/71	Orbit insertion 12/2/71 Lander crashed
Mariner 9	USA	05/30/71	Orbit insertion 11/14/71 7329 pictures
Mars 5	USSR	07/25/73	22 orbits; 60 pictures
Viking 1	USA	08/20/75	Orbiter 6/19/76–8/07/80 Lander 7/20/76–11/13/82 22.5°N 48.0°W
Viking 2	USA	09/05/75	Orbiter 8/07/76–7/25/78 Lander 9/03/76–4/11/80 48.0°N 225.7°W
Phobos 2	USSR	07/12/88	Orbited 01/29/89–03/27/89

Landers scooped up soil samples with their robotic arms and tested these samples for metabolic products indicative of life. The results of the *Viking* experiments were enigmatic, but most scientists think that the observed activity was the result of chemical reactions with the oxidized surface material and not due to organic activity. Thus, the *Viking* experiments most likely suggest that life does not currently exist on the martian surface. Future missions to Mars undoubtedly will search for fossil evidence of life in some of the older stratified rocks.

The Surface of Mars

Ten orbiter missions and two landers have successfully visited Mars (Table 2) and provided information about its surface characteristics. Observations from early flyby missions revealed a heavily cratered surface in the southern hemisphere. However, orbital images from *Mariner 9* showed that while 60 percent of the martian surface is heavily cratered, the remaining 40 percent of the surface has experienced recent geologic activity. Among the geologic units identified on Mars are cratered terrains, volcanic provinces, canyons, and water-carved channels. The channels in particular were a surprise to planetary scientists since current climatic conditions preclude the existence of liquid water on the martian surface.

The distribution of geologic units is not random across Mars (Figure 1). Much of the southern hemisphere is covered by heavily cratered terrain, indicating an ancient formation age for this material. Much of the northern hemisphere has fewer craters and is thus younger. In addition, the northern hemisphere is about 3 km lower than the southern hemisphere. The two hemispheres are separated

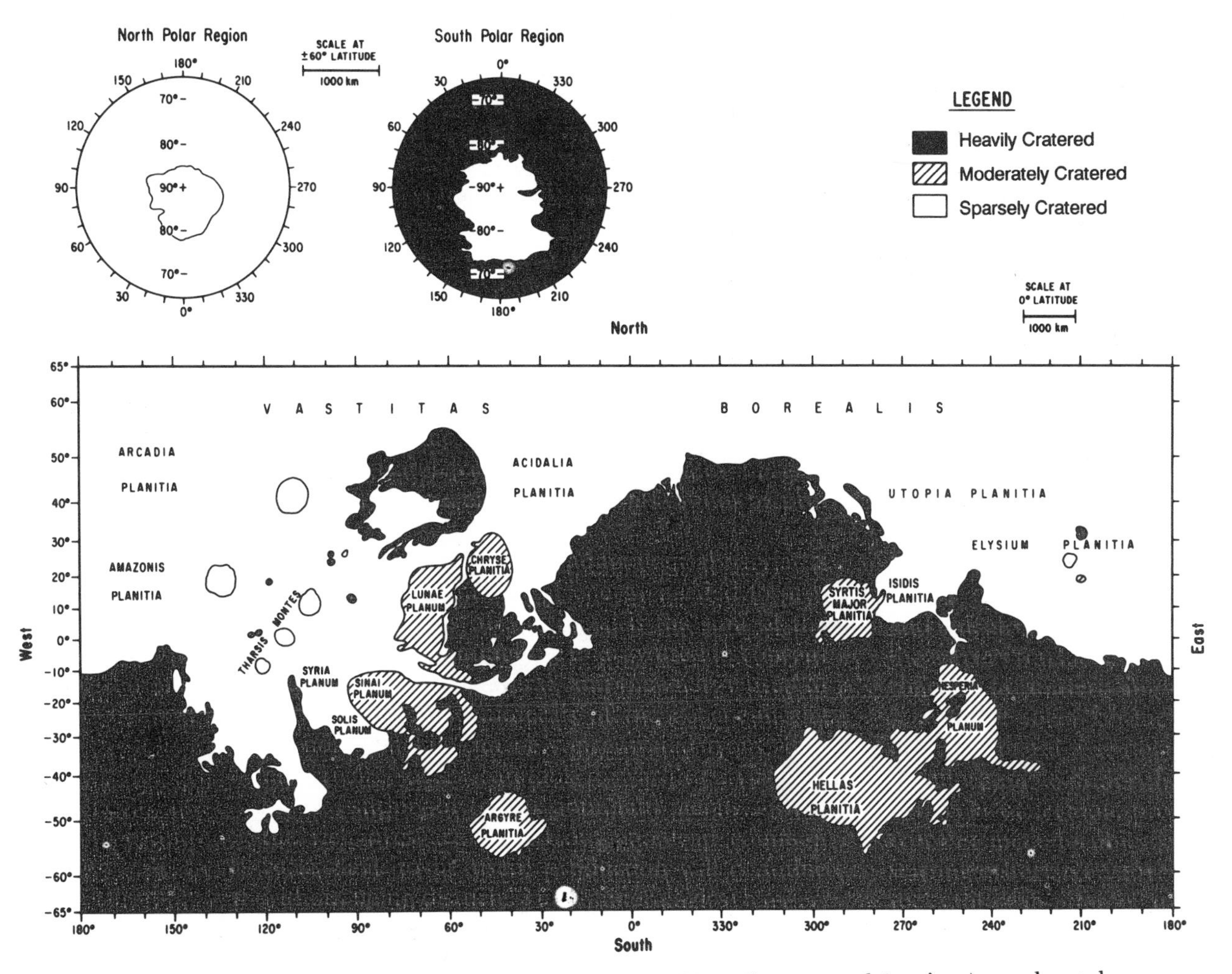

Figure 1. This map shows the nonuniform distribution of heavily cratered (ancient), moderately cratered (intermediate aged), and sparsely cratered (young) terrains on Mars. (From Barlow, 1987.)

in many places by a kilometer-high escarpment that extends around the planet tilted at a 35° angle from the equator. The cause of this hemispheric dichotomy is unknown—theories include a 7,000-km-diameter impact basin in the northern hemisphere and the arrangement of internal convection cells.

Impact craters on Mars are similar to impact craters elsewhere in the solar system, with a few important exceptions. Craters range in diameter from 50 m (projectiles forming smaller craters burn up entering the martian atmosphere) to over 2,000 km. Small craters (smaller than 6 km diameter) show a simple bowl-shaped appearance, whereas larger craters display more complex morphologies with terraced walls, central peaks or pits, and shallow floors. Ejecta blankets surrounding fresh martian craters in the 6- to 50-km-diameter range tend to be lobate in form (Figure 2) rather than radial as seen on the Moon. Either the atmosphere or vaporization of subsurface ice causes the fluidized emplacement of these features.

A number of volcanic features are evident in the *Mariner 9* and *Viking Orbiter* images, including large expanses of flood basalts, huge shield volcanoes, strange paterae (ancient volcanoes), and

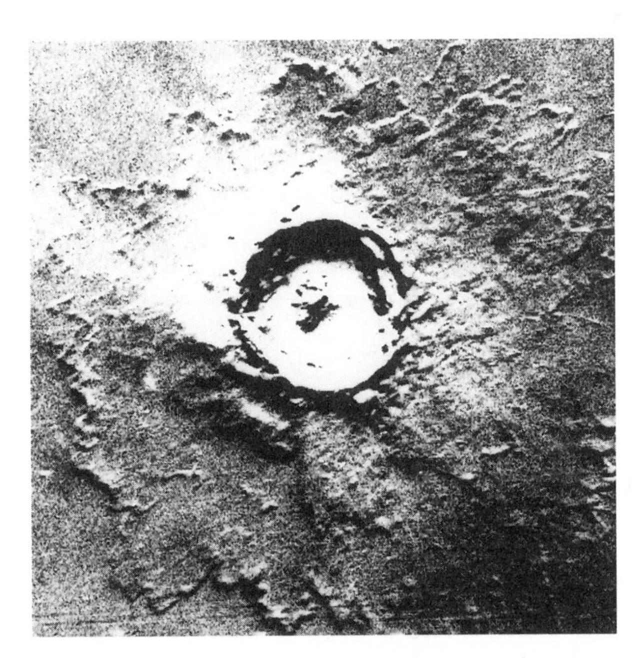

Figure 2. Fresh martian impact craters often show a lobate ejecta blanket, indicating formation by fluidized processes. This crater is approximately 15 km in diameter.

small cones and domes. The largest volcanic complex on Mars is located just north of the dichotomy escarpment in the Tharsis Province, where over ten huge volcanic constructs are found. Four shield volcanoes dominate the region, with the largest, Olympus Mons, rising over 25 km above the surrounding plains and extending over 600 km in diameter (Figure 3). These volcanoes were formed by effusive basaltic eruptions, similar to those that formed the Hawaiian Islands. Smaller domes dot the Tharsis landscape as well as the Elysium region in the eastern hemisphere of the planet. The highlands region around the Hellas Basin contains a number of huge, low-relief, heavily eroded, ancient volcanoes (Figure 4). No terrestrial analog is known for paterae, but most scientists believe they formed from explosive eruptions when magma encountered near-surface water reservoirs early in martian history. Volcanism on Mars seems to have reached a peak around 3×10^9 years ago and has declined in extent since then. It is unknown if young volcanoes like Olympus Mons are still active.

Tectonic activity is largely concentrated in the Tharsis region of Mars, where an extensive array of fractures have formed over the course of billions of years. The fractures are oriented in different directions, indicating changes in the stress fields over time, and are particularly well developed in the northern and southern areas of the Tharsis Province. Stretching over 4,000 km to the east of the Tharsis volcanic province is a huge canyon system called Valles Marineris (see Plate 27). The system consists of about ten canyon segments extending along the equatorial region between longitudes 30°W and 110°W. Individual canyons can be over 200 km wide and 7 km deep; many have huge landslides on their floors. The system has developed through a combination of faulting, mass wasting, and possibly water action.

The east end of the Valles Marineris canyon system debouches into large channels whose morphologies suggest formation by flowing water (Figure 5). These large channels, called outflow channels, originate in areas of disrupted terrain called chaos. Scientists believe that water burst through the surface in the regions, forming the outflow channel and leaving behind the collapsed surface material. Outflow channels are relatively young features and probably formed under present climatic conditions. Since liquid water is currently unstable on the surface, this implies that

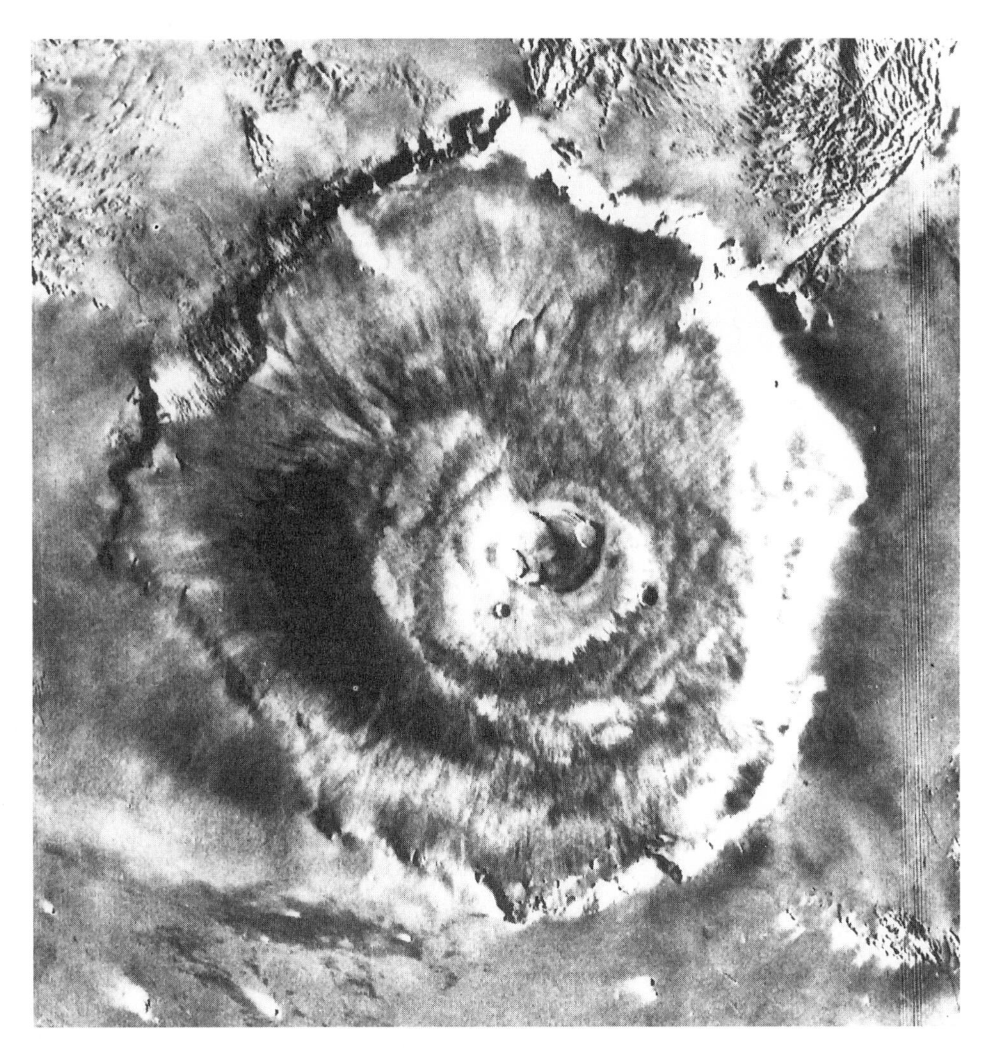

Figure 3. Olympus Mons is the largest volcano in the solar system. It is similar in shape to the Hawaiian shield volcanoes, but it is three times higher than Mauna Kea and is larger than the entire island of Hawaii.

each outflow channel formed in only a few days before the water evaporated.

Smaller channels, called valley networks, are found in the heavily cratered southern highlands and on the flanks of some of the volcanoes. These channels appear similar to terrestrial runoff channels and initial theories proposed a thicker ancient martian atmosphere which could have sustained rainfall. Scientists now believe these channels were formed by a process called sapping, where the removal of groundwater causes collapse of overlying terrain. Rainfall is therefore not necessary to explain these features.

The composition of the polar caps was a source of much controversy until the *Viking Orbiter* missions in the late 1970s. *Viking Orbiter* measurements indicated that the seasonally varying portion of each cap is composed of CO_2 ice. The remnant cap in the northern hemisphere is too warm to contain CO_2 ice in the summer so scientists believe it is composed of H_2O ice. The southern polar cap may or may not contain a remnant cap of H_2O.

Temperature differences in the thin martian atmosphere cause wind, which can transport dust and sand between locations. Several types of eolian features have been identified in both orbital and ground-based images of the planet, including sand dunes, drifts, ergs (vast areas covered by deep sand deposits and dunes), and wind streaks. Thick layered deposits surrounding the polar caps are be-

Figure 4. Tyrrhena Patera, located at 23°S 253°W, is an example of an ancient highlands patera. These features have no terrestrial analog but probably formed by explosive interaction between magma and water early in martian history. The broad channel emanating from the central caldera is about 5 km wide.

lieved to result from eolian deposition cycles caused by oscillations in the planet's orbital obliquity and eccentricity.

Composition of the Martian Surface

The lack of samples returned to Earth from either robotic or human missions to Mars precludes detailed analysis of the mineralogic composition of the planet's surface. However, a combination of Earth-based and Mars-orbit remote sensing techniques, *Viking* Lander results, and the SNC (shergottite, nakhlite, and chassignite) meteorites (*see* METEORITES FROM THE MOON AND STARS) have provided scientists with some clues as to the composition of the martian soil. The classic dark areas of the planet are believed to be composed primarily of pyroxenes, whereas the brighter regions are thought to consist of palagonite (basalt which has erupted in the presence of water). The existence of clays, carbonates, and crystalline minerals is still controversial, although water has been detected in the soil. Much weathering, both physical and chemical, has affected the size and composition of the martian surface material. The soil tends to be composed of fine-grained material that has been homogenized across the planet by the global dust storms. Clumping of the soil into clods called duricrust probably results from a high concentration of salts in the soil.

Interior Structure of Mars

Little is known about the internal structure of Mars due to the lack of seismic data for the planet. However, scientists expect that Mars is layered in shells in the same manner as Earth. The outer crust is only a few tens of kilometers thick, based on gravity and tectonic information, and is probably composed primarily of basaltic rocks (as indicated by the types of volcanic structures seen on the surface). Underlying the crust is a thick iron-rich mantle, and the planet's center probably consists of a small, high-density core composed of iron with minor amounts of nickel, sulfur, and oxygen.

Mars has a lower mean density than the other terrestrial planets: 3.9 g/cm^3 for Mars versus 5.2 g/cm^3 for the other terrestrial planets. Compression effects in the larger terrestrial planets like Earth and Venus can account for some of this difference in density, but composition probably also contributes. Scientists believe that the materials in the martian interior are more oxidized, and therefore less dense, than the materials comprising the interiors of the other terrestrial planets. The existence of higher concentrations of oxygen at the location of Mars is predicted by cosmochemical models of the distribution of elements at the time of planetary formation.

Future Exploration

Mars has figured prominently in the history of solar system astronomy and continues to be a major goal in present space exploration scenarios. With the loss of the *Mars Observer* mission just prior to orbit insertion around the planet in 1993, the United States is now discussing sending two *Mars Surveyor* missions to Mars to complete the geochemical and atmospheric studies that were to have been accomplished by *Mars Observer.* The Russians will emplace small landers and penetrators on the martian surface with their *Mars-96* mission, to be followed by the *Mars-98* mission that will

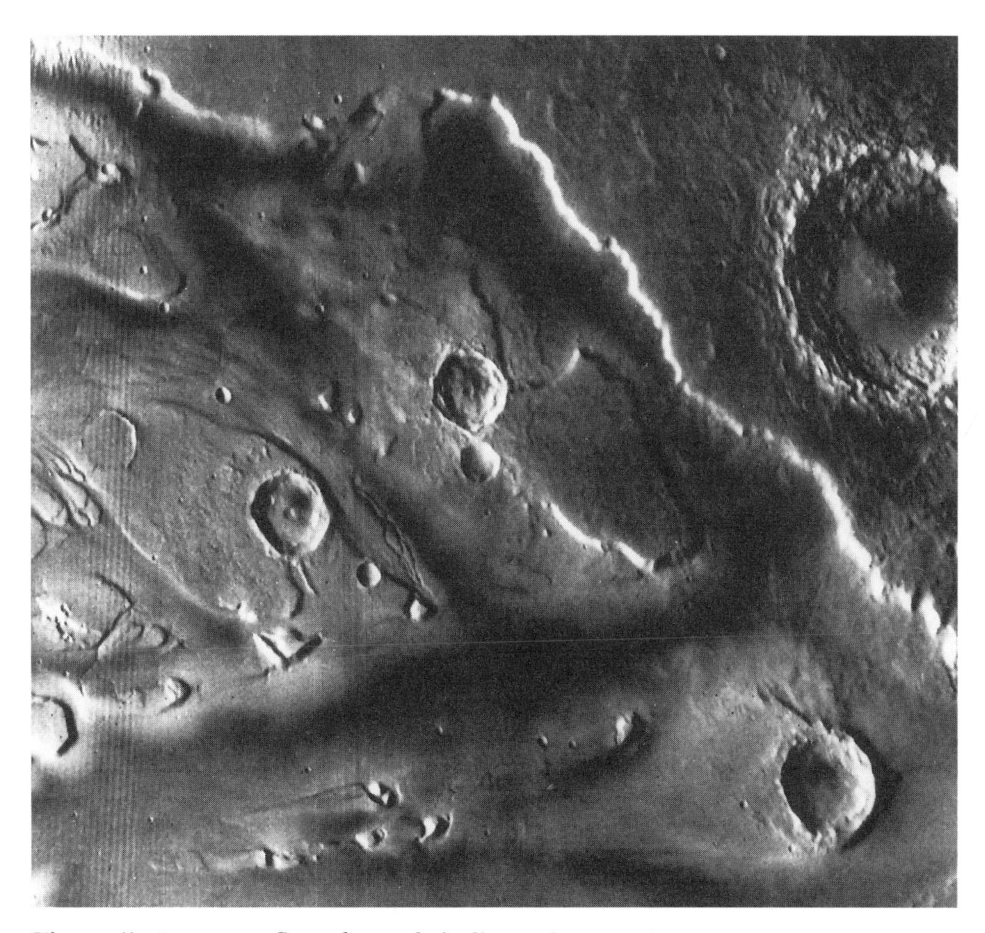

Figure 5. Large outflow channels indicate that massive floods of water once surged across the martian surface.

land rovers on the surface and send balloons into the atmosphere. The United States is developing plans for the *Mars Pathfinder,* which will land a rover on the martian surface and sample the chemical composition of soil and rocks within a few tens of kilometers from the landing site. Other countries, such as Japan and the members of the European Space Agency, also are interested in missions to Mars and may either fly spacecraft to the planet independently or perhaps jointly develop a network of small landers. Rovers, sample return missions, and eventually human exploration are future goals in our continuing study of the Red Planet.

Bibliography

CARR, M. H. *The Surface of Mars.* New Haven, CT, 1981.

———. "Mars." In *The New Solar System,* J. K. Beatty and A. Chaikin, eds. Cambridge, MA, 1990.

CATTERMOLE, P. *Mars: The Story of the Red Planet.* New York, 1992.

KIEFFER, H. H.; B. M. JAKOSKY; C. W. SNYDER; and M. S. MATTHEWS. *Mars.* Tucson, AZ, 1992.

NADINE G. BARLOW
JOSEPH BOYCE

MASS WASTING

See Landscape Evolution; Landslides and Rockfalls

MEINZER, OSCAR E.

Oscar Edward Meinzer (1876–1948) is known by scientists throughout the world as the father of groundwater science. When we think about the almost daily news about groundwater, either its supply or its contamination or conflicts concerning it, on an international basis or between groups within a country, it is difficult to realize that there was a time during this century when groundwater was little appreciated as a resource. Meinzer's major contributions included making the public, politicians, and policy makers aware that groundwater was a basic resource worthy of assessment, evaluation, and scientific study. He also recognized that aquifers are functional components of the hydrologic cycle and that the principles of occurrence, movement, and chemistry of groundwater must be incorporated into the concepts of earth science. He achieved these goals through his innate abilities as a creative scientist, an able administrator, and a proven leader.

Born on 28 November 1876 on a farm near Davis, Illinois, Meinzer was one of six children of William and Mary Julia Meinzer, German immigrants who sought to escape the oppressive rule of the Prussians. He graduated magna cum laude from Beloit College in Wisconsin in 1901. Meinzer served as principal of a public school at Frankfurt, South Dakota (1901–1903), and as a science teacher at Lenox College at Hopkinton, Iowa (1903–1905), where he met his future wife, Alice Breckinridge Crawford, whom he married in October 1906.

Meinzer joined the U.S. Geological Survey in 1907 and was appointed acting chief of the Division of Ground Water in 1912; he was made chief a year later and held that position until his retirement in 1946. He possessed a keen analytical mind and an unusually broad scientific outlook—the two traits that would see him make remarkable contributions during a new era marked by rapidly growing awareness of the value of natural resources and the challenge of developing and conserving them. During more than forty years of field investigations, research, and scientific leadership, he transformed the study of groundwater from a geologic and engineering sideline to a fully recognized science. He and his staff developed two perspectives on groundwater investigations—that of the geologist concerned primarily with description of the rocks and sediments and that of the engineer or physicist who focused primarily on the dynamics of fluid motion. He broadened the scope of groundwater investigations to take into account all facets of the hydrologic system that affected development and management of groundwater as a resource (*see* WATER QUALITY; WATER SUPPLY AND MANAGEMENT; and WATER USE).

Early in his professional career, he foresaw the importance of groundwater in the national economy and concluded that defining the areal extent, thickness, depth, geologic structure, and physical characteristics of aquifers would be major activities for groundwater scientists. He foresaw that these scientists also must learn about the chemical quality and the quantity of groundwater that would be available for beneficial use, and that they devise and test methods for determination of the quality and quantity of groundwater. He developed programs to measure water levels, pumpage from wells, and the discharge from springs for use in estimating the amount of water moving through a groundwater system. His "Outline of Methods for Estimating Ground-Water Supplies," first presented in 1920 but later revised and published in 1932, was the first clear statement of the various methods for estimating the yield of groundwater basins. His dissertation for the doctoral degree granted by the University of Chicago in 1923, "The Occurrence of Ground Water in the United States, with a Discussion of Principles," was, in effect, the first textbook on groundwater and still remains a major reference. He established a hydrologic laboratory to study the permeability and specific yield of rocks and the laws of fluid dynamics. He was the first to recognize the elasticity and compressibility of artesian aquifers and the significance of these characteristics in relation to the storage and release of artesian water (Meinzer, 1928). He led the way in the United States in the study of saltwater encroachment into coastal aquifers.

Although he was a shy and retiring man, he enjoyed people, especially young people, and was active in his church and a leader of the Boy Scouts of America movement nearly from its beginning. He had a keen orderly mind and an unusual gift for concentration that permitted him to work at a prodigious pace. He was a gifted writer and, although burdened with administrative and supervisory duties, he authored and coauthored about 110 reports and articles dealing with groundwater. He was a slow deliberate speaker and refused to indulge in any form of propaganda or self-aggrandizement. He was extremely active in national and international professional societies and served as an officer in many of them. For example, he was among the chief organizers of the Section of Hydrology of the American Geophysical Union and served as first chairman. He was president of the Society of Economic Geologists in 1945, president of the International Commission on Subterranean Water (1936–1948), and president of the American Geophysical Union at the time of his death. He was awarded the Bowie Medal of the American Geophysical Union in 1943, one of only a few hydrologists ever to receive this honor. He received an honorary doctorate from his alma mater, Beloit, in 1946, and was a member of the Cosmos Club of Washington, D.C.

In summary, Meinzer not only developed and defined the science of groundwater but also organized and trained a competent group of scientists and engineers who, working with him, were able to give the science practical value and would ensure its continuing development and use after he had finished his work. Meinzer's was a "monument of achievement which will endure" (Sayre, 1949). Up to the time of his death, he was mentally alert, apparently in good health and spirits. When preparing a treatise about groundwater and anticipating a scientific trip to Europe he died peacefully during an afternoon nap on 14 June 1948.

Bibliography

HACKETT, O. M. "The Father of Modern Ground-Water Hydrology." *Ground Water* 2, no. 2 (1964):2–5.

MEINZER, O. E. *The Occurrence of Ground Water in the United States, With a Discussion of Principles.* U.S. Geological Survey Water-Supply Paper 489. Washington, DC, 1923.

———. *Outline of Ground-Water Hydrology, with Definitions.* U.S. Geological Survey Water-Supply Paper 494. Washington, DC, 1923.

———. "Compressibility and Elasticity of Artesian Aquifers." *Economic Geology* 23 (1928):263–291.

———. *Outline of Methods for Estimating Ground-Water Supplies.* U.S. Geological Survey Water Supply Paper 638C. Washington, DC, 1932.

PARKER, G. G. "Early State of Hydrogeology in the United States, 1776 to 1912." *Water Resources Bulletin* 22, no. 5 (1986):701–716.

SAYRE, A. N. "Memorial to Oscar Edward Meinzer." In *Proceedings of the Geological Society of America.* Boulder, CO, 1949.

WILLIAM BACK

MERCURY

Mercury is the most elusive of the planets visible to the naked eye, partly because it is the closest planet to the Sun and partly because it is the second smallest planet in the solar system (only Pluto is smaller). Many people live their entire lives without ever seeing the planet—the famous astronomer Nicolaus Copernicus is said to have been one of those. Even in the space age, Mercury remains elusive, having been visited by just one spacecraft mission, which imaged only 45 percent of its surface. Many questions still remain about this small, enigmatic world.

Mercury is the innermost of the planets in our solar system, orbiting at an average distance of 57.9 million kilometers (km) from the Sun. Mercury's orbit, however, is one of the most elliptical of all the planetary orbits, so its actual distance from the Sun ranges from 46 million km to 70 million km. As such, the amount of sunlight that the surface receives varies considerably, with the light being about 2.3 times more intense when Mercury is closest to the Sun than when it is furthest away.

Mercury has a diameter of only 4,880 km. It has no moons circling it and has a mass that is only 5.5 percent that of Earth. However, Mercury has an amazingly large density of 5,400 kilograms/cubic meter (kg/m^3), similar to that of Earth, indicating that it is a rocky body with a large core and high concentration of heavy elements such as iron (Fe).

Studying Mercury is very difficult because of its small size and the fact that it never is seen very far from the Sun. The maximum angular distance between Mercury and the Sun is only 28°, less than one-third of the distance between the horizon and the zenith (point directly overhead). Thus, Mercury can never be observed in completely dark skies, only during twilight or during the daytime. In the late 1800s astronomers tried to unravel some of the mysteries of Mercury, including its rotation period. They believed they could identify faint markings on the surface of the planet and, by timing how long it took these features to reappear, they estimated that the rotation period of Mercury was the same as its orbital period of eighty-eight Earth days. This rotation period was accepted until 1965, when radar signals were bounced off the surface of the planet. Changes in frequency of the reflected radar signals can provide information about how fast the planet is rotating. The results indicated a rotation period of about fifty-nine Earth days.

Later observations further refined the rotation period to 58.65 days and the orbital period of the planet to 87.97 days. Astronomers quickly noticed that three rotations of Mercury exactly equal two orbits of the planet. This is known as a 3:2 spin-orbit coupling and is caused by tidal forces from the Sun acting on Mercury. This is a very stable configuration and Mercury is the only planet in the solar system to exhibit this type of spin-orbit coupling.

The only spacecraft that has visited Mercury was *Mariner 10* in 1974–1975. *Mariner 10* was a flyby mission of the planet, but the spacecraft went into an orbit around the Sun that permitted it to make three Mercury passes before its instruments ceased operating. The first encounter between *Mariner 10* and Mercury was in March 1974, the second was in September 1974, and the final encounter was in March 1975. During the three encounters, *Mariner 10* explored the space environment surrounding Mercury and took photos of the planet's surface. Most of the information we have about Mercury today is a result of this single mission.

Mariner 10 revealed that Mercury does not retain an atmosphere. This was not surprising since the planet is so small that it is easy for gases to reach escape velocity and disperse into space. Recent Earth-based observations of Mercury, however, show that a thin haze of hydrogen (M), helium (He), sodium (S), and potassium (K) atoms exist just above the planet's surface. The hydrogen and helium are solar wind particles captured by Mercury's gravity as the solar wind streams past the planet. They quickly escape to space. The sodium and potassium atoms are produced by the collision of solar wind particles with the mercurial surface, knocking these atoms off the surface rocks. These atoms remain in the mercurial atmosphere for only a short time before collapsing back onto the surface.

The lack of an atmosphere and the proximity of the planet to the Sun combine to give Mercury a very wide range of surface temperatures. When Mercury is closest to the Sun, its noontime temperature can reach 700 K. With no atmosphere, however, the heat quickly escapes to space, so nighttime temperatures get down as low as 100 K. The wide variation in surface temperature is the largest among the planets in the entire solar system.

Mercury's rotation axis is oriented perpendicular to its orbital plane, so Mercury does not experience seasons. This also implies that there are craters at the polar regions of the planet whose floors may be in permanent shadow. In recent years, Earth-based radar studies of the polar regions of

Mercury have revealed patches of a highly reflecting substance that may be ice lying in the permanently shadowed floors of polar impact craters. The ice may have been deposited during impacts of icy comets with the planet. Although controversy still exists about the material causing these strong radar reflections, it is intriguing to think of ice existing on the surface of the planet closest to the Sun! It is possible that similarly sheltered polar caps on the Moon (*see* MOON) also contain water ice. Water ice could provide, of course, resources for manned outposts on these bodies.

A magnetometer onboard *Mariner 10* revealed yet another surprise—Mercury has a weak magnetic field. Neither Mars, which is slightly larger than Mercury, nor the Moon, which is slightly smaller, have magnetic fields, so it was expected that such a small, slowly rotating planet as Mercury would not have one either. The field is very weak, only about 1 percent of the strength of Earth's field, but strong enough to form a bow-shaped shock wave, a magnetosheath, and a magnetosphere (*see* PLANETARY MAGNETIC FIELDS). It is a dipolar field, inclined about 11° from Mercury's rotation axis. The field is likely produced by an iron core in Mercury, but whether the field is being actively generated by motions in a liquid core or if it is a remnant field from a solid iron core is still open to debate.

Mariner 10 photos of the 45 percent of the mercurial surface that was imaged reveal a very heavily cratered body, similar in appearance to the lunar highlands (*see* MOON). The large number of impact craters indicates that the surface of Mercury formed very early in solar system history and has experienced little geologic activity since its formation. A few large impact basins are seen on the planet's surface, the largest being the 1,300-km-diameter multiple-ring structure called Caloris. Some lava flows are seen on the floor of the Caloris Basin, but all are very light in color and very ancient. No young flows comparable to the lunar maria are seen on Mercury.

Directly antipodal to the Caloris Basin is a region of broken up, hilly terrain. This region is officially called hilly and lineated terrain, although some scientists simply call it "weird" terrain. Scientists believe that when the impact that created the Caloris Basin occurred, most of the interior of Mercury was still hot and molten. Seismic waves from the impact propagated through the molten interior and were focused at the antipodal point, heaving up the surface in that area. The immense seismic forces caused the surface to break up into hills and valleys, creating the hilly and lineated terrain that we see today.

Mariner 10 images also revealed the presence of sinuous faults crossing the mercurial surface. These sinuous features, called lobate scarps, are thrust faults, caused by compression of the surface. The scarps vary in length from 20 km to over 500 km and have heights of up to 1 km. The presence of these scarps over the entire surface imaged by *Mariner 10* suggests they resulted from global-wide compression of the planet. Analysis of the size of the scarps indicates that Mercury has shrunk between 1 and 3 km in radius since the outer crust solidified.

What caused this dramatic shrinkage of Mercury? Scientists believe the key lies in the interior of the planet. The high density and the presence of a magnetic field suggest that Mercury contains a very large iron core. Models suggest that this core occupies about 40 percent of the entire volume of Mercury, compared to only 16 percent of the volume taken up by Earth's core (*see* CORE, STRUCTURE OF). Early in Mercury's history, the interior was very hot and the iron core was molten. The outer crust of Mercury cooled quickly and began retaining the scars of impact events shortly after the planet formed, while the interior stayed hot and molten. Over the past 4.5 billions years (Ga), the iron in Mercury's core had cooled and begun to solidify. As iron changes from a liquid to a solid, its volume decreases, so the interior of Mercury shrank. The crust of Mercury was thus exposed to compressive forces as the interior shrank, causing breaks in the crust that we see today as the lobate scarps. Volcanic activity on the surface was limited to early in the planet's history before the interior conduits for magma were closed off by shrinkage of the planet. Thus Mercury has been essentially a geologically dead world for about the past 4 Ga.

From the outside, Mercury looks very much like the heavily cratered lunar highlands. We now know, however, that the large iron core inside the planet has caused this small world to evolve very quickly and along a very different tectonic path than either the Moon or any other object within the solar system. We have learned a substantial amount about this world from having seen less than half of its surface. Who knows what analysis of the unseen 55 percent of the planet might reveal.

Bibliography

BEATTY, J. K., and A. CHAIKIN. *The New Solar System.* Cambridge, Eng., 1990.

MURRAY, B., M. C. MALIN, and R. GREELEY. *Earthlike Planets: Surfaces of Mercury, Venus, Earth, Moon, and Mars.* San Francisco, 1981.

STROM, R. G. *Mercury: The Elusive Planet.* Washington, DC, 1987.

VILAS, F., C. R. CHAPMAN, and M. S. MATTHEWS. *Mercury.* Tucson, AZ, 1988.

NADINE G. BARLOW

METALLIC MINERAL DEPOSITS, FORMATION OF

Approximately twenty metallic chemical elements are made widely available through the mining and processing of their ores—examples include iron (Fe), aluminum (Al), copper (Cu), zinc (Zn), and lead (Pb). Many additional metals are produced on a limited basis, principally as by-products—examples include cadmium (Cd), a by-product of zinc mining, and arsenic (As), a by-product of copper mining.

Metals can be separated into two groups, the geochemically abundant metals and the geochemically scarce metals. Among the abundant metals, six are of technological importance: aluminum, iron, magnesium (Mg), manganese (Mn), titanium (Ti), and silicon (Si); all have clarkes greater than 0.1 percent (the weight percentage of an element in the crust is termed its clarke). All of the other chemical elements with metallic properties—of which there are about forty—have clarkes below 0.1 percent and are said to be geochemically scarce.

Ore Deposits

A metallic ore deposit is a local concentration of one or more minerals that can be profitably mined in order to recover the contained valuable metal. The profitability of ore deposits of both the abundant and the scarce metals is determined by the same three factors:

1. The desired metal must be present in a mineral from which it can easily be recovered by smelting or some other inexpensive process. Most of the minerals in the earth's crust are the ubiquitous silicate minerals and are not suitable for smelting or other processing; most of the ore minerals, therefore, are less common minerals, particularly sulfides, oxides, hydroxides, native metals, and carbonates. The desired ore minerals must also be present in sufficiently large grains so that they can be inexpensively separated to prepare a pure concentrate ready for smelting.
2. The method content of the deposit, known as the grade, must be high enough to make mining and processing feasible.
3. The size of the deposit, its depth below the earth's surface, and its geographic location must be suitable for exploitation without entailing unreasonable expenses.

Ore Deposits of the Geochemically Abundant Metals

Deposits of the abundant metals are large—commonly they contain billions of tons of ore—and they are formed by the same processes that form common rocks; processes such as igneous differentiation, metamorphism, weathering, and sedimentation. Most of the iron and manganese ores, for example, are sedimentary rocks, while aluminum ores form as a result of weathering.

Because the clarkes of the abundant metals are high, the concentration factors needed to produce a deposit of sufficiently high grade are not great. For example, the clarke of iron is 5.80 percent; a very good iron ore has a grade of 55 to 60 percent iron by weight, and the enrichment needed is only 10 times the clarke. In some cases it is possible successfully to mine iron deposits that contain only 30 percent iron.

The important ore minerals of the geochemically abundant metals are listed in Table 1.

The two ore minerals of aluminum, gibbsite and diaspore, are both hydroxides that form from common aluminous silicate minerals such as feldspar and mica when warm tropical or semi-tropical rainfall leaches away silica and leaves an insoluble residue of aluminous hydroxides, called bauxite.

Table 1. Ore Minerals of the Geochemically Abundant Metals

Metal	*Ore Minerals*
Iron	Magnetite, Fe_3O_4; hematite, Fe_2O_3; limonite, FeO·OH; siderite, $FeCO_3$.
Aluminum	Diaspore, $HAlO_2$; gibbsite, H_3AlO_3.
Magnesium	Dolomite, $CaMg(CO_3)_2$; magnesite, $MgCO_3$.
Manganese	Pyrolusite, MnO_2; psilomelane, $Mn_2O_3 \cdot 2H_2O$.
Titanium	Rutile, TiO_2; ilmenite, $FeTiO_3$.
Silicon	Quartz, SiO_2.

Iron and manganese ores form as a result of sedimentary processes. The key to their formation rests in the chemical differences between the two common oxidation states of the elements—$Fe^{2+}$and Fe^{3+}, in the case of iron, and Mn^{2+} and Mn^{4+}, in the case of manganese. In their more reduced forms (Fe^{2+} and Mn^{2+}), both iron and manganese are quite soluble and can move readily in the aqueous environment. The more oxidized forms (Fe^{3+} and Mn^{4+}) are essentially insoluble so that when an iron- or manganese-rich water comes into contact with the atmosphere, or with an oxidizing solution, an iron or manganese mineral is precipitated. The extent to which sedimentary iron and manganese deposits have formed through geologic time has varied greatly—the largest deposits all formed 2 billion years (Ga) or more ago, suggesting that the early earth atmosphere was much less oxidizing than today's atmosphere.

Any quartz-bearing rock can serve as an ore of silicon, so many igneous, sedimentary, and metamorphic rocks are suitable ores. Similarly, any of the sedimentary carbonate rocks that contain the common mineral dolomite can serve as an ore of magnesium. Titanium ores are a somewhat different story. The two ore minerals of titanium, rutile and ilmenite, are widespread in igneous and metamorphic rocks but commonly they are only present in small amounts. However, both of the minerals are chemically and physically resistant to the weathering environment and so they tend to become concentrated in placers, particularly beach and marine placers. The main titanium ores are therefore clastic sediments.

Ore Deposits of the Geochemically Scarce Metals

Because the clarkes of the geochemically scarce elements are less than 0.1 percent, and in certain cases, such as gold, as low as 0.0000002 percent, large enrichment factors are needed to produce ore grades—factors of 10,000 times or more in some cases (Figure 1). Unlike the geochemically abundant metals, where ordinary rock-forming processes can produce the needed enrichment, the scarce metals are only concentrated by special, relatively rare geological processes. As a result ore deposits of the scarce metals tend to be small compared to those of the abundant metals.

There are three principal ways by which geochemically scarce metals can become sufficiently enriched to form an ore deposit—by magmatic segregation, by hydrothermal solution, or by placer concentration.

Magmatic segregation occurs in three forms. First, as certain mafic magmas cool, they separate into two immiscible liquids, one a molten silicate liquid, the other a molten sulfide liquid. The molten sulfide liquid is always iron-rich but also tends to concentrate metals such as nickel (Ni), copper, platinum (Pl), palladium (Pd), and other platinum-group metals. When the molten sulfide liquid cools, it crystallizes into a mixture of valuable ore minerals. The world's largest ore deposits of nickel and the platinum-group metals are all magmatic segregates: Sudbury, Ontario, and Noril'sk in Siberia are examples.

The second form of magmatic segregation happens when an ore mineral crystallizes from a cooling mafic magma and then settles out to form an essentially monomineralic layer in the igneous body. The world's major ore deposits of chromium (Cr) are magmatic segregates, formed by the settling of the mineral chromite (Fe, Mg) Cr_2O_4: the Bushveld Igneous Complex in South Africa and the Great Dike of Zimbabwe are examples.

A third form of magmatic segregation can occur in certain cooling magmas, such as granite, that contain a lot of dissolved H_2O. When the granitic magma crystallizes, the minerals that form are mostly anhydrous, so an increasingly H_2O-rich residual magma remains. Valuable metals such as cesium (Cs), lithium (Li), beryllium (Be), niobium (Nb), and tantalum (Ta) also become concentrated in the residual magma. When the H_2O-rich resi-

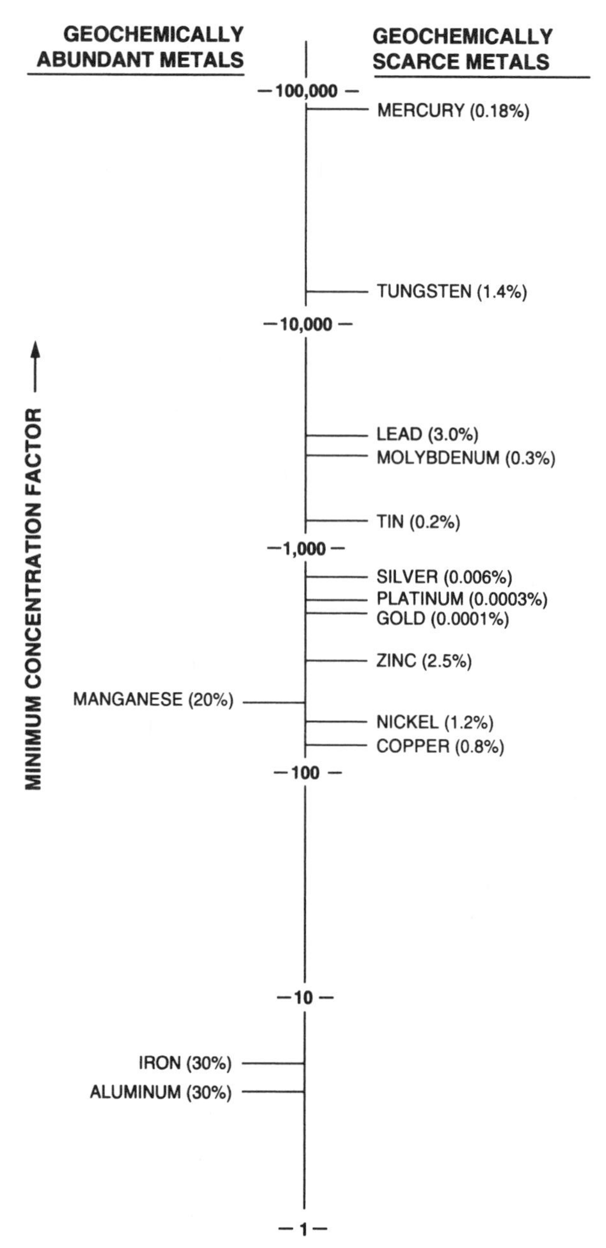

Figure 1. Minimum concentration factors needed to form ores of selected metallic elements. The clarkes of geochemically scarce metals are so low that very large enrichments are needed to make ore, while abundant metals require much lower enrichments. Numbers in parentheses represent the lowest feasible grade to mine for that metal.

due crystallizes, the result is a member of a very coarse-grained family or rocks called pegmatites, and in the pegmatites there are minerals of the concentrated scarce metals.

Hydrothermal solutions are the main collecting, concentrating, and transporting agents by which ores of the geochemically scarce metals are formed. Hydrothermal solutions, as the name implies, are heated waters that contain dissolved salts, especially sodium and calcium chlorides (NaCl and $CaCl_2$, respectively). Such halide-rich solutions have the capacity to dissolve small amounts of sulfide, native metal, and oxide ore minerals (see Table 2 for the ore minerals of the more important scarce metals). Hydrothermal solutions are ubiquitous in the crust—any deeply circulating water becomes heated and will dissolve minerals from the rocks it passes through. Not all solutions develop the capacity to form ores, but when a solution reacts with a suitable rock, the solution can extract a geochemically scarce metal from its ionic substitution trap. If such a solution then flows through a

Table 2. Ore Minerals of the Most Important Geochemically Scarce Metals

Metal	*Ore Minerals*
1. *Sulfide Minerals*	
Copper	covellite, CuS; chalcocite, Cu_2S; digenite, Cu_9S_5; chalcopyrite, $CuFeS_2$; bornite, Cu_5FeS_4; tetrahedrite, $Cu_{12}Sb_4S_{13}$.
Zinc	sphalerite, ZnS.
Lead	galena, PbS.
Nickel	pentlandite $(Ni, Fe)_9S_8$.
Molybdenum	molybdenite, MoS_2.
Antimony	stibnite, Sb_2S_3.
Mercury	cinnabar, HgS.
Silver	acanthite, Ag_2S.
2. *Oxide Minerals*	
Chromium	chromite, $(Fe, Mg)Cr_2O_4$.
Tin	cassiterite, SnO_2.
Niobium	columbite, $FeNb_2O_6$.
Tantalum	tantalite, $FeTa_2O_6$.
Tungsten	wolframite, $(Fe, Mn)WO_4$; scheelite, $CaWO_4$.
Uranium	uraninite, UO_2.

Metals not listed occur either as native metals (e.g., gold and platinum), or they are recovered as by-products.

defined channelway and undergoes a reaction such as boiling, rapid cooling, or chemical change by reaction with a totally different rock, the dissolved load of scarce metals can be deposited as ore minerals, and an ore deposit results. The geologic settings in which hydrothermal solutions form are varied. For example, solutions that form in sedimentary basins tend to migrate laterally through aquifers and form stratabound ores of copper, lead, and zinc. Solutions that form by convective flow around intrusive igneous bodies tend to form disseminated or vein deposits of copper, lead, zinc, molybdenum, tin (Sn), tungsten (W), mercury (Hg), silver (Ag), and gold (Au)—what is deposited depends on what is in the solution. Solutions that form when seawater reacts with volcanic rocks on the seafloor deposit ores of copper, zinc, and gold.

The third method by which geochemically scarce elements can become sufficiently enriched to form ores involves a step beyond magmatic segregation or hydrothermal solution deposition. When previously formed deposits are subjected to weathering, certain ore minerals are so tough they do not break down chemically or physically—examples include native gold and platinum, and cassiterite (the main ore mineral of tin). If the resistant minerals have high densities, they can become concentrated in streambeds or along shore/lines as deposits called placers. Much of the world's tin is recovered from placers in Malaysia, Thailand, and Indonesia, and at least half of the gold ever mined has been obtained from placers, particularly the ancient paleoplacers of the Witwatersrand in South Africa.

Bibliography

Craig, J. R., D. J. Vaughan, and B. J. Skinner. *Resources of the Earth.* Englewood Cliffs, NJ, 1988.

Kesler, S. E. *Mineral Resources, Economics, and the Environment.* New York, 1994.

Skinner, B. J. *Earth Resources,* 3rd ed. Englewood Cliffs, NJ, 1986.

BRIAN J. SKINNER

METALLIC MINERAL DEPOSITS FORMED BY SEDIMENTARY PROCESSES

A number of metals are enriched to ore grade by sedimentary processes. Among the more valuable are copper, lead, and zinc. Both copper and lead were known to the Romans, who used lead widely for water pipes. In fact, the English word "plumber" is derived from Latin—one who works with lead. Zinc was not known as a separate metal in antiquity but was widely used in mixtures with copper to make brass. In the New World, native Americans produced copper from deposits of metallic copper in northern Michigan, and traded copper artifacts widely. Today, copper and lead are still used widely in plumbing but also in electrical applications as copper wiring and lead storage batteries; zinc is used in protective coatings for iron and steel products.

In sedimentary systems, metals can either be deposited at the same time as the enclosing rocks, or can come in later, after the sediments have been converted to sedimentary rocks. Lead and zinc tend to occur together and can be of either early or late timing. In contrast, copper usually occurs alone and has a late timing of introduction to its host. The early style of lead and zinc deposition is well displayed by ore bodies in Alaska, in western Canada, and in Australia. The Alaskan deposits are particularly intriguing because they have evidence of fossil animals that lived in the sediment when the lead and zinc were depositing (Moore et al., 1986). These worm-like organisms are probably related to those found today growing around the hot springs that issue onto the seafloor at mid-ocean ridges (*see* METALLIC MINERAL RESOURCES FROM THE SEA). Here, cold seawater is drawn into cracks in the seafloor, where it reacts with the hot lava to extract metals, and then vents back to the seafloor as a plume of metal-laden hot water. Investigators who have watched these plumes from deep-diving submarines have referred to them as "black smokers." Worms, clams, and crabs feed on the nutrients provided by these hot water springs, forming a characteristic vent fauna. Hence the presence of similar vent faunas at deposits like Red Dog in Alaska leads us to believe that these deposits, too, formed by the venting onto the seafloor of hot water charged with metals.

When we see lead and zinc occurring as very fine-grained precipitates, delicately banded with sediments carrying signs of deposition in deep water, we ascribe to them an early, synsedimentary origin. When, however, the lead and zinc occur as large crystals that fill irregular spaces in limestones, we class these as late deposits that formed long after their host rocks became lithified. Deposits of this latter type are common in the upper Mississippi Valley in the United States and have been much studied. Their normal association with limestone, a moderately soluble rock, suggests that they form largely by replacement of preexisting rocks. It appears that deeply circulating saline groundwaters carried the lead and zinc from deep parts of the sedimentary basin to be deposited at higher levels (Bethke and Marshak, 1990).

Copper, when it is introduced late to its host rocks, tends to form by filling in the spaces between the grains of its host rather than by replacing the host. Also, the typical host is a sandstone or a shale rather than a limestone. These sandstones and shales show evidence of having formed on land under arid climatic conditions, producing a group of rocks often referred to as redbeds because of their distinctive color. Many redbeds have no copper, but the ones that do always form in rift structures. Rifts are areas like those seen today in east Africa where Earth's crust has pulled apart slightly, producing a down-dropped keystone-like feature. The down-dropped area usually contains a river, such as the Rio Grande of the United States, or a lake, such as Lake Tanganyika in Africa. Another ingredient should also be present if copper is to be enriched: abundant volcanic rocks of basaltic composition. When three conditions are met—a rift with volcanics, an arid climate setting, and lake deposits or similar fine-grained, organic rich sediments—the setting is good for the formation of a sedimentary copper deposit. The famous white pine deposits of northern Michigan fit this pattern, as does the Kupferschiefer of Germany and Poland. Both have produced copper abundantly for many years. In both cases, the copper is found in dark gray shales that overlie thick redbed sandstones. It appears that pyrite in the shales acted to precipitate copper that was being leached out of the underlying redbeds by saline groundwater (Brown, 1992).

Sedimentary processes account for much of the world's deposits of lead, zinc, and copper, and the metals can be introduced either early or late in the history of the enclosing sediments. Iron and manganese are two metals whose deposits are almost entirely sedimentary and almost always deposited with the host strata. These two metals have very similar chemical behaviors and hence they are concentrated by similar geological processes (Maynard, 1991; Force and Maynard, 1991). The key to the generation of large deposits of iron and manganese is the presence of a large reservoir of oxygen-free water. Both metals are much more soluble under reducing than under oxidizing conditions, so they tend to move from low-oxygen to high-oxygen areas. Water bodies like the Black Sea, which contains hydrogen sulfide instead of oxygen in its deeper waters, constitute large reservoirs of soluble manganese that is available to be moved into shallow water and precipitated as manganese oxide in a "bathtub ring" around the edges of the basin where the deep oxygen-free water meets the shallow well-oxygenated surface water.

Iron does not show the enrichment in shallow Black Sea sediments that manganese does, probably because modern seawater contains too much sulfur. Iron forms a very insoluble sulfide, pyrite, that immobilizes the iron in deep-water sediments, so iron deposits are not forming today, and although small deposits have formed over the past 500 million years (Ma), no large deposits have formed since the middle Precambrian. This time distribution is usually interpreted by geologists as resulting from a lower oxygen content in the atmosphere during the early part of Earth's history. Low atmospheric oxygen leads to oceans with no oxygen and low sulfur, so iron would have become mobile in large quantities in deep water, available for precipitation in shallow water as the giant banded iron formations that make up such a distinctive part of the Precambrian sedimentary record.

Bibliography

Bethke, C. M., and S. Marshak. "Brine Migrations Across North America—The Plate Tectonics of Groundwater." *Annual Review of Earth and Planetary Sciences* 18 (1990):287–315.

Brown, A. C. "Sediment-hosted Stratiform Copper Deposits." *Geoscience Canada* 19 (1992):125–141.

Force, E. R., and J. B. Maynard. "Manganese: Syngenetic Deposits on the Margins of Anoxic Basins." In *Sedimentary and Diagenetic Mineral Deposits: A Basin Analysis Approach to Exploration,* eds. E. R. Force, et al. El Paso, TX, 1991, pp. 147–159.

MAYNARD, J. B. "Iron: Syngenetic Deposition Controlled by the Evolving Atmosphere-Ocean System." In *Sedimentary and Diagenetic Mineral Deposits: A Basin Analysis Approach to Exploration*, eds. E. R. Force, et al. El Paso, TX, 1991, pp. 141–147.

MOORE D. W.; L. E. YOUNG; J. S. MODENE; and J. T. PLAHUTA. "Geologic Setting and Genesis of the Red Dog Zinc-Lead-Silver Deposit, Western Brooks Range, Alaska." *Economic Geology* 81 (1986):1696–1727.

J. BARRY MAYNARD

METALLIC MINERAL DEPOSITS FORMED BY WEATHERING

The environment at Earth's surface is characterized by abundant oxygen, the presence of water, generally acidic conditions, and fluctuating temperatures. These conditions are different from those under which most rocks are formed. Therefore, most rocks are unstable under surface conditions and decompose (or weather) to form new compounds that are stable.

In addition to decomposing rocks, the surface environment is important in concentrating several important metallic mineral deposits, especially aluminum, iron, and manganese. These three elements form extremely insoluble minerals in the weathering environment and tend to remain as residual accumulations near the surface when all other ingredients in the rocks have been removed in solution.

Aluminum is the third most abundant element (after oxygen and silicon) in rocks that make up Earth's crust. It makes up 8.13 weight (wt.) percent of crustal rocks, where it is especially important in the major rock-forming feldspar minerals.

However, virtually all commercial aluminum deposits form by chemical weathering. The aluminum minerals formed in the weathering environment, Gibbsite [$Al(OH)_3$], Diaspore (αAlO · OH], and Boehmite [γAlO · OH] are among the most insoluble substances on Earth. Aluminum deposits are typically mixtures of these three minerals and are called "Bauxite." They form residual deposits at Earth's surface where all of the more soluble elements have been dissolved out of the rocks by downward moving groundwaters.

Enrichment of aluminum occurs best in subtropical climates with alternating wet and dry seasons. In the weathering environment, elements such as potassium, sodium, calcium, and magnesium are dissolved out of the soil, leaving behind soils that are rich in iron oxides and clay minerals (aluminum silicates). These soils are referred to as "laterites" (or pedalfers). Prolonged leaching of the laterites tends to remove the silica from the aluminum silicates, leaving behind hydrous aluminum oxides and iron oxides.

Alumina and silica are both very insoluble in weathering environments. Silica is removed chemically from alumina during alternating wet and dry seasons. During dry seasons both alumina and silica are somewhat soluble. With the onset of the wet season the alumina and silica tend to be carried downward in the groundwater. However, the solubility of alumina decreases more abruptly than that of silica, so the silica is carried farther downward in the soil before it is precipitated. Alumina commonly precipitates in the soil as pea-sized, concentrically layered structures called pisolites, giving the bauxite ore a distinctive texture. Over long periods of time the silica is effectively separated from the alumina. Because chemical weathering is a very slow process bauxite deposits accumulate only in areas with little relief. In areas of high relief erosion removes the soil before meaningful accumulations of bauxite form. Bauxite deposits are mined on Cape York in northern Australia, in Jamaica, in Arkansas in the United States, in Venezuela, and in the countries of West Africa.

Iron is concentrated in the surface environment as a result of the presence of oxygen that converts iron from relatively soluble ferrous iron (Fe^{+2}) to extremely insoluble ferric iron (Fe^{+3}). Ferrous iron is far more common than ferric iron in rock-forming minerals. In the surface environment the iron in minerals is oxidized to Fe^{+3}, rendering it unstable. The ferric iron tends to combine with oxygen and some water to form hematite (Fe_2O_3), goethite [$FeO(OH)$], or limonite [$Feo(OH) \cdot nH_2O$]. These compounds rank among the least soluble substances known, and once they form, they are very stable and are not transported in waters moving downward through the rocks. Other ingredients in the rock, including calcium, magnesium, sodium, potassium, and silica are more soluble in ground water than the iron minerals and, with time, are carried away in solution.

Iron deposits formed by weathering that are

Figure 1. Natural iron ore mine on the Mesabi range in northern Minnesota. The mine is about 100 m deep and more than a kilometer wide.

rich enough to mine form mainly by enrichment of iron-rich sedimentary rocks (iron-formations) that contain little or no aluminum, because aluminum is undesirable in iron ores. Iron mines (Figure 1) are typically very large features in which millions of tons of iron oxides have accumulated.

Like iron, manganese forms extremely insoluble compounds with oxygen, and, therefore, is stable in the weathering environment. Manganese-oxide ores may form by prolonged chemical weathering of manganese-bearing rocks. The original rock may contain manganese silicates or carbonates disseminated in large volumes of valueless rock. Chemical weathering removes the valueless minerals in solution, leaving the manganese oxides as a mineable residue. Large deposits of this type are present at Morro da Mina, Brazil, where manganese-bearing sedimentary rocks have been enriched by weathering (Dorr et al., 1956) in West Africa and Mexico.

Bibliography

DORR, J. V. N., II; I. S. COELHO; and A. HOREN. *The Manganese Deposits of Minas Gerais, Brazil: Symp. sobre Yacimentos de Manganeso.* Twentieth International Geological Congress, Vol. 3. Mexico City, 1956.

GENE L. LA BERGE

METALLIC MINERAL RESOURCES FROM THE SEA

The oceans are an enormous reservoir of elements dissolved in seawater. With a total mass of nearly 1.35×10^{21} kg, the sea contains vast quantities of trace metals, including about 340 million tons of iron (Fe), 280 million tons of copper (Cu), 430 million tons of zinc (Zn), 100 million tons of manganese (Mn), and 715 million tons of nickel (Ni) (Whitfield and Turner, 1983). Such a readily accessible and abundant source of metals raised the notion in the late nineteenth century that seawater could be "mined" for profit. This hope was fueled by a report in 1872 that seawater contained as much as 65 parts per billion gold (Au). In the

1920s Fritz Haber carried out a comprehensive study of gold in seawater with the intent of developing a commercial process for its recovery. Instead, Haber concluded that the gold content of seawater was several orders of magnitude lower than suggested by earlier studies and finally reported an average concentration of only 0.004 parts per billion. With improved analytical techniques, several studies in the late 1980s found the gold content of seawater to be still lower, and the oceans are now estimated to contain "only" about 14,000 tons of gold in solutions. Although the oceans are a significant reservoir of dissolved metals, for the most part the concentrations of metals in seawater are far too small to make their recovery feasible. Actual concentrations of metals such as Cu, Fe, Zn, Ni, and Mn are typically less than a few hundred parts per trillion, and concentrations of rare metals such as gold are much lower (e.g., only a few parts per 100 trillion). By far the greatest concentrations of metals in the sea are found in the mineral deposits that occur on the ocean floor.

Metallic mineral deposits in the deep sea were first discovered in the early 1870s during an oceanographic survey by HMS *Challenger.* During this expedition, large quantities of manganese-rich nodules were recovered from the Pacific seafloor. Since their discovery, a wide variety of metal-rich deposits have been found on the ocean floor, including ferromanganese crusts and nodules (Fe, Mn, cobalt [Co], Cu, Ni, platinum [Pt]), deep-sea metalliferous sediments (Fe, Mn), and polymetallic massive sulfides (Fe, Cu, Zn, lead (Pb), silver (Ag), Au). Most of the metals in these deposits, especially the sulfide deposits, originate from hot springs that discharge into the oceans in areas of recent volcanic activity. Thermal energy from deep-sea volcanism causes the convective circulation of seawater deep within the oceanic crust, where the fluids gradually become heated to temperatures of 350–400°C. By reaction with the rocks, the heated seawater leaches metals and sulfur from the crust and transports them in solution through fractures and faults back to the seafloor, where they are discharged at hydrothermal vents known as "black smokers." Some of the less soluble metals (e.g., Cu and Zn) are precipitated as sulfide minerals around the active hot springs, and other more soluble metals (e.g., Fe and Mn) are dispersed into the open ocean and deposited in distal environments. Leaching of metals from the oceanic crust by hydrothermal circulation of seawater is now considered to be of major important as a source of metals in the oceans (Corliss, 1971) and plays a key role in the chemical buffering of seawater, along with other processes such as rock weathering, river input, and biological activity.

Polymetallic Sulfide Deposits

In the early 1960s, the discovery of large metalliferous deposits in the Red Sea demonstrated the significance of hot spring activity on the modern seafloor and confirmed that large quantities of metals are being deposited by hydrothermal vents in the oceans (Degens and Ross, 1969). On the floor of the Red Sea, hot springs are venting metal-rich brines into several deep basins at water depths of about 2,000 m. The minerals precipitated from the brines have accumulated as metalliferous muds that blanket the bottom of the basins in layers about 10 m thick. One such deposit is estimated to contain 90 million tons of metalliferous material, including 450,000 tons of Cu and 1,800,000 tons of Zn.

The first "black smoker" deposits were discovered in 1978 at a water depth of 2,600 m on a portion of the mid-ocean ridge system known as the East Pacific Rise. At "black smoker" vents, metals are precipitated as the rising hydrothermal fluids at 350°C are cooled by mixing with seawater at 1 or 2°C. The venting solutions become blackened by a fine-grained precipitate of metal sulfides that forms a turbulent cloud of "black smoke" above the hot springs. Chimney-like structures commonly grow around the hot springs as a result of mineral precipitation, and mounds of sulfide minerals gradually accumulate to form massive, polymetallic sulfide deposits (Figure 1). Hydrothermal precipitates may also form in veins and stockworks beneath the seafloor, along the fractures and faults that feed the hot springs. A corollary discovery of these amazing deposits is the existence of prolific life forms in the inky depths of the oceans, deriving energy not from photosynthesis, but chemically, from the heated waters (*see* LIFE, ORIGIN OF).

Massive, polymetallic sulfide deposits on the seafloor are composed mainly of the minerals pyrite, pyrrhotite, chalcopyrite, sphalerite, and ga-

Figure 1. Undersea photo of black smoker vent. (Courtesy of IFREMER, France.)

lena, together with silica, anhydrite, and barite. Economically interesting concentrations of a variety of metals are found in most of the deposits, including several weight percent of Cu and Zn, up to several hundred parts per million of cadmium (Cd), Co, molybdenum (Mo), arsenic (As), antimony (Sb), selenium (Se), and Ag, and up to several parts per million of Au. Most "black smoker" deposits have dimensions of a few tens of meters and contain a few thousand tonnes of massive sulfides. However, some of the largest deposits (e.g., the TAG Hydrothermal Field, Mid-Atlantic Ridge) are several hundreds of meters in diameter and tens of meters thick, with total accumulations of several million tonnes of massive sulfide. These larger examples are analogous to ancient deposits of Fe, Cu, and Zn that are mined on land today and belong to a group of ore deposits known as "volcanogenic massive sulfides." The term "volcanogenic" refers to the close spatial association of the deposits with volcanic rocks, like those found on the modern mid-ocean ridges. Exploration of the seafloor since the late 1970s has led to the discovery of widespread hydrothermal venting and associated metallic mineral deposits along the mid-ocean ridges, as well as on volcanic seamounts and in back-arc rifts (Figure 2).

Whereas "black smokers" are a product of focused, high-temperature discharge, much of the hot spring activity in volcanic areas of the seafloor is caused by hydrothermal circulation at much lower temperatures. Diffuse, warm springs at temperatures of less than about 50°C are widespread in areas away from high-temperature hydrothermal venting. The fluids in these warm springs are not hot enough to leach metals such as Cu and Zn from the underlying rock but can dissolve the more soluble and abundant metals such as Fe and

Mn. The Fe and Mn are precipitated around warm springs as oxide crusts on volcanic rocks close to the vent source or within unconsolidated marine sediments through which the fluids leak.

Deep-Sea Metalliferous Sediments

The clouds of "black smoke" discharged into the water column at high-temperature vents form a buoyant plume that is carried away from the hydrothermal center by ocean currents. Here, the iron and manganese are dispersed into the oceans by "black smoker" plumes and eventually settle to the seafloor as metal-oxide particles, becoming incorporated in normal marine sediments. As a result much of the sediment in the deep sea is slightly metalliferous and may contain thin layers rich in Fe- and Mn-oxides. Some of the Fe and Mn remain in solution and become part of the ocean's budget of dissolved metals.

Metalliferous layers have been found in deep marine sediments up to 2,000 km from the mid-ocean ridges. They were first discovered in holes drilled by the Deep Sea Drilling Project (DSDP) in the early 1960s. The metal-rich layers lie on top of the volcanic basement, beneath a thick cover of normal marine sediments. Although they are now

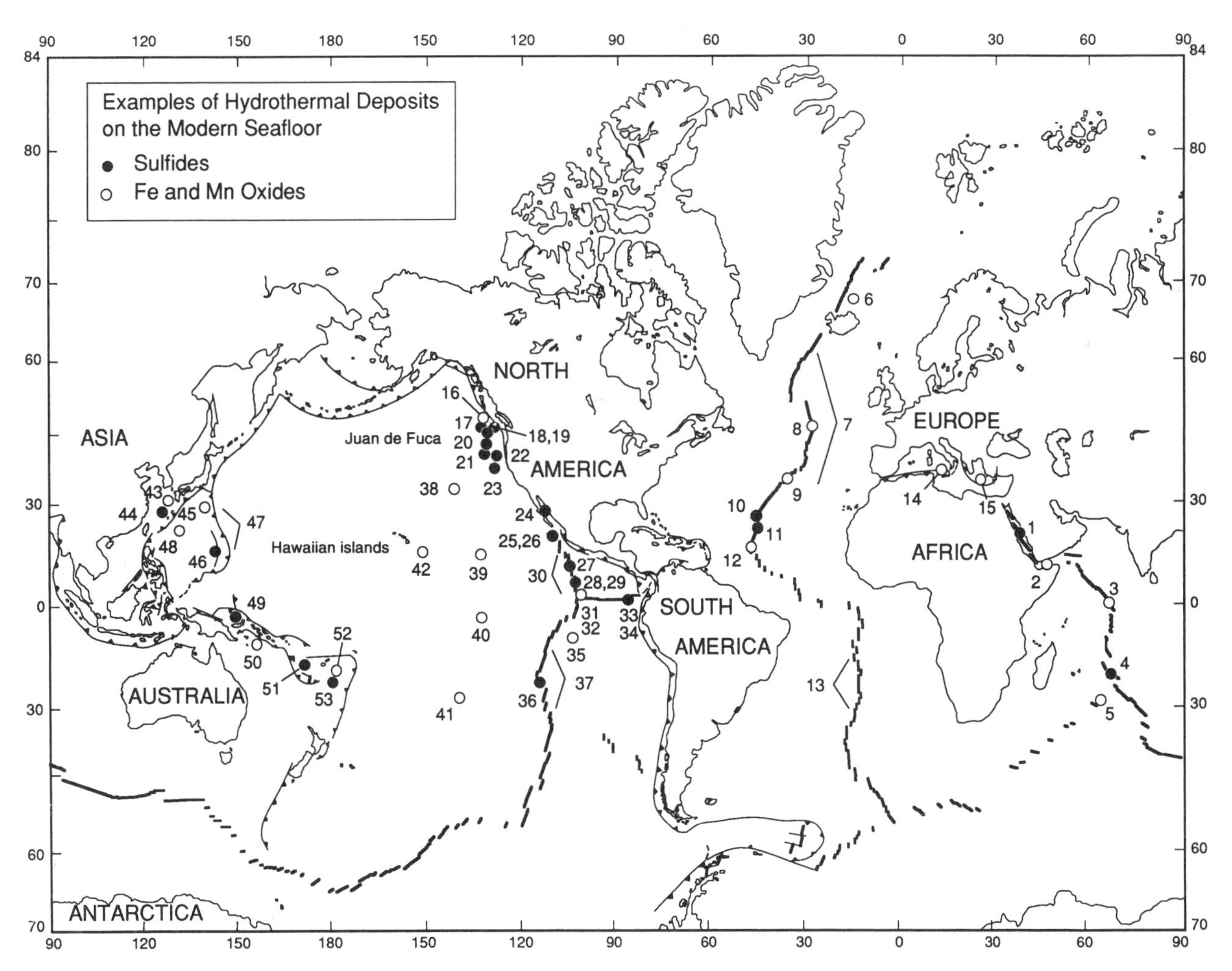

Figure 2. Examples of hydrothermal deposits on the modern seafloor.
● Sulfides
○ Fe and Mn oxides

buried, these metal-rich layers were once exposed at the seafloor on the flanks of mid-ocean ridges and have arrived at their present location by seafloor spreading. The deposits range from a few centimeters up to tens of meters in thickness and typically contain 25–35 wt. percent Fe and 5–10 wt. percent Mn. Their mineralogical and chemical characteristics can be related to such factors as proximity to the hydrothermal source and the chemical behavior of Fe and Mn in the hydrothermal plume. Deposits of deep-sea metalliferous sediments are now known to be common throughout large areas of the ocean basins.

Deep-Sea Ferromanganese Crusts and Manganese Nodules

The deep ocean basins also contain abundant deposits of Fe and Mn crusts (ferromanganese crusts) and Mn nodules. Unlike the low-temperature Fe and Mn deposits around active hydrothermal vents, most of the deep-sea ferromanganese deposits are too far removed from known hydrothermal sources to be considered direct precipitates from hot springs. Instead, deep-sea ferromanganese crusts and Mn nodules are formed mainly by adsorption of Fe, Mn, and other metals from ambient seawater or from interstitial waters in deep-sea sediments (so-called hydrogenetic deposits). In contrast to hydrothermal deposits that grow at rates of several millimeters to centimeters each year, deep-sea crusts and nodules accumulate at rates of only a few millimeters per million years. Despite their formation in the absence of direct hydrothermal activity, a principal source of Fe and Mn in deep-sea crusts and nodules is widely believed to be the input of metals into the oceans by hydrothermal venting along the mid-ocean ridges, a finding supported, for example, by the isotopic composition of their contained Pb.

Cobalt-rich ferromanganese crusts are commonly found on the flanks of seamounts and oceanic islands, especially in the equatorial Pacific, and typically at water depths between about 800 m and 2,400 m. The crusts consist of coatings of Fe- and Mn-oxides, a few millimeters up to 10 cm in thickness, on the surface of marine sediments or on exposed volcanic rocks. The deposits have compositions ranging from 7–18 wt. percent Fe, 15–30 wt. percent Mn, up to 2 wt. percent Co, 0.5 wt. percent Ni, and traces of Pt. Manganese nodules are spherical concretions composed of Mn-oxides, with lesser amounts of Fe-oxides. They range in size from a few millimeters to large concretions up to several tens of centimeters in diameter. Most nodules occur in vast fields or patches covering large areas of abyssal sediments at water depths of 4,000–5,000 m. High-grade nodules range in composition from 1–20 wt. percent Fe, 15–38 wt. percent Mn, and up to 2 wt. percent combined Cu, Co, and Ni. The size, shape, chemical characteristics, and distribution of manganese nodules are highly variable and appear to depend on a number of environmental factors, such as the rate and nature of sedimentation and local biological activity. Among the largest and highest-grade deposits is one that occurs in the eastern equatorial Pacific between the Clarion and Clipperton fracture zones. This deposit may contain between 600 and 4,500 million tonnes of potentially recoverable, high-grade nodules (Heath, 1981). One estimate suggests that the equatorial Pacific deposits may contain a total of 38 billion tons of nodules of all types (Glasby, 1977).

Ocean Mining

The possibility of mining in the sea on a commercial scale was first seriously considered by John L. Mero (1959, 1965), following his comprehensive assessment of the occurrence of marine Mn nodules. Pilot mining tests were conducted in the early 1980s, and the major Mn nodule belt in the central equatorial Pacific has been extensively surveyed for future mining operations. The nodules are considered principally as a source of Ni and to a lesser extent Cu, Co, and Mn. In the 1970s, a quantitative assessment of the contained metal in the Red Sea deposits was made, and a pilot project to test the recovery of Cu, Zn, Ag, and Au from the muds was successfully completed. A number of the polymetallic sulfide deposits on the ocean floor may be of sufficient size and metal content that they would be mined if they were found on land. However, the technology developed for the mining of Mn nodules and the metalliferous muds in the Red Sea is not applicable to the "hard-rock" mining conditions that would be encountered in sulfide deposits like those found on the mid-ocean ridges.

In more than a century of exploration less than about 1 percent of the world's ocean floor has been surveyed. In that time, more than one hundred occurrences of hydrothermal mineralization (including Fe and Mn crusts, metalliferous sediments, and polymetallic sulfide deposits) and extensive deposits of Mn nodules have been found. Vast unexplored portions of the ocean floor are likely to contain countless more examples of seafloor minerals. The long-term future of this resource will be determined by the continuing need for metals and by the limited supply of minerals on land. However, important questions remain about the economic viability of deep-sea mining, and complex political, legal, and environmental considerations for the use of the oceans are still being addressed (e.g., United Nations International Seabed Authority and Tribunal for the Law of the Sea).

Bibliography

CORLISS, J. B. "The Origin of Metal-Bearing Hydrothermal Solutions." *Journal of Geophysical Research* 76 (1971):8128–8138.

DEGENS, E. T., and D. A. ROSS, eds. *Hot Brines and Recent Heavy Metal Deposits in the Red Sea.* New York, 1969.

GLASBY, G. P., ed. *Marine Manganese Deposits.* Amsterdam, 1977.

HEATH, G. R. "Ferromanganese Nodules of the Deep Sea." In *Seventy-Fifth Anniversary Volume of Economic Geology,* ed. J. Skinner. New Haven, CT, 1981.

MERO, J. L. *The Mineral Resources of the Sea.* Amsterdam, 1965.

RONA, P. A., K. BOSTROM, L. LAUBIER, and K. L. SMITH, JR., eds. *Hydrothermal Processes at Seafloor Spreading Centers.* New York, 1983.

———. "Mineral Deposits from Sea-Floor Hot Springs." *Scientific American* (January 1986): 84–92.

WHITFIELD, M., and D. R. TURNER. "Chemical Periodicity and the Speciation and Cycling of the Elements." In *Trace Metals in Seawater,* eds. C. S. Wong et al. New York, 1983.

MARK D. HANNINGTON
PETER M. HERZIG

METALLOGENIC PROVINCES

The concept of metallogenic provinces was introduced by de Launay in 1900. Metallogenic provinces are areas of the Earth's crust characterized by an unusual abundance of ores of a particular metal or type of ore deposit (Rose, 1979). The definition implies nothing of the genesis or size of the province. Metallogenic provinces may form as a result of one or more extraction processes affecting areas unusually rich in a particular element(s) (geochemical provinces), extraordinarily effective extraction processes acting on average source materials, a combination of the two, or by other processes. Extraction is followed by concentration of the element(s) into ore deposits.

A metallotect (a term introduced by Lafitte et al. in 1965) is a geologic, lithologic, tectonic, or geochemical feature that may have been important in the concentration of element(s) into deposits within a metallogenic province. Metallogenic provinces are plentiful in the world and some classic examples are presented in Table 1.

Deposits of a particular type within a metallogenic province generally have many features in common. Deposits are usually hosted by the same rock type or rock suite, have similar characteristics, and formed in similar tectono-stratigraphic settings by the same ore-forming process(es). However, deposits differ in size, ore grade, amount of associated wall rock alteration, associated metals, and absolute ages of formation. The size and boundaries of provinces, number of deposits within a province, and whether provinces contain one or more deposit types also vary. Some provinces are unusual in that they crosscut tectonic boundaries and the deposits are of various ages. An example is the 1,500-km-long Asturian-Pyrenean-Alpine belt lead-zinc of (Pb-Zn) deposits that cuts across different tectonic provinces between northwestern Spain and northern Yugoslavia (Peters, 1978).

Acceptance of the plate tectonic theory in the 1960s revolutionized the thinking of geologists about both the distribution and genesis of ore deposit types and metallogenic provinces. De Launay (1913) had recognized that metallogenic provinces were related to particular tectonic settings. Lindgren (1909) introduced the term "metallogenic epoch," stating that these epochs coincided with

Table 1. Metallogenic Provinces

Commodity	*Ore Type*	*Age*	*Host Rocks*	*Tectonic Setting*
Au				
Greenstone-hosted gold (GSHG) Example: Motherlode, California	quartz veins	Cretaceous?	greenstones	convergent plate boundary
Witwatersrand, S. Africa	paleo placer	Proterozoic	conglomerates, sandstones	synolinal basin
Sn				
Bolivia	Sn-Ag veins	Tertiary	intermed. plutons, tuff	convergent plate boundary
Cu				
Andes, S. A.	Cu-Mo porphyry	Tertiary	intermed.-felsic plutons	convergent plate boundary
Belt Supergroup, NW MT, U.S.	diagenetic Cu-Ag disseminations	Proterozoic	clastic rocks	intra plate-extensional
Pb-Zn				
Mississippi Valley-Type (MVT)	veins, disseminations	Paleozoic?	carbonates, sandstones	cratonic
Mo				
Colorado	stockwork veinlets, fractures	Tertiary	felsic plutons	atectonic to rift-related

major orogenic events. Bilibin (1968) and other Russian geologists also recognized that certain ore deposit types formed during different stages of geosynclinal (orogenic) development. Presently, many ore types can be related to plate tectonic processes. Examples include copper-molybdenum (Cu = Mo)-type porphyry deposits and Kuroko-type deposits in Andean-type and island arc-type subduction zones, respectively. However, we do not know how far back the present style of plate tectonics has operated. If plate tectonics was operational only in the Phanerozoic and Proterozoic, what process(es) could have formed some of the very ancient deposits and metallogenic provinces? These and other questions continue to receive study as earth scientists learn more about the genesis of ore deposits, the crust, and the mantle.

In addition to plate tectonic–related origins, researchers have attributed metallogenic provinces to phenomena such as inhomogeneities in the crust and mantle, magma chemistry, earth evolution, changes in the earth's atmosphere, depth of erosion, and combinations of the above.

A relationship between metallogenic provinces and crustal rock composition was shown by Titley (1991), who found that Ag-to-Au ratios from epigenetic ore districts in parts of Arizona and New Mexico differ systematically. Ag-enriched ores occur in terranes consisting of Paleozoic marine rocks on Proterozoic clastic rocks. Au-enriched ores occur within or above Proterozoic mafic-felsic volcanic arc rocks.

Some metallogenic provinces are clearly related to the types of igneous rocks present. Examples include the association of platinum (Pt) group elements and chromite with ultramafic rocks, gold-tellurium (Au-Te) deposits with alkalic igneous provinces, and tin (Sn) with sulfur-type granites (granites formed by the melting or anatexis of sedimentary rocks).

Some metallogenic provinces may be related to the evolutionary development of Earth, but the story is complex because of our lack of knowledge of plate tectonic interactions in the Precambrian; for example, see Barley and Groves (1992).

The Lake Superior-type banded iron formation and uranium-bearing placer deposit metallogenic provinces of the Proterozoic (Table 1), however, are apparently related to the development of the earth's atmosphere. These deposits formed after photosynthesis by plants commenced, but apparently prior to the accumulation of oxygen in the atmosphere, which led to the formation of hematitic-bearing clastic subaerial rocks. Whether this concept can be applied to the abundant Late Proterozoic clastic-hosted copper-silver and copper-cobalt deposits and some other deposit types is still debated.

Some metallogenic provinces are probably a function of the amount of erosion of the rock package (Laznicka, 1973; Barley and Groves, 1992). Deposits that form on or near the earth's surface are generally rapidly eroded following formation. Examples are Tertiary and Cretaceous hot spring and epithermal precious metal and porphyry Cu-Mo-type deposits. Because igneous activity has always occurred on the earth, these types of deposits have probably also formed throughout most of earth history. Tectonic burial probably accounts for the rare occurrences of these types of deposits in ancient rocks.

Finally, some metallogenic provinces have witnessed more than one period of ore deposit formation. The region of Arizona southwest of the Colorado Plateau is an excellent example. It has been subjected to at least three periods of superposed mineralization, starting in the mid Proterozoic (Titley, 1991).

Bibliography

BARLEY, M. E., and D. I. GROVES. "Supercontinent Cycles and the Distribution of Metal Deposits Through Time." *Geology* 20 (1992):291–294.

BILIBIN, Y. A. "Metallogenic Provinces and Metallogenic Epochs." In *Geological Bulletin*. New York, 1968.

DE LAUNAY, L. "Sur les types régionaux de gîtes métallifères." In *Compte-rendus des séances de l'Académie des sciences*, 130 (1900):743–746.

LAFITTE, P., F. PERMINGEAT, and P. ROUTHIER. "Cartographie métallogénique, métallotect et géochimie régionale." *Société Française Minéralogique* 88 (1965):3–6.

LAZNICKA, P. "Development of Nonferrous Metal Deposits in Geological Time." *Canadian Journal of Earth Sciences* 10 (1973):18–25.

LINDGREN, W. "Metallogenic Epochs." *Econic Geology* 4 (1909):409–420.

PETERS, W. C. *Exploration and Mining Geology*. New York, 1978.

ROSE, A. W., H. E. HAWKES, and J. S. WEBB. *Geochemistry in Mineral Exploration*. London, 1979.

TITLEY, S. R. "Correspondence of Ores of Silver and Gold with Basement Terranes in the American Southwest." *Mineralium Deposita* 26 (1991):66–71.

IAN M. LANGE

METALS, GEOCHEMICALLY ABUNDANT AND GEOCHEMICALLY SCARCE

Eighty-eight naturally occurring chemical elements have been found in the earth's crust. Additional elements are known—some located spectroscopically in stars, the rest man-made—but all have isotopes with such short half-lives that evidence of their former presence in the crust can only be inferred from the presence of their stable isotopic daughters.

The most abundant element, accounting for 45.2 percent of the earth's crust by weight, is oxygen (Table 1). At the other end of the abundance scale are elements such as osmium (Os), rhenium (Re), rhodium (Rh), and ruthenium (Ru), which are present in the crust at levels below 0.00000002 percent. The weight percent of an element in the crust is called its clarke, a unit named for Frank Wigglesworth Clarke (1847–1931), one of the founders of the science of geochemistry. Clarke was a chemist who worked for the U.S. Geological Survey (USGS) in Washington, D.C. In 1908, Clarke published *The Data of Geochemistry*, a critical compilation of chemical analyses of rocks and

Table 1. Twelve Chemical Elements (the Geochemically Abundant Elements) Account for More than 99 percent of the Mass of the Crust by Weight

Chemical Element	*Amount in Crust (wt. %)*
Oxygen	45.20
Silicon	27.20
Aluminum	8.00
Iron	5.80
Calcium	5.06
Magnesium	2.77
Sodium	2.32
Potassium	1.68
Titanium	0.86
Hydrogen	0.14
Manganese	0.10
Phosphorus	0.10
Total	99.23

After Skinner and Porter, 1995, Table B-1.

minerals. Clarke eventually published five editions of *The Data of Geochemistry* and the information in the volumes is one of the bulwarks on which the science of geochemistry has been built.

There are great disparities among the clarkes of the eighty-eight naturally occurring elements. The twelve elements in Table 1 have clarkes of 0.1 percent or greater and as a group account for 99.23 percent of the mass of the crust; these elements are referred to as the geochemically abundant elements, and the six in the abundant list that are widely used metals—aluminum (Al), iron (Fe), magnesium (Mg), manganese (Mn), silicon (Si), and titanium (Ti)—are known as the geochemically abundant metals. All other elements, many of which are also metals, have clarkes below 0.1 percent and are said to be geochemically scarce elements and geochemically scarce metals, respectively. There are important differences between the abundant and scarce metals so far as their distribution in nature is concerned, and the differences influence the availability of the metals for industrial purposes.

Goldschmidt demonstrated that the three most important factors governing the way a chemical element occurs in the crust are the clarke of the element, the electronic charge, and the radius of the element in its ionic form or forms. All common rocks consist largely or entirely of minerals made up of the twelve geochemically abundant elements. The seventy-six geochemically scarce elements are present in at least tiny amounts in essentially all rocks, but they rarely form separate minerals. Goldschmidt demonstrated that the geochemically scarce elements are present by ionic substitution in the crystal lattices of minerals made up of geochemically abundant elements. For example, a mineral such as pyroxene, which contains Fe^{2+} and Mg^{2+}, two abundant elements with the same ionic charge and similar ionic radii, will invariably contain small amounts of elements such as Cu^{2+}, Ni^{2+}, and Zn^{2+}, substituting for the Fe^{2+} and Mg^{2+}.

Many consequences follow from the fact that the geochemically scarce elements are largely hidden in the crust by ionic substitution. One of the consequences concerns the long-term availability of supplies of metals.

Distribution of Chemical Elements in the Crust

Historically, the way the chemical elements are distributed in the crust was a scientific puzzle. The puzzle existed because of the difficulties in analyzing chemical elements with very low clarkes using traditional wet chemical methods of analysis such as titration. During the first half of the twentieth century the development of rapid and accurate analytical techniques, such as emission spectroscopy, X-ray fluorescence, and atomic absorption spectroscopy, made it possible for the first time to systematically analyze elements with low clarkes, and thereby to investigate the way chemical elements are distributed in the crust. The scientist who made many of the early analyses and who discovered the basic rules governing elemental distributions in the crust was VICTOR MORITZ GOLDSCHMIDT (1888–1947). Goldschmidt did most of his work in Norway and Germany, and his influential *Geochemistry*, published posthumously in 1954, is another of the bulwarks on which the science of geochemistry was founded (*see* GEOCHEMISTRY, HISTORY OF).

Geochemically Abundant Metals

The abundant metals form common minerals that are present in nearly all common rocks. Iron, for example, is present as either magnetite (Fe_3O_4), hematite (Fe_2O_3), or limonite ($FeO \cdot OH$) in most igneous, metamorphic, and sedimentary rocks. Each of the iron minerals is an important ore mineral because it can be readily reduced to metallic iron in a blast furnace. Of course, the most desirable feed to send to the blast furnace is a pure mineral concentrate. There are well-tested ways of making near-pure concentrates of a mineral so that in theory, and albeit at a high cost, even if a rock only contained small amounts of magnetite, hematite, or limonite, that rock could be the source of a pure mineral concentrate. A similar argument can be presented for each of the abundant metals. By extension, one can argue that supplies of geochemically abundant metals are practically unlimited, because ordinary rocks could, if necessary, serve as their ores. Without any needed technological change or discovery almost every common rock can be considered a potential ore for one or more of the abundant metals.

Geochemically Scarce Metals

Where the geochemically scarce metals are concerned, a very different picture emerges. Geochemically scare metals are present in the crust by ionic substitution—commonly only a few parts per billion (ppb)—for abundant metals in common minerals. A concentrate of such a concealed element can be no richer than the amount in the host mineral, and a concentrate of the host mineral would be an exceedingly lean material to send to the smelter if the sole purpose were to recover the scarce metal. In a few instances, where the host is an ore mineral of an abundant metal, a concealed element can be inexpensively recovered as a by-product. Such circumstances are rare and it is not possible to consider ordinary rocks as even potential sources for most geochemically scarce metals because no concentration step or by-product route is possible. To mine and produce geochemically scarce metals, prospectors seek the rare results of geological processes that have yielded ore deposits, wherein a separate mineral of a scarce metal has formed (*see* METALLIC MINERAL DEPOSITS, FORMATION OF).

There is a concentration level above which most scarce metals exceed the capacity of common minerals to serve as hosts by ionic substitution. The level varies by metal and host mineral but for many combinations of metals and minerals saturation is reached between 0.01 and 0.1 percent by weight. Above that level, which is known as the mineralogical barrier, scarce metals form separate minerals that can be concentrated for smelting. Below the mineralogical barrier rich concentrates of geochemically scarce element minerals cannot be prepared.

The significance of the mineralogical barrier is profound. The amount of energy needed to produce a unit mass of any metal is the sum of the energy used in mining, the energy used in concentration of the ore mineral, and the energy used in smelting. Because minerals of the geochemically abundant metals can be concentrated from common rocks, society can look forward to a virtually endless supply of the six abundant metals—when the rich ores are mined out and lean common rocks have to be used, production will continue without being disturbed by a mineralogical barrier. Minerals of the geochemically scarce metals cannot be concentrated below the mineralogical barrier, so at the barrier a major hiatus will be reached when all the scarce metal ore deposits have been discovered. What society will choose to do when that point is reached is an open question.

Ore deposits of the geochemically scarce metals are formed in the crust by processes such as the interaction of a hot brine (commonly called a hydrothermal solution) and a common rock. The brine can alter the minerals and extract the scarce metals hidden by ionic substitution in the process. Subsequent reactions, such as boiling, can cause the hydrothermal solution to deposit its dissolved load and form an ore deposit. Estimates of the fraction of the scarce metal in the crust that is concentrated into ore deposits are loosely constrained, but do not appear to exceed 0.01 percent of the amount present.

Bibliography

CRAIG, J. R., D. L. VAUGHAN, and B. J. SKINNER. *Resources of the Earth.* Englewood Cliffs, NJ, 1988.

SKINNER, B. J. "A Second Iron Age Ahead?" *American Scientist* 64 (1976):258–268.

SKINNER, B. J., and S. C. PORTER. *The Dynamic Earth,* 3rd ed. New York, 1995.

BRIAN J. SKINNER

METAMORPHIC PROCESSES

Many types of tectonic processes result in deep burial, deformation, and heating of sedimentary and igneous rocks that originally formed at or near the earth's surface. The rocks respond to these changes by undergoing metamorphism (literally "change of form"), which means recrystallization to new mineral assemblages and/or textures that are more stable under the new conditions. Different types of tectonic processes produce different types of metamorphic rocks, even from identical starting material. Features of ancient metamorphic rocks thus can often be used to infer the nature of tectonic processes operative in the past. This article looks at the processes whereby igneous and sedimentary protoliths (starting materials) are transformed into metamorphic rocks in different tectonic settings, and at the kinds of clues metamorphic rocks retain regarding the tectonic history of their transformation.

Metamorphic Transformations

Several variables control the transformation of a sedimentary or igneous rock into a metamorphic one. The most important of these are the composition of the original rock, temperature, pressure (depth of burial), the composition and amount of any fluid that migrates through the rock at depth, and the amount of time that the rock remains at depth before exposure at the earth's surface. Metamorphism is driven largely by changes in depth and temperature conditions such that an assemblage of minerals that is thermodynamically stable in one regime in the crust becomes unstable and undergoes chemical reaction to produce a new assemblage in the new regime. These reactions occur largely in the solid state (i.e., no melt phase is involved). For example, a silica-rich limestone contains the mineral assemblage calcite + quartz at 25°C and 1 bar, the conditions that prevail at Earth's surface. Burial and heating of the limestone to depths of a few kilometers and temperatures in excess of 600°C, however, causes the chemical reaction

$$\underset{\text{calcite}}{CaCO_3} + \underset{\text{quartz}}{SiO_2} \leftrightarrow \underset{\text{wollastonite}}{CaSiO_3} + \underset{\text{fluid}}{CO_2}$$

to proceed, producing the new mineral wollastonite at the expense of calcite and quartz and liberating CO_2 fluid. Depending upon the original proportions of calcite and quartz, the recrystallized limestone (now called marble) will consist of calcite + wollastonite, quartz + wollastonite, or wollastonite alone. The CO_2 released by the reaction may leave the site of reaction by migration along grain boundaries or through fractures, or it may remain trapped in local pore spaces. The double-headed arrow indicates that the reaction has the potential to proceed in the opposite direction during cooling, but if the CO_2 leaves the reaction site the reaction becomes irreversible. In this case, wollastonite will be preserved during uplift and cooling of the marble even at depth-temperature conditions at which it is thermodynamically unstable. Thus, the wollastonite-bearing assemblage may ultimately be observable at the earth's surface, allowing geologists to deduce its prior burial and heating history.

Rapid burial of the limestone to a depth in excess of 15 km at temperatures less than approximately 400°C would have resulted in a different chemical reaction

$$\underset{\text{calcite}}{CaCO_3} \leftrightarrow \underset{\text{aragonite}}{CaCO_3}$$

that produced a rock composed of aragonite + quartz rather than a wollastonite-bearing assemblage. (Note that although the mineral aragonite has the same chemical composition as calcite, it has a different crystal structure and greater density than calcite.) For any specific starting composition, the metamorphic mineral assemblage is a direct result of the changes in pressure and temperature that occur during the metamorphic event. Recognition of aragonite in a siliceous marble would alert an astute geologist to the high-pressure history experienced by the original limestone, whereas the presence of wollastonite would indicate that the rock had experienced a high-temperature metamorphic event.

We can examine the effect of bulk composition (chemistry of the protolith) by looking at a layer of shale interbedded with the original siliceous limestone. At conditions near the earth's surface, the shale will be composed largely of clay minerals, which are aluminum- and water-rich, and quartz. Because the shale layer does not contain calcite, it will not experience the same chemical reactions as the adjacent limestone layers. Rather, it will undergo a series of reactions that release H_2O rather than CO_2 during heating and which ultimately produce a rock composed of micas, quartz, and aluminous minerals such as andalusite, garnet, cordierite, or staurolite if metamorphosed to high temperatures at shallow depths; this rock will be referred to as either a schist or a hornfels, depending upon whether or not the minerals show a strong preferred orientation. At high pressures and low temperatures, the shale will develop micas, quartz, stilpnomelane, and such exotic phases as magnesian chloritoid or carpholite rather than the aragonite of the adjacent marble. It will have experienced exactly the same tectonic history as the marble, but its final mineral assemblage will differ from that of the marble because of the different original compositions and hence reaction histories of the two rock types.

Metamorphism can also be driven by the infiltration of fluids rather than, or in addition to, changes in pressure and temperature. This can be seen by examination of the reaction of calcite and quartz to produce wollastonite and CO_2: at a pressure of 2,000 bars (approximately 7 km depth in the earth), calcite, quartz, wollastonite, and pure CO_2 all coexist in thermodynamic equilibrium at a temperature of approximately 740°C. However, if water infiltrates into the marble from an adjacent shale, it will cause dilution of the CO_2 (i.e., the two fluids will mix to produce a single fluid phase with an intermediate composition). According to Le Chatelier's principle, which states that any perturbation to a system in chemical equilibrium will cause a change that tends to counteract that perturbation, dilution of the CO_2 by H_2O will result in further reaction between calcite and quartz as the system tries to produce a more CO_2-rich fluid. Eventually, all of the calcite and/or quartz will be consumed as the rock tries to reestablish a pure CO_2 fluid, even though the temperature appears to be too low to cause the reaction to go to completion. In this case, migration of water along the grain boundaries of the marble caused the metamorphic reaction to proceed. Numerous studies of metamorphic terrains have demonstrated the important role that fluid infiltration can play in driving metamorphic reactions, usually in conjunction with changes in temperature or pressure of the rocks.

One of the other key variables that controls the features of metamorphic rocks is time. Many metamorphic reactions proceed slowly relative to the rates of tectonic processes (i.e., heat transfer is faster than mass transfer) and chemical equilibrium is thus not fully established. The resultant preservation of disequilibrium mineral chemistries and reaction textures actually benefits geologists who attempt to reconstruct tectonic histories of ancient rocks. Because evidence of several old reactions may be preserved in an individual rock, the geologist is able to gain information on the progression of pressure-temperature states experienced by the rock during metamorphism; in contrast, if the rock were always perfectly equilibrated, all record of its early history would be eradicated by reaction and recrystallization. In general, a complete metamorphic cycle, from initial burial or heating of the rock to final exposure at Earth's surface, can take anywhere from tens of thousands of years adjacent to a granite pluton to tens of millions of years during a mountain-building event.

Contact Versus Regional Metamorphism

Metamorphic processes can be divided into two principle categories depending upon whether reactions occur mainly in response to changes in the thermal regime of the rocks at approximately constant depth (contact metamorphism), or whether pressure changes are as important as changes in temperature (regional metamorphism). The former case results mainly from heating caused by intrusion of hot magma into the crust, whereas the latter type is typical of areas undergoing large-scale crustal deformation.

Emplacement of a hot body of magma into a cooler sequence of rocks causes heat to flow from the intrusion into the surrounding country rocks. This heat flow results in transient heating of the surrounding rocks with very little accompanying deformation or change in depth of burial. As the surrounding rocks heat up, they undergo metamorphic reactions and grow new minerals that generally show little to no preferred orientation, reflecting the absence of directed stresses on the rocks. Rocks closest to the magma body will experience a greater degree of heating than those further away, and the metamorphic minerals that develop will reflect this thermal zonation. Subsequent erosion of the region following cooling of the magma and its surroundings will expose a roughly symmetric "contact aureole" around the pluton. Rocks close to the pluton will contain high-temperature metamorphic mineral assemblages, whereas those further away will show progressively lower temperature assemblages. All of the rocks will record metamorphism at similar depths of burial, with little indication of depth changes during the metamorphic cycle. The width of the contact aureole, which may range from a few meters to a couple of kilometers in radius, reflects both the original temperature difference between the hot magma and the surrounding rocks and the degree of fluid circulation driven by crystallization of the magma and devolatilization of the surrounding rocks. Large amounts of fluid circulation and intrusion of magma into rocks that are already at moderately elevated temperatures both tend to produce wider contact aureoles.

In contrast to most contact metamorphic rocks, regionally metamorphosed rocks usually show a strong preferred orientation of the metamorphic minerals, indicating that deformation accompanied the metamorphic recrystallization. In general, regional metamorphism results from large changes in depth of burial and temperature of the protoliths in response to events such as continent-continent collision or collision between oceanic and continental crust. Changes in depth of burial come about as a result of large-scale underthrusting, overthrusting, emplacement of nappes (large-scale recumbent folds), extensional faulting, and isostatic (gravitational) forces. Temperature increases with depth in the earth, and hence increasing the depth of burial of a rock usually causes heating to occur. However, pressure and temperature changes may occur at different rates, and regionally metamorphosed rocks typically reflect a complex series of changes in pressure and temperature over 10 to 100 million years (Ma). Most of the world's major mountain belts are cored by regionally metamorphosed rocks, and the continental interiors of Canada, India, Australia, and Greenland also expose large tracts of highly deformed, ancient metamorphic rocks.

Tectonic Settings

Regional metamorphism occurs at a variety of tectonic sites, with characteristic pressure and temperature regimes associated with different plate tectonic processes. In areas where oceanic crust is subducted beneath continental (or oceanic) crust, material is rapidly carried down to great depth and experiences a large increase in pressure before significant heating can occur. This material thus experiences high-pressure, low-temperature metamorphism, producing metamorphic rocks that are called blueschists after the characteristic blue amphiboles that develop in rocks with the composition of basalts. Some of this material may be tectonically exhumed and exposed at the earth's surface, as has occurred in much of the circum-Pacific region with spectacular examples in the Franciscan complex of California and the Sambagawa metamorphic belt of Japan. Because blueschist formation is restricted to this tectonic setting, recognition of blueschists in a metamorphic terrain provides immediate information on the paleotectonic setting of the region at the time the metamorphism occurred.

Continued subduction of oceanic material over a long period of time leads to melting of the overlying mantle and intrusion of magma into the crust in a zone that is approximately 100 km above the subducted slab (e.g., in the Andes, the Aleutians, Japan, and the Philippines at present). This movement of hot melts into the crust produces a zone of high-temperature metamorphism at relatively low pressures a few tens of kilometers to the continent side of the zone of blueschist metamorphism. The common association of high-pressure, low-temperature metamorphic rocks with high-temperature, lower-pressure rocks is referred to as a paired metamorphic belt. Recognition of such paired belts in the ancient rock record provides information on the location of ocean-continent collision zones and hence on the geography of continental margins in the past.

Continent-continent collision, which is the process that formed the Himalaya, the Alps, the Appalachian mountain belt, and the Scandinavian Caledonides, causes extreme overthickening of the continental crust from an equilibrium thickness of approximately 35 km to thicknesses in excess of 75 km. Regional metamorphism occurs throughout much of this overthickened crust over the entire time interval that collision is active. Continental collisions generally result in rapid and complex changes in the pressure and temperature regimes of individual rocks, and structures such as large faults may juxtapose rocks from a variety of different metamorphic regimes in the crust. The record of continent-continent collision is thus typically complicated, though medium-pressure, medium-temperature metamorphic rocks predominate in many eroded continent-continent collision zones. In many cases the reaction history that can be read from these rocks provides insight into the interactions between pressure-temperature changes and deformational processes, and permits a relatively detailed reconstruction to be made of the collisional event.

Bibliography

Daly, J. S., R. A. Cliff, and B. W. D. Yardley, eds. *Evolution of Metamorphic Belts*. Oxford, Eng., 1989.

ENGLAND, P. C., and A. B. THOMPSON. "Pressure-Temperature-Time Paths of Regional Metamorphism I. Heat Transfer during the Evolution of Regions of Thickened Continental Crust." *Journal of Petrology* 25 (1984):894–928.

GILLEN, C. *Metamorphic Geology: An Introduction to Tectonic and Metamorphic Processes.* London, 1982.

THOMPSON, A. B., and P. C. ENGLAND. "Pressure-Temperature-Time Paths of Regional Metamorphism II. Their Inference and Interpretation Using Mineral Assemblages in Metamorphic Rocks." *Journal of Petrology* 25 (1984):929–955.

YARDLEY, B. W. D. *An Introduction to Metamorphic Petrology.* New York, 1989.

JANE SELVERSTONE

METAMORPHIC ROCKS

Igneous rocks are those produced by Earth's internal processes, and sedimentary rocks are those produced by Earth's external processes. Metamorphic rocks, the third main group of rocks that comprise the earth, are produced by Earth's dynamism. That is, Earth, as a dynamic planet, has the capacity, through the action of lithospheric plate tectonics, to place already existing rocks into different tectonic environments. Different regimes of pressure and temperature characterize these environments. For example, at great depths in Earth's crust and mantle, both pressure and temperature are high, whereas near Earth's surface in proximity to a volcano, temperature may be quite high though pressure is not. When moved from one pressure-temperature regime to another, rocks change in response to the different pressure-temperature conditions of the new environment (*see* METAMORPHIC PROCESSES). In short, rocks respond to a changed environment.

In Greek, the word "meta-" means change and "morphe" means form; thus, metamorphism is the word used to describe the process by which rocks change in response to being moved from one environment to another. One may think of metamorphic rocks as "reincarnated rocks," that is, rocks that have previously existed in another form. Any rock—igneous, sedimentary, or metamorphic—may be metamorphosed to form a new metamorphic rock. A formal definition is as follows: metamorphism is all changes in mineral assemblage and rock texture that occur in the solid state in Earth's crust as a result of being placed in a new environment and therefore being in a new realm of pressure and temperature (Skinner and Porter, 1989). Metamorphism also includes weathering at Earth's surface.

In addition to heat (temperature) and pressure, two other factors are relevant in the development of metamorphic rocks: time and the presence of fluids. The formation of a metamorphic rock from a preexisting one (called a protolith) requires time on the order of millions of years, because in order for a metamorphic rock to form, the elements present in the minerals of the protolith must rearrange themselves to form crystal structures of the new metamorphic minerals. This is a fortunate circumstance since once a metamorphic rock forms, it would take a similarly long time for the rock to metamorphose again at Earth's surface. Thus, it is possible for metamorphic petrologists to study rocks exposed at Earth's surface in order to determine the pressures and temperatures of metamorphism and thereby infer their histories of formation. Water- or carbon dioxide-rich fluids, enhance metamorphism if present during the process. Dehydration or decarbonation reactions (reactions that liberate water and carbon dioxide, respectively) produce such fluids. These reactions take place during prograde metamorphism, which is metamorphism that occurs with successively higher pressures and temperatures. The liberated fluids occupy pore spaces in the rock and facilitate diffusion of ions between the minerals that form and those that break down during the metamorphic event.

Metamorphic rocks develop changes in their texture, mineral assemblage, or both in response to a changed physical environment. As temperature on a metamorphic protolith increases, the rock recrystallizes: the size of individual mineral grains becomes larger and the boundaries between the grains become straight and flat. If pressure also increases, platy minerals such as micas tend to arrange themselves parallel to one another with their maximum surface area perpendicular to the direction of pressure exerted on them. Such parallel arrangement of platy minerals causes the rock to exhibit a planar fabric called a schistosity or, more generally, a foliation (Yardley et al., 1990).

Changes in mineral assemblage, that is, the group of minerals that constitutes the rock, occur because particular minerals have particular stability ranges. The stability ranges are regions of pressure and temperature in the crust at which rocks "feel comfortable" and can therefore stably exist. Consequently, a rock that at great depths in Earth's crust consisted of the high-pressure mineral kyanite along with quartz, plagioclase, and biotite, for example, will develop a changed mineral assemblage if it is tectonically moved to regions of lesser pressure where kyanite is not stable. Such a rock might develop the mineral assemblage andalusite-quartz-plagioclase-biotite because andalusite is stable at lower pressures than kyanite. This example emphasizes the point that metamorphism, most often, is an isochemical process: though mineral assemblage and texture may change, the overall composition of the rock does not. An apt analogy for this process is bread-baking; one combines flour, yeast, honey, and salt and heats the mixture. After a few hours the combination has a completely different texture and taste even though it has the same composition as the initial mixture of flour, yeast, honey, and salt.

Types of Metamorphic Rocks

Metamorphic rocks may be categorized and named on the basis of texture or type of protolith. The most basic distinction one can make between metamorphic rocks is the presence or absence of a foliation. Rocks that have been metamorphosed in the presence of high pressure develop a foliation, a planar fabric that pervades the rock, whereas those that have been affected by temperature alone do not. The latter rocks are hornfelses and are characterized by randomly oriented, interlocking grains with straight-grain boundaries and the presence of minerals stable at high temperatures (Yardley, 1989).

Rocks that have been metamorphosed as a result of both increased temperature and directed pressure develop a distinct foliation that becomes more pronounced with greater degree, or grade, of metamorphism. Such foliated textures are best displayed by protoliths that contain clay minerals because the metamorphism of clay minerals produces phyllosilicates; phyllosilicates, with their pronounced single cleavage direction, when aligned, produce the planar fabric characteristic of metamorphism resulting from increasing pressure and temperature. Rocks formed from the metamorphism of clay-bearing sediments are named slate, phyllite, schist, or gneiss depending on the degree to which coarse grains and the parallel arrangement of phyllosilicates in the rock are developed. Slates are very fine grained and have well-developed "slaty cleavage" produced by the alignment of the small phyllosilicate grains in the rock. Slate has been used for roofing material because it cleaves so perfectly into thin plates. With increasing metamorphic grade, slates become more coarse-grained and the cleavage surfaces of the micas in the rock are more visible. Hence, the rock develops a sheen, breaks into less than perfect slabs and is called a phyllite. Schists are produced by higher grades of metamorphism and are characterized by the parallel arrangement of coarse grains of micas, typically biotite and muscovite, which are visible with the unaided eye. Amphiboles are also common in schists and form the schistosity along with biotite and muscovite. Gneisses are very coarse grained, foliated, and typically layered; coarse grains of light-colored minerals such as quartz, alkali feldspar, and plagioclase occur in distinct bands separated by concentrations of dark, platy, or elongated minerals such as biotite and hornblende (Yardley et al., 1990).

Since hornfels, slate, phyllite, schist, and gneiss as described above are textural terms that do not indicate mineralogic composition, other common metamorphic rock names derive from the composition of the protolith from which the metamorphic rock formed (Table 1).

Thus, a rock formed from the metamorphism of a clay-rich sediment is called a pelite. Because this name connotes nothing about texture, metamorphic petrologists often combine textural and compositional rock names. For example, a pelite that has large micas arranged in a planar fabric is called a pelitic schist. Furthermore, because metamorphism is generally isochemical, minerals characteristic of particular compositions consistently develop in each of the metamorphic rocks as defined by composition. Since clays are rich in aluminum, Al-rich minerals such as muscovite, garnet, the alumino-silicate polymorphs (andalusite, sillimanite, kyanite), chloritoid, and staurolite form in pelites. Therefore, we refer to a pelitic schist with well-developed garnet and sillimanite as a garnet-

Table 1. Common Metamorphic Rocks and Their Protoliths

Metamorphic Rock	*Protolith (original material)*
pelite	clay-rich sediment
psammite	sandy sediment
quartzite	quartz-rich sediment
marble	limestone
calc-silicate	calcite- or dolomite-rich sediment
metabasite	basalt
metasediment (very general name)	any sedimentary rock
meta-igneous rock (very general name)	any igneous rock

Source: Adapted from Yardley, 1989.

sillimanite-schist. Similarly, since calcite and dolomite contain abundant calcium and magnesium, calc-silicates regularly contain Ca- and Mg-bearing silicate minerals such as tremolite, diopside, and forsterite. The characteristic minerals that appear in any one of these groups of metamorphic rocks indicate the pressure and temperature at which the rock formed and are thus termed index minerals.

Environments of Metamorphism

Metamorphism occurs on both local, small scales, and over large areas on the scale of mountain belts. Metamorphism in a restricted area is typically caused by the intrusion of hot magmas (600 to 1,000°C) into crustal rocks. Because metamorphism is caused by the juxtaposition of hot material adjacent to colder surrounding rock, heat is the agent of metamorphism and the process is called thermal or contact metamorphism. The effect of such intrusion is the development of a contact aureole: concentric zones in the surrounding rock around the emplaced magma. Since heat is transmitted to the surrounding rock by conduction, the extent to which the intruded country rocks "feel" the emplacement of hot magma is a function of the size and temperature of the intrusion and the initial temperature difference between the intrusion and the surrounding rocks. The larger the intrusion, the more heat available for contact metamorphism; an intrusion as large as 5 km in diameter may produce an aureole more than a kilometer in width that may stay hot for several tens of thousands of years (Gillen, 1982). The resulting concentric zones contain minerals characteristic of a distinct temperature range. Surrounding rocks immediately adjacent to the intrusion will contain anhydrous metamorphic minerals that are comfortable at high temperatures. If the surrounding rock is pelitic, such anhydrous minerals might be spinel, sillimanite, and garnet, whereas if it is calcareous, olivine and pyroxene would be stable in that zone. Zones farther from contact with the intrusion will contain hydrous metamorphic minerals such as cordierite, if the surrounding rock is pelitic, and tremolite, if it is calcareous. In both cases, however, the textural name applied to these rocks will most likely be hornfels, for little if any pressure is involved in this type of metamorphic environment.

The most common metamorphic rocks of the continental crust form by metamorphism on the scale of hundreds and thousands of kilometers in a process known as regional metamorphism. The formation of mountain belts, such as the Himalayas, the Appalachians, or the Alps, involves regional metamorphism of surrounding rock as it is heated and deformed. This type of metamorphism is in marked contrast to contact metamorphism because with this type, pressure is as important a variable as temperature. Consequently, regional metamorphic rocks tend to be foliated and thus are the slates, phyllite, schists, and gneisses described previously. Also common in regional metamorphic terranes is a rock called migmatite, or mixed rock, that has the appearance of marble cake and forms at very high pressures, and temperatures at which the rock being metamorphosed becomes ductile and flows.

Regional metamorphism occurs over a wide range of temperatures and pressures, roughly 200 to 750°C and 2 kbar to 10 kbar (approximately 5–35 km depth) (Yardley, 1989). As with contact metamorphism, mineral assemblages develop that are characteristic of particular temperatures and pressures, and it is by reading the record left by the minerals that metamorphic petrologists work to reconstruct the histories of mountain belts.

Bibliography

GILLEN, C. *Metamorphic Geology*. London, 1982.

SKINNER, B. J., and S. PORTER. *The Dynamic Earth: An Introduction to Physial Geology*. New York, 1989.

YARDLEY, B. W. D. *An Introduction to Metamorphic Petrology.* New York, 1989.
YARDLEY, B. W. D., W. S. MACKENZIE, and C. GUILFORD. *Atlas of Metamorphic Rocks and Their Textures.* New York, 1990.

JILL S. SCHNEIDERMAN

METEORITES

Every year, about 1,000 million grams or g (about 1,000 English tons) of material enters Earth's atmosphere and makes its way to the surface. Shooting stars and meteor showers are the visible trails of dust to pea-sized fragments of rock and metal burning up in the atmosphere. Occasionally, a large chunk of space debris falls as a meteorite. Even larger events, those that form impact craters, are rare and none have been recorded in written history. [The Tunguska event of 1908, which occurred in a sparsely populated region of north central Russia, is a close analog; the object apparently detonated high in the atmosphere and flattened trees over 2,150 square kilometers (km^2) but no crater was formed.] This entry covers meteorites, those natural objects that impact Earth and are recovered. Technically, a meteorite is a recovered fragment of a small natural rocky or metallic body from interplanetary space that has survived transit through Earth's atmosphere.

There is also a class of objects known as interplanetary dust particles, or colloquially as "cosmic dust." These are tiny bits of space debris that are collected in the upper atmosphere by NASA, using high-altitude research planes. Slightly larger than cosmic dust are micrometeorites. These are meteorites that are too small to find with the unaided eye, a few millimeters in size. Interplanetary dust particles and micrometeorites are discussed in INTERPLANETARY MEDIUM, COSMIC DUST, AND MICROMETEORITES.

Traditionally, meteorites are divided into three broad categories based on their composition: stones, irons, and stony-irons. Stones are composed largely of silicate minerals, principally olivine, orthopyroxene, clinopyroxene, and plagioclase. Some of them can be very similar to Earth rocks, although most also contain some metallic-iron-nickel and are thus distinct. Irons, as the name indicates, are made of iron alloyed with nickel and cobalt, and they usually contain troilite (FeS). Irons are the most easily distinguished meteorites and are often quite large. The Willamette iron, found in Oregon, masses 12,800 kilograms or kg (over 14 English tons). The last group, the stony-irons, are composed of roughly equal mixtures of rocky material and metallic iron-nickel.

Meteorites are also divided into two groups based on the circumstances of their recovery. Those whose atmospheric passage is witnessed and are subsequently recovered are commonly referred to as falls. Those that are simply found lying in the fields (or commonly turned up by plows), mountains, ice caps, and so forth are called finds. This distinction may not seem important, but it is. Falls have a well-documented history while on Earth, so a scientist studying them can confidently exclude contamination by Earth materials or alteration by Earth processes as causes for any unusual properties.

The population of falls and finds are different. This is partly because some meteorite types withstand the degrading effects of Earth's surface environment better than others and thus have greater chances of recovery. Another reason is that some types of material, the irons and stony-irons particularly, are so distinct from Earth materials that they are easy to spot as being unusual and therefore worthy of collecting. Finds include material that fell on Earth in the past (up to about a million years ago for some meteorites recovered from Antarctica), and possibly this older population of material *was* different. This reason is generally discounted, but some researchers have suggested that subtle differences in meteorite populations recovered in Antarctica, as compared to modern falls, are indeed due to differences in the populations of objects in Earth-crossing orbits between now and a few hundred thousand years ago.

Meteorites can be either primitive or differentiated depending on whether their bulk chemical composition was largely established by processes acting in the early solar nebula, or as a result of processes acting on their parent body, respectively. Most stones are primitive, while most iron and stony-irons are differentiated.

Table 1 summarizes the current classification scheme of meteorites and gives statistics of falls. The observed falls represent our best guess as to the spectrum of meteorite types currently falling

Table 1. Classification of Meteorites Including Statistics of Falls; Meteorite Types with No Falls Are Known from Finds Only

Category	*Group*	*Falls*	*% Falls*
Meteorites total		969	
Iron total		46	4.7
	IAB	6	0.6
	IC	0	0.0
	IIAB	6	0.6
	IIC	0	0.0
	IID	3	0.3
	IIE	1	0.1
	IIF	1	0.1
	IIIAB	8	0.8
	IIICD	2	0.2
	IIIE	0	0.0
	IIIF	0	0.0
	IVA	4	0.4
	IVB	0	0.0
	Anomalous irons	5	0.5
	Unclassified irons	10	1.0
Stony-irons total		11	1.1
	Mesosiderites	6	0.6
	Pallasites	4	0.4
	Lodranites	1	0.1
Stones total		912	94.1
Achondrites total		71	7.3
	Angrites	1	0.1
	Aubrites	9	0.9
	Lunar	0	0.0
	Martian (SNC)	4	0.4
	Ureilites	4	0.4
HEDS or basaltic achondrites			
	Diogenites	10	1.0
	Eucrites	23	2.4
	Howardites	18	1.9
Primitive achondrites			
	Acapulcoites	1	0.1
	Brachinites	0	0.0
	Winonaites	1	0.1
Chondrites total		767	79.2
Enstatite chondrites		14	1.4
	EL	6	0.6
	EH	8	0.8
Ordinary chondrites		715	73.8
	H	299	30.9
	L	345	35.6
	LL	71	7.3
Carbonaceous chondrites		37	3.8
	CH	0	0.0
	CI	5	0.5
	CK	2	0.2
	CM	16	1.7
	CO	5	0.5
	CR	2	0.2
	CV	7	0.7
Rumurutri chondrites		1	0.1
Anomalous stones		1	0.1
Unclassified stones		73	7.5

on Earth. Some meteorite classes are not represented among observed falls. These are only known from finds, so obviously the fall statistics are not a complete inventory of the types of materials reaching Earth's surface. A discussion of the three broad categories of meteorites (stones, irons, and stony-irons) and the different meteorite groups in each of these categories follows.

Stones

The stony meteorites are the most common types of space debris falling on Earth today. Most stones are of a type known as chondrites. Most of them contain small, spherical silicate objects on the order of a millimeter in size that are known as chondrules, from which this general class of meteorites gets its name. There are three large chondrite associations that are grouped because of similarities in properties: the enstatite, ordinary, and carbonaceous chondrites. Chondrites are primitive meteorites; their chemical compositions were established in the solar nebula.

The enstatite chondrites share the property that their minerals are highly reduced. They contain no oxidized iron, a portion of the silicon is reduced to the metallic state, and calcium and manganese, normally bounded to oxygen in Earth rocks, are partially combined with sulfur as sulfides. The principal mineral in enstatite chondrites is the iron-free orthopyroxene, enstatite. The enstatite chondrites are further broken into two subgroups, named the EH and EL chondrites, respectively, for *h*igh iron and *l*ow iron varieties.

The most common type of chrondrite (74 percent of all falls) is the ordinary chondrite clan. This is broken up into the groups H, L, and LL chondrites for *h*igh iron, *l*ow iron, and *l*ow iron-*l*ow metallic iron, respectively. The ordinary chondrites are more oxidized than the E chondrites, and the degree of oxidation increases in the sequence H, L, LL. They are composed of olivine, iron-bearing orthopyroxene, clinopyroxene, and plagioclase as the major silicate minerals, and metallic iron-nickel and troilite.

The third major group of chrondrites is collectively known as the carbonaceous chondrites. These are among the most oxidized chondrites, with some members containing no reduced (metallic) iron; a portion of their iron is in the trivalent oxidation state. There are seven subgroups of carbonaceous chondrites: CH, CI, CK, CM, CO, CR, and CV. The second letter of the designation is derived from the name of a prominent member of the group. For example, the CI group is named for the Ivuna chondrite. There is more diversity in composition and mineralogy within the carbonaceous chondrite group than within other chondrite groups. Common features shared by the carbonaceous chondrites are higher contents of carbon and water. The CI chondrites do not contain chondrules, but are classified as chondrites because their bulk composition is a very close match to the nonvolatile components of the Sun. This strongly confirms that CI chondrites are primitive nebular materials like the other chondrite groups.

Stony meteorites that do not contain chondrules are called achondrites. Most of these meteorites are differentiated; they were formed by magmatic processes on their parent asteroid. Two types of achondrite meteorites are known or suspected to come from larger bodies in the solar system. There is a small group of meteorites that are fragments of the Moon, based on comparison with returned lunar samples. There is also a group of meteorites that are widely suspected to have come from Mars, based on limited measurements made on Mars by the American *Viking* landers (*see* METEORITES FROM THE MOON AND MARS).

The largest group of achondrites is the so-called HED meteorites or basaltic achondrites. The acronym stands for the three meteorite types: *h*owardites, *e*ucrites, and *d*iogenites. These meteorites are mostly brecciated igneous rocks that formed on an asteroid-sized body (*see* ASTEROIDS). The reflection spectra of these meteorites closely match those of the asteroid 4 Vesta, and it is widely suspected that 4 Vesta is the source of these meteorites. The eucrites are basalts composed of the low-calcium clinopyroxene, pigeonite, and calcium-rich plagioclase. Diogenites are orthopyroxenites, rocks composed of more than 90 percent orthopyroxene. The howardites are breccias composed mostly of fragments of eucrites and diogenites. The compositions of the HED meteorites indicate that they are crustal samples from a differentiated asteroid, one that underwent extensive igneous processes resulting in separated core, mantle, and crust. Hence, 4 Vesta is a mini-analog of the terrestrial planets.

The next largest group of achondrites is the ureilite group. These are unusual meteorites consisting of olivine and either pigeonite or augite as the dominant minerals, and are noteworthy for having substantial amounts of carbon, up to several percent. Most of the carbon is graphite, but many ureilites contain tiny diamonds as well. The ureilites are widely believed to be residues from partial melting on their parent asteroid; they have lost a basaltic component (through volcanism) and are similar to certain types of ultramafic rocks from Earth.

The aubrites, or enstatite achondrites, are the next largest group of achondrites. These rocks are composed almost entirely of the mineral enstatite. They are highly reduced meteorites like the enstatite chondrites. They are thought to be crystal accumulations formed from magmas on an asteroid whose composition was very similar to that of the enstatite chondrites.

There are a number of achondrites whose compositions are very nearly chondritic, but whose textures are nonchondritic. These are often called primitive achondrites. Most of these achondrite types are simply highly metamorphosed chondrites and really ought to be considered chondrites. Some, however, have undergone igneous processes and are true achondrites in every sense of the term. Possibly, the primitive achondrites represent material from asteroids that were more extensively heated than chondrite parent bodies, but not enough to undergo classic igenous differentiation such as happened on 4 Vesta.

Irons

The irons are composed of Fe, Ni metal occurring as two major minerals, kamacite and taenite. Kamacite is a low Ni alloy, while taenite is a high Ni alloy. Minor minerals are troilite and the Fe, Ni phosphide, schreibersite. Some irons contain graphite and/or iron carbide.

The textural relationships of the two Fe, Ni alloys have been used to classify irons into the textural types ataxites, hexahedrites, and octahedrites. The octahedrites are subdivided into fine, medium, coarse, and coarsest varieties. This textural classification is not widely used now because environmental (for example, cooling rate) as well as compositional differences define the textures. Hence, closely related iron meteorites can have different textures, and unrelated irons can have the same texture. Modern classification of irons is based on compositional parameters. Decreasing contents of gallium (Ga) and germanium (Ge) are used to subdivide the irons into four major types designated I through IV. Increasing Ni content is used to subdivide each major type into subtypes designated A, B, and so on. It is now known that some iron meteorite groups are part of a continuum. For example, the IIIA and IIIB irons represent a continuous sequence and are now referred to as the IIIAB irons.

The compositions of irons show that most iron meteorite groups represent the cores of their parent asteroids. Commonly, members of an iron meteorite group display a strong anti-correlation between Ni and iridium (Ir); low Ni members have high Ir contents and vice versa. This anti-correlation results from the crystallization of molten Fe, Ni in the cores of asteroids. The first Fe alloy to crystallize has a lower Ni and higher Ir content than the molten metal. Crystallization depletes the molten metal in Ir and enriches it in Ni, and the final solid to form is rich in Ni and poor in Ir. Some iron meteorite groups, the IAB and IIICD are examples, do not follow the trend expected for a crystallizing core. The IAB and IIICD irons contain chondritic silicate inclusions. These irons are believed to have formed by impact melting on their parent body, and they probably do not represent fragments of a core.

Stony-Irons

There are two major groups of stony-irons; the mesosiderites and the pallasites. The lodranites are a minor group, and there are anomalous stony-irons as well. The lodranites are closely related to the acapulcoites, a class of primitive achondrite.

The pallasites are composed of roughly 50 percent metal plus troilite and 50 percent olivine. The pallasites are logically what one would expect for the core-mantle boundary of a differentiated asteroid. The metal phase represents the outer portion of the core, while olivine is a plausible candidate for the lowest layer of the silicate mantle. Many pallasites seem to be closely related to the IIIAB irons based on comparison of their metal compositions.

The mesosiderites are composed of roughly 50 percent metal plus troilite and 50 percent complex silicates very similar to howardites in mineralogy. The origin of the mesosiderites has long been a puzzle because the silicates are clearly rocks from the surface of an asteroid, while the metal must have come from an asteroidal core. How these two disparate minerals were mixed is an enigma.

The Solar Nebula, Parent Bodies, and Processes

The reason we study meteorites, of course, is to learn about the earliest history of the solar system. The primitive meteorites contain clues about processes that occurred in the solar nebula and later processes that affected their parent asteroids, such as metamorphism, aqueous alteration, and impacts. The differentiated meteorites tell us something about the igneous activity that occurred on the smaller bodies in the solar system, and give us clues about the heat source that melted them.

Clues to pre-solar nebula events come largely from the various carbonaceous chondrite groups. The CV chondrites contain what are called calcium-aluminum-rich inclusions, CAIs for short, that are the most refractory components of solar system materials. These inclusions contain O, titanium (Ti), and other elements with anomalous isotopic abundances. These isotope anomalies indicate that the material that went into forming the CAIs was not well mixed with the general matter of the solar nebula and show that shortly before the solar nebula began to form the solar system, it received an influx of material from one or more nearby supernovas. More direct evidence for extra–solar system matter comes from micro-sized diamonds, graphite, and silicon carbide grains found in the CM and CV chondrites. These grains contain anomalous C, Si, and N that indicate that they were formed in the outer envelopes of red giant stars during novas.

Clues to solar nebula processes can be found in the variations of bulk compositions among the different chondrite groups. For example, the three ordinary chondrite groups show systematic increases in oxidized Fe/total Fe and decreases in total Fe/Si going from H to L to LL. This is evidence for two different nebular processes—increasing oxidation and fractionation of metal from silicate, in the sequence H, L, LL. Other chondrite types hint at still more compositional features of the nebula. The E chondrites contain no oxidized Fe, and elements that typically bond to oxygen, Ca and Mn, are bonded to sulfur. This demonstrates that there was a region of the nebula with extremely reducing conditions, which is thought to indicate a very high C/O ratio in the local nebular gas.

Both the primitive and differentiated meteorites yield clues about solar system processes. Most meteorites have undergone impact metamorphism. This is displayed in the brecciation of the rocks, the evidence for shock-damaged minerals, localized melting, and the juxtaposition of clearly unrelated materials. This process is general to the solar system; virtually all meteorite groups contain members that exhibit the effects of impacts that have occurred throughout solar system history.

Primitive meteorites have undergone metamorphism and/or aqueous alteration while metamorphism commonly affected differentiated meteorites. Evidence for parent body metamorphism is particularly well shown by the ordinary and enstatite chondrites. In these meteorite classes, there is a range of textural features from unmetamorphosed nebular matter to highly recrystallized rocks. The level of metamorphism is given a numerical designation from 3 to 6. Type H3 chondrites, for example, are composed mostly of chondrules in a fine-grained matrix formed in the solar nebula, and contain minerals with a wide range in composition. In H6 chondrites, the minerals have recrystallized, destroying the chondrules and matrix, and mineral compositions are homogeneous.

Aqueous alteration is evident primarily among some of the carbonaceous chondrite groups, but is also observed in the type 3 ordinary chondrites. The aqueous alteration has converted the primary anhydrous minerals, pyroxene and olivine, into hydrated, layer-lattice silicates of the serpentine group. Oxidation has also occurred, and sulfates and magnetite are present in the most altered meteorites, the CI chondrites.

Igneous differentiation on asteroidal-sized bodies is evidenced by the various achondrites, the stony-irons, and most iron meteorites. Many achondrites are basalts, rocks formed by quickly cooling a magma. Eucrites are the prime example. Other achondrites are cumulates, crystals that separated from a magma; the diogenites are such. Still others may represent the unmelted residues of magmatism. The lodranites are examples of this. Most iron meteorite groups are fragments of once totally molten cores of asteroids.

All of these various parent body processes happened very early in the solar system, about 4,560 million years ago (4.56 Ga). Age dating of the various meteorite groups shows that the difference in ages between differentiated meteorites, the most highly processed material, and the CAIs in CV chondrites is only of the order of a few million years. This indicates that whatever the energy source was that heated asteroids, it was very intense very early in solar system history and rapidly decayed away. What this heat source was is still a mystery, but three general possibilities have been suggested—heating by short-lived radioactivity, heating by an early intense active sun, and impact heating. Only continued study of the meteorites and theoretical calculations will allow us to ferret out the nature of the heat source.

Formation of the Planets

The primitive meteorites are the only samples we have available of the solar system that still retain direct information of the chemical and physical processes that occurred as the cloud of gas and dust condensed and accumulated into the rocky bodies. Hence, chondrites give us valuable clues about fractionation processes that operated on the building blocks of the planets. Estimates of the bulk compositions of the planets are therefore constrained by what we have learned from the study of the chondrites. For example, the ratio of moderately volatile elements, such as K, to refractory elements, such as U, is variable among the different chondrite groups. The K/U ratio measured in crustal rocks gives an estimate for the entire planet, which then gives an estimate of the volatile/refractory lithophile element fractionation that occurred during formation of the planet (*see* EARTH, COMPOSITION OF).

Summary

Meteorites have been referred to as the poor man's space probes. They are inexpensive to acquire and yield a great wealth of information on the history of the solar system. As more meteorites fall or are found, our knowledge and understanding of the earliest history of the solar system become ever more complete.

Bibliography

DODD, R. T. *Meterorites: A Petrologic-Chemical Synthesis.* Cambridge, Eng., 1981.

———. *Thunderstones and Shooting Stars. The Meaning of Meteorites: A Petrologic-Chemical Synthesis.* Cambridge, MA, 1986.

MCSWEEN, H. Y., JR. *Meteorites and Their Parent Planets.* Cambridge, MA, 1987.

WASSON, J. T. *Meteorites: Their Record of Early Solar-System History.* New York, 1985.

DAVID W. MITTLEFEHLDT

METEORITES FROM THE MOON AND MARS

Meteorites are pieces of rock and iron that fall from space onto Earth. Most meteorites come from the asteroid belt (between the orbits of Mars and Jupiter), where millions of small planetary bodies (asteroids), rocks, and boulders orbit the Sun (*see* METEORITES). A few meteorites, however, are from the Moon and Mars. Meteorites from the Moon are similar in most ways to rocks collected there by the Apollo astronauts and the Soviet Luna robotic spacecraft (*see* MOON). The idea of identifying meteorites as having come from Mars seems unlikely, because spacecraft have never returned Mars samples to Earth. Lacking such samples, we have nothing to compare to the meteorites. Nevertheless, the characteristics of a small group of meteorites match everything known about Mars from telescopic and spacecraft investigations, and these meteorites seem to have no other possible source (*see* MARS).

Through the middle ages, meteorites were thought to come from heaven, as messages or warnings. Before the Enlightenment had run its course, it was considered irrational to believe that stones could fall from the sky. Not until around 1800 were E. Chladni (in 1794) and J. Biot (in 1803) able to prove that rocks did indeed fall from the sky. Where they originated was unknown: was it Earth's atmosphere, the Moon (propelled to Earth by volcanoes), comets, or interstellar space? The fall of an iron meteorite during the Andromedid meteor storm of 1885 (related to the Beila comet) was taken as proof that meteorites come from comets. Since the mid-twentieth century there has been general agreement that most meteorites are fragments of asteroids: the paths of some meteorites have been tracked back to the asteroid belt. How asteroid fragments might come to Earth is generally understood and there are close matches between the optical properties of some meteorites and some asteroids. The meteorite fall during the 1885 meteor storm was merely a coincidence, a random piece of iron asteroid that happened to fall during a cometary meteor shower.

Martian Meteorites

The idea that meteorites could come from Mars appeared in the late 1970s and was generally accepted by about 1985. The fundamental facts are that almost all meteorites are very old, 4.5 billion years (Ga), while a few crystallized from basaltic lava within the last 1.3 Ga. These young meteorites must come from a planet-sized object, because only a planet could stay hot enough to make lava so recently (Jupiter's moon Io is an exception). There is convincing chemical evidence that the meteorites did not come from Earth. Mars became the likely source, because it has huge, relatively young volcanoes (*see* MARS).

At first, the theory that meteorites came from Mars was quite controversial. Meteorites from the Moon had not been found, even though the Moon is much closer to Earth and has much lower gravity than Mars. And no one knew how solid rocks could be ejected from Mars at the 5 km/s speed needed to leave its gravity. These questions were resolved in 1982–1983 with the first recognition of a meteorite that came from the Moon (see below), and the discovery that meteorite impacts could accelerate solid rocks to speeds high enough to leave Mars. The most convincing tie to Mars also was discovered in 1983: one of the meteorites contains a trace of the martian atmosphere, a unique mixture of gases that had been analyzed on Mars in 1976 by the *Viking* Lander spacecraft. That meteorite is EETA79001 (see Plate 28), from the Elephant Moraine area (EETA) of the Transantarctic Mountains of Antarctica; it was found in the winter of 1979–1980 (hence ". . . 79 . . ."), and was the first meteorite described from that season (hence ". . . 001"). Since then, traces of martian atmosphere have been found in most of the other martian meteorites.

Twelve martian meteorites have been recognized as of mid-1995 (Table 1). All of them are igneous rocks, solidified from molten lava (*see* MAGMA). All the martian meteorites formed from basaltic magma, the same general type that forms the ocean floor and most ocean islands (such as Hawaii) on Earth. Some of the martian meteorites are crystallized basalt magma, solidified essentially as they erupted (Table 1); they consist principally of pyroxene minerals (high-calcium and low-calcium varieties) and plagioclase feldspar. These martian basalt rocks are called *shergottites*, after the meteorite Shergotty (Table 1). The other martian meteorites come from rock that formed as crystals

Table 1. Martian Meteorites (also called SNCs)

Name	*Rock Type*	*Where Found*	*Fall Year*
Chassigny	Dunite (olivine-rich)	Haute Marne, France	1815
Shergotty	Basalt lava rock	India	1865
Nakhla	Pyroxenite (high-Ca pyroxene)	Egypt	1911
Lafayette	Pyroxenite (high-Ca pyroxene)	Indiana, USA	unknown
Governador Valadares	Pyroxenite (high-Ca pyroxene)	Brazil	~1958
Zagami	Basaltic lava rock	Nigeria	1962
ALHA77005	Lherzolite (olivine-pyroxene-plagioclase)	Antarctica, Trans-antarctic Mts.	unknown
EETA79001	Basalt lava rock	Antarctica, Trans-antarctic Mts.	unknown
LEW88516	Lherzolite (olivine-pyroxene-plagioclase)	Antarctica, Trans-antarctic Mts.	unknown
ALH84001	Pyroxenite (low-Ca pyroxene)	Antarctica, Trans antarctic Mts.	unknown
Y-793605	Lherzolite (olivine-pyroxene-plagioclase)	Antarctica, Yamato Mountains	unknown
QUE94201	Basaltic lava rock	Antarctica, Trans-antarctic Mts.	unknown

Meteorites are named for towns or places near where they fell; meteorites from Antarctica are generally given (nearly) arbitrary names.

of these minerals and of olivine settled out of basaltic magmas. Martian meteorites rich in high-calcium pyroxene are called *nakhlites* after the meteorite Nakhla; one martian meteorite, Chassigny, is rich in olivine. Together, the martian meteorites have been called the SNC group or SNCs (pronounced "snicks") after these three groups. This name is still used even though one martian meteorite is not "S", "N," or "C" (ALH84001; Table 1).

The martian meteorites have become critical in the study of Mars. The Mariner and Viking spacecraft missions to Mars provided complete imagery of its surface from orbit, but did only limited analyses of its surface, and did not return samples to Earth. Therefore, these meteorites provide the only martian samples that can be analyzed in the sophisticated laboratories available on Earth. Data from the martian meteorites have been essential in understanding the bulk composition and early history of Mars, the composition of its mantle, the composition of its crust (including why Mars is red), the history and abundance of water on Mars, and the evolution of its atmosphere.

Martian meteorites cannot tell us everything important about Mars, partly because we do not know where on Mars they came from, and also because they represent only a portion of Mars. The martian meteorites are not a representative sampling of the martian surface. First, they are much younger than most of the martian surface. Second, Mars' surface includes many rock types besides basalts: sandstone and conglomerate, chemical limestone or caliche, and impact breccias. Finally, the southern highlands of Mars are likely to be rich in the low-density minerals like feldspar and quartz (by analogy with Earth and the Moon); no martian meteorites have this characteristic.

Lunar Meteorites

The notion that meteorites came from the Moon has been advanced and repudiated many times. It was first suggested seriously in 1660, and again around 1800. With the recognition of cometary and asteroidal sources for meteorites, the idea of lunar meteorites lost favor. It was briefly revived in the 1960s during planning for the Apollo lunar missions, because some scientists thought the Moon was ancient and primitive like the chondrite meteorites (*see* METEORITES). But rocks returned from the Moon were unlike known meteorites. Basalt lava rocks from the Moon were different from the basalt meteorites from the asteroids (and Mars). Rocks of the lunar highlands, broken mixtures (breccias) rich in the feldspar mineral anorthite (*see* MOON), were completely unknown as meteorites.

Then, in the winter of 1981–1982, an American meteorite hunting expedition in Antarctica found a small (31 gram) sample of a broken rock mixture (a breccia) that looked like a lunar highland rock (see Plate 29). Intense analyses by many groups in 1982–1983 showed that everything about this meteorite, ALHA81005, was consistent with an origin on the Moon, and it was accepted as the first true lunar meteorite. Three lunar meteorites had been collected earlier (in 1979 by a Japanese expedition) but were not recognized as lunar until 1984–1987. The other lunar highlands meteorites that we know about were recognized immediately on examination, but the lunar mare basalt meteorites (especially EET87521) have been confused with asteroidal basalts. The only lunar meteorite found outside Antarctica, Calcalong Creek, was discovered in the south Australian desert (Table 2).

The lunar meteorites represent the same rock types collected on the Moon by the Apollo astronauts and the Luna robotic probes—basalt rocks from the mare regions, feldspar-rich breccia rocks from the highlands, and hardened soils from both regions. The lunar meteorites are not identical to any of the returned lunar samples, nor are they very different either. And the proportion of mare to highland lunar meteorites is approximately the same as the proportion of mare to highland areas on the Moon. This suggests that the lunar meteorites are a fairly representative, random sample of the whole lunar surface.

Lunar meteorites have been important for lunar science, but not so important as martian meteorites have been for Mars science. The Apollo and Luna missions had already returned Moon rocks to Earth, rocks from known sites and known local geology. There is currently no way of determining the precise origins of the lunar meteorites. Studies of lunar meteorites have not produced serious modifications to the history and geology of the Moon based on returned samples. Some interesting new rock types have been found, but more as variations on the themes established by the returned samples.

Table 2. Lunar Meteorites

Name	*Rock Type*	*Where Found*
ALHA81005	Highlands breccia	Antarctica, Transantarctic Mts.
Y-791197	Highlands breccia	Antarctica, Yamato Mountains
Y-793169	Basalt	Antarctica, Yamato Mountains
Y-793274	Basalt breccia	Antarctica, Yamato Mountains
Y-82192, Y-82193, Y-86032	Highlands breccia	Antarctica, Yamato Mountains
EET87521	Basalt breccia	Antarctica, Transantarctic Mts.
MAC88104, MAC88105	Highlands breccia	Antarctica, Transantarctic Mts.
Asuka-881757	Basalt	Antarctica, Yamato Mts. Area
Calcalong Creek	Highlands breccia	Nullarbor Plain, South Australia
QUE93069, QUE94269	Highlands breccia	Antarctica, Transantarctic Mts.
QUE94281	Basalt breccia	Antarctica, Transantarctic Mts.

Bibliography

DODD, R. T. *Thunderstones and Shooting Stars—The Meaning of Meteorites.* Cambridge, MA, 1986.

HUTCHISON, R., and A. GRAHAM. *Meteorites: The Key to Our Existence.* London, 1993.

MCSWEEN, H. Y. "What We Have Learned About Mars from SNC Meteorites." *Meteoritics* 29 (1994):757–779.

NORTON, O. R. *Rocks from Space.* Missoula, MT, 1994.

WARREN, P. H. "Lunar and Martian Meteorite Delivery Services." *Icarus* 111 (1995):338–363.

ALLAN TREIMAN

METEOROLOGY, HISTORY OF

Meteorology, or the observation and scientific study of the weather, is as old as human civilization. In the past weather phenomena were believed to be manifestations of a divine will. Texts such as the Bible and the Vedas are explicit in this regard. Other classical literature is more descriptive. The Babylonian *Epic of Gilgamesh* invoked astrometeorology, assuming a relationship exists between the configuration of the heavens and the weather. The Chinese *Book of Songs* contains weather proverbs and prognostications, and the epic poem *Mahabhrata* describes the onset of the summer monsoon and the dust storms of India.

Ancient Greek natural philosophers, including the Milesians and atomists, speculated on meteorological phenomena. Medical climatology began with Hippocrates of Kos who emphasized the effects of climate and other geographical factors on human health in his treatise *On Airs, Waters, and Places.* Aristotle's *Meteorologica* (ca. 350 B.C.E.), a systematic explanation of natural phenomena occurring "in the region nearest to the motion of the stars," was also a synthesis of contemporary meteorological knowledge. Aristotle explained such diverse phenomena as the Milky Way, comets, meteors, thunderbolts, whirlwinds, and earthquakes by combining his doctrine of four causes (formal, material, efficient, and final) with the theory of Empedocles—that the universe consists of four basic elements (air, earth, fire, and water), four primary qualities (hot, cold, moist, and dry), and two fundamental properties (opposition and affinity). Following the work of Parmenides of Elea (b.ca. 515 B.C.E.), Aristotle attempted to describe the various climates of the known world. He believed that climate—from the Greek term *klima,* meaning inclination—was due to the Sun's height above the horizon. Theophrastus continued the Aristotelian tradition with his treatises *On Weather Signs* and *On Winds.*

Seneca's *Naturales Quaestiones* contains both a comprehensive discussion of meteorological phenomena and a summary of earlier theories, as does book two of Pliny's *Natural History*. Both works date to the first century C.E. Claudius Ptolemy (second century) included astrological weather prognostications in his *Tetrabiblos*. The writings of Isidore of Seville (seventh century) and the Venerable Bede (eighth century) combine theological and meteorological speculation. In the eleventh century the great Arabic scientist Alhazen's *Opticae Thesaurus* treated atmospheric optical phenomena including refraction and the onset of twilight. Between the thirteenth and the mid seventeenth centuries, Aristotle's works, translated into Latin and the vernacular, again dominated Western thought. Notable commentators on his meteorology included Albertus Magnus, Roger Bacon, and Thomas Aquinas.

In the seventeenth century the continuing revolution in the practice of experimental natural philosophy provided a break from the authority of Aristotle and a scientific basis for weather studies. Francis Bacon advocated a new methodology for investigating nature based on observation, inductive reasoning, and the compilation of reliable natural histories, for example, of the winds. In 1637 René Descartes published his *Meteorology* along with *Discourse on Method* as an example of his new deductive philosophy and geometry-based physics. He explained rainbows using optical principles and other meteorological phenomena as the result of the motions and mechanical interactions of particles of matter.

Also in the seventeenth century new instruments—the thermometer, barometer, and hygrometer—were developed for measuring, weighing, and describing the atmosphere. Both Galileo Galilei and Santorio (Sanctorius) developed air thermometers; the sealed liquid-in-glass thermometer (1641) is attributed to Ferdinand II of Tuscany. Robert Boyle and Robert Hooke attempted to standardize the thermometer scale. Gabriel Fahrenheit, Anders Celsius, and René de Réaumur developed their thermometers and standard scales in the eighteenth century. The invention of the barometer (1644) is attributed to Evangelista Torricelli, with improvements by Boyle and Hooke. In the eighteenth century Jean André de Luc developed a reliable, portable siphon barometer. Devices for measuring humidity date to the fifteenth century, although Hooke's calibrated instrument (1667) is the best known. A century later, J. H. Lambert improved the hygrometer and conducted basic investigations of atmospheric humidity.

The new scientific societies of Europe promoted the collection, compilation, and dissemination of meteorological observations from remote locations and over widespread areas of the globe. The first documented system of uniform meteorological observations was that of the Accademia del Cimento in Florence, founded by Ferdinand II of Tuscany. Observations were taken by members of the Jesuit order with barometers, thermometers, and hygrometers from 1654 to 1670 at seven locations in Italy and four elsewhere in Europe. Some twenty years later Edme Mariotte in France attempted to delineate the weather over the region embraced by his numerous correspondents. His data allowed him to estimate the average annual rainfall in France and to advance a theory for the wind systems of Europe and the globe. The best documented observational project of the eighteenth century was the Societas Meteorologica Palatina (1781–1795), founded by Karl Theodore of Palatinate-Bavaria and managed by the court priest, Father Johann Jakob Hemmer. Fifty-seven stations, extending from Siberia to North America and southward to the Mediterranean received instruments, forms, and instructions and contributed observations to the Society's *Ephemerides*.

Using the growing amount of data, in 1686 Edmund Halley explained the trade winds and monsoonal circulations as the result of differential solar heating. Two years later he published the first chart of the world's winds, perhaps the earliest of all meteorological charts. George Hadley extended this work, explaining tropical wind systems (1735) as the combined result of solar heating and Earth's rotational motion. On the topic of innovative meteorological charts, Alexander Humboldt published the first isothermal charts in 1817; James Espy and Elias Loomis developed synoptic charts in the 1840s, and Urbain Le Verrier published weather maps based on telegraphic weather reports in 1863.

Other important issues that attracted the serious attention of early savants included the influence of weather on health, the influence of climate on culture, and the influence of the stars on the weather. Hippocratic medicine experienced a modest revival in the eighteenth century coinciding with the beginning of regular collections of vital statistics in England. In the 1730s John Arbuthnot popularized the notion that the conditions in the atmosphere—especially seasonal changes and rapid changes of temperature—were related to the recurrence and spread of disease. The notion that climate influenced culture and society was stimulated by European expansion and worldwide colonization. Theories of climatic determinism are found in the works of enlightenment philosophers, notably Jean-Baptiste Abbé Du Bos and his more famous disciple, the Baron Montesquieu. Astrologers searching for connections between the heavens and Earth were among the first to keep daily weather diaries. Popular almanacs, like the *Bauern-Praktik* (1508; with sixty editions), typically contained astrological tables and weather proverbs for the amusement of their readers. Almanacs also published seasonal and yearly weather forecasts for better planting and harvests. Some included blank pages for weather observations. At the turn of the seventeenth century the noted scientist Johannes Kepler conducted astrologically motivated weather research that was widely cited, if not imitated. Questions of lunar influence on the weather were pursued by serious researchers until the middle of the nineteenth century.

Early settlers in the New World found the climate harsher and the meteorological phenomena more violent than in the Old World. Many colonial Americans kept weather journals but, compared to European standards, few had adequate instruments. Benjamin Franklin's famous lightning studies and Thomas Jefferson's support for widespread and comparative observations are worthy of note. Jefferson's hopes for a national meteorological system did not materialize in his lifetime, although the U.S. Army Medical Department and other federal agencies established large-scale climatological observing programs early in the nineteenth century.

For over two decades William Redfield and James Espy argued over the nature and causes of storms and the proper way to investigate them. Redfield, supported by the work of William Reid in England and Henry Piddington in India, focused on hurricanes as circular whirlwinds; Espy emphasized centripetal wind flow and the release of latent "caloric" in updrafts. While it came to no clear intellectual resolution, the "American storm controversy" of the 1830s and 1840s stimulated the development of new observational efforts, notably by the Smithsonian Institution's meteorological project under the direction of Joseph Henry. The Smithsonian experimented with the telegraphic transmission of weather data and supported James Coffin's work on the winds of the globe. William Ferrel used this information to develop his mathematical theory of the general circulation of the atmosphere.

National weather services began in England with the Board of Trade (1854) and the Meteorological Office (1867); in France with the Paris Observatory (1864) and Central Meteorological Office (1878); in Germany with the Norddeutsche Seewarte (1868) and Deutsche Seewarte (1872); in the United States with the Army Signal Office (1870); and in Russia with the Hydrographic Department (1872). International conferences to standardize observational practices began in Brussels (1853). A regular series of meetings among the directors of national weather services started with meetings in Leipzig (1872) and Vienna (1873) and led to the formation of the International Meteorological Organization (IMO) and its successor, the World Meteorological Organization (WMO). The *Meteorologische Zeistschrift* was founded in 1884. Valuable references from this era include Cleveland Abbe's translations of articles on the mechanics of the Earth's atmosphere (1877, 1891, 1910) and the four-volume international *Bibliography of Meteorology* (1889–1891) published by the U.S. Army Signal Office.

Early in the twentieth century Max Margules worked on the inclination of frontal surfaces and the energetics of storms. New theories and methods for practical forecasting were developed by Vilhelm Bjerknes (*see* BJERKNES, J. A. B. AND V. F.) and the Bergen school of meteorology. Employing hydrodynamic theory developed by Herman von Helmholtz, graphical methods inspired by Heinrich Hertz, and thermodynamics, Bjerknes and his coworkers, Tor Bergeron, H. S. Solberg, CARL-GUSTAV A. ROSSBY, and Jacob Bjerknes, developed air mass analysis, the polar front theory, and a new cyclone model. These innovations were adopted worldwide and completely transformed meteorological theory and practice.

Although in Europe there were professors and institutes of meteorology in the late nineteenth century, the discipline developed rapidly in the twentieth century, especially after 1930, with the establishment of university departments, graduate education, well-defined career paths, and new specialized societies and journals. A large number of meteorologists received their training during World War II when a worldwide system of weather reports was established in support of military operations, especially aviation.

New tools and instruments became available to meteorologists in the twentieth century. Balloon-borne radiosonde flights were instituted in the 1930s and provided regular measurements of the upper atmosphere. After World War II surplus radar equipment and airplanes were used in storm studies, radioactive fallout from atmospheric nuclear tests provided worldwide tracers of upper air wind patterns, and weather modification on both small and large scales was attempted using silver iodide and other cloud seeding agents. At the Institute for Advanced Study in Princeton, New Jersey, John von Neumann and Jule Charney experimented with electronic computers for weather analysis and prediction. The first meteorological satellite, Tiros, was launched in 1960.

New interdisciplinary problems, approaches, and techniques now characterize the modern subdisciplines of the atmospheric sciences. Specialties in cloud physics, atmospheric chemistry, satellite meteorology, and climate dynamics have developed along with more traditional programs in weather analysis and prediction. The U.S. National Center for Atmospheric Research and many new departments of atmospheric science date from the 1960s.

Since the late 1960s atmospheric scientists have been increasingly involved in international projects such as the Global Atmospheric Research Programme. Fundamental contributions have been made by Ed Lorenz on the chaotic behavior of the atmosphere (1963) and by Sherwood Rowland and Mario Molina on chlorine-ozone reactions (1974). Since the 1980s interest in "global change" studies has been driven by environmental concerns such as stratospheric ozone depletion and global warming.

Bibliography

BRUSH, S. G., and H. E. LANDSBERG. *The History of Geophysics and Meteorology: An Annotated Bibliography.* New York, 1985.

FLEMING, J. R. *Meteorology in America, 1800–1870.* Baltimore, MD, 1990.

FLEMING, J. R., ed. *Meteorology Since 1919: Essays Commemorating the 75th Anniversary of the American Meteorological Society.* Boston, MA, 1995.

FRIEDMAN, R. M. *Appropriating the Weather: Vilhelm Bjerknes and the Construction of a Modern Meteorology.* Ithaca, NY, 1989.

FRISINGER, H. H. *The History of Meteorology to 1800.* New York, 1977.

HENINGER, S. K., JR. *A Handbook of Renaissance Meteorology.* Durham, NC, 1960.

KHRGIAN, A. KH. *Meteorology: A Historical Survey*, Vol. 1. Trans. from the Russian by R. Hardin. Jerusalem, 1970.

KUTZBACH, G. *The Thermal Theory of Cyclones: A History of Meteorological Thought in the 19th Century.* Boston, 1979.

SCHNEIDER-CARIUS, K. *Weather Science, Weather Research: History of Their Problems and Findings from Documents During Three Thousand Years.* Trans. from the German by Indian National Scientific Documentation Centre (INSDOC). New Delhi, 1975.

JAMES RODGER FLEMING

MICROMETEORITES

See Interplanetary Medium, Cosmic Dust, and Micrometeorites

MILKY WAY

See Galaxies

MINERAL DEPOSITS, EXPLORATION FOR

Mineral deposits must be discovered before mines can be developed, thus a mining operation begins with prospecting and exploration—stages with a long period of investment and a high risk of failure. Success in exploration ultimately determines survival of the mining industry.

Prospecting

Long before philosophers or scientists arrived at satisfactory explanations for occurrence and distribution of minerals in Earth's crust, the miner/prospector was successfully finding and developing mineral deposits. Before Aristotle proposed a classification of the elements, the mines of Laurium in Greece had been actively mining ores of lead and silver and accurately locating dislocated portions of the veins. Recognition of surface discoloration as a guide to mineralization dates back to early Babylonian and Anatolian cultures.

Recording of early search techniques did not come until the classical works of Georgius Agricola in Saxony and Vannoccio Biringuccio in Italy. Agricola emphasized that an important step in locating a deposit was through some agency that would strip off the soil and expose the ore to view—uprooted trees, landslides, farmers' plows, or forest fires. Erosion by streams not only uncovered deposits but also carried the minerals for long distances. This displacement called for skill in determining whether metallic nuggets were of local or remote origin. This is an approach still used by prospectors today.

Today's prospector often follows rule-of-thumb methods but has a keen eye for landscape, vegetation, and color variations as he or she systematically searches for ore. Few modern "discoveries" are made that do not have marks of earlier prospector activity. The prospector recognizes "friendly country" for certain kinds of ore. The exploration geologist might put this in technical terms of favorable host rock, localizing structures on wallrock alteration, but the prospector and the geologist are looking at the same guides to ore.

Armed with a gold pan, a metal detector, and a four-wheel-drive vehicle, today's prospectors continue to discover favorable indications of ore that they negotiate to mining companies. Today's prospectors are seldom the miners.

Exploration

Exploration is considered more sophisticated than prospecting. In 1952 the first major mineral exploration organization independent of mining operations was formed. Geologists recruited for this task were given special instruction in recognition and characteristics of various types of deposits and utilization of geochemical and geophysical search techniques. Success of this approach resulted in its adoption by most mining companies, and many exploration groups were formed. Fields of operation spread throughout the world and gave great stimulus to advancement of geochemical and geophysical methods. It changed the supply of most world mineral resources from shortage to surplus.

Modern exploration begins with defining objectives as to commodity, size, and economic factors. Search is based on conceptual models of probable target deposits that suggest likely location, structural controls, age and character of enclosing rocks, shape, mineralogy, grade, and techniques for identification.

Target deposits are generally encased in halos of altered rock that contains traces of metallic elements. These clues lead to discovery of buried deposits using field examination, geological mapping, and geochemical sampling of rocks, stream sediments, or ground waters for trace elements (Figure 1). Where the target body has physical characteristics different from enclosing rocks—differences in gravity, electrical properties, or radioactivity—geophysical methods can be used in the search (Table 1).

Most exploration programs focus on areas of decreasing size, using methods increasing in cost per unit area and declining risk of failure. These include: (1) literature research and reconnaissance with selection of geologically favorable localities; (2) coverage of the selected area with detailed geological mapping, geochemical and geophysical prospecting, and/or use of special techniques such as isotope studies; (3) sampling by pitting, trenching, drilling and/or underground exploration by shafts, drifts and crosscuts; and (4) metallurgical testing of the mineralized body and an economic feasibility study.

Geological Methods

Exploration begins with selection of a recognized metallogenic district or province such as the Mother Lode system of veins in California, the copper province of the southwestern United States, or the lead-zinc province of the Mississippi Valley area. Or the region may encompass a large area with few known deposits but with geological features similar to known mineralized areas. Reconnaissance entails screening of the entire area for some combination of exposed or indicated evi-

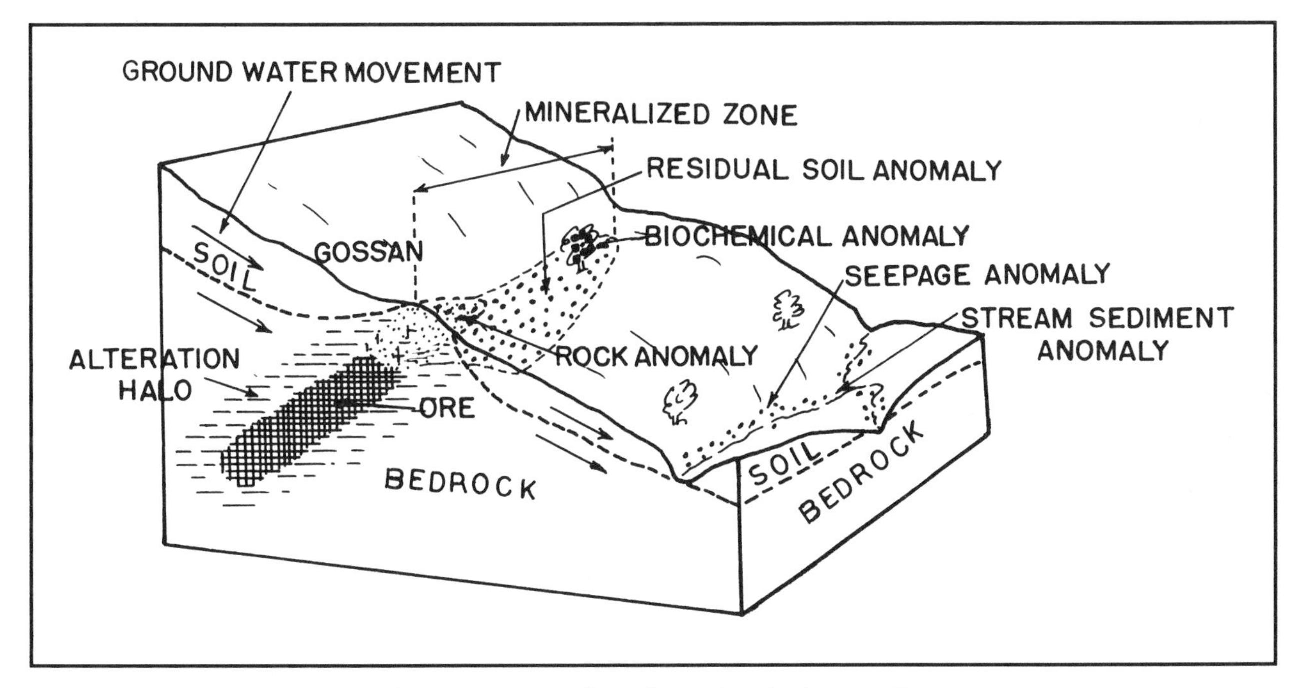

Figure 1. Formation of geochemical anomalies.

dence of the proposed model by field traversing—looking for favorable source and host rocks, and favorable structures and rock alteration as evidence of passage of mineralizing solutions. This screening is normally accompanied by testing of stream alluvium by panning or geochemical sampling, aerial geophysical surveys over covered areas, and analysis of satellite imagery and aerial photography. Large areas with little or no promise are eliminated. This phase is often referred to as prospecting, designed to identify at low cost an anomalous area with promise.

Table 1. Geophysical Methods

Method	*Name of Method*	*Physical Property*	*Type of Field*
Gravity	gravimetric	density	natural gravity
Magnetic	magnetometric	magnetic susceptibility natural remanent magnetism	natural magnetic
Electrical	telluric magnetotelluric self-potential (SP) applied potential resistivity induced polarization (IP) electromagnetic (EM) controlled source audio-frequency magnetotellurics (CSAMT) radar, microwave	electrical conductivity dielectric permitivity magnetic permeability	natural telluric currents natural electrochemical reactions artificially applied electric or electromagnetic
Seismic	reflection seismology refraction seismology	seismic wave velocity (elastic moduli and density)	artificially created seismic waves
Radiometric	radiometric	radioactive decay	natural radioactivity

Modified from Hartman and Sumner, 1992.

Specific target areas, generally less than 100 km^2, are examined in more detail after land acquisition, followed by geological mapping, geochemical sampling, ground geophysical testing if appropriate, and exploratory drilling in favorable locations.

Favorable areas that fit the conceptual model, generally less than 10 km^2, are selected for detailed pitting, trenching, drilling, and/or underground workings, accompanied by accurate sampling, petrologic studies of rocks and minerals, and detailed structural analysis and testing of rock's physical properties. At this stage environmental factors must be considered.

Geochemical Testing

Exploration geochemistry or geochemical prospecting entails the systematic measurement of one or more chemical elements contained in the deposit sought. It may entail analyses for specific chemical elements, minerals, or even bacterial counts. Rock, soil, stream sediment, surface water, groundwater, vegetation, microorganisms, animal tissue, particulates, or gases including air may be analyzed (Figure 1). Geochemical exploration identifies anomalous metal concentrations over and adjacent to mineral deposits.

Geophysical Testing

Geophysical prospecting depends upon physical differences between rock types or between ore and enclosing rocks. It does not specifically identify mineral components except in the case of radioactive minerals, but it identifies anomalous areas with greater gravity or hardness, higher or lower electrical conductivity or resistance, rocks or minerals with distinctive magnetic properties, or the presence of sulfide minerals. It identifies anomalies that must be interpreted in terms of geological and geochemical input, and on the basis of the conceptual model (Table 1).

Evaluation

Most deposits discovered today are found through a combination of exploration techniques, though in some portions of the world ore deposits may be exposed in outcrop and can be immediately identified and move directly into the evaluation stage.

Evaluation of a prospect discovered by an exploration program entails obtaining representative samples of the mineralized rock through pitting and trenching if the mineralization is exposed at the surface, or by drilling or underground workings (adits, shafts, crosscuts). Samples for analysis are selected at regular intervals as representative portions of material excavated—drill core or cuttings, or bulk or channel samples from sides and faces of underground workings.

Evaluation entails defining geometry, grade and distribution of values, metallurgical recoveries, suitable mining methods, environmental impact, and economic feasibility. Representative samples must be selected for metallurgical testing to determine the best method for extracting the valuable minerals and to determine the amount recoverable from the mineralized rock. Physical testing determines the crushing and grinding characteristics, blasting fragmentation, and mine support requirements.

Early studies determine effects of mining on water quality in surface water and groundwater, flora and fauna, historical and archaeological sites, scenic areas, and social cultures.

Probability and Cost of Discovery

It is estimated that from one thousand favorable prospects, only 4 percent or 40 will qualify as target areas, and only 0.4 percent or 4 may prove to be economically viable mines. Cost per discovery averages between $50 million and $100 million, depending on the locality.

Bibliography

BANATA, H. *Anatomy of a Mine.* USDA Forest Service, General Technical Report INT-35. Washington, DC, 1977.

HARTMAN, H. L., and J. C. SUMNER, eds. *SME Mining Engineering Handbook,* Vol. 1. Baltimore, MD, 1992.

WILLARD C. LACY

MINERAL DEPOSITS, FORMATION THROUGH GEOLOGIC TIME

Concentration of minerals or elements into ore deposits is not an accidental or random phenomenon in either time or in space. This notion of fundamental associations has been accorded increasing acceptance in the 1980s and 1990s as a result of a growing knowledge of the events of crustal evolution, from an increasing body of reliable radiometric age dates, and from our knowledge of where and how different kinds of ore bodies form.

The dating of ore deposits is based upon both rigorous determination from radiometric methods and from conventional methods employing interpretations from stratigraphic position. Methods from stratigraphy allow interpretations based upon the notion that the mineral deposit can be no older than the rocks that contain it, thus giving an oldest age, and for certain ores can be no younger than an unconformity that directly overlies them. Many kinds of ores whose age cannot be determined at this time from properties of radioactivity are still dated within such constraints. Widespread application of radiometric dating of ores has taken place only during the past half century employing measurements of potassium, argon, strontium, uranium, and lead in appropriate minerals and rocks. Some kinds of ore deposits such as igneous intrusions may be dated directly where they occur as parts of the evolution of datable rocks. Other ores may be dated from minerals formed during their origin.

Charles Meyer addressed many aspects of the occurrence of ore deposits across geologic time in a series of papers. Meyer reported the temporal variations in ore-forming processes (1981), the changing metal patterns in time (1985), and the correspondence of metal ores with geological events (1988). Data from his papers, especially his 1988 work, have been extracted and compiled in the accompanying figure for discussion here (Figure 1).

The temporal variation and evolution of ore deposit styles and their metals are both secular over the expanse of geologic time, and the recurrence of some kinds of deposits are cyclical within shorter (billion-year) periods. Our knowledge of such timing, however, is imperfect. The assignment of ages to ore deposits and the recognition of temporal patterns is constrained both by our capacity to reliably date certain kinds of ore deposits,

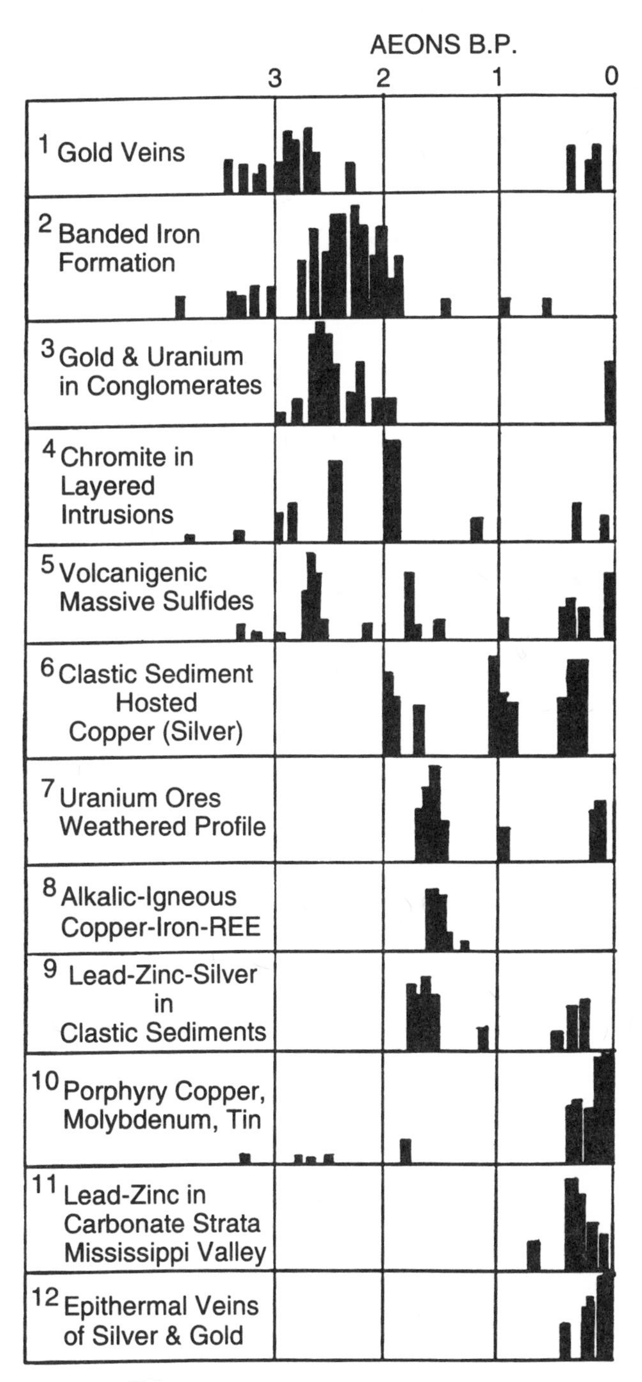

Figure 1. Adapted, modified, and augmented from the diagrams of Charles Meyer (1988). The figure shows a horizontal scale (AEONS B.P.) left to right, from about 3.5 Ga ago to the present time (0). Vertical bars are shown on approximate 50 million-year intervals. The height of each bar is an approximation of total percentage of ores of the type represented during the 50 million-year time interval shown.

and by geological accidents of emplacement and protection that influenced their survival through geologic time, or that leave them hidden from our inspection. For example, the first significant cycle of concentration of platinum group metals is with distinctive calcium-, iron-, and magnesium-rich, layered igneous rocks intruded at deep levels in the crust. Those we recognize are accidents of deep erosion. Platinum weathered and eroded from such rocks concentrates in modern placer deposits whose survival is tenuous because of their vulnerability to the same erosion that exposed their original hosts to weathering in the first place. There is abundant evidence that both emplacement of igneous rocks and their weathering are among the oldest and most enduring of geological processes, but their effects at or near the earth's surface are vulnerable to erosion, and evidence that such ores may have continuously evolved during crustal weathering has been erased. Thus, continuous modification of the crust plays a direct role in our capacity to accurately assess the existence of deeply buried deposits, and to assess the temporal importance of certain processes forming ore deposits.

Enhanced understanding of ores has led to the well-substantiated notion that their formation is a consequence of the interaction of specific geological processes with, or in, specific kinds of crustal environments. Moreover, the record of distribution of ores in time reveals contrasts in the dominance of concentration of certain elements in preference to others, in the sudden appearance or disappearance of some elements from certain ore deposits, and in temporally restricted dominance of particular styles or types of ore deposits in certain geological environments. Many such variations correspond with stages of crustal evolution, and some important kinds of ores appear to correspond with the changing nature of weathering and the atmosphere. Some styles of ore deposits—such as placer deposits, those associated with volcanic activity, and some iron ores—have remained unchanged in time, but reveal changes in metal endowment or mineralogy that are related to chemical changes in the crust and atmosphere during crustal evolution during the past 4 billion years (4.0 Ga). A consideration of ore deposits in a temporal framework based upon the distinction and separation of geologic time into the Archean, Proterozoic, and Phanerozoic is not a matter of mere convenience; those times, as well as time boundaries within them, correspond in very distinctive ways with formation of specific ore deposit styles.

The Archean, the period spanning more than 1.5 Ga of earth history up to about 2.5 Ga ago, was an interval during which the first primitive crust formed. The geologic style of this time is marked by the results of volcanism, chemical weathering in a reduced atmospheric environment, erosion and sedimentation, and by episodic metamorphism of the rocks of the evolving and thickening crust. The Proterozoic, from about 2.5 Ga to 0.6 Ga ago, was a period of thickening and maturing of the crust by episodic intrusion of felsic igneous rocks and widespread metamorphism; it was also a time during which the atmosphere became chemically oxidizing. The Phanerozoic, the period since about 0.6 Ga ago, has been a time during which life has flourished on land as well as in oceans, a time of continued volcanic activity, and a time during which the geological record reveals intermittent and episodic ocean ridge spreading, continental break-up, drift, assembly and reassembly of parts of the older Proterozoic landmass.

A sample of some common kinds of ore deposits, which reveal different traits of temporal occurrence, and some of whose origins may be related to stages evolution of Earth, are shown in Figure 1. The figure shows occurrence of ore deposits across the span of earth history since about 3.5 Ga ago; omitted because of their great antiquity are the oldest known concentrations of metal, bedded iron deposits of about 3.8 Ga in Archean rocks of Greenland. The following discussion follows the numbers on specific deposit types shown in Figure 1.

The oldest gold veins (1) occur in successions of volcanic rock complexes, the greenstone belts, and in metamorphic and magmatic belts and orogens of complex continental margins of the Phanerozoic, where they are recognized at levels of deep erosion as of high temperature in origin; they include those of the Mother Lode of California. They have originated from the metamorphism of volcanic and associated strata, and from fluid flow generated by intrusions of magma. Comparatively few occurrences of gold veins are known in the Proterozoic, especially the interval from about 2.0 Ga to about 1.5 Ga ago; the absence of gold during these times has no simple explanation, but the ore style represents one example of time-constrained metallogenesis.

The immense resource of Precambrian banded

iron formation (BIF) (2) is mostly Precambrian in age, when iron was precipitated in stable basins on old shields. The limiting youngest age of the BIF of about 1.8 Ga is believed to mark the advent of the oxidizing atmosphere after which it was no longer possible to transport significant amounts of dissolved iron in surface waters. The younger occurrences are those of mostly deep, chemically reducing, shelf environments of younger continental margins. Stable basins on shields were likewise repositories of gold and uranium concentrated as placer deposits in conglomerates (3); the diminishment of uranium in gold-bearing placers corresponds in time with the decline of Precambrian banded iron formation and may also reflect the change in atmospheric composition.

Layered intrusions (4) resulted from emplacement of magma at deep crustal levels, where slow cooling resulted in gravitationally separated layers of different igneous rocks. Different elements were concentrated at different levels of contrasting rock composition within the igneous mass and include platinum, iron, and vanadium, as well as chromium. To a greater degree than most other styles of ore occurrence, our knowledge of these systems has resulted from studies of bodies that are mostly accidents of exposure by deep erosion. Although the largest and most important are exposed by deep weathering of Precambrian old shields, sufficient numbers of younger intrusions are known to suggest that the process of their formation is not absolutely controlled by time.

Volcanigenic massive sulphide ores (5), although related to the episodic habit of volcanism, are nonetheless ores of all the ages spanning the whole range of age of ores shown in Figure 1. Composed chiefly of base metals with accessory precious metals, these ore bodies were formed in submarine volcanic environments such as are presently believed to exist, for example, in the Bonin arc south of Japan. They appear to be associated with centers of intrusion; from the fact that they are found in continental margin environments of the Phanerozoic, similar settings have been proposed for them during the Archean and Proterozoic. In the case of these ores of submarine volcanigenic ores, we see an example of the role that knowledge of ore deposit setting and timing may play in solving other problems of the geological evolution of Earth.

The Proterozoic was a time of cratonization (building of continental interiors), which together with the influence of an oxidizing atmosphere resulted in significant changes in the style of mineral deposit formation. Our knowledge of ores of the Proterozoic, however, is limited by both superjacent cover of Phanerozoic rocks and by a long period of weathering that may have removed deposits formed in shallow (less than 5 km) crustal environments. As are ores of the Archean, so are the ore deposits of the Proterozoic exposed in "windows" of basement rocks.

The oxidizing atmosphere resulted in solution of uranium minerals with their subsequent trapping and enrichment in the weathered profile (7), and the thickened and evolving crust attained a stage of mechanical competence that allowed the formation of long and deep rifts and rift valleys. These structural basins were sites of accumulation or precipitation of copper (6) and lead-zinc-silver ores in clastic strata (9). During the mid-Proterozoic, especially the interval between 1.7 Ga and 1.1 Ga, numerous bodies of magma evolved that were emplaced in the evolving crust. Authorities differ on the source of the magmas and the reasons for their generation. Importantly, however, the igneous intrusions of this time interval gave rise to their own genre of ore deposits. One example is shown for comparative purposes in Figure 1, the Alkali Granites and their uranium-gold-copper-rare earth element (REE) ores (8); others include tin with certain granites, titanium with anorthosite massifs, and copper and REE with unusual intrusions of carbonate minerals. A manifestation of deeply stored heat beneath the cratons during these times are many diamond pipes, vertical carrot-shaped bodies in the crust believed to have derived by explosive release of mantle-derived deep thermal energy. Ensuing geologic time was witness to the intermittent development of these same kinds of rocks and same associated metals, but never again of the size and at the scale as that seen in the Proterozoic.

The Phanerozoic, the time since about 0.6 Ga ago, witnessed the origin of hydrocarbon accumulations as well as ore deposits of many genetic types representing accumulations of a wide variety of metals, many of which have a long history of intermittent formation. Virtually all known resources of hydrocarbon in the crust are of this age and were formed from modification and maturing of the great amount of biomass that was deposited in sediment during this geologic time. Further, although there is evidence of accumulation of evaporite minerals (salt, gypsum) in basins of older eons, the greatest surviving accumulation of this group of minerals corresponds with formation of the Pangean supercontinent during the mid-Phanerozoic, a time when ocean levels were lower than present times and marginal basins received intermittent influx of ocean water that evaporated.

The history of the formation, motion, and breakup of continental plates and ocean crust formation is attended by consistent temporal and spatial correlations with some styles of ore deposits. Significant ores formed on both the craton and in continental margin regions. Relevant examples are shown on Figure 1. Most deposits of copper, molybdenum, and tin of these times are associated with the intrusion of magmas to shallow crustal levels (1 km) in continental margin regions when ocean floor and continents were converging at centimeters per year rates; the ore deposits are recognized as stages in the development of andesite-dominant or rhyolite-dominant volcanoes. These ores are alluded to as "porphyry" types (10) because of their association with shallow igneous intrusions having porphyritic textures owing to rapid cooling of magmas. Figure 1 (10) shows that although the preponderance of occurrence of these ores are of Phanerozoic age, there are temporally scattered occurrences back into the Archean where similar tectonic and lithologic settings may have existed and where the intrusion systems appear to have escaped subsequent erosion. These same volcanic settings are the sites of the epithermal (shallow but hot) precious metal veins (12); their paucity in older rocks is attributed to their erosion from topographically high and exposed locations.

Bibliography

MEYER, C. "Ore-forming Processes in Geologic History." *Economic Geology, 75th Anniversary Volume* (1981): 6–41.

———. "Ore Metals through Geologic History." *Science* 277(1985):1321–1328.

———. "Ore Deposits as Guides to Geologic History of the Earth." In *Annual Reviews of Earth and Planetary Science*, eds. G. W. Wetherill, A. A. Albee, and F. G. Stehli. Palo Alto, CA, 1988, pp. 147–171.

SPENCER R. TITLEY

MINERAL DEPOSITS, IGNEOUS

VICTOR MORITZ GOLDSCHMIDT (1954) classified elements into groups according to their geochemical affinities (*see* GEOCHEMISTRY, HISTORY OF). Siderophile elements including Co, Ni, Mo, Au, and the Pt group (Pt, Pd, Os, Ir, Rh, and Ru) occur with iron and are presumably concentrated in the core of the earth. Chalcophile elements S, Se, Te, As, Cu, Ag, Pb, Sb, Bi, Zn, Ag, and Mo are concentrated in sulfide minerals while lithophile elements O, Si, Li, Na, K, Cs, Al, Ti, Be, Zr, Ca, Mg, W, U, Sn, and Mn reside in the silicic crust of the earth. Elements such as the Cr and the Pt group are found commercially only in igneous rocks rich in Fe and Mg and low in SiO_2 (ultramafic rocks). Chromium deposits are classified as syngenetic- or magmatic segregation-type deposits because they form as the surrounding igneous magma crystallizes. Copper-bearing deposit types, in contrast, are not only numerous but occur within many different kinds of rocks. They may form during host rock formation or later. The deposits are classified as epigenetic if they develop after host rock formation.

A variety of magma types and sites of magma generation exist. Magmas may originate in the mantle, for example, ultramafics such as peridotite, and mafics such as in basalt, or the crust, that is, granite and rhyolite. Some igneous rocks that are of mantle derivation only attain their final composition by processes such as magmatic differentiation (separation within the magma chamber) and/or contamination by country rock (intruded host) material that occurs in the crust. Ore deposits of magmatic origin are found in both intrusive and extrusive rocks of different compositions.

The formation of ore magmas requires liquid immiscibility (the development of ore and silicate magmas that, like oil and water, do not mix). Unusually high oxygen or sulfur contents are probably required to form oxide and sulfide magmas, respectively. Finally, a spectrum of ore deposit types exists from purely magmatic to late magmatic stage hydrothermal (hot water) to hydrothermal types where metals and/or fluids are igneous derived. This entry discusses only syngenetic igneous-hosted mineral deposits.

Chromium has been mined from chromite mineral-bearing deposits found in ophiolite complexes (pieces of the ocean floor composed of ultramafic rocks and basalt, or sediments) commonly attached to upper mantle ultramafic rocks. These rock assemblages were tectonically raised onto land exposing two types of chromite deposits. The basal section of seafloor rocks contains sack or pod-shaped masses of chromite; tectonized mantle rock stratigraphically below may contain lenses and disseminations of chromite. Important deposits are found in Cuba, the Philippines, and from Cyprus and Greece eastward.

The world's greatest concentration of mineable Cr, however, is found together with Pt group metals in large, layered, mafic intrusions that formed in cratons. Examples of these intrusions, some of which contain Cr and Pt deposits, include the Bushveld, South Africa (67,340 km^2 in outcrop), Sudbury complex, Canada (1,342 km^2), Duluth complex, Minnesota (4,715 km^2) and the Stillwater complex, Montana (194 km^2). Within the Bushveld and Stillwater complexes, chromite crystals together with Pt group minerals crystallized early and accumulated into layers near the base of the very large magma chambers. Stratigraphically above the chromite layers are valuable, thin mafic pegmatite horizons (reefs) containing chromite and Pt group minerals. Both the Merensky reef in the Bushveld and the JM reef in the Stillwater complex are being mined.

The Sudbury, Stillwater, and Duluth complexes have basal zones of massive and/or disseminated Ni and Cu sulfide minerals. Hypotheses, but no consensus, exist to explain the origin of the intrusions, complex layering, and precious metal-bearing reefs. Some intrusions, such as Sudbury, possibly resulted from a meteorite impact (Dietz, 1964).

Kimberlite, a mica-bearing potassium-rich peridotite, is the primary source of diamonds. This rock type apparently originates in the mantle at depths of 100 to 200 km, and rises into the upper crust in pipe- to carrot-shaped structures in cratonic, non-orogenic areas. Southern Africa and Russia contain the most valuable deposits (Skinner and Clement, 1979).

Magmatic segregation deposits of ilmenite-magnetite (Ti and Fe ore minerals) are found in anorthosite massifs (large intrusive bodies of igneous rock composed of at least 90 percent Ca-rich plagioclase). These massifs are middle Proterozoic in age and hosted by metamorphic rocks. They apparently formed in anorogenic intracontinental rifting settings. The ilmenite-magnetite bodies probably crystallized from an oxide liquid that separated from an anorthosite-gabbro parent magma. Examples include the Lake Sanford and Allard Lake deposits in New York State and Quebec, Canada, respectively.

Carbonatite, an unusual igneous rock composed of carbonate minerals, is found generally as cylindrically shaped intrusions in continental rift zones. The rock is mined for Cu, apatite, vermiculite mica, and by-product Fe, U, Co, Zr, Ha, Ni, Au, Ag; for Pt group metals at Palabora, Republic of South Africa; for rare earths such as La at Mountain Pass, California; and for the industrial minerals apatite, vermiculite, and barite elsewhere (Guilbert and Park, 1986).

Large iron oxide (magnetite and/or hematite) deposits are associated with intrusive and extrusive rocks of intermediate to felsic composition. The Fe-rich magmas, containing 4–5 percent P, apparently separate from silicate parent magmas. Iron Mountain, Missouri, contains intrusive and extrusive deposits; the deposits at El Laco, Chile, and Cerro de Mercado, Mexico, are lava flows; and the deposits at Kiruna, Sweden, may be either intrusive or extrusive deposits (Parak, 1975).

Because of the large size and charge of ions of elements such as U, common silicate mineral structures cannot accommodate them. These ions then accumulate in the last vestiges of silicic melt and form minerals such as uraninite and/or complex U-bearing minerals. These minerals may be disseminated within the rock and/or form veinlets. The Rossing deposit of Namibia and Bokan Mountain in southeast Alaska are examples.

Finally, pegmatites (very coarse crystalline bodies of generally granitic composition) are "store houses" of many valuable minerals and elements that do not fit structurally into common silicate minerals. These pod-shaped bodies, up to hundreds of meters in length, are late-stage magmatic crystallization features composed primarily of quartz and feldspar. Commercial quantities of Li, Be, Nb, Ta, Rb, Cs, muscovite mica, feldspar, silica, and the gem stones topaz, tourmaline, and emerald are found in pegmatites together with lesser amounts of Zr, U, Th, rare earths (REE), Sn, P, Cl, F, B, Mo, and W.

Bibliography

Dietz, R. S. "Sudbury Structure as an Astrobleme." *Journal of Geology* 72 (1964):412–434.

Goldschmidt, V. M. *Geochemistry.* Oxford, Eng., 1954.

GUILBERT, J. M., and Paper 575-D. 1967, pp. 123–126. C. F. PARK, JR. *The Geology of Ore Deposits.* New York, 1986.

PAGE, N. J., and E. D. JACKSON, eds. "Preliminary Report on Sulfide and Platinum Group Minerals in the Chromitites of the Stillwater Complex, Montana." In *Geological Survey Research,* U. S. Geological Survey Professional Paper 575-D. 1967, pp. 123–126.

PARAK, T. "Kiruna Ores Are Not Intrusive-magmatic Ore of the Kiruna Type." *Economic Geology* 70 (1975): 1242–1258.

SKINNER, E. M. W., and C. R. CLEMENT. "Mineralogical Classification of Southern African Kimberlites." In *Kimberlites, Diatremes, and Diamonds:* (I) *Their Geology, Petrology, and Geochemistry,* eds. F. R. Boyd and H. O. A. Meyer. Washington, DC, 1979.

IAN M. LANGE

MINERAL DEPOSITS, METAMORPHIC

Metamorphic hydrothermal fluids are derived by dehydration or decarbonation of minerals at high temperature and pressure. During metamorphism, rocks and minerals recrystallize as they are heated and squeezed deep within Earth. This recrystallization results in the release of water, carbon dioxide, and other volatile constituents to produce a hydrothermal fluid. For example, at high temperature, the common mica, muscovite, converts into feldspar and quartz with the release of water, and the common minerals quartz and calcite react to form wollastonite with the release of carbon dioxide gas:

Muscovite →
K-feldspar + quartz + water (dehydration)
$KAl_3Si_3O_{10}(OH)_2 \rightarrow KAlSi_3O_8 + SiO_2 + H_2O$

Calcite + quartz →
wollastonite + carbon dioxide (decarbonation)
$CaCO_3 + SiO_2 \rightarrow CaSiO_3 + CO_2$

Any reaction that results in the release of liquid or gas is known as a devolatilization reaction. Other constituents released during metamorphic reactions include carbon monoxide (CO), methane (CH_4), and hydrogen sulfide (H_2S) gas, and trace metallic and nonmetallic ions.

Devolatilization reactions produce vapor-rich hydrothermal fluids. However, to carry metals in solution, a hydrothermal fluid must generally be saline because chloride ion is required to form soluble metal complexes. Metamorphic hydrothermal fluids are not highly saline, and therefore have not traditionally been considered important sources of ore-forming fluids. The notable exception to this generalization involves gold. Unlike many other metals, gold forms a highly soluble bisulfide complex [$Au(HS)_2^-$] over a relatively wide range of temperatures. Because metamorphic reactions commonly produce sulfur, most commonly as hydrogen sulfide (H_2S) gas, metamorphic fluids may be important carriers of dissolved gold. This gold may be precipitated when the bisulfide complex is destabilized by cooling, or by oxidation, acidication, or neurtralization reactions:

$Au(HS)_2^- + H^+ + 1/2H_2O \Leftrightarrow Au^\circ + 2H_2S + 1/4O_2$ (acidification)

$Au(HS)_2^- + 1/2H_2O \Leftrightarrow Au^\circ + 2HS^- + H^+ + 1/4O_2$ (neutralization)

$Au(HS)_2^- + 1/2H_2O + 15/4O_2 \Leftrightarrow Au^\circ + 2SO_4^= + 3H^+$ (oxidation)

Gold may also precipitate if the bisulfide complex reacts with iron in rocks to form iron sulfide minerals, such as pyrite (FeS_2) and pyrrhotite (FeS), a process known as sulfidation.

Metamorphic Hydrothermal Fluids

A famous example of a mineral deposit believed to have formed from metamorphic hydrothermal fluids is the so-called Mother Lode of central California, which fueled the great Gold Rush of 1849. These deposits, in which gold occurs within quartz-carbonate veins, are referred to as mesothermal because they formed from moderate temperature fluids (250–450°C) associated with regional metamorphism. Similar deposits occur worldwide in Archaean age rocks (greater than 2.5 billion years or Ga old), and include the famous Yellowknife and Porcupine-Timmins-Kirkland Lake districts in Canada, and the Golden Mile of the Kalgoorlie district in west Australia.

Mesothermal gold deposits are characterized by free gold or gold tellurides within milky white quartz veins. The veins also contain carbonate minerals, most commonly ankerite [$Ca(Mg,Fe)(CO_3)_2$] or dolomite [$Mg(CO_3$], and iron sulfides, but only minor amounts of other base metal sulfides such as stibnite (Sb_2S_3) and arsenopyrite (FeAsS). The veins vary in thicknesses from centimeters to meters, but they invariably occur near very large-scale regional structures, such as major faults and shear zones, tens or hundreds of meters wide and up to several hundreds of kilometers long. Most commonly, the host rocks are mafic or ultramafic igneous rocks, known as greenstones, that display a characteristic wall rock alteration to carbonate-chlorite-sericite-schist produced by hydration and carbonation of primary mafic minerals in the wall rock at low-moderate temperatures.

The characteristic alteration assemblage and the lack of abundant associated base metals indicate that mesothermal gold-quartz veins are precipitated from low salinity, CO_2-rich fluids that could be produced by regional metamorphism and devolatilization of greenstones. Metamorphogenic fluids would have been concentrated into and migrated along large-scale regional structures; gold would have been precipitated by cooling and changes in fluid chemistry caused by wall rock interaction during the late stages of regional metamorphism. However, although the occurrence of mesothermal gold deposits within regionally metamorphosed rocks is unequivocal, the origin of the gold-bearing hydrothermal fluids remains controversial, primarily because the association with crustal-scale structures suggests the possibility of tapping deeper sources of hydrothermal fluids. Some geologists have pointed out that low-salinity CO_2-rich fluids can also be produced by separation of a magmatic-hydrothermal fluid early in the crystallization history of a mantle-derived magma. Others have proposed a "granulitization" model, in which CO_2 released from the upper mantle (or from a thickened oceanic crust) mixed with water derived from dehydration reactions in the lower crust to form partial melts and CO_2-rich fluids; the latter leached gold and carried it upward along structures. Finally, some geologists have argued for the involvement of meteoric waters, which could have penetrated deeply into the crust along regional structures during the late stages of metamorphism (Keays and Skinner, 1988).

Metamorphic and Metamorphosed Ore Deposits

In addition to those mineral deposits formed from metamorphic hydrothermal fluids, some ore deposits consist of minerals formed directly by metamorphic mineral reactions. Examples include industrial minerals such as asbestos and talc, which form by metamorphism of ocean-floor basalts. Finally, some hydrothermal ore deposits, because of their geological context, invariably become metamorphosed as part of their geological history. For example, volcanogenic massive sulfides deposits, which form on the seafloor, are invariably metamorphosed during their subsequent incorporation into continental crust during plate tectonic collisions. The resulting metamorphism may increase the grade of the ore, but it may also obscure the origins of the deposit.

Perhaps the most famous example of a metamorphosed ore deposit is the lead-zinc-silver (Pb-Zn-Ag) district of Broken Hill, Australia, the origins of which were hotly debated in the early part of the twentieth century. The diverse and unusual minerals found there seemed to suggest a much greater range of temperatures of formation than normally expected in hydrothermal ore deposits. In the 1950s, geologists demonstrated that the ores were metamorphosed hydrothermal deposits, originally deposited in sedimentary rocks. Recrystallization had improved ore grades somewhat by eliminating impurities within the sulfides, but the overall metal grades and metal ratios remained essentially undisturbed (Stanton, 1972). Thus, at Broken Hill, metamorphism itself was not a significant cause of base metal sulfide mineralization. However, as indicated above, debate continues about the possibility of metamorphic remobilization of mineral components in other geological settings.

Bibliography

GUILBERT, J. M., and C. F. PARK, JR. *The Geology of Ore Deposits.* New York, 1986.

KEAYS, R. R., and B. J. SKINNER. "The Geology of Gold Deposits: The Perspective in 1988." *Economic Geology Monograph* 6 (1988):1–8.

STANTON, R. L. *Ore Petrology.* New York, 1972.

NAOMI ORESKES

MINERAL PHYSICS

Mineral physics encompasses the study both of Earth and planetary materials and of related substances. Minerals, traditionally defined as homogeneous, naturally occurring, inorganic crystals having well-defined chemical compositions, represent only a subset of the materials that are studied. In addition, melts and glasses (noncrystalline materials), biominerals and other substances of organic origin, polycrystalline aggregates such as rocks, and synthetic materials that are related to minerals in structure or properties all fall within the domain of mineral physics.

The approach of mineral physics is to understand the packing or spatial arrangement of atoms (crystalline and noncrystalline structures), the chemical bonding forces between atoms, and the consequent bulk properties of mineral-like materials. That is, a major aim is to explain macroscopic properties in terms of microscopic or atomic-scale phenomena. Both chemical and physical properties are of interest, so that "mineral physics" is really a short-hand term for studies of the chemistry and physics of mineral-like materials (*see* MINERALS AND THEIR STUDY).

As a discipline within the earth and planetary sciences, one of the primary goals of mineral physics is to provide constraints on the nature and geological evolution of planets through the understanding of mineral properties. Therefore, much effort is placed on studying minerals and related materials at the high pressures and temperatures characteristic of planetary interiors. More generally, mineral physics concerns itself with ceramics, metal alloys, and other materials that resemble minerals in various ways. It thus overlaps a wide variety of disciplines, ranging from condensed-matter physics and solid-state chemistry to materials science.

Over the past few decades, several discoveries associated with mineral physics have contributed significantly to the earth and planetary sciences (Table 1). One of the most important has been the recognition that the majority of crystalline substances known at Earth's surface transform to new, more densely packed structures when taken to the high pressures of the interior.

Perhaps the most famous example is the transformation of graphite to diamond at conditions existing approximately 150 kilometers (km) beneath the surface. Another important example is stishovite: a dense crystalline form of SiO_2 that was first discovered in high-pressure laboratory experiments and then found in nature at Arizona's Me-

Table 1. Selected Accomplishments Associated with Mineral Physics

- Minerals undergo crystallographic transformations at high pressures:
  - Confirms significance of meteorite impacts on Earth.
  - Explains metamorphic rocks observed at the surface.
  - Helps explain the seismological structure of the mantle and constrain the temperature at depth.
- Primary mineral of Earth's interior revealed: high-pressure silicate perovskite phase.
- Magmas undergo structural changes with depth; affects differentiation throughout bulk of planet.
- Chemical diffusivity measurements demonstrate that fluid–solid separation is the primary means of global geochemical differentiation.
- Pressure tends to suppress temperature-induced volatility:
  - Components that are "volatile" near Earth's surface can be involatile at greater depth.
  - Interior may be a major reservoir of H_2O and CO_2.
- Many elements change in chemical-bonding character deep inside planets:
  - Hydrogen is metallic inside giant planets.
  - Oxygen becomes a metallic alloying component inside large, terrestrial planetary cores.
- Minerals and rocks deform at geological strain rates when subjected to planetary interior stresses and temperatures:
  - Planetary materials are rheid (deform by viscous flow).
  - Movements of atomic-scale defects cause deformation.
- Thermal diffusivity measurements show that solid-state convection is the primary means of global heat transfer and tectonic deformation.
- Both pressure and temperature suppress brittle deformation.
- Minerals can retain reliable paleomagnetic information (both direction and intensity).
- Overlap with other disciplines:
  - Numerous organisms, past and present, synthesize magnetic minerals; biomagnetic activity confirmed in several cases.
  - Advanced materials, including CVD diamond and C_{60} fullerene, identified in nature.
  - Polytypism and modulation of crystallographic structures established on all scales.

teor Crater, where it formed during the transient high pressures caused by the iron meteorite impact that produced the crater. The discovery of stishovite and other high-pressure mineral phases at sites around the world has provided conclusive evidence that extraterrestrial impacts on Earth have been significant in the geological past. Impact is now viewed as one of the main processes by which planets are formed and modified early on.

All of the important minerals of Earth's crust and outermost mantle transform to new crystalline structures at high pressures, as is the case with many related synthetic materials (analog compounds). Thus, just as the minerals SiO_2 quartz and Mg_2SiO_4 olivine transform to stishovite and a spinel form, respectively, the analogous GeO_2 and Mg_2GeO_4 exhibit the same types of structural transformations. Indeed, germanates are attractive analog materials to study because they transform at much lower pressures, hence at more accessible conditions, than the corresponding silicates.

Much of the seismological structure of the mantle, consisting of variations in density and wave velocities with depth, is convincingly explained in terms of the high-pressure mineralogical transformations that have been documented in the laboratory. The observed depths of the seismological transitions can be translated directly into pressures. As a result, comparison with the experimentally measured transformation pressures, which depend on temperature, provides a first-order constraint on the temperature in the mantle: it is near 2,000 K at 650-km depth, for example.

Remarkably, the principal minerals of the upper mantle all transform to perovskite-structured (Mg, Fe)SiO_3 ($\pm$additional oxides) at conditions 600–700 km beneath the surface. This perovskite-type mineral is stable to the bottom of the mantle and is therefore considered the single most abundant material of the planet, making up over 40 percent of Earth's interior by volume. The physical properties of the high-pressure perovskite phase have no doubt played an important role in controlling the geological evolution of the planetary interior.

Just as crystalline structures are transformed, the atomic arrangements in molten materials are also found to transform with increasing pressure. The result is that melts become more tightly packed, rapidly increasing in density as pressure is raised. Near Earth's surface, magmas are typically less dense than surrounding rocks: the melts are buoyant and are consequently erupted volcanically. At depth, however, melts tend not to be so buoyant; they may even be denser than the rocks from which they form, implying that the melts would sink.

This is significant because the primary means by which the major regions of the planet become distinct—the geochemical differentiation by which the metallic core, rocky mantle and crust, and fluid atmosphere and hydrosphere separate from each other—is through processes involving the formation and movement of fluids. Measurements show that chemical diffusion is far too slow to account for global-scale transport of chemical species, so the only effective way for differentiation to occur is through partial melting (or vaporization). The resulting fluid typically differs greatly in composition from the surrounding rock.

If it is sufficiently buoyant, the fluid can move over large distances, thereby causing global-scale separation of chemically distinct regions (e.g., extraction of the atmosphere, hydrosphere, or crust from the underlying mantle). What the high-pressure experiments show, however, is that melt buoyancy becomes limited at depth. Hence, geochemical differentiation is likely to be different, and perhaps far less efficient, for the bulk of Earth's interior than is seen in the volcanic activity of the near-surface environment.

A related discovery bears on "volatile" species, such as H_2O and CO_2, which readily escape from rocks that are heated at low pressures. Yet this is no longer the case at elevated pressures. Laboratory studies establish that these normally volatile molecules can become strongly locked within mineral structures at conditions occurring just a few hundred kilometers into the mantle. In short, pressure strongly counteracts the effect of temperature in causing volatile release. Consequently, it is thought that the deep mantle may be far more enriched in volatile components than is observed for the outermost mantle. Indeed, the bulk of Earth's H_2O and CO_2 might reside at depth, within the solid planet, rather than in the atmosphere and hydrosphere.

It is not just the geometric packing of the atoms, but even the chemical bonding between atoms that can change radically with pressure. Hydrogen, for example, is theoretically expected to become a metal at the conditions existing inside giant planets (e.g., Jupiter and Saturn). Recent experiments are in accord with this expectation and support the view that metallic hydrogen is the primary constituent of the giant planets and the Sun.

For Earth, it is oxygen rather than hydrogen that is the predominant element; after all, rock is more than 50 percent oxygen. It is therefore significant that oxygen changes dramatically in chemical properties, becoming a metallic alloying component at deep-mantle conditions. Thus, oxygen is now thought to be a significant constituent of Earth's metallic core, where the geomagnetic field is produced, and the rocky mantle may even be slowly combining with (dissolving into) the underlying liquid core over geological time.

Recent studies have shown that materials can exhibit novel behavior at the conditions existing deep inside planets. However, it is over the exceptionally long time periods of geological history that material properties are found to be truly remarkable. Now the distinction between liquid and solid behavior becomes quite blurred, such that the crystalline rock of the mantle—brittle, solid material over human timescales—flows vigorously over periods of tens and hundreds of millions of years. A more familiar, and less extreme, example of this is in the way that glaciers made of "solid" ice flow down mountainsides.

For both cases, ice and rock, it is the movement of atomic-scale defects such as dislocations that provides the fundamental mechanism of large-scale deformation. And, it is the long-term flow of the mantle that ultimately causes plate tectonics, volcanic eruptions, earthquakes, and most of the other geological processes of Earth's crust. Indeed, measurements of the thermal conduction, optical properties, and deformation rates of minerals at high temperatures confirm that it is this large-scale tectonic motion, the fluid-like convection of the mantle, that is the primary means by which the deep interior is cooling down.

Although largely explained by plate-tectonic motions (*see* PLATE TECTONICS), one feature of the global pattern of earthquakes remains enigmatic. Experimental studies demonstrate that rock loses its brittle character at high pressures, as well as at high temperatures and over long time periods. It is therefore hard to understand why some earthquakes take place at great depths within subduction zones, down to about 700 km beneath the surface and far below the approximately 100–150 km thick lithosphere. Some experiments suggest that structural instabilities can take place in the subducted minerals at high pressures and might be the cause of the deep-focus earthquakes. However, the evidence remains incomplete, and further work will be required to fully understand the deepest seismicity of the mantle.

Historically, observations of magnetic "stripes" within the oceanic crust provided some of the first clear evidence for plate tectonics. Interpreted as a record of the geomagnetic-field orientations in the past, the magnetization pattern of the oceanic crust is like a tape recorder documentation of geological motions. It was thus a major accomplishment to show that many rocks and minerals can retain a reliable record of the orientation and even the intensity of Earth's magnetic field in the distinct geological past. One of the most profound results of paleomagnetism, this is the basis for deciphering the past history of the magnetic field, showing that it has flipped in orientation or polarity numerous times (a few times per million years, on average).

An unexpected outcome of studies in rock magnetism has been the discovery that a wide variety of living organisms contain magnetic minerals. Though typically small in abundance, biomagnetic minerals have been found in bacteria, bees, fish, birds, whales, and even humans, and can also be detected in ancient fossils. Perhaps most remarkable of all is the demonstration that the magnetic minerals are not only produced by the organisms, but also have biological functions for at least a few species.

This is but one example of how research related to mineral physics overlaps strongly with disciplines outside the earth sciences. Another is the recognition that some of the most advanced materials, first discovered in laboratory syntheses, have subsequently been found in nature, where they were created far in the geological past: diamond produced by chemical vapor deposition (CVD) and C_{60} fullerene are notable examples.

Even the concept of crystalline structure—the underlying atomic-scale description for most materials—has been dramatically broadened by studies of minerals. Specifically, a given crystal structure can exhibit "incommensurate" modulations (nonproportionate variations in structure) and different structures can be intermixed down to the smallest (unit-cell or atomic) scales, such that a complete spectrum is now recognized between ordered and disordered arrangements of atoms. These supra-crystalline structures are not only found in nature, notably among the biopyriboles (mineral grains that consist of intimate mixtures of mica, pyroxene, and amphibole crystal structures on the unit-cell scale), but are also characteristic of the new high-temperature superconducting oxides.

Bibliography

BROWN, G. C., and A. E. MUSSETT. *The Inaccessible Earth,* 2nd ed. New York, 1991.
PUTNIS, A. *Introduction to Mineral Sciences.* New York, 1992.

RAYMOND JEANLOZ

MINERALIZED MICROFOSSILS

Microfossils are any tiny remains of organisms ranging from bacteria to vertebrates. They include the microscopic shells or tests of whole organisms, or the microscopic parts of larger organisms. Microfossils are a heterogeneous assemblage without any phylogenetic relationship grouped together for the practical reason that they can be found through the microscopic study of sedimentary rocks. Their microscopic size makes their recovery from small rock samples likely; hence, they are valuable in geologic applications and have been much studied for over 150 years.

Not all microscopic organisms or parts of organisms fossilize easily because most do not have fossilizable parts (*see* FOSSILIZATION AND THE FOSSIL RECORD). The ability to biomineralize or otherwise construct skeletal parts makes some groups particularly abundant in the microfossil record. These groups use organic compounds or a wide variety of hard materials—not all are minerals by proper definition, but they are resistant and hence fossilize. The mineral-like materials include hydrated SiO_2, $CaCO_3$, PO_4, $CaPO_3$, and grains of rocks, minerals, or even other microfossil skeletons (Table 1). These are assembled or secreted into whole or parts of skeletons that can then be found in the rock record. Organic-walled microfossils, such as phytoplankton cysts, spores, and pollen, are also very common in the fossil record, but are not considered here (*see* PALYNOLOGY).

Because many microfossils are not only extremely abundant but represent particular environments or geologic times, they are extremely important in deciphering the geologic history of life and of the earth. They are useful in biostratigraphy, age dating, paleoenvironmental interpretation, paleoecology, paleoceanography, paleoclimatology, among others, and can be useful in the modern counterparts of these disciplines (*see* PALEOECOLOGY; PALEOCLIMATOLOGY; OCEANOGRAPHY, BIOLOGICAL). Indeed, some entire areas of earth history, for example, paleoceanography, are based almost entirely on data retrieved from microfossils. Microfossils have been utilized in industry for decades to aid in the exploration for petroleum worldwide, generating billions of dollars of productivity. In the diatomite industry, microfossils are the resource itself, and when processed, they are made into filters, abrasives, chemical and pigment carriers, and many other items in daily use.

Microfossils represent the first known life on Earth, found in rocks some 3.5 billion years old (Ga). These are the mineralized sheaths of cyanobacteria found in silicified stromatolites (bacterially constructed calcareous mounds) in Australia and South Africa. Other than cyanobacteria, which secrete $CaCO_3$ around themselves, mineralized microfossils are virtually absent from the fossil record until about 545 million years (Ma) ago. Other microfossils were present from about 1.6 Ga ago, but they are organic in composition. Only in the very latest Precambrian are tiny elliptical and circular siliceous plates, probably from some kind of unicellular algae, found.

Like larger animals, microfossils appear abundantly at the base of the Cambrian, the first period of the Paleozoic era (*see* HIGHER LIFE FORMS, EARLIEST EVIDENCE OF). These skeletons include many types of hard parts—silica, carbonate, phosphate, and agglutinated materials. Many are tiny parts of larger animals that appeared at that time, but others represent new appearances of mineralized single-celled protists themselves. While most microfossils derived from animals and some protists appear in the Cambrian, others do not occur until much later (Table 1).

The first appearance of mineralized microfossils in the Cambrian, however, probably does not mark the initial evolutionary development of the groups. DNA sequence data from some living representatives of those ancient forms suggest that they diverged much earlier in geologic time, perhaps more than 1 Ga ago. When microfossils are first found, they clearly represent advanced, divergent groups of organisms that must have had a significant earlier history. None of this, however, is

Table 1. Mineralized Microfossils

Mineralized Microfossil	*Description of Fossils*	*Size*	*Composition of Fossils*	*Habitat*	*Geologic Range*
Prokaryotes: Single-Celled Organisms with DNA Loosely Organized in the Cell					
Cyanobacteria	Unicells	<50 μm	Some with $CaCO_3$ sheaths	Shallow saline and normal marine waters	Archean-Recent
Eukaryotes: Single or Multicellular Organisms with DNA Organized in the Nucleus of the Cell					
Algae: Single-celled, photosynthetic eukaryotes:					
Chlorophytes	Unicells	100–400 μm	$SiO_2 + nH_2O$, calcite, aragonite	All	Proterozoic-Recent
Charophytes	Spheres, ovoids	0.5–3 mm	Calcite	Freshwater	Silurian-Recent
Dinoflagellates (endoskeletal)	Lattice of rods	5–150 μm	$SiO_2 + nH_2O$	Marine planktonic	
Ebridians	Mesh of solid rods	<200 μm	$SiO_2 + nH_2O$	Marine planktonic	Paleocene-Recent
Diatoms*	Circular, elongate, or irregular duel valves	<200 μm	$SiO_2 + nH_2O$	Nonmarine, marine, soil, ice	Jurassic-Recent
Chrysomonads	Spherical cysts	<100 μm	$SiO_2 + nH_2O$	Aquatic	Neoproterozoic-Recent
Silicoflagellates	Lattice of hollow rods	<100 μm	$SiO_2 + nH_2O$	Marine planktonic	Cretaceous-Recent
Calcareous nanno-fossils*	Plates, asters, rods	<50 μm	Calcite	Marine	Silurian (?), Pennsylvanian (?) Triassic-Recent
Protozoans: Single-celled heterotrophic eukaryotes:					
Arcellinids	Single-chambered tests	<150 μm	Agglutinated	Freshwater, damp litter	Mississippian, Tertiary-Recent
Foraminifera*	Single- or multi-chambered tests	0.01–100 mm	Agglutinated, calcite, aragonite, $SiO_2 + nH_2O$	Marine bethonic and planktonic	Cambrian-Recent

Radiolaria*	Spherical or conical mesh of solid rods	0.03–1.5 mm	$SiO_2 + nH_2O$	Marine planktonic	Cambrian-Recent
Acantharians	Spicules	<0.05 mm	$SrSO_4$	Marine planktonic	Eocene-Recent
Heliozoans	Spines, scales	<0.05 mm	$SiO_2 + nH_2O$	Marine, freshwater	Pleistocene-Recent
Tintinnids	Single-chambered lorica	<300 μm	Agglutinated, $CaCO_3$	Marine planktonic	Ordovician-Recent
Fungi: Mostly multicellular eukaryotes that absorb digested food:					
Fungi	Coccoid, filaments	<10 μm	Some calcite	All	Precambrian-Recent
Animals: Multicellular eukaryotes that ingest their food:					
Sponges*	Spicules	<100 mm	$SiO_2 + nH_2O$, $CaCO_3$	Marine and freshwater benthonic	Cambrian-Recent
Octocorallians	Spicules	1–3 mm	$CaCO_3$	Marine benthonic	Silurian-Recent
Brachiopods	Tiny shells	>0.1 mm	$CaPO_4$, calcite	Marine benthonic	Cambrian-Recent
Bryozoans	Zooids	<0.5 mm	$CaCO_3$	Marine, freshwater benthonic	Ordovician-Recent
Ostracods*	Bivalved carapace	0.1–5 mm	$CaCO_3$	Freshwater, marine	Cambrian-Recent
Echinoderms	Spicules, plates	0.05–1 mm	Calcite	Marine	Cambrian-Recent
Conodonts*	Toothed or flattened plates	0.1–3 mm	$CaPO_4$	Marine	Precambrian-Triassic
Mollusks	Univalves, mostly coiled, bivalves	>0.1 mm	Calcite	Marine, terrestrial	Cambrian-Recent
"Fish"	Bones, teeth, and scales	>0.1 mm	$Ca_5[PO_4]_3$	Marine, terrestrial	Cambrian-Recent
"Higher vertebrates"	Bones, teeth	>1 mm	$Ca_5[PO_4]_3$	Terrestrial, marine	Devonian-Recent

* Particularly abundant and useful microfossils.

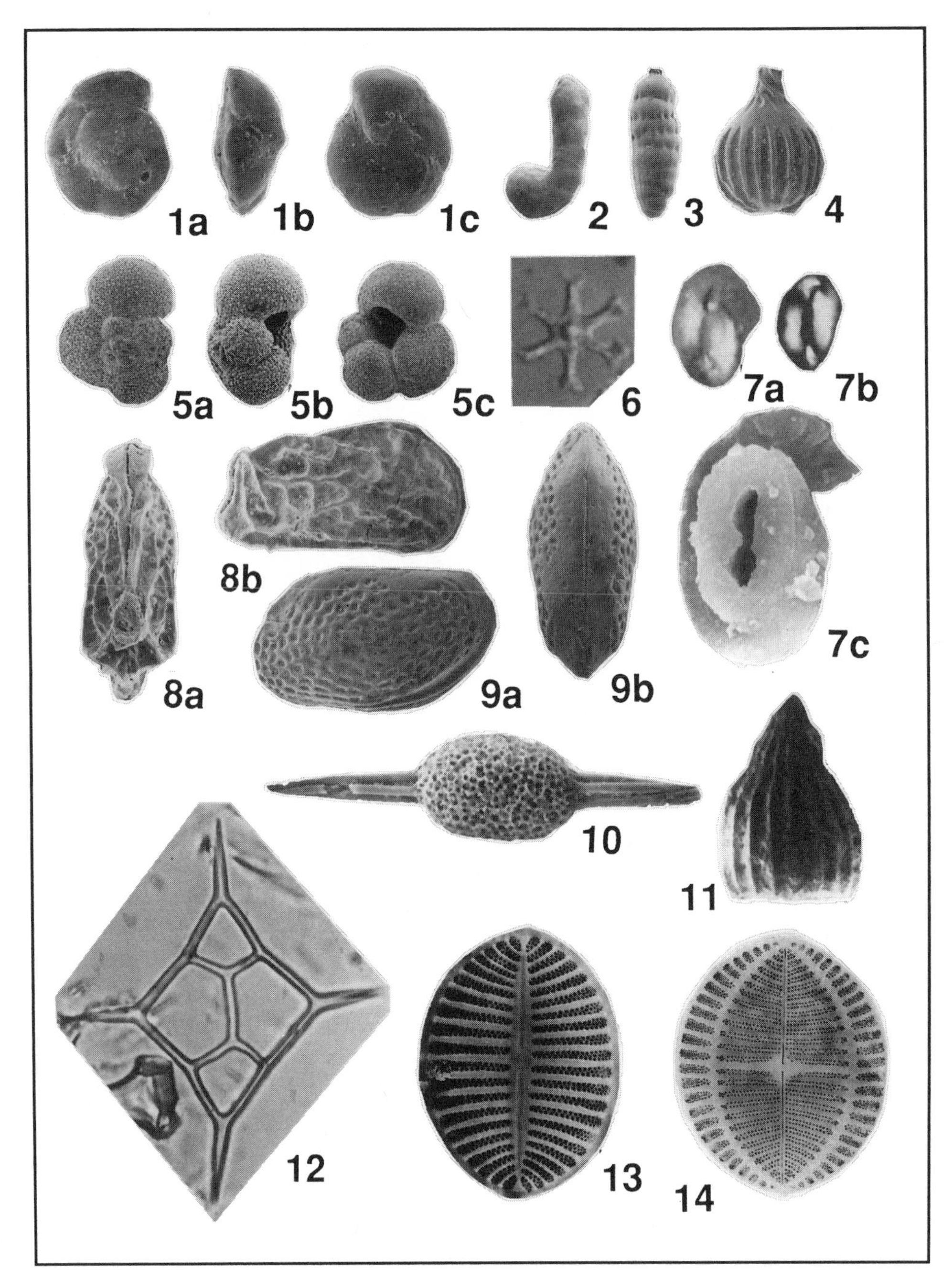

Figure 1. A selection of important kinds of mineralized microfossils found in the geologic record (see Table 1), including calcareous (1–9) and siliceous (10–14) kinds.

Calcareous Microfossils: Foraminifera (1–5), benthic and planktic protozoa, from the Miocene of central California (from Finger, Lipps, Weaver, and Miller, *Micropaleontology* 36 [1990]:1–55): 1. *Oridorsalis umbonata,* a deep-water benthic species: a. spiral view, b. edge view, c. umbilical view (× 134). 2. *Marginulinopsis* sp. a benthic species (× 28). 3. *Rectuvigerina loeblichi,* a probable infaunal benthic species (× 30). 4. *Lagena timmsana,* a benthic species (× 115). 5. *Globigerina bulloides,* a planktic species: a. spiral view, b. edge view, c. umbilical view (× 112). All illustrations are scanning electron micrographs.

Calcareous nannofossils (6, 7), planktic algae, from the Miocene of central California (from Finger, Lipps, Weaver, and Miller, *Micropaleontology*

known yet, as few, if any, mineralized microfossils have been found in older rocks. Even among microfossils that occur later in the fossil record, DNA evidence suggests a longer history than is represented by the fossils themselves.

After the initial appearance of microfossils in the Cambrian, their diversity increased among both benthic and planktic types, just as among larger fossils. Many invertebrate microfossils (see Table 1) diversified then, but agglutinated foraminifera and siliceous radiolaria are also first found with them. In addition, organic-walled microfossils and larger calcareous benthic algae became more diverse and complex. In the Ordovician and into the Silurian, all these groups radiated again with many more species of invertebrates, foraminifera, and radiolaria occurring, along with the first appearance of tintinnids. In the later Silurian, foraminifera with calcareous tests began to appear. Thus, all major kinds of organisms and mineralized types appeared between 545 and 400 Ma ago.

These initial waves of radiation populated the seas with familiar kinds of larger and smaller organisms. Although not exactly the same as modern kinds, nearly all can be placed in modern groups at some level. This similarity of form indicates that the general body plans not only of invertebrates but protists as well were established early in the Paleozoic. The radiation of body plans, species and higher taxa, biomineralization, habitat utilization, and feeding strategies further indicates that these were responses not to some internal biologic selection mechanism, but to factors in these early ecosystems. One mechanism that transcends all habitats, body plans, and skeletal types is trophic interactions—the flow of energy between the various kinds of organisms. The way in which this energy flowed, as carbon fixed by photosynthesizing algae eventually to carnivorous animals, became more complex.

In the later Paleozoic, the fossil record indicates that marine life underwent a series of extinctions and reradiations, each transforming the biota in certain ways. The microfossil record, of course, parallels that of larger fossils. Some particularly noteworthy developments were the great diversification of fusulinid foraminifera, especially in the Carboniferous, and of radiolaria throughout the later Paleozoic. These indicate that both benthic and planktic organisms were radiating at similar times.

At the end of the Paleozoic, a major extinction took place that reduced the marine biota by at least

36 [1990]:1–55): 6. A single star-shaped plate from *Discoaster variabilis* in transmitted light micrograph (× 2,200). 7. Plates from *Helicosphaera scissura:* a. Nomarski micrograph of a plate (× 2,150), b. polarized light micrograph of a plate (× 2,000), c. scanning electron micrograph of a single plate (× 4,500).

Ostracodes, benthic metazoans, from the Miocene of central California (from Finger, Lipps, Weaver, and Miller, *Micropaleontology* 36 [1990]:1–55): 8. *Ambostracon* sp.: a. dorsal view of paired valves, b. right lateral view (× 73). 9. *Loxoconcha* sp.: a. left lateral view, b. dorsal view of paired valves (× 110). All illustrations are scanning electron micrographs.

Siliceous Microfossils: Radiolaria, planktic protozoa. 10. *Archaeospongoprunum salumi* from the Cretaceous of Northern California (× 250). 11. *Thanarla* sp. from the Cretaceous of the deep sea, Pacific Ocean (× 250). Both illustrations are scanning electron micrographs courtesy of Michelle Silk.

Silicoflagellate, a planktic alga. 12. *Dictyocha fibula* from the Miocene of the Pacific Ocean, Chatham Rise, east of New Zealand (× 1,000) (from D. Bukry, *Initial Reports of the Deep Sea Drilling Project* 40 [1986]:925–937).

Diatoms, benthic and planktic algae, from Holocene fjord sediments in the Vestfold Hills, Antarctica (from Whitehead and McMinn, *Marine Micropaleontology*, 1996). 13. *Cocconeis costata* (× 1,300) and 14. *Cocconeis fasciolata* (× 1,250), both attached benthic diatoms. Illustrations are scanning electron micrographs.

80 percent (*see* EXTINCTIONS). All microfossils underwent this extinction as well. Although the general types of organisms did not go extinct, some important groups died out. Among microfossils, fusulinid foraminifera disappeared altogether, just like trilobites and others. Some other groups, like the microfossil conodonts and larger invertebrate ammonites, declined to a few species at the end of the Paleozoic, but did not disappear until later.

After the great extinction, the Mesozoic witnessed a reradiation of all groups. Among the microfossils, in particular, was the appearance of calcareous plankton and a radiation of additional siliceous plankton, as well as unmineralized organic plankton. Calcareous nannoplankton, a heterogeneous group bearing microscopic $CaCO_3$ plates and rods on a single cell, foraminifera, radiolaria, diatoms, and silicoflagellates became important constituents of the plankton, changing the deposition of carbonates and siliceous sediments largely from the continental regions to the deep sea. The calcareous nannoplankton and foraminifera began to extract $CaCO_3$ and diatoms, and to a much lesser extent silicoflagellates, began to extract silica from the surface waters of the world's oceans to make their skeletons, which were then deposited on the deep seafloor. This coincided with the breakup of the supercontinent of Pangaea and the formation of the modern ocean basins. Perhaps the invasion of the pelagic realm and diversification of so many kinds of plankton, including a number of larger animals as well, were related to increased complexity and intensity of circulation patterns in these oceans. At the end of the Mesozoic, a major extinction struck the plankton and shallow water benthos, especially reefs. This event reduced open-ocean planktic and certain benthic microfossil groups from many tens of species to just a few. One possible cause could have been the collision of an asteroid with Earth at this time—evidence now suggests that an impact crater of the right age is located in Yucatan, Mexico. Exactly how such an impact might affect shallow-water planktic and benthic organisms is still conjectural, but it must have caused great atmospheric and oceanographic reorganization that surely changed habitats significantly (*see* IMPACT CRATERING).

As with previous extinctions, planktic and benthic life reradiated into a wide variety of similar types, many represented by microfossils. Indeed, most of the groups that existed in the Mesozoic can be found in the following Cenozoic. The Cenozoic is characterized by a general cooling of high latitudes, especially in the last 25 Ma that culminated in the alternating glacial-interglacial periods of the last few million years. This cooling took place chiefly in high latitudes, resulting in oceanographic and climatic changes. Oceanographic circulation changed from warm deep water derived from dense, high salinity water originating in tropical seas to dense, cold water from polar regions. This created a vertical thermal gradient from top to bottom and a horizontal gradient from the poles to the equator that permitted evolutionary radiation of benthic and planktic microorganisms, as well as larger animals, into the new environments. Likewise, on land new habitats developed as climates changed, and the terrestrial biota changed too. In the past million years, microfossils recorded the fluctuating glacial changes through biogeographic changes and stable isotope shifts found in their skeletons. This glacial-interglacial record indicates that the current global climate is interglacial and that the earth can be expected to return to a glacial climate sometime in the next few thousand years.

While there are many kinds of microfossils that may contribute to our knowledge of this history, the most useful mineralized ones are foraminifera, radiolaria, calcareous nannofossils, diatoms, ostracodes, and conodonts (Figure 1). The last two are metazoans, while the first two are protozoans and the middle two are single-celled algae. The fossil record of all groups contributes enormously to earth and life history.

Bibliography

BOARDMAN, R. S., A. H. CHEETHAM, and A. J. ROWELL, eds. *Fossil Invertebrates.* Oxford, Eng., 1987.

CLARKSON, E. N. K. *Invertebrate Palaeontology and Evolution.* London, 1986.

HAQ, B. U., and A. BOERSMA. *Introduction to Marine Micropaleontology.* New York, 1978.

LIPPS, J. H. "What, If Anything, Is Micropaleontology?" *Paleobiology* 7 (1982):167–199.

LIPPS, J. H., ed. *Fossil Prokaryotes and Protists.* Oxford, Eng., and Boston, 1992.

TAPPAN, H. *The Paleobiology of Plant Protists.* San Francisco, 1980.

JERE H. LIPPS

MINERALOGY, HISTORY OF

The earliest interest in minerals undoubtedly sprang from their use as utilitarian and decorative materials. We find mention of numerous minerals in ancient writings, particularly their use in jewelry; minerals also hold a certain fascination because of reputed mystical properties. The earliest surviving study of minerals are the *Vedas* (1100 B.C.E.). The first scientific work concerning mineralogy is the *History of Stones* by Theophrastus (370–287), who, as a student of Aristotle, dealt with the physical properties of minerals rather than their mystical properties, as had earlier authors. Pliny the Elder (23–79 C.E.) documented the then current knowledge of all known minerals in an encyclopedic work titled *Historia Naturalis*. At least five volumes of this work dealt with practical aspects of minerals, including the preparation of gemstone fakes. This latter topic was apparently in vogue because the Roman Emperor Diocletian ordered the destruction of all books describing the faking of gems in approximately 300 C.E. George Bauer (latinized to Georgius Agricola) was the city physician for Joachimsthal, Bohemia, but his consuming interest in the local mining industry led him to accumulate an unprecedented amount of information concerning mining practices, smelting, and minerals. His books *De Natura Fossium* (1546) and *De Re Metallica* (1556) are considered the foundations of mineralogy. The latter book, in particular, contains fascinating accounts of contemporary mining practices. The former book set out the first classification of minerals by criteria other than simple color. His classification of minerals was based on such physical properties as color, luster, transparency, hardness, density, and cleavage, a practice that provided a significant advance in their study.

The modern study of the internal constituency and structure of minerals can probably be said to have originated with the great astronomer Johannes Kepler who, in 1611, documented the sixfold (hexagonal) symmetry of snowflakes, and reasoned that it was due to the internal cubic or hexagonal closest packing of spheres. The identity of these spheres could not then be determined. The Danish scientist Nicolaus Steno (Niels Stensen) demonstrated in 1669 that the interfacial angles of a particular mineral (he studied such common minerals as quartz and hematite) are constant, regardless of the overall crystal shape (Steno's law). He hypothesized that crystal growth was directional in nature, depending upon the addition of unit "particles." It is worth noting that this work was based upon relatively crude measurements with a contact goniometer. Verification of these measurements awaited invention of the far more accurate optical goniometer a century later.

The new science of crystallography provided the next major advance in mineralogy in 1784, when René-Just Haüy proposed in his *Traité de Crystallographie* that crystals are composed of "integral units" assembled in a uniform manner (Figure 1). This theory, that the external form of a crystal was a result of internal order, neatly explained Steno's law. Haüy was led to his theory by his studies of the uniform cleavage of calcite ($CaCO_3$). Apparent problems with Haüy's theory arose when it was observed that the common mineral aragonite, although chemically indistinguishable from calcite, had entirely different properties and crystal forms. A solution to this apparent paradox was proposed by Eilhard Mitsherlich (1821), who realized that it was perfectly possible for minerals with identical compositions to have entirely different crystal structures, a concept known as polymorphism. Haüy's concept of an integral unit survives in the modern concept of the "unit cell."

Through the first half of the nineteenth century the science of crystal forms, crystallography, was developed further. The six crystal systems (isometric, tetragonal, hexagonal, orthorhombic, monoclinic, and triclinic) were described by Christian Weiss and Fredrich Mohs. Mohs also developed the modern mineral hardness scale named for him.

Study of the symmetry elements rotation axes, mirror planes, and centers of inversion led Johann Hessel (1830) and Auguste Bravais (1848) to derive the thirty-two possible crystal classes, which are the symmetries exhibited by the outward forms of crystals. Between 1881 and 1890, working independently, E. S. Federov, Artur Schoenflies, and William Barlow combined the crystal classes with translational symmetry elements (glide planes and screw axes) to yield the 230 possible space groups, which are the internal symmetries of crystal structures. However, these theoretical investigations provided no actual knowledge of the atomic arrangements within minerals. This latter advance came in 1912. Barlow also proposed what ultimately turned out to be the correct crystal struc-

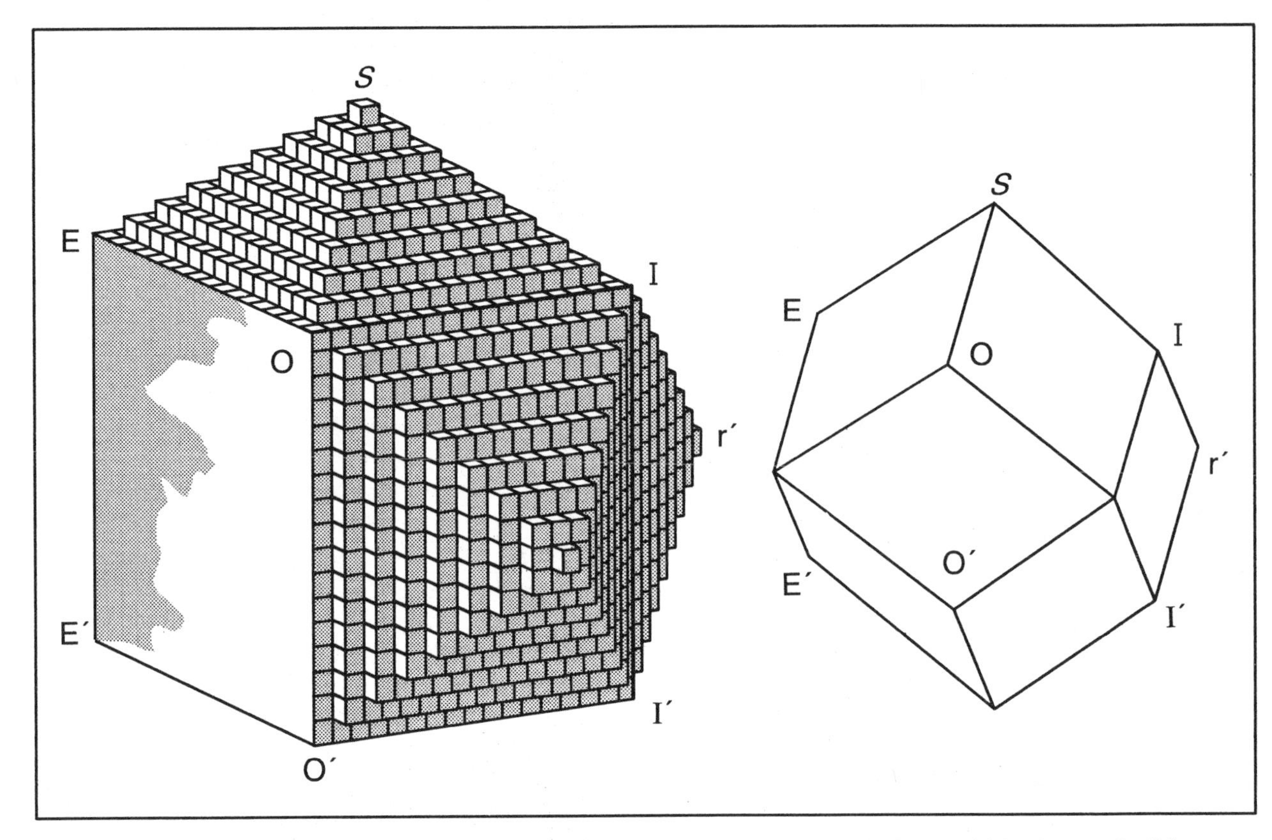

Figure 1. On the left is a diagram showing how an ordered stacking of identical blocks can build up a complex crystal form (a dodecahedron), after Haüy's *Traité de Crystallographie.* On the right is a drawing of a complete dodecahedron, a crystal form exhibited by the common mineral garnet.

ture for the mineral halite (NaCl), although this was not demonstrated for more than sixty years.

At the time the principal characterization of minerals was by observation of physical properties. These included:

1. Form, being principle morphology, or crystal shape
2. Color
3. Cleavage, parting, and fracture, which describe the morphology of freshly broken surfaces and are related to internal structure
4. Specific gravity, which is the weight of a specific volume of the mineral normalized to an equal volume of water
5. Hardness, referred to a relative scale ranging from talc (softest) to diamond (hardest)
6. Tenacity, or resistance to breaking
7. Opacity, ranging from transparent to opaque
8. Magnetic properties, if present

The French naturalist Pierre Cordier appears to have been the first scientist to have recorded by microscope observations of minerals immersed in water (1815); this technique was developed into a powerful tool when special immersion oils were utilized later. The microscope was further enhanced through addition of polarizing elements, done by William Nicol in 1828, which allowed the systematic examination of the behavior of light in minerals and aided in their identification. These polarizing elements are still called "nicols" in honor of this Scottish scientist. The completed development of the specialized petrographic microscope is generally attributed to Henry C. Sorby, in the 1850s; his basic design continues to give good service today. Sorby was also the first individual to report observations of fluid inclusions (trapped microscopic remnants of mineralizing solutions and melts) in minerals and meteorites in the microscope.

Up until this time, minerals were grouped according to physical properties, which although firmly grounded on observation, hindered deeper understanding of mineral genesis. However, the increasing capabilities of chemists soon provided a new view of minerals. John Dalton's work (1766–1844), in particular, establishing atomic theory and the rules of stoichiometry provided a foundation for new classification of minerals based on composition. Over the next century advances in analytical chemistry permitted initial classifications of minerals by composition to be proposed by ABRAHAM GOTTLOB WERNER and Jons Jakob Berzelius; the latter (1779–1848) was the first to group minerals according to the electronegative elements, yielding groups such as the oxides, sulfides, phosphates, nitrates, and so on. The modern classification of minerals was codified by the American mineralogist James Dwight Dana (1813–1895) in his *System of Mineralogy* (1837) and *Manual of Mineralogy* (1848) (*see* JAMES D. DANA). Dana used essentially Berzelius's classification, extending it to all known minerals (at that time numbering about 700). This is the classification in use today, although it has been expanded to include the approximately 3,300 minerals described since the 1840s.

The most celebrated geochemist in this century was the Norwegian VICTOR MORITZ GOLDSCHMIDT (1888–1947). His crusade was to improve the knowledge of the chemistry of minerals, the relationships between coexisting minerals in rocks, and thus better understand the origin and evolution of Earth. Goldschmidt and his collaborators succeeded in determining many basic rules governing the distributions of elements within minerals. Much of this work has continued principally through the experimental synthesis of minerals in the laboratory. This laboratory work thus provides a necessary complement to analytical work on minerals.

In 1912, at the suggestion of their thesis advisor, Max von Laue, students Walter Friedrich and Paul Knipping discovered that X rays would be diffracted by a crystal (originally copper sulfate), at once establishing that atoms and X rays were both on the order of 1 angstrom (Å) in diameter/wavelength, and that atoms are truly arrayed in crystals in a periodic fashion. The young British physicist William L. Bragg (*see* BRAGG, WILLIAM LAWRENCE AND WILLIAM HENRY) realized that this X-ray diffraction could be approximated by the reflection of X rays off planes of atoms, greatly simplifying interpretation of the results. In X-ray diffraction, the position of the diffracted X-ray beam is dependent upon the position and periodicity of the atomic planes in the mineral structure, while the intensity of the beam is a function of the number and identity of the atoms in the planes. Together with his father, William H. Bragg, they performed the first crystal structure analysis, that of halite (NaCl). The Braggs and coworkers continued to solve mineral structures at a rapid pace; by 1935 the structures of most major rock-forming minerals were known. Crystal structure analysis has provided great insight into the atomistic reasons for macroscopic physical and optical properties of minerals. The American scientist Linus Pauling was among the first to apply the newly developed quantum physics to chemistry and, based on his observations of mineral structures, derived a set of empirical rules governing the structures of minerals (1929). Originally, crystal structure analysis required the use of a single, relatively defect-free crystal. The development of X-ray powder diffraction, in 1917, permitted for the first time a rapid, unambiguous identification technique for mineralogists. Use of a powdered sample greatly facilitates collection of diffraction data from all possible atomic layers and use of simplified data reduction analysis. The development of specialized X-ray diffraction equipment has been rapid. In the 1970s the Gandolfi X-ray camera was developed to permit collection of essentially complete powder diffraction patterns from a single grain. In the 1980s the Rietveld method was developed to permit crystal structure determinations to be performed using high-resolution X-ray powder diffraction data, permitting analysis of minerals that fail to form sufficiently defect-free crystals.

Electron beam instruments have, since the 1960s, become indispensable tools for the mineralogist, providing both morphological, structural, and compositional information. In the last generation, the application of a plethora of new analytical techniques have yielded a new picture of minerals and the worlds they compose (*see also* MINERALS AND THEIR STUDY).

Bibliography

KLEIN, C., and C. S. HURLBUT, JR. *Manual of Mineralogy*, 20th ed. New York, 1985.

MICHAEL ZOLENSKY

MINERALS AND THEIR STUDY

A mineral is a naturally occurring, inorganic, crystalline material with a definite chemical composition. Mineralogy is the study of minerals, their origin, occurrence, crystal structure, and composition. At present approximately four thousand minerals are known, although only a couple hundred minerals are of major importance, and approximately fifty to one hundred new minerals are described each year. A few naturally occurring amorphous materials called mineraloids fall short of the qualifications for minerals. Examples of mineraloids are opal and some samples of turquoise.

Into the nineteenth century, the principal characterization of minerals was by observation of physical properties. These included:

1. Form, being principally morphology, or crystal shape
2. Color
3. Cleavage, parting and fracture, which describe the morphology of freshly broken surfaces, and are related to internal structure
4. Specific gravity, which is the weight of a specific volume of the mineral divided by the weight of an equal volume of water
5. Hardness, referring to a relative scale ranging from talc (softest) to diamond (hardest)
6. Tenacity, or resistance to breaking
7. Opacity, ranging from transparent to opaque
8. Magnetic properties, if present.

These properties continue today to be of use, particularly for the scientist in the field. However, the increased analytical developments of the past century have left their mark on mineralogy.

With very few exceptions, minerals possess an internal arrangement of atoms characterized by long-range order in three dimensions, called crystal structure. The science concerned with the study of these arrangements and the rules that govern them is called crystallography. Any crystal structure can be viewed as a simple motif unit of atoms that is repeated by geometrical operations (or symmetry operations) to fill up space. The symmetry operations include mirror planes, rotations, and a center of inversion, all of which govern the external symmetry of crystals. These previous symmetry operations can be combined with a small (unit) translation to produce additional symmetry operations called glide planes and screw axes. All of these symmetry operations, combined in the 230 different ways possible, provide the total symmetries exhibited by internal crystal structures. At the simplest level, all crystals and crystal structures can be referred to only six crystal systems. These are isometric (with the most symmetry), hexagonal, orthorhombic, tetragonal, monoclinic, and triclinic (least symmetry). Usually, the crystal symmetry of a mineral can be determined by inspection of well-formed crystals, but it is best characterized by examination in a petrographic microscope and by X ray and other diffraction techniques.

Optical properties of minerals are less easily determined than physical properties, but are important because of the information they offer concerning mineral identification, internal structure, and paragenesis. These properties are most readily determined using the petrographic microscope, where plane-polarized light is passed through a suitably thin section (typically 30 micrometers or μm thick) of a mineral or rock; loose grains may also be used. In crystalline minerals that are not of the isometric crystal group this light is broken into two polarized rays (the ordinary and extraordinary rays) vibrating in mutually perpendicular directions, whose speed and divergence depend upon the atomic structure and refractive indices of the mineral being examined. Careful study of the behavior of these two rays and their interaction can serve to identify the mineral, indicate the orientation of the principal crystallographic directions, give compositional data for some minerals, and yield other useful information. The optical properties of opaque minerals can also be studied through the use of reflected light. This latter technique is invaluable in the characterization of ore minerals, which are often opaque.

By the beginning of the nineteenth century analytical chemistry had become sufficiently advanced to permit accurate analyses to be made of minerals. Besides obviously providing much useful new information on the constituents of minerals, chemical data permitted the reclassifications of minerals on the basis of composition. Today, chemical analysis of minerals continues to have a central role in mineral characterization. Traditional wet-chemical techniques (dissolution and precipitation) have in most circumstances been replaced by other techniques as described below.

X-ray diffraction analysis, first developed in the second decade of the twentieth century (*see* BRAGG, WILLIAM LAWRENCE AND WILLIAM HENRY), continues to be one of the most powerful tools of the mineralogist. This method is based on the diffraction of X rays by the electrons in planes of atoms, being somewhat analogous to the reflection of light by a mirror. In X-ray diffraction, the direction (orientation) of the diffracted X-ray beam is dependent upon the relative position and periodicity of the atomic planes in the mineral structure, while the intensity of the diffracted beam is a function of the number and identity of the atoms in the planes. In actuality, the electrons surrounding the atoms are what scatter X rays. Since 1914, single crystal X-ray diffraction has been the principal means for elucidating the crystal structure of all materials, including such organic materials as DNA. Originally, crystal structure analysis required the use of a single, relatively defect-free crystal. The development of X-ray powder diffraction, in 1917, permitted for the first time a rapid, unambiguous identification technique for mineralogists, since no two minerals have exactly the same X-ray powder pattern. Use of a powdered sample greatly facilitates collection of diffraction data from all possible atomic layers and use of simplified data reduction analysis. This technique can use either an X-ray diffractometer or powder diffraction camera such as the Debye-Sherrer camera. The Gandolfi X-ray camera was later constructed to permit collection of essentially complete powder diffraction patterns from a single mineral grain, allowing careful, nondestructive study of ever smaller samples. In the 1980s the Reitveld method was developed to permit crystal structure determinations to be performed using high-resolution X-ray powder diffraction data, permitting analysis of minerals that fail to form sufficiently defect-free crystals. The availability of stable, reliable sources of neutron beams has permitted the development of neutron diffraction as a complementary technique to X-ray diffraction. While X-ray diffraction is sensitive to the distribution and scattering power of electrons, in neutron diffraction it is the nucleus that causes the scattering; thus neutron diffraction is considerably more revealing of the positions of light atoms, which, like hydrogen, have few electrons, in a mineral structure. In the past decade the availability of synchrotron beams, and associated high-intensity X-ray beams, has permitted the collection of X-ray diffraction data in a small fraction of one second. This capability now allows physicists to observe structural changes in minerals at the atomic level in real time, greatly expanding knowledge concerning mineral reaction paths and kinetics, structural transformations, and phase changes.

One of the first things that was learned from these structural studies was the sizes of atoms and ions, which are typically on the order of 1 Å (10^{-8} cm) in radius. In general, negatively charged ions are significantly larger than positively charged ions. This means that the former generally comprise the skeleton of mineral structures, and underscores the utility of Dana's mineral classification (*see* DANA, JAMES D.), which divides minerals into groups based on principal anions. X-ray crystal structure analysis thus provides great insight into the atomistic reasons for macroscopic physical and optical properties of minerals. The American scientist Linus Pauling was among the first to apply the newly developed quantum physics to chemistry and, based on his observations of mineral structures, derived a set of empirical rules governing the structures of minerals, which were further developed in his book *The Nature of the Chemical Bond* (1939). Although these rules were derived for use with minerals containing predominantly ionic bonds, it has recently been argued that they may be usefully applied to molecular-orbital bonded structures as well. Application of these principles has permitted many structures to be fairly accurately predicted, and today allows solid-state scientists to tailor design materials for specific uses. An example of this is the development of new zeolites to serve as molecular sieves, for example, in purifying water.

The elucidation of crystal structures has often shed light on mineral stabilities and relationships. The silicate minerals are the most important example. The silicates contain SiO_4 tetrahedra that, in different classes of silicates, are increasingly polymerized from single tetrahedra, through rings, chains, and sheets to three-dimensional frameworks (*see* MINERALS, SILICATES). The American petrologist NORMAN L. BOWEN realized that silicate melts form silicates with increasing degrees of silicate polymerization as temperatures decrease and the melt incrementally crystallizes. Thus mineral structures can have an important influence on petrologic and geochemical processes.

In the last thirty years electron beam instruments have become indispensable tools for the mineralogist. In all electron microscopes an elec-

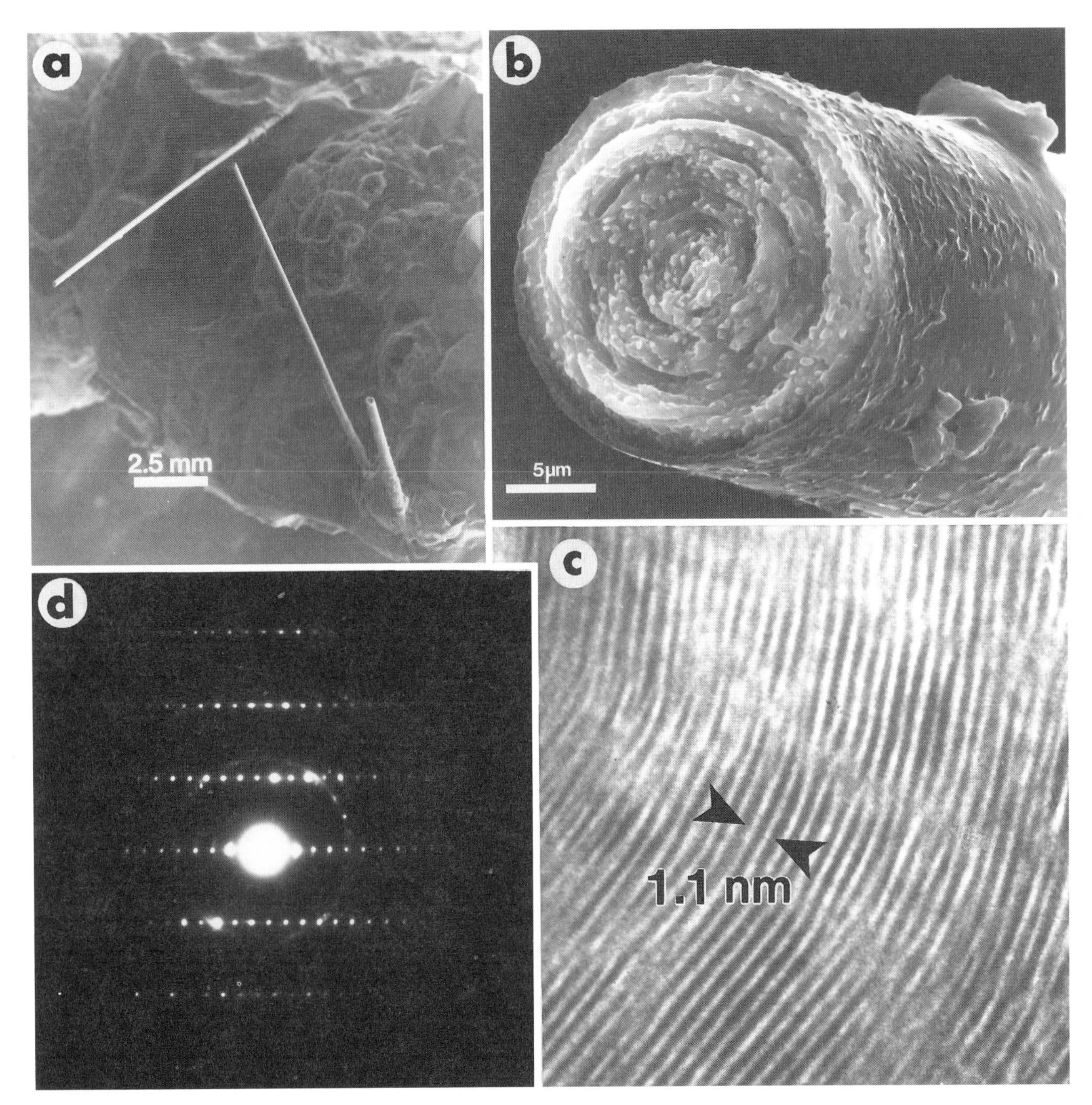

Figure 1. The study of a mineral specimen usually involves collecting detailed information concerning its form, structure and composition. As an example, examine these four successively more-detailed views of the unusual mineral tochilinite provided by electron microscopy. The structure of tochilinite consists of successive sheets of iron sulfide and magnesium-iron hydroxide. The scanning electron microscope (a) reveals that long crystals of tochilinite are in reality cylinders constructed of tightly rolled sheets (b). In the transmission electron microscope layers of atoms forming the sheets can be imaged, and the double sheets are revealed to have a total thickness of 1.1 nm (c). Additional information concerning the arrangement of atoms within tochilinite is provided by electron diffraction patterns (such as that shown) obtained with the TEM (d).

tron beam is directed down a column to interact with a suitably prepared sample. In the transmission electron microscope (TEM) the sample is thinned to approximately 1,000 Å thick, and the electron beam passes entirely through it; the exiting electron beam is imaged to reveal details of the sample's structure (Figure 1). Some incident electrons are diffracted during this process, and imaging of the diffracted electrons can reveal additional structural information. Another common application of the TEM is to detect the energies of X rays resulting from the collision of atoms in the sample with the electron beam, commonly called energy dispersive X-ray spectroscopy (EDX or EDS). These X rays can be used to yield quantitative compositional information on areas as small as a few nanometers in diameter. Thus the TEM is the most useful tool for the study of very small mineral grains, and has greatly advanced our understanding of crystal structures, structural defects, mineral transformations, and other reactions. In petrology (*see* PETROLOGY, HISTORY OF), TEM studies can reveal critical details concerning mineral paragenesis, and therefore the natural history of a rock or mineral deposit; this information can frequently not be acquired in any other way. High-resolution TEM can, in favorable circumstances, provide images of individual molecules or atoms, and this technique, together with electron diffraction, can permit crystal structure determination of mineral samples not suitable for X-ray diffraction analysis. Additional compositional and atomic bonding information can be provided in TEM by the application of electron energy loss spectroscopy (EELS), with this information being gleaned from the changes in energy of the electron beam as it passes through the thin sample. One recent application of this technique has been the discrimination of silicon in crystalline structures from that in amorphous structures by the energies of the transmitted electrons.

In the scanning electron microscope (SEM) the electron beam is rapidly scanned over a small region of a sample, and analysis is performed on the resulting secondary, backscattered, diffracted, and auger electrons, and X rays. If suitably imaged, the secondary and backscattered electrons provide morphological and limited structural information on the mineral. Because of the small wavelength of the electrons, great depth of field can be achieved in the high-magnification secondary electron images. X rays, originating when electrons are completely absorbed into an atom, have wavelengths and energies characteristic of the atomic number of the atom from which they sprang, and thus yield compositional information on the sample. Scanning coils can be added to the TEM column, resulting in scanning transmission electron microscopes (STEMs).

The electron microprobe permits the rapid chemical analysis of mineral grains on the scale of a few micrometers, through collection and analysis of the wavelengths of resultant X rays. The measurement of wavelengths of X rays can be made with considerably greater accuracy than that of X-ray energies. Thus compositional analyses by electron microprobe are inherently superior to those by any typical SEM. Such microfeatures as chemical zoning and exsolution can be examined by the electron microprobe, permitting an increased understanding of conditions attending mineral genesis and development.

Spectroscopic analyses provide a window onto the atomic nucleus and bonding, that is, short-range atomic order, and can be used to characterize such noncrystalline materials as glasses. Mossbauer spectroscopy deals with the resonant absorption and emission of gamma rays by certain atomic nuclei, most important of these being ^{57}Fe, ^{119}Sn, and ^{112}Sb. For example, using this technique, iron in its different valence states can be recognized and quantified. Recognition of differential bonding environments can also yield information on iron-containing mineral structures. Mossbauer spectroscopy also provides information concerning intervalence charge transfers, and magnetic properties of iron-containing minerals. Atomic vibrations in a crystal will absorb electromagnetic radiation if the radiation and vibration are of the same frequency. If the excited vibration results in a change in the dipole moment of the crystal, infrared absorption effects result whose analysis can yield valuable information on atomic structure and bonding. Similar information can be gleaned from observation of radiation scattered by a crystal, where vibrational energy is absorbed or imparted to the scattered radiation; this is termed the Raman effect. Nuclear magnetic resonance (NMR) spectroscopy involves the resonant absorption and subsequent emission of radio-frequency radiation by an atomic nucleus through the interaction of a nucleus' magnetic moment with an applied magnetic field. NMR spectroscopy can yield information concerning the degree of structural order in diamagnetic minerals. Each year sees the fruition of new spectroscopic techniques, providing a more detailed look at minerals. New techniques include electronic absorption spectroscopy, electron spin resonance spectroscopy and X-ray absorption spectroscopy. Of particular interest is secondary ion mass spectrometry (SIMS). In SIMS an ion beam is used to sputter ions away from a sample, whose masses and charges are measured with a mass spectrometer, permitting analysis of trace elements and isotopes with spatial resolution of a few micrometers. Lately, this tool has been extensively applied to the dating of individual mineral grains with great resolution. Since SIMS is a surface analysis technique involving the removal of ions, continued sputtering of a single spot results in depth profiles of the distributions of specific elements and isotopes.

There are several other powerful techniques used for the analysis of major, minor, and trace elements in submicrogram-sized amounts. The recent availability of high-intensity X rays from synchrotron beam lines has allowed the development of synchrotron X-ray fluorescence. This is an ideal trace-element analytical technique because of very high signal-to-noise ratios for many important elements. If a proton beam is focused on a surface, the resulting X-ray emission (called proton-induced X-ray emission or PIXE) is well suited to trace-element analysis, since the background radiation (bremsstrahlung) is much lower than for impinging electron beam X-ray emission. Artificial radioactivity can also be induced through high neutron fluxes during residence in the core of a reactor. Subsequent measurement of the gamma rays produced by the decaying radioactive elements (called instrumental neutron activation analysis or INAA) can be used to analyze many elements of interest, including the rare earth elements.

Thermal analytical techniques can aid in the identification of unknown minerals, through a variety of techniques. In differential thermal analysis (DTA), samples are heated in a standardized manner, and the energy absorbed or evolved by chemical and structural transformations are recorded, and can serve as a fingerprint for identification. Thermal gravimetric analysis (TGA) is similar, with sample weights being monitored as a function of temperature, yielding information concerning mineral transformations.

A recent development has been the determination of changes in mineral properties, structures, and stabilities with increasing static pressure. In the most extreme of these experiments the sample is held between opposed diamond "anvils," with pressure being applied hydrostatically (*see* PETROLOGIC TECHNIQUES). The transparency of the diamond anvils permits observation of the squeezed sample via X-ray diffraction or spectroscopic techniques. This technique is permitting the exploration of the interior of the planets, including Earth, in laboratory simulations.

Since the 1980s the scanning force microscope has been developed to permit direct imaging of individual atoms, revealing important details concerning mineral surfaces, and chemical reactions like catalysis and crystal growth.

The twentieth century has seen the continuous introduction of new analytical instruments and techniques to mineralogy. It is clear that these developments will continue, since minerals are the basic material of the physical universe.

Bibliography

KLEIN, C., and C. S. HURLBUT, JR. *Manual of Mineralogy* 20th ed. New York, 1985.

SINKANKAS, J. *Mineralogy.* New York, 1975.

MICHAEL ZOLENSKY

MINERALS, NONSILICATES

Minerals are arranged into classes according to their major anionic group. There is insufficient space here to describe more than a few of the 3,500 known minerals; the silicates are dealt with in a separate entry (*see* MINERALS, SILICATES). The important mineral classes are (in alphabetical order) antimonates, antimonites, arsenates, arsenides, borates, carbonates, chromates, halides, hydroxides, iodates, molybdates, native elements, nitrates, organics, oxides, phosphates, selenates, selenites, sulfarsenides, sulfates, sulfides, sulfosalts, tellurates, tellurides, tungstates, vanadates, and vanadium oxysalts. The most important of these groups are discussed below briefly.

Native Elements. Only about twenty elements are found in the solid state as minerals, and these can be divided into metals, semimetals, and nonmetals. Metals are subdivided on the basis of crystal structure: gold, silver, copper, and lead share the same crystal structure, as do platinum, palladium, iridium, and osmium. Kamacite (Fe) and taenite (Ni, Fe) have related structures. Mercury, tantalum, tin, and zinc have also been found in the native state. The native semimetals compose two structural groups: arsenic, antimony, and bismuth in the first and selenium and tellurium in the second. The most important nonmetals are sulfur and carbon, the latter existing in radically different crystal structures as graphite, diamond, and fullerenes (C_{60} and C_{70}).

Sulfides. Sulfides, together with sulfosalts, sulfarsenides, and arsenides, include the majority of ore minerals. Most sulfides are opaque and distinctly colored, many with metallic to submetallic luster due to metallic to covalent bonding (but a strong ionic bonding component can also be present). Many sulfides exhibit rampant polymorphism (different crystal structure for the same mineral compositions); for example, pyrite and marcasite (both FeS_2). Pyrite is the most abundant sulfide in Earth's crust and is also present on Mars and likely present on Venus; marcasite is far less common. Troilite (FeS) and pyrrhotite ($Fe_{1-x}S$) are the more common iron sulfides on asteroids and comets. Sphalerite and wurtzite (both ZnS) are the most important ores of zinc; wurtzite has recently been observed to form around volcanic vents on the ocean floor. Chalcopyrite ($CuFeS_2$), chalcocite (Cu_2S), and covellite (CuS) are widespread sulfides of the important element copper.

Oxides. Since oxygen is the most abundant element in the crust of the terrestrial planets, oxides naturally compose an important mineral group. The most important oxides are subdivided into the following structural groups: the hematite group (A_2O_3; A = Al, Cr, Fe, V), spinel group (AB_2O_4; A = Fe, Mg, Mn, Ni, Ti, Zn, Co, Cu, Ge; B = Fe, Al, Cr, Mg, Mn, Ti, V), and rutile group (AO_2; A = Ti, Mn, Sn, Te, Si, Ge, Mn). Hematite (Fe_2O_3) is an important terrestrial weathering product, often forming from the breakdown of magnetite (Fe_3O_4). The latter mineral, along with spinel ($MgAl_2O_4$) and chromite ($FeCr_2O_4$) are important components of many igneous and (following weathering) sedimentary rocks, as well as primitive meteorites. They are also among the more important oxides in Earth's interior.

Hydroxides. Minerals such as goethite [FeO (OH)], brucite [$Mg(OH)_2$], gibbsite [$Al(OH)_3$], and diaspore [AlO(OH)] are important weathering products on Earth, familiar to all persons as components of rust.

Halides. These minerals are characterized by the dominance of the electronegative halogen ions Cl^-, Br^-, F^-, and I^-, and ionic bonding, which explains their high solubility in water. The most common halide, halite (NaCl), is a critical substance for life, and fortunately it is present in large sedimentary deposits. Other common halides are sylvite (KCl) and fluorite (CaF_2).

Carbonates. These minerals contain the anionic complex $(CO_3)^{2-}$ and are important sinks for carbon at the surface of Earth, the outer belt asteroids and, probably, on Mars. The important carbonates are divided into three structural groups: the calcite group [$A(CO_3)$; A = Ca, Mg, Fe, Ni, Zn, Mn, Co, Cd], aragonite group [$A(CO_3)$; A = Ca, Sr, Pb, Ba], and dolomite group [$AB(CO_3)_2$; A = Ca, Ba; B = Mg, Fe, Mn, Zn]. Calcite ($CaCO_3$) is the most abundant carbonate (indeed, one of the most abundant minerals on Earth), being the dominant phase in limestone, marble, and carbonatites. Aragonite, a polymorph of calcite, is the dominant phase in the skeletons of many marine invertebrates. Dolomite [$CaMg(CO_3)_2$], which forms as a porous replacement of other carbonates, is an important host for many sulfide ores. Hydrous carbonates, such as malachite [$Cu_2CO_3(OH)_2$], are common weathering products of ore bodies.

Borates. The essential feature of borates are BO_3^{3-}, triangular planar units that can be polymerized like the silicate tetrahedra in silicates. Borates may also contain tetrahedral BO_4^{5-} groups, and even more complex molecules such as [$B_3O_3(OH)_5$]$^{2-}$, which consists of one triangle and two tetrahedra. Colemanite [$CaB_3O_4(OH)_3 \cdot H_2O$] contains chains of triangles and tetrahedra, and borax [$Na_2B_4O_5(OH)_4 \cdot 8H_2O$] features a complex ion consisting of two tetrahedra and two triangles. Ulexite [$NaCaB_5O_6(OH)_6 \cdot 5H_2O$] can occur as closely packed parallel fibers with fiber-optic properties. All three of these common borates occur in arid regions from evaporating playa lakes.

Sulfates. Sulfur coordinated by four oxygens, forming a tetrahedron, is the important structural feature of sulfates. The most widespread sulfates are gypsum [$CaSO_4 \cdot 2H_2O$], anhydrite [$CaSO_4$], and barite [$BaSO_4$]. All three are widespread sedimentary minerals; anhydrite can form directly (in hot environments) or through the dehydration of gypsum.

Phosphates, Arsenates, and Vanadates. These mineral groups are closely related in that the three anionic groups [$(PO_4)^{3-}$, $(AsO_4)^{3-}$, and $(VO_4)^{3-}$] all form tetrahedral units capable of substitution for one another. For example, the secondary minerals pyromorphite [$Pb_5(PO_4)_3Cl$], mimetite [$Pb_5(AsO_4)_3Cl$], and vanadinite [$Pb_5(VO_4)_3Cl$] are isostructural and exhibit complete solid solution. Apatite [$Ca_5(PO_4)_3(OH, Cl, F)$] is the most widespread phosphate, found in both igneous and sedimentary rocks, and, together with whitlockite [$Ca_9(Mg, Fe)H(PO_4)_7$], in many meteorites.

Bibliography

Klein, C., and C. S. Hurlbut, Jr. *Manual of Mineralogy*, 21st ed. New York, 1993.

Sinkankas, J. *Mineralogy*. New York, 1975.

MICHAEL ZOLENSKY

MINERALS, SILICATES

Silicates contain essential silicon and oxygen, and since these are the two most common elements in Earth's crust, silicates predictably make up the bulk (about 95 percent) of the crust and thus our geological environment (*see also* MINERALS, NONSILICATES). The same combination probably holds for the other terrestrial planets (Mercury, Mars, and Venus) and Earth's moon. In fact, approximately 25 percent of the 3,500 known mineral species are silicates. The essential feature of the silicates is the silicate tetrahedron (SiO_4^{4-}), with four oxygens lying at the corners and silicon at the heart (Figure 1). The diversity of the silicates is attributable to the myriad ways in which silicate tetrahedra can be polymerized into pairs, rings, chains, sheets, and three-dimensional networks. The classification of silicates is thus based upon this linkage, and six subclasses are generally recognized, which will be discussed in turn. Another important feature of silicates is the frequent partial substitution of Si^{4+} by Al^{3+} or, less commonly, Fe^{3+}, Be^{3+}, or Ti^{3+} in tetrahedra, which results in an additional degree of net negative charge on the silicate backbones that is satisfied by other cations coordinating the tetrahedra.

Nesosilicates. In the simplest silicates, silicate tetrahedra exist as independent units, cross-linked only through other cations or anion groups. Thus the silicon-to-oxygen ratio for the silicate portion of these minerals is 1:4. The most common nesosilicates are olivine, garnet, and zircon. Both olivine and garnet are family names for groups of silicates; within each family the crystal structure is basically unchanging while the composition varies (through solid solution). Olivines have the general formula $(Mg, Fe)_2SiO_4$ (Ni and Mn can also substitute for the cations) and are common constituents of many igneous and a few metamorphic rocks. The Mg and Fe endmember olivines are called forsterite and fayalite, respectively, and there exists complete solid solution between these phases although pure fayalite is very rare. Material with the olivine composition but a more densely packed structure (called majorite) is probably the principle component of Earth's mantle, and thus the most common material composing Earth. Olivine is also the principle component of most primitive meteorites and the asteroids from which they were derived. Garnets [$A_3B_2(ZO_4)_{3-x}(OH)_{4x}$; A = Ca, Fe, Mg, Mn; B = Al, Cr, Fe, Mn, Si, Ti, V, Zr; Z (tetrahedral site) = Si, Al, Fe] are common constituents of many types of rocks, most notably high-grade metamorphic rocks and ultramafic rocks from Earth's upper mantle. Zircon ($ZrSiO_4$), a common accessory phase in felsic igneous rocks, has particular importance due to its great resistance to chemical and physical weathering and to impurities of uranium that make it ideal for the dating of rocks. The oldest indigenous mineral grains (4.3 billion years, or Ga, old) yet discovered on Earth and its moon have been zircons.

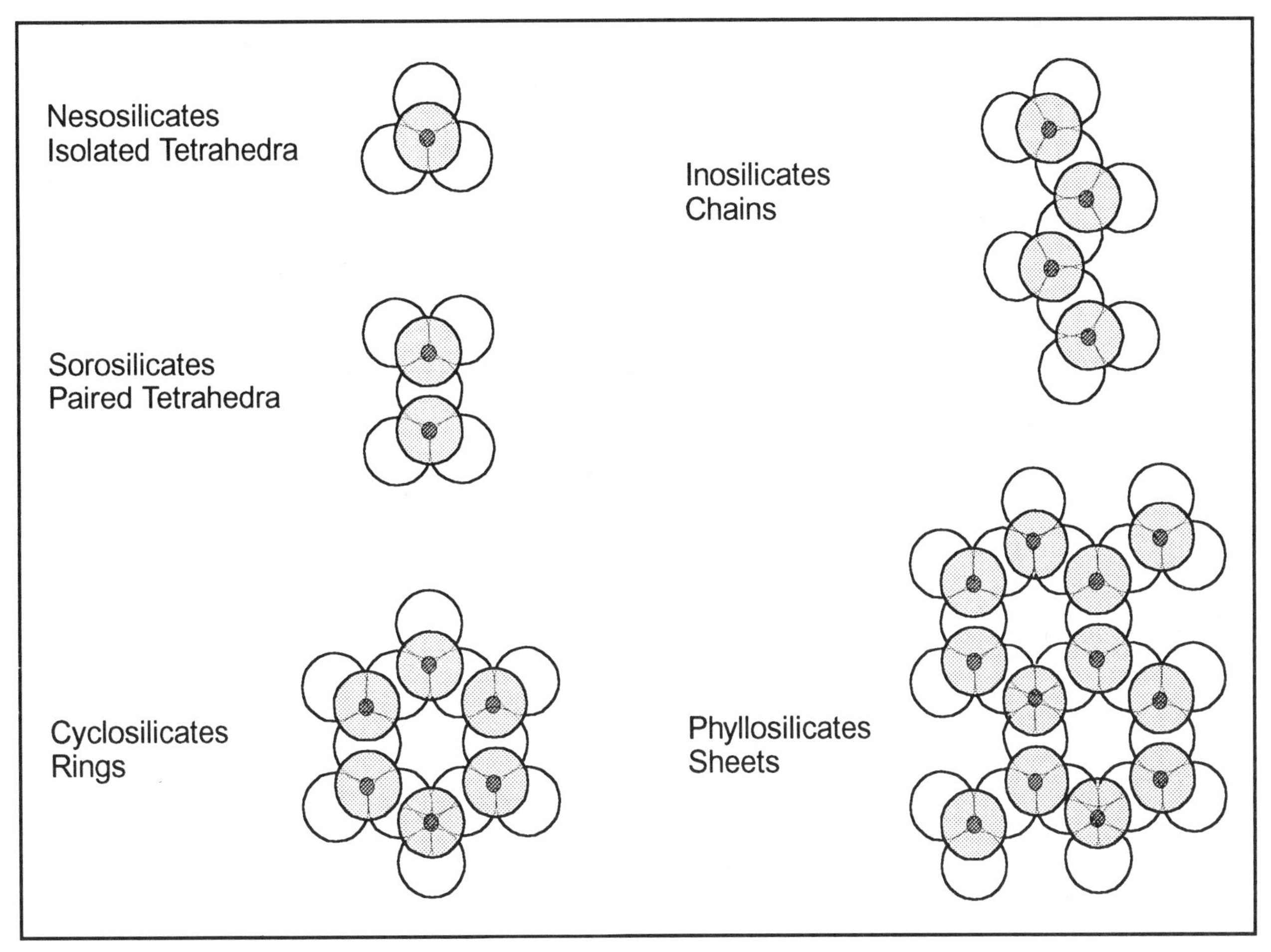

Figure 1. A sense of the polymerization of SiO_4 tetrahedra can be gained from this figure, which shows the basic silicate structures for the nesosilicates (isolated tetrahedra), sorosilicates (pairs), cyclosilicates (rings), inosilicates (chains; a single chain is shown), and phyllosilicates (sheets). Tectosilicates are not shown. Large circles represent oxygen atoms, shaded oxygens project above the plane of the other oxygens, and the smaller circles are silicon atoms nestled in between.

Sorosilicates. Pairs of silicate tetrahedra, sharing one oxygen atom, form the structural basis of these minerals, which thus have a silicon to oxygen ratio of 2:7. These pairs are therefore bonded to other cations and anion groups to complete the crystal structures. Common sorosilicates include lawsonite [$CaAl_2Si_2O_7(OH)_2 \cdot H_2O$] and hemimorphite [$Zn_4Si_2O_7(OH)_2 \cdot H_2O$]. Lawsonite is a common constituent of high-pressure, high-temperature, regional metamorphic rocks called blueschists; hemimorphite is a product of low-temperature aqueous alteration. Some sorosilicates also have independent silicate tetrahedra within the structure, for example, the common low-grade metamorphic phase epidote [$Ca_2(Al, Fe)Al_2O(SiO_4)(Si_2O_7)(OH)$].

Cyclosilicates. In these minerals three, four, or six silicate tetrahedra are joined into closed rings, producing a silicon-to-oxygen ratio of 1:3. The rings are linked in different ways by other cations or molecules, which sometimes also occupy the centers of rings. Important cyclosilicates include beryl [$BeAl_2(Si_6O_{18})$] and tourmaline [$(Na, Ca)(Li, Mg, Al)(Al, Fe, Mn)_6(BO_3)_3(Si_6O_{18})(OH)_4$]. Both of these minerals are found in pegmatitic and felsic igneous rocks; beryl can also be found in mica schists. Beryl is the principle source of beryllium, and when found in crystals of sufficient clarity and pleasing colors it is used in jewelry (emerald, aquamarine, morganite).

Inosilicates. Here silicate tetrahedra share oxygens to form single and double chains, which form the basis for the tabular to elongate crystal forms inosilicates take, and the obvious cleavage that runs parallel to them. Other cations cross-link the chains to complete the structures. The most important inosilicates are among the most important rock-forming mineral groups: pyroxenes (single chains: ABZ_2O_6; A = Ca, Fe, Li, Mg, Mn, Na, Zn; B = Mg, Fe, Al, Cr, Mn, Ti, V; Z = Si, Al) and amphiboles (double chains: $A_{0-1}B_2Y_5Z_8O_{22}(OH, F, Cl)_2$; A = Ca, Na, K, Pb; B = Ca, Fe, Mg, Li, Mn, Na; Y = Fe, Mg, Al, Cr, Mn, Ti; Z = Si, Al, Be, Ti). The lengthy list of different cations for each structural site testifies to the extensive solid solutions observed in these minerals. Single-chain inosilicates have silicon-to-oxygen ratios of 1:3; that of the double chain minerals is 4:11. Pyroxenes and amphiboles are essential constituents of most igneous and many metamorphic rocks; pyroxenes are also found in most stony meteorites and lunar basalts and, together with amphiboles, are present in meteorites from Mars (*see* METEORITES FROM THE MOON AND MARS). The most widespread of the inosilicates are those rich in Mg, Fe, and Ca, including the pyroxenes enstatite ($Mg_2Si_2O_6$), diopside ($CaMgSi_2O_6$), and augite [$Ca(Mg, Fe)Si_2O_6$], and the amphibole magnesiohornblende [$Ca_2(Mg, Fe, Al)_5(Si, Al)_8O_{22}(OH, F)_2$].

Phyllosilicates. The next logical step in the silicate polymerization scheme is sheets, which are composed of tetrahedra sharing three out of four oxygens with neighboring, coplanar tetrahedra. A second type of layer, consisting of other cations in six-fold coordination, constitutes an essential feature of phyllosilicate structures. Considerable complexity results from the different ways in which these tetrahedral and octahedral layers can be stacked, often with large interlayer cations or (OH) being present. Phyllosilicates exhibit perfect sheety cleavage, since generally only weak hydrogen bonding holds the layers together. The silicon-to-oxygen ratio of phyllosilicates is 2:5. Phyllosilicates compose a major portion of clays and most soils, and are important phases in most types of terrestrial rocks as well as some primitive meteorites. Important phyllosilicates include serpentine [$Mg_3Si_2O_5(OH)_4$], kaolinite [$Al_2Si_2O_5(OH)_4$], and the mica group [$XY_{2-3}Z_4O_{10}(OH, F)_2$; X = Ca, K, Na, Ba, ($H_3O$); Y = Al, Cr, Fe, Mg, Li, Mn, V, Zn; Z = Si, Al, Fe, Be]. Important micas include biotite [$K(Mg, Fe)_3(Si, Al, Fe)O_{10}(OH, F)_2$] and muscovite [$KAl_2(Si_3Al)O_{10}(OH, F)_2$], both essential components in most felsic igneous and pelitic metamorphic rocks, and phlogopite [$KMg_3(Si_3Al)O_{10}(OH, F)_2$], found in ultramafic igneous rocks where it is a major host of water in Earth's upper mantle. An additional complication to the silicate classification scheme is the observation that minerals with structures intermediate between double-chain inosilicates and phyllosilicates exist. These phases, called biopyriboles, are created during exceedingly slow cooling in metamorphic rocks.

Tectosilicates. The most common minerals in Earth's crust are tectosilicates, where all silicate tetrahedra share all oxygens with neighboring tetrahedra to complete a three-dimensional linkage. In this completely polymerized arrangement the silicon-to-oxygen ratio assumes its maximum value of 1:2. In most tectosilicates other cations occupy open spaces in the tetrahedral network, although the most common mineral in Earth's crust, quartz (SiO_2), consists only of the silicate tetrahedra arranged into three- and six-sided "springs." In the minerals of the feldspar group [XZ_4O_8; X = Ca, Na, K, Ba, Sr; Z = Si, Al], the silicate network takes the form of nested "staircases" with cations sitting on the steps. The feldspars are subdivided into plagioclase (a solid solution between albite [$Na(Si_3Al)O_8$] and anorthite [$Ca(Si_2Al_2)O_8$]), characteristic of mafic to intermediate igneous and pelitic (Al-rich) metamorphic terrestrial rocks, many meteorites, and lunar highlands rocks, and potassium feldspars [most notably orthoclase and microcline, both $K(Si_3Al)O_8$], which are found in felsic igneous rocks and their weathering products.

Bibliography

KLEIN, C., and C. S. HURLBUT, JR. *Manual of Mineralogy*, 21st ed. New York, 1993.

SINKANKAS, J. *Mineralogy*. New York, 1975.

MICHAEL E. ZOLENSKY

MINERAL STRUCTURE AND CRYSTAL CHEMISTRY

By definition, minerals are crystalline, which means that the arrangement of their constituent atoms is ordered and periodic in three dimensions (*see* MINERALS AND THEIR STUDY). Moreover, the chemical, physical, mechanical, and transport properties of minerals are largely controlled by their crystal structures. It is therefore important to understand how the atoms are arranged within the structure of a mineral and how the atoms are connected together by chemical bonds.

Since the earliest determination of a mineral structure, that of halite (NaCl) in 1914, the detailed crystal structures of thousands of minerals have been determined through X-ray diffraction and other experimental techniques. These structures range from simple to extremely complex. The chemical bonding may be metallic, ionic, covalent, or Van der Waals (weak, residual bonding), or even a combination of these types within the same structure. The type of bonding depends primarily upon the constituent atoms. In the native metallic elements, for example, the bonding is largely metallic, whereas mixed ionic and covalent bonding dominate in minerals containing metals and oxygen, such as the silicates.

Figure 2. The hexagonal closest-packed (HCP) structure, with stacking sequence (AB).

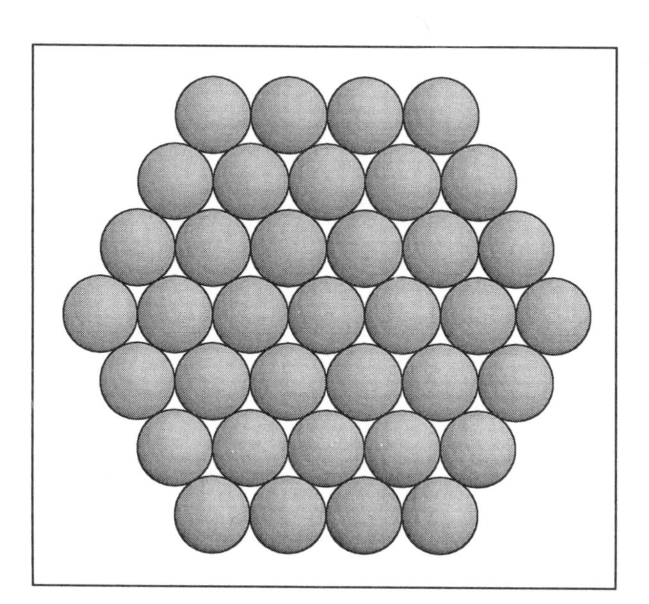

Figure 1. A closest-packed sheet of atoms.

Simple Structures Based on Closest Packing of Atoms

Many mineral structures can be described in terms of closest-packed, planar sheets of neutral atoms, anions (e.g., O^{2-}, F^-, Cl^-), or cations (e.g., Mg^{2+}, Al^{3+}). Such a sheet, in which the atoms are represented by equal-diameter spheres, is shown in Figure 1.* These sheets commonly are stacked upon each other in a characteristic sequence, to form three-dimensional, closest-packed arrays. The structures of some native elements are described by such arrays, but in a far greater number of mineral structures, one type of atom, such as oxygen, forms a closest-packed array, with other atoms,

* Figures showing crystal structures were prepared with the computer program ATOMS by Shape Software.

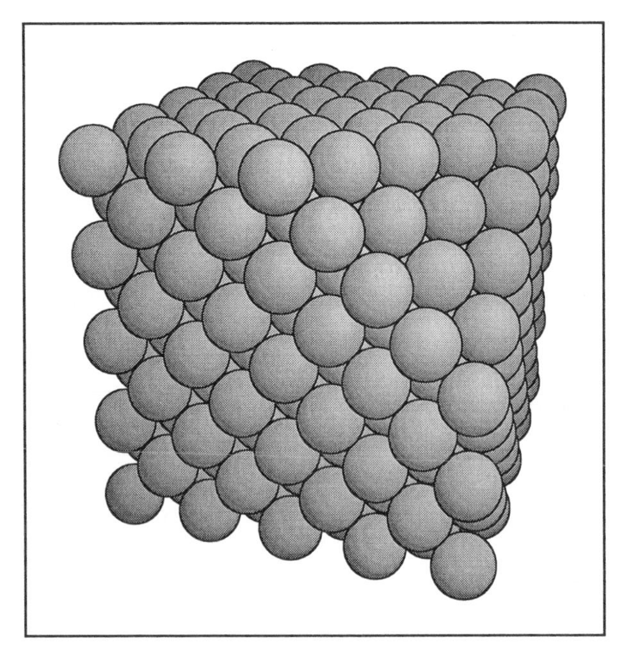

Figure 3. The face-centered cubic (FCC) structure. The nearest corner of the large cube has been removed to show the closest-packed sheet.

such as metals, filling spaces or voids within the array. In many cases, the closest-packed array is slightly distorted, but closest packing still provides a simple way of describing such structures.

In the simplest way of stacking closest-packed sheets, the second sheet is fitted onto the first sheet so that its atoms overlie the gap between three atoms of the first sheet; the third sheet is added so that its atoms directly overlie those of the first sheet; the fourth sheet is in the same position as the second; and so on. This "stacking sequence" can be referred to as ...ABABAB..., or simply (AB). The resulting structure, shown in Figure 2, is known as hexagonal closest packing, or the HCP structure. Although this structure does not occur among terrestrial minerals, it does occur in metals such as magnesium (Mg), and this packing type forms the structural basis for numerous compound minerals (those containing more than one chemical element).

In deriving the HCP structure, we chose to place the third closest-packed sheet in the same position, A, as the first sheet. However, we could have placed it in a third position, designated C, that overlies neither the first or second sheets. If repeated periodically, this stacking sequence becomes ...ABCABCABC..., or simply (ABC). The resulting structure, shown in Figure 3, is cubic, and the cube-shaped unit cells contain atoms not only at their corners, but also in the centers of all their faces. Hence, this is called the face-centered cubic, or FCC, structure. Several important native elements crystallize with the FCC structure, including copper (Cu), silver (Ag), gold (Au), platinum (Pt), and the nickel-iron (Ni-Fe) alloy taenite, which occurs as an important component of iron-nickel meteorites.

Examples of Halides, Sulfides, Oxides, and Carbonates. The FCC and HCP packing types form the basis for numerous other compound structures (*see* MINERALS, NONSILICATES). An example is the structure of halite (Figure 4), in which the chlorine (Cl) ions occupy the positions of cubic closest packing, and the sodium (Na) ions occur in spaces between the Cl ions. Each Na ion is bonded to six nearest-neighbor Cl ions, and each Cl ion is bonded to six surrounding Na ions. In the terminology of crystal chemistry, we say that the Na is in a six-coordinated crystallographic site, or in octahedral coordination. Both FCC and HCP arrays also contain four-coordinated, or tetrahedral sites.

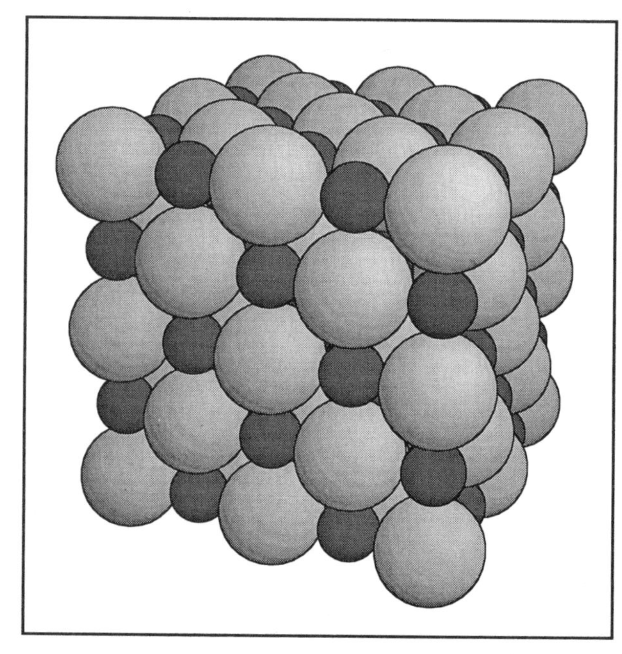

Figure 4. The halite (NaCl) structure. The large spheres represent Cl, the small spheres Na.

In addition to halite, several other simple minerals possess the NaCl structure. These include the halides sylvite (KCl) and cerargyrite (AgCl), the sulfide galena (PbS), and the oxide periclase (MgO). Pyrite (FeS_2) is a simple derivative of the NaCl structure, in which Fe occupies the Na positions, and the Cl ions are replaced by covalently bonded pairs of sulfur (S) atoms. Similarly, the rhombohedral carbonates calciate ($CaCO_3$), dolomite ($CaMg[CO_3]_2$), magnesite ($MgCO_3$), siderite ($FeCO_3$), and rhodochrosite ($MnCO_3$) all possess structures in which the metal ions occupy the positions of Na, and carbonate groups (CO_3^{2-}) the positions of Cl, in a distorted NaCl-type arrangement. In the spinel group of minerals, including spinel proper ($MgAl_2O_4$), magnetite ($Fe^{2+}Fe_2^{3+}O_4$), and chromite ($Fe^{2+}Cr_3O_4$), metal ions occupy both octahedral and tetrahedral sites within a cubic closest-packed array of oxygen ions.

Many simple compound minerals also can be described in terms of HCP arrays. The oxides corundum (Al_2O_3), hematite (Fe_2O_3), and ilmenite ($FeTiO_3$) have metal ions occupying two-thirds of the octahedral sites in a slightly distorted HCP array of oxygen. In the structure of the nonstoichiometric sulfide pyrrhotite ($Fe_{1-x}S$), iron atoms occupy octahedral voids in a distorted HCP array of sulfur atoms. In contrast, in the orthorhombic carbonates aragonite ($CaCO_3$), witherite ($BaCO_3$), strontianite ($SrCO_3$), and cerussite ($PbCO_3$), it is the cations that occupy the positions of a distorted HCP array, with the carbonate groups occupying complex positions between the metal ions.

Other Simple Structure Types. Many other simple structure types are not easily described by closest packing. Examples include the body-centered cubic structure, in which each cube-shaped unit cell contains atoms at the corners, with one additional atom at the center of the cube. This is the structure of Fe metal at room temperature, as well as the most abundant mineral of iron-nickel meteorites, the Fe, Ni alloy kamacite.

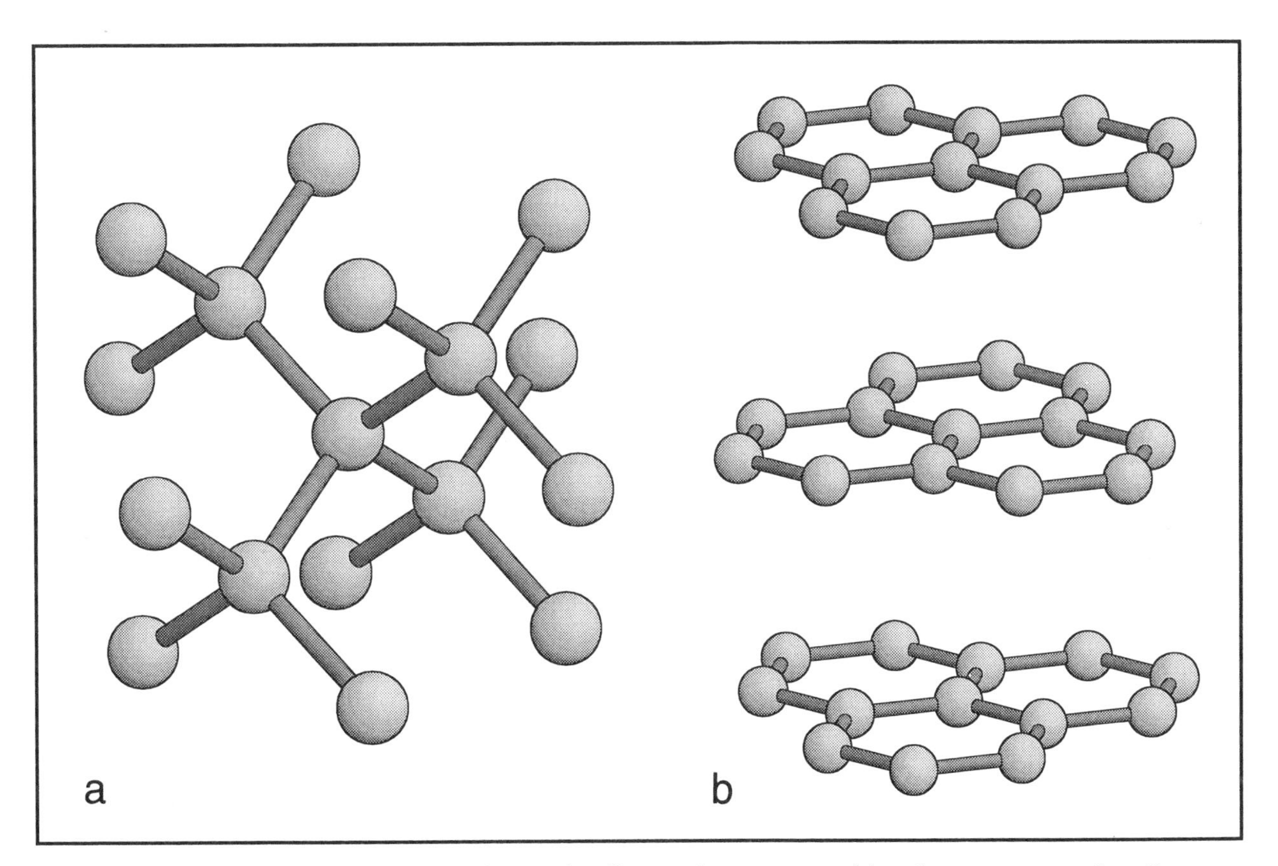

Figure 5. Structures of carbon minerals. a. The diamond structure, with spheres representing C atoms and sticks representing bonds. b. The graphite structure.

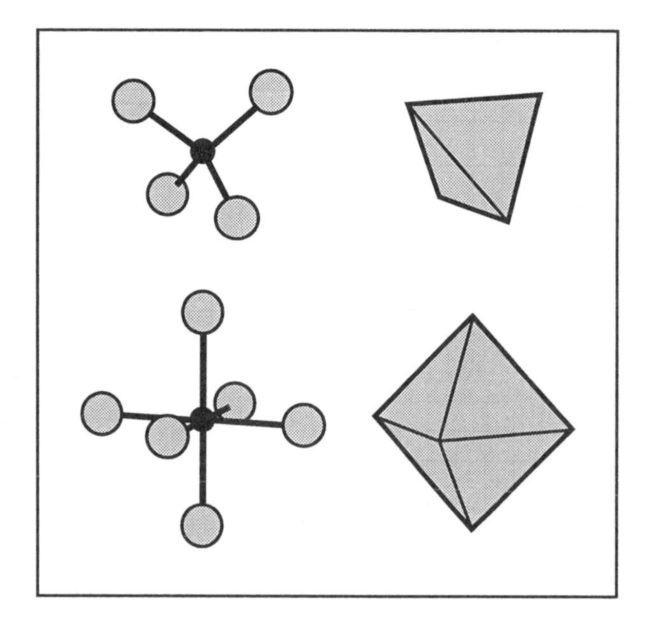

Figure 6. Fourfold (tetrahedral) and sixfold (octahedral) coordination. In the ball-and-stick representations (left), the black spheres are the cations, the gray spheres anions, and the sticks are the chemical bonds. In the polyhedral representations, the vertices of the polyhedra are anions, and the cations reside in the centers of the polyhedra.

The structure of diamond, in which each carbon atom is connected to four others by a three-dimensional network of strong covalent bonds, is illustrated in Figure 5a. Carbon also can crystallize in the graphite structure, Figure 5b, in which each atom is covalently bonded to three others to form planar sheets. These sheets are only weakly connected to each other through Van der Waals bonding. Whereas diamond is the hardest mineral, graphite is extremely soft, to the point that it is used as a lubricant and in pencil "lead," thus illustrating the profound influence that crystal structure can have on mechanical properties. Diamond and graphite also provide a classic example of a substance occurring in two different crystal structures. This phenomenon is known as polymorphism, and the different structures are known as polymorphs. The important semiconductor metals silicon (Si) and germanium (Ge) also crystallize in the diamond structure.

Just as the HCP and FCC structures provide the basis for many derivative structures, the diamond structure can be used to describe additional minerals. If alternating carbon (C) atoms in the diamond framework are replaced by zinc (Zn) and S atoms, we obtain the sphalerite (or zincblende) structure. Numerous important compound semiconductor materials, such as GaAs, crystallize with the sphalerite structure. In addition, if alternating layers of Zn are replaced by Cu and Fe, the result is the structure of the most important ore mineral of copper, chalcopyrite ($CuFeS_2$).

Silicates and Silica Minerals

Silicates are minerals that contain oxygen, silicon, and other metal atoms (*see* MINERALS, SILICATES). Together with the silica minerals (polymorphs of SiO_2), they make up over 95 percent of Earth's crust and mantle. Whereas simple structures are usually represented by packing models or drawings (e.g., Figures 1–4), or by ball-and-stick drawings (e.g., Figure 5), most silicate structures are so complex that such representations become confusing. Therefore, silicates are commonly illustrated using polyhedral drawings, in which a cation and all its coordinating anions are represented by a single, regular or distorted polyhedron, the corners of which are the anion positions and the center of

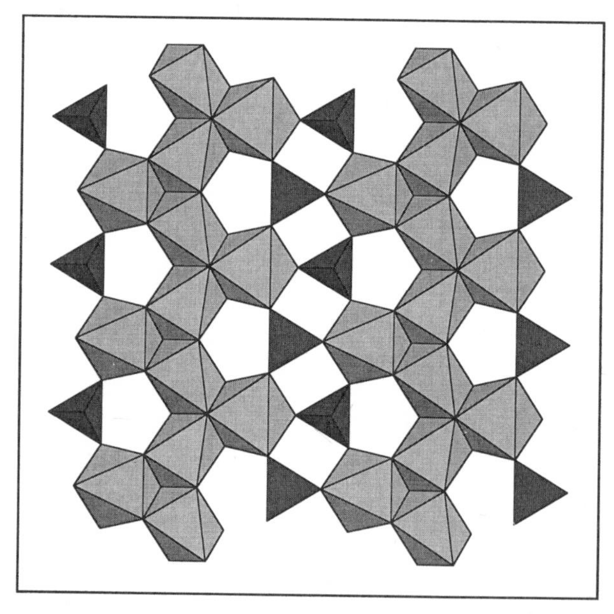

Figure 7. A layer of olivine structure. The tetrahedra (darker polyhedra) contain Si, and the zigzag chains of octahedra contain Mg and Fe.

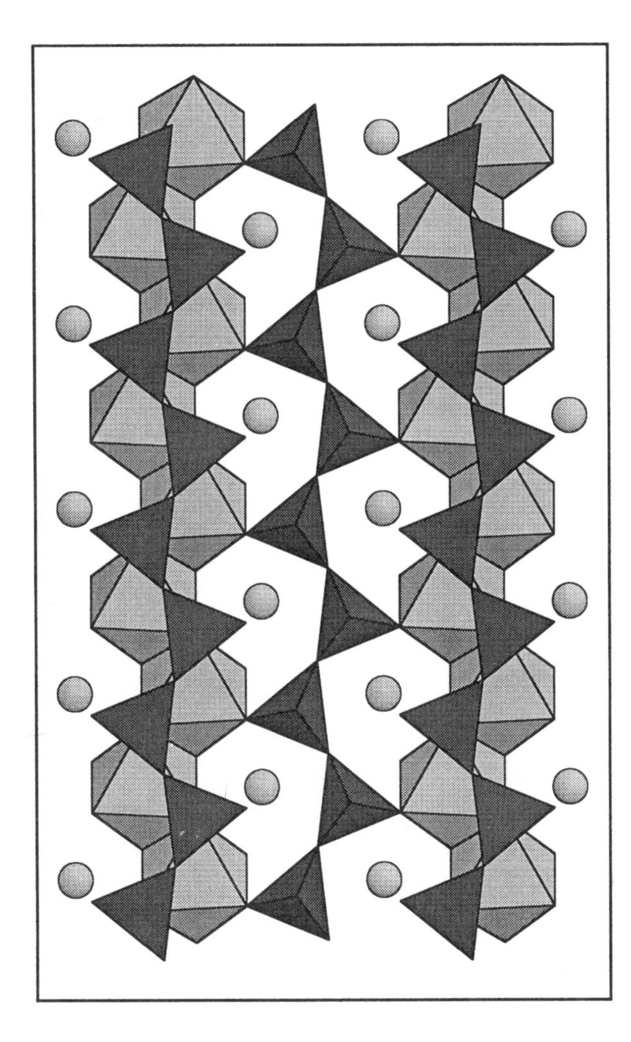

Figure 8. A layer of clinopyroxene structure, showing how the silicate chains are connected to the octahedral chains. The octahedra represent the M1 cations and the spheres the M2 sites.

which is occupied by the cation. Figure 6 shows both ball-and-stick and polyhedral representations of fourfold (tetrahedral) and sixfold (octahedral) coordinations. Generally, among the common rock-forming elements, Si, and sometimes Al, are tetrahedrally coordinated by oxygen, whereas cations such as Mg, aluminum (Al), titanium (Ti), manganese (Mn), and Fe are commonly found in octahedral coordination. The cations Ca and Na may be in sixfold coordination with oxygen or have higher coordination numbers, whereas potassium (K) typically is eight- to twelve-coordinated. Hydrogen (H) commonly is bonded to only one oxygen atom to form a hydroxyl ion (OH), but H also may participate in hydrogen bonding between two oxygen atoms.

Nesosilicates. The nesosilicates, also called orthosilicates or island silicates, contain isolated SiO_4 tetrahedra, which share no oxygen atoms with neighboring tetrahedra. Figure 7 shows a layer from the structure of olivine, $(Mg, Fe)_2SiO_4$, in which Si occupies the tetrahedra and Mg the zigzag chain of edge-sharing octahedra. Other important nesosilicates include the garnet group, the humite group, the aluminosilicate minerals (Al_2SiO_5), and zircon ($ZrSiO_4$).

The nesosilicates are the only silicate group in which all four-coordinated cations occur in isolated tetrahedra. All other silicate groups have structures in which oxygen ions are shared between adjacent tetrahedra, to form more complex anionic groups called polyanions. This phenomenon of connecting tetrahedra via shared corners is called polymerization. In addition to systematic structural variations among silicate groups, there is a systematic chemical variation: nesosilicates typically have a ratio of tetrahedral cations : oxygen of 1:4, and this ratio increases through the various groups, reaching 1:2 in the framework silicates and silica minerals.

Cyclosilicates. Also called ring silicates, cyclosilicates contain clusters of three, four, or six silicate tetrahedra that are polymerized to form closed rings. Representatives include beryl and tourmaline.

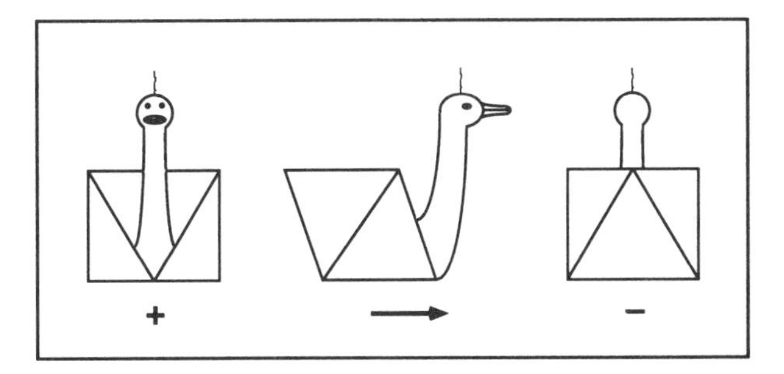

Figure 9. The duck convention used for visualizing the stacking sequences of chain silicates. The octahedral "duck" in the center is swimming to the right, that on the left is swimming toward us in the + orientation, and that on the right is swimming away from us in the – orientation. (After Thompson, 1981.)

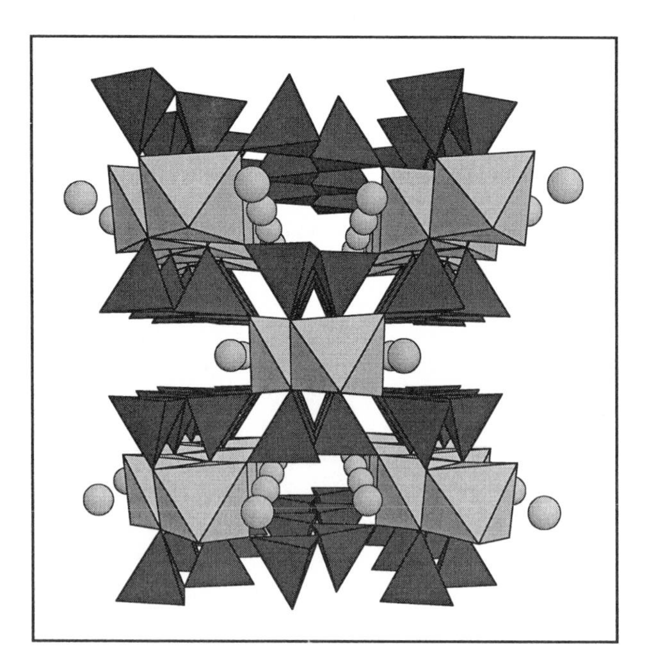

Figure 10. A perspective view of the clinopyroxene structure parallel to the silicate chains. All of the octahedra are oriented the same way, in the + orientation.

Inosilicates. Also called the chain silicates, inosilicates are characterized by the polymerization of tetrahedra into linear chains. In the pyroxene subgroup, there are single chains, as shown in Figure 8, which are attached to chains of edge-sharing octahedra. In most pyroxenes, there are two non-tetrahedral metal sites, a fairly regular octahedron known as M1, and a less-regular, six- to eight-coordinated site called M2 (represented by spheres in Figure 8). The M1 site typically contains Mg, Fe, and/or Al, whereas the M2 cation may be calcium (Ca), Na, Mg, and/or Fe.

Just as for the closest packed sheets in the HCP and FCC structures, the layers of chain silicates may be assembled with different stacking sequences. These sequences can best be seen using the duck convention (Thompson, 1981). In Figure 9, the octahedral "duck" in the center is swimming to the right. The duck on the left is swimming toward us (or is in the "+" orientation), whereas that on the right is swimming away from us (or is in the "−" orientation). The structure of monoclinic pyroxene (clinopyroxene), viewed parallel to the silicate chains, is shown in Figure 10. This figure shows that each octahedral strip in the chain silicate structures is connected to two silicate chains, and in this structure all of the ducks (octahedra) are swimming in the same direction (i.e., toward us), giving the stacking sequence ... +++ ... , or simply (+). Pyroxene minerals with the monoclinic structure include diopside, hedenbergite, clinoenstatite, clinoferrosilite, augite, pigeonite, jadeite, acmite, and omphacite.

The other common stacking sequence found in pyroxenes is that of orthorhombic pyroxene (the orthopyroxene structure), illustrated in Figure 11. From bottom to top of this figure, the octahedral layers are oriented, −, +, +, −, −, producing the sequence ... ++−−++−−++−− ... , or simply (++−−). The most abundant orthopyroxene minerals are solid solutions between enstatite and ferrosilite.

Figure 11. A perspective view of the orthopyroxene structure, which has the octahedral stacking sequence (++−−).

In addition to the pyroxenes, there is another group of single-chain silicates known as pyroxenoids. Rather than repeating every two tetrahedra (Figure 8), the chains in these minerals have periodicities of 3, 5, 7, or 9, producing such species as wollastonite, rhodonite, pyroxmangite, and ferrosilite III.

After pyroxenes, the most abundant chain silicates belong to the amphibole group, which is characterized by double silicate chains, as shown in Figure 12 (like two single chains, stuck together by sharing of oxygen atoms). Like pyroxenes, the amphiboles contain strips of edge-sharing octahedra known as M1, M2, and M3, which are chemically and structurally analogous to M1 in pyroxene. Similarly, they possess a six- to eight-coordinated site, called M4, that is analogous to the pyroxene M2 site. There is also an additional site, the A-site, between the back-to-back silicate chains in amphiboles, shown by the large spheres in Figures 12, 13, and 14. The A-site can either be empty or occupied by large cations such as K and Na. The large number of different crystallographic sites in the amphibole structures accounts for their extreme chemical complexity, which results in hundreds of mineral names that are used for different amphibole compositions.

There are two stacking sequences that appear in the common, rock-forming amphiboles, and they are precisely the same as those that occur in the common pyroxenes. Figure 13 shows the monoclinic amphibole (clinoamphibole) structure, viewed parallel to the chains, with the (+) stacking sequence. The common orthohombic amphibole (orthoamphibole) structure, with the stacking sequence (++−−), is shown in Figure 14.

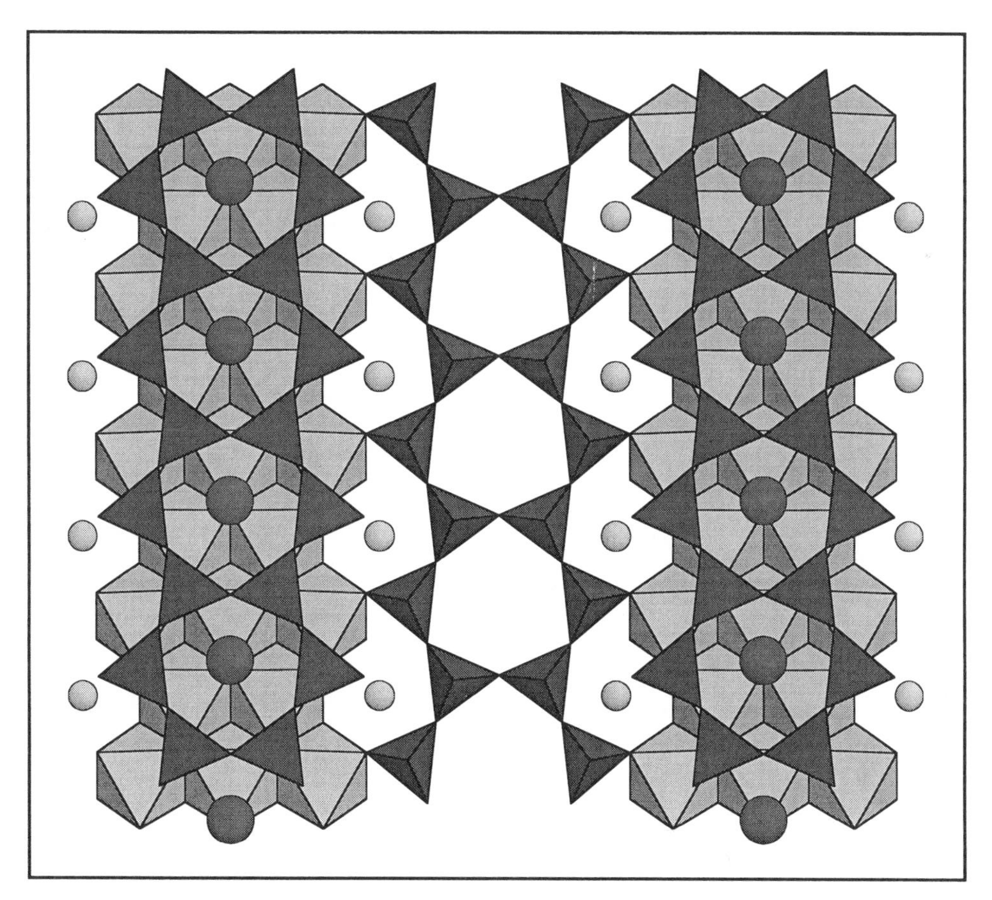

Figure 12. A layer of amphibole structure, showing the double silicate chains (darker polyhedra) and wide octahedral strips. The M1, M2, and M3 sites are represented by octahedra; the M4 site is shown as small spheres; the A-site is shown as larger, darker spheres.

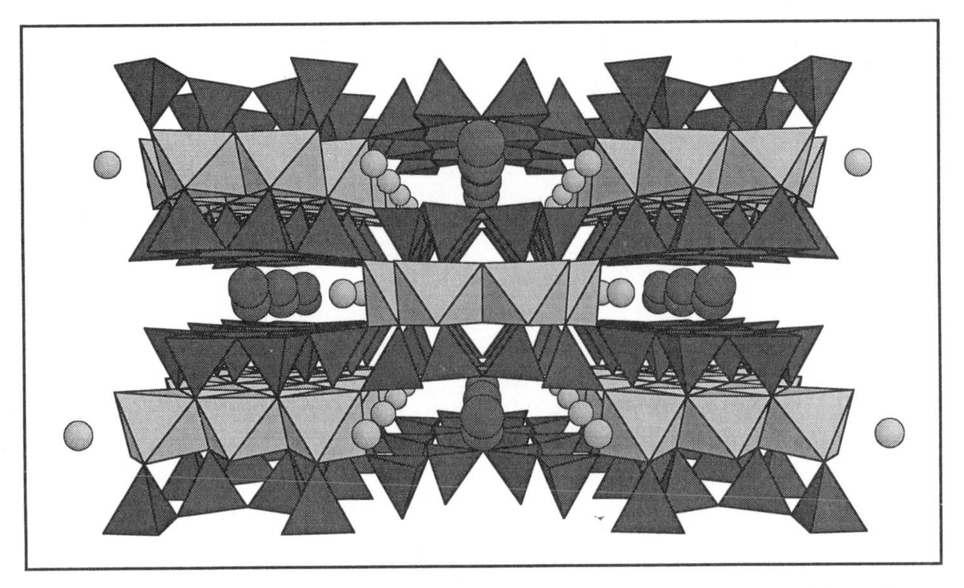

Figure 13. A perspective view of the clinopyroxene structure parallel to the silicate chains, showing the stacking sequence (+).

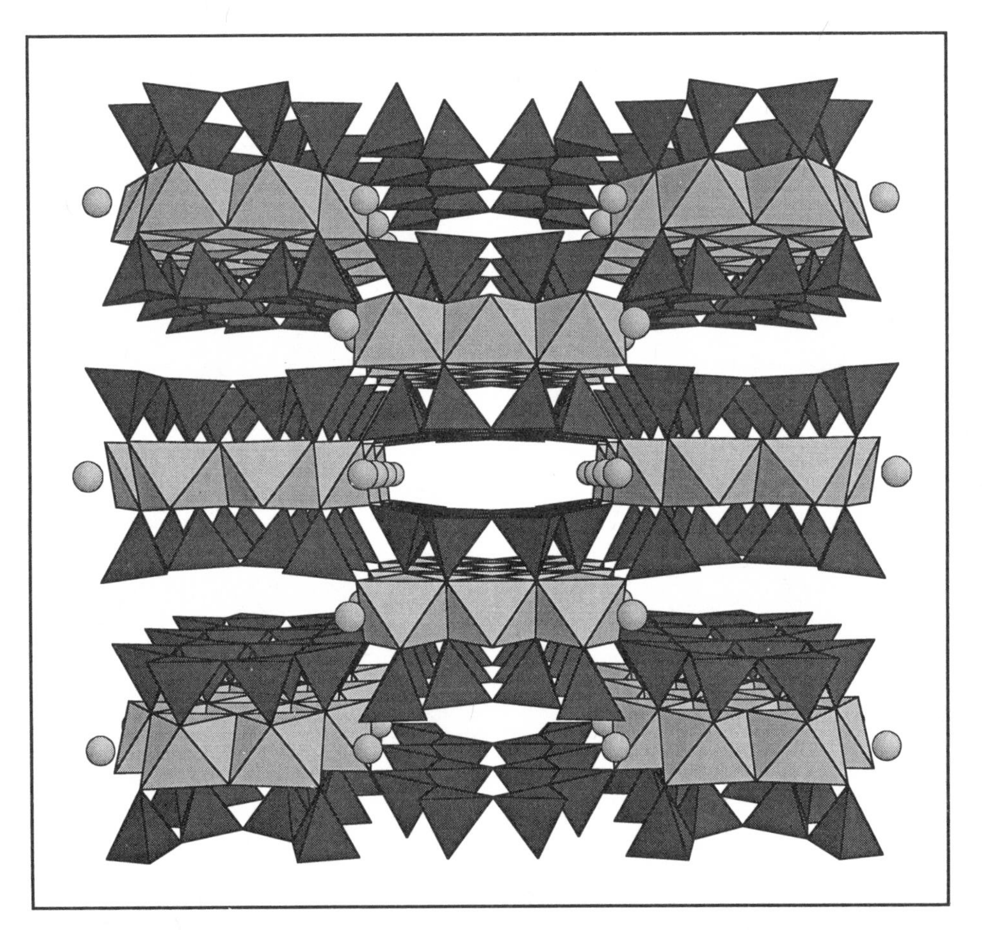

Figure 14. A perspective view of the orthoamphibole structure, which has the stacking sequence (++−−).

Phyllosilicates. Also called layer silicates or sheet silicates, phyllosilicates possess silicate sheets that are polymerized in two dimensions. As shown in Figures 15 and 16, these sheets are attached to sheets of edge-sharing octahedra. The type of octahedral sheet in Figure 15 is called a trioctahedral sheet, made up of octahedra that typically contain Mg or Fe. The trioctahedral sheets of phyllosilicates are similar to those found in the hydroxide mineral brucite, $Mg(OH)_2$. In the octahedral sheet of Figure 16, called a dioctahedral sheet, one-third of the octahedral sites are vacant. This type of sheet is typically occupied by Al and is similar to the sheets of the hydroxide mineral gibbsite, $Al(OH)_3$.

In the 1:1 subgroup of phyllosilicates, there is one tetrahedral sheet attached to each octahedral sheet, and the resulting layers are stacked atop one another, as shown in Figure 17. The 1:1 phyllosilicates include the serpentine minerals (trioctahedral) and the clay mineral kaolinite (dioctahedral). If two silicate sheets are attached to each octahedral sheet, the 2:1 phyllosilicates are derived. When the 2:1 layers are electrostatically neutral, the result is the talc structure (trioctahedral) or the pyrophyllite structure (dioctahedral). If, however, the 2:1 layers are charged, either monovalent cations, such as K^+ or Na^+, or divalent cations, such as Ca^{2+}, enter sites between the layers, producing the true micas and the brittle micas. Figure 18 shows the interlayer K^+ cations between the 2:1 layers of biotite (a trioctahedral, black mica). The common white mica minerals are dioctahedral, with the interlayer cation K^+ in muscovite and Na^+ in paragonite. One more common subgroup of phyllosilicates is represented by chlorite, which contains hydroxide sheets between 2:1 layers, rather than interlayer cations. Just as in the chain silicates, all of the phyllosilicates are subject to variations in the stacking of their layers, resulting in numerous structural variants known as polytypes.

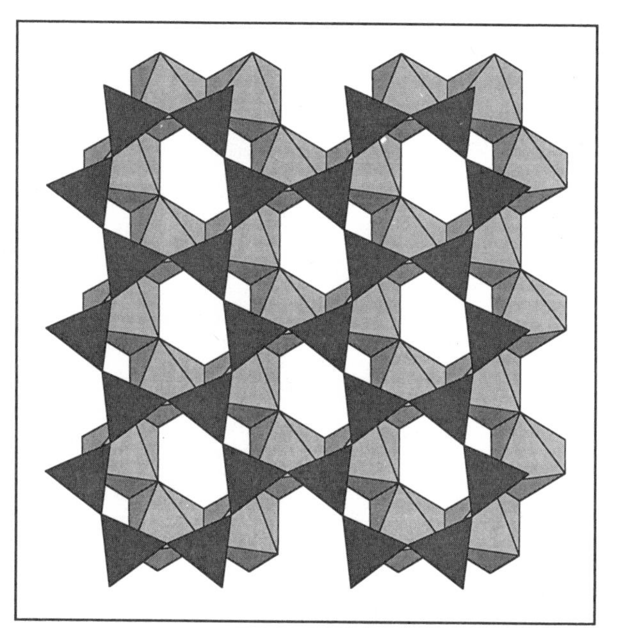

Figure 16. A layer from the kaolinite structure, showing a silicate sheet connected to a dioctahedral sheet. One third of the octahedral sites are vacant, compared to the trioctahedral sheet shown in Figure 15.

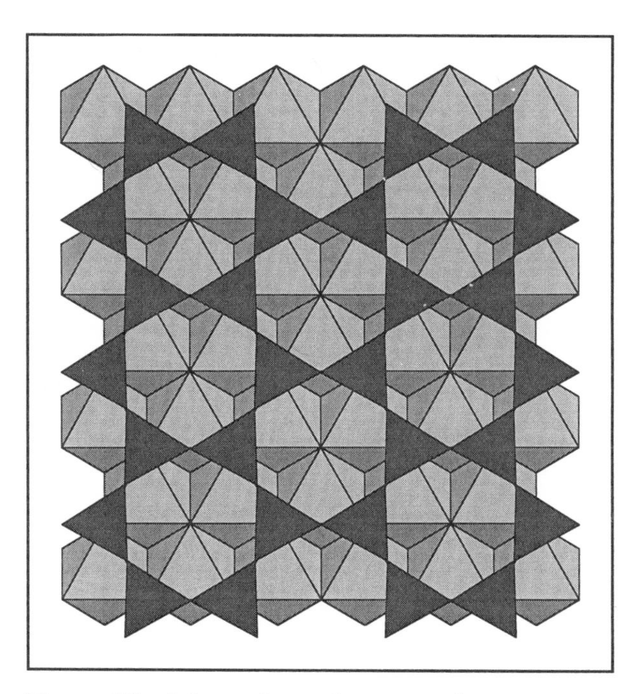

Figure 15. A layer from the serpentine structure, showing a two-dimensional silicate sheet connected to a trioctahedral sheet.

Tectosilicates and Silica Minerals. Also called framework silicates, tectosilicates and silica minerals are characterized by fully polymerized, three-dimensional frameworks of tetrahedra, and all four corners of each tetrahedron are shared by adjacent tetrahedra. The structure of quartz,

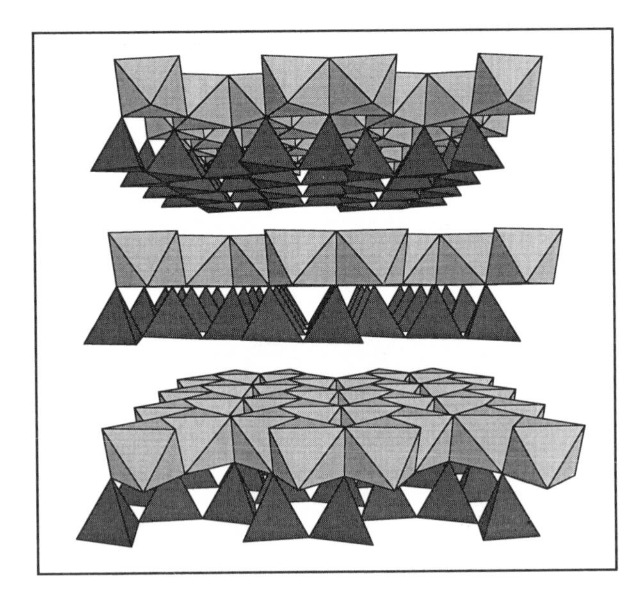

Figure 17. A perspective view parallel to the 1:1 layers of the kaolinite structure.

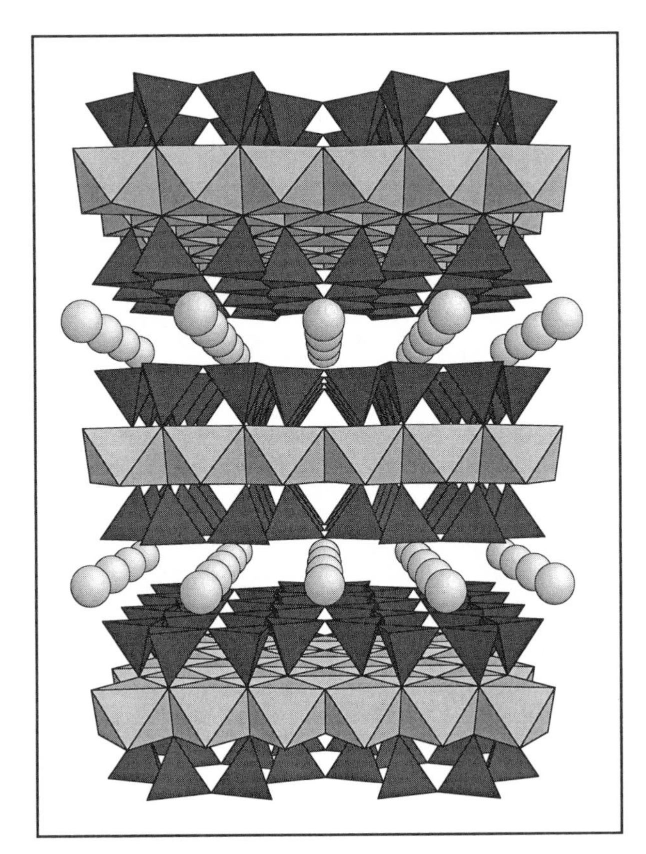

Figure 18. A perspective view parallel to the 2:1 layers of the trioctahedral mica biotite. The interlayer cations (K^+) are represented by spheres.

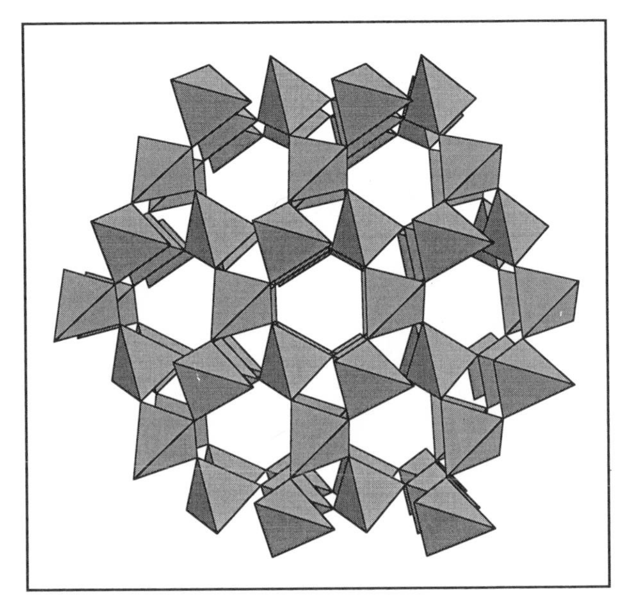

Figure 19. The structure of quartz, showing the linking of tunnels to form a three-dimensional framework.

which is by far the most abundant silica mineral (SiO_2) in Earth's crust, is illustrated in Figure 19. The linked SiO_4^{4-} tetrahedra occur as double helices that form narrow tunnels through the structure. These tunnels are cross-linked to form the three-dimensional framework. Other silica minerals include the high-temperature polymorphs tridymite and cristobalite, which have much less-dense, open frameworks, and the more densely packed, high-pressure form known as coesite.

Although not identical, the framework of coesite is similar to that of the feldspar minerals, shown in Figure 20. In the feldspar tetrahedral framework, some of the Si atoms are replaced by Al, and larger, charge-balancing cations occur in cavities within the framework: Ca in anorthite, Na in albite, and K in sanidine, orthoclase, and microcline. Feldspar crystal chemistry is extremely complicated because the Al and Si atoms may occur randomly (are disordered) on the tetrahedral sites of the framework at high temperatures, but they order into a variety of complex, nonrandom patterns at lower temperatures. At low temperatures, the rock-forming feldspars are triclinic, but at high temperatures, K-feldspar and albite become monoclinic.

Figure 20. The structure of the monoclinic K-feldspar sanidine. The tetrahedra of the framework are occupied by Si and Al, and the spheres representing the K ions occur in large cavities within the framework.

After feldspars, the most common rock-forming framework silicates are the feldspathoids, which have lower silica contents. The frameworks of some of these minerals are similar to those of tridymite and cristobalite, but with Al replacing part of the Si and cations such as Na and K stuffed into the large cavities within these frameworks. Also notable are the zeolite minerals, which have aluminosilicate frameworks with very large cavities or tunnels. These cavities can contain a number of different large cations and H_2O molecules. Ion exchange of these cations can occur rapidly, and certain zeolite frameworks can also act as catalysts for reactions involving other molecules that enter the structure. Because of their unique structures, zeolites are employed in industry as ion-exchange media, as molecular sieves, and as catalysts in processes such as petroleum cracking.

Mineral Structures at High Pressure

In all of the silicate structures we have examined, Si is tetrahedrally coordinated by oxygen. At pressures such as those in Earth's lower mantle, however, Si is forced into octahedral coordination, and silicates assume structures that are generally simpler than those of most crustal silicates. At these pressures, the simple pyroxene composition (Mg, Fe)SiO_3 assumes the perovskite structure, shown in Figure 21. The Si occupies the octahedral sites, and the Mg and Fe occupy a large, twelve-coordinated site in the octahedral framework (represented by a sphere in Figure 21). Although the mineral perovskite itself ($CaTiO_3$) is relatively rare, silicate perovskite may well be the most abundant mineral in our planet, given the huge volume of the lower mantle.

Another probable lower-mantle mineral is magnesiowüstite, $(Mg, Fe)_{1-x}O$, which possesses the NaCl structure. At very high pressures, the Si in silica itself forms SiO_6 octahedra, arranged in the rutile structure (Figure 22). At lower pressures, this structure occurs for a number of minerals, including rutile itself (TiO_2), pyrolusite (MnO_2), and cassiterite (SnO_2). Thus, at high pressure, the silicates tend to assume the structures of simple oxides that are observed at lower pressures for other compositions.

Figure 21. The ideal, undistorted structure of perovskite. In high-pressure silicate perovskite, the octahedral sites are occupied by Si, and the large, twelve-coordinated site shown as a sphere is occupied by Mg and Fe.

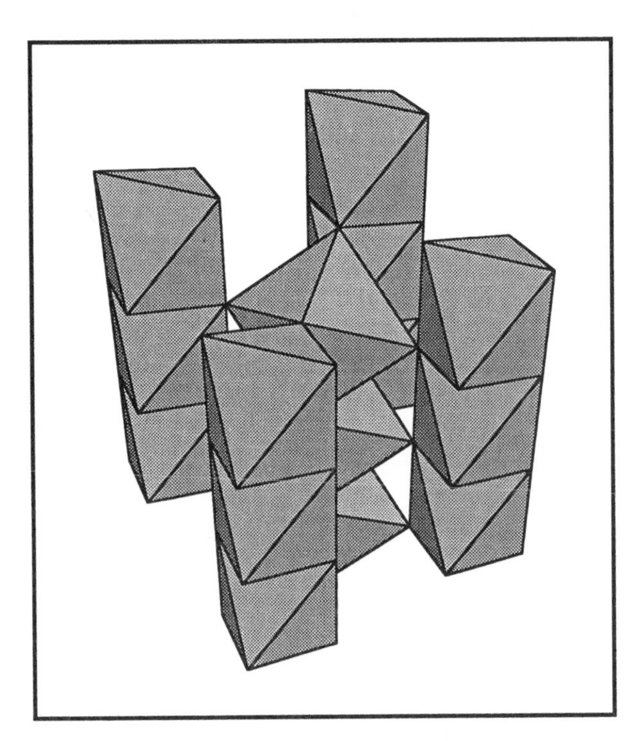

Figure 22. The structure of the SiO_2 polymorph stishovite. In this high-pressure form, Si occupies the six-coordinated sites in the crosslinked, edge-sharing octahedral chains of the rutile structure.

Bibliography

DEER, W. A., R. A. HOWIE, and J. ZUSSMAN. *Rock-Forming Minerals.* New York, 1963.

KLEIN, C., and C. S. HURLBUT, JR. *Manual of Mineralogy (after J. D. Dana),* 21st ed. New York, 1993.

LIEBAU, F. *Structural Chemistry of Silicates.* Berlin, 1985.

MINERALOGICAL SOCIETY OF AMERICA. *Reviews in Mineralogy* Washington, DC. Sulfides—Vol. 1; Feldspars—Vol. 2; Oxides—Vol. 3, 25; Zeolites—Vol. 4; Nesosilicates—Vol. 5; Pyroxenes—Vol. 7: Amphiboles—Vol. 9A; Carbonates—Vol. 11; Micas—Vol. 13; Phyllosilicates—Vol. 19; Aluminosilicates—Vol. 22; Silica—Vol. 29.

PAPIKE, J. J. "Chemistry of the Rock-forming Silicates: Ortho, Ring, and Single-Chain Structures." *Reviews of Geophysics* 25 (1987):1483–1526.

———. "Chemistry of the Rock-Forming Silicates: Multiple-Chain, Sheet, and Framework Structures." *Reviews of Geophysics* 26 (1988):407–444.

THOMPSON, J. B., JR. "An Introduction to the Mineralogy and Petrology of the Biopyriboles." In *Amphiboles and Other Hydrous Pyriboles—Mineralogy,* ed. D. R. Veblen. Washington, DC, 1981.

DAVID R. VEBLEN

MINERS, HISTORY OF

The earliest miners were hunters who sought rocks for weapons from stream beds and from the soil. In soft soil they made shallow excavations. As societies became more complex and more organized the task of obtaining supplies of required rock and flint was assigned to special groups.

Rock drawings from around 4000 B.C.E. at the Timna Mine in southern Israel suggest that early miners worked as free men, not as slaves, and that they contributed to the technology. In contrast, the Nubian gold fields, which furnished Middle East treasure for over two thousand years, was operated, at least in part, with slave labor. In about 1400 B.C.E., the Egyptian pharoah Amenhotep II sacked Nubia and returned to Egypt with more than fifty tons of copper, half a ton of gold, and many slaves.

About 1200 B.C.E. or earlier the Phoenicians, originating from what is now Lebanon, opened up trade routes throughout the Mediterranean world. They encouraged mining by supplying a market for the products. Their major contribution in mining included the opening of the copper/silver mines at Rio Tinto, Spain, the tin mines of Cornwall, England, and the lead/silver mines in Greece. They created a system by which prosperity and power could be measured in actual exchangeable wealth.

The mines of Laurium in Greece were worked by slaves procured for this purpose and rented at a rate of twenty-five grams of silver per day. Their life expectancy under the Athenian lash was only four years, but their labors produced the silver that sustained the Golden Age of Greece and the birth of democracy. Gold from Macedonia financed the initial campaign of Alexander the Great, and during his brief lifetime these mines produced more than 30,000 gold talents, about three quarters of a ton. Diodorus recorded: "There is absolutely no consideration or relaxation for sick or maimed, for aged man or weak woman. . . . all are forced to labor at their tasks until they die, worn out by misery among their toil."

The Romans were efficient administrators. During the reign of Hadrian (138 C.E.) the Roman Empire began to recognize advantages of a degree of individual ownership and permitted mining by freedmen in increasing numbers. One of their greatest contributions was introduction of mining law detailing the rights and obligations of miners. Gradually slave labor was replaced by skilled artisans. It was Roman custom, however, to bind the population of a mining district to the soil. Workers were forbidden to migrate from the vicinity of the mines to seek employment elsewhere. This was the beginning of the tradition of villenage, a feudal system from which the miner did not completely extricate himself for a thousand years.

During the Middle Ages little progress was made in mining in the Western world. Emergence from this period coincided with the discovery and development of rich silver deposits in Saxony, Bohemia, and the Tyrol by Saxon miners.

The migration of the Celts, a people from south-central Europe who settled in the metal-rich areas of central Germany, was probably the single most important factor in dissemination of mining technology and mining legal concepts throughout Europe. These Celts became the Saxons and carried a tradition of *bergbaufreiheit,* or the rights of the free miner, whereby the poorest villein could become his own master merely by marking his own mining claim and registering its boundaries after making a discovery—subject to a tribute or royalty paid to the royal landowner. The Saxon miner formed guilds from which later grew the labor unions. The guilds set rules for working conditions. By the sixteenth century Saxon "free miners" led the way not only in their own country but throughout all of Europe and Britain.

The Rammelsburg silver deposits found in 938 furnished Henry the Fowler, the Duke of Saxony, with money to fortify the duchy of Harz, provided a large part of the funds for his military campaigns and for those of his son, Otto the Great, and helped finance development of an empire. The free Saxon miner at Rammelsburg became the most advanced in mining technology.

The legal foundation for the rights of the free miner was recorded in a treaty initiated by the Bishop of Trent in 1185, where miners were invited to explore and mine the regions of northern Italy as free men with rights of discovery. This practice was copied throughout Europe and Britain. Rights were extended by various princes in the Germanic territories in 1209 and by Edward II of England in 1288. The right of ownership based on discovery by a free miner became the foundation for mining laws and was carried by the Saxon and Cornish miners to the Americas, Australia, and South Africa. As mining extended underground the free miner found that he had to cooperate with his companions in working a mine and partnerships formed. As the operations grew, other men were required, and this led to self-governing groups—"cost-book associations"—which later became stock companies.

In 1562 Queen Elizabeth sent to Germany for Saxon miners to introduce better mining practices to Britain. These Saxon-Celts who settled in Cornwall were tin miners, and during the gold rushes of the nineteenth century the Saxon and Cornish miners supplied mining technology and the basis for mining laws throughout the world.

Discovery of gold in California in 1848, the Australian discoveries at Ballarat and Bendigo in 1851, and discoveries in the South African Witwatersrand in 1886 coincided with a period of revolution, economic depression, misery, and famine in Europe and Britain. Gold mining offered the potential for millions of people to better themselves. Thousands left the Cornish and Saxon mines and migrated to the goldfields of America, Australia, and South Africa. These individuals formed the nucleus of skilled miners in the development of all three countries. The Chinese, in whose country mining was discouraged or prohibited, left their impoverished home and joined in the search for gold in California and Australia, often bonded to hong merchants until their earnings had more than repaid transportation costs. The Chinese worked primarily the low-grade placer deposits that had been abandoned by the Europeans. These miners eventually became pioneer farmers, engineers, and traders in these developing countries.

Modern mechanized mining has changed the economics of mining and the lives and prosperity of modern miners. Today's miner is a highly skilled worker receiving top wages. These developments can be attributed to D. C. Jackling, a struggling American mining engineer who in 1903 succeeded in interesting a group of investors in the idea of bulk mining of low-grade ores by open-pit methods. The bulk-mining concept spread to underground mining as well, with development of block caving and sublevel caving methods.

Bibliography

BLAINEY, G. *The Rush That Never Ended.* Melbourne, Australia, 1963.

HARTMAN, H. L., ed. *SME Mining Engineering Handbook.*

Society of Mining, Metallurgy and Exploration. Baltimore, 1992.

RAYMOND, R. *Out of the Fiery Furnace.* Melbourne, Australia, 1984.

WILLARD C. LACY

MINING TECHNIQUES, PAST AND PRESENT

About 300,000 years ago man began searching streams and loose soil for fragments of chert or flint, tough rocks that would not shatter when used as hammers, and for things of beauty—gems and gold. Deposits in the soil or soft rock led to excavations, and by the Neolithic period (around 3,500 B.C.E.) flint mines in soft chalk beds in France and England had well-developed systems of shafts and stopes carved with bone or horn picks and stone hammers. During the same period in Egypt, the mining that supported the stone industries was carried out largely through quarrying—the wedging loose of blocks and shaping with stone hammers.

Before 4,000 B.C.E. at Timna in southern Israel, malachite and azurite in soft white sandstone were mined using different methods: through excavation of loose material and stone quarrying; by tunneling and stoping of copper mineralized rock; and by sinking vertical circular shafts up to 20 m deep. Ventilation was provided by small openings above the tunnels, drilled as the tunnel advanced.

Early copper mines (about 2,700 B.C.E.) on the island of Cyprus and in the Danube valley were mined by trenches and rows of shafts along veins.

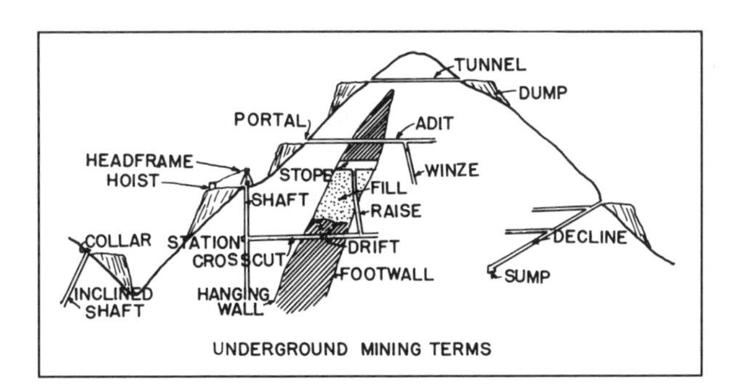

Figure 1. Underground mining terms.

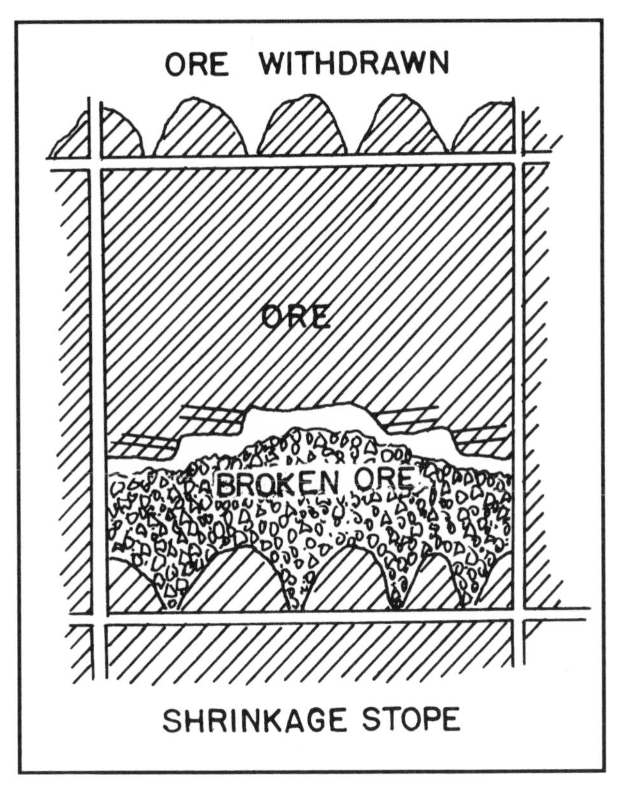

Figure 2. Shrinkage stope.

Rock was broken by fire-setting techniques—heating the face with fire and quenching with water. Underground workings of this period consisted of adits, inclined shafts, and stopes measured in hundreds of meters (Figure 1).

During the early Iron Age at the Hallstatt salt mines near Salzburg, Austria, miners used iron implements. Sloping shafts with timber supports were dug, and when masses of salt were encountered stopes were opened.

The Laurium mine in Greece was worked by Myceneans in the second millennium and then abandoned. Athenians renewed working there about 600 B.C.E. from open pits with short adits dug into oxidized ores. Later, more than two thousand rectangular vertical shafts were sunk in pairs and connected by stopes. Parts of pulley wheels and marks of axles at the top of shafts indicate a system for hoisting the ore. Ores were sorted underground and the better grade ores hoisted while poorer grade material was left as pillars or for filling old workings. Work was done with wrought iron hammers and picks with wooden handles, chisels, and wedges. Each miner had his own lamp and niches were cut on the sides of the workings

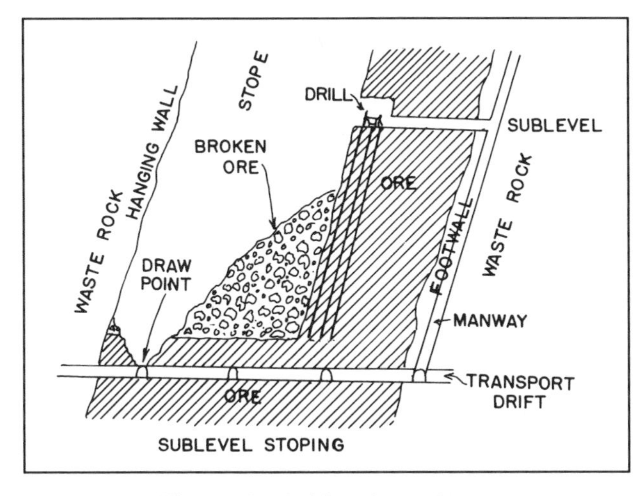

Figure 3. Sublevel stoping.

for oil lamps. Ventilation was controlled by underground doors and fires at the bases of certain shafts to create updraft.

The Romans developed ingenious methods for dewatering mines to allow them to penetrate to depths below the water table. At Cyprus and Rio Tinto, Spain, they introduced the "Egyptian Screw," a water wheel and chains of pots for lifting water, all powered by slaves. Fire-setting was seldom used to break rock underground because of ventilation problems. Picks, wedges, hammers, and battering rams weighing up to 60 kg were used to break up quartz veins. On the surface, water wheels were used for grinding and aqueducts were built to supply water for sluicing ores.

The practice of breaking rock underground by fire-setting gave way to the use of explosives, first used at Schemnitz, in former Czechoslovakia, in 1672. The Saxon miners devised new pumps, utilized water power, and devised the first fuel-operated power engine. The Savery steam engine, developed at Cornwall, England, in 1698 began the mechanization of pumps and drills, and the development of systems to transports ores and men from underground.

Illumination evolved from tallow or oil dip lamps used by the early Saxon miners, to candles fixed in clay on helmets, ladders, and rock faces by the eighteenth century Cornish miners, to the twentieth century illumination provided by acetylene lamps and electrical cap lamps.

China's contacts with the Middle East enabled early mining and metallurgical technology to remain abreast of, and in some areas surpass, that of the West. The Chinese have mined and smelted iron for at least 2,500 years, and gunpowder was first used by them in the third century for breaking rock. China was isolated during the industrial revolution, however, and fell behind in industrial technology.

In England, Cornwall was the center from which much of modern underground technology evolved. The steam engine of James Watt and the high-pressure steam engines of Richard Trevithick enabled mining in Cornwall and elsewhere to prosper during the first half of the nineteenth century. These engines powered the Cornish pump, the hoisting of ore and men, and the compression of air for ventilation and powering of drills. Black powder (gunpowder) was replaced by "giant powder" and then by Alfred Nobel's dynamite.

A variety of modern mining methods have evolved to take advantage of the size and shape of the ore bodies, overburden, strength of the ore and host rock, water conditions, value of the ore, and available technology. Contributions have come from the Saxon and Cornish miners, and from miners in the Americas, Australia, and South Africa.

Underground Self-supporting Mining Methods

Small ore bodies and narrow veins can be completely extracted, and where the rock is stable,

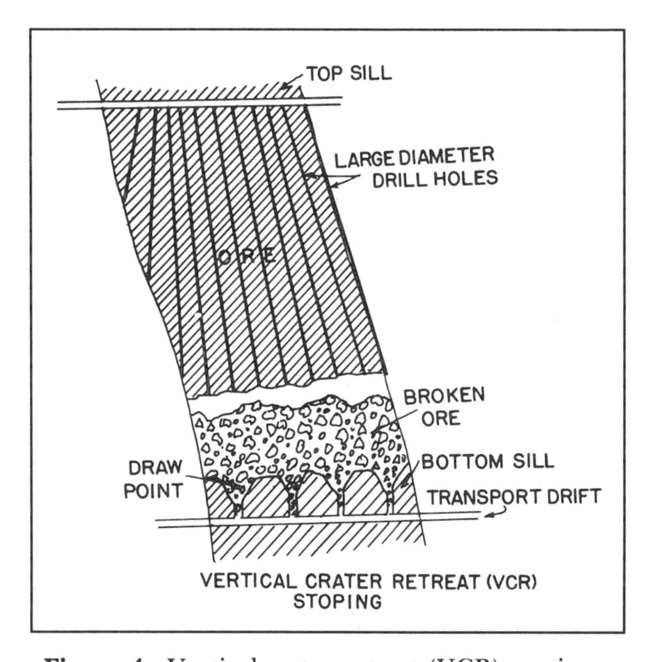

Figure 4. Vertical crater retreat (VCR) stoping.

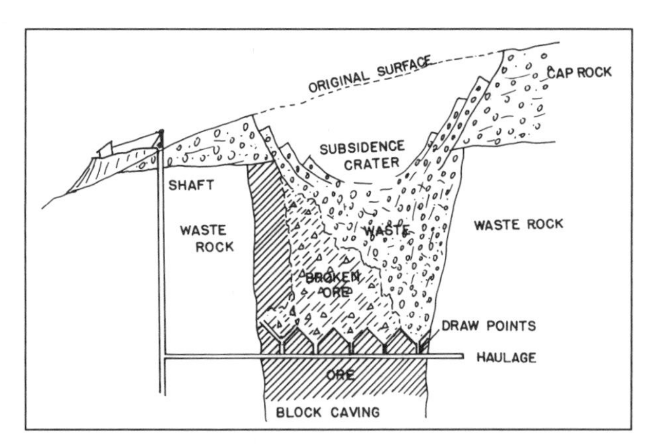

Figure 5. Block caving.

huge stopes (underground excavation formed by the extraction of ore) may stand open for many years.

In flat or gently dipping bedded ores, such as coal, salt, phosphate, and oil shale, room and pillar mining is used. This technique leaves some rock in place as support pillars. These are generally removed just before abandoning that portion of the mine. In some instances artificial supports are used for temporary support.

Shrinkage stoping (Figure 2) is used in narrow, steeply dipping ore bodies with stable walls, by breaking the ore from beneath and allowing the broken ore to support the walls. Broken ore is partially drawn from chutes at the bottom to supply space for the miner to drill overhead. After the stope is completed all the ore is drawn and the stope allowed to collapse.

Sublevel open stoping is a high-production, bulk-mining method applicable to large, steeply dipping, regular and competent ore bodies enclosed in stable rock. Stoping is carried out by blasting vertical slices of ore into expansion slots and drawing the broken ore from draw-points at the haulage level (Figure 3).

A variation of sublevel stoping, vertical crater retreat, uses charges in vertical drill holes to break the ore, making horizontal cuts into expansion space provided by drawn ore, and advancing upward. The broken ore provides support for the walls of the ore body until ore is drawn (Figure 4).

A block caving mining method is used in the mining of large ore bodies with appreciable vertical extent (over 100 m) that are too deep to mine by open pit methods. The entire ore block is undercut in panels that cave as a result of the crushing weight of overlying ore and overburden (Figure 5). Broken ore is drawn from raises and transported to the surface. A subsidence crater forms above the caved ore blocks that may later be used for open pit mining of peripheral ores or leaching of collapsed material.

Underground Supported Mining Methods

Cut and fill stoping is used where one or both walls of the ore are weak and would collapse and dilute the ore. It is similar to shrinkage stoping except that as each cut of ore is removed, a layer of waste rock or mill tailings is placed in the stope to serve as a platform on which the miners work. Ore is taken from the stopes as mined and dropped down timbered raises (Figure 6).

Where ore and enclosing rock are weak and the value of the ore is high, a square-set method of mining with timber support is used. A small block of the ore—roughly 2 m × 2 m × 2.5 m—is removed and replaced by a timber set. As mining progresses the timber sets are interlocked and filled with waste rock or sand fill. The high cost of

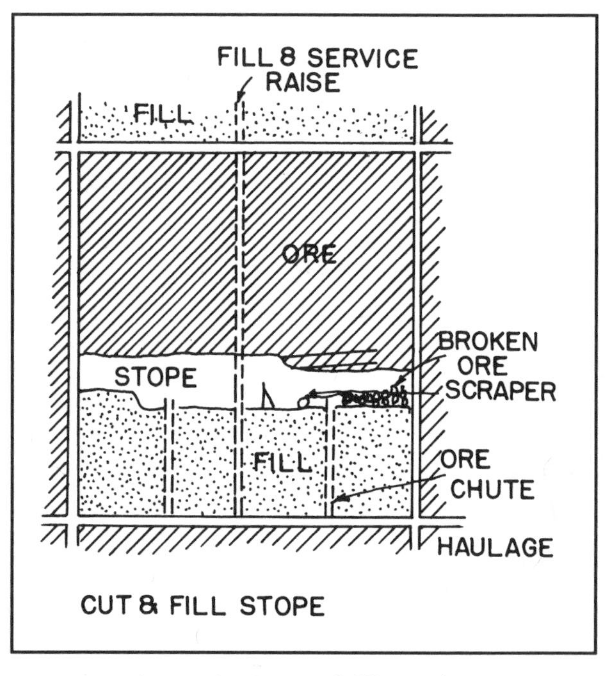

Figure 6. Cut and fill stoping.

timber and labor has made this mining method nearly obsolete in the United States.

Surface Mining Methods

Surface mining applies to extraction of minerals from water and sediments of lakes and rivers, seas and oceans, from groundwater (natural or introduced), and from rock and soil excavated from the surface.

Placer deposits are generally concentrations of heavy minerals such as gold, diamonds, and tin and tungsten minerals in alluvium. They are generally low-grade deposits and must be close to a supply of water, near the surface, and loosely consolidated. Most of the placer mining falls into three groups: panning and sluicing, hydraulicing, or dredging.

Where the deposit is near the surface, glory holing may be used. It involves a pit at the surface from which ore is dropped by gravity through raises connected beneath to haulageways along which ore is transported to the surface (Figure 7).

Open pit mining is applied to surface excavation of large, often low-grade ore bodies in which waste material is stored adjacent to the pit as excavation progresses. Mining occurs as sequential pushbacks of the pit slopes or by the lateral mining of sequential benches. Standard operation includes drilling, blasting, excavating, loading, and haulage of ore to concentrator and low-grade material to leach or waste dumps. Pit slopes range from 35° to 50° dependent upon rock conditions. Giant drills, shovels, and trucks break and remove the ore. Pits may extend to depths greater than 300 m. A future challenge is discovery of uses for the excavations and the shaping of dumps for future uses (Figure 8a).

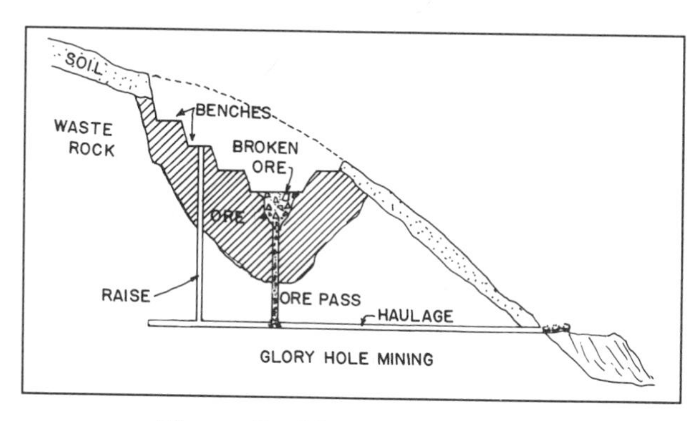

Figure 7. Glory hole mining.

Strip mining is similar to open pit mining but reclamation can be carried out simultaneously with extraction. It is used on flat-lying tabular deposits close to the surface. Mining progresses by removal first of a box cut and storage of the overburden waste material. Thereafter, overburden is broken, excavated, and hauled by truck, scraper, or conveyor to fill previously mined-out pits with little or no waste deposited outside the mine area. Mined areas can be readily landscaped following mining (Figure 8b).

Solution mining techniques are used in extraction of soluble ores such as potash, salt, and other minerals where conventional mining methods would not be economic. Sulfur is mined by the Frasch process where the sulfur is liquified by steam in order to bring it to the surface in drill holes. Copper oxide minerals are dissolved by acid solutions in mine dumps, cave areas over old mines, and fractured rocks. Uranium is leached by acid solutions in permeable sandstones, and gold is extracted by cyanide or thiourea solutions from dumps, fractured rocks, or alluvium by circulation of these solutions. Permeability often needs to be enhanced by blasting, caving, hydrofracturing, or actual mining and placing broken ore on leach pads. Leach solutions are applied through ponding, sprays, drip or injection methods.

Drilling and Blasting

Rock excavation in most non-coal underground mining is carried out by drill and blast techniques using percussive drills. In surface mines and large-scale caving methods large-diameter rotary drills are used.

An explosive or blasting agent is a compound or mixture of compounds that are capable of rapid decomposition when subjected to heat, impact, friction, or shock. When confined within a drill hole this decomposition releases tremendous amounts of heat and gas. A shock wave and energy of expanding gases fractures and displaces adjacent rock. Varieties of dynamite and gelatin dynamite are generally utilized in underground mining, whereas in surface mining or in large drill holes underground dry porous pellets (ANFO) saturated with diesel fuel are used.

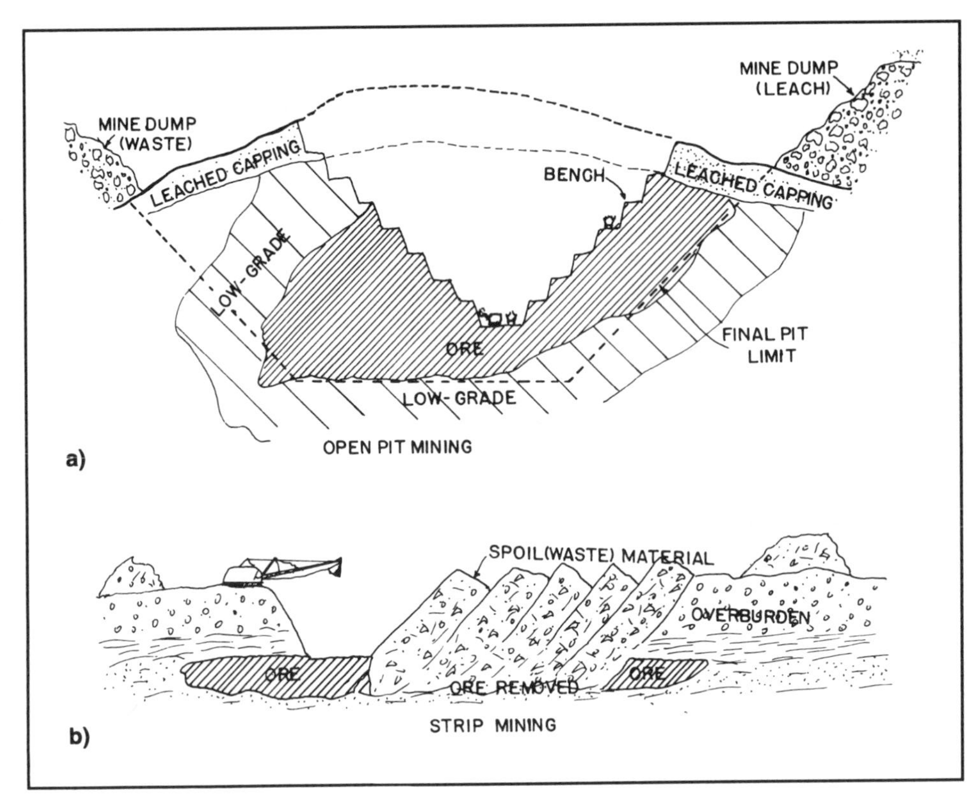

Figure 8. a. Open pit mining. b. Strip mining.

Loading and Transport

In underground mining transport of ore and waste is generally by rail, conveyor belt, or truck—loaded by gravity flow from ore chutes or front-end loaders. Surface mining uses large trucks (100- to 300-ton capacity), or by rail or conveyor belt, loaded by large shovels (to 35 m^3 capacity). In soft material continuous bucket-chain and bucket-wheel excavators can be used.

Automation and Robotics

Because of possible health and safety considerations, automation, remote control, and robotic applications in mining are under investigation.

Bibliography

Banata, H. *Anatomy of a Mine.* USDA Forest Service, General Technical Report INT-35. Washington, DC, 1977.

Hartman, H. L. ed. *SME Mining Engineering Handbook.* Baltimore, MD, 1992.

Raymond, R. *Out of the Fiery Furnace.* Brisbane, Australia, 1984.

WILLARD C. LACY

MINORITIES IN THE EARTH SCIENCES

The population of profession geoscientists includes a broad range of people: men and women, people with disabilities, and representatives of many minority groups. Minority groups, in particular, are proportionately underrepresented in the geosciences. Among the various minority groups represented in the overall U.S. population, African Americans, Hispanics, and Native Americans are grossly underrepresented in the sciences generally and the earth sciences in particular.

The primary factor responsible for the underparticipation of these groups in the profession has its origin in the low number of minorities enrolled in bachelor programs in geoscience. For example, during the 1993–1994 academic year, just 4.5 percent of the students enrolled in undergraduate programs in geoscience were members of minority groups. At the graduate level the number declines to 2.8 percent for the same period.

A 1995 survey of geoscience faculty shows that minority diversity (5.5 percent overall) is generally higher among lower-level positions, with the exception being at the instructor level, which is far less diverse than the assistant professor level (Claudy, 1995). Data on other employment sectors are limited but indicate that only 2.3 percent of employed geoscientists are members of underrepresented minority groups (Claudy and Kauffman, 1988). (See Tables 1 and 2.)

Historical Perspective

To better understand this imbalance, a historical perspective is useful. At the time the geosciences were forming as a profession in the United States in the late 1600s, nonwhites were legally barred from any educational training, including learning to read and write. Although a small number of African Americans and Native Americans learned informally, rarely were such individuals accorded professional status. One notable exception was Benjamin Banneker, an African American born in Maryland in 1731. Banneker was a farmer for much of his life, but he also cultivated an interest in astronomy and mathematics, which he learned through extensive reading. In 1791, Banneker's mathematical expertise earned him an appointment by then Secretary of State Thomas Jefferson to assist Major Andrew Ellicott in laying the plans for a federal city on the Potomac River. Banneker performed a major portion of the surveying for the District of Columbia, and in so doing became the first documented African American employed in the geosciences in America (Kaplan, 1973).

Table 1. Ethnic Minority Distribution in Faculty by Ethnicity

	Faculty (%)
African Americans	1.0
Hispanics	0.9
Native Americans	0.2

Table 2. Minority Distribution in Faculty by Level

	Faculty (%)
Professor	4.2
Associate Professor	5.6
Assistant Professor	9.9
Instructor	3.5

Source: Clandy, 1995.

Also in 1791, an expedition was mounted in Mexico, traveling from Acapulco to Mexico City. The scientific leader for the expedition was Antonio Pineda y Ramirez, whose mission was to make "a tedious comparative examination of the soil and the primary core of this Kingdom," including naturally occurring metals and water bodies. A consultant to the expedition party was José Antonio Alzate y Ramirez, whose interests included cartography. The scientific team examined waterworks, mines, and local geoscientific practices, which included meteorology (Engstrand, 1981). Another expedition commenced that year, led by José Longinos Martinez, who had developed a well-regarded natural history museum (including a mineral collection) in Mexico City. The Martinez expedition, which began in Mexico City and ended in Monterey, California, included an examination of the mines in Baja California (then operated by Native Americans) and the observation of "ore deposits of silver, gold, lead, antimony, magnetic iron, etc." in Saltillo, near Cabo San Lucas. Though these two expeditions were preceded by Spain's earlier Royal Scientific Expeditions, the reports from these journeys probably reflect some of the earliest documentation of geoscientific research by Hispanics in the New World.

The Civil War and the changes that it wrought on the U.S. Constitution, along with related postwar legislation, improved educational opportunities for nonwhites. The bill establishing the Bureau of Refugees, Freedmen, and Abandoned Lands (1865), for example, provided a basis for establishing a number of state agricultural and mechanical colleges, many of them created to educate freed slaves (Bennett, 1993). This latter group of colleges later became known as HBCUs, an acronym for historically black colleges and universities. HBCUs provided education to African Americans in chemistry and biology, but few provided training in the earth sciences.

The period extending from the 1950s through the mid–1970s is sometimes referred to as the "second reconstruction." Two factors during this period had particular importance for underrepresented groups: civil rights legislation and the race to land men on the moon. Civil Rights legislation passed in the late 1950s and early 1960s increased the number of minority Americans who voted and who enrolled in higher education. Further, the pivotal Supreme Court decision in the case of Brown vs. Board of Education (1954) theoretically invalidated the system of racially separate education. These actions, combined with the American space program objectives fueled by competition with the former Soviet Union's program, significantly increased the numbers of minority and female students matriculating toward science degrees and finding employment in the sciences, including geoscience.

Individual Successes

It was also during this same period that the first African Americans achieved doctoral degrees in the geosciences. One of these individuals was Mack Gipson, Jr. Gipson, a native of Trenton, South Carolina, received his bachelor's degree in natural science from Paine College (an HBCU) of Augusta, Georgia, in 1953, and his master's and doctoral degrees in geology from the University of Chicago in 1961 and 1963, respectively. Gipson was the first African American to receive a Ph.D. in geology from the University of Chicago, and his primary career was in higher education. Gipson founded and was instrumental in developing the geology department at Virginia State University at Petersburg (another HBCU), and later taught at the University of South Carolina at Columbia. Gipson also did a substantial amount of work in petroleum exploration, both domestic and international.

Another pioneering African American earth scientist is Randolph W. Bromery, a geophysicist who earned his bachelor's degree at Howard University in 1956. Bromery went on to earn his master's degree at American University (1962) and his doctoral degree in geophysics from the Johns Hopkins University (1968). Bromery embarked on a career in the federal government as a geophysicist with the U.S. Geological Survey (USGS), and later served as a faculty member and in administration in higher education at the University of Massachusetts at Amherst, where he served as chancellor. Bromery served as the president of the Geological Society of America (1989–1990), the first African American in that position. In the mid–1990s Bromery was employed in a consulting firm and served on several corporate boards (Exxon, Nynex), and provided administrative leadership in the Massachusetts higher education system as a member of the Board of Regents.

Gipson and Bromery both achieved rewarding careers, the origins of which can be traced to historically black colleges. Their access to graduate school and later employment opportunities were significantly enhanced by the Equal Opportunity/ Affirmative Action policies that were promulgated during the 1960s.

Charles Baskerville proved yet another example of an African American whose notable career in the earth sciences may have been based in part on "second reconstruction" opportunities. A geological engineer who earned his bachelor's degree at the City College of New York (1953), Baskerville went on to earn master's (1958) and doctoral (1965) degrees at New York University. Baskerville began his professional career with the New York State Department of Transportation. He subsequently joined the faculty of the City College of New York, and later worked at the U.S. Geological Survey, from which he retired in 1990. In the mid–1990s Baskerville taught at Central Connecticut University.

Latinos, such as Louis A. Fernandez, also benefited from the improved access for minorities in higher education. Fernandez, who was born in 1939 in Puerto Rico, earned his bachelor's degree at the City College of New York in 1962, his master's degree from University of Tulsa in 1964, and his doctoral degree from Syracuse University in 1969. Fernandez worked as a research geologist at Yale University in New Haven, Connecticut, from 1968 to 1971, then joined the geology faculty at the University of New Orleans in 1971 where he served in several positions, including chair of the department of geology and dean of the College of Sciences. In 1991 Fernandez became dean of the School of Natural Sciences at the California State University at San Bernardino, and in 1995 was elevated to the position of Vice President of Academic Affairs.

Another underrepresented group in the geosciences is individuals with physical disabilities. According to National Science Foundation data (1994), in 1990 people with disabilities accounted for only 11,100 people, or 2.6 percent, of the total science workforce compared to 22 million disabled in the American workforce overall (O'Brien, 1993). The employment of disabled Americans specifically in the geosciences has been tracked only minimally. A survey of college student enrollment in 1993–1994 shows only twenty-six disabled students enrolled as undergraduate geoscience majors (AGI, unpublished data). Legislation such as the Americans with Disabilities Act (1990) may provide increased opportunities for this group to engage in geoscience careers.

Portrait of Change over Time

Data on the number of doctoral degrees earned in the earth, atmospheric, and ocean sciences reflect the continuing pattern of underparticipation by some minority groups in the geosciences, despite a moderate increase between 1983 and 1991 (see Table 3).

Problems with Data

It is difficult to ascertain accurately the level of underrepresentation of minority groups as the available demographic data are limited. Frequently, surveys that address the gender and minority status of the student or professional population in the sciences are not reported. Also, such surveys tend to be discontinuous, limiting analysis to short-term changes rather than changes in participation over time. Further, the small size of the minority geoscience population makes any survey sample susceptible to inaccuracy—the omission of an individual may have a significant impact on the survey outcome.

For example, are Asian Americans to be considered a minority in the earth sciences when statistically they appear to be more numerous in the earth science workforce than in the U.S. population? The resolution of such debate lies partly in the design and implementation of the surveys and the methods for gathering data on participation.

Many issues have been identified as factors contributing to the paucity of some minorities in science, including poor preparation in science and mathematics in grades K through twelve; cultural elements, such as lack of family and peer support; educators' low expectations of minority students' academic performance; and the isolation or lack of collegiality among individual groups because of the low numbers of peers (Malcom, 1993). In the geosciences, these factors are exacerbated by the virtual absence of earth science curricula in high schools, leaving most students unaware of the discipline. Recent efforts to reform and improve science education have resulted in several new educational frameworks that include earth science as a discipline that all students need to know, and in equal portions to those disciplines usually thought of as traditional sciences: physics, chemistry, and biology (NRC, 1994; AAAS2, 1993; NSTA, 1992). Such changes are likely to increase opportunities for all students, including minorities, to pursue meaningful careers in the earth sciences.

Table 3. Doctoral Degrees in Geosciences

	African-American Students	*Hispanic Students*	*Native American Students*	*All Students*
1983	1	18	2	483
1984	2	2	–	474
1985	2	8	1	442
1986	–	9	2	422
1987	1	6	–	425
1988	2	4	2	511
1989	3	12	6	529
1990	3	22	1	535
1991	2	24	3	606
1992	6	26	1	526
1993	3	16	4	473

Source: NSF, 1994.

Bibliography

American Association for the Advancement of Science (AAAS). *Benchmarks for Scientific Literacy.* New York, 1993.

Bennett, Jr., L. *Before the Mayflower: A History of Black America,* 6th ed. New York, 1993.

Claudy, N. H. "Faculty Salaries Rise." *Geotimes* 40, no. 9 (1995):5.

Claudy, N., and M. E. Kauffman. *North American Survey of Geoscientists.* Alexandria, VA, 1988.

KAPLAN, S. *The Black Presence in the Era of the American Revolution, 1770–1800*, New York, 1973.

MALCOM, S. "Increasing the Participation of Black Women in Science and Engineering." In *The "Racial" Economy of Science: Toward a Democratic Future,* ed. S. Harding. Bloomington and Indianapolis, IN, 1993.

NATIONAL RESEARCH COUNCIL (NRC). *National Science Education Standards.* Washington, DC, 1994.

NATIONAL SCIENCE FOUNDATION (NSF). 1994, "Selected Data on Science and Engineering Doctorate Awards: 1993." NSF 94-318, Arlington, VA, 1994.

———. "Women, Minorities, and Persons with Disabilities in Science and Engineering: 1994." NSF 94-333, Arlington, VA, 1994.

NATIONAL SCIENCE TEACHERS ASSOCIATION (NSTA). *Scope, Sequence, and Coordination of Secondary School Science,* Vol. 1, *The Content Core.* Washington, DC, 1992.

SUITER, M. J. "Programs for Ethnic Minorities in the Geosciences." *GSA Today* 1, no. 9 (1991):187–191, and 1, no. 10 (1991):217–218.

———. "Tomorrow's Geoscientists: Recruiting and Keeping Them." *Geotimes* 36, no. 1 (1991):12–14.

MARILYN J. SUITER

MITCHELL, MARIA

America's first woman astronomer, Maria Mitchell, was born on Nantucket Island, forty-eight kilometers off the southern coast of Massachusetts on 1 August 1818. She was the third of ten children of Quaker parents. Maria's father, William, was a liberal, cultured gentleman, a teacher in the Nantucket schools, and from 1836 to 1861 cashier of the local bank. He was an ardent amateur astronomer and close confidant of the first two directors of Harvard College Observatory, William and George Bond. Maria received her basic training in astronomy from her father and at an early age became his assistant. At the age of twelve, during the annular solar eclipse in 1831, she timed the contacts for him. Cyrus Peirce, before he accepted the directorship of Horace Mann's first normal school in Lexington in 1839, taught at a school for young ladies on Nantucket. One of his students was Maria, whose aptitude for mathematics he recognized and encouraged. By means of stellar observations William Mitchell checked the accuracy of the rates of chronometers for whaling captains. One time, when William was not at home, a whaler brought his chronometer and Maria volunteered to check it, having learned by watching her father. The whaler was skeptical but let her try; to his amazement she accomplished the task perfectly.

In 1835, at age seventeen, Mitchell started a private school for girls age six and older. It was the first school on Nantucket to admit children of all races, color, or religious preference. (At that time the public school would not accept black children.) Unfortunately this experiment did not survive because the following year she was appointed librarian at the Atheneum, Nantucket's public library, a position she held for twenty years. Maria went to work early every morning to study some of the books under her care before the doors opened to the public. She surveyed Nantucket for her father for the purpose of making a map of the island. This was arduous work, as much of the land had previously been explored only long ago by native Indians. William Mitchell published their map in 1838.

On clear nights, after her work at the Atheneum, William and Maria scanned the skies for interesting objects, searching for comets, observing variable stars, planets, and their satellites, and timing lunar occultations. One memorable evening, on 1 October 1847, working alone with their 7.5 cm Dollond telescope, Maria discovered a new comet. For this discovery she received a gold medal established by the King of Denmark to be awarded to anyone who first sighted a telescopic comet (one invisible to the naked eye at the time of discovery). Maria Mitchell became the first woman and the first American to receive the medal, and this catapulted her to international fame. She declared, however, that she could not claim the comet as her own unless she could compute its orbit. Her earlier mathematical studies under Cyrus Peirce paid dividends. Her orbit for the comet indicated that it was traveling in a parabolic orbit and would not return—a conclusion confirmed much later by one of her Vassar students, Margareta Palmer, at Yale. In 1859, a group calling itself "The Women of America" presented Mitchell a 12.5 cm Alvan Clark telescope with which she might discover more comets. Indeed, she discovered three more, and computed their orbits, but others anticipated her in these discoveries. The telescope is still in active use at the Nantucket Maria Mitchell Obser-

erected in 1908 in her memory. In 1848 she became the first woman to be made a member of the American Academy of Arts and Sciences, indeed the only woman inducted for nearly a century, when five women were admitted in 1943.

While a librarian, she was given the opportunity to compute ephemerides of Venus for the *American Ephemeris and Nautical Almanac,* a task she continued along with her other duties from 1849 through 1868. Alexander Bache, director of the U.S. Coast Survey, loaned the Mitchells a transit instrument for them to contribute observations for the determination of the figure of Earth.

Vassar College for women was opened in 1865 with Maria Mitchell as the first American woman professor of astronomy, a post she held until 1888. In Nantucket she had lived in a community where a large percentage of the male population were whalers, spending years at a time at sea. Hence the women of the island experienced greater responsibilities and greater freedom of decision than most women at that time. But at Vassar she recognized discrimination for the first time when she discovered that the men in other departments, with responsibilities similar to her own, were being paid more. Henceforth she became an ardent advocate of women's rights, and in 1870 she was elected president of the American Association for the Advancement of Women.

Maria Mitchell proved to be an outstanding teacher. One of her ablest students, Mary Whitney, who became her successor, reflected that Mitchell's gift as a teacher was her ability to provide stimulus, not drills. Few teachers were more inspiring. Among her students were later prominent woman astronomers, including Antonia Maury (1866–1952) at Harvard, noted for her pioneering work in stellar spectroscopy, and Margareta Palmer (1862–1924), the first woman in the United States to earn a Ph.D. in astronomy, at Yale in 1894 (see Hoffleit, 1983).

Although Maria Mitchell was a diligent observer she published very little: one article in the *Astronomical Journal* (1856) on her observations of minima of the eclipsing star Algol; one in the *American Journal of Science* in 1863 on observations of thirty-six double stars and four articles between 1873 and 1879 on the satellites of Jupiter and Saturn; and a few less technical articles elsewhere, notably in Nantucket's weekly newspaper.

Although she was primarily an astronomer, Maria Mitchell's work on chronometry, celestial navigation, cartography, and her determination of the latitude and longitude of the Vassar College Observatory (Furness, 1934) constituted relevant contributions to earth sciences.

Bibliography

FURNESS, C. E. "The Longitude of the Vassar College Observatory. Determination by Maria Mitchell, 1865–77." *Publications of Vassar College Observatory* (1934):1–15.

HOFFLEIT, D. *Maria Mitchell's Famous Students.* Pamphlet published by the American Association of Variable Star Observers, in celebration of the seventy-fifth anniversary of the Maria Mitchell Observatory, 1983.

KENDALL, P. M. *Maria Mitchell: Life, Letters, and Journals.* Boston, 1896.

WRIGHT, H. *Sweeper in the Sky.* New York, 1949.

DORIT HOFFLEIT

MODERN ORE DEPOSITS

Many of today's rich mineral deposits were formed by hydrothermal fluids circulating in rocks close to Earth's surface. Where these heated solutions have escaped to the surface active hot springs have often developed. The minerals that are deposited around active hot springs today are recognized as modern ore deposits in the making. The essential components of these active ore-forming systems are:

1. Heat from igneous activity or Earth's natural geothermal gradient to drive the hydrothermal circulation of fluids
2. Structural pathways, such as faults or rifts, to allow infiltration of water into the crust and to focus the flow of heated solutions to the surface
3. A source of metals that may be provided by leaching of bedrock or from volcanic emanations (e.g., metal-enriched fluids evolved from magmas)
4. A means of transporting the metals (e.g., in solutions such as meteoric water, seawater, connate water trapped in sedimentary formations, magmatic fluids, or metamorphic fluids)

5. A mechanism for precipitating metals from the rising hydrothermal solutions (e.g., conductive cooling, reaction with wall rocks, mixing with near surface waters, or boiling)

Active ore-depositing geothermal systems occur in different geologic settings throughout the world. There is a concentration of hot spring activity around the margins of the Pacific that coincides with volcanic activity on island arcs (e.g., Japanese Islands, Philippine Islands, New Zealand). Hot springs are also common in continental volcanic environments, where heat is provided by the cooling magmas that once fed volcanos at the surface (e.g., Geysers—Clear Lake area, California). Major rifts in continental crust and the extensions of these rifts into the sea also focus hydrothermal circulation and have led to the formation of mineral deposits in the Salton Sea, the Gulf of California, and the Red Sea. The most productive of the modern ore-forming systems occur where new oceanic crust is being created by seafloor spreading (e.g., along the mid-ocean ridges at places like the East Pacific Rise, the Mid-Atlantic Ridge, and Iceland). Not all hydrothermal systems are related directly to igneous activity. An enhanced geothermal gradient and high heat flow during mountain building or deep burial of sedimentary basins can also lead to the movement of hydrothermal fluids and the formation of ore deposits. Some saline brines that are expelled from compacting sediments in subsiding sedimentary basins contain abundant metals, such as zinc (Zn) and lead (Pb), as well as hydrocarbons that are leached from the sediments (e.g., so-called oil field brines in the coastal areas of the Gulf of Mexico). The "oil field brines" are considered by many to be possible modern analogs for ore-fluids that produced some ancient sediment-hosted Pb-Zn deposits.

Modern ore deposits forming in active hot springs include those of gold (Au), silver (Ag), mercury (Hg), arsenic (As), and antimony (Sb)—known as "epithermal deposits"—as well as deposits of copper (Cu), Zn, and Pb. Although none of the actively-forming mineral deposits in the world are currently being mined, a few of the producing gold mines in parts of the western Pacific still have active hot springs. Some hot spring deposits on land may be the surface expressions of major mineral deposits (e.g., porphyry copper deposits) forming at depth.

Hot Spring Deposits on Land

Since the 1950s, extensive drilling of hot spring systems for geothermal power has exposed many occurrences of active mineral deposition. Most of these deposits are found within continental volcanic environments, and many are characterized by "geysers" produced by near-surface boiling of the hydrothermal fluids. Well-known examples of such deposits are found in the Geysers—Clear Lake area of northern California; the world's largest geothermal power producer, at Steamboat Springs in Nevada; and at the Broadlands geothermal field in New Zealand. These hot springs are heated to temperatures of up to 300°C at depths of up to a kilometer or more beneath the surface. Where the hydrothermal fluids discharge along faults, metals—such as silver, gold, arsenic, antimony, and mercury—are precipitated together with abundant opaline silica in deposits known as sinters. In the Geysers—Clear Lake area, several major deposits of Hg have formed from the recently active hot springs. At Steamboat Springs, the sinters contain up to 150 parts per million of Ag and 15 parts per million of Au. Such deposits are similar to many of the gold deposits being mined in the Great Western Basin of the United States. At Broadlands, New Zealand, fluids discharging into hot spring pools are precipating high-grade Au, As, Sb, Hg, and thallium (Tl) in sinters around their margins. It has been estimated that hot springs in the Broadlands system could deposit as much as 450,000 kg (or 4.5 tons) of gold in less than one thousand years, a relatively short time in geological terms. In many developed geothermal systems, minerals are deposited as scales in the pipe casing of deep wells, and these scales are sometimes rich in metals such as Ag, Au, Cu, Zn, and Pb.

The Salton Sea geothermal system in the Imperial Valley of California contains some of the most metal-rich hot springs on land. Hydrothermal fluids in this system have reacted extensively with sediments in the Imperial Valley and are much more saline than other hot springs on land. As a result, the Salton Sea brines carry abundant metals including iron (Fe), Cu, Zn, Pb, Ag, cadmium (Cd), and As. Ore minerals are being precipitated from the brines in veins, along fractures and faults, and as disseminations within the surrounding rocks at depths of up to a kilometer beneath the surface. The formation of a major ore deposit in the Salton

Sea geothermal system is prevented by a lack of sufficient sulfur to combine with all the metals and form sulfide minerals.

Hot Spring Deposits on the Ocean Floor

In 1977 the first deep-sea hot springs were discovered at a water depth of 2,600 m along the Galapagos Spreading Center (Corliss et al., 1979). In the following year, large deposits of metal sulfides were found in a similar setting on the East Pacific Rise near 21°N latitude (see Edmond and Von Damm, 1983). These deposits were formed by hot springs known as "Black smokers" venting hydrothermal fluids onto the seafloor at temperatures of up to 350°C. Such deposits are now known to be common along the entire length of the world's mid-ocean ridge system and in other areas of seafloor volcanic activity (*see* METALLIC MINERAL RESOURCES FROM THE SEA). As in geothermal systems on land, hydrothermal convection of seawater at temperatures of 350–400°C causes leaching of metals and sulfur from the ocean crust along the mid-ocean ridges. The heated seawater is similar to the Salton Sea brines and is able to carry abundant metals such as Fe, Cu, and Zn, but, unlike the Salton Sea, hot springs on the modern seafloor also carry enough dissolved sulfur to combine with all of the metals in solution. These metals accumulate as deposits of massive sulfides where the hydrothermal fluids vent onto the seafloor. "Black smoker" deposits are now recognized as being the modern analogs of ancient deposits of Cu and Zn that are mined on land today. One large deposit of this kind is currently forming in the Red Sea.

Bibliography

CORLISS, J. B.; J. DYMOND; L. I. GORDON; J. M. EDMOND; R. P. VON HERZEN; R. D. BALLARD; K. GREEN; D. WILLIAMS; A. BAINBRIDGE; K. CRANE; and T. H. VAN ANDEL. "Submarine Thermal Springs on the Galapagos Rift." *Science* 203 (1979):1073–1083.

EDMOND, J. M., and K. VON DAMM. "Hot Springs on the Ocean Floor." *Scientific American* (April 1983):78–93.

ELLIS, A. J., and W. A. J. MAHON. *Geochemistry and Geothermal Systems.* New York, 1977.

HENLEY, R. W., and A. J. ELLIS "Geothermal Systems Ancient and Modern: A Geochemical Review." *Earth Science Reviews* 19 (1983):1–50.

WEISSBERG, B. G., P. R. L. BROWNE, and T. M. SEWARD. "Ore Metals in Active Geothermal Systems." In *Geochemistry of Hydrothermal Ore Deposits,* ed. H. L. Barnes. New York, 1979.

WHITE, D. E. "Active Geothermal Systems and Hydrothermal Ore Deposits." In *Seventy-Fifth Anniversary Volume of Economic Geology,* ed. B. J. Skinner. New Haven, 1981.

MARK D. HANNINGTON
PETER M. HERZIG

MOLLUSKS

The phylum Mollusca comprises such morphologically diverse animals as snails, slugs, clams, oysters, mussels, squids, and octopuses. There is no single morphologic character that is shared by all members of the phylum, although most are free-living, unsegmented, have a head with a mouth and sense organs such as eyes, have a complete gut, and have a dorsal calcareous shell that provides support for a muscular foot and the internal organs. The shell grows by accretion and is secreted by a mantle, which is a fold of the body wall that surrounds the other soft parts. Most mollusks also have a radula, which is a hard, elongate, toothy structure that can be protruded from the mouth and is used to rasp the substrate while collecting food. The word Mollusca comes from Latin and literally means "soft bodied," an allusion to the soft parts of these animals. Mollusks are mostly marine animals, although some pelecypods (clams) and gastropods (snails) live in brackish water and fresh water, and some gastropods live in terrestrial, nonaqueous environments. The sexes are separate in most mollusks, although a small number of pelecypod and gastropod species are hermaphroditic, and a few gastropod species are unisexual. Various pelecypods, snails, and cephalopods are commercially sought after as food for humans. Buttons or jewelry are made from the shells of some pelecypods and gastropods; some oysters are cultivated for pearl production. A number of mollusks, such as garden slugs and zebra mussels, are considered to be pests.

Classification and Evolution of Mollusks

Mollusks (Figure 1) have been diverse and abundant since the Cambrian period and, because most

Figure 1. Shells of representative fossil mollusks, illustrating the morphological diversity of the phylum.
a. An internal mold of a monoplacophoran, *Bipulvina croftsae*, showing well-preserved, paired muscle scars; from the Ordovician of Missouri; ×1.25.
b. A rostroconch, *Conocardium cuneus;* from the Devonian of Ohio; ×1.
c. A gastropod, *Busycon echinatum*, showing well-preserved spines; from the Pleistocene of Florida; ×0.75.
d. A gastropod, *Fasciolaria scalarina*, viewed from the apertural side, showing a borehole drilled by a predatory gastropod; from the Pleistocene of Florida; ×0.75.
e. A pelecypod (oyster), *Gryphaea arcuata*, showing asymmetry of the two valves; from the Jurassic of France; ×0.75.
f. A gastropod, *Isonema humile*, showing the characteristic spiral shape and well-preserved growth lines on the shell; from the Devonian of Ohio; ×2.
g. A scaphopod, *Dentalium intercalatum;* from the Cretaceous of Tennessee; ×0.75.

Figure 1. Continued
h. A pelecypod (scallop), *Pecten fraternus*, showing asymmetry in the shell and a small borehole drilled by a predatory gastropod; from the Miocene of Maryland; ×0.75.
i. A pelecypod, *Eucrassatella* sp., showing a borehole drilled by a predatory gastropod; from the Miocene of Maryland; ×0.75.
j. A hyolith, *Haplophrentis reesei;* from the Cambrian of Utah; ×1.
k. An ammonoid cephalopod showing goniatitic sutures, *Imitoceras rotatorium;* from the Mississippian of Indiana; ×0.75.
l. An ammonoid cephalopod showing ceratitic sutures, *Ceratites compressus* (left), and a pelecypod, *Hoernesia socialis* (right); from the Triassic of Germany; ×0.75.
m. An ammonoid cephalopod showing complex ammonitic sutures; from the Cretaceous of Seymour Island, Antarctica; ×0.5.
All photos by Loren E. Babcock.

forms have a calcareous shell, they have generally left a good fossil record (*see* FOSSILIZATION AND THE FOSSIL RECORD). Currently, mollusks are second only to arthropods in species diversity. The number of living species of mollusks is estimated to be more than 100,000; at least 35,000 species have been described from fossils.

There are eight or nine classes of mollusks, of which all but two have living representatives. Most mollusk classes have fossil representatives dating from the Cambrian period. An unequivocal ancestor of all the mollusk classes is not known from the fossil record.

Aplacophorans (class Aplacophora) are small, anatomically simple, wormlike mollusks that lack a shell and are instead covered with a spicule-bearing cuticle. They have a complete gut, gills, and a radula. Aplacophorans have no known fossil record because they are not easily preserved.

Chitons (class Polyplacophora) are oval-shaped mollusks that seem to have developed from a spicule-bearing ancestor. They evolved eight dorsal shell valves and retained spicules only in the mantle girdle, which surrounds the shell. Chitons live in high-energy, shallow marine environments, where they graze on algae and bacteria.

Monoplacophorans (class Monoplacophora) have a cap-shaped or twisted shell. Members of this class have a series of paired muscles running lengthwise along the inside of the shell. Most early monoplacophorans probably lived in shallow water and grazed on algae and bacteria, but present-day ones live in deep water and feed on microscopic organisms in mud.

Gastropods or snails are members of a large class (Gastropoda) whose members have helically coiled shells; the apex of the shell points away from the animal's head. The viscera of a snail becomes twisted, or torted, in the early larval stage. This torsion allows the animal to squeeze a lot of body mass into a small amount of space, and brings most organs into a position above the head. A gastropod can retract the head and foot into the protective shell, and an operculum allows the animal to close off the open end of the shell when the head and foot are retracted. The gastropods have undergone great evolutionary diversification since the Cambrian period, in part because of their shell features and because of the presence of a radula, which allows them to easily break down food. Some gastropods made the transition from the marine environment to fresh water and land by the Pennsylvanian period. Most gastropods remained in the oceans, however, where some developed into voracious predators that could either use the radula to bore into the shells of other mollusks, or use poisoned radular teeth to subdue their prey. Other gastropods filtered microorganisms from water, whereas some became swimmers, burrowers, or parasites on other animals. A few gastropods (such as sea slugs and the terrestrial garden slugs) have lost their shells.

Cephalopods are members of a large class (Cephalopoda) whose members commonly have a bilateral symmetry and a shell that is partitioned into chambers by septa. Cephalopods, which include the living *Nautilus*, squids, cuttlefishes, and octopuses, have a well-developed head region that includes a brain, tentacles, and sense organs including eyes. Representatives of this group may have evolved from monoplacophorans, some of which have septa. Among the cephalopods are the largest living invertebrate animals, the giant squids. These animals have a body length of up to 12 m, and the two longest tentacles may stretch another 50 m. A key event in the evolution of cephalopods was the development of a siphuncle, which is a tube-like, buoyancy-control organ that pierces all of the septa. The siphuncle allowed cephalopods to live in the marine water column, where they were among the earliest swimming carnivorous animals. The oldest cephalopods, the nautiloids, appeared in the late Cambrian period, reached maximum diversity in the Silurian period, and declined from the Devonian to the Triassic period, when they became extinct.

One group of cephalopods, the ammonoids, is particularly noteworthy because rapid evolution in this group and a readily preservable shell make them excellent biostratigraphic tools for rocks of Devonian to Cretaceous age. During their evolution, the septa generally became more complexly folded. The position where a septum intersects the outer part of the shell is called a suture, and suture patterns are important for classifying ammonoids. Ammonoids underwent three great episodes of evolutionary radiation and extinction. Those with the simplest suture patterns (agoniatitic and goniatitic sutures) appeared during the middle Devonian period. Ammonoids having goniatitic sutures diversified quickly but the group dwindled to a few species by the end of the Permian period during an interval of widespread mass extinction. Ammonoids having ceratitic sutures appeared in the

Triassic period and quickly diversified but again ammonoids suffered near-total extinction at the end of the Triassic period. Finally, ammonoids having ammonitic sutures appeared in the Jurassic period, underwent rapid diversification, and became completely extinct, along with a number of other organisms, at the end of the Cretaceous period (*see* FOSSILIZATION AND THE FOSSIL RECORD).

Rostroconchs belong to a small class (Rostroconchia) whose members have a two-piece shell similar to that of a pelecypod. Unlike the shell of a pelecypod, though, the shell of a rostroconch could not open or close. Rostroconchs probably fed on microorganisms. They are known only from Paleozoic rocks, and are thought to have evolved from monoplacophorans.

Scaphopods or tusk shells (class Scaphopoda), which are known from the Ordovician period to the present, have a narrow tubular shell that is open at both ends. This group probably arose from rostroconchs. Scaphopods live within the sediment, and most feed on microorganisms.

Pelecypods, including clams, mussels, oysters, and scallops belong to a large class (Pelecypoda; also known as Bivalvia or Lamellibranchiata) whose representatives are characterized by a bivalved shell and a strong muscular foot that is used for locomotion. Each of the two valves of the shell is asymmetrical. Pelecypods probably evolved from rostroconchs. Filtering microorganisms from water has been the dominant means of obtaining food by representatives of this class ever since the Cambrian period. However, a few species, including the oldest known species, have used microorganisms in sediment as a food source. Some pelecypods that grow to enormous size (up to 1.5 m) have supplemented their nutrient intake by developing a symbiotic relationship with green algae that live within the tissues of the pelecypod. Early pelecypods mostly lived buried in sediment to shallow depths, but by the end of the Ordovician period, species had appeared that could live attached to hard substrates by threadlike structures; others could bore into hard substrates. Some pelecypods made the transition to fresh water by the middle Devonian period. Scallops, which are the only group of swimming pelecypods, have a fossil record extending from the Mississippian period. Beginning in the Mesozoic era, pelecypods underwent a great adaptive radiation following the development of siphons, or tubes, from part of the mantle. The siphons enhanced the animals' ability to burrow in sediment, even to great depths, because the siphons could reach to the water column, thus allowing the animals to respire and feed on suspended food particles. Also in the Mesozoic era, one group of reef-forming pelecypods, the rudists, appeared and underwent great evolutionary diversification. Rudist species were decimated during the late Cretaceous extinction event, although a few species survived into the early Tertiary period.

Hyoliths, which are not regarded as mollusks by all authorities, have a fossil record spanning virtually the entire Paleozoic era. Some authorities consider hyoliths to be an extinct phylum of animals that had a common ancestry with mollusks. Hyoliths have an elongate, cone-shaped shell that can be closed at the wider end by an operculum. Some hyoliths also have a pair of small, thin, blade-shaped shell parts that extended outward from the open end and whose function is speculative. Most hyoliths probably fed on microorganisms in the mud.

Bibliography

BARNES, R. D. *Invertebrate Zoology.* Philadelphia, 1987.

BOTTJER, D. J., C. S. HICKMAN, and P. D. WARD, eds. *Mollusks: Notes for a Short Course.* Knoxville, TN, 1985.

HICKMAN, C. P., JR., L. S. ROBERTS, and F. M. HICKMAN. *Biology of Animals.* St. Louis, MO, 1982.

MOORE, R. C., ed. *Treatise on Invertebrate Paleontology, Mollusca.* Lawrence, KS, 1957–1971. Parts I, K, L, and N.

MORTON, J. E. *Molluscs.* London, 1967.

POJETA, J., JR., B. RUNNEGAR, J. S. PEEL, and M. GORDON, JR. "Phylum Mollusca." In *Fossil Invertebrates,* eds. R. S. Boardman, A. H. Cheetham, and A. J. Rowell. Palo Alto, CA, 1987.

LOREN E. BABCOCK

MOON

Earth's moon has been a source of fascination and curiosity throughout history. For thousands of years, humans have wondered about the nature and origin of the Moon. The space age has brought profound changes in our knowledge of the Moon. Numerous robotic spacecraft have acquired high-resolution images and a wide variety of remote-sensing data. Twelve humans have conducted research on the lunar surface and the Apollo missions and Luna landers returned rock and soil from the Moon. Extensive analyses of the returned lunar samples and remote-sensing data have provided an understanding of the Moon that is deeper and broader than for any other solar system object except Earth.

Why do we study the Moon? First, the Moon provides a baseline for understanding the early history and evolution of the terrestrial (inner) planets. By studying the processes and evolution of this simplest and nearest solar system object, we obtain a deeper understanding of planetary geologic processes as well as a fuller appreciation of the much more complex histories of the terrestrial planets.

A second reason for investigating the Moon is to determine its origin. The Moon and Earth are closely related, and determining lunar origin could yield information concerning the dynamical and accretional processes that produced Earth.

A final reason for the continued investigation of the Moon is its proximity to Earth. Spacecraft travel times to the Moon are relatively short, and the Moon is routinely available for Earth-based telescopic observations. It seems likely that a lunar base will be established early in the next century and that local resources will be utilized. The Moon may well prove to be a stepping stone in the human exploration of the solar system.

The Moon always shows the same face toward Earth. This is the result of tidal forces from Earth acting on the Moon. The side facing Earth is called the nearside of the Moon. The side never seen from Earth is called the farside of the Moon. The farside can only be observed by spacecraft orbiting the Moon.

Even when seen with the naked eye, one recognized that the lunar surface consists of two major types of terrain: bright highlands and relatively darker plains commonly called "seas" (or maria in Latin). The dark maria cover about 17 percent of the lunar surface and are concentrated on the nearside of the Moon. In contrast, rugged highlands dominate the lunar farside.

Lunar highlands consist largely of cratered terrains. These overlapping impact craters range in size from small craters to large multiringed basins, which are hundreds or even thousands of kilometers in diameter. It is important to note that all of the basins and almost all of the craters were formed by meteorite impact very early in lunar history. Cratered terrain is characterized by high albedo and rugged relief and is generally thought to consist of brecciated and impact reworked material derived from the ancient lunar crust (Taylor, 1982; Greeley, 1994).

Lunar basins are defined as very large impact structures that are hundreds of kilometers in diameter and typically display concentric rings (Spudis, 1993; Greeley, 1994). Impact basins are ancient. All large basins formed prior to about 3.8 billion years (Ga) ago and the earlier structures have been largely obliterated. Impact basins and basin-related features and deposits dominate vast portions of the lunar surface. Basin formation created topographic lows that were sites for the accumulation of basaltic volcanism (mare basalts) and basin ejecta deposits cover large areas. Hence, many of the Apollo and Luna missions were specifically targeted to areas related to these giant impact features, to minimize topographic hazards in landing.

Orientale basin is the youngest large impact basin on the Moon. This Texas-size impact structure is located on the western limb of the Moon and exhibits at least four major rings. The Cordillera Mountains form the outer, more prominent ring, which is about 900 kilometers (km) in diameter. The inner and outer Rook Mountain rings are well defined, but the smallest ring is an unnamed inner bench. Many researchers consider the outer Rook ring to define the edge of the original Orientale impact crater. According to this model, the Cordillera Mountains were produced as a consequence of faulting inward toward the crater when the initial crater cavity collapsed. Rings interior to the original crater rim are generally thought to have been produced by post-impact rebound (Spudis, 1993; Greeley, 1994). Most researchers think that the initial crater cavity was excavated within a few minutes.

The interior of the Orientale basin is only partly flooded by mare material, and large expanses of highlands can be seen within the basin. Based on morphologic evidence, most researchers agree that large amounts of impact-melted rock is present on the basin interior. This impact melt was produced from rocks in the Orientale target site by the extremely high shock pressures associated with the impact event. The radially textured Orientale ejecta deposit termed the Hevelius Formation occurs outside of the Cordillera ring. This ejecta material was excavated from a variety of depths in the crust beneath the Orientale target site. At large distances from the basin, Orientale secondary crater chains and clusters are abundant (Wilhelms, 1987). These secondary craters were formed by the impact of giant blocks ejected by Orientale.

Numerous large impact basins are visible on the nearside of the Moon. These include the Humorum, Nubium, Nectaris, Crisium, Serenitatis, and Imbrium basins. The Imbrium basin is particularly important because it is the youngest impact structure on the central nearside and its deposits dominate the northern portion of the Moon's Earth-facing hemisphere. The Imbrium basin exhibits at least three major rings. The main ring is 1,140 km in diameter and consists of the Apennine Mountains, the Alps Mountains, the Carpathian Mountains, and other topographic highs.

Material excavated and emplaced as a result of the Imbrium impact event is present on large portions of the lunar nearside. These ejecta deposits are generally mapped as the Fra Mauro Formation. Because of its widespread distribution on the Moon's nearside, the Fra Mauro Formation is an important time marker (Wilhelms, 1987). Imbrium-related material was sampled by the *Apollo 14* and *15* missions. Radiometric age data obtained for the returned samples indicate that the Imbrium basin formed 3.85 Ga.

Light plains deposits are widespread in many areas of the lunar highlands. This unit is characterized by a moderate to high reflectance and a relatively smooth, flat surface. Prior to the Apollo missions, many lunar scientists argued that the light, highlands plains were emplaced by eruptions of volcanic ash or flows with possible silicic compositions. A highlands plains unit, the Cayley Formation, was sampled by the *Apollo 16* astronauts and its was determined that the unit was dominated by impact-generated breccias at this site. Soon after the *Apollo 16* mission, highlands plains previously interpreted to be of volcanic origin were reinterpreted as products of impact basin–forming processes. However, more recent studies have provided evidence that certain highlands plains units are indeed of volcanic origin (Hawke and Head, 1978; Spudis, 1993). A prime candidate is the Apennine Bench Formation, a highlands plains deposit southwest of the *Apollo 15* landing site. In addition, it now appears that at least some light plains were formed by the emplacement of highlands-rich material on top of ancient mare basalt units by large impacts.

What processes were responsible for the formation of the highlands crust? Intensive studies of the samples returned by the Apollo and Luna missions as well as the results of remote-sensing investigations provide a partial answer to this important question. Most lunar scientists agree that the Moon was surrounded by an ocean of molten rock or magma very soon after it formed about 4.55 Ga. As this waterless "magma ocean" slowly cooled, crystals of low-density plagioclase feldspar, a silicate rich in calcium and aluminum, would have risen upward after forming. In contrast, denser mafic minerals such as olivine and pyroxene (silicates rich in magnesium and iron) would have sunk to lower levels in the magma ocean. This differentiation process produced an early Moon with a low-density, plagioclase-rich crust that was tens of kilometers thick. The denser minerals olivine and pyroxene accumulated in the mantle beneath the crust. Later melting of this mafic mantle material is thought to be the source for mare basalts (Taylor, 1975; Spudis, 1990).

An unusual chemical component was identified during the analysis of the returned Apollo samples. This material is characterized by an enrichment in incompatible trace elements. These elements do not fit well into the structures of such common lunar minerals as pyroxene, olivine, and plagioclase feldspar as magma cools and crystallizes. This group of incompatible elements includes potassium (K), the rare-earth elements (REE), and phosphorus (P). Hence, lunar scientists refer to this chemical component as KREEP. The current consensus is that KREEP represents the last crystallization product of the global magma ocean and is present in a thin layer at the base of the crust. Some KREEP appears to have been emplaced on the lunar surface by volcanic and/or impact processes.

The highlands samples returned from the Moon can generally be identified as members of one of the two dominant highland rock classes. The first is the ferroan anorthosite suite. The members of this suite contain very large amounts of plagioclase feldspar and only minor quantities of mafic silicates. Most members of this class have been reworked by impact processes and appear to have formed by plagioclase floatation during the crystallization of the magma ocean, as discussed previously.

The second important highland rock class also contains abundant plagioclase feldspar, but it contains major amounts of pyroxene and olivine. This rock class is termed the magnesium suite. The members of the magnesium-rich suite exhibit a wide variety of crystallization ages and appear to have undergone the same intense impact processing as the ferroan anorthosites. There is now nearly unanimous agreement that the ferroan anorthosites and magnesium-rich suite rocks could not have crystallized from the same parent magma. Some researchers have proposed that magnesium-rich suite rocks were formed by the differentiation of plutons that were emplaced in the plagioclase-rich crust (Spudis, 1990).

Although the maria constitute only 17 percent of the lunar surface, they have generally received much more attention than the highlands. Analyses of the returned lunar samples have conclusively demonstrated that the maria are of volcanic origin. The rocks returned from the maria sites are basalts that exhibit a fine-grained texture. They are dominated by pyroxene with lesser amounts of plagioclase feldspar and accessory minerals such as olivine and ilmenite. The mare basalts vary in composition from place to place on the lunar surface. Basalts in Mare Tranquillitatis and portions of Mare Serenitatis are enriched in titanium (Ti).

Samples returned from the mare plains range in age from 3.1 to 3.8 Ga. Evidence from the study of mare basalt clasts in highlands breccias and remote-sensing data indicate that more volcanism may have occurred as long ago as 4.3 Ga. Some researchers have suggested that minor amounts of mare material were emplaced within the last billion years. The mare basalt magmas are thought to have originated hundreds of kilometers deep within the Moon. Heat generated by the decay of radioactive isotopes in the lunar mantle produced partial melts that moved upward and were emplaced on the surface (*see* HEAT BUDGET OF THE EARTH).

Sinuous channels, called rilles, occur in many mare regions. Prior to the Apollo missions, the origin of these features was very controversial. Now, most lunar scientists agree that sinuous rilles are former lava channels and/or collapsed lava tubes. Thermal and mechanical erosion appears to have played an important role in the formation of some lunar sinuous rilles.

In addition to sinuous rilles, other volcanic features have been identified on mare surfaces. Small volcanic domes are abundant in some areas. They are relatively low, flat-topped circular structures that commonly exhibit summit pits or craters. Dark-haloed craters of volcanic origin occur on the floor of Alphonsus Crater and in many other areas. These are generally interpreted to be pyroclastic vents formed by explosive eruptions. These small, localized pyroclastic deposits are quite distinct from the much more extensive regional dark mantle deposits of pyroclastic origin that blanket and subdue thousands of square kilometers of lunar terrain. The tiny orange and black spheres returned by the *Apollo 17* astronauts may be representative of the pyroclastic debris present in a nearby regional dark mantle deposit.

The relatively simple tectonic history of the Moon as compared with that of Venus or Mars does not lessen its importance. Three surface features dominate broad-scale lunar tectonic expression—wrinkle ridges, linear rilles, and scarps. The origin of wrinkle-like, mare ridges has been the subject of controversy. Some early studies suggested a volcanic origin or an origin directly related to the cooling of mare basalts. Research in the post-Apollo era produced evidence from a wide variety of sources for a structural origin. The complex nature and differences in appearance of wrinkle ridges may require a variety of origins, including thrust faulting due to crustal compression. Most wrinkle ridges are found in the interior of maria and probably represent compression caused by settling of the basins after the emplacement of the dense basaltic rock.

The many linear rilles that occur on the surface of the Moon offer striking evidence of the tectonic deformation of the lunar crust. Linear rilles typically take the form of steep-walled, flat-floored depressions or troughs. They are generally several kilometers in width and some extend for hundreds of kilometers across both highlands and mare terrain. There is general agreement that linear rilles are fault-bounded grabens formed by normal faulting due to crustal extension. Because most grabens were formed prior to about 3.6 Ga, and most wrinkle ridges continued to form well after 3.0 Ga, some researchers have proposed that a change in tectonic style from graben formation due to crustal extension to wrinkle ridge formation caused by compression may mark a shift in global stress patterns. This change may have occurred when the last linear rilles or grabens formed, about 3.6 Ga. Such a shift could have happened when the

Moon began to cool and contract after an earlier expansion caused by internal heating. The entire story may be much more complex. Relatively young scarps have been identified in the lunar highlands and some researchers have interpreted these as thrust faults produced by continued global contraction.

The surface of the Moon exhibits a large number of craters with a wide variety of sizes and ages. The majority of lunar craters are thought to be of impact origin. Detailed studies have demonstrated that fresh impact crater morphologies change as a function of crater size. Simple craters less than 15 to 20 km in diameter are typically circular and bowl-shaped and they exhibit sharp, raised rims. These simple craters are surrounded by hummocky ejecta blankets that extend from the rim crests to distances approximately equal to the diameter of the parent crater. Bright rays commonly extend many crater diameters beyond the edge of the ejecta blanket.

Impact craters larger than 15 to 20 km in diameter display morphologies different from smaller craters. The large, complex craters are proportionally more shallow than small craters. Terraces formed by slumping are present on the walls of large craters. Complex crater floors are generally broad and flat, and ponds and flows of impact melt can be identified on many large crater floors. In addition, large craters have one or more central peaks that rise abruptly from the floor. Most lunar scientists agree that central peaks were formed by the rapid rebound of compressed target material.

The continuous ejecta deposits around large, complex craters are somewhat different from those associated with small craters. Ponds and flows of impact melt are commonly seen on the rims of large craters. The outer portions of the continuous ejecta blankets of large craters display radial textures and grade into discontinuous ejecta deposits that are dominated by secondary crater chains and clusters.

Since the Moon essentially lacks an atmosphere and has never experienced liquid water on the surface, lunar degradational processes are dominated by mass wasting, including landslides and rockfalls, and impact cratering events (Greeley, 1994). Impact cratering is also responsible for the formation of lunar regolith (or soil) deposits. Surfaces on the Moon are constantly being fragmented and pulverized by impact cratering at a variety of scales. Meterorite and micrometeorite bombardment over hundreds of millions of years has produced a fine-grained regolith that consists of tiny rock fragments, mineral grains, impact glasses, and mixtures of these that are bound by impact-generated glass. The regolith's thickness at a given point on the lunar surface depends on the age of the underlying bedrock and thus on how long the surface has been exposed to meteoritic bombardment; regolith developed on mare units is 2 to 8 meters (m) thick, whereas in highlands areas its thickness may exceed 15 m (Spudis, 1990).

Our knowledge concerning the lunar interior comes largely from data returned by missions during the Apollo era as well as from photographs taken by the *Galileo* and *Clementine* spacecraft. Chief among these are the seismic measurements made at the Apollo landing sites. Using instruments called seismometers (*see* EARTHQUAKES AND SEISMICITY), the Apollo experiments measured the strength of "moonquakes" at various depths within the Moon as well as the signals from occasional meteorite (and man-made) impacts on the Moon. The results of the analysis of this data show that impacts have brecciated and mixed the upper portions of the lunar crust to a depth of at least 2 km and perhaps to depths of 10–20 km. The lunar crust has an average thickness of about 70 km, but it varies from a few tens of kilometers beneath the impact basins on the nearside to over 100 km in some portions of the lunar farside.

The Apollo seismic data indicate that at a depth of 60 to 100 km, a major increase in seismic velocity occurs. This seismic boundary is thought to mark the contact between the lunar crust and mantle. The mantle constitutes approximately 90 percent of the volume of the Moon and is thought to consist of olivine and pyroxene as well as minor amounts of other minerals. Most lunar researchers agree that the source regions for the mare basalts were located 200 to 400 km beneath the surface and that the composition of the mantle varies as a function of location and depth within the Moon.

A major unresolved question is whether or not the Moon has a metallic-iron core. As yet, the available evidence does not require an iron core, but permits a small one. However, if a core exists, data from a variety of sources place firm constraints on its size. Several geophysical measurements suggest, but do not prove, the existence of a lunar metallic core whose radius is in the range of 350 to 450 km. If we assume a dominantly iron composition, a core with a radius in this range would constitute 2 to 4 percent of the total lunar mass (Spudis, 1990). The Moon does not possess a magnetic field today, but some lunar rocks show evidence of having formed in the presence of a magnetic field 3.5 to 4.0×10^9 years ago.

Understanding the origin of the Moon is a major goal of lunar science. Traditionally, researchers have investigated three hypotheses of lunar origin. The simplest model, termed co-accretion, calls for the formation of the Moon by accretion of material in orbit around a still growing Earth. According to this hypothesis, Earth and the Moon formed together and have existed as a pair from the beginning. The capture hypothesis suggests that the Moon was formed elsewhere in the solar system and was later trapped in orbit by Earth's gravity when it strayed too near our planet. Strong objections have been raised to the capture hypothesis based on dynamical considerations. The third scenario maintains that Earth accreted without a satellite but later began to spin at such a high rate that a portion of its original mass was thrown off to form the Moon. This model is referred to as the fission hypothesis.

In the immediate post-Apollo era, the three primary hypotheses for lunar origins—capture by, fission from, and co-accretion with the Earth—were all seen by lunar scientists as seriously flawed, though each model had its hard-core advocates. Most objections to these hypotheses stem from inconsistencies with the Moon's composition, as well as from dynamical considerations.

Recently, a fourth hypothesis for the origin of the Moon has gained popularity (*see* EARTH-MOON SYSTEM, EARLIEST HISTORY OF). According to this model, a giant, Mars-sized object collided with the Earth almost 4.6 Ga and a mixture of terrestrial and projectile material was ejected into Earth orbit. The impactor had already segregated or differentiated into a metallic core and a rock mantle and crust. The ejected debris, which came mostly from the rocky portions of the impacting body, later coalesced to form the Moon. While the giant impact hypothesis appears to explain many of the salient features of the Moon (for example, its bulk chemical composition), additional research will be needed to confirm this model.

After many years of intensive study of the returned lunar samples as well as of spacecraft remote-sensing data and imagery, we now have a very general outline of the origin and evolution of the Moon. The following summary presents our current understanding of the Moon.

About 4.6 Ga, a Mars-sized object collided with the proto-Earth. This collision threw a cloud of vaporized debris into Earth orbit, and this material, which was derived primarily from the impacting body, quickly cooled and coalesced to form the Moon.

The rapid rate of lunar formation resulted in the release of large amounts of heat that produced melting to a depth of at least a few hundred kilometers. As this magma ocean cooled, minerals began to crystallize. Low-density, aluminum- and calcium-rich minerals (plagioclase feldspar) floated upward and formed a plagioclase feldspar-rich crust with an average thickness of about 70 km. High-density minerals rich in iron (Fe) and magnesium (Mg) (olivine and pyroxene) crystallized and sank to the lower levels of the magma ocean, forming the lunar mantle. The magma ocean solidified by about 4.4 to 4.3 Ga, when the last residues of the original magma ocean crystallized as the thin KREEP layer. After the formation of the crust, large bodies of magma were emplaced in some portions of the crust and crystallized to produce layered plutons.

During this time, violent impacts continued to overturn and mix the upper crustal materials thoroughly. Early surface features were destroyed and the large impacts created multiringed basins. Some scientists think that a brief surge in the cratering rate occurred at about 3.9 Ga. The Orientale and Imbrium basins represent the last major impacts on the Moon. The Imbrium impact occurred 3.85 Ga and Orientale basin probably formed about 3.8 Ga. The cratering rate was declining rapidly at this time and may have become relatively constant at about 3.0 Ga.

Some magmas produced by the partial melting of the mantle were erupted on the lunar surface as early as 4.3 Ga. These ancient lava flows were largely destroyed by later impact events. The vast bulk of the dark plains that dominate the lunar nearside are mare basalts that were emplaced between 3.8 and 3.0 Ga. Mare volcanism continued until at least 2.5 Ga and may have occurred as recently as 1 Ga, or even more recently.

The dominant process operating on the Moon for the last 3 Ga has been the ongoing peppering of the lunar surface by meteorites. Occasionally, a large crater such as Tycho forms. In most ways, however, the Moon is now geologically "dead."

Bibliography

Greeley, R. *Planetary Landscapes*, 2nd ed. New York, 1994.

Hamblin, W. K., and E. H. Christiansen. *Exploring the Planets*. New York, 1990.

Hawke, B. R., and J. W. Head. "Lunar KREEP Volcanism: Geologic Evidence for History and Mode of Emplacement." *Proceedings Lunar Planet Science Conference 9th* (1978):3285–3309.

Heiken, G. H., D. T. Vaniman, and B. M. French, eds. *The Lunar Sourcebook*. Cambridge, Eng., 1993.

Spudis, P. D. "The Moon." In *The New Solar System*, eds. K. Beatty and A. Chaikin. Cambridge, Eng., 1990.

———. *The Geology of Multi-Ring Impact Basins: The Moon and Other Planets*. Cambridge, Eng., 1993.

Taylor, S. R. *Lunar Science: A Post-Apollo View*. New York, 1975.

———. *Planetary Science: A Lunar Perspective*. Houston, 1982.

Wilhelms, D. E. *The Geologic History of the Moon*. U.S. Geological Survey Professional Paper 1348. Washington, DC, 1987.

B. RAY HAWKE

MOUNTAINS AND MOUNTAIN BUILDING

Mountains are a source of inspiration to poets and scientists alike. How do mountains reach such lofty heights? What do rocks found within mountains tell us about the history of the mountains and of Earth? (*see also* STRUCTURAL GEOLOGY, HISTORY OF)

Orogeny refers to the process of forming mountains. Early geologists developed the geosynclinal theory, which posited the formation of long, narrow troughs that subsided and were filled with sedimentary and volcanic strata. The deeply buried layers were subjected to such high temperatures that they were metamorphosed and expanded, much like a cake rising in an oven. The expansion caused deformation and uplift of Earth's surface as mountain ranges.

Vertical movements were an important aspect of geosynclinal theory; mountains resulted from materials moving up and down without large lateral displacements, which were thought unlikely (*see* DANA, JAMES D.). Observations that blocks of Earth's crust had, indeed, moved horizontally over long distances led to drastic revision of mountain building ideas. The plate tectonic theory (*see* PLATE TECTONICS), with horizontal motion as a major premise, explains the origins of mountain ranges. Geosynclinal theory has been discarded.

Mountains and Plate Tectonics

Most mountains are associated with processes occurring along lithospheric plate boundaries. Where plates diverge, the cold, upper crust rips apart, forming fault-block mountains in between down-dropped valleys. The lithosphere thins, so that hot, buoyant material of the asthenosphere rises, elevating continental rift zones and mid-ocean ridges (Figure 1). Where plates converge, rocks are scraped off the descending plate, compressed, and uplifted as mountains (Figure 2). One plate may descend so deeply that its crust gets very hot; the resulting magma forms a volcanic mountain chain on the overriding plate. Sometimes entire oceans close as plates converge, causing blocks of continental crust to collide; mountains form as the crust is compressed and thickened (Figure 3).

Although most mountains occur at diverging and converging plate boundaries, there are other ways mountains form. Where plates slide past one another along transform boundaries, such as in New Zealand or at the San Andreas Fault of California, mountains develop in a complex zone of sheared crustal blocks. Mountains in the interior of lithospheric plates have been related to various processes, including the passage of a plate over a mantle hotspot (Hawaiian Islands) and shallow subduction of a plate beneath a continent (Rocky Mountains of Colorado and Wyoming).

Diverging Plate Boundaries

When a continent rips apart, the two fragments drift away as parts of different lithospheric plates. If the process continues long enough, a new ocean separates the continental fragments. Mountains are thus associated with two types of diverging plate boundaries: continental rift zones and mid-ocean ridges.

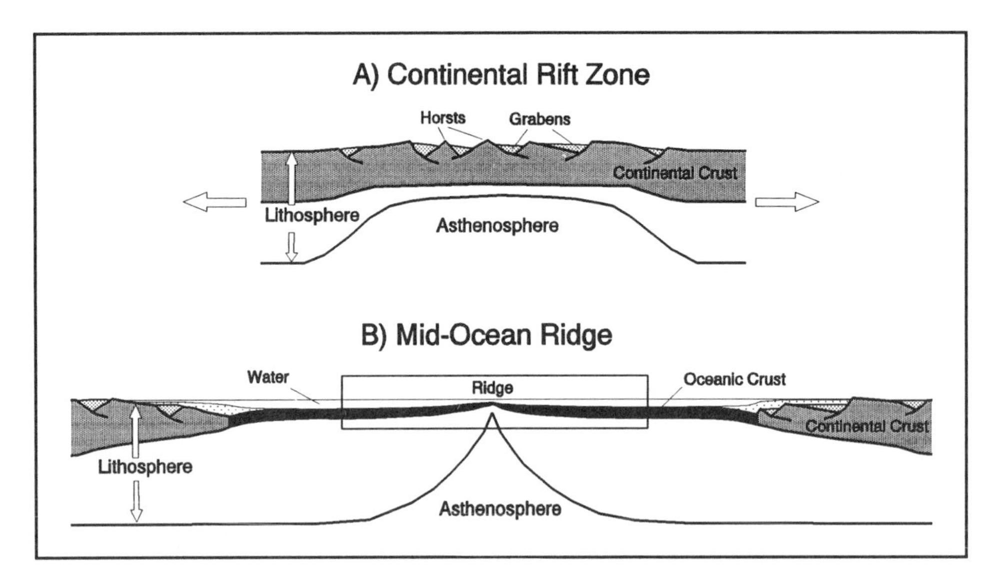

Figure 1. Formation of mountain ranges at diverging plate boundaries.
A. Continental Rift Zone. As continental lithosphere rips apart, hot asthenosphere uplifts a broad region, such as the Basin and Range Province in the western United States. Long, parallel ranges of mountains form as eroding horst blocks, separated by grabens that subside and fill with sedimentary and volcanic strata.
B. Mid-Ocean Ridge. When continental crust is completely breached by the asthenosphere, new oceanic lithosphere forms. The hot asthenosphere elevates mid-ocean ridges, the most extensive range of mountains on Earth.

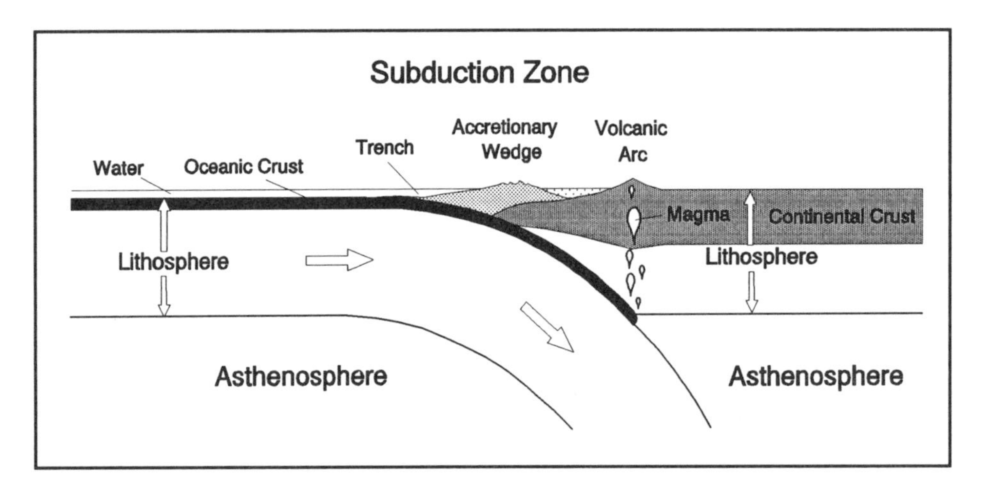

Figure 2. Mountain ranges formed at a subduction zone. Near the trench, sediments and oceanic crust are scraped off the downgoing plate. The material is compressed along thrust faults and folds, uplifting mountains of the accretionary wedge. Where the top of the plate reaches 100 to 150 km depth, hot fluids rise, melting high-silica minerals in their path. A chain of volcanic mountains forms on the overriding plate (volcanic arc).

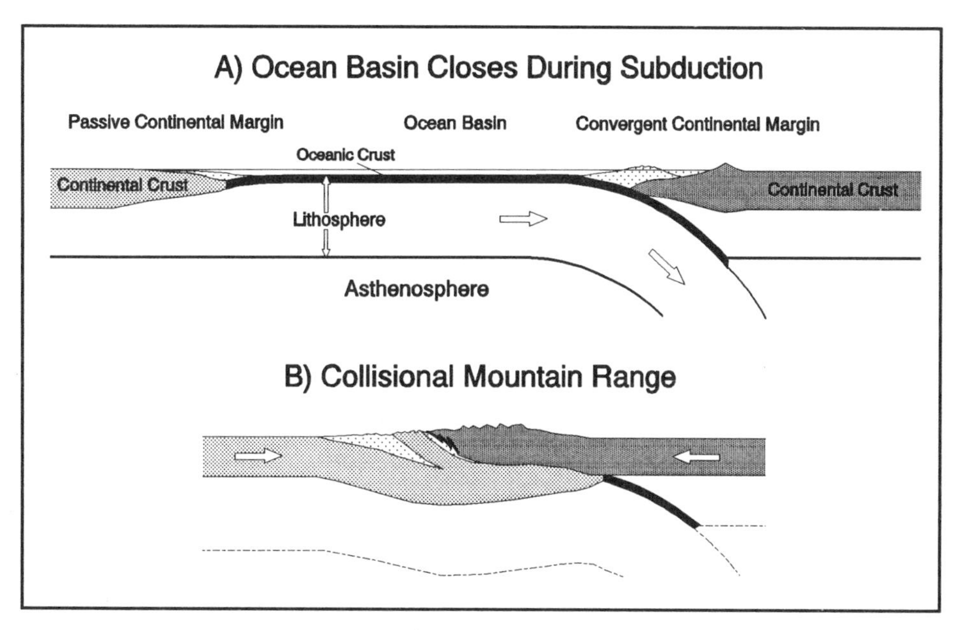

Figure 3. Formation of collisional mountain range.
A. Continental margins approach as ocean basin closes through subduction.
B. As the continents collide, compression deforms the oceanic crust and sediments, as well as continental crust of both plates. Continental crust is too buoyant to subduct; if convergence continues a broad region rises, isostatically, as the crust thickens. Individual mountains result from deformation and erosion of the uplifting region.

Continental Rift Zones. As a continent pulls apart, it stretches, thinning the crust and the entire lithosphere (Figure 1A) (*see* RIFTING OF THE CRUST). The region is commonly at high elevation because the underlying asthenosphere is hot and buoyant. The upper part of the crust deforms in a cold, rigid fashion, causing long, elevated ridges (*horst* in German) separated by down-dropped valleys (*graben* in German). The grabens fill with sedimentary and volcanic strata as they further subside, forming basins; the adjacent horst blocks are left high as mountain ranges. A region of fault block mountains in western North America—comprising all of Nevada and portions of Utah, Idaho, Oregon, California, Arizona, and northern Mexico—is thus called the Basin and Range Province (Figure 4). Other continental rift mountains are found in the Pannonian Basin of central Europe and the East African Rift zone.

Mid-Ocean Ridges. When continents completely rift apart, new oceanic lithosphere forms. The buoyant asthenosphere elevates a ridge of seafloor up to 4,000 kilometers (km) wide (Figure 1B). A chain of mid-ocean ridges extends in a more or less continuous fashion for over 50,000 km, including the mid-Atlantic Ridge, the East Pacific Rise, and the Indian Ocean Ridge. Though mostly submerged below water, the mid-ocean ridge system is the longer mountain range on the surface of Earth.

Converging Plate Boundaries

Mountains that formed when lithospheric plates converge are created due to compression, to magma generated by heating the descending plate, or to the buoyancy of thickened crust. Two types of converging boundaries are common, depending on if the descending plate is capped by thin (oceanic) or thick (continental) crust.

Subduction Zones. When the descending plate has oceanic crust, two chains of mountains form parallel to the deep-sea trench at the surface juncture of the plates (Figure 2). On the trench, where the top of the plate is shallow and cold, some of the

Figure 4. Fault block mountains in the Basin and Range Province. View south along Steens Mountain, a horst block in southeastern Oregon. Wildhorse Lake is on the right, resting on the edge of sediments that fill a graben; the Alvord desert occupies a continuation of the graben in the distance, upper left. (Photo by Joseph M. Licciardi.)

sediments and underlying rock are scraped off, deforming into a wedge-shape. These materials attach (or "accrete") to the overriding plate; portions of the accretionary wedge rise above sea level as the sediments and rock are compressed, folded, and faulted, forming long ridges and valleys. Accretionary wedge mountains include the coastal ranges of Washington, Oregon, and northern California, as well as the islands of Barbados and Timor.

Farther from the trench the top of the descending plate is 100 to 150 km deep, where it is so hot that fluids are driven from its crust. The fluids rise, melting silicate minerals from the mantle and crust of the overriding plate. The portion of magma that makes it to the surface erupts as a chain of volcanoes that may lie along straight or curved lines (volcanic arcs). Large mountains in the chain are steep-sided, composite volcanoes, although broad shields also form (*see* VOLCANISM). Smaller mountains in the chain may be cinder cones or ridges of hard lava flows. Examples of volcanic mountains formed by subduction include Japan, the Andes of South America, the Cascades of California, Oregon and Washington (Figure 5), and the Aleutians in Alaska. The Sierra Nevada mountains in southern California are the roots of a volcanic arc that was subsequently uplifted; the volcanoes have eroded away, exposing the hard rocks of former magma chambers.

Continental Collision Zones. At collisional mountain ranges, two plates that both have thick crust converge. Collision occurs when the oceanic part of the downgoing plate is lost to subduction (Figure 3A). When the continental crusts meet, both plates are too light (i.e., they have too much thick, "buoyant" crust) to subduct into the deeper mantle. The continents are deformed by compression, their rocks metamorphosed and uplifted. If

Figure 5. Cascade Mountains of central Oregon, looking north. The volcanoes follow the line, at 100 to 150 km depth, where the top of the subducting Juan de Fuca Plate extrudes hot fluids. The bare patch in the lower right is Belknap Crater, a cinder cone at the crest of a shield volcano. Mount Washington, at the center of the photo, and Three Fingered Jack, on the left-hand edge, are volcanoes deeply eroded by glaciers. Snow-capped Mount Jefferson and Mount Hood (to the right in the far distance) are large composite volcanoes. (Photo by Robert J. Lillie.)

convergence continues, one continent may plow under the other (as the Indian subcontinent extends beneath Asia today); the result is a broad region of high elevation (the Himalayas and Tibetan Plateau) due to isostatic uplift (*see* ISOSTASY) as crust thickens (Figure 3B).

The highest mountains on earth, the Himalayas, are part of a large chain of mountains that extends from south-central Asia to western Europe. The chain forms by convergence of Eurasia with other continental fragments, including India, Saudi Arabia, and smaller pieces of thick crust swept northward as Africa moves toward Europe. Other mountains in the collision zone are the Pamirs, Hindu Kush, Zagros, Caucasus, Carpathians, Alps, and Pyrenees. The Appalachian Mountains in the eastern United States and Canada formed about 300 million years (Ma) ago when Europe and Africa collided with the eastern edge of North America; the same collision formed the Caledonide Mountains in southern Greenland, the British Isles, and Scandinavia, which were separated from the Appalachians as the Atlantic Ocean opened.

Types of Rocks Associated with Mountains

Many types of rocks are found within mountains, owing to the variety of mountain building processes (*see* ROCKS AND THEIR STUDY). Generalizations can be made, however, because rocks form in systematic ways. Distinction of rocks formed before mountain building processes began (pre-orogenic) from those formed in association with the mountains (syn-orogenic) is particularly useful.

Diverging Plate Boundaries

Pre-orogenic rocks of continental rift zones occur in eroded horst blocks forming the ranges (Figure 1A). They are of a wide variety, including continental and oceanic sedimentary strata and underlying crystalline rocks. Syn-orogenic strata include sedimentary and volcanic deposits filling grabens. The sediments are mostly nonmarine clastic and evaporitic deposits (*see* SEDIMENTARY STRUCTURES), because the hot asthenosphere holds the region well above sea level. Volcanic strata are often of two types: high-silica rhyolite that tends to extrude early, followed by larger amounts of low-silica basalt as rifting progresses (*see* IGNEOUS ROCKS).

At mid-ocean ridges the entire crust is formed in association with the mountain building (i.e., ridge formation). The upper crust is basalt, forming dikes and pillow lavas on the ocean floor; the lower crust is gabbro. Sediments that fill down-dropped fault blocks and cover the ocean crust are mostly deep-sea chert, because the ridge is normally far from eroding landmasses (Figure 1B).

Converging Plate Boundaries

Pre-orogenic rocks in the accretionary wedge of a subduction zone are commonly oceanic basalts and sediments scraped off the downgoing plate (Figure 2). These rocks may be metamorphosed to blueschist (*see* METAMORPHIC ROCKS), owing to the high pressure and low temperature of the subduction zone near the trench. Deposits of thick sand and thin mud layers ("turbidites") fill the trench; these syn-orogenic strata may deform and intermingle with older rocks of the accretionary wedge in a complex folded and faulted mass (mélange).

Many types of volcanic, plutonic, and sedimentary rocks occur in volcanic arcs. Rising fluids promote melting in the mantle and crust of the overriding plate, producing magmas of varying composition. Erupting lavas may be basaltic, but not normally higher in silica (andesite to rhyolite). Interlayered with the lava flows are volcanic mudflows and material thrown into the air by eruptions (ash, cinders, and so on). Plutonic rocks include shallow sills and dikes of varying composition, representing magma that fed the summits and fissures on the sides of volcanoes (*see* INTRUSIVE ROCKS AND INTRUSIONS). Deeper chambers of high-silica magma can solidify into granitic batholiths.

Rocks and sediments of an ocean basin and its margins are trapped in the collision zone between continents, where they are compressed, deformed, and uplifted as mountains (Figure 3). The pre-orogenic materials include deep-water sediments, basalt, and gabbro from the ocean basin (ophiolites), as well as clastic and carbonate deposits and crystalline rocks from the continental margins. Syn-orogenic strata are of two varieties: course-grained sediments deposited in the trough between the converging continents (flysch) and shallow-marine to continental sediments eroded from the rising mountains as the continents collide (mollasse). Materials are subjected to high pressure and temperature as they are pushed deeper into the earth, metamorphosing to amphibolite and granulite. If the continental rocks are pushed so deep that they melt, granitic intrusions form. Continued compression or isostatic rebound may later expose the granite and metamorphic rocks at the surface.

Bibliography

HATCHER, R., H. WILLIAMS, and I. ZIETZ, eds. "Contributions to the Tectonics and Geophysics of Mountain Chains." Memoir 158, Geological Society of America. Boulder, CO, 1983.

HSÜ K., ed. *Mountain Building Processes.* New York, 1982.

MIYASHIRO, A., K. AKI, and A. SENGÖR. *Orogeny.* New York, 1982.

MOORES, Ed., ed. *Shaping the Earth: Tectonics of Continents and Oceans.* New York, 1990.

ROBERT J. LILLIE

MURCHISON, RODERICK

Roderick Murchison is best remembered by modern geologists for his lengthy battle with ADAM SEDGWICK (1785–1873), Woodwardian Professor at Cambridge University, over the boundary between Murchison's Silurian system and Sedgwick's Cambrian. This was only one of several major geological controversies in which the pugnacious Murchison was a protagonist, yet his work remains of fundamental importance to the development of stratigraphy. When Murchison began his studies, rock type was believed to hold the key to stratigraphy, but he demonstrated overwhelmingly that fossils were of far greater use in correlation. Moreover his dispute with Henry de la Beche (first director-general of the British Geological Survey) over the Devonian changed the word "system" from its original meaning of a group of rocks affected by the same tectonic event to the modern usage of a globally significant temporal unit (more accurately a time-rock unit) based on a distinctive assemblage of fossils within a package of rocks. The evolution of the concept of system was integral to the elaboration of the modern concept of the geologic time scale (*see* GEOLOGIC TIME).

Murchison was born to the Scottish landed gentry in Eastern Ross, Scotland, on 19 February 1792. Murchison's father died early in his life. The younger Murchison received a military education and served briefly as an officer during the Peninsular War. After leaving the army, Murchison married the heiress Charlotte Hugonin, an amateur naturalist, and established himself as a country gentleman of "independent means" (mostly those of his wife). He was evidently a man of boundless energy, and encouraged by his wife and friends, he chose geology, particularly stratigraphy, as his avocation; the demands of fieldwork combined easily with his love of hunting and riding. Sedgwick, seven years Murchison's senior, became the younger man's guide to the intricacies of geology, principally through their extensive field trips. Murchison joined the Geological Society of London in 1825 and became the foreign secretary by 1829. The Geological Society of London was the first learned society devoted exclusively to science and it was to have a major role in the development of the science during the nineteenth century.

In 1831 Murchison began working on a relatively undeformed and highly fossiliferous series of rocks that lay beneath the Old Red Sandstone in the Welsh Borderlands. He recognized that the distinctive fossils found in these rocks allowed him to trace the interval into South Wales. He named this unit the Silurian, and emphasized that it was the fossils, rather than the type of rock, which characterized the different intervals of geologic time (unlike the previously defined Carboniferous and Cretaceous, which were initially recognized as distinctive lithologies). Murchison's Silurian was characterized by an assemblage of distinctive marine invertebrate fossils but contained only a few, fragmentary fish remains and lacked any sign of terrestrial plants. After tracing this readily definable assemblage into other areas, Murchison finally published *The Silurian System* in 1839, in which he argued that the Silurian was a distinctive interval in the history of life, much as the Cenozoic was the age of mammals. To lend further support to his concept of systems, Murchison quickly became proselytizer for the concept of systems in general. He prevailed upon Sedgwick to describe the Cambrian system in 1835. In 1840–1841 Murchison made two visits to Russia with the French paleontologist Edouard de Verneuil in search of more Silurian rocks. He found these, but also a distinctive package of rocks between the Carboniferous and Triassic and described them as the Permian system. Many historians view Murchison's "paramilitary" approach to extending the writ of systems through stratigraphy as an important example of scientific imperialism. Certainly Murchison conducted his work much like a military campaign, attacking any opposition and seeking to destroy any foe.

The economic significance of the Silurian was also important to Murchison. He took an avowedly progressive view of the history of life and argued that there was no point in searching for coal below Silurian rocks, since land plants did not evolve until later. This stance led, in turn to his acrimonious debate with de la Beche over the reliability of fossils as opposed to lithology in stratigraphy. De la Beche argued that since none of the marine fossils he found in the ancient rocks of Devon were those of the Silurian, these rocks and their associated coal seams must represent a system which underlay the Silurian. This called into question both Murchison's progressivist view of the history of life as well as his view of the Silurian as the first system with a well-developed marine fauna. Murchison went forth into battle again, vanquishing de la Beche, and he and Sedgwick established the Devonian system intermediate between the Silurian and Carboniferous. Murchison viewed the Silurian as the oldest distinctive such assemblage, particularly since Sedgwick's Cambrian contained few distinctive fossils, and there was considerable stratigraphic overlap between the two systems as originally described. This led to a substantial dispute with Sedgwick involving stratigraphy, a debate over priority, theoretical import, and other issues that finally destroyed the friendship between them.

By middle age Murchison became increasingly interested in the public voice of science and contribution of science to the British Empire. He was a strong imperialist and rejuvenated the Royal Geographical Society after 1851 as a center for overseas exploration, imperial expansion, and the collection and dissemination of information. Following de la Beche's death in 1855 Murchison was appointed director-general of the British Geological Survey. Murchison's careful attention to his social status rebounded to the benefit of science, but there apparently is little doubt that he was overly self-satisfied with his scientific status. Murchison was knighted in 1846, Knight-Commander of the British Empire in 1863, and baronet in 1866. He received a medal from the emperor of Russia as well as the Wollaston, Copley, Brisbane, and Prix Cuvier medals, honorary degrees from Oxford and Cambridge and the honors of numerous scientific societies. Murchison died in London on 22 October 1871.

Bibliography

OLROYD, D. R. *The Highlands Controversy.* Chicago, 1990.

RUDWICK, M. J. S. *The Great Devonian Controversy—The Shaping of Scientific Knowledge among Gentlemanly Specialists.* Chicago, 1990.

STAFFORD, R. A. *Scientist of Empire: Sir Roger Murchison, Scientific Exploration and Victorian Imperialism.* Cambridge, Eng., 1989.

DOUGLAS H. ERWIN

N

NANSEN, FRIDTJOF

Though remembered today mainly for his achievements as an Arctic explorer and humanitarian, Fridtjof Nansen left an enduring mark on a wide field of science including anthropology, biology, geology, oceanography, and climatology. Born at Store-Fröen near Christiania (now Oslo), Norway, on 10 October 1861, to Baldur Fridtjof Nansen and Adelaide Johanne Isadore née Wedel-Jarlsberg, Nansen attended school in Christiania where, in 1880, he entered the university to study zoology. His choice of subject was strongly influenced by a hope that the fieldwork would enable him to indulge his passion for hunting and the outdoors. He was awarded his doctoral degree mainly on the basis of his highly innovative paper, "The Structure and Combination of Histological Elements of the Central Nervous System" (1887).

In 1882, while studying marine life along the eastern coast of Greenland, Nansen became intrigued by the possibility of crossing the ice cap and resolving some of the questions about glaciation that were currently being debated. At that time, the interior of an ice cap had not been seen, so the nature of continental glaciation was still a matter of speculation. According to a theory proposed by A. E. Nordenskiöld, ice was confined to the humid coastal regions, and the interior was an ice-free oasis.

Despite the opinions of "experts" who considered his plan suicidal, Nansen obtained the necessary financial support. After surmounting a series of obstacles and delays, he and his five companions started from the east coast on 15 August 1888 and reached the west coast on 26 September. Forced to winter at the settlement of Godthaab, Nansen took the opportunity to study the Eskimo community and gather material for his book *Eskimoliv* (1891; translated into English as *Eskimo Life,* 1893), still a major source of information on the life and customs of the people.

In this, the first crossing of an ice cap, Nansen established the great thickness of continental glaciers and from this concluded that the crust would be depressed isostatically under its weight. Together with his meteorological records from the interior of the ice cap, his geological observations launched a new surge of interest in continental glaciation. Though not a geologist by training, Nansen played an important role in this work. He saw an analogy between the present glaciation of Greenland and the Pleistocene glaciation of Scandinavia and recognized the glacial origin of fjords. He contributed support for the theory of isostatic rebound and was among the first to consider a possible link between polar wandering and climatic change.

Noting several lines of evidence that currents flowed westward from eastern Siberia into the North Atlantic, Nansen devised a plan to explore the polar regions by constructing a ship capable of drifting with the currents without being crushed by the ice. By this means he hoped to reach the unexplored central regions of the Arctic Ocean and determine whether or not a landmass existed near the pole. With a complement of thirteen men, his ship, the *Fram* ("Forward" in Norwegian), sailed from Christiania on 24 June 1893. Heading east along the coast of Siberia until the end of summer, Nansen turned north and on 22 September was caught in the ice at 78°50′N, 133°37′E. The ship resisted the pressure of the ice, and for the next year and a half drifted north and west. On 14 March 1895, having concluded that there was not a large polar landmass and that the *Fram* would continue to drift safely, Nansen and F. H. Johansen left the ship at 84°4′N, 102°27′E and started northward with dog sleds. Realizing that they had little chance of finding the ship on their way back, they took kayaks in order to cross the open water between the ice pack and Franz Josef Land. On 8 April, owing to limitations of food and their slow progress across the broken ice, they were forced to turn back at 86°14′N, the highest latitude yet reached by humans. They reached Franz Josef Land, where they passed the winter in a tiny hut while living on bear and walrus meat. Nansen and Johansen returned to Norway on 13 August 1896; the *Fram* reached Norway a week later, having drifted north to 85°57′ before emerging from the ice pack between Greenland and Spitzbergen. Nansen's two-volume account of the expedition, *Fram over Polavet* (English translation *Farthest North*), appeared in 1897.

During the period 1896–1917 Nansen held a post as professor at the University of Christiania, where he devoted most of his time and energy to oceanography. He compiled a six-volume report of the *Fram* expedition that still remains a major reference on the Arctic Ocean. He participated in the establishment of the International Council for the Exploration of the Sea and for some time directed the central laboratory in Christiania. He participated in cruises in the North Atlantic, the Barent and Kara seas, and the region of the Azores. His contributions to oceanography were both technical and theoretical. He was responsible for the design of several instruments that are still in common use today, for example, the "Nansen bottle," widely used to collect contamination-free samples of seawater at depth. He helped show how bottom waters are formed and, in collaboration with Vilhelm Bjerknes (*see* BJERKNES, J. A. B. AND V. F.) and V. W. Ekman, showed that the phenomenon of "dead water" is the result of internal waves at the interface between a surface layer of freshwater and denser seawater. He also worked with Ekman on wind-driven currents of the seas, showing that the force of the wind is not transmitted to deep water simply by viscous drag. Nansen had noted that ice tended to drift 20 to 40 degrees to the right of the wind direction and, attributing the effect to Earth's rotation, predicted that the deviation would increase with depth. This explanation was verified and elaborated by Ekman who proved that Nansen's deduction was correct and that the Coriolis effect is a basic element in the circulation of ocean currents.

After gaining fame as an explorer, Nansen became increasingly involved in public affairs. In 1917, during World War I, he led a Norwegian commission to the United States to negotiate an agreement with that nation to permit imports of food and other essential supplies through the naval blockade to Norway. He headed the Norwegian delegation to the first assembly of the League of Nations and in 1920 was appointed high commissioner for repatriation from Russia of prisoners of war from the former German and Austro-Hungarian armies. The Soviet government refused to recognize the League of Nations but negotiated with Nansen personally. In September 1922 he reported that his task was completed and that 427,886 prisoners of war had been repatriated at amazingly little cost.

During 1921 and 1922, Nansen directed an effort to bring relief to famine-stricken Russia. Because the League of Nations refused financial assistance, he resorted to public appeals and succeeded in raising the necessary funds. On 5 July 1922, as a result of his initiative, an international agreement was signed in Geneva introducing the identification card for displaced persons known as the "Nansen Passport." In 1922 Nansen was awarded the Nobel Peace Prize; he used the prize money for the furtherance of international relief work.

Nansen died at his home at Lysaker near Oslo on 13 May 1930.

Bibliography

HESTMARK, G. "Fridtjof Nansen and the Geology of the Arctic." *Earth Science History* 10 (1991):168–212.

NANSEN, F. *Eskimo Life.* London, 1893.

———. *The Norwegian North Polar Expedition.* 5 vols. Oslo, 1901–1905. Repr. New York, 1969.
SÖRENSEN, J. *The Saga of Fridtjof Nansen.* London, 1932.
WALKER, J. M. "Farthest North, Dead Water and the Ekman Spiral. Part II: Invisible Waves and a New Direction in Current Theory." *Weather* 46 (1991): 158–164.

ALEXANDER R. MC BIRNEY

NATURAL SELECTION

See Life, Evolution of

NEBULA

See Stars

NEPTUNE

See Uranus and Neptune

NEUTRON STAR

See Stars

NIER, ALFRED O. C.

Alfred Otto Carl Nier was a physicist whose contributions to the development of mass spectrometers and the measurement of natural isotopic variations produced major advances in the earth and planetary sciences. In the years 1936–1938, while a postdoctoral fellow, he put the uranium-lead method of dating uranium-bearing minerals on a sound basis by determining the isotopic composition of both the radioactive parent, uranium, and the radiogenic daughter, lead (Pb), ushering in the modern quantitative era of geochronology. In 1939, shortly after becoming an assistant professor at the University of Minnesota, he separated enough of the rare uranium isotope, ^{235}U, to allow its identification as the fissioning isotope of uranium, a discovery of great significance to the later development of the atomic bomb by the Manhattan Project (*see* ENERGY FROM THE ATOM). Also during the 1940s, Nier designed a mass spectrometer that became the prototype for nearly all later instruments applied to problems of geochronology and isotope geochemistry.

In the 1950s, Nier and his students determined the natural isotopic composition of many elements, several of which form the basis for methods of measurement of geologic time (*see* ISOTOPE TRACERS, RADIOGENIC), or for the study of natural processes in geologic and biologic systems (*see* ISOTOPE TRACERS, STABLE). Nier and E. A. Gulbransen first reported variations in the isotopic composition of carbon in nature, and, subsequently, Nier designed an "isotope ratio" mass spectrometer leading to the development of instruments for the precise determination of the isotopic compositions of hydrogen, carbon, oxygen, and sulfur in geological and biological materials. In later work, he applied mass spectrometry to problems in planetary science, including the compositions of the upper atmospheres of Earth and Mars, and the causes of isotopic variations in extraterrestrial samples. At the time of his death, he was investigating methods of distinguishing cosmic dust of asteroidal and cometary origin.

Because of his pioneering and fundamental contributions to modern geochronology and isotope geochemistry, Nier is considered by many to be the father of those fields. Nier died in Minneapolis, Minnesota on 16 May 1994 following an automobile accident.

Nier's reminiscences paint a picture of a boy growing up with a passion for mechanical and electrical things. Born on 28 May 1911 in St. Paul, Minnesota, Nier's early interest in electricity and radio directed him into the study of electrical engineering at the University of Minnesota. New jobs in that field were unavailable when he graduated in 1931 at the height of the Great Depression, and he returned to the university to earn a master's degree in electrical engineering, which he received in 1933. His interest was captured by exciting new developments in physics, and he began to work towards a Ph.D. in that field. He constructed a mass spectrometer based on the design of an instrument used by his advisor to study the ionization potential of rare gases and alkali metals. Nier applied his newly constructed instrument to show the existence of ^{38}Ar, the rarest of the argon isotopes, and to determine the relative abundances of the three argon isotopes, ^{36}Ar, ^{38}Ar, and ^{40}Ar. He also showed that potassium contained the rare isotope ^{40}K with an abundance of only 1/8600 that of ^{39}K, the principle isotope. ^{40}K was later shown to be responsible for the radioactivity of potassium.

After receiving his Ph.D. in 1936, Nier was awarded a National Research Council Fellowship to Harvard University. At Harvard he worked with the physicist K. T. Bainbridge, famous for his pioneering work in the mass spectrographic study of the masses and relative abundances of isotopes, with the chemist G. P. Baxter, who specialized in the determination of the atomic weight of elements by chemical means, and with the geologist Alfred Lane (Tufts College), who was chairman of the National Research Council Committee on Geologic Age and a source of geological samples for Nier's work. At that time, the ages of uranium-bearing minerals were calculated from their contents of uranium and thorium and of radiogenic lead from radioactive decay of those elements. The isotopic composition of the lead was inferred from its chemically determined atomic weight. Nier put the age calculations on a sound footing by measuring both the present-day isotopic composition of uranium, and the isotopic composition of the lead in the samples. In this work, the age of one uranium-rich sample was calculated to be 2.2 billion years (Ga), a value greater than the then assumed age of Earth. Following World War II, scientists such as ARTHUR HOLMES of the University of Edinburgh and F. G. Houtermans of the University of Berne recognized that the isotopic composition of common lead from the lead ores analyzed by Nier could be used to estimate the age of Earth. A flurry of activity to determine the most precise age followed, leading to initial estimates of Earth's age of about 2.9–3.3 Ga. The issue remained unsettled until 1953 when measurements of the isotopic composition of lead from meteorites were combined with those of lead from terrestrial oceanic sediments by C. C. Patterson to obtain the presently accepted value of 4.56 Ga (*see* OLDEST ROCKS IN THE SOLAR SYSTEM).

Following his two years at Harvard, Nier returned to the University of Minnesota. He took his Harvard mass spectrometer with him, leaving behind its two-ton magnet. He was soon back in operation with a replacement magnet, but recognized that the requirement for massive magnets severely limited the availability of mass spectrometers for research. In 1939, he designed and built an instrument requiring a much smaller wedge-shaped "sector" magnet. The design was refined following World War II, and Nier's machine shop at the University of Minnesota built dozens of mass spectrometers based on it, enabling many other scientists to participate in isotopic research. In the interim, uranium fission was discovered in 1939, and Nier was urged to attempt separating enough ^{235}U so that it could be verified as the fissioning isotope. He succeeded, using some of the uranium compound from his Harvard geological work for the ^{235}U separation. At the time, his was the only mass spectrometer in the world capable of this work. Also during this time, he and his students became interested in the isotopic composition of carbon, and they built a thermal diffusion column in which the $^{13}C/^{12}C$ ratio in methane could be enriched more than ten times the natural value. This led to a number of collaborations with colleagues in the biological sciences interested in using the enriched carbon as a tracer for biological processes.

Following the war, Nier left interpretation of his prewar uranium-lead data to others such as ARTHUR HOLMES (with whom he maintained a lively correspondence) and Houtermans. In 1948, he and L. T. Aldrich demonstrated the presence of ^{40}Ar in potassium-bearing minerals from the decay of ^{40}K, discovered earlier by Nier, and they suggested that this decay scheme also could be used to determine geological ages. Nier contributed to development of the rubidium-strontium method of dating rocks and minerals by determining the natural isotopic abundances of strontium and rubidium. The main emphasis of his research during the 1950s and 1960s, however, was the very precise determination of atomic masses and the application of mass spectrometry to space research. He and his colleagues measured the products of nuclear reactions produced by cosmic rays in meteorites, the composition of Earth's upper atmosphere using mass spectrometers carried on sounding rockets, and the composition of the Martian atmosphere using a mass spectrometer carried aboard the *Viking* spacecraft. He retired from the faculty of the University of Minnesota in 1980, but continued his research as a professor emeritus. During the 1980s he returned to the isotopic analysis of extraterrestrial samples. At the time of his death he was developing a means to distinguish interplanetary dust particles of cometary origin from those of more mundane asteroidal origin by the way they released cosmic-ray produced helium when they were heated.

"Al" Nier as he preferred to be called, derived great joy from creating new "gadgets." His friendly manner and his vast store of experiences made him an immensely enjoyable conversationalist. Once, near a display of construction toys, he reflected on the influence of mechanical toys on development of his own later interest in science. This was characteristic for a man who had a rare combination of humbleness and pride, who delighted in doing those things in mass spectrometry that others could not do, but was always willing to help a student or younger colleague. It is not surprising that he was returning home from his laboratory (at the age of 82) when he suffered the automobile accident which led to his death, or that from his hospital bed he asked how soon he would be able to return to his lab. He was an excellent teacher, and many of his former students have held leading positions in academia, government, and industry. Nier was an avid traveler, even in his later years, but also enjoyed working around his home or at his cabin on a lake in northern Minnesota.

Alfred Nier received many national and international honors and awards, including the Day Medal of the Geological Society of America (1956), the AEC award (1971), the NASA Medal of Exceptional Scientific Achievement (1977), the Goldschmidt Medal (*see* GOLDSCHMIDT, VICTOR MORITZ) of the Geochemical Society (1984), the Field and Franklin Award of the American Chemical Society (1985), the Thomson Medal of the International Mass Spectrometry Conference (1985), and the Bowie Medal of the American Geophysical Union (1992). He was a member of the National Academy of Sciences, the American Physical Society, the American Philosophical Society, the Geochemical Society, the Meteoritical Society, the American Association for the Advancement of Science, the Max-Planck Society (in Germany), and the Royal Swedish Academy of Sciences. He was a role model who had a great influence on those who followed him scientifically, not only by founding whole areas of research, but also by establishing the standards of excellence necessary to attain meaningful results from exacting measurements of the type he pioneered.

Bibliography

ALDRICH, L. T., and A. O. NIER. "Argon 40 in Potassium Minerals." *Physical Review* 74 (1948):876–877.

NIER, A. O. "A Mass Spectrometer for Routine Isotope Abundance Measurements." *Review of Scientific Instruments* 11 (1940):212–216.

———. "Some Reminiscences of Isotopes, Geochronology, and Mass Spectrometry." *Annual Review of Earth and Planetary Sciences* 9 (1981):1–17.

———. "The Isotopic Constitution of Radiogenic Leads and the Measurement of Geological Time. II." *Physical Review* 55 (1939):153–163.

NIER, A. O., and E. A. GULBRANSEN. "Variations in the Relative Abundances of the Carbon Isotopes." *Journal of the American Chemical Society* 61 (1939):697–698.

NIER, A. O., and M. B. MCELROY. "Composition and Structure of Mars' Upper Atmosphere: Results from the Neutral Mass Spectrometers on Viking 1 and 2." *Journal of Geophysical Research* 82 (1977):4341–4349.

PEPIN, R. O., and P. SIGNER. "Alfred O. Nier." *Meteorites* 29 (1994):747–749.

LAURENCE E. NYQUIST

NONFERROUS METALS

Humans obtain sustenance, the material goods required for survival, from the bountiful resources of the earth. Many of these critical resources are minerals—naturally occurring, predominantly inorganic solids that we take from the earth. We can think of obvious uses of mineral resources: shelter, tools, transportation, and energy. Other uses are subtle and may surprise you. Minerals are used as pigments, to purify water, in pharmaceutical and over-the-counter drugs, in making this page and the ink you are scanning. Nonferrous metals are a subset of these mineral resources.

Definitions

Minerals are subdivided into fuel and nonfuel resources. Fuels are extracted primarily for their energy content—coal, petroleum, and uranium. Nonfuel minerals are subdivided into metals and nonmetals on the basis of chemical and physical properties such as thermal conductivity, electrical resistance, ductility, opacity to various wavelengths of electromagnetic radiation, malleability, and fusibility. Most can also be mixed with one or more other metals in varying proportions to form alloys that possess unique properties. Many metals have high strength-to-weight ratios important in structural uses where weight limitations are important (e.g., the transportation industry and multistory construction).

Metals may be classified by their chemical behavior (as in the periodic table of the elements), abundance in the earth's crust, mineralogy, or end uses in modern industry. This latter, end-use classification is preferred by mineral economists and is shown in Table 1.

"Nonferrous" might lead one to expect that this group of metals includes all metals except iron (*see* IRON DEPOSITS). It is not that simple. In general, the nonferrous metals in Table 1 are those metals (1) whose primary end use is not in the ferro-alloy (steel) industry; (2) that are not used primarily for monetary or jewelry purposes; and (3) whose annual production/consumption is significantly large and critical to basic heavy industry. "Significant" production is a vague term. In general, the annual production of nonferrous metals is in the range of millions of tons of metal, as contrasted to tens of thousands of tons or less for the special metals. Cadmium (Cd) and mercury (Hg) are exceptions.

History

The nonferrous metals, with the exception of aluminum (Al) and cadmium, have been known since before recorded history. Copper (Cu) artifacts, often with gold (Au), have been found in prehistoric sites. Copper, lead (Pb), zinc (Zn), tin (Sn), and mercury were used in China and the eastern Mediterranean region by 2500–3000 B.C.E. The Bronze Age, from 2000–3000 B.C.E., is characterized by implements and weapons of copper-tin alloys. Cadmium, which behaves like zinc, was first discovered in 1817. Aluminum, common in modern society and the second most abundant metal in the earth's crust, was not isolated in metallic form until 1825 and remained a rare and expensive curiosity until development of commercial aluminum production in the late 1880s.

Medieval alchemists recognized the physical and chemical similarities of copper, lead, tin, and zinc with gold and silver (Ag). To them, the nonferrous metals were of lesser value than gold and silver. They were base metals in contrast to the precious or noble metals. Their fruitless efforts to convert nonferrous to noble metals laid the foundations for modern chemistry and metallurgy.

With the exception of aluminum and cadmium, the nonferrous metals (Table 1) have longstanding commercial and industrial end uses. All are critical to modern industrial economies.

Table 1. Classification of Metallic Mineral Resources

Ferrous metals	Iron, chromium, vanadium, cobalt, nickel, molybdenum, tungsten
Nonferrous metals	Copper, aluminum, lead, zinc, tin, mercury, cadmium
Precious metals	Gold, silver, and the six platinum group elements (i.e., platinum, palladium, rhodium, iridium, osmium, and ruthenium)
Special metals	Arsenic, antimony, beryllium, bismuth, cesium, gallium, germanium, hafnium, indium, niobium, rare earth elements,* selenium, tantalum, tellurium, zirconium

* The rare earths are the fifteen elements, atomic numbers 57 to 71, belonging to the lanthanide series.

Table 2. Important Physical Properties of the Nonferrous Metals

	Symbol and Atomic #	*Specific Gravity*	*Melting Point (°C)*	*Resistance (μohm-cm)*	*Thermal Conductivity (W cm^{-1} K^{-1} at 0°C)*
Aluminum	Al_{13}	2.70	660	2.65	2.37
Cadmium	Cd_{28}	8.65	321	6.83	0.97
Copper	Cu_{29}	8.96	1083	1.67	4.03
Lead	Pb_{82}	11.35	327	20.64	0.35
Mercury	Hg_{80}	13.55	−39	98.4	0.08
Tin	Sn_{50}	7.31	232	11.0	0.68
Zinc	Zn_{30}	7.13	420	5.92	1.17

μohm-cm stands for micro-ohm centimeters, W cm^{-1} K^{-1} means watts per centimeter-degree Kelvin at O°C.

Properties of the Nonferrous Metals

Table 2 lists the important physical properties of the nonferrous metals. The unique properties of each metal that are important in their end uses are:

Aluminum—light, corrosion-resistant, and durable. Aluminum oxides are inert, hard, and have high melting points.

Cadmium—compounds of cadmium have bright primary colors: reds, yellows, blues, and greens.

Copper—readily alloyed with tin to make bronze and with zinc to make brass. Both alloys are durable, easily cast, and corrosion-resistant.

Lead—soft, corrosion-resistant, effective radiation shield. Highly toxic.

Mercury—liquid at room temperature and alloys with other metals (amalgams). Highly toxic.

Tin—corrosion-resistant and alloys with many metals (pewter is tin and lead with variable amounts of other nonferrous metals). Many alloys (solders) have low melting points.

Zinc—corrosion-resistant.

Specific Uses of Nonferrous Metals

Nonferrous metals are basic to modern industrial economies. Table 3 lists the primary consumers of

Table 3. Principal End Uses of the Nonferrous Metals

Aluminum	containers, transportation, high-voltage transmission lines, heat exchangers, abrasives (alumina)
Cadmium	Ni/Cd batteries, corrosion-resistant platings, pigments, plastics
Copper	building and construction, electrical wiring, heat exchangers, industrial machinery, transportation
Lead	batteries, cable, sheet, and solder
Mercury	electric lighting and wiring devices, batteries
Tin	solder, tinplating, brass
Zinc	galvanized steel, die-cast alloys, brass, paints, alkaline batteries, sacrificial anodes

the nonferrous metals in modern industrial societies. In addition to these current uses of nonferrous metals, changing technologies, heightened environmental concerns, and the development of alternative materials will both increase and decrease demands for specific metals. The likelihood is that mercury use, in particular, will decrease in coming years due to its high toxicity and difficulty of handling.

While difficult to predict, technological advances suggest significant changes in demand for nonferrous metals in the following areas:

Communications and electronics: increasing demand in general and increased use of nonferrous metals in semiconductors and superconducting materials.

Energy storage: increasing demand for lead batteries and growth and development of new cadmium- and zinc-based batteries.

Materials: new ceramics and nonmetal-based composites may result in lower demand for metals in many construction and transportation applications.

Transportation: aluminum continues to replace heavier metals (including copper) and new methods of ion-implantation may develop high-strength aluminum alloys.

Although our focus is nonferrous metals, precious metals have extensive uses in industries where they directly compete with nonferrous metals. More than half the annual domestic consumption of gold and silver is in the industrial sector. The chemical stability and low electrical resistance of gold and silver make them important as electrical conductors, particularly in microelectronic applications where small amounts of low-resistance conductors are required. The platinum (Pt) group elements, long used in industrial catalysis, are increasingly used in electrical circuits where low resistance, chemical inertness, and high melting points are important.

World Production and Consumption

Nonferrous minerals are produced on all continents except Antarctica, where mining is banned. As elaborated below, economic nonferrous mineral deposits occur only in certain geological settings. Since metals are not homogeneously distributed in the earth's crust, the production of nonferrous minerals is the result of a complex interaction between geologic, economic, social, technical, historical, and political factors.

Table 4 lists the major producing nations of nonferrous metals. Table 5 shows the 1994 production and consumption statistics for the United States and the world. In 1992, the United States mined about 19 percent of the copper, 12 percent of the lead, 9 percent of the cadmium, 7 percent of the zinc, and 2 percent of mercury in the world. Neither aluminum nor tin are mined in the United States. As Table 5 demonstrates, the United States consumes more nonferrous metals than it produces, making the United States a net importer.

Levels of national consumption of nonferrous mineral resources are most often a function of standard of living. This correlation reflects the heavy use of nonferrous metals in the communications, consumer, and transportation sectors of industrial economies.

An increasingly important aspect of the nonferrous metal markets is recycling. In 1992, the estimated value of the United States' domestically recycled nonferrous metals was $7.6 billion, about one-and-a-half times the value of domestically produced nonferrous metals. Except for mercury, which is difficult to handle because of its liquid state and toxicity, the growth in the amount of nonferrous metal recycling in the early 1990s ranges from 1 percent (lead and zinc) to 30 percent (aluminum) per annum.

Production of minerals allows only one harvest. Once a mineral is extracted from the earth, that resource has been exhausted for all time on a human scale. Therefore, we must ask, How much nonferrous metal remains to be mined?

Estimates vary depending on the degree of certainty required. Mineral economists estimate a reserve base that includes known economic and marginal deposits. This base can change with market price, changing mining and mineral beneficiation technologies, and the general level of geological understanding of the earth. In Table 6 we estimate the world's nonferrous metals reserve base. One needs to be careful not to infer too much from these estimates. For example, the reserve base of lead is only twenty-two times current annual consumption. Will we run out of lead about 2020? No. No serious restrictions on lead supply for the foreseeable future are likely, barring significant changes in demand or in political and economic conditions. Significant lead resources remain to be

Table 4. Major Producing Nations of Nonferrous Metals in 1992 (in rank order, with percent of world's total 1992 production given)

Aluminum	Australia (36.7), Guinea (15.6), Jamaica (10.5), Brazil (9.4), Surinam and Guyana (4.9), India (4.4), Russia (4.4)
*Cadmium**	Japan (15.7), United States (8.6), Canada (8.2), Belgium (8.1), China (6.4), Australia (5.3). Kazakhstan (5.3)
Copper	Chile (18.4), United States (17.3), Canada (9.0), Russia (7.2), Zambia (5.0), Zaire (4.9), Poland (4.7), Peru (4.3)
Lead	Australia (15.5), Russia (12.8), United States (12.6), China (9.8), Canada (9.1), Peru (5.6), Mexico (5.1)
Mercury	China (31.5), Mexico (23.2), Algeria (14.1), Russia (13.3)
Tin	China (21.0), Indonesia (15.3), Brazil (14.9), Malaysia (10.5), Bolivia (8.5), Thailand (7.6), Russia (6.9)
Zinc	Canada (15.8), Australia (14.4), Russia (8.9), China (8.9), Peru (8.6), United States (7.5), Mexico (4.1)

* Since cadmium is a refinery by-product of zinc production, production statistics reflect the location of refinery capacity rather than mining.
Note: all data in Table 4 and subsequent tables are from the most current *Annual Report of the U.S. Bureau of Mines*, 1994. Typical publication of these data is delayed 1–2 years for domestic and 2–3 years for foreign statistics.

discovered and evaluated. It requires time and money to determine these additional reserves and there is no compelling reason for mining companies to identify reserves beyond ten to twenty years in the future.

Mining

Nonferrous metals are mined from rock. In general, the nonferrous metals are scarce and must be naturally concentrated before profitable mining is

Table 5. Production and Consumption of Nonferrous Metals (in thousand of metric tons, unless otherwise noted)

	Consumption		*Production*	
	U.S. (1992)	*World (1991)*	*U.S. (1992)*	*World (1992)*
Aluminum	5,725	17,194	0	19,219
Cadmium (tons)	3,515	20,000	1,620	18,750
Copper	2,311	10,714	1,765	9,290
Lead	1,237	5,342	398	3,242
Mercury (tons)	621	—	64	3,014
Tin	34.9	218	0	179
Zinc	1,276	6,993	523	7,137

Table 6. Estimated World Reserves and Resources of Nonferrous Metals (in approximate rank order, 1994)

	Est. World Reserve Base	*Principal Locations*
Aluminum	very large	Australia, Guinea, Jamaica, Brazil, Surinam and Guyana, India, Russia
Cadmium	not applicable	same as zinc
Copper	587,000,000 tons	Chile, U.S., Russia, Poland, Zambia, Zaire, Peru, Canada, Australia, Mexico
Lead	120,000,000 tons	Australia, United States, Canada
Mercury	not available	Russia, Spain, Algeria, China, Mexico
Tin	40,000,000 tons (est.)	SE Asia, Brazil, Bolivia, Zaire, Russia, U.K.
Zinc	325,000,000 tons	U.S., Canada, India, South Africa, Peru

possible. An economic concentration of valuable minerals is an ore deposit, the valuable rock is ore, and the minerals which contain the metals are called ore minerals.

Nonferrous ores are extracted by either underground mining or open-pit (quarry-like) operations. Both techniques rely on large earth-moving equipment and blasting technology to break and transport rock so that the ore minerals can be processed. In most cases, the ore minerals are metal sulfides or oxides and constitute only a small percentage of the rock. If that percentage is too low, the rock cannot be mined at a profit. The lowest concentration of a metal that can be profitably mined is called the cut-off grade.

Cut-off grades are not fixed. They are geologically controlled, and also dependent on nongeological factors such as mining methods, world mineral prices, competition from new mines, declining or rising world market demand, proximity to markets, and government policies. For example, prior to the development of large earth-moving equipment and modern techniques of concentrating minerals at the turn of the last century, the cut-off grade for copper was about 3.0 percent. Within a few years, driven by technological innovation alone, the cut-off grade dropped below 1.0 percent. Today, it is 0.5 percent. Domestic U.S. copper mines have operated with cut-off grades as low as 0.3 percent due to proximity to existing mills and smelters and government guaranteed start-up loans that allowed producers to capitalize mines at favorable discount rates. The cut-off grade of a lead deposit in Missouri will be significantly less than that for a deposit in the Yukon Territory, where transportation and labor costs are higher. In more than one instance, government policy in centrally managed economic systems has encouraged "nonprofitable" mines to operate at below free-market cut-off grades when the nonferrous metal being mined is an important export product.

After extraction of ore from the earth, it is normally crushed and the ore minerals concentrated by one or more physical processes. These processes include density separation and flotation (a technique that relies on the differential adherence of finely powdered minerals to bubbles in a froth of organic chemicals). Concentrates are then converted to pure metal variously by pyrometallurgy (roasting and melting in the presence of gases and fluxes to remove impurities), solvent extraction, and electrowinning (electroplating of metal from a melt or aqueous solution). In situ dissolution of nonferrous minerals and pumping of the aqueous metal-rich solutions to the surface for treatment are experimental in the early 1990s. Peters (1988) provides an excellent introduction to mining and related mineral technologies.

Mining of nonfuel mineral resources currently is limited to land and a few shallow submarine environments. Significant nonferrous metal resources may exist offshore on the continental shelves. In the deep oceans, manganese nodules containing several tenths of a percent of nonferrous metals are found. The majority of offshore and deep-sea nonferrous metals cannot be mined with current technology.

Origin of Deposits

As indicated, a nonferrous metal cannot be extracted from a rock economically unless some natural process concentrates that metal ten to several thousand times the crustal average. In the absence of such concentration, a miner would have to remove and process large volumes of rock and could not compete with other mines with richer ores.

Two natural processes concentrate nonferrous metals to make economic deposits. In the first, the valuable metal is retained in the rock while components of no economic value are removed. This model is applicable to economic deposits of aluminum that consist of aluminum oxides and hydroxides. These are insoluble in water. When a rock is chemically weathered, the water-soluble elements in the rock are taken into solution and removed from the rock. The resulting residue is a soil enriched in alumina called laterite. Bauxite is a laterite sufficiently rich in alumina to be ore. Thus, bauxite becomes ore by the extensive selective leaching of soluble constituents from rock.

The second common mechanism for concentrating nonferrous metals to economic concentrations involves the addition of metal to a sediment or rock. For nonferrous metals, the most common agents for mobilizing and carrying metals are hot, often saline, watery brines called hydrothermal solutions. These solutions can be: (1) hot spring emanations from cooling volcanic rocks usually beneath the ocean; (2) crustal waters circulated by the intrusion of magma into shallow levels of Earth's crust; and (3) waters trapped in sediments during deposition of marine sediments that are then heated and driven off by continuing sedimentary deposition, burial, and compaction.

The source of water for hydrothermal solutions is local surface or ground water, water contained in rocks known as connate waters, water generated during the devolatilization of a cooling magma body, or some combination of these. These fluids either migrate into surface environments of active sedimentation and deposit the metals or slowly percolate through permeable or fractured bodies of rocks where chemical and physical changes result in metal precipitation.

In the case listed of hot springs, hot, acid waters leach soluble metals from the rocks they encounter. The resulting metal-rich fluids then vent to the seafloor. On mixing with the cold neutral waters of the ocean, metal sulfides precipitate as nearly pure chemical sediments or as individual mineral grains finely disseminated in marine muds. Modern examples of this phenomenon are black smokers on the seafloor, which are like smoking chimneys that form as metal-laden brines bubble up from cracks in the rocky seafloor and precipitate fine-grained, black metal sulfides. These sulfide grains coalesce to build mounds and chimneys of pure metallic minerals. The end result is a massive deposit rich in iron, copper, lead, zinc, and silver sulfides (*see* MINERAL DEPOSITS, IGNEOUS).

Hydrothermal solutions also are formed when an intrusive magma body comes to rest in the earth's crust. As the stationary magma slowly cools, it heats large volumes of connate waters. Convection cells of hot water form above the cooling magma body and leach metals as they circulate in the shallow crust. Metal-rich, magma-derived waters from the magma body may also be introduced into the convection stream. The resulting metal-rich hydrothermal solution rises to (1) boil at shallow depths due to pressure release; (2) interact chemically with colder, reactive rocks marginal to the magma chamber; or (3) mix with cold and dilute shallow waters. Each or all of these processes may rapidly precipitate metal sulfides in a confined volume of rock. One example of such deposits occurs when the hydrothermal fluid is limited to depths of 2–10 kilometers (km) in the crust. Porphyry copper deposits are produced. These are large, many hundred million- to billion-ton deposits of copper ore that account for the majority of world copper production. In other cases, the circulating fluid may encounter highly reactive rocks like limestone. The rapid precipitation of metals that results produces an altered carbonate (CO_3)-rich ore known as a skarn deposit. Skarns host many tin deposits and a few copper deposits as well. Hydrothermal fluids may also reach near-surface environments where boiling or dilution by ground water results in the precipitation of metal sulfides at depths beneath the surface of no more than a few tens or hundred of meters—a common ore occurrence for mercury and gold (*see* MINERAL DEPOSITS, METAMORPHIC).

The third hydrothermal process involves hot, basinal brines that migrate from thick sequences of marine sediments during subsidence. These saline solutions leach nonferrous metals and carry them upward toward the basin margin. Since limestone and dolomite are common in most marine sequences, the metal-bearing brines often encounter reactive carbonate rocks and precipitate lead and zinc (with cadmium) sulfides. These carbonate-hosted lead-zinc deposits are the major source of these metals worldwide.

These models for ore-forming processes of nonferrous metals are simplified. Every individual deposit differs in some detail. There are deposits whose origins are more complex or even different from the generalized models presented above (see Guilbert and Park, 1986, for a detailed introduction to ore-forming processes). The guiding principle in all cases is that some set of physical or chemical processes must operate to concentrate scarce nonferrous metals. Recognizing the telltale signs that such processes have occurred is the challenge for the modern mineral prospector.

Bibliography

CRAIG, J. R., D. J. VAUGHAN, and B. J. SKINNER. *Resources of the Earth.* New York, 1988.

GUILBERT, J. M., and C. F. PARK. *The Geology of Ore Deposits.* New York, 1986.

PETERS, W. C. *Exploration and Mining Geology,* 2nd ed. New York, 1988.

U. S. DEPARTMENT OF THE INTERIOR. *Annual Report of the Bureau of Mines.* Washington, DC.

P. GEOFFREY FEISS

NONSILICATES

See Minerals, Nonsilicates

NORTHERN LIGHTS

See Auroras

NUCLEAR WASTE

See Hazardous Waste Disposal

NUCLEOSYNTHESIS AND THE ORIGIN OF THE ELEMENTS

The study of nucleosynthesis (the origins of atomic nuclei) is a fundamental discipline in astrophysics involving important interdisciplinary relationships with the earth sciences through the study of meteorites (*see* METEORITES). Nucleosynthesis theory is based on the development of nuclear-astrophysical models describing the formation of atomic nuclei, and can be tested and refined by astronomical and meteoritic observations. Overall, the aim is to understand all significant astrophysical mechanisms involving the formation, destruction, and distribution of atomic nuclei, and to model the time evolution of nuclide abundances across the diverse range of galaxy types, especially our own Milky Way spiral. Topics include "big bang" light element synthesis in the very early universe, and various galactic nucleosynthetic processes in stars, novas, and supernovas that have been active over most of the history of the universe and are ongoing today.

Nucleosynthesis of the major rock-forming elements in stars is a natural consequence of their principal mechanism of energy generation and proceeds by stages of fusion of lighter nuclei up to the region of the binding energy per nucleon maximum at atomic number (A) = 56 (= ^{56}Fe). Synthesis of nuclides heavier than in this "iron peak" region is largely a consequence of neutron captures in stars and supernovas. Stars contribute products of their interior nuclear reactions into the interstellar medium out of which new generations of stars form in an evolving cycle of galactic chemical evolution (*see* STARS). Understanding this cycle is a complex task. It requires well-dated stellar "fossil records" of the chemical and dynamical evolution of galaxies, and accurate modeling of nuclear processes in stars, novas, and supernovas across a wide range of compositions, masses, configurations, and evolutionary histories. Efforts to make increasingly precise laboratory determinations of critical hard-to-measure nuclear reaction rates, lifetimes, decays, and so on, play an important role in advancing the field, as do refinements in predictive nuclear theory. Nucleosynthesis studies necessarily draw from a wide range of scientific fields and contribute importantly to associated disciplines, including particle physics, cosmology, and early solar system studies.

Efforts to understand the origins of the solar abundance distribution of nuclides (Figure 1) have been a central preoccupation to which geochemistry has made important contributions, beginning with the pioneering meteorite abundance studies of VICTOR MORITZ GOLDSCHMIDT (1937) and EDUARD SUESS and HAROLD UREY (1956). Abundances in the terrestrial planets exhibit large chemical fractionations related to both high-temperature processing in the inner nebula and

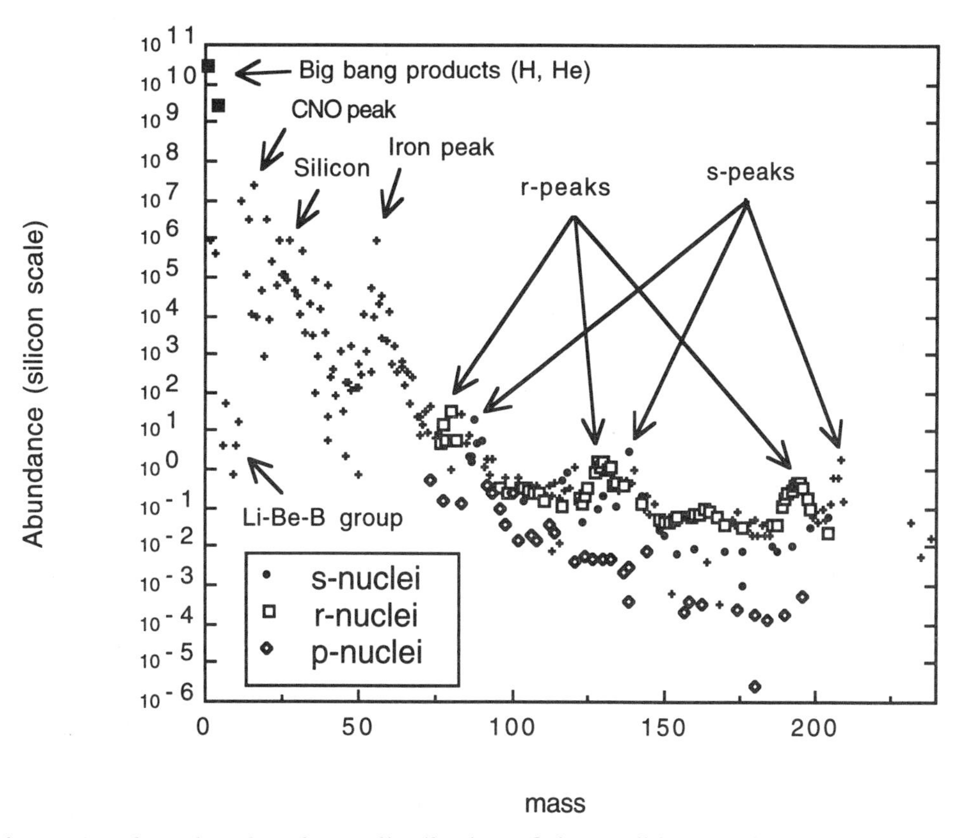

Figure 1. The solar abundance distribution of the nuclides as a function of atomic number (A) from Anders and Grevesse (1989). Abundances are given as number of atoms on a log scale normalized to Si = 10^6. Some nuclei are grouped or labeled according to the nucleosynthetic process that produced them. Note the triple peak structure apparent in both the *s*-only and *r*-only nuclei. This corresponds to neutron capture processes in the vicinity of closed neutron shells at N = 50, 82, and 126, where destruction rates are small with respect to production rates, so that an abundance buildup results at equilibrium.

segregation of chalcophile and siderophile elements into planetary cores (*see* CORE, STRUCTURE OF). Only a few rare primitive meteorite types record unfractionated nonvolatile element abundances representative of the bulk solar system. Spectroscopic studies of the solar photosphere provide a link to hydrogen (H), helium (He), and other noncondensing volatiles in the Sun (*see* SUN). Suess and Urey's study stimulated foundational advances in nucleosynthesis theory published by Cameron and the group of Burbridge, Burbridge, Fowler, and Hoyle in 1957. Discoveries in the 1960s of isotopically anomalous xenon (Xe) in meteorites attributable to the early decay of extinct radionuclides of iodine (I^{129}) and plutonium (Pu^{244}) demonstrated that nucleosynthesis had been an extended process in the Galaxy prior to the formation of the Sun 4.57 billion years (Ga) ago. The discovery of evidence of decay of shorter-lived aluminum-26 (Al^{26}) in meteorites in the 1970s demonstrated that nucleosynthesis had been ongoing near or within the protosolar cloud right up until the solar system's formation 4.6 Ga ago. Meteorites continue to provide seminal insights via the study of a wide range of stable and radiogenic isotope anomalies (unusual isotopic compositions) with increasingly sophisticated mass spectrometers. A new form of "isotopic astronomy" based on the study of meteorites has attracted wide-scale interest, especially since the discovery of isotopically anomalous inclusions and presolar stardust in primitive meteorites. Isotope studies of these and other meteoritic materials yield uniquely precise data for evaluation of nucleosynthetic models.

Big Bang Nucleosynthesis of Light Elements

Hydrogen and He are the dominant elements in galaxies, and comprise about 70 and 28 percent, respectively, of the mass of the outer portion of the Sun. The H and He^4 mass fractions of astrophysical regions are commonly denoted by "X" and "Y." "Z" is used to denote the mass fraction of elements heavier than He, such that $X + Y + Z = 1$. The sum of the Z-elements is described as the "metallicity," and is about 2 percent in the Sun. It is a consequence of stellar evolution in the galaxy prior to the formation of the Sun. Most of the H and He in galaxies (and all of the deuterium, D, and some of the lithium, Li) was synthesized in the first few minutes of the big bang by assembly of subatomic particles during cooling of the expanding primordial universe at a temperature of about one billion degrees K (*see* COSMOLOGY, BIG BANG AND THE STANDARD MODEL). This view is strongly supported by evidence from the cosmic microwave background radiation and excellent agreement of the predictions of the standard hot big bang model with observations of the relative abundances of H, D, He^3, He^4, and Li^7 in low-metallicity regions. The existence of stars with Z less than about 10^{-4} of the solar value indicates that heavy elements were not significantly produced in the big bang. Heavy element abundances began to grow only during the epoch of galaxy formation and appear in highly redshifted quasar spectra at inferred ages of approximately 1 Ga after the big bang.

Nucleosynthesis and Chemical Evolution in Galaxies

On the large scale, the chemical evolution of galaxies is affected by several factors, including galactic masses and morphologies, accretion timescales, and galactic encounters and/or collisions (*see* GALAXIES). Disk galaxies are gas accretion structures and possess ellipsoidal halo populations of very old stars preserving the shape of their early precollapse configurations. Important insights into nucleosynthetic sites can be gained from studies of heavy element abundances in populations of very old low-mass, low-metallicity stars, both in the halo and the disk. Low-mass stars preserve records of their initial abundances because their nuclear evolution is slow on cosmic timescales and occurs in their deep interiors. The Sun, for example, has preserved its initial surface abundances over the past 4.6 Ga with exceptions only for fragile nuclei such as D, $Li^{6,7}$, boron-10, boron-11, and beryllium-9. Significant clues from studies of old low-mass stars include the growth of oxygen (O/H) before iron (Fe/H), and europium (Eu/H) before barium (Ba/H). These trends indicate that O and Eu are predominantly produced by massive (fast-evolving) stars, whereas Ba and Fe are predominantly products of less massive (longer-lived) stars.

Most massive stars end their relatively brief galactic lives (less than about 30 million years, or Ma) as Type II supernovas, ejecting large quantities (greater than about 1 solar mass) of O, but relatively small quantities of Fe (about 0.1 solar masses) (*see* STARS). Another less frequent class, Type Ia supernovas, is thought to occur in lower-mass binary systems by mass-transfer accretion onto a CO-rich white dwarf to the explosive limit of about 1.4 solar masses. Systems ending in Type Ia explosions have very long evolutionary timescales (greater than about 1 Ga). Their eventual explosion typically burns about 50 percent of the exploding star's mass into Fe, with very little survival of O. Thus the high O/Fe ratios of very old stars reflect the shorter evolutionary timescale of O-producing Type II supernovas relative to Fe-producing Type Ia supernovas. The latter events began contributing their nucleosynthetic products to the interstellar medium later in the early history of the galaxy after early generations of massive stars had formed and exploded.

Rates of chemical evolution in galaxies are controlled by the initial stellar mass distribution of newly formed stars and the rate of star formation. Galaxies stop forming stars and effectively cease to evolve chemically when they exhaust their gas supplies, either by supernova-driven outflow (low-mass dwarf galaxies) or efficient star formation (some elliptical galaxies and eventually all disk galaxies). The evolution of the star formation rate in the solar neighborhood (8.5 kiloparsec, or kpc, from the galactic center) is recorded in the age distribution of low-mass dwarf stars. This record suggests early growth in the first few billion years followed by a roughly constant rate of star formation until the present. This is in agreement with slow inside-to-outside accretion models for the formation of disk galaxies. Age determinations of disk and halo stars suggest that our galactic disk began to accrete at the solar galactocentric radius (8.5 kpc) about 12 Ga ago, about 3 or 4 Ga after the initial formation of the galactic halo. Heavy elements in the solar system were synthesized over about 7 Ga of presolar history in the disk.

Specific Processes

Galactic nucleosynthesis is largely a process of the progressive gravity-driven nuclear burning of H and He into heavier elements. Roughly twelve distinct processes in stars, novas, supernovas, and interstellar medium can account for most of the solar distribution of nuclides heavier than He. Major nuclear reactions controlling some of these processes are given in Table 1 using the in/out shorthand form where, for example, $^{12}C(\alpha, \gamma)^{16}O$ represents the reaction: $^{12}C + \alpha \rightarrow {}^{16}O + \gamma$. The symbols α and γ represent alpha particles (^4He) and gamma rays, respectively.

Cosmic Ray Spallation. In the interstellar medium over galactic history this is a significant nucleosynthetic process only for the rare light elements Li, Be, and B. These elements are overabundant in cosmic rays by a factor of about a million relative to other elements. Thus, a roughly one millionth part of stopped cosmic rays is sufficient to account for these rare elements, some of which are also made in the big bang and supernovas.

Hydrogen Burning. Occurs in stars of 1.2 solar masses or less by the proton-proton chain, and in more massive stars additionally by proton captures on C, N, and O involving proton-in, alpha particle-out (p, α) steps in a cycling reaction network known as the CNO cycle. In stars containing neon (Ne) and magnesium (Mg), from galactic chemical evolution prior to their formation, additional NeNa and MgAl cycles produce ^{22}Ne, ^{23}Na, and $^{26,27}Al$ during H-burning. When the hydrogen fuel in the stellar core becomes exhausted, the resulting He-enriched core contracts until sufficiently hot and dense conditions are present to initiate the next stage of nuclear burning.

Helium Burning. Progresses by the triple-α reaction and alpha captures. In lower-mass stars, this process occurs during the red giant phase in which a He-burning zone migrates outward in a thin shell located on top of an expanding pile of He-burning ashes (a CO core).

The slow neutron capture *s*-process is driven by the neutron-producing reactions $C^{13}(\alpha, n)O^{16}$ and $Ne^{22}(\alpha, n)Mg^{25}$, which occur during He-burning in shell regions in red giants as well as in the cores of massive stars. Neutrons (n) are captured initially on Fe with continued n-capture reactions (n, γ), producing increasingly heavier elements up the valley of nuclear stability all the way to Bi^{209}. This is shown for a region of the rare earth elements in Figures 2 and 3. Temperature and neutron-density-dependent "branch points" occur along the *s*-process path at unstable nuclides with lifetimes comparable to neutron capture timescales (typically days to months). Yields near these points put strong constraints on *s*-process conditions. Epi-

Table 1. Main Nuclear Burning Reactions in Stars and Supernovas

Symbols:	p = proton (^{1}H)
	D = deuterium (^{2}H)
	e^+ = positron
	$\beta-$ = beta particle (electron)
	ν = neutrino
	γ = gamma ray
	α = alpha particle (^{4}He)
	$^{12}C(\alpha, \gamma)^{16}O$ is short hand for: $^{12}C + \alpha \rightarrow {}^{16}O + \gamma$
Hydrogen burning:	
p-p chain:	$p(p, e^+ + \nu)D$
	$D(p, \gamma)^{3}He$
	$^{3}He(D, p)^{4}He$
	$^{3}He(^{3}He, 2p)^{4}He$
CNO cycle:	$^{15}N(p, \alpha)^{12}C$
	$^{17}O(p, \alpha)^{14}N$
	$^{18}O(p, \alpha)^{15}N$
	$^{13}C(p, \gamma)^{14}N$
	$^{15}N(p, \gamma)^{16}O$
	$^{12}C(p, \gamma)^{13}N(e^+, \nu)^{13}C$
	$^{16}O(p, \gamma)^{17}F(e^+, \nu)^{17}O$
	$^{14}N(p, \gamma)^{15}O(e^+, \nu)^{15}N$
Helium burning:	
Triple-α:	$\alpha + \alpha \leftrightarrow {}^{8}Be(\alpha, \gamma)^{12}C$
α-captures:	$^{12}C(\alpha, \gamma)^{16}O$
	$^{16}O(\alpha, \gamma)^{20}Ne$
	$^{20}Ne(\alpha, \gamma)^{24}Mg$
	$^{14}N(\alpha, \gamma)^{18}F \rightarrow {}^{18}O(\alpha, \gamma)^{22}Ne$
Carbon burning:	
C-fusion:	$^{12}C(^{12}C, \alpha)^{20}Ne$
	$^{12}C(^{12}C, p)^{23}Na$
	$^{12}C(^{12}C, \gamma)^{24}Mg$
Ne-photodissociation:	$^{20}Ne(\gamma, \alpha)^{16}O$
Oxygen burning:	
O-fusion:	$^{16}O(^{16}O, p)^{31}P$
	$^{16}O(^{16}O, \alpha)^{28}Si$
	$^{16}O(^{16}O, \gamma)^{32}S$
Silicon burning:	
Si-photodissociation:	$^{28}Si + \gamma \leftrightarrow {}^{24}Mg + {}^{4}He$
$(\alpha, \gamma) \leftrightarrow (\gamma, \alpha)$ "e-process":	$^{28}Si(\alpha, \gamma)^{32}S \leftrightarrow {}^{32}S(\gamma, \alpha)^{28}Si$
	$^{32}(\alpha, \gamma)^{36}Ar \leftrightarrow {}^{36}Ar(\gamma, \alpha)^{32}S$
	$^{36}Ar(\alpha, \gamma)^{40}Ca \leftrightarrow {}^{40}Ca(\gamma, \alpha)^{36}Ar$
	. . . $^{56}Ni\text{-}e^+ \rightarrow {}^{56}Co\text{-}e^+ \rightarrow {}^{56}Fe$

sodes of convection in red giants mix He-burning and associated *s*-process products up to the stellar surface, where the resulting heavy element enrichments and isotopic structures have been observed in stellar spectra. A rich record of these conditions is also preserved in silicon carbide stardust in primitive meteorites (Figure 4).

Most stars do not evolve beyond the He-burning stage and cool as CO dwarf stars after ejection of their metallicity-enriched envelopes into the inter-

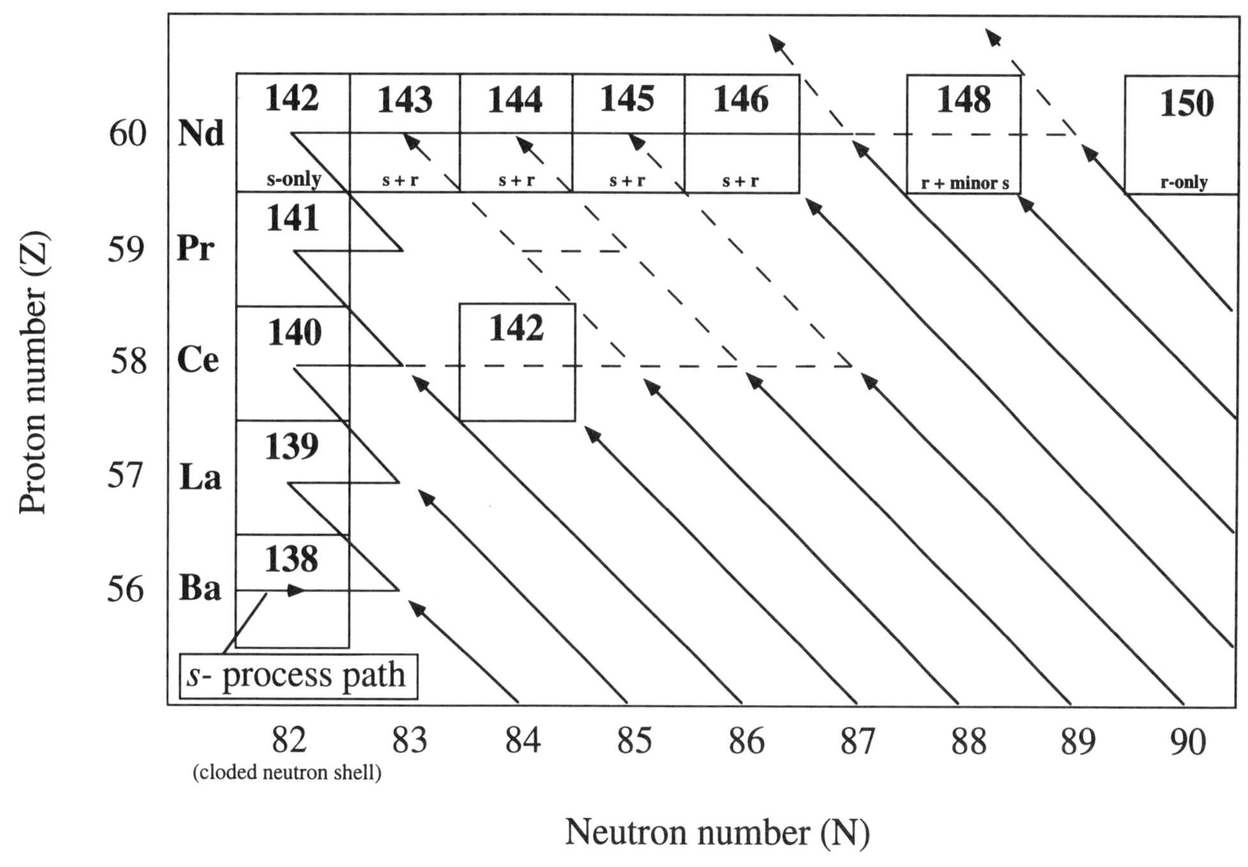

Figure 2. The neutron versus proton number (N-Z) plane in the region of the rare earth element neodymium (Nd). Only the stable nuclei are shown, indicating the location of the nuclear "valley of stability" in this region. Branch points across unstable nuclei with lifetimes in the range of the neutron capture time in the *s*-process are shown by dashed lines. The *s*-process path at N = 82 is at a closed neutron shell in which the destruction reactions are at a minimum. This produces an abundance peak at ^{138}Ba, which has the lowest neutron capture cross section of the *s*-nuclei. ^{142}Nd is an *s*-only species due to shielding from the *r*-process by Ce^{142}. There are several branchings in this region, including a minor and diagnostically important one across Nd^{147} (half-life: 11 days) to stable Nd^{148}. The most neutron-rich isotope, Nd^{150}, is an *r*-only species because Nd^{149} is too short-lived for branching. Decay to stability of unstable neutron-rich nuclei generated in the *r*-process follows the diagonal arrows. Flow in the *r*-process is far to the neutron-rich side of stability. The location of the *r*-process abundance peak at ^{128,130}Te and ^{129}Xe, as shown in Figures 1 and 3, indicates that it passes through the N = 82 closed shell at ^{128}Pd (Z = 46), ^{129}Ag (Z = 47), and Cd^{130} (Z = 48).

stellar medium as red giant winds and "planetary nebulas."

Carbon Burning. Follows He-burning in stars with initial masses greater than about eight solar masses where gravitational contraction drives central temperatures high enough to fuse C^{12}. Further α and p captures synthesize additional nuclei up to A = 28.

C-burning is followed by further contraction of the stellar core and a brief O-producing episode of neon photodissociation.

Oxygen Burning. Ensues after further contrac-

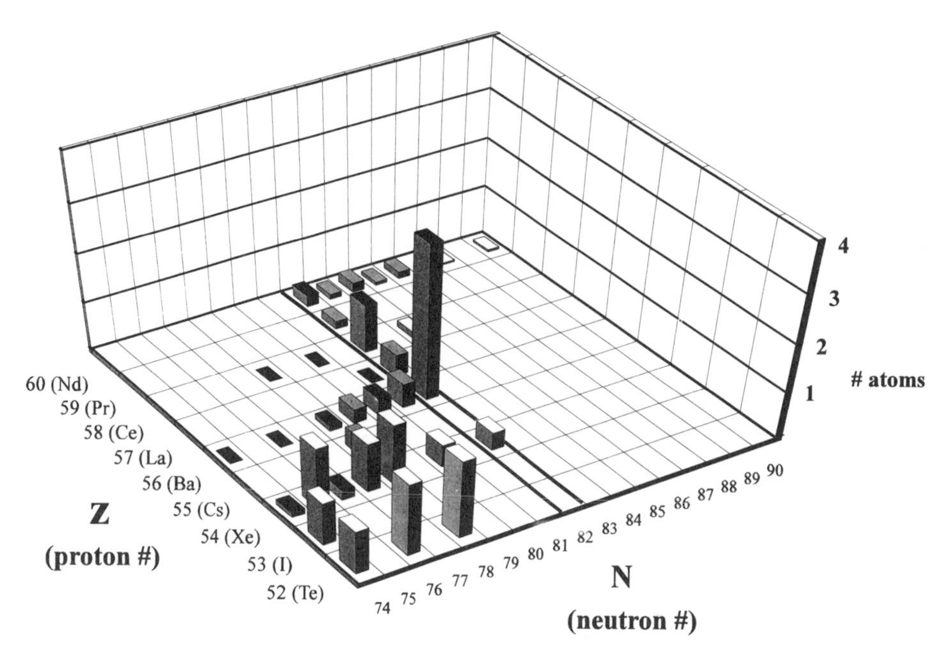

Figure 3. Here the N-Z plane in the Te-Nd region is shown with a vertical scale indicating the solar atom abundances (on a linear scale normalized to Si = 10^6). The tops of the "buildings" are differently shaded to indicate nucleosynthetic origin. White indicates *r*-only nuclei or nuclei with abundances dominated by the *r*-process component. Black indicates the same for the *s*-process. Dark grey indicates p-only nuclei. Mixed-source nuclei are shown in light grey. The *s*-process abundance peak seen in Figure 1 at mass 138 can be seen to be dominated by ^{138}Ba on the closed N = 82 neutron shell. This isotope has a very small neutron capture cross section, which limits its destruction in the *s*-process and is produced by neutron capture on ^{137}Ba, which has a much higher cross section. The peak in *r*-process abundances in this region is offset from N = 82 by 8–10 mass units (to ^{128}Te-^{129}Xe-^{130}Te) as a consequence of the *r*-process nucleosynthesis path being far to the neutron-rich side of stability. For comparison, the (lower level) *s*-process abundances in this region are visible at the shielded *s*-only nuclei ^{128,130}Xe and ^{134,136}Ba. Maximum production in the *r*-process occurs also on the N = 82 closed neutron shell (where neutron capture cross sections are relatively small and half-lives against beta decay are relatively high), but far to the neutron-rich side of stability. The abundance peak corresponds to production of the unstable precursor isotopes ^{128}Pd-^{129}Ag-^{130}Cd (which beta decay to ^{128}Te-^{129}Xe-^{130}Te). Neutron densities high enough to drive neutron capture nucleosynthesis this far from stability are believed to occur deep inside supernovas.

tion of massive stars, with alpha and proton captures synthesizing additional nuclei up to A = 40.

Silicon Burning. Is the final nuclear burning stage in stars. It commences at approximately 3 billion degrees Kelvin with the photo disintegration of nuclei in the A = 28 region and capture of the resulting protons, neutrons, and alpha particles by residual nuclei in a process of nuclear statistical equilibrium. The products of Si-burning exhibit a prominent abundance peak at Fe^{56} in the solar abundance distribution.

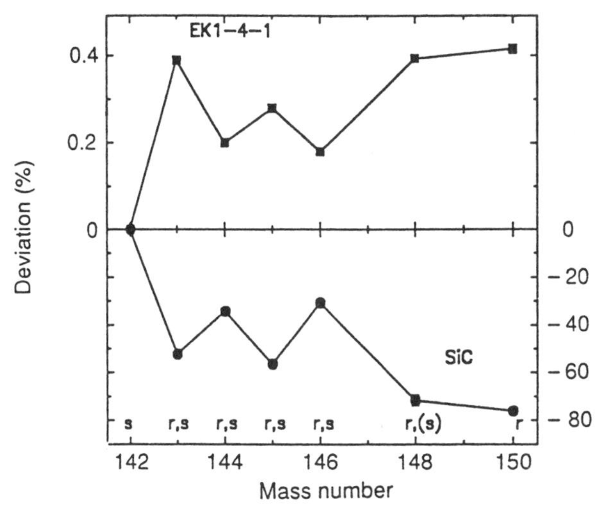

Figure 4. Mirror isotopic anomalies in meteoritic materials for neodymium obtained by mass spectrometry. These are expressed as deviations with respect to normal solar system Nd and indexed to mass-142, which is produced uniquely in the *s*-process. Mass-150 is produced uniquely in the *r*-process. All other isotopes have a mixed derivation. The top pattern is for a high-temperature refractory inclusion (EK1-4-1) from the Allende meteorite, and exhibits a small *r*-process enrichment pattern. The bottom pattern was obtained for silicon carbide (SiC) stardust separated from the Murchison meteorite and shows a nearly pure *s*-process composition. These observations empirically confirm theoretical model decompositions of the heavy nuclei into distinct *r*- and *s*-process components. The SiC stardust is believed to have condensed around red giants and was incorporated as part of the presolar "cosmic sediment" preserved in primitive meteorites. The chemical history of the inclusion EK1-4-1 is not well understood, but implies some environment or process that did not participate in the efficient homogenization of presolar components reflected in the uniformity of heavy isotopic compositions in most planetary materials. The figure is reproduced from Ott (1993).

Mass-accreting white dwarf stars approaching the Chandrasekhar mass are the biggest powder kegs in the universe and explode as Type Ia supernovas after ignition by initiation of a "spark" of core C-burning. A nuclear flame proceeds through the star on a timescale of seconds driving C-, O-, and Si-burning, and generating about 0.7 solar masses of radioactive Ni^{56} ($T_{1/2}$ = 6.1 d). Ni^{56} decays to Fe^{56} via Co^{56} (77.3 d) and powers the supernova light curve with a characteristic exponential decay law. These events are believed to be responsible for roughly half of the iron in the solar system.

Type II supernovas are the terminal result of sequential nuclear burning stages that leave behind a layered structure of ashes in the presupernova star: (H)-(He)-(C, O)-(O, Ne, Mg)-(O, Mg, Si)-(iron peak nuclei). Once Si is burned to exhaustion in the core (in weeks to days), no further energy supply is available to inhibit further core collapse and a catastrophic core collapse implosion squashes the Fe-group ashes into a compact neutron star or black hole. Conversion of gravitational potential energy into neutrino luminosity probably drives the supernova explosion of the overlying star. Compression during passage of the resulting shock wave initiates explosive burning stages in the overlying onion-structured star, with explosive Si-burning merging into a high-temperature nuclear statistical equilibrium (*e*-process) synthesis of a new suite of Fe peak elements. These are ejected into the interstellar medium along with further products of explosive burning reactions at higher levels as well as unburned ashes from the preexplosive burning stages. Modeling of the evolution of massive stars into Type II supernovas can explain the relative abundances of most of the major rock-forming elements in the solar system, as shown in Figure 5.

Isotopic anomalies has been observed in meteorite inclusions in a suite of neutron-rich nuclides crossing the iron peak (Ca^{48}, Ti^{50}, Cr^{54}, Fe^{58}), and in three heavier nuclei ($Ni^{62,64}$ and Zn^{66}). This pattern of anomalies is the result of excesses of material produced in neutron-rich nuclear statistical equilibrium in explosive nucleosynthesis in supernovas.

The synthesis of elements heavier than iron in supernovae involves four additional processes. Nuclei that can be attributed uniquely to a specific process are described by an identifying letter representing that process, for example, as an *r*-only species in the case of the *r*-process. The least abundant heavy nuclei are a consequence of the *p*- or γ-process and the ν-process (ν = neutrino), both of which involve mass loss from preexisting heavy nu-

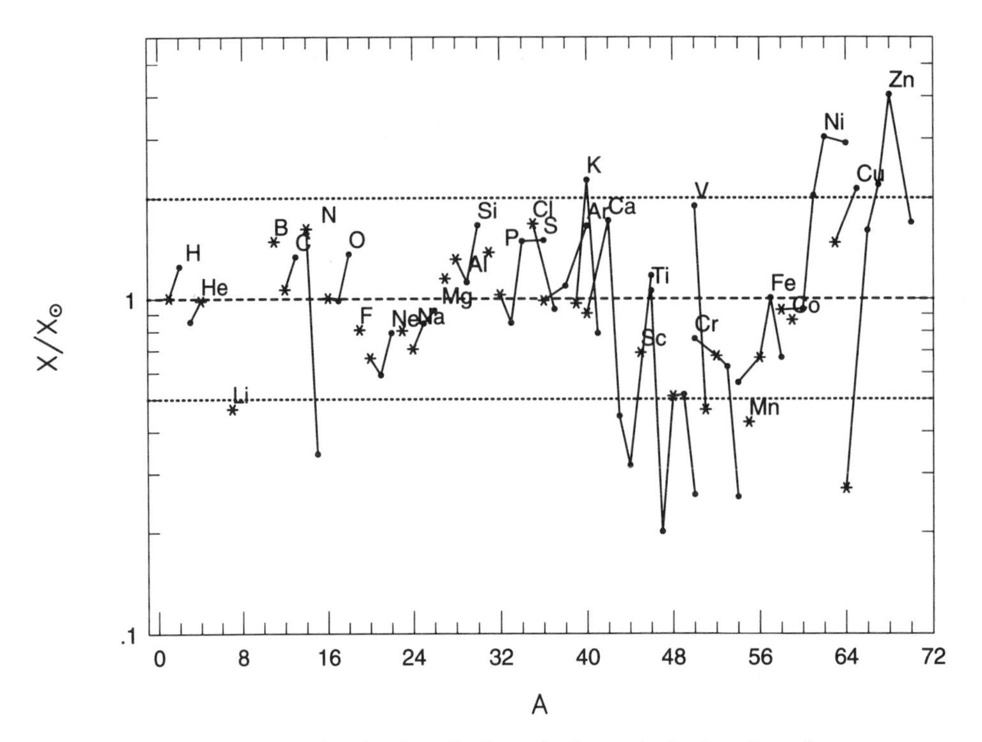

Figure 5. Results of a galactic chemical evolution calculation based on modeling of nucleosynthesis in an integrated range of core collapse supernovas representative of the mass distribution of massive stars formed in the galaxy. Abundances normalized to solar are plotted on the y-axis against atomic mass (A) on the x-axis. Isotopes of each element are joined by a line, and the most abundant isotope is indicated by an asterisk. The dashed horizontal line indicates a fit to the solar distribution. It can be seen that the model fit is quite good for most nuclei to within a factor of two. Thus, most of the major rock-forming elements are attributable to nucleosynthesis in core collapse supernovas. The figure is reproduced from F. X. Timmes, S. E. Woosley, and T. E. Weaver, "Galactic Chemical Evolution: Hydrogen through Zinc," *Astrophysical Journal Supplement* 98 (1995):617–658.

clei, by photo disintegration [(γ, n), (γ, p), (γ, α)], and neutrino spallation [(ν, n + ν'), (ν, p + ν'), and (ν, α + ν')], respectively. The ν-process occurs in core collapse supernovae (which have very high neutrino luminosity), and may be the predominant source of Li^7, B^{11}, F^{19}, La^{138}, and Ta^{180} in the solar system. It may also produce significant amounts of B^{10}, N^{15}, Al^{26}, P^{31}, Cl^{35}, $K^{39,40,41}$, Sc^{45}, $Ti^{47,49}$, $V^{50,51}$, Mn^{55}, Co^{59}, and Cu^{63}, and be responsible for reducing the odd-even effect in the rapid neutron capture process. The γ- or *p*-process necessarily occurs in all supernovas by photo disintegration of preexisting nuclei within a limited range of temperatures and pressures. The *p*-process contribution to the solar abundance distribution is small and significant only in the *p*-only nuclei (which are shielded from beta decay by stable nuclei and do not lie on the *s*-process path). Meteorite studies, however, have identified two important *p*-process cosmochronometers, Nb^{92g} and Sm^{146}.

A rapid neutron capture *r*-process is required to explain a component present in the nuclei that are unshielded with respect to beta decay. The distribution of *r*-nuclei can be seen in the *r*-only nuclei shown in Figure 1. Synthesis of these species requires neutron captures at extremely high neutron densities sufficient to generate enough sequential captures on seed nuclei to produce actinides up to at least mass 244. Accurate modeling of the *r*-process is critical for dating the galaxy by nuclear cosmochronology using the unstable *r*-only nuclei Th^{232}, U^{235}, U^{238}, and Pu^{244}. The pattern of the *r*-only nuclei exhibits a triple peak structure at Se^{80}, Te^{130}, and Pt^{195}. This exhibits an offset to lower masses relative to the three *s*-process abundance peaks. Both sets of peaks correspond to nuclear reactions occurring at closed neutron shells in the nucleus at neutron numbers of 50, 82, and 126. In the *s*-process these reactions produce stable nuclei at these closed shell locations on the *s*-process path. The offset in the case of the *r*-nuclei indicates that the *r*-process synthesis path passed through closed neutron shells far from stability through the unstable but relatively long-lived species Zn^{80}, Cd^{130}, and Yb^{195} in a $(n, \gamma) \leftrightarrow (\gamma, n)$ equilibrium. This indicates neutron densities in excess of 10^{20} cubic centimeters (cm^{-3}). Such conditions can only exist in extreme explosive conditions as in the deep interiors of supernovas. A promising site for the *r*-process is associated with rapid α-capture synthesis of elements above the Fe peak. Under conditions which may exist in Type II supernovas, the α-process is expected to be neutron-rich and merges into an *r*-process in "winds" blown off of the surface of the proto-neutron star by neutrino pressure. There has been no observational confirmation of this model, but a source in Type II supernovas would be consistent with the early growth trend of Eu relative to Ba in very old stars. (Eu is predominantly an *r*-element, whereas Ba is predominantly an *s*-element.)

A New Astronomy from Meteoritic Grains

Meteorite data can contribute importantly to the refinement of nucleosynthetic models. For example, some meteorite inclusions possess *r*-process anomalies (Figure 4), which correlate with neutron-rich *e*-process isotopic anomalies in elements in the Fe-peak region. This suggests an association between the *r*-process and *e*-process sites, as predicted in core collapse supernova models. However, it is by no means clear that the anomalies in large objects represent a spatially associated sampling of in-mixed material from a single supernova. Such associations more likely might be the result simply of similar chemical properties in presolar grain populations deriving from distinct formation sites. The solution to this problem is found in the study of individual supernova and circumstellar condensate grains isolated from meteorites. The discovery of such grains and their analysis by ion microprobe mass spectrometry and other microtechniques has opened up a new window in astronomy with profound implications for the science of nucleosynthesis. Though technically demanding, this is an exciting new research field in which the earth sciences will continue to contribute important discoveries enhancing our understanding of nucleosynthesis and the origins of the elements.

Bibliography

ANDERS, E., and N. GREVESSE. "Abundances of the Elements: Meteoritic and Solar." *Geochimica et Cosmochimica Acta* 53 (1989):197–214.

ANDERS, E., and E. ZINNER. "Interstellar Grains in Primitive Meteorites." *Meteoritics* 28 (1993):490–514.

BURBRIDGE, G. R., E. M. BURBRIDGE, W. A. FOWLER, and F. HOYLE. "Synthesis of Elements in Stars." *Reviews in Modern Physics* 29 (1957):547–650.

CLAYTON, D. D. *Principles of Stellar Evolution and Nucleosynthesis.* Chicago, IL, 1984.

FOWLER, W. A. "The Quest for the Origin of the Elements." *Science* 226 (1984):922–935.

GILMORE, G., I. R. KING, and P. C. VAN DER KRUIT. *The Milky Way as a Galaxy.* Mill Valley, CA, 1990.

KIRSHNER, R. P. "The Earth's Elements." *Scientific American* (October 1994):59–65.

OTT, U. "Interstellar Grains in Meteorites." *Nature* 364 (1993):25–33.

PEEBLES, P. J. E. *Principles of Physical Cosmology.* Princeton, NJ, 1993.

PENZIAS, A. A. "The Origin of the Elements." *Science* 205 (1979):494–554.

ROLFS, C. E., and W. S. RODNEY. *Cauldrons in the Cosmos: Nuclear Astrophysics.* Chicago, IL, 1988.

WOOSLEY, S. E., and T. A. WEAVER. "The Great Supernova of 1987." *Scientific American* (August 1989):32–40.

CHARLES L. HARPER, JR.

OBSERVATIONAL ASTRONOMY

A clear mild evening with a sparkling sky is one of life's most cherished happenings. It is also one of life's most ignored events, for as the starry sky shines outside, we remain inside. Observing the sky is one of the easiest things to do, whether with naked eye, a pair of binoculars, or a small telescope. It requires no more effort than spending a quiet evening outside getting acquainted with our immediate neighborhood of the Moon and the bright planets, and then looking outward to see the city of stars we call the Milky Way galaxy.

Observing with the Naked Eye

The evening sky is a panorama where things happen at varying distances. You might see a "shooting star"—properly called a meteor, whose flash of light is the interaction of a tiny particle with Earth's upper atmosphere some 60 kilometers (km) above us. There will probably be an artificial satellite wandering slowly across the sky more than a hundred km up. After that we take a leap into space for the next closest object, the Moon, which is some 380,000 km out, and Jupiter, the largest planet, which is some 800 million km away.

Curiously enough, the stars—the farthest things we usually see—are the easiest to follow. They form patterns in the sky (called constellations) that do not change historically, so that constellation figures dreamed up five thousand years ago can be visualized by a child. Even under a bright city sky, these simple figures are easy to see: Orion the hunter in the first part of the year, Leo the lion in March and April, the "Summer Triangle" in June and July, and the Great Square of Pegasus in October. For Northern Hemisphere dwellers, the Big Dipper is usually in the sky, as is the Southern Cross for people in the Southern Hemisphere.

Comets and Meteors

It is not often that a comet can be seen in the sky without the aid of a telescope. In recent years there have been only a few: Ikeya-Seki in 1965, Bennett in 1970, West in 1976, Halley in 1986, and Levy in 1990. But on some nights you can see the residue from comets as showers of meteors. The best of the showers are shown in Table 1.

Binoculars

A simple pair of binoculars allows an observer to see the sky in much greater detail. To get the most out of using binoculars, you should hold them comfortably while sitting in a lawn chair or some other relaxed position. Then, the Milky Way, a band of milky light to the naked eye, becomes a rich panorama of faint stars, star clusters, and faint

Table 1. Meteor Showers and Associated Comets

Name of Shower	*Best Evening*	*Associated Comet*
Lyrids	21 April	
Eta Aquarids	4 May	Halley
Delta Aquarids	30 July	
Perseids	11 August	Swift-Tuttle
Orionids	20 October	Halley
Taurids	31 October	Encke
Leonids	16 November	Tempel-Tuttle
Geminids	13 December	(asteroid) Phaethon

clouds called nebulae. And when you look at Jupiter, you might see the four moons that Galileo discovered in 1610.

Telescopes

A well-made small telescope will open huge vistas. But there are some precautions: do not buy a telescope on a wobbly mounting, and do not pay attention to the manufacturer's claims of high power. Magnification is not as important in a small telescope as its ability to gather light and focus its image clearly; in fact, telescopes should not use more than 60 power per inch of aperture; thus a 10-cm-diameter telescope does not need magnifications higher than 240.

The two basic kinds of telescopes are the refractor and the reflector. A refractor uses an objective lens that gathers light from the sky and bends it to a focus where an eyepiece magnifies the light. A reflector uses a concave mirror that reflects the light back through the tube to a smaller mirror, and then to the eyepiece. The Schmidt–Cassegrain is a popular type of a reflector that uses both a lens and a mirror to gather light, which is then reflected to a small mirror and then sent back through a hole in the main mirror to an eyepiece.

Magnitudes

The first thing you notice on looking skyward on a clear night is that the stars are not all the same brightness. Stars are fainter the farther away they are, and some stars are intrinsically more luminous than other stars. Our system of magnitudes (a quantitative measure of brightness) dates back to the second century B.C.E. Greek astronomer Hipparchus, who divided the stars into six brightness sets with the 20 brightest stars being called first magnitude and the faintest stars sixth magnitude. A first magnitude star is defined as being 100 times brighter than a sixth magnitude star, which is traditionally the faintest star visible without a telescope. A second magnitude star is 2.5 times fainter than a first magnitude star, a third magnitude star, in turn, is 2.5 times fainter again.

Vega, the brightest star in the summer triangle, is a zero magnitude star. Polaris, the north star, as well as most of the stars in the Big Dipper, is second magnitude.

Limiting magnitude is the brightness of the faintest star that one can see. On a clear night away from city lights, the limiting magnitude for the naked eye is about six. With binoculars, the limiting magnitude is typically eight, although that depends on the power of the binoculars, the condition of the sky, and the acuteness of the observer's eyes. The typical limiting magnitude in a suburb is three or four. In the middle of a big city, only the brightest first magnitude stars might be visible. Limiting magnitudes with telescopes depend on the telescope's size. Traditionally, with a 20-cm-diameter telescope you should be able to see fourteenth magnitude stars, though good observers can see fainter.

When a star is rising or setting, it fades as its light shines through a thicker amount of atmosphere than when it is high in the sky. Even the Sun fades when it is near the horizon. This effect is called atmospheric extinction.

Observing the Planets

Each month astronomical magazines like *Sky and Telescope* and *Astronomy* publish the positions of the bright major planets. Since these planets move across the background sky, they cannot be plotted on permanent star charts. The planets trace paths among the stars that you can easily follow over a few weeks. As Mars, Jupiter, and Saturn orbit the Sun, they appear to move eastward across the sky. But as Earth overtakes these planets as it travels about the Sun, they slow down, and then reverse course for a few months. We call this effect retrograde motion.

Since Mercury and Venus are closer to the Sun than we are, they travel different types of paths through the sky. When they are visible in the sky they appear close to the Sun, never in the midnight sky. Mercury's greatest elongation is 28 degrees, and Venus's greatest elongation is 47 degrees from the Sun (*see* MERCURY; VENUS).

Jupiter is the most rewarding planet to see through a telescope. Its two brownish equatorial belts are visible through almost any-sized telescope, and more and more details, including its Great Red Spot, become evident through larger instruments (*see* JUPITER AND SATURN).

In 1993, Gene and Carolyn Shoemaker and David Levy discovered a comet that Jupiter's gravity had split apart into twenty-one pieces. In July of 1994, these fragments collided with Jupiter, producing the biggest explosions ever witnessed on another planet. Dark spots larger than Earth were visible using small telescopes, and a dark band was still present a year later (*see* COMETS).

Saturn is a joy to observe, its enormous ring a highlight through any telescope. However, because the angle of the rings changes as Saturn orbits the Sun, they are almost invisible for a few months every fourteen years (*see* JUPITER AND SATURN).

Mars is at its best only when its orbit and the Earth's orbit bring the two planets within 80 million km of each other. Then you can see dark and bright markings called albedo features, with exotic names like Syrtis Major (a dark patch that resembles the Indian continent), and Solis Lacus (the "eye of Mars"). Finally, Venus and Mercury are worth a look through a telescope, if only to see their phases (*see* MARS).

Observing Deep Sky Objects

Double stars, clusters of stars, nebulae, and distant galaxies are beautiful objects that await the gaze of a patient observer (*see* STARS; GALAXIES). Double stars can be striking through small telescopes. Albireo, the star opposite Deneb in Cygnus, the Swan, is a beautiful pair of stars. Although the two stars of Albireo are not physically related, they are so close together in the sky that they appear as a set of brightly colored stars.

There are two types of star clusters. Open clusters appear as densely populated regions of stars, and globular clusters look like small round hazes of light. The clouds of gas and dust that we call nebulae look like irregular patches of fuzziness. Finally, there are many distant galaxies in space. Although most of these look like faint fuzzy spots, some, like the Andromeda galaxy, are bright enough to be seen through binoculars. The best of these objects were catalogued by Charles Messier, an eighteenth century comet hunter. Spread over most of the sky, these showpieces give rewarding views through small telescopes.

Improving Your Observing Ability: Getting Dark-Adapted

Going from a well-lit room to a dark sky involves an adjustment that your eyes cannot do immediately. On going outside you may first see a bright sky with very few stars, but within a few minutes, as your eyes adapt to the darkness, you will see more of the fainter stars. Give yourself time to enjoy an observing session.

Transparency and Seeing

These two factors affect what you see in the sky. Transparency is simply how clear the sky is: a crystal clear night right after a cold front passes through is more transparent than a night with haze or cirrus clouds. Seeing refers to how steady the sky is. When the stars twinkle, that usually means that the seeing is poor, resulting in fuzzy views of planets and stars at the telescope.

Light Pollution

Although a few marvels of the deep sky are visible from city skies, we recommend a dark sky away from city lights for the best views. As a result of poor design and lots of glare from urban lights, it is impossible to see any but the sky's brightest stars, the Moon, and some bright planets from most downtown areas.

The problem with city lighting is not the light fixtures themselves, but where they are pointed, and the blinding glare they produce. Some cities, particularly Tucson, Arizona, now have lighting ordinances that specify that lights must be shielded from direct glare, an idea that results in safer lighting as well as better views of the night sky for skywatchers.

Whether you observe from a brightly lit city or from a pristine site in the country; whether you have a large telescope or just a pair of binoculars, give the night sky a chance. It might reward you with a lifetime of learning and fun.

Bibliography

LEVY, D. H. *The Sky: A User's Guide.* Cambridge, Eng., 1991.

RIDPATH, I. *Norton's 2000.0.* New York, 1989.

DAVID H. LEVY

OCCULTATIONS

See Eclipses and Occultations

OCEAN BASINS, EVOLUTION OF

Ocean basins evolve as a direct and inevitable consequence of the global plate tectonic cycle. The outer shell of the earth, the lithosphere, behaves in an essentially rigid manner on short to intermediate timescales. It is divided into a number of large units known as plates. These plates move slowly about the face of the earth and interact almost exclusively at their boundaries. Earthquakes result almost entirely from plate boundary interactions and form bands that outline the edges of the major plates. Ocean basins form at one type of plate boundary (constructive boundaries) and are consumed or destroyed at others (destructive boundaries).

Constructive plate boundaries are the centers of seafloor spreading that occur in a near-central location in most of the world's ocean basins (see Plate 30). It is here that the material that floors the ocean basins is created. Consider the boundary that lies almost exactly in the center of the South Atlantic Ocean. To the west is the South American plate, to the east is the African plate. In a relative sense these two plates are moving away from one another in a general east–west sense: the South American plate moves west, the African plate moves east. Because the plates are rigid, a potential void must be created in the boundary region where they form. This region is the spreading center.

In fact, no void is created. In a continuum process, the horizontal plate motion gives rise to a vertical ascent of the less rigid mantle material beneath. In effect, the "void" is continuously filled with rising mantle material. It is this process that creates new ocean basins. As the mantle rises, it begins to melt. Melting is not complete as only part of the mantle rock actually becomes liquid, but that liquid separates from the solid matrix and moves toward the surface because it is much more buoyant than the solid from which it melted. In fact, the solid does not allow easy flow of the liquid component and is retained enough to add its buoyancy affect to the upward motion of the mantle, enhancing the flow. Additionally, by a process that is not yet completely understood, the flow of melt toward the surface becomes highly concentrated and reaches the surface in a very narrow zone less than 1 km wide, known as the neo-volcanic zone of the spreading center. It is here that the new ocean floor is created.

When the melt reaches near the surface, it fills a small magma chamber where it resides for a short while. These magma chambers can be "seen" by a geophysical technique known as seismic reflection imaging. The chamber is the source region for magma that shoots upward to form dikes and extrusives, and also the source of a large body of plutonic rock that slowly cools in place. These three components form the basic structure of oceanic crust, a rock layer about 6 km thick that is the floor of ocean basins.

The crust is created at a narrowly focused region of magmatism and deformation at the spreading center. The region is relatively elevated and forms the mid-ocean ridges that are the central mountain ranges of most ocean basins. Apart from the regions immediately adjacent to the continents, the mid-ocean ridges are the shallowest parts of the ocean basins, a discovery made in the early part of the twentieth century when ships were laying telephone cables on the seafloor between the United States and Europe. The depth of ocean basins increases away from the ridges in a gentle, systematic way that is a result of the cooling of the underlying layers as they move away from the spreading centers. In fact, the deepening is not so much related to distance from the spreading axis but to the time since formation of the element of seafloor. The depth of the seafloor is proportional to the square root of the time since formation.

On an Earth of finite size, the creation of new ocean basins by seafloor spreading must be balanced by some process that consumes an equivalent amount of seafloor. This process, called subduction, occurs when one of the plates dives beneath another; in effect, the plate reenters the mantle from which it was originally derived, where it is then reassimilated. Subduction is far from a simple process and strong interaction of plates occurs as they slide against one another. Parts of the plate being consumed are scraped off, the leading edge of the overriding plate is highly deformed, and magma is derived from melting of the wedge behind the subduction zone, causing massive volcanism along the plate edge. The Andean Mountains along the western coast of South America result from subduction processes associated with the consumption of the Nazca plate beneath the South American plate. Occasionally, the collisional deformation is so extreme that a slice of the oceanic crust of the seafloor is thrust up onto the land to form a structure known as an ophiolite. These rate structures (examples occur in Cyprus, Oman, and Newfoundland) give earth scientists special insight into the process of seafloor creation.

Rarer still are glimpses into the process by which a new ocean basin comes into creation. Ocean basins form by the rifting apart of two continental fragments. South America and Africa were once joined, and the South Atlantic ocean basin formed by the initiation of seafloor spreading along the mid-ocean ridge in what is now the center of that basin. This rifting and initiation of spreading must have happened numerous times in the history of Earth but are taking place in very few places on Earth today. One place is off the eastern end of Papua New Guinea, where data collected in the last several years have shown the process of initiation of spreading to involve a complex pattern of propagation (see Plate 31). The Woodlark Basin is progressively growing in a westerly direction, unzipping the very thick crust of the Papuan Peninsula. There is evidence to suggest that rifting and the initiation of a new ocean basin may commonly be associated with propagation of basin formation. Such propagation may be required as the continental lithosphere is not weak and does not rupture easily. Hence a clean synchronous breakage along thousands of kilometers as would be required to break open the South Atlantic, for instance, would take forces much greater than are available. In a propagation process, stresses are concentrated at the tip of the rift, making breakage much more likely.

Bibliography

Mutter, J. C., and J. A. Karsan. "Structural Processes at Slow-Spreading Ridges." *Science,* no. 257 (1992): 627–634.

Taylor, B., A. Goodliffe, F. Martinez, and R. Hey. "Continental Rifting and Initial Sea-Floor Spreading in the Woodlark Basin." *Nature,* no. 374 (1995):534–572.

JOHN MUTTER

OCEANIC CRUST, STRUCTURE OF

The bottoms of the oceans are covered nearly everywhere by sediments; oceanic crust directly underlies these sediments. Oceanic crust consists of sedimentary rocks, basalts, diabases, gabbroic rocks, ultramafic rocks, and their metamorphic equivalents (metabasalts, metagabbros, amphibolites, and serpentinites; *see* IGNEOUS ROCKS). The oceanic crust averages about 7 kilometers (km) in thickness, but locally can be much thinner (less than 2 km thick) due to tectonic activity. Oceanic crust directly overlies Earth's mantle.

Oceanic crust is created at mid-ocean ridge spreading centers, which form a 65,000-km-long network of immense mountain ranges on the ocean floor. The most well known of these mountain ranges are the Mid-Atlantic Ridge, which runs north to south along the center of the Atlantic Ocean, and the East Pacific Rise in the eastern part of the equatorial Pacific Ocean. At these spreading centers, new oceanic crust is continuously created by seafloor spreading in which the two oceanic plates move apart at the rate of 5–20 centimeters (cm)/year. Because the crust moves away from the mid-ocean ridge as it forms, the age of the oceanic crust is youngest (zero age) at the mid-ocean ridges and becomes progressively older with increasing distance from the mid-ocean ridge. The oldest oceanic crust is Jurassic (180–190 million years old or Ma) and is located in the western Pacific Ocean. Although oceanic crust has been produced by seafloor spreading since early in Earth history, most has been subducted into the mantle and is no longer on Earth's surface (*see* OCEAN BASINS, EVOLUTION OF).

The known structure of the oceanic crust is constrained by geophysical studies of the crust and by studies of rocks recovered from the seafloor by drilling, dredging, or manned submersible programs. An additional major influence on most studies of the structure of oceanic crust has been observations obtained from ophiolites, which are fragments of ocean-floor-like crust emplaced in mountain ranges.

Geophysical studies of the structure of oceanic crust most commonly involve seismic reflection and refraction studies in which the velocity of seismic waves being transmitted through the crust is used to constrain its structure. In geophysical studies, the ocean crust is divided into several major layers, which are numbered from top to bottom. The different layers, defined on the basis of geophysical studies, can be linked to specific rock types on the basis of rocks recovered from the ocean basins. Studies of the ocean floor by dredging, drilling, and manned submersibles have shown that the most common types of rocks within the oceanic crust are basalts, diabases, gabbroic rocks, and ultramafic rocks (including their metamorphic equivalents). The knowledge of distribution of these rock types within the oceanic crust and their origin is constrained by studies from ophiolites and the oceans.

At the top of the oceanic crust is Layer 1, which has low seismic velocities (1.5–3.4 km/s) and is considered to be composed mostly of oceanic sediments and sedimentary rocks. Layer 1 is commonly absent where new oceanic crust forms at mid-ocean ridges and thickens with distance from the mid-ocean ridge as sediments are progressively deposited on the oceanic crust. Beneath these sediments and sedimentary rocks are basaltic rocks, which occur as pillowed lavas or lava flows that erupt onto the floor of the ocean, flow laterally, and are cooled quickly by the adjacent seawater. Diabases occur as isolated dikes or, more commonly, as sheeted dikes that form by seafloor spreading. These basaltic and diabasic rocks form Layer 2, which has intermediate seismic velocities (3.5–6.0 km/s). Layer 3 has higher seismic velocities (6.5–7.8 km/s) and consists of gabbroic rocks, possibly with some cumulate ultramafic rocks. The gabbroic rocks are coarse-grained mixtures of the minerals olivine, plagioclase, and pyroxene (*see* IGNEOUS ROCKS). The gabbroic rocks are formed by the crystallization of basaltic liquids in magma chambers located within the oceanic crust. The bottom of Layer 3 is the base of the oceanic crust; beneath Layer 3 is Earth's mantle, which typically has seismic velocities greater than 8.1 km/s and consists of ultramafic rocks. The ultramafic rocks, often called abyssal peridotites, are coarse-grained mixtures of olivine, pyroxene, and spinel. These rocks are upper-mantle samples that represent residues of the partial melting processes that produce basaltic liquids.

The formation of oceanic crust occurs at mid-ocean ridges as a consequence of melting of Earth's mantle to produce magmas. In the context of plate tectonic processes, convection in Earth's mantle results in areas where the mantle is rising and is locally at a higher temperature than the surrounding mantle. As the convecting mantle rises beneath mid-ocean ridges, it begins to melt due to decompression at about 100 km below the seafloor and continues melting during ascent up to approximately 20 km below the seafloor. Melting of the mantle produces high-temperature (1,300–1,450°C) primary basaltic liquids, which accumulate at various depths within the mantle and then rise to form the oceanic crust.

The continuous upwelling of mantle beneath mid-ocean ridges produces a semicontinuous supply of basaltic magma that forms the oceanic crust. Basaltic magmas rise from the mantle and accumulate in the oceanic crust to form magma chambers a few kilometers beneath the seafloor. Crystallization of basaltic liquids in these magma chambers occurs because of the lower temperature, thereby producing cumulate gabbroic and ultramafic rocks. It is the episodic eruption of residual basaltic magmas from these crustal magma chambers that produces the diabase dikes, pillowed lavas, and lava flows that form the upper part of the oceanic crust. The formation of oceanic crust, therefore, is the end result of the differentiation of mantle-derived primary basaltic liquids in which crystallization produces cumulate rocks that form Layer 3, with the residual basaltic liquids that remain after crystallization erupted from the top of the magma chamber to form Layer 2.

The amount of oceanic crust at a mid-ocean ridge spreading center and the composition of the crust are clearly linked to the composition of the mantle and the average depth at which melting occurs. If the average depth of melting is approximately 100 km beneath the seafloor, the composition of the oceanic crust in that locality would be that of a basalt with 15–18 percent MgO (see Table 1, column 1). If the average depth of melting is approximately 30 km beneath the seafloor, however, the composition of the oceanic crust in that locality would be similar to that of a basalt with 10–12 percent MgO (see Table 1, column 2). The composition of the crust at a specific locality could be determined by adding up the compositions of the basaltic, diabasic, and gabbroic rocks at that locality, but such information is not available because systematic sampling of the deeper parts of the gabbroic rocks has not been very successful. Because the compositions of mid-ocean ridge basalts differ from one location to another, however, it is likely that the composition of the crust differs significantly from one location to another.

In spite of these uncertainties in the composition of the oceanic crust, it is clear that the oceanic crust composition differs substantially from the continental crust composition. Estimates of continental crustal composition (see Table 1, columns 3–5) have high SiO_2 and Al_2O_3. The continental crust is often refered to as "sial" because of high abundances of the elements silicon and aluminum. Similarly, the oceanic crust is often referred to as "sima" because of the high abundances of silicon and magnesium.

Superimposed on the magnetic processes involved in creating oceanic crust is the formation of active hydrothermal circulation cells, particularly within the upper part of the oceanic crust. Tectonic activity at mid-ocean ridges and the strong temperature difference between hot oceanic crust and cold ocean water cause the oceanic crust to become pervasively cracked so that seawater actively circulates throughout the upper parts of the oceanic crust. This circulation of hydrothermal fluids occurs even in deeper levels of the oceanic crust, but the amount of water circulating is greatly diminished. The circulation of seawater through the oceanic crust causes the crust to cool, changes the composition and mineralogy of the crust, and causes active hydrothermal venting onto the seafloor.

Table 1. Estimated Compositions of Ocean Crust and Continental Crust

	1	*2*	*3*	*4*	*5*
SiO_2	47.7	49.5	58.0	64.8	57.3
TiO_2	0.85	0.82	0.8	0.51	0.9
Al_2O_3	13.4	15.2	18.0	16.1	15.9
FeO	9.58	8.15	7.5	4.8	9.1
MgO	16.9	10.7	3.5	2.7	5.3
CaO	9.94	12.2	7.5	4.6	7.4
Na_2O	1.07	1.94	3.5	4.4	3.1
K_2O	0.05	0.16	1.5	2.0	1.1
Total	99.5	98.6	100.3	99.9	100.1

Column 1: Composition of high-MgO basalt formed at 100 km depth that could differentiate to produce oceanic crust.
Column 2: Composition of low-MgO basalt formed at 30 km depth that could differentiate to produce oceanic crust.
Columns 3–5: Estimated compositions of the continental crust.

Bibliography

NICOLAS, A. *Structure of Ophiolites, and Dynamics of Oceanic Lithosphere.* Dordrecht, Netherlands, 1989.

SINTON, J. M., and R. S. DETRICK. "Mid-Ocean Ridge Magma Chambers." *Journal of Geophysical Research* (1992):197–216.

DON ELTHON

OCEANOGRAPHIC EXPEDITIONS

The average depth of the world's oceans is approximately 4.5 kilometers (km), posing a great challenge for scientists wishing to study the seafloor. Oceanographers have had to devise clever ways to sample and survey this vast and remote terrain. This entry provides highlights of several memorable oceanographic expeditions and the technologies scientists utilized to explore the ocean's depths (Kennett, 1982; Hsü, 1992).

Seafloor Depth and Shape

Early students of the oceans were unaware of their great depths because rope, thrown over the side of the ship as a measuring device, was inevitably too short to reach the ocean floor. One of the first significant ocean basin measurements came in 1840, during an Antarctic expedition of the HMS *Erebus* and *Terror*, when Sir James Ross measured a bottom depth of almost 4.44 km. During the 1840s, the needs of the prosperous shipping industry prompted the federal government to make the first bathymetric maps of the inner part of the U.S. continental shelf. Shortly thereafter, an influential navy lieutenant, Matthew Fountain Maury, convinced the navy to equip its ships with 10,000-fathom (18,000-meter) reels of bailing twine and 32-kilogram (kg) cannon balls for use as sinkers on the sounding lines. Lieutenant Maury's survey resulted in the first published deep-sea bathymetric chart of the Atlantic (52°N to 10°S), and served as the basis for laying the first transatlantic telegraph cables.

Questions about the deep ocean environment, including the type and location of seafloor-dwelling organisms, spurred the next round of contributions to marine geology. From 1872 to 1876, the HMS *Challenger* sailed all the world's oceans, measuring their depth, taking bottom samples, and making other measurements. Sir John Murray was responsible for the publication of the volumes of reports and results of the *Challenger* expedition, which remained the major source of knowledge of the ocean floor until the 1930s. The *Challenger* expedition established the general morphology, or shape, of the seafloor, showing that far from being a vast, flat plain, it also contains mountains, plateaus, and valleys.

The next great advance in the study of the ocean floor occurred in the years just before World War II. The crucial technological breakthrough was the echo sounder, which works by measuring the time it takes for a "ping" of sound sent by a ship to bounce off a submerged object and return to the ship where it is recorded. The echo sounder was originally developed to detect submarines, but in the 1930s the German ship *Meteor* used it to show that a rugged ridge extended the length of the middle of the Atlantic Ocean.

During and after World War II, the echo sounding technique was refined. By the 1950s, scientists could continuously record the depth of the seafloor beneath the ship to an accuracy of 1 m in 5,000 m of water. This precision depth-recording capability was used by an expanding oceanographic fleet, allowing Bruce Heezen and Marie Tharp to synthesize thousands of kilometers of new ocean depth data into detailed physiographic maps of the ocean floor. During these exploratory oceanographic expeditions in the 1950s, Heezen and WILLIAM MAURICE EWING first realized that the Mid-Atlantic Ridge mapped by the *Meteor* was only one part of a mid-ocean ridge that circled the globe.

Seafloor Sediments

In addition to the wealth of new information about seafloor topography provided by the HMS *Challenger* from 1872 to 1876, samples dredged from the seafloor during this expedition also showed, for the first time, the incredible diversity of sediment and rock found at depth. Clays, silts, pebbles, and even black volcanic rock were recovered. Muds, or "oozes," made up of carbonate or silica microfossil shells covered much of the deep seafloor. Seafloor sediment samples taken by the American ship *Albatross* from 1888 to 1920 added greatly to the basic knowledge of deep-sea sediments provided by *Challenger*. One insight gained from study of *Albatross* samples was that seafloor currents keep large areas of the seafloor free of sediment cover over long time periods (Shepard, 1963).

Tubes of sediment called cores yielded even more valuable data and discoveries. In the 1920s and 1930s, the German ship *Meteor* and the Dutch ship *Snellius* recovered the first cores. The scientists aboard *Meteor* used a simple gravity core while the *Snellius* crew used an explosive coring device called a piggot gun. Wolfgang Schott studied *Meteor's* 1-m-long gravity cores of the equatorial Atlantic, demonstrating for the first time that deep-sea sediments record continental glacial stages. A Swedish deep-sea expedition (1947–1948) used Borge Kullenberg's new piston corer, which provided scientists with deep-sea sediment cores from 7 to 20 m long. By studying changes in the sediments and fossils within the cores, scientists could study geologic history over longer intervals. Even so, conventional piston cores still provide only the most recent history of sediment deposition in the ocean.

To obtain even longer cores and extend the study of ocean history back even further requires a drilling vessel. During the Deep Sea Drilling Project (DSDP) from 1968 to 1982, the drillship *Glomar Challenger* recovered 97 km of core for scientific study at 635 sites in the world's oceans (see Hsü for an excellent history of DSDP). The availability of undisturbed, continuous sedimentary cores spanning the last 200 million years (Ma) of Earth history spurred the birth of a whole new discipline, paleoceanography, the study of the development of ocean systems. Since the end of DSDP, the Ocean Drilling Program's drillship, the *JOIDES Resolution*, has continued to take core samples in the world's oceans, filling in the details of early ocean circulation and climate history (see "25 Years of Ocean Drilling" in the 1993/1994 issue of *Oceanus* for several articles on this subject). Scientific study of drilling cores has shown that many periods in geologic history have experienced rapid climate change, shedding light on questions of contemporary climate change.

Seafloor Spreading

As early as 1620, Francis Bacon recognized that the shorelines of South America and Africa might once have been joined, but it was not until the early 1900s that ALFRED WEGENER built a strong case for the theory of continental drift. In the mid–1960s the theory of seafloor spreading replaced that of continental drift, and the oceans provided the critical data. While the continental drift theory required that the continents move freely over the earth's surface to change position, the seafloor spreading theory stated that new oceanic crust is created at the axis of the mid-ocean ridge, then it moves passively away from the mid-ocean ridge as if on a conveyor belt toward deep-sea trenches, where it is consumed.

In the early 1950s, Arthur D. Raff developed a shipborne magnetometer which provided seafloor magnetic measurements that built on Wegener's theory of continental drift. In 1955, aboard the U.S. Coast Guard vessel *Pioneer*, Raff and R. G. Mason used this new tool to map the intensity of magnetization of the northeast Pacific seafloor. They found that the magnetic "anomalies" formed linear stripes on the seafloor. In the early 1960s, Drummond Matthews collected magnetometer surveys of the Carlsberg Ridge in the Indian Ocean taken from the HMS *Owen*. Fred Vine studied these data and showed that the magnetic stripes on the Indian Ocean seafloor were the result of the seafloor having been reversely magnetized at times in the past. Together Vine and Matthews predicted that there should be symmetry of these magnetic stripes about the axis of the mid-ocean ridge as a result of magnetization of the basaltic seafloor crust being "locked in" at the time it formed along the ridge. Walter Pitman verified this conclusion in 1965 during the *Eltanin* cruises to the South Pacific when he showed the perfect bilateral symmetry of the magnetic profiles he collected. The theory of seafloor spreading was gaining momentum.

Analysis of sediment cores retrieved by the *Glomar Challenger's* earliest "legs" (as drillship cruises are called) provided the critical documentation to support the theory of seafloor spreading. By dating microfossil assemblages in sediments that immediately overlay the basaltic "basement," scientists found that the age of the seafloor increased with distance from the mid-ocean ridge. Deep drilling also proved that the oldest seafloor is only about 170 Ma, a fraction of the earth's age, supporting the theory that oceanic crust is consumed and thrust under adjacent crust at the trenches that form the "Ring of Fire" around the Pacific Ocean (*see* PLATE TECTONICS).

While the seagoing expeditions of the nineteenth century first discovered the mountain chain in the middle of the Atlantic Ocean, it was not until 1974 during the French-American Mid-Ocean Undersea Study (FAMOUS) that scientists got their first up-close look at the Mid-Atlantic Ridge using the U.S. submersible *Alvin,* and French submersibles *Archimede* and *Cyana.* During this project, the three-person crews in these subs collected 1,350 kg of rock and took 100,000 photos (Deep Submergence Vehicle *Alvin* brochure, Woods Hole Oceanographic Institution). Scientists could now see the volcanic forms making up the Mid-Atlantic Ridge and the deep faults and fissures that slice it up.

In 1977, scientists diving in *Alvin,* in search of heated waters and metalliferous sediments, discovered the spectacular biology and warm water vents of the Galapagos Spreading Center in the equatorial Pacific Ocean. Since there is no light at mid-ocean ridge depths of about 2 km, studies found that the unusual animals who reside there derive their life-sustaining energy from chemosynthesis rather than photosynthesis. During dives in 1979, at another location along the Pacific mid-ocean ridge, black smokers spewing 350°C waters (650°F) were first discovered. The "black smoke" is the particulate matter that precipitates out of the lava-heated water from depth when it mixes with the cold (near zero degrees centigrade) seafloor water.

At times *Alvin* has also been used as an archeological tool. In 1986, using *Alvin* and the remotely operated vehicle *Jason Jr.*, Robert Ballard found and explored the wreckage of the sunken ship *Titanic.*

Conclusion

In the 1990s, advances in satellite communications technologies continue to revolutionize the way science is done at sea. Data from ships can be transmitted instantaneously to shore laboratories for analysis, and scientists can conduct "oceanographic expeditions" from shore by remotely operating seafloor instruments. Like their predecessors who used rope and buckets to establish the fundamentals of seafloor topography and sediment distribution, future scientists will continue to push the frontiers of science as far as their imaginations and technical innovations can take them.

Bibliography

HSÜ, K. J. *Challenger at Sea: A Ship That Revolutionized Earth Science.* Princeton, NJ, 1992.

KENNETT, J. *Marine Geology.* Englewood Cliffs, NJ, 1982.

"25 Years of Ocean Drilling." *Oceanus* 36 (Winter 1993/1994):49–61.

SHEPARD, F. P. *Submarine Geology,* 2nd ed. New York, 1963.

ELLEN S. KAPPEL

OCEANOGRAPHY, BIOLOGICAL

Biological oceanography, the science of the study of the living organisms of the oceans and seas, seeks to define the identity, distributions, abundance, and interrelationships of the broad assemblages of living organisms inhabiting the oceans. Biological oceanographic studies also focus on how these complex living assemblages are affected by the physical and chemical properties of the sea, such as the currents, tides, ocean circulation, temperature, salinity, as well as how the ocean inhabitants themselves alter the characteristics of the ocean such as chemical composition, gas exchange, and color. Increasingly, biological oceanographic studies focus on large-scale problems, at the regional or ecosystem level, which address how changes in the global environment affect marine ecosystems and, in turn, how marine ecosystems contribute to and control the important global elemental cycles, such as those of carbon, nitrogen, oxygen, and sulfur.

There are several institutions throughout the United States and in other nations that offer advanced degrees (M.S., Ph.D.) in biological oceanography. However, scientists trained in a broad array of disciplines ranging from molecular biology through nearly all subdisciplines of the biological sciences (systematics, microbiology, cell biology, physiology, ecology, marine ecology) as well as biochemistry, natural-products chemistry, micropaleontology, mathematical modeling and bioengineering, participate in and contribute to the field of biological oceanography. Biological oceanographic researchers actively collaborate with specialists in the three other major oceanographic disciplines of physical oceanography—chemical oceanography, marine geology, and geophysics—and participate in joint projects and programs. (*See* OCEANOGRAPHY, CHEMICAL; OCEANOGRAPHY, PHYSICAL; OCEANOGRAPHY, GEOLOGICAL.)

Methods of Research

Biological oceanography involves making observations and measurements, collecting discrete samples, undertaking experiments in a variety of settings, making predictions based on various hypotheses or models, and utilizing an array of platforms and collecting gear. Research studies are undertaken by biological oceanographers in their home laboratories; at remote marine laboratories and field sites around the world; and aboard small and large research vessels, utilizing manned and remote underwater vehicles, working in underwater habitats, and via remotely sensed data downloaded from various aircraft and orbiting satellites (*see* OCEANOGRAPHIC EXPEDITIONS). Large oceanographic research vessels capable of traversing the world's oceans are still the main observational, collecting, and research platforms of the profession, but rapid advances in remote-sensing techniques and automated instrumention are occurring. Sampling gear ranges from simple plastic buckets to specialized collecting bottles, large net arrays that can be opened and closed at specified depths, grab samplers, and various coring devices; to remotely operated "landers" (free vehicles that descend to the ocean bottom to make various measurements and analyses) with specialized chambers for holding the samples under "near" natural conditions; and to the rapidly developing suite of moored and towed acoustic and optical samplers.

Biological oceanography involves the study of a broad array of living forms ranging from microscopic viruses, bacteria, and plants, to whales, the world's largest mammals. Living organisms are believed to have originated in the ocean realm (*see* ORIGIN OF LIFE), and thus the oceans represent an important repository of planet Earth's biodiversity. Representatives of all known taxonomic groups of animals are found in the sea; some, like the sea urchins and starfishes, are found only in marine waters (*see* ECHINODERMS).

The range of plants inhabiting the oceans is much more limited and consists primarily of the primitive plants referred to as marine algae, as well as some flowering plants, such as the seagrasses found in shallow coastal waters. Marine algae, such as the brown kelps and green algae found along rocky coasts, may be large and visible (macroscopic) but the predominant plants of the sea are very small (2–200 μm) microscopic forms. Diatoms and dinoflagellates, as well as the recently discovered prochlorophytes, generate organic (carbon) matter through photosynthesis and are referred to as the primary producers upon which most other ocean organisms are dependent (*see* MINERALIZED MICROFOSSILS). Cyanobacteria (sometimes referred to as blue-green algae) and chemosynthetic bacteria are also important primary producers in the ocean. Chemosynthetic bacteria, like photosynthetic organisms, generate organic matter from carbon dioxide and water. However, rather than using light energy, they use the oxidation of energy-rich compounds such as sulfides. Other microorganisms, such as viruses, bacteria, and fungi, are important biological components of the sea and play key roles in the recycling of organic matter in the oceans.

Classification System

The living organisms of the oceans are often separated into three primary groups depending on their mode of life: if they drift with the movement of the water, they are called plankton (bacterioplankton, phytoplankton for plants, zooplankton for animals, icthyoplankton for larval fish); if they are able to swim, they are called nekton; and if they live on or in the bottom of the ocean, they are referred to as benthic. Some organisms, primarily those in shallow waters, may have both planktonic and benthic stages in their life cycles. Marine organisms are also classified based on: how they obtain their carbon (autotrophs use CO_2, heterotrophs use preformed organic carbon); energy (phototrophs or chemotrophs); the distance they live from the land (coastal, neritic, oceanic); the depth of the water that they inhabit (upper zone or epipelagic, midwaters or mesopelagic, deep sea or abyssopelagic/abyssobenthic); their actual physical dimensions (size) such as pico- (less than 2 μm), nano- (2–20 μm), micro- (20–200 μm), meso- (20–2,000 μm), or macroplankton (greater than 2,000 μm); or the temperature region of the globe they inhabit, such as tropical, temperate, boreal, and polar.

Biological oceanography studies may focus on distinct taxa, such as diatoms or copepods, or on whole assemblages, such as phytoplankton or zooplankton, or on whole ecosystems like coral reef communities or hydrothermal vent communities. Alternatively, they may concentrate on regional features such as Gulf Stream rings or the Peru upwelling, or processes such as carbon flux along a sector of a major ocean basin.

Marine life is not homogeneously dispersed throughout the oceans either in space or time but rather occurs in patches that can range from a few centimeters to hundreds of kilometers in extent. Oceanic life is dilute in certain regions and concentrated in others. Biological oceanography studies have a major interest in understanding the processes, or interacting factors, both abiotic (waves, currents, general circulation patterns) and biotic (life history, behavior, food chain interactions), that regulate the distribution of life in the sea, and in learning how these factors may be altered with changing climate.

The major surface currents of the ocean are similar in all three of the major ocean basins and are an important factor dividing the oceans into distinct biological provinces. The basins are divided by the prevailing surface currents into circular systems called gyres, which have their own distinct assemblage of organisms. Each north and south ocean basin of the Atlantic and Pacific, and the single basin of the Indian Ocean, has a subpolar gyre, a subtropical gyre, and an equatorial zone that results in three definable biological provinces. The subpolar province is characterized by a limited number of species in large concentrations, while the subtropical province has a large number of species with limited abundance, and the equatorial province is characterized by both high species composition and large biomass. The three ocean basins are interconnected in the southern hemisphere by the circumpolar Antarctic convergence current, which circulates around the Antarctic continent. Large concentrations of marine life are usually found in the waters of coastal regions of all the world's oceans, particularly where upwelling of nutrient-rich deep waters occurs.

Ocean Food Chain

The upper open ocean is the realm of microscopic phytoplankton. These phytoplankton are limited to this upper region, where light is present to fuel photosynthesis, upon which the majority of other ocean organisms are dependent. The microscopic phytoplankton form the base of the oceanic food chains and food webs. They are fed upon by a variety of zooplankton, usually passing through several sizes of zooplankton, and on to fishes and even to certain species of whales. These planktonic plants and animals often pass through the guts of their consumers and their remains are packaged in fecal pellets that sink to the bottom of the ocean. Sometimes they are not consumed at all and sink directly to the ocean bottom. These sinking pellets and organisms are not only a major source of food for deeper dwelling organisms both in the water column and on the ocean floor, but they also comprise an important part of deep-sea sediments.

Deep-sea Species and Sediments

Hydrothermal deep-vent communities, first discovered in the late 1970s, are associated with spreading mid-ocean ridges and continental margins, where hot water rich in minerals, particularly sulfide, vent to the bottom of the oceans (*see* HYDROTHERMAL ALTERATION AND HYDROTHERMAL MINERAL DEPOSITS). They are unusual and prolific oases in the otherwise sparsely inhabited deep sea. Vent communities are some 10,000- to 100,000-fold richer in organic matter than most deep-sea benthic communities, which are dependent on the settling of organic matter produced kilometers above in the upper photic regions of the ocean. Chemosynthetic bacteria that are associated with many of the animals found at the vents are the key to these remarkable deep-sea communities, producing the organic matter that supports these rich and diverse deep-sea communities.

Biogenic particles from the skeletons and shells of marine organisms are a major contributor to deep-sea sediments. Biogenous sediments (defined as more than 30 percent shells by volume) and deep-sea muds (less than 30 percent biological material by volume) cover a large area of the ocean floor. These biological constituents include the siliceous diatoms and radiolarians, and the calcareous Foraminifera, coccoliths, and Pteropods, all of which are important microfossils. The presence or absence of certain species of microfossils within these biogenic sediments, in conjunction with the amount of specific radionuclides or stable isotopes present, serve as important stratigraphic recorders and indicators of the ages of the deep-sea sediments (*see* ISOTOPE TRACES, STABLE).

The microfossil record found on the deep ocean floor is potentially one of the best undisturbed sources of information on evolutionary changes that occurred in these planktonic organisms and how they responded to various environmental factors (*see* PALEOCLIMATOLOGY). Analyses of fossil species with biological studies on the ecology of their modern counterparts provide a powerful tool in forecasting the biological effects of current and predicted global climate change.

Bibliography

DUXBURY, A. C., and A. DUXBURY. *An Introduction to the World's Oceans.* Reading, MA, 1984.

GROSS, G. M. *Oceanography: A View of the Earth,* 5th ed. Englewood Cliffs, NJ, 1990.

MANN, K. H., and J. R. N. LAZIER. *Dynamics of Marine Ecosystems—Biological-Physical Interactions in the Oceans.* Cambridge, MA, 1991.

NATIONAL RESEARCH COUNCIL. *Oceanography in the Next Decade—Building New Partnerships.* Washington, DC, 1992.

PARSONS, T. R., M. TAKAHASHI, and B. HARGRAVE. *Biological Oceanographic Processes,* 3rd ed. Oxford, Eng., 1984.

SIEBURTH, J. M. *Sea Microbes.* New York, 1979.

LINDA E. DUGUAY

OCEANOGRAPHY, CHEMICAL

The seawater that covers most of Earth is salty; it contains about 35 grams (g) of dissolved salt per kilogram (kg). Because most of the water on land has a much lower content of dissolved salts, it is natural to wonder where the salt in the ocean came from and how it got there. The study of the ocean's chemistry is conducted by chemical oceanographers working closely with geologists, biologists, and physicists who also study the ocean. U.S. research on the chemistry of the ocean is supported mainly by the National Science Foundation, the Office of Naval Research the National Oceanic and Atmospheric Administration (NOAA), the Department of Energy (DOE), and other federal agencies.

The largest amount of dissolved material in the ocean is composed of the substances listed in Table 1. The elements in Table 1 are listed with positive or negative charges (cations and anions). Adding up the total molar equivalents of charge results in zero, which means that the ocean is electrically neutral. Two of the substances, bicarbonate (HCO_3^-) and sulfate (SO_4^{-2}), consist of several elements bound together so strongly that they behave as easily identifiable units; such substances are referred to as complex ions (*see* SEAWATER, PHYSICAL AND CHEMICAL CHARACTERISTICS OF).

The composition of seawater differs slightly from place to place in the ocean because of areas where evaporation exceeds rainfall, or vice versa. However, the ratio of the major dissolved ions to Cl^- (chloride ion) is constant everywhere in the ocean. Elements with this characteristic are referred to as conservative elements and include the major elements as well as many trace constituents of seawater.

In the case of the conservative elements, oceanographers assume that the ocean is at steady state; that is, the rates of addition and removal are equal and constant over time. If the amount of an element, say, sodium (Na), in the ocean is divided by the rate at which it is added to or removed from the ocean, the result is a length of time that is called the residence time. This time represents the average length of time an atom remains dissolved in seawater before being removed. For very soluble elements, residence times are typically very long, for example, 80 million years (Ma) for Na^+. But for less soluble elements, residence times may be very short. Iron (Fe), which is barely detectable in seawater, has a residence time of fifty years.

The salinity of seawater is determined by reference to the Practical Salinity Scale. Before the 1960s salt content was commonly measured by titrating (combining) the Cl^- ion with silver nitrate and reporting the result as chlorinity, or grams of Cl^- per kilogram of seawater (19.4 parts per thousand or ppt chlorinity for average seawater). Modern studies use electrical conductivity to measure the salinity of seawater. The electrical conductivity

Table 1. Chemistry of Seawater (salinity S = 35.000)

Major Constituents

Constituent	*g/kg (ppt)*	*mmol/kg*	*Residence Time (10^6 years)*
Na^+	10.781	469.0	80
K^+	0.400	10.21	12
Mg^{+2}	1.284	52.83	13
Ca^{+2}	0.412	10.28	1
Sr^{+2}	0.008	0.091	5
Cl^-	19.353	545.9	
SO_4^{-2}	2.712	28.23	10
HCO_3^-	0.126	2.06	
Br	0.844	130	

Some Trace Constituents

	mmol/kg	*Residence Time (years)*
Al	~0.03	600
Fe	$\sim 1 \times 10^{-3}$	50
Co	$\sim 3 \times 10^{-5}$	340
Ni	$\sim 8 \times 10^{-3}$	8200
Cu	$\sim 4 \times 10^{-3}$	970
Zn	$\sim 6 \times 10^{-3}$	500

Nutrient Concentration Ranges

	mmol/kg
NO_3^-	0–45
PO_4^{-3}	0–4
H_4SiO_4	0–200

Dissolved Gases

	Partial Pressure, Dry Air (atm)	*Equilibrium Concentration, Surface Seawater (ml/l)*	
		0°C	*24°C*
N_2	0.781	14	9.0
O_2	0.209	8.8	5.5
Ar	9.3×10^{-3}	0.36	0.22
CO_2	3.2×10^{-4}	0.36	0.23

mmol = millimol. atm = atmosphere, standard.

of a seawater sample is compared to the conductivity of a standard KCl solution chosen to be equal to that of average seawater with a salinity of S = 35.000. The conductivity ratio between the sample and the standard can be measured with a precision greater than 1 part in 40,000.

Interactions of Chemistry, Biology, and Physics in the Ocean

Trace amounts of other elements also occur in seawater, in concentrations less than 1 part per million (ppm). Trace elements in seawater have a wide range of chemical behaviors and may have residence times from years to millions of years. Included in the list of trace substances are nutrients, metals, dissolved gases, and dissolved organic carbon. Nutrient elements are among the most significant trace elements in seawater. These substances include nitrogen (as NO_3^-, NO_2^-, and NH_4^+) and phosphorous (as PO_4^{3-}), which along with carbon comprise the bulk of the tissue of phytoplankton; these elements occur in the proportions of C : N : P equal to 106 : 16 : 1, based on averaged analyses of plankton. The ratio is know as the Redfield–Richards ratio in honor of the scientists who conducted the first major survey of plankton composition. The photosynthesis reaction by which phytoplankton grow can be written using this ratio in the following expression:

$$106CO_2 + 122H_2O + 16HNO_3 + H_3PO_4 \rightarrow (CH_2O)\ 106(NH_3)\ 16H_3PO_4 + 138O_2$$

In this equation CO_2, NO_3^-, and PO_4^{-3} are used during photosynthesis to create phytoplankton biomass with dissolved oxygen as a by-product. Silica (H_4SiO_4) is also an important nutrient for diatoms and other phytoplankton with skeletal material of amorphous silica.

The productivity of these organisms depends on the availability of the nutrient elements as well as the presence of sunlight. This productivity is effective enough to impoverish most surface waters of the nutrient elements; the consumption of the phytoplankton by larger zooplankton results in the formation of fecal pellets that sink below the surface layer of the ocean, removing nutrients to deeper water. There the slow processes of solution and bacterial metabolism result in higher concentrations of the nutrient elements at depth; further productivity then depends on the occurrence of upwelling of deep water to return some of the nutrients to the surface to reenter this cycle. Nutrient elements have a characteristic vertical profile in the ocean, rapidly increasing from near zero at the surface for the first 1,000 meters (m), then remaining roughly constant to the seafloor. Some of the trace elements that do not directly affect the productivity of plankton at the surface have very similar vertical profiles in seawater because their chemical behavior is similar to that of a nutrient. Germanium (Ge), which is not necessary for the formation of the H_4SiO_4 tests (skeletal parts) of diatoms, nevertheless has a profile shape very similar to that of H_4SiO_4 because it tends accidentally to be included in minor amounts in the cycles that influence H_4SiO_4.

Removal of nutrients from surface waters and remineralization in deeper water, combined with the overall circulation pattern of the ocean, results in a horizontal segregation of the nutrient elements. Cold, dense water formed in the North Atlantic sinks toward the seafloor and travels to the South Atlantic, then eastward into the Pacific Ocean, and finally toward the North Pacific Ocean. Along the geographic path of the deep water, nutrient elements are removed from surface water and added to deep water, increasing amounts of dissolved nutrients in the deep water. Thus North Pacific deep water has a much higher nutrient content than North Atlantic deep water.

Dissolved gases in seawater are an important type of trace constituent. Nitrogen and oxygen are the two most abundant atmospheric gases: both of these are dissolved in surface water and carried throughout the ocean. Oxygen is consumed during respiration so that the O_2 content of seawater decreases away from contact with the atmosphere. Decay of large blooms of plankton sometimes can locally remove nearly all available O_2. CO_2 reacts chemically with water after being dissolved and forms other dissolved species:

$$CO_2 + H_2O \rightarrow HCO_3^- + H^+$$

This reaction removes part of the dissolved CO_2 from the water, in turn causing more atmospheric CO_2 to be removed into the ocean surface. CO_2 is also removed from the atmosphere by photosynthesis during phytoplankton growth. Formation of $CaCO_3$ shells of some of the plankton also removes HCO_3^-, and thus CO_2, from seawater. These effects result in the flux of atmospheric CO_2 into the surface of the ocean. The Department of Energy and NOAA are sponsoring research on this aspect of chemical oceanography; the National Science Foundation funds the Joint Global Ocean Flux Study (JGOFS) to study the vertical transport of carbon downward from the surface of the ocean to the sediment.

Increasing use of fossil fuels and deforestation is elevating levels of CO_2 in the atmosphere, where this "greenhouse" gas appears to be causing a gradual warming of Earth's climate (*see* GLOBAL ENVIRONMENTAL CHANGES, NATURAL). There has been great interest in the possibility of the ocean acting as a CO_2 sink, thus protecting the atmosphere from climate change (*see* EARTH'S ATMOSPHERE, CHEMICAL COMPOSITION OF).

Cycling of Materials Through the Ocean

Dissolved and solid materials are constantly added to the ocean by rivers, by wind-blown dust, by anthropogenic activities, by submarine hot springs at mid-ocean ridge systems, and by exchange of gases with the atmosphere. Water added to the ocean by rain and rivers is removed by evaporation, but dissolved salts are not. Why does the sea not get saltier? The answer lies in the cycling of materials through the ocean. In order for cycles of these materials to exist, there must be removal processes to balance out the additions.

A varied group of removal processes complete the cycles of elements through the ocean. Evaporation of water out of isolated bodies of seawater may locally remove all dissolved salts from a part of the ocean as it dries up. Phytoplankton and zooplankton living at the ocean surface remove such elements as Ca and H_4SiO_4 from seawater to make hard parts, and nitrate, phosphate, and carbon to generate the soft tissues. These small organisms become food for larger ones, and the remains are packaged into fecal pellets that fall to the ocean floor, removing the constituent elements to the sediments. Amorphous masses of organic matter called marine snow move slowly toward the ocean floor. As these particles fall, interactions continue to affect the chemistry of seawater. Bacterial decomposition of the organic matter uses up O_2 from seawater and helps to dissolve or remineralize the soft tissue constituents back into the water. At the same time CO_2 is added to deep water as a byproduct of bacterial respiration. Many trace elements are only slightly soluble in seawater; they tend to be adsorbed to the surfaces of all kinds of particles in seawater and removed very effectively to the ocean floor when the particles sink. At mid-ocean spreading centers, seawater circulates deep into Earth's crust and upper mantle where it is heated and chemically changed before it returns to the seafloor as hydrothermal springs. The seawater loses all its Mg^{2+} and SO_4^{2-}, and its dissolved O_2, returning to the seafloor hotter, more acid, and enriched in Ca, Mn, H_4SiO_4, and other elements. Huge plumes of dark manganese (Mn) and Fe oxides and white H_4SiO_4 precipitate from the submarine springs when the hot fluids mix with cold, oxygenated normal seawater.

Interactions between seawater and seafloor sediment continue to take place after the sediment has been deposited. Any organic matter not destroyed in transit to the sediment continues to be food for a variety of organisms and bacteria living in the mud. Metabolic processes use up the O_2 in the pore water between the sediment grains. In response to the gradient of decreasing O_2, more O_2 diffuses downward into the sediment from the overlying seawater. Beyond the reach of the diffusing O_2, other chemical processes enable bacteria to metabolize their food. Nitrate (NO_3^-) is stripped of its oxygen; when the nitrate is used up, SO_4^{-2} is reduced, resulting in sulfide (S^-) in the pore water. More SO_4^{-2} then diffuses into the seafloor in response to the gradient that is set up. Over long periods of time changes in the prevalence of such reactions on the seafloor may affect the geochemical cycles of sulfur and carbon in the ocean.

Not all of the interaction between seawater and sediment involves bacterial intervention. H_4SiO_4 may continue to dissolve from the skeletal debris in the sediment and from some types of aluminosilicate minerals derived from land. H_4SiO_4 concentrations in pore water are commonly rather high, and the H_4SiO_4 continues to diffuse out of the sediment into the seawater. Over long periods of time mineralogical changes in carbonate minerals and aluminosilicate minerals take place deep in the sediment stack. In response to these changes Ca^{+2} and K^+ diffuse out of the seafloor into the ocean and Mg moves in the opposite direction.

The composition of seawater has been relatively constant for hundreds of millions of years. The complex biogeochemical cycles that control seawater chemistry maintain the system close to a steady-state condition.

Bibliography

BROECKER, W. S., and T. H. PENG. *Tracers in the Sea.* Palisades, NY, 1982.

LIBES, S. M. *An Introduction to Marine Biogeochemistry.* New York, 1992.

MILLERO, F. J., and M. L. SOHN. *Chemical Oceanography.* Ann Arbor, MI, 1992.

MARTHA SCOTT

OCEANOGRAPHY, GEOLOGICAL

Geological oceanography is sometimes contrasted with marine geology, depending on whether the emphasis is upon studying the ocean itself using geological techniques, or upon studying geology that happens to be under salt water. For the purpose of this article we shall consider them equivalent. The subject can be divided into three broad areas: sediments, rocks, and geophysics, including tectonics and structure.

Geophysics, Structure and Tectonics

The methods used by marine geophysicists are covered in the entry on GEOPHYSICAL TECHNIQUES. They include studies of gravity, magnetism, heat flow, and seismic analysis, using sensors towed behind the ship on long cables, or instruments deployed on the seafloor. Variations in gravity and in seismic velocity are used to infer differences in density or thickness of the layers of ocean crust and the upper mantle. Seismic techniques can determine the subsurface structure such as location of fault zones, buried seamounts, and even the top of the mantle. They permit us to develop a three-dimensional image of the crustal layer cake. Intensity and direction of magnetization in rocks of different ages provide insight into variations in Earth's magnetic field. In addition, the reversals in direction of the field over varying lengths of time produce a characteristic pattern of "normal" and "reversed" magnetization in the rocks, some intervals broader than others, rather like bar-codes used in stores. These patterns can be used to determine the age of ocean crust and changes in the direction of seafloor spreading. Heat flow varies with the age and thickness of ocean crust and is used to detect regions of recent volcanic activity.

The topics studied by marine geophysicists are described in the following entries: OCEAN BASINS, EVOLUTION OF; OCEANIC CRUST, STRUCTURE OF; EARTH AS A DYNAMIC SYSTEM; PLATE TECTONICS; and EARTHQUAKES AND SEISMICITY. Some principal topics of study are: (1) The mid-ocean ridge system and its interaction with Earth's mantle. The Mid-Atlantic Ridge spreads slowly, has a narrow steep valley along the axis with small magma chambers beneath it, low heat flow, and high earthquake activity; the East Pacific Rise spreads three times as fast and has a broad shallow valley, large magma chambers with high heat flow, and few earthquakes. What causes these differences? Has the growth pattern of either ridge system changed over time and if so why? What stresses cause the faulting of the plates as they move away from the ridge? (2) The trenches and island arcs where ocean crust is colliding with other plates (active margins). If oceanic crust collides with a continent (for example, off Peru and Chile), the ocean crust bends downward and slides beneath the continent. The frictional stresses and heating as the mass of crust and overlying sediment is forced beneath the continent lead to large earthquakes and active volcanoes. At what depth are earthquakes triggered and is it different for different trenches? How far under the continent does the ocean material travel before it is melted into the mantle? Especially complicated systems occur in the western Pacific and Caribbean, where collision is occurring between island volcano systems, continents, and oceanic crust; often there are small basins behind the volcanic systems which are spreading at the same time that collision is occurring on the other side (for example, the Sea of Japan or the Philippine Sea). It is a major challenge to sort out the different sources of stress, heat, and magma supply in these areas, and to predict the occurrence of earthquakes and volcanism. (3) Passive continental margins where the continents are separating from each other. About 300 million years ago (300 Ma) the Americas, Europe, and Africa were a single giant continent. Heating and rupture lead to formation of a rift valley system that finally opened up to form a narrow sea and then the Atlantic Ocean, with the rift system along the middle. Why did the supercontinent split where and when it did? What sort of stresses were placed upon the new continental margins, causing them to flex downward? Where is the true boundary between continent and ocean crust in these regions, buried under kilometers of sediment?

Rocks and Hydrothermal Deposits

Rocks from the ocean crust and upper mantle are collected by dredging or by a drilling rig mounted on a ship, and are studied by petrologists and geochemists. The association of minerals in a rock and the ratios of different elements and isotopes in the minerals can be used to infer the temperature and pressure conditions under which the rock formed, as well as the original nature of the magma and modifications it may have undergone before it solidified. Major topics of study include: (1) Variations in the source and supply of magma from the mantle. The mantle is believed to be chemically variable in different parts of Earth: the "plum pudding" model. Why do these differences exist, and do they have any systematic relationship to the different ridge and volcano systems? As the magma rises from the mantle into chambers within the crust, it may undergo a sort of distillation process, with some minerals crystallizing out at depth (fractional crystallization). These in turn may later be dissolved into a fresh batch of magma. What effect does this have on the final mineral composition and chemistry of the rocks that get erupted at the seafloor? What is the variation in size, depth, and temperature of the magma chamber beneath different segments of ridges or different volcanic regions? Why do rocks from different parts of the same ridge have different chemistry? Does this represent different mantle sources or just different degrees of fractional crystallization from a single source? Why do volcanic islands have a different chemistry from the ridges? What does the composition of the rocks tell us about layering and convection in the mantle? (2) Alteration of the rocks as they age. Prolonged exposure to seawater gradually removes some elements and precipitates new minerals along fractures in the rock. How does this change the chemistry of the rocks and of seawater itself? What effect does it have on physical properties of the rock, such as magnetic intensity? A special aspect of this topic involves the interaction of seawater with very hot rocks and magma, referred to as hydrothermal alteration. At the ocean ridges cold seawater enters the crust through fractures and penetrates toward the magma chamber. As it heats up, it interacts with the nearby rock and eventually is forced upward again through a different set of fractures. When the hot, mineral-laden water is ejected, unusual structures are precipitated that are rich in metals. What controls the chemistry of these deposits, which may be analogs of some metal deposits on land? How fast do they grow and how long do they last? How do they change with time, and is there a relationship between fluid composition and the community of organisms found near these vents? Why are they found only in certain parts of the ridge system?

Marine Sediments

Marine sediments can be divided into organic particles (fragments of coral or the microscopic fossils of single-celled plankton) and inorganic particles (sand, silt, and clay derived from erosion of continental or island rocks). A special class of inorganic sediments are those rich in metals that are precipitated from seawater involved in hydrothermal alteration of magma (*see* METALLIC MINERAL DEPOSITS, FORMATION OF). Sediments derived from erosion occur mostly near the continents on the shelves and slopes, or near ocean islands where they are mostly thick layers of volcanic ash. Organic (or "biogenic") sediments occur in shallow tropical regions where reefs are common, and in the deep ocean basins, especially under areas that are rich in plankton such as around Antarctica and in the tropics. The plankton organisms live in the surface waters, and when they die, their external shells settle to the ocean floor. Sediments composed purely of microfossils may accumulate as much as a kilometer of thickness over tens of millions of years.

In coastal zones the study of sediment transport and deposition is extremely important for understanding processes such as beach erosion and the silting-up of harbors. Important questions include: Are short, intense events like storms more important than long-term conditions in controlling sediment transport? What are the effects of waves and currents, respectively, on erosion and transport of sediments? What is the ultimate fate of the vast amount of sediment carried by rivers? Does it pile up on the shelf, as in the vast delta systems of major rivers, or is it gradually carried off the shelf and down the continental slope? How do changes in sea level affect the transport and deposition of sediments?

Because deep ocean sediments accumulate very slowly, they record thousands or even millions of years of microfossil production in the surface water. Studies of chemical changes in the fossils, and of the abundance of different species, are used to examine how the surface circulation and chemistry of the oceans have changed. For instance, some species are known to have distinct temperature restrictions; their presence in the sediment at any location gives us some idea of the surface water temperature at times in the past. The ratio of stable isotopes of oxygen in the shells is affected by water temperature and also depends on the total amount of water locked up in ice on the continents because light isotopes are preferentially evaporated from the ocean to form snow and ice, leaving behind the heavier isotopes. Carbon isotopes can be used to date the sediment (with the radioactive carbon isotope ^{14}C) and also may provide insight on the rates of growth and utilization of carbon dioxide. Many of these parameters show cyclic changes corresponding to the timing of the ice ages. Research is focused on questions like: What causes climate change, and how does the ocean change during an ice age? For instance, does it get colder everywhere, or only in the polar regions? Is the ocean more or less productive during a glacial time, and how might that affect the amount of carbon dioxide in the atmosphere? How fast can changes occur? At the end of the last Ice Age, the ice sheets melted very rapidly. What was the effect of vast volumes of cold freshwater flooding into the ocean? On a longer timescale, how were the oceans affected by the event that caused extinction of the dinosaurs and marine invertebrates?

The dissolved ions in the water associated with sediments (pore water) interact with the different kinds of sediment, causing some compounds to dissolve and others to precipitate. This process of diagenesis is involved in the transformation of sediment to rock. Chalk and limestone are formed from coral fragments and from some of the planktonic microfossils; chert (flint) is formed from other types of microfossils. What are the relative effects of temperature, pressure, and porosity upon the rate and intensity of diagenesis? Why are chert beds or nodules found at certain layers within chalks? Does this reflect original layering of different assemblages of microfossils, or is it an effect of migration of pore waters with precipitation concentrated in certain intervals? An aspect of diagenesis that is receiving much attention is the nature of pore fluids in the mass of sediment on the continental slopes, which can be several kilometers thick. As the sediment pile grows, pore waters are squeezed along certain layers. When the fluids are expelled from the sediment, how does their chemistry affect the chemistry of the seawater? When the sediments are buried in ocean trenches, how do the fluids and sediments contribute to the explosive volcanism nearby?

Geological oceanography thus addresses a wide range of topics in earth sciences. Almost all of the subdisciplines are represented: paleontology, sedimentology, stratigraphy, petrology, geochemistry, geophysics. In essence, it consists of applying the techniques and concepts of these classical fields to the 75 percent of Earth that lies beneath the sea.

Bibliography

KENNETT, J. *Marine Geology*. Englewood Cliffs, NJ, 1982.

CONSTANCE SANCETTA

OCEANOGRAPHY, HISTORY OF

Oceanography—the scientific study of Earth's ocean—is a modern activity but built on ancient roots. Prehistoric peoples sailed across narrow parts of the ocean to settle Australia and across thousands of kilometers to settle the island of the South Pacific. No written records of these migrations survive, but the people must have been skillful sailors. For example, fish bones and shells in refuse piles near ancient coastal villages show that fishing and shellfish gathering provided protein to their diets. Bones of deep-sea animals in the piles suggest that substantial fishing occurred offshore, necessarily involving boats.

As ships and navigation improved, sailors and traders explored the ocean, looking for trade routes to ply, or products to buy and sell. Their discoveries were too valuable to share and thus are rarely recorded. Most early sailors were illiterate, but efficiently transmitted their knowledge of the sea by telling stories of exploration and discovery. Most of their knowledge of the ocean was lost to modern oceanographers.

European Ocean Exploration

The great age of European ocean exploration began early in the fifteenth century, stimulated by knowledge gained from Greek and Arab geography in the Islamic libraries. These ideas were "rediscovered" when Christian armies recaptured southern Spain from the Islamic kingdoms. Publication in 1410 of a Latin edition of Ptolemy's (ca. 140 C.E.) map of the world greatly influenced thinking and exploration. Along with Ptolemy's maps came two ideas well known to the Greeks: (1) Earth is a sphere, and (2) the ocean is navigable. These ideas came to Europe from the manuscripts recovered from the Islamic libraries.

Perhaps the most influential individual in Portugal's early ocean exploration was Prince Henry the Navigator (1392–1460), who established a center for seafaring in southern Portugal. He brought Europe's most learned people to teach Portuguese sailors the latest techniques in navigation. In the voyages that followed, Portuguese ships explored the west African coast; their ships reached the tip of South Africa in 1488. Vasco da Gama (1460–1524) reached India in May 1498, laying the foundation for Portugal's empire and its lucrative spice trade with India and southeast Asia.

The impetus for ocean exploration came from the need to replace closed caravan routes to China and India. For centuries these caravan routes, known as the Silk Road, had supplied Europe with Asian luxuries and cooking and medicinal spices. The decline of the central Asian Mongol Empire and the Turkish conquest of Constantinople in 1453 cut off these overland trade routes from Europe. Thus, developing new ocean trading routes was the only solution.

New technology helped. Cannons carried aboard newly developed, highly maneuverable European ships were formidable weapon systems. This led to European domination of the ocean and to later conquests and colonization of coastal areas. Eventually western European nations dominated both land and ocean for more than four hundred years. During these centuries of exploration, most advances in knowledge about the ocean came from working on practical problems. For instance, improvements in navigation resulted in improved maps of ocean shorelines. France (1770) and England (1795) established national offices for ocean mapping.

Captain James Cook (1728–1779) made three exploring voyages to map the Pacific islands, using the latest navigational instruments, including newly developed chronometers, to determine longitude. His maps were the most accurate made up to that time and are still used in some remote islands.

Individual Scientific Inquiries

The first efforts to establish oceanography as a science came from individuals who compiled ocean observations. Benjamin Franklin (1706–1790) published an early map of the Gulf Stream—the strong current system along the U.S. Atlantic coast. Franklin, the colonial postmaster, noted that mail ships coming from England to America took much longer to make the voyage than did American ships heading in the other direction. His cousin, Timothy Folger, a Nantucket whaler, had told Franklin about the Gulf Stream. American whalers and seamen apparently knew the current well, but Franklin's Gulf Stream map and his instructions for locating its warm waters allowed other ships to locate the current. Thus, they learned to use Gulf Stream currents to speed eastbound voyages and to avoid bucking strong current on westbound voyages.

Despite the benefits obtained from such advances, there was no systematic exploration or study of the ocean until the mid-nineteenth century.

American Ocean Exploration

In the United States, government support for oceanography in the nineteenth century grew out of practical concerns. Among these was the need to insure safety of passengers and goods aboard U.S. ships, maintain adequate coastal defenses, and protect fisheries. In 1830, the U.S. Navy created a Depot of Charts and Instruments (later the U.S. Naval Observatory) to supervise use of navigational instruments on navy vessels. Matthew Fontaine Maury (1806–1874), a naval officer, became superintendent in 1842.

To make ocean transportation safer and speedier, Maury compiled observations on winds, currents, and weather recorded from ships' logbooks. Maury also furnished ships' captains with blank charts for recording weather and ocean conditions in data-scarce areas. Wind and current charts compiled in this way revolutionized navigation and cut weeks off transoceanic runs by clipper ships, the fastest ships of their time. The general world charts of ocean surface currents still in use come primarily from Maury's compilations.

In 1853 Maury organized the first meteorological conference, which led to international cooperation in collecting weather information at sea. Maury wrote *The Physical Geography of the Sea* (1855), the first major oceanographic book in English.

Challenger Expedition

By the 1860s, scientific interest was building in England for studying the deep ocean. In the late 1860s, the Royal Society of London provided financial support and the British Admiralty provided two ships for North Atlantic deep-sea studies. Finally, the Royal Society agreed to sponsor an ambitious ocean exploring project. The expedition aboard HMS *Challenger* (1872–1876) was the first large-scale, government-supported research project; these have since become an important component of oceanography. Before returning to England, *Challenger* traveled 109,000 kilometers (km) collecting water samples, ocean-bottom rocks and sediments, and taking soundings of water depths. The expedition sampled all ocean basins except the Arctic.

The British government established the *Challenger* Expedition Commission to analyze the samples and to publish the results. Headed by Sir John Murray (1841–1914), a Canadian-born geographer and naturalist, the commission produced a twenty-three-volume report that laid the foundations for modern oceanography.

Oceanography continued during the early part of the twentieth century, dominated primarily by fisheries concerns. Large fluctuations in fish catches and concerns about the survival of fish stocks led to the formation in 1902 of the International Council for the Exploration of the Sea (ICES). ICES is still an influential organization for European and North American oceanographers.

Following the *Challenger* Expedition, small oceanographic expeditions occurred. Notable among them were those sponsored by Prince Albert I of Monaco, a wealthy man with a lifelong interest in the ocean. He supplied his private yachts for expeditions in the Atlantic, and organized the Oceanographic Museum in Monaco.

Wartime Ocean Studies

Research supported by the military during World War II greatly stimulated oceanography, improving our understanding of the ocean. For example, improvements in predicting wave conditions were necessary for the amphibious invasions of Europe and the Pacific islands. Mapping ocean basin features, such as their magnetic fields, helped detect submarines. These results also contributed to scientific studies after the war ended. For instance, HARRY HESS (1906–1969), while on duty in the U.S. Navy, mapped ancient seabed volcanoes that had been eroded at the ocean surface but later sunk to depths of several kilometers. This finding provided clues that finally led to the theory of plate tectonics.

Ocean studies to improve detection of submarines continued until the end of the cold war in the early 1990s. After the end of the cold war, newly declassified systems used by the U.S. Navy for tracking Soviet submarines were turned to detecting undersea earthquakes and tracking movements of whales, offering scientists new opportunities to observe the ocean and marine life.

International Studies

Ocean studies after 1950 required more ships and scientists than any one country could provide, and cooperative projects involving ships and other resources from many countries became the norm. In 1957–1958, the International Geophysical Year (IGY) accelerated this trend. IGY projects involved scientists from sixty-seven nations to study Earth features and ocean phenomena during a short period of intensive observations. The International Indian Ocean Expedition (1959–1965) began the modern phase of joint ocean studies with many nations contributing ships, scientists, and other support. The International Decade of Ocean Exploration (1970–1980) involved fifty-two countries around the world. Its purpose was to learn more about the ocean as a source of food, fuel, and minerals to replace depleted supplies on land.

These scientific studies provided the background for the United Nations Conference on the Law of the Sea, which in 1982 finished a comprehensive treaty. The new Law of the Sea set out new rules governing the ocean. It redefined maritime boundaries and established Exclusive Economic Zones. Finally, it set up a regime to govern exploitation of deep-sea mineral resources. After sixty countries ratified the treaty in 1994, it went into effect.

Undersea Exploration

Exploration under the sea surface began in ancient times. Alexander the Great (356–323 B.C.E.) reportedly used a glass barrel to observe below the sea surface. Various other devices used over the years did not give the mobility or visibility required for scientific studies. The first breakthrough came with the invention of the scuba (self-contained underwater breathing apparatus) in 1943 by Jacques Cousteau and Emil Gagnan. Scuba uses pressurized gas tanks to supply air for breathing underwater. The device uses a pressure regulator and mouthpiece to allow swimmers to remain underwater for an hour or more free from cables or hoses. When used first in the late 1940s, scuba opened the undersea world for both scientists and commercial divers. It is still widely used in surface waters (100 m or less) and on the continental shelf. Various new combinations of gases now allow divers to remain underwater much longer and to diver deeper.

The next advance was the bathyscaphe, developed by the balloonist, physicist, and engineer Auguste Piccard (1884–1962). A gasoline-filled float provided buoyancy (like a balloon). A steel ball vessel under the float protected the two divers from the extreme pressures in the ocean depths. The bathyscaphe *Trieste,* purchased by the U.S. Navy, dove on 23 January 1960 12,000 m to the bottom of the Challenger Deep, near the Mariana Islands in the western North Pacific Ocean. Unfortunately the balloon-like *Trieste* had limited mobility and no capability to take samples and was soon retired.

Perhaps the most famous submersible was *Alvin,* built in 1964 by the U.S. Navy for the Woods Hole Oceanographic Institution. Initially used for deep-sea engineering studies, *Alvin* was later used by scientists supported by the National Science Foundation (NSF), the National Oceanic and Atmospheric Administration (NOAA), and the Office of Naval Research.

Because of Cousteau's early work, the French were very active in undersea research and built two submersibles, *Archimede* and *Cyana.* These two submersibles joined with *Alvin* during 1973–1974 in the French-American Mid-Ocean Undersea Study (FAMOUS) examining the Mid-Atlantic Ridge. This was the first scientific use of submersibles and again demonstrated the benefits of international studies. In the 1980s the United States and France built deep-diving (6,000 m) submersibles, followed in the early 1990s by the Japanese, who built a submersible with a 6,500-m capability. The Soviet Union built two deep-diving (6,000 m) submersibles, later operated by Russia.

In the mid–1970s, *Alvin* and U.S. scientists explored a rift zone near the Galapagos in the eastern equatorial Pacific. There they discovered entirely unknown communities of fast-growing organisms living on the dark ocean bottom. These organisms and the bacteria that live in them obtain their energy from poisonous hydrogen sulfide and metals, not from the Sun. These materials came from waters superheated (400°C) while flowing through recently erupted lava on the sea bottom.

Remotely operated vehicles (ROVs), combining highly maneuverable platforms with cameras and manipulators operated from surface ships, began to supplant manned submersibles. Lacking the constraints of a crew, ROVs can stay submerged for much longer periods and can descend much deeper. In early 1994, the Japanese announced that their ROV had returned to the Challenger Deep and approached but did not exceed the previous depth record set by *Trieste.*

Satellite Oceanography

In the late 1950s, satellites were first launched carrying cameras that transmitted images of Earth's surface. Observations of clouds and atmospheric features were first obtained. *Seasat,* launched 7 July 1978, was the first ocean-observing satellite; it remained in orbit only 102 days. The mass of data from *Seasat* showed the utility of satellites to provide global ocean observations every few days. Next was the Coastal Zone Color Scanner, a spectrometer that sensed different colors in surface sea waters, which was launched in 1978 on *Nimbus*-7. CZCS observed ocean surface color until 1988. This ten-year record provided the first global picture of the worldwide distribution and abundance of chlorophyll in the ocean, and thus the first picture of global ocean productivity. Now satellite observations are indispensable for ocean research and for making predictions of ocean conditions.

Many countries have launched ocean-observing satellites. The European Space Agency's *ERS-1* carried a variety of ocean-observing sensors. The U.S. and France launched *Topex/Poseidon,* which observed subtle variations in ocean-surface topography that indicate locations and strength of ocean currents. Japan and several other countries have established ocean-observing satellites. Some have been launched by commercial companies and sell their data directly to users. These observations are also used for ocean predictions and commercial services.

Field Experiments

After 1960, field experiments became the dominant procedure. Whereas earlier expeditions obtained data and samples for later study ashore, field experiments study oceanic processes as they occur at sea.

Such experiments now require ships, aircraft, satellites, and other support, usually supplied by several countries, all working together. Samples taken from ships are analyzed while the experiment is underway. As aircraft and satellites survey the area, they transmit data to ships and to laboratories. This allows scientists to change sample locations and timing as they learn more about the processes involved. Finally, scientists can share results, using electronic networks, so that the data are available nearly simultaneously to investigators anywhere in the world. These field experiments can study large areas of the ocean, which is now necessary to understand ocean processes.

Ocean Drilling

In the 1960s scientists began drilling into the ocean bottom. Initially the intent was to drill through Earth's outer layer, the crust, to study the rocks below; this was called the Mohole Project. Preliminary work demonstrated that commercial drilling ships could drill into the ocean floor to sample sediment deposits and crustal rocks. Deep-drilling capabilities to penetrate the crust required building large and expensive ships. Because of the high costs, the deep-drilling project was cancelled but shallow drilling survived. In 1968, the U.S. National Science Foundation began the Deep Sea Drilling Project (DSDP), and the *Glomar Challenger* was converted for scientific drilling. In 1975, France, Germany, Great Britain, Japan, and the Soviet Union were invited to join the United States, forming the International Phase of Ocean Drilling. In 1983, the *Glomar Challenger* was retired.

The successor Ocean Drilling Program (ODP) began operations in 1985 after converting a commercial drillship, the *JOIDES Resolution*. ODP was also an international effort, with the original partners participating plus Canada, Australia, and several small European nations. Scientific ocean drilling is planned to continue through the end of the century.

Bibliography

DEACON, M. *Scientists and the Sea: 1650–1900.* New York, 1971.

GROSS, M. G. *Oceanography: A View of Earth.* Englewood Cliffs, NJ, 1993.

HSU, K. J. *Challenger at Sea: A Ship That Revolutionized Earth Science.* Princeton, NJ, 1992.

KAHARL, V. A. *Water Baby.* New York, 1990.

SEARS, M., and D. MERRIMAN, eds. *Oceanography: The Past.* New York, 1980.

M. GRANT GROSS

OCEANOGRAPHY, PHYSICAL

Physical oceanography is the study of the motion and physical properties of water in the oceans, marginal seas, and nearly enclosed seas. Physical oceanography covers a large range of space and timescales, from the smallest surface ripples to climate dynamics that equally involves the atmosphere. It is a subdiscipline of fluid mechanics, which is the study of the physics of fluids, and a subdiscipline of oceanography, which is the study of all phenomena in the ocean (*see* OCEANOGRAPHY, CHEMICAL; OCEANOGRAPHY, GEOLOGICAL; and OCEANOGRAPHY, BIOLOGICAL).

Forcing Mechanisms

The open ocean averages about 4,500 meters (m) in depth, and is bordered by continental slopes that rise to continental shelves and then to the shoreline. The forces that drive seawater motions are frictional stresses, gravitational attraction, and pressure differences. For periods longer than one day, the rotation period of Earth, the Coriolis and centrifugal accelerations are also important. In a nonrotating reference frame fixed to the stars, unforced motion must be in a straight line. However, in our rotating frame of reference, the centrifugal force appears to fling objects, seawater, and the atmosphere outward; this outward pseudo-force is countered by Earth's gravity, which holds objects on Earth. Also in our rotating frame, the Coriolis acceleration turns moving objects to the right in the Northern Hemisphere and to the left in the Southern Hemisphere, at the rate of changing the direction of motion completely around in one day, without changing the speed.

The gravitational attraction between Earth and the Moon and the Sun causes the tides. Winds blow across the sea surface driving surface waves that range from centimeter-long capillary waves to long ocean swells; the restoring force for the surface waves is Earth's gravity. Winds cause turbulence and mixing in the upper 50 to 100 m of the ocean, creating a "mixed layer." Turbulence at the base of the mixed layer can force internal waves, which are waves pulled down by gravity, and which ride the stratification of the interior ocean just as surface waves ride the stratification between water and air. Finally, winds acting longer than a day cause flow in the upper 100 m to the right of the wind in the Northern Hemisphere and to the left in the Southern Hemisphere; this is called the "Ekman transport" (*see* EKMAN, VAGN VALFRID). Where the Ekman transport is convergent, water is pushed downward and forces the large-scale ocean circulation to flow toward the equator; where it is divergent, large-scale flow is poleward. This forces the wind-driven general circulation, between the surface and about 2,500-m depth; the circulation consists of ocean-wide clockwise and counterclockwise "gyres." Both equatorward- and poleward-flowing waters return to their origin in strong currents located at the western boundary. The deep ocean circulation is forced by sinking of dense water in small regions at high latitudes. The compensating deep upwelling elsewhere in the ocean drives the deep water poleward; the circulation is closed by strong western boundary currents.

The pressure differences causing motion on timescales longer than a day and horizontal scales larger than the ocean depth are parallel to Earth's surface, or more precisely along surfaces of constant geopotential. The pressure at depth depends on the total mass of seawater above that depth, and so depends on the sea surface height and the density of the seawater lying above the point. The strongest surface currents are about 1 meter per second (m/s) and the sea surface height difference across them is about 1 m over 100 kilometers (km); in comparison, wind-forced surface waves can reach 20 m in height. Seawater density varies with position and depends on pressure, temperature, and salinity. The pressure dependence is not important for the general circulation. Seawater temperature and salinity can be changed by heating/cooling at the sea surface, by heating from below Earth's crust, and by evaporation and precipitation at the surface.

Properties of Seawater

The temperature range in the ocean is −1.8°C to 30°C. The freezing point is lower than that of freshwater due to the presence of dissolved salts. The highest temperatures occur at the sea surface in the tropics; higher temperatures are not attained because the air is heated from below by the ocean, rises, and produces clouds, which keep the Sun from further warming the sea surface. The ocean is warm at the surface and cold below; most of the ocean is colder than 2°C, indicating that the waters below about 1,000 m originate at the sea surface at high latitudes.

Salinity is the number of grams of dissolved matter in 1 kilogram (kg) of seawater; in practice it is measured using a proxy, the seawater conductivity, which depends on both temperature and salinity. Sea salts originate from continental weathering as well as from weathering of submarine rocks and marine inputs of volcanic fluids. The different constituents of sea salt are present in remarkably uniform proportions because the overall mixing time of the oceans, about 1,000 years, is much shorter than that of weathering, and the residence times of the major elements dissolved in seawater (*see* WEATHERING AND EROSION). Salinity in the open ocean ranges from 32 to 37. Lowest open ocean salinities are found at high latitudes where precipitation is larger than evaporation, and where melting sea ice floods the surface with relatively freshwater. (Sea ice extrudes salt as it freezes; thus the salinity of the sea ice is much lower than the salinity of the original seawater.) Highest salinities are found near the sea surface at low, but not equatorial, latitudes, where evaporation dominates over precipitation.

Water retains heat much more efficiently than the atmosphere. The physical quantity describing this is the heat capacity, which is the change in heat content for a given temperature change. The ocean is thus a major moderating influence in weather and climate at all timescales.

Ocean Circulation

The ocean circulation is clockwise around high-pressure centers in the Northern Hemisphere and counterclockwise around highs in the Southern Hemisphere, due to the Coriolis force. The gyres are asymmetric, with much stronger currents at the western boundary returning slow interior flow that is forced equatorward or poleward by Ekman layer convergence/divergence or deep upwelling. The strong western boundary current best known to North Americans is the Gulf Stream, which returns the flow of the North Atlantic's subtropical gyre. The other subtropical western boundary currents are the Kuroshio (North Pacific), the Brazil Current (South Atlantic), the Agulhas Current (Indian), and the weaker East Australian Current (South Pacific). The western boundary currents for the subpolar gyres are the Labrador Current (North Atlantic) and the Oyashio (North Pacific). Because the ocean is open all around Antarctica, eastward winds force a strong current all around the continent, called the Antarctic Circumpolar Current.

The ocean circulation changes with increasing depth. The wind-driven subtropical gyres shrink poleward and toward the western boundary; most evidence of the wind-driven circulation disappears below about 2,000 m depth. However, the western boundary currents and strong recirculating currents just next to them may reach to the ocean bottom, as does the Antarctic Circumpolar Current. The global overturning circulation, sometimes referred to as the conveyor belt, involves sinking of cold, dense water in the northern North Atlantic/Arctic Ocean and around Antarctica. Deep water is not formed in the northern North Pacific. The sinking water requires upwelling throughout the oceans, which drives the deep circulation; the global distribution of upwelling is not well known. The upwelling vertical velocities are on the order of 0.00001 centimeters per second (cm/s), whereas the horizontal circulation that they drive is on the order of 1–10 cm/s.

In the Atlantic and Pacific Oceans, the overturning cell consists of deep water between about 1,000 and 3,000 m flowing southward to the Antarctic, and bottom water below and intermediate water above flowing northward from the Antarctic. In the North Atlantic and North Pacific, the low salinity intermediate water of Antarctic origin loses its identity, and intermediate water formed locally is more apparent. The Indian Ocean has no high northern latitudes, and its overturning cell consists of northward flow of deep and bottom waters, and southward flow of upper ocean waters.

North Atlantic deep water (NADW) is a mixture of overflow of cold, oxygenated water from the Arctic, outflow from the Labrador and Mediterranean Seas, and inflow of South Atlantic bottom and intermediate water. Because some sources of NADW are at the sea surface, its properties in the North Atlantic indicate recent renewal. The timescale for deep water renewal in the northern North Atlantic is only about twenty years. NADW flows southward into the South Atlantic, then eastward around Africa into the Indian Ocean, and then past Australia into the Pacific Ocean. Its eastward flow is part of the Antarctic Circumpolar Current, where it is characterized by high salinity and high oxygen. As it mixes with the much older deep water exiting the Pacific Ocean, its properties are modified and it becomes known as Circumpolar Deep Water.

The bottom water flowing northward in all oceans is formed in the Weddell and Ross Seas south of the Antarctic Circumpolar Current. The Pacific Ocean's deep water is formed as a mixture of the inflowing intermediate, deep, and bottom waters; because the surface water in the northern North Pacific cannot be made dense enough to sink to great depths, the only sources of Pacific Deep Water are the Atlantic and Antarctic; hence it is the oldest water in the world ocean. The age of Pacific Deep Water (determined by ^{14}C analysis; *see* ISOTOPE TRACERS, RADIOGENIC) in the northern North Pacific is 500 to 1,000 years, which is thus the timescale for the complete global overturning cell.

Shallow and weak boundary currents are found at the eastern boundaries of the oceans. These are forced by large-scale winds that veer to blow alongshore as they approach the continents. When the wind blows toward the equator, it forces Ekman layer flow offshore; this results in upwelling from about 200 to 300 m below the sea surface. Because the upwelled water originates below the euphotic (sunlight) zone, it is cold and undepleted in nutrients, and therefore sustains richly productive fisheries. An alongshore current toward the equator is also created by the pressure difference arising from the offshore Ekman transport. The eastern boundary currents and upwelling regimes are: the California Current (North Pacific); the Peru or Humboldt Current (South Pacific); the Canary Current (North Atlantic); and the Benguela Current (South Atlantic). The Indian Ocean does not have a counterpart; instead the Leeuwin Current flows poleward along the entire west coast of Australia.

At the equator in the Pacific and Atlantic Oceans, the trade winds blow westward, causing both Ekman transport away from the equator, and westward flow right along the equator where the Coriolis effect is negligible. The first results in upwelling at the equator, so the surface water on the equator is relatively cold. The westward flow creates a warm surface water pool in the west; this creates a pressure gradient that pushes water toward the east just below the surface. The resulting eastward flow, centered at 150 m depth, is the Equatorial Undercurrent, whose speed rivals the Gulf Stream, but which is only several hundred meters thick. Below the Equatorial Undercurrent there are many reversing jets flowing along the equator; in general the circulation within several degrees of the equator is strongly east–west and complex.

Climate Variations and the Ocean

The ocean moderates weather and climate on all timescales because of its large heat capacity relative to air. Examples are: the daily sea and land breeze alternation common along the coast; the seasonal moderation of coastal climates; the strong seasonal cycle of the monsoons in the northwestern Indian Ocean; changes over several years to decades, such as El Niño, that are related to the timescale of circulation in the upper ocean; and changes over hundreds of years that are related to the timescale of the global overturning circulation.

El Niño was originally described as a devastation of the Peruvian coastal fisheries occurring every three to seven years, with its onset late in the year—near Christmas, hence its name. El Niño is now known to be a coupled atmosphere-ocean phenomenon that involves the entire tropical Pacific and eastern Indian Ocean. Normally, surface trade winds blow from high pressure in the eastern Pacific to low pressure in the western Pacific, with rising air over Indonesia resulting in large precipitation, and sinking air over the eastern Pacific and western South America causing desert conditions. The trade winds sustain the equatorial circulation described in the previous paragraph. The cold surface water in the eastern Pacific sustains the high atmospheric pressure there, and the warm surface water in the western Pacific sustains the low pressure and rising, cloud-forming air there. During an El Niño, the trade winds slacken, and the warm water in the west sloshes back toward the east, further weakening the trades. The center of rising air moves eastward and weakens, causing drought in Indonesia and Australia, and much higher precipitation than normal in the central and eastern tropical Pacific. The warm surface water in the eastern equatorial Pacific flows southward along South America beneath the northward-flowing Peru Current; the water that is upwelled into the Peru Current is thus nutrient-depleted and cannot sustain the normal fisheries yield. Normal conditions return when a region of rising air and precipitation moves into the western Pacific from the eastern Indian Ocean.

The length of time between El Niño episodes is about three to seven years. This time is related to the length of time it takes water to flow across the Pacific from west to east. Advances in modeling and observations as a result of the intensive Tropical Ocean–Global Atmosphere (TOGA) program mean that an occurrence can be predicted with skill six months to a year in advance. One legacy of TOGA is a reporting network consisting of island and coastal measurements and a large array of instruments moored in the tropical Pacific, measuring subsurface currents and temperature, and surface meteorological conditions.

Climate changes over periods of decades to centuries are much less well measured and understood than El Niño. It has become apparent that there is a ten-year oscillation in the North Atlantic that is important for weather conditions on both sides of the ocean; this is comparable to the timescale for circulation around the northern North Atlantic. Climate changes on glacial timescales of 20,000 to 100,000 years are mainly driven by variations in Earth's orbit (Milankovitch cycles; *see* GLOBAL ENVIRONMENTAL CHANGES, NATURAL). Deep-sea sediment cores are used to track deep water properties and suggest that changes in climate during these cycles actually occur over about 1,000 years; this shorter timescale matches that of the ocean's global overturning circulation, suggesting its importance in modifying the climate.

The World Ocean Circulation Experiment (WOCE) of the 1990s seeks to understand today's circulation as clearly as possible, including analysis of existing time series and startup of new series in regions other than the tropical Pacific. Understanding what controls the overturning circulation, including North Atlantic Deep Water and Antarctic Bottom Water formation, is a goal of both WOCE and the Atlantic Climate Change Program (ACCP). Measurements include: the seawater properties that determine density and from which the pressure gradients that drive the circulation can be estimated; natural and man-made tracers such as oxygen, nutrients, chlorofluorocarbon (freon), tritium, helium-3, carbon dioxide, and radiocarbon from which the pathways of circulation can be observed; repeated profiling of temperature across the oceans from ships of the merchant marine to determine changes in temperature and circulation; and velocity from surface drifters and subsurface floats deployed throughout the oceans. Satellites provide important new tools for ocean measurements: for example, the altimeter which measures sea surface height to within centimeters of Earth's geoid; sea surface temperature; and surface color that is related to chlorophyll and hence ocean productivity.

Modeling improves continuously as computer power expands. It is now possible to model today's global ocean circulation with a resolution of 15 km in the horizontal and 100 m in the vertical. Present modeling of long-term changes reveals the great importance of salinity in determining the occurrence of deep overturning; oscillations in the global overturning circulation on the order of thousands of years are evident.

Unraveling the natural climate cycle requires understanding how the ocean circulates, overturns, and interacts with the atmosphere. Manmade climate change is a concern because of the potential for changing the advent of glaciations, rate of desertification, and sea level rise. One goal of climate research, which involves the coupled ocean-atmosphere system, is to understand the overall response to increasing amounts of greenhouse gases, such as carbon dioxide and chlorofluorocarbons. Oceanographic measurements to monitor long-term shifts in average temperature include traditional time series near islands, analysis of long coastal records collected for fisheries and navigational purposes, merchant ship measurements of surface and subsurface properties, and a new effort, Acoustic Thermometry for Ocean Climate (ATOC), to measure temperature change acoustically using several sound sources and many receivers spread around the world's oceans. The coupled ocean-atmosphere system is being modeled in a number of laboratories, focused on the response to a doubling of greenhouse gases in the atmosphere.

Bibliography

DUXBURY, A. C., and A. B. DUXBURY. *An Introduction to the World's Oceans*, 4th ed. Dubuque, IA, 1994.

GILL, A. E. *Atmosphere-Ocean Dynamics*. San Diego, CA, 1982.

PICKARD, G. L., and W. J. EMERY. *Descriptive Physical Oceanography*, 5th ed. Tarrytown, NY, 1990.

TOMCZAK, M., and J. S. GODFREY. *Regional Oceanography: An Introduction*. Tarrytown, NY, 1994.

LYNNE D. TALLEY

OIL AND GAS, IMPROVED RECOVERY OF

Improved recovery methods refer to all processes and treatments used for the recovery of oil left in the reservoirs at the exhaustion of the natural drive energy of the reservoir, that is, at the termination of the primary recovery process. A summary of conventional and improved oil recovery techniques is given in Figure 1 (Donaldson et al., 1989; Leonard, 1986; Gregory, 1994).

Oil from underground oil-bearing reservoirs is produced through a series of wells drilled into the reservoirs (*see* OIL AND GAS, RESERVES AND RESOURCES OF). A schematic of a typical system of reservoir and wells referred to as a five-spot arrangement is shown in Figure 2. Reservoir oil flows to the production wells due to the difference between the fluid pressure in the reservoir and in the production wells.

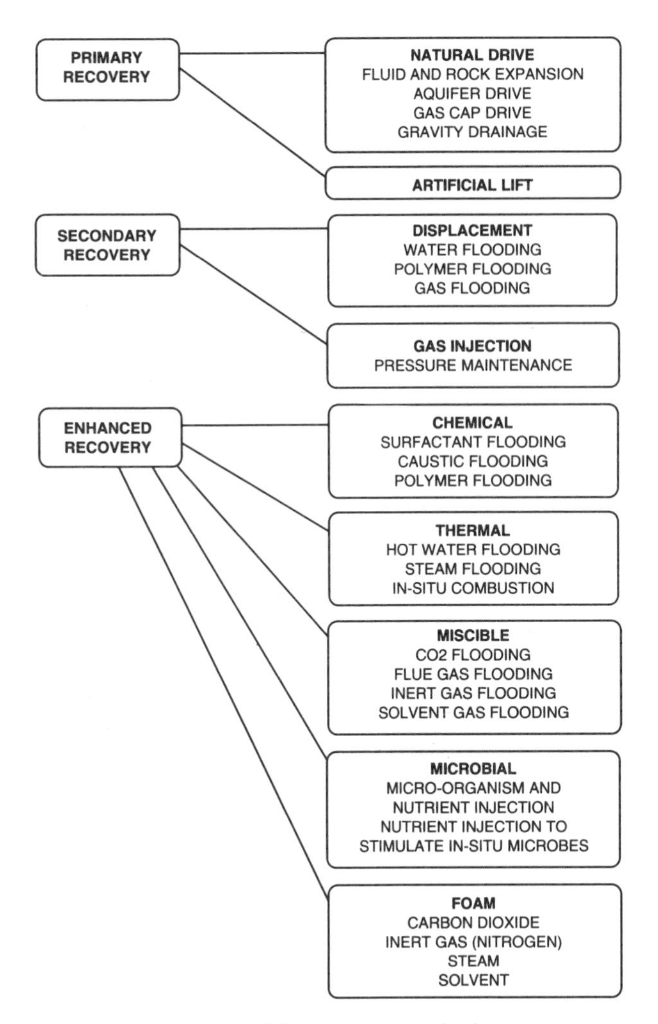

Figure 1. Oil recovery techniques.

Primary Recovery

During the primary recovery of oil, which begins when newly completed wells are opened for production, the energy driving the oil production is provided primarily by the expansion of oil, water, oil-bearing rock, and any free gas liberated from oil and water (Tiab and Civan, 1989). Under favorable conditions the gravity force provides additional drive energy (Gregory, 1994). The amount of oil that may be recovered from an oil reservoir depends on the properties of the oil-bearing formations, fluids contained in these formations, and the development and operational management strategies followed during the oil production. When the natural energy becomes insufficient to drive the oil to the surface, it may be supplemented by applying artificial lifting methods to lift the oil

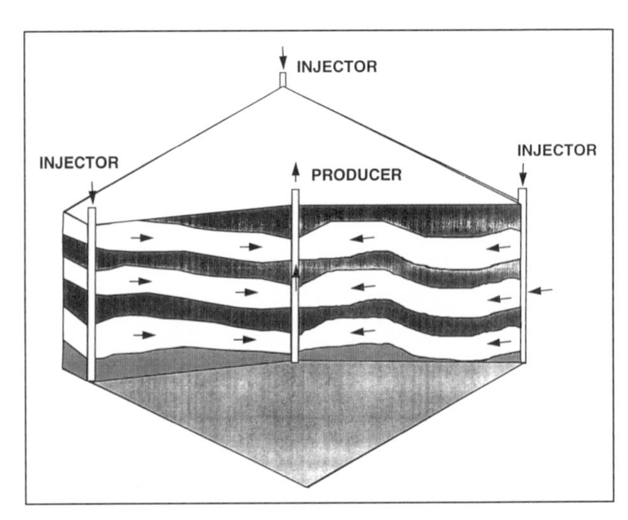

Figure 2. Typical reservoir and wells system.

from the producing wells to the surface (Leonard, 1986).

The primary recovery of oil terminates upon the exhaustion of the natural reservoir energy. Therefore, improved recovery techniques must be used to recover additional oil. The improved recovery techniques are generally classified as the secondary or supplementary recovery techniques and the tertiary or enhanced recovery techniques.

Secondary Recovery

In the secondary recovery stage, some production wells are converted to injection wells in order to inject fluids such as water, gas, or polymer solutions for supplementing the energy necessary to displace the oil toward the producing wells. The fluids used for the secondary recovery processes do not react with the reservoir fluids and, therefore, do not cause any alteration of the fluids in the reservoir (Gregory, 1994).

Gas Flooding. Gas injection methods are applied for pressure restoration, pressure maintenance, and/or gas-drive purposes based on the way the gas is injected into a reservoir. In the pressure restoration method, the gas is injected into the reservoir through some wells while other wells are closed until the desired pressure is reached. Restoration of pressure may require a period of over a year. Then the production wells are reopened to start the production of oil.

In the pressure maintenance method, the gas from producing wells is injected into the selected wells to prevent the loss of the reservoir pressure. In this method some wells are operated as injection wells while the others are operated as the production wells.

In the gas-drive method, the gas is injected into the reservoir under pressure and a continuous gas flow is maintained from the injection wells to the production wells. The injected gas pushes the oil bank ahead of the gas front toward the production wells.

Water Flooding. In the water flooding process, water is injected into the reservoir, through injection wells, to drive oil toward the production wells. To improve the efficiency of the water flooding method, the viscosity of the water is adjusted by adding chemicals such as polymers, molasses, and glycerin. These additives help achieve better sweep of the reservoir oil by decreasing the tendency of the injected water to finger through and bypass the oil zone.

In some cases, the recovery efficiency can be maximized by applying a combination of these methods, an example of which is the method known as the water alternating gas injection (Boberg, 1988).

Enhanced Recovery

Experts estimate that about 20 to 50 percent of oil in reservoirs can be recovered by the primary and secondary recovery techniques depending on the characteristics of the oil and the reservoir formation (Donaldson et al., 1989). To recover the oil remaining after the secondary recovery, enhanced oil recovery techniques are used. Enhanced oil recovery involves processes such as chemical, thermal, miscible, and microbial recovery. Enhanced oil recovery techniques involve reactive processes that may alter the physicochemical properties of the fluids and the oil-bearing rock. Further, enhanced oil recovery techniques are usually more expensive and often may not be technically, economically, and environmentally justifiable. For example, the loss of fluid conductivity of oil-bearing formation resulting from various rock-fluid interactions can severely affect the performance of the enhanced oil recovery processes.

Chemical Recovery. Chemical recovery can be accomplished by injecting reactive fluid systems into reservoirs such as surfactant, polymer, or caustic solutions. Surfactant flooding is aimed at lowering the interfacial tension between the water and oil for the purpose of releasing and mobilizing the oil trapped in rock-matrix. Surfactant flooding is technically feasible for low- to medium-viscosity oils, but requires extensive expertise for field applications.

Water-soluble polymer flooding is aimed at the mobility control of the oil–water front in order to improve the sweep efficiency of water flooding. The performance of polymer flooding is adversely affected by decomposition due to thermal, chemical, and/or biological processes, and by injectivity decline due to plugging of the rock-matrix of the near wellbore formation.

Caustic flooding is aimed at reducing the interfacial tension between water and oil by in situ formation of surfactants due to the reaction of the caustic solution with the acidic constituents in oil. Caustic solution often undergoes chemical reactions with the petroleum-bearing rock resulting in severe conductivity loss which diminishes the advantage of caustic flooding.

Thermal Recovery. Thermal recovery processes are aimed at improving the recovery of viscous oil and tar by heating to reduce viscosity. This process is achieved by means of hot water or steam injection and in situ combustion. Reduced viscosity allows oil to flow through the flow passages more easily. Thermal recovery methods are among the most successful enhanced oil recovery techniques.

Miscible Flooding. The goal of miscible flooding is to reduce the interfacial tension between the water or gas and the oil by dissolving oil in injected fluids such as compressed carbon dioxide, nitrogen, and light hydrocarbons. Miscible flooding is practical only for low- and medium-viscosity oils. One of the major deficiencies of the miscible flooding process is the poor sweep efficiency resulting from the fingering and gravity overriding problems associated with the injected fluids. However, some of these difficulties can be overcome by using cyclic injection of gas and water or by using foams and polymers.

Microbial Methods. The microbial enhanced oil recovery process utilizes microbes to recover oil at low cost. This method facilitates multiple recovery mechanisms that participate simultaneously in the oil recovery process. This is accomplished by introducing or stimulating viable microorganisms in an oil reservoir to produce a variety of gases, acids, surfactants, and biopolymers. These products improve the overall oil displacement efficiency. Field implementation of microbial enhanced oil recovery techniques has been rather limited compared to the other techniques due to the lack of understanding of the reliability and effectiveness of these techniques.

It is apparent that successful application of the improved oil recovery techniques requires strong cooperation between science, engineering, and technology for optimal selection and design of processes and the reservoir management strategies suitable for the characteristics of the subsurface oil-bearing formations.

Bibliography

BOBERG, T. C. "Thermal Methods of Oil Recovery." In *An Exxon Monograph.* New York, 1988.

DONALDSON, E. C., ed. "Microbial Enhancement of Oil Recovery—Recent Advances." In *Developments in Petroleum Science* 31, ed. G. V. Chilingarian. Amsterdam, 1991.

DONALDSON, E. C., G. V. CHILINGARIAN, and T. F. YEN. "Enhanced Oil Recovery, II, Processes and Operations." In *Developments in Petroleum Science* 17B, ed. G. V. Chilingarian. Amsterdam, 1989.

GREGORY, A. T. "DTI's Improved Oil Recovery Strategy." In *Chemical Engineering Research and Design, Part A: Transactions of the Institution of Chemical Engineers* 72, part A (March 1994):137–143.

LAKE, L. W. *Enhanced Oil Recovery.* Englewood Cliffs, NJ, 1989.

LEONARD, J. "Increased Rate of EOR Brightens Outlook." *Oil & Gas Journal* (April 14, 1986):71–161.

NATIONAL SCIENCE FOUNDATION—EXPERIMENTAL PROGRAM FOR STIMULATION OF COMPETITIVE RESEARCH. R11-8610676 Project Final Report. Washington, DC, February 1989.

PETZET, G. A., and B. WILLIAMS. "Operators Trim Basic EOR Research." *Oil & Gas Journal* (February 10, 1986):41–46.

TIAB, D., and F. CIVAN. "Thermodynamic Analysis of Fluid Flow in Porous Media."

FARUK CIVAN
ANUJ GUPTA

OIL AND GAS, PHYSICAL AND CHEMICAL CHARACTERISTICS OF

The terms oil and gas, or more specifically crude oil and natural gas, are commonly used to define the liquid and gaseous phases of petroleum. It is a widespread misconception that oil and gas are essentially separate and unrelated entities, each having a relatively simple and constant composition. In reality, these two substances constitute a continuous spectrum of related chemical compounds numbering into the thousands. They are often, but not always, found together in the same subsurface reservoir, from which they are extracted simultaneously and physically separated after reaching the surface. The petroleum accumulated in some reservoirs is entirely in the gas phase, but an appreciable portion may condense to liquid upon encountering cooler temperature at the surface. This liquid product usually is called condensate.

Crude oil, being an extremely complex mixture of chemical compounds, is highly variable in composition and physical properties. It ranges from a light (low-density), almost colorless liquid to a tarry black substance that barely flows. Some oils that are liquid in the reservoir will actually solidify after cooling to surface temperature because they contain a high proportion of waxes. Oils of this type are common in the Uinta Basin of northern Utah.

In contrast to crude oil, natural gas is a much simpler and less variable mixture. Some accumulations consist very predominantly of methane and are defined by production engineers as "dry gas"; others may contain significant quantities of condensate, and are termed "wet gas." Natural gas usually contains no more than 15–20 components in measurable amount.

Composition of Crude Oil

To understand the reasons for crude oil variations and their significance in commercial uses of an oil, a rudimentary knowledge of its chemical composition is necessary. Although there are exceptions, as will be discussed later, a crude oil is dominated by hydrocarbons. As the name implies, these compounds contain only the elements carbon (C) and hydrogen (H) in their molecular structures. Hydrocarbons consist of these general types:

1. Alkanes—open-chain molecules with only single bonds between carbon atoms. They may occur

as either a straight chain or a branched chain, and are often called "paraffins" by organic chemists, petroleum scientists, and engineers.

2. Cycloalkanes—special forms of alkanes in which the carbon atoms form one or more rings, almost always with either five or six atoms per ring. They are frequently called "cycloparaffins" or "naphthenes" by petroleum technologists. Alkanes and cycloalkanes are the dominant hydrocarbon types in a vast majority of oils.
3. Alkenes—hydrocarbons with one or more double bonds between carbon atoms. Alkenes, also commonly called "olefins," are rarely found in crude oil.
4. Aromatics—hydrocarbons that contain one or more benzene rings in the molecule. They are normally much less abundant than alkanes, but on rare occasions will constitute more than half of an oil deposit.

Crude oil may also contain nonhydrocarbon components, defined as those in which at least one atom of an element other than H and C is present. In crude oil, these elements are almost always nitrogen, sulfur, or oxygen (N, S, O). Sulfur-containing compounds are usually much more abundant than those with N or O (Table 1).

Effect of Composition on Physical Properties

Variations in relative proportions of the above chemical types among oils are responsible for much of the variability observed in physical properties such as density, viscosity, and "freezing point" (pour point). These properties are especially important to oil producers, who are responsible for pumping the oil out of the ground and transporting it to refineries.

Table 1. Average Composition of Crude Oil

Elemental:		
Carbon	84.5%	(by weight)
Hydrogen	13.0%	
Sulfur	1.5%	
Nitrogen	0.1%	
Oxygen	0.1%	
Molecular Type:		
Alkanes	25%	(by weight)
Cycloalkanes	50%	
Aromatics	17%	
Asphaltics	8%	

Hunt, J. M. *Petroleum Geochemistry and Geology*, 1979.

Density of crude oil is almost always less than 1.0 (the density of water). On very rare occasions a thick tarry oil will be heavier than water. Oil producers and refiners measure crude oil density as degrees of American Petroleum Institute (API) gravity. On this scale, water has a value of 10°, and gravity increases with decreasing density. An oil's density (API gravity) is primarily a function of the relative amounts of various compound types it contains, although molecular size also is significant. Among components of comparable molecular size, API gravity decreases in the order straight-chain alkanes > branched-chain alkanes > cycloalkanes > aromatics > nonhydrocarbons. For example, an oil with high aromatics content will have lower gravity (more dense) than one containing mostly alkanes, and an oil with high sulfur content will have the lowest gravity of all. Producible oils generally range between about 10–50° API, although condensates will usually exceed 50° because their average molecular size is much smaller than that of crude oil.

Viscosity generally parallels density (varies inversely with API gravity). Alkanes are less viscous than aromatics or nonhydrocarbons, but molecular size becomes more important in determining viscosity than it is for density.

A property called pour point is another element of special importance to oil producers. It is the temperature at which an oil ceases to flow as it is slowly cooled. Pour point can be considered as the freezing point of an oil, although there actually is no specific freezing point for a crude oil because it contains so many components. In general, the more viscous low-gravity oils will have a higher pour point, but there are many important exceptions. Most notable are the waxy oils from the Uinta Basin, mentioned above, that have high pour points because the waxes are dominated by compounds with straight carbon chains. Straight-chain components typically freeze at higher temperatures than do branched and cyclic components. These same oils, however, have higher-than-average gravities and lower viscosities at temperatures above the pour point.

Reasons for Crude Oil Variability

There is no single cause for the wide variations observed in oil composition. Many factors enter in, but most of them can be grouped into two general categories: (1) nature of the source material from which crude oil is derived; and (2) processes that can alter the composition of an oil after it has accumulated in a reservoir.

Variations in Source. The type of organic matter that generates crude oil is determined primarily by two interrelated factors, that is, the kinds of organisms originally deposited in a source sediment and the environment in which they are deposited. Depositional environment actually dictates to a great extent what species of organisms will be dominant at any particular subaqueous location. It includes important considerations such as water salinity, depth, oxygen supply, input from rivers, and so on. Petroleum geochemists can now successfully determine the depositional environment of an oil's source from analysis of the oil.

Oils from open marine sources tend to have a dominance of cycloalkanes and aromatic hydrocarbons, plus moderate amounts of nonhydrocarbons. Those formed in deltas and coastal swamps usually are high in paraffins but low in cyclic components and nonhydrocarbons, primarily because of input from land plant remains. Freshwater lake (lacustrine) oils generally contain small amounts of nonhydrocarbons and will often have high concentrations of paraffin waxes (e.g., Uinta Basin oils). Exclusion of oxygen from a sediment (reducing environment) often causes an oil derived from the organic matter in that sediment to have much lower API gravity, a greater proportion of aromatic hydrocarbons, and higher sulfur content.

Alteration of Oil Composition in the Reservoir.

1. Effects of temperature—As an oil accumulation becomes subjected to increasingly higher temperatures with deeper burial, it undergoes appreciable change in composition. Larger molecules, especially aromatic hydrocarbons and those containing N, S, and O, are lost from the oil as a result of polymerization reactions that form substances so complex they are no longer soluble. At the same time, larger alkane molecules are converted to smaller ones by thermal cracking reactions. The net result is a much higher gravity oil with greatly reduced amounts of N-S-O components.

2. *Biologic activity*—Microorganisms are the cause of some specific changes in oil composition. Many species of aerobic microbes have the capability of selectively consuming certain component types from an oil. Initially, microbes preferentially remove straight-chain alkanes, but once these components are nearly gone, branched alkanes become the next item on the microbial menu. With even more severe microbial activity, cycloalkanes and aromatic hydrocarbons will also be appreciably affected. Thus the action of microorganisms will cause a crude oil to become much lower in API gravity (from selective loss of high-gravity compound types), and have a much higher concentration of N-S-O components because they are relatively resistant to microbial attack. This kind of oil alteration always occurs in relatively shallow reservoirs, usually less than 1,500 m in depth, because the microorganisms are not active at the higher temperatures of deeper reservoirs.

3. *Water washing*—Movement of a water stratum beneath an oil accumulation can selectively remove those components having the greatest water solubility. Because aromatic hydrocarbons have higher solubility than alkanes, they will be lost to a greater extent. The net effect of water washing is a slight decrease in API gravity.

Effects of Composition on Commercial Value of an Oil. When an oil enters a refinery, it usually is first distilled into fractions that are subsequently processed further to obtain the products familiar to most everyone, such as gasoline, diesel fuel, jet fuel, heating oil, and lubricating oil. Heavy, nondistillable material becomes asphalt. The commercial value of a crude oil is determined to a great extent by its yield of those products that are the most profitable for the refiner. Gasoline, jet fuel, and diesel fuel are the principal moneymakers simply because they are consumed in greatest quantity; they are derived primarily from the lighter (higher-gravity) fractions of an oil. It therefore follows that an oil refiner will be willing to pay an appreciably higher price for a high-gravity oil than for one with low gravity because it will provide greater quantities of high-demand products.

Crude oils containing a high percentage of either sulfur or nitrogen have less commercial value than low-sulfur and low-nitrogen oils. In addition to causing odor and air pollution problems, high concentrations of either S or N interfere with many refinery processes and must be removed from initially obtained distillation fractions. Such a procedure is rather expensive to carry out, so the value of high-S and high-N oils to a refiner will necessarily be reduced.

It is evident that microbial activity is generally somewhat detrimental to the value of an oil, as it depletes the more desirable components and increases the concentration of sulfur and nitrogen. On the other hand, thermal alteration of an oil increases its value by increasing its gravity and removing S and N. The relative proportions of alkanes and cycloalkanes do not significantly affect an oil's commercial value, but might necessitate different refining strategies for maximizing particular products.

Bibliography

HUNT, J. M. *Petroleum Geochemistry and Geology*. San Francisco, 1979.

TISSOT, B. P., and D. H. WELTE. *Petroleum Formation and Occurrence*, 2nd ed. Berlin, 1984.

JACK A. WILLIAMS

OIL AND GAS, PROSPECTING FOR

Prospecting for oil and gas has evolved from a basic trial-and-error drilling to the application of sophisticated geophysical techniques to predict the best locations for drilling. This entry begins with a brief description of the terminology for oil and gas environments. Next, using this terminology, the more popular methods of exploration are described along with a brief history of their early use. Finally, this entry summarizes how a prospective oil or gas well is evaluated once a potential drilling location has been identified.

Basics of Oil and Gas Traps

Oil and gas are usually associated with sedimentary rocks (*see* SEDIMENTS AND SEDIMENTARY ROCKS, CHEMICAL AND ORGANIC). The three basic types of sedimentary rocks are shales, sands, and carbonates. In general, the shales are the sources of the hydrocarbons while the sands and carbonates act as the conduits and/or the containers.

Source Rocks. Shales often are deposited with some organic matter, for example, microscopic plant or animal matter, which become part of the rock. When the shales are squeezed under the pressure of overlying rocks and heated by the natural heat flowing from the earth, oil and gas may be formed depending on the type of organic matter involved and the temperature to which the source rock is raised. If a rock is capable of producing oil or gas, it is referred to as a source rock since it is a source for the hydrocarbons. An important element of modern exploration involves making certain that a new area being explored has source rocks capable of generating hydrocarbons. The petroleum industry relies on organic geochemists to evaluate the source rocks in new exploration environments. If the source rocks have not been heated enough or do not have enough organic matter, geochemists may recommend that a company try another area.

Reservoir Rocks. Once the potential for hydrocarbons has been verified in a region, explorationists look for reservoir rocks that potentially contain the hydrocarbons. Reservoir rocks are typically comprised of sands or carbonates. Sands are present at most beaches, so it is not surprising that a great deal of petroleum has been found in sands that were once part of an ancient beach. Other environments responsible for sandstones include rivers, river deltas, and submarine fans. An example of a carbonate environment is a coral reef, much like the reef located off the coast of Florida. The sediment deposited on reefs consists of the skeletons of tiny sea creatures made up primarily of calcium carbonate. The standstones and carbonates become a type of underground pipeline for transporting the oil and gas from the source rocks. These reservoir rocks have pore spaces (porosity) and the ability to transmit fluids (permeability). A rock with a high porosity and high permeability is considered a desirable reservoir rock because it can not only store more petroleum but the petroleum will flow easily when being produced from the reservoir. Finding reservoir rocks is not always easy and much of the effort of exploration is devoted to locating them.

Traps. Once the petroleum has been generated and squeezed from a source rock into a neighboring sand or carbonate, the petroleum is free to move through the porous and permeable formations until it reaches the surface or meets a place beyond which the petroleum can no longer flow. When the petroleum flows to the surface, the hydrocarbons are referred to as seeps. When the hydrocarbons are prevented from reaching the surface, they are said to be trapped. Traps are divided into two basic categories: structural and stratigraphic. Structural traps are caused by a deformation of the earth that shapes the strata of reservoir rocks into a geometry, which allows the hydrocarbons to flow into the structure but prevents their easy escape. Stratigraphic traps are the results of lateral variations in layered sedimentary rocks so that porous rocks grade into nonporous rocks, which will not allow the petroleum to move any further. The easiest traps to find have been structural traps, because deformation affects all strata in a pile of rocks. The stratigraphic traps are the most difficult to find because they occur without affecting much of the surrounding geology.

Geological Mapping for Oil and Gas

Geologists can be thought of as the historians of the earth. As discussed above, it is important to know when the hydrocarbons were generated and where and when the oil and gas traps were created. Thus history is important to exploration success. When a well is drilled, a number of devices are lowered down the well to log or identify the different formations that have been penetrated. The geologist uses the logs from many different wells to put together an interpretation or structure map of the depth of sedimentary rock formations beneath the surface of Earth. In addition to being historians, geologists must solve geometric puzzles because they use the bits and pieces of information about rock layers observed from the well logs in order to form a complete picture of the earth. They can, for example, compare well logs from two adjacent wells and determine if a fault cuts through one of the bore holes. The interpretation for modern geologists has been more challenging because many oil and gas wells purposely deviate from the vertical. Modern geologists employ numerous geometric tricks to assemble complicated 3-D interpretations of the earth and to identify where the reservoir rocks are located.

Early Geophysical Methods (1900s to 1950s)

Geophysical methods are used to obtain information on the geology away from the wells where the location of oil- and gas-producing formations is known. In the early 1900s, the use of geophysics to find structural traps was accomplished by employing seismic and gravity methods. Seismic methods were first used in World War I as a means of locating cannons. Influenced by this idea, seismic methods later were introduced to oil and gas exploration. In 1919–1921, J. C. Karcher mapped the depths to producing formations in Oklahoma by setting off explosions and recording the sound waves reflected from the different layers of rock. Since the depth to the top of the oil and gas formations could be mapped with this technique, the seismic reflection method was the first technique used to identify structural traps. Gravity methods were the earliest geophysical method used for oil and gas exploration, starting with the torsion balance developed by Roland von Eotvos of Hungary. Sensitive gravity instruments measure the changes in the pull of Earth's gravity at different locations due to density variations in the different types of rocks. Since salt is lighter than most sedimentary rocks, gravity measurements were a popular tool for finding reservoirs associated with salt domes. Many other geophysical methods were attempted, but the gravity and seismic methods have survived the test of time. Some early work on magnetic measurements were accomplished, but magnetic surveys did not become popular until more sophisticated magnetometers were developed after World War II. When Conrad and Marcel Schlumberger joined forces with Henri Doll in the 1920s, the first geophysical logging of wells was accomplished. The electric logs and other logs they developed played a major role in the early history of oil and gas exploration. The company they formed, Schlumberger Limited, has been an active contributor to the technological advances discussed below.

Explosion of Technology

In the 1950s the petroleum industry experienced an explosion of technology, a trend that continues today. One of the contributors in this effort was Harry Mayne, who developed common depth-point stacking, a method of adding seismic signals together. The resulting stacked signals were then plotted in the form of a picture called a seismic section. The seismic sections, which appear as cross-sections of the earth, were used initially to find structural traps. The creation of the stacked seismic sections benefited from digital recording methods that could be used to accurately record the amplitudes of seismic signals. True amplitude recording led to the discovery that the presence of hydrocarbons, especially gas, caused high-amplitude seismic reflections. These reflections, which appeared as bright spots on the seismic sections, enabled researchers to distinguish hydrocarbons from the surrounding rocks. Thus, in the 1970s bright-spot technology was developed by the petroleum industry to directly detect hydrocarbons from the surface. Shell Oil Company was the first company to use this new technology in the offshore region of the Gulf of Mexico, lowering the risk of falsely identifying oil and gas reservoirs. Unfortunately, other geological circumstances can create bright reflectors and some of these features were drilled.

As a result of these limitations in identifying stratigraphic traps, the industry sought a better understanding of stratigraphy. An important contribution in using seismic methods to map stratigraphy was developed. This improved method shaped the waveform leaving a seismic source into a shorter duration signal (the wave form is said to have been deconvolved). The deconvolution methods were introduced to the industry by Enders Robinson, who had been a part of radar research group at MIT during World War II. Robinson applied some of the same radar technology to compress the seismic wavelets so that small changes in the stratigraphy could be mapped.

Seismic stratigraphy, originally developed by Peter Vail and his associates at Exxon, was another important contribution to industry's ability to unravel the stratigraphy of the earth, that is, the geometry of the layering. As the industry pushed to find even more information about stratigraphic variations, methods were developed to use information about the shear-wave properties of rock as well as the compressional-wave velocities which had been the primary tool for seismic exploration. Shear waves or shear-related methods of exploration, such as the seismic method known as amplitude versus offset, or AVO, are used by the industry to identify changes in lithology (e.g., from sandstones to shales) to evaluate bright spots or other types of reflection amplitude anomalies, and to find rocks that have been fractured, thus making them into better reservoir rocks. Today, the petroleum industry has advanced to sophisticated seismic imaging and analysis techniques called inversion and migration (migration here refers to the proper placement of the reflectors rather than the migration of oil as discussed above). In the early 1970s Jon Claerbout at Stanford University introduced an important seismic imaging principle that is used to migrate seismic data accurately.

An important part of using surface seismic data is the identification of the oil- and gas-bearing zones on the seismic data. Vertical seismic profiles (VSPs) are often recorded with a source on the surface and with receivers down the well. In this way, the travel time to the reservoir can be measured along with other information relating the well data to the surface seismic data.

Another type of technology that has influenced oil and gas exploration is the aquisition of 3-D surface seismic data. This technology produces so much information that modern interpreters have been forced to use computers just to view their data. The 3-D seismic surveys are very much like a solid section of the earth. An interpreter can sit at a work station and literally slice through and view the section in almost anyway he or she wishes. Maps that might have taken months to produce in the early days of seismic mapping can now be accomplished in minutes. Computers are used not only for seismic processing and interpretation but for geological modeling and more accurate reservoir modeling. Integrated teams of scientists consisting of geologists, geophysicists, and petroleum engineers are pooling their talents to construct detailed 3-D models of reservoirs.

Exploration Today

Based upon the above discussion, we can now put together a picture of how modern oil and gas exploration is accomplished. First, if the area being explored is a new one, the company has to make an assessment of the source rock potential. Next, the harder problem of finding the reservoir rock and the trap are addressed. Sophisticated seismic methods, including 3-D seismic surveys and VSPs, are used to create a picture of where the oil and gas traps are located, when the traps were formed, and the amount of oil and/or gas they contain. Since financial risk is a major element of exploration, putting together a map showing where to drill is only part of the exercise. Someone has to be convinced to put money into the effort to drill a well. Large companies go through multiple computer runs to evaluate the risks of drilling at a prospective site. Smaller companies tend to use a formula in which they multiply the probability of success times the current value of the oil that will be obtained if the well is successful. Next, they subtract from this product the probability of failure times the cost of drilling a dry well. If the difference is a positive number, the well is judged to be a potentially profitable risk. Once a reservoir is found, a team of engineers, geologists, and geophysicists is formed to characterize the reservoir. The integrated team of scientists then pool their computer modeling efforts to maximize the production, and hence the profit, of the reservoir.

Bibliography

ALLAUD, L. A. *Schlumberger: The History of a Technique.* New York, 1977.

CURTIS, D., P. W. DICKERSON, D. M. GRAY, H. M. KLEIN, and E. W. MOODY. *How to Try to Find an Oil Field.* Tulsa, OK, 1989.

MAYNE, H. W. *Fifty Years of Geophysical Ideas.* Tulsa, OK, 1989.

MCCASLIN, J. C. *Petroleum Exploration Worldwide: A History of Advances Since 1950 and a Look at Future Targets.* Tulsa, OK, 1983.

NETTLETON, L. L. *Geophysical Prospecting for Oil.* New York, 1940.

SCHLUMBERGER, A. G. *The Schlumberger Adventure.* New York, 1982.

RAYMON L. BROWN

OIL AND GAS, RESERVES AND RESOURCES OF

Oil and natural gas are nonrenewable resources found within sedimentary rocks. Examples of these rocks include sandstones, siltstones, and limestones, which contain organic remains of living creatures from the geologic past. These rock contain microscopic pores that form a three-dimensional network that is normally filled with water. Under geologic conditions of elevated temperature and pressure, the organic remains decay and, over time, generate oil and natural gas. The generated molecules of fluid move through the network of pores displacing the water until their movement is hindered and stopped. The molecules aggregate in these localized traps to form oil and gas deposits. These traps can be structural, stratigraphic, or a combination of the two. Structural traps prevent the movement of oil and gas by geologic features such as faults, folds, or salt intrusions. Stratigraphic trapping is caused by changes in the rock or rock pores that prevent further migration. The rocks in which accumulations of oil and natural gas occur are called petroleum reservoirs.

Petroleum reservoirs contain the oil and natural gas resources of the world. They are not large, open pools of fluids but rather porous rocks that contain reservoir fluids between microscopic grains of the rock. Due to the nature of the rock and fluids, a large portion of the oil and gas cannot be extracted from the earth.

This entry discusses the natural resources of oil and gas and distinguishes between reserves and resources. The methodology used to estimate oil and natural gas resources is discussed and estimates of these important resources are reviewed.

Definition of Resources and Reserves

Before we can discuss reserves and resources of oil and natural gas, we must understand that reserves and resources are not synonymous words when used to describe the quantities of oil and gas in the world. Resources are those quantities of oil and natural gas that have been generated by geologic processes and are currently stored in the earth. This includes quantities that are known to exist as well as quantities that are believed to exist even though they have not been located, identified, or confirmed.

Reserves, on the other hand, are known quantities of oil and gas resources that remain to be economically recovered from the earth. This definition of reserves implies three conditions that must be met. First, the oil and natural gas must be physically identified. Second, technology must exist that allows us to produce the oil and gas from the earth.

Finally, the oil and natural gas must be economically produced. The recovery of the oil and gas must be done in such a manner that the costs of extracting and processing the oil and natural gas are less than the market value of the product. If this element of profit is not present, the oil and gas resource will not be exploited.

Oil and gas reserves can be thought of as resources that remain to be economically produced while resources are the quantities of oil and gas occurring in the earth. Thus, estimates are affected by both technical and economic uncertainty. Figure 1 demonstrates this relationship. The box represents the oil and natural gas resources of the world. The horizontal axis represents the technical uncertainty associated with the estimate. Technical uncertainty includes geologic and engineering uncertainty. Geologic uncertainty relates to whether the hydrocarbon volumes estimated by geologists and petroleum engineers do exist in the earth. Engineering uncertainty is related to the technology available to produce the oil and natural gas from the reservoirs. Therefore, as one moves to the right on the axis, resource volumes become increasingly uncertain and may or may not be present or producible. The left half of Figure 1 represents the identified oil and natural gas resources in the world.

The vertical axis represents the economic feasibility of the resource volumes and captures the economic uncertainty associated with the reserve estimates. This uncertainty includes product prices dictated by market demand and the effects on those prices by taxes, environmental regulations, political instability, and global concerns. At a given market price, only a portion of the resources in the earth are producible, and as the market price increases, additional resources will be recoverable since economic confidence increases. If one moves down the vertical axis, indicating improved economic confidence, economic reserves will increase. Under these conditions the upper portion of the diagram represents the economically producible oil and natural gas resources.

The volumes of oil and natural gas represented by the intersection of identified resources with economic resources are the oil and gas reserves. Reserves can be further categorized as shown in the diagram. Measured reserves are those reserves that are currently producing with sufficient geo-

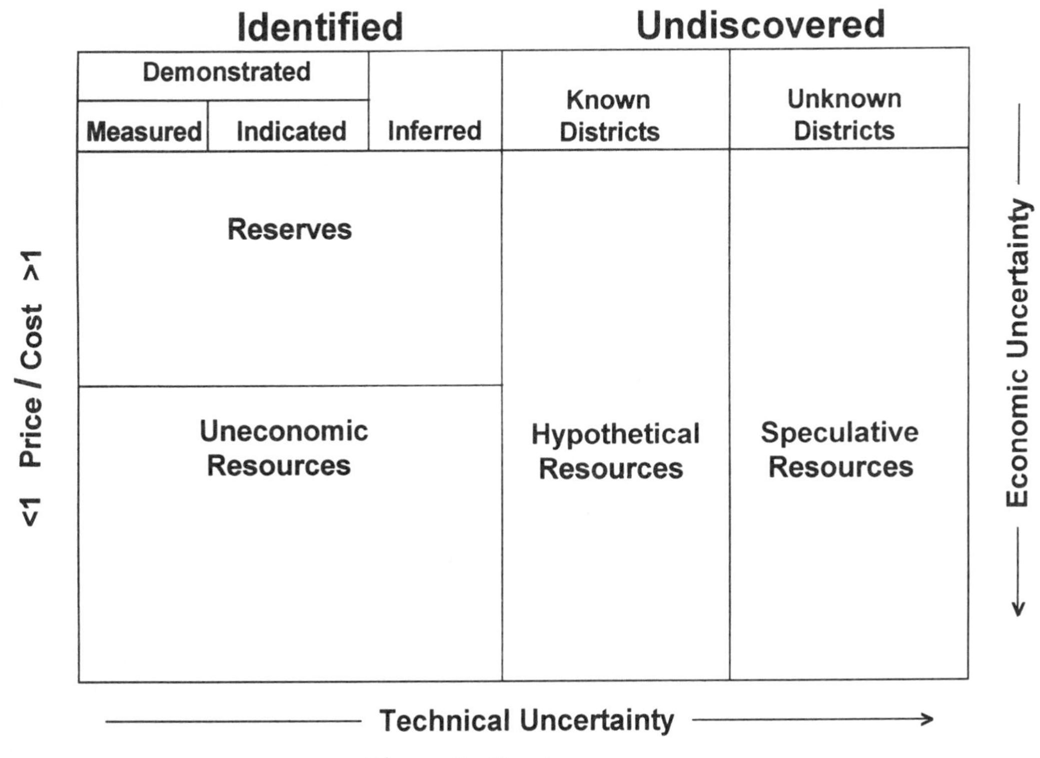

Figure 1. Total resources.

logic and engineering data available to make confident estimates. Indicated reserves are those reserves that can be estimated based upon known data but are not currently being produced. Examples of these reserves include reservoirs that have been penetrated by wells but are currently producing other reservoirs, reserves that will result from the application of an improved recovery technique, or reserves that could be produced by drilling additional wells. Inferred reserves are those that represent undeveloped reserves based upon geologic and engineering data that have less certainty than the category of measured and indicated reserves. Reserves that could be developed by drilling wells outside the current field area are examples of these reserves. While there is no universal agreement on what parameters dictate these categories, one should realize each category has increasing uncertainty associated with them.

Estimates of oil and gas reserves and resources have a timeline associated with them. As time passes and reserves are produced, the resource and reserve base decline due to the removal of the produced quantities of oil and gas. During the same period, additional resources may be discovered by drilling and exploration activities that increase the resource and reserve base. Consequently, resource and reserve estimates are made for a given point in time and are updated on a regular basis to account for new discoveries and produced volumes.

To allow the comparison of estimates at different periods of time, we often utilize the term "ultimate reserves" or "ultimate resources." Ultimate reserves is the sum of cumulative production plus identified reserves. Ultimate resources is the sum of cumulative production, identified reserves, uneconomic resources, and undiscovered resources. We should keep in mind that estimates of oil and gas resources and reserves are just that, estimates, and subject to a large degree of uncertainty.

Estimating Reserves and Resources

Estimating oil and natural gas reserves and resources is an important responsibility of the geologist and petroleum engineer. These estimates give an indication of the worth of these resources to a company, country, or region. They also indicate the ability of a particular region of the world to meet its own energy demand.

Reserve Estimates. Reserves are those quantities of oil and gas that are economically producible and are represented by the upper left-hand portion of Figure 1. This estimate represents a portion of the known or identified quantities of oil and gas in the world.

Estimates of these resources are made after they have been identified by drilling into oil and gas reservoirs. Once these resources are known to exist, reserves can be estimated based upon information about the reservoir. There are essentially five methods for estimating these reserves:

1. Analogy
2. Volumetrics
3. Material balance
4. Decline curves
5. Reservoir simulation

Each method has its advantages and disadvantages and can be used independently of the other methods. Each method is suitable for different phases in the life of a reservoir and requires different data to arrive at estimates of reserves.

Analogy is often used early in the life of a newly discovered reservoir. When little data are available, the method allows reserves to be estimated based upon comparison to nearby and geologically similar reservoirs. These estimates are subject to a large degree of uncertainty as we are expecting different reservoirs to behave similarly. This does not always occur.

Volumetrics also allow reserve estimates to be made early in the life of a reservoir with limited data. This method is based upon the pore volume of the reservoir being studied. If we estimate the area and height of the reservoir from geologic data, a total or bulk volume of the reservoir can be estimated. This bulk volume is composed of rock and fluid. The fluid is located in the microscopic pore space between the rock grains and can be water, oil, gas, or any combination of the three fluids. With data from subsurface well logs, the percentage of the bulk volume that is actually pore space can be estimated. The well logs will also indicate the amount of the pore space filled by oil, gas, and water. These values are then used to make a volumetric estimate of the oil and gas resource located in the reservoir. Reserves are then estimated based upon a recovery factor that is an indication of the economically producible oil and gas in the reservoir.

The material balance method of reserve estimation is based upon the conservation of mass. Simply stated, the mass produced equals the mass originally in place less the mass currently in place. This method requires that we have additional information about the reservoir such as production and pressure history, fluid properties, and relative permeability data. Since the method requires historical pressure and production data, the reservoir must have been producing sufficiently long that the average reservoir pressure has declined prior to the application of this method. The technique allows one to estimate reserves and resources in a reservoir independently of the size of the reservoir.

Decline curves are used to estimate reserves based solely upon production history. This method requires no knowledge of reservoir size, pressure behavior, or fluid content. In this method, plots of production history versus time are extrapolated to estimate reserves and, due to their simplicity, can be strikingly deceptive in the proper analysis of the data. Several months, and sometimes even years, of production history are required before this method is suitable for estimating reserves.

The last method used to estimate reserves is reservoir simulation. While reservoir simulation is not primarily used for reserve analysis, its results can be used to estimate resources and reserves in a reservoir. This method requires enormous amounts of data to describe the geology, fluid distribution, and the rock and fluid properties of the reservoir. In this approach one simulates, mathematically, the behavior of the reservoir by matching its past history and forecasting its future behavior. In doing so, the engineer can estimate reserves. In general, this is a very expensive and time-consuming method of estimating reserves.

It is common for reserve estimates to improve with time. When the estimates are first made, the geologist and engineer know very little about the reservoir. As the reservoir is developed and produced, they gain additional knowledge to describe the reservoir and, consequently, the uncertainty surrounding their reserve estimates decreases.

This process of estimating reserves is repeated over and over on a regular basis for all the reservoirs in the world by geologists and petroleum engineers. By adding together these values, estimates of the world oil and natural gas reserves are developed.

Resource Estimates. Resource estimates are made to yield an idea of how much oil and natural gas are stored in the earth. These estimates have a large degree of uncertainty since they include volumes that have not been discovered or identified. As discussed earlier, the reserve estimates are for identified, economically producible volumes of oil and natural gas. In developing these reserve estimates, the total volume of oil and gas present in a reservoir is also estimated and generally serves as the basis for all resource estimates.

The next step in estimating oil and natural gas resources is to speculate on what parts of the world have undiscovered resources. This is accomplished by using surface and subsurface data to estimate the volumes of sedimentary basins around the world. This volume is an indication of how much of the earth appears suitable for the accumulation of oil and natural gas. Based partly on analogy to the producing regions of the world and the production histories of similar regions, an estimate of resources can be developed for the world.

Estimates of undiscovered resources are made for both the producing and nonproducing regions of the world. As shown in Figure 1, estimates for nonproducing regions are more uncertain than those estimates for producing regions. This process is carried out for the entire world and the sum of regional estimates yields an indication of worldwide resources. Resource estimates are not done as often or as methodically as reserve estimates.

Historical Estimates of Resources and Reserves

Oil production began in the United States with the discovery of oil in Pennsylvania in 1859. World oil production grew slowly from 1860 until 1900 when production rates began to increase rapidly. The increasing world production and increasing rates of discoveries led some people to envision an almost limitless supply of oil and natural gas for the world. Though most realized that the world oil and gas resource was limited, supply still far exceeded consumption. However, energy consumption began to grow very rapidly in the industrialized nations during the 1920s and placed an ever growing demand upon the oil and gas resources of the world.

In 1937, the American Petroleum Institute began to make systematic, annual estimates of oil reserves in the United States and by 1948 were making reserve estimates for the world. DeGolyer and MacNaughton began publishing estimates of world reserves in 1945. The U.S. Geological Survey (USGS) has been estimating the oil and gas resources of the world since 1965. These sources yield considerable information on worldwide production, reserve, and resource estimates. They also allow us to develop a perspective on the historical estimates of oil and gas resources.

Early estimates of worldwide reserves were made based upon historical data regarding oil and gas production in the United States. This was due in part to the maturity of exploration and production in the United States that served as the best analogy for extrapolation to world reserves. In 1948, L. G. Weeks estimated world producible resources to be 610 billion barrels of oil (610 BBO), with 110 BBO attributed to the United States. From 1948 to the 1960s, the estimates of both U.S. and world producible resources continued to grow due to additional production worldwide and major discoveries throughout the world. The production and discoveries yielded valuable additional information in which better estimates of oil and gas resources in the producing regions of the world could be analyzed. Table 1 summarizes oil and gas reserve estimates for the period of 1948 through 1994.

In 1956, M. KING HUBBERT surprised the United States by identifying a time at which U.S. production would reach its maximum output and begin to decline. His bold prediction was based on the historical data of production and discoveries in the United States. Using a logistics curve to fit the oil data with an estimate of 150 BBO of ultimate producible resources, he estimated that a maximum oil output of 2.75 BBO per year in the United States would occur in the mid–1960s, a brief ten years away. He further concluded that even if the ultimate producible resources were closer to 200 BBO, the year of peak production would only be moved ahead by several years to approximately 1970. His estimates of course excluded the major Alaska Prudhoe Bay discovery in 1968.

Table 1. Historical Estimates of World Oil and Gas Reserves

	United States		*Total World*	
Year	*Oil, BBO*	*Gas, TCFG*	*Oil, BBO*	*Gas, TCFG*
1948	21.49	165.9	68.20	
1950	24.65	179.4	76.45	
1960	31.72	261.2	290.04	
1970	29.63	275.1	530.53	1,491.3
1980	29.81	194.9	644.93	2,574.8
1990	26.50	167.1	1,002.21	3,991.2
1994	23.75	165.0	999.12	5,016.2

Sources: American Petroleum Institute, *Basic Petroleum Data Book*, May 1994. *Oil & Gas Journal*, 27 December 1993. Reserve estimates exclude the inferred reserves of Figure 1.

Table 2. World Estimate of Ultimate Oil Resources, 1 January 1990

Region	*Cumulative Production BBO*	*Identified Reserves BBO*	*Undiscovered Resources BBO*	*Ultimate Resources BBO*
North America	187.0	103.5	97.8	388.3
South America	59.3	72.4	56.4	188.1
Western Europe	17.2	35.9	17.4	70.5
Eastern Europe/CIS	111.5	85.2	93.7	290.4
Africa	48.7	71.1	34.9	154.7
Middle East	166.4	624.2	118.3	908.9
Asia Pacific	39.3	60.3	58.3	157.9
Total World	629.3	1,052.7	489.3	2,171.3

Source: U.S. Geological Survey, Thirteenth World Petroleum Congress.

Table 3. World Estimate of Ultimate Gas Resources, 1 January 1990

Region	*Cumulative Production TCFG*	*Identified Reserves TCFG*	*Undiscovered Resources TCFG*	*Ultimate Resources TCFG*
North America	850.5	466.2	816.7	2,133.4
South America	38.7	178.6	254.0	471.3
Western Europe	127.1	203.6	225.7	556.4
Eastern Europe/CIS	406.9	1,565.0	1,256.4	3,228.3
Africa	32.6	261.6	328.3	622.5
Middle East	57.5	1,479.2	943.4	2,480.1
Asia Pacific	81.4	345.7	486.3	913.4
Total World	1,594.6	4,499.8	4,417.3	10,511.7

Source: U.S. Geological Survey, Thirteenth World Petroleum Congress.

Using information for the world, Hubbert extended his analysis to an evaluation of the world petroleum resources. Applying his methodology and an ultimate producible resource estimate of 1,250 BBO, he predicted that world oil output would peak at 12.5 BBO annually in the year 2000. He predicted that natural gas production would peak in 1970 at 13.8 trillion cubic feet (13.8 TCFG) in the United States assuming ultimate resources of 850 TCFG.

From a historical viewpoint, annual U.S. crude oil production peaked at 3.3 BBO in 1970 and remained above 3.0 BBO through 1985 and is now declining while gas production reached 22.65 TCFG in 1973. World oil production is slowly increasing and was estimated to be 22.0 BBO in 1993. Hubbert's estimates, while controversial, did underline that fact that the petroleum resources of the world were finite and were in fact being consumed at a very rapid pace.

Table 4. Estimated World Oil and Gas Reserves, 1 January 1994

Region	*Oil Reserves BBO*	*Gas Reserves TCFG*
North America	80.08	330.9
South America	73.76	197.4
Western Europe	16.64	191.1
Eastern Europe/CIS	59.17	2,017.8
Africa	61.96	343.5
Middle East	662.87	1,581.0
Asia Pacific	44.65	354.5
Total World	999.12	5,016.2

Source: *Oil & Gas Journal,* 27 December 1993. Reserve estimates exclude inferred reserves.

Current Estimates of Oil and Gas Resources

Tables 2, 3, and 4 indicate current estimates of world oil and gas resources and reserves. Tables 2 and 3 show resource estimates prepared by the USGS for the Thirteenth World Petroleum Congress. As of 1 January 1990, the resources of the world were estimated to be 2,171 BBO and 10,512 TCFG. The majority of these resources are located in the Middle East, Eastern Europe, and North America. These tables also indicate cumulative resource production to date. These values give an indication of the percentage of the resources we have already consumed.

Table 4 is a 1 January 1994 estimate of reserves as prepared by the *Oil & Gas Journal.* Estimated oil reserves are 999.1 BBO, while gas reserves are 5,016 TCFG. Approximately two-thirds of the current oil reserves lie in the Middle East. The majority of the gas reserves are found in Eastern Europe and the Middle East.

Bibliography

AMERICAN PETROLEUM INSTITUTE. *Basic Petroleum Data Book, Petroleum Industry Statistics.* Washington, DC. Updated three times per year.

DEGOLYER and MACNAUGHTON. *Twentieth Century Petroleum Statistics.* Dallas, TX. Updated annually.

HUBBERT, M. K. "Nuclear Energy and the Fossil Fuels." *API Drilling and Production Practice* (1956):7–25.

Masters, C. D., D. H. Root, and E. D. Attanasi. "World Resources of Crude Oil and Natural Gas." In Vol. 4, *Proceedings Thirteenth World Petroleum Congress*, Buenos Aires, Argentina, 1991, pp. 51–64.

"Worldwide Production Report." *Oil & Gas Journal*. Tulsa, OK. Prepared annually.

MICHAEL L. WIGGINS
RONALD D. EVANS

OIL SHALE

During the early days of their exploitation, oil shales were commonly given provincial names such as torbanite, tasmanite, kerosene shale, bituminous shale, kukersite, albertite, cannel coal, and boghead coal, to name a few. As a group, these rocks were enigmatic and little was known about their composition or the sources of their organic matter. Since about 1970 geological and geochemical studies using modern analytical tools, especially fluorescence microscopy, have led to a clearer understanding of the origin and classification of oil shales as outlined by Cook, Hutton, and Sherwood (1981) and Hutton (1987).

Oil shale is a fine-grained sedimentary rock containing abundant organic matter that is rich in hydrogen, and that will yield substantial quantities of oil and combustible gas upon destructive distillation. Little or no oil can be extracted from most oil shales by petroleum solvents. Oil shales range in age from Cambrian to Tertiary and are found on all continents (*see* TARS, TAR SANDS, AND EXTRA HEAVY OILS).

Oil shales usually originate under anoxic conditions in many different depositional environments, including marine basins, epicontinental shelves, freshwater to saline lakes, and stagnant ponds or shallow lakes in coal-forming swamps and marshlands. The precursors of the organic matter in most oil shales are predominantly algae and associated microorganisms such as sulfate-reducing bacteria. Other precursors include subordinate amounts of spores, pollen, leaves, and woody material of terrestrial plants derived from lands marginal to the depositional basin.

Most oil shales are brown to black in color, fissile to massive in aspect, and finely laminated or possess streaked and blebby sedimentary structures. Typically, potentially commercial grades of oil shale contain, by weight, about 10 to 20 percent organic matter and 80 to 90 percent inorganic gangue (matrix)—chiefly silicate and carbonate minerals. Such oil shale can yield between 75 and 150 liters of oil per metric ton of rock. Some oil shales also contain potentially recoverable phosphate and alum. The oil shale of the Eocene Green River Formation in Colorado contains, in addition, abundant nahcolite ($NaHCO_3$) and dawsonite [$NaAl(OH)_2CO_3$], which are potential sources of soda ash and alumina. Trace metals, including uranium, vanadium, copper, lead, and zinc, found in some marine oil shales, add potential by-product value to the deposit.

The world resources of oil shale are enormous, exceeding by far the known reserves of crude oil. One of the world's largest deposits is found in Eocene lake beds of the Green River Formation in Colorado, Utah, and Wyoming. Green River oil shale is estimated to contain 1.7 trillion barrels (about 291 $\times$ 10^9 metric tons) of in-place shale oil in rocks that average 15 or more gallons of shale oil per ton (about 57 liters per metric ton). About three-quarters of this resource is in western Colorado (Pitman, Pierce, and Grundy, 1989).

A large region in the eastern United States, including parts of Indiana, Kentucky, Ohio, Tennessee, and adjacent states, is underlain by marine oil shale of Devonian and Mississippian age. These rocks contain an estimated 423 billion barrels (about 72 $\times$ 10^9 metric tons) of potentially recoverable shale oil in near-surface deposits averaging 10 or more gallons of oil per ton (about 42 liters per metric ton) of rock (Matthews, 1983). The organic matter in this oil shale contains less hydrogen and consequently yields less oil on retorting than Green River oil shale. However, the oil yield of eastern oil shale can be doubled by retorting the rock under pressure in a hydrogen atmosphere. To be of economic value, such oil shale should be in near-surface deposits amenable to mining by open-pit, room-and-pillar, or in situ methods.

Synthetic crude oil and combustible gas are extracted from oil shale by retorting. In this process, crushed oil shale is heated in a closed vessel to a temperature of about 500°C. At this temperature, the organic matter thermally decomposes into oil, gas, and char. Retorting may be carried out in surface facilities, or by in situ techniques whereby the oil shale is retorted underground in chambers mined in the deposit. The shale oil is then pumped to the surface (Allred, 1982).

Hydrocarbon vapors from the retort are condensed into shale oil, upgraded with hydrogen, then sent to the refinery for conversion into gasoline, jet fuels, lubricants, and petrochemicals. Combustible gas from the retort (largely methane) can be recycled for use in an oil shale facility or marketed. The char remaining on the retorted shale can be burned for use in the oil shale plant. Oil shale can also be burned directly for electrical generation.

Depending upon the method used to recover the energy in oil shale, the environmental cost can be high. Large amounts of crushed waste shale are produced from surface retorts. This waste rock is susceptible to leaching of noxious elements and other deleterious substances that could enter surface streams and ground waters. Some countries that have had a sustained oil shale industry have serious problems of groundwater contamination and air pollution from the waste products created by mining and retorting oil shale.

To date, shale oil has been unable to compete with lower-priced crude oil in the western industrialized nations because of the costs of mining, retorting, and upgrading. However, in countries under centrally controlled economies, such as the former Estonian Soviet Socialist Republic and China, oil shale has been used for many years for generation of electric power and for the manufacture of transportation fuels and petrochemicals. A total of more than 1 billion barrels of oil or oil-equivalent energy has been produced from oil shale worldwide.

Large-scale use of oil shale will probably not occur until world supplies of conventional crude oil diminish substantially and the price of crude oil increases severalfold. New technologies that provide less costly methods of recovering synthetic crude from oil shale could accelerate the development of an oil shale industry. In today's world, new methods of mining and retorting oil shale will also need to include provisions for a clean environment with minimum generation of water and air pollutants.

Bibliography

Allred, V. D., ed. *Oil Shale Processing Technology.* East Brunswick, NJ, 1982.

Cook, A. C., A. C. Hutton, and N. R. Sherwood. "Classification of Oil Shales." *Bulletin du Centre de Recherche Exploration-Production Elf-Aquitaine* (1981):353–381.

Colorado School of Mines. *Proceedings of Oil Shale Symposia for the Period from 1964 to 1992.* Golden, CO.

Hutton, A. C. "Petrographic Classification of Oil Shales." *International Journal of Coal Petrology* (1987): 203–231.

Matthews, R. D. "The Devonian-Mississippian Oil Shale Resource of the Eastern United States." In *Sixteenth Oil Shale Symposium Proceedings,* ed. J. H. Gary. Golden, CO, 1983.

Pitman, J. K., F. W. Pierce, and W. D. Grundy. *Thickness, Oil Yield, and Kriged Resource Estimates for the Eocene Green River Formation, Piceance Creek Basin, Colorado.* U.S. Geological Survey Oil & Gas Investigations Chart OC-132. Washington, DC, 1989.

Russell, Paul L. *Oil Shales of the World, Their Origin, Occurrence, and Exploitation.* New York, 1990.

University of Kentucky. *Proceedings of the Eastern Oil Shale Symposia for the Period from 1981 to 1993.* Lexington, KY.

JOHN R. DYNI

OLDEST ROCKS IN THE SOLAR SYSTEM

Few subjects in the earth sciences are more intriguing or fundamental than those involving time. What is the age of Earth? When did life arise on Earth? What are the oldest rocks in the solar system? How old is our sun, our galaxy, the universe?

This entry will address primarily information from experiments that indicate the time of formation of the oldest rocks and minerals available for laboratory analysis—rocks collected at or near Earth's surface, rocks returned from the Moon during Apollo spaceflights, and meteorites—Rosetta stones from space because of their seminal use in the understanding of Earth and the solar system (see also Dalrymple, 1991, and Faure, 1986).

Time can be discussed from different viewpoints. Not provided here is a discussion of time as a theoretical or cosmological subject (*see* COSMOLOGY, BIG BANG, AND THE STANDARD MODEL). In

Earth history, time is measured either in a "relative" sense or in an "absolute" sense. Time that can be measured quantitatively, that is, assigned a number—years or billions of years—is termed "absolute"; rocks and geologic events whose ages can be described only by terms such as "older than" or "younger than" have "relative" age determinations, established, for example, by field relations (position in a stratigraphically layered sequence, or by cross-cutting relationships such as faulting and intrusive penetration by igneous rocks), chronologic use of fossils, or by impacting of planetary surfaces by asteroids, meteors, or comets. Absolute ages thus are determined by counting—annual tree rings or annual layers of lake deposits (varves), for example. The most useful absolute ages are those calculated by determining the abundance of naturally occurring but unstable elemental species (radioactive or "parent" isotopes), their rates of decay to stable elemental species (radiogenic or "daughter" isotopes), and the abundance of their radiogenic products.

Meteorites

Most meteorites appear to have originated on asteroid-sized bodies, such as those in the asteroid belt between Mars and Jupiter. Numerous absolute (isotopic) dates indicate that most meteorites were formed very close to 4.56×10^9 years, or 4.56 billion years (Ga) ago. Additionally, studies of short-lived isotopic systems, such as those of iodine-xenon and samarium-neodymium, demonstrate that meteorites were formed during a very short time span early in the solar system's history. Because asteroids and the planets and their satellites are assumed to have formed at essentially the same time, the formation age of meteorites is accepted as the age of formation of Earth and all other parts of the solar system, excepting the slightly older Sun. [An indirect, or hypothesis-dependent, isotopic age ("model age") can also be assigned to Earth through the use of the isotopes of lead, discussed below.] A few meteorites yield ages significantly younger than 4.56 Ga; these rocks are assumed either to have been altered, for example, by melting or by impacts, or to have originated on other satellites or planets (the Moon, Mars) and been thrown into Earth's gravitational field by very large impacts on these planets (*see* METEORITES FROM THE MOON AND MARS).

Moon Rocks

Aside from meteorites, the only sizable extraterrestrial specimens so far available for age and other direct analysis are the more than 300 kilograms (kg) of rock returned from the Moon by the *Apollo* astronauts and small amounts of soil from the *Luna* missions of the former Soviet Union (*see* APOLLO ASTRONAUTS). Many age determinations have been made on these samples; they range from about 3.0 Ga to about 4.42 Ga.

The two main geologic divisions of the Moon are the dark maria, which are large impact basins floored by extrusive (volcanic) basaltic rock, and the highlands, which contain a variety of instrusive (plutonic) igneous rocks, including anorthosite, the rock type that has yielded the oldest, unequivocal age for Moon rocks, about 4.42 Ga. Because no lunar rocks have been found to be younger than 3.0 Ga, it is assumed that, unlike that of Earth, the Moon's internal heat supply, the source of extensive volcanism, neared exhaustion about this time. However, photographs of regions of the Moon not yet sampled show evidence suggesting volcanism more recently than 3.0 Ga ago, perhaps as recently as 800 million years (Ma) ago, or even less.

More pertinent to this entry are the oldest directly dated lunar rocks. A few volcanic rocks (basalts) from the highlands have extrusion ages near 4.3 Ga, only 120 Ma younger than the age of the anorthosite. Thus, available evidence shows that, in the short geologic time span of about 140 Ma after the accretion of the Moon, crustal anorthosite rocks formed, and, 120 Ma later, the earliest known volcanic rocks were extruded (*see* EARTH-MOON SYSTEM, EARLIEST HISTORY OF).

A compelling theory of the Moon's origin, proposed in 1975 and accepted by a significant number of scientists since 1984 (Hartmann, Phillips, and Taylor, 1986), is that Earth was impacted by a very large asteroid early in its history (soon after Earth's core formed). This threw part of the Earth's mantle material into orbit, and it coalesced to form the Moon. If this theory is correct, the Moon is of course somewhat younger than Earth. Some scientists believe that this rather small difference (perhaps 200 Ma) is discernible in certain isotopic studies.

Earth's Oldest Rocks

Since the 1950s and, increasingly, since the advent of modern mass spectrometry in the 1980s and 1990s, many thousands of isotopic ages have been determined for crustal rocks from the continents and the ocean basins, Earth's major geologic and topographic divisions. The rocks that floor the world's oceans are geologically young, with ages of formation extending from the present only to about 200 Ma (*see* EARTH AS A DYNAMIC SYSTEM). Continents, especially the geologically stable interiors of continents known as cratons or shields, have long been known to contain Earth's oldest rocks—the Precambrian rocks.

Hadean, Archean, and Proterozoic (collectively the Precambrian) are terms used to describe successively younger intervals of Earth's earliest history. Rocks formed during this time, many of them igneous and metamorphic, generally underlie thick sequences of sedimentary rocks, which contain abundant fossils—those of the Phanerozoic (visible life) Eon, less than 600 Ma in age.

Owing to Earth's dynamic history, rocks and the isotopic systems recording their ages of formation may be continually affected by processes such as fracturing, melting, metamorphism, and weathering. Consequently, older rocks are less likely to have survived intact than younger rocks. Prior to the availability of absolute age determinations, these oldest rocks were "dated" in a relative sense, in some cases with errors of billions of years, on the basis of field relations and rock type. Field relations include "superposition" (in an undisturbed sequence of layered rocks, the oldest are at the bottom) and cross-cutting relations (rocks that have been folded, faulted, or intruded must be older than the structural events or the intrusive rocks). With respect to rock type, it was reasoned that the oldest rocks would have the greatest opportunity to have been highly metamorphosed; in some cases rocks, such as granite gneiss, were correlated across great distances owing simply to a similarity of rock type. With absolute dating, it quickly became obvious that high-grade metamorphic rocks may be geologically young and that very old Precambrian sedimentary rocks may be essentially unaltered, thus appearing "young."

Numerous isotopic age determinations performed beginning in the 1950s quickly established that the Precambrian-Phanerozoic boundary was about 500–600 Ma. Precambrian rocks with ages greater than about 2.8 Ga, some up to perhaps 3.2 Ga, were few in number and were exposed in limited continental areas. In the 1960s, age dating of the Morton Gneiss, a granitic metamorphic rock exposed in Minnesota, established the record for the oldest documented rock, about 3.4 Ga (later increased to about 3.6 Ga), a record that held for many years. In the late 1970s, isotopic dating established even older rocks, again metamorphic, in eastern Greenland. The well-documented age of the oldest of these rocks, the Isua Formation, is close to 3.8 Ga.

Through the use of further advances in mass spectrometry along with careful geologic study of the samples collected for analysis, rocks with isotopic ages of formation of 3.9 to 4.0 Ga have been documented on several continents. In addition to the craton of eastern Greenland, areas containing rocks about 4.0 Ga are located in the Canadian shield near Lake Superior in Canada, the Transvaal Province of South Africa, and cratonic rocks, of western Australia. Isotopic ages near 4.0 Ga, however, are being reported with increasing frequency from other areas of Precambrian outcrops.

A significant and well-dated though younger (near 3.5 Ga) exposure of rocks occurs at North Pole, Australia. This exposure is noteworthy in that, unlike the more complex metamorphic rocks previously described, it includes sedimentary rocks that are but slightly altered and contain the oldest fossil evidence of life on Earth in the form of stromatolites (layered algae mats) and bacteria. Both stromatolites and bacteria, distinctly evolved life forms, require an even earlier period of evolution, of course.

A technique in mass spectrometry developed in the 1980s, the ion probe, allows accurate and precise age determinations to be made on tiny regions of individual mineral grains in rocks. Through the use of this technique, detrital (clastic) zircon crystals from sandstone in western Australia have yielded a range of very old ages. A very few of these minerals have been dated at about 4.3 Ga, the oldest dates recorded for Earth materials. The rock from which these crystals weathered has not been found. However, zircons are typical minerals of continental crystal rocks; this work thus indicates that Earth had a continental crust less than a quarter of a million years after it accreted.

Age of Earth

The determination of Earth's age is a fabulous story that can be treated only briefly here (see Dalrymple, 1991, for an excellent summary). If the materials of the solar system originated from a well-mixed medium and later segregated into the inner planets (including the rocks of Earth), the asteroids, and the outer planets, there should be linear relationships among certain radioactive-radiogenic isotope pairs in rocks of these different offspring. Claire Patterson (1956) found such a relation among the isotopes of lead (two of which result from the decay of two isotopes of uranium) in stone and iron meteorites, indicating a starting time of about 4.5 Ga for meteorites (recall that the presently accepted value is about 4.56 Ga!). Further work by Patterson and other scientists found that lead from different types of Earth and Moon rocks also fit this relationship, indicating (but not proving) that all are about 4.5 Ga. This type of isotopic analysis results in model, or hypothesis-dependent, ages. It is noteworthy that the meteorites thought to have originated from Mars also have *model* ages of about 4.5 Ga (*see* METEORITES FROM THE MOON AND MARS).

Rocks that may have formed during Earth's earliest history—from 4.56 to about 4.0 Ga—have not been found. This is not surprising given Earth's violent early history of impact and the immense subsequent time allowing for metamorphosing, weather, and other destructive Earth processes. Subtle and unexpected variations in certain stable isotope abundances from Earth's oldest rocks, however, indicate to some scientists that they represent the radiogenic products of radioactive isotopes that have since disappeared owing to their short half-lives. These isotopic "footprints" may be a chemical record of even earlier Earth rocks (*see* NUCLEOSYNTHESIS AND THE ORIGIN OF THE ELEMENTS).

(*See also* GEOLOGIC TIME, MEASUREMENT OF; SOLAR SYSTEM; MOON; METEORITES; and METEORITES FROM THE MOON AND MARS.)

Bibliography

DALRYMPLE, B. *The Age of the Earth.* Stanford, CA, 1991.

FAURE, G. *Principles of Isotope Geology,* 2nd ed. New York, 1986.

HARTMANN, W. K.; K. R. PHILLIPS; and G. TAYLOR, eds. *Origin of the Moon.* Houston, 1986.

PATTERSON, C. C. "Age of Meteorites and the Earth." *Geochemica et Cosmochimica Acta* 10 (1956):230–237.

E. JULIUS DASCH

ORE

Ore is an assemblage of minerals or rocks that has formed by geologic processes and can be exploited for use by society. The term is most commonly used for geologic materials from which metals are extracted, but it applies to nonmetallic mineral resources as well. An ore deposit is an accumulation of ore that can be exploited profitably using known technology and under present economic conditions. In contrast, a mineral deposit is an accumulation of potentially usable geologic material that at present may or may not be amenable to profitable exploitation.

An ore may be used as extracted from the ground without much further processing, as is the case for some nonmetallic resources like high-grade phosphate rock or potassium-bearing salts used as fertilizers. Generally, however, the valuable constituents are contained in ore minerals that have to be separated from the rock after mining, and then further processed or smelted to extract the valuable metals or nonmetallic constituents. The ore minerals occur in nature in several major chemical groups, among them the native metals (e.g., gold, mercury), sulfides (e.g., chalcopyrite for copper, galena for lead, or sphalerite for zinc), oxides (e.g., magnetite and hematite for iron), and carbonates (e.g., rhodochrosite for manganese). The noneconomic minerals that dilute the ore are called gangue minerals, for example, quartz and calcite associated with ore minerals in many metallic mineral deposits. The concentration of the metal or nonmetallic constituent in the ore, the ore grade, is generally expressed as a percentage (e.g., 0.8 percent copper in a typical, large-scale copper deposit), or as a unit established by old tradition (e.g., 0.3 troy ounces of gold per ton of ore in a high-grade gold deposit).

Whether a mineral deposit can be profitably exploited depends on a variety of geologic, technologic, economic, and geographic factors: On the positive or revenue side are a high value of the commodity per unit (e.g., the high price per troy ounce of gold or platinum), a high concentration of the ore constituents in the rock (i.e., a high ore grade), and a large tonnage (i.e., a large ore reserve); on the cost side are the expenses of mining, processing, and smelting of the ore, including any costs of transportation and environmental protection. The feasibility of exploitation further depends on the quality of the infrastructure, the local political and economic climate, and the availability of financial resources for the investment. To achieve an optimal balance between short-term profit and long-term maximization of recovery of the metal or mineral contained in the ore once extraction has begun, high-grade, high-revenue ore is generally mixed with low-grade, low-revenue ore. So-called high-grading of a deposit will yield maximum profit in the short term, but will leave the lower-grade material in the ground and unexploitable in the future because of its low value.

Mineral resource assessments on a global or national scale are generally based on degree of geologic assurance and economic/technologic feasibility of exploitation of the resource. In individual ore deposits, exploitable ore reserves are categorized by geologic assurance of their presence: "proven ore" represents measured quantities of ore; "probable ore" is ore indicated by geologic criteria and preliminary exploration; "possible ore" is ore postulated on the basis of general geologic considerations, but without the benefit of direct observation or measurement. Ore reserve estimates must be revised continually, to account for the balance between tonnage of ore extracted and ore added to reserves by geologic exploration, and to adjust for the economic factors that influence the profitability of extraction, namely fluctuations in commodity prices and in costs of extraction.

Bibliography

GOCHT, W. R., H. ZANTOP, and R. G. EGGERT. *International Mineral Economics.* Heidelberg, 1988.

PETERS, W. C. *Exploration and Mining Geology,* 2nd ed. New York, 1987.

HALF ZANTOP

ORGANIC GEOCHEMISTRY

Organic geochemistry is a relatively new subdiscipline in the earth sciences, combining elements of organic chemistry and geology. It was firmly established in the 1930s with the pioneering work of Alfred Treibs in Germany. He clearly demonstrated how biologically derived organic molecules (biomolecules) alter over geologic time to become geological molecules or molecular fossils (also called biomarkers because they retain structural features that relate to the original biomolecules). The first molecular fossils to be studied were pigments derived from chlorophyll, which is the basic organic molecule of photosynthesis, the process that converts light energy to chemical energy.

Photosynthesis is the way most new organic molecules are created on Earth, and it is the ultimate source for nearly all living organic matter and for most dead organic materials, including molecular fossils, that are buried in sediment and rocks. The process of photosynthesis in its simplest terms can be expressed as a chemical reaction as follows:

$$2H_2O + CO_2 \xrightarrow{\text{light energy}} H_2O + (CH_2O) + O_2$$

In this equation of life, water (H_2O) and carbon dioxide (CO_2) react during photosynthesis, utilizing mainly the energy from the Sun. Oxygen (O_2) is split from water, and organic matter, expressed as CH_2O, is created. Through living organisms, this organic matter takes the form of a vast array of organic molecules, including lipids, carbohydrates, and proteins. When death occurs, many of these organic molecules cannot be sustained and begin to undergo reactions such as hydrolysis (loss of water), decarboxylation (loss of CO_2), oxidation (when oxygen is present), and reduction (when oxygen is absent). If not protected in some way, the dead organic matter reacts in the reverse direction of the above chemical reaction, being oxidized by O_2 to yield carbon dioxide and water. Fortunately, a small amount (usually less than 1 percent) of the organic matter is protected by being buried during sedimentation. This organic matter enters the realm of geology, and some of it eventually becomes deposits of crude oil, coal, and natural gas.

The geological processes that affect organic matter are themselves complex (Figure 1). Where the initial organic matter is mainly from algae and

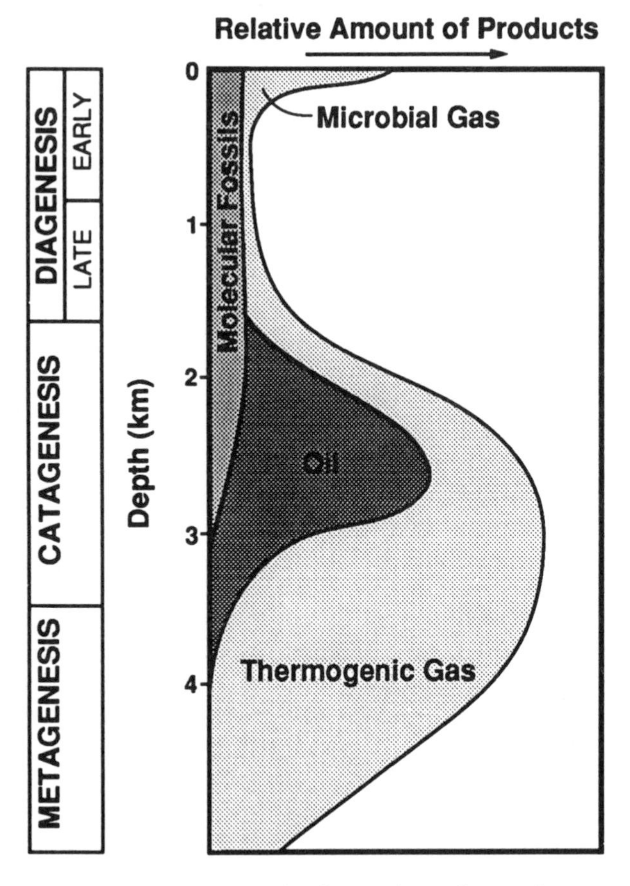

Figure 1. Stages in the formation of petroleum.

bacteria (both marine and nonmarine), the processes can lead to the formation of crude oil as well as natural gas. In contrast, where the initial organic matter comes mainly from terrestrial plants, the processes can lead to coal and also natural gas. The following description applies to organic matter that alters to petroleum (crude oil and natural gas). First the buried organic matter undergoes early diagenesis, that is, the biological, chemical, and physical alterations of organic matter in sediments prior to changes that result from the effects of heat generated within Earth. With deeper burial and rising temperatures, organic matter undergoes late diagenesis when more intense chemical and physical alterations result in the formation of some bitumen (dispersed, solvent-soluble organic matter), with much of the organic matter becoming a complex polymeric material called kerogen. Kerogen is very stable and insoluble in most common organic and inorganic solvents. Most organic matter in sedimentary rocks is in the form of kerogen. With increasing burial depths (greater than about 1 kilometer or km) and at higher temperatures (greater than about 80°C) some of the kerogen as well as some of the bitumen alters to petroleum in a process called catagenesis.

Where the original organic matter is terrigenous trees and other plants that are buried in swamps, the resulting product is coal of varying kinds (peat, lignite, subbituminous, bituminous, and anthracite) depending on the intensity of the diagenetic and catagenetic processes. At temperatures in the range of 150° to 200°C, the geological process affecting organic matter is called metagenesis and involves cracking of any remaining organic matter to its simplest forms, graphite (pure carbon) and methane (CH_4).

But metagenesis is not the only process by which methane forms geologically. From the earliest stages of organic matter alteration, methane has been produced, first by biological processes, that is, by microbial reactions of fermentation and reduction of CO_2, and then by diagenesis, catagenesis, and metagenesis. Methane resulting from all of these processes becomes the principal component in natural gas. Methane generated from biological processes and during early diagenesis is very pure, accompanied by only minor amounts of heavier hydrocarbon gases such as ethane and propane. In contrast, methane formed during late diagenesis and catagenesis has increasing amounts of heavier hydrocarbon gases. However, in metagenesis the methane produced is again very pure. Differentiation of the processes by which methane forms can often be determined by its carbon and hydrogen isotopic compositions (*see* ISOTOPE TRACERS, STABLE).

Molecular fossils also undergo important changes during the geological processes of diagenesis, catagenesis, and metagenesis. The molecular fossils entering diagenesis strongly reflect their biological precursors. During diagenesis, chemical reactions take place that cause the loss of functional groups, such as OH, COOH, and NH_2, with the result that the molecular fossils become hydrocarbons, that is, compounds composed only of carbon and hydrogen atoms. As molecular fossils move from the diagenetic to the catagenetic realm during continued burial, important changes take place in their stereochemical configurations (spatial relationship between atoms in the molecule). These changes can be measured to determine the extent of change (the thermal maturity) of the molecular fossil.

Such molecular information becomes important in the study of petroleum source rocks. These are the rocks from which petroleum originated. Petroleum migrates from source rocks into reservoir rocks and collects in traps formed by some impermeable barrier. If sufficient petroleum enters the trap, a commercial petroleum accumulation may form. It is a principal business of the petroleum industry to search for, find, drill, and produce these accumulations to provide energy fuel, plastics, and lubricants for society.

It is interesting to note that energy is produced from fossil fuels (crude oil, coal, and natural gas) using the reverse of the photosynthesis reaction, that is,

$$(CH_2O) + O_2 + H_2O \xrightarrow{\text{combustion}} 2H_2O + CO_2 + \text{energy}$$

Thus the energy from the Sun that was captured in the organic matter through photosynthesis is converted to fossil fuel energy (heat) by the reverse process of combustion.

Some of the CO_2 resulting from combustion is available to continue the cycle by reacting with H_2O through photosynthesis to produce new organic matter. Some of the remaining CO_2 is dissolved in the oceans to form reefs and later carbonate rocks. However, at present there is an excess of CO_2 produced by the combustion of fossil fuels. Since the beginning of the industrial revolution, atmospheric CO_2 levels have been increasing, and now the rate of increase is about 0.5 percent per year. Because CO_2 is a "greenhouse" gas, it absorbs radiant heat energy from Earth's surface. The increasing amounts of CO_2 in the atmosphere thus appear to contribute to global warming trends (*see* GLOBAL CLIMACTIC CHANGES, HUMAN INTERVENTION).

Not all molecular fossils end up in crude oil, coal, and natural gas. Organic geochemistry is concerned with more than the geochemical processes leading to fossil fuels. Rather it deals in its broadest sense with the occurrence, distribution, and fate of organic matter in general, not only on Earth but also in the cosmos. The following are other examples of some of the applications of organic geochemistry to aspects of earth sciences:

1. Paleoceanography—long-chain, unsaturated alkenones to determine paleotemperatures of the ocean.
2. Paleontology—distributions of amino acids for taxonomy of fossils.
3. Quaternary Geochronology—ratios of amino acid stereoisomers (different spatial arrangements of constituent atoms) for geologic age assessment.
4. Paleo- and Geothermometry—ratios of amino acid stereoisomers for prediction of temperatures of depositional environments; methane hydrate locations to ascertain geothermal gradients and heat flow.
5. Cosmochemistry—fatty acids, amino acids, and hydrocarbons in meteorites to understand processes of extraterrestrial synthesis.
6. Ecology—aromatic hydrocarbons are natural and anthropogenic pollutants in sedimentary environments.
7. Paleobotony—molecular fossil records extend back in time beyond 600 million years providing evidence for the origin and evolution of early life.
8. Stratigraphy and Sedimentation—fossil hydrocarbons used for correlation of geologic units.
9. Climatology—methane from geologic sources affects the balance of "greenhouse" gases in the atmosphere and is of concern in global warming.
10. Hydrology—soluble organic molecules as tracers in surface and ground water movements.
11. Isotope Geology—isotopic compositions of organic matter and organic molecules provide details of bio-geo-organic processes.

From this list it is apparent that organic geochemistry involves many different scientific disciplines, providing a strong link between the atmosphere, hydrosphere, biosphere, and geosphere.

Bibliography

HUNT, J. M. *Petroleum Geochemistry and Geology.* San Francisco, 1979.

PETERS, K. E., and J. M. MOLDOWAN. *The Biomarker Guide.* Englewood Cliffs, NJ, 1993.

TISSOT, B. P., and D. H. WELTE. *Petroleum Formation and Occurrence*, 2nd ed. Berlin, 1984.

WAPLES, D. W., and T. MACHIHARA. *Biomarkers for Geologists.* AAPG Methods in Exploration, no. 9. Tulsa, OK, 1991.

KEITH A. KVENVOLDEN

ORIGIN OF THE ELEMENTS

See Nucleosynthesis and the Origin of the Elements

OWEN, SIR RICHARD

In the middle of the nineteenth century, Richard Owen was acclaimed as one of Britain's greatest natural scientists. By the time of his death in London on 18 December 1892, his fame as a zoologist had been eclipsed by Darwin (*see* DARWIN, CHARLES), and Owen was characterized as the remnant of a reactionary, idealistic morphological tradition opposed to Darwinian evolution. Only in recent decades has his importance as a scientific thinker been reassessed.

Owen was born in Lancaster, England, on 20 July 1804, the son of Richard Owen, a West India Company merchant, and Catherine Parrin Owen. After completing primary schooling in 1818, he served three indentures to Lancaster surgeons between 1820 and 1825. Owen matriculated in 1824 as a medical student at the University of Edinburgh, where he studied under many of the same teachers and experienced the same curriculum as would Darwin the next year. At Edinburgh he came under the strong influence of the anatomist Dr. John Barclay (1758–1826), the owner of a private anatomical school in Edinburgh. In April 1825, with Barclay's recommendation, he moved to London to apprentice to the prestigious London surgeon John Abernethy (1764–1831). Although Owen was licensed as a surgeon in August 1826, his practice seems to have been limited to a brief period of hospital work in London and Birmingham.

Through Abernethy's influence, Owen was appointed in 1827 assistant to William Clift (1775–1849), curator of the Hunterian Museum of the College of Surgeons, one of the great anatomical museums of the world. In 1842, Owen succeeded Clift as curator, a position he held until 1856. Owen's long collaboration with Clift, resulting in the multivolume illustrated catalogs of the Hunterian Collection (1830–1856), forms the background for understanding his subsequent anatomical and paleontological work. Owen married Clift's daughter Caroline on 20 July 1835. Their only child, William (b. 6 October 1837), died by apparent suicide in 1886.

In March of 1827, Owen attended the Hunterian Lectures in Comparative Anatomy of Joseph Henry Green (1791–1863). There he was first acquainted with Green's efforts to synthesize German-derived natural philosophy with the anatomical theories of GEORGES CUVIER (1769–1832) and JEAN-BAPTISTE LAMARCK (1744–1829). In July of 1830 Cuvier visited the Hunterian Museum and was conducted through the collection by Owen. The following summer Owen visited Cuvier's Gallery of Comparative Anatomy in Paris and met many of the leading figures of Parisian life science. The Paris experience was important for Owen's entry into the field of professional comparative anatomy. After his return he began a steady output of papers on the comparative anatomy of various mammals and invertebrates that numbered in the hundreds.

In 1836 Owen received the prestigious appointment to the Hunterian Lectureship in Comparative Anatomy at the College of Surgeons. Beginning in May of 1837, he delivered these lectures annually until 1856, covering the vertebrate and invertebrate collections of the Hunterian Museum. Owen's later public career included lectureships on paleontology at the Royal School of Mines in London (1857–1861), and as Fullerian Professor of Comparative Anatomy at the Royal Institution (1859–1861).

In 1837 Owen was commissioned to describe the vertebrate fossil materials collected by Darwin aboard the HMS *Beagle* (1831–1836), which was published as *Fossil Mammalia*, the first volume of the *Zoology of the Voyage of H.M.S. "Beagle"* (1840). The proportion of paleontological studies in his published work grew steadily after this date. His two-part "Report on British Fossil Reptiles," delivered to the British Association in 1839 and 1841, introduced the order *Dinosauria* into the literature. His papers on the fossil reptiles were collected in the four-volume *History of British Fossil Reptiles* (1849–1884). The most general overview of his paleontological work is his popular text *Paleontology* (1860).

Two interrelated issues united his anatomical and paleontological work of the 1840s. In the *Lectures on Comparative Anatomy of the Vertebrate Animals* (1846), his Hunterian lectures for 1844–1845, he first laid the groundwork for his theory of the archetypal vertebrate plan, which was first presented to the public at the British Association Meetings in Southampton in September 1846, and published in the British Association Reports as "Report on the Archetype and Homologies of the Vertebrate Skeleton" in early 1847. This was subsequently issued in revised form with extensive plates in 1848. In this landmark study Owen proposed a fundamental reform of anatomical nomenclature, bringing under one system of terms the various designations of anatomical structures found in the British, French, and German literature of comparative anatomy. A complex and often misunderstood concept (Rupke, 1993, 1994), Owen's archetype was an ideal plan that unified the competing claims of Cuvier and the rival claims of Étienne Geoffroy St. Hilaire (1772–1844). Cuvier emphasized the principle of the functional adaptation of animal structure to conditions of life. St. Hilaire developed the claims for a common anatomical plan, independent of conditions of life, underlying both vertebrate and invertebrate structure. Owen's ideal structural form functioned both as an ideal archetype by which forms were created, and also as an "inner law" operating as a secondary cause in nature. On the basis of this concept, Owen was able to enunciate a clear distinction in 1846 between relations of "analogy," designating purely functional (i.e., Cuvierian) relations of parts, and those of "homology" that designated relationships of individual and general structures to the ideal archetype. This principle would become a staple of anatomical discussion in the subsequent tradition.

The paleontological implications of the concept of the archetype were also important. As an "inner law" of nature, the successive appearances of more complex vertebrate forms in the fossil record—fishes, amphibians, reptiles, mammals—could be accepted as a real historical sequence, a point denied by such illustrious contemporaries as CHARLES LYELL (1797–1875). For Owen, the history of vertebrate life displayed historical divergences away from the ideal archetypal plan.

In this version of the "unity of type" principle, it was an easy step for Darwin to claim in the *Origin of Species* (1859) that the principle of homology indicated the literal historical derivation of forms from primordial common ancestors with homological relationships demonstrating this historical descent. Thus reinterpreted, Owen's principle became a mainstay of Darwinian interpretations of animal relationship. Owen never, however, accepted these reinterpretations as correct understandings of his concept. Controversies with the Darwinian party over such issues, particularly with Thomas Henry Huxley (1825–1895), clouded much of Owen's reputation after 1860.

After leaving the Hunterian Museum in 1856, Owen's scientific career shifted increasingly to administrative duties. He was appointed superintendent of the burgeoning galleries of zoological and fossil materials housed in the British Museum in Bloomsbury. Following long negotiations, his dream of a new national museum of natural history was finally realized with the construction of the Natural History Museum in South Kensington (1873–1881), which officially opened in 1881. Owen served as director of this institution until his retirement at the close of 1883.

Owen's role in metropolitan London science was recognized by knighthood in January 1884, and he was the recipient of numerous other international honors, including election to the most prestigious international scientific societies of his day. In Britain he received, among other honors, the Copley Medal of the Royal Society of London (1851), the Gold Medal of the Linnean Society (1888), and honorary degrees from Oxford, Cambridge, and the University of Dublin. Extensive unpublished materials are held at the British Museum of Natural History and the Royal College of Surgeons, both in London (Gruber and Thackray, 1992).

Bibliography

DESMOND, A. *Archetypes and Ancestors: Paleontology in Victorian London.* London, 1982; Chicago, 1984.

———. *The Politics of Evolution: Morphology, Medicine, and Reform in Radical London.* Chicago, 1989.

GRUBER, J., and J. C. THACKRAY. *Richard Owen Commemoration: Three Studies.* London, 1992.
MACLEOD, R. M. "Evolutionism and Richard Owen, 1830–1868." *Isis* 56 (1965):259–280.
OSPOVAT, D. *The Development of Darwin's Theory: Natural History, Natural Theology, and Natural Selection, 1838–1859.* Cambridge and New York, 1981.
OWEN, R. S. *The Life of Richard Owen,* 2 vols. London, 1894; repr. Westmead, Eng., 1970.
RUPKE, N. "Richard Owen's Vertebrate Archetype." *Isis* 84 (1993):231–251.
———. *Richard Owen: Victorian Naturalist.* New Haven, CT, and London, 1994.
SLOAN, P. R., ed. *Richard Owen's Hunterian Lectures: May and June 1837.* London and Chicago, 1992.

PHILLIP R. SLOAN

OZONE HOLE

See Pollution of Atmosphere

P

PALEOBOTANY

Paleobotany is the subdiscipline of paleontology dealing with the scientific study of fossil plants. Paleobotany involves the careful description and interpretation of plant remains from any stage of geologic time and the application of this information to evolutionary, ecological, floristic, geographic, or geological issues. Paleobotanical studies address such questions as: When did plants originate? What are the major events in the evolution of plants? What are the relationships between ancient plant lineages and modern ones? When did plant-animal or plant-fungus interactions first arise and are they different from those observed today? How did modern floras, such as temperate woodlands, prairies, or tropical forests, come into being? Why do we see certain biogeographical distributions in some plant groups? What information can the plant fossil record provide concerning ecological responses of ancient plants to their environment, plant migrations, or past climates? How is knowledge of fossil plants and their preservation useful in the search for oil or coal or in predicting future climate? Because plants seldom are preserved in their entirety and not all fossilized plants have yet been discovered, the story of their past history still is unfolding.

Paleobotany includes in its study not only the predominantly land-dwelling, multicellular, autotrophic (can make their own food) organisms presently classified in the Kingdom Plantae, but also fungi, some protists such as the single-celled, colonial, or filamentous, autotrophic freshwater and marine algae (Kingdom Protista), and primitive autotrophs called cyanobacteria (Kingdom Monera). This is appropriate because the fossils being found in Precambrian and early Paleozoic sediments represent likely progenitors of plants. Fungi often are found in or on plant tissues and merit study in their own right and as possible sources of alteration or decay of plant tissues.

Fossil representatives of the major extant plant groups (liverworts, hornworts, mosses, ferns; fern allies such as club mosses, spike mosses, and horsetails; seed-producing plants such as conifers, ginkgoes, cycads, or sago palms; and flowering plants) are documented starting in the Silurian. Evidence thus far suggests that club mosses can be recognized as a distinct group from the rest of the tracheophytes (plants with specialized water-conducting cells) since the Late Silurian or Early Devonian; that ferns and seed plants have existed since latest Devonian times; that conifers first appeared about 50 million years later at the end of the Carboniferous; and that flowering plants appeared in significant numbers in the Early Cretaceous. Debate continues over the possibility that both land plants and flowering plants arose much earlier

Figure 1. A reconstruction of Renalia hueberi, a simple land plant of Early Devonian age. (All figures courtesy of Patricia G. Gensel.)

Figure 2. First tree, Archaeopteris.

than presently documented by megafossils (Figures 1 and 2).

Many now-extinct plant lineages recognized as a result of paleobotanical studies expand the total diversity of the plant kingdom. These include at least ten distinct lineages of early land plants (particularly important lineages include the leafless rhyniophytes, trimerophytes, and zosterophylls, and the shrubby to tree-sized progymnosperms), and at least nine lineages of extinct seed plants (Table 1). Comparative data suggest that seed plants arose from within the progymnosperms and that gymnosperms (naked-seeded plants) were

Table 1. Major Groups of Now Totally Extinct Plants

Non-seed Plants	*Seed Plants*
Rhyniophytes	Calamopityaleans
Trimerophytes	Lyginopteridaleans
Zosterophyllophytes	Medullosaleans
Barinophytes	Callistophytaleans
Progymnosperms	Cordaitaleans
Cladoxylaleans	Bennettitaleans
Iridopterids	Glossopteridaleans
Pre-ferns	Caytonialeans
Coenopteridaleans	Pentoxylaleans
	Czekanowskialeans

much more abundant and diverse in the Mesozoic than they are today.

Microscopic plant remains, such as spores and pollen grains or fragments of resistant plant tissues, are studied along with megafossils or extracted from sediments that lack megafossils and analyzed for biostratigraphic or evolutionary purposes (*see* PALYNOLOGY). Combining megafossil and microfossil approaches to address problems of plant evolution often has resulted in major advances in knowledge; for example, microscopic-sized plant fragments and spores indicate the existence of land plants much earlier in time (Ordovician) than can be deduced from megafossils (Gray and Shear, 1992) and study of pollen ultrastructure of modern and fossil seed plants has considerably improved recognition of the earliest flowering plants (Doyle, 1978).

What Paleobotanists Do

Paleobotanists collect fossils and use a variety of techniques, some adapted from ones used to study modern plants and some derived from geological techniques, to extract the maximum possible information from them. A major task of the paleobotanist is to examine the disarticulated plant parts and put them together, thus reconstructing whole plants. Plants or plant parts are identified and their structures, functions, and variation are interpreted to the extent the information permits. Paleobotanical studies are improved by incorporating geological information, especially analyses of taphonomy (events during transport, deposition, and fossilization), sedimentary environments, and inferred paleoenvironment. Paleobotanists develop a broad knowledge of geology and botany and employ a uniformitarian (the present is the key to the past) approach but continually test the hypothesis that ancient organisms may have differed in significant ways from modern ones.

The kind of information that can be obtained from plant megafossils depends on conditions and events during deposition and preservation and ranges from the whole plant or an imprint of it to cell and tissue-level details of component parts. If only external form (impressions) is preserved, this is exposed and documented (Figure 3). Some fossils consist of altered organic matter (compressions); these are exposed more fully by microexcavation of rock matrix with needles or treated chemically with acids to digest the enclosing sediment and release the organic remains. Resistant plant parts typically are the waxy cuticle, lignified conducting cells (tracheids, vessels), and sporopollenin-coated spores. These entities are extracted, oxidized, and studied with light or electron microscopy, either directly or after sectioning. Chemical analyses to determine the composition of the resistant materials, such as cutin or cutan in cuticles or sporopollenin in spore walls, can be conducted on preserved organic remains and compared with living counterparts. If cells or tissues are impregnated with mineral (permineralizations), thin sections can be prepared, or the mineral removed and plant pieces studied microscopically. The data thus obtained, appropriate illustrations, and interpretations are communicated to others via papers, books, talks, or museum exhibits. Specimens and photos are maintained in collections.

Figure 3. Leaf of the earliest known cycad ("sago palm") called Leptocycas, from Late Triassic of North Carolina.

Certain attributes of plants influence the type and amount of information available from plant fossils. Plants grow throughout their lifetime and many shed parts regularly. These features, plus events during transport and burial, explain the abundance of fragmentary remains. Events during the development of a plant are readily affected so that the form of mature organs can vary in response to environmental differences, making assignment to a single taxon challenging. Mosaic evolution, namely different parts of an organism evolving at a different rate than other parts, occurs in plants and must be considered when interpreting structures. With adequate collection of all possible variations and as preservation permits, it is possible to depict, from fragments, what the whole plant probably looked like and to infer ways in which it functioned. Even so, aspects of rooting structures, habit, and the spacing and distribution of plants in communities presently are poorly known.

Certain stages in plant life cycles, such as haploid sexual structures (gametophytes, gametangia, or gametes) of non-seed plants, lack resistant coverings and are not found as fossils. As an exception, several gametophytes with gametangia containing either sperm or eggs recently were described from the Early Devonian Rhynie Chert flora of Scotland that offer information about haploid phases of some early land plants (Remy et al., 1993). Flowers are delicate and until recently were seldom found; however, in the past decade a large number of Cretaceous and Tertiary flowers have been documented (Dilcher and Crane, 1984).

Sub-cellular features are preserved only in exceptional conditions, such as the possible chromosomes in spores of a permineralized Carboniferous lycopsid and the remains of organelles shown in well-preserved early Tertiary leaves. Most biochemical features (enzymes, metabolites) are unknown or uncertain. Recent attempts to isolate and amplify deoxyribonucleic acid (DNA) from chloroplasts of exceptionally well-preserved fossils have produced some data that were compared cladistically to modern groups (Golenberg et al., 1990; Soltis et al., 1992). If additional molecular data can be obtained, comparisons to extant counterparts and evaluation of divergence times based on estimated molecular clocks can be tested for many plant types.

Contributions to Evolutionary Studies

The study of plant fossils has provided considerable information about plant diversity and evolution, including the origin(s), duration, and possible extinction of plant groups, changes in plants or plant structures through time, and rates of evolution. Through comparative analysis of structures, fossil plant data allow for rigorous postulation of genealogies, that is, how various lineages relate to one another and to modern plants or plant groups. For example, recognition of the progymnosperms as a lineage of plants that combine fern-like reproduction and conifer anatomy led Charles Beck and colleagues to regard them as the group from which seed plants probably arose (Beck and Wight, 1988). Studies have documented the first occurrence of plant tissues (tracheids, the cambium) or organs (leaves, flowers, and roots), provided evidence of how some structures evolved (conifer female cones), and documented changes in their construction and function through time. Floristic analysis of plant assemblages, some of which represent types of plant communities no longer present in the biosphere, suggests ways communities may have evolved.

Geologic Importance

Fossil plants or their reproductive diaspores (spores or pollen grains) are increasingly useful in determining the age of a deposit or in correlating rock units from one place to another, although biostratigraphy is based mainly on animal (particularly invertebrate) remains. Plants are the source of coal, and their various forms of preservation improve understanding of coal formation and use of coal as a fuel. Similarly, plants along with animals are involved in the formation of oil deposits and are useful in estimating appropriate sites for exploration or drilling. Distribution of plant types, for example, the leaves of *Glossopteris,* is used to support ideas about continental drift or confirm past positions of continental plates. Plants are used to infer climate (see next section).

Paleogeographic Importance

The fossil record may demonstrate changes in floral composition in a given area or more globally through time and contribute to understanding how modern floras came into being. Comparison of floras from different geographic regions within a specific time interval is useful in demonstrating the extent to which certain plants were localized or widely distributed. The most ancient floras appear to have been largely cosmopolitan, but phytogeographic regions are clearly evident by the Carboniferous. More study is needed because there are some regions of the world for which fossil floras are unknown or incompletely known.

Based on assumptions that plants exhibited the same environmental tolerances to climate in the past as they do today, ancient floras or individual plants or plant structures have been used to postulate climatic conditions in past time periods. Currently, it is possible to use aspects of leaf morphology or wood construction to determine climatic parameters such as temperature or moisture conditions. By tracing changes in communities and climate through time, we can gain a much better understanding of the impact of biotic and abiotic events on long-term climatic conditions.

Paleoecological Aspects

As assemblages or lineages of plants become better known, it is possible to investigate both fossil and associated geological data to determine what the environments were like in which certain plants lived and to infer how plants may have responded to or altered their environments. It also is becoming possible to judge the extent to which environment may have influenced evolutionary changes and to distinguish changes in components of communities caused by ecology rather than evolution. Further study may help address questions such as: Did plant communities in the past function and respond to environmental conditions and each other in the same way as modern plants do? What were plant–animal and plant–environment interactions like among the earliest plants to live on the land, among early seed plants, or among the first flowering plants?

Important Floras

Of the many informative fossil-bearing sites worldwide, a selected few are listed that have had major impact on some aspect of plant history. The oldest land plant megafossils are best known from the late Silurian of Wales, United Kingdom, and if dated correctly, from Victoria, Australia. The famous silica-impregnated plants from the Rhynie Chert, Scotland, document internal structure of early land plants. The Early Carboniferous floras of France and Scotland provide evidence of early diversification of ferns and seed plants. Coal-producing swamp vegetation is represented in deposits in the central United States, England, Western Europe, and the Donets basin, Ukraine. One example of well-preserved impression and compression remains of this period are the fossils of the Mazon Creek flora.

The distinctive Gondwana flora is well documented by the Permian *Glossopteris* flora of Australia, India, Africa, South America, and Antarctica. The Petrified Forest of Arizona and fossils of the Chinle Formation document ferns and gymnosperms of the Triassic age. Ferns and diverse seed plant types are known from the Jurassic flora of Yorkshire. Earliest flowering plants are found in the Early-Middle Cretaceous Potomac Group and the Dakota Group (United States), and from Portugal and Sweden. Pronounced floristic changes are documented at the Cretaceous-Tertiary boundary in the Raton Basin, New Mexico. Numerous Tertiary floras show modernization of plants and floras, for example, the Clarno Nut Beds and associated plants, Oregon, the London Clay flora (United Kingdom), and Messel deposits of Germany document the tropical vegetation typical of the Early Tertiary. Cooler climates are inferred on the basis of Oligocene floras, such as the Bridge Creek flora of the John Day Formation of Oregon and Brandon Lignite flora of Vermont. Late Tertiary floras from the western United States and Europe have been extensively studied and document pre-glaciation floras.

Bibliography

ANDREWS, H. N. *Ancient Plants and the World They Lived In.* Ithaca, NY, and London, 1947.

BECK, C. B., and D. C. WIGHT. "Progymnosperms." In *Origin and Evolution of Gymnosperms,* ed. C. B. Beck. New York, 1988.

BEHRENSMEYER, A. K., J. D. DAMUTH, W. A. DIMICHELE, R. POTTS, H.-D. SUES, and S. L. WING. *Terrestrial Ecosystems Through Time.* Chicago.

DILCHER, D. L. and P. R. CRANE. "In Pursuit of the First Flower." *Natural History* 3(1984):56–61.

DOYLE, J. "Origin of Angiosperms." *Annual Review of Ecology Systematics* 9 (1978):365–392.

GOLENBERG, E. M., D. E. GIANNASI, M. T. CLEGG, C. J. SMILEY, M. DURBIN, D. HENDERSON, and G. ZURAWSKI. "Chloroplast DNA Sequence from a Miocene Magnolia Species." *Nature* 344 (1990):656–658.

GRAY, J., and W. SHEAR. "Early Life on Land." *American Scientist* 80 (1992):444–456.

REMY, W., P. G. GENSEL, and H. HASS. "The Gametophyte Generation of Some Early Devonian Land Plants." *International Journal of Plant Sciences* 154 (1993):35–58.

SOLTIS, P. S., D. E. SOLTIS, and C. J. SMILEY. "An rbcL Sequence from a Miocene Taxodium (Bald Cypress)." *Proceedings of the National Academy of Sciences* 89 (1992):449–451.

STEWART, W. N., and G. ROTHWELL. *Paleobotany and the Evolution of Plants.* New York, 1993.

TAYLOR, T. N., and E. L. TAYLOR. *The Biology and Evolution of Fossil Plants.* Englewood Cliffs, NJ, 1993.

THOMAS, B. A. *The Evolution of Plants and Flowers.* London, 1981.

THOMAS, B. A., and R. A. SPICER. *The Evolution and Paleobiology of Land Plants.* London, 1987.

PATRICIA G. GENSEL

PALEOCLIMATOLOGY

Paleoclimatology is the study of ancient climates, those in existence prior to instrumental records. The results of paleoclimatic studies may take the form of general descriptions of past condition(s) of Earth's climate, or may be more detailed and quantitative reconstructions of a region at some point in time (e.g., the CLIMAP project; see below).

Climate has influenced many aspects of Earth history, just as it impacts the present-day planet. Through geologic time, climate elements have affected the evolution of life, biotic distributions across Earth, the nature of surface environments of Earth, and the nature of sediment deposition. It is through geologic evidence of ancient surface processes and conditions—fossils and sedimentary rocks—that indirect evidence of climatic characteristics can be derived. Thus, fossils and sedimentary rocks are the major tools of paleoclimatology (*see* PALEOGEOGRAPHY; PALEONTOLOGY; SEDIMENTS AND SEDIMENTARY ROCKS, CHEMICAL AND ORGANIC; SEDIMENTS AND SEDIMENTARY ROCKS, TERRIGENOUS (CLASTIC); SEDIMENTARY STRUCTURES; TAPHONOMY).

Origin and Evolution of Paleoclimatology

As a field of study, paleoclimatology has developed in conjunction with geology, especially with the fields of paleontology and sedimentary geology. Once fossils were understood to represent the remains of ancient organisms, and therefore, that they represented past conditions, the concept of changing climates and environments through Earth history was established. The understanding that certain kinds of rocks, primarily sedimentary rocks, formed under specific environmental conditions, added an important element necessary for deciphering and understanding past climatic and surficial conditions of Earth.

Historically, Leonardo Da Vinci, as well as others before and after him, observed fossil seashells at inland and uplifted areas and surmised that ancient seas had to have existed in those regions at some distant time to explain their presence. Centuries later, English and French geologists and surveyors, in their quest to construct canals and catalog regional strata, noticed hints of ancient environments and life-forms that were different from those of the present. Later, geologists and geographers, most notably Wladimir Koppen and ALFRED WEGENER, applied modern climatology zones and processes to geologic records of the ancient world in perhaps the first systematic approach, reconstructing ancient configurations of continents based upon distributions of fossils and rocks that indicated latitudinal trends of wet or dry conditions. In another landmark advance in early paleoclimatology, JEAN LOUIS RODOLPHE AGASSIZ realized that the Quaternary rock record could be read as a chronology of waxing and waning ice ages. All of these efforts contributed to the field that is now known as paleoclimatology.

Major Tools of Paleoclimatology

Paleoclimatic indicators, also known as proxy climate indicators, can be categorized as being primarily chemical, biological, or physical in nature (although some indicators may be considered to belong to more than one of those classifications). Chemical indicators of past climatic conditions include stable isotope ratios (oxygen being the most widely used), which are most commonly recovered from ice cores, marine microfossils (foraminifera, primarily), carbonate rocks, and corals (*see* ISOTOPE TRACERS, STABLE). The isotropic ratio of two isotopes of the same element contained in a substance is fixed, or locked in, at the time of formation for these materials. For example, the growth of shell material on a marine organism will incorporate oxygen from the surrounding seawater. The ratio in the shell of isotopes Oxygen-18 to Oxygen-16 will be a function of the ambient climatic conditions (primarily temperature) of that water. This ratio can be measured in the laboratory. A major advantage of this technique is the widespread availability of potential samples in both the marine and terrestrial realms. A major limitation of this application is the possibility of alteration of the original material, which must be investigated prior to interpretation of the records.

Biological indicators of ancient climatic conditions can be either macro- or microfossils. Two approaches are possible in the interpretation of fossils. In one case, the ancient forms are related to modern forms and their associated climatic preferences. In the other case, morphological features of fossil organisms are used to provide some indication of climatic conditions under which the organism exist. The method that relies on the nearest living relative becomes increasingly uncertain as the time period studied becomes older because relationships to existing forms become less direct. Morphologic assessments are less uncertain in this respect because it is the form of the plant or animal that is important, and the association has little dependence upon modern relatives. However, this method is most useful when an entire population or assemblage of fossil organisms are examined together. Plant and animal macrofossils are important terrestrial paleoclimate proxies, that is, substitutes, and plant pollen is a valuable and abundant paleoclimate microfossil. Marine microfossils are abundant and extremely useful for reconstructing past ocean temperatures and water mass characteristics.

Another type of biologic proxy is any record of regular (often annual) growth that responds to ambient conditions. These records include tree rings, coral growth bands, or bone growth. Analyzing these proxies assumes that optimum growth conditions will result in larger growth bands; optimum growth conditions will be different for corals and trees, for example, and this must be considered in these analyses.

Physical proxy climate indicators are those types of geologic evidence for which processes of formation provide some inorganic indication of climatic conditions during formation. Examples of physical proxy climate indicators include sand dunes, glacial evidence such as morianes and scour features, volcanic ash lobes, ripple marks, and sedimentary rock types such as coals and evaporites.

Using Paleoclimate Evidence to Reconstruct Ancient Climates

Most types of paleoclimate indicators can provide at least a general sense of wet or dry, or cold or warm, paleoclimate conditions. For example, evaporites can be used to indicate arid conditions, and coals and bauxites to indicate high precipitation. Evidence of glaciers also provides specific paleoclimatic details, of cold climates, and possibly direction of ice movement. Remains of corals and other temperature-sensitive flora and fauna can be used to provide minimum temperature estimates. Some evidence, such as oxygen isotope ratios (*see* ISOTOPE TRACERS, STABLE), may be used to derive quantitative paleoclimate elements, such as sea surface temperature or mean annual land temperature. All paleoclimate evidence is of most value when the evidence is located in its ancient paleogeographic (latitudinal) position (*see* PALEOGEOGRAPHY). For example, corals are located in regions where ocean surface temperatures are above 25°C in the modern world, and this limitation can be used to relate the presence of ancient corals to the latitudinal extent of warm ocean surface temperatures in ancient paleogeographic positions.

Major Theories of Climate Change Through Geologic Time

Climatic changes on different timescales throughout Earth history have been the subject of much study and debate among geologists. Some of the major topics are: (1) tectonic forcing of climate, including land/sea configuration and the impact of continental distribution on global climate through time, via processes of albedo and ocean current controls; (2) forcing of global climate by changes in atmospheric trace gas concentrations; (3) the "Faint Young Sun Hypothesis," explaining how the Precambrian Earth had liquid water (based on geologic evidence) when the young Sun was less bright than today; and (4) the issue of periodic oscillations in glacial advances and retreats over the past several million years, commonly known as Milankovitch Effects, a hypothesis which postulates that periodic glaciations were caused by solar insolation variations as the product of Earth's changing orbital motions (*see also* GLOBAL CLIMATIC CHANGES, HUMAN INTERVENTION; GLOBAL ENVIRONMENTAL CHANGES, NATURAL).

Bibliography

BRADLEY, R. S. *Quaternary Paleoclimatology: Methods of Paleoclimatic Reconstruction.* 1985.

CLIMAP PROJECT MEMBERS. "Seasonal Reconstruction of the Earth's Surface at the Last Glacial Maximum." Geological Society of America, Map and Chart Series no. MC-36. Boulder, CO, 1981.

CROWLEY, T. J., and G. R. NORTH. *Paleoclimatology.* New York, Boston, 1991.

FRAKES, L. A. *Climates through Geologic Time.* Amsterdam, 1978.

LISA C. SLOAN

PALEOCLIMATOLOGY, HISTORY OF

The science of paleoclimatology attempts to describe and explain the history of past climates. The scope of the subject covers changes before development of a relatively widespread instrumental observing system in the middle of the nineteenth century. Thus the subject spans a range of topics from climate change on an annual timescale over the past few hundred years, to the ice ages of the Pleistocene (last 1.6 million years or Ma, with the most recent Ice Age ending about 14,000 years ago), and to climates when continents or continental barriers were in different positions. The latter topic involves everything from climates of the early earth (3–4 billion years ago or Ga) to climates of the more "recent" past (pre-Pleistocene). For example, certain intervals in the Pliocene (approximately 1.6–5.0 Ma) were warmer than the present. Paleoclimatologists attempt to describe and explain the causes of the warmth during that time interval. Part of the description involves an accurate understanding of continental positions, and since some of the mountains of western North America had not yet reached their present elevation, and the central American isthmus was partially open between the Atlantic and the Pacific, then study of the Pliocene legitimately falls into the older category. In contrast, solid-earth boundary conditions were virtually identical to the present during the Pleistocene ice ages.

Methodologies

To reconstruct climates of the past, it is necessary to examine a whole range of geological evidence. For example, as the climate has changed, the distribution of plants and animals has varied, with warmth-loving organisms migrating toward the poles (higher latitudes) during warm periods and organisms preferring cooler climates migrating toward the equator (lower latitudes) during cold periods. Thus the presence or lack of such indicators can be evidence for warmer or colder climate. Similarly, as air and sea temperatures change, chemical ratios in fossils also change, so geochemical measurements can be used to infer climate change. These data are called "proxy climate indices" (proxy because the indices are considered as a stand-in for actual measurements of temperature, precipitation, etc.). This relatively simple concept requires application of advanced statistical methods to assess the relationship between organisms and environmental variables. Obtaining data involves examining everything from deposits left by the vast ice sheets of the Pleistocene, to pollen deposited in bogs and lakes, to deep-sea expeditions to retrieve samples from the ocean floor, and to cores from the Greenland and Antarctic ice sheets. Dozens of different approaches have been developed, and all require intercomparison and cross-checking to develop the most accurate possible picture of past climate variations.

An accurate chronology of past events is required to place all data in a correct time sequence. Placing events in time requires both a relative and an absolute timescale (*see* STRATIGRAPHY). The relative timescale involves determining whether one event happened before or after another event. The absolute timescale requires knowing when these events occurred in absolute time. The accuracy of timescale resolution depends on the nature of the process studied. For samples covering the last few hundred years (e.g., tree rings, corals, and sometimes ice cores), information is often accurate to within one year. For example, very large acidity spikes in polar ice cores can be accurately dated and related to massive volcanic eruptions of Indonesia's Tambora (1815) and possibly El Chichon (1259) in central America. The best temporal resolution attainable for the Pleistocene is usually about one thousand years, although occasionally an accuracy of a few hundred years can be obtained. On longer (tectonic) timescales, resolution decreases considerably, with a resolution accurate to within around 10,000 years occasionally obtainable, although it is usually closer to 100,000, a million, or for older deposits even several million years. Despite this loss of resolution, valuable conclusions can still be derived from such deposits, but the conclusions require couching with caveats concerning timescale uncertainties.

Pleistocene Ice Ages

Probably the most important conclusion to be drawn from observations of past climate change is that at various times in Earth history the climate was either substantially colder or warmer than at present. The magnitude of the climate changes is quite impressive. During the Pleistocene ice ages, great ice sheets extended to about 40–45°N latitude in North America and Europe, with virtually all of Canada and the northern half of the United States (approximately north of St. Louis) covered by an ice sheet 2–3 kilometers (km) thick. Most of Europe north of the Alps also had an ice sheet. Sea ice extended to approximately the same latitude of 40–45°N in the North Atlantic and about 50°S latitude around Antarctica. The volume of water contained by these ice sheets was about 50×10^6 km^3, which lowered sea level about 110 meters (m) and moved the shoreline to the edge of the continental shelves.

Other changes occurring during the Ice Age include air and sea surface temperature decreases of several degrees centigrade, massive displacements of plants and animals, a generally drier climate, and probably significant reductions in tropical rain forest extent. One of the more remarkable results obtained from studies of past climates involves ice core measurements of ancient air samples indicating that levels of carbon dioxide (CO_2) and methane (CH_4) decreased during the ice ages. Because these gases tend to trap outgoing radiation reemitted from Earth's surface, they are greenhouse gases (*see* GLOBAL CLIMATIC CHANGES, HUMAN INTERVENTION and GLOBAL ENVIRONMENTAL CHANGES, NATURAL). Their lower Ice Age concentrations indicate that they contributed to the overall cooling inferred for the Pleistocene (about 4°C global average temperature decrease).

Although trace gas changes probably amplified the changes occurring during the Pleistocene, it is generally agreed that the most important factor responsible for triggering the ice ages involves orbitally driven changes in solar energy received at the top of the atmosphere. The "wobbles" in Earth's axis due to the gravitational pulls of primarily the Sun and Jupiter affect the amount of radiation seasonally incident on any particular latitude. Although globally averaged mean annual changes are virtually negligible, seasonal changes in insolation received can approach 10 percent at some latitudes. The basic idea behind the orbital hypothesis is that decreases in summer insolation in critical northern high latitudes allowed development of permanent ice sheets, and increases in summer insolation caused the ice sheets to melt.

The above hypothesis was advanced in near-present form by the Serbian astronomer Milutin Milankovitch in 1930 and 1941. Although challenged for many years, convincing evidence for its role was published in 1976 (Hays et al.) and subsequently many papers have verified those results. The present goal of research is to better understand how the orbital "signal" was transmitted through the land-sea-air-ice system in order to construct ice sheets of the dimensions inferred. As discussed earlier, CO_2 and CH_4 changes probably amplified the orbital signal. Changes in the ocean circulation and perhaps atmospheric dust levels (due to drier climates) may also be important for understanding the numerous oscillations of Ice-Age glaciers in the Pleistocene (Figure 1).

Climate Change on Tectonic Timescales

Orbital theory predicts that Milankovitch fluctuations have occurred continuously through Earth history, yet ice ages comprise only about 25 percent of the climate record of the last 600 Ma (Figure 2). During these past warm time periods subtropical fauna and flora migrated at least ten to twenty degrees poleward of their present extent. There is very little evidence for polar ice sheets, and the waters of the deep ocean, presently about

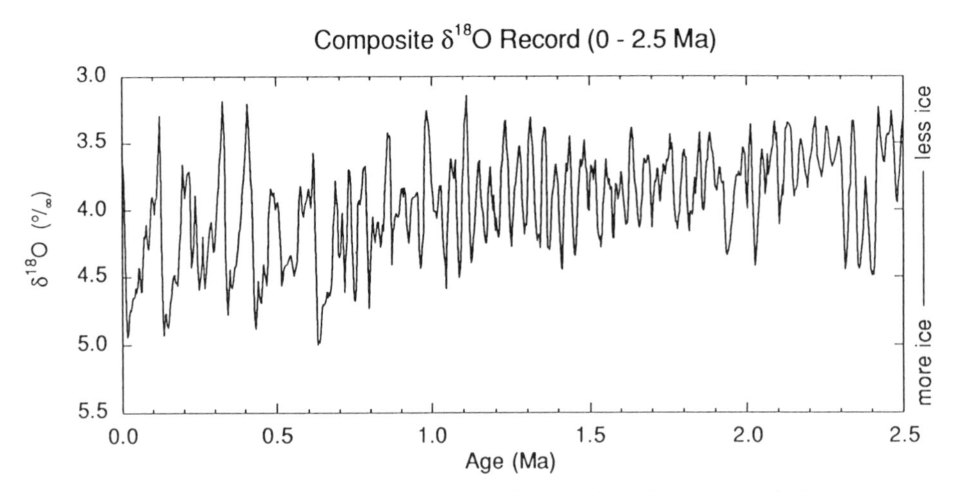

Figure 1. Variations of global ice volume for the last 2.5 Ma, as inferred from oxygen isotope measurements of single-celled organisms living on the seafloor. The many oscillations are related to changes in forcing driven by orbital insolation variations. (Based on data in Raymo et al., 1990.)

2°C, were about 15°C. Some of the most dramatic indications of greater warmth during these times involve Late Cretaceous (70 Ma) evidence for dinosaurs on the North Slope of Alaska and Eocene (55 Ma) evidence for alligators on Ellesmere Island (78°N paleolatitude) and 50-m trees with diameters of 1.5 m on nearby Axel Heiberg Island. Global temperature may have been about 6 to 8°C greater than present during these warm time periods.

Initially it was thought that changes in continental position may have been responsible for the long-term climate changes observed on tectonic timescales. However, a number of climate modeling experiments suggest that such changes, although important, are not large enough to explain all the changes that have been inferred. These results are consistent with geochemical models and proxy observations suggesting that past changes in CO_2 levels of the atmosphere may have also been primarily responsible for past warm time periods. The CO_2 levels are controlled by outgassing from volcanoes, and weathering and carbon storage on land. There is good agreement between estimated CO_2 changes and glaciation over the last 600 Ma except during the Late Ordovician (440 Ma), when a unique paleogeographic configuration may have permitted coexistence of glaciers with higher CO_2 (the generally high CO_2 values of the early Paleozoic reflect absence of terrestrial plant cover).

Relevance to Greenhouse Warming Studies

Studies of past climates and climates of other planets (*see* VENUS) have also helped climatologists develop a frame of reference for interpreting future changes due to a greenhouse climate warming. Greenhouse gas levels may reach several times present values in the next few centuries. The geologic record supports the general idea of the role of greenhouse gases in that climate change varies with CO_2 during most fluctuations of the last 600 Ma. Although there is considerable uncertainty in greenhouse predictions, there are some indications that temperatures in the next few centuries could be as warm as any time in the last 600 Ma. However, because ice sheets respond very slowly to temperature changes, we may well have a situation where "nonglacial" air temperatures coexist with polar ice sheets. This configuration may represent a unique climate in Earth history. However, geologic data indicate that tropical sea surface temperatures did not increase as predicted by climate models. Continued study of these and other climate fluctuations will enable climatologists to better understand causes for past climate change and result in more confident use of past changes as a yardstick for assessing the consequences of future climate change caused by anthropogenic activities.

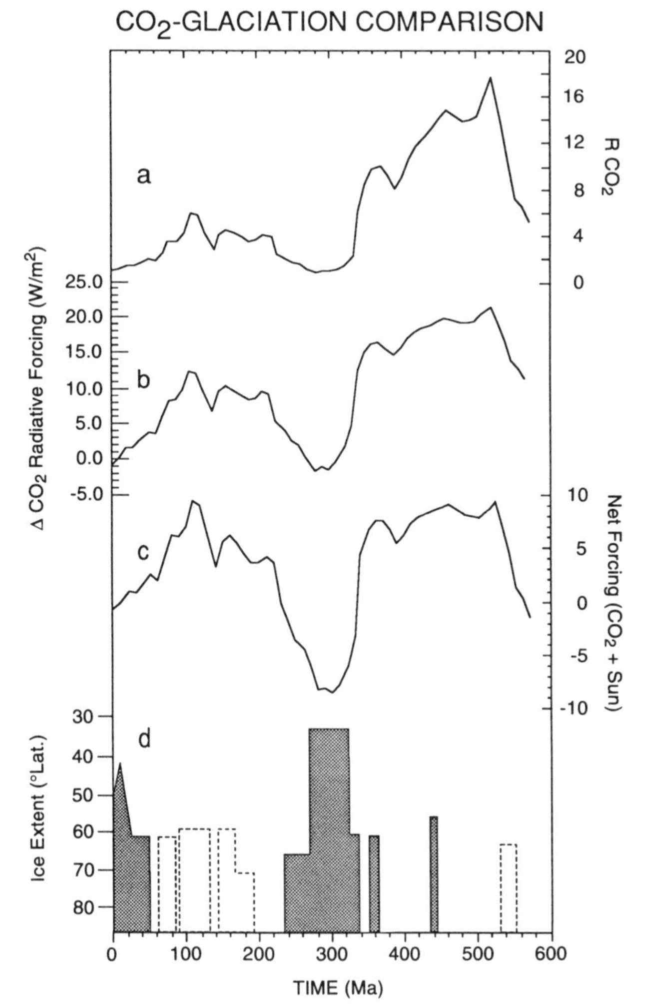

Figure 2. a. Phanerozoic CO_2 variations as calculated by a geochemical model (from Berner, 1991); R refers to ratio of past CO_2 values to present levels. b. Radiative departures from present of part a; this conversion is necessary to consider because CO_2 is a less effective greenhouse gas at high concentrations. c. Combined effect of radiative forcing and solar luminosity variations versus time; solar variability is necessary to consider because the output of the Sun was significantly less than present during earlier stages of Earth history. d. Latitudinal extent of glacial deposits versus time as estimated by Frakes and Francis (1988). Dashed areas represent evidence for ice but not necessarily continental glaciation; filled-in areas represent times of continental glaciation. Note the inverse relation between "net CO_2 + Sun" and glaciation for all of the last 600 Ma except the Late Ordovician (440 Ma) glaciation. (After Crowley, 1993.)

Bibliography

BERNER, R. A. "Models for Carbon and Sulfur Cycles and Atmospheric Oxygen: Application to Paleozoic Geologic History." *American Journal of Science* 287 (1987):177–196.

CROWLEY, T. J. "Geological Assessment of the Greenhouse Effect." *Bulletin of the American Meteorology Society* 74 (1993):2363–2373.

CROWLEY, T. J., and G. R. NORTH. *Paleoclimatology*. New York, 1991.

FRAKES, L. A., and J. E. FRANCIS. "A Guide to Phanerozoic Cold Polar Climates from High-latitude Ice-rafting in the Cretaceous." *Nature* 333 (1988):547–549.

HAYS, J. D., J. IMBRIE, and N. J. SHACKLETON. "Variations in the Earth's Orbit: Pacemaker of the Ice Ages." *Science* 194 (1976):1121–1132.

MILANKOVITCH, M. *Canon of Insolation and the Ice Age Problem*. Trans. by the Israel Program for Scientific Translations. Jerusalem, 1970.

RAYMO, M. E., W. F. RUDDIMAN, N. J. SHACKLETON, and D. W. OPPO. "Evolution of Atlantic-Pacific $\delta^{13}C$ Gradients Over the Last 2.5 m.y." *Earth Planetary Science Letters* 97 (1990):353–368.

THOMAS J. CROWLEY

PALEOECOLOGY

Paleoecology, like ecology, is the study of interrelationships between organisms and their environments. But the realm of paleoecology is the fossil remains of organisms, be they marine or nonmarine, single-celled fungi, algae or foraminifera, or highly developed, complex, multicellular plants and animals, and environments of the past. Thus, paleoecology can be viewed as historical ecology because it deals with organisms and environments of the geologic record. On the one hand, paleoecology is concerned with what fossils can reveal about the physical habitats in which organisms lived; on the other, with what fossils can reveal about the once-living organisms themselves.

Fossils are among the principal windows to physical environments for the last 600 million years (Ma) or so of geologic time. Fossils of terrestrial plants and animals may provide specific details about temperature, amount and seasonality of rainfall, and elevation of habitats of once-living organisms. Since the Jurassic, for example, fossil crocodilians have indicated warm-temperate to tropical climate. Fossils of freshwater organisms may be used to infer the size, depth, turbulence, and substrate characters of a waterbody, and the oxygen content, chemical composition, and amounts of suspended sediments of its water, and whether it represented a lake or a small, temporary pond. Since the Carboniferous or even the Devonian, for example, fossil lungfishes have been seen as a guide to climatic seasonality and waterbodies that seasonally have dried up. Fossils of marine organisms provide clues to environments of intertidal, subtidal, and estuarine habitats, and to water and substrate parameters. The brachiopod *Lingula,* for example, commonly indicates a nearshore environment, as does the Jurassic to Recent trace fossil *Ophiomorpha,* which represents the burrow of a ghost shrimp. Such information enables paleoecologists to determine short-term environments of the geologic past, in special circumstances measured in terms of a few years time or even a single year. Such information also enables them to determine long-term environmental changes throughout the history of life that show how past and present environments differ in climatic and physical factors. It enables them to learn what climatic fluctuations have occurred, their frequence and severity, and how long they lasted, and to determine changing surface relations between land and sea and changes in land elevation. Knowing the differences and changes in climate and physical environments over time is essential for an understanding of the past distribution of organisms and their evolution and extinction.

Fossils are the only direct sources of information about the ecology of the organism they represent. Paleoecology of individual fossil organisms is called paleoautecology; ecologists refer to similar studies with living organisms as autecology. Paleosynecology is the study of populations, or assemblages or communities of fossil organisms. It is concerned with relationships that existed between organisms whose remains are found together in rock deposits. Ecologists refer to similar studies with living organisms as synecology.

Paleoecology provides the means to flesh out the bones, teeth, and shells of fossil animals and to reconstruct plants from their leaves and wood—to understand fossils as once living organisms and as part of an assemblage of other organisms. By considering fossils within the context of modern living organisms, it is possible to determine how they functioned, the significance of their adaptations, their evolution, and the evolution of the communities of which they were a part.

Types of Evidence

Paleoecologists draw conclusions from all evidence of fossils. In some circumstances, noted below, they may also infer environmental information from the sedimentary rocks in which the fossils are buried. Under exceptional circumstances, such as the famous amber fossils and the frozen Siberian mammoths, whole organisms, including soft-body parts, may be preserved. Most fossils are rarely so complete (*see* FOSSILIZATION AND THE FOSSIL RECORD; TAPHONOMY). Invertebrates are generally represented by mineralized shells, tests, or skeletons deposited in life—the whole organism excluding soft-body parts, although mineralized structures sometimes show the imprint of some of an animal's soft-body anatomy. Some organisms are represented merely by an organism's impression in rock, an outline, without preservation of either mineralized or organic remains. Impressions may include the outline of soft-body parts and impressions of skin or other surface body features, such as wings, that may help paleoecologists understand how an organism functioned. Much of the paleoecologist's record is based on parts of organisms, such as pollen grains, spores, leaves, and wood of plants; the jaws, teeth, and bones of vertebrates; or the durable reproductive structures produced by some unmineralized freshwater invertebrates.

Organisms lacking mineralized or preservable organic structures provide no direct evidence of their presence and hence no direct ecological information. Some such organisms leave paleoecologically useful trace fossils, indirect evidence of an organism's presence useful in determining aspects of its behavior, life style, and feeding habits (*see* TRACE FOSSILS). Trace fossils are "marks" made by organisms as they move, rest on the substrate, or feed. Footprints, burrows, or bite marks made in leaves by herbivorous insects are common trace fossils. Other trace fossils are fossilized feces (excrement or dung) and phytoliths, silicate bodies formed in leaves and fruits of some plants. Feces distinctive to the organisms that produced them may contain pollen or leaf fragments of the plants on which an organism browsed or bones and teeth of animals on which they preyed. Phytoliths may be characteristic of plants and provide the only evidence for their presence or of communities of which they were a part. Community paleoecology also enables paleoecologists to infer the presence of soft-bodied organisms in fossil assemblages, if they were regular members of associations, whose fossilizable members are buried together. Thus trace fossils and community paleoecology enable paleoecologists to gain ecological information even for organisms that themselves lack fossil records. Community paleoecology is best studied among bottom-dwelling aquatic organisms commonly buried and fossilized where they lived together. Fully terrestrial organisms most often have their remains transported by wind or water currents, or by other organisms, variable distances before burial, both scattering and separating body parts and separating their remains from those of their common biotic associates in life. This significantly complicates reconstruction of living communities.

Examining the Evidence

How paleoecologists gather information depends on the relationship between fossil and living organisms. Many fossils represent organisms without close living relatives or ones that are even wholly extinct. An indirect, deductive approach is used in the study of these fossils. Others have closely related living relatives and may even represent living organisms. Many genera, for example, have fossil records for the last 60 Ma, and living species are common in rocks only a few million years old. A direct, inductive approach is possible where there is a close relationship between fossil and living organisms.

The direct approach applies ecological information about modern organisms and their environmental tolerances to fossils often independent of any physical aspects of their burial and fossilization. Fossil organisms that are the same as living species, or closely related to living species, provide information about their environments based on the ecology of living relatives, their biological associates, and their physical environments. Biologically determined information inferred from the ecology of living relatives permits often precise and detailed conclusions about the paleoecology of fossil organisms limited only by imprecise or incomplete knowledge of the modern organism's ecology and biogeography.

In the indirect approach, it is rarely possible to use the ecology of living organisms to provide useful information about fossils, their biological associates, or their physical environments. Paleoecological conclusions are based most often on circumstantial information related to the physical environment in which the fossil is found. In limited circumstances, information about living organisms that appear to be similar in form or may appear to have functioned similarly can provide some useful information. Generally, it is necessary to employ evidence drawn from local and regional geological data. These data may include the sediments in which fossils are buried and regional geological evidence that provide climatic information independent of the fossils. For some fossils, ecological conclusions may be inferred from the form (morphology) of the structure preserved in the fossil record independent of biological relationship. Size, shape, texture, and marginal characteristics of fossil leaves, for example, provide generalized climatic information because trees in different major climatic regions (tropics, temperate, desert) have distinctive leaves. Annual rings in fossil wood, even of wood from wholly extinct trees, have been used to infer climatic seasonality. Shapes and enamel characters of fossil vertebrate teeth, among the most commonly preserved vertebrate remains, are closely correlated with their feeding habits and food, thus providing autecological information. Specifically, for fossil mammal teeth, wear patterns and microscopic abrasion marks are useful for inferring diet. In the case of herbivores, dietary information also enables paleoecologists to infer something about the vegetation and, in turn, the prevailing climate in the animal's habitat. In marine and nonmarine mollusks, shell shape and thickness may sometimes be used to infer water turbulence and whether the animals were burrowers, lived on the substrate surface, or attached to rocks. The razor clam, *Solen,* is a burrowing organism; clams with similarly shaped shells in the fossil record are interpreted as burrowers, even if they are not related to *Solen.* Cap-shaped limpet gastropods usually live on rocks. The mangrove oyster that tightly attaches itself to mangroves may bear the impression of the tree's bark on its shell. Fossil mangrove oysters with bark imprints may be used to infer the presence of mangrove, even in the absence of remains of the mangrove itself, and that these fossil oysters occupied the same substrate as modern mangrove oysters.

Interpretation

The information that paleoecologists can skillfully tease from the fossil record is formidable despite the variable and sometimes unpredictable loss of information between the living organism or living community and the paleoecologists's fossil assemblage. Common sources of information loss are the complete absence of soft-bodied organisms except under the most unusual circumstances of burial; the absence of many dry-ground organisms that live too far from burial sites in lakes and streams (only exceptionally is the ocean a major burial site for terrestrial/freshwater organisms) to ever become a part of the fossil record; and the effects of burial that may diminish fossil assemblages or bias them in unpredictable ways. An additional complicating factor is the absence of close relatives among living organisms for many fossils. The field of taphonomy is now developing principles that will enable paleoecologists to attempt to correct for biases built into the fossil record because of these and other limitations in interpreting fossils and fossil assemblages.

To apply to fossils the ecological information gleaned from living organisms and living relatives of fossils, it is also necessary to assume that the physiological tolerances of organisms have remained unchanged over time and that organisms have functioned similarly regardless of relationship. If changes in tolerances have occurred, they are more apt to follow the rule that the more remote the relationship between fossil and modern organisms, the more ancient the fossils. These limitations in concert, and singly, mean that even the best paleoecological information obtained from the fossil record is limited compared with what is possible from both empirical study of, and experimentation with, living organisms.

Bibliography

BOUCOT, A. J. *Principles of Benthic Paleoecology.* New York, 1981.

KRASILOV, V. A. *Paleoecology of Terrestrial Plants.* New York, 1975.

JANE GRAY
ARTHUR BOUCOT

PALEOGEOGRAPHY

Paleogeography, the term used to describe the geography of the ancient past, is different from the study of modern geography because the earth's surface is in constant motion due to plate tectonics (*see* PLATE TECTONICS). The continents move at rates of 2–10 centimeters per year (cm/yr). Though this may seem slow, over millions of years continents can travel thousands of kilometers. The goal of paleogeography is to map both the past positions of the continents and changing distribution of mountains, lowlands, shallow seas, and deep ocean basins through time. The best way to illustrate the earth's changing paleogeography is through a series of maps.

Paleogeographic maps are useful because they show us how the earth has changed through time. Accurate paleogeographic maps are necessary to understand global climatic change, migration routes, oceanic circulation, mountain building, and the formation of many of the earth's natural resources, such as oil and gas. The paleogeographic maps shown in Figures 1–3 illustrate both the movement of the continents and the changing distribution of mountains (dark stipple), land (medium stipple) shallow sea (light stipple), and deep sea (no stipple) during the last 650 million years (Ma). Present-day geographic regions are labeled, and the modern coastlines are shown as black outlines. Past plate boundaries are also shown. The jagged lines in the center of ocean basins are mid-ocean ridges; the curving lines along the edges of continents are deep-sea trenches; and the triangular teeth point in the direction of subduction.

Late Precambrian, 1100–545 Ma

To produce paleogeographic maps, it is necessary to be able to accurately date the rock layers (*see* GEOLOGIC TIME, MEASUREMENT OF) and paleomagnetic information in order to determine the past positions of the continents. The absence of distinctive fossils makes it difficult to determine the age of Precambrian strata, and the paucity of reliable paleomagnetic data makes it difficult to produce paleogeographic maps for the Precambrian. With available data, 650 Ma is about as far back as we can go with any accuracy.

The climate was cold during the late Precambrian. Evidence of glaciation is found in the rocks of nearly every continent. Why cold conditions were so widespread during the late Precambrian has long puzzled geologists. As Figure 1 illustrates, the mystery of why evidence of this late Precambrian ice age is so widespread can be explained by the fact that during the late Precambrian many continents just happened to be at high latitudes. Ice-rafted debris and glacial deposits formed on these continents as they traveled near the north and south poles.

Early and Middle Paleozoic, 545–360 Ma

A new ocean, the Iapetus Ocean, widened between the ancient continents of Laurentia (North America), Baltica (northern Europe), and Siberia. Gondwana, the largest continent, stretched from the equator to the South Pole. During the Ordovician period, warm water deposits, such as limestones and salt, are found in the equatorial regions of Gondwana (Australia, India, China, and Antarctica), while glacial deposits and ice-rafted debris occur in the south polar areas of Gondwana (Africa and South America).

By middle Paleozoic time, approximately 400 Ma ago, the Iapetus Ocean had closed, bringing Laurentia and Baltica crashing together. This continental collision resulted in the formation of the Caledonide mountains in Scandinavia, northern Great Britain, and Greenland, and the northern Appalachian mountains along the eastern seaboard of North America.

Late Paleozoic, 360–245 Ma

By the end of the Paleozoic era, most of the oceans that had opened earlier were consumed as the continents collided to form the supecontinent of Pangaea. Centered on the equator, Pangaea stretched from the South Pole to the North Pole and separated the Paleo-Tethys Ocean to the east, from the Panthalassic Ocean to the west. During the late Carboniferous and early Permian the southern regions of Pangaea (southern South America and southern Africa, Antarctica, India, southern India, and Australia) lay under a thick sheet of ice that extended as far north as southern Arabia. There is evidence of a smaller north polar ice cap in eastern Siberia during the last Permian (Figure 2).

The term "Pangaea" means "all land." Although the supercontinent that formed at the end of the Paleozoic era is called Pangaea, this supercontinent probably did not include all the landmasses that existed at that time. In the eastern hemisphere, on either side of the Paleo-Tethys Oceans, there were

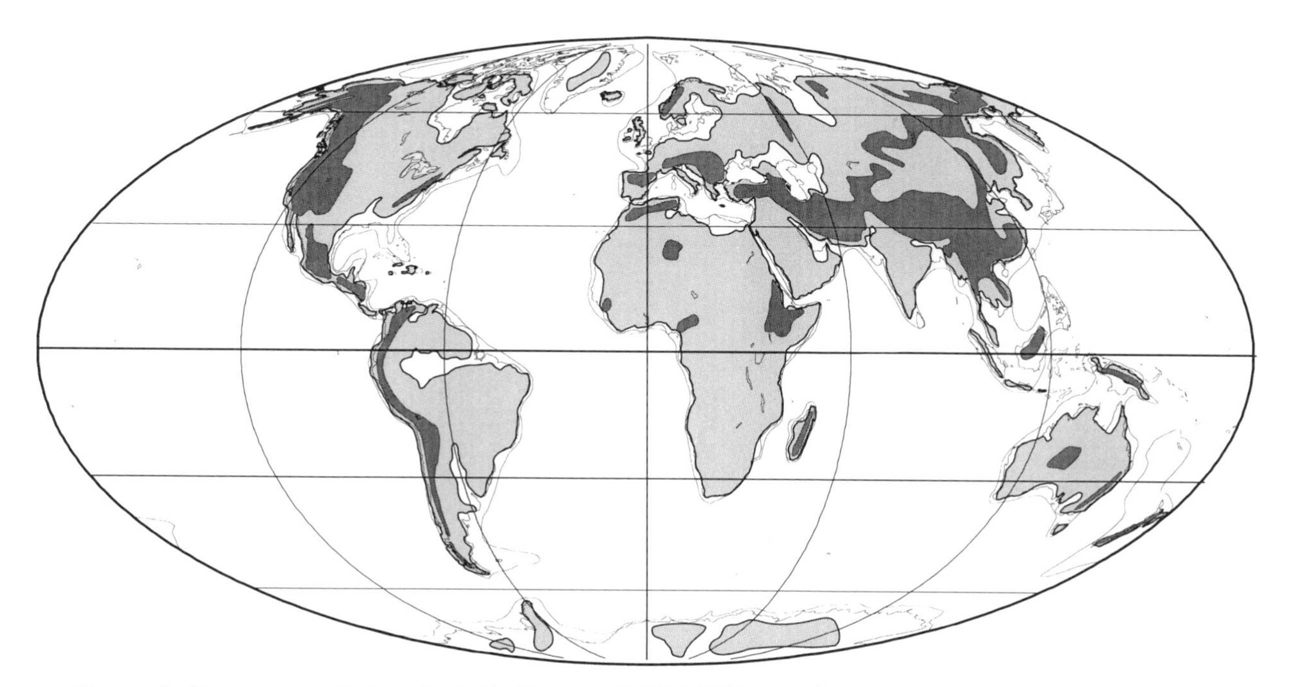

Figure 1. Late precambrian. © C. R. Scotese, PALEOMAP project, University of Texas Arlington.

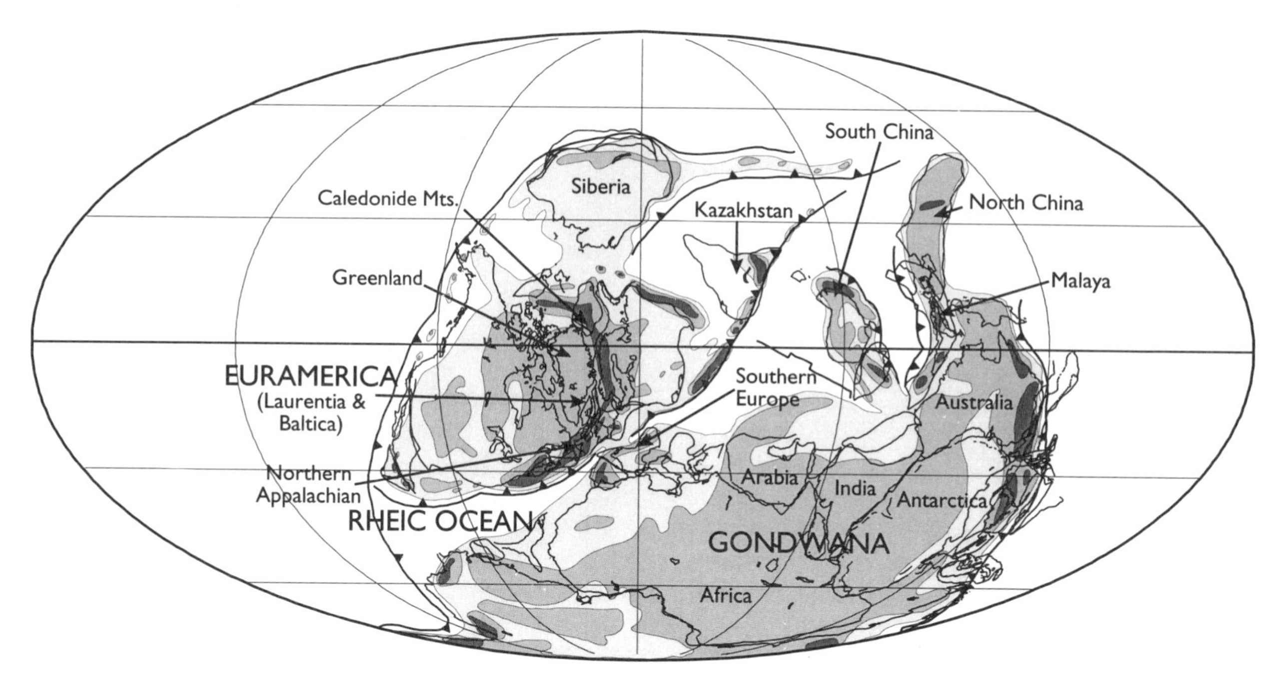

Figure 2. Late Permian. © C.R. Scotese, PALEOMAP project, University of Texas Arlington.

continents that were separated from Pangaea. These continents were North and South China, and a long irregularly shaped continent known as Cimmeria. Cimmeria consisted of parts of Turkey, Iran, Afghanistan, Tibet, Indochina, and Malaya.

Early Mesozoic, 245–144 Ma

The continental collisions that led to the formation of Pangaea began in the Devonian and continued through the late Triassic. In a similar fashion, the supercontinent of Pangaea did not rift apart all at once, but rather was subdivided into smaller continental blocks in three main episodes.

During the Mesozoic, North America and Eurasia were one landmass, called Laurasia. As the central Atlantic Ocean opened, Laurasia rotated clockwise, sending North America northward and Eurasia southward. Coals, which were abundant in eastern Asia during the early Jurassic, were replaced by deserts and salt deposits during the late Jurassic as Asia moved from the wet temperate belt to the dry subtropics. This clockwise, seesaw motion of Laurasia also led to the closure of the wide V-shaped ocean, Tethys, that separated Laurasia from the fragmenting southern supercontinent, Gondwana.

Late Mesozoic, 144–166 Ma

The second phase in the breakup of Pangaea began in the early Cretaceous, about 140 Ma ago. Gondwana continued to fragment as South America separated from Africa opening the South Atlantic, and India together with Madagascar rifted away from Antarctica and the western margin of Australia opening the eastern Indian Ocean.

Globally, the climate during the Cretaceous period, like the Jurassic and Triassic, was much warmer than today. Dinosaurs and palm trees were present north of the Arctic Circle and in southern Australia. Though ice and snow may have seasonally occurred in polar regions, there were no large ice caps at any time during the Mesozoic era.

These mild climatic conditions were due in part to the effect of shallow seaways that covered the continents during the Cretaceous. These seaways allowed warm water to be transported northward, warming the polar regions. These seaways also tended to make local climates milder, much like the modern Mediterranean Sea, which has an ameliorating effect on the climate of Europe.

Shallow seaways covered the continents because sea level was 100–120 meters (m) higher than today. Higher sea level was due, in part, to the creation of new rifts in the ocean basins that displaced water onto the continents. The Cretaceous was also a time of rapid seafloor spreading. Because of their broad profile, rapidly spreading mid-ocean ridges displace more water than do slow spreading mid-ocean ridges. Consequently, during times of rapid seafloor spreading, like the Cretaceous, sea level will rise.

Cenozoic Era, 66–0 Ma

The third and final phase in the breakup of Pangaea took place during the early Cenozoic. North America and Greenland split away from Europe, and Antarctica released Australia. Australia, like India some 50 Ma earlier, moved rapidly northward on a collision course with Southeast Asia. Additional, important rifting events have taken place during the last 20 Ma of the Cenozoic era and continue today, including: the rifting of Arabia away from Africa opening the Red Sea; the creation of the east African Rift System; the opening of the Sea of Japan; and the northward motion of California and northern Mexico away from North America.

Though several new oceans have opened during the Cenozoic, the last 66 Ma of Earth history are better characterized as a time of intense continental collision. The most significant of these collisions has been the collision between India and Eurasia, which began about 50 Ma ago. During the late Cretaceous, India approached Eurasia at rates of 15–20 cm/yr—a plate tectonic speed record. After colliding with marginal island arcs in the late Cretaceous, the northern part of India, called Greater India, began to be subducted beneath Eurasia, raising the Tibetan Plateau.

On a global scale, the area of the ocean basins has increased slightly during the Cenozoic, at the expense of the continents. Because the ocean basins are larger, they can hold more water. As a result, sea level has fallen during the last 66 Ma. In general, sea level is lower during times of continental collision such as the early Devonian, late Carboniferous, Permian, and Triassic periods.

During times of low sea level the continents are emergent, land faunas flourish, migration routes between continents open up, the climate becomes more seasonal, and most important, the global climate tends to cool off. This is largely due to the fact that land tends to reflect the Sun's energy back to outer space, while the oceans absorb the Sun's energy. Also, landmasses permit the growth of

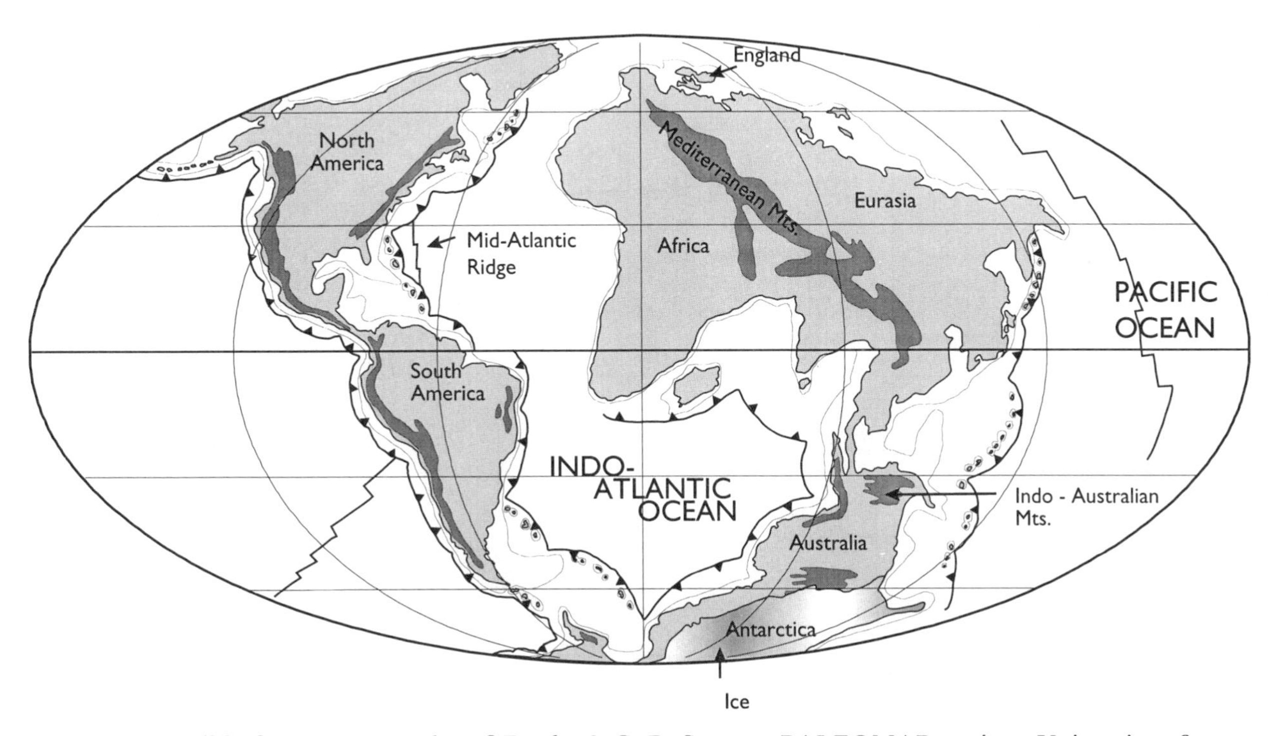

Figure 3. Possible future geography of Earth. © C. R. Scotese, PALEOMAP project, University of Texas Arlington.

permanent ice sheets, which because they are white reflect even more energy back to space. The formation of ice on the continents, of course, lowers sea level even further, which results in more land, which cools the earth, forming more ice, which lowers sea level, which results in more land, which cools the earth, and so on. The lesson to be learned is: once the earth begins to cool (or warm up), positive feedback mechanisms push the earth's climate system toward greater and greater cooling (or heating). During the last half of the Cenozoic, the earth began to cool off. Ice sheets formed first on Antarctica and then spread to the Northern Hemisphere. For the last 5 Ma the earth has been in a major Ice Age. There have been only a few times in the earth's history when it has been as cold as it has been during the last 5 Ma (late Precambrian, late Ordovician, and late Carboniferous–early Permian).

The World, 0–250 Ma in Future

Though there is no way of knowing what the future geography of the earth will be, it is possible to project current plate motions into the future and make an educated guess (Figure 3). The map presented here shows the earth 250 Ma in the future. In general the Atlantic and Indian oceans continue to widen until new subduction zones recycle the ocean floor in these ocean basins and bring the continents back together in a new Pangaean configuration some 250 Ma in the future.

Bibliography

DALZIEL, I. W. D. "Pacific Margins of Laurentia and East Antarctica–Australia as a Conjugate Rift Pair: Evidence and Implications for the Eocambrian supercontinent." *Geology* 19 (1991):598–601.

McKERROW, W. S., and C. R. SCOTESE. *Palaeozoic Palaeogeography and Biogeography.* Geological Society of London, Memoir 12. London, Eng., 1990.

SCOTESE, C. R., R. K. BAMBACH, C. BARTON, R. VAN DER VOO, and A. M. ZIEGLER. "Paleozoic Base Maps." *Journal of Geology* (1979):217–277.

SCOTESE, C. R., and W. W. SAGER, eds. *Mesozoic and Cenozoic Plate Reconstructions. Tectonophysics,* Vol. 155, no. 1–4. Amsterdam, 1989.

SMITH, A. G., D. G., SMITH, and B. M. FUNNELL. *Atlas of Mesozoic and Cenozoic Coastlines.* Cambridge, Eng., 1994.

VAN DER VOO, R. *Paleomagnetism of the Atlantic, Tethys, and Iapetus Oceans.* Cambridge, Eng., 1993.

ZIEGLER, A. M., C. R. SCOTESE, and S. F. BARRETT. "Mesozoic and Cenozoic Paleogeographic Maps." In *Tidal Friction and the Earth's Rotation II,* eds. P. Borsche and J. Sundermann. Berlin, 1983.

CHRISTOPHER R. SCOTESE

PALEONTOLOGY

The word "paleontology" derives from the Greek, literally translated as "ancient life study." The term refers to the study of ancient life as revealed by fossils. The field encompasses numerous subdisciplines, defined on the basis of: (1) a particular fossil group; (2) a process; or (3) study of fossil distribution in time and space. The major fossil group subdisciplines are: invertebrate paleontology (the study of macroscopic invertebrate fossil organisms; *see* ARTHROPODS, MOLLUSKS, BRACHIOPODS, ECHINODERMS, and COLONIAL INVERTEBRATE FOSSILS); micropaleontology (the study of microscopic marine organisms; *see* MINERALIZED MICROFOSSILS); paleobotany (the study of fossil plants; *see* GYMNOSPERMS, ANGIOSPERMS, PTERIDOPHYTES, and PALEOBOTANY); palynology (the study of fossil pollen; *see* PALYNOLOGY); and ichnology (the study of tracks, trails, and burrows; *see* TRACE FOSSILS). Process-oriented subdisciplines include evolution, extinction, paleoecology, and taphonomy (see separate entries on each of these subjects). Biostratigraphy (*see* STRATIGRAPHY) and paleobiogeography (*see* PALEOGEOGRAPHY) are based on fossil distribution data.

The fossil record is our window to reconstructing the history of life on earth. It documents organic change through time and by this establishes the facts of evolution and extinction. It provides physical data on which our understanding of evolutionary theory rests. Paleontologists address fundamental questions of our existence: When did life first appear on earth? What was the nature of this early life, and how did we get here (to our species, Homo sapiens) from there?

The answers to such big questions may come from surprisingly modest sources. For example, their respective work on trilobites and snails led paleontologists Niles Eldredge and Stephen Jay Gould to challenge Charles Darwin's widely accepted gradualistic view of evolution (*see* LIFE, EVOLUTION OF and DARWIN, CHARLES). No one person would be able to study the entire fossil record, especially at the level of detail required to answer the questions posed above. Traditionally, individual paleontologists specialize in a single group of organisms, studying their morphology (physical appearance), geographic distribution, and geologic placement (i.e., position within the vertical sequence of sedimentary rock or stratigraphic record). The fundamental data of paleontological research are species-level descriptions, that is, detailed descriptions of the different basic types of fossils. Taxonomy is the description and naming of species. From these individual, species-level studies higher-level questions can be addressed, for example, how are different species related? Species by species, organisms are grouped into higher taxonomic categories (species into genera, genera into families, families into orders, orders into classes, classes into phyla, and phyla into kingdoms). Phylogeny is the study of the interrelatedness of organisms. Once sufficient data are described, higher and higher order relationships can be pursued.

Paleontological research is by its very nature historical. Constructing phylogenies, documenting evolution, reconstructing paleoecological relationships, or making biostratigraphic correlations rely on identifying patterns in morphological characteristics, geographic, or geologic distribution. From patterns come hypotheses of the underlying cause (or process). Once hypotheses are constructed, they are amenable to testing.

Paleontological is a specimen-based science (as is archaeology), which is exciting in that a new discovery may overturn existing hypotheses. Unfortunately, specimen-based disciplines are subject to abuse. The history of paleontological research is marred (or made more colorful, depending on your point of view) by frauds (*see* PALEONTOLOGY, HISTORY OF) perpetrated to gain wealth, fame, revenge, or to advocate a personal agenda. Fortunately, modern scientific procedure includes requirements that make data available for peers to examine, and thus makes such abuses harder to perpetrate.

Nature of the Fossil Record

The term "fossil" comes from the Latin word meaning "dug up." Originally the concept included unusual rocks, minerals, ores, artifacts; basically anything hard that came from the ground. Today, the term is restricted to objects dug from the ground that have an organic origin. The most familiar fossils are body fossils, the petrified remains of the actual organism (e.g., brachiopod shell). Common in some groups, especially mollusks, is preservation as an internal mold or impression (*see* FOSSILIZATION AND THE FOSSIL RECORD). The relatively recent appreciation of the true nature of trace fossils (some of which were previously interpreted as fossil plants) expanded the definition of fossil to include indirect evidence of life, evidence of an organisms's activity or behavior.

The concept of what constitutes the fossil record (i.e., the history of life as recorded in sedimentary strata by fossil data) continues to expand. Recent discoveries of fossil amino acids, including DNA, opened a new field of inquiry. Studies of the isotopic composition of fossil material (e.g., vertebrate bones) is another promising new avenue for paleontological research. New technologies produce new ways of examining fossil material and provide new tools for constructing phylogenetic relationships and ancient ecological relationships.

Limitations of Fossil Data

The chief limitation of the fossil record is its incompleteness. Only a small proportion of once-living things left a fossil record, prompting the adage, "fossilization is the exception and not the rule." Not all organisms face equal chances of fossilization; therefore, the record is biased toward some groups over others. The major factors that affect fossilization potential are the morphology of the organism, the environment in which the organism died, and the geologic age of the fossil (*see* FOSSILIZATION AND THE FOSSIL RECORD). For example, hard-shelled invertebrates (e.g., brachiopods) living in the marine environment have a higher fossilization potential than a sparrow nesting in the backyard. The brachiopod is likely to be buried by sediment carried to the shallow marine environment (via rivers or storms), and its robust shell is a good candidate for fossilization. Coversely, upon its demise the sparrow likely would be preyed upon by the neighborhood cat, leaving little but feathers as potential fossils in an environment that does not experience rapid deposition of protective sediment.

Geologically young fossils (70 million years or Ma old or younger) may be better preserved than their ancient relations because the shell material has not experienced as long an interval of potential destruction by post-fossilization processes (erosion of the fossiliferous sedimentary unit or dissolution of the skeletal material). Many fossil taxa from geologically young strata have living descendents, making paleobiological interpretations easier. We may never fully understand the biology and ecological relationships of extinct organisms not closely related to modern living organisms (e.g., trilobites).

Perfectly preserved fossils are relatively uncommon; most suffer some degree of breakage, abrasion, or other alteration (*see* TAPHONOMY). Exceptions include spectacular assemblages of well-preserved fossils, term *lagerstätten* (German for "mother lode"). Well-known fossil *lagerstätten* include fish and other aquatic vertebrates from the Eocene-age Green River Formation of Wyoming; insects preserved in amber (various localities); Pleistocene-age vertebrates, including saber-toothed cats from California's LaBrea tar pits; mysterious soft-bodied organisms from the Pennsylvanian Period Mazon Creek locality of Illinois; and *Archaeopteryx* from Germany's Jurassic Solenhofen Limestone. *Lagerstätten* occur in a variety of geological and geographical settings and through a wide range of geologic time. Although the temporal and physical differences between these deposits are striking, they do share basic characteristics of preservation style—all were produced by entombment of organisms in a biologically inert environment.

Such well-preserved, "museum quality" fossils provide important morphological information basic to taxonomic research, but less-well-preserved fossils may reveal other aspects of a fossil organism's biology. For example, repetitive patterns of breakage in skeletal material may indicate a pattern of predation, and thus give information about fossil behavior and ecological relationships. Hypotheses of predator/prey relationships are sometimes confirmed by the fortuitous discovery of fossils in which prey remains are identifiable in the body cavity of the predator, or a predator's tooth marks match damage to the prey's remains. The ever-present possibility of future exceptional discoveries adds a measure of drama to paleontological research.

Paleontology and Other Disciplines

Paleontology is intimately linked to biology. Both fields use similar data, specifically, the morphology of the organism, its life habit (ecology), and its behavior. The obvious difference between how research is pursued in the two fields is that the biologist has access to more morphological information than the paleontologist (information on soft tissue, DNA fingerprint, etc.) and can directly observe the behavior and ecological relationships of modern organisms. Paleontologists often must use comparison by analogy to modern organisms to infer paleoecological relationships of fossil groups. Thus, biology has much to offer paleontology, but paleontology in turn offers a unique and important contribution to biology: The fossil record offers biological data that reach back to about 3.5 billion years (Ga), "only" 1 Ga after the origin of Earth. Fossils provide the evidence of organic change through time. The concepts of evolution and extinction are owed largely to the existence of fossils.

Paleontology is also closely allied with geology in that the fossil data come from geologic material (usually sedimentary strata). Mapping the vertical (stratigraphic) distribution of fossils led to a fundamental principle of sedimentary geology, the principle of faunal succession. This is the observation that fossils tend to appear in a definite stratigraphic order (*see* STRATIGRAPHY). This principle enabled correlation of strata (demonstrating temporal equivalency) and compilation of the first geologic maps. The corollary to faunal succession is that a sequence of geologic time can be recognized by its characteristic fossils. This observation led to the development of the geologic timescale (*see* GEOLOGIC TIME).

Paleontological data also provided early evidence for continental drift (later, plate tectonics). For example, fossils of the terrestrial plant *Glossopteris* occur on the now widely separated landmasses of South America, Antarctica, Australia, India, and Africa. ALFRED WEGENER argued that these landmasses must have been joined at that time to account for this distribution.

Biostratigraphic data, detailed information on the vertical distribution of fossil species, are an important tool in oil exploration. The fossil content of a rock unit serves as a valuable fingerprint in identifying economically important sedimentary layers.

Paleontological Research

All research in paleontology utlimately originates in the field with the collection of fossil specimens. The nature of the paleontologist's field area is variable; it may be roadside outcrop (exposure of fossiliferous rock at Earth's surface) or an offshore oil platform, where drill core from the seafloor is examined for microfossils. Not all paleontologists have the opportunity to collect for themselves the material they study. Some work in the laboratories of oil companies, poring over washed drill samples, looking for diagnostic microfossils. Other paleontologists work in museums, where they may make new discoveries from specimens collected years earlier.

The typical paleontological study begins in the library, with checking historical reports and maps to identify the best geographic/geologic localities for the job at hand. Practical constraints on what follows (location of the field area relative to the researcher's home base, length of field season, type of equipment used) depend on the level of funding available to the researcher. The paleontologic field season begins with a reconnaissance visit, identifying the more promising localities for more intensive study. During reconnaissance, fossils are collected by surface collecting, which is simply gathering material exposed at Earth's surface. Once the promising localities are identified, detailed collection and, if necessary, excavation begin. The goal of paleontological fieldwork is to retrieve the fossil material while retaining as much information as possible. Typical field observations include recording the exact geographic and stratigraphic location, the nature of the surrounding sediment, and the identity, abundance, condition, and orientation of the fossils. Paleontologists commonly use a grid system in field collections, marking the location of fossils, and noting relative abundance or presence/absence of different species. The tools used in field excavation of fossils may include bulldozers and backhoes (to strip away overburden—unfossiliferous overlying sediments); jackhammers, diamond-edged rock saws, hammers, and chisels to extract blocks of fossiliferous material from the outcrop; or whisk brooms, putty knives, and dental implements for removing delicate fossil material (e.g., vertebrate bones) from soft sediment. Choice of tools depends on the nature of the host rock and the objectives of the research.

Final fossil preparation takes place in a laboratory. The first step is to separate the fossils from the matrix that entombs them. This may be accomplished by chemical or physical means, for example, an acid bath to dissolve soluble matrix from insoluble fossils, or cleaning with a miniature sandblaster unit. New technologies from other disciplines have been adapted for paleontological work. X ray, CAT-scan, and NMR (nuclear magnetic resonance) have been used when nondestructive exploration of fossil material (e.g., dinosaur eggs) is desired.

After the fossil material has been cleaned, the next step is the description, measurement, and quantification of specific characteristics chosen for study. In measuring and describing fossils paleontologists employ tools ranging from calipers to computer imaging techniques. Modern paleontological data analysis often reduces fossil organisms to a numerical quantity, the better to analyze the data using a wide variety of mathematical models. Despite the attempt to objectify the specimen description process by using sophisticated computer hardware and software, the researcher must make decisions on which characters are potentially the best (i.e., most informative or appropriate) to answer the proposed questions. The ability to make these decisions comes with experience. Enhanced technology (e.g., computer techniques) may speed the process, but it does not necessarily make for better scientific results. If the underlying assumptions are not valid or the data are flawed (e.g., error in measurement), the conclusions may be invalid. To paraphrase the distinguished paleontologist and geologist Derek Ager, the best paleontologist is the one who has seen the most fossils.

Future of Paleontological Research

Advances in technology have made it possible to expand paleontological research to previously unimaginable venues. The National Aeronautics and Space Administration (NASA) sponsors a program in exobiology, that is, biology outside of Earth. The search for life on other planets falls within the scope of paleontology. For example, scientists agree that Mars is currently devoid of life, but erosional surficial features on the red planet suggest its surface was sculpted by running water in the distant past. If Mars once supported water, would not life have been present also? If life were present on Mars in the past, the evidence, if preserved, would be paleontological, that is, fossilized direct or indirect evidence of life. Paleontologists participating in NASA's exobiology program are charged with the task of predicting what this potential martian fossil record would look like so that a mission to Mars would be prepared to recognize such evidence.

Technological advances may change the tools and techniques of paleontologists but the fundamental data of paleontology—the fossils—persist largely unchanged through the ages. The questions raised by fossils and the scientists who study them persist also because of their fundamental importance. After all, what paleontology offers is an awareness of who we are, where we came from; an understanding of our place in what Aristotle called the *scala naturae,* the "great chain of being."

Bibliography

LAPORTE, L. F., ed. *The Fossil Record and Evolution.* San Francisco, 1982.

RUDWICK, M. J. S. *The Meaning of Fossils.* Chicago, 1985.

SIMPSON, G. G. *Fossils and the History of Life.* New York, 1983.

SKINNER, B. J., ed. *Paleontology and Paleoenvironments.* 1981.

STEARN, C., and R. CARROLL. *Paleontology: The Record of Life.* New York, 1989.

DANITA S. BRANDT

PALEONTOLOGY, HISTORY OF

Although fossils have interested humans for centuries, the understanding of those objects as the remains of once living organisms was a topic of considerable debate until the eighteenth century. Naturalists of the sixteenth and seventeenth centuries who studied fossil objects interpreted them in the context of prevailing religious and philosophical views. In the framework of Renaissance Aristotelianism and Neoplatonism, fossils were defined as organic objects that "grew" inside the earth. It did not necessarily follow, however, that they were the remains of living organisms. Religious views emphasized the divine creation of the earth and its inhabitants and militated against the idea that God would have allowed any organism to become extinct. Questions concerning the positions and modes of preservation of fossils presented additional interpretive difficulties. Gradually, detailed observations, closer attention to chronology, and recognition of distinct strata favored an organic interpretation of fossils.

The confirmation of extinction and the establishment of paleontology as a scientific field came with the work of GEORGES CUVIER (1769–1832). A French scientist who arrived at the Paris Natural History Museum in 1795, Cuvier was well versed in comparative anatomy. Focusing on the study of large fossil vertebrates, particularly mammoth bones that had been a lively topic of discussion since their discovery in 1739, Cuvier produced a detailed anatomical analysis demonstrating that the remains of a Siberian mammoth differed markedly from those of any living elephant and belonged to a species that had become extinct. Cuvier's additional studies, culminating in his four-volume work, *Recherches sur les ossemens fossiles* (1812), did more than provide documentary evidence for extinction; it established the theoretical structure for further research. Working at much the same time as the Englishman William Smith, Cuvier and ALEXANDRE BRONGNIART conducted studies in the Paris Basin that demonstrated that the vertical arrangement of fossils could be used to determine the age of formations. Cuvier defined the succession of fossils in progressive terms. Rejecting evolution, Cuvier explained earth history as a series of catastrophes produced by physical laws that had resulted in extinctions (Appel, 1987; Rudwick, 1976).

Cuvier's work had a powerful impact on subsequent research. Not only did the interest in and discovery of fossils increase, but geologists interpreted remains along Cuvierian lines. By the 1830s European scientists were describing the sequence of the fossil record in terms of the ages of fishes, reptiles, and mammals, a scheme that emphasized progression. In contrast to Cuvier, a committed Newtonian, William Buckland and JEAN LOUIS RODOLPHE AGASSIZ defined catastrophes and subsequent separate creations as the work of God (Bowler, 1976; Rupke, 1983). By the 1830s geologists were also employing fossils, rather than rocks, as the means for determining the relative age of the earth. Efforts to establish the temporal and spatial features of the Silurian, Devonian, and other age divisions produced heated scientific battles, but by the 1860s much of the geological timescale was in place (Rudwick, 1985; Secord, 1986; *see* GEOLOGIC TIME).

Paleontology also attracted public attention. The discovery of large, extinct animals had fascinated naturalists and scholars in the seventeeth and eighteenth centuries, many of whom believed such curiosities possessed medicinal or magical qualities. By the early nineteenth century such specimens had gained popular attention. The interest in fossils led American artist Charles Willson Peale to develop one of the first exhibits of the mastodon at his Philadelphia Museum in 1801. By the 1810s Peale's son and other publicists regularly included such displays in their traveling shows. Dinosaurs, first discovered in the 1820s, created even more excitement. At England's Crystal Palace Exhibition in 1851 the anatomist Richard Owen and the artist Benjamin Waterhouse Hawkins created life-size restorations of dinosaurs. Hawkins' 1868 skeletal reconstruction of a dinosaur drew such large crowds that Philadelphia's Academy of Natural Sciences had to start charging admission to limit attendance.

In the United States paleontology also attracted government support. Beginning in the 1820s, state surveys served as an important resource for scientific employment as well as paleontological research. Agencies such as the New York State Survey supported James Hall's extensive work on fossil invertebrates, while the interest in coal in Pennsylvania and Ohio led those states to sponsor important paleobotanical research by Léo Lesquereux and John Strong Newberry. Following the Civil War, the federal government's interest in the trans-Mississippi West led to the creation of large-scale surveys that promoted paleontology. Through the Department of the Interior, Ferdinand V. Hayden's Geological Survey of the Territories supported considerable research by a host of paleontologists including Joseph Leidy, Edward Drinker Cope, and Fielding Bradford Meek. In 1879, with the creation of the U.S. Geological Survey, CLARENCE RIVERS KING and later JOHN WESLEY POWELL helped sponsor the publication of O. C. Marsh's discoveries of dinosaurs, birds with teeth, and large fossil mammals. Government sponsorship of fieldwork led to the discovery of hundreds of new fossils specimens and helped push America to the forefront of vertebrate paleontology (Goetzmann, 1967; Manning, 1967).

Paleontology gained additional importance following the publication of CHARLES DARWIN's *Origin of Species* (1859). Darwin's work suggested that the fossil record would provide documentary evidence for evolution, and scientists in Europe and America actively sought to find intermediate forms and to construct evolutionary sequences. Newly discovered individual specimens, such as *Archeopteryx*, served as important links between major taxonomic groups. Such dramatic findings were rare; more frequently scientists constructed evolutionary sequences linking older and more recent forms. Discoveries of extinct species of horses and tapirs enabled Albert Gaudry, Ludwig Rütimeyer, and Othniel Charles Marsh to construct classic vertebrate phylogenies. In the 1870s and 1880s Melchior Neumayr, Wilhelm Waagen, and Leopold Würtenberger did the same for fossil invertebrates (Rudwick, 1976).

Most paleontologists of the late nineteenth century continued to interpret fossil sequences in progressive terms. Biologists and geologists embraced Ernst Haeckel's doctrine that ontogeny recapitulates phylogeny. (Individual embryological development, or ontogeny, constituted a brief and rapid recapitulation of evolutionary history, or phylogeny.) Neo-Lamarckism, an alternative promoted by the American paleontologists Edward Drinker Cope and Alpheus Hyatt, emphasized environmental influence and the inheritance of acquired characteristics as the principal causal mechanisms of evolution. Others turned to orthogenesis, the theory that an internal force drove evolution along predetermined, linear paths.

While paleontology constituted an important area of biological and geological study throughout the nineteenth century, it subsequently lost ground to other fields. Many paleontologists did not embrace the methods of experimental biology and the findings of genetics that became prominent in the early twentieth century. In the United States a split occurred between natural history and experimental biology that led many laboratory scientists to debunk the study of the fossil record. As experimental biology gained a foothold in university departments, vertebrate paleontology found a home in museums supported by governments or private philanthropies. At major new natural history museums in London, New York, and Washington, D.C., expeditions, displays, and debates over human fossils continued to generate public interest and attention. But increasing specialization and reliance on experimental methods that yielded reproducible, predictable results left vertebrate paleontology as a small, peripheral field of inquiry (Rainger, 1991).

Invertebrate paleontology also experienced significant change. While some invertebrate paleontologists remained interested in evolutionary problems, by the early twentieth century the field had become closely tied to the oil industry, especially in the United States. Studies by Joseph Cushman and John J. Galloway identified certain foraminifera as excellent indicators of the possible occurrence of petroleum deposits, and invertebrate paleontology became a valuable asset to oil companies.

Since the 1940s both vertebrate and invertebrate paleontology have undergone important changes. In that decade, GEORGE GAYLORD SIMPSON, one of the architects of the modern synthesis, helped bring paleontology into the mainstream of biology by interpreting the data of the fossil record in terms of population genetics and evolution by natural selection. His work highlighted the importance of genetics and statistical methods for the study of fossil remains. Biologists readily accepted Simpson's work, but it did not significantly change vertebrate paleontology. Studies still emphasized morphological problems, and the field remained centered in museums. Since the 1970s new interpretations of dinosaur anatomy and physiology have given rise to new investigations and dynamic new exhibits, resulting in the tremendous popularity of dinosaurs. Luis Alvarez's 1979 claim for an extraterrestrial explanation of dinosaur extinction (*see* IMPACT CRATERING AND THE BOMBARDMENT RECORD) has produced a flourishing of interdisciplinary research, with chemists and astrophysicists joining paleontologists in examining the causes, processes, and patterns of mass extinction (Glen, 1994).

Since the 1940s important developments have occurred in invertebrate paleontology. In the 1940s and 1950s, influenced by the evolutionary synthesis, Norman Newell at Columbia and Bernard Kummel at Harvard emphasized the importance of biological questions for paleontologists. In the 1970s Newell's students, Niles Eldredge and Stephen Jay Gould, advanced a new evolutionary interpretation: punctuated equilibrium. In contrast to Darwinian theory, which emphasized gradual and continuous change, proponents of punctuated equilibrium interpreted the fossil record in terms of long periods of stasis interrupted by brief episodes of rapid evolutionary change. Still a topic of controversy, that theory nonetheless sparked considerable investigation of the fossil record and opened up a wide range of questions concerning evolutionary processes. Paleontologists have been at the forefront of cladistics, a new taxonomic system allowing classification of organisms on the basis of shared characteristics rather than chronological relationships. Statistical methods did not quickly catch on among vertebrate paleontologists, but in the late 1950s Martin Rudwick and David Raup employed biometric techniques and computer-generated analyses to examine the size, shape, and other dimensions of invertebrate fossils. Such studies have since become integral to paleontology. Equally important has been the emergence of paleoecology and taphonomy (*see* PALEOECOLOGY; TAPHONOMY). While Rudolph Richter and Johannes Weigelt studied such problems in the 1920s, the interest in sedimentation as related to the formation of petroleum truly fostered paleoecology. In the United States, W. H. Twenhofel was among the leaders in studying how sediments influenced fossil deposition and accumulation. By the 1950s he and others were examining the processes of fossil burial and transport and how they relate to paleoenvironmental reconstruction and interpretation. Subsequently, as Western scientists learned of the work of Soviet paleontologist I. A. Efremov, taphonomy became an important dimension of paleontology. While the health of the oil industry continues to affect employment opportunities in invertebrate paleontology, those scientists are very much concerned with biological questions and their work has significantly reshaped the study of evolution, systematics, and ecology.

Bibliography

APPEL, T. A. *The Cuvier-Geoffroy Debate: French Biology in the Decades Before Darwin.* New York, 1987.

BOWLER, P. J. *Fossils and Progress: Paleontology and the Idea of Progressive Evolution in the Nineteenth Century.* New York, 1976.

———. *The Eclipse of Darwinism: Anti-Darwinian Evolution Theories in the Decades Around 1900.* Baltimore, 1983.

DESMOND, A. *Archetypes and Ancestors: Paleontology in Victorian London, 1850–1875.* Chicago, 1984.

GLEN, W. *The Mass Extinction Debates: How Science Works in a Controversy.* Stanford, 1994.
GOETZMANN, W. *Exploration and Empire: The Explorer and Scientist in the Winning of the American West.* New York, 1967.
MANNING, T. G. *Government in Science: The U.S. Geological Survey, 1867–1894.* Lexington, KY, 1967.
RAINGER, R. *An Agenda for Antiquity: Henry Fairfield Osborn and Vertebrate Paleontology at the American Museum of Natural History, 1890–1935.* Tuscaloosa, AL, 1991.
RUDWICK, M. J. S. *The Meaning of Fossils: Episodes in the History of Palaeontology,* 2nd ed. New York, 1976.
RUPKE, N. *The Great Chain of History: William Buckland and the English School of Geology (1814–1849).* Oxford, Eng., 1983.
SECORD, J. A. *Controversy in Victorian Geology: The Cambrian-Silurian Dispute.* Princeton, NJ, 1986.

RONALD RAINGER

PALEOTHERMOMETRY

See Isotope Tracers, Stable

PALYNOLOGY

Palynology is the study of modern and fossil pollen grains and spores produced by terrestrial plants, and of other tiny fossils (microfossils) that can only be seen and studied with powerful optical or electron microscopes. These diverse fossils, from wholly unrelated organisms, are commonly referred to as "palynomorphs." Sometimes, the term "paleopalynology" is used to distinguish the study of fossil palynomorphs from palynology, which includes the study of pollen and spores of both living and fossil plants.

Palynomorphs most commonly are single-celled organisms, or single-celled body parts of multicellular organisms, or sometimes multicellular, microscopic structures. They represent organisms from marine, terrestrial, and freshwater habitats that form the record of organic or organic-walled microfossils. Organic microfossils are composed of compounds whose main components are carbon, hydrogen, oxygen, and nitrogen, elements that make up most of the bodies of all living things except for their mineralized parts. In contrast, most fossil organisms are represented by mineralized shells or tests of calcium carbonate or silica, or bones and teeth of calcium phosphate. The chemical composition of most organic microfossils has been little studied, but many react in recovery from rocks as if their organic matter was in part composed of sporopollenin, a substance first investigated in connection with pollen and spore preservation. Less commonly, some seem to be chitin, a complex carbohydrate, or some chitin-like compound. Even those apparently composed of sporopollenin are unlikely to have identical compositions since some evidence suggests that sporopollenin represents a class of compounds rather than a single compound. Organic compnents of palynomorphs are among the most durable and inert known in organisms and frequently the fossils retain, unaltered, the composition that they had in life, unlike many mineralized fossils. This durability accounts for the fact that organic microfossils are perhaps the commonest of all fossils. Additionally, many are common organisms in life or, in the case of organic microfossils that are parts of organisms, such as pollen and spores, produced in huge quantities and readily spread over wide areas by wind and water currents.

Palynology thus deals with a variety of different structures from unrelated organisms of different kingdoms, united by their organic composition (and hence by the techniques of recovering them from rocks), their tiny size, measured in units called micrometers (1 micron or μm = 0.001 millimeter or mm), and their often extreme abundance.

Relationship, Structure, and Function

Palynomorphs include structures that can be assigned to familiar living or fossil organisms as well as structures whose relationship to any living or fossil organisms is unknown. In some, but not all cases, it is possible to determine the function of palynomorphs. For example, the major group of palynomorphs produced by terrestrial organisms are pollen grains of seed-bearing plants such as angiosperms and gymnosperms, and spores of ferns and mosses and their relatives (*see* ANGIOSPERMS; GYMNOSPERMS; PTERIDOPHYTES). We can identify these tiny fossils with the major plant groups that produced them, and we know that they functioned as reproductive structures. Pollen and spores, usually measuring between 10 and 250 μm, are extremely common in fine-grained sedimentary rocks, both marine and nonmarine. Spores and hyphae (tubular, thread-like filaments) of fungus mostly from terrestial habitats are also common palynomorphs. Some freshwater green algae leave microscopic remains, both resting spores and vegetative remains, that may be common in some rock deposits.

Marine single-celled algae, such as dinoflagellates, produce resting cysts that are common organic microfossils in some rocks. We know the function of these cysts and can identify them as dinoflagellates, but most fossil cysts cannot be related to a particular dinoflagellate that produced them. Two important marine palynomorph groups of uncertain relationship to any known modern organisms are chitinozoans and acritarchs. Both are extremely common in marine deposits in some parts of the geological column. Most acritarchs are probably algal, but we have no idea what their relationship is to either modern or described fossil algae. Chitinozoans are extinct, confined to the Paleozoic, and of uncertain function and relationship. It is not even known whether they represent plants or animals! Scolecodonts, jaws of marine worms, may also be included among palynomorphs.

Study and Application

To study these tiny objects, palynologists recover them from rocks by using acids and bases to dissolve the rock's minerals and organic debris less resistant than the palynomorphs. These elaborate techniques free and concentrate palynomorphs from the rock matrix so that they may be mounted on microscope slides, identified, described, illustrated, counted, and sometimes subjected to statistical techniques.

The uses of palynomorphs and limitations to their study vary from group to group. Nevertheless, as a whole they have some specific advantages. They are extremely common fossils, even compared with other microfossils, often numbering in the millions in a few grams of rock, such as might be obtained in oil-well core samples. Their general resistence to biochemical and physical degradation means that they are commonly found in rocks where remains of other organisms will not normally be preserved or where they will be destroyed before recovery. The ease by which palynomorphs are transported by wind or water currents means that they can be found in rocks that lack other fossils, and their occurrence may be independent of local environments. In ecological and geological applications this dispersal potential may be both advantageous and disadvantageous—a mixed blessing. Organisms may come from far away and have little to do with local environments, thus complicating their interpretation. Understanding of regional environments may, however, be enhanced. On the other hand, wide dispersal may be useful in geology in determining ages and stratigraphic relations of rocks over wide areas. It is often possible to estimate ages of rocks within 3–4 million (Ma) years by this means. With regard to ecological applications, a principal limitation in the study of some palynomorphs is their uncertain biological affinities. Because of their extreme durability palynomorphs are also readily recycled: redeposited from older rocks, as they are eroded, into younger ones. Without extreme care this may confuse the stratigraphic ranges and, in some cases, the significance of the fossils in environmental interpretation since the rocks into which they are redeposited may not represent the same environment or environments as the one from which the palynomorphs originated.

Palynomorphs as a group are among the most versatile fossils in a wide context of applications throughout much of geologic time. Because of their varied biological origins and relationships and the diversity of environments represented, palynomorphs have proved useful in geology, biology, paleobotany, paleontology, paleoecology, paleogeography, climatology, anthropology, and archaeology. Some applications depend on identifying fossils (at some level) with modern organisms whose ecology and biogeography are known; others depend mainly on information about environments (marine/nonmarine; terrestrial/aquatic) that the fossils represent; still others depend on being able to recognize repeated, distinct assemblages of the same microfossils.

For geologists, palynomorphs have contributed important information in dating rocks, and stratigraphic correlation (regional and global) of rocks with similar assemblages of fossils. By charting the changing ratios of a sequence of marine and nonmarine palynomorphs in a rock unit, geologists and paleobiogeographers are able to recognize changes in sea level and environments of the rocks over time and locate the position of ancient shorelines. From the fact that palynomorphs are readily redeposited as a result of erosion of older rocks, it is possible to determine the source rocks of the newer deposits and sometimes determine the directions of water currents that carried sediment into the depositional basin. Different assemblages of palynomorphs, such as acritarchs, may also indicate different water depths and relative distance from ancient shorelines. Palynomorphs have also proved useful in paleogeography, through studies of when the landmasses had a different configuration than they do now.

Pollen and spores of terrestrial plants have been important in charting vegetational changes since the early Paleozoic and correlating such changes with those determined from leaves, wood, and other structures. Some of these changes mark adaptive radiations of other structures. Some of these changes mark adaptive radiations of newly evolved plant lineages and replacement of old lineages, others reflect plant migrations due to climatic changes. Spores of land plants are now providing significant evidence regarding the earliest adaptive radiation of land plants, the beginnings of the colonization of the land by plants that paved the way for the first land animals, and the evolution of the terrestrial ecosystem. Pollen has also shed light on intriguing questions that have plagued botanists since Darwin's time about the evolution of angiospermy and spread of the flowering plants throughout the globe, both questions of paramount importance to biologists and paleobotanists.

For the last 60 Ma or so, when pollen grains and spores can be identified as living genera and even species, they are particularly important in documenting vegetational changes and determining their regional and global environmental causes. Detailed climatic information about the last 2 to 3 Ma has been inferred from vegetational changes registered in pollen and spore assemblages. This information concerns rainfall and its seasonality, and fluctuations in temperatures. Pollen and spores have been a major source of information about climatic and vegetational changes associated with the last global glaciation.

Pollen records, including those of agricultural crop plants and agricultural weeds commonly associated with human settlement, have permitted tracing introduction of agriculture into various regions of the globe, and the spread of various agricultural plants from their centers of domestication. Pollen has also been useful, particularly in some regions such as northern Europe, for identifying land use patterns, and change in land use, and even sometimes specific agricultural practices. In North America it has been possible to trace deforestation from pollen records in lakes as European farmers and miners moved into the west.

Archaeologists have even been interested in the pollen/spore content of human and animal feces during the last several thousand years as a source of information about food habits and sometimes in providing evidences of parasitic disease. In some human habitats it has been possible to indicate how rooms were used (food storage, ritual, livestock habitat) from pollen and spores. Pollen and spores associated with archaeological artifacts have also helped to identify the use to which the artifact may have been put and occasionally have helped to put dates on such artifacts.

Bibliography

FAEGRI, K., and J. IVERSEN. *Textbook of Pollen Analysis,* 4th ed. New York, 1981.

GRAY, J. "Pollen Stratigraphy." *Encyclopaedia Britannica,* 15th ed. Chicago, IL, 1974.

MOORE, P. D., J. A. WEBB, and M. E. COLLINSON. *Pollen Analysis,* 2nd ed. Cambridge, MA, 1991.

JANE GRAY

PERMANENTLY FROZEN GROUND

Permafrost is defined as a thickness of permanently frozen surficial sediment or bedrock below the surface of the earth where the temperature has remained below freezing for two or more years, with or without the presence of moisture. The term "permafrost" and its definition, proposed by Siemon W. Muller of the U.S. Geological Survey in 1943, have now gained general acceptance.

Permafrost is allowed to develop when the mean annual air temperature is low enough to maintain a situation whereby the depth of winter freezing of the ground exceeds the depth of the summer thaw. The depth of downward winter freezing is counterbalanced by the upward flow of geothermal heat toward the surface. The balance of these opposing factors controls the presence, distribution, and thickness of permafrost through time (Lachenbruch, 1968; Washburn, 1980).

While geothermal heat flow essentially remains constant in a given area, surface temperatures can vary in response to geographic location, nature of the sediments or bedrock at the surface, seasonal vegetation, annual variations in precipitation and cloud cover, and drainage.

Although the near surface may thaw rapidly on a seasonal basis, the deeper permafrost may remain for thousands of years after a general atmospheric warming because of the very slow downward rate of warmth penetration, therefore leaving fossil or residual permafrost, primarily reflecting surface air temperatures that prevailed in much earlier times, perhaps as far back as thousands of years during the last Ice Age (Pleistocene epoch), which began 2.5 million years (Ma) ago.

In general, summers are significantly warmer than winters and, therefore, the ground above the perennial permafrost melts in the summer and refreezes in the winter. This zone is called the active layer, and its thickness varies through time with fluctuations that control the permafrost itself.

The top of the permafrost, as measured in the summer, is called the permafrost table. Depending upon seasonal and annual variations, the depth of freezing of the active layer on any given year may not reach the permafrost table, leaving an unfrozen layer, called the talik, between the two frozen sections. This talik sometimes contains water under pressure which often causes problems for engineering projects.

Permafrost underlies as much as 5 percent of Earth's surface, primarily in the Northern Hemisphere; however, it does exist in the limited areas of land exposure in Antarctica. In the exposed land areas generally surrounding the Arctic Ocean, for example, permafrost depths of up to 700 meters (m) have been reported from Alaska and up to 1,800 m from Siberia (Washburn, 1980).

Permafrost thins and becomes patchy as it merges with areas where seasonally frozen ground exists but permafrost is not present. Because of this, permafrost is classified into two categories, continuous and discontinuous. In general, continuous permafrost is present in areas where the mean annual air temperature is below about −8°C; and discontinuous permafrost is present in areas where the mean annual air temperature is between approximately −4°C to −8°C. In North America the continuous permafrost is generally present in northernmost Alaska and Canada, including the Elizabeth Island archipelago, above the latitude of approximately 65°N in Alaska and 60°N in central Canada. In Siberia, the continuous permafrost lies north of 65°N latitude in general (Troll, 1944; Washburn, 1980).

Ice is commonly present in the upper part of the permafrost layer as large and small masses of variously shaped ground ice, often occupying up to half the volume in the upper 3 m of permafrost. The occurrence of this ice takes the form of individual grains, veins and lenses, and downward-penetrating ice wedges of various sizes.

The landscape features in regions of permafrost are largely dominated by conditions related to them. In the simplest form, angular, fresh rock debris, derived predominantly from local bedrock, covers large areas of polar and mountainous regions. This is produced by the prying apart of bedrock by the freezing of water in crack systems. On horizontal surfaces the debris tends to remain where produced; however, it tends to move downslope to form "rock rivers" on gentle slopes, or talus cones, aprons, or rock glaciers moving out from the base of cliffs.

The melting of ice and the formation of the seasonal active zone produce a water-saturated sediment layer that moves downslope by gravity as a fluid flow. This process is called solifluction and is exemplified by lobate sheets of sediment moving downslope. These sheets are generally smooth, up to several meters thick, with larger fragments segregated and often concentrated around the lobate margin. Many sheets may be superimposed on a slope.

In addition to creating and moving sediments, frost action also produces characteristic structures. Frost cracks are caused by thermal contraction and are often filled with ice which, in turn, causes the expansion of the crack. Pingos are essentially small circular hills caused by growing subsurface lenses of ice that expand and dome up the overlying sediments as the ice lenses increase in volume due to addition of water. Patterned ground is composed of a group of features, including polygons, stripes, and circles, which individually or collectively form distinct patterns on the landscape. Although the details of the formation of some of these features are not clear, the primary processes involved are frost heaving, frost cracking, and variations in water content and hydrostatic pressure. Although some of these features may form in areas of seasonally frozen ground, the larger examples of these features are found in areas of permafrost.

Permafrost and frozen ground present few engineering and construction problems as long as they are undisturbed; however, any human-induced changes in the surface environment that lead to the thawing of the permafrost can produce serious problems in the human uses of such areas. Human disturbances include road construction, removing insulation by clearing vegetation, for example, and engineering construction, which in turn can result in road failure, building failure, and problems related to destabilizing the ground in such a manner as to produce landslides, slumping, and subsidence.

Available geological data on the presence and distribution of relict permafrost features now present in the warm midlatitudes allow a reconstruction of former cold climates and their distributions. However, former temperatures can only be generally approximated because the lack of firm understanding of the origins of the features does not permit inferences of specific paleotemperatures.

Bibliography

LACHENBRUCH, A. H. "Permafrost." In *The Encyclopedia of Geomorphology,* ed. R. W. Fairbridge. New York, 1968.

TROLL, C. "Strukturböden, Solifluktion and Frostklimate der Erde." *Geologische Rundschau* 34 (1944):545–694.

WASHBURN, A. L. *Geocryology.* New York, 1980.

HAROLD W. BORNS, JR.

PETROLEUM

Crude oil and natural gas, which together constitute petroleum, have been major sources of energy for most of the twentieth century. Being such an important part of everyone's life, these substances have evoked considerable curiosity as to how they came into existence. Gaining a knowledge of petroleum origin, however, represents more than just the simple satisfaction of explaining a scientific curiosity; it has considerable practical importance to those who explore for petroleum. If a good perspective on the history of petroleum accumulations is to be developed, several pertinent questions need to be answered. By what kind of process did petroleum originate? What were the starting materials from which it was made? When did it originate? Were petroleum deposits formed in their present locations or, if not, how did they get there?

Because of the complexity and variability of geologic conditions, there are no simple, universal answers to any of these questions. Specific modes of origin and subsequent histories of any two petroleum deposits will certainly not be identical. As a result, many aspects of petroleum origin do not retain a strong consensus among geoscientists as to specific processes involved. Fortunately, there is now general agreement as to the overall scheme. Most scientific debate on the subject of petroleum origin now centers on details of the processes rather than on the general theory of its origin.

Organic Versus Inorganic Origin

For many years there was considerable dispute over the question of whether petroleum had an organic or an inorganic origin, or more specifically, whether its origin was biogenic or abiogenic. Even though petroleum, particularly crude oil, consists almost entirely of organic substances, a group of noted scientists long maintained that both oil and gas had originated abiogenically. Among the more prominent of that group was the famous Russian chemist Dimitri Mendeleev, who originated the periodic table. Theories of abiogenic origin seemed to fall into one of two general schemes: (1) Methane and possibly other gaseous hydrocarbons were formed by chemical reduction of primordial carbon or its oxidized forms (carbonates) at high temperatures deep in Earth's crust. The resulting hydrocarbon gases polymerized to form liquid hydrocarbons, then both gases and liquids migrated to shallower strata and accumulated in the reservoirs where they are now found. (2) A postulated methane atmosphere early in Earth's history provided the starting material for formation of petroleum, which soaked into the subsurface and was preserved.

During the last half of the twentieth century, development of more sophisticated and precise analytical techniques has helped clarify this question of biogenic versus abiogenic origin. Evidence from crude oil composition itself is so overwhelmingly indicative of origin from biologic species that essentially all geoscientists are now in full agreement. Specificities in component distribution, low stable carbon isotope ratios, and the presence of significant optical activity (not possible in natural abiologic systems) all clearly point to a biogenic origin. In the application of petroleum geochemistry data and methods by explorers, abiogenic origin theories are no longer even considered. For all practical purposes, they have been relegated to the status of historical relics, although there remains a possibility that a small portion of the methane in natural gas accumulations was formed abiogenically.

Petroleum Generation Processes

Remains of once-living organisms dispersed in a sedimentary deposit represent the starting material from which petroleum is formed. The variability of species contributing to a sediment has made the determination of individual precursors a monumental problem, although progress is steadily being made. What is known to date about precursor species and compounds will be summarized in a later section.

For crude oil to be ultimately generated by sedimentary biologic remains, it is essential that the sediment be protected from a highly oxidizing environment after deposition, or its capability to generate oil will be practically eliminated. Gases can still be formed from oxidized organic materials, but even the generation of natural gas hydrocarbons will be somewhat impaired. Exclusion of oxygen from a deposited sediment requires that it remain in a quiescent subaqueous environment, whether in the ocean or a freshwater body.

Throughout early stages of burial, the dispersed organic matter undergoes a series of transformations involving both microbial and chemical processes. These conversions result in considerable alteration of chemical composition and a practical elimination of cell morphology. During this phase, commonly called diagenesis, the organic matter remains in a predominantly solvent-insoluble state. After diagenesis is completed, usually at temperatures below about 50°C, the resulting insoluble organic mass is defined as kerogen, while the relatively small portion that is soluble in organic solvents is termed bitumen.

As burial of the sediment increases in depth throughout geologic time, accompanied by a temperature increase, the process of catagenesis begins. In this stage, the higher temperatures bring about additional changes in the kerogen, mostly by thermal breakdown into smaller molecules. The process continues until an appreciable portion of the kerogen has been converted into bitumen. The proportion converted is usually between 20 and 50 percent, depending on original kerogen composition. Bitumen generated by this thermal cracking process represents crude oil generated by that particular section of sediment. The sediment is then designated as a source bed, or perhaps more appropriately as a source rock, as it will usually have undergone lithification by compaction at that stage of burial. While oil is being generated, considerable quantities of hydrocarbon gases are also formed by the same thermal cracking process. The relative proportion of oil and gas thus formed will vary depending on the composition of organic material originally deposited and the extent of oxidation it has undergone during its early burial history.

With still further burial and temperature elevation, additional methane is generated, although in a quantity less than that formed during liquid hydrocarbon generation. Actually, more methane is derived from thermal cracking of bitumen remaining in the source rock than is generated by further thermal cracking of the kerogen. This process represents the initial stage of metagenesis, or organic metamorphism, that eventually can culminate with all original organic matter having been converted to methane and graphite.

For a sediment to be effective in generating significant amounts of petroleum, it must necessarily be organic-rich. However, there is not general agreement on the percentage of organic matter that would be considered a minimum requirement. In the past, values as low as 0.05 weight percent organic carbon (the most accurate method for measuring organic matter content in rocks) have been put forward by some geochemists as a practical minimum, but this figure is now generally considered to be too low. By 1990, a majority of geochemists had reached the conclusion that at least 1.0 weight percent organic carbon is needed.

Temperatures required for the petroleum generation process are fairly well understood, but must be stated as a rather wide range because of the exchange between temperature and time inherent in most chemical reactions. The end of the diagenetic stage and the onset of catagenesis is believed to occur at about 50°C. The temperature at which bitumen generation reaches a maximum rate will then depend to a great extent on the length of time involved. On the average, maximum generation will occur around 90–100°C. If an extremely slow increase in burial depth occurs and a long period of time, say, 200 million years (Ma), elapses before maximum generation rate is reached, a temperature of 70–80°C will be sufficient. On the other hand, with rapid burial and a corresponding rapid temperature increase, only a few million years at a temperature of about 120–130°C will accomplish petroleum generation.

The question of when petroleum was formed, insofar as geologic ages are concerned, can hardly be debated. It very likely has gone on continuously since life forms originated on Earth, perhaps up to 3 billion years (Ga) ago. The oldest accumulations found, however, are in the range of 1–1.5 Ga old, and they are not large. From a practical commercial standpoint, essentially all existing petroleum originated since the beginning of the Paleozoic era about 600 Ma ago. Although more oil and gas were formed in some geologic periods than in others, major source beds can be attributed to each period throughout the Paleozoic, Mesozoic, and Cenozoic eras.

Migration

With rare exceptions, oil and gas accumulations occur in rocks other than source rocks, so migration has been a major factor in the history of petroleum deposits. Movement out of the fine-grained source into more porous and permeable beds is termed primary migration. Further migration through these porous strata to a final reservoir location is called secondary migration.

Primary Migration. The actual mechanism(s) by which primary migration takes place represents the least understood of all processes that crude oil undergoes. Small gaseous molecules can readily move through even the fine-grained source rocks, either by diffusion or dissolved in water expelled during compaction, so there is no problem with gas migration. Explaining the migration of larger components in crude oil is much more difficult. Many theories have been offered to explain primary oil migration, and there is yet no consensus as to which, if any, of them have been operative. For each postulate that has been presented, there seem to be legitimate objections as to why it will not work. It is highly unlikely that any single primary migration mechanism has accounted for all movement out of source rocks. In any given situation, more than one of them might well be operative, and their relative contributions to the migration process will undoubtedly vary among source rocks. It is also possible that mechanisms not yet considered will later be proposed and demonstrated to be feasible.

Secondary Migration and Accumulation. As mentioned above, secondary migration takes place in the more porous and permeable strata (carrier beds). Thus there is much less restriction to migration because of molecular size relative to pores through which the oil moves. Oil is essentially free to migrate as a separate phase, moved along by pressure differentials and buoyancy on the water phase during hydrodynamic movement. Secondary migration will proceed until the migrating fluids encounter any kind of barrier that prevents further movement. In some instances no barrier will be encountered, so the migrating oil will proceed to a surface outcrop where it becomes a seep. Migrating fluids do not always follow a simple path through a single carrier bed. They sometimes find avenues of escape through faults to a shallower, geologically younger formation where further migration might occur. In some instances, migrating petroleum can move across a fault into another carrier bed that might be much older, creating a situation where the oil will be accumulated in a reservoir older than the source where it originated.

Whenever a barrier to migration is encountered, the petroleum begins to accumulate behind it. A barrier may be any one of several types. In some instances, it is simply a matter of decrease in porosity and permeability in the same stratum through which migration takes place. Sometimes a fault is encountered where an impermeable section has been juxtaposed to the carrier bed. Very often the migrating fluids move into a domal structure where there is no avenue for exit. The petroleum accumulations usually referred to as oil and gas fields are thus formed.

Changes in phase relationships among oil, gas, and water take place during secondary migration and continue through the accumulation process. As migrating fluids move updip (uphill) through carrier beds, lower temperatures and pressures are encountered, resulting in less mutual solubility and a clearer distinction of separate phases. This process tends to continue after accumulation in a trap has been completed. The resulting effect is that a separate gas phase (gas cap) will occur in the reservoir, while water will be found beneath the oil phase with a distinct separating interface. Sometimes a gas cap will not be present in an oil reservoir because the overlying stratum (seal) cannot prevent small gas molecules from escaping while oil is retained. In other instances, gas may escape through the seal but is replenished by additional gas migrating from the source bed at a rate sufficient to maintain a separate gas phase in the reservoir.

A frequently observed phenomenon, especially in very old reservoir formations, is the occurrence of remigration. Common geologic events such as faulting or reversal of dip (change in direction of slope) will permit the oil and gas to move again until it encounters another trapping situation or possibly continues to a surface outcrop. Another type of reservoir destruction often takes place when erosion of overlying strata removes enough of the seal to let all of the trapped petroleum escape to the surface.

Petroleum Precursors

It has already been mentioned that oil and gas are formed primarily from the remains of living organisms. The question then remains as to which types of organisms were the original source of most petroleum, and, more specifically, what individual organic compounds present in biologic species represent precursors of some of the crude oil components. Answers to this question were unknown or at best uncertain until about 1970. Prior to that time, only a small portion of the thousands of crude oil components had been identified. The development of more sophisticated analytical techniques, particularly the gas chromatography–mass spectrometry combination (GC–MS), completely revolutionized the geochemist's ability to identify crude oil components and relate them to molecular species in precursor organisms.

Since petroleum has been in the process of formation for such a great extent of geologic time, it means that a wide variety of precursors may have been involved. The extinction of so many of those organisms adds some measure of difficulty to the problem of identifying which ones were most prominent in the petroleum generating process. The general approach has been to study modern organisms and infer the compositions of their ancient counterparts. Fortunately, it turns out that the kinds of organisms responsible for much of our petroleum are those that seem to have undergone the least change in composition over geologic time.

An idea that received commercial but not scientific promotion in the early part of this century attributed petroleum to the remains of dinosaurs. This concept of petroleum origin is still popular among many in the general public who are greatly intrigued by those immense creatures, but it has no scientific basis whatsoever. Much petroleum was being formed long before and long after the dinosaurs' approximately 150 Ma existence, precluding any major contribution on their part. Also, despite their huge size, they could not have contributed sufficiently to the sedimentary biomass to account for all petroleum generated during the Jurassic and Cretaceous periods when dinosaurs lived.

Many types of organisms contribute to the sedimentary organic matter from which petroleum is formed, so relative contributions to the process must be quite variable among effective source rocks. In general, most organic input to sediments comes from minute aquatic plant and animal species such as phytoplankton and zooplankton, plus a significant contribution from microorganisms that rework the organic remains during diagenesis. In a marine setting, phytoplankton are by far the most abundant biologic species. When deposition occurs near shore, particularly in a river delta, input from terrestrial plants becomes more prominent. The organic matter contained in source rocks deposited in a deltaic environment will often be dominated by terrestrial material.

In general, organisms containing the highest

proportion of lipids contribute more heavily to oil generation. Lipids are soluble in most organic solvents and include substances such as fats, oils, and waxes. Of all the components in living organisms, lipids possess chemical structures most similar to those of hydrocarbons in crude oil, and thus require less conversion during catagenesis. The actual route undergone by biological lipids to become crude oil hydrocarbons is still poorly understood. During diagenesis, much of the lipid material is either trapped in or weakly bonded to the insoluble, complex organic matter that becomes kerogen and ultimately regenerates hydrocarbons. Although no direct link to lipids can yet be conclusively proved, it seems almost certain that they are the primary precursors of the acyclic saturated hydrocarbons (alkanes) that comprise more than half of most crude oils.

Identification of specific precursors for cyclic hydrocarbons in crude oil is more difficult. Cycloalkanes and aromatic components may have several common precursors, and the latter may actually be derived from the former by a simple loss of hydrogen. Some biologic compounds such as steroids and pigments, also included in lipids, contain ring structures that can be related to those in some crude oil components, and are considered likely precursors. It is also possible for cyclization reactions to occur among acyclic compounds, either the hydrocarbons themselves or their precursors. Some aromatic compounds in crude oil may have originated in the lignins common in woody plants.

The origin of complex petroleum constituents such as resins and asphaltenes is highly speculative. Some geochemists postulate that they are formed by polymerization of smaller crude oil components, while others offer the theory that resins and asphaltenes are simply small fragments of kerogen that were split off during the generation process. The extreme complexity and current limited knowledge of the true molecular structure of these substances prevent any conclusive proof as to their mode of origin.

Since about 1970, numerous minor components of crude oil with moderately complex molecular structures have been shown to possess close similarity to specific compounds of known biological origin. These compounds are defined as biological markers, or more simply as biomarkers, and include molecular types such as steranes, triterpanes, isoprenoids, and porphyrins. The biomarkers have been a subject of intense research since the 1980s, and have been the source of much vital information about petroleum origin and its history. Analysis of biomarkers in an oil can reveal not only the kinds of organisms from which the oil was derived, but also the type of environment in which the source rock was deposited. In addition, biomarkers can be used to help identify the source rock in which a particular oil originated, and provide information about its thermal history.

Determining specific precursors for natural gas may never be possible. Gas consists of a relatively small number of simple molecules, none of which have specific structural counterparts in living systems, so there is no basis for comparing its composition to that of the organic matter in source rocks. Given the mobility that gas enjoys, it is very likely that most accumulations of it consist of components derived from several sources.

It is now recognized that appreciable quantities of methane can result from the action of some species of anaerobic bacteria on deposited organic matter at relatively shallow depths. Some major gas accumulations around the world, especially those in northwest Siberia, are now known to have originated in this manner.

Bibliography

Hunt, J. M. *Petroleum Geochemistry and Geology*. San Francisco, 1979.

Tissot, B. P., and D. H. Welte. *Petroleum Formation and Occurrence*, 2nd ed. Berlin, 1984.

JACK A. WILLIAMS

PETROLOGIC TECHNIQUES

Petrologists are geologists who study the origins and properties of rocks. However, geologists cannot directly observe most rocks forming because they originate deep within the earth and are only later brought to the surface through tectonic or volcanic activity. Some petrologists (experimental petrologists) use specialized experimental techniques to infer some of the information they need to study the origins and properties of rocks. To develop a description of the plate in the earth where a rock formed, experimental petrologists assume that if a particular feature of a natural rock can be recreated consistently by imposing specific parameters (e.g., time, temperature, pressure, chemical composition), then they can use fundamental scientific principles to extrapolate those laboratory parameters to geologically relevant conditions and, therewith, to describe the natural process that formed the rock.

Because conditions in Earth and terrestrial planets are extremely diverse, and because each set of conditions may create a different type of rock, experimental petrologists tend to specialize. For example, many specialize in "mapping" the conditions under which specific equilibrium phase assemblages (minerals + liquid + gas) form from rock chemical compositions. Others concentrate on how minor or trace elements distribute themselves among phases under different conditions. Still others may concentrate on processes found in a single type of geologic environment (e.g., Earth's upper mantle, rocks in subduction zones) or a single topic (e.g., ore deposit formation, meteorite genesis). However, experimental petrologists typically control parameters such as temperature, pressure, initial ("parent") composition, and cooling history, as well as the concentration of oxygen, water, and other volatiles. Moreover, several fundamental problems are ubiquitous. These problems, and the ways in which they are overcome, are discussed below.

Reproducing Geologic Conditions

Geologic Time. One major problem encountered by all experimental petrologists is the large difference between geologic time (measured in millions, sometimes billions, of years) and the time available for experiments in the laboratory (measured in hours or days). For this reason most experimental petrologists investigate situations where time and petrogenetic history are not a factor, that is, chemical equilibrium. NORMAN L. BOWEN began laying the basis for this type of work in the late 1920s when he effectively demonstrated that igneous rocks (rocks produced by the melting of other rocks) can be described using the fundamental principles of high-temperature inorganic chemistry. However, achieving chemical equilibrium is not a simple task. Experimental techniques used to approach chemical equilibrium include: (1) using very fine-grained powders, so that the large amount of (reactive) surface will expedite the chemical reactions; (2) running otherwise identical experiments for different lengths of time to demonstrate that the system's chemistry and mineralogy do not change with time; and (3) reversing experiments (e.g., if ice melts to water at 0°C with increasing temperature, then water should solidify into ice at 0°C with decreasing temperature; that is, the temperature measured for the phase transformation is independent of how the experiment was run). Establishing that chemical equilibrium has been obtained is paramount in any phase equilibrium study.

Of course, not all rocks were at chemical equilibrium when they formed. For example, chemical disequilibrium in the form of zoned crystals is common, even in igneous systems where high temperatures allow elements to move relatively quickly. Accordingly, some experimental petrologists study time-dependent (kinetic) problems, such as the rate at which foreign crystals can be incorporated into molten rock (magma), or the rate at which rocks can be altered in hydrothermal systems (e.g., mid-ocean rifts or sulfide mineral deposits). Here the issue of the difference between geologic and laboratory timescales rises again. Luckily, rate problems are generally both time- and temperature-dependent. While the experimental petrologist has little control over time, temperature is easily controlled. Increasing the temperature of a system decreases the time necessary to complete a reaction or to diffuse an element into a crystal. Therefore, most kinetic studies are done at temperatures much higher than those in nature, and the results must then be extrapolated to more geologically reasonable conditions. Even so, especially in low-temperature systems, experiments with durations of several months are not uncommon.

Some geological phenomena, particularly crystal textures in lava flows and lakes, lend themselves to experimental time-temperature studies because they occur relatively quickly for geologic processes. A lava flow may crystallize in a matter of hours or days. To model these type of processes, cooling rates must be duplicated. Since the advent of microprocessors, temperature controllers can be found on most furnaces, which can duplicate almost any kind of cooling history at atmospheric pressure.

Temperature. A variety of commercial furnaces can be used to take experimental charges (samples) up to, and well beyond, most geologically reasonable terrestrial igneous temperatures (around 1,200C). Furnaces come in various designs. Some furnaces have a square or rectangular heated chamber lined with refractory bricks, while others are made from ceramic tubes that allow access from either end and have a distinct "hot zone" in the center of the tube. Tube furnaces may be constructed with the tube in either the vertical or horizontal position, and may have ports to allow controlled amounts of gases or fluids to be injected, or samples of materials or gases to be removed and analyzed during the course of an experiment.

Because they are relatively inexpensive to purchase and easy to maintian, the most commonly used furnaces are resistance-heated. The heat is generated by passing electric current through a furnace element (e.g., a coil, wire, or bar) similar to those found in an electric oven. For an individual furnace, the temperature is proportional to the electric current applied, and the maximum temperature limit is constrained by the material out of which the furnace element is fabricated. For example, a furnace made with platinum or silicon carbide heating elements can routinely operate at 1,450°C; a furnace with molybdenum disilicide elements can operate at 1,600°C; a furnace with graphite or ZrO_2 elements can operate above 2,000°C. The higher the maximum temperature that a furnace can attain, the more expensive it becomes to purchase, maintain, and operate.

Two problems prevail while performing experiments at high temperatures. First, it is important to have a sample container (or support) that does not react with the sample. Usually, noble metals such as gold, palladium, platinum, or rhodium are used in order of increasing temperature, respectively. However, even these usually inert metals will react with experimental charges under some conditions. For example, when iron is a component of the sample, a ubiquitous experimental problem is iron loss from the sample to the noble metal sample container or holder. Some experimentalists circumvent this problem by pre-saturating the container with iron, while others add extra iron to their experimental charge to compensate for the iron loss.

The second problem is quenching (i.e., "freezing in") the sample rapidly to preserve the textures and phases that were present at high temperature so they can be analyzed. Solutions to this problem depend upon the type of equipment being employed and the chemistry of the experimental charge. For example, silicate glasses that contain in excess of 65 percent SiO_2 do not crystallize rapidly and may be easily quenched by removing the sample from the furnace and blowing on it with a stream of compressed air, while other compositions such as basalts—which are low in SiO_2 and crystallize rapidly—may require being run in a vertical tube furnace so that the sample may be drop-quenched. Drop-quenching is done by hanging the experimental charge in the furnace by a single wire, which can be electrically melted. Then the sample free-falls into a water or mercury bath at the bottom end of the furnace tube.

Pressure. Many rocks form under high pressures and temperatures deep in the interior of the earth; however, duplicating these high pressure and temperature conditions in the laboratory can be rather difficult. While no single piece of equipment can be used to attain the entire range of geological pressures, several pieces of equipment have been developed that, together, are commonly used by experimental petrologists to span the entire range of temperatures and pressures likely to be found in the earth's crust and upper mantle.

For geologic processes that occur at relatively shallow depths in the crust of the earth (low pressures) and moderate temperatures (such as low- to medium-grade metamorphism) externally heated cold-seal pressure vessels are most commonly used. These vessels consist of a high-strength steel alloy tube (about 200 mm long, 37.5 mm wide, with a 6.25 mm hole running down the center), which is closed on one end. The other end is sealed using a screw-on cap through which a small steel capillary links the inside of the vessel with a source of compressed gas. The experimental charge is encapsulated in a precious metal, placed inside the vessel, which is then sealed and pressurized with compressed gas. The end of the vessel holding the experimental charges is then inserted into a furnace, which is slowly heated to the desired temperature. This high-pressure technique is limited by the strength of the tube and is generally used for geologically low pressures and temperatures, that is, pressures up to 2 kilobars (kb) and temperatures to 800°C.

To emulate moderately high temperatures and pressure ambient during most igneous processes in the earth's crust, an internally heated, gas pressure vessel is commonly used. The internally heated vessel is a large steel tube, generally about 300 to 450 mm in diameter, which has a 25 to 37.5 mm hole through the center portion. Threaded steel plugs and special gaskets are used to seal the ends. As with the externally heated vessel, the sample is sealed in a noble metal. But instead of being placed directly into the vessel, the sample container is placed inside a miniature resistance furnace that is sealed inside the steel vessel. Electric leads extend through one of the steel plugs to the small furnace, and current flowing through the miniature furnace heats the sample. The other steel plug has a capillary tube that is connected to compressed gas, which provides pressure to the sample. Because of its thick steel body—more massive than the externally heated vessels—the internally heated gas pressure vessel can obtain maximum pressures of 10 kbars and temperatures to 1,200°C.

Duplicating the higher temperature and pressure conditions found at the base of the crust and upper mantle of the earth requires a different type of apparatus. Instead of using a compressed gas to pressurize the vessels, a solid pressure medium is used. The advantage of using a solid pressure apparatus is that solid media (usually NaCl or talc) are not very compressible, so little volume change is required to generate the high pressures; conversely, gas is very compressible so a large volume of gas must be compressed to generate high pressure. This low compressibility makes solid media apparatus much less dangerous at high pressures than a gas apparatus, which may explode.

The most common solid medium pressure design is a piston-cylinder apparatus. This apparatus consists of a tungsten carbide cylinder (which has a inside diameter of generally either 12.5 mm or 18.75 mm, and an outside diameter of about 50 mm) surrounded by a series of three thick steel rings: these rings are interference (pressure) fit around the carbide so that they form a single unit. The experimental charge to be studied is placed in a small sample container, which most commonly is made of graphite, and resembles a cup with a lid. This sample container is placed inside an assembly that consists of (1) an outer sleeve composed of the solid pressure media, and (2) an inner graphite sleeve. During the experiment, the graphite sleeve functions as a miniature resistance furnace; it generates heat when an electric current is passed through it. The sample container is centered in the assembly by "filler plugs" (one on each side of the sample container) made of pressure-transmitting media. The loaded assembly is placed into the carbide cylinder. A tungsten carbide piston is then driven into the cylinder with a hydraulic press compressing the assembly to the desired pressure. Besides acting as a pressure medium, the outer sleeve is also an insulator that allows the electric current to flow only through the graphite heater. The piston cylinder apparatus is capable of pressures up to 50 kbar and temperatures to 1,800°C. Because the geologic sample in this apparatus is surrounded by a large mass of metal components, the sample quenches rapidly when the power to the graphite heater is disconnected.

Although most geologically interesting pressures and temperature can be attained with the equipment described previously, several other solid-pressure media devices, such as multi-anvil and diamond-anvil, are used to achieve even higher pressures and temperatures.

Other Parameters. Pressure, temperature, and bulk composition are not the only factors that must be controlled to mimic natural conditions. Experimental petrologists often go to great lengths to duplicate more subtle parameters of importance to their systems, for example, hydrothermal, aqueous fluids and their pH, or any of a variety of gases. The most commonly controlled parameter is the partial pressure (fugacity) of oxygen.

Controlling the fugacity of oxygen is very important in systems containing iron, not only because it has several valence states, but because the dominant valence state may be a function of geologic environment. Iron in different valence states will produce different mineral assemblages. This latter concept can be seen by looking at "natural" iron oxide minerals. At the surface of the earth, the equilibrium valence is iron Fe^{+3}, and iron metal heated in air (say, on an engine) will eventually form a red coating, or rust, if it is not continually protected from the atmosphere. On the other hand, in most igneous terrestrial systems, the dominant valence state for iron is actually a combination of Fe^{+3} and Fe^{+2}. This fact can be seen in the ubiquitous nature of the iron-oxide mineral magnetite, or loadstone, which is Fe_3O_4 or $Fe^{3+}(Fe^{2+}\ Fe^{3+})O_4$. Conversely, in extraterrestrial systems, often the iron is a combination of Fe^{+2} (dispersed in silicate or sulfide minerals) and metal (Fe^0).

Because the valence state of iron is very sensitive to small changes in the amount of oxygen present, the oxygen pressure in geologic systems cannot be controlled directly. Rather, it is controlled using reactions with equilibrium constants that are known as functions of pressure and/or temperature. Often, solid buffers are used. For example, if a chunk of iron metal is placed in an evacuated silica-glass tube and heated, any oxygen remaining in the tube will oxidize the iron. If the silica tube remains mostly intact, and if both iron metal and iron silicate are present at the end of the experiment, then the ambient partial pressure of oxygen that must have been present during the experiment can be calculated. The reaction becomes:

$$SiO_2 + 2Fe^0 + O_2 = Fe_2SiO_4$$

and all of the reactants and products have thermodynamic values that are well known as functions of pressure and temperature.

In one-atmosphere pressure systems, oxygen partial pressures can be controlled by mixing gases that react to form small amounts of oxygen. For example,

$$CO + 1/2O_2 = CO_2$$

or

$$CO + H_2 + O_2 = CO_2 + 2H_2O$$

By calculating the equilibrium constant for these reactions as a function of temperature, the amount of oxygen can be controlled with extreme precision. For example, at 1,200°C, a mixture of CO and CO_2 gas that is 90.8 percent CO_2 by volume gives an oxygen partial pressure of $10^{-9.0}$ atmospheres, while a gas mix that is 81.5 percent CO_2 by volume gives $10^{-9.7}$ atmospheres of O_2.

Summary

Through a variety of techniques based on inorganic chemistry, engineering, and intimate knowledge of geology, scientists have developed several means of approximating natural conditions in the laboratory. Through study of the behavior of either simulated or actual natural materials in the laboratory, the experimental petrologist makes an informed guess as to how natural rocks that show characteristics similar to those made in the laboratory sample might have formed. It is up to the individual scientist to decide which of many possible parameters influence the problem under study, and to devise a means of controlling each of them during laboratory experiments. Moreover, they must devise experiments that run for a reasonable length of time. Accordingly, new techniques are always being devised, and there are probably as many individual procedures as there are problems that can be solved through experimental petrology, and scientists who work on those problems.

Bibliography

BOWEN, N. L. *The Evolution of the Igenous Rocks.* Princeton, NJ, 1928.

DEINES, P., R. H. NAFZIGER, and G. C. ULMER. "Eduard Woermann, Temperature-Oxygen Fugacity Tables for Selected Gas Mixtures in the System C-H-O at One Atmosphere Total Pressure." *Bulletin of the Earth and Mineral Sciences Experiment Station* 88. University Park, PA, 1974.

EDGAR, A. D. *Experimental Petrology: Basic Principles and Techniques.* Oxford, Eng., 1973.

ULMER, G. C., and H. L. BARNES. *Hydrothermal Experimental Techniques.* New York, 1987.

ULMER, G. C., ed. *Research Techniques for High Pressure and Temperature.* Berlin, 1971.

AMY J. G. JUREWICZ
STEPHEN R. JUREWICZ

PETROLOGY

See Rocks and Their Study

PETROLOGY, HISTORY OF

The concept of a rock, consisting of naturally formed materials in the earth, as a geological unit distinct from a stratigraphic or layered formation, is a relatively recent idea even though minerals were first described by Theophrastus about 300 B.C.E. The identification of a rock, an assemblage of one or more minerals with some measure of chemical and mineralogic consistence, was initially dependent on visual recognition of the minerals. For example, the properties of color, texture, hardness, form, and occurrence of the constituent minerals were used to characterize rocks. The first treatise on rocks in which the word "petralogy [*sic*]" appears was by J. Pinkerton (1811). Petrology developed as an exact quantitative science as the related fields of mineralogy and geology evolved. When the common occurrence of specific rock types was recognized, the nature of their origin in terms of the physical, chemical, and geological processes that produced them became of paramount interest.

Proposed Origins

Three schools of thought arose initially in regard to the origin of specific rocks: Vulcanist, Plutonist, and Neptunist. Jean E. Guettard (1752) of the Vulcanist school believed the Auvergne hills in France were extinct volcanoes. His interpretation was well supported by the observations of Strabo as early as 7 C.E. that the extrusion of lava from an active volcano cooled to form a common rock called basalt. Additional support came from Nicolas Desmarest (1763), who thought basalt was produced by the fusion of granite. That idea was tested by H. B. de Saussure (1779) when he tried to partially melt granite, the first experimental approach to a petrological problem. He concluded that basalt, now known to be the most common rock on the surface of the earth, cannot be derived by the fusion of granite. (It would take another 150 years before it was discovered that the inverse is true: granite is a potential partial melting product of basalt.) A major conceptual view that molten rock results from the internal heat of the earth was presented by JAMES HUTTON (1785). He also noted that granite had invaded a rock called schist essentially as a liquid. These views became associated with the Plutonist school, but were rejected by ABRAHAM GOTTLOB WERNER (1787), a prominent and persuasive teacher who headed the Neptunist school of thought. Werner adhered to the biblical account of the Flood, claiming that granites and basalts were crystallized from the primeval ocean. In his view no basalt was volcanic in spite of the direct observations of Strabo. Werner's students eventually abandoned the Neptunist school, and the belief that all rocks were of low-temperature aqueous origin was abandoned for all rocks except those later described as sedimentary. Detrital sediments, mixtures of minerals accumulated in water, air, or ice, were eventually recognized as a separate class of rocks.

Cyclical Processes and Catastrophism

Hutton (1795) introduced another major concept in which he viewed geological events as cyclical; that is, the same processes operating today occurred in the past. This view of geological events has dominated the thinking of many geologists. Others, GEORGES CUVIER (1812) for example, thought these cycles were separated by catastrophic events that resulted in abrupt changes in the fossil record, frequency of volcanic eruption, and in structural changes that led to mountain building. The abrupt and worldwide effects of impact by an extraterrestrial object were added to the list of catastrophic events in 1962. The debate between Hutton's view, uniformitarianism, and that of Cuvier, catastrophism, continues today, but with greater understanding of the physics and chemistry of the processes.

Diversity of Rocks Recognized

Petrology became an attractive and exciting field of study with the invention of HENRY C. SORBY (1851) of a method for making thin sections of rocks whereby the constituent minerals could be identified with a microscope. The field of study grew rapidly and was called petrography. The great diversity of rocks was soon recognized and the new debate centered on their relationships and origins. In a few years H. Coquand (1857) divided rocks into igneous, metamorphic, and sedimentary, a classification that remains useful today. For the igneous rocks a plethora of major ideas arose to account for their diversity. For example, Leopold von Buch (1825) noted that crystal setting in magma (= liquid + crystals) would yield different rock types. He reported its experimental demonstration by de Drée in the same paper. CHARLES DARWIN (1844) noted that the floating of crystals, leading to the formation of the rock trachyte, was a similar process contributing to the diversity of igneous rocks. These ideas were the first hints of a major physiochemical process known as crystal fractionation, described below, that was characterized some fifty years later.

Two Versus One Parental Magma

A significant idea was presented by Robert W. E. von Bunsen (1851), who started with two extreme magma types, one relatively low in silica and the other more silicic, that were mixed to give rocks that were compositionally intermediate. Joseph Durocher (1857) claimed the same idea adding the notion that such intermediates could be described as "hybrids." The magma invaded the surrounding rock, making room by literally dissolving the country rock, defined by Theodor Kjerulf (1855) as "assimilation," or by the sinking of the overlying blocks. Bunsen's idea of two magmas was rejected by Joseph B. Jukes (1857), who argued for a single parental magma. In his view, the refractory rocks were at depth, the more fusible portions having been squeezed out toward the surface. The relationship of mineral fusibility and sequence of crystallization of minerals was raised by Johann N. von Fuchs (1837), but it required many more years before the exact physicochemical principles could be established. These relationships were important in accounting for the sequence or succession of rock types.

Magma as a Solution

A major breakthrough came with Bunsen's (1861) recognition of magma as a complex solution and that the proportions of the constituents determined the order of crystallization of the minerals. These ideas led Ferdinand P. W. von Richthofen (1865) to the identification of an eruption sequence with basalt first, followed successively by andesite-dacite-trachyte-rhyolite. That alleged "law of periodic succession" has been highly debated because of the many exceptions around the world. The compilation of chemically analyzed rocks by Justus L. A. Roth (1861) contributed to the recognition of the interrelationships of the various igneous rocks.

Transformed Rocks

The metamorphic rocks, literally transformed rocks, were attributed to a different set of processes. The term "metamorphism" was invented by Ami Boué (1819) to describe those changes in sedimentary rocks that led to slaty or fissile cleavage, a significant change in structure and appearance of those rocks. In addition, Carl F. Nauman (1826) defined rocks that resulted from the introduction of material from external sources as metasomatic. The introduction of water, for example, produced profound changes in mineralogy. The transformation of sediments into granite or syenite was described as "grantification" or "granitization" by Balthazar M. Keilhau (1838).

Thermodynamic Basis Established

All of these petrological processes were eventually put on a firm theoretical base when the principles of thermodynamics were deduced and described by Josiah W. Gibbs (1876). These principles provided a rigorous and quantitative framework for relating magmas, crystallization behavior, and the reaction relations in both the liquid and solid state. Ten years passed before these powerful principles were adopted by Bakhius Roozeboom (1887) and practical applications made to rocks. They serve as the physicochemical basis for most petrological theory. Consistent with theory, Frederick Guthrie (1884) proposed that granites arose at the beginning of melting of a system containing its constituent minerals. The notion was extended to all major rock types, each arising at the beginning of melting of its appropriate chemical system. The primary cause of melting was believed by CLARENCE RIVERS KING (1878) to be due to the local release of pressure on the existing crystalline rocks at elevated temperatures.

Others developed the idea that there was a continuum of magmas from a single parent. By plotting various oxide constituents, such as CaO, MgO, FeO, and Na_2O, against silica, described as a variation diagram, Joseph P. Iddings (1890) deduced that the parent was at the low-silica, basic end of the chemical series. How a derivative magma was produced from such a basic parent was worked out by George F. Becker (1897), who applied the old idea, known since the time of Aristotle, of fractional crystallization, the differentiation of magma by the subtraction of early formed crystals. By extracting early formed crystals the diverse igneous rocks occurring in a region with common chemical and mineralogical characteristics could be related. Such associations defined a "petrographical province" according to the earlier observations of John W. Judd (1886).

Experimental Petrology

As these ideas were being formulated, laboratories were being established in Europe to test in a reproducible way the nature of the melting of natural rocks and of simple mineral systems that closely approached natural rocks. For examples, the University of Yuryev (Dorpat), Polytechnic Institute of St. Petersburg, University of Warsaw, University of Paris, University of Gratz, and University of Göttingen maintained laboratories for such research in the period 1892–1904. In 1905 the Carnegie Institution of Washington established its Geophysical Laboratory for carrying out high-temperature and high-pressure research on rock systems, including ore-bearing assemblages. The laboratory contributed and continues to contribute a vast number of studies bearing on the melting relation of minerals, especially defining their stability regions (a descriptive term first used by August Doelter, 1905) at the conditions of pressure and temperature relevant to the earth's crust. Its emphasis on accuracy and precision by an integrated, multidisciplinary group of investigators set the standard for at least the following sixty years. The first published study (Arthur L. Day, E. T. Allen, and Joseph P. Iddings, 1905) was on the feldspar system, albite and anorthite, which form a complete solid solution series as the mineral plagioclase and make up about 60 percent of the earth's crust. The application of pressure developed slowly, but the depth factor (pressure) was recognized by Joseph P. Iddings (1892) when he observed two different mineral assemblages with the same bulk composition in Yellowstone National Park.

Oxides Versus Minerals

It was not difficult to select important mineral systems to study, but a method for translating the chemical composition of a rock, containing usually about ten important oxides, into a set of simple minerals was needed. Fortunately, this "translation" has been provided by C. W. Cross, J. P. Iddings, L. V. Pirsson, and H. S. Washington in 1902. The scheme of calculation, called the CIPW normative system after the first letter of the last name of the authors, converted the oxides into proportions of ideal equivalents of the common rock-forming minerals. In this way a rock could be reduced to a set of minerals even though part of the rock was glassy. The scheme produces an ideal mineral set that is remarkably close to the actual minerals observed in the rock; hence, the CIPW system is widely used throughout the world.

Productive Period, 1905–1915

By 1915, there were a sufficient number of laboratory studies available for NORMAN L. BOWEN to summarize the results and apply them to major petrological problems. Fractional crystallization dominated the thinking of petrologists at that time, but further work was held up until after World War I. Other significant advances were made in petrology prior to World War I. For example, the deformation of rocks and the relative viscosities of melted natural rocks were studied quantitively, and calorimetric determination of the heat of melting, rock compressibility, and mineral age calculations were carried out. In addition, the pressure-temperature curve for a metamorphic reaction was calculated. The heat conduction equation was used to explain the variation of grain size in a dike, and a general solution for the cooling of a lava was presented.

These measurements of physical and chemical properties were complemented by advances in theory. The imaginative REGINALD A. DALY (1903) promoted the idea that basaltic magmas originated in a single vitreous basaltic layer below the crust. Friedrich Becke (1903) emphasized the role of pressure by noting that eclogite is the high-pressure form of gabbro. Gottfried G. Steinmann (1905) described the ophiolite assemblage (diabase, pillow lavas, radiolarian chert) as typical of the ocean floor sequence. The metamorphic rocks were also given detailed attention by George Barrow (1912), who defined the boundaries of metamorphic zones by noting the appearance of certain minerals in a region of progressive metamorphism. Soon afterward, PENTTI ESKOLA (1915) outlined the concept of metamorphic facies in which suites of minerals were identified as diagnostic of specific conditions of pressure and temperature. The role of volatiles in metamorphism and in some igneous processes (e.g., explosive volcanism, gaseous transfer) was defined, and the groundwork laid for understanding the theory of systems containing volatile components.

Post-World War I Resurgence

In the period after World War I, petrology prospered again and several major concepts appeared. Alfred Harker (1919) proposed that some minerals in metamorphic rocks resulted from a stress environment, whereas other minerals would not grow under stress and were confined to undeformed rocks. None of the alleged "stress minerals" could be produced in the laboratory on investigation some twenty years later (E. S. Larsen and P. W. Bridgman, 1938), but there was reason to believe that stress had a catalytic effect (that is, initiated or assisted) on mineral growth. In 1920 Waldemar C. Brögger proposed an unusual origin for calcite from the refusion of limestone. Although Ebenezer Emmons (1842) had first suggested that some carbonate rocks were igneous in origin, it was not until 1960 that an active carbonatite volcano was observed (Dawson, 1962). The diversity of rocks was addressed independently by Bowen (1922) and VICTOR MORITZ GOLDSCHMIDT (1922). They prepared a converging series of reactions of the mafic or dark minerals on the one hand and the progressive change of the feldspars, the most abundant mineral in Earth's crust, on the other hand. Objections to this characterization of the genetic lineage of the igneous rocks were raised by Daly (1925) and Jakob J. Sederholm (1925), who cited the paucity of intermediate rocks and the failure to provide a method for concentrating potassium in the residual magmas. Unfortunately, the laboratory demonstration of the complex interrelationships of the two series has not as yet been attempted.

Influential Text Book

A significant change in the course of petrology occurred with the appearance in 1928 of Bowen's book *The Evolution of the Igneous Rocks*. His simple thesis was that basaltic magmas were derived from the partial melting of peridotite and subsequently crystallized by the process of fractional differentiation, thereby producing a sequence of rocks increasing in silica content to yield finally a granite, that is, a silica-enrichment trend. His application of experimental data to specific field problems was especially persuasive. Some of his ideas have been modified, but his philosophy of approach has endured. It consists of recognizing a set of related field observations, simplifying those relations into a set of experiments that are carried out under conditions simulating those in nature, executing the experiments in an unambiguous manner reproducible by others, interpreting the results in the light of the specific field occurrences, and finally returning to the field to test the interpretation with new observations. The process is reiterated until a satisfactory interpretation of the observed facts can be achieved. No other book has had a greater impact in petrology during the twentieth century, nor has it survived without criticism. Bowen's colleague Clarence N. Fenner (1929) held that magmas followed an iron-enrichment trend as first proposed by Jethro J. H. Teall (1897) and not a silica-enrichment trend. Later it was demonstrated by Osborn (1959) that both trends were possible provided due consideration was given to the effective oxygen pressure. Major support for fractional crystallization as a dominant process was obtained by Lawrence R. Wager and William A. Deer (1939) in their very detailed study of the layered gabbro intrusion of Skaergaard, East Greenland. In addition, as for the occurrence of an unusual series of iron-rich rocks, they emphasized the role of convection currents in the development of the layered rocks.

During World War II

In addition to the intensive mapping programs involving ore petrology during World War II, some new ideas were published. For example, Bowen (1940) deduced the sequence of reactions for the progressive metamorphism of a siliceous dolomite, a paper that served as a model for future field and experimental studies. The results of a very large number of experiments on the melting relations of a four-component oxide system were summarized by J. Frank Schairer (1942) in a "flow sheet" that portrayed the invariant points, that is, the beginning of melting points, for assemblages of many important rock-forming minerals. He emphasized the incompatibility of certain assemblages, and the important role of invariant points in magma generation was thereby revived. The system investigated by Schairer contained mostly ferrous iron in equilibrium with a metallic iron crucible. An alternative method for dealing with iron-bearing systems was developed by Lawrence S. Darken and Robert W. Gury (1945), who used a gas-mixing technique to control the oxygen pressure over the system, a very important variable in rock formation.

Rise in Experimental Investigation

There was a great increase in petrologic research after World War II mainly because of increased funding for research, new materials developed during the war for apparatus, and an array of measuring devices with greater precision. The petrologic community expressed the need for more data on volatile-bearing systems. The granites became the focus of attention in an effort to resolve the nature of their origin: igneous, metasomatic, or metamorphic. The hydrothermal studies of Bowen and O. Frank Tuttle (1958) provided several interpretations that appealed to those supporting an igneous or metamorphic origin. Granitic magma might arise through the partial melting of existing rocks or as an end product of magma fractionation. A major study on the origin of the various basalt magmas by Hatten S. Yoder, Jr. and Cecil E. Tilley (1962) supported Bowen's view that garnet peridotite was the parental material and a "flow sheet" was derived relating the various basalts, some of which were separated by thermal divides, high-temperature barriers, that could be breached at high pressures. The role of pressure and water was established as important factors in relating basalts to eclogites and amphibolites, respectively. An apparatus was developed by Loring Coes (1953) and adapted to petrologic studies by Francis R. Boyd, Jr. and Joseph L. England (1960) to investigate the behavior of minerals and rocks under the conditions of the upper mantle. Interest then turned to the nodules in alkali basalts and kimberlite pipes because they were perceived to be samples of the upper mantle from depths as great as 450 kilometers (km). The diamond-anvil, high-pressure cell (Charles E. Weir et al., 1959) was applied to petrological problems and the conditions near the center of the earth have been attained, calibrated, and maintained (Ho-kwang Mao et al., 1990). The mineralogy and petrology to the center of the earth were thereby amenable to investigation.

Calibration of Natural Rocks

These investigations were undertaken in part as a result of the calibration of the various geothermometers, geobarometers, and geospeedometers found in the common rocks. The partitioning of an element between two or more minerals is a function of temperature, pressure, and bulk composition including volatiles. For this reason careful studies were undertaken to define how iron and magnesium, for example, are partitioned between an olivine and a pyroxene, or between two different pyroxenes. With the use of two or more elements both the temperature and pressure last endured by a rock could be established. The rate of formation could be deduced from the zoning or exsolution of a mineral. Geochemistry began to assume a large role in petrology as isotopes and trace elements were used to ascertain age, as progress variables of a process, for identification of potential source materials, and in characterizing interrelationships. With the temperature and pressure of formation of a rock defined, it was possible to deduce their variation with time in a given metamorphic region (e.g., Peter C. England and Alan B. Thompson, 1984). As the stability regions of minerals and their assemblages were outlined, the quantitative data could be combined with available thermodynamic data to determine the limits of other assemblages. In this way, a database was accumulated and the network of stability regions for a wide array of rocks could be established by calculation.

Petrology in Plate Tectonics

With the advent of the plate-tectonic hypothesis in the 1960s, attention turned to the region of magma generation at the mid-ocean ridge and at convergent boundaries, such as the Andes. At mid-ocean regions the models of upwelling of magma tended to ignore the large displacements of the source by the deep-seated and numerous transform faults. The presence of an all-pervasive magma source has been characterized by tomographic seismic methods, but the depth and distribution of the magma remain controversial. Another area within the plate-tectonic hypothesis critical to petrologists was the proposal that the subducted plate was a source of magma. It has not as yet been demonstrated how the coldest part of the mantle, the subducted plate, yields magma even though the presence of a small amount of residual water is alleged to lower the melting temperature of the overlying rocks. Water released as a result of the metamorphism of the subducted plate would be absorbed by the overlying hotter wedge, and would be contained in newly formed hydrous phases or in a greatly undersaturated liquid phase when formed. Whereas the plate-tectonic hypothesis has provided a broad framework for understanding the structure of the crust and mantle, it has also provided many new challenges for petrology.

Major Change in Direction

With the growing social needs for natural materials, attention turned toward understanding the processes that concentrate elements. Where the emphasis in the past had been on closed systems at equilibrium conditions, which is a close approximation for many common rocks, the need is for data on the open systems and nonequilibrium conditions pertinent to such problems as ore deposits and pollution problems. The flow of fluids usually dominates such processes and the isotopes of carbon (C), hydrogen (H), oxygen (O), nitrogen (N), and sulfur (S) are used to track the flow through rocks. It is evident that the boundaries of petrology are becoming diffused with mineralogy, geochemistry, geophysics, and geology as a consequence of the integrated approach to solve pertinent social problems.

Bibliography

For text references, see bibliographical reference to Yoder, 1993.

BAILEY, E. B. "Some Aspects of Igneous Geology." *Transactions of the Geological Society of Glasgow* 23 (1958):29–52.

BASCOM, F. "Fifty Years of Progress in Petrography and Petrology." *The Johns Hopkins University Studies in Geology* 8 (1927):33–82. Edited by E. B. Mathews.

BERKEY, C. P. "The New Petrology." *Eighteenth Report of the Director of the State Museum and Science Department, New York State Museum Bulletin* 251 (1924):105–118.

CROOK, T. "The Rise of Petrology." In *History of the Theory of Ore Deposits.* London, England, 1933, pp. 133–161.

CROSS, C. W. "The Development of Systematic Petrography in the Nineteenth Century." *Journal of Geology* 10 (1902):331–376 and 451–499.

DEER, W. A. "Trends in Petrology." *Memoirs and Proceedings of the Manchester Literary & Philosophical Society* 94 (1953):63–92.

KNOPF, A. "Petrology." *Geological Society of America Memoir*, Fiftieth Anniversary Volume (1941):333–363.

LOEWINSON-LESSING, F. I. *A Historical Survey of Petrology*, trans. S. I. Tomkeieff. Edinburgh, Scotland, 1954.

PIRSSON, L. V. "The Rise of Petrology as a Science." *American Journal of Science* 46 (1918):222–239.

TEALL, J. J. H. "The Evolution of Petrological Ideas." *Proceedings of the Geological Society, London* 57 (1901):62–86.

YODER, H. S., JR. "Timetable of Petrology." *Journal of Geological Education* 41 (1993):447–489.

ZITTLE, K. A. *History of Geology and Palaeontology to the End of the Nineteenth Century*, Ch. 4, trans. M. M. Ogilvie-Gordon. London, England, 1901, pp. 324–362.

H. S. YODER, JR.

PHYSICAL PROPERTIES OF MAGMAS

Magmas are typically generated under extreme and variable conditions of temperature, pressure, and differential stresses. Their formation, ascent toward the surface, and eventual eruption as volcanic centers or, failing that, their intrusive emplacement within the earth's crust, are physiochemical processes that have fundamentally influenced the geochemical and geophysical evolution of the earth through time and are largely responsible for the gross geochemical stratification of the earth and the terrestrial planets today. The links between the intensive parameters reigning in the earth where magmas are generated and the resultant behavior of magmatic systems are the physical and chemical properties of the magma and its host rocks.

This entry addresses the physical properties of magmas: the identification of the most important physical properties; their dependence on pressure, temperature, and magma state and composition; and how they influence volcanic activity, the surface expression of magmatism.

The most important magma properties can be divided into four types: (1) the pressure-volume-temperature equation of state; (2) the thermodynamic properties; (3) the kinetic or transport properties; and (4) the surface properties.

Real magmas are complex, multiphase mixtures. Thus the physical properties must also be divided into those of the constituent phases (minerals, melts, fluids) and the properties of the mixture. The first two categories of properties, the pressure-volume-temperature equation of state and the thermodynamic properties, are dominated by the contributions of the individual phases that can be summed together to obtain the magma property.

The relevant volumetric properties are the density and its temperature-, pressure-, and composition-dependence. The specific volumes of the mineral phases and fluids such as H_2O and CO_2 are combined linearly with the volume of the melt to obtain the volume or density of the magma. Densities of the phases generally decrease in the order mineral > melt > fluid. The densities of the individual phases vary quite independently with temperature and pressure, with the expansivities and compressibilities generally increasing in the order mineral < melt < fluid. Due to such differences, the relative densities of the phases can change, for example, crystals become less dense than the melt. Most of the phases present in a magma are subject to variation in their chemical composition during the evolution of the magmatic system. Thus the composition-dependence of the density of the individual phases is quite important in determining magma density and the relative buoyancy of phases.

Similarly, the calorimetric properties of the magma are obtained from the calorimetric data for the pressure-, temperature-, and composition-dependence of the internal energy of the individual phases. In calculations involving magmatic processes the thermodynamic state of the system is usually described in terms of the enthalpy and entropy of the system. The nature of the phases present in the system may be predicted by solving for a minimum in the Gibbs free energy. (Gibbs free energy is the work energy available to the system from chemical reactions.) Thermodynamic equations of state for the individual phases supply the information required to generalize the thermodynamic data for individual phases.

The kinetic or transport properties of magmas include the viscosity, the thermal conductivity, and the self- and chemical diffusivities (the rate chemical species diffuse through the system) of components and combinations of components of the individual phases. The transport properties of the system are, in general, dominated by the transport properties of the throughgoing fluid phase, the melt. The viscosity (rheology) of the phases decreases drastically in the order mineral > melt > fluid. The creep and deformation of crystals play only a minor role in the transport of magma once the melt phase is fully throughgoing. Crystal rheology is important, however, under certain pre-magmatic conditions of partial melting and segregation of partial melts. The fluid phase can be considered to have a negligible value of viscosity for most purposes. The viscosity of silicate melts is a very strong function of temperature and composition and a weak function of pressure. The viscosity of silicate melts increases at eruptive temperatures from a minimum of approximately 10^1 pascals (Pa) for an ultrabasic melt to a maximum of 10^9 Pa for a highly silicic calc-alkaline melt. Many melts do not crystallize upon eruption and the practical limit of magmatic processes in such magmas is the transition of the melt from liquid-like behavior to glassy or solid-like behavior. This occurs over a range of temperatures in nature from 800–700°C for dry basaltic, andesitic, and calc-alkaline rhyolitic systems down to 400°C for water-rich or peralkaline silicic systems. The temperature-dependence of the viscosity of the melt phase for most natural compositions is very large. For water-poor calc-alkaline silicic melts the temperature-dependence is approximately Arrhenian (i.e., the viscosity varies as an exponential function of the temperature). For all others it is strongly non-Arrhenian. The pressure-dependence of the viscosity of silicate melts illustrates the interesting behavior, anomalous compared with most other liquids, that the viscosities of highly silicic compositions have a negative pressure-dependence. Highly silicic in this context includes most dry natural melt compositions. Viscosities of hydrous silicic melts appear to have little dependence on pressure. The mixture of phases in a magma can lead to drastic consequences for the rheology of the system. The solid phases (minerals) do not contribute internally to the deformation of the system, so their presence as rigid particles therefore results in an increase in the viscosity. Several models of the effects of crystals on melts have been proposed to deal with complexity of crystal size, shape, size distribution, and volume fraction in magmas. The situation with fluid-filled vesicles is more complex. Here the vesicles may deform due to the negligible viscosity of the fluid phase. The vesicle deformation requires a finite relaxation time to be overcome. When deformation occurs faster than this vesicle shape relaxation time, the vesicles act as rigid particles and the viscosity rises as in the case for crystal suspensions. If the deformation of the vesicular magma is slow enough, however, the bubbles deform their shape and take up strain more efficiently than the equivalent amount of melt would do. Thus the viscosity decreases with the vesicularity.

There is a critical crystal fraction under set conditions of stress, temperature, and pressure above which the viscosity of the crystal suspension drastically increases and the magma may even cease to behave ductilely during deformation. Beyond the above considerations of the rheology of magma exist a set of effects contributing to the non-Newtonian—a non-Newtonian fluid is one for which the coefficient of viscosity is not constant, in contrast to a Newtonian fluid—and irreversible rheology of multiphase systems that are related to the temporal evolution of the texture or structure of the material. Reorientation and clotting of crystals and coalescence or necking down of bubbles, together with the growth or dissolution of either phase, with time, in a magma contribute to irreversible changes in the rheology. Such effects can dominate the temporal variation of the rheology of a lava flow.

Even silicate melts without crystals and vesicles can exhibit non-Newtonian rheology. This occurs when the stresses driving viscous flow are high enough, for a given viscosity, that the strain rate resulting from the Newtonian flow is very close to the microscopic strain rate generated by the self-diffusion of Si. Thus in volcanic systems where the stress is externally applied and the melt-dominated magma is relatively viscous, non-Newtonian flow occurs by the phenomenon of shear thinning. The viscosity appears to decrease with continued strain until a point is reached where the structure of the melt is so distorted by the flow that the stress exceeds the shear strength of the melt. The result is brittle failure of the melt.

The brittle strength of magmas is not well constrained. Estimates range from 1–100 megapascals (MPa). The possibility of magma fracturing brittlely during deformation is very poorly constrained experimentally but could have major consequences for transport mechanisms in the deep crust. It is clear that magma responds brittlely in the late eruptive stage of silicic volcanism. The pyroclasts that are produced by brittle failure of the magma may be subsequently viscously deformed if the temperature remains high enough for an extended period of time.

The diffusivities of chemical components of the phases decrease in the order minerals < melts < fluids. Thus for a typical magma, the melt diffusivities dominate in melt-fluid diffusion-controlled exchanges (including crystal growth) because they are rate-limiting, as is demonstrated by the presence of disequilibrium zoning within crystals. The flux of each component is influenced not only by its own chemical potential gradient, but also by those of other components. Since diffusivities are also strong functions of composition and self-diffusivities of different components are quite different, chemical diffusivities are substantially different than trace diffusivities (rate of diffusion of trace components through the system).

The self-diffusivities of components in silicate melts vary over a very wide range of values. In general the diffusivities converge with increasing temperature. Most of the generalities of the self-diffusivities of cations and anions can be simplified by considering (1) the relationship to the diffusivity of the solvent species Si, and (2) the properties (charge and radius) of the diffusing species. The diffusivity of Si, the major component of the melt phase, is fixed by the Stokes–Einstein relationship to be inversely proportional to the viscosity of the melt. All other components exhibit diffusivities that are either equal to or higher (faster) than that of Si. As a result, the diffusion of Si is often the rate-limiting step in processes of crystal growth and dissolution where Si transport is required. The fastest cations in silicate melts typically exhibit a low but positive temperature-dependence of diffusivity and a positive pressure-dependence. Cations with diffusivities similar to that of Si typically exhibit a high temperature-dependence of the diffusivity and a negative pressure-dependence. Thus two regimes can be identified, one where Si-like values of diffusivity are obtained and very different cations have similar diffusivity values dominated by the solvent cation (Si) diffusivity (liquid state or extrinsic diffusivity), and one where the cation diffusivities are much faster than Si and very cation-specific (glass state or intrinsic diffusivity). The intrinsic diffusivities exhibit a cation-specific variation in diffusion coefficient such that the diffusivity decreases with increasing cation size and charge. In the intrinsic diffusion regime the diffusivities of cations are only weakly dependent on melt composition, whereas in the extrinsic regime where viscosity plays a central role, the cation diffusivities vary strongly with composition. Volatile species (H_2O, CO_2) generally exhibit much higher diffusivities than the species that control viscous flow and relaxation in silicic melts but not in basic melts, due to their lower viscosities.

Although very few data are available, the thermal conductivity of silicate melts yields a thermal diffusivity that is much higher than the chemical diffusivities of most components of melts. A strong compositional dependence has not been observed.

The surface of a magma, in contact with another condensed phase such as a crystal, can strongly influence the textural development of partially molten material before a magma is developed as a transportable entity. The surface tension of the liquid–solid interface varies with composition of the melt phase as well as the structure and composition of the crystal phase. Additionally, the free energy of the surface is dependent on the crystallographic orientation of the mineral surface. This complexity leads to a significant composition-dependence of the equilibrium distribution of the melt phase within the partially molten rock at low melt fraction and can control the degree of partial melting required before the melt phase has achieved connectivity throughout the sample. If the melt wetting the interface between two surfaces is sufficiently thin, then the properties of the melt phase are so significantly affected by the presence of the surface that the bulk properties of the melt no longer describe adequately the properties of the melt film.

Of all the melt properties discussed above, the most obvious influence of melt properties on the behavior of magmas in nature is the influence of the rheology on the eruptive style of volcanism. The frequent, effusive flows of basaltic lavas in oceanic islands and continental hot spots are a direct consequence of the low viscosity of the magma. The gas contents of such magmas are readily lost due to relatively efficient degassing processes within the magma chambers at depth, and the eruptions are only of risk to property in the immediate path of the flowing lava and in areas affected by it. In stark contrast, the infrequent but highly explosive behavior of andesitic stratovolcanoes and of dacitic to rhyolitic domes is related to the high viscosity of the melts, sometimes augmented by the high crystal fractions present in the erupting lava and the common presence of a volatile phase located in pressurized vesicles in the volcanic edifices that serve as a driving force for some eruptions.

The risks associated with such eruptions are immense and very difficult to predict. At the close of the twentieth century much effort is being expended by researchers worldwide to reduce the risk to millions of people situated near such active volcanoes scattered around the globe. A major part of that task is characterizing the properties of magmas under eruptive conditions.

Bibliography

CHAKRABORTY, S. "Diffusion in Silicate Melts." *Reviews in Mineralogy* 32 (1995):411–504.

DINGWELL, D. B. "Viscosity and Anelasticity of Melts and Glasses." In *Mineral Physics and Crystallography: A Handbook of Physical Constants AGU Reference Shelf 2,* ed. T. Ahrens. Washington, DC, 1995.

———. "Relaxation in Silicate Melts: Some Applications in Petrology." *Reviews in Mineralogy* 32 (1995).

LANGE, R. A., and I. S. E. CARMICHAEL. "Thermodynamic Properties of Silicate Liquids with Emphasis on Density, Thermal Expansion and Compressibility." *Reviews in Mineralogy* 24 (1990):25–64.

MYSEN, B. O. *Structure and Properties of Silicate Melts.* New York, 1988.

NAVROTSKY, A. "Energetics of Silicate Melts." *Reviews in Mineralogy* 32 (1995):121–143.

RICHET, P., and Y. BOTTINGA. "Thermochemical Properties of Silicate Glasses and Liquids." *Reviews of Geophysics* 24 (1986):1–26.

———. "Rheology and Configuration Entropy of Silicate Melts." *Reviews in Mineralogy* 32 (1995):67–93.

RIVERS, M. L., and I. S. E. CARMICHAEL. "Ultrasonic Studies of Silicate Melts." *Journal of Geophysical Research* 92 (1987):9247–9270.

WEBB, S. L., and D. B. DINGWELL. "Viscoelasticity." *Reviews in Mineralogy* 32 (1995):94–119.

DONALD B. DINGWELL

PHYSICAL PROPERTIES OF ROCKS

Because most of the earth is not accessible for direct observation, it is important to understand as much as possible about the physical properties of rocks. In many geological settings or studies (e.g., those concerning oil exploration, crustal structure, mantle convection), the only information we have comes from indirect observations of gravity and magnetic anomalies and/or seismic velocities. In order to make an assessment of the composition or structure of the subsurface for inaccessible regions, we need to know as much as possible about the physical properties of the rocks that compose the subsurface. The rock properties most often studied include the density, elastic constants (bulk modulus, shear modulus, Poisson's ratio), thermal conductivity, coefficient of thermal expansion, magnetic susceptibility, electrical conductivity, electrical resistivity, and viscosity. A detailed discussion of all these properties is beyond the scope of this entry, so we will focus on the elastic constants, which can be directly related to the seismic velocity, density, coefficient of thermal expansion, and viscosity.

Elastic Constant and Seismic Velocities

An isotropic solid is one in which the response of the solid to an imposed stress is independent of the orientation of the solid. Only cubic minerals are truly isotropic. Because of the nature of the crystal structure of minerals, almost all minerals are slightly anisotropic, with as much as a 10 percent difference in their elastic constant depending on the orientation. However, an aggregate of crystals in various random orientations is often assumed to be isotropic. We begin our discussion of elastic constants with the isotropic case.

For an isotropic, homogeneous solid there is a linear relation between stress and strain, under the limit of infinitesimal strains. The limit of infinitesimal strains is a reasonable approximation for the deformation caused by seismic waves as they travel through the earth. For an isotropic system, the number of independent elastic constants reduces to two. We can express the relationship between stress (σ) and strain (ε) as

$$\sigma_{ij} = \lambda\delta_{ij} \sum_{k} \varepsilon_{kk} + 2\mu\varepsilon_{ij} \qquad (1)$$

where δ_{ij} is equal to 1 if $i \neq j$ and to zero if $i \neq j$, and λ and μ are the two independent Lamé constants. The elastic properties of an isotropic material can be described by elastic moduli: the shear modulus, μ, and the bulk modulus or incompressibility, K, defined as

$$K = \frac{3\lambda + 2\mu}{3} \qquad (2)$$

We can measure the bulk modulus with the following experiment. Given a uniform cube of an isotropic material with a volume of V, at a starting pressure P, we increase the pressure to a value of $P + \Delta P$ and measure the change in volume ΔV. The bulk modulus is given by

$$K = -V \cdot \frac{\Delta P}{\Delta V} = \frac{dP}{d \ln V} \qquad (3)$$

Another useful parameter is Poisson's ratio, ν. We can measure Poisson's ratio with the following simple experiment. We take a long, cylindrical rod and subject it to a uniform stress along its axis of symmetry. Poisson's ratio is defined as minus the ratio of the strain normal to the stress axis to the strain along the stress axis (i.e., ratio of thinning to elongation or thickening to contraction):

$$\nu = -\frac{\varepsilon_{22}}{\varepsilon_{11}} \qquad (4)$$

Poisson's ratio is useful in earth science because it can be expressed as a function of the ratio v_p/v_s of the velocities of pressure (v_p) and shear (v_s) seismic waves. We can write the seismic velocities in terms of the elastic moduli or the Lamé parameters:

$$v_p = \left(\frac{\lambda + 2\mu}{\rho}\right)^{1/2} = \left(\frac{K + 4\mu/3}{\rho}\right)^{1/2} \qquad (5)$$

$$v_s = \left(\frac{\mu}{\rho}\right)^{1/2} \qquad (6)$$

Hence,

$$\nu = \frac{(v_p/v_s)^2 - 2}{2[(v_p/v_s)^2 - 1]} \qquad (7)$$

In the crust, we often find $v_p = \sqrt{3}\, v_s$, which corresponds to $\nu = 0.25$. In this case, $\lambda = \mu$, so there is only one independent elastic modulus. Such a material is called a Cauchy solid.

Since the mid–1960s, the seismic velocity of a larger number of rocks and minerals has been measured. These have provided a basic understanding of many factors that influence seismic velocity and attenuation of rocks believed to be abundant constituents of the continental lithosphere.

The effect of confining pressure on velocities has been reported in a number of investigations. An example of data for a typical crystalline rock (Twin Sisters Dunite, an olivine-rich igneous rock) is shown in Figure 1. The characteristic shape of the curve of velocity versus pressure is attributed to the closure of microcracks. As can be seen from Figure 1, much of the closure takes place over the first 100 megapascals (MPa). Velocities measured in crystalline rocks at pressures up to 3,000 MPa demonstrate that changes in velocity with pressure do not approach those of single crystals until the confining pressure is above 1,000 MPa. Even at these high pressures, solid contact between the mineral components is probably only approximate because some porosity has originated from anistropic thermal contraction of the minerals.

Fewer data are available on the influence of temperature of rock velocities. It has been well known that the application of temperature to a rock at atmospheric pressure results in the creation of cracks that often permanently damage the rock. Thus, reliable measurements of the temperature derivatives of velocity must be obtained at confining pressures high enough to prevent crack formation. In general, pressures of 200 MPa are sufficient for temperature measurements to 300°C.

A wide variety of techniques have been employed to measure the influence of temperature on rock velocities. An example of data showing the influence of temperature on velocities is shown in

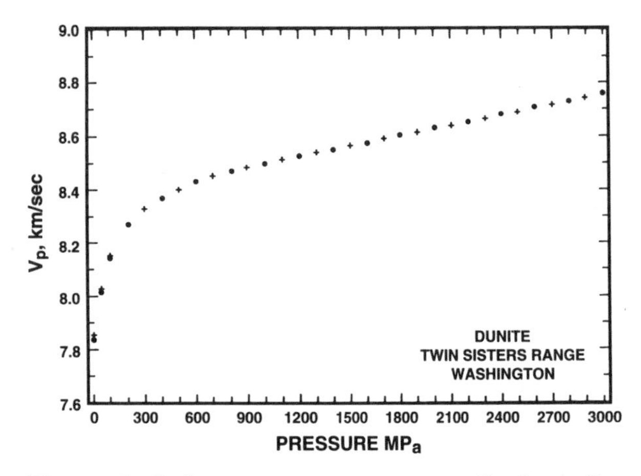

Figure 1. Laboratory measurement of seismic P velocity as a function of pressure for Twin Sisters Dunite (olivine-rich igneous rock).

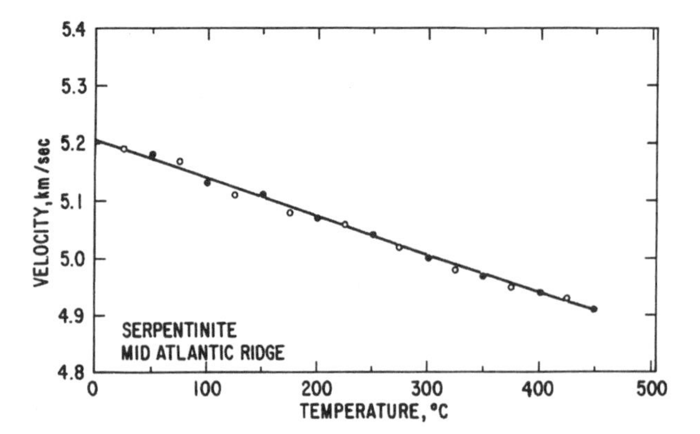

Figure 2. Laboratory measurement of seismic velocity as a function of temperature for serpentinite (metamorphic basalt or gabbro) from the Mid-Atlantic Ridge.

Figure 2. Increasing temperature decreases velocities, whereas increasing pressure increases velocities. Thus, in a homogeneous crustal region, velocity gradients depend primarily on the geothermal gradient. The change of velocity with depth is given by

$$\frac{dV}{dz} = \left(\frac{\partial V}{\partial P}\right)_T \frac{dP}{dz} + \left(\frac{\partial V}{\partial T}\right)_P \frac{dT}{dz} \tag{8}$$

where z is depth, T is temperature, and P is pressure. For regions with normal geothermal gradients (25–40°C/km), the change in compressional velocity with depth dV_p/dz is close to zero. However, in the high heat-flow regions, crustal velocity reversals are expected if compositional changes with depth are minimal.

Amplitudes of seismic waves decrease with increasing distance from their source. This property is called seismic wave attenuation. Seismic wave attenuation has great potential as a tool to yield a better understanding of the anelastic properties, and hence the physical state, of rocks in the earth's interior.

The three parameters most often reported as the attenuation are the seismic quality factor Q, also referred to as the specific attenuation Q^{-1}, the attenuation coefficient, α, and the logarithmic decrement, δ. These are related for low-loss materials ($Q > 10$) by

$$\frac{1}{Q} = \frac{\alpha V}{\pi f} = \frac{\delta}{\pi} \tag{9}$$

where V is the phase velocity and f is the frequency. In both the field and laboratory, difficulties arise in separating the intrinsic dissipation of the rock, that is, processes by which seismic energy is converted into heat, from geometric spreading, transmission losses, scattering, and other factors. Nevertheless, the utilization of laboratory attenuation measurements to tie seismic data to the anelastic properties of rocks is promising, and the refinement of laboratory techniques and the theory concerning the mechanisms involved has yielded and will continue to supply valuable insights into the structure and composition of the continental crust and upper mantle.

All investigations have found that Q increases with increasing confining pressure. Laboratory measurements show a sharp increase in Q at low pressures, which then levels off at high pressures, a response similar to that observed for velocities. The form of the Q versus P curve is generally attributed, therefore, to the closure of microcracks. As with velocity measurements, few researchers have studied attenuation as a function of temperature for rocks of the lithosphere. At temperatures below the boiling point of the rock's volatiles, Q appears to be temperature-independent, and above this Q increases, indicating outgassing of pore fluids and/or thermal cracking. At the onset of partial melting, Q decreases.

Seismic velocities are the most often used properties of rocks. Seismologists have produced one-dimensional (radial) models of seismic velocity from the surface to the center of the earth (Figure 3). While seismic velocities generally increase gradually as a function of pressure (depth), an abrupt jump in seismic velocity over a small pressure (depth) range usually indicates a change in chemical composition or a change in the solid phase of the material. For example, the jump in v_p and v_s at approximately 400-km depth corresponds with the pressure and temperatures of the olivine [$(Mg,Fe)_2SiO_4$] to wadslayite phase change observed in the laboratory. Seismic velocity models place an important constraint on the composition of the interior of the earth. Because the seismic velocities of minerals such as olivine, pyroxene, garnet, and perovskite behave differently as a function of depth, one can attempt to match the seismic velocity curve with simple models of mantle composition. At 2,900-km depth, there is an abrupt change in both density and seismic velocity. This marks the chemical boundary

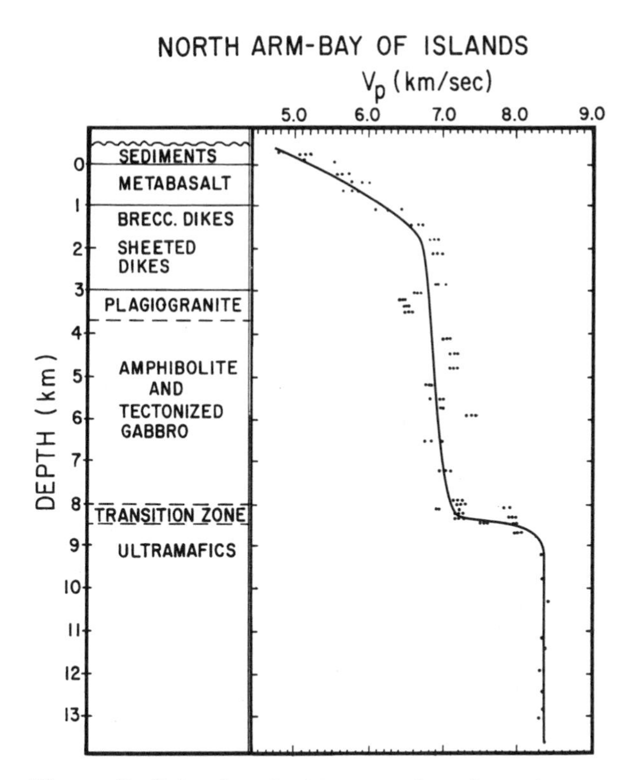

Figure 3. Seismic velocities as a function of depth from the Preliminary Reference Earth Model (PREM).

between the silicate mantle and the liquid iron outer core.

In the 1990s, seismic velocities have played an important role in understanding the structure and composition of the earth's crust. An excellent example is offered by combining oceanic crustal seismic studies with laboratory measurements of the velocity structure of ophiolites. Ophiolites, on land exposures of oceanic crust and upper mantle, contain a stratified sequence of rocks which, from top to bottom, consists of marine sediments, pillow basalts, dikes, gabbros, and peridotites. The laboratory-determined velocity structure of ophiolites (see Plate 32) matches very well with field measurements of velocities in oceanic basins. The initial rapid increase of velocity extending to depths of 3 km originates from a decrease in porosity with depth. At depths between 4 and 7 km, velocities are fairly constant. This region contains mainly gabbro and metagabbroic rocks. The rapid increase in velocity encountered at 7 km is similar to field observations of velocity changes at the Mohorovičič discontinuity. The ultramafic sections of ophiolites show 6–8 percent anisotropy originating from preferred orientation of olivine and pyroxene. This anisotropy correlates well with upper mantle seismic anisotropy measured with marine seismic surveys.

Velocities in single crystals of the common rock-forming minerals vary significantly with propagation direction. In general, for a given propagation direction in anisotropic media such as single crystals, there are three waves, one compressional and two shear. Their vibration directions form an orthogonal set, which usually are not parallel or perpendicular to the propagation direction. The propagation of waves is related to the single-crystal elastic constants through the Christoffel equation, which gives the three velocities for each direction as roots of a cubic equation. Details of wave propagation are related to the crystal symmetry. Most metamorphic rocks and some cumulate igneous rocks have preferred mineral orientations that are usually related to cleavage, foliation, or banding. It follows that many rocks are seismically anisotropic in a manner similar to single crystals. Compressional wave velocities vary with propagation direction and two shear waves travel in a given direction through the rock with different velocities. This latter property of anisotropic rocks, termed shear wave splitting, has recently been observed in several crustal and upper mantle regions.

Density and the Coefficient of Thermal Expansion

The density of rock is important for understanding gravity anomalies. Density depends on the chemical composition of the rock, the mineral structure, and the void space between minerals. Gravity anomalies are the result of mass anomalies at the surface of the earth (for example, changes in elevation) as well as mass anomalies within the earth (for example, mineral deposits or sedimentary basins). The relationship between a mass anomaly, dm, and the resulting gravity anomaly, dg, is

$$dg = G\,dm/r^2 = G d\rho\, V/r^2 \qquad (10)$$

where V is the volume of the mass anomaly and $d\rho$ is its density difference from the surrounding rocks, G is the universal gravitational constant (6.67×10^{-11} $m^3/kg\ s^2$), and r is the distance between the mass anomaly and the point where gravity is being measured. If we knew the distribution of mass anomalies within the earth exactly, we could integrate equation (10) over the whole earth and calculate the observed gravity everywhere. In routine gravity surveys, it is not uncommon to measure anomalies as small as 0.00001 percent of the average earth gravity to detect subsurface structure. From equation (10) it is clear that there is a trade-off between the size of the body (V) and the density difference ($d\rho$). Thus, the better we know the density of the subsurface material, the more accurately we can estimate the size of the body of interest.

The influence of temperature on density is also important in understanding crustal and mantle dynamics. For all earth materials, the density of a rock or mineral increases with increasing pressure and the density decreases with increasing temperature. The relationship between density and temperature is measured by the coefficient of thermal expansion, α, which is defined as

$$\rho(T) = \rho_0(1 - \alpha T) \qquad (11)$$

The coefficient of thermal expansion for the mineral olivine is approximately 3×10^{-5}/°C. Thus, a temperature change of 1,000° changes the density of the material by 3 percent. This density difference provides the force that drives mantle convection and, ultimately, plate tectonics. Density is also affected by pressure (mainly at mantle depths), changes in phase, and changes in composition. Thus, convection in the mantle is more complicated than a simple, uniform fluid as in most tank experiments of convection in the laboratory.

Viscosity

While earthquakes provide evidence that the earth behaves elastically on short timescales, on the timescales of postglacial rebound or mantle convection, the earth behaves like a viscous fluid. The viscosity of the mantle is one of the most important properties for understanding mantle flow, but it is also one of the most poorly constrained properties.

Laboratory measurements of deformation indicate that the viscosity of upper mantle minerals such as olivine is a strong function of temperature, pressure, grain size, and stress (Karato and Wu, 1993). For a temperature increase of 100 K, the viscosity decreases by an order of magnitude at constant stress. An increase of deviatoric stress by a factor of two decreases the viscosity by an order of magnitude. Other factors, such as partial pressure of oxygen and water content, may also have important effects, but are less well studied.

Because of the large difference between time and space scales in the laboratory and the mantle, estimates of mantle viscosity based on modeling large-scale geophysical observations (e.g., postglacial rebound) play an important role in our understanding of mantle viscosity structure; however, viscosity models deduced from these observations are not unique and require additional assumptions. For example, in modeling postglacial rebound, the thickness and extent of the ice sheet through time cannot be determined directly. The uncertainty in the ice sheet model adds to the complexity of the problem. In addition, the theoretical models are often greatly simplified to keep them mathematically tractable.

From these studies, several classes of viscosity models appear, one with essentially a uniform viscosity throughout the mantle and one with a low viscosity channel beneath the lithosphere and a higher viscosity in the lower mantle. At present, it seems that the observational constraints are not strong enough to exclude one of these models. Perhaps more worrysome, however, is that even less is known about the effect of lateral viscosity variations on surface observables.

Bibliography

ANDERSON, D. L. *Theory of the Earth.* Boston 1989.

KARATO, S. I., and P. WU. "Rheology of the Upper Mantle: A Synthesis." *Science* 2670 (1993):771–778.

SCOTT KING
NIKOLAS CHRISTENSEN

PHYSICS OF THE EARTH

The contributions of physics to our understanding of the structure, properties, and behavior of the earth, and the processes by which it reached its present state, are termed "physics of the earth." The expression has a subtly different meaning from "geophysics," which encompasses these things but refers also to the physics-based methods of exploration for oil and minerals. This article is restricted to fundamental questions about the earth and the experiments and theories that are used to find the answers to them.

For most of its early history, physics of Earth was dominated by its connection with astronomy. Long before there was any detailed information about its interior, the shape of Earth, the variation in gravity over the surface, and its motion in space were well measured. This branch of geophysics is known as geodesy (*see* EARTH, SHAPE OF). In its modern form geodesy relies heavily on measurements, using satellites that give truly global coverage as well as high accuracy in many types of observation.

Precise satellite measurements of Earth's ellipticity established that the equatorial bulge is slightly greater than the equilibrium value, which Earth would have if it behaved as a fluid. The ellipticity is the result of a balance between gravity, which pulls Earth toward a spherical shape, and the centrifugal effect of rotation. The equilibrium figure of Earth would give it an equatorial radius 21,287 meters (m) greater than the polar radius; the observed difference is 100 m more than this. The principal reason for the excess is believed to be that the polar regions were depressed by massive ice loads during the most recent ice ages, and that they are still springing back toward equilibrium. The rate at which the ellipticity is decreasing has been measured now, and it is the best indication we have of the viscosity of the deep interior.

Post-glacial rebound on a smaller scale has been studied for many years, first in the land surrounding the Baltic Sea (Fennoscandia), and then also in eastern Canada. These observations give a measure of the viscosity of shallower regions of Earth's mantle. Taken together, the two kinds of observation reveal a general increase in viscosity from the asthenosphere (weak layer), at depths of 100 to 300 kilometers (km), to the deep mantle.

The principle that Earth's surface tends toward an equilibrium shape is known as isostasy (*see* ISOSTASY). Satellite measurements of Earth's gravity show that isostasy operates on a global scale. The departure of Earth from equilibrium is nowhere greater than about 100 m, although extreme differences in surface elevation between mountain tops and the deep ocean floor exceed 10,000 m.

Moreover, maps of the geoid, the surface of constant gravitational potential closest to mean sea level, show features that are not related to the positions of the continents. Thus the continents must be isostatically quite precisely balanced. The geoid anomalies are attributed to deeper density variations, evidently connected with convection in the mantle (*see* MANTLE CONVECTION AND PLUMES).

Our knowledge of the structure of the interior of Earth is derived almost entirely from the study of seismology. Earthquakes or seisms are rapid displacements between adjacent blocks in the crust or upper mantle (down to 700 km depth) that send waves throughout the earth. Systematic timing of the arrivals of these waves at remote stations reveals the variation of wave speed with depth (Figure 1). There are two distinct kinds of waves that pass through Earth. The faster, referred to as P or Primary waves because they arrive first, are alternating compressions and rarefactions, propagating in the same way as sound waves in air. The slower S or Secondary waves are alternating deformations or shear waves in which the motions of individual particles of material are perpendicular to the wave motion, as with waves produced by shaking a rope. S waves can travel only through solids (*see* EARTHQUAKES AND SEISMICITY).

There is a general increase in wave speed with depth due to pressure, but as Figure 1 shows, there are major departures from a uniform Earth having properties varying only with compression. The most dramatic change in wave speeds occurs at a depth of about 2,900 km, which marks the boundary between the outer core and mantle. S waves cannot propagate at all in the outer core, which is liquid, and the P-wave speed there is less than in the overlying mantle. There is, however, a small solid inner core (*see* LEHMANN, INGE), the existence of which is believed to be important to the mechanism by which Earth's magnetic field is generated. Within the upper mantle there are several changes in properties, mostly appearing sharp in Figure 1, at depths down to about 670 km. The region below this appears homogeneous and is referred to as the lower mantle. Overlying the mantle is the thin crust, typically 35 km thick in continental areas but only 5 km thick in oceanic areas, in which wave speeds are clearly lower than in the mantle (*see* SEISMIC LINES, COCORP, AND RELATED GEOPHYSICAL TECHNIQUES).

The seismic wave speeds outline the major regions of Earth's interior, but additional information is needed to deduce the densities of the various layers and thereby infer their compositions. Gravity, g, at the surface (radius r) gives the total mass, M, of the earth since $g = GM/r^2$, where $G = 6.673 \times 10^{-11}$ m^3 kg^{-1} s^{-2} is the gravitational constant appearing in Isaac Newton's law of gravity.

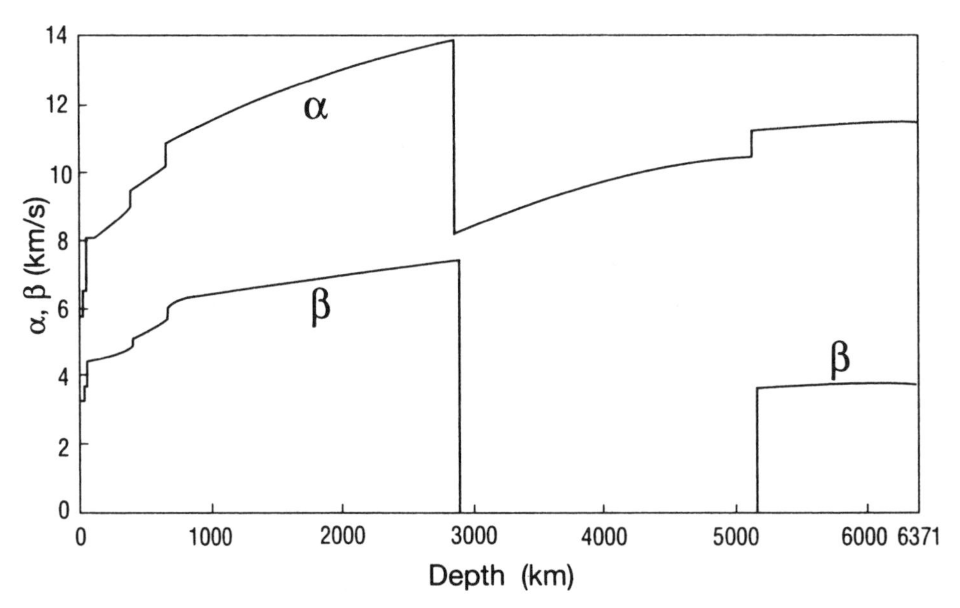

Figure 1. Iasp 91, a seismic wave velocity model of Earth, derived from the travel times of waves from numerous earthquakes to seismometers around Earth. α is the P-wave speed and β is the S-wave speed.

The average density, $\bar{\rho}$, is given by $M = \frac{4}{3} \pi r^3 \bar{\rho}$, from which $\bar{\rho} = 5515$ kg m^{-3}, approximately twice the average density of the crust. Compression of the interior accounts for only a small part of the difference. Another important measure of the density structure of Earth is its moment of inertia, which represents the degree of concentration of mass toward the center. This is a quantity that has been measured astronomically by observing the precession of Earth, a slow change in orientation of the axis of rotation caused by gravitational action of the Moon and Sun on the equatorial bulge, that is, the slight departure of Earth from perfectly spherical form (*see* EARTH, MOTIONS OF). Using this information, detailed "models" of Earth's density structure were calculated in the 1930s by K. E. Bullen (*see* GEOPHYSICS, HISTORY OF).

Improvements in Bullen's Earth models became possible in the 1960s when it was discovered that free oscillations of Earth could be observed (*see* EARTH, HARMONIC MOTIONS OF). After a large earthquake, seismic waves travel throughout Earth and the longer wavelengths persist for several days as standing waves. The earth vibrates as a whole. There are many distinct modes of vibration, each with its own period of oscillation, and more than one thousand of them have been identified. The longest period, fifty-four minutes, is due to the simplest of the spheroidal modes, in which Earth's shape oscillates, in this case between prolate and oblate ellipsoidal deformations. Every fifty-four minutes Earth becomes slightly elongated along a particular axis and between these times it is flattened on the same axis. There is another class of oscillations, the torsional modes, in which Earth is twisted, but its external shape is unaffected. The simplest of the torsional modes is an alternating twist between two halves.

The periods of free oscillation depend upon the internal structure of Earth and therefore give information about the structure independently of that obtained from seismic wave travel times. Modern Earth models (*see* EARTH, STRUCTURE OF) are adjustments of the earlier models calculated to fit the free oscillation periods as nearly as possible. The mathematical process is known as inversion; direct calculation gives the mode periods for any particular model and the inverse process of obtaining an accurate model involves adjusting a starting model to bring the calculated periods into line with those observed.

A result of modeling studies is a profile of density with depth (Figure 2). This shows the layers apparent in Figure 1, but is more useful in suggesting their compositions. For this purpose it is convenient to estimate the densities at zero pressure, using a theory of strong compressions (finite strain theory; see, for example, Stacey, 1992, Sec. 5.5). We can compare the densities of candidate materials with the values given by the broken line in Figure 2. We find that, in spite of the breaks in Figures 1 and 2, the whole of the mantle can be explained by a common composition determined by the mineral olivine, (Mg, Fe) SiO_3, which undergoes a series of transitions to denser crystal structures with increasing pressure. The difference between the upper and lower regions of the mantle is not attributed to composition but to different crystal structures (*see* EARTH, COMPOSITION OF).

On the other hand, the core must have a quite different composition as no combination of magnesium (Mg), iron (Fe), and silicon (Si) oxides could have the observed core density. The abundance of iron in other well-studied members of the solar system, meteorites and the atmosphere of the Sun, points to metallic iron as the dominant component. Certainly the core must be metallic as no nonmetal could carry the electric currents that must be responsible for the earth's magnetic field. Therefore, the outer core is identified as liquid iron. But it is not pure iron, which is 8 to 10 percent more dense than the core. Lighter elements must be dissolved in it, and there is a long history of attempts to deduce what they are. The strongest present contender is oxygen, but sulfur (S), silicon, carbon (C), and perhaps even some hydrogen (H) are probably also present. The solid inner core is also believed to be composed predominantly of iron but with less of the impurities.

The most widely used Earth model, known by its acronym PREM (Preliminary Reference Earth Model), was published in 1981 by A. M. Dziewonski of Harvard and D. L. Anderson of the California Institute of Technology. It has proved to be not as preliminary as its name implies. Lateral variations in the real Earth now attract more attention than further refinements of a spherically averaged model such as PREM. It is the purpose of SEISMIC TOMOGRAPHY to seek such lateral variations. In the crust the lateral variations are very obvious (continents and oceans are structurally so different that a spherically averaged Earth model has little meaning for the crust). Much smaller variations extend through the upper mantle, where they are accessi-

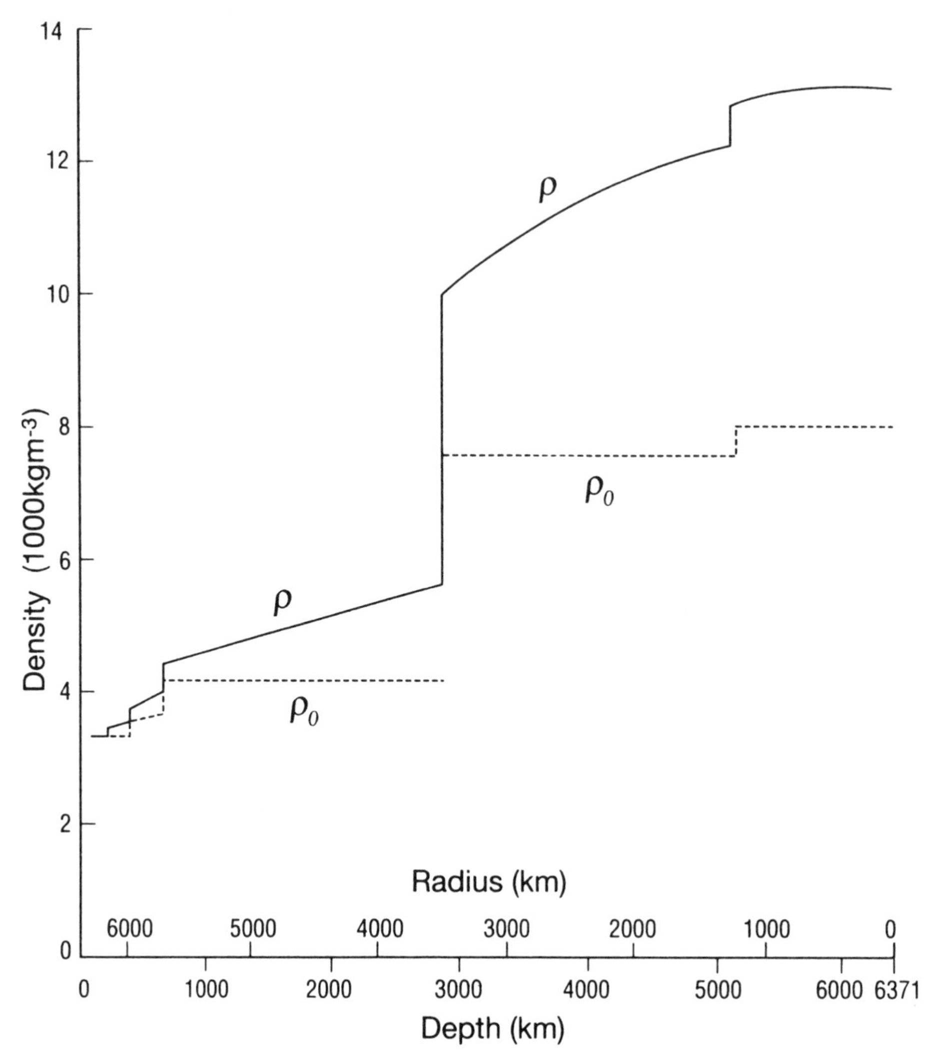

Figure 2. Density of Earth (solid line) and estimated zero-pressure densities of the deep Earth materials (broken line). (Reproduced by permission from Stacey, F. D., *Physics of the Earth*, 3rd ed., Brisbane, Australia, 1992.)

ble to analysis by seismic surface waves. These are waves guided by the layering near the surface. Lateral variations in the lower mantle are smaller still and more difficult to resolve because they must be "seen" through the heterogeneous upper mantle, but strong lateral variations in the lowest few hundred kilometers are well documented. This layer at the base of the mantle is known as D″ (dee double primed), following the original Earth model nomenclature of K. E. Bullen, who gave alphabetical identifications to the various layers but later found that he needed to subdivide some of them. The lateral variations in D″ appear important to interactions of the mantle with the core that influence the geomagnetic field.

Until the discovery of Earth's core by seismology at the turn of the century, a satisfactory explanation of Earth's magnetic field was impossible. Many early students of geomagnetism supposed that it was an extraterrestrial influence, an idea apparently supported by the well-observed correlations of geomagnetic storms with auroras and with sunspots. Although his ideas do not appear to have been highly regarded, Edmund Halley, a contemporary of Isaac Newton, came closer to the correct interpretation in the late seventeenth century with his suggestion that Earth contained concentric shells with magnets embedded in them and that the shells rotated slowly with respect to one another, giving the slow variation in features of the field (the secular variation). The internal origin of the field was finally proved in the 1830s by the German mathematician and physicist Carl Friedrich Gauss. Even Gauss did not believe that Earth's magnetic field was caused by electric currents, and the beginning of a satisfactory theory waited for another eighty years.

In 1919 Joseph Larmor proposed that internal motions of the fluid, metallic-conducting core would drive self-exciting dynamo action. Although it is now universally recognized as the only plausible mechanism, Larmor's idea was slow to gain acceptance. It was given a secure foundation in the 1940s and 1950s by the work of WALTER M. ELSASSER and Edward Crisp Bullard and is the basis of all modern work on the subject. The essential ingredients of the geodynamo are a large fluid core of high electrical conductivity driven in convective motion combined with rotation. The general principles are illustrated in Figure 3. According to Lenz's law of electromagnetic induction, when a magnetic field is moved through an electri-

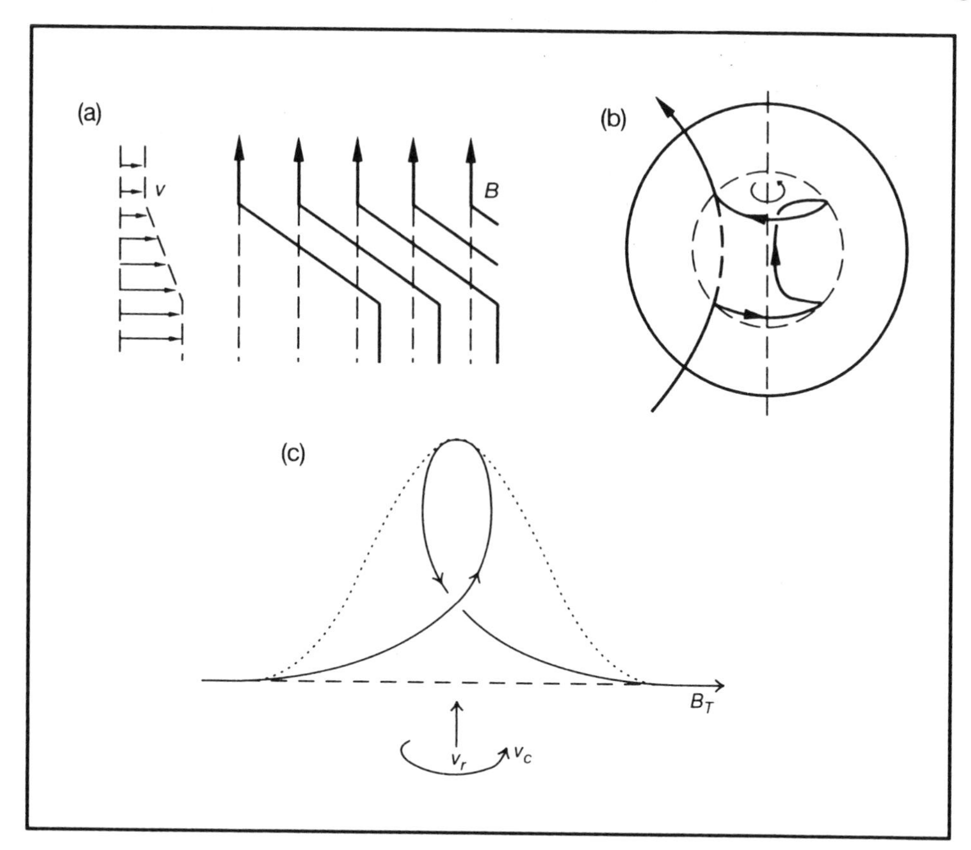

Figure 3. Basic physics of the dynamo mechanism. a. A velocity shear (v) perpendicular to a magnetic field (B) in a fluid conductor deforms and intensifies the field. Field strength is represented by closeness of the field lines. b. The different rotational speeds of the inner and outer parts of the core draw out the field lines of a dipole-type field to produce an additional field circulating within the zone of shear. This is known as a toroidal field. c. A toroidal field is drawn out by convective upwelling and twisted by the rotation to produce a field component reinforcing the dipole-type field in (b). V_r is the radial velocity of convective fluid motion, and V_c is the velocity of rotation of the inner region of the core relative to the outer part. (Reproduced by permission from Stacey, F. D., *Physics of the Earth*, 3rd ed., Brisbane, Australia, 1992.)

cal conductor it generates currents that oppose the motion of the field. Thus a magnetic field in a flowing fluid conductor tends to be carried along with the fluid and only diffuses out of it slowly. If the flow pattern is such as to intensify the field and is fast enough, then the field is self-maintained. Figure 3a shows how shearing motion in a conductor may intensify a field, and applications of this principle to Earth's core are illustrated in Figures 3b and 3c.

We need also to explain how the fluid motion is driven. The required energy must be provided by a convective mechanism. There is no evidence of heat sources (such as radioactivity) in the core but the core is slowly cooling (*see* HEAT BUDGET OF THE EARTH). This means that the solid inner core is growing slowly at the expense of the fluid outer core. The solidification process rejects some of the light solutes in the liquid iron, which remain in the fluid at the inner core boundary, producing a layer of buoyant fluid enriched in the light components. This is convectively unstable and mixes with the rest of the outer core, driving radial (up and down) motion of the fluid. This process is known as compositional convection. Its importance in the core was first stated by S. I. Braginsky and confirmed in detail by D. E. Loper. Because each element of fluid tends to conserve its angular momentum, the radial motion also establishes a differential rotation within the core. The inner part rotates faster than the outer part. Thus the convective motion coupled with rotation provides the two kinds of motion needed for self-regeneration of the field, as in Figures 3b and 3c. The gravitational energy released by the mixing ultimately appears as ohmic heat due to the electric currents.

William Gilbert, physician to Elizabeth I and a lifelong student of magnetism, noticed that the magnetic field of the earth resembled that of a uniformly magnetized sphere of lodestone (iron oxide) with which he experimented. Such a field is known as a dipole field and could be produced by a physically small but strong bar magnet at the center. About 80 percent of the earth's magnetic field, as observed at the surface, is accounted for by a dipole field. The best-fitting dipole field is inclined at 11° to the geographic axis, and when this is subtracted from the observed field, we are left with an interesting nondipole component (Figure 4). Observations over a long period have shown that the features of the nondipole field grow, decay, and drift about, with a generally westward drift dominating for several centuries. There is a slower variation in strength and direction of the dipole field.

The historical record gives only limited insight into the behavior of the geomagnetic field. To see how it behaves on a longer timescale we must turn to paleomagnetism, the record of the field preserved in rocks. Many rocks and archeological specimens have retained the magnetizations they acquired when they were formed or last heated. Archaeological samples from Britain yielded a 2,000-year record of the field direction there (Figure 5), and records from lake sediments have extended the record back a further 10,000 years. When the whole record is plotted, it resembles a tangled bird's nest centered on the star in Figure 5. The star is the local direction of a dipole field centered at Earth's center and with its axis parallel to Earth's axis. This illustrates the axial dipole principle: averaged over 10,000 years or more the geomagnetic field is very nearly that of a dipole aligned with Earth's axis. The secular variation is an apparently random wandering of the field about this direction, and the actual direction at any one time has no special significance. This observation, well confirmed by many longer records, led to the scientific revolution that culminated in plate tectonic theory (*see* PLATE TECTONICS).

If we ignore magnetic reversals (referred to below) and plot the apparent position of the time-averaged magnetic pole, without regard to its sign (i.e., magnetic north and south), then by the axial dipole principle we are also plotting the geographic pole. Within the uncertainties of the magnetic measurements (a few degrees), the pole has been in its present position for about 20 million years (Ma). But when the record is extended further back in time, discrepancies appear. The position of the pole follows a path across the globe. More important, the pole paths from different continents diverge (Figure 6). Polar wander, a shift of Earth's surface features relative to the axis of rotation, is not a sufficient explanation: as first demonstrated in the 1950s and subsequently confirmed with increasing precision, these observations demonstrated continental drift and paved the way for the theory of plate tectonics.

Another observation of paleomagnetism played an important role in this scientific revolution. Although the averaged magnetic axis is seen to coincide with the geographic axis, when several million years of records are considered, the magnetic field is seen to have repeatedly reversed, with the North

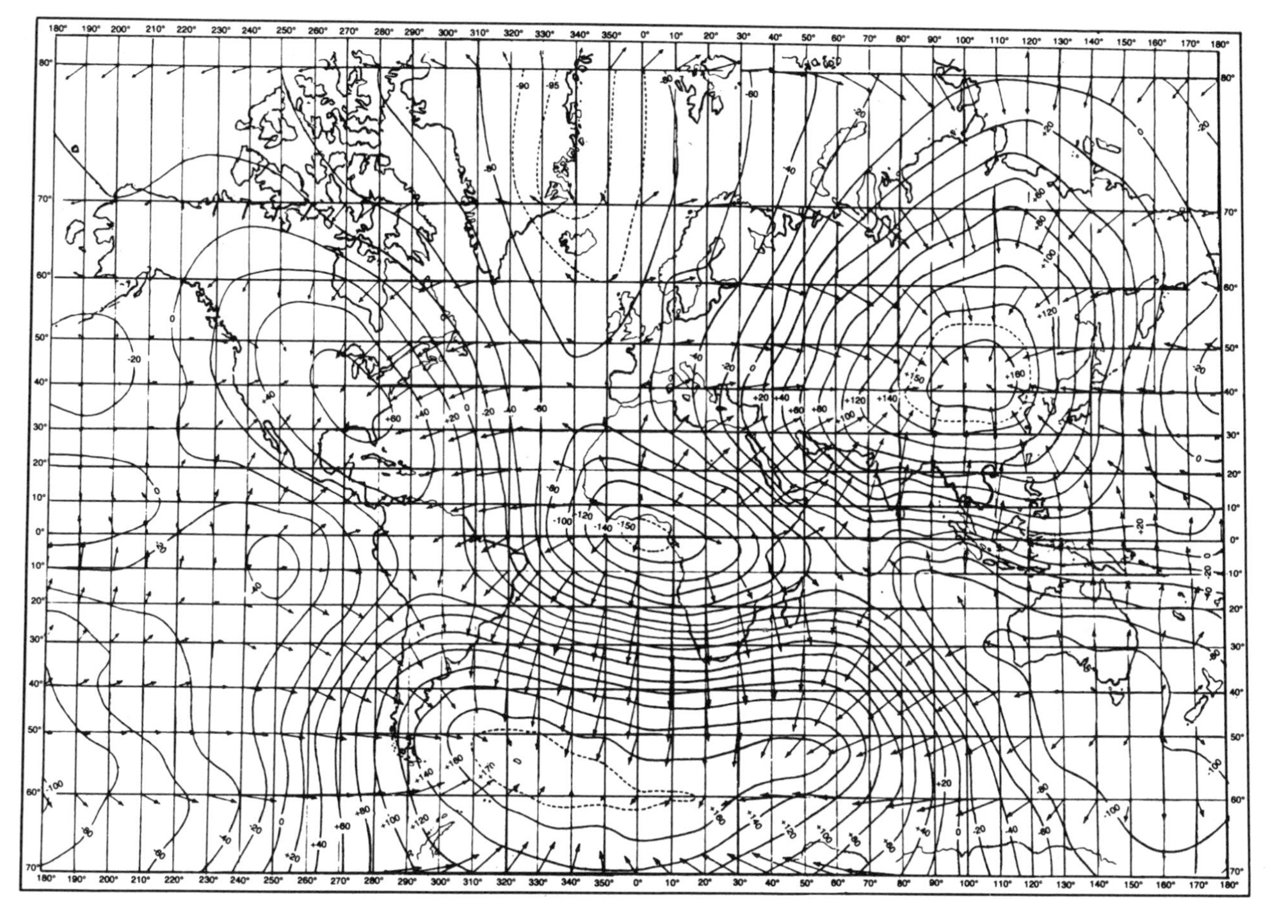

Figure 4. The nondipole field in 1945, as plotted by Bullard et al. (1950). Contours give the strength of the vertical component and arrows are vectors representing the horizontal component. The pattern is suggestive of eddies in the fluid core.

Pole replacing the South Pole and vice-versa. The reversal process itself is of fundamental interest to the theory of the dynamo. Over hundreds of millions of years (Ma) there has been a slow variation in reversal frequency from 5 per Ma for the last 20 Ma to none at all for 20 Ma to 100 Ma ago. The 100 Ma timescale is indicative of mantle influence because this is the time required for mantle features to move. But the discovery of reversals had a much more dramatic influence on geophysical thinking because it explained the stripes of alternating magnetic polarity seen in the igneous rocks of the ocean floors (*see* PLATE TECTONICS). As fresh igneous rock cools at an ocean ridge it is magnetized in the direction of the field at the time and carries this magnetization with it as it moves away from the ridge in the seafloor spreading process. Later rock may see a reversed field and so on repeatedly, producing a sort of tape recording of the field direction. The seafloor magnetic stripes provided not only convincing evidence of seafloor spreading but a direct measure of the rate of spreading.

The surface pattern of plate tectonics is well observed but details of the underlying convective pattern in the mantle are less obvious. Inclined planes of earthquake foci extend into the mantle from the subduction zones, where cooled plates are reentering the mantle, and show that at least several of them reach 700 km depth. At this level there is evidently a hiatus in the convective pattern and debate continues about whether or how the plates are subducted to greater depths. This debate really hinges on the question, What is the immediate driving mechanism of plate tectonics? To answer

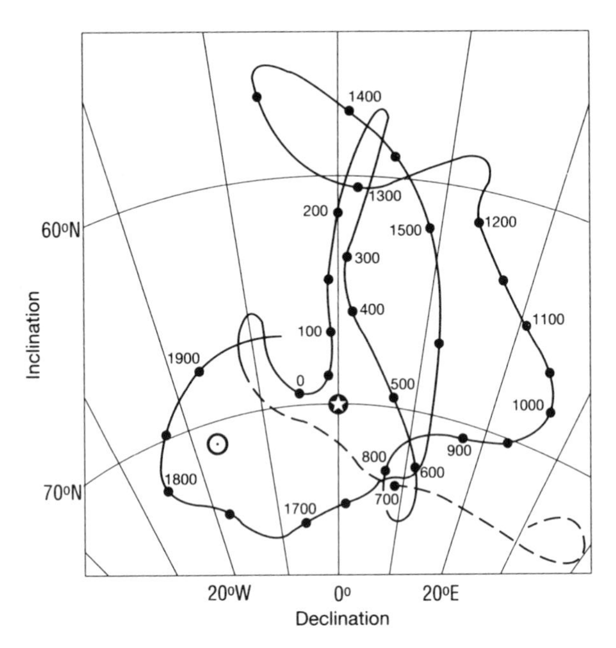

Figure 5. Direction of the magnetic field in Britain for the past 2,000 years. The record of direct observations from about 1600 is extended back by archeomagnetic data. Numbers on the curve give dates. The star gives the direction of an axial dipole field and the open circle is the direction of the field of the present inclined dipole component. (Reproduced from Stacey, F. D., *Physics of the Earth,* 3rd ed., Brisbane, Australia, 1992, using data by Tarling, 1989.)

this question we can use the value of the mechanical energy generated by thermal convection of the whole mantle, 4.8×10^{12} watts (W) (*see* HEAT BUDGET OF THE EARTH) and compare it with the gravitational energy released by sinking plates.

As the lithospheric plates cool on the ocean floors, they shrink thermally, and the oceans become progressively deeper away from the ridge sources of fresh ocean floor. Allowing for the further depression caused by the added weight of ocean water, the amount of shrinkage averages 2.4 km. This contraction represents an increase in density that provides a negative buoyancy force, causing the plates to sink into the mantle. Approximately 3 km^2 of fresh ocean floor is produced and destroyed (subducted) annually, and we can therefore calculate the energy of subduction of the slabs to 700 km depth as the gravitational energy released by transferring 7.2 km^3 of mantle material from the surface to 700 km annually. Allowing for the decrease in thermal expansion coefficient with pressure, we obtain 3×10^{12} W.

Thus we find that 60 percent of the convective energy of whole mantle convection is accounted for by subduction of the plates to 700 km, a quarter of the mantle depth. If we calculate the energy available from upper mantle convection alone, it is less than the subduction energy within the upper mantle. Thus subduction is faster than could be accounted for only by upper mantle convection. We must conclude that convection is driven by subduction of the plates and that the lower mantle cannot be convecting independently of the upper mantle. This argument confirms the picture of mantle convection advocated by Loper (1985), in which subducting slabs tend to break up at about 700 km depth and find their way into the lower mantle more slowly than their penetration of the upper mantle. The ocean rises and production of fresh ocean floor are passive consequences of subduction-driven seafloor spreading.

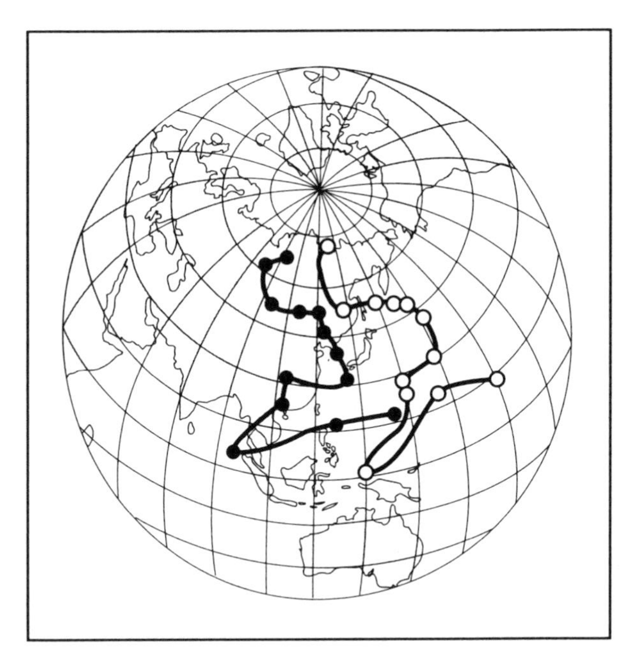

Figure 6. Pole paths for Europe and North America between 500 and 200 Ma ago. These paths would coincide if the continents were moved to close up the Atlantic Ocean. (From van der Voo, 1990, as redrawn by Stacey, 1992, Figure 7.15.)

As pointed out above, the geomagnetic dynamo is almost certainly driven by compositional convection and not by thermal convection. However, it is necessary to the process that the core be losing heat, estimated to be 3×10^{12} W, slightly less than 10 percent of the heat loss from Earth as a whole. The core heat must escape into the mantle, and the only way that it can do this is by conduction across the core-mantle boundary. It therefore heats up a layer at the base of the mantle, which becomes a boundary layer of hot and softened rock. It is also buoyant because it is thermally expanded and so rises convectively through the mantle. But the hot, soft boundary later at the bottom behaves quite differently from the cold, rigid boundary layer at the top. Whereas the cool material plunges down as rigid slabs, the hot material rises as narrow plumes. These have surface expressions in isolated hot spots of volcanism, the most vigorous of which are at Hawaii, Reunion Island in the Indian Ocean, Iceland, and Central Africa.

Earth has been compared to a clockwork toy with a spring that is unwinding, leading to the impression that it is heading for a state of inactivity and geological death. But the timescale for changes is so long that this is not the picture that should be painted. We can expect not only plate tectonics but all the familiar geological and geophysical phenomena, including the geomagnetic dynamo, to persist for a further 5 billion years (Ga). After that the evolution of the Sun will decide the fate of Earth.

Bibliography

BULLARD, E. C., C. FREEDMAN, H. GELLMAN, and J. NIXON. "The Westward Drift of the Earth's Magnetic Field." *Philosophical Transactions of the Royal Society of London* A258 (1950):41–51.

DZIEWONSKI, A. M., and D. L. ANDERSON. "Preliminary Reference Earth Model." *Physics of the Earth and Planetary Interiors* 25 (1981):297–356.

KENNETT, B. L. N., and E. R. ENGDAHL. "Travel Times for Global Earthquake Location and Phase Identification." *Geophysical Journal International* 105 (1991): 429–465.

LOPER, D. E. "A Simple Model of Whole Mantle Convection." *Journal of Geophysical Research* 90 (1985):1809–1836.

STACEY, F. D. *Physics of the Earth,* 3rd ed. Brisbane, Australia, 1992.

TARLING, D. H. "Archaeomagnetism." In *The Encyclopedia of Solid Earth Geophysics,* ed. D. E. James. New York, 1989.

VAN DER VOO, R. "Planerozoic Paleomagnetic Poles from Europe and North America and Comparisons with Continental Reconstructions." *Reviews of Geophysics* 28 (1989):167–206.

FRANK D. STACEY

PLACER DEPOSITS

Placer deposits are mechanical concentrations of dense mineral grains produced by fluid flow at the earth's surface. The main fluids involved are water and air. The concentration of dense grains appears simple at first glance, but actually in detail is complex. The term "placer" also has legal and other definitions that are not treated here.

Placer deposits supply considerable fractions of the world's gold, tin, titanium, diamond, platinum, and other commodities. Some important placer minerals are listed in Table 1. Placer mining was among the earliest of human mining activities.

Table 1. Economic Minerals Recovered from Placer Deposits (in approximate order of importance)

Mineral	*Density*	*Deposit Types*
Gold	15–19	Mostly wet alluvial fan to fluvial, close to source
Cassiterite	6.8–7.1	Mostly wet alluvial fan to fluvial
Ilmenite	4.7	
Rutile	4.2	Mostly shoreline
Diamond	3.5	
Other gems	Mostly 2.8–4	Fluvial and shoreline
Platinum	14–19	Mostly fluvial close to source
Uraninite	9	Wet alluvial fan to fluvial, Archean deposits
Zircon	4.7	Mostly shoreline
Monazite	5	Shoreline and fluvial
Columbite-tantalite	5–8	Mostly fluvial

Enrichment Processes

Studies of placer enrichment processes have only recently been put on a reliable scientific footing, mostly by Rudy Slingerland, Paul Komar, and their colleagues (summarized by Slingerland and Smith, 1986; Komar, 1989). Earlier authors (especially Rubey, 1933) had shown that sediment accumulation in equilibrium with a given flow produces a sorting among mineral grains such that a denser grain will generally be smaller than its equivalent neighbors. Slingerland and Komar showed that the relation varies with different flow types. For example, the behavior of grains is different in turbulent and sheet flow.

Placer enrichment generally occurs where an equilibrium established with one flow type is periodically disturbed by another flow type; the sediment can maintain equilibrium with both flow types only by becoming enriched in small dense minerals. Most commonly, an equilibrium based on equal settling velocity, which produces a population of smaller dense grains and larger less-dense grains, is disturbed by a flow that removes the largest grains. Figure 1 illustrates the process on a beach face—turbulent flow associated with breaking waves alternating with erosive sheet flow in the backwash.

Such interactions between flow types occur at all scales, from individual sediment ripples a few centimeters across to entire sedimentary basins many kilometers wide. Slingerland and Smith (1986) describe such hierarchies of enrichment scales within fluvial environments. Unusual events such as floods or storms can produce extreme enrichments.

Because the last stage of enrichment is quasi-erosional, the preservation of a concentration is unusual but necessary if a deposit is to form. Quick burial is the most common type of preservation. Concentration followed by burial can result from fluctuations in sediment balance; where net accu-

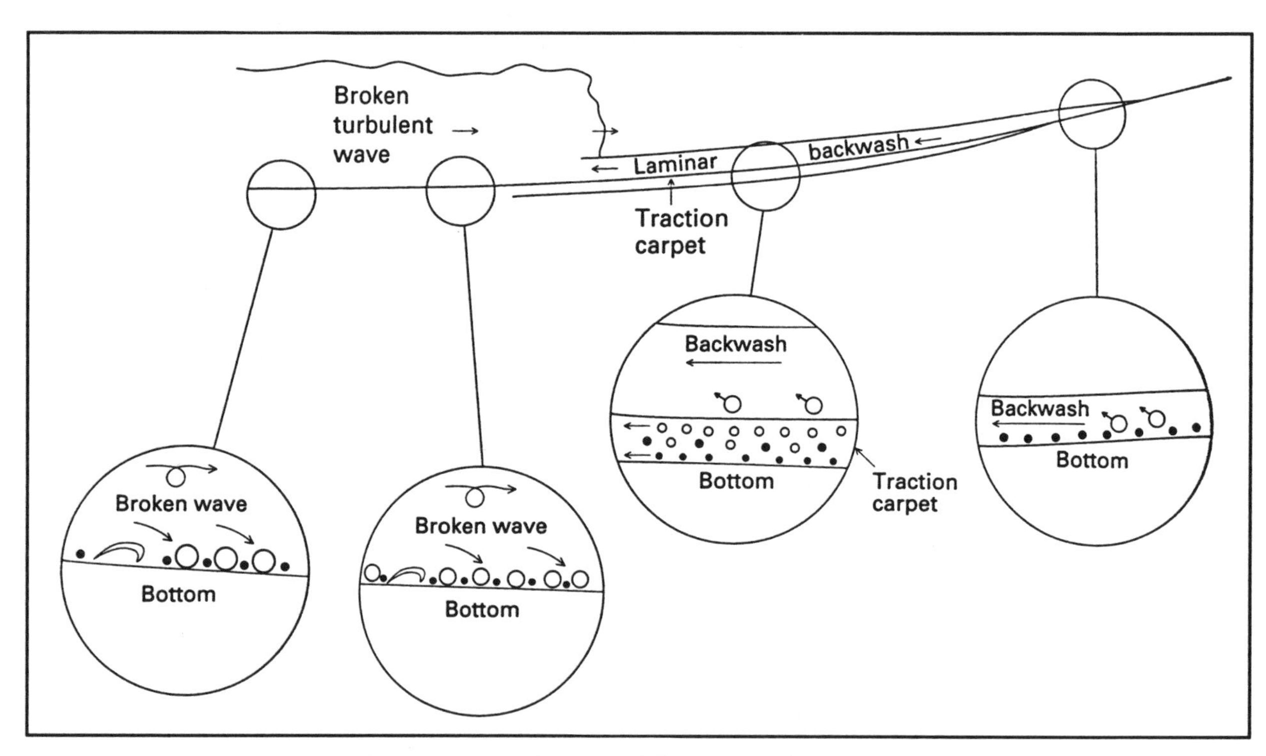

Figure 1. Diagram of successive stages in placer enrichment on the swash zone of a beach. A breaking wave is decelerating from left to right, whereas the accelerating backwash of the previous wave is moving from right to left. Insets show two stages of deposition from the turbulent breaking wave by settling from suspension (on the left) and two stages of placer enrichment in the sheet flow of backwash (on the right) by removal of large, less-dense grains, and slower motion of smaller dense grains (in black) (from Force, 1991).

mulation is slight, dilution of the concentrate is minimal. Figure 2 shows how a beach system does this. Alternatively, the concentrate may be removed from the concentrating environment, as by the wind into eolian dunes.

Weathering also produces (residual) enrichment in placer deposits. Most placer minerals (Table 1) are relatively inert and resistant to chemical decomposition.

Placer deposits can be of any age. Those deposits of great antiquity are commonly referred to as "paleoplacers." By contrast, some placers are presently being formed. Many others are sufficiently young so that elements of their original physiographic environment are preserved. Many more, slightly to much older, are buried or even lithified and deformed. Some are so old that conditions on the earth's surface were fundamentally different from those of today.

Types and Environments

The term placer deposit is applied to a range of types in which the importance of fluid concentration varies. This range includes weathered colluvial-slope deposits, through alluvial-fan and fluvial deposits, to shoreline deposits including related eolian deposits. The densest minerals tend to form near-source deposits in the alluvian fan and fluvial environments, whereas medium-density minerals are more common in shoreline and eolian deposits (Table 1).

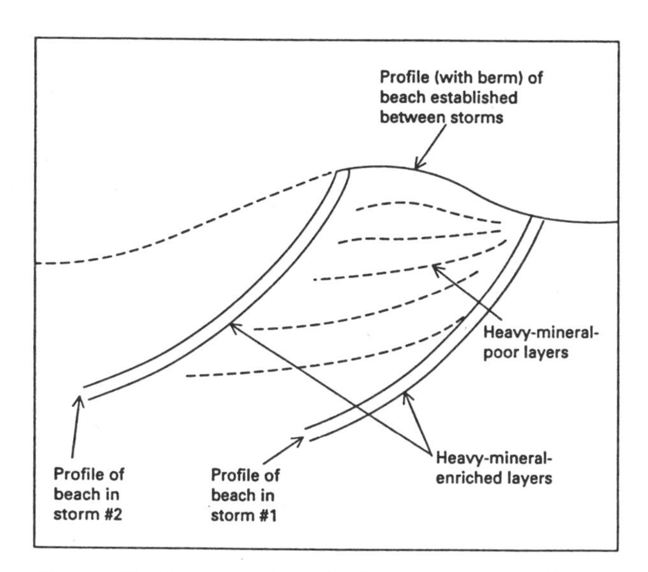

Figure 2. Preservation of placer concentrations (heavy mineral layers) formed during storms on beaches (from Force, 1991).

Deposit History

A valuable placer deposit typically represents a number of factors not apparent at the depositional site: (1) a source area containing the minerals of value; (2) a history of intense weathering of the constituent grains in soil over source rocks, during transport, and in the deposit itself; and (3) a transport mechanism that supplies favorable detritus with minimal dilution. Force (1991) describes such transport mechanisms for both fluvial and shoreline environments. The total transport distance is typically greater for the medium-density placer mineral assemblages.

Heavy Mineral Deposits

Placer concentrations of the minerals rutile, ilmenite, zircon, monazite, and some other minerals, generally with specific gravities ranging from about 3.2 to 7 (Table 1), are commonly called "heavy mineral deposits." Exploited deposits are typically young geologically (Miocene to Holocene), and form on streambeds and beaches. Such concentrations can be observed forming today as the result of floods or of storms at sea.

Archean Gold-Uraninite Placers

Gold-uraninite deposits of the Witwatersrand (South Africa) type are so important and so unusual that a separate description is warranted. A modified placer origin (i.e., placer concentration with grain morphology and composition locally modified by later processes) is generally accepted, (Minter et al., 1993), though a large body of contentious literature debates the question. Concentration occurred more than 2.7 billion years (Ga) ago, when Earth's atmosphere was apparently more reducing, and detrital pyrite and uraninite were stable. Concentration occurred on erosion surfaces separating different stages of the evolution and growth of wet alluvial fans. The concentrate of fine gold, uraninite, and pyrite was trapped in conglomerates forming the beds of broad, shallow channels, and in organic mats.

Bibliography

FORCE, E. R. "Placer Deposits." In *Sedimentary and Diagenetic Mineral Deposits: A Basin Analysis Approach to Exploration*, eds. E. R. Force, J. J. Eidel, and J. B. Maynard. Reviews in Economic Geology, vol. 5. El Paso, TX, 1991, pp. 131–140.

KOMAR, P. D. "Physical Processes of Waves and Currents and the Formation of Marine Placers." *CRC Critical Reviews in Aquatic Sciences* 1 (1989):393–423.

MINTER, W. E. L., M. GOEDHART, J. KNIGHT, and H. E. FRIMMEL. "Morphology of Witwatersrand Gold Grains from the Basal Reef: Evidence for Their Detrital Origin." *Economic Geology* 88 (1993):237–248.

RUBEY, W. W. "The Size Distribution of Heavy Minerals Within a Waterlaid Sandstone." *Journal of Sedimentary Petrology* 3 (1933):3–29.

SLINGERLAND, R., and N. D. SMITH. *Occurrence and Formation of Water-laid Placers: Annual Review of Earth and Planetary Science* 14 (1986):113–147.

ERIC R. FORCE

PLANETARY GEOSCIENCE, HISTORY OF

Planetary geoscience is a term that first came into widespread use in the 1960s. It denotes applying the principles of earth science to the study of other planets. Although early objections were raised to use of the root "geo," indicating "Earth" with reference to other bodies, geoscientists adopted it as a unifying terminology for all planetary bodies of a rocky nature. These include the three other "terrestrial" (earthlike) planets—Mercury, Venus, and Mars—Earth's moon, the rocky moons of other planets, asteroids, and meteorites. The large gas-rich, outer planets—Jupiter, Saturn, Uranus, and Neptune—and the tiny, anomalous planet Pluto cannot be understood by direct comparisons with the solid Earth. Strictly speaking, their study falls in the realm of planetology rather than of planetary geoscience.

Geoscience Studies of the Moon

Looking backward, we may date the beginnings of planetary geoscientific observations to 1610, when Galileo Galilei (1564–1642) turned his small, unmounted telescope toward the Moon and saw that, far from being the smooth sphere thought by philosophers, it has a rugged surface with mountains, deep valleys, and circular craters. He remarked that the brighter areas might represent land and the darker regions water. To this day, we call the bright lunar highlands "terrae" (lands) and the dark plains "maria" (seas), although we know the "seas" are not water but volcanic lavas. Galileo published the first maps of the Moon showing some of its more striking basins and craters. Not until the latter part of the twentieth century would we understand the origin of these features.

The earliest experimental studies of lunar crater formation were carried out in 1665 by Robert Hooke (1635–1703) in England. He duplicated the forms of certain craters by dropping balls into damp clay. Hooke could not, however, conceive of any source for projectiles that might strike the Moon. Consequently, he turned to experiments with boiling alabaster and concluded that the lunar craters are collapsed blisters in a once-molten lunar crust. Strong support for Hooke's rejection of impacting projectiles came in 1704 when Isaac Newton (1642–1727) declared that no small bodies exist in outer space beyond the Moon. As a result, volcanism, which is by far the most familiar crater-forming process on Earth, remained the most favored explanation for lunar craters until the 1970s.

An alternative origin by meteorite impact was suggested in 1873 by the English astronomer Richard A. Proctor (1837–1888), and was strongly advocated in 1893 by the American geologist, GROVE KARL GILBERT (1843–1918). The idea gained few adherents, however, until after an American astronomer, Ralph B. Baldwin, published a small book entitled *The Face of the Moon* in 1949. Baldwin showed that a smooth curve connects the depth-to-diameter ratios of point-source explosion craters, excavated by bombs and shells, with those of lunar craters and of four suspected meteorite craters on Earth. Baldwin's curve implied a common mode of origin: namely, explosive impact. Volcanic craters and calderas fell far off the line. *The Face of the Moon* convinced a number of scientists of the importance of impact as a geologic process.

Among them was the geologist Eugene M. Shoemaker, who persuaded the U.S. Geological Survey (USGS) to institute the first systematic geologic mapping program of another planetary body. Telescopic photographs of the Earth-facing portion of the Moon were separated into forty-four quandrangles on which relative ages of geologic formations were judged by cross-cutting and overlapping relationships, by the sharpness of features such as mountain peaks and crater rims, and by densities of craters that were taken as a measure of the time a given area was exposed to impacting meteorites. The resulting maps established a relative timescale for the geologic history of the Moon.

The geosciences became truly interplanetary in July 1969, when the *Apollo 11* astronauts made the first six geological field excursions to the Moon. To the surprise of many, the samples they collected contained no traces of water or of primitive forms of life, past or present. As a further surprise, samples from the highlands proved to be impact breccias instead of volcanic rocks. Radiometric dating showed that these breccias, consisting of broken and reassembled fragments of feldspar-rich crustal rocks, formed during a massive bombardment that excavated huge basins in the lunar surface beginning about 4.2 billion years (Ga) ago. Volcanic lavas began to fill the basins about 3.9 Ga ago, and continued to do so, creating the dark plains of the maria, until the Moon lost most of its internal heat about 3.0 Ga ago. The brecciated highland rocks turned the tide of opinion decisively toward meteorite impact as the chief basin and crater-forming process on the Moon. At the same time, fly-by and orbital missions were showing cratered surfaces on so many planets and satellites that in 1977 Shoemaker declared impact to be the most fundamental of all processes that have taken place on the terrestrial planets. Today, the most widely-favored hypothesis, formulated in 1985, postulates that the Moon itself originated as the result of a collision between Earth and a Mars-sized object that ripped material from the planet, demolished the impactor, and spun out a filament of mixed debris that finally collected into Earth's unique satellite (*see* EARTH-MOON SYSTEM, EARLIEST HISTORY OF).

The evidence from space prompted many scientists to consider the geological effects of impacts on Earth. Beginning in the 1970s, this subject, which previously had been pursued by a small number of specialists, became of central importance to the geosciences. By the mid–1990s more than one hundred impact structures had been identified on Earth, and there was increasing evidence that some impacts may have been powerful enough to trigger mass extinctions (*see* CATASTROPHIC IMPACT PROCESSES).

Geoscience Studies of Mars, Mercury, and Venus

Mars. We begin with Mars because, with its red color, polar caps, and variable markings visible from Earth, it has the longest record of observations by scientists. Numerous nineteenth-century observers believed Mars could be inhabited, and this idea gained support in 1877 when Giovanni V. Schiaparelli (1835–1910) reported sighting "canali" (channels) on Mars. Unfortunately, some English-speaking astronomers, particularly Percival Lowell (1855–1916) in America, mistranslated "canali" as "canals." In 1894, Lowell founded an observatory at Flagstaff, Arizona, primarily to study Mars, and there followed decades of observations focused largely on Martian engineering works.

The first close-up views of Mars were obtained in 1965 by the *Mariner 4* mission that flew past a moon-like surface covered by impact craters. In late 1971, *Mariner 9* imaged a different terrain with four towering volcanoes, including Olympus Mons, the largest volcano in the solar system. *Mariner 9* also revealed enormous canyons and smaller, branching valleys that look like terrestrial stream channels. It found no Martian canals, however, nor did any of the actual channels match the patterns of canals on early maps. No sign of life has been detected although diligent searches for it were carried out by the Viking missions, which landed on Mars in 1976 and analyzed its soils and atmosphere. Soviet spacecraft visited Mars in 1971 and 1974 and they, too, failed to find any evidence of life (*see* MARS).

Mercury. When the *Mariner 10* spacecraft paid three visits to Mercury in 1974 and 1975, it imaged a cratered surface with a history of impact events stretching back more than 4 Ga. Evidence suggests that a heavy bombardment excavated an abundance of large basins and craters on Mercury about 4.2 Ga ago. By 4.0 Ga ago, a plains unit had covered much of the surface and obliterated most of the craters smaller than about 500 kilometers (km) in diameter. Whether the plains unit consisted of volcanic lavas or of fine-grained impact debris remains unknown. The nature of certain younger, smoother plains on Mercury also is unknown, but a lack of evidence for basaltic compositions suggests that the plains are impact-generated. Mercury's unique crustal features are lobate cliffs about 1 km high and hundreds of kilometers long. The cliffs appear to result from stresses due to the cooling and shrinking of Mercury's interior. Although Mercury is a small planet with a mass equal to only 5.5 percent of the mass of Earth, the mean densities of the two planets are nearly equal (approximately 5.5 grams per cubic centimeter, or gm/cm^3). Consequently, it seems that an iron-rich core must occupy about half of Mercury's interior. That core must be at least partially molten because, like Earth, Mercury has an internally generated, dipolar magnetic field. Inasmuch as metallic iron would have solidified long ago in so small a planet, the core must contain a component, such as iron sulfide (FeS), with a lower melting point than iron. The balance of the evidence suggests that Mercury melted and differentiated very early into a body with an iron-rich core and a silicate mantle. Shortly thereafter a high-energy collision with a smaller body stripped away Mercury's crust and half of its mantle, leaving it with the highest proportion of iron-to-silicate rock of any body in the solar system (see MERCURY).

Venus. Earth's nearest neighbor and "sister planet," which has almost the same size, mass, and density as Earth, is swathed in so thick an atmosphere that no one ever saw its surface until spacecraft, launched from the United States and the former Soviet Union, began to investigate Venus in the early 1960s. The Russian Venera missions of the 1970s were the first to penetrate Venus's atmosphere, consisting of carbon dioxide and swirling clouds of sulfuric acid droplets, and to land on its surface. The landers sent back images for a few minutes before they succumbed to the enormously high temperature (470°C) and pressure (100 bars). A series of missions followed, culminating in the U.S. Magellan radar-mapping missions of 1989 and the early 1990s (*see* VENUS).

The Magellan images showed that a long history of volcanism has covered the entire surface of Venus with basaltic lavas. Thousands of small shield volcanoes dot the landscape, as do rounded domes and other circular features suggestive of hot spots with rising mantle plumes. Relatively fresh meteorite craters also occur on Venus. Their density suggests that the surface is well under a billion years old. Much of Venus consists of low plains and rolling hills, but two continent-sized highlands and several smaller ones rise abruptly to great heights. One plateau area, called Ishtar Terra, is as high as Tibet and includes spectacular mountain ranges with at least one peak higher than Mt. Everest. The highlands are basaltic terrains that have been intensely folded and faulted and pushed upward by tectonic forces that appear to be extensional near the equator and compressional in higher latitudes. Unlike the granitic continents and basaltic ocean basins of Earth, the highlands and lowlands of Venus show no discernible contrasts in composition or density. Clearly, Venus is a geologically active planet but painstaking searches for spreading ridges and subduction zones have yielded no positive evidence that its crust ever was split into moving plates like those on Earth. Despite their basic similarities, Venus differs from Earth in numerous ways. It has no magnetic field, no moon, no oceans, and no river systems. There has been neither stream erosion on Venus nor any means of removing CO_2 from the atmosphere and fixing it in limestones as has occurred on a vast scale on Earth. Most important of all, Venus, with its hostile environment, shows no sign of the life, which thrives so bounteously on Earth.

Meteorites: Samples of Asteroids, the Moon, and Mars

Ancient and medieval manuscripts report that, on rare occasions, fragments of stone or iron were seen plunging out of the sky, often accompanied by brilliant fireballs and sudden explosions. During the eighteenth century, however, philosophers adopted a skeptical attitude toward such stories and either rejected outright the idea that solid bodies could fall from the sky or argued that the objects were ejected by distant volcanoes or formed in the atmosphere by the combustion of dust and gases.

The first serious investigation of bodies falling from the sky was made in Germany by the physicist Ernst F. F. Chladni (1756–1827), who compiled reports from widely scattered sources and found them to be so startlingly similar that he concluded the witnesses spoke the truth. In 1794, Chladni wrote that meteors, exploding fireballs, fallen stones, and isolated masses of native iron all are closely related, and they result when small bodies orbiting the Sun are captured by Earth's gravitational field. Despite Isaac Newton's dictum that small bodies do not exist in space, Chladni proposed that meteorites are small masses of material left over from the formation of planets or pieces of planets disrupted by explosions or collisions.

At first, Chladni's ideas wielded little influence. Within the next nine years, however, seven meteorites were observed falling—some singly and some in showers of fragments from the break-up of a single body—and pieces were saved from six of them. In 1802 and 1803, chemists in London, Paris, and Berlin analyzed fallen stones and irons. They found the newly discovered element nickel, which is absent from ordinary rocks, in the irons and in metal grains of the stones. The analysts also reported mineralogical and textural features including chondrules (spherical silicate bodies about one millimeter in diameter) that set the fallen stones apart from rocks of Earth's crust. Ultimately, the last doubts about falling bodies were banished when a great shower of more than two thousand stones pelted down at L'Aigle, France, north of Paris, at one o'clock in the afternoon of 26 April 1803. Following that event, scientists agreed that stones could fall from the skies, though nearly six decades would pass before Chladni's idea of their origin beyond the Moon would be widely accepted (*see* METEORITES).

The first small body orbiting the Sun was discovered by Giuseppi Piazzi (1746–1826) on the night of 1 January 1801. A year later, Wilhelm Olbers (1758–1840) found a second one and noted that both bodies orbited between Mars and Jupiter. Two more small bodies were found in the same space within the next five years. Their presence led to a popular but erroneous hypothesis that they were pieces of a large planet that had exploded or been struck by a comet. Today, thousands of so-called asteroids or minor planets (mostly less than 500 km in diameter) have been found populating a wide belt in that region. In the 1960s scientists concluded from several lines of evidence that asteroids originated as small bodies rather than being fragments of any larger planet (*see* ASTEROIDS).

Most asteroids follow their "race-track" orbits in an orderly fashion, but some of them collide and shatter into fragments that are flung into elliptical orbits crossing that of Earth. Every day several asteroidal fragments, large enough to survive passage through the atmosphere, fall upon Earth as meteorites. In the nineteenth century, scientists began using iron, stony iron, and stony meteorites to model the interior layers of Earth as well as those of hypothetical meteorite parent bodies.

In 1888, the first small, rare diamonds were discovered in meteorites. Diamonds come from deep within Earth and, by analogy, meteoritic diamonds were expected to originate at depth in large bodies with high internal temperatures and pressures. Meteorites contain no other minerals indicative of high pressures, however, and the great majority of stony meteorites never have been heated to melting temperatures. The carbonaceous chondrites, in particular, contain water and amino acids, of non-biologic origin, indicative of low-temperature environments. In the early 1960s, experimental and textural evidence showed that meteoritic diamonds were formed not under high confining pressures but by impact-generated shock waves during collisions in space or with Earth. This removed the need for large bodies and led to the conclusion that the meteorite parent bodies were of asteroidal size from the beginning.

Not all meteorites come from asteroids. In 1982, a U.S. field party in Antarctica discovered a meteorite with the mineralogy and isotopic signatures of rocks from the lunar highlands. Since then, additional meteorites, from both the lunar highlands and maria, have been collected in Antarctica, and one has been found in Australia. Clearly, these meteorites were ejected by high-energy impacts on the surface of the Moon.

Once the authenticity of lunar meteorites was accepted, scientists began reconsidering problems posed by rare stony meteorites called shergottites, nakhlites, and chassingites (SNCs). In contrast to the great majority of meteorites, which formed 4.6 Ga ago, most of the SNCs crystallized from igneous melts only 1.3 Ga ago, and some appear to be even younger. Such meteorites could come only from a planet large enough to have maintained volcanic heat for nearly 2 Ga after Earth's moon cooled. Mars, with its huge volcanic cones, is the only likely source. More direct evidence is provided by analyses showing that the SNC meteorites are similar in bulk chemical composition to the Martian soils, and one of them contains gases similar to those in the Martian atmosphere, as measured by the *Viking* landing craft. On the basis of these and other lines of evidence, most meteoriticists now view SNC meteorites as samples projected to Earth by high-energy impacts on the surface of Mars (*see also* METEORITES FROM THE MOON AND MARS).

The Outer Solar System: *Voyager 1* and *2* Missions

When spacecraft began exploring the outer solar system, they detected a major change in the character of planetary bodies as they passed the center of the asteroid belt. All of the familiar gray, brown, and red bodies of silicate rock were left behind. As they entered colder regions, the craft found that more and more asteroids were black bodies of the types that are rich in carbonaceous matter and water. Then came the two giant gaseous planets, Jupiter and Saturn, and beyond them the immense icy planets, Uranus and Neptune. Pluto, alone, remains unvisited by spacecraft. Although earlier missions had traveled as far as Saturn, nothing compared with the excitement of scientists as they watched *Voyagers 1* and *2* approach the outer planets, beginning with Jupiter in 1979 and ending with Neptune in 1989. Even at close range, the surfaces of these four planets remained hidden by their thick atmospheres, but the *Voyagers* imaged their satellites and discovered numerous new ones. They also discovered that all four outer planets are orbited by rings, and they imaged the fine structure of Saturn's rings in astonishing detail.

The Voyager missions showed that every outer planet and satellite is decidedly different from every other, as is true of the bodies of the inner solar system. At Jupiter, the *Voyagers* found the moon, Io, in full volcanic eruption, emitting plumes of sulfurous gases hundreds of kilometers high and continually renewing its surface. At Saturn, they found that the frigid moon, Enceladus, acts as a water volcano erupting water from a warmer interior onto its icy surface. Triton, near Neptune, has four erupting geysers of nitrogen gas eight kilomets tall. Most of the sixty satellites now known in the solar system are heavily cratered by impacts. Some, such as Jupiter's Ganymede, consist of dark, cratered terrain girdled with bright, icy extensional grooves. Miranda, at Uranus, is so oddly patterned that it appears to have been completely broken apart by impact and reassembled in unmatched pieces. Nearly every body of the solar system from Mercury to Pluto is exotic; each one has developed its unique character due largely to chance collisions with other bodies. Today, after revealing to us the realm of the outer planets, *Voyagers 1* and *2* are headed out of the solar system into interstellar space.

Planetary geosciences have revealed the presence on Earth of samples from asteroids, the Moon, and Mars, and have revolutionized geology by transforming it from a science that focused exclusively on Earth to one that views Earth in context with the other planetary bodies in the solar system.

Bibliography

BALDWIN, R. B. *The Face of the Moon.* Chicago, 1949.

BEATTY, J. K., and A. CHAIKIN, eds. *The New Solar System.* Cambridge, MA, 1990.

MARVIN, U. B. "Meteorites, the Moon and the History of Geology." *Journal of Geological Education* 34 (1986): 140–165.

MCSWEEN, H. Y., JR. *Meteorites and Their Parent Planets.* Cambridge, Eng., 1987.

TAYLOR, S. R. *Solar System Evolution, a New Perspective.* Cambridge, Eng., 1992.

URSULA B. MARVIN

PLANETARY MAGNETIC FIELDS

The magnetic fields of the planets provide us with information about their present internal structures and dynamics, and in the case of Earth, Moon, and Mars, something about their past. They also exert a dominant influence on their near-space environments and thereby influence both the structure and evolution of their atmospheres. Spacecraft experiments have now measured the magnetic fields around all of the major planets with varying degrees of completeness. This entry describes those basic findings and the internal (planetary) and external (space) implications.

Requirements for a Planetary Magnetic Field

The first requirement for the generation of a planetary magnetic field is an electrically conducting fluid core composed of molten metallic material. For the inner planets, the core radius is usually about half that of the solid planet. For Earth and Mercury this core is probably an iron-nickel mixture to explain the observed density (*see* CORE, STRUCTURE OF). For Mars the core may contain iron sulfide. The Moon's core is clearly smaller than this, and Mercury's core is larger. Some planets, like Earth, probably have a solid central core. In this situation the fluid surrounding it is the important region for magnetic field generation. The largest two giant planets, Jupiter and Saturn, appear to have small rocky cores surrounded by metallic hydrogen that becomes electrically conducting at the great pressures of their interiors. In contrast, the fields of Uranus and Neptune are believed to be generated in water-ice mantles between their rocky cores and hydrogen atmospheres.

The second requirement for the generation of a planetary magnetic field is that the conducting fluid circulates in a pattern that can amplify and maintain the field through a "dynamo" action in which currents are generated by the motion of the conducting fluid. Rotation of a planet combined with internally driven "stirring" of the core fluid from any of a variety of energy sources can be effective for dynamo generation. The necessary stirring motions can result from the gravitational settling of cooled and solidified core material to the center of the planet or from chemical or radioactive reactions that generate heat and thus thermal convection. These fluid patterns generally create a field that externally is primarily dipolar in appearance, as illustrated by Figure 1. The field "lines" shown in Figure 1 can be thought of as indicating the pattern that iron filings would assume in the presence of a magnetic dipole. For a dipole, this pattern is the same in all planes containing the dipole axis. Higher "order" fields, such as quadrupole and octupole fields, are also generated at the same time. In most cases near the planet's surface these are weaker than the dipolar part of the field, and so it is customary to compare planetary magnetic fields simply in terms of their dipole "moments" (a moment is a vector pointing from the south to the north pole of the planet whose magnitude equals the product of the equatorial surface field and the radius cubed), as given in Table 1. However, in the generation region the field is complex, with all orders contributing about equally to the energy of the magnetic field.

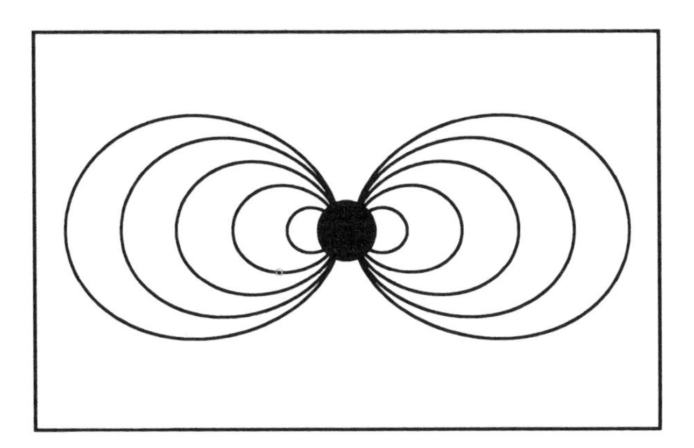

Figure 1. Dipole field "lines."

Table 1. Comparison of Dipole Moments

Planet	*Angle Between Rotation and Dipole Axes (degrees)*	*Dipole Moment Relative to Earth's Dipole Moment*
Mercury	Unknown	.0004
Earth	11	1.0
Jupiter	10	18,000
Saturn	0	580
Uranus	59	50
Neptune	47	24

The theory of magnetic field generation in a moving conducting fluid is called dynamo theory. The mathematical treatments of dynamo theory show how motions in a conducting fluid can distort a weak interior "seed" field to make it grow in strength. One popular dynamo theory predicts that the size of the dipole moment of a planet, and thus the strength of its dipole field, are proportional to the product of the core radius to the fourth power, the square root of the core density, and the angular rotation rate. The measured dipole moments of the planets are compared to this quantity in Figure 2. Some uncertainty exists in the predictions because the theoretical core sizes of the planets are only weakly constrained by our knowledge of their gravitational fields, but nevertheless, the theory seems to work quite well except for Mars and Venus. Mars and Venus are thought to have smaller dipole moments because the source of energy to drive core convection is either missing or very weak. The field of Pluto, if Pluto has a molten core of any consequence, is unknown, as is the field of all planetary satellites except the Moon. The fields of asteroids are also unknown, although meteorites appear to be weakly magnetized, as if their

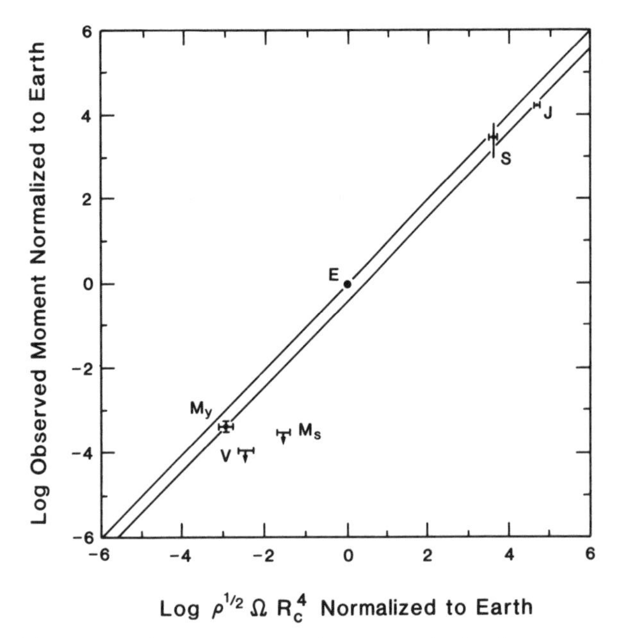

Figure 2. Comparison of measured dipole moments.

parent bodies had active internal dynamos at one time.

Styles of Planetary Magnetism

In most cases, the dipole magnetic fields of planets have moments that lie at a small angle to the axis of rotation. Table 1 gives the relative angles between the two for cases where spacecraft measurements are available. The large angles between these axes for Uranus and Neptune are the major exceptions. They may reflect the location of the dynamo generation at rather high levels in the interior of the planet. In some ways these dynamos may resemble the solar dynamo. It is important to appreciate that planetary magnetic fields change with time, albeit slowly on the timescale of human life. As Earth's dipole magnetic field was varied in strength and orientation over geologic and recorded history, so, most likely, have those of the planets.

One quite different type of magnetism is found on the Moon, where the internal dynamo generation of field appears to have ceased. Rocks in the lunar mantle and crust are magnetized. The manner in which they became magnetized is still a matter of debate. A likely interpretation is that the Moon had an active internal dynamo several billion years ago. During its subsequent history, the Moon's crust and mantle were rearranged by large meteor impacts and crustal movements that caused lava flows. Wherever there was significant heating the crust would lose its magnetization. The spacecraft that visited the Moon found the resultant patchwork of localized crustal magnetic dipoles. This patchy "remanent" magnetism has so far been observed only at the Moon, although some scientists believe it may also be present at Mars. A Martian dynamo is thought to have been active at a minimum just after the planet formed, because internal heat left over from the planetary accretion process would have kept the fluid in the new core moving for at least a few hundred million years. The same scenario may have occurred at Venus, but its surface is too hot because of the "greenhouse" effect to have retained remanent fields. Rock magnetization is preserved only if the temperature remains below a certain upper limit called the "blocking temperature."

Planetary Magnetospheres

The "magnetosphere" is the term used to describe the space around a planet where the magnetic field of the planet can be detected. Although a dipole field in a vacuum extends to infinity, the presence of the supersonically outflowing solar wind, the highly ionized outer atmosphere of the Sun, confines the planetary magnetic fields inside volumes that are roughly bullet-shaped, as illustrated in Figure 3. This confinement occurs because one of the properties of a highly ionized gas such as the solar wind is that it can exert a pressure on a magnetic field and can confine it to a cavity. The bullet-shaped boundary between the solar wind and the planetary magnetic field or magnetosphere is called the "magnetopause."

The positions and shapes of the planetary magnetopauses are determined by the balance between the pressure of the oncoming solar wind and the pressure associated with the planetary magnetic field. Thus, the stronger dipole fields make the biggest magnetospheres. Solar wind pressure also decreases with distance from the Sun due to radial expansion, thereby further increasing the sizes of the most remote magnetospheres. The locations of the subsolar or "nose" positions of the magnetopauses for planets with significant dipole moments are given in Table 2. Inside the magnetopause, the dipolar magnetic field of a planet is

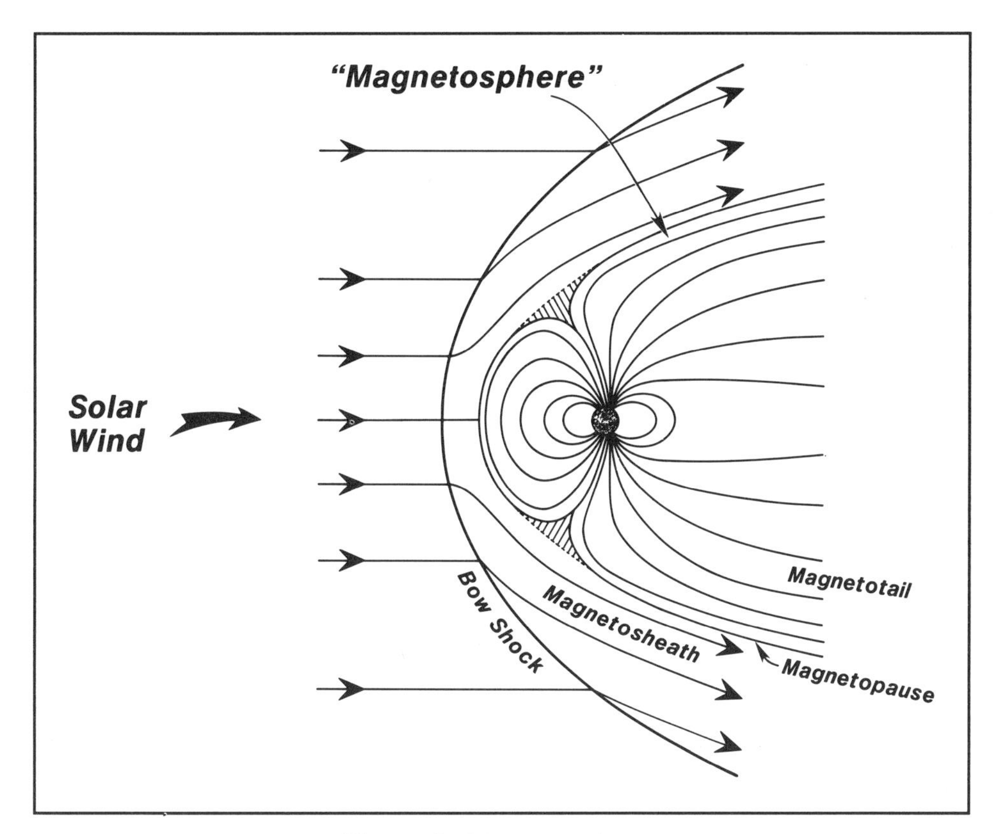

Figure 3. Magnetosphere.

distorted from its original configuration. At the "nose" of the magnetosphere it is compressed, while in the wake the field is highly stretched out in the direction away from the Sun. This latter region is called the "magnetotail." The full range of magnetosphere sizes is illustrated in Figure 4. The largest planetary magnetosphere is that of Jupiter. If it were visible, the magnetopause cross section of Jupiter as seen from Earth would be larger than the full Moon. The smallest known magnetosphere is that of Mercury, with a cross section that is little more than planet-sized. The magnetospheres all vary somewhat in size in response to changes that occur in the solar wind pressure. One of the most unusual magnetospheres is that of Neptune. Because of the large (47°) angle between the magnetic dipole axis and the rotation axis, the planetary rotation periodically places the magnetic poles at the nose of the magnetosphere. (This was not seen at Uranus in spite of the large difference between its rotation and dipole axes because the Uranian spin axis was nearly pointing at the Sun at the time its magnetosphere was probed.)

Table 2. Locations of Planetary Magnetopauses

Planet	*Typical Distance of Magnetopause (in planetary radii from center of planet)*	*Planet Radius (km)*
Mercury	1.5	2,440
Earth	10	6,370
Jupiter	60	71,400
Saturn	20	60,000
Uranus	25	25,600
Neptune	26	24,800

Another general feature of planetary magnetospheres seen in Figures 3 and 4 is a "bow" shock wave in the solar wind upstream of the magnetopause. The bow shock is analogous to the shock that forms in front of a supersonic aircraft. In the region between the bow shock and the magnetopause, the solar wind is diverted around the magnetopause. This region is called the "magnetosheath" because it covers the magnetopause. The

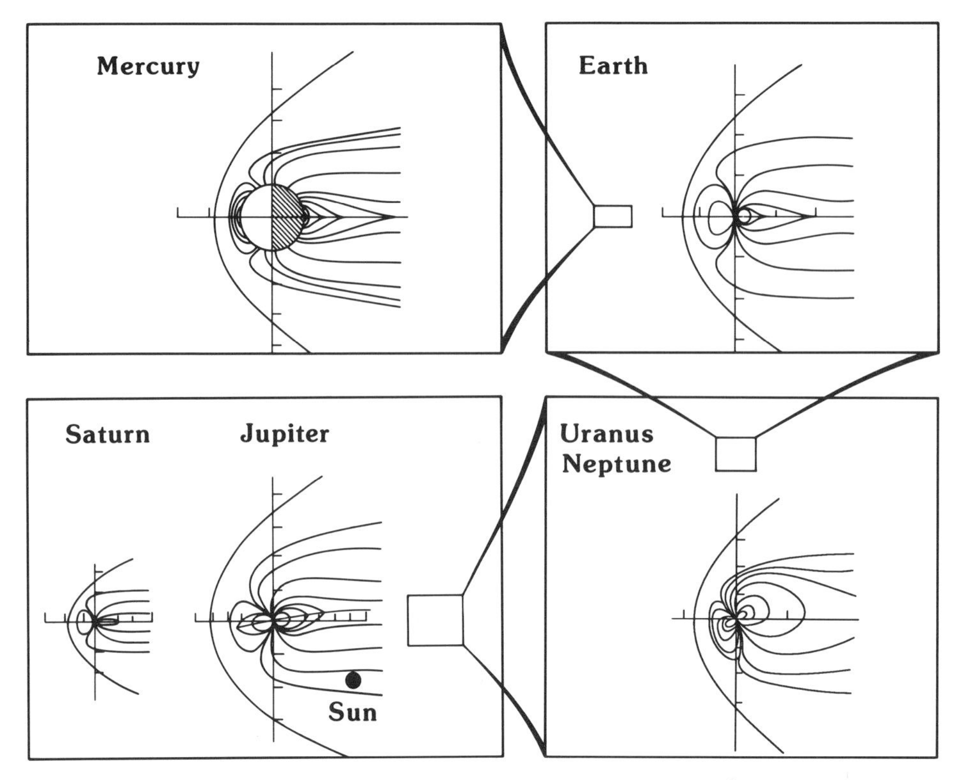

Figure 4. Relative sizes of planetary magnetospheres.

magnetosheath contains the magnetic field that is carried from the Sun in the solar wind.

The nonmagnetic planets Venus and Mars also have magnetosheaths and structures resembling magnetotails in their wakes. In these cases, however, the magnetopause is replaced by the boundary of the highly conducting ionosphere. The magnetotail is produced by the solar wind magnetic fields that drape over the ionosphere. This type of magnetotail is often called "induced," and the entire solar wind disturbance caused by the planet in the interplanetary field is called an "induced magnetosphere" to distinguish it from a magnetosphere of intrinsic planetary origin.

Other Consequences of Planetary Magnetic Fields

The presence of a planetary dipole magnetic field generally leads to trapping of energetic charged particles in radiation belts like Earth's Van Allen belts. These particles may come from an outside source, as cosmic rays, or from the internal source of the planet's ionosphere. Sometimes, as with Jupiter and Saturn, planetary satellites with orbits situated inside the magnetosphere provide an additional internal source of charged particles through ionization of their own atmospheres and "sputtering" of their atmospheres and surfaces by the trapped energetic particles. Sputtering is a process in which energetic particles collide with other atoms, allowing these atoms to gain energy and escape from the body. These internal sources of particles can modify the appearance of a magnetosphere through their pressure or momentum. (For example, the Io source inflates and flattens Jupiter's magnetosphere in its equatorial plane, as illustrated in Figure 4.) At the same time, planetary dipole fields shield the atmospheres and ionospheres of the magnetic planets from scavenging processes associated with direct interaction with the solar wind. Finally, physical processes associated with the interactions of the planetary ionospheres, the planetary fields, and the unsteady solar wind cause phenomena such as "magnetic substorms" and planetary auroras. These magnetic substorms are disturbances in the magnetic magnetospheres due to the sudden injection of energy from the planetary magnetotail into the night-time magnetosphere.

Bibliography

PARKER, E. N. "Magnetic Fields in the Cosmos." *Scientific American* (August 1983):44–54.

RUSSELL, C. T. "Planetary Magnetism." In *Geomagnetism*, Vol. 2, ed. J. Jacobs, New York, 1987.

JANET G. LUHMANN
CHRISTOPHER T. RUSSELL

PLANETARY MISSIONS

In a brief thirty years, we have explored the solar system from Mercury to Neptune—nearly fifty U.S. spacecraft and perhaps as many Soviet ones, with two from Europe and one from Japan as well. A total of about one hundred spacecraft have left Earth to explore neighboring worlds.

The earliest missions were flybys, which perform a reconnaissance of planets, moons, and some comets and asteroids. After reconnaissance the next step is exploration—systematic mapping and investigation of global characteristics of the other worlds. This is usually done with orbiters or with multiple flybys and perhaps multiple landers. It will also be done on planetary surface with mobile vehicles (rovers).

The third step is specific investigations or utilization of some aspect discovered at another world. We haven't reached this stage yet, although *Viking* on Mars and *Veneras* on Venus were highly capable landers beginning some specific investigations.

The Moon has been the most extensively visited place. In addition to dozens of robot vehicles, eighteen people have visited there and twelve have worked on its surface. In addition to eight planets and the Moon, we have visited close-up the satellites of Mars, Jupiter, Saturn, Uranus, and Neptune, as well as several asteroids and comets. We have orbited and landed only on three terrestrial planets—Moon, Mars, and Venus—although in 1995 an orbiter was en route to Jupiter.

In addition to the planets, dozens of new moons have been discovered or explored at Jupiter, Saturn, Uranus, and Neptune, together with rings and a magnetosphere at each. Only Pluto stands alone as an unvisited planet in our (known) solar system, although many more objects—asteroids, comets, and moons—await their first visit as well.

These worlds have been remarkable in the surprises they have offered scientists, revealing images of planets different than those suspected. Table 1 gives a comprehensive listing of all lunar and planetary missions, and the list below highlights the solar system as revealed by these missions. (It should be noted that any such listing is subjective and incomplete and will be subject to change as exploration continues.) Some highlight of planetary exploration include:

- The heavy, poisonous Venus atmosphere causing a runaway greenhouse effect in the atmosphere that makes the surface extremely hot (450°C) (discovered by *Venera* and *Pioneer*)
- The evidence of running water on Mars some billions of years ago—indicating an early Mars as warm and wet as early Earth (discovered by *Mariner* and *Viking*)
- The thousands of rings of Saturn, with strange dynamical behavior caused by little moons in and near these rings (discovered by *Voyager*)
- The mini solar system of Io, Europa, Ganymeade, and Callisto around Jupiter, images of which provide a stunningly simple description for the complex processes of planetary evolution (discovered by the *Voyagers*)
- The sulfur-covered surface of Io and continuously erupting volcanoes (discovered by *Voyager*)
- The enormous and complex magnetosphere of Jupiter and the diversity of the magnetospheres of Jupiter, Saturn, Uranus, and Neptune (discovered by *Pioneers* 10 and 11 and the *Voyagers*)
- The still unexplained big blue spots of Neptune's atmosphere (discovered by *Voyager*)
- The black-as-asphalt crusted snow and rock of comet Halley's nucleus (discovered by *VEGA* and *Giotto*)

The list of highlights should grow significantly in the next decade or so. The following missions are planned. In 1996 *Galileo*'s probe will enter Jupiter's atmosphere and the orbiter will make repeated close flybys of the *Galilean* satellite. In 2004 *Cassini* will orbit Saturn and its *Huyghens* probe will enter Titan's atmosphere and possibly land on its surface; in 1997 *Lunar A* is scheduled to orbit the Moon; in 1999 the *Near Earth Asteroid Rendezvous* will encounter asteroid Eros; in 2008 *Rosetta* will rendezvous with a comet; in 1997–1998 *Pathfinder* will land on Mars and *Mars 96* will send small sta-

Table 1. Notable Unmanned Lunar and Interplanetary Probes

Spacecraft	*Launch Date*	*Destination*	*Remarks*
Pioneer 3 (U.S.)	Dec. 6, 1958	Moon	Max. alt.: 107,246 km. Discovered outer Van Allen layer.
Lunik 1 (U.S.S.R.)	Sept. 12, 1959	Moon	Landed in area of Mare Serenitatis.
Mariner 2 (U.S.)	Aug. 27, 1962	Venus	Venus probe. Successful mid-course correction. Passed 34,831 km from Venus Dec. 14, 1962. Reported 800°F, surface temp. Contact lost Jan. 3, 1963 at 87 million km.
Ranger 7 (U.S.)	July 28, 1964	Moon	Impacted near Crater Guericke 68.5 h after launch. Sent 4,316 pictures during last 15 min of flight as close as 305 m above lunar surface.
Mariner 4 (U.S.)	Nov. 28, 1964	Mars	After mid-course correction, passed behind Mars July 14, 1965, taking 22 pictures from about 9,650 km.
Zond 3 (U.S.S.R.)	July 18, 1965	Moon	Sent close-ups of 4,827,000 sq km of Moon. Now in solar orbit.
Luna 9 (U.S.S.R.)	Jan. 31, 1966	Moon	1,560-kg instrument capsule of 100 kg soft-landed Feb. 3, 1966. Sent back about 30 pictures.
Surveyor 1 (U.S.)	May 30, 1966	Moon	Landed June 2, 1966. Sent almost 10,400 pictures, a number after surviving the 14-day lunar night.
Lunar Orbiter 1 (U.S.)	Aug. 10, 1966	Moon	Orbited Moon Aug. 14. 21 pictures sent.
Surveyor 3 (U.S.)	April 17, 1967	Moon	Soft-landed 65 h after launch on Oceanus Procellarum. Scooped and tested lunar soil.
Venera 4 (U.S.S.R.)	June 12, 1967	Venus	Arrived Oct. 17. Instrument capsule sent temperature and chemical data.
Surveyor 5 (U.S.)	Sept. 8, 1967	Moon	Landed near lunar equator Sept. 10. Radiological analysis of lunar soil. Mechanical claw for digging soil.
Surveyor 7 (U.S.)	Jan. 6, 1968	Moon	Landed near Crater Tycho Jan. 10. Soil analysis. Sent 3,343 pictures.
Pioneer 9 (U.S.)	Nov. 8, 1968	Sun orbit	Achieved orbit. Six experiments returned solar radiation data.
Apollo 8 (U.S.)	Dec. 21–27, 1968	Moon	First manned spacecraft in circumlunar orbit. TV transmission from this orbit.
Venera 5 (U.S.S.R.)	Jan. 5, 1969	Venus	Landed May 16, 1969. Returned atmospheric data.
Mariner 6 (U.S.)	Feb. 24, 1969	Mars	Came within 3,218 km of Mars July 31, 1969. Sent back data and TV pictures.
Apollo 10 (U.S.)	May 18–26, 1969	Moon	Manned descent to within 15 km of moon's surface.
Apollo 11 (U.S.)	July 16–24, 1969	Moon	First manned landing and EVA on Moon; soil and rock samples collected; experiments left on lunar surface.
Apollo 12 (U.S.)	Nov. 14–24, 1969	Moon	Manned lunar landing mission; investigated *Surveyor 3* spacecraft; collected lunar samples. EVA time 15 to 30 min.
Apollo 13 (U.S.)	April 11–17, 1970	Moon	Third manned lunar landing attempt; aborted in transit due to pressure loss in liquid oxygen in service module and failure of fuel cells. Flew around the moon.
Venera 7 (U.S.S.R.)	Aug. 17, 1970	Venus	Reached Venus Dec. 15, 1970. Sent data, apparently from surface, for 58 min.
Luna 16 (U.S.S.R.)	Sept. 12, 1970	Moon	Soft-landed Sept. 20, scooped up rock, returned to Earth Sept. 24.

Table 1. Continued

Spacecraft	*Launch Date*	*Destination*	*Remarks*
Luna 17 (U.S.S.R.)	Nov. 10, 1970	Moon	Soft-landed on Sea of Rains Nov. 17. *Lunokhod 1*, self-propelled vehicle, used for first time. Sent TV photos, made soil analysis, etc.
Apollo 14 (U.S.)	Jan. 31–Feb. 9, 1971	Moon	Third manned lunar landing; returned largest amount of lunar material.
Mariner 9 (U.S.)	May 30, 1971	Mars	First craft to orbit Mars, Nov. 13. 7,300 pictures, 1st closeups of Mars' moon. Transmission ended Oct. 27, 1972.
Luna 19 (U.S.S.R.)	Sept. 28, 1971	Moon	Orbited Moon, making measurements and taking photos. Soft-landed Feb. 21 in Sea of Fertility. Returned Feb. 25 with rock samples.
Pioneer 10 (U.S.)	March 3, 1972	Jupiter	998-million-km flight path through asteroid belt passed Jupiter Dec. 3, 1973, to give humans first closeup of planet. In 1986, it became first man-made object to escape solar system.
Apollo 15 (U.S.)	July 26–Aug. 7, 1972	Moon	Fourth manned lunar landing; first use of lunar rover; first live pictures of LM lift-off from Moon; exploration time 18 h.
Apollo 16 (U.S.)	April 16–27, 1972	Moon	Fifth manned lunar landing; second use of lunar rover. Total exploration time on the Moon 20 h, 14 min, setting new record.
Apollo 17 (U.S.)	Dec. 7–19, 1972	Moon	Sixth and last manned lunar landing; third to carry lunar rover. H. Schmitt—first scientist on another world.
Luna 21 (U.S.S.R.)	Jan. 8, 1973	Moon	Soft-landed Jan. 16. *Lunokhod 2* (moon-car) scooped up soil samples, returned them to Earth Jan. 27.
Mars 4 (U.S.S.R.)	July 21, 1973	Mars	Arrived Feb. 1974, briefly sending back photos.
Mariner 10 (U.S.)	Nov. 3, 1973	Venus, Mercury	Passed Venus Feb. 5, 1974. Arrived Mercury March 29, 1974, for humans' first closeup look at planet. First time gravity of one planet (Venus) used to whip spacecraft toward another (Mercury).
Venera 9 (U.S.S.R.)	June 8, 1975	Venus	Soft-landed Oct. 25, 1976. Photographed surface of planet.
Viking 1 (U.S.)	Aug. 20, 1975	Mars	Carrying life-detection labs. Landed July 20, 1976, for detailed scientific research, including pictures. Designed to work for only 90 days, it operated for almost 6 1/2 years before it went silent in Nov. 1982.
Viking 2 (U.S.)	Sept. 9, 1975	Mars	Like *Viking 1*. Landed Sept. 3, 1976. Functioned 3 1/2 years.
Luna 24 (U.S.S.R.)	Aug. 9, 1976	Moon	Soft-landed Aug. 18, 1976. Returned soil samples Aug. 22, 1976.
Voyager 1 (U.S.)	Sept. 5, 1977	Jupiter, Saturn, Uranus	Flyby mission. Reached Jupiter in March 1979; passed Saturn Nov. 1980; passed Uranus 1986.
Voyager 2 (U.S.)	Sept. 20, 1977	Jupiter, Saturn, Uranus	Like *Voyager 1*. Encountered Jupiter in July 1979; flew by Saturn Aug. 1981; passed Uranus January 1986; to pass Neptune 1989.
Pioneer Venus 1 (U.S.)	May 20, 1978	Venus	Arrived Dec. 4 and orbited Venus, photographing surface and atmosphere.
Pioneer Venus 2 (U.S.)	Aug. 8, 1978	Venus	Four-part multi-probe, landed Dec. 9.
Venera 11 (U.S.S.R.)	Sept. 9, 1978	Venus	Soft-landed Dec. 25, 1978. Transmitted data for 95 min.

Table 1. Continued

Spacecraft	*Launch Date*	*Destination*	*Remarks*
Venera 13 (U.S.S.R.)	Oct. 30, 1981	Venus	Landed March 1, 1982. Took first X-ray fluorescence analysis of the planet's surface. Transmitted data 2 h, 7 min.
VEGA 1 (U.S.S.R.)	Deployed on Venus, June 10, 1985	Encounter with Halley's Comet	In flyby over Venus while enroute to encounter with Halley's Comet, *VEGA 1* and *2* dropped scientific capsules onto Venus to study atmosphere and surface material. Encountered Halley's Comet on March 6 and March 9, 1986. Took TV pictures and studied comet's dust particles.
VEGA 2	Deployed on Venus, June 14, 1985		
Suisei (Japan)	Encountered Halley's Comet March 8, 1986	Halley's Comet	Spacecraft made flyby of comet and studied atmosphere with ultraviolet camera. Observed rotation nucleus.
Sakigake (Japan)	Encountered Halley's Comet March 10, 1986	Halley's Comet	Spacecraft made flyby to study solar wind and magnetic fields. Detected plasma waves.
Giotto (ESA)	Encountered Halley's Comet March 13, 1986	Halley's Comet	European Space Agency spacecraft made closest approach to comet. Studied atmosphere and magnetic fields. Sent back best pictures of nucleus.
Phobos 1 (U.S.S.R.)	July 7, 1988	Mars	Failed early on interplanetary trajectory due to command error.
Phobos 2 (U.S.S.R.)	July 20, 1988	Mars	Succeeded to Mars orbit to transmit scientific data; failed before closeup rendezvous with Phobos and final orbit of Mars was achieved when spacecraft lost attitude control.
Mars Observer (U.S.)	Sept. 25, 1992	Mars	Failed just prior to orbit insertion at Mars with cause unknown.

tions and penetrators to Mars, and an orbiter about it; in 1999–2000 *Mars 98* will place a rover and balloon on the Mars surface and *Mars Surveyor* will send an orbiter and lander there. *Planet B* will also orbit Mars in those years. *Galileo* is now on its interplanetary trajectory, while the other missions mentioned are in development.

Clementine, Galileo, Pathfinder, Cassini, and *Surveyor* are U.S. missions. *Huyghens* and *Rosetta* are European (from the European Space Agency), the *Mars Balloon* is French, the *Mars 96* and *98* are Russian, and *Planet B* and *Lunar A* are Japanese. This diversity emphasizes the international nature of and widespread interest in planetary exploration by planet Earth's inhabitants. Most of these missions are being planned cooperatively with participants from many nations.

Bibliography

BEATTY, K., and A. CHANKIN. *The New Solar System.* New York, 1990.

LANG, K., and C. WHITNEY. *Wanderers in Space.* New York, 1991.

SAGAN, C. *A Pale Blue Dot.* New York, 1994.

LOUIS FRIEDMAN

PLANETARY RINGS

Until 1977, rings were considered to be rare occurrences around planets. The only planet known to have rings was Saturn, whose rings were first seen by Galileo in 1610 as "bulges" on the side of the planet and later identified as rings by Huygens (1659). But on 10 March 1977, Saturn lost its distinction as the only planet with rings. On that night astronomers were preparing for a rare event, the occultation of a star by the planet Uranus. Stellar occultations are of interest to astronomers because they provide important information about the atmospheric structure of the planet that occults the star.

On the night of 10 March 1977, a group of astronomers led by James Elliot of the Massachusetts Institute of Technology were aboard the Kuiper Airborne Observatory (KAO), ready to observe the occultation from southwest of Perth, Australia. They began observations of Uranus 41 minutes before the predicted occultation time (the timing of stellar occultations is often quite uncertain). They could see both Uranus and the star through the 0.9 meter (m) telescope aboard the KAO. But before the star could be occulted by Uranus, the group detected a drop in the star's brightness. At first they thought that perhaps thin clouds were getting in the way, but the sky proved clear. By the end of the evening, the astronomers had observed a number of drops in the star's brightness and soon realized that the cause of the observations was not clouds but rather the existence of a series of thin rings around Uranus. Soon it was determined that Uranus was surrounded by nine thin rings. Saturn was no longer the only ringed planet in the solar system!

A ring around Jupiter was subsequently discovered by the *Voyager 1* spacecraft in 1979 (*see* JUPITER AND SATURN). Now with three of the four Jovian planets known to possess rings, astronomers began detailed observations of stellar occultations by Neptune to see if the outermost Jovian planet also was surrounded. Observations in 1981 and 1983 proved inconclusive, with the existence of small moons or "ring arcs" proposed to explain the enigmatic data. The *Voyager 2* flyby of Neptune in 1989 finally resolved the question by showing that Neptune is surrounded by three rings. Thus, all four Jovian planets are now known to possess ring systems, although that of Saturn remains the only one easily visible from Earth.

Early observers of Saturn's rings believed the rings to be solid. Even such prominent astronomers as Sir William Herschel (*see* HERSCHEL FAMILY) believed the rings to be a solid halo that had solidified in place out of an early ring of dust and gas. By 1675, Cassini had detected a dark division between the rings (now called the Cassini Division), which most astronomers believed was a gap of empty space between the two solid rings (designated A and B). Not until the discovery of a third faint ring, the C ring, in 1850, did astronomers begin to realize that the rings were not solid but rather composed of a myriad of particles. Based on radar observations, astronomers now believe the rings of Saturn are composed of billions of water-ice particles ranging in size from dust grains to boulders a few tens of meters in size. The faint rings of the other three Jovian planets are composed of darker, smaller dust-sized particles.

After the discovery of the C ring in 1850, the number of bright rings and dark divisions reportedly increased dramatically. In 1867, Kirkwood proposed that the structure of Saturn's rings could be explained by resonances with the planet's moons. This was well illustrated by the determination that a particle orbiting within the Cassini Division would have an orbital period equal to half that of the satellite Mimas, about a third that of Enceladus, a quarter that of Tethys, and a sixth that of Dione. Such an orbit would be unstable due to the gravitational tugs exerted by the moons, especially nearby Mimas. Kirkwood thus proposed that the Cassini Division was a permanently empty gap. Not until *Voyager 1* and *2* flew past Saturn in 1980–1981 did high-resolution pictures show that the Cassini Division does contain some material, just lower concentrations than the brighter rings.

Voyager revealed that Saturn's rings were not comprised of just the well-known major rings (A, B, C, D, E, and F) and divisions, but rather each ring and division were composed of thousands of individual ringlets (Figure 1). As camera resolution increased, so did the number of ringlets. The brighter rings contained more material and ringlets than did the fainter rings and the divisions, but very few areas were revealed to be completely free of material.

Another surprise revealed by the *Voyager* spacecraft was the presence of dark wedge-shaped features seen against Saturn's bright B ring. These wedges, called spokes, were found to rotate with the planet's magnetic field and were composed of small dust-sized grains of material. These small particles, produced by collision within the rings, acquire an electric charge and are levitated above the rings by Saturn's magnetic field. Because of their small size, they scatter little light in the forward direction, thus appearing dark when seen against the bright B ring.

What keeps the rings in place? This has been a long-standing question among planetary scientists studying ring dynamics. The discovery of narrow rings around Jupiter, Uranus, and Neptune, to-

Figure 1. This *Voyager 1* mosaic of Saturn's rings was taken at a distance of 8 million km (5 million miles) from Saturn in November 1980. The image shows approximately 95 individual ringlets comprising the rings.

gether with the discovery that Saturn's rings are comprised of narrow ringlets, complicated the original idea that resonances with the satellites were keeping the rings in place. At first scientists proposed that the narrow ringlets were maintained by small "shepherding" moons on either side of the ring. This idea received support when Saturn's tenuous F ring was discovered to be maintained by two small satellites, one inside the ring and the other outside. However, searches for shepherding moons around other rings have been unsuccessful—either the moons are too small to be detected by current techniques or, more likely, the rings are maintained by some other mechanism such as gravitational density waves (which are believed to also maintain the arms of spiral galaxies).

Many questions remain about planetary rings. Scientists are still uncertain how rings even form. Do they form along with their parent planet out of the original cloud of gas and dust? Do rings originally start out as moons that are torn apart by gravitational forces when they come too close to their parent planet? Or are they remnants of moons that have been destroyed by collisions with other large particles? All three of these theories have their pros and cons.

Regardless of how rings form, they should be short-lived phenomena because of the influence of gravity on larger particles and drag effects on smaller particles. Both effects cause the particles comprising the rings to be pulled into the atmospheres of the giant planets. So does this mean we are living at a special time when rings just happen to be abundant around the outer planets? Scientists do not like theories that suggest we live in a special time, so they continue to look at mechanisms that can maintain the rings we see, such as satellite resonances, shepherding satellites, and density waves. Now that we know that Saturn is not alone in possessing a ring system, scientists are finding that the abundance of riches leads to even more unanswered questions about these enigmatic but beautiful features of our solar system.

Bibliography

ELLIOT, J., and R. KERR. *Rings: Discoveries from Galileo to Voyager.* Cambridge, MA, 1984.

GREENBERG, R., and A. BRAHIC. *Planetary Rings.* Tucson, AZ, 1984.

SMITH, B. A., et al. "Voyager 2 at Neptune: Imaging Science Results." *Science* 246 (1989):1422–1449.

NADINE G. BARLOW

PLANETARY SYSTEMS, OTHER

The Polish astronomer Nicolaus Copernicus changed the way most of humanity viewed the universe by suggesting that Earth was not the center of the universe and that it revolved around the Sun as did the other planets. This Copernican revolution in thought has progressed to the point that we now know that the Sun is but one of roughly a billion stars in the Milky Way galaxy, and it resides in the outer regions of the Milky Way. Modern astronomy has also shown that our galaxy is just one of billions of other galaxies in the observable universe. One celestial grouping has so far remained unique, at least as far as observations have been able to show, namely, our planetary system. While our models for how the solar system formed suggest that other planetary systems should occur frequently in nature, there is at present no confirmed detection of another planetary system. Recognizing the importance of evidence for or against the presence of other planetary systems, a coordinated effort has begun to search for and characterize other planetary systems. This entry will explore the possible nature of other planetary systems and examine various observational methods that can be used to search for such systems.

It is now generally believed that the Sun and its retinue of planetary companions formed together some 4.6 billion years (Ga) ago, and that the solar system is the likely outcome of the gravitational collapse of slowly rotating, dense regions in giant molecular clouds (e.g., Black and Matthews, 1984; Levy and Lunine, 1993; Wetherill, 1989). The unexceptional nature of this formation process suggests that the conditions necessary for the formation of planetary systems may well occur every time a star is born. If this is indeed the case, planetary systems should exist in association with many, if not all, single stars.

The statement above refers to single stars, stars like the Sun that have no stellar companion or companions. This may be important because nature prefers to make stars in multiple, mainly binary, systems. While the exact statistics are not known because of the possible incompleteness of the survey, data available as of the early 1990s suggest that at most, only one out of every three stars is single (Mayor et al., 1992, McAlister et al., 1987). Stated somewhat differently, at least two out of every three stars is a member of a multiple star system.

Does the prevalence of binary star systems mean that we should expect to find planets revolving around no more than a third of the stars? We do not know. What we do know is that studies of orbital stability (Black, 1982, Graziani and Black, 1981) indicate that if planets can form in binary systems, there are several possibilities for stable orbits. One such configuration would have the planets revolving at great distance around both of the stars, and the other two possible configurations would have planets revolving around the individual stars just as the planets in the solar system revolve around the Sun. In the former case, the planets would have to be at a distance that is greater than roughly five times the separation between the stars, and in the latter cases the planets must be at distances from their parent star that are no greater than about one-fifth of the stellar separation. Given these facts, it seems clear that we should search binary stars as well as single stars for the presence of planetary companions.

The paradigm referred to above suggests that a planetary system forms from a rotating disk of gas and dust that is formed in association with the birth of a star. The formation from a disk accounts for the fact that our planetary system is nearly coplanar and that all of the planets move in the same direction as they revolve about the Sun. A more subtle aspect of the paradigm is that it implies that we find giant planets (such as Jupiter and Saturn) where they are because the temperature in the disk was cold enough for icy material to be present in condensed form, giving rise to a relatively rapid buildup of a core of material that could accrete the vast amount of gas that characterizes these planets.

All of these aspects of the paradigm should, in some general manner, hold for any planetary system. That is, we would expect that almost all planetary systems would be roughly similar to our own; they should be nearly coplanar, and they should have "rocky" planets, broadly similar to Mars or Earth, and be near the star, and giant volatile-rich planets at greater distances from the star. While simple in detail, this top-level prediction of the paradigm is important because it is amenable to observational testing.

Efforts have been made to search for other planetary systems. These have relied mainly on what are referred to as indirect detection methods, that is, methods of detection where one infers the presence of planetary companions to a star by virtue of some observable effect that they have on that star. The most studied effect is that of the gravitational perturbation on the motion of a star arising from an orbiting planet or planets. The observational techniques used are astrometry (to detect the projection of stellar motion in the plane of the sky) and spectroscopy (to detect the projection of stellar motion along the line of sight to the observer). Each of these techniques has a rich history.

Perhaps the best-known astrometric study was that by Peter van de Kamp, who studied a number of stars over several decades. The most famous of his studies dealt with a star called Barnard's Star, a relatively close, rapidly moving, small star, about one-seventh the mass of the Sun. Van de Kamp concluded in 1982 that Barnard's Star had two roughly Jupiter-mass companions. However, more accurate studies by George Gatewood at the Allegheny Observatory and, independently, by Robert Harrington at the U.S. Naval Observatory have shown that van de Kamp's data are in error and that the putative planets are not there. This does not yet rule out the possibility that there are small planets in orbit around Barnard's Star; such a conclusion can only come from much more accurate observations than are currently available.

Spectroscopic searches have proved to be quite interesting. The star HD114762 was observed (Latham et al., 1989) to undergo periodic motion due to the presence of an unseen companion. The orbital motion had a period of roughly ninety days, which for this solarlike star meant that the companion was in an orbit comparable to that of the planet Mercury. Studies by W. D. Cochran et al. (1991) confirm the discovery but suggest that the companion is either a brown dwarf or perhaps a low-mass (and therefore quite dim) star; it is not a planet. ("Brown dwarf" is a term used to describe objects that form by the same process as a star but that are of too low a mass to ignite nuclear burning of hydrogen in their interior. This means that they are less massive than 0.08 solar masses, roughly 80 times the mass of Jupiter. A lower limit to the mass of a brown dwarf has been suggested to be about 0.02 solar masses. The major distinction between a brown dwarf and a giant planet such as Jupiter is their mode of formation—the former arising from a large-scale gravitational collapse and the latter arising principally from the process of accretion of small objects into progressively larger objects.) Other spectroscopic studies have turned up hints of perturbations that could be due to planetary companions, but unlike the case for HD114762, none of these have been confirmed.

A major surprise was announced in 1992, when astronomers studying millisecond pulsars discovered a periodic effect in the time of arrival of pulses from these exotic astronomical objects.

Upon careful analysis, the discoverers (Wolszczan and Frail, 1992) concluded that the periodic effect was due to the presence of two companions, each having a lower mass limit of around 3 Earth masses. The orbital periods for these companions were determined to be approximately ninety-eight days and sixty-six days. Combining these orbital periods with an assumed mass for the pulsar of a little under 1.5 solar masses, the orbits must be roughly 0.36 and 0.47 times the size of Earth's orbit around the Sun. This discovery was very exciting, but it remains to be shown that the periodic effect is really due to companions. It has been suggested (Malhotra et al., 1992) that if companions are present, their mutual gravitational interactions should affect their orbits sufficiently to be observed by the late 1990s, thereby providing some verification of the interpretation of this exciting discovery.

If confirmed, is this system a planetary system? The masses of the companions would be consistent with those of planets, but they were formed under quite different circumstances than were the planets in the solar system. Pulsars represent a phase near the end of a star's life, and these companions must have been formed at that stage in the star's evolution. In contrast, the solar system was formed at the beginning of the Sun's life. So, while confirmation of the existence of these companions will provide us with valuable insight, there are many reasons to believe that these pulsar systems are "genetically" distinct from planetary systems like our own, in much the same way that apes and human beings are distinct in several key details.

The future will see the use of an even more diverse set of instruments, some in space, to search for other planetary systems. These systems are intrinsically difficult to detect, but the new generation of instruments should be able to detect objects as small as Uranus or Neptune revolving around virtually any of the more than one thousand stars within 30 light-years of the Sun. A thorough search program will take between one and two decades and is advocated by the National Aeronautics and Space Administration (NASA). The foundation of NASA's program is outlined in a report entitled "TOPS: Toward Other Planetary Systems." At the end of that time we will be able to say with some confidence whether the revolution in thought started by Copernicus some five centuries ago is complete. No matter what the outcome, the search will have a profound and irreversible effect on the perception of humanity's place in the universe.

Bibliography

BLACK, D. C. "A Simple Criterion for Determining the Dynamical Stability of Three-Body Systems." *Astronomical Journal* 87 (1982):1333–1337.

BLACK, D. C., and M. S. MATTHEWS, eds. *Protostars and Planets II.* Tucson, AZ, 1984.

COCHRAN, W. D., A. P. HATZES, and T. J. HANCOCK. "Constraints on the Companion Object to HD114762." *Astrophysical Journal Letters* 380 (1991):L35.

GRAZIANI, F., and D. C. BLACK. "Orbital Stability Constraints on the Nature of Planetary Systems." *Astrophysical Journal* 251 (1981):337–341.

LATHAM, D. W., T. MAZEH, R. P. STEFANIK, M. MAYOR, and G. BURKI. "The Unseen Companion of HD114762: A Probable Brown Dwarf." *Nature* 339 (1989):38.

LEVY, E. H., and J. I. LUNINE, eds. *Protostars and Planets III.* Tucson, AZ, 1993.

McALISTER, H. A., W. I. HARTKOPF, D. J. HUTTER, M. M. SHARA, and O. G. FRANZ. "ICCD Speckle Observations of Binary Stars." *Astronomical Journal* 92 (1987):183.

MAYOR, M., A. DUQUENNOY, J.-L. HALBWACHS, and J.-C. MERMILLIOD. "CORAVEL Surveys to Study Binaries of Different Masses and Ages." In *Proceedings of IAU Colloquium 135 on Complementary Approaches to Double and Multiple Star Research,* eds. H. A. McAlister and W. I. Hartkopf. ASP Conference Series, vol. 32, 1992.

WETHERILL, G. W. "The Formation of the Solar System: Consensus, Alternatives, and Missing Factors." In *The Formation and Evolution of Planetary Systems,* eds. H. A. Weaver and L. Danly. Cambridge, Eng., 1989.

DAVID C. BLACK

PLANETOLOGY, COMPARATIVE

In the fictional voyages of the television series *Star Trek,* planets similar to Earth are referred to as class M planets. Classification is an ultimate goal of all science and it is exciting to contemplate a future where planets, with all of their complex interrelated systems of geoscience, atmosphere, and biosphere, may be organized into an elegant classification system. That future classification system, if it is ever invented, will be the end result of the first attempts at comparative planetology that began in the 1960s with the beginning of planetary exploration by spacecraft.

The science of geology began only two hundred years ago with observation and interpretation of local rock layers in England (*see* GEOLOGY, HISTORY OF). Understanding of the underlying geological processes and geologic evolution of local rocks in many different areas was acquired slowly and each area was pieced together, like the squares of a quilt, until we began to think about geology and geological processes in a more regional sense. Since the early 1970s, with the acceptance of the paradigm of plate tectonics and the synoptic view provided by orbiting spacecraft cameras and other instruments, we have expanded our understanding of geology to planetary-wide processes, and we have begun to think about our home planet as an interrelated geological system.

Nevertheless, fundamental characteristics cannot be defined on the basis of a single example. By way of analogy, you might consider cats. If you had seen only one cat in your entire life, you might assume that all cats had pointed ears and whiskers and purred—and you would be right. But you might also assume that all cats were very small with white fur on their bodies and dark fur on their ears, paws, and long slender tail—and you would be wrong. You might suspect that certain characteristics would occur in all cats and that other characteristics might vary, but you could not be certain. This is the situation that geologists have faced with respect to Earth. For most of the two hundred years of the development of geology as a science, we have had only one datum—Earth. We do not know if Earth is unique as a planet; we do not know the basic similarities or differences, or general trends in geological evolution, of planets; we do not know how planets work geologically—but we are beginning to find out. By looking at the other planets of our solar system and, ultimately, at planets in other solar systems, we can bring that knowledge home to render a better understanding of our own planet. This is the goal of comparative planetology.

Basic Building Blocks of Planets

Within just a few decades of space exploration, the planets of our solar system have become geological, rather than astronomical, entities. Our understanding of the geology of these planets, however, begins with basic astronomical observations and assumptions concerning the solar nebula out of which the planets of our solar system "accreted," or formed (*see* SOLAR SYSTEM). Generally speaking, the bulk chemistry and major-element mineralogy of each planet theoretically should reflect the composition of solid particles formed at the highest temperature experienced at that location in the solar nebula. This means that, while all of the planets of our solar system are siblings, there is some diversity in their basic elemental composition or in their relative percentages of elements, and this diversity occurs in a graduated sequence with distance from the center of the solar nebula. The main building materials of the solar system can be grouped into rock (formed primarily by iron [Fe], sulfur [S], magnesium [Mg], silica [Si], and oxygen [O]), gases (such as hydrogen [H] and helium [He]), and ices (water, ammonia, and methane). In general terms, rock-rich planets (Earth-like, solid-surface planets) formed close to the Sun, and gas- and ice-dominated (volatile-rich) planets formed farther from the Sun. The change in planetary composition with distance from the center appears to have been reproduced within the mini-system of Jupiter and its satellites (*see* GALILEAN SATELLITES). Compositional mixing and changes over time in the amount of volatiles accreted by the planets clearly occurred, so the entire story is not simple. And some of the natural satellites (moons) in our system may not have formed in place, but instead may represent secondary planets formed from the material of their primary planets as current theories suggest for Earth's moon, or they may represent objects formed elsewhere and captured by gravitational attraction, like the moons of Mars. However they were formed, the planets in our solar system began their geological lives with present elemental abundances. This bulk composition is directly related to planetary density and mass. Therefore, if we can calculate a planet's density and mass, we then can make assumptions concerning the planet's composition and interior structure, even if we have never directly observed the geology of the planet. A first attempt at a general classification of planets, therefore, might incorporate the following data: (1) distance from the Sun; (2) density and mass of the planet; and (3) implied bulk composition.

But this is only the beginning—the planets have also been changed by dynamic processes. Very early in solar system history, a period of heavy bombardment of debris fragments left over from initial planet formation formed major impact basins and craters on the surface of the solid-surface (also known as "rocky" or "terrestrial") planets and may have affected interior dynamic processes within these planets (*see* IMPACT CRATERING AND BOMBARDMENT RECORD). Planetary atmospheres, either ephemeral or permanent, appear to be derived early in a planet's evolution due to outgassing of volatile elements, and possibly due to contributions of volatiles from the impact of comets and other small bodies. However, initial atmospheric composition and density can evolve and change with time, as has been the case with Earth (*see* ATMOSPHERES, PLANETARY). The presence of a permanent atmosphere is related to the mass, and therefore gravity, of the planet and the distance of the planet from the Sun. Also evolving very early, the interior structure of each planet appears to have been derived and organized through one or more of the following processes: degassing of volatile elements; chemical differentiation due to internal melting; dynamic processes such as core formation; and compositional stratification by a variety of mechanisms (*see* EARTH-MOON SYSTEM, EARLIEST HISTORY OF). Our proposed general classification scheme for all planets might now be expanded to (1) distance from the Sun; (2) density and mass of the planet; (3) implied bulk composition; (4) internal structure and presence/type of compositional stratification; and (5) presence (density/composition) of atmosphere.

Planetary Geological Processes

Solid-surface planets are of particular interest in comparative planetology because we seek comparisons with the geology of our own planet. Geology is a process-oriented science; planetary surfaces are acted upon by internal and external dynamic processes so that they evolve geologically with time. Earth is a water-dominated planet, with 70 percent of the surface covered by water and with a strongly interactive hydrologic cycle. Geologically speaking, Earth is a "water-damaged" planet, since water is a dominant agent of erosion and deposition on the surface of our world. The entire subdiscipline of sedimentary geology has been derived in order to study the processes and rock types produced primarily by water and secondarily by wind. These processes are extremely important in understanding the geological evolution of Earth's surface, but is that the case for all planets?

In general, we have discovered that there are four major types of geological processes that have acted upon all of the solid-surface planets we have observed so far. These are (1) impact cratering; (2) volcanism; (3) tectonism; and (4) gradation. Impact cratering and volcanism have altered *all* of the solid-surface planets in our solar system. Tectonism and gradation vary in type and importance from planet to planet.

Impact Cratering

Prior to our geological observations of moons and other planets, impact cratering was not considered important to geology, but our first look at our neighboring worlds demonstrated its importance as a continuing geological planetary process on all bodies, including Earth (*see* IMPACT CRATERING). The second planet to be explored geologically was, of course, the Moon. Geologically, Earth's moon represents a snapshot of the surface of the early Earth hanging in our night sky, a missing portion of Earth's geological history that has been "erased" and "overprinted" by the continuation of the dynamic geologic processes which alter Earth's surface. If not for the continuation of these processes, Earth would look like the Moon, with a record of 4.6 billion years (Ga) of impact cratering written on its surface. Study of the impact craters of the Moon directly led to two major breakthroughs in our understanding of planets. First, it led to increased awareness of the presence and importance of impact craters on Earth. Although impact cratering is a short-time process, unusual in geology, we now understand that the formation of an impact crater can significantly affect the geology, atmosphere, and biosphere of our world. It has been theorized that individual impact craters on Earth were responsible for creating some ore deposits or for causing major life-form extinctions (*see* EXTINC-

TIONS). Second, studies of impact craters led to a fundamental technique in planetary geology that is known as "crater counting."

The basic idea of crater counting is that the longer a surface exists on a planet, the more impact craters it will accumulate (*see* IMPACT CRATERING AND THE BOMBARDMENT RECORD). We are not dealing with the age of the creation of rocks (*see* RESEARCH IN THE EARTH SCIENCES), but rather with the age of the actual surface. For example, if a road with many potholes is resurfaced with asphalt, the age of the original road remains the same but the surface of the road is younger. Planetary surfaces can similarly be resurfaced by a number of geological processes. By actually counting the number and measuring the diameter of impact craters on a specific surface, it is possible to gain information on the relative age (older or younger) of this surface compared to other surfaces. On planets where we have no direct information about the age or type of rocks, we can make inferences about the average age or range of ages of the surface and therefore the timing, abundance, and even type of geological activity of the planet, by using the technique of crater counting. If we knew nothing else about Earth's geology, the low number of impact craters on its surface would tell us that we were dealing with a geologically active planet. Venus has a similarly low number of impact craters on its surface, an indication that it is also a geologically active planet, although volcanism and tectonism appear to be more important as resurfacing agents than erosion and deposition.

Volcanism

The recognition that most volcanism on Earth is related to global plate tectonics, and is the process by which new crust is continually formed at the mid-ocean ridges, occurred at about the same time as the discovery of the first volcanic edifice on another planet (*see* PLATE TECTONICS). That volcano, Olympus Mons on Mars, is now considered to be the largest volcano in the solar system (it would cover the entire state of New Mexico), but it appears to have been formed in exactly the same way as much smaller, but otherwise identical, shield-type volcanoes on Earth. Volcanic flows with Earth-like basaltic compositions have filled the large impact basins and craters of the Moon (producing the bright "man in the moon" features visible from Earth) and volcanoes or volcanic flows have been discovered or inferred on Mercury, Mars, and Venus. The recognition of volcanism as an important planetary process was an outgrowth of our analysis of other planets and our increasingly global view of Earth (*see* VOLCANISM, PLANETARY). The presence of volcanoes or volcanic flows on a planet tells us about the thermal evolution and heat flow of the planet; the type, number, and location of volcanoes on a planetary surface tell us the way in which that planet loses its internal heat and other related geologic information. For example, the long, linear chains of explosive volcanoes that surround the Pacific Ocean are a direct result of global plate movements, and the type and composition of this volcanism are primarily due to the presence of water on Earth. Thus far, linear chains of explosive volcanoes appear to be unique to Earth, but basaltic effusive (lava flow) volcanism is a fundamental process by which planets lose heat and create new crust, and volcanoes can be used to infer information about the interior crustal thickness, and resurfacing rate, of a planet.

Volcanism produced by processes similar to volcanism on Earth, but with materials or chemistry other than silicate rock, has been identified on some of the outer planet satellites. Continuous volcanic eruptions that are believed to erupt sulfur and silicates on the surface of Jupiter's moon, Io, appear to be due to strain and frictional heating of the interior of Io by "tidal bulging" due to its proximity to Jupiter and the other near-Jupiter satellites. Ice "volcanism" has been identified on Oberon and Titania, two moons of Uranus. Internal heat apparently melted ice that erupted and flowed onto the surface and then refroze, erasing or filling impact craters and acting as a planetary resurfacing process. Triton, the largest satellite of Neptune, shows evidence of active geyser-like eruptions believed to be of nitrogen gas and dust particles. The processes and implications of volcanism on a planetary surface remain the same, although the materials may differ (*see* SATELLITES, MIDSIZED; SATELLITES, SMALL).

Tectonism

The processes of tectonism produce deformation of a planet's crust. Depending upon the strength of its crust (which in turn depends primarily on composition and surface temperature), the surface of the planet will respond to tectonic processes by flowing, buckling, folding, fracturing, or a combination of all of the above. On the solid-surface planets of our system we have identified tectonic landforms such as long linear valleys, large scarps, grooves or ridges, and folded terrain (*see* TECTONISM, PLANETARY). Some of these features appear to represent global compression or extension that occurred early in the planet's history (features on Mercury and Earth's moon are examples), but others have been interpreted to represent regional or local resurfacing by deformation due to active tectonism of long duration. The surfaces of Earth, Venus, Europa, Ganymede, Enceladus, and Triton all appear to have been greatly altered by tectonic processes.

The global paradigm of plate tectonics provides a framework by which to study and understand the record of tectonism on Earth. Prior to this global tectonics framework, it was the vertical movement of most mountains, valleys, and faults that was considered to be important. Now we believe that the vertical movements of Earth's crust are simply the end result of long-term horizontal movements of multiple lithospheric plates, and are caused by plate boundary interactions as plates spread apart or collide. But is this the way all planets work?

So far, although some regionally or globally organized tectonic processes appear to have operated on Mars and are certainly still operating on Venus, no other planet has as yet been found to have Earth-type plate tectonics. Mars, Mercury, and the Moon are often referred to as one-plate planets. Venus is currently being studied for evidence of its style of tectonism, but the indications are that it may have its own unique type of regionally or globally interconnected tectonism. However, we do not yet fully understand Earth-type plate tectonics. We may not yet be able to recognize other, similar or dissimilar, types of planetary tectonism. In fact, it is disquieting to realize that we have generally described all tectonic features on other planets as due to dominantly vertical movements—just as we did on Earth before the discovery of plate tectonics. It is possible that, even if they occur, horizontal movements are extremely difficult to recognize on a planetary surface.

Just as with volcanism, tectonism is driven by the internal heat of a planet. Small planets (like the Moon and Mercury) generally lose their internal heat early, that is, during the first 2 Ga of their existence. There is no record of continuing volcanism or tectonism on the surface of these planets. Mars, almost intermediate in size to Mercury and Earth, appears to have been volcanically active up to a few hundred million years ago—and perhaps may still experience a "marsquake" now and then. Earth, of course, continues to be extremely volcanologically and tectonically active. Therefore, in our beginning studies of comparative planetology, we have developed a working hypothesis that appears to be generally correct and may someday become the first "Law of Planetary Geology"—the size of a rocky planet is related to the length of time that the planet is geologically active. In the future, seismic stations on newly discovered planets may provide an indication of the level and scale of geological activity of a planet. The planet can then be assigned to a scale of "geological liveliness" that would have further implications for its internal heat flow, thermal evolution, and interior structure.

Gradation

Gradation is a term that can be used to describe both erosional (degradational) and depositional (aggradational) planetary processes (*see* WEATHERING AND EROSION, PLANETARY). These resurfacing processes operate to break down, transport, and redeposit material on a planetary surface. Some gradation mechanisms operate on all planetary surfaces whether they have an atmosphere or not (such as gravitational mass wasting and obliteration of surface features due to impact cratering). Glacial landforms can occur on any ice planet (with ice of any composition). But gradation, including mass-wasting (gravity-driven), aeolian (wind), aqueous (water), and glacial (ice of any composition) processes, occurs with greater effect on planets with atmospheres. An abundance of both atmospheric and surface liquid water over a long time period appears to be unique to Earth's geologic history and is the dominant process of gradation on Earth. The surface of Mars also shows evidence of deposition and erosion due to transport of material by water. Current models postulate relatively short periods of large-scale flooding, flowing water, and even lakes or oceans in various locations of Mars, although currently the atmosphere of Mars cannot sustain water in a liquid state. Mars has polar ice caps and shows features that have been interpreted to be due to glacial (carbon dioxide and water ice) erosion and deposition. Aeolian modification has occurred on both Mars and Venus. And an unusual type of gradation occurs on the surface of planets with no atmosphere, due to impact cratering. This process has been called "gardening" on the Moon, where a loosely consolidated regolith (a soil-type material without the organic matter) has been produced on the surface.

Although evidence for gradation can be observed and recognized on other planets, it can be difficult to tell the difference between erosional and depositional surfaces. For example, during analysis of images or other remotely sensed data, we may be unable to determine whether an area is a wind-blown desert or an old lake bed—two geological landforms with similar surfaces but very different implications for understanding the geological history of that particular region. One of our first methods for planetary analysis is determining the composition of rocks on the surface; however, the composition (or even the age of material) cannot help us when we are dealing with gradation because that material may have been transported many times in many different ways since its initial formation. It is the process of transportation itself that we must attempt to recognize on another planet, and this will require more experience.

Comparing Planets

For solid-surface planets, our proposed general classification scheme can now be expanded to (1) distance from the Sun; (2) density and mass of the planet; (3) implied bulk composition; (4) internal structure and presence/type of compositional stratification; (5) presence (density/composition) of an atmosphere; and (6) type and relative importance of the four primary geological processes that have modified the planet's surface.

To assign a solid-surface planet to number 6 of our proposed classification scheme, we must be able to link a geologic process to an end result (a modification of that planet) that can be imaged or observed by other spacecraft instruments. We can do this by analog studies (comparison with observations and knowledge from other planets including Earth) or by theoretical studies (comparison to experimental or numerical modeling). Analog studies are useful when dealing with complex modifications, due to multiple interrelated geological processes. A theoretical approach can be used to understand those planetary phenomena that cannot be directly measured or observed, or to predict changes in the results of geological processes under varying planetary conditions. Examples include planetary gravity data that can be used to model the interior of planets; numerical modeling that can predict the resultant landforms of various styles of volcanism on planets with different gravity and atmospheric density; or surface features on the ice moons of the outer planets in our solar system that can be interpreted because the strength and deformation of ice at that temperature can be calculated. Understanding the geology of another planet requires a combination of observation, acquisition of many different data sets, and theoretical modeling.

On our own planet, our understanding of geology evolved from small-scale, local problems to a global synoptic view of interrelated geological processes. On other planets, the situation is reversed. We begin analysis with a global view and then attempt to study the geological details using the best image resolution and other data sets available. This means that the scale (global, regional, local) of our geological knowledge may vary from planet to planet. In the case of the Moon, we have local geological "ground truth" for six landing sites. This has enabled lunar geologists to link global geology with local rock types and ages and to put together a true geological time-rock stratigraphy for the Moon. However, for most planets, a true geologic map, one that includes age and type of rock, is impossible. Currently when we map the geology of other planets, we are mapping physiographic/surficial or terrain-type units. Interpretations can be made of their probable rock type, origin, and relative age in comparison to other units; however, the interpretations of the units may change as more data are acquired. Also on other planets we have little or no information concerning the subsurface. Impact craters provide us with one tool to "see" into the subsurface of tectonically inactive planets, since the cratering process brings material from depth in the center of the crater and deposits it as part of the ejecta blanket. This is one of the reasons that the Apollo astronauts visited and collected samples from the ejecta blankets and rims of many impact craters during their traverses on the Moon.

Since the 1960s, we have attempted to gain as much information as possible about the other bodies in our system through (1) Earth-based observations (if possible); (2) flyby spacecraft observations and data collection; (3) orbiting spacecraft with multiple instruments; (4) probes or landers with multiple instruments, (5) return of samples to Earth; or (6) human exploration. The advantage of the last two methods is flexibility, the ability to use many methods of analysis and to take advantage of unforseen discoveries. So far, only our moon, out of all of the other bodies in our system, has been studied by all of the above mission types. In our first attempts at comparative planetology, we did not know what information was critical to understanding planets, so we tried to accumulate as much data as possible (constrained by technological and financial considerations). Thus far, we have found the fundamental characteristics outlined above. However, it is also true that each planet we have explored in our solar system has some unique feature, process, or combination of processes.

Biosphere

In looking at our planet as an interrelated system, we must include the effects of life on the evolution of our atmosphere and on the redistribution of elements and materials in the crust. For example, the geological process of limestone formation would not be possible without marine life, and this process may have significantly altered the distribution of carbon dioxide on Earth. As we learn more about the environmental geology of Earth, we may discover other similar interrelated effects. In our fictional *Star Trek* classification scheme, class M planets, in addition to being Earth-like in other respects, are capable of supporting humanoid life. Earth conceivably represents only one potential style of carbon-based life requiring an atmosphere of oxygen/nitrogen/carbon dioxide, water, and solar radiation of a specific type and intensity. Estimates of the "Earth-style habitable" range around stars are dependent upon the size and type of the star. There may be other habitable types as well. Some scientists have postulated that life may have existed in the past on Mars. For the future exploration of Mars, and for other planets in other systems, exopaleontology may become important in understanding the geologic evolution of planets.

Our proposed general classification scheme for comparative planetology can now be expanded to a final—but still very preliminary—form: (1) size and type of star in the system; (2) distance of the planet from the star; (3) density and mass of the planet; (4) implied bulk composition; (5) internal structure and presence/type of compositional stratification; (6) presence (density/composition) of an atmosphere; (7) type and relative importance of the four primary geological processes that have modified the planet's solid surface; and (8) presence/absence of a biosphere either in the present or the past. Many of these are, of course, interrelated in complex ways. Some of these may be interrelated in ways we do not yet understand. For example, one might post the question, Is plate tectonics a planetary requirement for life? As the science of comparative planetology continues to develop, other diagnostic indicators will be recognized, other classification systems will be invented, and our knowledge of our own planet will be enhanced.

Bibliography

Greeley, R. *Planetary Landscapes.* Boston, 1987.

Hamblin, W. K., and E. H. Christiansen. *Exploring the Planets.* New York, 1990.

Head, J. W. "Surfaces of the Terrestrial Planets." In *The New Solar System,* eds. J. K. Beatty, B. O'Leary, and A. Chaikin. Cambridge, Eng., 1981, pp. 45–46.

Lewis, J. S. "Putting It All Together." In *The New Solar System,* eds. J. K. Beatty, B. O'Leary, and A. Chaikin. Cambridge, Eng., 1981, pp. 205–212.

Moore, P., and G. Hunt, eds. *Atlas of the Solar System.* Chicago, 1984.

JAYNE AUBELE

PLATE TECTONICS

Earth is a dynamic planet. Its rigid, outermost layer, the lithosphere, has been in motion since early Precambrian times. The theory of plate tectonics describes how the lithosphere moves.

The lithosphere, unlike the interior of the earth, which is warm and deforms plastically, is cool and fractures when it is under stress (*see* MANTLE, STRUCTURE OF). Earthquakes represent the energy that is released when the lithosphere fractures. The lithosphere is thin, 100–150 kilometers (km) thick beneath the ocean basins, though it may be nearly twice as thick beneath continents. An analogy can be made between the thickness of the lithosphere and the rind of an orange. If Earth were shrunk to the size of an orange, the lithosphere would be about as thick as the rind.

The lithosphere on Venus and Mars appears to be one continuous layer. On Earth, however, the lithosphere is fractured into large areas called "plates." There are seven large plates—North American, South American, African, Eurasian, Indo-Australian, Pacific, and Antarctic—and an equal number of smaller plates (Figure 1). The largest plate, the Pacific plate, is made entirely of oceanic lithosphere and is one of the fastest moving plates (10 centimeters per year or cm/yr). One of the smallest plates is the Juan de Fuca plate, which is located off the northwest coast of Oregon and Washington. The Eurasian plate, which is primarily made up of continental lithosphere, has a deep root that penetrates far into the interior of the earth, and consequently, it is one of the slowest moving plates (2–3 cm/yr). Other plates, such as the North American plate, South American plate, African plate, and Indo-Australian plate, are made up of both continental and oceanic lithosphere and move at speeds that range from 3–6 cm/yr (or about as fast as your fingernails grow).

Though the theory of plate tectonics was formulated during the mid–1960s, the idea that the continents are mobile dates back to the early 1900s. "Continental drift" is the term usually applied to these earlier hypotheses describing continental movement. One of the earliest proponents of continental drift was German meteorologist ALFRED WEGENER. In 1915, Wegener published *Die Entstehung der Kontinente und Ozeane* (*The Origin of the Continents and Oceans*), which presented convincing geological evidence that all the continents had once been joined together in a supercontinent that he called Pangaea. He proposed that the modern ocean basins and continents formed when this primordial supercontinent broke apart.

Wegener's evidence for Pangaea included the fact that many of the continental margins fit back together much like pieces of a jigsaw puzzle. He also noted that when the continents are put back together, the geology and trends of mountain belts match from one continent to another. Furthermore, animals that could not cross wide oceans were often present on both sides of the ocean separating continents, suggesting that the landmasses were once connected. The movement of continents north and south across climatic belts also explained why ancient glacial deposits exist in the Sahara Desert and how fossil coral reefs appear north of the Arctic Circle.

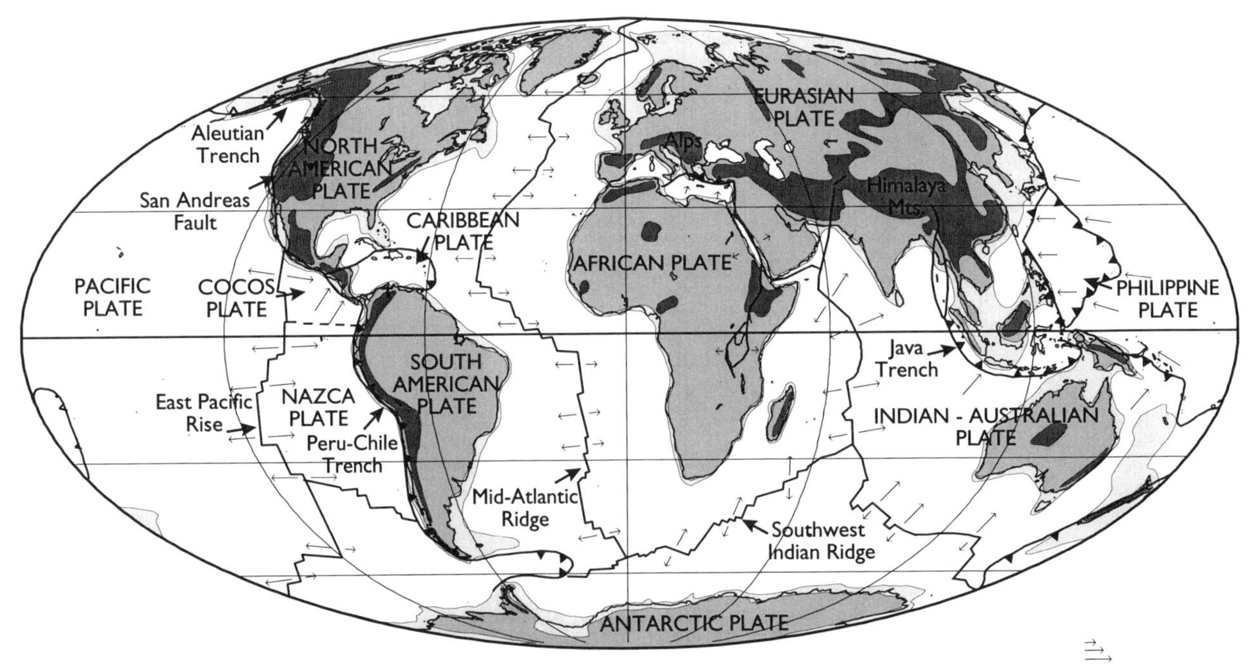

Figure 1. Plates of earth's lithosphere. © C. R. Scotese, PALEOMAP project, University of Texas Arlington.

Although Wegener's theory of continental drift was attractive to many geologists and paleontologists, it had one serious flaw. Wegener could not satisfactorily explain why Pangaea had broken apart or what caused the continents to move across the globe. The mechanism that Wegener proposed, called "pole-fleeing force," suggested that forces tied to the earth's rotation caused the continents to move away from the north and south poles. Calculations showed that this force was far too weak to move continents, and scientists objected to the notion that continents could somehow plow through the solid crust that formed the ocean basins like ships sailing through the seas. As a result, though Wegener's theory of continental drift explained many geological mysteries, it was not generally accepted, especially by scientists in the United States.

Acceptance of the idea of continental drift had to wait until new information was obtained from the ocean floor and from the interior of the earth. Prior to World War II little was known about the deep ocean basins. It was assumed that if the ocean basins were ancient features of the earth, they would contain a relatively undisturbed record of sediments extending back hundreds of millions, possibly billions, of years. Also it was assumed that if the oceans were filled with sediments, the ocean floor would be relatively flat. Both of these assumptions turned out to be incorrect. Deep-sea drilling of the ocean basins has revealed that, on average, only a thin layer of sediments covers the ocean floor (*see* DRILLING FOR SCIENTIFIC RESEARCH). The oldest sediments drilled are just over 100 million years (Ma) old and represent less than 3 percent of earth history. What has happened to all the older sediments?

It turns out that the ocean basins are not ancient, but are relatively young. The oldest known oceanic lithosphere is approximately 150 Ma old. As Wegener suggested, the Atlantic and Indian oceans formed when the supercontinent of Pangaea broke apart 150 Ma ago during the Jurassic period of the Mesozoic era.

The invention of sonar and echo-sounding techniques developed during World War II resulted in new, detailed maps of the ocean floor. It turns out that the ocean basins are far from flat. Bathymetric maps of the oceans indicate that the world's longest continuous mountain chain runs down the center of the Atlantic and Indian oceans, between Australia and Antarctica, and into the Pacific Ocean (Figure 1). Like the seam on a giant baseball, the mid-ocean ridge system girdles the earth and forms the boundary between several lithospheric plates. Another prominent bathymetric feature, the deep ocean trenches, is the second kind of plate boundary. While the mid-ocean ridges stand about 2,000 meters (m) above the ocean floor, the deep-sea trenches commonly extend to depths of 6,000 m. The deepest trench bottoms out at more than 11 km.

The mid-ocean ridges and the deep-sea trenches are the two principal kinds of plate boundaries. New lithosphere is produced at mid-ocean ridges and then moves symmetrically away from the ridge. This process is called seafloor spreading. Because new material is continuously being created at mid-ocean ridges and their continental equivalents, such as the East African Rift, these boundaries are termed constructive or divergent plate boundaries. Most divergent plate boundaries begin as sinuous chains of intra-continental rifts. Because the lithosphere is thinner and hotter beneath the rift, the rift stands high above the surrounding land surface and the rift valleys are often filled with water, forming a linear chain of lakes. As the rift grows, the continent breaks apart and new oceanic lithosphere is formed. The Red Sea is an example of a newly formed ocean basin. As the plates continue to move apart, the ocean basin widens, and the previously uplifted rift shoulders subside as the lithosphere cools.

The new lithosphere that is produced at the mid-ocean ridges cools, moves away from the ridge, and finally returns to the interior of the earth via the deep-sea trenches. The deep-sea trenches are the surface expression of a curtainlike slab of lithosphere that is continuously injected into the mantle. The descending portion of the lithosphere is called the subduction zone. Earthquakes in the descending slab indicate that the subduction zone extends to depths of at least 650 km. Though no longer seismically active, geochemical evidence suggests that the subducting lithosphere may continue to plunge deeper into the mantle and may eventually pile up at the boundary between the core and mantle. This subduction zone graveyard may give rise to superheated mantle plumes, or "hot spots." The volcanic islands that make up the Hawaiian–Emperor Island chain were produced by a mantle hot spot.

Because oceanic lithosphere moves toward the trench, down the subduction zone, and is eventually recycled back into the earth's interior, this kind of plate boundary is known as a convergent or destructive plate boundary. On average it takes about 200 Ma to recycle all of the world's oceanic lithosphere. Convergent plate boundaries are also the site of extensive mountain building and volcanic activity. When oceanic lithosphere is subducted beneath the margin of the continent, a linear belt of mountains, such as the Andes mountains of western South America, is formed along the edge of the continent. This mountain belt is also volcanically active because magma continually rises from the melting subducting slab. This magma erupts at the surface of the earth producing a linear chain of highly explosive volcanoes. Mount Saint Helens in Washington state is a typical example of an explosive Andean volcano.

When oceanic lithosphere is subducted beneath oceanic lithosphere, as is the case in the western Pacific, an arcuate chain of volcanic islands, called an island arc, is produced. The Aleutian Islands, south of Alaska, or the Mariana Islands, east of the Philippines, are examples of island arcs produced above active subduction zones.

Once a subduction zone starts, it is difficult to stop. Oceanic lithosphere subducts because it is cooler, hence slightly denser, than the hot mantle rock from which is was derived. Because the oceanic lithosphere is heavier than the underlying mantle, it sinks, pulling the rest of the plate with it. Imagine a tablecloth on a polished wooden table. If enough of the tablecloth hangs over the edge of the table, it will pull the rest of the tablecloth off the table and onto the floor. In a similar fashion, the subducting slab continually pulls the oceanic lithosphere back into the mantle. This is the primary reason why the plates move. This powerful driving mechanism of plate tectonics is called "slab pull."

The energy that powers the plates ultimately comes from the earth's great heat engine. The earth loses this heat through convection (*see* HEAT BUDGET OF THE EARTH; MANTLE CONVECTION AND PLUMES). The cool, subducting slab, together with the hot rising material beneath the ridge and in mantle plumes, are the two components of the earth's convective system.

Because plate motion is driven by the tendency for the cool oceanic lithosphere to sink back into the mantle, the only way to stop plate motion is to jam a subduction zone by consuming all the oceanic lithosphere in an ocean basin and bringing together continents that were once widely separated. Because it is made of lighter elements, continental lithosphere does not easily subduct, so subduction must stop. Although oceanic lithosphere recycles, continental lithosphere does not easily recycle. That is why the continents are much older than the ocean basins.

When continents collide, they form the highest and most extensive mountain ranges known. In the Mediterranean region, a chain of mountains has been produced as a result of the collision of Africa with Europe. These mountains are the Pyrenees in Spain and France, the Alps in France and Italy, the Dinarides in the Balkan states, and the Helenides in Greece. The most colossal mountains of Earth, however, are the Himalayan Mountains and Tibetan Plateau. This mountain range, which includes all ten of the world's highest peaks, is a result of the collision of the Indian plate with Eurasia. This collision began nearly 50 Ma ago and is continuing today. This collision was so violent that a large chunk of continental lithosphere that was once part of India now lies subducted beneath the Tibetan Plateau.

If you were to measure the length of all the plate boundaries, 55 percent would be mid-ocean ridges, 35 percent would be trenches, and the remaining 10 percent would be the third type of plate boundary, which is neither divergent nor convergent. This type of plate boundary is called a strike-slip or transform boundary. Movement along this type of plate boundary does not create or destroy lithosphere; rather, the two lithospheric plates simply slide past one another. On land, the best example of a strike-slip boundary is the San Andreas Fault system of California, which is the boundary between the Pacific and North Americans plates. Much of western California is actually part of the Pacific plate and is moving northward toward Alaska. Contrary to popular belief, California will not fall off into the ocean, but rather will slide northward so that in approximately 10 Ma, Los Angeles will become a suburb of San Francisco. In the ocean, strike-slip boundaries connect segments of the mid-ocean ridge. This special class of strike-slip boundary is called a transform fault.

The creation of ocean floor at mid-oceans ridges, its destruction below the deep-sea trenches, and sideways motion of the lithosphere along strike-slip faults, all release tremendous amounts of energy. Part of this energy is expressed in the form of earthquakes (*see* EARTHQUAKES AND SEIS-

MICITY). Shallow earthquakes occur at the crests of mid-ocean ridges and along the transform faults that join ridge segments. Shallow earthquakes also occur along strike-slip faults like the San Andreas fault and where the oceanic lithosphere bends as it enters the deep-sea trenches. Intermediate depth and deep earthquakes also occur along the length of the subducting slab. The deepest earthquakes occur in subduction zones at depths of approximately 650 km.

During the last thirty years, the boundaries of the plates have been mapped in detail by bathymetric surveys and by the precise location of earthquakes. Recently, a new technique has provided high-resolution maps of the ocean floor. This technique, called satellite altimetry, uses satellites equipped with radar transmitters to precisely measure the height of the ocean. Over short periods of time, changes in the height of the ocean are due to storms, tides, and ocean currents; however, over long periods of time, differences in the height of the ocean are due to local variations in the force of gravity. The force of gravity varies from place to place, depending on the topography of the ocean floor. A submerged island, which represents an excess of mass, will produce a gravity high. Gravity measurements over deep-sea trenches will be low, because the trench is essentially a mass deficit. Because the gravity signature directly correlates to the bathymetry of the ocean floor, it is possible to use this gravity measurement to map plate tectonic features.

Satellites using radar altimetry have mapped in great detail plate tectonic features such as mid-ocean ridges, fracture zones, and deep-sea trenches. In areas like the oceans around Antarctica, where ship traffic is light and bathymetric surveys are few and far between, satellite maps of the ocean floor have brought to light numerous, new plate tectonic features. These features include extinct mid-ocean ridges and fracture zone patterns revealing subtle changes in plate motion. These satellite surveys of the ocean floor have nearly doubled the number of mapped submarine islands. These submerged islands, or seamounts, are often missed by marine surveys because they are small and they fall between the survey tracks.

Bathymetric maps of the ocean floor, the location of earthquakes, and new satellite images clearly illustrate the modern geometry of the tectonic plates. But what about past plate configurations? How do we map the past positions of the continents and the shape of the ocean basins? When did plate tectonics begin? How far back in time can we reconstruct the positions of the continents? These are all interesting questions. Let's first start by discussing how we can map the past positions of the continents and ocean basins.

The key to mapping the past is Earth's magnetic field. Earth, like the Sun, has a magnetic field. It is as if there is a bar magnet stuck in the center of the earth (*see* EARTH'S MAGNETIC FIELD). We routinely use magnetic field lines emanating from the earth's core to tell north from south. However, these field lines also provide information about our location on the globe. This is because the magnetic field lines pierce the earth's surface at different angles depending on the latitude, or distance, from the North Pole. At the North Pole the field lines come straight out of the earth. The field lines bend over, becoming more and more horizontal toward the equator.

When some rocks containing iron form, they permanently record the orientation of the magnetic field. Laboratory measurements can determine whether the remanent magnetization is steeply inclined, indicating the rock was formed near the pole, or whether the remanent magnetization is more horizontal, indicating that the rock was formed near the equator. This technique of reading the earth's ancient magnetic field is called paleomagnetism. Paleomagnetism allows us to reconstruct the past position of the continents and determine how far they have moved in a north–south direction. Accurate paleomagnetic measurements have been made on rocks billions of years old. Using paleomagnetism, we can plot the paths of the continents back into the Precambrian.

Earth's magnetic field has another important property. Like the Sun's magnetic field, Earth's magnetic field "flips" or reverses polarity. During a magnetic reversal the north magnetic pole becomes the south magnetic pole, and vice-versa. On the Sun, these magnetic reversals occur in a twenty-two-year cycle known as the sun-spot cycle. On Earth, the turbulent motions in the core are much less vigorous than the motions in the interior of the Sun, so the interval between magnetic reversals is much longer. The period between magnetic reversals varies between hundreds of thousands of years to tens of millions of years.

When basalt, an iron-rich rock that makes up much of the oceanic lithosphere, is erupted at the crest of the mid-oceans ridges, it quickly cools and retains a record of the prevailing polarity of the earth's magnetic field. When the earth's magnetic field flips, the basalt forming at mid-ocean ridges is magnetized in the opposite, or reverse, direction. Ships towing magnetometers measuring the intensity of Earth's magnetic field can detect these polarity reversals. Fluctuations, or "anomalies," in the intensity of the magnetic field occur at the boundaries between normally magnetized seafloor and the seafloor magnetized in the "reverse" direction.

The age of the magnetic anomalies can be determined using fossil evidence and radiometric age determinations. Maps of these magnetic anomalies have been made for all the ocean basins. These maps show that the magnetic anomalies form linear patterns that parallel the crests of the mid-ocean ridges. These linear magnetic anomalies indicate that seafloor spreading has been nearly symmetric about the mid-ocean ridges for millions of years. The past positions of the continents and the shape of ocean basins can be directly reconstructed by superimposing linear magnetic anomalies of the same age. In this manner we can accurately reconstruct the past shapes of the ocean basins and reposition the continents in the locations they occupied millions of years ago. Obviously this technique can only be used on present-day ocean basins, which means it is restricted to about the last 200 Ma of Earth's history.

Plate tectonics is more than a description of how the lithosphere moves; it is a tool that helps us understand how the earth has changed through time. These changes include not only the movement of continents and the growth of ocean basins, but also the creation of the mountain ranges, eruption of volcanoes, and flooding of the continents by broad shallow seas. These changes have had profound effects on the geological record, the climate of the earth, and even the evolution of life.

The earth's heat engine is still vigorous. We can expect that plate tectonics will continue for hundreds of million, if not billions of, years into the future. During this time new supercontinents will form and break apart as the face of the earth continues to change and evolve. (*See also* MOUNTAINS AND MOUNTAIN BUILDING; PALEGEOGRAPHY; PALEOCLIMATOLOGY; RIFTING OF THE CRUST.)

Bibliography

ALLEGRE, C. *The Behaviour of the Earth: Continental and Seafloor Mobility.* Cambridge, MA, 1988.

COX, A., and R. N. HART. *Plate Tectonics: How It Works.* Palo Alto, CA, 1986.

SANDWELL, D. T., and W. F. SMITH. "Global Marine Gravity from ERS-1, Geosat and Seasat Reveal New Tectonic Fabric." *EOS Transactions of the American Geophysical Union* 73, no. 43 (1992):113.

UYEDA, S. *The New View of the Earth: Moving Continents and Moving Oceans.* San Francisco, 1971.

WEGENER, A. *The Origin of the Continents and Oceans.* Trans. John Biram. New York, 1915.

WYLLIE, P. J. *The Way the Earth Works: An Introduction to the New Global Geology and Its Revolutionary Development.* New York, 1976.

CHRISTOPHER R. SCOTESE

PLATINUM-GROUP ELEMENTS

Historically, platinum has received more active research for attention by scientists than any other element since its identification in 1748. Though there is no reference to platinum in the numerous books on metallurgy, assaying, and chemistry that began to appear in Europe in the sixteenth century, the earliest use of platinum is thought to have been about 700 B.C.E., when a single forged grain of platinum was used as a hieroglyphic inscription to decorate an etui from Thebes, Egypt. It has since beeen found that natural alloys of several of the platinum-group elements (PGE) occur as tiny inclusions in ancient jewelry from Egyptian (Twelfth Dynasty onward), Roman, and Byzantine times (Ogden, 1976). The high specific gravity of PGE alloys, similar to that of gold, caused them to be mechanically concentrated unknowingly from gold- and platinum-bearing placers, and thus inadvertently incorporated into the goldwork.

Later, several hundred years prior to the exploration of the Americas by Europeans, the Indians of Ecuador and Colombia made crude articles of platinum and platinum-gold alloys. The Spanish conquistadores, late in the seventeenth century, actively exploited gold placers in an area in the south central part of the Chocó region of Colombia, a long narrow strip between the main Cordillera of the Andes and the Pacific Ocean. They found a white metal that they called *platina*, a derogatory diminutive of *plata*, their word for silver. It was considered to be a nuisance because it concentrated in the gold washings, accompanied by black magnetic sands, which demanded a good deal of labor for separation. However, it was not until 1748 in Madrid, after publication of the journal of the voyage of Don Antonio de Ulloa, that reference to a new metal quickly aroused the interest of so many European scientists. This Spanish naval officer, astronomer, and mathematician, was a member of a French expedition to Ecuador between 1736 and 1743.

The identification, naming, and characterization of the six PGE (i.e., the light PGE palladium, rhodium, and ruthenium and the heavy PGE platinum, iridium, and osmium) spanned about one hundred years. Platinum was first identified in 1750 (C. Wood, W. Brownrigg, and W. Watson), followed by palladium and rhodium in 1802–1804 (W. H. Wollaston), iridium and osmium in 1805 (S. Tennant), and ruthenium in 1844 (K. K. Klaus) (McDonald and Hunt, 1982). (Iridium has played a major role in the identification of extraterrestrial impacts on Earth as a source of major geologic and biologic processes such as extinction—*see* IMPACT CRATERING.)

Table 1. Platinum Supply and Demand, Western World

Pt Supply (kg)	*1981*	*1987*	*1993*
South Africa	55,986	78,381	108,028
Russia (sales)	11,508	12,441	21,863
Canada*	4,043	4,354	6,752
Others	933	1,244	4,180
Total	72,470	96,420	140,822
Pt Demand (kg)	*1981*	*1987*	*1993*
Japan	35,769	51,321	63,016
North America	21,772	27,993	24,596
Western Europe	13,063	17,418	28,775
Rest of Western World	4,977	5,599	13,504
Western sales to China	933	933	0
Movements in stocks	(4,043)	(6,843)	10,931
Total	72,471	96,421	140,822

*Includes U.S. production after 1987.
Source: Johnson Matthey, PLC.

Production and Uses of the PGE

Production of platinum and palladium has exceeded that of the other four PGE and has been dominated in recent years by South Africa and Russia. World demand for both platinum and palladium has been led by Japan, followed by North America, as shown in Tables 1 and 2. In 1994 the price of platinum, calculated in U.S. dollars per troy ounce, averaged $406.19 (range $380/428) and for palladium the average was $157.50 (range $124/165). Western world demand for the other four PGE in 1993 was as follows: rhodium (12,971 kilograms or kg), ruthenium (6,295 kg), iridium (1,141 kg), and osmium (less than 172 kg) (Johnson Matthey, 1994). The price of rhodium, in U.S. dollars per troy ounce, has been the most volatile over the period between 1990 and 1993, ranging from a high of $7,000 to a low of $850. The principal uses of platinum in 1993 were jewelry (40%) and autocatalysts (35%) (Figure 1). For palladium, the principal uses were electrical (46%), dental (29%), and for autocatalysts (15%). In 1993 the bulk of the rhodium was used in autocatalysts (90%), with minor amounts used in the chemical (3%), electrical (2%), and glass industries (1%). Demand for ruthenium in 1993 was mainly distributed between the electrical (68%) and chemical industries (30%). Use of iridium in 1993 was largely in the chemical industry (62%); iridium is also in demand for crucibles (6%), and the remainder (32%) is used in a range of other applications (electrodes for biomedical devices, such as irradiated platinum-sheathed iridium wire for treating some tumors, in spark plugs, in fountain pen nibs, and

Table 2. Palladium Supply and Demand, Western World

Pd Supply (kg)	*1981*	*1987*	*1993*
South Africa	23,304	33,903	44,851
Russia (sales)	44,478	55,675	73,948
Canada*	4,977	5,910	11,896
Others	2,177	2,799	2,251
Total	74,936	98,287	132,945
Pd Demand (kg)	*1981*	*1987*	*1993*
Japan	25,505	44,167	62,213
North America	25,505	32,970	43,243
Western Europe	9,331	17,262	22,184
Rest of Western World	4,666	5,443	9,324
Movements in stocks	14,930	(1,555)	(4,019)
Total	79,937	98,287	132,945

*Includes U.S. production after 1987.
Source: Johnson Matthey, PLC.

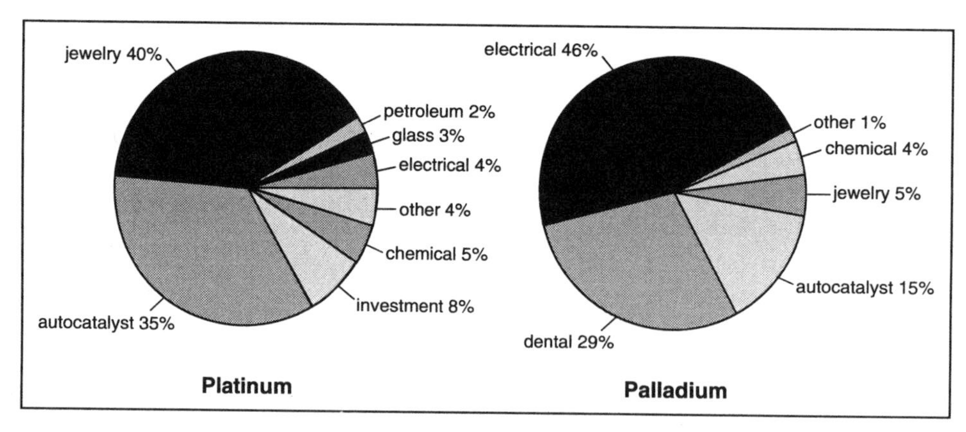

Figure 1. Principal uses of platinum and palladium in 1993 (after Johnson Matthey, 1994).

in platinum jewelry). Osmium is both the most scarce and least useful of the PGE, and it is not usually recovered from PGE deposits. It is used mainly as a biological staining agent and as a catalyst for drug synthesis.

Modern History of Platinum Mining

Until the discovery in 1824 of rich platinum-bearing placers in rivers draining Russia's Ural mountains of Siberia, Colombia was the only source of PGE, where mining of the placers began in the eighteenth century. In the 1920s the former Soviet Union was the leading producer of these metals, at which time by-product recovery of PGE from the Cu-Ni sulfide ore deposits of the Sudbury district, Canada, increased in importance, to the extent that Canada had become the leading producer by 1936. However, the cessation of supplies of crude platinum from Russia was a serious problem for platinum refiners such as Johnson Matthey (England) during the 1920s. Thus, the discovery and rapid exploitation of the world's largest resource of the PGE in the 1920s, in a layer of the Bushveld Igneous Complex (Transvaal, South Africa), now known as the Merensky Reef, enabled South Africa to become the largest producer by 1956–1957. The Noril'sk Ni-Cu sulfide ore deposits, in northwestern Siberia, also rich in PGE, were discovered in 1919 by geologists while prospecting for coal, but it was not until 1935 that action was taken to exploit these deposits. Thus, by the 1940s, the decline in supply of PGE from Russian placers was replaced by increased recovery from the new Ni-Cu mines from Noril'sk and the Kola Peninsula.

Mineralogy and Geology of PGE Deposits

The low crustal abundance of the PGE and the usual occurrence of their minerals as small inclusions (less than 1 millimeter or mm) in other minerals hindered knowledge of PGE mineralogy until the development in the 1960s of microanalytical tools such as the electron microprobe, which provided quantitative nondestructive analyses of tiny grains (approximately 10 micrometers or μm). By 1981 there were seventy-five recognized platinum-group minerals (PGM) (Cabri, 1981), the majority being minerals of palladium, followed by platinum. Since then, three have been reclassified and an additional fifteen new PGM species had been named and characterized, for a total of eighty-seven accepted PGM by 1993. The PGE form alloys and a variety of mineral compounds with elements of groups IV (Sn, Pb), V (As, Sb, Bi), and VI (S, Se, Te). The PGE also occur as a solid solution in common minerals that may also be alloys, tellurides, selenides, arsenides, sulfarsenides, sufides, and perhaps oxides and silicates. More recent developments of newer techniques for microanalysis, even more sensitive than the electron microprobe (ion- and proton-microprobes), have provided much new data on the distribution of PGE. There are also probably over two hundred poorly characterized minerals containing major quantities of PGE that occur in grains too small (less than 50 μm) for complete characterization and confirmation as new PGM. Improvements in microanalytical methods in the future for the full characterization of such small grains will improve our understanding of the mineralogy of the PGE. Some of these poorly understood PGM are oxides, hydroxides, and chlorides, but these are rare and not fully characterized.

PGE occur in a wide variety of geological environments, but it has been estimated that more than 99 percent of primary PGE production comes from sulfide ores of magmatic origin (Naldrett and Duke, 1980). Most of the world's resources and production are found in large stratiform bodies of basaltic rocks, usually intruded into continental rocks (Figure 2). Layering is prominent in these rocks and the PGE are associated with Fe-Ni-Cu sulfides, sometimes with chromite, in relatively thin layers (10–200 centimeters or cm thick). In most cases, the PGE-bearing "reefs" occur several hundred meters above the emergence of cumulus plagioclase, in their respective intrusion. The grade in these deposits is variable, ranging from about 3–22 grams per ton (g/ton), and these are currently exploited in southern Africa and in the U.S. state of Montana. Most of current world production comes from mines located in the Bushveld Igneous Complex of southern Africa, which mine either the Merensky or the UG2 reefs. The principal mines belong to Johannesburg Consolidated Investments (Rustenburg), Impala Platinum, Western Platinum Limited (Lonhro S.A. Ltd.), and Northam Platinum Limited (Gold Fields of South Africa Ltd.). A new mine, operated by BHP Minerals, began producing PGE concentrates in 1995 from the Great Dyke, about 80 kilometers (km) from Harare, Zimbabwe. The sulfide mineralization at the Great Dyke occurs in massive bronzitite rock, *below* the first appearance of cumulus plagioclase. The Stillwater Mining Company began producing PGE concentrates from the Stillwater Complex, Montana, in March 1987. In contrast to most of the African deposits, the ore in Montana contains more palladium than platinum. More than 95 percent of the PGE produced from Russia are derived from the Noril'sk-Talnakh Ni-Cu sulfide mines in Siberia, operated by the Noril'sk Nickel Combine. In Canada, most of the PGE are produced as a by-product from the Cu-Ni sulfide ores belonging to INCO Limited and Falconbridge Limited, associated with the Sudbury Irruptive. Some PGE are also recovered from deposits in Manitoba: INCO'S Thompson mines and the Namew Lake mine, owned by Hudson Bay Mining & Smelting Co. and Outokumpu Mines Ltd. Some PGE concentrates are also produced from the palladium-rich mineralization at Lac des Iles, Ontario.

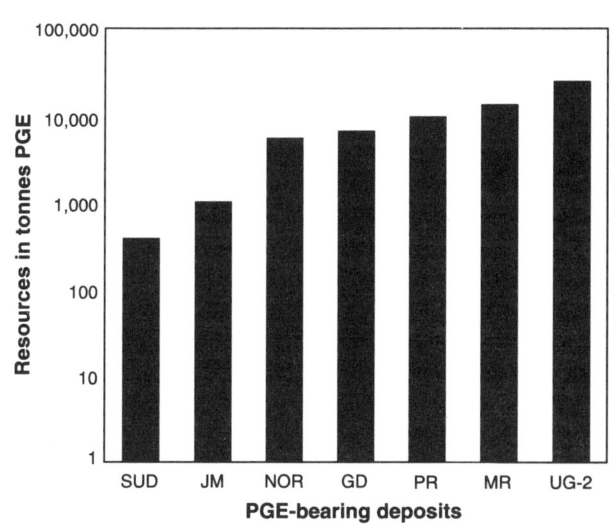

Figure 2. Identified resources of PGE in metric tonnes in different deposits (from Hulbert et al., 1988). Abbreviations: SUD = Sudbury (Canada), JM = J-M Reef (U.S.), NOR = Noril'sk-Talnakh (Russia), GD = Great Dyke (Zimbabwe), PR = Platreef (South Africa), MR = Merensky Reef (South Africa), UG-2 = UG-2 Reef (South Africa).

Bibliography

Cabri, L. J. *Platinum-Group Elements: Mineralogy, Geology, Recovery*. The Canadian Institute of Mining and Metallurgy, CIM Special Volume 23. Ste-Anne de Bellevue, Quebec, Canada, 1981.

Hulbert, L. J., J. M. Duke, J. W. Lydon, R. F. J. Scoates, L. J. Cabri, and T. N. Irivine. *Geological Environments of the Platinum Group Elements*. Geological Survey of Canada, Open File 1440. Ottawa, Canada, 1988.

Johnson Matthey Public Limited Company. *Platinum 1994*. London, 1994.

McDonald, D., and L. B. Hunt. *A History of Platinum and Its Allied Metals*. London, 1982.

Naldrett, A. J., and J. M. Duke. "Platinum Metals in Magmatic Sulfide Ores." *Science* 208(1980):1417–1424.

Ogden, J. M. "The So-called 'Platinum' Inclusions in Egyptian Goldwork." *Journal of Egyptian Archaeology* 62(1976):138–144.

Louis J. Cabri

PLAYFAIR, JOHN

John Playfair was a Scottish mathematician, geologist, and foremost advocate of the Huttonian concepts of uniformitarianism and plutonism. He is most noted for his book *Illustrations of the Huttonian Theory of the Earth* (1802) and his Law of Accordant Junctions also known as "Playfair's law."

Playfair was born in Benvie, Forfarshire, on 10 March 1748. He received his education from his father, a minister, and then entered the ministry at the University of St. Andrews, remaining there until 1769. In 1773, he was appointed minister at Benvie and Liff in succession to his father, but after nine years moved to Edinburgh. Playfair was talented in mathematics and as a result was quickly accepted into Edinburgh's scientific community. He became a professor of mathematics in 1785 and of natural philosophy in 1805 at the University of Edinburgh.

The late 1700s was a time of great controversy in geology and geomorphology. Prior to this time, Neptunism dominated the geological thinking of the period. The "Neptunists" (*see* WERNER, ABRAHAM GOTTLOB) maintained that the earth was once covered by a primeval, Noachian ocean created by the great biblical flood. The Neptunists classified and interpreted rocks as having formed either by primary deposition of chemical precipitates in this ocean or by sculpturing and subsequent deposition during retreat of this ocean from the land. Similarly during this period, geomorphologists interpreted landforms as relict features sculpted by the great biblical flood, with modern processes such as stream erosion having no significant effect on landform development. The "diluvialists" or "catastrophists," as they were called, viewed the earth as young, no more than a few thousand years old. This biblical, catastrophic view of earth history is not to be confused with the twentieth century recognition of catastrophic processes (e.g., impact cratering and outburst flooding from glacially dammed lakes) in the geologic record.

Playfair's predecessor and colleague JAMES HUTTON (1726–1797) challenged the Neptunist's doctrine with his theory of plutonism. In this theory, the earth is viewed as a dynamic heat engine where igneous sills and dikes are evidence of the planet's driving forces. His theory acknowledged the ability of streams to erode continents and to subsequently transport sediment to the oceans. With this last view, Hutton also challenged the catastrophist's orthodoxy by stating that landforms are created gradually, over long periods of geological history, and that modern geological processes are similar to those that operated in the past. These ideas would eventually give rise to "uniformitarianism," a concept that became largely synonymous with the phrase "the present is the key to the past." Although Hutton had great insight, his primary work, *Theory of the Earth*, was difficult to read and many of his concepts lacked clear example. The selling of his theory to the scientific community would be undertaken by Hutton's foremost advocate, John Playfair, who wrote eloquently.

Playfair became interested in geology from Hutton, who was also at the University of Edinburgh. After Hutton's death in 1797, Playfair started to write a memoir of his friend's life and work. Because of the continued attacks on Hutton's ideas and because of the "confused and repetitive" style in which it was written, Playfair decided to undertake a major rewriting of Hutton's two-volume *Theory of the Earth.* Playfair skillfully distilled Hutton's writings to a fraction of the original length and added copious illustrative notes. His summary and commentary of Hutton's original work was published as a book in 1802 and entitled *Illustrations of the Huttonian Theory of the Earth.* Although Playfair's book did little to advance his own ideas, it did much to acquaint nineteenth century geologists with Huttonian theory and became an unrivaled geomorphic text during this time (Davies, 1969).

Although Playfair's own geologic perspective generally did not diverge greatly from Hutton's, he did make significant advances in proving the fluvial origin of valleys. Playfair argued against those who held that valleys are crustal fissures formed during earthquakes or had been cut by diluvial currents (from the great biblical flood) or submarine currents (Neptunist's view). He formulated five "proofs" or arguments (see Davies, 1969) in support of the fluvialist doctrine. One of these arguments was based on a sketch map of a drainage system in plan view. He noted that where a higher elevation tributary joins a lower elevation tributary, an obtuse angle is always formed on the descending side. This observation led Playfair to recognize the dendritic pattern formed by most drainage systems. He pointed out that it is improbable that such a regular pattern could have formed by crustal fissures resulting from earthquakes.

Three other of Playfair's proofs were based on his field studies. He noted that in many regions, valleys form a radial pattern or drain down two sides of a ridge and that in each case, valley depth increases in the downstream direction. Playfair attributed this increase in valley depth to an increase in degradational processes, a phenomenon that he felt could only be explained by the action of streams. The next proof is similar to the last in which he states that it is ridiculous to invoke the action of a single debacle as being capable of eroding complex valley systems containing streams that are oriented and flow in all possible directions. Still another argument he invoked against the diluvialist theory is based on what he called "longitudinal valleys" where valleys are arranged at right-angles to each other and have several broadside openings by which water is discharged. Again, he noted that this phenomenon would be difficult to explain by a single debacle.

The most well known of Playfair's "proofs," however, is his Law of Accordant Junctions, which has become a standard way of appraising a terrain's history, as well as classic geological prose. Playfair wrote, "Every river appears to consist of a main trunk, fed from a variety of branches, each running in a valley proportioned to its size, and all of them together forming a system of vallies, communicating with one another, and having such a nice adjustment of their declivities, that none of them join the principal valley, either on too high or too low a level; a circumstance which would be infinitely improbable, if each of these vallies were not the work of the stream that flows in it." This law states that a river cuts its own valley with size and gradient adjusted to the resistance of the terrain in which it flows; tributaries to the main trunk of a river are similarly adjusted to form a mutually integrated valley system. This law describes an ideal equilibrium state within a fluvial system where streams are neither actively aggrading nor degrading their valley floors. This state most often occurs in terrains that have been subjected to long-term stream erosion. In geomorphology, the development of a fluvial system is commonly interpreted with respect to this ideal state. Fairbridge (1968) lists three instances where fluvial systems may deviate from the ideal pattern: (1) regions where glacial erosion has left tributary streams "hanging" above the floor of the main valley; (2) regions where a change in climate or regime has resulted in an underfit or misfit stream, for example, a small stream might flow in a broad, deep valley; and (3) coastal regions where the distal reaches of Pleistocene-age streams (formed during the last major glacial period) once extended across the continental shelf and are now "drowned" due to post-glacial sea level rise.

Playfair had planned to write a sequel to *Illustrations* based on his numerous field excursions in the British Isles. He had also hoped to travel to the continent to expand his geomorphological writings but was frustrated by the Napoleonic War. Finally in 1816, at the age of sixty-eight, he journeyed to the Alps but was unable to finish the second edition of his book before he died on 20 July 1819.

It is unfortunate that Playfair was not able to finish his second book, because his wide field experience, his ability to interpret landforms objectively, and his unusually lucid literary style would have added much to the field of geomorphology. However, the clarity and persuasiveness with which Playfair presented the concepts of plutonism and uniformitarianism in *Illustrations of the Huttonian Theory* not only helped to gain general acceptance of Hutton's theories, but also helped to address the cyclical nature of geological processes, and to establish the fluvial origin of valleys.

Bibliography

DAVIES, G. L. *The Earth in Decay*. New York, 1969.

FAIRBRIDGE, R. W. "Playfair's Law." In *Encyclopedia of Geomorphology*, ed. R. W. Fairbridge. New York, 1968.

HALLAM, A. *Great Geological Controversies*. New York, 1983.

TINKLER, K. J. *A Short History of Geomorphology*. Totowa, NJ, 1985.

WHITE, G. W. *Facsimile Reprint of Illustrations of the Huttonian Theory of the Earth by John Playfair*. Urbana, IL, 1956.

VIRGINIA C. GULICK

PLUMES

See Mantle Convection and Plumes

PLUTO AND CHARON

Speculation regarding the existence of a ninth planet dates back to the late 1800s. Largely encouraged by the successful mathematical prediction of Neptune's position by Adams and Leverrier, several people set out to find the trans-Neptunian planet. Prominent among these were William Pickering and Percival Lowell. Prior to his death in 1916, Lowell pursued the construction of a wide-field telescopic camera dedicated to the search. After numerous delays, in 1929 the camera was put into service at Lowell's observatory in Flagstaff, Arizona. A young astronomer, Clyde W. Tombaugh, had been hired by the observatory staff to undertake the search. Despite overwhelmingly daunting odds, at 4:45 p.m. on the afternoon of 18 February 1930, Tombaugh walked into the Observatory Director's office and announced, "Dr. Slipher, I have found your Planet X."

Even the largest telescopes on Earth have been unable to resolve details of the surface of Pluto, the outermost planet of the solar system, owing to its large distance from Earth, small size, and the blurring effects of Earth's atmosphere. In 1978 the discovery of Pluto's satellite, Charon, taxed the ability of Earth-based telescopes to the limit (Christy and Harrington, 1980). It provided, however, the "key" to unlock many of the secrets of the ninth planet, secrets it had retained since its discovery in 1930.

Almost immediately it was realized that if Pluto had a satellite, there must be two intervals during its 248-year orbit around the Sun when Earth passes through the satellite's orbital plane around Pluto (Andersson, 1978). During these times, we view the orbit "edge-on." Charon alternately is hidden by (occults) the disk of Pluto and passes directly in front of (eclipses) Pluto's disk exactly half an orbit later. These occultations and eclipses are termed mutual events (Figure 1). One such mutual-event series lasted from 1985 to 1990 (Binzel, 1989).

Although Pluto and Charon cannot be separately resolved from Earth's surface, it is a relatively easy task to measure the total light, or flux, that Pluto and Charon reflect back toward Earth. Most of the time we see the sum of Pluto and Charon's flux contribution (Pluto's flux + Charon's flux). When Charon is totally obscured by Pluto, the flux seen is just Pluto's contribution. Subtraction of Pluto's flux from the total flux yields Charon's contribution.

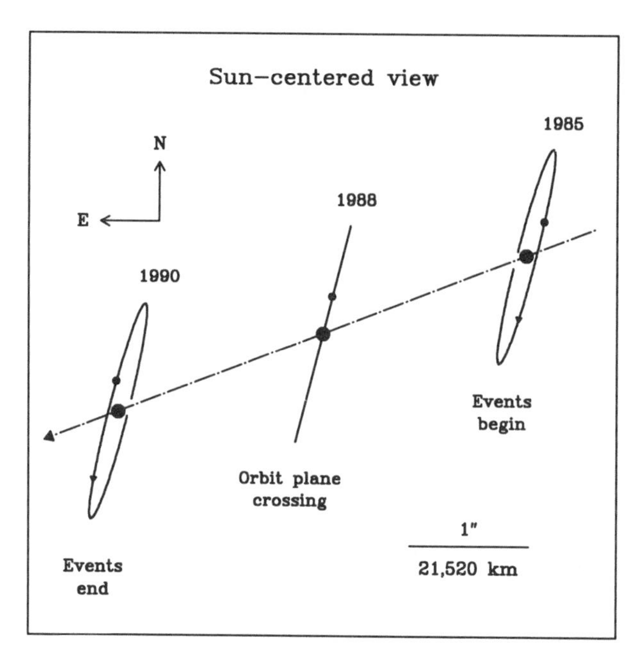

Figure 1. Mutual events.

Johannes Kepler's laws of planetary motion relate the size of a satellite's orbit to how long it takes to orbit its primary planet and the combined mass of the two. Therefore, precise timing of how long a mutual event lasts, multiplied by the velocity of the satellite, yields estimates for the sizes of Pluto and Charon. Simultaneous analysis of many such events also yields the orbital parameters of Charon around Pluto, and the reflectivities, or albedos, for each body. Analysis of several dozen mutual events monitored by one set of observers in Hawaii has resulted in the solution summarized in Table 1 (Tholen and Buie, 1990).

Charon's radius is almost exactly half that of Pluto. Most satellites (with Earth's moon the notable exception) are only a few percent the size of their primary planet. It should be noted that Pluto and Charon are not very large bodies—if you were to stack them one on top of the other, their combined diameter would be just about that of Earth's moon. Pluto, not Mercury, is the smallest of the Sun's major planets.

When Charon passes in front of Pluto, it covers selected subregions of the planet. Detailed analyses of the short-term brightness variations during these "transit" events are expected to yield an albedo (reflectivity) map for much of the surface of

Table 1. Orbital and Physical Parameters of the Pluto-Charon System

Semiaxis major	19,640	±	320	km
Eccentricity	0.00020	±	0.00021	
Inclination*	98.9	±	1.0	deg
Ascending node*	222.407	±	0.024	deg
Argument of periapsis*	210±	31	deg	
Mean anomaly†	259.96	±	0.08	deg
Epoch	JDE 2,446,600.5	=	19 June 1986	
Period	6.387246	±	0.000011	days
Pluto radius	1151	±	6	km
Charon radius	593	±	13	km
Pluto blue geometric albedo	0.44–0.61			
Charon blue geometric albedo	0.378	±	0.015	
Mean density	2.029	±	0.032	g cm^{-3}

* Referred to the mean equator and equinox of 1950.0.
† Measured from the ascending node.

Pluto. A few preliminary attempts at modeling Pluto's albedo have been made (e.g., Young and Binzel, 1993), but there are uniqueness problems with the results of these models; more than one acceptable solution may exist. This intriguing problem in cartography is an active area of research throughout the 1990s (Drish et al., 1994), and information gathered by the Hubble Space Telescope should aid in its resolution (Figure 2).

A mutual event can be monitored simultaneously at several wavelengths to obtain separate spectra of Pluto and Charon (Marcialis et al., 1992). If these wavelengths (colors) are chosen judiciously, information about the surface compositions can be retrieved. Each surface "mineral" has its own spectrum that can be used for identification, in the same way as human fingerprints. When this experiment was performed, it was found that Pluto's spectrum looked like that of methane (CH_4) frost, while Charon's could be matched only by water (H_2O) ice. Theoretical calculations show that at their distance from the Sun, Pluto and Charon are just in the size range where Pluto's gravity can retain a thin methane atmosphere, while smaller Charon's cannot. Thus, even if Charon once had a methane frost veneer, it would have rapidly "leaked off" into space. Loss of many meters of methane, but retention of the dark imbedded "impurities," explains why Charon's albedo is substantially darker than Pluto's (Marcialis et al., 1987).

"Mean density" is the most useful parameter for determination of the interior structure and composition of a planet. Accurate estimates for the sizes of Pluto and Charon from the mutual events, combined with the total mass of the system derived from Kepler's third law, show that Pluto's estimated mean density is about twice that of water. This means that Pluto has a substantially higher fraction of rock in its interior, about 50 percent, than indicated by theoretical models from the 1970s. Being denser than ice, this rock probably has sunk to form a core (820–920 kilometers, or km, in radius). On top of this core lies a rigid "mantle" of water ice, some 210–310 km thick. A methane "crust" forms the topmost layer of the planet, anywhere from 10–30 km in thickness (McKinnon and Mueller, 1988; Simonelli and Reynolds, 1989). Pluto's surface gravity is only 1/29 that of Earth's.

A theoretical model of the methane layer shows that it is structurally weak (Marcialis, 1990). Even though Pluto's surface temperature is only about 50 K, this is a substantial fraction of the melting temperature of methane. Topographic features greater than about 20 km in lateral extent should "relax" away in only a few million years. Therefore, Pluto's surface is assumed to be geologically very young, with few craters, much like the surface of Neptune's satellite Triton, which was imaged by *Voyager 2* in 1989.

On 9 June 1988 Pluto (but not Charon) passed directly in front of a star (Elliot et al., 1989, and Hubbard et al., 1988). Rather than "winking out" in a few seconds as would be expected if the star

Figure 2. Pluto and Charon as photographed by the Hubble Space Telescope faint object camera.

were covered by the limb of an airless body, the star gradually dimmed. Its reappearance about a minute later also was gradual. As of the mid–1990s, this observation is the strongest evidence for a thin atmosphere surrounding Pluto and supersedes previous indirect evidence. Gradual refraction by the gases in the atmosphere has been modeled and fit to the observations. It appears that Pluto's atmosphere extends nearly 1,000 km above the planet's surface, has a temperature of between 50 K and 60 K, a scale height (altitude change for pressure to fall by a factor of $1/e$) of about 46–60 km, and a temperature-to-mean molecular weight ratio (T/μ) of 4.2 ± 0.4 K (Hubbard et al., 1990). The occultation also showed that the "yardstick" of the Pluto-Charon system may be up to 4 percent larger than the mutual event solution in Table 1.

Spectral absorptions due to ices of carbon monoxide (CO), carbon dioxide (CO_2), and nitrogen (N_2) have been detected from a telescope in Hawaii (Owen et al., 1992). Considering the vapor pressures of these additional constituents, and their plausible relative abundances, it seems that nitrogen, and not the more spectrally active species methane, is the dominant constituent of Pluto's atmosphere. Still, the probable surface pressure is tiny, on the order of 40 microbars (1 μbar = 10^{-6} × the density of Earth's atmosphere at sea level).

The dynamics of Pluto's atmosphere are under study. It has been suggested that Pluto's atmosphere will condense as surface frost as Pluto recedes from the Sun (perihelion was in 1989) over the next few decades (Stern and Trafton, 1984). Throughout the long plutonian year, the bright, most volatile surface frosts actually may migrate from pole to pole and back again (Binzel, 1993, Binzel et al., 1991), similar to what is observed with the Martian polar caps and what is presumed to happen on Triton. Much additional work, both theoretical and observational, needs to be done in this area.

Plate 27. The Valles Marineris canyon system is the 2,000-km-long gash seen along the martian equator in this *Viking 1 Orbiter* image.

Plate 28. Sawed surface of the martian meteorite EETA79001 found in the Transantarctic Mountains of Antarctica. The cube (with letter S) in the lowermost left is 1 cm on a side. The vertical lines are saw marks, and the small lighter patches show the structure of this basalt rock. The black patches are glass that contains traces of martian atmosphere. (Photo courtesy of Office of the Curator, NASA Johnson Space Center.)

Plate 29. Broken surface of ALHA81005, the first recognized lunar meteorite. The cube to the left is 1 cm on a side. The white areas are pieces of anorthosite, rock rich in feldspar, and are essentially identical to anorthosites found by the Apollo missions. The dark material is hardened lunar soil. (Photo courtesy of Office of the Curator, NASA Johnson Space Center.)

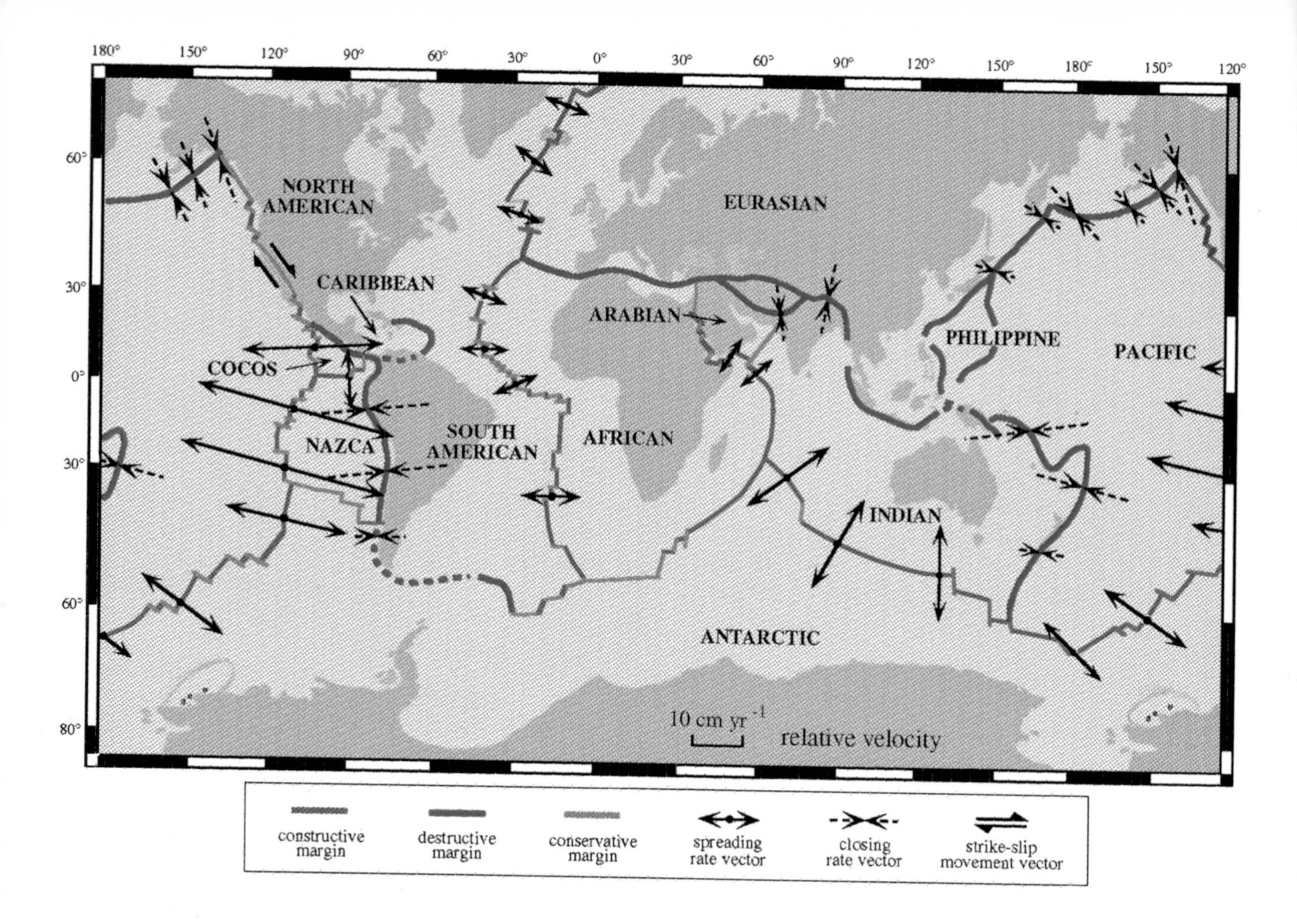

Plate 30. The outer surface of the planet is divided into a relatively small number of "plates" that move slowly about Earth's surface. They interact at their boundaries, or margins. Where they move apart, as in the center of the Atlantic Ocean, new seafloor is created by the process of seafloor spreading. Most of the world's great destructive earthquakes occur where plates impinge on one another and are destroyed. In regions such as the Aleutian Island chain in the northernmost Pacific, the ocean basin is sliding beneath the islands and being consumed back into Earth's mantle.

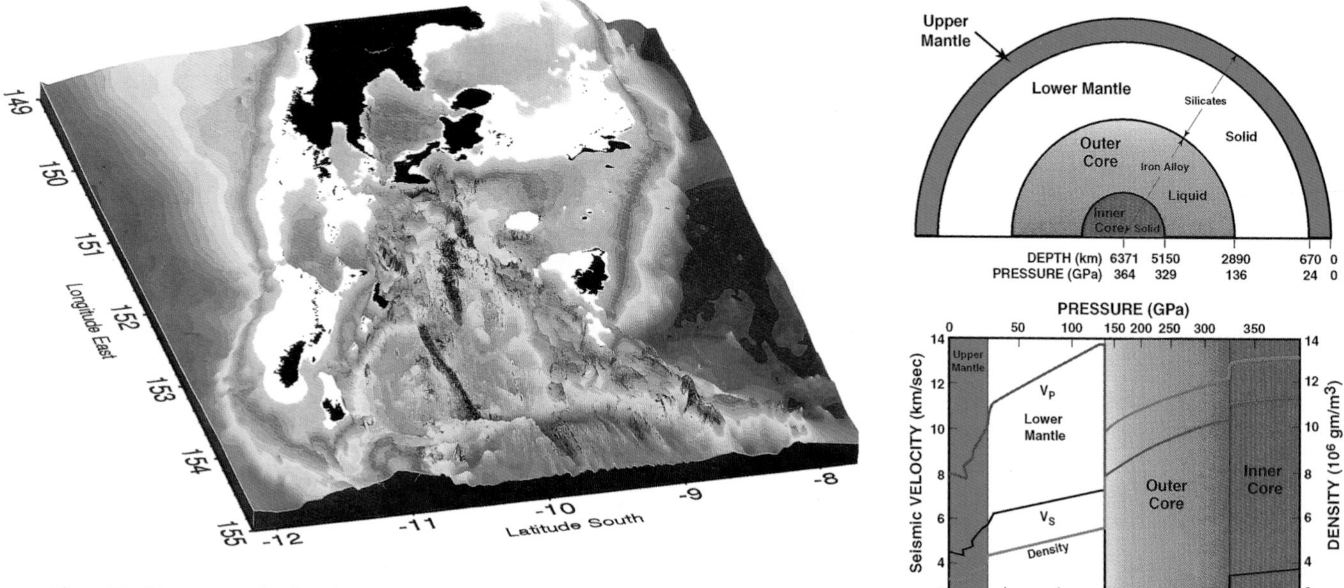

Plate 31. New ocean basins form when continents split apart. This process is active today off Papua, New Guinea where this rendering of seafloor topography (blue areas are deep, yellow and red are shallow) shows a distinct V-shaped depression oriented away from the viewer. In this depression a new ocean basin is being produced by seafloor spreading and the continental crust is being torn apart in an action not unlike an opening zipper. (Image provided by B. Taylor, University of Hawaii.)

Plate 32. Laboratory measurements of seismic velocity of ophiolites compared with field measurements in ocean basins.

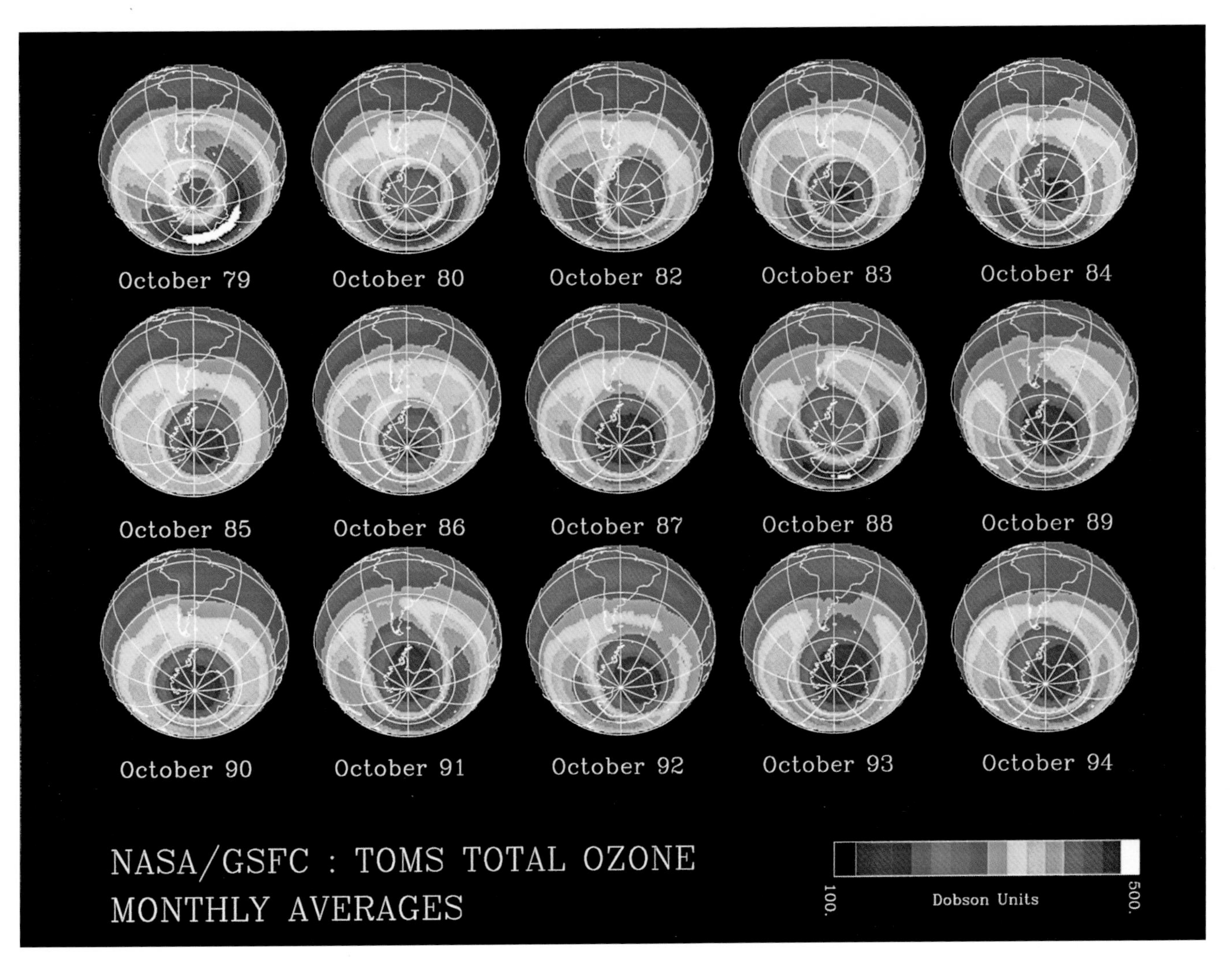

Plate 33. Monthly averages of TOMs.

Plate 34. Photograph of basalt dikes (light-colored due to alteration by acid-rich hot water) cutting sedimentary rocks in the Burns Basin gold mine, Nevada. Rigs in the foreground drill holes used for blasting to break the rocks for mining. Geologists use the cross-cutting relationships of dikes to help determine the geologic history.

Plate 35. River Andes (Huaraz), Peru. (Photo courtesy Ellen E. Wohl.)

Plate 36. This 3D plot of shear velocity anomalies in the mantle is viewed from an altitude of 37,000 km. The arc length of each side of the cut's triangle is 10,000 km. The bottom of the cut maps the velocity anomalies at the core's mantle boundary. A thin black line marks the 660-km discontinuity, and plate boundaries are in yellow. The scale range is ±1.5 percent; the slowest velocities are in red, the fastest are in blue. There is considerable scale saturation in the upper mantle. The velocity model is from the work of Su, Woodward, and Dziewonski.

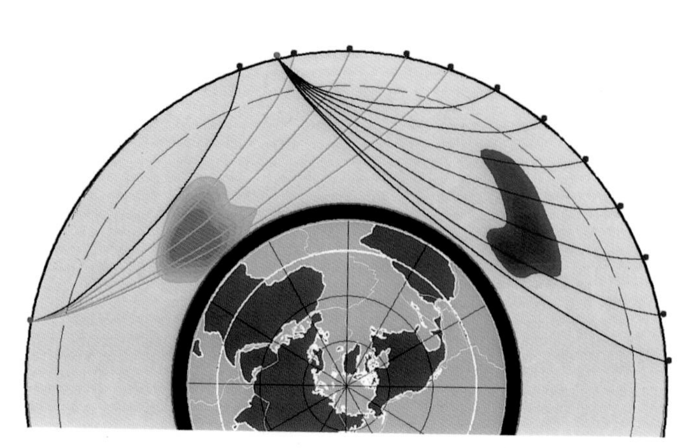

Plate 37. Seismic tomography locates a positive (blue) or negative velocity anomaly in the mantle by combining information on the travel time of many seismic rays from earthquakes (red and green dots) to seismic stations along criss-crossing paths. Waves that miss the anomaly show the normal travel times (black rays); waves that penetrate the anomaly are speeded up (blue rays) or slowed down (orange rays). A dense mesh of intersecting rays can be used to define the anomaly and to measure the velocity with which it transmits waves. The structure of the anomaly must account for the observed deviations in travel times of all the rays that traverse it. These are actual anomalies obtained in the equatorial cross section; see Plate 40's caption for an explanation of the details.

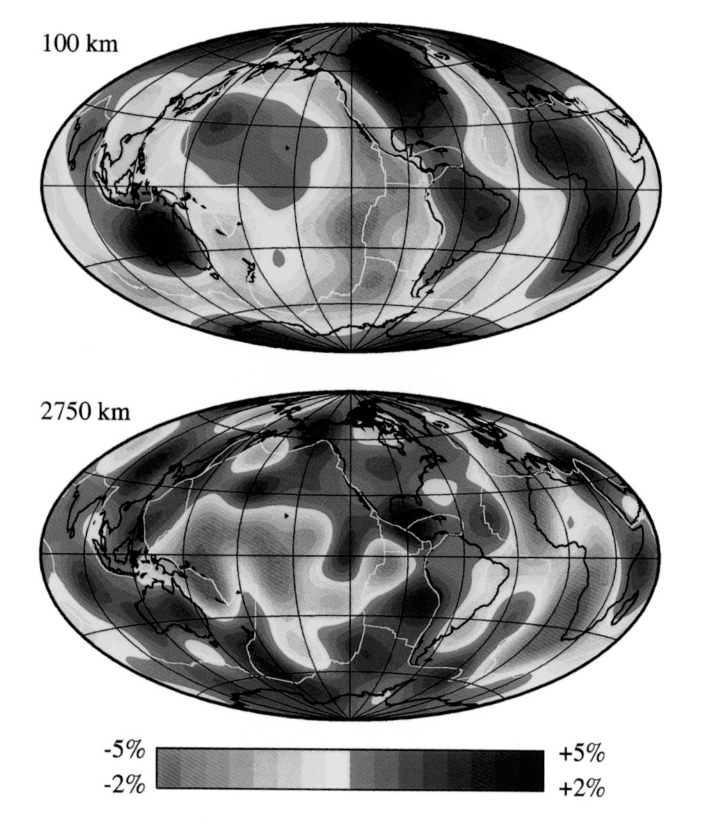

Plate 38. These two maps reveal shear velocity anomalies at a depth of 100 km (top: plate boundaries are shown in yellow) and 2,750 km (bottom) derived from a model of Su, Woodward, and Dziewonski. Comparing the lower map with Plates 36 and 38 indicates that many of the large-scale features seen near the core-mantle boundary extend throughout the lower mantle. They can be considered mega-structures of the earth's interior. Some are named: high-velocity structures such as "Pangea Trough," extending through much of the lower mantle under North and South America, "Tethys Trough" from the Macquarie triple junction to the Straits of Gibraltar, "China High," and "North Pacific High." Low-velocity anomalies include the "Equatorial Pacific Plume Group" and the "African Plume Group."

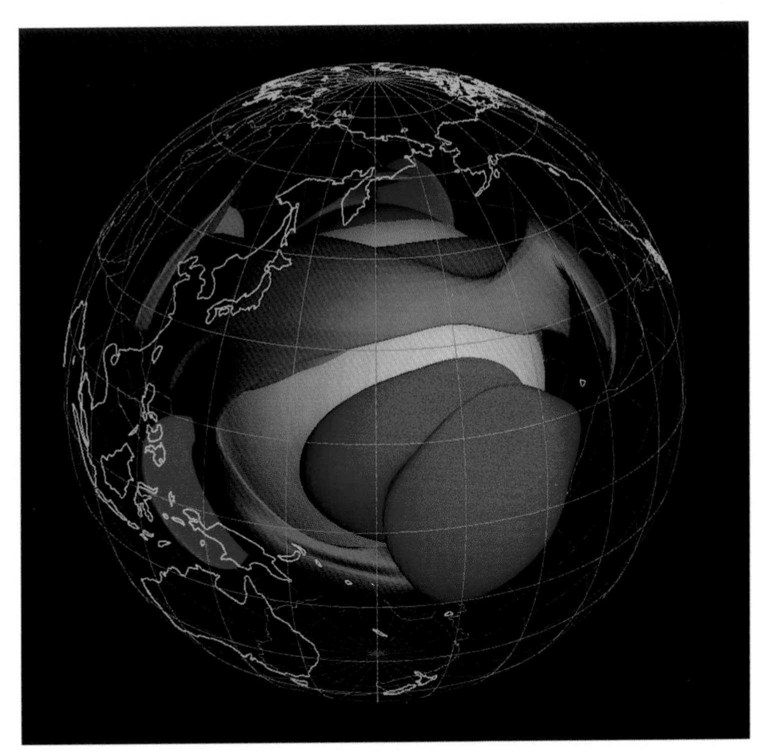

Plate 39. 3D view of the +0.4 percent (blue) and -0.4 percent (red) iso-surfaces of the shear velocity anomalies from Su, Woodward, and Dziewonski. The structure in the upper mantle has been removed to allow the unobstructed view of the large-scale pattern in the lower mantle. The ring of high velocities circumscribing the Pacific Ocean is very distinct, as is the mega-upwelling in the center of the Pacific. The other upwelling, under Africa, can be seen at the edge of the figure.

Plate 40. A polar great circle cross section through the shear velocity model of Su, Woodward, and Dziewonski. The cross section is made along the heavy white line in the inset map and passes through the center of the earth. The outermost ring is closest to the earth's surface; the innermost corresponds to the core-mantle boundary. The depth of the 660-km discontinuity is indicated by a dashed line. The scale ranges from -1.5 percent (red) to +1.5 percent (blue). Significant saturation of the scale is possible in the upper mantle.

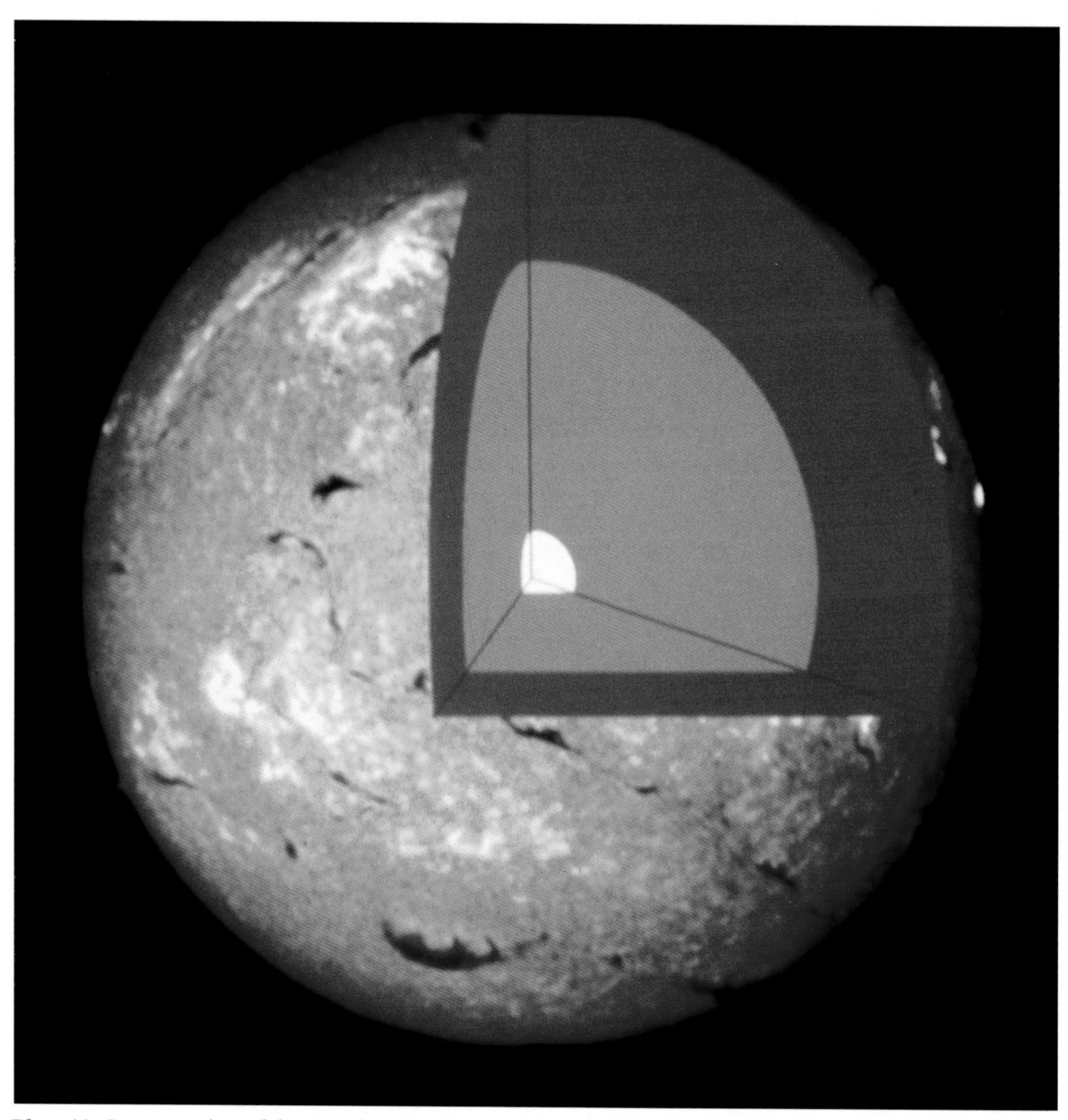

Plate 41. Cut-away view of the Sun showing the core, photosphere, chromosphere, sunspots, and the corona.

Plate 42. Photo of Uranus as compiled by *Voyager 2* on 10 January 1986, when the NASA spacecraft was 18 million km from the planet. The picture has been processed to show Uranus as human eyes would see it from the spacecraft.

Plate 43. Neptune, photographed by the *Voyager 2* narrow-angle camera. (Photo courtesy of NASA.)

Plate 44. Apollo Earth view. (Courtesy of NASA.)

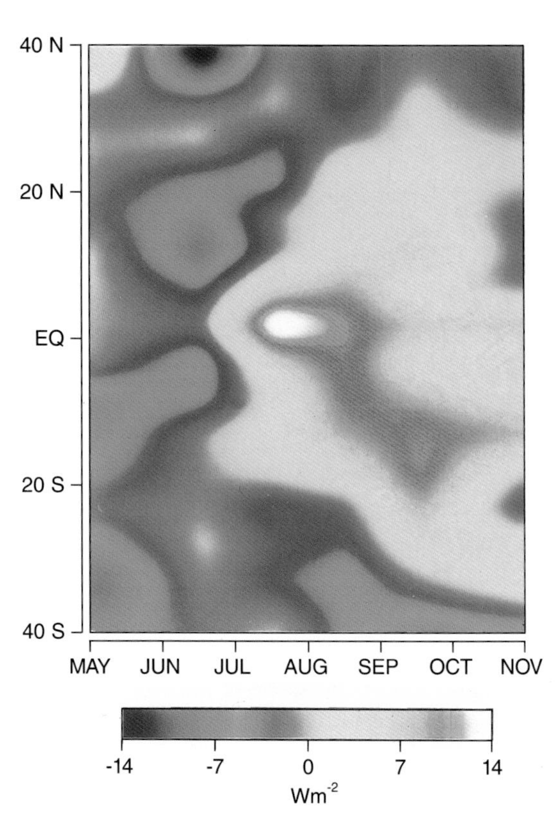

Plate 46. Earth Radiation Budget Experiment (ERBE) data on Mount Pinatubo aerosols. (Courtesy of NASA.)

Plate 45. Camera view of the Sinai Peninsula. (Courtesy of NASA.)

Bibliography

ANDERSSON, L. E. "Eclipse Phenomena of Pluto and Its Satellite." *Bulletin of the American Astronomical Society* 10 (1978):586.

BINZEL, R. P. "Pluto-Charon Mutual Events." *Geophysical Research Letters* 16 (1989):1205–1208.

———. "1991 Urey Prize Lecture: Physical Evolution in the Solar System—Present Observations as a Key to the Past." *Icarus* 100 (1993):274–287.

BINZEL, R. P., E. F. YOUNG, and E. S. DITCHBURN. "Insolation History on Pluto: Implications for Frost Models." *Bulletin of the American Astronomical Society* 23 (1991):1216–1217.

CHRISTY, J. W., and R. S. HARRINGTON. "The Discovery and Orbit of Charon." *Icarus* 44 (1980):38–40.

DRISH, W. F., JR., R. HARMON, W. J. WILD, and R. L. MARCIALIS. "Images of Pluto Generated by Matrix Lightcurve Inversion." *Icarus* 112 (1995).

ELLIOT, J. E., E. W. DUNHAM, A. S. BOSH, S. M. SLIVAN, L. A. YOUNG, L. H. WASSERMAN, and R. L. MILLIS. "Pluto's Atmosphere." *Icarus* 77 (1989):148–170.

HUBBARD, W. B., D. M. HUNTEN, S. W. DIETERS, K. M. HILL, and R. D. WATSON. "Occultation Evidence for an Atmosphere on Pluto." *Nature* 336 (1988):452–454.

HUBBARD, W. B., R. V. YELLE, and J. I. LUNINE. "Nonisothermal Pluto Atmosphere Models." *Icarus* 84 (1990):1–11.

MARCIALIS, R. L. "The Pluto-Charon System as Revealed During the Mutual Events." Ph.D. dissertation, University of Arizona, 1990.

MARCIALIS, R. L., G. H. RIEKE, and L. A. LEBOFSKY. "The Surface Composition of Charon: Tentative Identification of Water Ice." *Science* 237 (1987):1349–1351.

MARCIALIS, R. L., L. A. LEBOFSKY, M. S. DISANTI, U. FINK, E. F. TEDESCO, and J. AFRICANO. "The Albedos of Pluto and Charon: Wavelength Dependence." *Astronomical Journal* 103 (1992):1389–1394.

MCKINNON, W. B., and S. MUELLER. "Pluto's Structure and Composition Suggest Origin in the Solar, Not a Planetary, Nebula." *Nature* 335 (1988):240–243.

OWEN, T., T. GEBALLE, C. DE BERGH; L. YOUNG, J. ELLIOT, and D. CRUIKSHANK. "Pluto." *International Astronomical Union Circular 5532* (1992).

SIMONELLI, D. P., and R. T. REYNOLDS. "The Interiors of Pluto and Charon: Structure, Composition, and Implications." *Geophysical Research Letters* 16 (1989): 1205–1212.

STERN, S. A., and L. TRAFTON. "Constraints on Bulk Composition, Seasonal Variation, and Global Dynamics of Pluto's Atmosphere." *Icarus* 57 (1984):231–240.

THOLEN, D. J., and M. W. BUIE. "Further Analysis of Pluto-Charon Mutual Event Observations—1990." *Bulletin of the American Astronomical Society* 22 (1990):1129.

YOUNG, E. F., and R. P. BINZEL. "Comparative Mapping of Pluto's Sub-Charon Hemisphere: Three Least Squares Models Based on Mutual Event Lightcurves." *Icarus* 102 (1993):134–149.

ROBERT L. MARCIALIS

PLUTONISM

See Igneous Processes; Intrusive Rocks and Intrusions

POLAR EARTH SCIENCE

The polar regions encompass the land and ocean areas geographically north of the Arctic Circle and south of the Antarctic Circle. Despite common physical characteristics—rigorous cold, ice-covered oceans, ice sheets and glaciers, and alternating six months of daylight and darkness—the two regions are distinctly different. A deep, ice-covered ocean basin, surrounded by land except for the Bering Straits and the North Atlantic Ocean, dominates the Arctic. While glaciers are common in the Arctic, only the 18 million square kilometer (km^2) Greenland Ice Sheet remains. In contrast, Antarctica is a perennially ice-covered continent, surrounded by the southernmost Atlantic, Pacific, and Indian oceans. The massive East Antarctic and West Antarctic Ice Sheets cover all but 2 percent of the 14 million km^2 continent.

Arctic

Three parallel ridges—the Lomonosov, Nansen (an extension of the Mid-Atlantic Ridge), and Alpha Ridges—cross the deep basin of the 14 million km^2 Arctic Ocean. Unusually shallow continental shelves surround the basin and occupy more than half its area. These shelves extend outward from the mainland for distances ranging from about 100 km between Greenland and Barrow, Alaska, to an average of 800 km in the Chukchi, East Siberian,

Laptev, and Kara seas. Although the Arctic has been ice-covered from 5–15 million years (Ma) ago, the ocean has been warmer and ice-free. During the last few million years, twenty-six glaciations have affected northern polar seas and lands.

Three major physiographic groupings make up the terrestrial environment:

- Rugged uplands of igneous Precambrian shield, many of which were overrun by continental ice sheets that left behind scoured, rocky surfaces and deeply cut fjords.
- Flat plains and plateaus covered by deep glacial, alluvial, and marine deposits.
- Folded mountains that range from the high peaks of the Canadian Rockies to the older, rounded slopes of the Ural Mountains.

Permafrost—soil that remains continuously below 0°C for more than two years—covers more than half of Russia and Canada, 85 percent of Alaska, and 20 percent of China, as well as the polar ocean floor. Permafrost develops where the mean annual air temperature is about −3°C. In the cold, high latitudes of the Arctic, permafrost extends more than 700 meters (m) deep, with a maximum age between one and a half and 2 Ma.

In July 1993, U.S. glaciologists finished drilling through Greenland's ice sheet to bedrock, producing an ice core (the longest from the Northern Hemisphere) that records more than 100,000 years of regular climate oscillations and climatic interactions between the northern and southern hemispheres. Gradually compressed into ice, the accumulating snow trapped atmospheric gases, chemicals, and dust. The core's unbroken record reveals that, during the last glacial period, concentrations of calcium, chloride, potassium, magnesium, sodium, and sulfate ions rose and fell. These ions trace the dynamics of the Northern Hemisphere's polar atmospheric circulation and indicate that in colder periods, the polar atmosphere shifted southward and was better mixed. Researchers, using these data to trace a pattern of climate history, have discerned periods of climate change on scales ranging from decades to tens of thousand of years.

Other researchers, studying shallow ice cores from the same area, believe that they have evidence that an asteroid plummeted to Earth in 1908, exploded with an estimated energy of 15 megatons, and flattened trees for hundreds of square kilometers in Siberia's Tunguska River region (*see* METEORITES; ASTEROIDS). Early results show four- to twenty-fold leaps in iridium (Ir) concentration in Greenland ice from 1908 (*see* IMPACT CRATERING). While high iridium levels indicate a meteorite impact, glaciologists must distinguish the meteorite's signature from that of a volcanic eruption, but so far they have found no volcanic fingerprint in the 1908 ice.

Past volcanic eruptions are evident in the ice core's record and help show how particular eruptions affect climate. Examining ice over the past 9,000 years, glaciologists found that eruptions were larger and more frequent in the millennium just after the last deglaciation (*see* EARTH'S GLACIERS AND FROZEN WATERS). These data suggest that the melting of the ice sheets may have spurred volcanism by removing a great weight from Earth's crust. In the ice core record, a prominent sulfate spike, spanning several years, may mark the eruption, 68,000–75,000 years ago, of Toba, Sumatra—a long-lived event that could have altered the temperature and significantly affected global climate.

Antarctic

Antarctica, representing nine percent of Earth's continental crust, has been in a near-polar position for about 100 Ma. East Antarctic is an old Precambrian shield lying south of Australia, India, and Africa, while geologically young West Antarctica is part of the "Pacific ring of fire." The Transantarctic Mountains mark the boundary between the two blocks. The narrow, steep Antarctic Continental Margin has an average width of 30 km compared to a worldwide average of 70 km, with depths ranging from an average of 400–600 m to the deepest recorded drop of 800 m.

The continental ice sheets average about 3 km thick and cover 98 percent of the continent. In East Antarctica, the ice sheet flows from the high central plateau toward the coast, gradually thinning from about 4,000 m above sea level in the interior. In West Antarctica, six parallel ice streams, 50–100 km wide and 500 km long, drain the slow-moving ice from the interior. Flowing rapidly through slow or stagnant ice, the streams merge into the floating Ross Ice Shelf that fans out across the Ross Sea and buffers the inland ice from the ocean's destructive power. The behavior of these ice streams and their interaction with under-ice geology determine the stability of the West Antarctic Ice Sheet, Earth's last ice sheet grounded on bedrock well below set level.

Discovery in Antarctica of the fossil fern *Glossopteris* provided early paleontologic evidence that the continent, along with Australia, India, and South America, formed the supercontinent Gondwanaland and helped to support continental drift theory (*see* CONTINENTS, EVOLUTION OF; PLATE TECTONICS). In 1967 researchers discovered in the Transantarctic Mountains the fossils of *Lystrosaurus*, a freshwater reptile that could not have migrated to Antarctica unless the continents were connected. This discovery is considered one of the most significant proofs of continental drift.

In 1992 geologists made a more startling announcement concerning Antarctica and its relationship to the other continents. Data from the 150 km long Shackleton Range near the Weddell Sea and from North America suggest that 550 Ma ago North America's eastern margin was contiguous with South America, forming a pie-shaped wedge between Antarctica-Australia and South America. The ancestral Appalachian Mountains were thrust up when eastern North America collided with the western margin of South America. When the continents rifted apart, the Appalachians split, abruptly truncating the mountain range in Georgia and in the Andes near the Chilean-Peruvian border.

Until the 1980s most geologists agreed that the East Antarctic Ice Sheet had been in place for at least 15 Ma. In 1983 paleontologists, reexamining rock samples collected earlier at Transantarctic Mountain sites 2,500 m above sea level, found marine microfossils 2–5 Ma old. The age of these fossils began a debate about the stability of the East Antarctic Ice Sheet. On one side, the "dynamists" believe that the ice sheet expands and decays periodically and about 3 Ma ago it melted, leaving behind tree-covered islands. Their theory rests on evidence in the fossil record. Besides the marine microfossils, they have found fossilized beech trees that may have grown along the ice-free coasts of these small islands.

The opposing group (or "stablists") contends that the ice sheet first extended over East Antarctica 15–20 Ma ago and has remained stable. In 1995 this group found ice 8 Ma old in a region that the dynamists believe was clear of ice 3 Ma ago. Because a layer of volcanic ash that covered the glacial ice is about 8 Ma old, they assert that the ice is as old or older. The dynamists propose that temperatures may not have been warm enough to melt all the ice sheet, while other geologists suggest that a glacier may have pushed the volcanic ash over younger ice.

The marine-based West Antarctic Ice Sheet has also fluctuated in size and may have disintegrated at least once during the last 600,000 years, perhaps as recently as 125,000 years ago. Since the 1980s researchers have discovered that the thickness, velocity, and surface characteristics of the modern ice sheet are rapidly changing and may be tending toward collapse. In 1993 U.S. geologists reported the discovery of an active volcano beneath ice streams flowing through the ice sheet. Beneath the ice, a peak, registering a magnetic signature common to active volcanoes, rises 650 m high, atop a 23 km diameter caldera or crater. Their data show a distinct hold in the ice, 6 km across and 50 m deep, where ice flows downward and melts into a hot spot.

This discovery implies that local geology affects the ice sheet's dynamics, independent of climate change. The heat from subglacial volcanoes melts the ice from beneath, saturating sub-ice sediments and triggering the flow of the ice streams over 7 m of sediments. A change in geothermal flux or rock type beneath the ice could cause the streams to flow more rapidly. If an ice stream's downstream terminus retreats to the geological boundary at the active edge of the rift system, the inland ice sheet would no longer be protected and could fall prey to rapid razing by the sea.

Bibliography

Antarctic Journal of the United States, ed. W. Reuning. Office of Polar Programs, National Science Foundation, Washington, DC.

Arctic Research of the United States, ed. C. Myers. Office of Polar Programs, National Science Foundation. Washington, DC.

WINIFRED REUNING

POLE FLATTENING

See Earth, Shape of

POLLUTION OF ATMOSPHERE

The atmosphere of Earth, when viewed from space, appears as a fragile blue blanket enveloping the water planet. This air is a curious mixture of gases and particulate material. The nature of the mixture is highly variable depending on the altitude, geographic location, and time of year (*see* EARTH'S ATMOSPHERE). The troposphere is that part of the atmosphere that extends about 10–15 kilometers (km) above Earth's surface. Weather events and volcanic eruptions occur in the troposphere. The troposphere is where emissions from industrial sources, such as factories, automobiles, and power plants, are released. Another source of emissions comes not from industry but rather from the burning of forests, grasslands, and agricultural debris. Are the activities of the human inhabitants of Earth deleteriously changing the composition of the air?

This entry will address the key sources of atmospheric pollution. The sources and transport of the emissions, especially for sulfur dioxide, carbon monoxide, ozone, and nitrogen dioxide, will be described. An assessment of the changes in the Antarctic ozone hole will be presented.

Air Quality

Air pollution has been a major concern of the United States since the early 1980s. The Clean Air Act Amendment of 1990 continues to be updated, most recently in July 1994, and now lists sulfur dioxide, suspended particulates, carbon monoxide, ozone, non-methane hydrocarbons, nitrogen dioxide, and lead as the top seven pollutants to be monitored. Each of these pollutants has standard maximum concentrations above which the levels are considered to be unsafe. As a consequence of monitoring conducted by the U.S. Environmental Protection Agency, we know that the air quality has been gradually improving: lead levels in the air have dropped by 89 percent since 1982 as a result of the removal of lead from automobile gasoline; carbon monoxide levels are lower by 30 percent because of improvements in combustion efficiencies in power plants and automobiles; nitrogen oxides have been reduced by 6 percent, ozone by 8 percent, and sulfur dioxide by 20 percent. Even levels of fine particulates have shown 10 percent reductions since 1987.

But even with these general improvements in air quality, urban smog remains the most common air pollution problem. Tropospheric ozone, formed from hydrocarbon and nitrogen oxide emissions, mainly from motor vehicles, factories, and power plants, is a primary constituent in urban smog. High levels of ozone in the lower atmosphere may cause severe respiratory discomfort to sensitive people and may cause damage to forests and crops. Releases of carbon monoxide and particulates exceed federal safe guidelines in some areas during the cooler months; large annual releases of sulfur dioxide and nitrogen oxides contribute to a variety of deleterious conditions including acid rain, reduced visibility, and respiratory illness (Curtis and Walsh, 1993).

Air motions recognize no international boundaries and transit for hundreds to thousands of kilometers without constraint, save that of gravity. As the global community embraces newer technologies and consequently releases more emissions of potentially toxic materials, the need for global awareness and monitoring of air pollution will increase. The possibility for mankind to alter the global climate is becoming a reality. Global emissions of greenhouse gases continue to increase, although the rate of increase has declined to only 1–2 percent during the 1990s (Novelli and Rossen, 1995).

Sulfur Dioxide and Acid Rain

Sulfur dioxide emissions are estimated globally at 200 million tons per year. The use of coal for energy is a major source of sulfur dioxide emissions to the atmosphere. The sulfur dioxide is actually created by the desulfurization of the coal before it is used; then it is released in the flue gases to the atmosphere. Untreated coal may release even more SO_2 than that released by the desulfurization process. The sulfur dioxide then travels away from the release point by the prevailing winds, usually between 300 meters (m) and 1.5 km above the ground. When this air is swept into a weather system that produces rain, the sulfur dioxide chemically combines with water and ozone to produce sulfuric acid. A similar pathway exists for nitrogen dioxide, transforming it into nitric acid. Acid rain is the removal of these acids from the atmosphere by precipitation; this process requires about 100 hours in average conditions. The effect of the acid rain can be detected not only in the pH level (acidity) of the rainfall, but also in the pH of surface water and soil, in the health of the ecosystems and people, and in damaging effects on buildings and historic monuments, like the Parthenon in Greece, the Taj Mahal in India, and the Colosseum in Rome. Higher concentrations of suspended sulfates (6–10 micrograms per cubic meter or $\mu g/m^3$), sulfur dioxide (50 $\mu g/m^3$), and sulfuric acid (10 $\mu g/m^3$) can lead to increases in the incidence of chronic bronchitis, emphysema, allergies, asthma, and lung cancer (*see* PUBLIC HEALTH AND EARTH SCIENCE). In land ecosystems, plants react in an indirect way as a result of the nutrient breakdown in the soil from the washout of aluminum (Al) and magnesium (Mg) by acid rain. Often plant growth and photosynthesis decline from the penetration of the acids through the stomata of the leaves. Lichens die when exposed to atmospheric concentrations of sulfuric acid at 0.01–0.03 milligrams per gram (mg/g). Coniferous trees die at concentrations of 0.07–0.08 mg/g. Bird eggs suffer nearly 80 percent breakages when acid rain falls. Aquatic ecosystems are damaged by the runoff of the acid rain and from the washout of the metals from the soil matrix. Polluted snow can represent a hazard to freshwater ecosystems, where fish eggs and amphibians are particularly vulnerable, throughout the melting period. Clearly, the negative effects from anthropogenic emissions of sulfur dioxide and the acid rain that forms from it warrant continued global attention. International cooperation in reducing the emissions of sulfur dioxide, for example, by the use of smokeless fuel (switch from coal to gas) or by increasing the use of nuclear or solar power, is critical in order to significantly reduce the effects of global pollution and acid rain (Izrael, 1983).

Carbon Monoxide and Carbon Dioxide

The most well known of greenhouse gases, carbon dioxide, is released into the atmosphere by fossil fuel combustion and deforestation (*see* GLOBAL CLIMATIC CHANGES, HUMAN INTERVENTION). The seasonal fluctuations in carbon dioxide are largely controlled by the respiration of plants in the Northern Hemisphere. The average annual carbon dioxide concentration is about 350 mg/g and is increasing by roughly 15 mg/g per year. Chemical reactions in the troposphere also contribute to the global abundance of carbon dioxide because all of the carbon species, carbon monoxide and methane in particular, oxidize to carbon dioxide.

Carbon monoxide is a colorless, odorless, tasteless gas that is formed during incomplete or inefficient combustion. Carbon monoxide is an important trace gas in the troposphere because it is the primary reactant with the hydroxyl radical. The hydroxyl radical is more rapidly consumed in areas of enhanced carbon monoxide. This situation gives rise to apparent increases in other radiatively important greenhouse gases, such as methane. Peak concentrations of carbon monoxide of about 200 parts per billion (nanograms per gram or ng/g) occur in the Northern Hemisphere in the late winter with minimum concentrations in the Southern Hemisphere of about 50 ng/g. The concentrations are believed to be increasing by about 1 percent per year. The total global emission from natural and anthropogenic sources is estimated to range from 600–1,700 teragrams (10^{12}) of carbon per year (Wuebbles and Edmonds, 1991). Natural sources and anthropogenic sources are thought to contribute 50 percent each to the global emissions. Because carbon monoxide is highly reactive, the global source and sink relationships are very uncertain. Because the lifetime is relatively short, on the order of a few weeks to a few months, the global distribution of carbon monoxide exhibits significant variability in response to the strength and locations of the sources for the gas and to changes in the weather patterns. Thus, space-based measurements of carbon monoxide represent the best sampling strategy to determine the global distribution of this highly variable trace gas.

During 1994, the Measurement of Air Pollution from Satellites (MAPS) experiment flew on the space shuttle during April and October. The MAPS carbon monoxide measurements are shown in Figure 1. The MAPS data represent the average carbon monoxide mixing ratio between 2–12 km above the earth's surface, with the greatest sensitivity to the carbon monoxide present at about 8 km. In April, the highest mixing ratios (120 parts per billion by volume, or ppbv) were located over the northern high latitudes (in dark shades of gray). The mixing ratios decreased to about 50 ppbv over the high southern latitudes (in light shades of gray). This gradient can be seen in Figure 2 where the average carbon monoxide as measured by MAPS from 57°N to 57°S is plotted with independent measurements of carbon monoxide at various

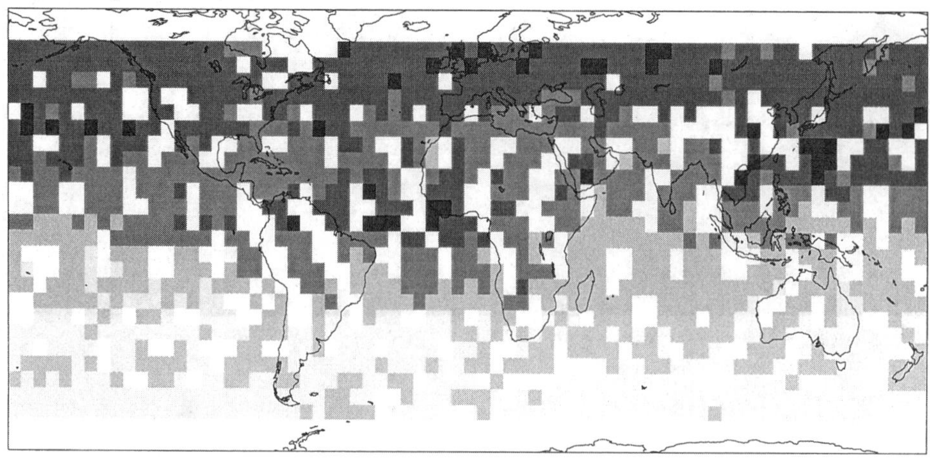

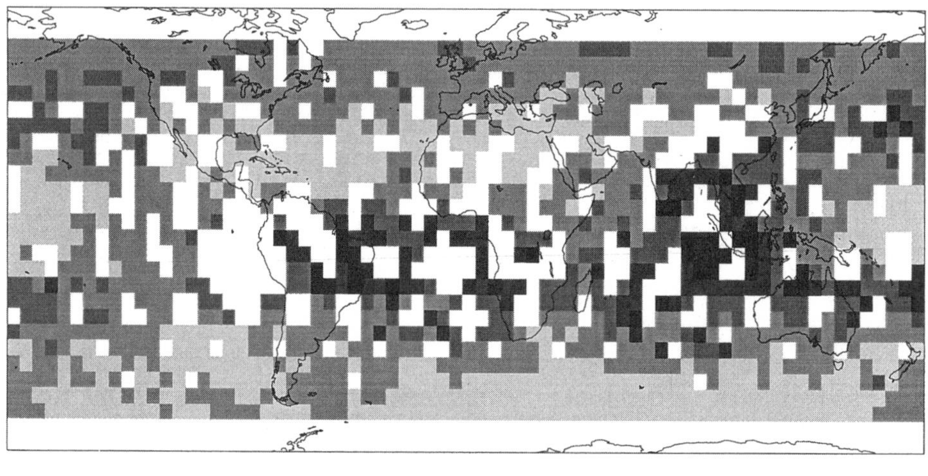

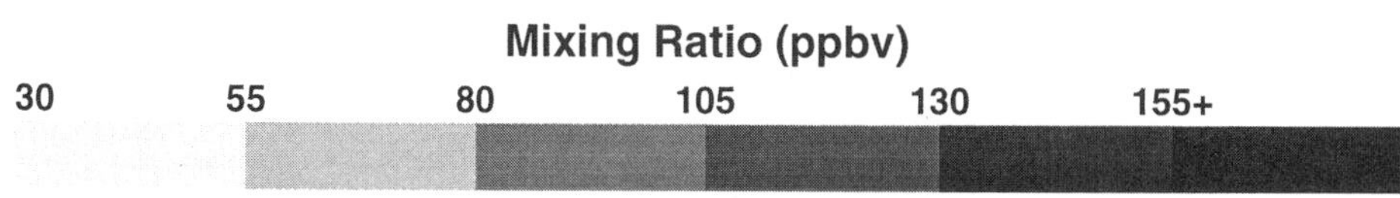

Figure 1. Measurement of air pollution from satellites: carbon monoxide mixing ratios in middle troposphere during April and October 1994. (NASA Langley Research Center/Atmospheric Sciences Division.)

locations. The gray shading shows one standard deviation in the measurements across each degree of latitude. Again the classic late winter/early springtime distribution is reflected in the MAPS data. In October, the photochemical destruction in summer and early fall has decreased the mixing ratios in the northern high latitudes to about 100 ppbv. The highest mixing ratios (greater than 130 ppbv) are located in the Tropics near or downstream from areas experiencing biomass burning. The mixing ratios south of 25°S decrease to a background value of about 70 ppbv. The zonal average of CO in October is shown in Figure 3 where the peak values lie clearly between the equator and 30°S. Two ten-day missions only gave a hint of what is really occurring in the atmosphere. Many unanswered questions remain about the nature of the shift in the global patterns from one season to the next, of the relative impact of the southern hemispheric burning season as compared to the

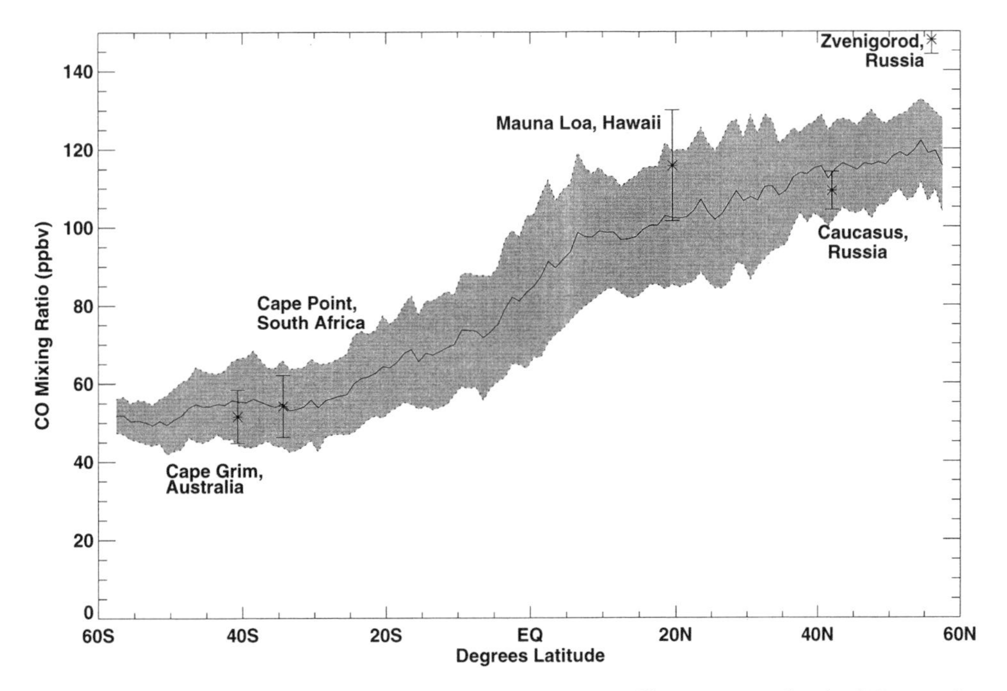

Figure 2. Measurement of air pollution from satellites tropospheric CO zonal averages, 9–19 April 1994.

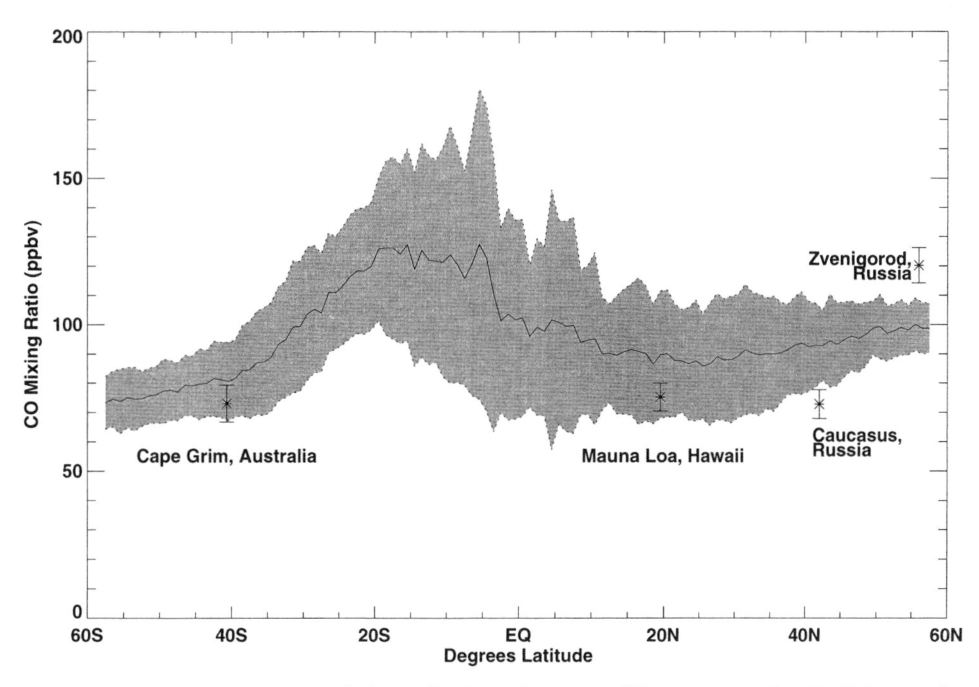

Figure 3. Measurement of air pollution from satellites tropospheric CO zonal averages, 30 September–11 October 1994.

technological emissions of the northern hemisphere, and of the relationship between CO emissions and long-range transport. New CO measurement systems are being readied for launch into space: the MicroMaps instrument on the Clark Spacecraft in early summer 1996; a modified MAPS instrument on the Russian space station *Mir* in late summer 1996; the Measurement of Pollution in the Troposphere (MOPITT) experiment on the first Earth Observing Platform in 1998; and the Tropospheric Emission Spectrometer (TES) on the second Earth Observing Platform in 2002. These longer-duration experiments, lasting from six months to three years, will hopefully address these and other critical questions on the nature of carbon monoxide and other trace gases.

Ozone

Between 15 and 30 km above the earth, ozone is produced naturally by the photodissociation and recombination of molecular oxygen. Photodissociation is a process by which absorbed photons provide the energy needed to break molecular bonds. The resulting monotonic oxygen combines with diatomic oxygen to form ozone, O_3. For O_2, the photons must be at wavelengths shorter than 242 nanometers (nm). Usually these short wavelengths of light are totally absorbed only in the stratosphere, and so, before the 1940s, there was no known mechanism for the production of tropospheric ozone. Yet, following World War II, with cars being driven in southern California in ever-increasing numbers, a new air quality condition was recognized and named photochemical "smog," a word combination of smoke and fog. The primary constituent of smog is tropospheric ozone.

Tropospheric ozone is formed over and especially downwind of large urban areas where the emissions of nitric oxide and nitrogen dioxide with volatile organic compounds in the presence of visible sunlight (wavelengths less than 420 nm) react to form ambient ozone concentrations up to three times the World Health Organization's standard of 120 ng/g over a one-hour period. The unpleasant and unhealthy effects of the formation of tropospheric ozone have spurred extensive research by the international scientific community. The global distribution of tropospheric ozone can be estimated by using the ozone measurements from two different satellites: the Total Ozone Mapping Spectrometer (TOMS) and the Stratospheric Aerosol and Gas Experiment (SAGE). The total ozone levels are described by a special unit of measure, the dobson unit (DU), which is 2.69×10^{16} molecules of ozone per cubic centimeter (cm^3). A typical amount of ozone in the stratosphere is 300 DU. Tropospheric ozone values are on the order 30 DU. In the middle and high latitudes the ozone concentrations can range from about 225–500 DU. Near the Antarctic ozone hole, values as low as 100 DU have been observed. Figure 4 shows the average global distribution of tropospheric ozone based on data collected between 1979 and 1991. The highest concentrations of ozone pollution are between tropical regions of South America and Africa and downwind from North America, Europe, and Asia. The long-range transport of locally produced ozone and its precursors is an important aspect of which areas will be affected by this air pollutant. If tropospheric ozone is the "bad" ozone, then stratospheric ozone is the "good" ozone. Stratospheric ozone is responsible for filtering and removing the ultraviolet wavelengths from the incoming radiation from the sun. During the long and dark winter months over the South Pole, the air becomes cold and the mixture includes some compounds not produced in the natural environment. Chlorofluorocarbons (CFCs), CFC-11 and CFC-12 in particular, have concentrations of about 0.260 and 0.445 ng/g, respectively, as measured at Mauna Loa in 1989, and lifetimes of 65 and 120 years. They are strong absorbers in the infrared region of the spectrum, which means they are also "greenhouse gases." A molecule of CFC-12 has about a 16,000 times the radiative forcing impact (warming effect) of a molecule of carbon dioxide. In 1991 CFCs were increasing at a rate of about 4 percent per year; in 1994 the growth rate was determined to be still positive but only about 2 percent per year. This slowing down of the CFC emissions is the result of the planned phaseout in the use of CFCs by manufacturing industries through international agreements; however, CFCs will continue to be present in the Antarctic stratosphere until the end of this century.

In October, when spring begins over the South Pole, the sunlight reaches the cold atmospheric mixture and the chemical reactions that result in stratospheric destruction of ozone begin. Fifteen years of TOMS data are shown in Plate 33. The area in deep blue and purple shows the location of the ozone hole. In the early 1980s the stratospheric

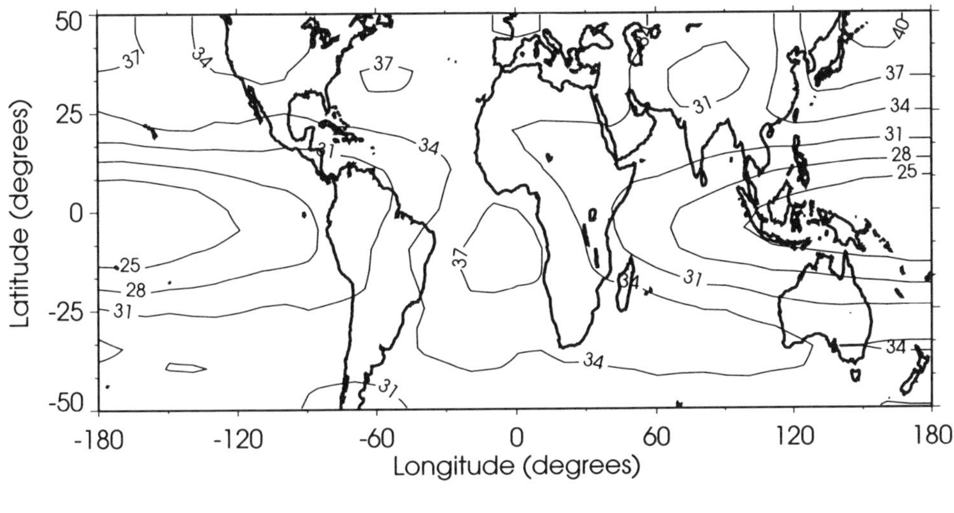

Figure 4. Tropospheric residual ozone (January–December).

concentrations were roughly 200 DU, but in the late 1980s and consistently in the 1990s, the stratospheric losses were even more severe, with the ozone concentrations falling to 100 DU in 1993.

Nitrogen Dioxide

Nitrogen oxide and nitrogen dioxide, often described together as NO_x, are one of the important precursors of tropospheric ozone. Sources for NO_x include the microbial activity in soils, lightning, biomass burning (particularly in the Tropics), aircraft emissions, and fossil fuel combustion. Fossil fuel combustion is the largest source of NO_x at about 24 teragrams of nitrogen per year. Since 1970, the emissions of NO_x from fossil fuel combustion have increased nearly 25 percent. The removal processes of NO_x include the oxidation of NO_x and the dry deposition of NO_2. While the removal processes are generally well understood, the global distribution of the amount of NO_x is not.

By developing extensive monitoring strategies and new measurement systems, the process by which air pollution forms and the extent to which the populations of the earth are affected can be understood. Clearly, air pollution will be a continuing concern in the years ahead. The NASA Earth Observing System will begin a new era of atmospheric monitoring from space in 1998. A better and more complete understanding of the changes in our atmospheric composition will be possible with the new suites of instruments viewing the earth from space and communicating these atmospheric observations directly to the international community of scientists.

Bibliography

Albritton, D. R., Watson, and P. Aucamp. *Scientific Assessment of Ozone Depletion: 1994.* World Meteorological Organization Global Ozone Research and Monitoring Project, Report No. 37. Geneva, 1995.

Bakwin, P., P. Tans, and P. Novelli. "Carbon Monoxide Budget in the Northern Hemisphere: Impact of Emission Controls." *Geophysical Research Letters* 26 (1994):433–436.

Connors, V., M. Flood, T. Jones, B. Gormsen, S. Nolf, and H. Reichle, Jr. "Global Distribution of Biomass Burning and Carbon Monoxide in the Middle Troposphere During Early April and October 1994. In *Biomass Burning and Global Change.* Cambridge, MA, in press.

Curtis, D., and B. Walsh, eds. *Environmental Quality.* Washington, DC, 1993.

Izrael, J. "Acid Rain and Its Environmental Effects." In *Long-Range Transport of Sulfur in the Atmosphere and Acid Rain,* ed. A Wiin-Nielsen. WMO Report No. 603. Geneva, 1983.

Logan, J., M. Prather, S. Wofsy, and M. McElroy.

"Tropospheric Chemistry: A Global Perspective." *Journal of Geophysical Research* 86 (1981):7210–7254.

MELLANBY, K., ed. *Air Pollution, Acid Rain and the Environment.* Watt Committee on Energy Report No. 18. London, 1988.

NOVELLI, P. "Recent Changes in Atmospheric Carbon Monoxide." *Science* 263 (1994):1587–1590.

NOVELLI, P., and R. ROSSON, eds. *Report of the World Meteorological Organization Meeting of Experts on Global Carbon Monoxide Measurements.* Boulder, CO, 1995.

WUEBBLES, D., and J. EDMONDS. *Primer on Greenhouse Gases.* Chelsea, MI, 1991.

VICKIE S. CONNORS

POLLUTION OF GROUNDWATER AQUIFERS

Groundwater pumped from below the land surface is an important source of drinking water throughout the world (*see also* GROUNDWATER). Groundwater is used extensively even in areas where there is plenty of surface water in lakes and streams. The preference for groundwater stems from the fact that it is generally assumed to be less susceptible to recognized threats to surface water quality, such as releases of untreated sewage, industrial process water, and agricultural runoff. While the assumption that groundwater is generally less vulnerable to pollution than surface water is correct, this does not mean that groundwater is always free of contaminants. On the contrary, pollution of groundwater resulting from human activities is neither a new nor an uncommon environmental problem. This entry will describe sources of groundwater contaminants, the ways in which pollutants enter and move in groundwater aquifers, and the options available to deal with the environmental problems posed by these pollutants.

Groundwater pollution has posed risks to human health since people began to settle in villages and locate wells near areas used for disposal of wastes. Only in the last few centuries, however, have the links between polluted groundwater and disease been clearly recognized. A cholera outbreak in London in the 1800s was possibly the first epidemic to be directly attributed to a contaminated well. Outbreaks of disease due to bacteria or viruses in groundwater continue to occur where well or spring water is untreated or inadequately disinfected. One study identified over 40,000 cases of illness in the United States between 1971 and 1982 that were caused by biological contamination of groundwater (Patrick et al., 1987).

Other pollutants found in groundwater are the result of natural processes, in some cases accentuated by human activities. Isotopes of radium and radon arising from the radioactive decay of uranium in rocks making up the aquifer can reach concentrations that exceed recommended drinking water standards. High concentrations of arsenic, dissolved from minerals in the aquifer, are often found in groundwater from volcanic rocks and in geothermal areas. Selenium, an essential element for human or animal growth but toxic in high concentrations, can dissolve from soils during irrigation. Aquifers in coastal areas may contain both fresh and salt water. A relatively stable boundary between the two water types can exist because salt water is denser than freshwater, allowing the freshwater to float above it. This boundary will shift if excessive pumping causes salt water to flow inland, a process called salt water intrusion. Once the salt water has moved inland, it cannot be easily flushed out, and affected wells must be abandoned because they no longer yield water that is suitable for human consumption.

While naturally occurring and biological pollutants in groundwater are still of concern today, even greater risks are posed in many industrialized nations by a variety of natural and synthetic chemicals. Again this is not a new problem. Petroleum production in Indiana during the early 1900s generated oil and salt water (oil field brine) that entered a shallow aquifer and polluted several wells. Acidic water draining through coal mines was also recognized in the early 1900s as a source of both surface and groundwater pollution. The extent of chemical contamination has increased dramatically with the increased production and use of industrial and agricultural chemicals since the 1940s. These chemicals can pollute groundwater as a result of accidental spills, leaks of pipelines and storage tanks, intentional disposal in ponds or landfills, and fertilizer and pesticide applications to farmland and suburban lawns.

Pollutants enter groundwater by a variety of pathways and processes. Waste spilled or disposed of at the land surface as a liquid can infiltrate through soil to the water table. Melted snow that has dissolved road salt and water from polluted lakes and streams can also infiltrate and serve as a source of groundwater contamination. Some liquids will mix with and dissolve in groundwater at the water table. Once mixed with groundwater, these pollutants move in the direction of groundwater flow. Other chemicals, such as gasoline and diesel fuel, do not mix readily with water and can form a separate "non-aqueous phase liquid," or NAPL, that is lighter than water and spreads along the water table. A light NAPL can move into and float on the water surface in a well. Still other chemicals, such as organic solvents used for dry cleaning and a variety of metal finishing processes, form dense NAPLs that sink through the saturated zone until they encounter a low permeability confining unit. As it sinks through the saturated zone, some of the NAPL may become trapped in pores as small, irregularly shaped blobs. The remainder accumulates at the base of the aquifer. Although most of the NAPL remains as a separate phase that does not move readily into a well, chemicals in the NAPL will also dissolve slowly as groundwater flows past. The NAPL therefore becomes a long-term subsurface source of pollutants.

Fertilizers and pesticides deposited as solids at the land surface and solid wastes buried in landfills can dissolve in rainwater that infiltrates through the soil to become groundwater recharge. Once the infiltrating water reaches the water table, the fate of these pollutants is similar to that described above for liquids that mix with groundwater. Other pollutants are added directly to groundwater through injection wells used to dispose of oil field brines and chemical wastes. Although injection systems are usually designed to force the wastes into a deep aquifer containing water that is already too salty to be used as a water supply, wastes can enter shallower aquifers due to leaks.

Systematic investigations to determine the extent of groundwater contamination from chemical sources in the United States began in the 1980s, following the discovery of a number of abandoned disposal sites such as that at Love Canal in Niagara Falls, New York. Many chemicals of concern pose potential health threats at concentrations that could not even be measured in the early 1980s. Thus, as investigations continue and methods for detecting pollutants become more sensitive, the extent of known contamination will increase. Despite the large number of sites at which pollutants have been identified, the overall quality of groundwater is still very good. A 1986 survey in the United States (U.S. Geological Survey, 1988) found, in every state, that most of the groundwater meets drinking water standards. Estimates of the volume of polluted groundwater are generally less than 2 percent (Patrick et al., 1987). Unfortunately, the small amount of groundwater that is contaminated tends to be at shallow depths and near population centers where groundwater is heavily used.

Once pollutants have entered an aquifer, they are very difficult to remove. Because of its slow velocity, groundwater may take hundreds to tens of thousands of years to flow from its recharge area to a discharge area under natural conditions. Groundwater velocities increase during pumping, but even under ideal conditions it may be necessary to pump for many years to remove dissolved contaminants. A pumping well may pull in groundwater from the zone of contamination, or plume, as well as additional clean water from nearby portions of the aquifer. As a result, the volume of groundwater that must be pumped is generally much greater than the actual volume of the plume. Some dissolved chemicals, including many organics, metals, and radionuclides, may also interact with the sediments or rock in the aquifer. This causes them to move even more slowly than the water itself, sometimes over 1,000 times more slowly. Finally, as noted above, NAPLs may serve as long-term hidden sources of pollutants. Very large volumes of water are required to completely dissolve only a small volume of NAPL chemicals. Less than one sixty-gallon drum of organic solvents can create a contaminant plume several kilometers long (Mackay and Cherry, 1989).

The most frequently used technology for treating polluted aquifers, called "pump-and-treat," involves removing groundwater from the ground by pumping and then treating it to remove pollutants before it is used. This strategy can be effective in preventing the further spread of pollutants, but it will not restore the aquifer to pristine conditions in the short term, particularly if NAPLs are present. In some cases where pump-and-treat systems initially appeared to have cleaned an aquifer, concentrations of pollutants later returned to their original levels as a NAPL continued to dissolve once the pumps were turned off. Development of methods to enhance the removal of groundwater contaminants and to treat them within the aquifer, either by chemical or biological means, is an active area of current research.

Bibliography

MACKAY, D. M., and J. A. CHERRY. "Groundwater Contamination: Pump-and-Treat Remediation." *Environmental Science and Technology* 23 (1989):630–636.

PATRICK, R., E. FORD, and J. QUARLES. *Groundwater Contamination in the United States*, 2nd ed. Philadelphia, 1987.

U.S. GEOLOGICAL SURVEY. *National Water Summary 1986—Hydrologic Events and Ground-Water Quality.* USGS Water Supply Paper 2325. Washington, DC, 1988.

JEAN M. BAHR

POLLUTION OF LAKES AND STREAMS

The hydrologic system is highly susceptible to human influence. Pollutants introduced to the hydrologic system eventually spread and enter the stream networks. All natural waters contain dissolved chemicals; however, some of those substances, often resulting from human activity, may and do cause contamination of stream and lake water. As the examples below illustrate, all too often in the past, the needs of development have ignored the environmental impact.

Ohio's Cuyahoga River winds through Cleveland and Akron. A large number of chemical factories line its once wooded banks. By the year 1959, chemical waste discharges to the Cuyahoga River from these industrial sites had reached nearly 100 tons per day. In the summer of that year, the river became so loaded with volatile organic chemicals that it actually caught fire and burned for eight days. In the fall of 1963, residents along the Mississippi River were startled to find thousands of dead fish floating in the river. Over five million fish had been killed by the release of the pesticide endrin into the river. In the late 1950s and early 1960s, native fish production in Lake Erie had fallen dramatically. The lake was receiving so much sewage, ladened with phosphates and detergents, that it was literally dying (Scheffer, 1991).

These incidents contributed to a growing concern regarding the potential impact of chemicals on the environment (Carson, 1962; *see* CARSON, RACHEL). In the late 1960s, the public learned of the dangers of DDT, dioxin, and agent orange (2,4-D and 2,4,5-T). These chemicals, known to be harmful to the environment, were now being found in some lakes and streams. In the early 1970s, a large number of toxic waste dump sites were discovered, and in 1976 major contamination was discovered at Love Canal, on the Niagara River in New York. Various industries had used the canal to dispose of chemical waste since the late 1940s. It is estimated that approximately 20,000 metric tons of chemical wastes had been buried in the canal (Miller and Miller, 1991). The area where much of the waste had been buried was now a residential area and exposure to leaking chemical waste had resulted in disease and injury to some of the residents.

The Love Canal incident led to the passage of the Clean Water Act of 1972, specifically designed to return lakes and streams to a more pristine state. Since the passage of the Clean Water Act, significant progress has been made toward that goal. However, the threat to our waterways still exists and their protection requires constant vigilance. Chemicals are an important part of our industrial- and technologically based society. They are used in a wide range of manufacturing processes, in agriculture, in medicine, in many common businesses, and in the household. Over seventy thousand chemicals are now used regularly, with hundreds added every year; the potential toxicity of many of these is poorly known (Global Tomorrow Coalition, 1990).

An additional threat to our waterways is wastewater from our growing population. Only since the passage of the Clean Water Act has the loading of raw municipal sewage into streams and lakes been curtailed. Primary treatment of municipal wastewater has significantly reduced the biological oxygen demand in our streams. Still a problem in many areas, however, is nutrient loading, particularly of nitrogen and phosphorous.

A regulatory framework is in place to reduce the contaminant loading of lakes and streams. However, it is important that we learn where the water in our lakes and streams comes from so that we can better understand how it can become contaminated.

Contaminant Sources

Surface water, of course, can become contaminated as a result of runoff through our across areas that have been contaminated or affected by human activity, or through the direct introduction of wastewater to the lake or stream. However, two additional less obvious pathways occur: by precipitation originating from or passing through polluted atmosphere and through the subsurface, as a result of seepage of contaminated groundwater into the stream or lake.

Atmospheric Sources. The atmosphere contains a number of pollutants (*see* POLLUTION OF ATMOSHPERE) including gases, hydrocarbons, and particulates (Global Tomorrow Coalition, 1990) consisting of dust, various chemicals, and metals such as lead. Precipitation typically removes some of these contaminants. Those particulates that serve as condensation nuclei (Miller et al., 1983) will fall with the resulting raindrop or snowflake. Gases in the atmosphere can dissolve within the condensed water.

A significant threat to watersheds is acid rain, resulting from the introduction of sulfur dioxide (SO_2) and oxides of nitrogen into the atmosphere. The presence of SO_2 is primarily from the burning of fossil fuels, whereas the nitrogen compounds are largely from automobile emissions (Drever, 1982). These compounds combine with water in the atmosphere to form sulfuric acid (H_2SO_4) and nitric acid (HNO_3). A pH of precipitation less than 4.5 is common in areas of high emissions (Press and Siever, 1994).

The impact of acid rain varies from accelerated weathering of buildings made of soluble rock such as limestone, to loss of vegetation and reduction of freshwater fish populations (Drever, 1982; Press and Siever, 1993). Damage resulting from acid rain is severe in Scandinavia, Central Europe, and in eastern North America. Fish have disappeared from four thousand lakes in Sweden, while 80 percent of Norway's lakes are either technically "dead" or have been seriously damaged; thousands of lakes on the eastern seaboard of the United States have levels of acidity making them incapable of supporting native fish (Global Tomorrow Coalition, 1990). In England there has been significant damage to forests, and in North America, tree damage extends from eastern Canada to Georgia (Global Tomorrow Coalition, 1990).

Efforts to offset the impact of acid rain vary from direct neutralization of acids in individual lakes and soils to more long-term solutions of reductions of contaminant emissions. In 1991, the U.S. Congress passed the Clean Air Act requiring a reduction by ten million tons the amount of sulfur emitted annually by coal-burning plants by the year 2000.

Surface Sources. Surface contaminant sources are described in terms of their localization as either "point" or "nonpoint" sources (Domenico and Schwartz, 1990). Point sources are characterized by a readily identifiable localized source, for example, a leaking underground storage tank, direct discharge of municipal or industrial wastewater, a contaminant spill, mine drainage, and so on. Nonpoint source contamination results from an area-wide practice: the contaminants cannot be traced back to a single location. Examples include the widespread application of agricultural chemicals, areas with individual septic tanks, stormwater runoff, or sediments derived from development and logging.

The Clean Water Act and its amendments provide a basis for controlling contaminant releases to streams and lakes. Certain point source discharges, such as wastewater discharges, are regulated through the National Pollutant Discharge Elimination System (NPDES) permit program (USEPA, 1993). Facilities gain a NPDES permit by demonstrating that their releases will not have adverse impacts on the water source. A number of best management practices (BMPs) are commonly used in order to reduce contaminant loading (USEPA, 1991; USEPA, 1993). Monitoring of discharge is required under the NPDES permit system.

Municipal wastewater discharges to waterways may contribute pathogenic organisms, such as bacteria, viruses, and parasites, as well as nutrients, particularly nitrogen (N) and phosphorous (P) (see agricultrual sources below). Nutrient loading, that is, the addition of N and P to surface water, may lead to algal blooms (Viessman and Welty, 1985). Subsequent die-off and decay of these organisms may result in eutrophication (depletion of dissolved oxygen) in the water body. Oxygen depletion in lakes and streams poses a serious threat to aquatic life.

A requirement of the Clean Water Act is the monitoring of releases in terms of total maximum daily loads (TMDLs) for a given stream. In the TMDL process, contributions from both point and nonpoint sources to a specific watershed are limited to what can be used or assimilated by the water body on a daily basis.

Industrial discharges, through direct conveyance, spills, leaking storage tanks and lagoons, or through underground injection wells, pose a risk to surface water sources. Contaminants, which include fuels, solvents, and heavy metals, may gain access to the surface water source directly, through discharge lines or overland flow, or indirectly by contaminating groundwater that is enroute through the aquifer to the river or lake.

Mine drainage occurs when water drains through mine debris that contains leachable minerals. Pyrite, a common iron sulfide mineral, reacts with water to form sulfuric acid. If this water reaches streams, vegetation may die, aquatic organisms are affected, and iron is precipitated along the stream (Drever, 1982). Acid mine drainage impacts have been observed from the coal mines of the eastern United States to metal mines in the West.

Nonpoint sources such as agriculture may contribute a range of contaminants to a watershed: nutrients from fertilizer and animal wastes, pesticides, and sediment. The potential impact of agricultural chemicals on lakes and streams can be minimized through the use of BMPs. Application rates of these chemicals should not exceed what can be used by crops in the field. Application should take into account the chemical's persistence, its tendency to be leached from the surface or to be carried away as runoff. Contamination of surface water (and groundwater) may occur if poor irrigation management practices are followed. Excess irrigation, or irrigation immediately after chemical application, may lead to transport of the chemical to waterways.

Storm water runoff may deliver nutrients, bacteria, petroleum products, and heavy metals to surface water. Contributions from runoff during storm events may exceed most other sources (Robbins et al., 1991). Runoff varies from approximately 10 percent on natural cover to over 50 percent when the percentage of paved surfaces exceeds 75 percent of the total surface area (MPCA, 1989). Stormwater NPDES permits are designed to minimize the potential impact of this contaminant source. A number of BMPs—such as grass filter strips, detention ponds, constructed wetlands, infiltration basins—coupled with source controls are being used (USEPA, 1993).

Sedimentation results from erosion, which increases when vegetation is removed from the watershed. A relationship is observed between certain land uses and erosion. Higher sediment yields are observed as a result of clear-cutting, conversion to farmland, and urban construction; the use of off-road vehicles has also led to increased sediment yield as a result of damage to vegetation (Keller, 1988). Best exploitation practices in forest management (e.g., selective logging and careful extraction techniques), in agriculture (e.g., land contouring and strip farming), and in stream management (e.g., riparian area maintenance) can minimize siltation problems (Global Tomorrow Coalition, 1990).

One of the pathways by which water is supplied to a stream or lake is the underground flow of groundwater. This supply of groundwater to a stream, which maintains stream flow between periods of rainfall, is referred to as the stream's base flow (Freeze and Cherry, 1979). If groundwater has become polluted as a result of contaminants percolating down from the surface to the aquifer, it is possible for those same contaminants to be transported to the surface water. An example of such a situation might be a case where shallow groundwater is affected by numerous individual septic systems or by the leaching of fertilizers from farmland. Nutrients, particularly nitrogen, may be delivered to a stream through base flow.

Watershed Protection

Watersheds can best be protected through a combination of contaminant source control (e.g., NPDES permits) and watershed protection strategies (Robbins et al., 1991). Watershed protection consists of several important elements. First of all, the watershed must be identified and sources of water (surface water and groundwater) that contribute to the lakes and streams be recognized. Next, the potential sources of contaminants within the watershed must be inventoried and, if possible, the relative risks they pose determined. Third, management strategies designed to minimize risk from these sources must be developed and implemented. One of these strategies should be public education aimed at providing information to individuals who live and work in the watershed on how their activities may affect water quality. Finally, because watershed protection must be an ongoing program, a monitoring scheme designed to evaluate the effectiveness of the program should be put into place.

Bibliography

CARSON, R. *Silent Spring*. Boston, 1962.

DOMENICO, P. A., and F. W. SCHWARTZ. *Physical and Chemical Hydrogeology*. New York, 1990.

DREVER, J. I. *The Geochemistry of Natural Waters*. Englewood Cliffs, NJ, 1982.

FREEZE, R. A., and J. A. CHERRY. *Groundwater*. Englewood Cliffs, NJ, 1979.

GLOBAL TOMORROW COALITION. "Fresh Water." In *The Global Ecology Handbook*, ed. W. H. Corson. Boston, 1990.

HEM, J. D. *Study and Interpretation of the Chemical Characteristics of Natural Water*. U.S. Geological Survey Water-Supply Paper 2254. Washington, DC, 1985.

KELLER, E. A. *Environmental Geology*, 5th ed. New York, 1988.

MILLER, E. W., and R. M. MILLER. *Environmental Hazards: Toxic Waste and Hazardous Material*. Santa Barbara, CA, 1991.

MILLER, A., J. C. THOMPSON, R. E. PETERSON, and D. R. HARAGAN. *Elements of Meteorology*, 4th ed. Columbus, OH, 1983.

MINNESOTA POLLUTION CONTROL AUTHORITY (MPCA). *Protecting Water Quality in Urban Areas: Best Management Practices for Minnesota*. St. Paul, MN, 1989.

PRESS, F., and R. SIEVER. *Understanding Earth*. New York, 1994.

ROBBINS, R. W., J. L. GLICKER, D. M. BLOEM, and B. M. NISS. *Effective Watershed Management for Surface Water Supplies*. Denver, 1991.

SCHEFFER, V. B. *The Shaping of Environmentalism in America*. Seattle, 1991.

USEPA. *Nonpoint Source Watershed Workshop*. U.S. Environmental Protection Agency, Seminar Publication No. EPA/625/4-91/027. Washington, DC, 1991.

———. *Urban Runoff Pollution Prevention and Control Planning*. U.S. Environmental Protection Agency Handbook No. EPA/625/R-93/004. Washington, DC, 1993.

VIESSMAN, W., JR., and C. WELTY. *Water Management: Technology and Institutions*. New York, 1985.

DENNIS O. NELSON

POWELL, JOHN WESLEY

As an explorer, geologist, and ethnologist, John Wesley Powell (1834–1902) contributed significantly to federal science and its management during America's Gilded Age. His work led to a better understanding of the arid lands, native peoples, and water resources of the American West.

Born in Mount Morris, New York, on 24 March 1834, Powell was the second son of Joseph and Mary Dean Powell. Powell's father, a Wesleyan preacher, moved his family to the Midwest. John Wesley Powell left farming in 1854 to teach and pursue natural history. Teaching in county schools supported his intermittent education at colleges in Illinois and Ohio, but he earned no degree.

Powell enlisted as an infantry private in April 1861, on the day after Fort Sumter surrendered, and was commissioned in June. A self-taught military engineer, he served on Ulysses Grant's staff and then led a battery of field artillery. In April 1862, during the struggle by Grant's army to hold the "Hornet's Nest" at Shiloh, a rifle ball destroyed Captain Powell's right arm. Nevertheless, returning to duty in February 1863, he saw combat again in the Vicksburg and Nashville campaigns. Powell commanded the artillery of the Seventeenth Corps before resigning with the rank of major in January 1865.

Powell soon began teaching geology and natural history at Wesleyan University, Illinois, and resumed museum curation for the State's Natural History Society. Summers spent with his wife and cousin Emma Dean, friends, and students in the central Rockies led to real exploration. On 24 May 1869, Powell and his team began a daring reconnaissance by boat down the Green and Colorado rivers that ended successfully on 30 August below the Grand Canyon. With funds from Congress, Powell made a second, more scientific voyage downriver in 1871. His account of these adventures, published in 1875, merged events from both trips.

Federal Surveyor

During the 1870s, Powell expanded his river reconnaissance into a survey of areas on and near the Colorado Plateau, first under the Smithsonian Institution and then within the Interior Department. From 1877, Powell's U.S. Geographical and Geological Survey of the Rocky Mountain Region operated south of the forty-second parallel as part of the Interior Department's work to assess and map the West by triangulation surveys that used a rectangular system, contour topography, and a uniform scale. By 1879, Powell and his colleagues had made significant contributions to understanding the development of mountain and valley landforms, to ethnology, and to classifying lands by their nature and use. Powell's interpretations of canyons as "antecedent," "consequent," or "superimposed" reflected the dynamic relations between rivers and the geologic structures they crossed. Powell also urged a more rational use of the West's lands and limited waters, and a wiser policy toward its native peoples.

In the debates after 1873 about improving the federal scientific surveys of the public lands in the West, Powell favored consolidation under civilian control. In March 1879, Congress and President Rutherford Hayes terminated the western surveys led by Powell, Ferdinand Hayden, and George Wheeler. To aid the nation's struggling economy, to advance a more reasoned use of its nonliving resources, and to improve its civil service, they established the U.S. Geological Survey (USGS) to classify scientifically the public lands and examine the geology, minerals, and products of the national domain. Powell supported Yale-educated CLARENCE RIVERS KING for USGS director. King had just completed his U.S. Geological Exploration of the Fortieth Parallel, the model for the other three surveys, and helped to shape the USGS.

Chief of the Bureau of Ethnology and USGS Director

The law that established the USGS also founded the Bureau of Ethnology (BE), and Powell moved to the Smithsonian to lead the BE. Powell served with King on the Public Lands Commission and, informally, as head of the USGS' General Geology Division, which worked to complete his and related studies in the West aimed at King's goal of producing a reliable geologic map of the United States.

King set high standards for the USGS. He organized a program of USGS applied and basic studies to support the mandated work on mineral resources and to provide a greater knowledge of Earth and its history. He gave the scientific staff freedom to accomplish these assigned mission-oriented tasks. When King resigned in March 1881, he recommended Powell as his successor..

Powell's goals and methods in science and its administration differed widely from those of King. Powell, representing an older, less-specialized tradition in science in America, believed that everything possible must be learned about a subject before the information gained could be applied to solving problems. This view paralleled his looser approach to management, one that allowed the USGS staff to choose their own subjects for study.

King and the other reformers did not succeed in establishing a separate agency for surveys of measurement and position. To advance the national geologic map, in 1882 Congress authorized USGS activities nationwide. Under this rubric, Powell quickly remade the agency into a bureau of topographic mapping and basic research in geology—but at the expense of the mandated studies of mineral resources.

USGS appropriations increased steadily in the 1880s under Powell's leadership, until Congress grew dissatisfied with the paucity of results useful to meeting national needs. Although Congress pressed for reform in 1887 by asking the USGS to itemize its requests for funds, the legislators also enabled Powell to pursue his longstanding goal of reforming land and water use in the West when they authorized and funded the Irrigation Survey (IS) within the USGS in 1888. When Powell refused to recommend sites whose selection would have reopened the public lands to entry and released dowry lands to six new states, Congress terminated the IS (1890). When the USGS could not respond to a renewed monetary crisis in 1890–1892, Congress selectively slashed the agency's statutory staff and its operating expenses. When Powell responded with a reduction-in-force that eliminated the USGS' best economic geologists and tried to transfer the agency from the Interior to the Agriculture Department, Congress encouraged him to leave by reducing his salary. Powell resigned in 1894.

CHARLES WALCOTT, the third director (1894–1907), restored Congress' confidence in and support of the USGS. His program balanced and expanded the agency's applied and basic research. Walcott reemphasized work in mineral resources and geology, reorganized other geological work, professionalized mapping, and began successful studies of water resources and land reclamation (*see also* GEOLOGICAL SURVEYS).

Last Years

After resigning as USGS Director, Powell underwent an operation on the stump of his arm and he remained chief of the BE. He continued to record his philosophy and to promote reclaiming arid lands by wise irrigation and land use. Powell died at his retreat in Haven, Maine, on 3 September 1902, less than three months after the Newlands Act established the Reclamation Service.

Bibliography

DARRAH, W. C. *Powell of the Colorado.* Princeton, NJ, 1951.

DUPREE, A. *Science in the Federal Government: A History of Policies and Activities to 1940.* Cambridge, MA, 1957.

GOETZMANN, W. H. *Exploration and Empire: The Explorer and the Scientist in the Winning of the American West.* New York, 1966.

MANNING, T. G. *Government in Science: The U.S. Geological Survey 1867[sic]–1894.* Lexington, KY, 1967.

RABBITT, M. C. *John Wesley Powell: Pioneer Statesman of Federal Science.* U.S. Geological Survey Professional Paper 669-A. Washington, DC, 1969.

RABBITT, M. C. *Minerals, Lands, and Geology for the Common Defence and General Welfare.* Vol. 1, *Before 1879;* Vol. 2, *1879–1904.* Washington, DC, 1979–1980.

STEGNER, W. E. *Beyond the Hundredth Meridian: John Wesley Powell and the Second Opening of the West.* Boston, 1954.

STRONG, D. H. *Dreamers and Defenders: American Conservationists.* Lincoln, NE, 1988.

WILKINS, T. *Clarence King: A Biography.* Albuquerque, NM, 1988.

CLIFFORD M. NELSON

PRECIOUS METALS

See Gems and Gem Minerals

PRIOR, GEORGE THURLAND

George Thurland Prior made contributions to mineral chemistry, igneous petrology, and, above all, to our understanding of the interrelationships among meteorites.

He was born in Oxford on 16 December 1862. At Magdalen College, Oxford, Prior took first class honors in chemistry (1885) and physics (1886). Prior went on to spend a few months studying chemistry under A. Classen, professor of inorganic chemistry at Aachen Technical High School, Germany, and in 1887 was appointed second class assistant in the Mineralogy Department of the British Museum (Natural History). He stayed there for the remainder of his career and was Keeper of Minerals (head of department) from 1909 until his retirement in 1927.

Prior was responsible for the museum's chemical laboratory and analyzed minerals in collaborative studies with his colleagues L. Fletcher, H. A. Miers, and L. J. Spencer. With philosophical and technical inventiveness the three mineralogists determined the optical properties of minerals and the angles between the faces of crystals, a diagnostic property. (Fletcher conceived the "optical indicatrix," a graphical representation of the variation in the velocity of light along different directions in a crystal. Miers invented a stage goniometer for measuring the angles between crystal faces under the microscope.) Prior was meticulous in insisting that material for analysis came from the exact specimen of which the physical properties had been determined. He was renowned for his manipulation and needed only a few milligrams of sample for an analysis, no mean feat with the techniques then available. New mineral deposits were being exploited worldwide, and Prior worked on a variety of substances sent to the museum from abroad, such as sulfarsenates (compounds of a metal, such as copper, with arsenic and sulfur), and copper and silver iodides. He and his collaborators sought order among minerals. Prior cautioned "against over-estimation of the exactness of mineral analyses when calculating formulae" and suggested, for example, that the rare earth element-bearing hydrated phosphates hamlinite, svanbergite, plumbogummite, beudanite, and florencite "form a natural group of rhombohedral minerals" (Smith, 1982). Although some of the mineral names have since been discredited, Prior's formulae are largely correct. For this mineralogical work, Prior was awarded the Wollaston Fund of the Geological Society of London in 1900.

Prior became interested in igneous petrology, so he learned optical petrography. He analyzed rocks returned from Antarctica by several expeditions, but more important, he studied suites of metamorphic and igneous rocks collected by others in Kenya and Uganda. Prior (1903) presented analyses and petrographic descriptions of the alkali-rich rocks of the Rift Valley, and data on related rocks from other igneous provinces. From his observations he concluded that the Atlantic, European, and East African volcanic chains "are characterized by the association of basalts and alkali-rich phonolitic rocks, whereas in the two other great Pacific chains . . . andesites are the prevailing lavas." This probably was the first recognition of what became known as the Atlantic and Pacific Provinces of igneous rocks. "Melilitite," the name he gave to a lava with over 10 percent of the mineral melilite, and analogous to the more common sodic rock, "nephelinite," has stood the test of time.

Like his predecessors, Story-Maskelyne and Fletcher, when Prior became Keeper in 1909, he began to work on meteorites. His analytical ability is exemplified by his analyses of the achondrite meteorite (*see* METEORITES) Nakhla (an igneous rock, possibly from Mars) that fell in 1911 (*see* METEORITES FROM THE MOON AND MARS). Prior obtained an analysis of the dominant mineral, calcium-pyroxene ("diopside"), by subtracting the contribution of acid-soluble olivine from his bulk analysis and recalculating to 100 weight percent. Prior skillfully maximized the data obtainable from each chemical analysis to provide mineralogical information. Most chondrites, the commonest meteorites, contain different proportions of metal, sulfide, and silicate. For an analysis, about 7 to 10 grams (g) of a chondrite were crushed and the metal extracted using weakly magnetic combs that he had devised. "Prior's combs" are kept in the chemical laboratory and are still used for analyzing chondrites. He analyzed the "attracted" (metal-rich) and "unattracted" (silicate-rich) portions separately, and used preferential acid attack to remove olivine from "unattracted" silicate. The bulk composition finally was assembled from all of the data.

From his own and other analyses Prior (1916) formulated two rules: "The less the amount of nickel-iron in chondritic stones, the richer it is in nickel and the richer in iron are the magnesium silicates". Metallic iron may oxidize to enter silicate, causing the abundance of metal to decrease, but the proportion of nickel increases, because it does not readily oxidize. He proposed "that meteorites have separated from a single magma which has passed through successive stages of progressive oxidation" (1920) and suggested that the most primitive meteorites are the enstatite chondrites and enstatite achondrites ("aubrites"), which contain negligible oxidized iron. Prior's classification scheme (1920) discounts secondary features such as color, veining, and brecciation (fragmentation and welding) and, correctly, relies on a primary feature, chemical composition. Although in advance of earlier schemes, it fails to accommodate the properties of most achondrites, irons, and stony-irons. More detailed chemical and oxygen isotopic analyses indicate that the chondrites are not part of a series based on redox state alone (Larimer and Wasson, 1988). Prior's rules may, however, hold within some individual chemical groups of chondrites. They also presaged, to some extent, the modern use of the compositions of meteorites in identifying chemical fractionations that resulted from processes that occurred in the embryonic solar system.

In 1923 Prior published a *Catalogue of Meteorites*, which established within the British Museum (Natural History) the world database on all known meteorites. The database is still maintained and disseminated.

Prior was Secretary (1909–1927) and President (1927–1930) of the Mineralogical Society of Great Britain and Ireland. He was awarded a D.Sc. by Oxford University and in 1912 he was elected a Fellow of the Royal Society. In recognition of his work he received the Murchison Medal (*see* MURCHISON, RODERICK) of the Geological Society of London (1927). In 1914 Prior married E. L. A. Cole of Cork and they had two daughters. He died on 8 March 1936 in Hatch End, northwest of London.

Bibliography

LARIMER, J. W., and J. T. WASSON. "Siderophile Element Fractionation." In *Meteorites and the Early Solar System*, eds. J. F. Kerridge and M. S. Matthews. Tucson, AZ, 1988.

PRIOR, G. T. "Contributions to the Petrology of British East Africa." *Mineralogy Magazine* 13 (1903):228–263.

———. "The Classification of Meteorites." *Mineralogy Magazine* 19 (1920):51–63.

———. *Catalogue of Meteorites*. London, 1923.

SMITH, W. C. "Seventy Years of Research in Mineralogy and Crystallography in the Department of Mineralogy, British Museum (Natural History), under the Keepership of Story-Maskelyne, Fletcher and Prior: 1857–1927." *Bulletin of the British Museum (Natural History) (Historical Series)* 10, no. 2 (1982):45–74.

SPENCER, L. J. "Biographical Notices of Mineralogists Recently Deceased." *Mineralogy Magazine* 24 (1936): 277–306.

ROBERT HUTCHISON

PTERIDOPHYTES

Earliest Land Plants

Pteridophyte, as commonly used, refers to all of the lower vascular (possessing specialized water-conducting cells in the xylem and food-conducting cells in the phloem) plants that reproduce by dispersing spores rather than seeds. There are four modern groups of plants that are all very distinct and only distantly related, and there were many other groups of these plants that are now extinct. (Further general references are provided for additional information.) Two groups of pteridophytes were the most likely first vascular land plants: the Zosterophyllophyta and the Rhyniophyta.

Rhyniophytes. A genus called *Cooksonia,* which was only a few centimeters tall and had forking branches and rounded sporangia at the tips that dispersed spores of one size (homosporous), is the first proven vascular plant of the Silurian period (Figure 1). There were no leaves, stems, or roots as we define those terms today, but related plants (*Rhynia*) were anchored by cellular rhizoids. Evidence from younger plants shows two stages in the life cycle which both existed independently: the diploid sporophyte, which produced the spores for dispersal, and the haploid gametophyte, which produced the sperm and egg for sexual reproduction, a life cycle characteristic of most of the pteridophytes. The gametophyte stage is only rarely found in the fossil record, and most of the information on pteridophytes is known from the sporophytes. There is increasing evidence, however, that some of the early plants had sporophytes and gametophytes of equal size and form.

Zosterophyllophytes and Lycophytes. The second group of early pteridophytes (Zosterophyllophyta) had multicellular flaps of tissue (enations) along the sides of their axes that are believed to

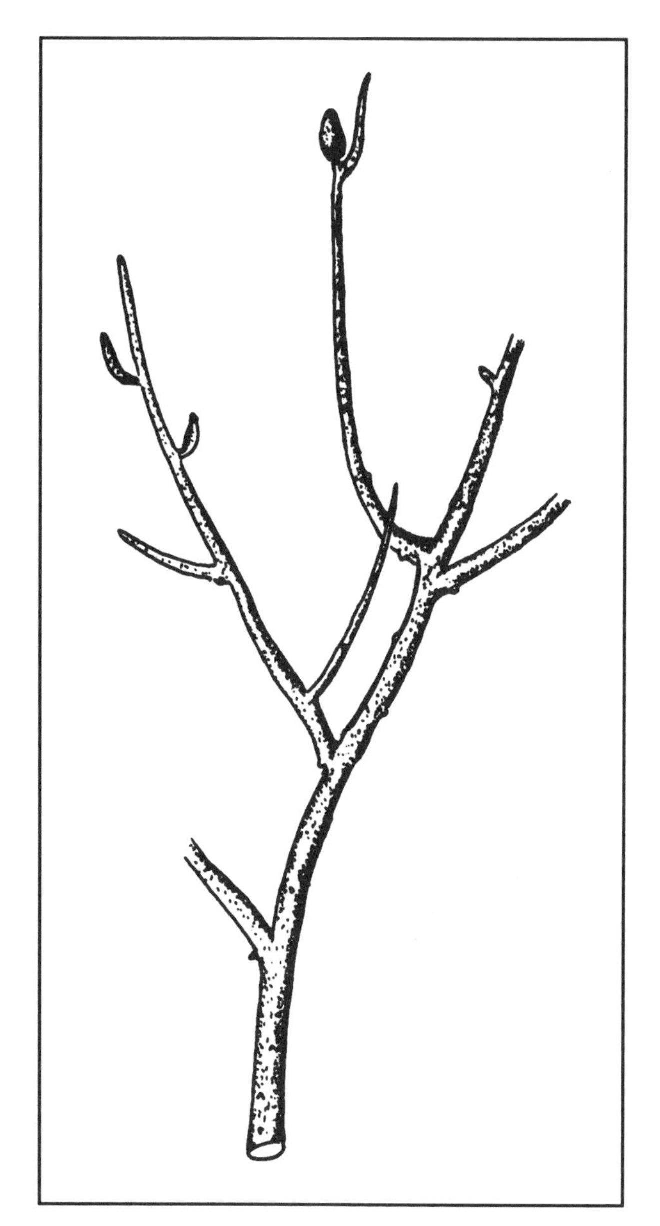

Figure 1. Rhyniophyta: *Rhynia gwynne-vaughanii.* (All illustrations courtesy of Andrew Sloane.)

have increased their food-producing ability by increasing the area for photosynthesis (Figure 2). These enations probably gave rise to the small leaves (microphylls) known in a group called Lycopodiophyta, the oldest known group of vascular plants that still occurs. Both these groups of plants have simple stems and roots and produce kidney-shaped sporangia that open into two flaps and are arranged along the sides of the stem. From evidence of the fossils and new molecular evidence from the modern representatives, the lycophyte

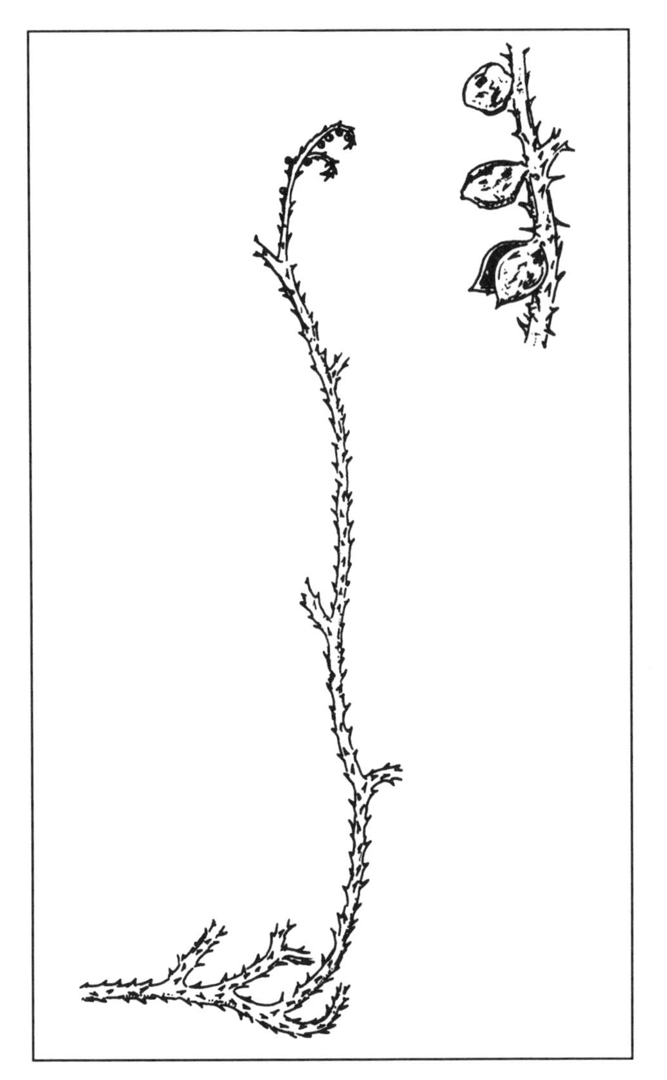

Figure 2. Zosterophyllophyta: *Sawdonia ornata.* (Sporangia enlarged at upper right.)

line was a sister group of all the other vascular plants (Figure 3). It appeared in the Middle Devonian and might have appeared earlier during the Silurian. This group of plants includes at least three separate major plant groups that persisted through time: the three extant orders are the Lydopodiales, the Selaginellales, and the Isoetales. The giant lycopod trees of the Carboniferous were the Lepidodendrales, dominant trees of the Carboniferous swamp forests. Some of them were more than 30 meters (m) tall, and up to 1 m in diameter. The leaves were produced on cushions of tissue on the trunk, which left distinctive scars when the leaves fell off, thus giving rise to the common name, scale trees. Strobili, large clusters of sporangia-bearing leaves, were produced at the ends of branches. Some of these plants produced two different sizes of spores (heterosporous), one of which produced the male gametophyte and the other the female, and several of them were adapted for a dispersal method similar to the seed plants of today, but the protective layers of the female structure opened to allow access by the sperm. In true seeds the coverings remain closed and specialized male structures are formed (i.e., a pollen tube; *see also* GYMNOSPERMS and ANGIOSPERMS). The growth of the lepidodendrids is also distinct because the large size was not attained solely by the production of wood (secondary xylem) but through an outer tissue formed separately. In addition, only the secondary xylem was

Figure 3. Lycopodiophyta: *Drepanophycus spinaeformis.* (Sporangia enlarged at upper left.)

produced by the vascular cambium and no secondary phloem. These large trees disappeared by the Mesozoic time, and only small aquatic forms remain today (Isoetales) in this line of plants.

Trimerophytes. In the middle of the Devonian period another group of pteridophytes that are descendants of the rhyniophytes appeared. These plants, the Trimerophyta, were larger, bushier, and had clusters of sporangia at the ends of the lateral branches. As evidence concerning these plants accumulates, we are realizing that there was much diversity within this group and that several lines of evolution were probably represented here. Most of the plants we see today are probably descended from this group of plants (Figure 4).

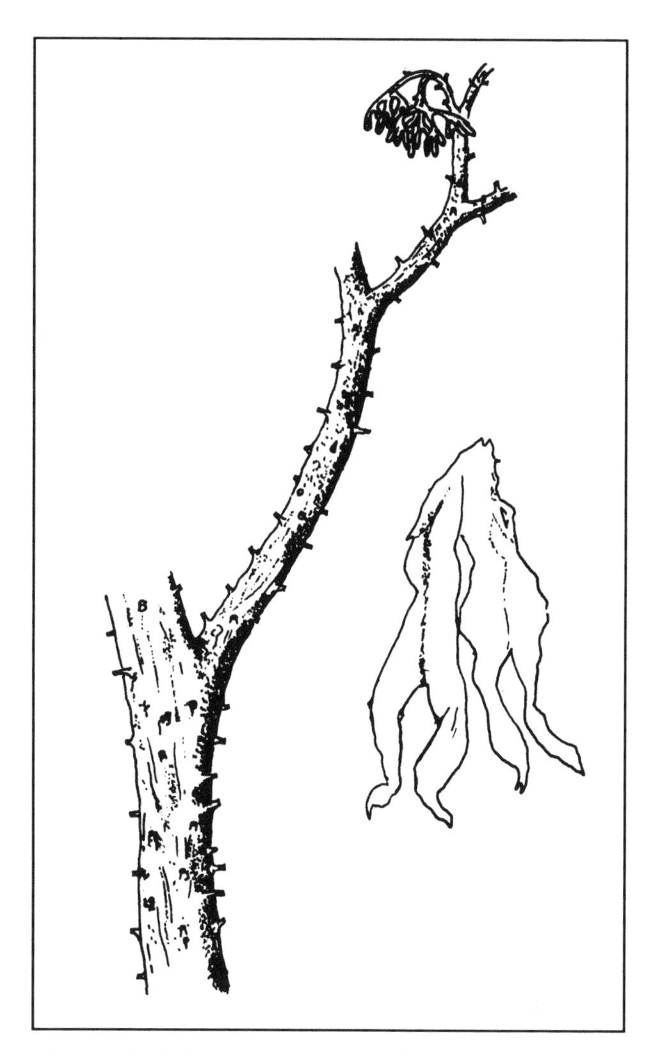

Figure 4. Trimerophyta: *Psilophyton princeps.* (Sporangia enlarged at lower right.)

Rise of Modern Pteridophytes

During the Devonian, other groups of pteridophytes also appeared, some of which formed leaves by modifications of branches (megaphylls). These plants gave rise to the modern horsetails (Equisetophyta), ferns (Polypodiophyta), and Psilotophyta. This last group of plants has no fossil record, but the morphology of the modern plants closely resembles the morphology of the Rhyniophyta (i.e., *Renalia*), with regard to the branching pattern, the growth habit, and the sporangial shape. The difference is that the modern plants have three fused sporangia (synangia) on the sides of the stem. However, the synangium may represent three branches greatly reduced and fused together. It must be emphasized, however, that there is a 350-million-year gap between these plants and no record of similar fossil forms have been found. This decreases the likelihood of any relationship between the two groups, and most botanists today accept the Psilotophyta as a separate highly modified group of plants adapted to specialized environments.

The Equisetophyta, the horsetails, had two Carboniferous representatives, Calamitales, that were larger during the Carboniferous, and smaller shrublike or vinelike Sphenophyllales. These plants also grew near or in the swamps but in a more disturbed environment. These plants persist into the Permian but by later Mesozoic time had disappeared, leaving behind only a single extant genus, *Equisetum.* Representatives of this group were characterized by whorled leaves and branches, sporangia borne on specialized axes called sporangiophores, and were homosporous or heterosporous. The internal structure of the stem shows a number of canals and cavities. The secondary growth produced by the arborescent forms was more similar to the type of secondary growth known today, that is, a vascular cambium producing both secondary xylem cells to the inner side and secondary phloem cells to the outer side. These plants were probably never major dominants in the environment, but were able to withstand slightly drier conditions than the lepidodendrids and thus survived longer. *Equisetum* plants of today grow in both wet areas and somewhat drier, sandier regions.

The Ferns. The ferns, Polypodiophyta, are a mixed group of plants today. All have large leaves and form their sporangia on the surface of the leaf. The modern ferns are divided into two main groups of plants: those that form their sporangia from several cells (eusporangiate) as do most other plants, and those that form their sporangia from one cell (leptosporangiate). The eusporangiate ferns undoubtedly appeared first in the fossil record, and one order of them, the Marattiales, appeared during the Carboniferous near swamps. In the uppermost Carboniferous in a drier interval of time between the wetter periods of swamp expansion, one of these plants, *Psaronius,* grew quite tall and became dominant. Its arborescent habit was accomplished by a mantle of roots that enveloped the stem and supported the treelike growth. Thus a very different mode of growth achieved the arborescent habit, and this mode of growth can still be seen in tree ferns of today (although these modern plants are not related to *Psaronius*). This group of ferns, which was abundant during the Carboniferous, has been reduced to only six genera in the modern flora, and these modern plants are still growing in tropical, moist areas. It is a good example of a plant group that succeeded in a particular environment and has maintained its success in competition without succumbing to other groups of plants. The other group of eusporangiate ferns is the Ophioglossales, whose fossil record goes back only to the Tertiary (Eocene) and whose precise relationship among the pteridophytes is still an area of active investigation.

The recent leptosporangiate ferns are probably the descendants of several other groups of ferns from the Carboniferous time: the plants that used to be grouped together under the order Coenopteridales and are now being divided up into more useful classification categories as we learn more about them. They often had creeping stems, growing as lianas (woody vines) or along the ground, had clusters of sporangia, had mainly primary growth patterns, and were homosporous. These plants also lasted through the Carboniferous and then disappeared by the end of that period. During the Permian and into the early Mesozoic families of ferns that are extant began to appear. Some of these are the Osmundaceae, Schizaeaceae, Gleicheniaceae, and Dicksoniaceae, which now grow in many places on Earth and have several genera and species. These plants appeared as forest undergrowth of the newly appearing conifer forests. They also moved out into the open areas away from the conifer forests, and probably were the ground cover for the savanna areas of that time. Most likely there were vast thickets of these ferns, some of which spread rapidly by rhizomes (underground creeping stems).

The next major time of change for the pteridophytes (late Cretaceous and early Tertiary) was during and after the appearance of the angiosperms, when the more modern families of ferns evolved. Again they appear to have been plants that survived best as understory plants in the changing environment of the angiosperm forests. By the late Tertiary the ferns were mostly modern in aspect. Also during the Cretaceous a group of ferns adapted to an aquatic environment and became heterosporous, represented now by only a few genera. Evidence from the past suggests that there were more representatives during the early Tertiary that are now extinct.

Evolutionary Trends

Evolutionary trends seen among the pteridophytes are examples of the trends seen among all plants. The uniqueness of the pteridophytes is the adaptation that has occurred with the two separate stages of the life cycle living independently. However, the ferns particularly have exploited many environments and developed reproductive strategies that are unknown in other groups of plants. Polyploidy is rampant among the ferns, as well as unique ways of asexual reproduction. Both the ferns and the lycopods are speciating and actively competing among the plants on Earth. These two groups are certainly not typical of older groups becoming more genetically stable through time. There is much being learned from the modern ferns that will help us to understand their past history and relationships, and much from the fossil record that will shed new light on the modern groups and their interactions within the environment and between themselves.

Bibliography

GENSEL, P. G., and H. N. ANDREWS. *Plant Life in the Devonian.* New York, 1984.

GIFFORD, E. M., and A. S. FOSTER. *Comparative Morphology of Vascular Plants,* 3rd ed. San Francisco, 1989.

MEYEN, S. V. *Fundamentals of Paleobotany.* London, 1987.

NIKLAS, K. J., ed. *Paleobotany, Paleoecology and Evolution.* New York, 1981.
STEWART, W., N. ROTHWELL, and G. W. ROTHWELL. *Paleobotany and the Evolution of Plants,* 2nd ed. Cambridge, Eng., 1993.
TAYLOR, T. N., and E. TAYLOR. *The Biology and Evolution of Fossil Plants.* Englewood Cliffs, NJ, 1993.
TIFFNEY, B. H., ed. *Geological Factors and the Evolution of Plants.* New Haven, CT, 1985.
TRYON, A., and B. LUGARDON. *Spores of the Pteridophyta.* New York, 1991.
TRYON, R., and A. TRYON. *Ferns and Allied Plants.* New York, 1982.

JUDITH E. SKOG

PUBLIC HEALTH AND EARTH SCIENCE

Probably the earliest connection between earth science and public health was the identification of the disease known as goiter. Murals from ancient Egypt and the earliest Chinese writings described the symptom: mild to severe throat swelling. In fact on every continent a portion of the populace once exhibited such reactions with "no distinction of race, nationality, color, creed or class" (Kelly and Snedden, 1960, p. 27), and in some geographic areas cretinism was also rife. High mountainous regions around the world were notorious goiter centers, but so was the area of the United States surrounding the Great Lakes. Modern medicine identified a cure and an effective prevention. Iodine provided as iodized salt or oil in human diets satisfies the necessary missing element required for a normally functioning thyroid.

There is an indelible link between humankind and the earth on which we live and from which we derive our nourishment. Our health and well-being are intertwined with whatever portion of Earth's surface we choose to call "home." Sometimes the interactions are obvious and overt, deliberate acts that juxtapose people with the environment, such as siting towns and farms, to maximize the production and distribution of adequate food and utilize water and waterways. Other interactions between geosciences and public health are effectively "silent," unplanned and unknown until a health problem develops. The low levels of iodine in certain soils leading to disease is an example, but the opposite situation, where naturally elevated concentrations of an element occur, may also prove disastrous.

The amounts of fluorine in the water in parts of Oklahoma, for example, may produce discolored or mottled teeth, and very high levels, as in parts of India, lead to osteopetrosis, a disease characterized by bone-like calcification elsewhere in the skeleton, resulting in gross deformity. Fluorine may be present in excess of human need in these environments, but on the other hand, the drinking water in many parts of the United States is now supplemented with 1 part per million (ppm) fluorine. This is because it has been scientifically demonstrated that adding a minuscule amount of fluorine is efficacious in preventing caries (tooth decay) and probably will ameliorate osteoporosis, a disease of old age in which bone mineral mass gradually disappears, leading to debilitating fractures. Fluoridation is thus an example of information provided by earth scientists being extracted and redirected by health scientists and applied to benefit public health.

Over the course of time humans have of necessity explored and learned to adapt to, or to use, the earth environment. Traditionally the earth sciences played a rather passive, albeit basic, role by providing accurate documentation of the topography, the rock varieties, and their structural relationships. The contiguous United States is now mapped at a scale of 1:250,000, and selected areas are known to 1:25,000, providing essential information for people ranging from agriculturists to town planners.

Since the 1920s, more detailed investigations of Earth's surface have been undertaken. Studies of the chemical composition of the rocks, soils, rivers, lakes, and oceans (geochemistry) have provided a greater appreciation of the natural sources and depositories of specific elements, information on the dynamic Earth system that can be applied to our advantage. Geochemical studies determine the distribution of elements, assure that excesses in contaminants can be avoided or that necessary trace elements may be added or treatments undertaken. Some of the beneficial results already demonstrated from such studies include marginal areas that become adequate pastureland with the addition of appropriate amount of cobalt (Co) or copper (Cu), or river systems and aquifers whose waters remain potable through chlorination or filtration. Geophysicists study the whole-Earth systems, including the oceans and the atmosphere, and have an obvious direct impact on public health when they predict impending earthquakes or volcanic eruptions.

It has become increasingly obvious in the 1990s that utilization of the land and its natural products by an ever expanding population will require not only stewardship and maintenance of the earth environment but also a much more detailed and integrated understanding of our planet and its reactions—if we are to maintain a healthy populace. It has become clear that the atmosphere is changing in response to the burning of fossil fuels, that our woodlands and jungles, if destroyed, probably will interfere with the future global response, let alone restrict biological diversity. It is important that geoscientists and health scientists pool their expertise and come together to work on these interdisciplinary problems. A little (partial) knowledge, can however, be a dangerous thing, and as illustrated below in two cases, can also be very expensive.

Radon, a colorless, odorless, radioactive gas, is emitted in the normal decay of the radioactive elements uranium (U) and thorium (Th), common constituents of some minerals found in rocks and soils. Pitchblende (uranium ore) miners exposed to large quantities of radioactive materials have a high risk of lung cancer. The granite bedrock of the Reading Prong in the northeastern United States, and of Devon and Cornwall in England, contains minerals with U and Th as part of their crystal structure. These are sites where we choose to live and build energy-efficient, highly insulated houses. Granites produce relatively more radon than other rock types, but the amount is several orders of magnitude lower than that experienced by the miners. Based on extrapolation from the uranium ore miners' occupational exposure, a radon hazard has been proclaimed for the populace inhabiting the granitic areas, a projection considered scientifically questionable.

Although radon may cause a physiological reaction at elevated levels, its potential as a primary cause of lung cancer in the general populace is debatable (Bowie and Bowie, 1991). There is only a small amount of data available on the actual level and duration of exposure for the few uranium miners and virtually no data on the low-dose effects in the case of the public. Any danger to the public can only be evaluated with data that provide statistically significant details that cover both aspects of the problem. Specifically, the amount and intensity of radon emitted over a period of time are needed to establish potential exposure levels, and the amount and intensity of disease in individuals exposed in mines, in the granitic and nongranitic areas, must be correlated and analyzed. Some data may become available on uranium miners in the former East Germany (Kahn, 1993) and for low-dose exposures data are accumulating from an epidemiological study in Connecticut, a state with patches of granitic rocks. In the meanwhile the public has heard that radon emission should be below 4 picocuries per liter (pCi/l) (an ultra-low level). Some well water radon measurements are considerably higher. The relationship between the counts of radon reported on the waters with the potential of radon-induced disease has frightened many families who, to protect themselves, choose to leave a house, and an area, probably incurring, in addition to their worries about personal health effects, a financial loss.

Another "hazardous" natural product that has been brought to the public's attention is asbestos. As with radon the identification of these fibrous materials as major public health problems was based on occupational exposures, where dust levels were documented as very high in the past. Over twenty-five years and many studies at mines and manufacturing plants, and on the workers and affected individuals, the level of occupational fiber exposure has been reduced. The pulmonary disease asbestosis has decreased markedly over that time and will probably essentially disappear by the year 2005. Data on lung cancer or mesothelioma, a cancer of the lining of the lung, in asbestos workers documented over that same time span (Skinner et al., 1988) led the hypothesis that casual exposure even to one asbestos fiber might be hazardous. This extrapolation to the low exposure led the Environmental Protection Agency (EPA) to promulgate the Asbestos Hazards Emergency Response Act (AHERA) of 1986, which mandated monitoring and encapsulation or removal of any asbestos (usually the insulation products) found in buildings (homes, churches, workplaces, and schools). The ensuing panic created a new industry—asbestos removal, which has an annual cost of hundreds of millions of dollars. Geologists pointed out that asbestos and other fibrous materials are the normal constituents of certain rocks in many sites over the globe. Though disease may be induced by high exposure in some occupations, there was no excess sickness even in the very areas and towns where asbestos was being mined. AHERA appears to have been an overreaction and is now being rethought (HEI-AR, 1991, 1993). Today health professionals in general suggest that low-dose asbestos exposures in air (inhalation) or in water (ingestion) are not public health problems. Removal of asbestos from buildings is usually considered unnecessary and indeed should be avoided (Mossman and Gee, 1990). Asbestos, once a marvelous, important and cheap natural product that made the populace comfortable and safe, became a public health hazard, and its future is in doubt. What is not in doubt is that the public accustomed to modern living requires insulation, and substitutes (fiberglass) are being used.

The many disciplines within earth science have much to contribute toward solving problems that confront health professionals. Geoscientists that actively seek to contribute would go a long way to insuring a cooperative focus not only on the prevention of disease, but for the betterment of the human condition. By bringing together individuals with scientific expertise of Earth systems with those focused on the human system, disease induction and potential treatments that have global implications, and may require remediation of a habitat, would be more easily effected. Endemic disease requires a multidisciplinary approach. Only a fraction of the globe has been researched with the detail necessary to address public health issues. The future has many challenges.

Bibliography

Bowie, C., and S. T. U. Bowie. "Radon and Health." *Lancet* 337 (1991):409–413.

Health Effects Institute—Asbestos Research (HEI-AR). *Asbestos in Public and Commercial Buildings.* Cambridge, MA, 1991, 1993.

Kahn, P. A. "Grisley Archive of Key Cancer Data." *Science* 259 (1993):448–451.

Kelly, F. C., and W. W. Snedden. *Prevalence and Geographical Distribution of Enemic Goitre.* World Health Organization. Geneva, Switzerland, 1960.

Skinner, H. C., W. M. Ross, and C. Frondel. *Asbestos and Other Fibrous Materials: Mineralogy, Crystal Chemistry and Health Effects.* New York, 1988.

H. CATHERINE W. SKINNER

PUBLIC POLICY

See Geology and Public Policy

PULSAR

See Stars

Q R

QUASAR

See Galaxies

RED GIANTS

See Stars

REFRACTORY MATERIALS

Refractory materials are those that will withstand high temperatures sufficiently to permit their use in a furnace lining or other location exposed to severe heat. The official definition of the National Bureau of Standards requires that a refractory material shall withstand temperatures of 1500°C. A measure of refractoriness that is commonly used is pyrometric cone equivalent (PCE), which indicates the fusion point. The fusion point is the temperature at which a particular specimen under a definite stress becomes sufficiently fluid to flow at a specified rate.

The value of a fusion point determination is that it shows whether or not a material is unsuitable above a certain temperature. For example, if a refractory material is needed to make a firebrick for use with furnace temperatures of 1600°C, the fusion point test would eliminate those with a fusion point at or below that temperature. Standard pyrometric cones with known fusion points are commercially available. Thus the bending characteristics of the sample can be compared with those of standard cones (Norton, 1968). Table 1 shows the end or bending point of Orton pyrometric cones.

The fusion point measurements are made on specimens formed into slender tetrahedrons similar in shape to the standard pyrometric cone. The fusion point of a specimen is when the top of the cone touches the base, as shown in Figure 1.

Only a few elements in Earth's crust are both refractory enough to form stable refractory compounds and abundant enough to be used in industry. These are silicon (Si), aluminum (Al), magnesium (Mg), calcium (Ca), chromium (Cr), zirconium (Zr), and carbon (C). These elements combine with oxygen and other elements in nature to form useable refractory minerals or synthetic materials. Typical refractory materials are high alumina clay such a kaolinite ($Al_2Si_2O_5[OH]_4$), silica (SiO_2), magnesite ($MgCO_3$), chromite ($[Mg,Fe]Cr_2O_4$), kyanite (Al_2SiO_5), dolomite ($Ca,Mg(CO_3)_2$), bauxite (principally diaspore $[Al_2O_3 \cdot 3H_2]$), olivine ($[Mg,Fe]_2S_iO_4$), and graphite (C).

Refractory Raw Materials

High alumina clays contain both alumina and silica; they are used to make firebricks that are classified as low duty (PCE fifteen), medium duty (PCE

Table 1. End Points of Orton Pyrometric Cones

Cone Number	*End Point °C*
12	1337
13	1349
14	1398
15	1430
16	1491
17	1512
18	1522
19	1541
20	1564
23	1590
26	1605
27	1627
28	1638
29	1645
30	1654
31	1679
31½	1699
32	1717
32½	1730
33	1741
34	1759
35	1784
36	1796
37	1830
38	1850
39	1865
40	1885
41	1970
42	2015

twenty-nine), high duty (PCE thirty-one and one half), and super duty (PCE thirty-four). The clays that are used for refractories are flint fireclays, plastic fireclays, and kaolins. Flint fireclays have PCEs of thirty-three to thirty-five and a low shrinkage. Because they have little or no plasticity, they are bonded with plastic fireclays or refractory kaolins. Flint clays are mined in eastern Kentucky and central Missouri in the United States; in England and Germany in Europe; in China in Asia; and in Australia. Plastic fireclays, which by definition have a PCE of fifteen or higher, are found in West Virginia, Pennsylvania, Kentucky, Ohio, Illinois, and Missouri, where they occur under coal beds. They also occur in England, Germany, Russia, China, and Australia. Refractory kaolins are found in Georgia and South Carolina in the United States, in Brazil in South America, and in England. Alumina-silica materials, as noted above, vary in refractoriness; the higher the alumina content, the greater the refractoriness. They exhibit a wide melting temperature range up to a hundred degrees centigrade, with the temperature at which melting commences separate from that at which the entire mass is molten. High alumina brick is made from highly aluminous materials such as diaspore or bauxite in combination with fireclays. Kaolinite melts at 1850°C, bauxite at 1800°C, and diaspore at 2035°C.

Silica bricks that are classified as super duty are made from crushed silica pebbles or a silica-rich rock called quartzite. Silica pebbles are separated from quartz pebble conglomerates and crushed and the high purity quartzites are crushed and screened prior to use. Silica materials must be exceptionally pure with a SiO_2 content of 99.5 percent or higher. Silica for refractory use is mined in West Virginia, Ohio, Wisconsin, Pennsylvania, and California. In Europe refractory silica deposits are mined in England, Germany, and France; and in Asia, they are mined in Japan. Silica bricks have excellent resistance to deformation under load at temperatures close to their melting points. About 2 percent lime is added to bond the silica grains together. The mineral in the silica pebbles and the quartize is quartz, which changes to cristobalite at high temperature. Silica bricks exhibit greater refractoriness than fireclays and expand rather than shrink at high temperature. They also have a good resistance to attack by most fluxes but are liable to severe spalling (breaking off in layers) during rapid temperature fluctuations below about 500°C.

Basic refractories comprise the products made from magnesite, chromite, and dolomite. They are capable of withstanding very high temperature and are chemically resistant. As a group, basic refractories, in contrast to the aluminum-silica and silica groups, exhibit much greater refractoriness and better resistance to chemical attack of slag and metallic oxides.

Magnesite is the magnesium end member of an isomorphous series of carbonates involving calcium, iron, and magnesium. Pure magnesite is rarely found, but when it occurs in nature, it is a replacement mineral in carbonate rocks, a vein-filling material, an alteration product in ultramatic rock, or a sedimentary rock. Magnesite is also synthetically produced from magnesium in seawater. Magnesite is calcined at temperatures of 1500°C or higher, which expels carbon dioxide and converts

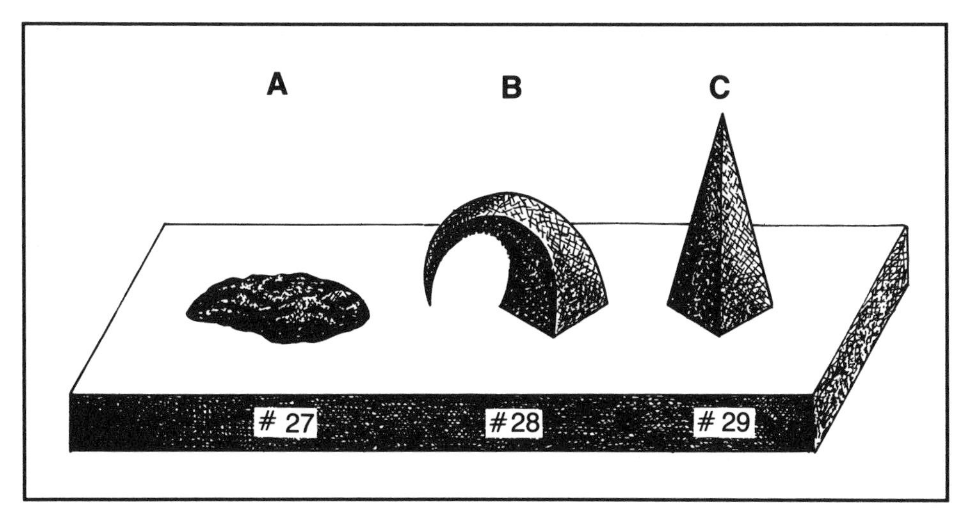

Figure 1. A. Pyrometric cone 27 is completely fused as the end point (1627°C) has been surpassed; B. the tip of pyrometric cone 28 is touching the base, indicating that the fusion point (1638°C) has been reached; C. pyrometric cone 29 is upright, indicating that the end point (1645°C) has not been reached.

the residue MgO to a crystalline form known as periclase. Natural magnesite is produced in China, North Korea, Russia, the former Czechoslovakia, Austria, Turkey, and Greece. Synthetic magnesite is produced from seawater in the United States, Mexico, and Israel.

Dolomite is a calcium magnesium carbonate and occurs most commonly as replacement of limestone. Dolomite is calcined at about 1500°C to form dead, burned $CaO{\cdot}MgO$, which has a melting point of 2600°C. Dolomite refractories are used in lining metallurgical furnaces where a basic brick is needed. Dolomite is rather common throughout the world but high-purity dolomite is rare. In the United States dolomite is mined for refractory purposes in Pennsylvania and Ohio.

Chromite is the major source of chromium and is a spinel with the formula $(Mg,Fe)Cr_2O_4$. For refractory uses chromite must have 30–40 percent Cr_2O_3 with a Cr:Fe and Mg ratio of 2–2.5 to 1, and with a maximum silica impurity of 6 percent. Chromite melts at 1990°C and is generally mixed with basic refractory materials, particularly magnesia, to form chromite and magnesium composite bricks. Chromite has excellent thermal conductivity, a low rate of thermal expansion, and a high resistance to thermal shock. The major areas where chromite is produced in the world are South Africa and Russia. Other producers include Albania, Turkey, India, Finland, Zimbabwe, and the Philippines. Chromite occurs almost exclusively in rocks like peridotite, dunite, and pyroxenite.

Kyanite is a naturally occurring aluminum silicate with a melting point of 1820°C. It is classified as midway between the silica and the basic magnesia refractories. Commercial grades of kyanite must contain a minimum of 56 percent Al_2O_3 and 42 percent SiO_2 with less than 1 percent acid-soluble Fe_2O_3, 1.2 percent TiO_2, and 0.1 percent each of CaO and MgO (Bennett and Castle, 1983). Kyanite converts to mullite at about 1100°C and expands in volume by about 18 percent. Therefore, kyanite is calcined before use to convert it to mullite, which is extremely refractory, has a small coefficient of expansion, and withstands abrasion and slag erosion. Kyanite typically occurs in aluminous, regionally metamorphosed rocks and their weathered derivatives (Harben and Bates, 1990). The major producers of kyanite are the United States, Russia, and India. Andalusite and sillimanite are minerals with the same formula as kyanite and, like kyanite, can be converted to mullite by calcining. Andalusite is formed through contact metamorphism of aluminous shales by granite intrusions. The major producers of andalusite are South Africa and France.

Bauxite is a mixture of aluminous minerals and is the principal ore of aluminum metal. A small percentage of high quality bauxite is used in refractories. Refractory-grade bauxite that is calcined at 1600°C to remove all the water converts any kaolinite present to mellite and gibbsite to corundum. A typical specification for refractory grade bauxite (calcined) is shown in Table 2.

Bauxite is a residual weathering product formed in tropical or subtropical climates. Although metallurgical grade bauxite to produce aluminum is abundant in many tropical countries, refractory grade bauxites are relatively rare. Refractory bauxites are produced in Guyana, Brazil, and China.

Olivine is a mineral that has two compositional end members, forsterite, which is Mg_2SiO_4, and fayalite, which is Fe_2SiO_4. There is complete isomorphous substitution of Mg and Fe, from forsterite to fayalite. The refractory grade olivine is the magnesium-rich variety that has a melting point of 1800°C. Olivine bricks have excellent heat retention and are used in ladle linings and linings of glass tank furnaces. Olivine occurs commonly as an accessory mineral in mafic igneous rocks. Economic deposits are restricted essentially to an igneous rock called dunite. Olivine for refractory use is produced in the United States in North Carolina and Washington. The largest production in the world comes from Norway. Other countries producing olivine are Sweden, Austria, Spain, and Japan.

Zircon ($ZrSiO_4$) is used extensively in refractories because of its very high melting point (2500°C), low thermal expansion, excellent thermal diffusivity, and chemical stability. By far the largest producer of zircon is Australia, which is responsible for over 60 percent of the world's production. Other areas where zircon is produced are South Africa, Russia, and the United States. Commercial concentrations of zircon occur in both primary rock deposits and in placer deposits of mineral sands on modern and ancient beaches (Harben and Bates, 1990).

Table 2. Refractory Grade Bauxite (calcined) Specification

Al_2O3	min. 84.5%
SiO_2	max. 7.5%
Fe_2O_3	max. 2.5%
TiO_2	max. 4%

Carbon in the form of graphite is used extensively as a refractory material because of its high melting point (3700°C), high conductivity, and excellent resistance to thermal stress. Generally, graphite is bonded with refractory clay to form crucibles for handling molten steel and other metals. Organic matter in sediments may be converted to graphite by regional or contact metamorphism. True graphite probably forms above 400°C (Landis, 1971). Graphite occurs dispersed in schists and gneisses, and in veins. The largest producer of graphite is China, followed by Korea, Russia, Austria, and Mexico.

Refractory materials are essential to our metals, glass, cement, lime, and many other industries. The tendency over the years has been the demand for higher purity materials. This requires better raw materials and better and more efficient processing.

Bibliography

BENNETT, P. J., and J. E. CASTLE. "Kyanite and Related Minerals." In *Industrial Minerals and Rocks*, ed. S. J. Leford. 5th ed. New York, 1983, pp. 799–808.

HARBEN, P. W., and R. H. BATES. *Industrial Minerals—Geology and World Deposits.* Metal Bulletin Plc. London, 1990.

NORTON, F. H. *Refractories,* 4th ed. New York, 1968.

LANDIS, C. A. "Graphitization of Dispersed Carbonaceous Material in Metamorphic Rocks." *Contributions to Mineralogy and Petroleum* 30 (1971):34–45.

HAYDN H. MURRAY

REPTILES

See Amphibians and Reptiles

RESEARCH IN THE EARTH SCIENCES

Research in the earth sciences is exciting and active, in large part because understanding processes that operate on and in Earth can have tremendous implications for our abilities to live safely and comfortably and to protect the environment. Earth scientists conduct research to better understand and protect us against natural hazards, such as earthquakes, volcanoes, landslides, floods, wildfires, and violent storms. Earth scientists investigate processes that help us find, conserve, and responsibly exploit resources, including energy (oil, gas, coal, hydroelectric, nuclear, wind, solar, biomass), mineral (metals, construction materials, chemical feed stocks), and water (groundwater, streams, rivers, lakes, and oceans). Earth scientists also look at records from the geologic past to help predict the ability of our planet to adjust to the impacts of human activity.

The earth sciences, or geosciences, encompass many disciplines that necessarily overlap with one another and with other branches of the sciences and engineering. The earth sciences include, among other disciplines, geology (which, in a traditional sense, includes the other disciplines), geophysics and meteorology (allied with physics), geochemistry (allied with chemistry), mineralogy (allied with material science, metallurgy, and physical chemistry), engineering geology (related to civil and geotechnical engineering), economic geology (related to mining engineering, petroleum geology and engineering, and economics), paleontology and soil science (allied with biology). The problems that we face increasingly require expertise from different fields. For example, complete understanding of processes in the fields of climatology, groundwater hydrology, and oceanography, requires interplay between the traditional disciplines of physics, chemistry, biology, and geology. Advances in mathematics are readily applied in the earth sciences. Even the field of medicine interacts with geology, geochemistry, and mineralogy in such areas as health hazards from naturally occurring pollutants.

This article draws heavily from reports by the National Research Council, which is the operating arm of the National Academy of Sciences, the National Academy of Engineering, and the Institute of Medicine, three private honor societies recognizing leaders in the fields of science, engineering, and medicine. The federal government frequently seeks guidance from the National Research Council regarding issues of science, technology, and policy. The National Research Council also reports on the status and future of scientific disciplines, including the hydrologic sciences (1991), oceanography (1992), and the solid-earth sciences (1993).

Uniqueness of Geological Research

An earth scientist's laboratory is the whole Earth (and other planets and moons, whenever we have an opportunity to look closely at them or obtain samples from them). Investigations range in scale from imaging surfaces of minerals at the atomic level to seismic experiments involving the entire planet. What is unique about geologic research is the concept of deep time, that geological processes often take place at timescales of thousands to millions of years. Rocks on Earth and the Moon record events that have taken place nearly as far back in time as the origin of the planet itself, about 4.55 billion years (Ga) ago.

Some processes, such as uplifts of mountains, movements of continental plates, and sculpting of the landscape, take place over extremely long periods of time, often millions of years. Other processes, such as earthquakes and volcanic explosions, take place in matters of seconds or less. Geologists have developed procedures and techniques for the study of deep time. These range from field techniques, such as geologic mapping, which helps to unravel questions concerning the layering, tilting, faulting, folding, and other disruptions of rocks, to laboratory techniques, such as measurements of concentrations of radioactive isotopes and their decay products in rocks to determine ages.

Geology is a field-based science. Although many advances are made in the laboratory, either through sophisticated examination and analysis of samples collected in the field or through computer simulation, interpretation of laboratory results and models hinges on relationships that the geologist observes in the field (*see* MAPPING, GEOLOGIC).

Geologic research is analogous to detective work. The geologist pieces together the events of the past from bits of evidence. At an outcrop in the field the geologist can often determine the relative ages of rocks and general aspects of how they formed and how they were transported to their present location.

As an example, imagine an outcrop with sandstone in contact with a basalt (Figure 1). On the basis of comparisons with fossils found elsewhere in the world, the geologist or paleontologist can estimate the age at which the sandstone formed

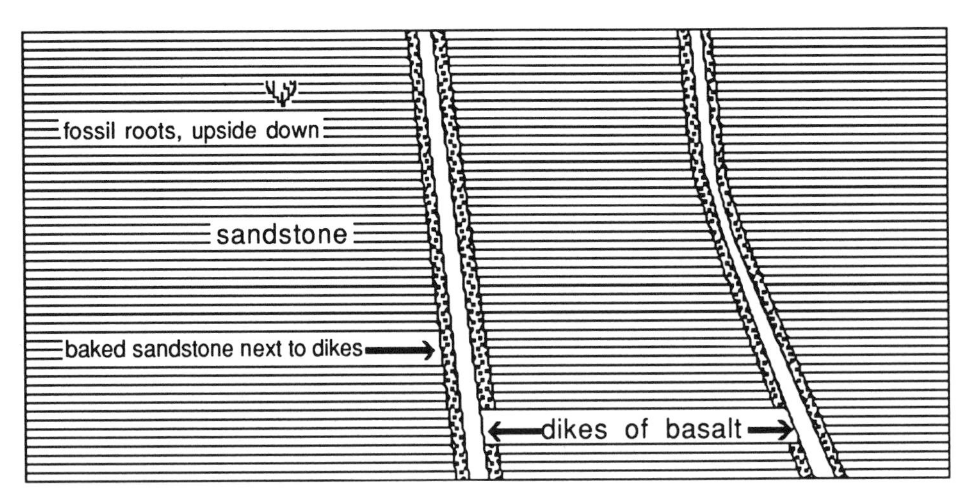

Figure 1. Cross-sectional view of an outcrop of sandstone and basalt. The geologist, aided by both field and laboratory investigations, can determine the history of the area from such outcrops.

(see Plate 34). Fossil root zones that are now upside down can tell the geologist that the rocks have somehow been turned over. Mapping of the layering in the sandstone can tell the geologist that the rocks were folded in a former mountain-building episode. From the baked contacts where the sandstone meets the basalt, and from the cross-cutting nature of the dikes of basalt, the geologist determines that the basalt formed after the sandstone. From laboratory measurements of radioactive elements and their decay products in the rocks, and from the physicist's knowledge of half-lives of radioactive isotopes, the geochemist can estimate the age at which the basalt solidified from a magma and cooled. Other laboratory measurements of orientations of magnetic minerals in the basalt can indicate that the basalt itself was not turned upside down, thereby bracketing the age of mountain building.

Such detective work is necessary to unravel the history of Earth and to understand the natural processes that have occurred in the past. But such detective work is also necessary to predict what the natural processes will do in the future. For example, the geologist can map the distribution of faults that moved during past earthquakes. By digging trenches across a fault, geologists can expose buried soil horizons and layers of sediment that were deposited along the fault after movements during ancient earthquakes (Figure 2). Geologists can collect samples of charcoal, which formed during ancient forest fires, and carbon-rich material from the ancient soil. Laboratory measurements of the carbon-14 ages of the soil and charcoal fragments can bracket the ages of last movement on the fault, that is, when the last earthquake occurred. With this information, geologists can better predict when the next earthquake is likely to occur—not to the day or even year, but certainly within the next few hundred to thousands of years. Such information is vital for earthquake preparedness.

Exciting Areas of Earth Science Research

The National Research Council's 1993 report titled *Solid-Earth Sciences and Society* identified many areas of research needed to better understand the processes that take place on the surface and within the interior of Earth. The following table of research topics provides a sense of the breadth of issues and concerns covered by earth scientists:

Research Opportunities in the Earth Sciences

A. To understand how the oceans, atmosphere, and life have evolved and how Earth will respond to environmental changes in the future, research is needed on:
   - processes taking place in the past 2.5 million years, that part of the geologic past that is most likely to correspond to changes in the near future;

Research Opportunities in the Earth Sciences Continued

- new techniques for dating soil development and for assessing soil contamination;
- biochemical and microbial processes taking place in soils;
- climate change deduced from layered ice in glaciers and from layers of sediments in ponds, lakes, oceans, and caves;
- past configurations of the continents, ocean circulation, and the distribution of life;
- sea-level change in the past;
- processes that control the ecology of coastal areas;
- factors that control environmental change and thresholds between states of apparent equilibrium in Earth systems;
- causes of abrupt and catastrophic changes as opposed to gradual changes;
- history of life deduced from the fossil record, from living species, and from chemical indicators;
- linkages between the atmosphere and the ocean;
- evolution of Earth's crust throughout geologic history and its effect on the chemistry of the oceans and atmosphere;
- chemical fluxes from the interior of Earth to the oceans and atmosphere, and vice versa;
- mathematical modeling of biogeochemical systems;
- ecosystems at deep-sea hydrothermal vents and hydrocarbon seeps, and their implications for the origin of life.

B. To understand how fluids have moved within and on Earth, leading to the formation of land forms, volcanoes, groundwater resources, mineral resources, and energy resources, more needs to be known about:
- history of drainage basins, including river flooding as it relates to past changes in climate;
- chemistry and physics of mineral–water interactions;
- role of fluids in tectonic processes, such as generation of earthquakes and eruption of volcanoes;
- generation and movement of magma in the crust and mantle.

C. To understand how Earth's crust and mantle formed and evolved and how they interact to concentrate volcanoes and earthquakes at the margins of continental-scale crustal plates, studies are needed on:
- evolution of Earth's crust from the mantle;
- responses of land forms to change in climate and tectonic forces;
- relations between tectonics and climate;
- dynamics of sedimentary basins;
- global comparisons of sedimentary sequences;
- details of the origin of new crust at ocean ridges;
- mathematical and computer modeling of land-form changes;
- mathematical modeling of continental-scale processes;
- recrystallization and chemical changes in the lower crust and upper mantle through time;
- details of the composition, temperature, and other physical properties of the lower crust;
- history of uplift of mountain ranges;
- quantitative understanding of earthquake ruptures;
- determining rates of geological processes;
- real-time measurements of plate motions and other processes, using the satellite-supported global positioning system (GPS) for highly accurate measurements, and using remote sensing;
- refining and quantifying geologic predictions;
- producing modern geologic maps, incorporating new theoretical concepts, new cartographic techniques, and more information, including geophysical, geochemical, and remotely sensed data.

D. To understand how the deep interior of Earth operates, earth scientists will be investigating:
- origin of Earth's magnetic field and why it changes with time;
- complexities of the core-mantle boundary;
- imaging Earth's interior using seismic and other geophysical techniques;
- experimental approaches to understanding processes at high pressures and temperatures;
- coupled geochemistry and geophysics of Earth's interior;
- modeling of the dynamics of processes in the interior of Earth, incorporating evidence from experimental geochemistry, samples brought to the surface by nature, seismic and gravity measurements, and other areas.

E. To sustain sufficient supplies of natural resources, including water, energy, and minerals needed in everyday life, more must be known about:
- kinetics of water–rock interaction;
- water quality and contamination;
- microbiology and soils, including new techniques for dating;
- quantitative modeling of water flow and chemical changes during flow;
- how mineral and energy deposits have differed in type, size, and grade throughout geologic time;
- organic geochemistry and the origin of petroleum;
- in situ mineral resource extraction;
- modeling of hydrocarbon sources, transport, and accumulation;
- numerical modeling of ore deposition;
- prediction of undiscovered mineral and energy resources;
- refined techniques for mineral exploration, including airborne geophysics and other approaches to find concealed deposits;
- integrated approaches to petroleum exploration;
- new techniques for discovering concealed geothermal resources;
- relationships of ore deposits to details of plate tectonics;
- advanced petroleum production and recovery methods;
- availability and accessibility of coal resources;
- coal quality;
- complexities of fluids in the crust.

F. To prepare for and mitigate geologic hazards, research is necessary on:
- paleoseismic studies of the effects of past earthquakes;
- seismic safety of dams and water reservoirs;
- earthquake prediction;
- predicting volcanic eruptions;
- geologic mapping of volcanoes to determine past effects;
- tectonics in the last 1.6 million years;
- landslide susceptibility maps as derivatives of geologic maps;
- landslide prevention;
- new techniques for dating geologic events;
- satellite-based surveying systems and remote sensing;
- real-time measurements of geologic processes;
- systems approaches to studying land-form changes;
- extreme events, including floods and storms;
- geographic information systems in land-use and urban planning;
- difficulties of building on volume-changing soils;
- bearing capacity of different rock types;
- use of underground space;
- subsurface geophysical exploration and underground void detection;
- detection of neotectonic features.

G. To minimize environmental damage and adjust to inevitable natural and human-induced global and local environmental change, earth scientists will contribute to:
- interdisciplinary earth science, materials, and medical research;
- radioactive waste isolation;
- groundwater protection;
- control of organic contaminants;
- waste management for municipal and hazardous waste;
- waste cleanup;
- waste disposal from industrial activity;
- evaluating environmental consequences of mining and burning coal;
- determining past global change deduced from the geological record;
- predicting future global change;
- building and interpreting global geochemical and geophysical databases;
- understanding abrupt and catastrophic changes in the past;
- evaluating climatic effects of volcanic emissions.

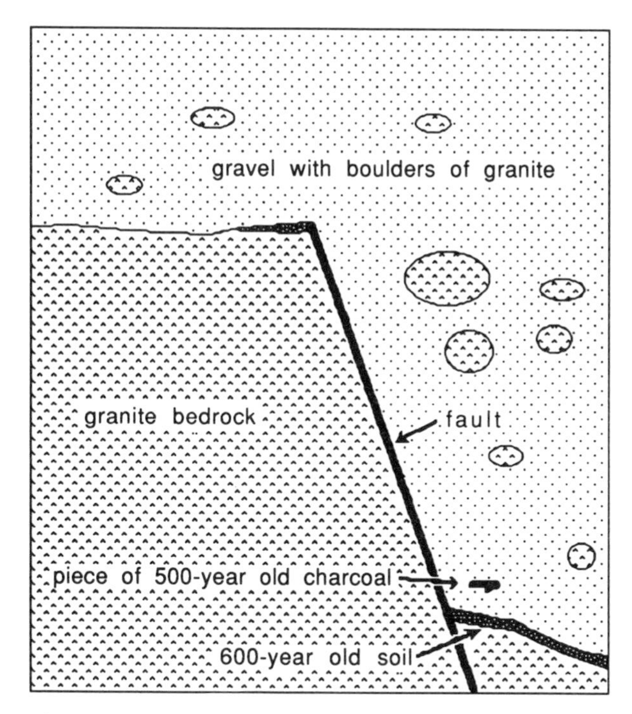

Figure 2. Sketch of details from a trench dug across an active fault. Geologists conduct detective work to determine that the last major earthquake on this fault was between about 500 and 600 years ago.

In recent years research in the earth sciences has led to major advances, including the theory of plate tectonics, which has given us understanding of many related processes taking place in and on Earth; and new evidence for causes of mass extinctions, which has given us better appreciation of catastrophic events, such as major episodes of volcanism, impacts of huge meteorites and comets, and extraordinary floods, all of which are parts of the geologic record.

Technological advances have given impetus to earth science research in many areas and will continue to do so. Satellite observations and satellite-assisted precise measurements of distances have given us abilities to make global observations of the oceans, continents, and atmosphere. Development of new instruments to analyze minute concentrations of elements and isotopes in small amounts of crystals and rocks has given us new techniques for dating rocks and unraveling details of their geologic histories. Faster computers with greater memory have given us the tools to more accurately determine the three-dimensional structure of Earth's interior at scales ranging from petroleum reservoirs to the whole Earth. These faster computers also have given us the ability to simulate natural processes, check the simulations against what the geologic record indicates actually took place in the past, and make predictions about the future.

Funding for Research

In part because of the broad spectrum of applied and basic research opportunities, funding for research in the earth sciences comes from a variety of sources, including government, industry, private foundations, and academia. A distinction between basic and applied research is often difficult to make in the earth sciences, because so much of what geoscientists do in their research can be directly or indirectly applied to issues of hazards, resources, or the environment. Many geoscientists are employed not in research but in the applications of geology and other earth sciences to such issues. Increasingly the problems that earth scientists are being asked to help solve are ones that require multidisciplinary teams, often incorporating not only natural scientists from other disciplines but also engineers, social scientists, policy makers, other professionals, and the public.

Bibliography

National Research Council, Committee on Opportunities in the Hydrological Sciences. *Opportunities in the Hydrological Sciences.* Washington, DC, 1991.

National Research Council, Committee on Status and Research Objectives in the Solid-Earth Sciences. *A Critical Assessment. Solid-Earth Sciences and Society.* Washington, DC, 1993.

National Research Council, Ocean Studies Board. *Oceanography in the Next Decade.* Washington, DC, 1992.

National Research Council, Panel on Effects of Past Global Change on Life. *Effects of Past Global Change on Life.* Washington, DC, 1995.

National Research Council, Panel on Global Surficial Geofluxes. *Material Fluxes on the Surface of the Earth.* Washington, DC, 1994.

JONATHAN G. PRICE
THOMAS M. USSELMAN

RESOURCE USE, HISTORY OF

The origins of mineral resource use are lost in the unrecorded prehistory of the human race, but it is likely that the two mineral resources first used were water and salt because both are essential for human survival. The daily need for water dictated that settlements and travel routes could never be far from it. Salt was adequately supplied in meat-rich diets but became a necessary separate commodity as cereal-based diets became widespread. Furthermore, salt was the cheapest way to enhance taste and to preserve meats from spoiling; hence salt became a prized and much traded commodity. More than one million years ago our ancestors began using stone implements, first crude and then increasingly better shaped by chipping, grinding, and cutting. The earliest of the tools were little more than blunt clubs to help in food gathering and game capture and sharp-edged scrapers that helped in cleaning animal skins. There was no doubt much rapid information transfer when new implements and techniques of manufacture were developed, but it is also likely that parallel discoveries occurred in distant places many times. The first rocks used were probably of many types, but the early stone workers soon recognized that obsidian, flint, and chert were especially useful because they were relatively easy to work, were very tough, and took and held very sharp edges.

Although stone working was raised to a fine art in many areas, great technological advancements occurred prior to 9000 B.C.E., when fired pottery developed. This represented the first synthesis of mineral resources, gave the creator total control over the shape of the product, and provided a better means of storage and transport of food and water. This discovery led to the rapid development of the ceramic arts of brickmaking, glazing, the making of mineral pigment paints, and even glassmaking by about 3500.

More than other metals, gold and copper occur in the native or uncombined state, and it was their accidental discovery long before 4000 that ushered in the ages of metals—copper, bronze, and iron. The greater density, different luster, and malleability of copper and gold led to their initial consideration as curiosities and for adornment. The ability to shape these metals into useful and desirable forms developed rapidly. By 4000 B.C.E. primitive copper-smelting techniques using charcoal as a fuel were widespread around the Mediterranean—the age of copper had begun. Within another 1,000 years or so, tin for making bronze, and also silver, lead, and zinc, were being extracted from sulfide and oxide minerals and several types of alloys were routinely prepared.

Iron, one thousand times more abundant in Earth's crust than copper, is much more difficult to extract from its natural oxides and sulfides because of its high melting point; hence its use lagged behind those of copper and gold. In fact, it appears that much of the earliest iron came from meteorites, where it occurred as a free metal. Its strength and hardness made it superior to copper alloys and led to its widespread use. Myths spread about the magical powers of iron.

Earth-derived iron appears in tools and wheel rims before 2000 B.C.E. This early iron was apparently derived by roasting iron oxide ores with charcoal in very hot fires. The fires were not hot enough to melt iron but would reduce the oxides to iron masses that could be pounded, shaped, and fused by repeated heating. This wrought iron, though widely prized, remained in short supply because of the difficulty and slow process of manufacture. The development of crude blast furnaces led to the greater availability of iron and made it the most used metal. The rapid flow of preheated air allowed for temperatures high enough to melt the iron and thus direct casting into desired shapes.

The period of global exploration and European colonialism began on a major scale at the end of the fifteenth century with the exploration of the Americas. The discovery of gold in the possession of the Native Americans provided an extra incentive to the Spanish and spurred rapid expansion of their influence in South and Central America. The return of 181 metric tons of gold and 16,000 metric tons of silver to Spain from the New World by 1650 enhanced its wealth and power and cemented Spain's hold and influence in the Western Hemisphere. Colonialism also opened other parts of the world that might provide raw materials and potential markets—North America to the English, French, and Dutch; Southeast Asia to the English and French; and Africa to the English, French, Germans, Dutch, and Belgians. Most of these parts of the world remained sources of mineral commodities for the Europeans for nearly two hundred years and some have only received independence since the middle of the twentieth century.

A major step in the use of mineral resources came with the industrial revolution, which swept across Europe in the 1700s and 1800s. The development of the iron and coal industries in turn stimulated a vast increase in the use of these resources and many others. The steam engine, requiring steel for construction and coal for fuel, necessitated development of an infrastructure to transport raw materials and finished products. This led to the construction of new roads, canal systems, and railroads. Similar developments occurred later in the Americas and other parts of the world.

Since the industrial revolution, there has been much expansion of, and sharper delineation in, the use of mineral resources. Many new chemical elements were discovered, advancing technology provided countless new uses for them, and world economies became more interdependent. Consequently, it is easier to trace concisely historical developments of Earth resource usage since the industrial revolution along a few broad lines.

Construction Materials

Rocks of almost every type have been used as building materials by humankind from the earliest of times in the construction of buildings, roads, bridges, dams, and so on. Though perhaps the least modified and processed material, rock is nevertheless used the most. In contrast to metals, fuels, and many other mineral resources, locally obtainable construction materials have generally been used because of their widespread availability, low cost, and large weight. Consequently, few worldwide production data exist. It is apparent that our ancestors made extensive use of natural shelters, such as caves, and that early in history they began to construct walls and houses by piling up angular rocks. Natural or crudely shaped rock fragments were supplemented by the use of mud (to fill crevices), sod, and earthen blocks, and by the use of crude, abode-type sun-dried mud bricks. Over time stone masonry advanced to a fine art in many parts of the world and brickmaking—made by firing clays—provided synthetic rocks. Cement, the most widely used "instant rock," was first used by the Greeks and Romans, who found that a moistened mixture of heated lime and finely ground volcanic ash hardened into a very stable, rocklike material. With the decline of the Roman Empire, the art of cement making was lost and not rediscovered until 1756, when an English engineer named John Smeaton translated the recipe from Latin writings. Since that time, the use of cement, especially when mixed with sand and gravel to form concrete, has become a primary building material of modern times. Crushed stone, often employed in concrete or as a foundation material beneath concrete, has grown to become the largest, most utilized mineral resource (more than 1.1 billion metric tons in 1990 in the United States) in most industrialized countries.

Natural stone of many types is still widely used in a myriad of shapes for building facings, monuments, and roads; but new ceramics, plastics, and polymers, many of which contain mineral matter, are receiving increasingly wide usage.

Metals

Humankind has made increasing use of metals since the early Bronze Age. The initial use of metals that were found in the native or uncombined state (e.g., copper, gold, silver) gave way to copper-based alloys (bronze-copper and tin; brass-copper and zinc) and ultimately to iron and steel. Mining for copper, gold, and silver developed around the Mediterranean after about 4000 B.C.E., with the ores and metals being transported through the known world primarily by ships. Metal production and technology grew slowly through the periods of the great Greek and Roman civilizations but then settled into quiescence during the Dark Ages.

Europe's awakening during the Renaissance, followed by the growth of the coal industry, the industrial revolution, and the advent of modern chemistry in the mid-eighteenth century, brought about a great increase in the demand for iron and the development of carbon steel. In rapid succession, new metals such as zinc, nickel, manganese, and titanium, among others, were discovered. Chemistry identified the new metals and their properties, and metallurgy rapidly found new uses and synthesized new alloys. Demand increased rapidly and new mines in the Old World and on the frontiers of the New World provided the raw materials. Table 1 provides some insight into the introduction of new metals and the world's expanding use of the ancient ones, such as gold, copper, silver, and tin. These data illustrate the rapid growth of metals during the 1800s and the ever increasing use of several of the metals during the 1900s. Mercury is the only one of the listed metals that appears to have peaked and been on the decline by the late twentieth century. The reduction in the production of mercury is a reflection of our knowledge of its toxicity. Growing concerns over

Table 1. World Average Annual Production in Thousands of Metric Tons (except gold in metric tons)

	1700–1724	*1725–1749*	*1750–1774*	*1775–1799*	*1800–1824*	*1825–1849*	*1850–1874*	*1875–1899*	*1900–1924*	*1925–1949*	*1950–1974*	*1986–1990*
Aluminum	—	—	—	—	—	—	—	0.8	63.9	661.3	6,236.0	16,886
Antimony	—	—	—	—	—	—	—	—	24.6	34.0	58.7	63.1
Chromium	—	—	—	—	—	—	—	16.8	76.2	400.3	1,840.0	11,548
Copper	—	2.4	4.3	6.5	16.2	36.1	85.3	254.9	899.5	1,865.0	4,894.0	8,555
Gold	11.9	18.2	23.6	17.2	12.8	37.0	173.3	213.1	583.0	769.2	1,048.7	1,812
Lead	—	—	—	—	—	94.5	227.7	524.2	1,051.0	1,495.0	2,689.0	3,372
Magnesium	—	—	—	—	—	—	—	—	—	40.6	151.5	336
Manganese	—	—	—	—	—	—	—	61.0	440.0	1,462.0	6,296.0	23,877
Mercury	—	—	0.8	1.2	0.8	1.1	2.3	4.0	3.6	4.7	8.1	5.9
Molybdenum	—	—	—	—	—	—	—	—	0.3	10.6	48.1	100.2
Nickel	—	—	—	—	—	—	0.4	2.6	24.0	97.5	422.6	854.6
Silver	0.4	0.4	0.5	0.8	0.7	0.6	1.2	3.6	6.0	6.7	7.6	14.0
Steel	—	—	—	—	—	—	518.0	8,350.0	59,968.0	111,915.0	404,022.0	758,760.0
Tin	1.8	2.3	4.5	4.4	9.5	14.7	23.7	62.9	116.1	132.4	183.8	199.4
Tungsten	—	—	—	—	—	—	—	—	13.1	16.0	38.1	41.9
Zinc	—	—	—	—	—	19.8	111.0	320.0	842.5	1,668.0	4,094.0	7,082.8

Data from C. J. Schmitz (1979) and U.S. Bureau of Mines, "World Non-Ferrous Metal Production and Prices 1700–1976."

potential toxic effects of antimony, lead, and zinc have limited their use and could result in a decline in their production.

Fuels

The forms and amounts of energy used by humankind mirror changes in technology and population growth. Prior to about 1700 C.E., world population had increased slowly and the primary sources of energy had been slaves, animals, firewood, and, in some areas, wind or water. The burning of wood for warmth and cooking was paralleled by the burning of various forms of biomass, including peat, compacted organic matter that is the precursor to coal. The significant use of coal, first derived from natural erosional exposures, appears to have begun in England in the ninth century. Its use remained subsidiary to that of wood but grew steadily, reaching 200,000 metric tons per year by 1550. The demand for increased energy with the emergence of the industrial revolution, combined with a shortage of wood resulting from the extensive deforestation of England, resulted in a rapid rise in the use of coal. As the industrial revolution spread across Europe, coal emerged as a major fuel in many other countries. In North America the industrial revolution occurred in the second half of the nineteenth century, just as many of the major American coalfields were being discovered. Hence, it was not until about 1900 that coal surpassed wood as the major fuel source in America.

Natural gas had been known and exploited in a few sites from ancient times, but its first commercial use occurred in the 1820s in Fredonia, New York. The absence of any significant pipeline network restricted major gas development until after World War II. Since that time the construction of massive networks, aggressive advertising, and concerns about acid rain and rising CO_2 levels have made natural gas a major fuel for domestic, industrial, and commercial use.

Petroleum's emergence as a major fuel is traced to a well drilled in Titusville in northwestern Pennsylvania by Edwin Drake in 1859. His backers sought petroleum primarily as a source of lamp oil but recognized its potential for the manufacture of solvents and lubricants. It did not emerge as a major fuel source until the development of the internal combustion engine around 1900. Even so, it did not surpass coal as the major energy source in the world until after 1950. Today, oil production rates exceed 60 million barrels (159 liters per barrel) per day. Although it is produced from a vast number of fields around the world, the largest concentrations occur in the Middle East, where approximately two-thirds of the world's nearly 1 trillion barrels of reserves are located. The Organization of Petroleum Exporting Countries (OPEC) has emerged as a major force in oil politics and pricing and accounts for more than three-quarters of the world's oil reserves.

Nuclear power emerged in the mid–1950s and was expected to replace most other mineral resource fuels by the twenty-first century. Its steady growth through the 1960s and 1970s matched expectations, but in 1979 a partial meltdown of the Three Mile Island plant in southern Pennsylvania ignited fears of nuclear power and led to a halt in new construction in the United States and elsewhere. This reaction was compounded by the massive explosion, fire, and radiation release at the Chernobyl nuclear plant near Kiev in the former Soviet Union in 1986. Some countries, such as France and Japan, have moved forward with new nuclear plants but many, like the United States, have seen nuclear growth slow or stop altogether. Some countries, such as Austria, have decided against any nuclear power, while others, such as Sweden, have begun to phase out existing nuclear power.

The report of room temperature fusion in 1991 ignited hopes that the use of heavy hydrogen (deuterium) from the oceans might replace the need for fossil fuels. The process allegedly produced heat energy as a result of hydrogen fusing to form helium when electricity was passed through a palladium wire immersed in water. Unfortunately, subsequent experiments have failed to reconfirm initial claims, and thus world energy production remains dominated by petroleum, coal, and natural gas and will likely remain so for at least the next fifty years.

Fertilizers

Humankind's need for fertilizers began as soon as our ancestors shifted from hunting-gathering to agriculture. It became apparent that the ashes from slash and burn techniques, which supply potassium, and manures from animals, which supply nitrogen and phosphorus, promoted plant growth. The use of ash and manures became widespread in many cultures but remained local until major exploitation of the large deposits of guano—manure from bird or bat droppings—began off the coast of Peru in the early 1800s. Between 1840 and 1880 more than 4 million metric tons of guano were shipped to England.

The discovery of rich nitrate deposits in what is today northern Chile resulted in a second major fertilizer trade from South America to Europe. Competition for the deposits and the revenues they brought resulted in the War of the Pacific from 1879–1883. Little blood was shed, but Chile took possession of Peru's southern province and Bolivia's western flank and only link to the sea. The industry grew until the Germans developed technologies to fix atmospheric nitrogen during World War I. These developments made Chilean nitrates unnecessary, and the mining of natural nitrates has all but vanished.

The modern phosphate fertilizer industry has its roots in Europe of the 1600s, when farmers found that ground animal bone promoted plant growth. In 1840, a German chemist found that the dissolution of the bones in sulfuric acid formed a much more soluble form of phosphate that provided far better plant growth. The same technology was applied to phosphate rock deposits when they were discovered in France in 1846, in England in 1847, in Canada in 1863, and in the United States in 1867. These discoveries led to the modern phosphate mining industry, which now exceeds 150 million metric tons per year.

The potassium fertilizer industry evolved from the recognition that potassium salts, concentrated in the ash from hardwood trees, promoted plant growth as well as serving in the manufacture of glass and soap. The discovery of potassium chloride-bearing evaporite deposits in Germany in 1857, and subsequently in France and the United States, converted the potassium fertilizer industry from one of plant ash treatment to the hard rock mining that exists today.

Chemical and Industrial Minerals

The history and growth of the chemical and industrial minerals industries reflect innovative technological developments and the growth of human population. The most widely used of the chemical minerals, halite or rock salt (NaCl), has been sought, traded, and used for millennia. In fact, the Latin "sal" for salt, which was paid as part of a soldier's wages, is the basis for our word "salary."

The earliest uses were as food additives and preservatives, but the advent of modern chemistry produced thousands of compounds needing sodium and chlorine as components. Halite is now a basic industrial raw material for sodium, chlorine, soda ash (Na_2CO_3), hydrochloric acid (HCl), and hundreds of other compounds.

The earliest sources of salt were flat, arid regions, where salt accumulated from natural evaporation. The discovery of salty brines from some springs, and later of thick evaporite beds, led to large industries that today extract salt from the brines or mine it in large underground operations.

Sulfur has nearly as many uses as salt and has been sought and used by humans for thousands of years. The first sulfur was extracted from volcanic vents where sulfur vapors crystallized on cooling. The rise of modern chemistry brought with it a great need for sulfuric acid because of its strength and versatility. Chemists discovered that by roasting pyrite, or "fool's gold" (FeS_2), sulfur oxide gases were produced, and when bubbled into water sulfuric acid was created. Until 1894 pyrite was mined widely to manufacture sulfuric acid. This method was replaced by the Frasch process, which used hot water to extract native sulfur from large deposits associated with evaporites. Pyrite mining for sulfur has nearly ceased, but the Frasch process remains a major source of sulfur. It has, however, been surpassed as a source of sulfur by the petroleum refining industry, which removes vast quantities of sulfur in order to provide clean burning fuels.

There are myriads of other chemical and industrial mineral resources. The reader is referred to the references below for that supplementary information.

Bibliography

CRAIG, J. R., D. J. VAUGHAN, and B. J. SKINNER. *Resources of the Earth.* Englewood Cliffs, NJ, 1988.

FLAWN, P.T. *Mineral Resources.* New York, 1966.

TYLECOTE, R. F. *A History of Metallurgy.* London, 1976.

U.S. BUREAU OF MINES. Mineral Commodity Summaries, issued annually.

WARREN, K. *Mineral Resources.* New York, 1973.

JAMES R. CRAIG

RESOURCES, RENEWABLE AND NONRENEWABLE

Depending on the context in which it is used, the word resource may mean anything from a source of supply or help to physical, emotional, or financial disposition. In the context of earth sciences, the term refers to the naturally occurring resources of the earth—the natural resources—that have some value to individuals and/or to society. The sustenance of life on Earth depends on the availability and consumption of natural resources, either in the form in which they occur in nature or in an appropriately processed form through human endeavor. An obvious example of a natural resource is water. Freshwater occurring in streams and lakes on the surface of Earth, or as groundwater in the subsurface soils and rocks, is used in irrigation, in all kinds of industrial processing operations, and, with some treatment, for drinking. Even the ocean water is a natural resource because of its recoverable salt and because it can be desalinated to yield drinking water. It may be possible someday to recover appreciable amount of energy utilizing the temperature differences of ocean water at different depths. Other familiar examples of natural resources include air, forests, and mineral deposits (*see* ENERGY, GEOTHERMAL; ENERGY FROM STREAMS AND OCEANS; ENERGY FROM THE SUN; ENERGY FROM THE WIND; MINERAL DEPOSITS, IGNEOUS; MINERAL DEPOSITS, METAMORPHIC; OIL AND GAS, RESERVES AND RESOURCES OF; and WATER SUPPLY AND MANAGEMENT).

Renewable and Nonrenewable Resources

From the perspective of availability relative to consumption, natural resources fall into two broad groups: (a) renewable resources, which either are replenished naturally or can be replenished through human intervention on a short timescale of a few months or years; and (b) nonrenewable resources, which cannot be replenished on a short timescale and, therefore, have a finite supply. The most obvious examples of renewable resources are agricultural crops, water, and solar energy. Crops can be grown again and again on the same piece of land; water is continuously being lost from the surface of the earth through evaporation, but only to return in the form of precipitation; and the nuclear reactions that are responsible for the Sun's heat source will continue for many more billion years.

Other resources, such as mineral deposits on which we depend for metals and other useful mineral substances or fossil fuels which supply most of our energy requirements at present, are nonrenewable resources. Although new deposits of oil, natural gas, and minerals are being formed today in appropriate geological environments, the formative processes are so slow that it may take tens of millions of years for the accumulations to be of any economic significance. Because the rate at which these resources are consumed by our society is orders of magnitude faster than the rate of their formation in the earth, for all practical purposes they should be considered as nonrenewable resources.

The distinction between renewable and nonrenewable resources is of critical importance in assessing the future supply of mineral substances and energy for our consumption. For a renewable resource, the limiting factor is the rate of consumption, so that the resource base can be maintained indefinitely if the rate of consumption is not allowed to exceed the rate at which it can be regenerated. For example, groundwater is a renewable resource because we can continue to draw groundwater from an aquifer as long as the withdrawal rate keeps pace with the rate of replenishment by infiltration of surface water. Solar energy is also a renewable resource, because we can count on having sunlight for an infinitely long time; the only theoretical limitation is that the rate of utilization of energy from this source cannot exceed the rate at which it arrives at the surface of the earth.

The limiting factors for a nonrenewable resource, such as a mineral commodity, are its abundance and availability. No matter how large the resource base, a nonrenewable resource is bound to be exhausted eventually. How long the resource will last, or its life expectancy, is a function of both the resource base and the rate of consumption.

Mineral and Energy Resources

Natural resources are of two broad types: mineral resources and energy resources (Table 1). Mineral resources refer to concentrations of useful mineral substances (*see* USEFUL MINERAL SUBSTANCES) from which economic extraction of a mineral commodity is currently or potentially feasible. Energy resources are those from which energy in usable form can be extracted. For convenience of description, energy-related mineral substances (fossil fuels and nuclear fuels) are commonly included under energy resources as they are used exclusively for production of energy through chemical or nuclear reactions. Examples of energy resources that are not mineral resources include solar energy, geo-

Table 1. Classification of Mineral and Energy Resources

Type of Resource	*Examples*
Mineral Resources	
A. Metallic mineral resources	Precious metals, ferrous (ferro-alloy) metals, nonferrous (base) metals, special metals (see Table 1 in USEFUL MINERAL SUBSTANCES)
B. Nonmetallic mineral resources	Precious stones and gemstones, building materials, fertilizer and chemical materials, abrasive materials, other industrial minerals, surface water and groundwater, soil (see Table 1 in USEFUL MINERAL SUBSTANCES)
Energy Resources	
A. Fossil fuels	Coal, crude oil, natural gas, oil shales, tar sands, methane contained in coal beds
B. Nuclear fuels	Uranium, thorium, hydrogen
C. Geothermal energy	Thermal energy contained in geysers, hot springs, hot water and steam at depth, hot rocks and molten lava, volcanic gases
D. Solar energy	Thermal energy contained in sunlight and ocean currents
E. Hydropower	Kinetic energy contained in running water
F. Miscellaneous	Kinetic energy contained in ocean tides, wind; biomass

thermal energy, tidal energy, and so on. Water is a versatile substance that qualifies both as a nonmetallic mineral resource and as a source of energy (e.g., hydroelectricity).

Resources Versus Reserves

Two fundamental questions regarding any mineral or energy resource are: (a) how much of it exists in or on Earth? and (b) how much of it can be exploited legally and commercially at any given time? This concept is the basis of a classification scheme, devised by the U.S. Geological Survey (USGS) that classifies resources into various categories depending on the degree of certainty of their existence and the economic viability of their exploitation (Table 2). The total resource includes not only identified resources but also less certain, undiscovered resources that, on geological grounds, can be reasonably expected to exist. Identified resources are labeled as measured where the quantity (tonnage) and quality (grade) estimates are based on sufficient geologic data, as indicated where such estimates are less certain, and as inferred where the estimates involve considerable extrapolation beyond the actual database. Economic resources are mineable for profit under the current conditions of price and technology; profitable mining of marginally economic and subeconomic resources would require increasingly higher price of the commodity, or lower cost of production through improved technology, or both. The most important category of the resource base of a commodity at a given time is its reserves, the resource component that is identified, measured, and economically and legally exploitable. All the other boxes in Table 2 collectively constitute potential resources, which are not commercially exploitable at the present time. Nevertheless, the potential sources of a mineral or energy commodity are much larger than the reserves and should be taken into account for long-term commercial and public planning.

The reserves of various commodities at any time are fairly well-defined quantities, but they change continually with time. Mining depletes reserves, whereas new discoveries through exploration add reserves. Reserves of a commodity increase with an increase in its market value or technological advances; they decrease when the market value falls because of excess supply or a slump in demand.

Life Expectancy of Mineral Commodities

On a global scale and over the time frame of a few years, the consumption of a mineral commodity is approximately the same as its production. Countries with large reserves of a mineral commodity generally account for most of its prodution in the raw material form (ore), but not necessarily of its consumption, the latter being largely a function of a country's state of industrialization. For example, the United States is a major producer of copper ore, almost all of which is consumed domestically, whereas much of the large amounts of copper ore produced by Chile and Peru are exported. The highly developed steel industry in Japan is almost entirely based on iron ores imported from a large number of countries around the world.

The world consumption of almost all mineral commodities has registered a significant overall in-

Table 2. Classification of Resources

	Identified Resources			*Undiscovered Resources*	
	Demonstrated		*Inferred*	*Hypothetical*	*Speculative*
Cumulative Production	*Measured*	*Indicated*			
Economic	Reserves		Inferred reserves		
Marginally economic	Marginal reserves		Inferred marginal reserves		
Sub-economic	Demonstrated subeconomic resources		Inferred subeconomic resources		

crease since the industrial revolution, although not with an uniform rate of growth, as is illustrated in Figure 1 for a few selected metals. This pattern of a general increase in world consumption of mineral commodities, which reflects the combined effects of the exponential growth in world population and the ever-increasing per capita consumption of mineral commodities, especially in the developing countries, has raised legitimate concerns about the future availability of mineral resources. The same is also true for nonrenewable energy resources.

How many more years the supply of a nonrenewable resource will last is impossible to predict because of the unpredictability of the variables involved. A minimum estimate of its life expectancy is simply the ratio of a resource's reserves to its current annual consumption. Such calculations, for example, would give a life expectancy of about two hundred years for iron ore and bauxite (ore material of aluminum) of which we have huge reserves, a modest thirty-five years for copper ore, but less than twenty years for either lead and zinc ores, the reserves of which are very limited. The estimates will be somewhat lower if a realistic annual rate of growth in the consumption, based on the historical record and projected population growth, is assumed for the calculations. The actual figures, however, are expected to be much higher because of the likelihood of additional reserves through new discoveries, or technological advances and price increases that would permit the exploitation of resources that are currently marginally economic or subeconomic.

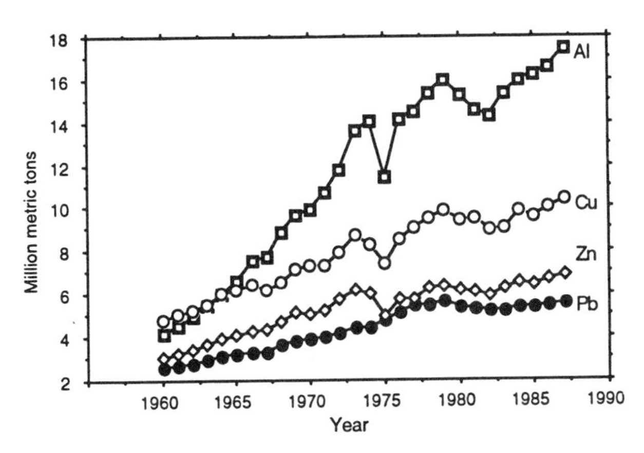

Figure 1. World consumption of selected metals: aluminum (Al), copper (Cu), zinc (Zn), and lead (Pb) from 1960 to 1987.

The life expectancy of a particular resource may also be increased by conservation measures, such as substitution and recycling, which effectively reduce the rate of consumption of the primary material. Some examples of substitution are plastics and composite materials for iron and steel in automobiles and for aluminum in airplanes, and optical fibers for copper in long-distance communication systems. However, substitution merely shifts the burden from one commodity to another, and not necessarily from a less abundant resource to a more abundant one. A much better conservation approach is recycling, especially for metals that are used in pure form and in sizeable pieces--copper in pipes and wiring, lead in batteries, aluminum in beverage cans—and it carries the additional appeal of reducing industrial and municipal waste. Some metals are extensively recycled, at least in the United States. During the period 1985–1992, recycled scrap metal accounted for 25–40 percent of the aluminum (Al), chromium (Cr), cobalt (Co), copper (Cu), Nickel (Ni), and zinc (Zn), and 60 percent of the lead (Pb) consumed in the United States.

Bibliography

CRAIG, J. R., D. J. VAUGHAN, and B. J. SKINNER. *Resources of the Earth.* Englewood Cliffs, NJ, 1988.

MISRA, K. C. *Mineral and Energy Resources: Current Status and Future Trends.* University of Tennessee Studies in Geology 14. Knoxville, TN, 1986.

MONTGOMERY, C. W. *Environmental Geology, 4th ed.* Dubuque, IA, 1995.

TILTON, J. E., ed. *World Metal Demand: Trends and Prospects. A Project of Resources for the Future and the Colorado School of Mines.* Washington, DC, 1990.

WORLD RESOURCES INSTITUTE. *World Resources. 1994–95.* New York, 1994.

KULA C. MISRA

RICHTER, CHARLES

Charles F. Richter was born in Hamilton, Ohio, on 26 April 1900. He earned a bachelor's degree in physics at Stanford in 1920, enrolled at the California Institute of Technology (Caltech) for graduate study, and received his Ph.D. in theoretical physics in 1928. Richter began working at the Seismological Laboratory of the Carnegie Institution of Washington a year before receiving his Ph.D. degree. The Seismological Laboratory was then located in Pasadena, California, and jointly operated by the Carnegie Institution and Caltech. Richter worked first as an employee of the Carnegie Institution and later as a member of Caltech when the laboratory was transferred to Caltech in 1936. He continued to work at Caltech as professor of seismology until his retirement in 1970. He taught seismology in Japan as a Fulbright scholar during 1959–1960. He served as president of the Seismological Society of America from 1959 to 1960 and was the second recipient of its medal in 1977. Richter married Lillian Brand in 1928. The Richters had no children.

In the 1930s, the Seismological Laboratory was preparing to issue regular reports of earthquakes occurring in southern California. While tabulating 200 to 300 earthquakes a year, Richter wanted to devise some means of grading them on a quantitative basis (the systems then in use were extremely qualitative). He measured the amplitudes of seismic waves recorded with the Wood–Anderson seismographs that just had been deployed in southern California. Using this data set, Richter introduced a local magnitude scale, M_L, to represent the physical size of earthquakes in southern California. M_L is determined by the logarithm of the amplitude A measured at a distance Δ from the earthquake's epicenter (the point on the earth's surface directly above the focus of an earthquake): $M_L = \log A + f(\Delta)$. Here $f(\Delta)$, called the amplitude attenuation function, is a function of Δ, but does not depend on A. Thus, at a given distance, M_L increases by one unit as the amplitude of seismic waves increases 10 times. $f(\Delta)$ is adjusted such that $M_L = 3$, if $A = 1$ mm at $\Delta = 100$ km. This scale, though developed empirically with data from only a few stations, turned out to be extremely useful not only for earthquake reporting purposes but also for various scientific and engineering studies. For this reason, Richter's paper on the local magnitude published in 1935 is generally regarded as a milestone in seismology. Later analyses using many more earthquakes and seismic stations confirmed that the function $f(\Delta)$ Richter determined is very accurate.

Although M_L was originally introduced to measure earthquakes only in southern California, the concept—the Richter scale—was later extended to earthquakes all over the world by BENO GUTENBERG (then director of the Seismological Laboratory of Caltech), Richter, and subsequently by many other seismologists. With the introduction of the magnitude scale, seismologists could begin to study earthquakes quantitatively, and earthquake seismology, which had been a somewhat descriptive science, became a modern quantitative science.

Another major contribution by Richter was his 1958 book *Elementary Seismology*. This book is in a way an encyclopedia of seismology; it touches on practically every aspect of earthquakes with a strong emphasis on field observations. It includes not only the conventional subjects like descriptions of important earthquakes, seismic waves, magnitude, intensity, seismographs, and distribution of earthquakes, but also sections on the effects of earthquakes on ground and surface water, insurance, seismic zoning, and earthquake-resistant construction. It is remarkable that even today not only students but also professional seismologists often refer to this thirty-five-year-old text. Although Richter did not have many students from his own laboratory, through this book he probably has more students than any other seismologist. Richter's descriptions of earthquakes is always very detailed and accurate. Very often when the original account was written in a foreign language he had someone translate it before writing his own account.

Another important contribution is the book *Seismicity of the Earthquake*, which Gutenberg and Richter published in 1949. This is not merely a catalog of earthquakes but a monumental work on tectonics of Earth as viewed from a seismological point of view.

Richter also coauthored with Gutenberg a series of monographs, *On Seismic Waves* (1934, 1935, 1936, and 1939). Interpretation of seismograms is one of the most important tasks in seismology. Many of the important features of Earth's interior and of earthquakes are determined by the interpretation of very subtle features in seismograms. The descriptions of seismic waves and their theoretical interpretations described in *On Seismic Waves* helped many seismologists to extract useful information from seismograms, which eventually led to many important discoveries.

Although Richter's works are generally on the observational aspects of seismology and appear descriptive, he was also interested in basic theories in seismology, as is evidenced by his paper on mathematical questions in seismology.

For many years until his retirement, Richter was in charge of the measuring room of the Seismological Laboratory. He examined literally thousands of seismograms to study every detail of earthquakes. In this capacity, Richter played an important role in educating the public about earthquakes. Whenever a major earthquake occurred, not only in southern California but also in other seismic regions in the world, reporters from the news media called Richter at his home or at the laboratory for his comments. Richter responded to these requests day and night. Talking to the press about earthquakes was apparently one of Richter's joys and pleasures. Richter was a man with remarkable memory and his account of an earthquake that just occurred in relation to the historical background of the area intrigued not only the general public but also professional seismologists.

Richter was known to be a man of good humor, but whenever he saw something that ran counter to his scientific or professional beliefs, he expressed his opinion in candid words, which can be seen in some of his writings.

Richter died on 30 September 1985 in Altadena, but the tradition he established of making detailed analysis of earthquakes in southern California and reporting the results to the public is maintained at the Seismological Laboratory.

Bibliography

GUTENBERG, B., and C. RICHTER. *Seismicity of the Earth.* Princeton, NJ, 1949.

———. "On Seismic Waves." In *Beiträge zur Geophysik* 43 (1934):56–133; 45 (1935):280–360; 47 (1936):73–131; 54 (1939):94–136.

RICHTER, C. F. "An Instrumental Earthquake Magnitude Scale." *Bulletin of the Seismological Society of America* 25 (1935):1–32.

———. *Elementary Seismology.* San Francisco, 1958.

HIROO KANAMURI

RICHTER SCALE

See Earthquakes and Seismicity

RIFTING OF THE CRUST

Rifting is the process by which tectonic plates, usually continents, are stretched and broken. A sedimentary basin is created when a continent is stretched a little. A new ocean is created when a continent is stretched beyond the point of breaking. When a continent rifts apart, the asthenosphere below the continent rises to fill the gap. As it rises, it partially melts to form new oceanic lithosphere. The South Atlantic Ocean, the South American continent, and the African continent formed in this way when the supercontinent of Pangaea rifted about 135 to 110 Ma.

Rifts in continents normally form elongate depressions bounded on one or both sides by normal faults. They are generally tens to hundreds of kilometers long and tens of kilometers wide. The flanks of the rift basin may be uplifted, the crust under the basin is usually thinner than unrifted continental crust, and there may be volcanism in or near the rift.

The largest continental rift active today is the East African Rift. It extends from the southern end of the Red Sea in Ethiopia to the Indian Ocean in Mozambique. The rift depression contains long narrow lakes where there is adequate water, and lies below sea level where there is not. Volcanic activity associated with the rift accounts for the highest mountains in Africa, Mount Kilimanjaro and Mount Kenya, both higher than 5,000 m. The volume of volcanic rocks in Kenya and Ethiopia alone is estimated to be 500,000 km^3. That is enough lava to cover the state of New York to a depth of 4 km!

The ultimate source of the forces that cause rifting is heat. The earth moves heat from the deep mantle to the surface by convection in the asthenosphere, which entrains parts of the lithosphere. Rifting can occur where the convecting asthenosphere causes the lithosphere to be in tension. The mechanism by which the convecting system produces tension in a continental plate remains the subject of some controversy. "Active" rifting models suggest that upwelling asthenosphere is the direct cause of continental rifting. In these models the lithosphere is hit from below by a plume of hot asthenosphere that lifts and then splits it apart. In active rifting the inferred sequence of events is doming—volcanism—extension. "Passive" rifting models suggest that horizontal tension in the lithosphere pulls the lithosphere apart, allowing the passive upwelling of asthenosphere to fill the gap. The most likely source for the horizontal tension in the lithosphere is the "slab pull" force at a nearby subduction zone. In passive rifting the inferred sequence of events is extension—doming—volcanism. When applied to the study of rifts, both these simplistic models are found wanting. It is likely that real rifts form by some varying combination of active and passive mechanisms.

Regardless of the overall rifting mechanism, active or passive, the exact location of a rift is controlled by the heterogeneity of the preexisting lithosphere. On the largest scale, rifts form along preexisting weaknesses. The most profound linear weaknesses in continental lithosphere are continental sutures and associated mountain belts. Therefore, continents tend to rift along the lines of previous continental collisions, and new oceans tend to form in roughly the same position as extinct oceans. Lineated weaknesses will only be activated if they are oriented roughly normal to the applied tension. On a finer scale and shallower in the crust, preexisting faults are commonly reactivated as graben bounding normal faults. These may have been older normal faults, but most commonly they are reverse faults that last moved during an earlier continental collision.

The final stage of "successful" continental rifting is the initiation of seafloor spreading and the creation of new oceanic lithosphere. The complete process of rifting, rift initiation to seafloor spreading, generally takes between 10 and 50 m.y. The initiation of seafloor spreading is not usually synchronous in a connected series of rifts. The South Atlantic Ocean, for example, rifted open from south to north. Locally, there are usually rift segments where seafloor spreading begins sooner, called "nucleation points," and rift segments where seafloor spreading begins later, called "locked zones." Seafloor spreading propagates from nucleation points into locked zones. Seafloor spreading does not necessarily begin along the center axis of a rift, leaving behind symmetric rifted margin basins. A more typical pattern is for seafloor spreading to begin on one side of the rift, leaving a wide margin basin on one side of the new ocean and a narrow margin basin in the conjugate position on the other side. It is also sometimes observed that the polarity of the margin, that is, which side is wide, reverses at an interval of 20–500 km along the rift axis. Along the U.S. Atlantic margin, the Georges Bank and Blake Plateau Basins are examples of wide margin basins and the Carolina and Baltimore Canyon Troughs are examples of narrow margin basins.

In some rifts the extension ceases before seafloor spreading begins, forming so-called failed rifts. Rift-rift-rift triple junction systems often have two successful arms and one failed arm. The failed arm extends into one of the continents and is called an aulacogen. The mouths of large rivers are often coincident with aulacogens.

Long rift axes, such as the East African Rift and the rift that led to the formation of the Atlantic Ocean, generally consist of a periodic system of half-graben basins with alternating polarity. A half-graben basin is formed by extending crust containing a set of parallel, steeply dipping (about 60°), normal faults. As the crust is extended the faults slip and the blocks that they bound are rotated in the direction that flattens the faults. This is like a set of books on a bookshelf. If you tilt the whole set of books, there will be normal slip between books, the books will rotate, and the horizontal surface area of the top of the books will increase. The polarity of a half-graben system can be defined in terms of the dip direction of the faults. The surface trace of the half-graben faults in a single polarity system is generally arcuate, concave in the fault dip direction. In seismic reflection profiles the faults are seen to flatten with depth. These are referred to as listric faults. The polarity of half-graben systems tends to reverse along the axis of a rift. The region where the polarity of the system reverses is known as a transfer or accommodation zone. The geometry of the faults in the subsurface of an accommodation zone is poorly known.

Rifts are characterized by diffuse, shallow, and relatively small earthquakes. They are diffuse because rifting involves many subparallel anastomosing normal faults in a region tens of kilometers wide. Most of these faults penetrate only to mid-crustal depth (10–20 km). Deformation deeper in the crust and mantle is accommodated by ductile flow rather than by earthquakes. Fault plane solutions for rifting earthquakes generally show them to occur on steeply dipping normal faults. This observation apparently conflicts with seismic reflection imaging of normal faults, which suggests that they flatten with depth and that most crustal extension is accommodated by slip on very low-angle portions of the faults. Some earthquakes in rift zones show transform and even reverse mechanisms. These are often associated with heterogeneities in the rifting crust or with accommodation zones.

Rifting increases heat flow. As lithosphere thins during rifting, the geothermal gradient, and hence the vertical heat flow, is increased. If the lithosphere is thinned by one half, then the immediate effect on the surface heat flow is an increase by a factor of two. This anomalous heat flow decays exponentially with time returning to roughly its original value over about 50 m.y.

As continental lithosphere is thinned during rifting, the surface generally subsides creating a water- or sediment-filled basin. This basin can be up to about 2.5 km deep if filled with water alone or up to about 7.5 km deep if filled entirely with sediment. Note that due to isostasy it takes a greater thickness of a denser material to fill a hole. This subsidence, referred to as "initial" or "synrift" subsidence, is due to the replacement of continental crust, with average density of about 2.8 g/cm^3, by hot mantle rock, with average density of about 3.2 g/cm^3. After rifting stops, the surface generally continues to subside slowly for tens of millions of years. This subsidence, referred to as "thermal" or "postrift" subsidence, is due to the cooling and contraction of the hot mantle rock emplaced under the rift basin. When this subsidence is complete and the effects of initial subsidence are included, the basin can be up to about 6 km deep if filled with water alone or up to about 18 km deep if filled entirely with sediment.

The sedimentary basins that result from continental rifting contain most of the world's oil and natural gas reserves. The formation of oil and natural gas requires a source of organic matter that is rapidly buried by sediment and cooked at temperatures of 100–200°C for tens of millions of years. Rifting creates the hole in which organic matter and sediments are rapidly buried and the combination of elevated heat flow and establishment of a normal geothermal gradient in the thickening pile of sediments provides the heat. It is common for young rift basins to accumulate significant layers of evaporites. This occurs when basins are isolated from oceans and the only way for water to leave the basin is by evaporation. The presence of salt and salt diapirs in mature rift basins provides the opportunity for oil and natural gas to migrate and accumulate in economically exploitable reservoirs.

The gravity signature of a rift is due to the combination of the effects of low-density sediment and water filling the rift basin, of high density upper mantle rock upwelling adjacent to continental crust, and of lower density, hot, asthenosphere rock upwelling adjacent to upper mantle rock. This produces a negative gravity anomaly localized over the surface manifestation of the rift, superimposed on an intermediate wavelength gravity high, over an even broader gravity low. The amplitudes of the three competing signals vary among rifts. Therefore there is no completely diagnostic rift gravity signature.

The magnetic signature over rifted crust is due almost entirely to the magnetic character of the crust that is being rifted and to the depth of the top of the rifted basement. Where the basement surface is deeply buried by sediment, the magnetic anomalies due to sources in the shallow crust will be broader and more subdued. Where the shallow crust is near the surface, the anomalies will be shorter wavelength and larger.

The volcanism during rifting is the result of decompression melting of the upwelling asthenosphere beneath the rift. Note that in either active or passive rifting the asthenosphere is rising. Asthenosphere rocks are solid at their equilibrium depth and temperature. They can be caused to melt either by raising the temperature or by decreasing the pressure. The pressure on a rock is controlled by its depth. The asthenosphere rocks undergo "partial" melting during their ascent. Typically 5–25 percent of the rock in the upper asthenosphere melts. The basaltic melt rises quickly through the denser mantle. The melt often accumulates at the base of the crust or at discrete levels within the crust. At this point fractional crystallization of the basaltic magma and/or melting of crustal rocks into the magma may occur. These processes account for some of the great diversity in magma types erupted during rifting. The magma probably makes its way to the surface along faults, but the mechanisms of magma movement in the crust are not well understood. In general, rift zone volcanism is highly explosive, and in some places pyroclastic rocks dominate the volcanic sequence.

Basalts erupted in continental rift zones tend to be alkalic, enriched in volatiles, and enriched in large ion lithophile elements (LILE), suggesting derivation from enriched mantle sources. Where rifting has been extreme and rapid, transitional basalts predominate. These are thought to be associated with a higher degree of partial melting. Another factor thought to have a significant affect on the volume of volcanics and their petrologic character is the prerifting temperature of the asthenosphere. The asthenosphere temperature is thought to be elevated, by up to 200°C in the vicinity of a hotspot or mantle plume. If rifting occurs in such a region, the amount of rift magmatism is greatly increased over what would normally be expected. There are rifts where the thickness of basalt is known to be at least 20 km. Asthenosphere temperature is probably one principal factor in explaining the occurrence of "wet" rifts (those with anomalously large amounts of volcanism) and "dry" rifts (those with little or no volcanism).

Bibliography

Bosworth, W. "Geometry of Propagating Continental Rifts." *Nature* 316 (1985):625–627.

Girdler, R. W. "Processes of Planetary Rifting as Seen in the Rifting and Breakup of Africa." *Tectonophysics* 94 (1983):241–252.

Wilson, M. *Igneous Petrogenesis: A Global Tectonic Approach.* London, 1989.

DALE SAWYER

RINGWOOD, A. E. (TED)

Alfred Edward (Ted) Ringwood was an outstanding theoretical and experimental geochemist who received many international honors for his contributions to the understanding of the nature and processes of the earth's mantle, the origin of the Moon, and the origin and evolution of the solar system. He applied his deep knowledge of mineral chemistry to devise and patent processes for the encapsulation of high-level nuclear waste (the SYNROC concept) and for the manufacture of new ultra-hard materials.

Ted Ringwood was born on 19 April, 1930, the only child of Alfred Edward, a World War I veteran, and Ena Grace Ringwood. Ringwood grew up in inner Melbourne, Australia, in a family of very limited means during the Great Depression. His mother and wider family encouraged the academic ability evident initially at Hawthorn Central School. Ringwood received a scholarship to Geelong Grammar School (one of Australia's leading private schools) and later a scholarship to study science at Melbourne University. Between 1948 and 1956 he obtained bachelor's, master of science, and doctor's degrees from Melbourne University. His master's thesis was on the geology and petrology of granites and acid volcanics in the remote and heavily forested Snowy River area of northeastern Victoria. It was during his Ph.D. research, initially begun as an experimental study relevant to economic geology, that he became interested in crystal chemistry and the applications of VICTOR M. GOLDSCHMIDT's principles to prediction of the effects of pressure and temperature on minerals in the earth's mantle. He also published classical papers on the partitioning of trace elements in magmatic crystallization, emphasizing the importance of electronegativity in controlling trace element behavior.

Ringwood pursued a postdoctoral fellowship (1957–1958) with Francis Birch at Harvard University, where he began the experimental studies of high-pressure phase equilibria that became one of his major scientific contributions. He used germanates (germanium analogs of silicate materials, for example, Mg_2GeO_4 as an analog for olivine, Mg_2SiO_4) to study high-pressure equilibria because the germanates will undergo similar phase transitions at lower pressures, making them more easily studied in the laboratory. In 1958 he was recruited by J. C. Jaeger to join the new department of geophysics at the Australian National University (ANU). He was appointed to a personal professorship in 1963 and as professor of geochemistry in 1967. Ringwood led the efforts that established the new Research School of Earth Sciences (RSES) at ANU in 1972 and was its Director from 1978 to 1983. Throughout his working career he remained committed to the ideal of the Australian National University as a distinctively Australian center of research operating at the highest international standards. Ringwood's own work did much to fulfill this vision.

Ringwood began his experimental work at ANU with diamond-anvil syntheses of high-pressure polymorphs of germanates and silicates. This was followed in 1963 by his collaboration in experimental studies using the piston-cylinder apparatus and electron microprobe to explore relationships between the earth's crust and mantle through the gabbro to eclogite reactions, the mineralogy of peridotite, and the origin of basalts as partial melts from peridotitic mantle. In his later work he used the multi-anvil device and laser-heated diamond anvil to directly synthesize mantle minerals under Transition Zone pressures (400–900 km depth) and then to pressures approaching those of the core-mantle boundary (2,900 km). In this later work, he explored the roles of mineral reactions in the subducted slab in determining questions of whole-mantle versus upper-mantle convection.

Ted Ringwood sought to address the important issues in earth sciences and worked effectively across the boundaries of geophysics, geochemistry, and geology. He was extraordinarily creative, formulating new ideas and new syntheses of data, but, being primarily an experimentalist, he also sought to test his ideas by real-earth data and by experiment.

Ringwood's high-pressure studies took him into the design and demonstration of ultra-hard materials based on diamond aggregates and on cubic boron nitride aggregates. He also turned his knowledge and insights of mineral chemistry to an issue of global concern—the safe disposal of high-level nuclear waste. He devised and demonstrated the SYNROC (synthetic rock) concept of containing radioactive elements in stable mineral assemblages as an approach superior to that of encapsulation in metastable glass.

Ringwood extended his crystallochemical interests to studies of meteorites and of the origin and evolution of the solar system. The return of lunar samples by the Apollo missions enabled him to apply the experimental techniques developed in unravelling the mysteries of terrestrial basalts to lunar basalts. He contributed in a major way to understanding the nature and diversity of lunar rock compositions and their implications for lunar origin and Earth-Moon relationships. The studies of planetary evolution included speculations on the nature of the light element(s) within the earth's core and, characteristically, he devised experimental approaches to test his models and their implications for the geochemical features of the earth's core and mantle.

Ted Ringwood's outstanding characteristic was that he made original, often controversial, contributions to knowledge across a wide spectrum of earth sciences. He presented his work with clarity and vigor in both written and verbal forms and engaged in scientific debate with conviction and independence. He presented speculative hypotheses but followed up with experimental tests or sought real-earth observations to confirm or deny the models presented. Although some of his vigorously argued ideas are now discarded, many others are cornerstones for lines of research stretching into the twenty-first century. His work profoundly influenced international geophysics and geochemistry from 1960 to 1993, and his leadership and stimulus at the Research School of Earth Sciences at the Australian National University was a major factor in its rise to an international leadership position in earth sciences. Ringwood published two very influential books, *The Composition and Petrology of the Earth's Mantle* (1975) and *The Origin of the Earth and Moon* (1979), and he authored or coauthored over three hundred research papers.

Ted Ringwood was one of the most internationally honored earth scientists, receiving thirty medals or prizes and being elected to many of the world's most prestigious learned academies. He was elected a Fellow of the Australian Academy of Science in 1966, of the American Geophysical Union in 1969, of the Royal Society of London in 1972, as a Foreign Associate of the U.S. National Academy of Science in 1975, and a Fellow of the Australian Academy of Technological Sciences in 1991. Ted Ringwood died prematurely on 12 November, 1993 following a long battle against lymphoma.

Bibliography

Green, D. H., and Ringwood, A. E. "The Genesis of Basaltic Magmas." *Contributions to Mineralogy and Petrology* 15 (1967):103–190.

Kesson, S. E., and Ringwood, A. E. "Mare Basalt Petrogenesis in a Dynamic Moon." *Earth and Planetary Science Letters* 30 (1976):155–163.

Ringwood, A. E. "The Principles Governing Trace Element Distribution During Magmatic Crystallization. Part I. The Influence of Electro-negativity." *Geochimica et Cosmochimica Acta* 7 (1955):189–202.

———. "Mineralogical Constitution of the Deep Mantle." *Journal of Geophysical Research* 67 (1962):4005–4010.

———. *Composition and Petrology of the Earth's Mantle.* McGraw-Hill, 1975.

———. *Origin of the Earth and Moon.* New York, 1979.

———. "The Earth's Core: Its Composition, Formation and Bearing Upon the Origin of the Earth." Bakerian Lecture. *Proceedings of the Royal Society* A395 (1984):1–46.

———. "Disposal of High-level Nuclear Wastes: A Geological Perspective" (Hallimond Lecture 1983). *Mineralogical Magazine* 49 (1985):159–176.

———. "Role of the Transition Zone and 660 km Discontinuity in Mantle Dynamics." *Physics of the Earth and Planetary Interiors* 86 (1994):5–24.

DAVID H. GREEN

RIVERS, GEOMORPHOLOGY OF

When water flows down a slope, it tends to concentrate in small-scale irregularities on the surface. This concentration increases the water's erosive capacity so that the irregularities are enlarged into rills and eventually small channels. These channels may be arranged on the surface in a variety of patterns, depending on the topography and resistance to erosion of the surface. The spatial pattern of channels is usually described at the scale of the drainage basin. A drainage basin includes the entire surface area that drains toward a given point. The boundaries of the basin are the drainage divides, which may be dramatic features, such as the Continental Divide of the western United States, or much more subtle high points, such as the gentle slope between the northern border of the Mississippi River and the channels that drain north to Canada. In either case, the drainage divide forms a line or point of higher elevation than the surrounding landscape; precipitation falling on one side of the divide flows into one drainage basin, while that falling on the other side of the divide flows into an adjacent basin. For example, precipitation on the western slope of the Colorado Rocky Mountains contributes to flow in the Colorado River, while precipitation falling on the eastern slope eventually joins the Missouri River.

Drainage Patterns

Within a drainage basin, the drainage pattern may be classified by spatial arrangement or by temporal sequence with respect to the structure of the underlying rocks. A dendritic drainage pattern is one of the most common categories in a spatial classification. Dendritic patterns form on homogeneous rocks with fairly gentle slopes and are named for their resemblance to the veins of a leaf. Trellis or rectangular patterns are characterized by channels that join at right angles. These patterns form where the underlying rocks have a strong orthogonal joint system. Radial patterns form where a series of channels flow out and downward from a single topographic high point, such as a volcanic cinder cone. Parallel drainage patterns form on steeply sloping, homogeneous surfaces where many long channels form parallel to one another.

Channels classified with respect to structure may be consequent or subsequent. Consequent channels are so named because their pattern is a consequence of the underlying structure. These channels form the expected patterns, flowing down inclined slopes, for example. Subsequent channels form by headward erosion along zones of weaker rock. The idea that channels could precede present-day structure and topography was developed during the late-nineteenth-century scientific expeditions to the American West. When geologists saw rivers like those of the Colorado Plateau region, where deep canyons are incised into broad uplifts, they realized that the rivers must be older than the uplifts. They coined the terms "superimposed" and "antecedent" to describe a situation where: (1) a river gradually incised down to a buried structure, maintaining its course as it eroded through the structure (superimposed); and (2) a preexisting river continued to incise as the surrounding land was uplifted (antecedent).

Within a drainage basin, channels may also be described in terms of stream order and drainage density. Stream order involves a numbering system designed to facilitate comparison among basins by means of dimensionless ratios. In the most commonly used system of stream ordering, a first-order channel is one that has no tributaries. Where two first-order channels join, they form a second-order channel. Similarly, two second-order channels form a third-order channel, and so on. A basin with a low stream order thus has many first- and second-order channels, which usually implies high sediment yields, flash floods, and relatively steep, short channels. Drainage density is a measure of the spatial density of channels; it is the total length of channels within the basin, divided by basin area. A high drainage density implies an older drainage pattern and relatively efficient transport of water and sediment through the basin.

Drainage basins may also be characterized by their vertical dimensions, using such measures as the relief ratio or the hypsometric integral. Relief ratio is simply the difference in elevation between

Figure 1. Upper Amazon Basin, Ecuador. (Photo courtesy of Ellen E. Wohl.)

the highest and lowest points in the basin. A high-relief ratio implies steep slopes, and the associated high sediment yields and flash floods. The hypsometric integral of a basin describes the distribution of mass within the basin. The original basin is assumed to have vertical sides rising from a horizontal plane passing beneath the basin at the level of the basin mouth. The integral is a plot of proportion of total basin area versus proportion of total basin height, and expresses the volume of the original basin that remains. Values for most natural basins range from 20 to 80 percent, with the higher values indicating that large areas of the original basin have not been altered into slopes. Low values imply that much of the basin stands at low elevation relative to the area of the original upland surface.

The vertical characteristics of an individual channel are indicated by a longitudinal profile, which is a plot of elevation versus downstream distance from the drainage divide. Channels in balance with external controls have a smooth, concave-upward longitudinal profile. Such a profile characterizes the ideal "graded stream," which is adjusted so as to provide just the velocity required for the transport of the sediment load supplied from the basin. The graded stream is referred to as being in equilibrium, with no net change through time unless one of the external controlling parameters (such as climate) changes. In contrast, a channel with a straight or convex-upward longitudinal profile may still be incising rapidly in response to a recent tectonic uplift. Layers of more resistant rock that cross a channel's course may produce irregularities, known as knickpoints, in the channel's longitudinal profile. The most dramatic knickpoints form huge waterfalls such as Niagara Falls.

Channel Classification

Individual channels within a drainage basin may be classified in terms of their substrate, surface pattern, flow characteristics, bedforms, or vertical stability. Channel substrate is generally divided into bedrock versus alluvial channels, although the dominant grain size of channel alluvium may also provide a basis for classification, as in gravel-bed or sand-bed channels. The three most common channel classes on the basis of surface pattern are braided, meandering, and straight. Braided channels have multiple flow paths, for the main channel is subdivided into many secondary channels by numerous bars and islands. These secondary channels shift back and forth relatively rapidly (over a period of days to years, depending on the river's size), and braided channels are generally characterized by unstable banks, steep channel slopes, large sediment loads, and high variability in the quantities of water and sediment moving along the channel. Channels are classified as meandering if their sinuosity, or ratio of channel length to straight-line valley length, is greater than 1.5. Meandering channels generally have lower channel slopes and smaller sediment loads than braided channels. Channels with a sinuosity less than 1.5 and a single flow path are classified as straight.

Channels may also be classified in terms of their flow characteristics; specifically, how often water flows in the channel, or the climatic scenarios that produce high flows. Under the former criterion, ephemeral channels flow only for short time periods—usually after a heavy rainfall. Intermittent channels may have surface flow along some reaches of the channel, but only subsurface flow along other reaches. Perennial channels flow throughout the year along the entire length of the channel. Considering the climatic controls on flow, channels may be snowmelt-dominated, or their flow may be produced by rainfall or rain-on-snow events. Flow in snowmelt-dominated channels rises gradually to a maximum volume during spring snowmelt, is sustained for a period of days to weeks, and then gradually declines. In contrast, rainfall- or rain-on-snow-dominated channels tend to have dramatic rises in flow that may last only a few hours or days, followed by a swift decrease in flow.

Some channel classifications focus on bedforms, the regularly repeating structures sculpted by water and sediment moving along the channel bed. Common bedforms include pool-riffle sequences, step-pool sequences, dunes, and the plane bed. Pools are topographic low points in the channel bed and riffles are topographic high points; these features alternate with each other in a downstream direction, producing an undulating channel bed. Step-pool sequences, as the name implies, are steps alternating with small pools in a downstream direction. Dunes are asymmetrical waves of sediment perpendicular to flow that move down the channel with a tractor-tread type of motion, while a plane-bed channel has a relatively flat, featureless bed.

Finally, channel classifications that focus on vertical stability may designate channels as: (1) eroding/incising, if the level of the bed is being consistently lowered over a period of years to decades; (2) stable, if there is no net change in channel-bed elevation; and (3) aggrading, if there is a net accumulation of sediment in the channel.

There is no single, universally accepted system of channel classification. Many of the systems described above are not mutually exclusive, so that a channel could be alluvial, braided, ephemeral, and aggrading, for example. The classification(s) used will be governed by the channel characteristics deemed most important in a particular situation.

Channel Hydrology and Hydraulics

Channels may also be described by focusing on the water and sediment moving through them. Discharge is a measure of volume flowing past a measuring point per unit time, and may be applied to either water or sediment. Water discharge, which is usually measured in cubic feet per second or cubic meters per second, is the product of channel width, flow depth, and velocity. Velocity is a vector quantity that describes the speed and direction of water flow. It is reported in feet per second or meters per second.

A network of gaging stations that measure river discharge in the United States is maintained by the U.S. Geological Survey (USGS). Depending on the type of station, discharge may be measured continuously, or in time increments ranging from fifteen minutes to once a year. Changes in discharge through time may be described by means of hydrographs, flood-frequency curves, or flow-duration curves. A hydrograph is a plot of discharge versus time. Such a plot may cover the interval of a single rainfall-generated flood (a few hours to days), or it may be an annual hydrograph that summarizes seasonal changes in discharge. A flood hydrograph has two components: baseflow and runoff. Baseflow is water that enters the channel from the subsurface, having moved relatively slowly downslope through subsurface mechanisms. Baseflow represents discharge in a channel before and after a flood. The flood peak is produced by runoff that quickly reaches the main channel through surface paths, such as sheetflow over slopes or small channels. A flood-frequency curve is a plot of flood discharge versus either exceedence probability (the probability that the corresponding discharge is equalled or exceeded in any one year) or the recurrence interval (the average time interval between years in which the annual peak discharge equals or exceeds the corresponding discharge). Larger discharges generally have lower exceedence probabilities and higher recurrence intervals. Exceedence probability and recurrence interval are often used to name floods. People speak of the hundred-year flood, for example, and often interpret this to mean that a flood of this magnitude only occurs once every one hundred years. However, although the exceedence probability for such a flood is 0.01, and the average recurrence interval is one hundred years, it is entirely possible that a "hundred-year flood" could occur in two successive years. Such an occurrence may not be probable, but it is possible. Flood-frequency curves may use either partial duration or annual maxima flood series. A partial duration series includes all of the discharges above a specified threshold, while an annual maxima series includes only the largest flood in each year of record. Flow-duration curves, which plot discharge versus percent of time that discharge is equalled or exceeded, include a range of lower discharges as well as floods. As in a flood-frequency curve, large discharges are seldom exceeded.

Flow velocity is generally measured at several points within a channel cross section, and then reported as a cross-sectional average. Velocity fluctuates widely both in time and space. If you have ever stood in flowing water, you have undoubtedly felt the pulses that are produced by short-term fluctuations in velocity at a point. These fluctuations result from turbulence as the water flows past the rough channel boundaries. Because of this flow resistance, water flowing at the sides and bed of the channel moves more slowly than water in the central portion of the channel. Velocity is thus generally highest at the top and center of the channel, where frictional resistance is lowest. It also tends to increase downstream. As discharge increases in a downstream direction, channel width and depth and velocity also increase correspondingly.

The relations between the controlling variable discharge and the dependent channel variables of width, depth, and velocity are described through hydraulic geometry. Hydraulic geometry refers to a series of simple power functions, such as:

$$\text{width} = a\,(\text{discharge})^{b}$$
$$\text{depth} = c\,(\text{discharge})^{f}$$
$$\text{velocity} = k\,(\text{discharge})^{m}$$

because discharge = (width)(depth)(velocity) = $a(\text{discharge})^{b}\ c(\text{discharge})^{f}\ k(\text{discharge})^{m}$, then $a \cdot c \cdot k = 1$, and $b + f + m = 1$ (Ritter, 1986). In other words, if discharge changes, width, depth, and velocity will change in a corresponding manner. The hydraulic geometry relations allow us to predict the direction of channel change in discharge.

The movement of water through a channel exerts drag and lift forces on the channel boundaries. These forces are often expressed as boundary shear stress and stream power per unit area. Boundary shear stress is most simply expressed as the product of flow depth, channel gradient, and

Figure 2. Big Box Canyon, Nevada. (Photo courtesy of Ellen E. Wohl.)

the specific weight of water; it is the shear force exerted on the channel bed, by the flowing water. Stream power per unit area is the product of shear stress and velocity; it is the power (or rate of doing work) per unit area of stream bed (Knighton, 1984). When the shear stress in a channel exceeds the resisting forces of grains on the bed (the submerged weight of the grains, and any frictional resistance due to neighboring grains), the grains will be entrained by the flow and begin to move. It takes more energy to initiate motion than to continue motion, and once a grain has been dislodged from the channel bed it is likely to move a distance many times its diameter before being redeposited (*see* GEOLOGIC WORK BY STREAMS). This transport may be quantitatively expressed in terms of stream power.

Bibliography

KNIGHTON, D. *Fluvial Forms and Processes.* London, 1984.
RITTER, D. F. *Process Geomorphology,* 2nd ed. Dubuque, IA, 1986.
SCHUMM, S. A. *The Fluvial System.* New York, 1977.

ELLEN E. WOHL

ROCKFALLS

See Landslides and Rockfalls

ROCKS AND THEIR STUDY

How rocks are studied, and what can be learned from them, depend entirely on the goals of the study. Rocks can be studied in their relation to humanity (as resources or hazards), for clues to processes and events outside human experience (past, future, inside the earth, or off the earth), or for wonder and curiosity. These goals are not mutually exclusive; a limestone could be important in finding oil, might preserve the history of an ancient ocean, and might contain beautiful fossils.

Rocks can have a significant impact on humankind, and the money to study rocks usually comes from a concern with human interests. Rocks may contain (or be) important natural resources, such as gold, oil, limestone, or road metal. Some rocks are used as gemstones, such as diamond, emerald, rock crystal (quartz), and agate. Less obvious rock resources include stable foundations for buildings, or impermeable sites for dams or for storage of radioactive waste. Rocks can also represent direct hazards to humanity (like volcanoes, earthquakes, or landslides) or be the sources of environmental contaminants (*see* PUBLIC HEALTH AND EARTH SCIENCE).

Rocks also bear witness to events in places we cannot go, guided by an understanding of present-day processes, and by simulations of natural processes in the laboratory. Looked at with a proper understanding, rocks can tell us the ages of ancient mountains and seas, the structure of the earth's interior, the sources of molten lava, and how the solar system began. Perhaps with a proper appreciation of the past, we can predict some of the future: the next earthquake, the next ice age, the odds of an asteroid or comet destroying New York City.

Rocks can be fascinating of themselves, without any potential use or application; a pretty stone can excite curiosity into the workings of the world, or an appreciation of form and texture. Curiosity leads to questions such as why rocks are just so, and whether there really are gray-green greasy-looking rocks near the banks of the Limpopo River (which there are, called charnockites; Mason, 1983).

What Is Rock?

Rock is the solid material of the earth (or elsewhere), made solid by inorganic processes. The "solidness" of rock is relative, and sometimes illogical. Rock salt can flow under its own weight to produce salt glaciers. Ice is not considered rock, even though it is solid. But shale is considered rock, although it may crumble in the hand and turn to mud in the rain. Living organisms do not commonly make rocks (limestone reefs may be considered exceptions); seashells and wood, for instance, are not rocks. But these materials can become parts of rocks if they are solidified together by inorganic processes; limestone and coal, formed from seashells and wood, are rocks. A good introductory guide to rocks and minerals is Chesterman (1978), and many others are available.

The word "rock" has other related but distinct meanings. A rock is an object, a piece of rock (the solid material) a few centimeters across or larger. And in planetary sciences, rock is used to mean material of rocklike density, 2–4 grams per cubic centimeter (g/cm^3), as opposed to ices (e.g., frozen water or methane), which have densities near 1 g/cm^3.

There are many varieties of rock or rock-types, which are commonly divided into igneous, sedimentary, or metamorphic, according to processes that yielded their present forms. Igneous rocks are products of magmatism, melting of solid rock inside Earth to a liquid, magma, or lava (*see* MAGMA). Igneous rocks form by solidification of magma or lava to glass or an aggregate of mineral grains. Volcanic igneous rocks form by eruption of lava onto the earth's surface; plutonic igneous rocks form when magma solidifies inside the earth.

Sedimentary rocks are solidified from sediment, material transported by physical processes on the surface of a planetary body as solid fragments or by chemical processes in water solution. Pieces of rock or single mineral grains can be transported mechanically as fragments (clasts) to form rocks such as sandstone, shale, or breccia; these are clastic sedimentary rocks. Rock can form directly from water solutions, as in rock salt deposits, some limestones, hot spring deposits (like travertine at Yellowstone National Park), and quartz veins (which may contain gold); these are chemical sedimentary rocks. Rock can also form from pieces of living organisms, as in coal, some limestones, and guano; these are organic sedimentary rocks.

Finally, metamorphic rocks develop from other rock types through chemical and physical transformations, collectively called metamorphism. For instance, if a shale (solidified mud) is buried in the earth and subjected to increasing temperatures and pressures, it changes progressively to the metamorphic rock types slate, phyllite, schist, gneiss, and granulite. If gneiss or granulite is heated more, it may melt to form a granite magma. In this progression, the rock's chemical composition may not change (except for loss of water), but its physical form and minerals change repeatedly. Metamorphic rocks can also form by extensive chemical replacements and reactions, called metasomatism. For instance, petrified wood is a metasomatic rock. Water-rich solutions dissolved the original wood (mostly cellulose) and replaced it with agate (fine-grained quartz) while preserving the shapes of the cell walls, rings, twigs, worm holes, and so on.

Rocks may form by many of these processes, so it is often impossible to classify a rock uniquely. A rock formed of volcanic ash and bombs fallen directly from a volcano is both igneous and sedimentary; it is called pyroclastic (Greek for "fiery fragments"). Inside a large body of molten rock (magma), solid crystals may settle to the bottom to yield igneous sediments (cumulates). Coal, formed of plant fragments that fell into a swamp, is both an organic and clastic sedimentary rock. Metamorphic rocks commonly preserve compositions or textures of their original sedimentary and igneous rocks. And the chemical compositions of igneous rocks can usually indicate what kind of rock they melted from.

Among these rock types and rock-forming processes, there tends to be an overall sequence of events, the rock cycle (Figure 1). Starting from magma in the earth, the first type of rock to form is igneous, either basaltic or granitic, volcanic or plutonic. If igneous rock becomes exposed at the earth's surface (by eruption or through erosion), it will be weathered and eroded, dissolved and broken down into constituent mineral grains or chemical compounds by water, wind, ice, and heat. These grains and chemicals will be transported as sediments and solidify to sedimentary rocks. Sedi-

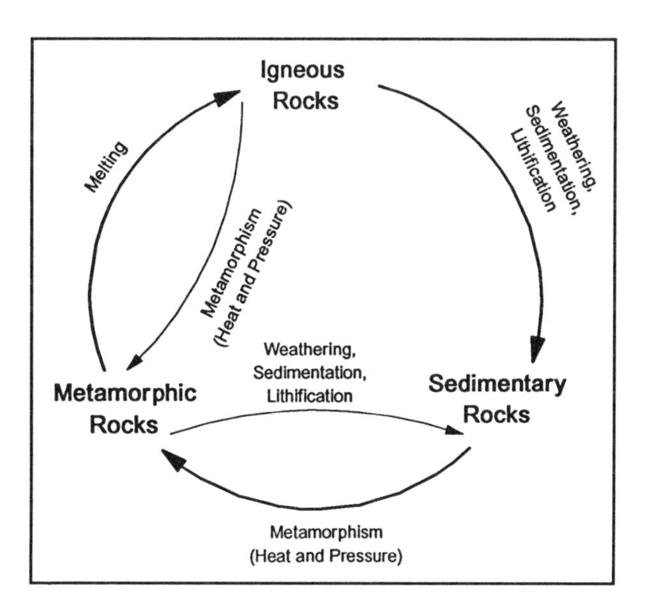

Figure 1. The rock cycle.

mentary and igneous rocks may become buried deep in the earth and metamorphosed (by heat and pressure) to yield metamorphic rocks. If heat and pressure continue, the metamorphic rocks may melt to magma, yielding new igneous rocks. Otherwise, the metamorphic rocks may become exposed at the earth's surface and yield sediments, which then go to form new sedimentary rocks.

Petrology: The Science of Rock

Petrology is the science and natural history of rock (*see* PETROLOGY, HISTORY OF). The goal of petrology is understanding the characteristics, distributions, origins, and fates of the great variety of rocks on and in Earth (and other planetary bodies). In its broadest definition, petrology includes all aspects of rock: description, identification, classification, occurrence in space and time, occurrence in relation to geologic events, minerals and chemical compositions, and theories of origins (Holmes, 1935; Philpotts, 1990). This definition is unworkably broad, and many aspects of petrology have split off into separate disciplines. Remaining in the core of petrology are two topics, petrography and petrogenesis, which tend to emphasize mineralogy, chemistry, and textures as tools for understanding rocks.

Petrography is the descriptive science of rock, and includes description itself (macroscopic and microscopic), identification, and classification; a classic reference book is Williams et al. (1954). Although rocks had been described and studied since ancient time, the modern science of petrology began in the 1800s as an inventory of the natural world. The turning point of petrography was HENRY C. SORBY's discovery in 1858 that rocks could be thinned (by grinding and sanding) and examined with an optical microscope. This new microscopic view of rocks, their minerals, and their textures (spatial relationships of grains) allowed identification of previously undetected mineral grains and textural relationships, and the optical microscope has become one of the most important tools in geology. Almost all studies of rock begin with petrographic descriptions, and almost all use petrography to dictate, or at least bolster, their conclusions.

Descriptive petrography as a science has languished in the last half-century, as it became overshadowed by new concepts in geology (e.g., plate tectonics) and advances in chemical instrumentation. Electron microscopy has permitted significant advances in understanding structures and textures at lengths less than a micrometer, but most of this work is in mineralogy, not petrography. Petrography may be headed for a rebirth through the development of new imaging systems, computer-assisted image analysis, and quantitative methods for interpreting textures.

Rock classification grew with rock description, but less easily. Arranging rocks by genus and species (as Linneaus had done with living organisms) failed because rocks' compositions, mineral proportions, and structures intergrade extensively. The great variety of rock types on Earth offered little order or structure to the fin-de-siècle petrologist. Metamorphic and sedimentary petrology responded by using a limited number of rock names with modifiers (e.g., garnet schist, sandy limestone); this scheme has proved adaptable to shifting ideas of petrogenesis and philosophies of classification. However, igneous petrology responded by giving each variant rock type a new name, commonly commemorating an obscure locality with a name difficult to pronounce except in the local dialect. Only with the advent of experimental petrology has it become clear which igneous rock types are of fundamental importance, finally permitting classification schemes similar to those of sedimentary and metamorphic rocks.

Petrogenesis is the science of the origins of rock, of interpreting what rocks are to discover how they formed. Petrogenesis includes most of what is called petrology today, although the word "petrogenesis" is now used almost exclusively for igneous rocks (Hess, 1989, exemplifies a modern view of igneous petrogenesis). Petrogenesis tends to emphasize chemistry and chemical processes, while physical processes are commonly delegated to related fields, like volcanology, sedimentology, rock mechanics, and tectonics.

Deciphering the origins of rocks is often difficult and ambiguous because most geological processes are hidden and occurred long ago and deep in the earth. The effects and processes must be inferred from clues they leave in the rock; often, the clues have been twisted or obscured by later processes. The petrologist has two additional sources of information: present-day observations of geological processes, and simulations of natural processes, including laboratory experiments and mathematical models.

A tenet of geology since the early nineteenth century is that Earth processes in the past are similar to those that act today—"The present is the key to the past." So, for instance, the petrologist observes active volcanoes to understand the origins of volcanic rocks. This source of information has limits in space, in time, and in frequency of events. We cannot observe the melting of rock to produce magma (although it is happening now) because we cannot go so deep inside the earth. Except under the special circumstances of the laboratory, we cannot observe the conversion of sand to sandstone because we do not have the time that this process likely takes. And we cannot observe the effects of a giant meteorite impact on the earth (nor might we want to) because there has been none recently (the last was 65 million years, or Ma, ago, the next preceding was 2 billion years, or Ga, ago).

Where rock-forming processes cannot be observed directly, the petrologist must simulate the real world, most commonly through laboratory experiments and mathematical models. In laboratory experiments, the petrologist tries to duplicate natural conditions as closely as possible in a controlled, reproducible environment. In effect, laboratory experiments create synthetic rocks that can be investigated with the same techniques as natural rocks. In mathematical modeling, the petrologist tries to describe geological processes by quantitative relationships, and then deduces the effects of the processes in the real world. The validity of a model can be tested by comparing its predictions with observations of real rocks. Most mathematical modeling is done numerically, using computers to calculate the effects of geologic processes.

Geologic time and distance are extremely difficult to simulate in laboratory experiments. To overcome limitations on time, experimental results must be examined in minute detail for the smallest trace effects of geological processes, and then these traces must be extrapolated to geological times. This extrapolation is huge, since an experiment can run for days to months, and the geological process might act over thousands to millions of years. Limitations of size can be as important as limitations on time because some processes will not operate in laboratory-sized environments. These limitations are not important in mathematical (numerical) models, which can be constructed to simulate any time and size desired. But mathematical models are only as good as the petrologist's understanding of all processes that may be important; a faulty understanding will lead to an unrealistic model.

Questions in igneous petrogenesis tend to focus first on the origins of magmas, where they formed and what was melted (Hess, 1989), and second on the evolution of magmas until they solidify as igneous rocks. The origins of igneous rocks can be attacked best through geochemistry and laboratory experiments; the former helps define what was melted, and the latter provides estimates for the physical conditions (temperature and pressure) of melting. Understanding the evolution of igneous rocks requires petrography, laboratory experiments, and numerical modeling. Crystals grown in a magma (now seen in an igneous rock) can preserve a history of events in the magma, much like tree rings preserve a history of a tree and its environment. Laboratory experiments can provide quantitative estimates of temperatures and pressures for the magmatic events recorded in the crystals. And numerical modeling can suggest which physical processes caused the chemical and thermal events.

Questions in sedimentary petrogenesis tend to focus first on the nature of the sediments involved, and second on how the sediments became rock (lithification). Study of sediment itself, the types of clasts or fragments involved and their textures, relies heavily on petrography and (to a lesser extent) a knowledge of sediments from present-day environments and from experiments. The study of lithification also relies heavily on petrography, but chemistry provides important clues about the nature and sources of the cements that hold sediments together. Mathematical modeling has become important in lithification studies, as the timescales and lengthscales involved are commonly beyond those of the laboratory. The study of chemical sediments is much like that of igneous rocks, relying equally on petrography, chemistry, and laboratory experiments.

Questions of metamorphic petrogenesis tend to focus on the physical conditions (pressure, temperature, and deformation) of metamorphism, and the causes of metamorphism. The physical conditions of metamorphism have been studied extensively in laboratory experiments so that it is commonly possible to determine a rock's maximum temperature of metamorphism to ±20°C. Mineral grains growing during metamorphism can preserve a history of changing pressure and temperature (much like tree rings), so the careful (lucky) petrologist may be able to determine how pressure and temperature changed during metamorphism. The combination of metamorphic pressures, temperatures, and times may suggest tectonic events that caused metamorphism, and permit a reconstruction of plate movements in the long-distant past.

Bibliography

CHESTERMAN, C. W. *The Audubon Society Field Guide to North American Rocks and Minerals.* New York, 1978.

HESS, P. C. *Origins of Igneous Rocks.* Cambridge, MA, 1989.

HOLMES, R. *Principles of Physical Geology,* 2nd ed. New York, 1935.

MASON, R. "The Limpopo Mobile Belt—Southern Africa." *Philosophical Transactions of the Royal Society of London* A 273 (1983):643–685.

PHILPOTTS, A. R. *Principles of Igneous and Metamorphic Petrology.* Englewood Cliffs, NJ, 1990.

WILLIAMS, H., F. J. TURNER, and C. M. GILBERT. *Petrography.* San Francisco, 1954.

ALLAN TREIMAN

ROSSBY, CARL-GUSTAF A.

Carl-Gustaf Rossby was born on 28 December 1898, in Stockholm, Sweden, the eldest of four brothers and a sister born to Arvid Rossby, a construction engineer, and Alma Marelius. Carl was precocious and had a wide range of interests including science and music. He pursued these interests in the company of a coterie of young intellectuals where he was exceptional for combining a jovial lighthearted spirit with a deductive mind that tended toward debate. This convivial nature in concert with a penchant for scientific argument was a hallmark of his life.

Rossby was educated at Stockholm University where he received the bachelor's degree (*filosofie kandidat*) in 1918 and the Swedish *licentiat* degree (corresponding to a Ph.D.) in applied mathematics in 1925. He studied under world-renowned mathematician Ivar Fredholm and was encouraged to gain practical experience at the Bergen School of Meteorology (Geophysical Institute at Bergen, Norway) prior to receipt of his *licentiat* degree. He spent the period from winter 1918/1919 through summer 1920 at Bergen and was inclined toward meteorology as a result of the invigorating intellectual environment that permeated the institute, which was headed by Vilhelm Bjerknes (*see* BJERKNES, J. A. B. AND V. F. K.).

Rossby came to the United States in 1925 on an American–Scandinavian Foundation fellowship and spent two years at the U.S. Weather Bureau. He advocated the use of the conceptual models related to air masses and fronts that were central approaches at the Bergen School, but the bureau was steeped in an empirical tradition of weather prediction and was generally unreceptive to these suggestions. At about the time his fellowship expired, he secured a position with the Daniel Guggenheim Foundation for the Promotion of Aeronautics. This foundation supported a project aimed at designing an ideal weather reporting network for commercial air carriers in California. The rugged terrain, valleys, and oceanic/land contrasts in California presented a host of challenging problems to weather forecasters. Rossby was eminently successful in establishing a reliable reporting network that proved especially valuable to Western Air Lines. When the Guggenheim Foundation decided to fund the first graduate program in meteorology at Massachusetts Institute of Technology (MIT) in 1928, Rossby was chosen to lead this program. Shortly after accepting his academic appointment at MIT, he married Harriet Alexander of Boston.

Rossby established a curriculum along the lines of the Bergen School and launched the program into a vigorous fluid dynamical approach to study the atmospheric and oceanic circulation. In these initial research efforts and others to come, he consistently mixed theory with practice, that is, the observational and experimental sides were deemed as important as theory. Upper-air observations from instrumented balloons became available in the early to mid–1930s, and meteorolographs attached to the wings of planes also recorded data up to the 6,096-m level. Nearby Woods Hole Oceanographic Institution had also initiated an active field program under the direction of Henry Bigelow and Columbus O'D. Iselin, and observations from the Gulf Stream were made available to Rossby.

In 1936–1937, Rossby made a series of contributions to understanding the dynamics of the Gulf Stream. He proposed that many of the features associated with the Gulf Stream might be explicable by analogy with the spreading of a turbulent wake stream as previously studied for engineering applications. He combined mathematical modeling with laboratory simulations conducted by his protégé Athelstan Spilhaus to explain the counter-current on the western side of the Gulf Stream, the entrainment of water mass into the Gulf Stream, and discharge of eddies along the main axis of flow. These investigations led Rossby to develop the theory of geostrophic adjustment, the tendency for large-scale circulations to come into geostrophic balance.

On the atmospheric side of the ledger, Rossby was intrigued by the upper-level waves that accompanied the surface cyclones that had been conceptually modeled by the Norwegians. Rossby was intent on finding a quantitative description of these upper-air waves. In one of the most daring and bold uses of analytical dynamics, Rossby viewed the atmosphere as a thin gaseous envelope attached to the rotating earth and was able to simplify the governing equations for fluid flow to such an extent that a formula for the wave speed could be extracted. This has become known as the Rossby-Wave Formula and it embodies the principle of conservation of vorticity, an extension of nineteenth-century work by Hermann von Helmholtz.

By the early 1940s Rossby had assumed leadership of the Institute of Meteorology at the University of Chicago and, when the United States entered World War II, he became scientific advisor to the military on issues related to meteorology. The Army Air Force and Navy projected that they would have 65,000 airplanes involved in the war effort and they would need about 10,000 weather forecasters in support of military aviation. Rossby was put in charge of the University Meteorological Committee that was responsible for the recruitment and training of forecasters. A nine-month course was established at five civilian universities (Caltech, UCLA, University of Chicago, MIT, and New York University) and at the Air Force Technological Center in Grand Rapids, Michigan. Rossby recruited the top teachers and researchers in meteorology as instructors. Over 6,000 forecasters completed the program and served in all theaters of war.

Following the war, Rossby had a vision to bring the top meteorologists together at an international institute. The cold war environment in the United States undermined his plans, so he decided to set up the International Institute of Meteorology at the University of Stockholm. This was accomplished over a period of about five-years (1946–1951). During this transition period, he worked alongside John von Neumann in planning numerical weather prediction (NWP) experiments using the self-programming digital computers that were under development. Success was achieved by several of his protégés in 1952, and this led the way to operational NWP that started in late 1953 at the International Institute in Stockholm.

On 19 August, 1957, while coordinating the multitude of activities at the International Institute and spearheading a program in atmospheric chemistry, Carl Rossby died at his desk of a heart attack. In almost forty years of work in meteorology and oceanography, he had set a standard of excellence in the pursuit of fundamental understanding of the dynamics of the ocean-atmosphere system. He brought the rigor of fluid dynamics together with his insight, and this allowed him to simplify the governing equations and extract the essence from complicated geophysical phenomena. In the best traditions of mentorship, he recognized the talent of his students and helped them realize their research potential. These people have admirably carried Rossby's lofty goals and traditions forward.

Bibliography

BERGERON, T. "The Young Carl-Gustaf Rossby." In *The Atmosphere and Sea in Motion (Rossby Memorial Volume)*, ed. B. Bolin. New York, 1959.

BYERS, H. R. "Carl-Gustaf Arvid Rossby." *Biographical Memoirs, National Academy of Sciences* 34, no. 11 (1960):249–270.

KUTZBACH, G. "Carl-Gustaf Arvid Rossby." In *Dictionary of Scientific Biography*, vol. 11, ed. C. C. Gillispie. New York, 1975.

LEWIS, J. M. "Carl-Gustaf Rossby: A Study in Mentorship." *Bulletin of the American Meteorological Society* 73 (1992):1425–1438.

PLATZMAN, G. W. "The Rossby Wave." *Quarterly Journal of the Royal Meterological Society* 94 (1968):225–248.

JOHN MAURICE LEWIS

RUBEY, WILLIAM W.

William Walden Rubey was born in Moberly, Missouri, on 19 December 1898, the son of Ambrose Burnside Rubey and Alva Beatrice (Walden) Rubey. After attending high school in Moberly, he enrolled in the University of Missouri and received the A.B. degree in geology in 1920. On graduating with highest honors, he continued academic studies at The Johns Hopkins University (1921–1922) and Yale University (1922–1924) for portions of the next four years, overlapping these with his work for the U.S. Geological Survey (USGS).

This latter association began in 1920 and extended practically unbroken for fifty-four years. Over this exceptionally long period, he exerted a strongly positive influence on the nature and thrust of research undertaken by the Survey, both through his own research and as a consequence of his widely sought judgment in scientific policy and in personnel matters. During World War II he served as the USGS liaison with the armed forces and contributed substantially to the entrance of both the Water Resources and Geologic Divisions into various phases of military geology.

He left full-time employment with the USGS to accept a professorial appointment at UCLA in 1960 but continued USGS-supported field studies in western Wyoming until his death. In addition, he became the first director of the Lunar Science Institute in Houston (1968–1971), serving during the period of the return of Apollo mission samples and the subsequent management of the most intensive scientific scrutiny yet received by natural materials.

Rubey's scientific contributions have received international recognition and include: the systematics of stream hydrology; sedimentation, stratigraphy, compaction, and origin of sedimentary rocks; areal geologic mapping of parts of the midcontinent, the Great Plains, and the northern Rocky Mountains (especially the Black Hills and western Wyoming); the origin of the atmosphere, seawater, and chemical differentiation of the earth; mechanisms of overthrust faulting and mountain building; and the factors influencing the release of seismic energy during earthquakes. He was a brilliant generalist, studying the interconnections of geologic phenomena in virtually all the problems he addressed; he brought rigor and quantification to bear on many subjects that hitherto had seemed intractable.

Rubey contributed early papers involving the petroleum geology of parts of Arkansas, Oklahoma, and Kansas, and in general discussed the relationships among porosity, compaction, stratigraphy, and structure. His early recognition of the magnitude of the Amarillo helium field influenced the government's decision to control its production. As a geomorphologist, he studied stream capture (the appropriation of the headwaters of one stream by another), the evolution of badland topography in the Great Plains, and factors influencing river channel development in Illinois. His sedimentological and stratigraphic investigations in the Black Hills of South Dakota and in western Wyoming led to: the systematic description of the Colorado Group strata of Late Cretaceous age; a discussion of the origin of the siliceous Mowry Shale; recognition of economic concentrations of vanadium and uranium in black shales and phosphorites (chemical sediments rich in phosphorus-bearing minerals); and to a major role in the drafting of the first American Stratigraphic Code in 1933. Sedimentological studies involved the quantitative elucidation of settling velocities and resultant size distribution of clastic grains deposited from flowing water. Other hydrologic investigations were directed toward the interdependence of flow regime, suspended particulate load, and stream bed topography.

Rubey's 1951 presidential address to the Geological Society of America (GSA) dealt with the origin of seawater, and it was subsequently amplified in GSA Special Paper 62 (1955); in this article, he demonstrated that the atmosphere and hydrosphere have accumulated gradually near Earth's surface over the course of geologic time as a consence of the outgassing of the deep interior. This is perhaps his best-known and most frequently cited work, although the papers coauthored with M. KING HUBBERT on the importance of aqueous fluid pressures in allowing slip along major, low-angle thrust faults may be equally famous. The solution to this latter perplexing problem grew out of Rubey's concern for the origin of complexly faulted structures such as he was mapping in western Wyoming; it shows how he applied his considerable understanding of physics to a general, first-order problem of tectonics. A related study involved the significance of the correlation observed by others of microseismic (earthquake) activity in the Denver, Colorado, area with fluid injection at the Rocky Mountain Arsenal. As fluids were injected underground, swarms of small earthquakes took place. The results of this study introduced a possible method of controlling release of earthquake energy in tectonically active areas.

This brief summary of some of Rubey's more substantial geologic contributions gives an indication of his scientific breadth. However, geology was but one aspect of his interests in the world around him. Rubey was a perceptive, scholarly, and dedicated naturalist, as aware of the beauty to be found in the living environment as he was of geologic features. His fieldwork included considerable attention to botanical and zoological observations. Like many of his USGS peers, he was an avid birdwatcher, and he was an associate of the Ornithological Union. His bibliography includes two articles dealing with barred owls and ravens, respectively. But while Rubey's written contributions were in the natural sciences, he was both interested in and well informed about politics, social problems, and history.

Rubey received many honors. He was elected to membership in the National Academy of Sciences (NAS) in 1945 and the American Philosophical Society in 1952, was awarded the U.S. Department of the Interior Award of Excellence in 1943 and the Distinguished Service Award in 1958; he received the National Medal of Science from President Lyndon B. Johnson in 1965, was president of the GSA from 1949 to 1950 and received that society's highest honor, the Penrose Medal, in 1963. He was president of the Geological Society of Washington in 1948 and of the Washington Academy of Sciences in 1957. Rubey was a member, fellow, or councillor of more than twenty learned societies. He was a fifty-year member of the American Association for the Advancement of Science and of the American Association of Petroleum Geologists. He received honorary degrees of D.Sc. from Yale University, the University of Missouri, and Villanova University, and an LL.D. from UCLA.

Bill Rubey's counsel was held in exceptionally high esteem as reflected by his appointment to many official advisory committees and panels, and as testified to by the innumerable requests from scientific colleagues for his thoughts on broadly ranging subjects. Invariably, his advice was wisely and generously, but modestly and tactfully, given. Rubey, however, regarded himself primarily as a field man and as a general geologist, and was happiest surrounded by the numerous field data on a mountaintop or in the desert. In his hesitating and modest manner, he shared these joys first with his contemporaries and later with the several generations of geologists he trained. He instilled the spirit of scientific inquiry without preaching it. By his own example of precise field investigation, he inspired others to study the rocks for the critical data before developing or accepting a working hypothesis. Then he would devise further field tests. But perhaps his personality, as much as his scientific integrity, influenced both older and younger associates; here clearly was a man whose respect was worth earning.

Rubey thus exerted a profound influence on the development of earth sciences over five decades in four specific ways: (1) through his own brilliant scientific works; (2) through his scientific advisory services to governmental agencies, to private foundations, and to universities; (3) through his stimulating, catalytic role as convener of informal but productive discussion groups, notably in Washington, DC, and at UCLA, but also in countless other places in an even more informal but no less significant fashion; and (4) through the enthusiasm he generated in his colleagues and associates. His accomplishments in any one of these areas alone would be considered singularly distinguished. That he was able to contribute significantly in all four areas attests to the remarkable intellect, quality, and drive of this man.

Bibliography

Rubey, W. W. "Origins of the Siliceous Mowry Shale of the Black Hills Region." U.S. Geological Survey Professional Paper no. 154, 1928, pp. 153–170.

———. "Geological History of Seawater: An Attempt to State the Problems." *Geological Society of America Bulletin* 62 (1951):1111–1147.

———. "Geological Map of the Afton Quadrangle and Part of the Big Piney Quadrangle, Lincoln and Sublette Counties, Wyoming." U.S. Geological Survey Miscellaneous Investigations, Map I-686, 1973.

W. GARY ERNST

RULES OF REACTION

See Thermodynamics and Kinetics

S

SALT WATER ENCROACHMENT

See Pollution of Groundwater Aquifers

SATELLITE LASER RANGING AND VERY LONG BASELINE INTERFEROMETRY

The first successful satellite laser ranging (SLR) experiments to an artificial satellite (Beacon Explorer B) were carried out by NASA's Goddard Space Flight Center in 1964. Since that time, the precision of the range measurement has improved by roughly an order of magnitude per decade—from a few meters in 1964 to a few millimeters in 1994. During the same thirty-year period, the number of active scientific SLR stations worldwide grew from one to about forty-five.

The Very Long Baseline Interferometry (VLBI) technique was first developed in the late 1960s by the radio astronomy community for the study of compact extragalactic radio sources such as "quasars," located at the edges of the known universe fifteen billion light years away. Because of their great distance from Earth, quasars act as stationary point sources for precise geodetic experiments. In 1993, there were approximately thirty active VLBI stations worldwide, and many intersite vectors are measured with a precision of few millimeters.

In 1979, the National Aeronautics and Space Administration (NASA) created the Crustal Dynamics Project for the purpose of applying these unique space-age technologies to the measurement of contemporary global tectonic plate motions, crustal deformation near plate boundaries, Earth's gravity field, and changes in the orientation of Earth's rotation axis and spin rate. The agency was rapidly joined in its efforts by a host of domestic and international partners.

Principles of the Techniques

Satellite Laser Ranging. The satellite laser ranging technique is illustrated in Figure 1. An ultrashort optical pulse is generated by the laser. The outgoing pulse is sampled by an optical detector that in turn starts a sophisticated electronic "stop watch" called a time interval unit or event timer. The remainder of the pulse travels through the atmosphere to the satellite and is reflected back toward the station by special "retroreflectors" mounted on the satellite. These retroreflectors (or "cube corners") have the useful property of reflecting incoming light precisely back in the direction from which it came. The light returning from the

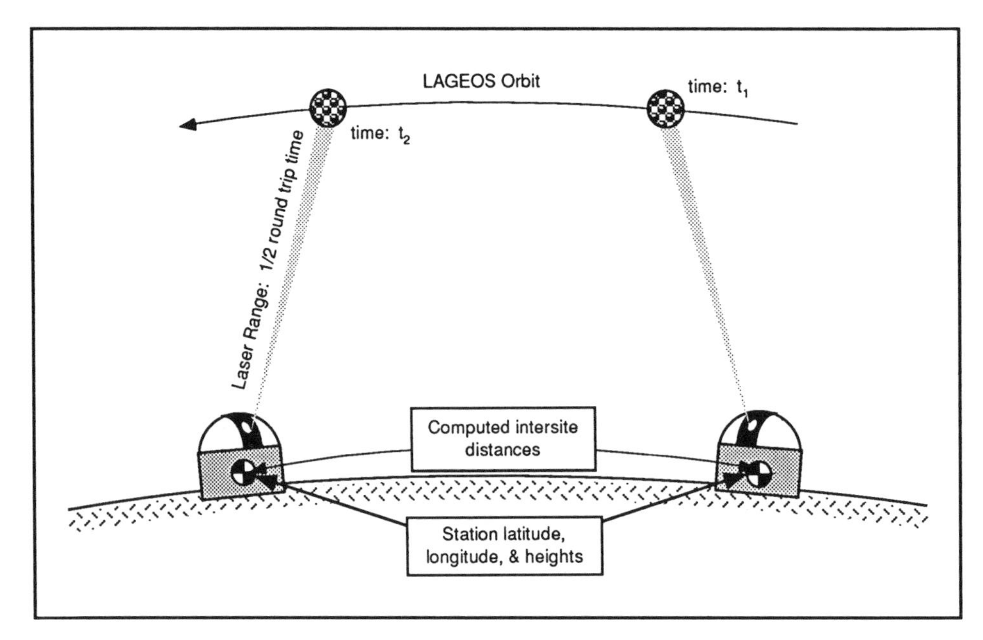

Figure 1. Principles of satellite laser ranging (SLR).

satellite is collected by a telescope and focused onto a second detector, the output of which stops the "stop watch." In modern systems, the roundtrip pulse time-of-flight (TOF) is typically measured with a precision of ten to twenty trillionths of a second.

To convert the measured TOF to a geometric range between the station and the satellite, one must correct for several systematic errors. Each SLR station is unique with its own set of optical and electronic delays that contribute to the measured TOF. An evaluation of these "system delays" is made possible by ranging to a nearby calibration target whose distance can be measured by conventional surveying techniques with millimeter precision. The atmospheric propagation delay, which is an error source common to all space geodetic techniques, will be discussed later.

To obtain global intersite distances (baselines) from SLR, the monthly data from the global network are first collected and fit to an orbit based on the best available satellite force models. For maximum accuracy, this force model must include not only a detailed model of Earth's static and dynamic gravitational field but also small nonconservative forces acting on the satellite. The latter include atmospheric and charged particle drag, radiation pressure from the Sun and Earth albedo (the percentage of the incoming radiation that is reflected by a natural surface such as the ground, ice, snow, water, clouds, or particulates in the atmosphere), and thermally induced forces caused by differential solar heating of the satellite.

Passive geodetic satellites, such as the Laser Geodynamics Satellite (LAGEOS), are constructed to have a high mass-to-area ratio and are placed in high orbits (6,000 to 20,000 kilometers or km) to minimize atmospheric drag and sensitivity to higher order components of the gravity field. Their spherically symmetric shape provides an omnidirectional and quasi-uniform response to laser irradiation and simplifies satellite center of gravity and nonconservative force modeling. Once the orbit is determined, station positions can be adjusted to give the best global fit.

In the data analysis process, various a priori station motion models associated with the orientation of Earth's pole, spin rate, a nominal plate motion model based on the geologic record, solar and lunar tidal forces acting on Earth's crust, and ocean loading at coastal sites must be taken into account. Atmospheric and relativistic light propagation delays are also modeled.

Very Long Baseline Interferometry. In a geodetic VLBI experiment (Figure 2), the weak microwave radiation from a quasar is simultaneously observed by two or more large radio telescopes equipped with special receivers. The incoming radiation is detected in several frequency bands, digitized, "timetagged" by onsite atomic clocks (hydrogen masers), and recorded by high-speed, high-density tape recorders. The observable being determined is the delay between the signal arrival times at the two stations. Multiplying this quantity by the speed of light gives the component of the baseline vector that lies in the direction of the quasar line of sight. To observe the full intersite vector, it is necessary to view approximately thirty quasars distributed fairly widely over the hemispherical sky. A typical VLBI experiment lasts twenty-four hours and involves about five stations, although as many as nine have participated in a single session. By including individual stations within different groupings, a global polyhedron of VLBI-determined baselines can be formed.

Following the experiment session, the station tapes are sent to a correlator facility. The function of the correlator is to determine the aforementioned time delays by effectively "sliding" one station's signals in time until it overlays that observed at another station. This process is not as straightforward as it might seem since the time delay is affected by the rotation of Earth, clock offsets, frequency drifts, and other instrumental delays. While VLBI analysis need not concern itself with satellite orbits and their errors, all of the other station motion and radiative propagation effects discussed for SLR must be taken into account.

Scientific Applications

Terrestrial Reference Frames. Since SLR is a satellite-based technique, station coordinates are determined in a "geocentric" reference frame, with Earth's center of mass (CofM) at its origin. Modern SLR measurements of the Earth CofM fluctuate at roughly the 1 centimeter (cm) level. VLBI, on the other hand, precisely orients Earth with respect to a field of "stationary" extragalactic quasars (celestial reference frame). The global polyhedron formed by VLBI-determined baselines has no sensitivity to CofM location but can be referenced to the geocentric frame through station collocation with SLR. As of this writing, global positions derived from these two very different techniques agree in all three axes at the 1 to 2 cm level. SLR and VLBI solutions are routinely submitted to the International Earth Rotation Service (IERS) in Paris, France, which maintains the International Terrestrial Reference Frame (ITRF).

Global Baselines and Tectonic Plate Motion. Repeating SLR or VLBI baseline measure-

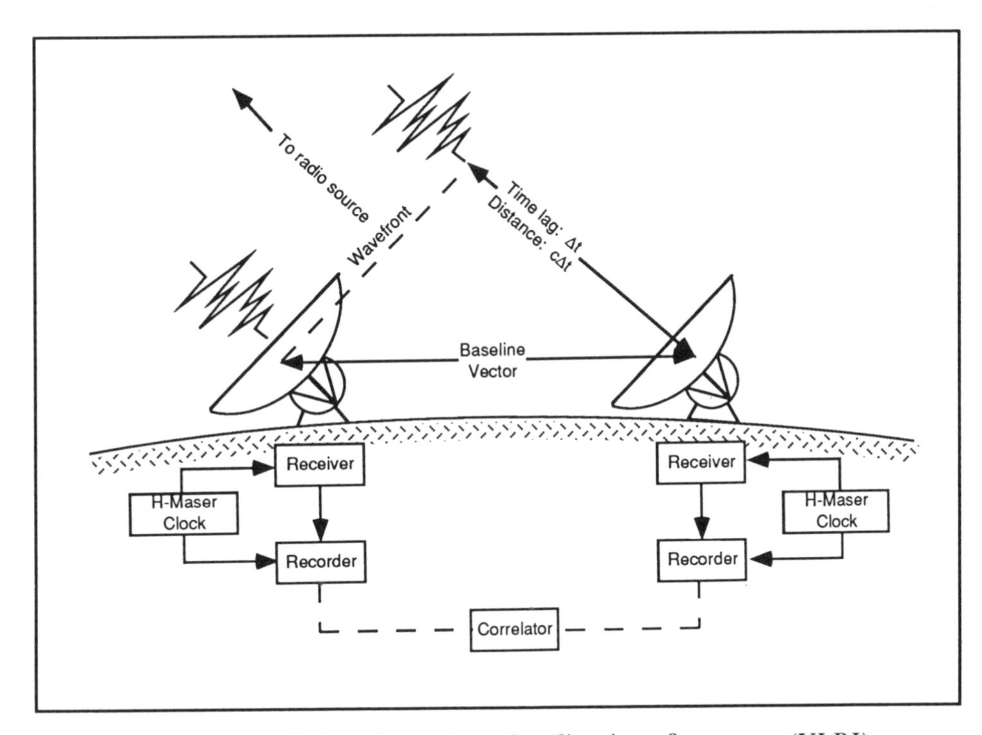

Figure 2. Principles of very long baseline interferometry (VLBI).

ments over a period of time yields the baseline rates. Since stations located deep within the interior of a tectonic plate travel at a velocity commensurate with that of the plate itself, their motions relative to interior stations on other plates yield the vector plate velocities on a global scale.

Space geodesy has unequivocably demonstrated that the plates are in constant motion. The contemporary baseline rates and tectonic plate motions established by SLR and VLBI are accurate to about 2 millimeters per year (mm/yr) and agree to better than 6 percent with long-term averages independently derived from the geologic record. Figure 3 illustrates the combined SLR/VLBI solution for global geodesic rates (time rate of change of the arc length between two sites) expressed in mm/yr.

Regional Plate Deformation. Forces acting between two tectonic plates will deform the interacting plates near the boundary and produce complicated site motions. Stress and strain in the boundary region increase until the rock elastic limit is reached and an earthquake occurs. In southern California, the earthquake-prone San Andreas Fault forms part of the boundary between the North American and Pacific plates. Figure 4 shows the measured vector velocities of SLR and VLBI sites as viewed by an observer on the North American Plate. Looking east to west across the plate boundary, our North American observer would see site velocities that increase monotonically (within a few hundred km) from zero to the full relative Pacific plate velocity. Similar SLR/VLBI studies of complex boundary kinematics have also been carried out in southern Europe where the Eurasian, African, and Arabian plates all collide.

Earth Gravity Field. High precision SLR data predominate in determining Earth's gravitational field and defining the fundamental constant GM (gravitational constant times Earth mass). The Joint Gravity Model 2 (JGM-2), developed jointly in 1993 by NASA's Goddard Space Flight Center and the University of Texas Center for Space Research, utilized laser and microwave tracking data from over thirty satellites, satellite altimetry data, and ground-based measurements of gravity. The model is expressed as a series of spherical harmonics complete through degree and order seventy.

Earth Orientation Parameters. Since 1980, SLR and VLBI have routinely monitored submilliarcsecond changes in the orientation of the instantaneous spin axis in Earth's body-fixed frame ("polar motion") and variations in Earth's rate of

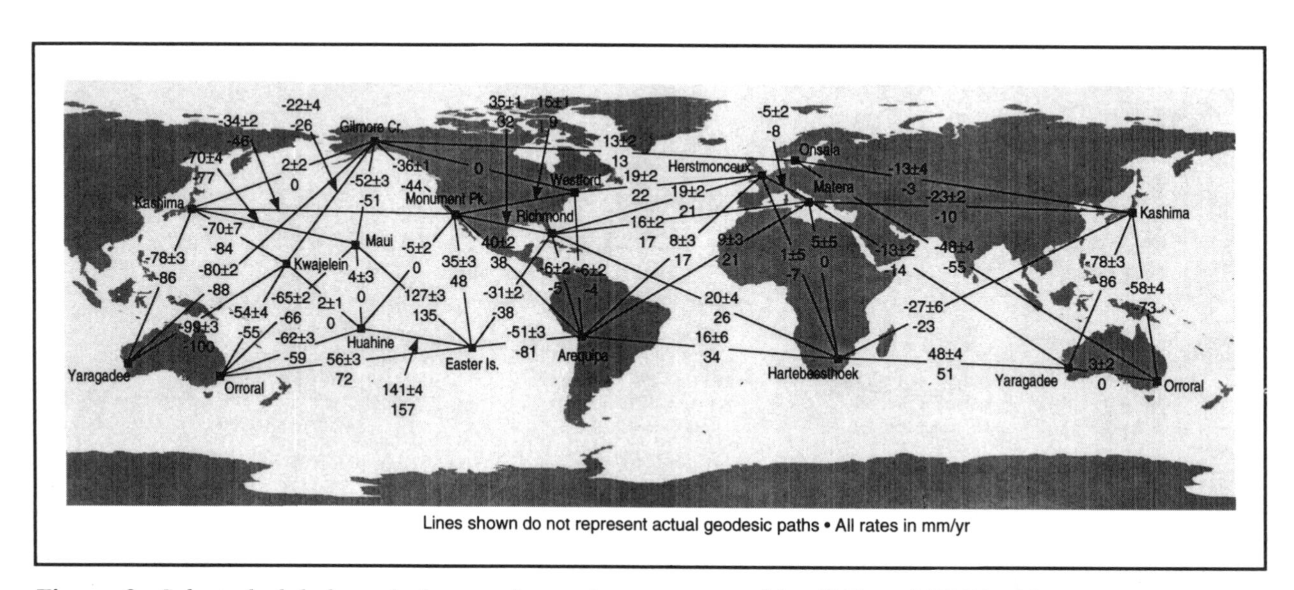

Figure 3. Selected global geodesic rates in mm/yr as measured by SLR and VLBI. The top numbers on each geodesic signify contemporary rates and their accuracy as measured by SLR or VLBI; the bottom numbers represent long-term averages as obtained from the geologic record (NASA/GSFC).

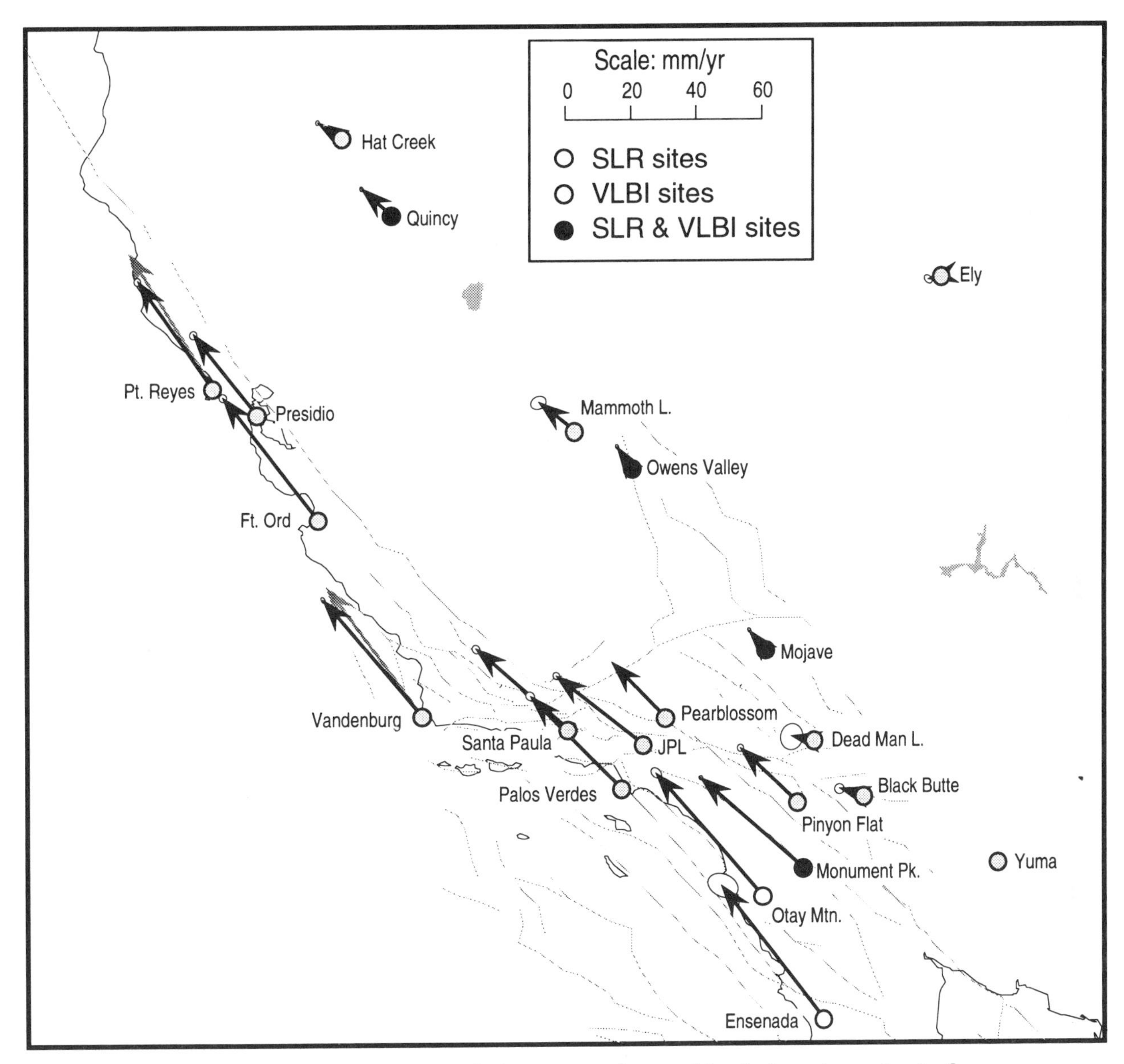

Figure 4. Vector velocities (relative to a stationary North American plate) of SLR/VLBI sites near the San Andreas Fault in southern California (NASA/GSFC).

spin (length of day). These are caused by: (1) angular momentum exchanges between solid Earth and the Moon, oceans, or atmosphere; and (2) mass redistributions that change Earth's moment of inertia tensor. Associated physical mechanisms include postglacial rebound, core-mantle interactions, tectonic processes, changes in ocean or atmospheric circulation patterns, melting of the polar ice caps, and so on. Redistributions of mass produce both seasonal and long-term variations in the gravity field, and these are also sensed by SLR.

Ocean and Ice Topography. The U.S./French TOPEX/Poseidon oceanographic mission, launched in August 1992, is producing sea surface topography and global ocean circulation models of unprecedented accuracy. Sea surface topography (SST) is defined as the height of the instantaneous sea surface relative to the marine geoid (a gravity equipotential surface approximating the mean sea level). Global ocean circulation is obtained by computing the two-dimensional slope of the sea surface topography.

The primary instrument for space-based oceanography is the microwave altimeter, which measures the range r from the host satellite to the ocean surface (Figure 5). In order to compute the SST, t, one also needs the satellite position relative to the geocenter (R + N + t + r) and the distance of the local geoid from the geocenter (R + N). As proven centimeter accuracy technique for precise orbit determination, SLR was chosen as the primary tracking network for the TOPEX/Poseidon mission. Because of its close tie to gravity, the marine geoid is also largely determined by SLR (at least on large spatial scales on the order of several hundred kilometers or more). SLR performs a third service by periodically calibrating the onboard altimeter. Without such calibration, long-term drifts in altimeter bias could be misinterpreted as changes in mean sea level.

The European ERS-1 spacecraft launched in 1991 also carries an altimeter and is tracked by laser. Due to its higher inclination orbit, ERS-1 can measure the topography of the polar ice sheets as well as ocean features.

Current Research Areas

The Atmosphere as Limiting Error Source. The varying index of refraction in Earth's atmosphere limits the absolute accuracy of SLR and VLBI geodetic measurements. The dominant effect is a variation in the group velocity of an electromagnetic wave with atmospheric density, that is, the wave speeds up as it travels from the ground station into the vacuum of space while the reverse occurs on the return trip. A second, smaller effect is a consequence of Snell's law of refraction, which predicts that an electromagnetic beam will follow a curved, rather than straight, path as it moves between atmospheric "layers" of differing refractive index.

In SLR these few meter level atmospheric corrections are made, with roughly 1-cm accuracy, via a model augmented by surface measurements of pressure, temperature, and humidity at the site. However, future SLR systems will use two laser wavelengths to measure directly the total integrated delay and produce geometric ranges with an accuracy of 2 mm.

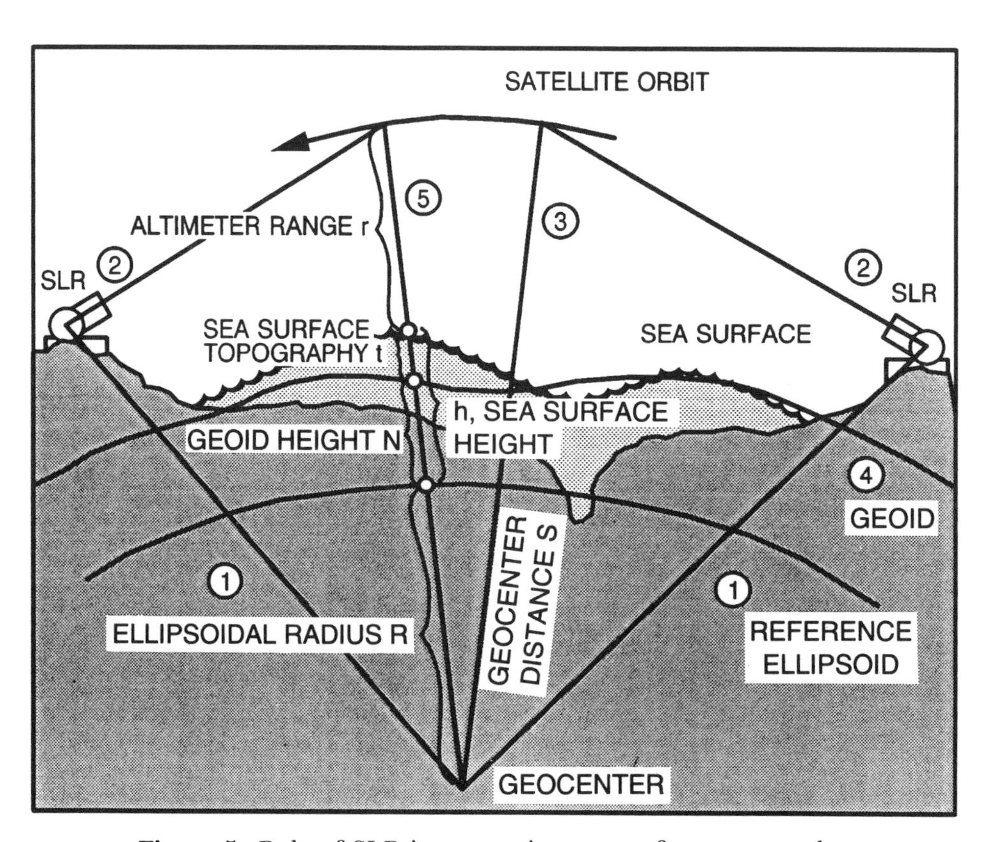

Figure 5. Role of SLR in measuring sea surface topography.

VLBI operates at much lower electromagnetic frequencies than SLR and is much more sensitive to propagation delays induced by the ionosphere and atmospheric water vapor. Since the ionospheric delay is highly frequency-dependent, it is well-estimated (at the subcentimeter level) by performing VLBI experiments in two widely separated frequency bands, that is, S-band (approximately 2.4 GHz) and X-band (approximately 8.4 GHz). The more difficult problem of water vapor delay is presently handled in routine VLBI data analysis through the application of stochastic estimation techniques.

Technology Trends. In SLR, the principal challenges are: (1) improving the performance and geographic distribution of the global network; (2) developing and implementing two color stations to reduce the uncertainty caused by atmospheric refraction; (3) continuing to improve the orbital and geodetic analysis; and (4) developing totally automated stations that are small, eyesafe, and inexpensive to construct and operate.

In VLBI, the principal emphasis is on the development of low noise, wide bandwidth microwave components and tape recorders to improve the signal-to-noise ratio (SNR) and the development of improved atmospheric models. Higher SNR will enable few millimeter precision from smaller (10-m class) radio telescopes as well as improved position resolution from the existing network.

In recent years, special Global Positioning System (GPS) (*see* GLOBAL POSITIONING SATELLITES, GEOGRAPHICAL INFORMATION SYSTEMS AND AUTOMATED MAPPING) geodetic receivers have begun to perform many of the same measurements with comparable precisions. Regular and frequent intercomparisons of the various space geodetic techniques on a global scale will be essential to understanding residual error sources and to the achievement of millimeter-absolute accuracies over the next decade.

Bibliography

COHEN, S. C., R. W. KING, R. KOLENKIEWICZ, R. D. ROSEN, and B. E. SCHUTZ, eds. Special Issue on LAGEOS Scientific Results. *Journal of Geophysical Research* 90, no. B11 (1985).

LAMBECK, K. *Geophysical Geodesy: The Slow Deformation of the Earth.* Oxford, 1988.

NEREM, R. S., B. H. PUTNEY, J. A. MARSHALL, F. J. LERCH, E. C. PAVLIS, S. M. KLOSKO, S. B. LUTHCKE, G. B. PATEL, R. G. WILLIAMSON, and N. P. ZELENSKY. "Expected Orbit Determination Performance for the TOPEX/Poseidon Mission." *IEEE Transactions on Geoscience and Remote Sensing* 31 (1993):333–355.

SMITH, D. E., and D. L. TURCOTTE, eds. *Contributions of Space Geodesy to Geodynamics.* Geodynamics Series, Vol. 24–26, American Geophysical Union. Washington, DC, 1993.

WALTER, L. S., ed. Special Issue on Satellite Geodynamics. *IEEE Transactions on Geoscience and Remote Sensing* GE-23 (1985).

JOHN J. DEGNAN

SATELLITES, MIDSIZED

A midsized satellite is any one of the celestial bodies in orbit around Saturn, Uranus, Neptune, or Pluto, intermediate in size between the large planetary satellites (the Galilean satellites of Jupiter, Earth's moon, and Titan), and the small irregular satellites. They range from large, planetlike, geologically active worlds with significant atmospheres, such as Triton, to geologically inert bodies with heavily cratered surfaces, such as Mimas.

These bodies are all believed to have as major components some type of frozen volatile, primarily water ice, but also methane, ammonia, nitrogen, carbon monoxide, carbon dioxide, or sulfur dioxide existing alone or in combination with other volatiles. Many show evidence of a wide range of tectonic and volcanic activitiy. Triton, the largest satellite of Neptune, was observed by *Voyager 2* to have active volcanic plumes. The Saturnian satellite Enceladus may also be currently active. Several of these satellites, including Phoebe, Triton, and possibly Charon, may be captured bodies. Our view of the midsized planetary satellites has been revolutionized by the *Voyager 1* and *2* encounters: the Saturnian system in 1980 and 1981, the Uranian system by *Voyager 2* in 1986, and the Neptunian system by *Voyager 2* in 1989. Before these encounters, all direct knowledge of the satellites was obtained from ground-based telescopic observations. Ground-based spectroscopic and photometric measurements continue to be important for understanding the composition and possible time-variable phenomena on these bodies. Table 1

Table 1. Summary of the Properties of the Medium-Sized Planetary Satellites

Satellite	*Distance from Primary (10^3 km)*	*Revolution Period (days) R = Retrograde*	*Orbital Eccentricity*	*Orbital Inclination (degrees)*	*Radius (km)*	*Density (g/cm^3)*	*Visual Geometric Albedo*	*Discoverer*	*Year of Discovery*
Saturn									
S1 Mimas	186	0.94	0.020	1.5	199	1.4	0.8	Herschel	1789
S2 Enceladus	238	1.37	0.004	0.0	249	1.2	1.0	Herschel	1789
S3 Tethys	295	1.89	0.000	1.1	523	1.2	0.8	Cassini	1684
S4 Dione	377	2.74	0.002	0.0	560	1.4	0.55	Cassini	1684
S5 Rhea	527	4.52	0.001	0.4	764	1.3	0.65	Cassini	1672
S7 Hyperion	1480	21.28	0.104	0.4	205 × 130 × 110		0.3	Bond and Lassell	1848
S8 Iapetus	3560	79.33	0.028	14.7	718	1.2	0.4–0.08	Cassini	1671
S9 Phoebe	12,950	550.4R	0.163	150	110		0.06	Pickering	1898
Uranus									
U5 Miranda	130	1.41	0.017	3.4	236	1.2	0.35	Kuiper	1948
U1 Ariel	191	2.52	0.003	0.0	579	1.6	0.36	Lassell	1851
U2 Umbriel	266	4.14	0.003	0.0	585	1.5	0.20	Lassell	1851
U3 Titania	436	8.71	0.002	0.0	789	1.7	0.30	Herschel	1787
U4 Oberon	583	13.46	0.001	0.0	761	1.6	0.22	Herschel	1787
Neptune									
N3 Proteus	117.6	1.12	0.0004	0.039	200		0.06	*Voyager 2*	1989
N1 Triton	354.8	5.875R	0.000015	157	1350	2.08	0.73	Lassell	1846
N2 Nereid	5509	360.1	0.753	6.7	170		0.14	Kuiper	1949
Pluto									
P1 Charon	19.6	6.39R	0?	99	593	1.4	0.40	Christy	1978

is a summary of currently known facts for the midsized satellites.

Summary of Characteristics

Discovery. None of the midsized satellites were known before the invention of the telescope. The first set of objects to be discovered were four midsized satellites of Saturn, by Giovanni Cassini in the latter half of the seventeenth century (Table 1). It was not until over one hundred years later that the next midsized satellite discoveries were made—the Uranian satellites Titania and Oberon and two smaller moons of Saturn. As telescopes acquired more resolving power in the nineteenth century, the family of satellites grew. Charon, the only known satellite of Pluto, was discovered in 1978. The last of these bodies to be discovered was the Neptunian satellite Proteus, by the *Voyager 2* spacecraft in 1989.

Planetary satellites are named after figures in classical Greek and Roman mythology who were associated with the namesakes of their primaries. They are also designated by the first letter of their primary planet and an Arabic number assigned in order of discovery: Mimas is S1, Enceladus S2, and so on. Geologic features on the satellites' surfaces are named after characters or locations from Western and Eastern mythologies. Official names for all satellite names and features are assigned by the International Astronomical Union.

Physical and Dynamical Properties. The orbit of a satellite is said to be regular if it is in the same sense of direction (the prograde sense) as that determined by the rotation of the primary, and if its eccentricity and inclination are both small. The orbit of a satellite is irregular if its motion is in the opposite (or retrograde) sense of motion, if it is eccentric, or if it has a high angle of inclination. The majority of the midsize satellites move in regular, prograde orbits. Exceptions are Iapetus, Phoebe, and Triton. Many of the satellites that move in irregular orbits are believed to be captured objects, or objects that have experienced a cataclysmic event such as a major impact.

All of the midsized satellites except Hyperion and possibly Nereid present the same hemisphere toward their primaries, a situation that is the result of tidal evolution. When two celestial bodies orbit each other, the gravitational force exerted on the near side is greater than that exerted on the far side. The result is an elongation of each body to form tidal bulges consisting of either solid, liquid, or gaseous (atmospheric) material. The primary body exerts a force on the satellite's tidal bulge to lock its longest axis onto the primary-satellite line. The satellite, which is said to be in a state of synchronous rotation, subsequently keeps the same face toward the primary. This despun state occurs rapidly, probably within a few million years.

Several of the satellites, including the Saturnian satellite Phoebe and areas of the Uranian satellites, are covered with C-type material, the dark, unprocessed, carbon-rich material found on the C class of asteroids (*see* ASTEROIDS). The surfaces of other satellites, such as Hyperion and the dark side of Iapetus, contain D-type primordial matter (named after the D class of asteroids), which is spectrally red and believed to be rich in organic compounds. Both D- and C-type material are common in the outer solar system. Because these primitive materials represent the material from which the solar system formed, an understanding of their occurrence and origin will yield information on the early state and evolution of the solar system. Iapetus presents a particular enigma: one hemisphere reflects ten times more light than the other.

Before these bodies were explored by spacecraft, scientists expected them to be geologically dead worlds. They assumed that heat sources were not sufficient to have melted their mantles to provide a source of liquid or semi-liquid ice or ice-silicate slurries. Reconnaissance of these bodies by the two *Voyager* spacecraft uncovered a wide range of geologic processes, including currently active vulcanism on Triton and possibly Enceladus. Several of the other medium-sized satellites of Saturn and Uranus, including Mimas, Ariel, and Titania, are large enough to have undergone internal melting with subsequent differentiation and resurfacing. It has been hypothesized that one object (Miranda) was split apart by a large impact and subsequently reaccreted.

Scientists now believe that heat provided by the decay of radioisotopes (radiogenic heating) could provide sufficient energy to melt Tethys, Dione, Rhea, and Iapetus. Recent work on the importance of tidal interactions and subsequent heating has provided the theoretical foundation to explain the existence of widespread activity on some of the smaller satellites. Another factor is the presence of non-ice components, such as ammonia (NH_3H_2O) hydrate or methanol (CH_3OH), which lower the melting point of near-surface materials. Partial melts of water ice and various contaminants—each with their own melting point and viscosity—provide material for a wide range of geologic activity. Conversely, the types of features observed on the surfaces provide clues to the likely composition of the satellites' interiors.

The prevalence of geologic activity on the midsized satellites has led planetary scientists to think in terms of unified geologic processes that function throughout the solar system. For example, partial melts of water ice with various contaminants could provide flows of liquid or partially molten slurries that in many ways mimic terrestrial or lunar lava flows formed by the partial melting of mixtures of silicate rocks. The ridged and grooved terrains on satellites such as Enceladus, Tethys, and Miranda may all have resulted from similar tectonic activities. The migration of ice-silicate slurries into large surficial cracks and subsequent flooding appears to have occurred on Enceladus, Tethys, Dione, Rhea, Miranda, Ariel, Titania, and possibly Umbriel. Finally, explosive volcanic eruptions occurring on Triton, and possibly Enceladus, may result from the escape of volatiles released as the pressure in upward-moving liquids decreases.

Formation and Evolution of Midsized Satellites

The sun and planets formed about 4.6 billion years (Ga) from a disk-shaped rotating cloud of gas and dust known as the protosolar nebula (*see* SOLAR SYSTEM). When the temperature in the nebula cooled sufficiently, small grains began to condense. The difference in solidification temperatures of the constituents of the protosolar nebula accounts for the major compositional differences of the midsized satellites. Since there was a temperature gradient as a function of distance from the center of the nebula, only those materials with high melting temperatures (silicates, iron, aluminum, titanium, and calcium, etc.) solidified in the central, hotter portion of the nebula. These materials are depleted in the outer solar system, whereas volatile materials are enriched. Beyond the orbit of Mars, carbon (C), in combination with silicates and organic molecules, condensed to form the carbonaceous and D-type material found on many of the midsized satellites. Beyond the outer region of the asteroid belt, formation temperatures were sufficiently cold to allow water ice to condense and remain stable. For the Saturnian and Uranian satellites, these materials are joined by methane (CH_4) and ammonia (NH_3), and their hydrated forms. For the satellites of Neptune and Pluto, formation temperatures were low enough for other volatiles, such as nitrogen (N), carbon monoxide (CO), and carbon dioxide (CO_2), to exist in solid form.

After small grains of material condensed from the proto-solar nebula, electrostatic forces caused them to stick together. Collisions between these larger aggregates caused meter-sized particles, or planetesimals to be accreted. Finally, gravitational collapse occurred to form larger, kilometer-sized planetesimals. The largest of these bodies swept up much of the remaining material to create the proto-planets and their companion satellite systems. One important concept of satellite formation is that a body cannot accrete within Roche's limit, the distance at which the tidal forces of the primary become greater than the internal cohesive forces of the satellite. Retrograde satellites are probably captured asteroids or large planetesimals left over from the period of planet formation. Triton is the only midsized satellite large enough to possess a gravitational field sufficiently strong to retain an appreciable atmosphere.

Soon after the satellites accreted, they began to heat up from the release of gravitational potential energy. An additional heat source was provided by the release of mechanical energy during the heavy bombardment of their surfaces by remaining debris. Mimas and Tethys both have impact craters caused by bodies that were nearly large enough to break them apart; probably such events did occur to other satellites. The decay of radioactive elements found in silicate materials provided another major source of heat, a source that still exists today. The heat produced in the larger satellites was sufficient to cause melting and subsequent chemical fractionation; the dense material, such as silicates and iron, went to the center of the satellite to form a core, while ice and other volatiles remained in the crust. A fourth source of heat is provided by tidal interactions. When a satellite is being tidally despun, the resulting frictional energy is dissipated as heat. Because this process happens very quickly for most of the midsized satellites (approximately 10 million years or Ma), another mechanism involving orbital resonances among satellites is believed to cause the heat production required for more recent resurfacing events. Gravitational interactions tend to make the orbital periods of the satellites within a system multiples of each other, a state known as a resonance. The result is that the satellites meet each other at the same point in their orbits. The resulting flexing of the tidal bulge induced on the bodies by their mutual gravitational attraction causes significant heat production in some cases.

Some of the midsized satellites, such as Tethys, Miranda, Ariel, and possibly Umbriel, underwent periods of melting and active geological processes within a billion years of their formation and then became quiescent. Others, such as Triton, and possibly Enceladus, are currently geologically active. For nearly a billion years after their formation, the satellites all underwent intense bombardment and cratering. The bombardment tapered off to a slower rate and presently continues. By counting the number of craters on a satellite's surface and by making certain assumptions about the flux of impacting material, geologists are able to estimate when a specific portion of a satellite's surface was formed. Continual bombardment of satellites by meteorites, including microsized bodies, causes the pulverization of both rocky and icy surfaces to form a covering of fine material known as a regolith.

Prior to the Voyager missions, most scientists believed that topographic features, such as craters formed on the outer planets' satellites, would have disappeared due to viscous relaxation. The two *Voyager* spacecraft revealed surfaces covered with craters which in many cases had morphological similarities to those found in the inner solar system, including central pits, large ejecta blankets, and well-formed outer walls. Scientists now believe that silicate mineral contaminants or other impurities in the ice provide the extra strength required to sustain impact structures.

Planetary scientists classify the erosional processes affecting satellites into two major categories: endogenic, which includes all internally produced geologic activity, and exogenic, which encompasses the changes brought by outside agents. The latter category includes the following processes: (1) meteoritic bombardment and resulting gardening (mixing) and impact volatization; (2) magnetospheric interactions by planetary magnetic fields and, in some cases, the solar wind, including sputtering of molecules into smaller fragments and implantation of the resulting energetic particles; (3) alteration by high energy ultraviolet photons; and (4) accretion of particles from sources such as planetary rings. These processes work throughout the eons to alter the optical properties of satellites' surfaces. In general, these processes darken surfaces and redden them. However, accretion of ring particles may brighten the inner midsized Saturnian satellites, and in some cases meteoritic bombardment can also brighten a surface by excavating fresh ice underneath.

Individual Satellites

The Satellites of Saturn. The six largest satellites of Saturn—Rhea, Dione, Tethys, Mimas, Enceladus, and Iapetus—are smaller than the four large Galilean satellites of Jupiter but still sizable. As such, they represent a unique class of icy satellite. Earth-based telescopic measurements showed the spectral signature of ice for Tethys, Rhea, and Iapetus; Mimas and Enceladus are close to Saturn and difficult to observe because of scattered light from the planet. The satellites' low densities and high albedos (Table 1) imply that their bulk composition is largely water ice, probably combined with ammonia or other volatiles. Resurfacing has occurred on several of the satellites. Figure 1 shows the six medium-sized icy satellites to scale.

Mimas, the innermost medium-sized satellite, is covered with craters, including one (named Arthur), which is as large as a third of the satellite's diameter (upper left of Figure 1). The impacting body was nearly large enough to break Mimas apart. Surficial grooves that appear on Mimas may be features caused by the impact. The craters on Mimas tend to be high-rimmed, bowl-shaped pits; apparently surface gravity is not sufficient to have caused slumping.

The next satellite outward from Saturn is Enceladus, an object known from ground-based measurements to reflect nearly 100 percent of the visible radiation incident on it (for comparison, the Moon reflects only about 10 percent). The only known composition consistent with this observation is almost pure water ice, or other highly reflective volatile. *Voyager* images of Enceladus show a surface that has been subjected, in the recent geologic past, to extensive resurfacing; grooved formations similar to those on the Jovian satellite Ganymede were evident (Figure 2). The relative lack of impact craters on this terrane is consistent with an age less than a billion years. Fresh ice on

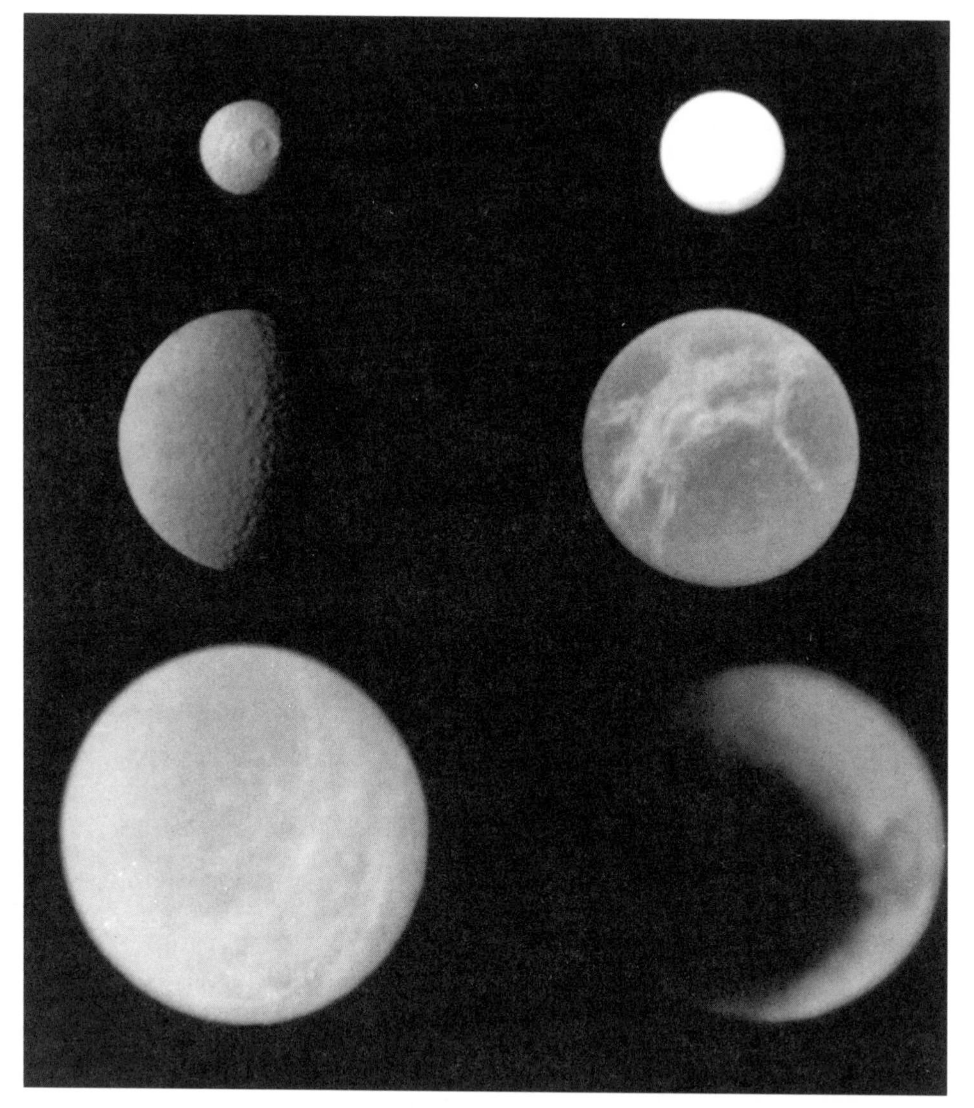

Figure 1. *Voyager* images of the six medium-sized icy Saturnian satellites. From the upper left: Mimas, Enceladus, Tethys, Dione, Rhea, and Iapetus.

the surface suggests that some form of ice volcanism is presently active on Enceladus, fueled by tidal heating caused by orbital interactions with Dione. About half of the surface observed by *Voyager* is extensively cratered and dates from the end of the period of heavy bombardment nearly 4 Ga ago.

Enceladus may be responsible for the formation of the E-ring of Saturn, a tenuous, distended collection of icy particles that extends from inside the orbit of Enceladus to past the orbit of Dione. The position of maximum thickness of the ring coincides with Enceladus's orbit. Currently active volcanic plumes could provide a source of particles for the E-ring.

Tethys is covered with impact craters, including Odysseus, the largest known impact structure in the solar system (approximately 420 kilometers or km in diameter). The craters tend to be flatter than those on Mimas or the Moon, probably because of viscous relaxation and flow due to Tethys' stronger gravitational field. Evidence for episodes of resurfacing is seen in regions that have fewer craters and higher albedos. There is a large trench formation, the Ithaca Chasma, which may be a degraded form of the grooves found on Enceladus.

Dione, about the same size as Tethys, exhibits a wide range of surface morphology. Most of the surface is heavily cratered, but indications that gradations in crater density occur suggest several periods of resurfacing during the first billion years of its existence. One hemisphere of the satellite is about 25 percent brighter than the other, due possibly to more intensive micrometeoritic bombardment. Bright wispy streaks (Figure 1) are believed to be the result of igneous activity and subsequent emplacement of erupting material.

Rhea appears to be superficially similar to Dione (Figure 1). Bright wispy streaks cover one hemisphere. There is, however, little evidence for resurfacing events in its history. There appears to be a bifurcation between crater sizes—some regions lack large craters, while other regions have a preponderance of such impacts. This dichotomy may be due to an event that obliterated small craters. However, the large craters may simply be due to a population of larger debris more prevalent during an earlier episode of collisions.

Shortly after Cassini discovered Iapetus in 1672, he noticed that at one point in its orbit around Saturn it was very bright, but on the opposite side of the orbit it nearly disappeared. He hypothesized that one hemisphere is composed of highly reflective material, while the other side is much darker. *Voyager* images show that the bright side, which reflects nearly 50 percent of the incident radiation, is typical of a heavily cratered icy satellite. The other side, which is centered on the direction of motion, is coated with a dark material reflecting only about 3 to 4 percent of visible radiation.

Two classes of theories have emerged on the origin of the dark material: one involves the deposition of the dark material from an exogenic source

Figure 2. A *Voyager 2* photomosaic of Enceladus. Old, heavily cratered terrane and recently resurfaced areas are both visible.

while the other involves internal geologic (endogenic) deposits. One scenario for the exogenic deposit of material entails dark particles being ejected from Phoebe and drifting inward to coat Iapetus. The major problem with this model is that the dark material on Iapetus is redder than Phoebe, although the material could have undergone chemical and coloration changes after its expulsion from Phoebe. One observation lending credence to an internal origin is the concentration of material on crater floors, which implies an infilling mechanism.

Other aspects of Iapetus are unusual. It is the only large Saturnian satellite in a highly inclined orbit. It is less dense than objects of similar albedo; this fact implies a higher fraction of ice, or possibly methane or ammonia, in its interior.

Hyperion is an irregular satellite in chaotic rotation around Saturn: both these facts suggest it may have been subjected to a collision relatively recently. Telescopic observations reveal water ice absorption bands and an albedo of 0.30, suggesting the existence of more contaminants on its surface than the brighter midsized Saturnian satellites, including D-type material.

Saturn's outermost satellite, Phoebe, a dark object (Table 1) with a surface composition apparently similar to that of C-type asteroids, moves in a highly inclined, retrograde orbit, suggesting it is a captured object. *Voyager* images show definite variegations consisting of dark and bright (presumably icy) patches on the surface. Although it is smaller than Hyperion, Phoebe has a nearly spherical shape.

The Satellites of Uranus. The medium-sized satellites of Uranus are Miranda, Ariel, Umbriel, Titania, and Oberon. Models of the formation of the solar system suggest that these satellites are composed of water ice, possibly in the form of methane clathrates (CH_4H_2O) or ammonia hydrates (NH_3H_2O), and silicate rock. Water ice has been detected spectroscopically on all five satellites. Their relatively dark visual albedos, ranging from 0.20 for Umbriel to 0.36 for Ariel (Table 1), and gray spectra indicate their surfaces are contaminated by a dark C-type component. Melting and differentiation have occurred on Miranda and Ariel, and possibly some of the other satellites. Tidal interactions may provide an important heat source in the case of Ariel.

The *Voyager 2* reconaissance of the Uranian system in 1986 revealed satellites that have undergone melting and resurfacing (Figure 3). Two features on Miranda, known as "coronae," consist of a series of ridges and valleys ranging in height from 0.5 to 5 km in height (Figure 5). The origin of these features is uncertain: some geologists favor a compressional folding interpretation, whereas others invoke a volcanic origin. Both Ariel, which is the geologically youngest of the five satellites, and Titania are covered with cratered terrane transected by grabens, which are fault-bounded valleys. Umbriel is heavily cratered and is the darkest of the satellites. Both of these facts suggest its surface is very old, although the moderate resolution images obtained by *Voyager 2* cannot rule out melting or other geologic activity. Oberon is similarly covered with craters, some of which have very

Figure 3. *Voyager 2* images of the five major satellites of Uranus, shown to relative size. They are, from left, Miranda, Ariel, Umbriel, Titania, Oberon.

dark deposits on their floors. On its surface are situated faults or rifts suggesting resurfacing events. In general, the Uranian satellites appear to have exhibited more geological activity than the midsized Saturnian satellites, possibly because of the presence of methane, ammonia, nitrogen, or additional volatiles that lowered the melting point of magmas.

Umbriel and Oberon, and certain regions of the other satellites, appear to contain D-type material, the organic-rich primordial constituent that seems to be ubiquitous in the outer solar system.

The Satellites of Neptune. Neptune has three midsized satellites: two are the satellites Triton and Nereid, known from telescopic observations, and the third, Proteus, was discovered during the *Voyager 2* encounter in 1989.

Triton moves in a highly inclined retrograde orbit, suggesting it is a captured body. Its bulk composition is water ice and silicates; methane, nitrogen, carbon dioxide, and carbon monoxide have been detected on its surface or atmosphere. Triton has a reddish color, possibly due to organic compounds forming from photochemical reactions on its surface. The satellite is probably differentiated with a rocky core and possibly liquid water—ammonia mantle. Because of its high inclination, Triton exhibits complex seasonal changes over the Neptunian year (165 Earth years). It appears to have a southern seasonal polar cap of nitrogen and methane, in addition to a permanent one; color changes that have occurred over the past several decades support this view.

The scarcity of craters on Triton means it has had episodes of resurfacing. The so-called cantaloupe terrane consists of pitlike depressions and crisscrossing ridges. Frozen lakes with layered morphologies suggest successive episodes of flooding and refreezing. Scores of dark, young (less than one thousand years old) plumes are found in the polar regions (Figure 6). Several actively erupting volcanoes or geysers were found on *Voyager 2* images. The features are typically 8 km high and disperse in horizontal wind-entrained plumes of more than 100 km. The geysers may be explosive eruptions of sun-heated nitrogen admixed with dark organics.

The surface temperature of Triton is 38 K, lower than any other known body in the solar system. It has a thin atmosphere with a surface pressure of 16 microbars, consisting primarily of nitrogen and a small amount (0.01 percent) of methane. Triton also has a photochemical haze layer and ionosphere.

Neptune's satellite Nereid moves in an eccentric orbit bringing it from 57 to 385 planetary radii from Neptune: this orbit is by far the most eccentric of any known natural satellite. Although its rotational state is not known, *Voyager* images showed that it is not in synchronous rotation. Its albedo is consistent with a surface composition of ices and silicates.

Proteus is the largest known irregular satellite in the solar system. The satellite has probably not been subjected to viscous relaxation; rather its mechanical properties have been determined by the physics of water ice, with an internal temperature below 110 K. Proteus shows signs of having been heavily cratered, including an impact basin. Because its albedo is significantly lower than that of Nereid, it may have carbonaceous material on its surface.

The Pluto-Charon System. Pluto's satellite Charon was discovered from photographic images of Pluto in 1978. Although the body appears as an ill-defined blob in Earth-based images, the Hubble Space Telescope has resolved it as a separate disk. Charon's mass is about 9 percent of Pluto's, a higher fraction than any other satellite in relation to its primary. Its density is 1.4, compared to Pluto's 2.1. Between 1985 and 1990, Pluto and Charon underwent a series of mutual eclipses that have allowed scientists to learn more about its physical character (*see* PLUTO AND CHARON). One hemisphere is redder than the other: this dichotomy implies compositional variegations. Spectral observations show water ice absorption bands, but there is no evidence for methane, which is found on Pluto's surface. Apparently, all methane has thermally escaped from Charon's weaker gravitational field. Charon probably has a rocky core and water-ammonia ice mantle.

Bibliography

BEATY, J. K., B. O'LEARY, and A. CHAIKIN, eds. *The New Solar System*, 3rd ed. Cambridge, MA, 1990.

BERGSTRALH, J., and E. MINER, eds. *Uranus*. Tucson, AZ, 1991.

BURATTI, B. J. "Planetary Satellites, Natural." In *Encyclopedia of Physical Science and Technology*. San Diego, CA, 1992.

BURNS, J., and M. MATTHEWS, eds. *Satellites*. Tucson, AZ, 1986.

GEHRELS, T., ed. *Saturn.* Tucson, AZ, 1984.
HARTMANN, W. K. *Moons and Planets*, 2nd ed. Belmont, CA, 1983.
MORRISON, D., ed. *The Satellites of Jupiter.* Tucson, AZ, 1982.
STONE, E., and the VOYAGER SCIENCE TEAMS. *Science* 246 (1989):1417–1501.

BONNIE J. BURATTI

SATELLITES, SMALL

Satellites in the solar system can be conveniently divided into large and ellipsoidal, and smaller, irregularly shaped satellites. Small satellites are known to orbit Mars and the four gas giant planets. Most information on them comes from spacecraft images.

Table 1 lists the known small satellites. They occur in three main settings: close to the planet in nearly circular orbits; in orbits shared with larger satellites; and in inclined, eccentric, and distant orbits. The small satellites that orbit close to the gas giant planets have dynamical and possibly compositional relationships to the ring systems. Many of the satellites act to confine gravitationally the ring particle orbits, and some of the satellites may supply particles to the rings from impact processes.

The satellites that orbit close to their primary are all tidally locked into rotation periods synchronous with their orbital period, with their long axes pointing toward the planet. Those in very distant orbits are not tidally locked into synchronous rotation, and Hyperion may have a chaotic rotation period.

Compositions

The composition of satellites is investigated using spectral properties, and in some instances, and where available, measurements of mean density. The visible spectra of Phobos and Deimos show that they are similar to C-type asteroids (*see* ASTEROIDS), and their mean densities of about 1.9 grams/cubic centimeter (g/cm^3) are consistent with this composition.

The composition of the inner satellites of Jupiter is unknown, but it is probably rocky for all of them. Spectrally, Amalthea appears to have a surface contaminated by sulfur and other materials ejected from the volcanic moon Io. The outer Jovian satellites have spectra that suggest they are similar to C-type asteroids.

All of the small satellites in the Saturnian system except Phoebe appear to have surfaces largely composed of water ice. The satellites Janus and Epimetheus have mean densities that suggest they are made of porous water ice with very little rocky material. Phoebe is largely covered with dark material common on many outer solar system objects.

The small Uranian satellites have low albedos (the fraction of incident radiation that is radiated by a surface or body) (Table 1), but little is known of their spectra. The small Neptune satellites likewise have albedos of about 6–7 percent, but reliable spectra are lacking. The outer Neptune satellite Nereid has a brighter surface very distinct from the inner small satellites.

Surface Features

Because small satellites lack significant internal heat sources, most surface features result from external influences, largely impacts of asteroids, comets, and other material orbiting the planet. All small satellites that have been well imaged are heavily cratered; some have craters with diameters more than half the mean diameter of the satellite.

Although the small satellites have very low surface gravities, the ejecta from these impact craters is largely retained on their surfaces. Some ejecta land close to the crater, but the rest that can escape the satellite go into orbit about the planet, and eventually reimpact the satellite. Phobos has well over 100 meters (m) of loose material (regolith) that has been generated by impact craters. The surfaces of both Phobos and Deimos have blocks up to 100 m across that can be associated with specific source craters.

Phobos's surface has distinctive linear depressions, termed "grooves," a few meters to 100 m deep, up to 1 kilometer (km) wide, and up to 15 km long. They appear to be formed in loose material and most theories of their origins suggest disturbance of the regolith by deeper fractures or invoke effects of ejecta from large craters.

Tapered, bright markings on Deimos show that the loose material on its surface moves downslope under the influence of its gravity, which is only

Table 1. Small Satellites

Primary Satellite	*Semimajor Axis of Orbit* ($\times 10^3$ *km*)	*Satellite Radii a,b,c (km)*	*Albedo*	*Most Likely Bulk Composition*
Mars				
Phobos	9.4	13.4, 11.2, 9.2	0.06	Carbonaceous
Deimos	23.5	7.5, 6.1, 5.2	0.07	Carbonaceous
Jupiter				
Metis	128.0	20, __, 20	0.05–0.1	Rock
Adrastea	129.0	12, 10, 8	0.05–0.1	Rock
Amalthea	181.3	131, 73, 67	0.06	Rock
Thebe	221.9	55, __, 45	0.05–0.1	Rock
Leda	11,110	5	0.05–0.1	Rock
Himalia	11,470	90	—	Rock
Lysithea	11,710	10	0.03?	Rock
Elara	11,740	40	—	Rock
Ananke	20,700	10	0.03?	Rock
Carme	22,350	15	—	Rock
Pasiphae	23,300	20	—	Rock
Sinope	23,700	15	—	Rock
Saturn				
Pan	133.6	10	0.5?	Ice
Atlas	137.7	18, __, 14	0.5	Ice
Prometheus	139.4	74, 50, 34	0.5	Ice
Pandora	141.7	55, 44, 31	0.5	Ice
Janus	151.4	97, 95, 77	0.5	Ice
Epimetheus	151.5	69, 55, 55	0.5	Ice
Telesto	294.7	__, 12, 11	0.6	Ice
Calypso	294.7	15, 8, 8	0.9	Ice
Helene	378.1	18, __, 14	0.6	Ice
Hyperion	3,560	170, 120, 100	0.25	Ice
Phoebe	13,210	115, 110, 105	0.06	Rock
Uranus				
Cordelia	49.8	13	0.065	Rock
Ophelia	53.8	16	0.07	Rock
Bianca	59.2	22	0.07?	Rock
Cressida	61.8	33	0.07?	Rock
Desdemona	62.7	29	0.07?	Rock
Juliet	64.4	42	0.07?	Rock
Portia	66.1	55	0.07?	Rock
Rosalind	66.9	29	0.07?	Rock
Belinda	75.2	34	0.07?	Rock
Puck	86.0	77	0.074	Rock
Neptune				
Naiad	48.22	29		Rock
Thalassa	50.07	40		Rock
Despina	52.53	79	0.063	Rock
Galatea	61.95	74	0.059	Rock
Larissa	73.54	104, __, 89	0.056	Rock
Proteus	117.65	218, 207, 201	0.061	Rock
Nereid	5511.2	170	(.16–.20)	Ice

about 1/3000th that on Earth. Thermal expansion and contraction or micrometeorite bombardment probably provide the initial impetus for the downslope movement. Amalthea shows evidence of similar features.

Shaping of Small Satellites

The small satellites are irregularly shaped because of two primary effects. First, they are small enough to experience collisions that can spall off large pieces or completely fragment them. Second, there is insufficient internal heat to raise temperatures to the point of viscous relaxation for the rocky satellites or even for the icy ones. Therefore, the satellites do not deform in response to their gravity and thus retain the irregular outlines caused by impacts and fragmentation.

Origins

The various orbits of the small satellites suggest a few different modes of origin. Those that have circular orbits close to their primary probably formed from the circumplanetary material that also condensed to form the larger satellites.

The compositions of Phobos and Deimos suggest that they are captured asteroids, but their orbital characteristics imply that they formed in orbit around Mars.

The inner satellites of Jupiter are probably rocky objects by extrapolation from the Galilean satellites that show an increase in rock content toward Jupiter. This gradient probably represents temperature effects on condensation of material in orbit about the forming planet. The outer Jovian small satellites are probably captured asteroids, based on their very eccentric, and for one group, retrograde orbits. The clustering of these satellites into two restricted sets of orbits further suggests they are derived from collisional fragmentation of two larger objects captured into Jupiter's orbit.

The inner small satellites of Saturn probably formed from the largely icy materials condensing near the planet. They are in a region that should encounter very high rates of impact and, as a result, these small objects may have been catastrophically disrupted and reaccreted several times. Some of the debris from such fragmentation events may become ring material.

The small objects that share orbits of Tethys and Dione (Calypso, Telesto, and Helene) might be debris ejected from them in very large collisions, but the dynamics of these objects' origins have not been solved.

Hyperion is something of an anomaly as it orbits between two substantial satellites, Titan and Iapetus. It appears to have undergone massive fragmentation (inferred from its shape) and could be a remnant of a somewhat larger object. Reaccretion of material to Hyperion from orbit may be difficult because of resonances with Titan.

Phoebe, orbiting Saturn in an inclined, retrograde orbit, is most likely a captured object.

The inner satellites of Uranus are all dark and probably have surfaces similar to the planet's ring material. They probably condensed in orbit about Uranus, but have likely been fragmented many times by impacts.

The inner satellites of Neptune likewise seem similar to the planetary ring material. These objects may have had a particularly complex history due to the likelihood that Triton was captured into orbit and should have disrupted the orbits of objects within five Neptune radii, causing many mutual collisions. The outer Neptune satellite, Nereid, could have formed with Neptune and also had its orbit disrupted by Triton's capture.

Bibliography

BURNS, J. A. "The Evolution of Satellite Orbits." In *Satellites*, eds. J. A. Burns and M. S. Matthews. Tucson, AZ, 1986, pp. 117–158.

THOMAS, P., J. VEVERKA, and S. DERMOTT. "Small Satellites." In *Satellites*, eds. J. A. Burns and M. S. Matthews. Tucson, AZ, 1986, pp. 802–835.

PETER C. THOMAS

SATELLITES, SOLAR POWER

The objective of solar power satellites (SPS) is to convert solar energy in space for use on Earth. Solar cell arrays would convert solar radiation directly into electricity and feed the electricity into microwave generators. Using a phased-array transmitting antenna, the satellite would direct the low-power density microwave beam at 2.45 gigahertz (GHz) to a receiving antenna on Earth. The microwave would safely and effectively be reconverted into electricity and feed into ground networks.

The SPS was first conceived as being in geostationary orbit (36,000 kilometers or km) from Earth; later, others proposed building the SPS on the Moon. Based on 1970s technology, scientists estimated that each SPS would produce 5 gigawatts (GW) of electric energy and each satellite would have an overall dimension of $10 \times 5 \times 0.5$ km. The peak power density of the receiving antenna would be 23 milliwatts/square centimeter (mW/cm^2) and at the exclusion edge fence of the receiving antenna 0.1 mW/cm^2; the environmentally acceptable level is 0.5 mW/cm^2. The overall dimension of each satellite would be nearly the equivalent of the size of Manhattan Island, New York.

The concept of wireless power transmission was first proposed by Nikola Tesla, the inventor of the induction motor and the polyphase power transmission, alternate current (AC). Construction of a demonstration tower was started on Long Island, New York, in 1906. This project was never completed. In 1968, Dr. Peter Glasser, scientist and Vice President of Arthur D. Little, first proposed the SPS concept (Figure 1). In 1975, The National Aeronautic and Space Administration (NASA) demonstrated microwave power transmission and initiated systems definition activities. A Department of Energy DoE/NASA joint study of SPS was conducted from 1977 to 1980, with independent reviews by the National Academy of Sciences and Office of Technology Assessment.

The objective of the DoE/NASA joint study was "to develop an initial understanding of the technical possibility, economic practicality, and the social and environmental acceptability of the SPS concept." The three-year study found no insurmountable problems but recommended future research into possible microwave effects, new materials, improved photovoltaics, robotics, and new launch capabilities.

Since 1980, many international symposia have been held, discussing all aspects of the SPS as a wireless power transmission method including new designs and data as to various orbits, size, environmental impact, and cost estimates. SPS Rio 92–Space Power Symposium was held in conjunction with the 1992 United Nations Conference for the

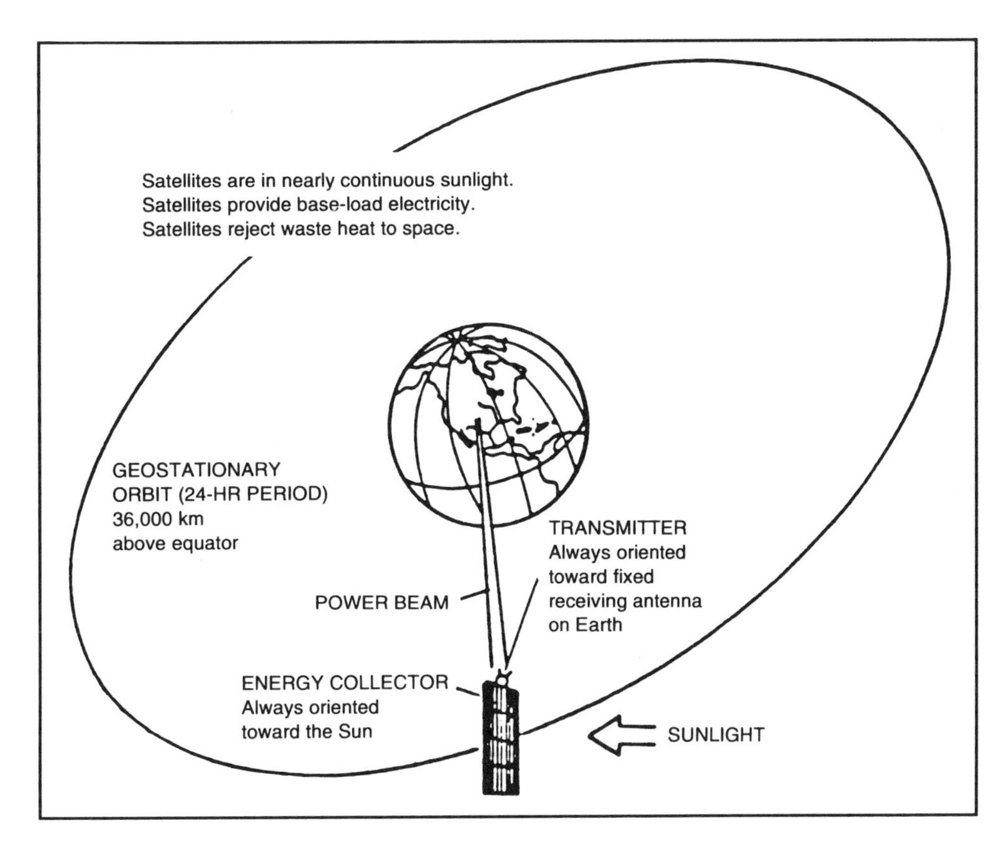

Figure 1. Reference system concept.

Environment and Development in Rio de Janeiro. Technologies of photovoltaics, robotics, and material sciences have made significant improvements since 1980. New space transportation systems are in the development stage, such as the SSTO (Single Stage to Orbit). The growing world population and increased demands for energy will require new energy sources in the twenty-first century. In 1995, the Japanese Ministry of International Trade and Industry (MITI) began a feasibility study concerning the realization of space solar power generation. This is an extension of Japan's SPS-2000 studies. Russian and European groups are also studying the application of wireless power transmission. In 1994, Dr. Mary L. Good, Undersecretary for Technology of the U.S. Department of Commerce, hosted a workshop on Wireless Power Transmission (WPT), bringing together representatives of utilities, industry, unions, and government to consider WPT possibilities in the global economy.

Bibliography

Glaser, P. E., F. P. Davidson, and K. I. Csigi, eds., "The Emerging Energy Options." In *Solar Power Satellites.* Chichester, Eng., 1993.

"Space Power Generation Feasibility Study." *Sanker Shimbun,* 1 February 1994, evening edition.

FREDERICK A. KOOMANOFF

SATURN

See Jupiter and Saturn

SCHUCHERT, CHARLES

Charles Schuchert was America's leading historical geologist and paleontologist in the first few decades of the twentieth century. Author or coauthor of over two hundred scientific articles, monographs, and textbooks, Schuchert forged a scientific career based on self-education and apprenticeship. His numerous textbooks on paleontology and historical geology were widely used throughout the early part of the twentieth century, and his studied opposition to continental drift had a significant impact on American thinking on this question.

Schuchert was born in Cincinnati, Ohio, on 3 July 1858, at a time when there were few academic geologists in America and only Yale offered formal graduate education in geology. Extensive scientific work was being done in state geological surveys and regional natural history societies, however, and it was through one of the latter that Schuchert found entry into science. Schuchert's parents were impoverished immigrants and Charles was forced to leave school at the age of thirteen to work in his father's tiny furniture factory. Throughout his childhood, Charles had pursued a hobby of fossil-collecting and had taught himself to identify the brachiopods he found near his home. After leaving school, Schuchert took night classes in lithography and began developing his skills as a scientific illustrator, primarily of his own still-growing collection. In 1878, Schuchert joined the Cincinnati Society of Natural History, where he met E. O. Ulrich, curator of the society's collections. Ulrich's position at the society was unpaid, but Schuchert convinced him that, with the aid of lithography, he could earn at least part of his living preparing illustrations for state geological surveys. In 1884, Schuchert abandoned the furniture shop to become Ulrich's paid assistant.

Schuchert worked with Ulrich for three years, and his work came to the attention of James Hall, State Geologist of New York. Hall hired Schuchert to assist in the production of *Paleontology of New York, Vol. VIII* (1889–1891). In Hall's orbit, Schuchert met many of the leading geologists of the period and was exposed to the problems of stratigraphy, mountain-building, and the relation of fossil evolution to Earth history. In 1893, Schuchert was offered his first professional position, as assistant paleontologist at the U.S. Geological Survey. One year later, he became Assistant Curator of Stratigraphic Paleontology at the U.S. National Museum in Washington, D.C. Schuchert soon established his reputation as an invertebrate paleontologist and curator of the first rank. His pamphlet *Directions for Collecting and Preserving Fossils* (1895) became a curatorial standard in the decades that followed. Meanwhile, he mounted a number of international collecting expeditions through which he broadened his expertise to encompass the global distributions of fossil species and their relation to historical geology.

In 1904, Schuchert was offered a position at Yale University, where the demands of teaching led him into the area for which he ultimately became most famous: paleogeography. Having never taken a college course himself, Schuchert was vexed by the difficulty of conveying to students the massive amounts of geological detail needed to understand the distribution of fossil species in relation to Earth history. He began to compile maps illustrating the distribution of fossils with respect to global lithology and stratigraphy. These maps became increasingly synthetic and interpretive: ultimately, the data of stratigraphy, paleontology, and structural geology were used to reconstruct the limits of land and sea in each succeeding geological period. The paleogeographic maps led Schuchert to write his magnum opus, the three-volume *Historical Geology of North America* (1935–1943). They also led to his involvement in the debate over continental drift.

Like most paleontologists, Schuchert was aware that the distribution of fossil species, particularly in parts of the late Paleozoic and early Mesozoic, strongly suggested that major portions of Earth's crust were previously contiguous. European geologists had accounted for this by the theory of Gondwana—a Paleozoic supercontinent that had fragmented sometime in the mid-late Mesozoic in response to terrestrial contraction caused by secular cooling. As the earth shrunk, portions of Gondwana had sunk beneath the ocean depths, and continental areas that were previously connected were sundered.

Although the contraction theory was widely accepted in Europe, most American geologists followed the views of James Hall and JAMES D. DANA and believed that the broad outlines of the continental blocks and ocean basins were fixed. In 1903, the discovery of radiogenic heat cast doubt upon the assumption of a cooling earth, and seemed to confirm the American view. Many American geologists considered the matter closed. However, Schuchert, with his broad knowledge of paleontology and paleogeography, realized that the prevailing American viewpoint was inadequate to account for the distribution of fossil species. When ALFRED WEGENER's theory of continental drift began to be widely discussed in the United States in the early 1920s, Schuchert recognized it as a possible solution to the problems of paleogeography. For more than a decade, Schuchert engaged in an extensive correspondence with geologists and paleontologists in America and Europe on the subject of continental drift and related theories. In the end, Schuchert decided against continental drift, primarily because the distribution of marine invertebrates did not appear to be consistent with the radical changes in climatic patterns implied by the theory. Together with Stanford geologist Bailey Willis, Schuchert developed the alternative concept of isthmian links—narrow land connections like the isthmus of Panama—which would permit faunal migration without invoking major climatic disturbances. Schuchert's solution was widely accepted by American geologists and contributed to the American rejection of the idea of moving continents until the debate was reopened by geophysical evidence in the late 1950s.

Among scientists, Schuchert is most well known for his *Outlines of Historical Geology* (1931) and *A Textbook of Geology (1929)*, coauthored with Carl Dunbar, and for his bibliographic catalogue, *Brachiopoda* (1929), coauthored with his long-time assistant, Clara Mae LeVene. However, Schuchert is also important historically as a leader of early-twentieth-century American science, having served in a large variety of leadership roles, including as the first president of the Paleontological Society (1910), as president of the Geological Society of America (1922), and as a vice president of the American Association for the Advancement of Science (1927). Ironically, the organizations he presided over helped to consolidate the professionalization of American science that ultimately excluded self-taught men like himself. By the end of Schuchert's life, graduate programs in geology were widespread in American universities and no one could expect academic employment without an appropriate degree. Schuchert's biography thus highlights the tremendous changes that occurred in American science near the turn of the century: beginning in an era where few scientists had formal training, and ending in a period where none did not. Charles Schuchert was one of the last professional scientists in America with no formal training in any scientific field. He died on 20 November 1942, in New Haven, Connecticut.

Bibliography

KNOPF, A. *National Academy of Sciences Biographical Memoirs* 27 (1952):363–437.

ORESKES, N. *The Rejection of Continental Drift. Historical Studies in Physical Sciences* 18 (1988):311–348.

SCHUCHERT, C. *Outlines of Historical Geology*. New York, 1931; 3rd rev. ed. with C. O. Dunbar. London, 1937.

———. *A Text-Book of Geology, Part II. Historical Geology*. New York, 1924; 2nd ed. with C. O. Dunbar. London, 1929.

———. *Historical Geology of North America*. New York, 1935–1943.

———. Papers, Yale University Archives. New Haven, CT.

NAOMI ORESKES

SCIENTIFIC CREATIONISM

"Creation science" has as its major tenets the sudden creation of the universe and life from nothing, a "relatively recently" created Earth whose geological features are the result of a catastrophic worldwide flood, the insufficiency of natural causes (i.e., mutation, natural selection) to significantly change, through evolution, the supernaturally created "kinds" of organisms, and, finally, a separate ancestry of humans and apes. Thus, creation science sets itself firmly in opposition to natural science that, since the earliest (sixth century B.C.E.) Greek physicalists, Thales and Anaximander, has persistently sought and substituted physical causes for supernatural explanations.

In style, too, scientific creationism is opposed to ordinary scientific practice. Science puts forth theories that are tested repeatedly against observations of the physical world, theories that are tentatively held and rejected as falsified if in conflict with a number of observations. In contrast, the cornerstones of creation science—special creation, a recent Earth less than 20,000 years old, and universal flood catastrophism—have been tested repeatedly since the early nineteenth century against geologic and fossil evidence and shown to conflict with Earth history. Yet, these empirically falsified theories continue to be dogmatically held and forcefully advocated by creationists.

Why, then, do some people inappropriately couple the term "science" to a religious concept, supernatural "creation"? Prior to the mid-twentieth century, the teaching of evolution was widely suppressed in American schools. Religious conservatives coerced many textbook publishers and teachers into avoiding mention of Charles Darwin's theory of natural selection and evolution in general—although these are the central principles of biology. Several states passed laws that made it a crime to teach evolution. It was not until the 1960s that the last of these laws (the Rotenberry Act of Arkansas) was declared unconstitutional. Following the Soviet launch of Sputnik in 1957, there were concerted efforts to strengthen the teaching of science and evolution throughout the United States. The term "scientific creationism" gained currency in the 1960s as Christian Fundamentalist organizations sought to gain equal time with evolution in public school science courses for their belief in creation ex nihilo as set forth in the Bible's first book, Genesis.

However, the First and Fourteenth Amendments to the U.S. Constitution clearly forbid the advancement of sectarian religion in public schools. Accordingly, the two mutually contradictory biblical creation myths (in Genesis 1, humans are created after the animals, but in Genesis 2, humans are created before the animals), together with the Genesis story of the deluge of Noah, were stripped of explicit religious imagery and names, reworded somewhat as in the beginning of this article, and repackaged as creation science. In Arkansas in 1981, a "Balanced Treatment of Creation-Science and Evolution-Science Act" was signed into law. The law was declared unconstitutional on 5 January 1982 in a U.S. District Court. Judge William R. Overton, who presided, concluded that creation science, although it does serve to advance a religion, has no scientific merit or educational value.

Bibliography

KITCHER, P. *Abusing Science: The Case against Creationism*. Cambridge, MA, 1983.

U.S. District Court's Memorandum Opinion. Rev. Bill McLean et al. *vs.* The Arkansas Board of Education et al., No. LR C 81 322, filed 5 January 1982 in the U.S. District Court, Eastern District of Arkansas, Western Division.

PAUL A. ROBERTS

SEAWATER CONVERSION

See Desalination

SEAWATER, PHYSICAL AND CHEMICAL CHARACTERISTICS OF

Water is one of the most remarkable compounds found in nature. Much of the behavior of water that is considered normal, such as ice floating on water or capillary action, is unique in comparison to other similar substances. Most of water's unusual characteristics are the result of slight electrostatic charges distributed around the water molecule. An oxygen atom has four pairs of electrons in its valence shell in a tetrahedral shape with an angle of 109°C between electron pairs. When hydrogen is bonded to a strongly electronegative atom, such as oxygen, a residual positive charge remains on the hydrogen atom and the unbonded electron pairs become slightly negatively charged. The repulsion and attractive forces between the hydrogen nuclei and lone electron pairs reduce the H—O—H bond angle to 104.5°. The water molecule is thus polar in nature with a positive side (two hydrogen nuclei) and a negative side (two lone electron pairs). Electrostatic bonds, or hydrogen bonds, form between the hydrogen nuclei of one water molecule and the electron pairs of another. Some of water's unique properties, such as a high surface tension and cohesion, are the direct result of hydrogen bonds (Table 1).

Hydrogen bonds are responsible for the unusually high freezing point (the temperature at which liquid becomes solid) and boiling point (the temperature at which a liquid becomes a gas) of water. For example, hydrogen sulfide (H_2S) and hydro-

Table 1. Unusual Physical Properties of Water

Property	*Comparison with Other Substances*	*Importance in the Ocean and Life*
Heat capacity	Highest of all solids and liquids with the exception of liquid NH_3	Prevents extreme ranges in temperature, allows large transfer of heat by currents; important in maintaining uniform body temperature
Latent heat of fusion	Highest except NH_3	Maintains temperature at freezing point in response to absorption or release of latent heat
Latent heat of evaporation	Highest of all substances	Very important in transfer of heat and water between the ocean and atmosphere
Thermal expansion	Temperature of maximum density decreases with increasing salinity and is 4°C for pure water	Temperature of maximum density for freshwater and dilute seawater is above the freezing point and at the freezing point of normal seawater, important in controlling temperature distribution and vertical circulation in lakes
Surface tension	Highest of all liquids	Controls surface phenomena such as drop formation and capillary action
Dissolving power	Dissolves more substances and in greater quantity than nearly any other liquid	Maintains salinity and nutrients essential for life
Transparency	High relative to most liquids	Absorption of radiant energy is large in infrared and ultraviolet wavelengths. There is little selective absorption of visible wavelengths and, as a result, pure water appears colorless in small volumes
Conduction of heat	Highest of all liquids	Important only on a small scale, as in living cells, outweighed by turbulent diffusion on a large scale

Modified from Sverdrup et al., 1942.

gen telluride (H_2Te), compounds very similar in structure to water, are gases at room temperature because they do not form hydrogen bonds. Generally, the boiling and freezing points of compounds increase with increasing molecular weight. The freezing point of H_2S and H_2Te are −82°C and −51°C, respectively, which suggests that the even lighter water molecule should freeze at −90°C instead of the observed 0°C. The higher freezing point of water is the result of the additional energy required to be removed to overcome hydrogen bonds upon freezing.

Because of hydrogen bonds water has a great tendency to resist changes in state. Water has the highest heat capacity of nearly all other compounds and because of this is used as a reference standard to compare other materials against. Heat capacity is the amount of heat required to raise 1 gram (g) of any substance 1°C. The unit of measure for heat capacity is the calorie, which is defined as the amount of heat required to raise 1 g of water 1°C. The high heat capacity of water is responsible for maintaining water in a liquid state when it is dropped onto hot pavement, whereas liquid mercury similarly treated would vaporize almost instantaneously. Because of its high heat capacity ocean water is able to release a large quantity of heat before freezing and this property is very important in warming high latitude coastal areas, such as Scandanavia, whereas continental areas at similar latitudes are nearly inhospitably cold.

To freeze water, an additional 80 calories of heat must be removed once temperatures have decreased to 0°C. Heat loss that can be measured as water cools is termed sensible heat. The latent heat of fusion is the additional heat lost at 0°C, without a change in temperature, that is needed to change the state of water from a liquid to a solid. The heat lost, without a change in temperature, is released during the formation of hydrogen bonds as ice forms. The same amount of heat must be added to ice to cause melting as hydrogen bonds are broken in the converse of this process, which is termed the latent heat of melting. This phenomenon is what makes ice such an effective refrigerant in coolers. A great deal of heat must be added to ice to effect a change in state and during this process cool temperatures are maintained. Not all of the hydrogen bonds are broken during melting, but all must be broken during vaporization. For this reason water has the highest latent heat of vaporization of any substance, 585 calories/g, and this figure represents the amount of heat needed to change liquid water at 100°C to a gas.

A familiar property of water, although unusual in comparison to almost all natural materials, is that solid water, ice, is less dense than liquid water. Density is defined as mass per unit volume and is commonly expressed in grams/cubic centimeter (g/cm^3). Upon cooling, molecules lose the thermal energy that prevents them from crystallizing. Loosely assorted molecules in a liquid are tightly packed within a crystal upon freezing. Because more molecules can be packed within a specific volume in a crystal than within a liquid, crystals are usually more dense. Water behaves differently and, once again, its unique behavior is a function of hydrogen bonding. Water cooling from 20°C to 4°C exhibits a trend of increasing density, from 0.9982 g/cm^3 to 1.0000 g/cm^3, owing to the loss of thermal energy that keeps molecules apart. However, below 4°C hydrogen bonds become more pronounced and water begins to form hexagonal ice crystals consisting of six widely spaced water molecules bonded together. These ice crystals occupy more space than liquid water and, therefore, water progressively cooled below 4°C actually decreases in density. Upon freezing all of the water molecules reside in hexagonal crystals. Solid ice has a density of 0.9170 g/cm^3 and is considerably less dense than liquid water.

The polar nature of water allows many compounds to be readily dissolved within it and for this reason water is often termed the "universal solvent." The charged nature of the water molecule serves to reduce the electrostatic attraction between atoms bonded by ionic bonds by as much as 1/80th of their dry values. Thus the attraction between sodium (Na) and chloride (Cl) that forms salt crystals is markedly reduced. Once the attraction is minimized, these atoms may become dissociated and positive ends of water molecules cluster around the negatively charged chloride ions and negative ends cluster around positively charged sodium ions. In this manner, a molecule of sodium chloride (NaCl) is dissociated and hydrated.

Salinity is a measure of the amount of total dissolved solids in water—for example, in seawater—and is commonly expressed in terms of parts per thousand (ppt) and given the symbol ‰ (or per million, ppm). Typical seawater has a salinity of 35‰, which means that 1 kilogram (kg) contains approximately 35 g of dissolved salts. Since the early 1980s it has been common to express salinity in terms of practical salinity (s) and dispense with the ‰ symbol. Practical salinity is based on the ratio of the conductivity of seawater to that of a standard KCl solution and adjusted to provide values nearly identical to ppt. Practical salinity is unitless because it is based on a ratio and is used because conductivity has become the predominant method for the determination of salinity.

Nearly every element in the periodic table can be found dissolved in seawater; however, only eleven ions account for 99.9 percent of the total ions found in seawater. These eleven ions are known as major constituents and are, in decreasing order of abundance, Cl^-, Na^+, So_4^{2-}, Mg^+, Ca^{2+}, K^+, HCO_3^-, Br^+, $H_2BO_3^-$, Sr^{2+}, and F^-. Sodium and chloride together account for 86 percent of all the dissolved salts. All other elements are present in concentrations of less than 1 ppm and are termed trace elements (Table 2).

Chemists in the 1800s observed that the proportions of the major constituents in seawater from various localities were constant despite variations in salinity. In other words, the ratio of one major ion to another is the same whether the salinity is 18‰ or 40‰. This observation has been termed the principle of constant proportions. Ions that obey the principle of constant proportions are said to behave conservatively. Deviations from the principle of constant proportions are rare but have been observed in certain areas within which river water dilutes ocean water to very low salinities. Trace elements do not follow this principle because they are often involved in the life cycles of living organisms and are said to exhibit nonconservative behavior.

Table 2. Average Concentrations of Major Ions in Seawater (g/kg) Normalized to 35‰

Ion	*Average Value*
Chloride	19.353
Sodium	10.775
Magnesium	1.295
Sulfate	2.712
Calcium	0.4121
Potassium	0.399
Bicarbonate	0.145
Bromide	0.0673
Boron	0.0046
Strontium	0.0080
Fluoride	0.0013

Modified from Kennish, 1989.

The reason that the principle of constant proportions is maintained in the oceans is that these ions are in steady state, that is, the input of ions to the ocean is equal to their losses. Dissolved solids enter the ocean from hydrothermal vents, the atmosphere, and rivers. Removal of dissolved solids occurs due to biological uptake, chemical precipitation, or absorption. The average time that an ion remains dissolved in seawater before being removed is termed its residence time. Residence time is calculated by dividing the amount of an element in the ocean by the rate at which that element is added or removed from the ocean. Ions in the oceans have residence times that vary dramatically from 80 million years (Ma) for chloride to as little as one hundred years or less for iron and aluminum that are involved in biological cycles. (Mixing, or turnover, rates for the oceans, contrastingly, are about 1,000 years; thus, conservative ions are well mixed in the oceans.) Because steady-state conditions have been maintained, it is estimated that the salinity of the oceans has remained constant for at least 600 Ma. Human activity has done little to alter the salinity of the ocean but has resulted in nuisance algal blooms in some coastal waters due to the input of biologically active compounds.

The temperature at which water reaches a maximum density is lowered by increasing the amount of dissolved salts and pressure. High pressures and the presence of hydrated salts inhibit the formation of ice crystals. For salinities greater than 24.7‰ density increases with decreasing temperature to the freezing point. For salinities less than 24.7‰ a temperature of maximum density is reached prior to the freezing point. At 24.7‰ the temperature of maximum density and freezing point coincide at −1.33°C. As salinities decrease below this level, the temperature of maximum density increases and is 4°C for pure freshwater.

The density of seawater is controlled by temperature, salinity, and pressure, and it generally increases the depth in the ocean. If this were not the case, as occasionally occurs, denser water at the surface would quickly settle to the bottom. In tropical regions surface waters are warmed extensively from solar heating and there exists a marked temperature change from surface to deeper waters commonly at a depth of 100 meters (m). The region within which temperature changes dramatically with depth is termed a thermocline. Heating of surface waters in high-latitude regions is not nearly as extensive as in equatorial waters and thermoclines are weak or absent. There often exists, however, a significant increase in salinity with depth owing to the dilution of surface waters with freshwater from melting sea ice or rain. The region in which salinity increases markedly with depth is termed a halocline. Both changes in temperature and salinity influence density and therefore the thermocline and halocline are often accompanied by a pycnocline, which is a region of marked density change with depth.

The density of open ocean water typically ranges between 1.02200 to 1.03000 g/cm^3. Oceanographers frequently express density in terms of in situ density or σ_t. The in situ density is determined from the temperature and salinity of seawater in place, or as measured at depth. In situ density is calculated by subtracting one from the specific gravity and multiplying by 1,000. Specific gravity is simply the density of a seawater sample divided by the density of pure water (1.0000 g/cm^3). Therefore, typical in situ densities in the ocean vary between 22.00 and 30.00. It is sometimes desirable to remove adiabatic effects of pressure on temperature measurements. The term "adiabatic" refers to an increase in temperature that accompanies the compression of water at depth from high pressures. Density calculated from in situ temperatures and salinities at great depths will sometimes yield lower in situ densities at the bottom than above. The potential density removes the adiabatic effect and provides more meaningful depth trends in density.

Meteorologists interested in studying climate define large air masses on the basis of temperature, pressure, and humidity. In much the same manner, oceanographers characterize water masses, which can move horizontally and vertically, on the basis of temperature and salinity. These water masses have unique salinities and temperatures that allow oceanographers to track their movements over considerable distances. The movement and mixing of water masses are assessed with the aid of temperature-salinity (T-S) diagrams. In Figure 1 a hypothetical T-S diagram for a station lo-

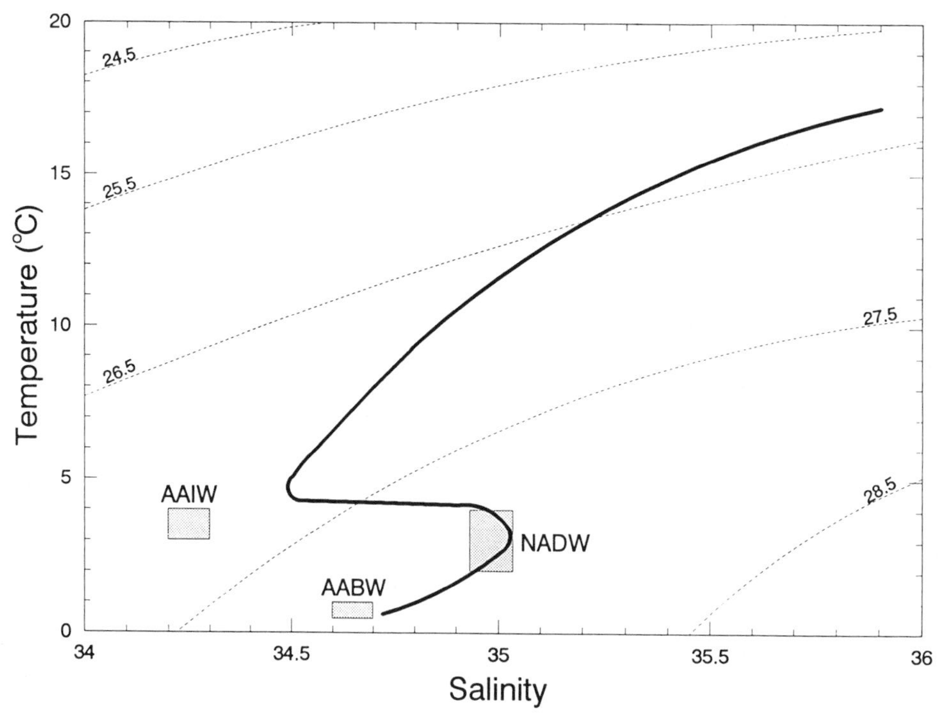

Figure 1. A T-S diagram for the equatorial South Atlantic Ocean. Regions of constant density are indicated by dashed lines. Major ocean water masses are indicated by boxes. AAIW = Anarctic Intermediate Water. AABW = Antarctic Bottom Water. NADW = North Atlantic Deep Water.

cated near the equator in the south Atlantic Ocean is shown. Shaded boxes illustrate typical salinities and temperatures of Antarctic Bottom Water (AABW), Antarctic Intermediate Water (AAIW), and North Atlantic Deep Water (NADW). Warmer waters are clearly a mixture of surface water and AAIW. The influences of NADW and AABW are apparent despite the fact that both of these water masses have traveled from polar regions.

The deep ocean floor is extensively covered with authigenic minerals, deposits that form in place from ions dissolved in seawater or leached out of sediments. Zeolites and clays are authigenic minerals that may form from volcanic or other material that has been chemically altered in the sea. Phosphorite nodules are authigenic minerals composed primarily of P_2O_5 often found in sediments below regions of intense upwelling activity. Perhaps the best known authigenic minerals are manganese nodules (*see* MANGANESE DEPOSITS; METALLIC MINERAL RESOURCES FROM THE SEA). Manganese nodules are commonly found on the seafloor and consist of approximately 65 percent manganese oxide, 35 percent iron oxide, less than 5 percent of nickel, copper, and cobalt oxides, and trace amounts of other minerals. Growth rates of manganese nodules are exeedingly slow and vary between 1 to 200 millimeters (mm) per million years. The processes forming manganese nodules are still uncertain; however, microorganisms may be the determining factor in their origin. *See also:* DESALINATION; DIAGENESIS; EARTH'S FRESHWATERS; EARTH'S HYDROSPHERE; ENERGY FROM STREAMS AND OCEANS; GROUNDWATER; OCEANOGRAPHY, BIOLOGICAL; OCEANOGRAPHY, CHEMICAL; OCEANOGRAPHY, GEOLOGICAL; OCEANOGRAPHY, HISTORY OF; OCEANOGRAPHY, PHYSICAL.

Bibliography

KENNISH, M. J. *Practical Handbook of Marine Science.* Boca Raton, FL, 1989.

MACINTYRE, F. "Why the Sea Is Salt." *Scientific American* 223 (1970):104–115.

MILLERO, F. J., and M. L. SOHN. *Chemical Oceanography.* Boca Raton, FL, 1992.

SVERDRUP, H. U., M. W. JOHNSON, and R. H. FLEMING. *The Oceans.* New York, 1942.

NATHANIEL E. OSTROM
DAVID T. LONG

SEDGWICK, ADAM

Adam Sedgwick was born in Dent, Yorkshire, England, on 22 March 1785. The son of a clergyman, Sedgwick himself was an active Anglican clergyman for most of his life. He studied theology and mathematics at Cambridge, graduating in 1808. In 1809 he was made an assistant demonstrator (a teaching position) at Trinity College, Cambridge, and in 1818 he succeeded John Hailstone as the Woodwardian Professor of Geology at the University of Cambridge, a position he retained for the remainder of his career.

Apparently Sedgwick knew relatively little about geology when he was first appointed to the chair in the subject at Cambridge, but he quickly made up for lost time. Sedgwick was known as a dynamic and energetic researcher and lecturer. He taught during the winters and undertook geological expeditions during the summers. He also became an active member of the Geological Society of London, eventually serving as the society's president.

Sedgwick is primarily remembered for his stratigraphic investigations of the Paleozoic rocks found in Wales, Yorkshire, the Lake District, and Scotland. At the time these were the oldest known fossil-bearing rocks, and on the basis of his studies Sedgwick defined the Cambrian system (named after an ancient term for the Welsh area). At approximately the same time, Roderick I. Murchison was working in the Welsh Borderland (east of where Sedgwick was concentrating his fieldwork) and there Murchison defined the Silurian system, named after an ancient tribe that had lived in the area during Roman times. (*See* MURCHISON, RODERICK.) Together Sedgwick and Murchison studied the rocks of the Devonshire area, distinguishing the Devonian system. Later in their careers, however, Murchison and Sedgwick argued over the boundary between Murchison's Silurian and Sedgick's Cambrian. Murchison increasingly insisted that first part, and then virtually all, of Sedgwick's "Cambrian" was actually part of the Silurian. This resulted in a protracted debate that did not effectively end until about 1879 (after both Murchison and Sedgwick had died) with Charles Lapworth's recognition of the Ordovician as a system intermediate stratigraphically between the Cambrian and the Silurian. Sedgwick died on 27 January 1873 in Cambridge, England.

Sedgwick had a reputation of being a straightforward, plain Yorkshireman with a rural background. He was characterized by high moral principles and he exhibited genuine piety in his church work. As a lecturer in geology, he was known for the combination of enthusiasm and humor he used in his teaching. He founded a geological museum at Cambridge, and he was the recipient of such prestigious awards as the Wollaston (1851) and Copley (1863) medals of the Royal Society, as well as an honorary degree from Oxford in 1860.

Sedgwick lived through the Darwinian revolution in paleontology and biology, yet remained opposed to the concept of evolution. This is somewhat ironic in that Charles Darwin had gained an interest in geology due in part to the direct influence of Sedgwick (as a young man, Darwin attended a field trip to North Wales with Sedgwick). Furthermore, it was Sedgwick's friend and student, John Henslow, who arranged for Darwin to be the naturalist on the epoch-making voyage of HMS *Beagle* (*see* DARWIN, CHARLES).

Bibliography

FAUL, H., and C. FAUL. *It Began with a Stone: A History of Geology from the Stone Age to the Age of Plate Tectonics.* New York, 1983.

RUDWICK, M. J. S. *The Great Devonian Controversy: The Shaping of Scientific Knowledge Among Gentlemanly Specialists.* Chicago, 1985.

ROBERT M. SCHOCH

SEDIMENTARY STRUCTURES

In our day-to-day lives we may witness or hear about environmental changes. We may see a forested area leveled to make way for a shopping center, or we may read about a river being dammed to create a reservoir for drinking water. It is hard to imagine, however, the sweeping changes that Earth's environments have undergone over its history. It seems incredible, for example, that 500 million years (Ma) ago North America was almost entirely covered by a shallow, tropical sea, that 300 Ma ago an immense mountain range comparable to the Himalayas extended along the eastern margin of our continent, and that 90 Ma ago dinosaurs roamed a swampy coastal plain in what is now the state of Utah. How do we know about these ancient environments? In large part from evidence preserved in sedimentary rocks. Sedimentary particles are derived from the weathering of previously existing rocks, and they are transported to and deposited in a wide range of environments on Earth's surface (*see* SEDIMENTS AND SEDIMENTARY ROCKS, CHEMICAL AND ORGANIC; SEDIMENTS AND SEDIMENTARY ROCKS, TERRIGENEOUS). The composition and texture (size, size variation, shape) of the particles in sedimentary rocks reveal some aspects of the conditions of sedimentary transport and deposition. Equally, if not more revealing, however, is the arrangement of the particles into sedimentary structures. Sedimentary structures form during or relatively soon after deposition of sedimentary particles, before the sediments are compacted and cemented together to form sedimentary rocks. This entry discusses some of the major types of sedimentary structures and what they tell us about ancient environments (Allen, 1982, and Collinson and Thompson, 1982).

The most common sedimentary structure, present in virtually all sedimentary rocks, is layering or stratification. If the individual layers are less than 1 centimeter (cm) thick, they are called laminae (singular lamina), and if they are thicker, they are called beds. Laminae and beds reflect periods of time in an environment when sediment accumulates under relatively steady physical conditions. Each lamina or bed is characterized by sediment with a composition, texture, and structure that distinguish it from the layers below and above. It is separated from these adjacent layers by a surface called a bedding plane. The changes in sedimentary properties across a bedding plane may be gradual or sharp. A gradational bedding plane forms when conditions in an environment change relatively gradually. For example, during the spring, melting snow may cause a mountain stream to flow gradually faster, and with time to carry progressively larger particles into a downstream lake. On the lake bottom a lamina of mud, deposited in winter and early spring, may be gradationally overlain by a lamina of sand, deposited during the later spring. Sharp bedding planes, on the other hand, may form when conditions of sedimentation change abruptly, or when periods of nondeposition or erosion of sediment are followed by renewed accumulation. These planes separate sedimentary rocks that are significantly different, and in some cases they may represent relatively long periods of time. For example, at water depths greater than a few tens of meters on the continental shelves, currents and waves are relatively weak during most of the year, and muddy sediments accumulate. When a storm affects a shelf area, however, the currents and waves are strengthened. The muddy seabed is commonly eroded, and sandy sediments may be transported from the coastal zone. At the end of the storm, a layer of sand will be found on the shelf, separated from the underlying mud by a sharp, erosional bedding plane.

The flow of wind and water currents and the back and forth movement of waves over loose beds of sediment may generate numerous types of sedimentary structures. Where such flows occur over sand-covered surfaces, for example, loose sand grains are commonly molded into regularly spaced, undulating forms called ripples and dunes. Ripples and dunes, collectively known as bedforms, produce several kinds of sedimentary structures. Ripple marks are sedimentary structures that are found on sandstone bedding planes. Viewed from the top, the peaks or crests of the ripples form low, long, parallel ridges separated by narrow troughs (Figures 1 and 2). The crests and troughs are oriented perpendicularly to the direction of ancient winds, currents, and wave motions. In side view, ripple marks formed by wind or water currents are shaped like asymmetrical triangles (Figure 1). The side of the ripple that originally faced upwind or upcurrent is known as the stoss side and it has a relatively shallow inclination. The side of the ripple that faced downwind or down current is called the lee side and it is more steeply inclined. The asymmetrical shape of wind and current ripples makes them particularly useful for determining the direction of ancient flows, or paleocurrents. Ripple marks made by waves have symmetrical shapes, with two sides inclined at

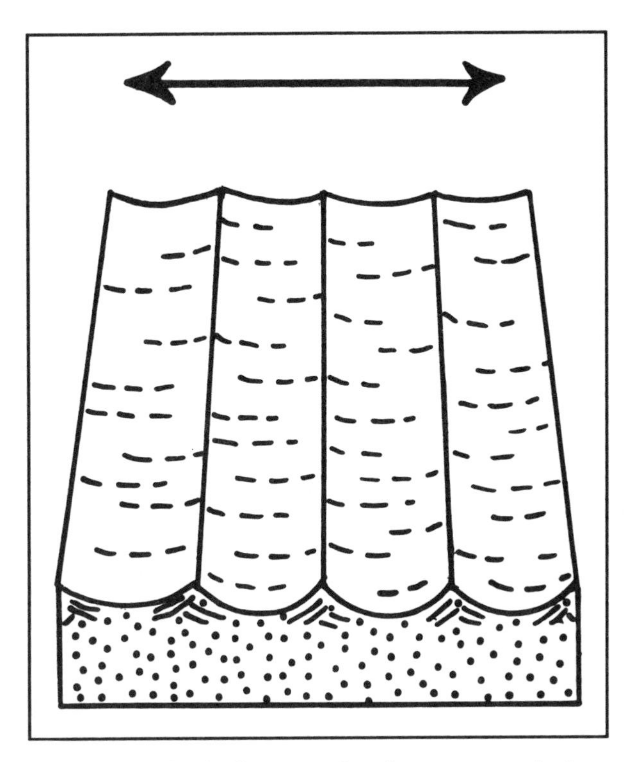

Figure 2. Block diagram showing symmetrical ripples on a sand bed beneath waves. The crests of the ripples are aligned perpendicularly to the direction of back and forth motion of the waves, shown by the arrow. Note that the cross laminae visible on the front of the block are tilted in two different directions.

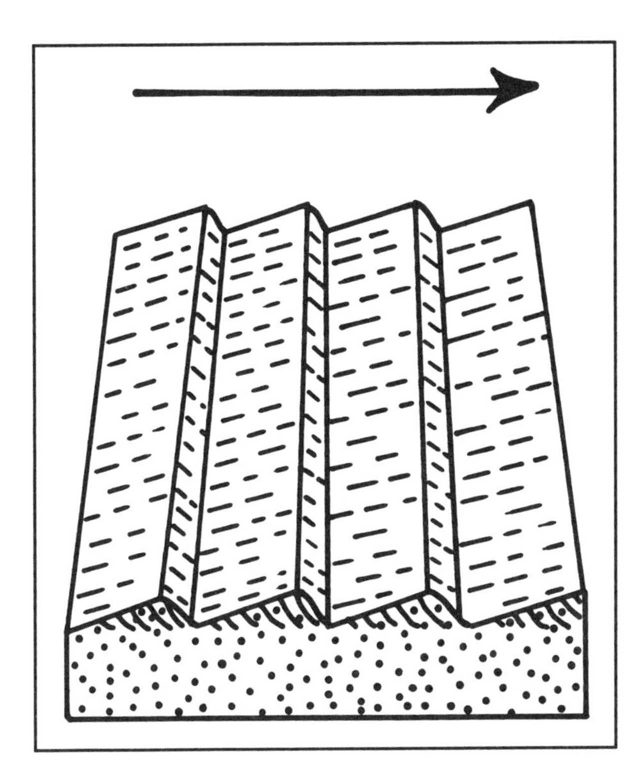

Figure 1. Block diagram showing asymmetric ripples formed on a sand bed by wind or water currents. Note the cross laminae visible on the front of the block, which are tilted downward in the direction of the flow, indicated by the arrow.

equal angles (Figure 2). Because the crests of wave ripples are usually sharp and the troughs are more broad and shallow, wave ripples are also useful for determining the original orientation of sedimentary beds that have been tilted or overturned. Such "way-up" indicators are important for geologists attempting to reconstruct the correct order of geologic events recorded in sequences of sedimentary beds.

Cross beds and cross laminae are sedimentary structures that form because ripples and dunes naturally migrate in the direction of winds and currents. As these bedforms migrate, sand grains roll or hop up the stoss sides toward the crests of the ripples and dunes, and then avalanche down the lee sides. The avalanching causes layers of sand to be deposited that are tilted downward in the downwind or down current direction (Figure 1). These layers, called cross beds or cross laminae, are commonly preserved within sandstone beds, and they are commonly used as paleocurrent indicators. Cross beds and cross laminae may also serve as way-up indicators because they are typically less steeply tilted near the bases of bedforms than near the crests.

Ripple marks, cross beds, and cross laminae are all examples of sedimentary structures that form when wind, water currents, and waves cause loose sediment to be deposited in regularly spaced, migrating bedforms. Sometimes, however, erosion of beds rather than deposition is responsible for producing sedimentary structures. When a very fast moving current flows over a bed of mud, for example, it may locally scoop out an elongate depression in the mud. Such depressions, known as scour marks, generally have rounded "noses" at their upstream ends. Downstream from these noses, the sides of the scour marks flare gradually outward and the bottoms of the marks become more shallow (Figure 3). Scour marks are common sedimentary structures in environments where conditions vary between the quiet conditions under which fine, muddy sediments usually accumulate, and the energetic conditions under which currents will scour the muddy beds. Scour marks, like cross beds and cross laminae, are good indicators of paleocurrents. Because they form in muddy beds and are usually subsequently filled in with coarser sediments, they serve as useful way-up indicators.

One environment in which scour marks commonly form is the deep sea, where quiet conditions are episodically interrupted by swiftly moving turbidity currents. Turbidity currents are underwater flows composed of mixtures of sediment and water. The mixtures have densities that are greater than sediment-free water and, under the influence of gravity, they move rapidly down slopes to the bottom of the oceans. Initially, as they flow over the ocean bed, turbidity currents are generally very erosive, and scour marks may form. As they flow over the flat ocean bottom, however, turbidity currents tend to slow down and the sedimentary particles that they carry are deposited. The largest particles settle from the turbidity current first; they are followed by progressively smaller particles. This process produces a type of sedimentary structure known as a graded bed (Figure 3). Graded beds are commonly used to determine the original way-up directions of tilted sedimentary deposits.

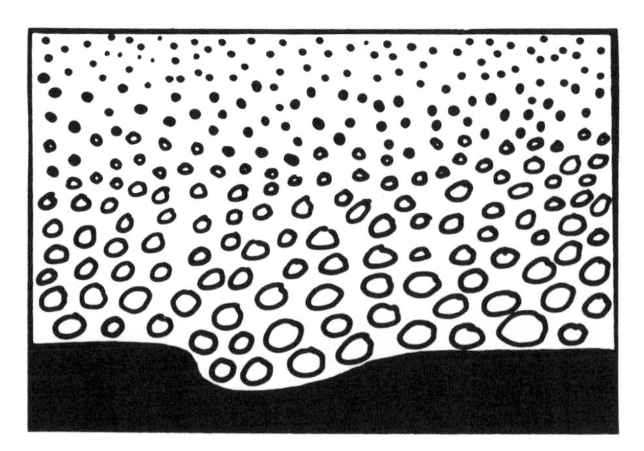

Figure 3. Side view showing a mudstone bed (colored black) with a scour mark, overlain by a graded bed.

A variety of sedimentary structures are formed when sediments are disturbed after they are deposited. Desiccation cracks are one such structure that forms when wet, muddy sediments dry up and crack, typically forming polygons of mud with edges that curl upward. The cracks may form, for example, on the bottoms of dried up ponds or lakes, on river flood plains, or on tidal flats. Commonly, the open cracks become filled with coarser-grained sediment, such as sand. In sedimentary rocks, the polygonal pattern of desiccation cracks is commonly viewed on bedding planes. In side view, desiccation cracks appear as downward tapering wedges, typically filled with sandstone. The wedges may be used as way-up indicators. The presence of desiccation cracks in sedimentary rocks indicates an environment that was alternately wet and dry.

Load casts and flame structures are two additional types of sedimentary structures that form by disturbance of sediments (Figure 4). These structures, which are inevitably found together, form where alternating beds of sand and mud have been deposited relatively rapidly from water currents. Because these sediments have been deposited rapidly, the water between the particles does not initially escape and the sediment beds are thus somewhat "soupy" and weak. In such cases the layers tend to undergo gravitational readjustment. The sand beds, which are denser than mud layers, begin to sink downward. Generally, the sand forms rounded, irregular lobes of various sizes. These lobes are known as load casts. Between the load casts the underlying muddy sediment projects upward in small, commonly tapering fingers. The appearance of these projections is aptly described by the name "flame structures." Load casts and flame structures may be used as way-up indicators. The presence of these structures in sedimentary rocks indicates an environment where both mud and sand were deposited rapidly. These conditions are commonly met, for example, in river deltas.

Biogenic sedimentary structures are a group of sedimentary structures produced by living organisms. Such structures include tracks, trails, and burrows. Biogenic sedimentary structures, also known as TRACE FOSSILS, are very useful in the interpretation of ancient environments from sedimentary rocks. They not only reveal something about the types of animals that lived in an environment and how they behaved, but give some information about the physical characteristics of the environment. Biogenic structures may give information, for example, about the relative energy of currents and waves in an environment, the firmness of the sediment bed, and the rate and constancy of sediment deposition.

Figure 4. Side view showing layers of sandstone (stippled pattern) with load casts at their bases, and overlying layers of mudstone (colored black) with flame structures at their tops.

There is thus a wide variety of sedimentary structures that tell us much about conditions during and soon after the deposition of sediments. By integrating studies of sedimentary structures with data on the composition, texture, and fossil content of sedimentary rocks, geologists can determine the types of environments that existed at various places on Earth's surface at particular times in Earth's history. Sedimentary structures can in particular give us detailed information about the roles and intensities of currents and waves in an environment, the directions that currents flowed, and the rates at which sediments were deposited. This information can be considered valuable simply because we are curious about the history of the planet on which we live. Such information is also valuable, however, because it may help us to find the natural resources on which our society depends. In searching for petroleum resources, for example, geologists seek to find porous and permeable types of sedimentary rocks, such as sandstone, which serve as reservoirs for oil, surrounded by impermeable rocks, such as mudstone, which seal the oil in. Such conditions may be met, for example, where ancient sandy river channels are flanked and overlain by flood plain mudstones, or where ancient barrier island sandstones are isolated amidst mudstone that accumulated in back-barrier lagoons or on continental shelves. Through interpretation of sedimentary structures and other aspects of sedimentary rocks, geologists are able to predict the presence and orientation of such "sand bodies" and therefore to guide exploration for and recovery of petroleum resources.

Bibliography

ALLEN, J. R. L. *Sedimentary Structures: Their Classification and Physical Basis,* Vols. 1 and 2. Amsterdam, 1982.

COLLINSON, J. D., and J. D. THOMPSON. *Sedimentary Structures.* London, 1982.

ELANA L. LEITHOLD

SEDIMENTOLOGY

Sedimentary rocks make up nearly three-fourths of Earth's surface crustal rocks; therefore, their study has great significance in interpreting Earth history. Sedimentology is the scientific study of the physical, chemical, and biological properties, classification, and origin of sedimentary rocks. Just as detectives search for clues that allow reconstruction of the scene of a crime, sedimentologists search for clues in sedimentary rocks that provide insight into Earth's history. Through sedimentological study, past climates, ocean environments and ecosystems, configurations of ancient landmasses, and the locations and compositions of long vanished mountain systems can all be reconstructed. Sedimentology is an integral part of the more specialized sciences of paleontology (study of ancient life forms on the basis of their fossil remains; *see* PALEONTOLOGY), stratigraphy (study of layered rocks; *see* STRATIGRAPHY), and sedimentary petrology (study of the composition and origin of sedimentary rocks).

Weathering, Soils, Sediments, and Sedimentary Rocks

The complex process of forming sedimentary rocks begins with the weathering of preexisting rocks, exposed on land to moisture and atmospheric gases (*see* WEATHERING AND EROSION). Rocks are composed of minerals that have different degrees of resistance to chemical weathering processes. Thus, some minerals, such as calcium-rich plagioclase feldspar, break down fairly readily during weathering and their constituent elements (e.g., calcium, sodium, silicon) go into solution. Other minerals, such as quartz and potassium-rich feldspars, are more resistant to chemical dissolution and thus become concentrated as weathering residues. Generation of these residues is aided by physical weathering processes, such as repeated freezing and thawing, that cause physical disruption of larger pieces of rock. Some of the chemical elements released from minerals during weathering are removed from the weathering site in surface waters and eventually make their way by rivers and streams to the ocean. On the other hand, some chemical elements recombine at the weathering site to form new minerals, such as iron oxides and clay minerals.

Chemical and physical weathering processes create a mantle of soil on top of weathered bedrock. This soil consists of resistant minerals concentrated by chemical weathering processes; small, broken pieces of the parent rock generated by physical weathering processes; and new minerals formed at the weathering site by chemical recombination. Some ancient soils, presumably those formed in more low-lying areas, managed to escape erosion and become preserved in the sedimentary record. These ancient soils are called paleosols or "fossil soil" (*see* FOSSIL SOILS). Most of the weathered soil residues generated in the geologic past were removed by erosion. Once weathering detritus has been eroded and transported by water, ice, gravity, or wind, we refer to it as siliciclastic sediment, that is, broken fragments composed dominantly of silicate minerals. Siliciclastic sediment (clay, silt, sand, gravel) is transported into and deposited in a variety of depositional environments ranging from river valleys and lakes on land to marine deltas, continental shelves, and the deep ocean floor. Upon burial, siliciclastic sediment is converted by burial pressure and cementation into sedimentary rock—common sandstone, shale, and conglomerate (*see* SEDIMENTS AND SEDIMENTARY ROCKS, TERRIGENOUS).

The chemical elements dissolved in surface waters at the weathering site are eventually transported by rivers and streams to the ocean, where additional chemical elements may be added by submarine alteration of volcanic rocks. Along spreading ridges, seawater seeps downward into fractures where it comes in contact with hot basalt below the ocean floor. Heated water, chemically charged with elements leached from the basalt, then rises to the seafloor where it is discharged at temperatures that may exceed 300°C. This superheated water issues from vents as bubbling clouds called black smokers (containing suspended, dark mineral grains) and white smokers (without dark grains). Some dissolved mineral matter precipitates rapidly around the vents to build tall "chimneys" as high as 10 meters (m). The remaining dissolved elements mix with the ocean bottom water to increase the concentration of "salts" in the ocean.

When the concentration of dissolved elements in ocean water reaches the saturation point, they are removed by chemical precipitation processes and/or biogenic activity to form so-called chemical/biochemical sediment (*see* SEDIMENTS AND SEDIMENTARY ROCKS, CHEMICAL AND ORGANIC). One of the most common and important kinds of chemical/biochemical sediment is calcium carbonate ($CaCO_3$). Calcium carbonate sediment becomes consolidated upon burial to form limestones. Other chemical/biochemical sediment includes evaporite deposits such as halite (salt) and gypsum, siliceous sediment (chert), iron-rich sediment (which is the source of most of our iron ores), and phosphorites (phosphorus-rich sediment). Most chemical/biochemical sediment accumulates on the shallow continental shelves of the ocean; however, chert is believed to form mainly in deeper water.

Methods of Studying Sedimentary Rocks

Sedimentologists use a variety of methods and analytical procedures for carrying out sedimentological studies. Study begins in the field, where investigators do geologic mapping, measure the thickness of rock units, and describe the field characteristics of the rocks (e.g., bedding, grain size, sedimentary structures such as cross beds, vertical and lateral stratigraphic relationships). At this time, rock specimens are collected for further study in the laboratory.

In the laboratory, techniques are available by which the grain size and shape of sediment grains can be measured and the mineral and chemical composition analyzed. For example, the grain size of sediment can be measured by sieving a disaggregated sample through a nested set of wire screens, or it may be determined indirectly by measuring the fall velocity of the grains as they settle through a column of water. Grain shape may be roughly estimated by techniques as simple as visually comparing the shape of a grain to the shapes of model grain images, or it may be determined more accurately by Fourier shape analysis, which requires application of complex mathematical routines and computer manipulation of data and images.

The mineralogy of very fine-grained sediments or sedimentary rocks is commonly determined by X-ray diffraction techniques that involve directing a divergent X-ray beam at a finely powdered sample and measuring diffraction of the minerals by the X rays. Coarser-grained rocks such as sandstones are studied with a petrographic (polarizing) microscope that allows the optical properties of mineral grains to be determined with precision (*see* PETROLOGY, HISTORY OF). More sophisticated tools, such as the scanning electron microscope (SEM), electron probe microanalyzer, and cathodoluminescence microscope, provide additional research capability. The SEM permits visual observation of grains at extremely high magnifications as well as semiquantitative chemical analysis of individual mineral grains. The electron probe microanalyzer is particularly useful for making chemical analyses of very small areas (1–10 micrometers, or μm) within mineral grains, and the luminoscope permits sedimentologists to study mineral cements and distinguish between clastic (original) mineral grains and secondary grains formed during sediment burial.

What Sedimentologists Do

Sedimentologists are geologists specially trained in the techniques of studying sedimentary rocks. They work in industry positions, government research laboratories, and university geoscience departments. Most industry sedimentologists work for petroleum companies, where they play an extremely important role in the search for petroleum reserves by studying and identifying the characteristics of sedimentary rocks that govern the migration and accumulation of petroleum. Sedimentologists also participate in industry activities that concern the movement, storage, and production of groundwater, as well as in other environmental studies that involve fluid movements (e.g., toxic waste management).

Sedimentologists who work for government research laboratories and universities commonly focus more closely on basic sedimentological research than on applied research. Such studies are generally oriented toward developing a deeper and more comprehensive understanding of the processes by which sedimentary rocks form, and improving our ability to understand the origin of these rocks. In addition, sedimentology professors in university departments participate in the training and supervision of sedimentology and other geoscience students.

Major Concepts in Sedimentology

Although sedimentology can trace its scientific roots back to Leonardo da Vinci (1500 B.C.E.), it is still a vigorous, growing science. New concepts continue to emerge, many of which have been shaped and made possible by the development of sophisticated new analytical techniques and experimental methods. For example, laboratory flume studies have provided enormous insight into the mechanisms of sediment transport and the origin of sedimentary structures such as ripples and cross-stratification. Another research milestone is the development of depositional facies models for sedimentary rocks. Facies models form the basis for interpretation of depositional environments in terms of hydrodynamic conditions and serve as predictors of new geological situations. Sedimentologists have now constructed facies models for all of the major depositional environments, making environmental interpretation much more reliable. An important development in facies modeling is the growing understanding that sea-level changes through time have had a major impact on the sedimentary record.

Significant improvements have been made also in our understanding of sediment origin or provenance (including interpretation of the probable tectonic setting of source rocks within a plate-tectonics framework) and the mechanisms of diagenesis—the physical and chemical changes that take place in sediment during burial. The availability of sophisticated analytical tools makes it possible to study physical and chemical changes on a micron scale, thus enabling researchers to relate diagenetic textures, mineralogy, and mineral chemistry to the variables of overburden pressure, temperature, and changed pore-water composition in the burial environment. These examples are but a few of the many that could be cited to illustrate major concepts of sedimentology that are currently shaping research directions in this exciting field of earth science.

Bibliography

BOGGS, S., JR. *Principles of Sedimentology and Stratigraphy,* 2nd ed. New York, 1994.

CHAMLEY, H. *Sedimentology.* Berlin, 1990.

FRIEDMAN, G. M., J. E. SANDERS, and D. C. KOPASKA-MERKEL. *Principles of Sedimentary Deposits: Stratigraphy and Sedimentology.* New York, 1992.

SAM BOGGS, JR.

SEDIMENTOLOGY, HISTORY OF

Sedimentology, a term first used by A. C. Trowbridge (1885–1971) and proposed in 1932 by H. A. Wadell (1895–1962), is most simply defined as the scientific study of sediments and sedimentary rocks. The term "sediment" refers to regolith that has been transported. Sediment is a word derived from the Latin *sedimentum,* which means "a settling." Not stated, but implied in the minds of many users of the term, is the idea that the settling occurred through air or water. Sedimentology is the scientific study of modern sediments and ancient sedimentary rocks. Sedimentology includes sedimentation, the scientific study of sedimentary processes, and sedimentary petrology, the petrographic study and genetic interpretation of sedimentary rocks. The interested reader is referred herein to: SEDIMENTOLOGY; SEDIMENTARY STRUCTURES; SEDIMENTS AND SEDIMENTARY ROCKS, CHEMICAL AND ORGANIC; SEDIMENTS AND SEDIMENTARY ROCKS, TERRIGENOUS (CLASTIC).

In the 1960s and 1970s, as defined by Friedman and Sanders (1979), sedimentology was described in the widest possible sense as the geology of sedimentary deposits. Thus, it included the study of sedimentary materials, sedimentary processes, and all aspects of the study of the products of sedimentation. This definition of sedimentology thus is about the same as some broad definitions of stratigraphy, such as those expressed by A. W. Grabau (1870–1946), by C. O. Dunbar (1891–1979) and J. Rodgers (b. 1914), and by J. M. Weller (1899–1976).

According to this broad view, the goal of sedimentology is to enable a geologist to interpret the vertical and lateral relationships of sedimentary strata that are now the subject of stratigraphy.

In the 1960s the term "megasedimentology" was coined as the scientific study of the sedimentology of vast regions, including entire sedimentary basins; lithosphere plates; or fold belts, such as the Appalachian or Alpine mountain chains, a study now considered as basin analysis.

Today the term "sedimentary geology" encompasses both sedimentology and stratigraphy, but in the 1960s and 1970s all sedimentary geologists called themselves "sedimentologists"; in fact, for practical purposes stratigraphy had become part of sedimentology. This period was the golden age of sedimentology. In the 1970s stratigraphy split off from sedimentology largely thanks to Exxon's development of seismic stratigraphy.

An interest in sedimentology, including the origin of sedimentary deposits, dates back to the very beginnings of geology. Modern geology begins with JAMES HUTTON (1727–1797) and his wide-ranging and revolutionary concept of the great geological cycle. The significance to geology of Hutton's cycle idea has been compared to the significance to astronomy and physics of Newton's law of gravity.

Hutton argued that the history of Earth could be deciphered from study of its layers of stratified bedrock, and that these ancient stratified rocks had formed in ways comparable to those now forming modern sediments. CHARLES LYELL (1797–1875) wrote his classic three-volume book, *Principles of Geology*, the first volume of which appeared in 1830, and the third, in 1833. It was Lyell who reestablished Hutton. Lyell went to extreme lengths to establish his case for the importance of what he termed "existing causes" as a basis for analyzing the geologic record. Many of Lyell's arguments came from his studies of modern sediments and comparisons with ancient sedimentary bedrock. Lyell's *Principles of Geology* contain many pages devoted to themes of sedimentology, especially the study of the processes and products of the environments in which modern sediments are being deposited as a comparative basis for paleogeography, paleoecology, and paleoclimatic reconstructions.

One of the principles first recognized by James Hutton, but most effectively presented by Lyell, was the doctrine of uniformitarianism, according to which the formation of all sedimentary strata can be explained on the basis of processes that are now in operation or experimental results. Thus, a study of modern sediments can provide the keys to understanding sedimentary bedrock. Sir Archibald Geikie (1835–1924) coined in 1905 what has become the motto of geology: "The present is the key to the past." This principle has also become known as the principle of "actualism," a term more recently favored by sedimentologists.

One of the earliest and most remarkable applications of actualism was that of JEAN LOUIS RODOLPHE AGASSIZ (1807–1873), who inferred in the 1840s that within the past tens of thousands of years, much of North America had been covered by ice sheets. Agassiz had grown up in Switzerland, and there became familiar with such glacial features as erratic boulders, glacial striae on bedrock, and various kinds of moraines. Later, after he moved to Cambridge, Massachusetts, he found that in nearly every place he visited in eastern North America, such erratic boulders, striated bedrock, and moraines are present. Moreover, these appeared to be identical with comparable features he had studied in his native Switzerland. Yet, in North America, no present-day glaciers exist. He insisted that glaciers had existed in North America at one time, but had vanished; without the former glaciers, the features noticed would not have been possible. By inferring the former presence of glaciers during a time interval named the Pleistocene epoch, he concluded that the climate in the recent past must have been colder than at present. Thus, on the basis of the principle of actualism, he could (1) reconstruct the paleogeography during the Pleistocene epoch; and (2) infer that the Pleistocene climate had been much colder than at present. Similarly, using the principles of actualism, other products in the rock record can be used to infer former climates: reefs mean a warm climate; evaporites, a dry climate; and coal, a moist climate.

One of the most effective pioneers in making the principle of actualism useful as a sedimentologic tool for a better understanding of the rock record was the German geologist Johannes Walther (1860–1937). Walther developed the practice of studying modern sedimentary environments, such as modern continental deserts, modern carbonate shelves, or modern rivers, and depositional processes as keys to the interpretation of the rock record. Walther traveled all over the world, and his fascinating and penetrating observations and writings on modern environments make fascinating reading today. These writings present some of the first real data for use in the interpretation of the origin of sedimentary strata in the bedrock. Some of Walther's observations form the cornerstone of modern sedimentology.

Two disciples continued Walther's tradition and thus his influence among later generations of geologists. The first was Grabau, whose textbook *Principles of Stratigraphy* (1913), a classic far ahead of its time, followed in the steps set by Walther. The pioneer American sedimentologist W. H. Twenhofel (1875–1957) continued the tradition of Walther and Grabau. The philosophy of Twenhofel's influential books, *Treatise on Sedimentation* (1925, 1932) and *Principles of Sedimentation* (1939, 1950), assured the continuing influence of Walther and Grabau. In his books, Twenhofel methodically presented the characteristics of modern environments in which sediments accumulate, thus providing a tool for interpreting strata in the rock record.

Since 1859, when HENRY C. SORBY (1826–1908) began to use the petrographic microscope for studying thin sections cut from sedimentary rocks, such study has become an essential part of sedimentology. Although not the first to recognize the importance of the petrographic microscope, Sorby admonished his colleagues "to avail ourselves of all the resources of polarized light."

The importance of sedimentology to geology is matched by the importance to sedimentology of knowledge about the oceans. This connection was expressed in the title of a paper published in 1958 by Ph. H. Kuenen (1902–1976): "No Geology Without Marine Geology."

The history of understanding of the oceans can be subdivided into four periods. These are (1) pre-*Challenger* expedition (prior to 1872); (2) the *Challenger* expedition (1872–1876) results; (3) post-*Challenger* and pre-World War II; and (4) post-World War II.

Useful knowledge about the oceans that had accumulated prior to 1872 dealt mostly with the coasts and shallow parts, but did include some important concepts based on the understanding of water circulation and the impact of circulation on bottom sediments. Biologic and biogeographic study of marine invertebrates, including coral reefs, provided Lyell and CHARLES DARWIN (1809–1882) with much of the background with which they interpreted the geologic record that they encountered during their travels.

The surveys for transatlantic submarine telegraph cables, carried out in the middle of the nineteenth century, aroused great interest in learning more about the deep sea. In 1872, organized exploration of the deep sea began in earnest. The Royal Society of London and the Royal Navy combined their efforts in a three-and-one-half-year expedition. They selected and outfitted a party of scientists aboard the wooden-hulled corvette HMS *Challenger*. The expedition, which set off on 21 December 1872, circumnavigated the globe.

The *Challenger* scientists measured the temperature of the water from the surface and at various depths, collected water samples, towed nets to collect marine animals and plants, sent small dredges to the bottom to bring up specimens of the sediment, and sounded the depth.

The *Challenger* scientists assembled not only a vast collection of specimens, but also, where available, samples from every other expedition that had taken place previously. Fifty monumental volumes were eventually published. The *Challenger* reports formed the foundation on which modern knowledge of the oceans has been built.

Since World War II marine sedimentology has been revolutionized as a result of information gained from the precision recording echo sounder; from the Ewing piston corer; from visual images of the seafloor recorded on photographic film, made directly by geologists using SCUBA or riding in research submersibles, or transmitted by underwater television camera; and from records made using side-scanning sonar.

Despite an august history of 150 years, sedimentology as a science advanced most rapidly since about 1950. This rapid advance resulted from a change of sedimentology from a pure to an applied science. Economic incentives, particularly in the exploration for petroleum (oil and gas), spurred prodigious expansion and rapid advances in sedimentology. Exploration personnel of major American oil companies began to realize that sedimentology was the key to success in exploration. Beginning with this recognition in the late 1940s and early 1950s, the first large-scale sedimentological research projects materialized. There had been large-scale research projects before, such as the boring of the atoll of Funafuti in the Pacific Ocean at the close of the nineteenth century, but such early efforts were isolated. The 1947 report of the Research Committee of the American Association of Petroleum Geologists, under the leadership of Shepard W. Lowman (1899–1967), stated that research in sedimentology is the most urgent need in petroleum geology. Project 51 of the American Petroleum Institute led to a methodical and detailed study of modern depositional environments on a scale not previously attempted. With the aid of research vessels and research teams, modern marine and deltaic depositional environments were explored. Much of the background of this largest-of-all projects of the American Petroleum Institute was prepared by Lowman, who first conceived the idea. A classic book that emerged from this team effort was published as a special volume of the American Association of Petroleum Geologists: *Recent Sediments, Northwest Gulf of Mexico.* It was coedited by FRANCIS P. SHEPARD (1897–1985), F. B. Phleger (b. 1909), and T. H. van Andel (b. 1923).

In the research laboratories of the major oil companies, eminent team leaders gave modern sedimentology a boost that led to rapid breakthroughs and advances. Most notable among this group were H. N. Fisk (1908–1964) of Exxon (formerly Humble), a student of deltas; H. A. Bernard (1915–1964) and R. J. LeBlanc (b. 1917) of Shell, pioneers in the study of rivers; and R. N. Ginsburg (b. 1925) of Shell, a researcher in carbonate sediments. Sedimentology owes a great debt of gratitude to the major oil companies for pioneering in large-scale research projects and for making this information available to the profession at large through the regular channels of publication. From oil-company research the depositional or facies model developed as a norm for comparison, genetic interpretation, and prediction. Later academic workers expanded these models.

Of profound importance in helping modern sedimentology was the publication in 1931 of the *Journal of Sedimentary Petrology* of the Society of Economic Paleontologists and Mineralogists, now SEPM (Society for Sedimentary Geology), and, in 1962, the journal *Sedimentology* of the International Association of Sedimentologists.

Bibliography

FRIEDMAN, G. M., and J. E. SANDERS. *Principles of Sedimentology*. New York, 1978.

FRIEDMAN, G. M., J. E. SANDERS, and D. C. KOPASKA-MERKEL. *Principles of Sedimentary Geology: Stratigraphy and Sedimentology*. New York, 1992.

GEIKIE, SIR A. *The Founders of Geology*. New York, 1905.

SHEPARD, F. P., F. B. PHLEGER, and T. H. VAN ANDEL, eds. *Recent Sediments, Northwest Gulf of Mexico*. American Association of Petroleum Geologists. Tulsa, OK, 1960.

GERALD M. FRIEDMAN

SEDIMENTS AND SEDIMENTARY ROCKS, CHEMICAL AND ORGANIC

Sediments and sedimentary rocks may be divided into two groups, intrabasinal and extrabasinal. Intrabasinal sediments and sedimentary rocks are those whose constituent particles were derived from within the basin of deposition. Most carbonate sediments are secreted or were precipitated and are preserved within a basin of deposition. By contrast the terrigenous variety belongs to the group designated extrabasinal, which includes sediments and sedimentary rocks composed of particles derived from outside the basin of deposition, such as sandstones and shales (*see* SEDIMENTS AND SEDIMENTARY ROCKS, TERRIGENOUS). Intrabasinal sediments and rocks are further classified into (1) carbonate sediments and rocks; (2) authigenic sediments and rocks (minerals that have grown in place subsequent to the formation of the sediment or rock of which they constitute a part); and (3) carbonaceous sediments and rocks.

Carbonate Sediments and Rocks

Carbonate sediments and rocks may be composed of carbonate particles. Particles that grew as solids in the depositional basin include skeletal and nonskeletal calcium carbonate materials. Much carbonate skeletal debris includes whole skeletons of calcium carbonate-secreting organisms, such as foraminifers (*see* MINERALIZED MICROFOSSILS) and mollusks (*see* MOLLUSKS), as well as broken pieces of the hard parts secreted by these and other organisms (*see* ARTHROPODS; BRACHIOPODS; COLONIAL INVERTEBRATE FOSSILS; and ECHINODERMS).

Ordinarily, organisms must die to contribute their skeletal material to sediments. Although death is involved in the origin of most fossil skeletal material, several exceptions are known. Ostracodes and trilobites discard or discarded their shells during molting and coccoliths shed plates during life.

The silt- and clay-size components of carbonate sediments are collectively designated as lime mud. Most lime mud consists of tiny needles and platelets of carbonate crystals. Organisms are responsible for secreting the tiny solids that compose most lime mud. Much lime mud forms by the accumulation of tiny skeletal components secreted by algae. During the Cretaceous, pelagic algae, known as coccoliths, were abundant enough to form an ooze that, when lithified, became chalk. Hence chalk is almost entirely confined to the Cretaceous. Some lime mud results from the physical breaking down of sand-size and larger skeletal material.

In the modern ocean, carbonate skeletal particles are composed of the minerals aragonite and calcite. Although the chemical formulas of both aragonite and calcite are the same, $CaCO_3$, these minerals differ in crystallographic arrangement and in their contents of trace elements. Aragonite crystallizes in the orthorhombic crystal system, and consists of essentially pure calcium carbonate. Calcite crystallizes in the hexagonal system, and contains various amounts of magnesium. One variety of calcite has been termed low-magnesian calcite and the other high-magnesian calcite. Modern shallow water skeletal carbonate particles consist mostly of high-magnesian calcite and aragonite.

Sand-size spherical or ellipsoidal particles of calcium carbonate are designated as pellets. Internally, pellets are generally homogeneous and devoid of structure. Most pellets are formed by deposit-feeding organisms that eat the mud. In modern carbonate sediments, pellets are probably the most common single kind of particle. Because a few individual organisms can excrete thousands of pellets, the abundance of pellets is easy to understand. A useful term, peloid, has been introduced for particles that resemble pellets, but for which no particular origin is implied.

Particles known as ooids derive their name from the Greek word that means egg or egglike because these particles resemble the roe of fish. Ooids generally consist of aragonite. They usually are spherical or elliptical. In sections that include their centers, ooids can be seen to consist of a central nucleus surrounded by a rim displaying a concentric or radial structure. The nucleus may be another calcareous particle, such as a skeletal particle, a foraminifer, or a pellet, and even a noncalcareous fragment, such as a particle of quartz. The rim of the ooid consists of one or more layers.

The ooids in cyanobacterial mats may be precipitated by the microbial organisms themselves, yet the origin of the majority of the ooids is still subject to speculation. Most modern ooids occur in or close to the intertidal zone, where the waves break and pound. Carbon dioxide presumably is removed from the calcium bicarbonate (the water is saturated with respect to calcium bicarbonate), and a rim of calcium carbonate is precipitated as a coating on existing particles.

The term "intraclast" refers to sand-size or larger particles, texturally analogous to a rock fragment broken from consolidated or hardened materials, which are accumulating within the basin of deposition. In this context "intra" means within the basin, and "clasts" denotes broken particles. Examples of intraclasts are recycled fragments of coherent carbonate sediment. Intraclasts are of various sizes and shapes. Many are angular and their diameters exceed 2 millimeters (mm).

Spherical or elliptical particles that exeed 2 mm in diameter are known as pisolites. The division between pisolites and ooids is one of size; ooids are smaller than 2 mm.

By definition, carbonate rocks are those containing more than 50 percent carbonate minerals. Two broad compositional divisions are limestones and dolostones; limestones are composed of the mineral calcite ($CaCO_3$) and dolostones of dolomite $CaMg(CO_3)_2$. Except for reefs, which framework builders construct, or for consolidated lime mud, the basic constituents of most limestones are the sand-size particles previously discussed. The space between these particles is occupied by (1) a matrix of lime mud, whose lithified equivalent is known as micrite; (2) a calcite cement or spar; and (3) void space.

Authigenic Rocks

Authigenic rocks are chemically and mineralogically diverse; their only common feature is their authigenic origin. Authigenic is a term that designates minerals that have grown in place subsequent to the formation of the sediment or rock of which they constitute a part. The most important kinds of authigenic rocks are (1) chert, (2) phosphate rock, (3) sedimentary iron ores, and (4) evaporites.

Chert. A tough, brittle siliceous rock, chert exhibits a splintery to conchoidal fracture and a vitreous luster. The silica of chert may be an alteration product of volcanic rock, such as smectite and volcanic glass, or a precipitate derived from the dissolution of tests of siliceous organisms. The most probable source of silica in the cherts of the central Pacific Ocean is dissolution of radiolarian tests. The silica in most oceanic cherts probably is of biogenic origin.

The tests of siliceous organisms, such as those of radiolarians and diatoms, as well as the spicules of sponges consist of opal. Likewise, siliceous shells in cherts of Tertiary age are mostly opaline. In contrast, nearly all silica in Paleozoic cherts occurs as quartz and chalcedony. Other silica minerals present in chert include quartzine and lutecine. Although most cherts are probably the end product of the reaction biogenic opal → cristobalite or opal-cristobalite → quartz, the possibility of direct crystallization of quartz should not be excluded. Most cherts in the record formed by replacement of carbonate sediment or of gypsum ($CaSO_4 \cdot 2H_2O$) or anhydrite ($CaSO_4$.); of skeletal particles, such as those of bryozoans, echinoderms, or brachiopods; of wood; or as nodules in carbonate deposits.

Phosphate Rocks

Sedimentary phosphate deposits usually consist of a carbonate fluorapatite $Ca_5(PO_4, CO_3)_3(F, OH)$; a variety of apatite known as francolite is the basic raw material for many phosphorous-containing compounds on which modern technological developments depend. When the apatite-like phase cannot be identified, the name collophane is commonly applied. Such phosphates, or phosphorites (the two terms are synonymous), are referred to as phosphate rock.

Phosphate rock occurs on the modern sea bottom and in ancient deposits. Phosphate-rich sediments are present on many modern continental shelves. Phosphate rock forms nodules on the modern sea bottom, extending from shallow to great water depths. In some nodules the concentration of phosphate reaches 96 percent, but in others abundant impurities are included, such as organic matter, siliceous tests of organisms, calcareous shells, some of which are partly phosphatized, sharks' teeth, and various terrigenous particles.

The chemical reaction that results in the precipitation of phosphorite is still not known. Phosphorites are thought to form where deep phosphate-rich waters upwell adjacent to a shallow shelf. However, upwelling by itself may not be enough to cause precipitation. In places, phosphorites have replaced carbonate sediments or rocks and are composed of skeletal particles, ooids, and pellets. In this reaction apatite replaced $CaCO_3$, but the geochemical setting, the timing, and the depositional and diagenetic conditions are not known.

Iron-Rich Sedimentary Rocks

Iron-rich sedimentary rocks have been defined as sedimentary iron ores. In this usage the term is a general rock name rather than a value term based on economics. Although all sedimentary rocks contain readily detectable amounts of iron, the term iron-rich deposits designates those consisting predominantly of iron minerals, notably oxides, hydroxides, carbonates, silicates, and sulfides. The oxides are hematite and magnetite; the hydroxide mineral is goethite (including limonite); the carbonate mineral is siderite; the silicate minerals are chamosite, greenalite, and glauconite; and the sulfide minerals are pyrite and pyrrhotite. One of the remarkable features of sedimentary iron ores is that the mineral compositions in those of Precambrian age contrast with those of younger (Phanerozoic) ages. In Precambrian deposits hematite typically is interbedded with chert. By contrast, in Phanerozoic deposits hematite replaces ooids and fossil fragments or forms an earthy matrix. This change in the distribution of hematite with time is probably related to the biologically driven buildup of oxygen in the earth's atmosphere during the late Precambrian. As a result, oxidized iron in the form of hematite is thought to have been precipitated on a massive scale.

The sedimentary iron ores are interpreted as products of nearshore, restricted, shallow-marine environments. The source of the iron for these deposits is thought to have been deeply weathered continental rocks. Deep weathering releases this iron from the rocks and various iron salts form.

After the iron has thus been released from the bedrock, streams carry away these iron salts in suspension or in solution to the sea, where ferrous iron is oxidized and deposited. Alternative or additional sources of iron may be submarine volcanoes or upwelling currents carrying iron derived from deeper parts of the ocean floor.

Evaporites

Evaporites form by precipitation from brines whose salinity values have been greatly increased by evaporation. Evaporation may take place in closed or in open systems in numerous environmental settings.

Although almost forty different precipitate-type minerals have been recorded from evaporite deposits, only about twenty are present in more than trace amounts. Of these, only two kinds are common, sulfates and halides, both of which form extensive deposits of sedimentary rock. Two kinds of sulfate minerals are common in sedimentary rocks:

gypsum ($CaSO_4 \cdot 2H_2O$) and anhydrite ($CaSO_4$). Although anhydrite forms in the modern sea-marginal flats of Abu Dhabi in the Persian Gulf, for all practical purposes, gypsum, as the hydrated sulfate, occurs at the earth's surface, and anhydrite, the anhydrous sulfate, in the subsurface. The change from one kind of sulfate to the other depends on the geothermal gradiant as well as on the salinity of the subsurface brine.

Among the halides, halite rock (rock salt) is the most common. It forms successions up to 1,000 meters (m) thick.

Carbonaceous Sediments and Rocks

Carbonaceous sediments and rocks, primarily coal, have been classified in two ways: by rank (content of carbon and caloric value) and by petrographic characteristics. In order of increasing rank, the classes of coal are (1) lignite; (2) subbituminous coal; (3) bituminous coal; and (4) anthracite (see *Coal*). The progression of carbonaceous material through the continuous series of lignite (or its precursor, peat) through bituminous coal to anthracite is known as coalification. During coalification the color of the products darkens and their luster increases. Also during this progression, their relative contents of carbon and caloric value gradually rise, and their contents of moisture, volatile matter, and oxygen decrease. Except in anthracite (in which hydrogen decreases), the proportion of hydrogen remains roughly constant through this series. Coalification is accomplished in two stages: biochemical and geochemical. In the biochemical stage, bacteria convert plant material to peat. During the geochemical stage, both chemical and physical changes result from the action of temperature and pressure. The increase in temperature is derived chiefly from the geothermal gradient. The pressure results from deep burial and from deformation. With application of even greater heat and pressure, anthracite can be converted to graphite. In essence, the rank of coal is a product of the temperature history. Older coals are likely to have been more deeply buried, and hence exposed to greater geothermal temperatures than younger coals.

Coals have been recorded in strata ranging in age from Precambrian to Holocene. However, only after the Late Silurian Period, when land plants had become established, was it possible for plant material to accumulate on a scale to form large deposits of coal. In the history of the earth, two particularly rich coal-forming periods are known, the Carboniferous (principally Pennsylvanian) and Permian periods; and the Late Cretaceous and Early Tertiary Periods. Most coals form from accumulated plant debris in sea-marginal swamps or in closed fluvial basins.

Bibliography

FRIEDMAN, G. M., J. E. SANDERS, and D. C. KOPASKA-MERKEL. *Principles of Sedimentary Deposits: Stratigraphy and Sedimentology*. New York, 1992.

GERALD M. FRIEDMAN

SEDIMENTS AND SEDIMENTARY ROCKS, TERRIGENOUS (CLASTIC)

Terrigenous sediments and rocks are sedimentary materials composed mostly of mineral particles and aggregates (detrital grains) derived from breakdown of older rocks during chemical and physical weathering (*see* WEATHERING AND EROSION). The term "clastic," meaning broken, is another term applied to this group of materials, but which, strictly speaking, also applies to sediments and rocks made of broken pieces of calcareous biogenic material (limestones). Around 85 to 90 percent of sedimentary rocks are terrigenous.

Some mineral particles in terrigenous sediments and rocks originate as chemical precipitates during reaction of unstable minerals with water and the atmosphere, as is the case for many clay minerals. Other detrital grains are fragments of older rocks. A fragment may be a single mineral crystal, an aggregate of several crystals of a single mineral, or an aggregate containing crystals of more than one mineral. Minerals that survive to become particles in terrigenous sediments are mostly ones that are more stable in the low-temperature aqueous environment of Earth's surface (for example, quartz and feldspar are common in many sandstones but olivine is not) (*see* WEATHERING AND EROSION). Polycrystalline aggregates are known as lithic fragments. Some lithic fragments can be readily identified and clearly reveal the nature of the parent rock. Identification of lithic types, together with examination of the proportion of lithic fragments, quartz, and feldspars and other minor mineral particles, can tell us a great deal about the nature of the rocks that eroded to produce a sediment. Deciphering the source of sediment in this way is known as the study of provenance and has provided many details about the history of Earth's crust, including information about rocks that once existed but otherwise have been lost to observation.

After terrigenous sediment is produced, it may be transported by gravity, wind, ice, or water. Transport may be very local (a rockfall deposit beneath a cliff), across continental-scale distances (river sediment carried thousands of kilometers from its source), or even across global-scale distances (fine-grained, wind-transported particles) (*see* GEOLOGIC WORK BY WIND; GEOLOGIC WORK BY STREAMS). Ultimate deposition of terrigenous sediment can take place in environments ranging from mountain slopes and valleys to abyssal depths in the sea. Thus, it is not surprising that about 60 percent of Earth's continental surfaces are covered by terrigenous sediments and rocks, or soils derived from such materials; much of the seafloor is also covered with terrigenous sediment. Knowledge about the origin and distribution of the terrigenous rocks is integral to understanding many facets of the crustal environments that humans inhabit and utilize.

Classification

Subcategories of terrigenous sediment and rocks are based on the size of the particles, the degree to which particles have become stuck together or lithified, and composition.

Classification Based on Grain Size and Lithification. The size of particles produced at Earth's surface ranges over several orders of magnitude (Table 1) and provides a useful basis for dividing terrigenous sediments and rocks into several distinctive groups (Figure 1). Mudrocks and claystones constitute about two-thirds of all sedimentary rocks; the sedimentary fill in some deep basins has an even higher proportion of fine-grained terrigenous sediment and rock. Despite their volumetric significance, finer-grained terrigenous materials have received little study compared to sandstones.

Table 1. Grain Size Classification Applied to Terrigeneous Sediments

*Particle Diameter**	*Size Class*	
>256	boulder	gravel
>64	cobble	
>4	pebble	
>2	granule	
>1	very coarse	sand
>0.5	coarse	
>0.25	medium	
>0.125	fine	
>0.0625	very fine	
>0.0039		silt
<0.0039		clay

* Millimeters.
Modified from Folk (1980).

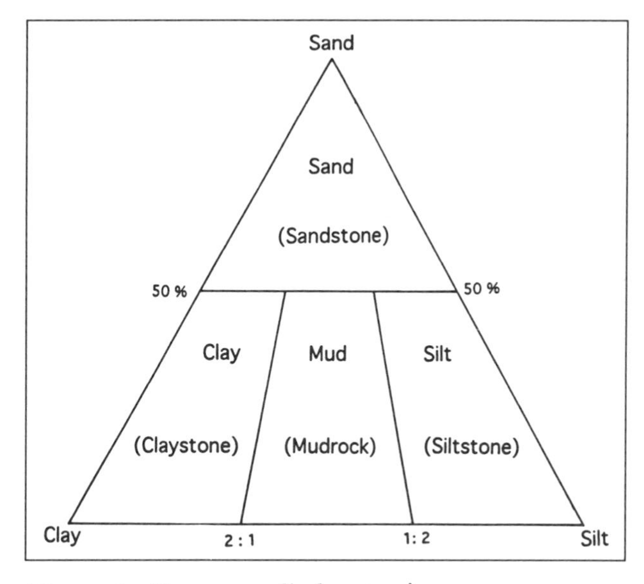

Figure 1. Names applied to terrigenous sediments and rocks (in parentheses) are based on proportions of different sizes (see Table 1) among the detrital grains. In addition to the names shown, samples with greater than about 30 percent gravel-size material are designated as gravels (or conglomerates). Mudrocks and claystones that readily break into thin sheets parallel to the bedding are known as shales. (Modified from Folk, 1980.)

The term "clay" is confusing in that it refers to both a range of particle size (generally less than 4 micrometers, or μm) and to a group of minerals with a distinctive sheet structure (clay minerals). In terrigenous sediments and rocks most clay minerals are also clay-size. In Table 1 clay refers strictly to the size of the particle. Size is also prominent control on the occurrence of lithic fragments (dominant among boulders, rare in silt, almost unknown in clay) and other minerals.

Divisions based upon the degree of lithification are somewhat subjective. For example, a mud may be relatively soft when wet, and hard, and thus a mudrock, when dry, with a continuous gradation of lithification states at the various intermediate stages of dehydration. Geologists generally apply the terms for lithified materials for cases in which the material can be handled without falling apart into its constituent grains. The processes by which terrigenous sediments turn into terrigenous rocks will be examined below (*see* DIAGENESIS).

Compositional Classification. Several classification schemes for sand-size materials have been based on the quantitative proportions of quartz, feldspar, and lithic fragments. Figure 2 shows one such classification. Single mineral grains other than quartz and feldspar constitute less than a few percent (by volume) of the typical sandstone. Important minor grain types include micas and a wide variety of "heavy" minerals such as zircon, tourmaline, garnet, and magnetite (so named because their specific gravity is greater than that of more common quartz and feldspar). Although they are not used in classification, these minor components provide clues to provenance and are more important in understanding the origin of terrigenous sediment than their volume might suggest.

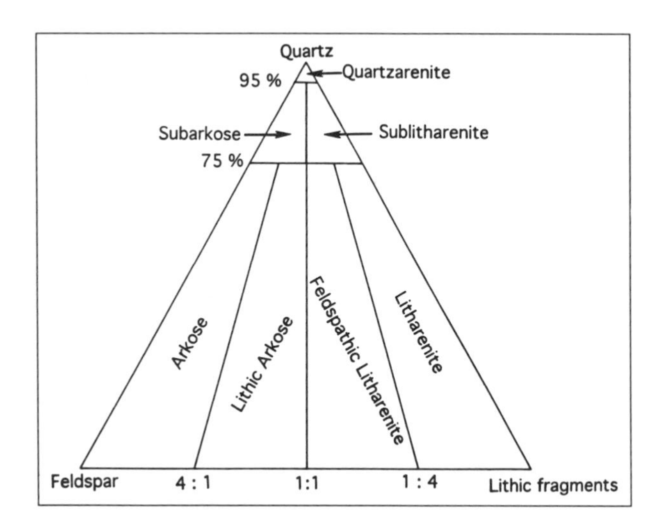

Figure 2. An example of a classification scheme for sands and sandstones. In this classification, authigenic minerals and grains other than quartz, feldspar, and lithic fragments are not considered. Such a classification is useful both for descriptive purposes and for deciphering provenance. (Modified from Folk, 1980.)

Other important grains not generally used for classification of terrigenous materials are particles formed in or near the depositional environment. Skeletal fragments from carbonate-secreting organisms are the most common of these and with increasing abundance of such particles terrigenous sediments and rocks grade into carbonate sediments and limestones (*see* SEDIMENTS AND SEDIMENTARY ROCKS, CHEMICAL AND ORGANIC). One arbitrary choice for a boundary between carbonate sediments and rocks versus terrigenous materials is at 50 percent (by volume) carbonate biogenic debris among the detrital particles (Folk, 1980). Other typically minor particles include glauconite formed in the marine environment, phosphatic bone fragments, and organic matter of marine or terrestrial origin.

Assignment of a sand or sandstone to a particular compositional category is usually accomplished through observation of grains in a thin section using a petrographic microscope. A common method is to determine the proportion of quartz, feldspar, and lithic grains through identification of many individual grains randomly selected using a spatial grid, a procedure known as point-counting.

Diagenesis and Gradation into Metamorphic Rocks

How does sand in a river or, say, mud in a bay, become transformed into sandstone or mudrock? Both physical and chemical processes play a role in lithification of terrigenous materials, as is also true for limestones.

After deposition, considerable space exists between the grains (porosity). In sandy sediments porosity at deposition is typically around 40 percent; muddy sediments are even more porous. As sediments are buried by accumulation of younger deposits, the weight of the overlying material causes grains to progressively shift closer together, thus reducing the porosity in a process known as compaction. Particle rearrangement can take the form of simple grain rearrangement, deformation of "squashing" of softer lithic particles, or actual breakage of more rigid grains. In rocks with many soft particles compaction can reduce porosity to less than a percent or so (*see* DIAGENESIS).

Chemical processes also have an important effect on the degree of lithification. Cementation is precipitation of minerals from aqueous solution into the pore spaces between the detrital grains. The presence of even small amounts of cement can cause a pronounced enhancement of lithification. Common cements in terrigenous sediments are quartz and calcite.

Some cements are irregularly distributed within terrigenous rocks, even over the scale of a few centimeters. Weathering often reveals discrete bodies of rock cemented to a much greater degree than the surrounding materials. Known as concretions, many of these bodies have dramatically geometric shapes such as spheres, oblate spheroids, disks, or highly elongate rods or tubes that attract the attention of collectors, both professional and amateur. Controls on the distribution and shape of concretions are a poorly understood aspect of terrigenous rocks.

Another important chemical process in terrigenous sediments is dissolution of grains or cements by an aqueous solution in the surrounding pore spaces. In some cases, material from the dissolved mineral is carried away by the solution leaving what is known as a secondary pore. Subsequent mineral precipitation may occur from aqueous solutions within the secondary pore and in such cases the grain or cement is said to be replaced.

As in other sedimentary rock types, cements and replacements that form chemically after deposition are collectively known as authigenic minerals. Some terrigenous sediments undergo so much diagenesis through chemical reaction with fluids that they are made mostly of authigenic minerals. Evidence of provenance can be partially lost in the process. Highly altered terrigenous rocks have typically undergone deep burial (greater than 3 to 4 kilometers, or km) or experienced elevated temperatures (greater than 100°C) and enhanced fluid flow through their proximity to igneous intrusions or to faults that have served as conduits for the flow of hot water (Figure 3).

Figure 3. Terrigenous clastic sedimentary rocks of volcanic derivation, island of San Miguel, Azores. Many clastic rocks display abrupt lateral variations in lithology at the outcrop scale. (Photograph by K. M. Marsaglia.)

Choosing a firm dividing line between highly cemented and replaced terrigenous rocks and their metamorphosed equivalents (e.g., slates, phyllites, and schists) is not easy. Such a division could be based on composition as well as texture. A commonly used compositional distinction is based on the existence of chemical equilibrium between different minerals in the rock, which is generally true for metamorphic rocks, but not sedimentary ones, which retain minerals of diverse origins (detrital grains from different sources together with authigenic minerals). Another possible divide between metamorphic and terrigenous rocks is the preservation of sedimentary fabrics (grains, pores, cements) in terrigenous rocks versus the destruction of such fabrics in metamorphic ones. Practicing scientists recognize the overlap that is inherent in the gradation between diagenesis and metamorphism and some terrigenous rocks are studied both by scientists who identify themselves as sedimentologists or sedimentary petrographers and also by those who think in terms of metamorphic petrology. It should be noted that textures such as banding, caused both by sedimentary and metamorphic processes, may be difficult to distinguish in some rocks.

Importance to Humans

As mentioned previously, terrigenous materials are significant to humans simply by virtue of their abundance at the surface. The grain size and mineralogy of terrigenous sediments or rocks can have a prominent controlling effect on the character of soil formed from them, and hence upon the distribution of both natural vegetation and agriculture. The most significant economic use of terrigenous materials is as aggregate—sand and gravel for direct use as fill or as components in building and road materials. More specialized uses include clay minerals mined as refractory materials, for ceramics, and as food additives. Glass manufacture represents another important specialized use for these materials. Very few sand deposits qualify for use in the production of glass, very quartz-rich compositions being required.

Interestingly, some of the most significant economic and environmental aspects of terrigenous sediments are related not to the portion of the rock composed of minerals, but rather to a portion of the rock that is not composed of minerals—the porosity (see various titles on petroleum, gas, and water resources). Much of the freshwater supply available for human use is in the form of subsurface aquifers and most of these are in terrigenous sediments. Documenting the nature and distribution of porosity in sandstones is thus an important aspect of managing water use in many areas. Other materials that may occupy pores in subsurface terrigenous sediments and rocks include brines, petroleum, natural gas (methane), and other gases such as carbon dioxide and hydrogen sulfide. Many important oil-producing sedimentary basins have reservoirs that are found dominantly in terrigenous sediments and rocks. Geologists who explore for these resources in terrigenous rocks try to document the 3-D geometry of the porous parts of terrigenous rock deposits. Documenting the chemical and compactional history that affects the nature and amount of pores in subsurface terrigenous rocks is also an important aspect of the search for oil and gas.

Figure 4. Particle size variation on a modern beach, island of San Miguel, Azores. Particles on this beach range from sand size to small boulders and are largely of volcanic provenance. (Photograph by K. M. Marsaglia.)

Bibliography

Blatt, H., G. Middleton, and R. Murray. *Origin of Sedimentary Rocks.* Englewood Cliffs, NJ, 1980.

Folk, R. L. *Petrology of Sedimentary Rocks.* Austin, TX, 1980.

Pettijohn, F. J., P. E. Potter, and R. Siever. *Sand and Sandstone.* New York, 1987.

KITTY L. MILLIKEN

SEISMICITY

See Earthquakes and Seismicity

SEISMIC LINES, COCORP, AND RELATED GEOPHYSICAL TECHNIQUES

The exploration of Earth in all its fascinating and perplexing complexity ranks as one of the ultimate human adventures. And the pace seems to be accelerating as such traditional means for exploration as ships, planes, wagons, and dogsleds are supplemented by the tools and methods of modern science. This article is about a major frontier of current Earth exploration, the huge subterranean portion of the crust of the continents, i.e., the uppermost 40 kilometers (km) or so of rocks that lie buried beneath the surface of the large land-covered areas of Earth.

Most of this vast and often unnoticed frontier that is hidden from, yet so close to, the bustle of human activity at the surface is not readily accessible for personal examination. However, we almost certainly will come to know it better, for we already have a variety of geological, geochemical, and geophysical techniques for sensing it remotely. Only a tiny fraction of the total volume of this vast frontier has been observed in detail, but preliminary results demonstrate unequivocally that we shall find there a dazzling array of buried features, features that once existed at the surface as mountain ranges, continental margins, deep sea trenches, rift valleys, and island arcs, or that were once formed in, and have since remained a part of, the interior. Those features are there waiting to be discovered, delineated, comprehended, named, and relegated to their proper role in the history of evolution of Earth. The frontier is there, the tools are there; we need only mount the effort to make the observations, and then, as cleverly as we can, deduce the meaning of those observations and so reveal the makeup of Earth and its overall geological history. What more promising opportunity, what finer challenge could be offered a young scientist aspiring to a career of discovery in the style of the great explorers of the past?

Currently the potential for discovery in the deep crustal unknown is highly favorable, but such potential is hardly unprecedented in the history of Earth exploration. Columbus, and his contemporaries and followers, enjoyed a parallel opportunity as they probed the geographic frontier of the New World during the great Age of Discovery just a few centuries ago. Somewhat later, from the late eighteenth century to the present in yet another interval of rapid revelation about Earth, geologists observed and mapped the exposed rocks of the surfaces of the continents, recording a fascinating variety of rock types in complex and unanticipated spatial patterns. These observations reveal Earth to be dynamic and ever changing in the long-term geological perspective, in sharp contrast to the serene and static image Earth presents to a single individual over a human lifetime. Mastery of Earth's geography and geology, and the derived knowledge of the resources contained therein, are, of course, critical to the development of civilization as we know it. To prosper and live wisely on their natural spacecraft, humans need to understand their planet as thoroughly as possible.

During the two decades following World War II, another great frontier of Earth, the deep seafloor, was explored in some detail. Once sketchily known, submarine features like mountain ranges and abyssal plains were revealed, defined, and shown to be parts of globe-encircling patterns that led scientists, during the 1960s, to the recognition of the huge global geodynamic system that goes by the strange name of plate tectonics, and that incorporates the concepts of spreading seas and drifting and colliding continents. In some ways, the sensing of the eerie, drowned features of the seafloor, which are never seen in all their grandeur by humans, is comparable to the exploration of the buried features of the continents. The subterranean features are also never seen, and their rock–rock interfaces may be considered analogous to the rock–water interface of the seafloor or the rock–air interface of the land surface.

One convenient way to describe and view the exploration of the buried crust of the continents is as a search for the third, or depth, dimension of geology at the surface. This characterization is somewhat more profound than it first appears, for it implies that the range of the qualitative yet powerful and productive style of geological reasoning that prevails at the surface can be, and will be, extended to considerably greater depths. The third, or depth, dimension of continental geology is somewhat known to us already, of course. Various techniques of geophysics based on gravity, magnetics, electromagnetics, seismic refraction and seismic surface waves, and various geochemical and geological methods have provided some knowledge, mostly of a reconnaissance nature, of the continental crust. One particular part of the crust, the sedimentary basins, is already known in astonishing detail in places. The sedimentary basins are the locale of hydrocarbons, and the enormous societal values of hydrocarbons have engendered a huge and expansive effort to explore the geology of the sedimentary basins and to find and produce hydrocarbons from them. Hydrocarbon production, at present at least, is essentially limited to the upper 10 km of those basins.

Hydrocarbon exploration has produced, as a sort of by-product, two techniques that are and will be especially useful for exploring the remaining, underexplored, portions of the continental crust. One is drilling, which permits sampling of rocks at depth and the use of various kinds of logging techniques for making observations within the hole. The deepest hole so far drilled into the crust (in the Kola Peninsula of the former Soviet Union) penetrated 12 km, and modern techniques may be capable of still greater penetration, perhaps ultimately the entire 40 km thickness of the crust. A program to explore the entire crust must surely include numerous deep drill holes, but the cost of drilling is so great and the area sampled so limited that it is not realistic to expect that the entire crust can be fully explored in this manner.

The other major technique, and the one of special interest here, is sounding based on the reflection of seismic waves from buried rock–rock interfaces. This technique is often called "seismic reflection profiling," particularly when measurements are made along a line at the surface, but other terms, such as "3-D seismics," are sometimes used when a more elaborate form of surveying is followed. Though more complex in practice, seismic profiling is analogous in essence to echo sounding at sea. In seismic reflection profiling, seismic waves are generated at a particular location by either a buried explosion, or by an array of truck-mounted vibrators that shake the ground in radio-controlled synchronism (the Vibroseis technique of Continental Oil Co.). The returning echoes from buried features are typically sensed by an array of thousands of detectors, perhaps spread over a distance of many kilometers, and recorded in digital form. After measurement at one source point is completed, the entire operation is effectively moved along the line, like a giant caterpillar, for perhaps a few tens of feet, and then another observation is made. This process is repeated over and over again for the entire path of the study, which might be tens, hundreds, or thousands of kilometers in length. The signals so recorded are processed intensively and in very sophisticated fashion by computer to produce a sort of cross section of the rocks beneath the line. The cross section is then interpreted with other data in geologic fashion to provide the best estimate of buried geology in that section. An exciting prospect is that, once sufficient data are in hand, it will be feasible to provide virtual reality tours of the geology of the subterranean, somewhat like the Jet Propulsion Laboratory video flyby of the surface topography of Venus.

Most of the world's seismic reflection profiling is carried out by the petroleum industry as part of the search for hydrocarbons in sedimentary basins. In the mid–1970s, however, as a consequence of technique development and a few scattered and limited observations in West Germany, Australia, Canada, and the United States, it became evident that the industrial seismic reflection method for basin exploration could be applied on a much larger scale to explore, more or less systematically, the entire continental crust. This line of reasoning led to the formation of Consortium for Continental Reflection Profiling (COCORP), a National Science Foundation-sponsored activity that is a principal focus of this article. COCORP is headquartered at Cornell University in Ithaca, New York, but it includes numerous participants from other universities and government laboratories, and it enjoys close communication and some support from the petroleum industry. The chief purpose of COCORP is to explore the entire continental crust in the United States and elsewhere using state-of-the-art, industrial seismic reflection techniques, modified as appropriate for this specific task. COCORP develops strategies and tactics, plans and contracts with industry for seismic surveys, maintains a modern data processing facility, does routine and nonroutine processing, interprets the seismic results in conjunction with related geophysical and geological information, and distributes data and results to other interested parties. As of 1992, COCORP had collected some 12,000 km of seismic reflection data in the United States, including substantial fractions of two continuous transcontinental profiles.

COCORP results are so diverse and plentiful that they cannot be fully summarized in this brief article. Almost all COCORP surveys have produced new, important, and commonly surprising information on the buried crust. For example, COCORP observations resolved a long-standing controversy over the fundamental structure of the Appalachian Mountains of the eastern United States, revealing the former margin of the North American continent before it was hidden in the great collision with Africa that bulldozed rocks onto that margin and buried it. COCORP traced faults from outcrops at the surface to great depths in the Basin and Range province of the west, and also revealed there the layered and reflective lower crust and the nature of the Moho, or crust-mantle boundary. COCORP has contributed to the understanding of the Laramide uplifts of the west, detected mid-crustal bodies of molten rocks in regions of extensional geology, and found huge volumes of deeply buried and layered Proterozoic rocks in the mid-continent. At a number of places in the United States, COCORP has found zones of dipping reflectors that penetrate the entire crust and seem to mark the sites of ancient sutures, the scars of great collisions hundreds, or thousands, of millions of years old. COCORP has also begun to map the spatial pattern of variability of the Moho, a type of information certain to be important in understanding the crust. And in a few cases, though rarely, COCORP has found reflectors in the mantle beneath the Moho, normally a part of the earth far less reflective than the crust. One can well imagine the delight that will result when these enigmatic components of Earth that have only been sketchily resolved at present are identified and understood. For those enamored by the thrill of discovery, the days of Columbus, Mercator, and Ortelius, when only a few landfalls in the New World were known, seem a parallel to the present level of deep crustal exploration.

COCORP's early successes, beginning in the mid–1970s, have stimulated related activities in some twenty-five to thirty other countries, and in other organizations in the United States. British Institutions Reflection Profiling Syndicate (BIRPS) in Britain, Deutsches Kontinentales Reflexionesseismisches Program (DEKORP) in Germany, Etude de la Croûte Continentale et Océanique par Réflexion et Refraction Sismiques (ECORS) in France, Australian Continental Reflection Profiling (ACORP) in Australia, and LITHOPROBE (Canadian program for the study of the continental lithosphere) in Canada are some examples of non-U.S. efforts, each similar to COCORP in its mission, but each with its own distinctive style and specific goals. LITHOPROBE, for example, involves a variety of other techniques in addition to reflection profiling for surveying the crust. BIRPS is unusual in that it employs shipborne equipment for studying the continental crust beneath shallow seas, a situation characteristic of the water surrounding the British Isles. Early success by BIRPS has stimulated deep seismic studies in other water-covered areas, sometimes in places where the anomalous crust of continental margins, or island arcs, can be surveyed.

A current trend in deep seismic profiling concerns joint international projects with scientists and support from two or more countries. Cornell scientists who are normally involved in COCORP surveys in the United States have, for example, recently, and with National Science Foundation support, joined with scientists from Stanford, Syracuse, and Columbia and from the Chinese Ministry of Geology and Resources to conduct a survey to test deep seismic reflection profiling in Tibet, where the seismic crew must operate at an elevation of about 5 km. The Tibetan plateau, like the Himalayas a product of the great collision between India and Asia, presents one of the great problems of modern global tectonic research. The question is, how did the collision produce this vast plateau, the largest high region on Earth? The preliminary results of that first survey are somewhat spectacular for they seem to reveal very deep and hitherto unsurveyed scars of the collision that must be clues to its nature. Cornell scientists are also in the early stages of comparable surveys in the Andes, the Urals, and parts of the Middle East, all locales of classic geological problems, and scientists of other countries have also banded together to probe areas of major geological problems elsewhere.

Finally, if past experience with exploration of unknown parts of the earth is a guide, the exploration of the crust seems certain to produce information that will be beneficial to society in the future—perhaps the sites of mineral resources or hydrocarbons, or the sources of geothermal heat, or the causes of earthquakes and volcanoes, or the patterns of fluid flow, or any of a variety of other kinds of useful information. COCORP-style projects occupy a niche that fosters scientific discovery and that also complements, supports, and stimulates the related activities of industry, government, and academia that seek to provide for the society of the future.

Bibliography

BROWN, L. D. "A New Map of Crustal 'Terranes' in the United States from COCORP Deep Seismic Reflection Profiling." Ithaca, NY, 1991.

NELSON, K. D. "A Unified View of Craton Evolution Motivated by Recent Deep Seismic Reflection and Retraction Results." *Geophysics Journal International* 105 (1991):25–35.

ZAO, W., K. D. NELSON, J. CHE, J. GUO, D. LU, C. WU, X. LIU, L. D. BROWN, M. L. HAUCK, J. T. KUO, S. KLEMPERER, and Y. MAKOVSKY. "Deep Seismic Reflection Evidence for Continental Underthrusting Beneath Southern Tibet." *Nature* 366 (1993):557–559.

JACK OLIVER

SEISMIC TOMOGRAPHY

The term "seismic tomography" means mapping in two or three dimensions (3D) the departures of the actual speed of seismic waves from a reference (initial) model, which is most often one-dimensional. Seismic tomography is somewhat similar to medical tomography (computer-aided tomography or CAT), but instead of measuring the absorption of X rays, the seismologist measures the travel times of seismic waves generated mostly by earthquakes (*see* GEOPHYSICAL TECHNIQUES; PHYSICS OF THE EARTH).

Seismic tomography is practiced on very different scales, from exploration for oil, where the dimensions of the mapped region are measured in kilometers, to the planetary scale (global seismic tomography), in which the whole Earth is scanned. There is an intermediate scale in which a large region of particular interest, say, 4,000 by 4,000 km in horizontal dimensions and 1,000 km in depth, is mapped because we wish to image the anomalous velocities associated with subducted slabs, for example. This article will deal with mapping of the whole Earth. A popular introduction to the subject can be found in Anderson and Dziewonski (1984) or Dziewonski and Woodhouse (1987).

Plate 36 is an example of a model obtained using global seismic tomography. It shows a triangular cut into a recent earth model of the shear velocity anomalies in Earth's mantle. The surface is the top of the mantle (Mohorovičić discontinuity), and the bottom the core-mantle boundary. Seismic wave speeds higher than average are shown using green to blue colors, and slower than average are represented by yellow to red.

If the internal properties of Earth changed only with depth and not geographic location, our planet would be tectonically dead. Both short (earthquakes, volcanoes) and long timescale (mountain building, seafloor spreading) observations indicate that this is not the case. This dynamic behavior must be driven by lateral differences in temperature and density (*see* PHYSICAL PROPERTIES OF ROCKS). Unfortunately, the internal distribution of these parameters cannot be uniquely inferred from observations at the surface.

Seismic velocities decrease with increasing temperature: the inference is that the red areas are hotter than average and the blue, colder. Seismic wave speeds also vary with chemical composition, but there are strong indications that the thermal effect is dominant. Thus our picture can be thought to represent a snapshot of the temperature pattern in the convecting earth's mantle. Density is also a function of temperature. Material hotter than average is lighter and, in a viscous earth, will tend to float to the surface; colder material is denser and will tend to sink.

In addition to small-scale features in Plate 36, there is a distinct large-scale difference between the anomalous wave speeds under South America and the middle of the Pacific Ocean. The material under South America tends to be cooler than average at nearly all depths, while the mantle under the Pacific Ocean is hotter than normal. This may be indicative of the large-scale flow of the material in the mantle. There has been subduction under the western coast of South America for more than 100 million years (Ma), meaning that some 10,000 km of a cold lithosphere has been subducted into the mantle (*see* PLATE TECTONICS). The hot upwelling under the Pacific Ocean may be responsible for many of the anomalous properties of this region.

A 3D model of seismic velocities is obtained by solving an "inverse problem." The data are the observed travel time anomalies of seismic phases and the parameters describing the location and size of wave speed anomalies are the unknowns. For the solution to be informative, there must be a sufficient coverage of the mantle volume with criss-crossing paths. This is necessary to locate the sources of the observed travel time anomalies. If the data sampling a particular region exist only for one azimuth, the observed anomaly could arise anywhere along the ray path. Plate 37 shows "fast" and "slow" anomalies at some depth in the mantle; the rays that travel through it arrive earlier or later than normal; those that miss it have normal travel times. From such observations, it is possible to locate the anomaly. In actual applications, there are as many as several million observations for rays criss-crossing all the parts of the mantle.

To obtain a solution, the model must be represented by a finite number of parameters. In the example shown in Plate 36, this is achieved by introduction of basis functions that vary smoothly with geographical location and depth. The model is obtained by summation of many such functions with the coefficients determined by solving a system of equations with, often, several thousand unknowns. A frequently used alternative approach is to divide the medium into a 3D array of cells, with a constant velocity perturbation within each cell.

The reason that either of these two approaches yields stable solutions is that the lateral heterogeneity in the earth is strongly dominated by very large wavelength features. Thus, the bias that might be caused by ignoring the short-wavelength perturbations is not a serious problem.

Plate 38 contains two maps derived from a 3D model of the shear velocity anomalies obtained by Su, Woodward, and Dziewonski. The first map shows anomalies at a depth of 100 km. At this depth there is a strong relationship between the velocity anomalies and the tectonic features at the surface. The continents are faster than average by as much as 5 percent; this is expected because they are old and have had a long time to cool.

The mid-ocean ridges, such as the East Pacific Rise, are very slow, 4 percent below normal. The velocities increase with age of the oceanic crust, because of the effect of cooling on the seismic velocities. The depth to which the velocity anomalies associated with the mid-oceanic ridges continue is an important and, perhaps, a controversial issue. Several tomographic models indicate the minimum depth of 200 km, twice the thickness of the lithosphere obtained in the cooling plate model.

The transition zone (410–660 km depth) is dominated by a large wavelength signal, which may be associated with accumulation of the subducted material above the upper–lower mantle boundary—possibly a chemical or rheological obstacle.

The structure in the middle mantle is known the least; the amplitude of the variations is low and they appear to be rather evenly distributed as a function of the wavenumber. At about 2,000 km this changes, with progressively more and more power shifting toward the very large wavelengths. This is illustrated in the other map in Plate 38 showing anomalies near the core-mantle boundary. There is a distinct ring of higher than normal velocities circumscribing the Pacific Ocean and two very large regions of intense negative perturbations: one in the Pacific, the other under Africa. The positive anomalies have been associated with the past subduction; thus, it is possible that the material subducted in the ocean trenches finds itself at the bottom of the mantle after tens of millions of years.

The pattern of variations of the anomalies with depth is in qualitative agreement with the predictions of a mantle convection model that considers the effect of phase transformations in the upper mantle. This model predicts the "avalanche" effect—a sudden flux of cold subducted material accumulated in the transition zone, which occurs once its mass is large enough for it to penetrate the chemical boundary layer.

Plate 39 shows the very largest scale features in a 3D view centered on the Pacific Ocean. The large upwelling in its center is the dominating feature. The ring of high velocities around the Pacific basin is also distinctly seen. It is difficult not to believe that this large-scale thermal structure has a profound influence on the overall dynamics of the earth. Yet, even though its existence has been known since the mid–1980s, there is no explanation of its possible role in terms of modeling the mantle convection.

There is an important distinction between global seismic tomography and medical tomography, for example, in which the location of sources and receivers can be chosen according to the objective. In global seismic tomography we are limited by the distribution of globally detected earthquakes and by the location of the seismographic stations. There is not much that can be done about the distribution of seismicity, except that now and then an earthquake occurs in an unusual place, so the coverage is expected to improve with time. Generally, the earthquake distribution is more even in the northern hemisphere.

Our ability to refine the 3D models of the mantle depends strongly on the coverage of the earth with receivers. At this time, there are no permanent ocean bottom seismographic stations, so the locations of the receivers are limited to the continents and islands. Significant progress has been made in recent years in the deployment on land of state-of-the-art seismographic instrumentation. A concept of the Ocean Seismographic Network has been under development since the early 1980s. Several countries are building instrumentation and carrying out testing of the noise levels at the ocean floor and in holes drilled by the international Ocean Drilling Project.

It has been believed since the time of publication of the first large-scale global seismic tomography study that 3D images of lateral heterogeneity will be an essential tool in solving some of the fundamental problems of geodynamics. The results accumulated since then confirm that statement. In what follows, examples are given of the application of global seismic tomography to various problems in earth sciences.

Mantle Convection

Distribution of seismic anomalies represents the current configuration of the thermal and compositional heterogeneity advected by mantle flow and it imposes a complex set of constraints on the possible modes of convection in the mantle. Plate 40 shows a polar cross section through the same earth model as in Plates 36 and 38. Clearly, the dominating features from the core-mantle boundary to mid mantle are the two slow and two fast regions indicative of a very large-scale convection pattern.

Mineral Physics

The in situ ratio of the relative perturbations in the shear and compressional velocities inferred for the lower mantle from global seismic tomography was much higher than determined, at relatively low pressures, in the laboratory. Only recently a generally accepted explanation of this fact has been provided.

Gravity and Rheology

Assuming seismic anomalies are proportional to density perturbations (a reasonable assumption), the seismic anomalies provide constraints on the modeling of large wavelength gravity anomalies, the viscosity distribution in the mantle, and the ratio of relative perturbations in density and seismic velocity.

Petrology and Geochemistry

The global seismic tomography models have the potential to provide integral constrains on petrological and thermal models of the ridge systems. Velocity anomalies associated with the continental shields confirm the hypothesis of "continental roots" (*see* CONTINENTAL CRUST, STRUCTURE OF). The cross sections through the upper mantle of North America, Africa, and Europe (Plates 36 and 40) show that positive velocity anomalies extend to 400 km or so.

Correlation has been noted between low velocity anomalies near the core-mantle boundary, such as seen in Plate 38 (bottom), and the occurrence of large-scale isotopic anomalies.

Geomagnetism

Coincidence of the regions with the high rate of secular variations in the magnetic field and slow velocity anomalies (higher than average temperature) provide constraints on the thermal and mechanical coupling between the mantle and the core.

Recent studies have pointed out that the virtual geomagnetic pole paths coincide with the two high-velocity regions circumscribing the Pacific and, effectively, connecting the north and south poles (*see* Plate 38, bottom).

Geodesy

Global seismic tomography models of velocity anomalies and the topography of major discontinuities in the earth's structure (the core-mantle boundary; 660 km discontinuity, for example) can be compared with the data on the earth's rotation obtained by very-long-baseline interferometry.

Knowledge of the 3D variations in seismic velocities allows more refined studies of the deepest parts of the earth's interior because corrections can be made for the distorting effect of the mantle structure. For example, Morelli, Dziewonski, and Woodhouse discovered in 1986 that even after correcting for the effects of mantle heterogeneity on the travel times of the waves traversing the inner core, there remained residuals as large as ±2 seconds. The indication was that the waves, which traverse the inner core in the direction parallel to the rotation axis, travel faster than those propagating in the equatorial plane.

The researchers failed to explain these residuals by lateral variations in the inner core, and after exhausting all alternatives, they concluded that the material in the inner core must be anisotropic. That is, the waves propagating in different directions travel with different speeds. This is known to occur in many crystals. The form of anisotropy inferred for the inner core exists in crystals with the hexagonal symmetry. This symmetry is predicted for iron, the main constituent of the inner core, at the pressure appropriate for this region (*see* CORE, STRUCTURE OF). A mystery remains concerning what physical process caused the crystals of iron to align, such that the inner core, with a radius of 1,220 km, appears to be a single crystal.

Bibliography

ANDERSON, D. L., and A. M. DZIEWONSKI. "Seismic Tomography." *Scientific American* 251, no. 4 (1984):60–68.

DZIEWONSKI, A. M., and J. H. WOODHOUSE. "Global Images of the Earth's Interior." *Science* 236 (1987):37–48.

DZIEWONSKI, A. M., B. H. HAGER, and R. J. O'CONNELL. "Large Scale Heterogeneity in the Lower Mantle." *Journal of Geophysical Research* 82 (1977):239–255.

MORELLI, A., A. M. DZIEWONSKI, and J. H. WOODHOUSE. "Anisotropy of the Inner Core Inferred from PKIKP Travel Times." *Geophysical Research Letters* 13 (1986):1545–1548.

ADAM M. DZIEWONSKI

SHEPARD, FRANCIS P.

Francis Parker Shepard, considered one of the fathers of marine geology, worked for fifty years at the Scripps Institution of Oceanography, now part of the University of California at San Diego. His teaching, his research on continental shelf sedimentation and submarine canyons, and his books, particularly *Submarine Geology* (1973), played an important role in elevating geological oceanography to parity with its sibling disciplines, physical, chemical, and biological oceanography. This parity passes unnoticed today, yet the first great summary of oceanography, *The Oceans* (Sverdrup et al., 1942), gives short shrift to the seafloor and the processes that shape it.

Shepard's careful studies of continental shelves and their sediments showed the influence of changes in sea level due to the waxing and waning of Ice Age glaciers on shallow-water geological processes and deposits. His early work dispelled the notion that sediments become progressively finer grained with distance from the coast. His efforts culminated in the famous "API-51" project, in which support from the American Petroleum Institute allowed Shepard and his associates and students to study the modern sediments of the northern Gulf of Mexico from 1951 to 1958. The results (Shepard et al., 1960) have helped geologists understand the oceanographic processes that formed ancient sedimentary rocks which today can be seen on land.

The development of echo sounders during the 1930s, and their marked enhancement during World War II, provided the essential tool that Shepard needed to study submarine canyons. The largest of these features, such as the Monterey Canyon off central California, dwarf the Grand Canyon. Shepard spent most of his career from1936 on studying the size and shape of canyons, their origins, and the way that they funnel sediment from the shoreline into the deep sea. His work resulted in numerous articles and books (such as Shepard and Dill, 1966) that still represent much of our knowledge of these features.

Submarine Geology, initially published in 1948, was the first attempt to provide a comprehensive description of the geology of the seafloor from the shoreline to the deep-sea trenches. A competing text, *Marine Geology* by Phillip Henry Kuenen (1950), appeared a short time later. These two books formed the basis for most geological oceanography classes during the subsequent thirty years. Shepard revised his text in 1963 and 1973.

Shepard was born on 10 May 1897 in Brookline, Massachusetts. He earned his B.A. at Harvard in 1919 and his Ph.D. in structural geology at the

University of Chicago in 1922. He began his career at the University of Illinois, but spent most of his time at the Scripps Institution from 1935 on. From 1941 to 1945, he worked for the University of California's Division of War Research. Following the war, he joined Scripps full time, where he taught until his retirement in 1964.

This pioneering scientist, who died in San Diego, California, on 25 April 1985, can fairly be described as the first American marine geologist. He trained many of the second generation "stars" of the field. Robert Sinclair Dietz, Kenneth Orris Emery, and Joseph R. Curray were among the students who carried on his work.

Bibliography

KUENEN, P. H. *Marine Geology.* New York, 1950.

MILLER, R. L. *Papers in Marine Geology; Shepard Commemorative Volume.* New York, 1964.

SHEPARD, F. P. *Submarine Geology,* 3rd ed. New York, 1973.

SHEPARD, F. P., and R. F. DILL. *Submarine Canyons and Other Sea Valleys.* Chicago, 1966.

SHEPARD, F. P., F. B. PHLEGER, and T. H. VAN ANDEL. *Recent Sediments, Northwest Gulf of Mexico.* American Association of Petroleum Geologists. Tulsa, OK, 1960.

SVERDRUP, H. O., M. W. JOHNSON, and R. H. FLEMING. *The Oceans.* New York, 1942.

G. ROSS HEATH

SILICATES

See Minerals, Silicates

SILVER

Silver is a bright, shiny, white metal, with the atomic number 47, and an atomic weight of 107.88; the chemical symbol is Ag, from the Latin *argentum.* It has a density of 10.5 grams/cubic centimeter (g/cm^3), and a melting temperature of 960°C. The crystal structure of silver is face-centered cubic, identical with that of gold, and at room temperature and above the two metals are miscible in all proportions. Because silver is durable, attractive, relatively rare, and easily worked, it is usually classed as a precious metal.

Uses

Silver was one of the first metals known and used by mankind. In small amounts the element can be found in its native, or metallic, form. Native metal fragments were worked into ornaments and utensils as early as seven thousand years ago. Silver also has a long history of use as a monetary medium. It has been used for trading and as the basis of wealth for more than four thousand years, but in the later 1960s the United States and most other countries stopped minting silver coins, thus bringing silver's long-time monetary use to an end, except for occasional commemorative coins. The cessation of silver coinage was occasioned by the combined coinage, industrial, and technical demands for silver exceeding the rate of new production.

Silver is an extremely valuable, even essential, industrial metal because it has the highest electrical and thermal conductivities of all metals and is therefore important in the electronic and electrical industries. Silver is also photosensitive, forming the basis of much of the photographic industry; in addition it is corrosion-resistant, readily electroplated, and an important alloying metal. Besides its many technical and industrial uses, silver is widely used in jewelry, cutlery, and many household goods. Approximately 15,000 tons of new silver are mined each year.

Occurrence

As previously noted, silver is completely miscible with gold; it is not common for either metal to be found pure in their native metal form. In fact, only a small fraction of mined silver occurs in the native form; considerably more is recovered from the minerals argentite (Ag_2S), proustite (Ag_3AsS_3), and pyrargyrite (Ag_3SbS_3). The most important silver-bearing minerals are members of the tennantite-tetrahedrite solid solution series ($Cu_{12}As_4S_{13}$ and $Cu_{12}Sb_4S_{13}$, respectively) in which silver replaces some of the copper in the atomic structure.

Deposits mined largely or entirely for silver are mostly epithermal veins. Such deposits consist of quartz-rich veins, veinlets, and disseminations, usually emplaced in volcanic rocks, formed by deposition from hydrothermal solutions, generally at temperatures below about 275°C. The source of the silver in epithermal veins and the hydrothermal solutions is shallow bodies of magma of felsic-to-intermediate composition. The Andean volcanic province, together with the Mexican and western North American volcanic provinces, hosts the world's largest and richest epithermal silver deposits.

A predominant amount of newly mined silver is recovered today as a by-product. Silver-bearing tennantite and tetrahedrite are present in small amounts in many copper, lead, and zinc deposits. Because large quantities of copper, lead, and zinc are mined, the amount of by-product silver is also large even though the silver content may be no more than a gram per ton. By-product silver is produced in large amounts from the mining of many porphyry copper deposits, as in Bingham, Utah, from the great stratiform lead-zinc deposits, as in Broken Hill and Mount Isa in Australia, and at most massive sulfide deposits, such as Kidd Creek in Ontario.

The average silver content of the earth's crust is low, only 0.000008 percent by weight, but even so silver ore deposits are widespread—fifty-five countries reported the production of newly mined silver in 1993. The largest producers, in order, are Mexico, the United States, Peru, Canada, Australia, the former Soviet Union, and Poland; each country produces in excess of 800 tons of silver per year and collectively the seven countries account for 75 percent of the annual world production.

Silver was mined by many early civilizations, especially by the Greeks and Romans, but production was always low and silver was a rare metal until the European invasion of the Americas. Spanish conquistadores looted the vast silver stocks of the Incas and Aztecs and then forced the conquered peoples to mine new silver at an increasing rate. The flow of silver from the Americas brought previously unimagined wealth to the courts of Europe. Silver production rose steadily from about the year 1520 to 1800, then declined slowly until the rich discoveries of the western United States in the 1860s once again boosted production. Until the beginning of the twentieth century most silver was produced as a primary product, but throughout the twentieth century, by-product silver production has steadily increased. It is likely that in the future by-product silver will become ever more dominant.

Bibliography

GUILBERT, J. M., and C. F. PARK, JR. *The Geology of Ore Deposits.* New York, 1985.

KESLER, S. E. *Mineral Resources, Economics and the Environment.* New York, 1994.

BRIAN J. SKINNER

SIMPSON, GEORGE GAYLORD

The American paleontologist George Gaylord Simpson was born in Chicago, Illinois, on 16 June 1902. Simpson was educated at the University of Colorado and Yale University, receiving his Ph.D. from the latter institution in 1926 for a dissertation on Mesozoic mammals. He joined the staff of the American Museum of Natural History in New York in 1927, and also held a concurrent faculty position at Columbia University. His initial appointment was as an assistant curator, but by 1944 he was chairman of the Department of Geology and Paleontology at the American Museum. In 1959 he moved to Harvard University as Agassiz Professor of vertebrate paleontology at the Museum of Comparative Zoology, a position he retained until 1970. In 1967 Simpson was appointed a professor of zoology at the University of Arizona in Tucson, a position he resigned from in 1982, two years before he died.

The core of Simpson's lifetime work was research on the evolution and classification of fossil mammals, although he did occasionally delve into other topics such as the evolutionary history of penguins. He pursued museum and fieldwork throughout North America and other parts of the world, including Brazil and Mongolia. An early trip to Patagonia formed the basis for a popular travelogue, *Attending Marvels: A Patagonian Journal* (1934). Simpson used his detailed research on fossil organisms to help establish the Neo-Darwinian orthodoxy of the 1930s and 1940s, which, for the first time, brought together and synthesized the then somewhat disparate fields of population genetics, chromosomal studies, evolutionary theory, and paleontology.

Simpson arrived at the American Museum of Natural History in the mid–1920s, just as the highly influential vertebrate paleontologist Henry Fairfield Osborn was ending his career there. But while Osborn had promulgated non-Darwinian theories of evolution, Simpson studied with and assisted the Darwinian vertebrate paleontologist William Diller Matthew. Simpson was also heavily influenced by the work being pursued by the geneticists at Columbia University and elsewhere in the 1930s. Early in his career Simpson established his reputation by making many fundamental contributions to early mammalian paleontology, from basic systematic analysis and faunal descriptions, to sweeping large-scale studies of evolutionary change through time. But perhaps more important, he also addressed larger issues in paleontology and evolution. With Simpson, and greatly influenced by contemporary evolutionists such as Ernst Mayr, Julian Huxley, and Theodosius Dobzhansky, vertebrate paleontology was brought fully into the mainstream of Neo-Darwinian evolutionary thinking.

Simpson was a strict Darwinian (*see* DARWIN, CHARLES) and wrote many works integrating Darwinian natural selection with the larger-scale evolutionary patterns observed in the vertebrate paleontological record. Among his classic works along these lines are *Tempo and Mode in Evolution* (1944; rewritten as *The Major Features of Evolution*, 1953), *The Meaning of Evolution* (1949), and the popular case study *Horses: The Story of the Horse Family in the Modern World and through Sixty Million Years of History* (1951). Simpson was extremely interested in the problem of variation within and among species, and the need to pursue statistical and biometrical analysis of organisms. This interest resulted in, among other works, *Quantitative Zoology* (1939), coauthored with his wife, Anne Roe. (The book was revised and added Richard Lewontin as a third author in 1960.) Simpson was also deeply concerned with the principles of classification, a necessary component to reconstructing evolutionary lineages (see, for instance, his *Principles of Classification and a Classification of Mammals*, 1945), as well as problems of functional morphology, paleoecology, and historical biogeography (examples of which appear in his collected volume of papers entitled *Why and How: Some Problems and Methods of Historical Biology*, 1980).

Perhaps Simpson's single most important contribution was to demonstrate that the vertebrate fossil record was compatible with the Neo-Darwinian evolutionary theory that was developing in the 1930s and 1940s (*see* LIFE, EVOLUTION OF). Simpson was a driving force in promoting cross-fertilization between the fields of paleontology, genetics, and evolutionary theory based on living organisms. A seminal contribution to such interdisciplinary studies was the volume of collected papers edited by Glenn L. Jepsen, Ernst Mayr, and Simpson entitled *Genetics, Paleontology and Evolution* (1949). Simpson was closely associated with the development of the Neo-Darwinian "synthetic theory" of evolution of the 1940s (see in particular Julian Huxley's 1942 book *Evolution: The Modern Synthesis*). With Simpson, vertebrate paleontology and evolutionary theory came of age.

Simpson, however, was not infallible. For instance, for two decades Simpson used paleobiogeographical evidence to argue against the direct continental connections that were postulated by early continental drift theorists, such as Alfred Wegener and his followers (*see* WEGENER, ALFRED). In the late 1960s and early 1970s, however, Simpson did finally acknowledge that "continental drift" (now in the form of plate tectonics) was real (*see* PLATE TECTONICS). To give one other example, at the end of his life Simpson took a somewhat adversarial position against the new "phylogenetic systematics" (cladistics) that revolutionized phylogeny reconstruction and organismal classification among paleontologists. Simpson died on 6 October 1984 in Tucson, Arizona.

Simpson's forte was the written word. It was through publications that he primarily communicated with his colleagues. He was not known as a good teacher. He did not particularly like to lecture, or even speak about serious or scientific subjects in an informal conversational setting. Throughout his career, however, he received many honors and awards for his contributions to science, including the Penrose Medal of the Geological Society of America in 1952.

Bibliography

LAPORTE, L. F. "Wrong for the Right Reasons: G. G. Simpson and Continental Drift." In *Geologists and Ideas: A History of North American Geology*, eds. E. T.

Drake and W. M. Jordan. Geological Society of America Centennial Special Volume 1. Boulder, CO, 1985, pp. 273–285.

OLSON, E. C. "Memorial to George Gaylord Simpson, 1902–1984." *Geological Society of America Memorials* 16 (1986):1–6.

SIMPSON, G. G. *Tempo and Mode in Evolution.* New York, 1944.

———. *The Major Features of Evolution.* New York, 1953.

———. *Concession to the Improbable: An Unconventional Autobiography.* New Haven, CT, 1978.

———. *Fossils and the History of Life.* New York, 1983.

ROBERT M. SCHOCH

SINKHOLES

See Caves and Karst Topography

SOIL DEGRADATION

Soils are the major medium of plant growth on the world's land surfaces, illustrating their importance in terrestrial ecosystems and as a critical natural resource. Soils form at the interface of the lithosphere, atmosphere, hydrosphere, and biosphere through a large number of physical processes, such as shrinkage and swelling, and chemical processes, such as chemical weathering. With time, these processes ultimately promote the formation of a soil profile, which consists of a sequence of soil "horizons," layers that differ biologically, mineralogically, texturally, and chemically from the soil parent material (the earth material in which the soil initially began to develop, e.g., granite, dune sand, alluvium).

Some Causes and Processes of Soil Degradation

Soil-forming processes (*see* SOILS, FORMATION OF; SOIL TYPES AND LAND USE) are countered by many processes of soil degradation. As with soil-forming processes, degradational processes are also physical and chemical. Some are naturally caused, such as erosion associated with rivers. Many others, however, are largely induced by humans (anthropogenic). Widespread removal of vegetation through clearcut timbering or by human-caused fires may greatly increase surface runoff and soil erosion as well as diminish soil organic matter content. It is occasionally difficult to separate natural from anthropogenic causes of soil degradation. Changes in stream behavior, exemplified by changes in meander pattern and migration or by extensive arroyo formation, have been attributed to changes in the pattern of stream discharge linked, for example, to urbanization. Changes in the frequency and patterns of forest fires have also been related to human activities. Other researchers, however, see such changes as being related to relatively recent changes in climate documented through detailed analysis of meteorologic and other climate proxy records.

Some processes of soil degradation are unambiguously anthropogenic. Examples include soil salinization related to irrigation in arid and semiarid regions, acidification due to acid rainfall, long-term nitrogen fertilizer additions, chemical spills and mine tailings runoff, improper practices of dumping, transport, or storage of liquid and solid waste, and changes in vegetation that alter soil chemistry. Increasing urbanization causes soil degradation by increasing contaminated runoff into soils, removing soils through excavation or building of parking lots and highways. Improper agricultural practices or heavy grazing pressure in sensitive soil-landscapes can promote erosion and compaction of the soil A horizon (the typical surface horizon of soils), which in turn increasingly exposes the often clay-enriched soil B horizon. Because the B horizon usually favors slower water infiltration compared to the A horizon, its near surface exposure typically causes increased surface runoff and erosion via sheetwash and rilling. As more of the A horizon is eroded, even more surface runoff occurs, accelerating erosion of the B horizon. This process constitutes a type of negative system feedback response, which once initiated, can be reversed only with great difficulty.

Estimates of soil losses or decreases in productivity vary significantly. Differences in estimates vary, partly related to differences in methods of measurement (e.g., total suspended load of rivers, or erosion-pin analysis). Nevertheless, regardless of the methods used, soil losses may be profound. For example, soil erosion from some agricultural areas may be as large as 300 tons per hectare per year (Hausenbuiller, 1985). Such depletion of the soil resource through erosion or alteration is associated with huge economic losses and exacerbates economic and geopolitcal problems. U.S. Department of Agriculture studies indicate that the economic costs of soil erosion are as high as 10 billion dollars annually; studies in Canada suggest annual economic costs approaching 1.3 billion dollars. Soil degradation is especially harmful in third world countries where increasingly larger populations result in overutilization of the soil resource or degradation associated with removal of vegetation by slash-and-burn methods. Also, insufficient capital is available to support costly soil rehabilitation. Upsetting the delicate balance in the soil-landscape-vegetation association in regions with semiarid climates has promoted widespread desertification and soil loss. This has significantly affected Subsaharan Africa, but is also occurring in the western United States.

From the earth scientist's perspective, the long-term impacts and consequences of these soil losses may be even more significant than presumed by many urban planners, politicians, and economists. Key soil horizons in arable, nutrient-rich soils include (1) certain types of A horizons, such as the mollic horizon (dark, organic-matter-rich and base-rich mineral surface horizons) of the soils of world's areal extensive grasslands of the plains and steppes; (2) B horizons, such as the argillic (clay-rich) horizons of soils that primarily form in regions subject to strongly seasonal climate that typifies many of the most agriculturally productive regions. These are the parts of soils that have chemical and mineralogical composition and structure that promote agricultural productivity. Important plant nutrients are available as exchangeable cations easily taken up by roots from exchange sites of organic humic compounds in the A horizon or silicate clay minerals of the A and B horizons. Soil structure promotes deep, even infiltration as clay and organic matter constituents act to bind soil particles into aggregates, or granular or subangular peds. The presence of such structure facilitates soil moisture movement, storage, and aeration. Evaluating the overall consequences of degradation of productive soils requires not only evaluation of the costs entailed in mitigating effects of soil degradation associated with contamination or salinization, but also an understanding of how long it will take for such soils to reform subsequent to their total degradation. Loss of nutrients present only in the organic-matter-rich upper horizons of tropical soils, for example, is essentially irreversible. An understanding of how fast and through what processes such horizons form is critical. Knowing this ultimately requires an appreciation of soil-landscape relationships.

Geologic Studies of Soil Development and Soil Degradation

If soils are lost through degradation, a few critical questions must be addressed. How long does it take to form soils? How is the rate of soil development influenced by the climate in which the soil is forming? These questions are not easily answered. Materials are rarely found in soils that allow direct determination of their age. More important, a soil is a highly complex system, as noted above, characterized by complex interactions between numerous and diverse processes acting simultaneously that, collectively, create the soil. It is difficult, if not wholly impossible, on the basis of even detailed, rigorous examination of a given soil, to determine how a given factor, such as climate, has influenced certain soil-forming processes over its time of formation.

A solution to this problem is provided by an approach mostly developed by Hans Jenny (1941; 1980). One of the foremost pedologists of the twentieth century, Jenny relied on pedologic research and ideas of eighteenth- and nineteenth-century Russian soil scientists, such as Lomonsov and Dokuchaev, proposing a method through which the relationship between a given soil-forming factor, such as parent material or climate, and certain aspects of soil development (e.g., clay content, horizon thickness) could be developed. This method entails selection of a group of soils in circumstances deliberately selected to isolate the impacts or effects of a given factor and to minimize influences of the remaining factors. The major factors—climate (cl), organisms (o, biotic factors), relief (r, topography, aspect), parent material (p), and time (t, soil age)—control or influence the rates and magnitudes of most soil-forming processes. For example, the amount of clay movement (illuviation) in a soil is primarily related to climate as the process of illuviation is driven largely by infiltrating soil water. Clay illuviation is also related to clay mineral type, which reflects another factor, parent material. An equation, referred to as Jenny's Fundamental Equation of Soil Development, can be written that relates properties or the overall state of soil development, S, to these factors: $S = f(cl, o, r, p, t)$.

A simple example shows how the factorial approach works. A group of soils on different parts of glacial moraine slopes exhibit differing soil properties. These differences should have little or nothing to do with factors such as climate, parent material, biota, or soil age as all of the soils must have been similarly influenced by these factors subsequent to soil development that began at the time of glacial retreat and moraine abandonment. Therefore, the observed change in soil properties must be attributable to slope position, and accordingly a quantifiable "topofunction" from this soil toposequence is derived. So, studies of these soils (Birkeland and Burke, 1988) can now focus on how soil properties might be influenced by slope position in a landscape (e.g., windward side or lee side, shoulder or midslope of moraine).

There are certain limitations to the factorial approach, and it has been strongly criticized by some researchers. Yet, the approach has supplied a philosophical methodology that provides a framework for investigating and formulating hypotheses for genesis of complex soils in diverse circumstances (Harden and Singer, 1994). Many studies that have utilized this approach have had great success, for example, in demonstrating how and why soil development varies in different regions and has been impacted by climatic change through geologic time. Primarily through soil chronosequence studies, the factorial approach has had an especially large impact in geopedological and geomorphic research. These studies have enabled the use of soils as a powerful tool to determine landscape ages, study landform evolution, and correlate land surfaces and stratigraphic units in both recent and ancient terrestrial sedimentary deposits.

Insights into Soil-Forming Processes from Chronosequence Studies

The most favorable circumstances for chronosequence studies occur in landscapes in which flights of abandoned terraces of different ages have formed. Such landscapes form when a trend of overall lowering of regional base level has occurred during recent geologic time, punctuated occasionally by periods of channel aggradation. Landscapes throughout much of the western United States have formed in this manner. In some cases, as many as ten fluvial terraces can be identified that may range in age from only decades to over a million years. In such landscapes, soils formed on terrace surfaces in much the same kind of parent material and the same topography. Differences in present vegetation cover are typically minimal (although vegetation may have varied in the past owing to global climatic changes). Thus variation in soil development on the terraces is not related strongly to climate and biotic factors, but chiefly reflects soil age.

Systematic temporally dependent increases in soil development on older terraces are shown especially well by increases in reddening (associated with increasing ferric oxide content) and total soil clay content. For example, soil chronosequence studies in the deserts of southern New Mexico show conclusively that the morphology and content of soil calcium carbonate-bearing horizons are closely related to soil age, and that the source of virtually all of the calcium in the pedogenic carbonate is incorporated eolian dust (Gile and Grossman, 1979). Many subsequent studies, based on mass-balance modeling and isotopic analysis of trace elements in soil-formed materials, also conclude that dust is a significant source of many important soil materials, even in strongly chemically weathered soils of tropical climates. Chronosequence studies are being greatly enhanced by the development of several new numerical dating techniques (i.e., cosmogenic surface age determination) that will enable age determination of many more soils and provide calibration of rates of soil development in diverse geographic settings. Better age determinations for soils and their associated landforms and deposits will help to distinguish soil property changes related to soil age from those associated with past changes in climate during Quaternary.

Chronosequence research demonstrates that many properties of soils form at different rates in different climates, a conclusion that has important implications for studies of soil degradation. In most circumstances, on geomorphically stable surfaces, at least several centuries are required to form thick, dark A horizons, and a minimum of thousands of years usually are required to form B horizons. In addition, on such stable surfaces soil formation is a strongly accumulative process over the long-term, not erosional. Processes of dust incorporation, accompanied in many cases by deposition from sheetwash, slowly add mass to soil, causing slow upward growth of soils. This process implies that the increasingly thinner A horizons observed on many old geomorphic surfaces are not due to erosional thinning (long-term degradation). Instead, the A horizon's decreasing thickness reflects its gradual replacement by an increasingly thicker, less permeable, reddish clay-rich B horizon.

Soils are present over extensive areas of landscapes on geomorphic surfaces that did not exist before 10,000 years ago, prior to the end of the last Ice Age, the Pleistocene epoch. In some landscapes, such as the midwestern United States, postglacial, fine-grained eolian deposits, called loess, blanket virtually all of the landscape and are several meters thick. In other more arid regions of the United States, eolian sand dunes blanketed landscapes during the end of, or following the end of, the last glacial period. This extensive loess and sand deposition occurred directly in response to the changes in climate that occurred at the end of the Pleistocene. Soils supporting extensive grasslands formed once deposition ceased and stable surfaces formed. Also, in other parts of the United States, but especially throughout much of the west, postglacial climatic changes caused the transportation of huge amounts of sediment onto the surfaces of adjacent piedmonts and intermontane basins (Bull, 1991). After abandonment of these fans and terraces by active river channels, geomorphic stability promoted soil development. On landscapes older than three to four thousand years in diverse parent materials and climates, moderately well developed soils with thick A horizons and at least moderately well developed B horizons have formed across widespread areas of North America. These geologically young, typically base-enriched soils are generally quite fertile and constitute the majority of agriculturally utilized landscapes in many regions.

Conserving the Soil Resource

Many processes degrade soils—some are natural and can occur quite slowly or rapidly. Many more, however, are anthropogenic and can very rapidly degrade the soil in a period of decades, years, or even shorter periods of time. Most moderately developed soils in the landscape require at least a few millennia to form, so once degraded significantly, they will require millennia to reform. Also, conversion of once geomorphically stable landscapes that were conducive to soil development to rilled or deeply channeled landscapes will further increase the time required to reform soils. Rehabilitation measures and environmental restoration are costly, are often relatively inefficient, and may be unfeasible in some circumstances. Moreover, distribution of many plant communities is strongly associated with soils, and thus degradation of many soils may impede reestablishment of certain species, or perhaps even seriously impact some ecosystem's biological diversity at a local level, and perhaps ultimately globally. Complicating matters, significant "greenhouse warming" of Earth's climate, which many climatologists predict will occur in the next century, will surely impact certain soil-forming processes in currently agriculturally important regions (*see* GLOBAL CLIMATIC CHANGES, HUMAN INTERVENTION). Will these impacts be negative, causing soil degradation? How fast will processes of soil degradation, if they occur, operate? These questions demonstrate the importance of increasing our understanding of soil-forming processes, which, as discussed above, can be aided significantly through geomorphic studies of soil–landscape–climate linkages. Finally, such research will provide a better understanding of how to manage and conserve Earth's remaining soil resources. More efficient and effective conservation of this critical resource may be as important as any other strategy to ensure the sufficiency of global soil resources to meet humankind's future needs.

Bibliography

BIRKELAND, P. W., and R. M. BURKE. "Soil Catena Chronosequences on Eastern Sierra Nevada Moraines, California, U.S.A." *Arctic and Alpine Research* 20 (1988):473–484.

BULL, W. B. *Geomorphological Responses to Climatic Change.* New York, 1991.

GILE, L. H., and R. B. GROSSMAN. *The Desert Soil Project Monograph.* U.S. Department of Agriculture, Soil Conservation Service. Washington, DC, 1979.

HARDEN, J., and M. SINGER. *Factors of Soil Formation: A Fiftieth Anniversary Retrospective.* Soil Science Society of America Special Publication 33. Madison, WI, 1994.

HAUSENBUILLER, R. L. *Soil Science.* Dubuque, IA, 1985.
JENNY, H. *Factors of Soil Formation.* New York, 1941.
———. *The Soil Resource—Origin and Behavior.* New York, 1980.

LESLIE D. MC FADDEN

SOIL TYPES

A record of how regolith has been transformed to soil becomes visible, in most places, when a pit is dug to a depth of 3 or 4 meters (m) or when the face of a road cut is cleaned with a spade. Here a sequence of several soil layers can be distinguished. This sequence is the soil profile and the individual layers are soil horizons. Together they form the basis for soil classification. Horizons are usually, but not always, more or less parallel with the surface of the ground. Where erosion or some other disturbance has occurred, soil horizons may become discontinuous or mixed with each other.

The horizon sequence commonly begins with a layer of organic matter on the surface that is called the O horizon. In older literature this is called the litter layer. The O horizon is composed of biomass that has fallen to the surface from the aboveground ecosystem, both plant and animal. This horizon is extremely important in wild ecosystems because it is the seedbed for successional plants that follow; it moderates and regulates the microclimate of the soil profile, especially with respect to moisture and temperature, and the nature of the O horizon is closely related to the species and population density of heterotrophic decomposers in the profile. The forest floor layer in many coniferous forests, some broadleaf forests, and most heathlands is a well-defined, unincorporated matted layer that rests on the mineral soil below it. A high proportion of the soil animal activity is focused in this layer.

The O horizon is subdivided into three zones: Oi, Oe, and Oa. Corresponding layer designations in the older literature are L, F, and H. The Oi layer is composed of fallen biomass that is easily recognizable with respect to origin. Decomposition has just begun. In the Oe layer the origin of the biomass is partly identifiable but decomposition is well under way. In the Oa layer the biomass is an amorphous, dark-colored residue of biomass resting on the surface of the mineral soil.

In temperate-zone forests composed of deciduous tree species and in many tropical forests, litter and mineral soil are mixed by soil animal activity. In this case the forest floor consists of scattered litter with exposed areas of dark-colored mineral soil high in organic matter. Here earthworms and other large heterotrophic decomposers are abundant. Usually the litter involved is higher in basic cations such as calcium and magnesium and, as a result, the upper few centimeters of the profile are less acidic than the upper part of the profile where the O horizon is present. In grassland soils the dark color is associated with the depth of root penetration and characteristic biological activity.

Below the O horizon, if all mineral soil layers are present, are the A, E, B, and C horizons in order of increasing soil depth. Boundaries of mineral soil horizons are not always distinct. Separation of one horizon from the one above or below may require a great deal of experience and close scrutiny. For this reason each of the mineral horizons may be further subdivided to describe subordinate distinctions.

The A horizon is the uppermost mineral soil horizon. It contains varying amounts of humified organic matter intimately mixed with the mineral soil fraction. This mixture is usually darker in color than the lower horizons. In wild soils not disturbed by human activity the lower boundary of this horizon is frequently wavy, showing considerable depth variation due to differential penetration of water caused by surface topographic features. Where this is the case, incorporation and mixing of organic matter with mineral soil are ascribed to a combination of soil animal activity and/or variation in depth of root penetration, especially in grasslands. In cultivated soils the lower boundary of the A horizon is more nearly a straight line that marks the deepest penetration of cultivation activity, and where this feature is present, it is designated as an Ap horizon.

The E horizon is characterized by maximum leaching loss (eluvial) of basic cations, clay, and iron and aluminum oxides. This results in higher concentrations of quartz and other resistant minerals in the sand and silt-size fractions of the soil. This horizon marks the zone of the most intensive biogeochemical weathering in the profile. It is differentiated from the A horizon by its lighter, ashy-grey color, reduced aggregation, and generally coarser texture.

The B horizon is the illuvial layer in which much of the material, including silicate clays, iron, aluminum, and humic substances, alone or in combination with other leached material, has accumulated. These substances move deeper into the profile from the overlying E horizon in solution or suspension, where they encounter a different biological or chemical environment that causes deposition of dissolved or suspended substances. Sesquioxides, especially of iron, frequently give this horizon a strong reddish color that may be used as a diagnostic tool. Silicate clays carried down from upper layers or synthesized in place from dissolved or suspended components form granular, blocky, or columnar prismatic structures in many B horizon soils. Transitional horizons between A and B, E and B, or B and C horizons are fairly common and are designated as A/B, E/B, or B/C horizons. In grassland soils formed in climatic zones with less total precipitation and consequently less leaching, calcium carbonate or sulfate deposits may appear in the B horizon. In grassland soils, the color is usually darker than profiles where more iron is leached from surface layers. Forested soils subject to severe leaching, or where groundwater is present at the proper depth, may form indurated layers that restrict continued downward movement of soil solution. Forests on such soils are usually shallow-rooted and susceptible to uprooting by windstorms that disrupt and mix soil horizons in that area.

The C horizon has been least affected by processes that formed the overlying A, E, and B horizons. In the absence of any obvious discontinuities in the regolith, it can be assumed that the C horizon represents the substrate from which overlying horizons have been formed. This is not precisely true because it is very likely that changes in the regolith have occurred due to substances that were carried beyond the B horizon by percolating water. In arid climates the C horizon sometimes marks the deepest penetration of water with light-colored salt deposits.

The R horizon is the designation given to consolidated bedrock, whatever its nature, when that bedrock appears close enough to the ground surface to be included in the profile description.

Other features that may be recorded as subscripts to the master horizons (O, A, E, B, and C) include buried horizons, frozen soils, mottled colors caused by groundwater saturation and plowing or other disturbance. It should also be recognized that not all horizons listed will always be present in a given profile. When wild soils are cultivated, all horizons in the top 15 to 20 centimeters (cm) will be obliterated and homogenized. The resulting mixture is designated Ap. In areas subjected to severe erosion the B horizon may be exposed at the soil surface, and crops grown in that horizon. The same designation, Ap, is still used in these circumstances even though the A and E horizons are missing and the B horizon has been truncated.

Most early soil classifications focused on the influence of soils and soil characteristics on production of agricultural crops based on trial and error experience. These systems emphasized soil genesis or assumed soil genesis that sometimes led to controversy. In 1975, a classification system called Soil Taxonomy was published in the United States. In this system soils are classified on the basis of characteristics such as precisely defined diagnostic horizons (surface horizons are called epipedons and subsurface layers are called horizons), soil temperature regimes, and soil moisture regimes, which can be quantitatively measured in the field or laboratory. This permits verification by other soil scientists. Also, nomenclature used to describe soils and their characteristics is rooted in Latin or Greek words in somewhat the same way that plants are classified. Soil Taxonomy is now the primary system used in the United States, and scientists in other countries also use it. Many features of Soil Taxonomy have been incorporated in the Food and Agriculture Organization of the United Nations Educational, Scientific, and Cultural Organization (FAO/UNESCO) Soil Map of the World. Some countries continue to follow their own classification systems because they provide advantages for particular needs.

Because soils occur in nature as part of a continuum of landscapes (more correctly soilscapes), for practical reasons, a smaller, representative sample for each landscape must be chosen for analysis. This unit must be the smallest volume that encompasses the morphological, physical, and chemical variability found in that part of the landscape represented by the sample. In Soil Taxonomy the sample unit is called a pedon and all contiguous pedons that are closely related in their characteristics are called polypedons. Polypedons correspond approximately to the concept of soil series that is the basic unit of classification in the genesis-based system that preceded Soil Taxonomy in the United States. Thus the new is linked with the old.

In Soil Taxonomy, soils are placed in categories that range from differentiation on a very broad scale to increasingly precise limited descriptions of groups of soils that are separated for practical, interpretive applications. Categories of classification and their numbers of representatives in the United States are Orders (10), Suborders (47), Great groups (230), Subgroups (1,200), Families (6,000), and Series (13,000). Only the orders are discussed here. For information on the other categories, readers are encouraged to consult the additional reading sources listed below.

Orders are named and briefly defined as follows:

1. Entisols—little profile development, weakly developed mineral soils
2. Inceptisols—embryonic soils of humid regions, usually with an altered horizon, frequently forested
3. Mollisols—dark-colored surface horizon high in basic cations and organic carbon, dark soils of grasslands and steppes
4. Alfisols—presence of an illuvial (B horizon) layer high in clay, medium to high content of basic cations, sometimes sodium, fertile for-

Table 1. Approximate Land Areas of the Ten Soil Orders in the United States and in the World with Representative Locations of Occurrence

	Percent of Ice-free Land Area		*Representative Geographic Location*	
Soil Order	*U.S.*	*World*	*U.S.*	*World*
Entisols	8.0	12.0	Rocky Mountains, Florida, Georgia, Alabama, Nebraska	Sahara Desert, Saudi Arabia, Africa, Australia, Siberia, Tibet
Inceptisols	18.2	20.0	Oregon, Washington, New York, Ohio, Pennsylvania	Ecuador, Colombia, Spain, France, China, Germany, Chile, Africa
Mollisols	25.0	9.0	Great Plains, Utah, Idaho, Oregon, Washington	Former Soviet Union, China, Mongolia, Argentina, Uruguay, Paraguay
Alfisols	13.5	14.0	Ohio, Indiana, Michigan, New York, Wisconsin, Minnesota, Pennsylvania, Texas, California, New Mexico	Baltic States, China, Central Europe, Africa, Western Russia, Brazil, India, Asia
Ultisols	12.8	10.0	Southeastern U.S., Hawaii, Oregon, California, Washington	Australia, Asia, China, Brazil, Paraguay
Oxisols	0.012	9.0	Hawaii, Puerto Rico	South America, Africa
Vertisols	1.0	3.0	Texas	India, Ethiopia, Sudan, Australia, Africa, Mexico, Venezuela, Paraguay, Bolivia
Aridisols	11.6	18.0	California, Nevada, Arizona, New Mexico	Africa, Australia, Argentina, Turkestan, Middle East
Spodosols	4.8	4.0	Northeastern U.S., Michigan, Wisconsin, Minnesota	Canada, Northern Europe, Siberia
Histosols	0.5	1.0	Alaska, Lake states, Washington, Indiana, Massachusetts, New York, New Jersey	Former Soviet Union, Canada, Finland, Scandinavia

ested soil frequently found near forest-grassland boundaries

5. Ultisols—presence of an illuvial (B horizon) layer high in clay, low content of basic cations, highly leached, frequently in forested, tropical, or subtropical climates
6. Oxisols—presence of a strongly altered horizon consisting of a mixture of hydrated oxides of iron and aluminum with variable amounts of highly weathered clay, usually deep red in color, occurring in tropical or subtropical climates
7. Vertisols—soils high in clay that shrink when dry, developing deep cracks that frequently fill with eroded soil particles; generally unstable soils
8. Aridisols—soils of deserts or semideserts with sparse xerophytic vegetation, sometimes with an illuvial (B horizon) layer high in clay and basic cations, including sodium
9. Spodosols—presence of a massive horizon composed of organic matter, iron, and aluminum; highly leached forested soil, sometimes with groundwater intersecting profile
10. Histosol—peat or bog soils with more than 20 percent organic matter, frequently forested and frequently saturated with water for periods ranging from a few days to six months or more

Land and representative distribution of each soil order based on information from Brady (1984) are found in Table 1 on page 995.

Theoretically, all soils that exist on the surface of the earth can be included in the Soil Order category. For management purposes, however, more precise differentiations are required. When soil surveys are conducted, the surveyor uses mapping units that fit the requirements of the management plan for which soil information is needed. In most cases the soil series (polypedon) is the unit of choice. The surveyor uses soil pits or a soil auger to examine soil horizon development and characteristics of the soil profile at a number of places on the landscape being mapped. The number of locations sampled is determined by observing landforms and topographic features. In this determination the experience of the surveyor ultimately defines the accuracy of a map showing distribution of mapping units and their boundaries. Boundaries of mapping units are drawn on aerial photographs of the study area. The resulting field map is verified and produced.

In addition to the map an interpretive bulletin that presents additional relevant information on field observations made by the surveyor, and on laboratory analyses of soil collected at the various field locations, is assembled. These maps and their accompanying bulletins are used by agriculturists, engineers, hydrologists, and other scientists. They also form the basis for certain regulations governing land use on the municipal, state, and federal level. In some circumstances remote sensing techniques are used to provide additional information on soils being classified.

Bibliography

BRADY, N. C. *The Nature and Properties of Soils*, 9th ed. New York, 1984.

BRIDGES, E. M., and D. A. DAVIDSON. *Principles and Applications of Soil Geography*. New York, 1982.

BUOL, S. W., F. D. HOLE, and R. J. MCCRACKEN. *Soil Genesis and Classification*, 2nd ed. Ames, IA, 1980.

OLSON, G. W. *Soils and the Environment: A Guide to Soil Surveys and Their Applications*. New York, 1981.

SOIL SURVEY STAFF. *Soil Taxonomy*. Agricultural Handbook 436, U.S. Department of Agriculture. Washington, DC, 1975.

GARTH K. VOIGT

SOIL TYPES AND LAND USE

Society has many uses for land. Their effects on soils are directly related to the degree to which the contemplated use departs from conditions that interacted to form the original soil (*see* SOILS, FORMATION OF). Very generally, land uses include production of food and fiber, watershed protection, living space, industry and business, transportation, recreation, parks and preserves, mining, and waste disposal. A rapidly expanding population puts all of these uses in direct competition with each other and, in many cases, the consequences of a decision to use resources that time and circumstances have placed there are so disruptive that future alternative uses are practically excluded forever. Soil is a primary factor in some of these enterprises, such as agriculture, forestry or watershed protection, but in highway or airport runway construction or in urban development it is the nature of the regolith that relates to the contemplated use. The soil is pushed aside merely to level topography.

Agriculture

Origins of agriculture are obscure and controversial but what is evident, as one views the record from the dawn of history to the present, is an ever-increasing intensity of land use. By the time humankind had achieved the ability to domesticate plants and animals to increase food production, these ancient people had undoubtedly also learned to cultivate, fertilize, irrigate, and otherwise manipulate soils to their advantage. Thus, wild plants and wild soils in self-sustaining ecosystems have been gradually transformed to modern agriculture. This process certainly has been one of the most significant advances made by humans. In recent years, however, many agricultural practices have become less sustaining and more extractive. While yields of some crops have been doubled, other crops appear to have reached yield plateaus. We do not yet know all of the environmental consequences of this trend. Continuing excessive soil erosion losses, soil compaction due to heavy machinery used in large-scale high-intensity agriculture, and extensive use of pesticides are among causes for concern.

Agricultural soil should provide a favorable growth environment for roots of domesticated crop plants. This includes a stable available supply of soil water, pore space for adequate gas exchange with the outer atmosphere, a balanced supply of nutrient elements essential for plant growth, mechanical support to keep the entire plant upright, and absence of toxic substances or conditions.

A fertile, silt loam surface soil provides nearly optimum levels of all of these requirements. On a volume basis, such a soil is composed of about 50 percent solid matter and 50 percent pore space. The solid matter portion is composed of 45 percent mineral matter and 5 percent organic matter. Pore space, due largely to soil animal activity, is divided into two categories of pore diameter. Micropores have a sufficiently small diameter (about 0.06 millimeters or mm) to hold water against the force of gravity because of capillary and surface attraction. Macropores are those larger in diameter than 0.06 mm. These are unable to retain significant amounts of water against gravity, allowing relatively fast free movement of water and air through the soil. Compared with the well-balanced pore diameter distribution of the silt loam soil, sandy soils have a higher proportion of macropore volume, whereas clay soils frequently have a much higher proportion of micropore volume.

When water falls on a soil surface at a rate sufficient to saturate the upper few centimeters, all pores in the saturated zone are, by definition, filled with water. Macropores are emptied rapidly by gravity. Water moves essentially downward as far as the supply of water permits. With all micropores in the wetted zone filled and at equilibrium with capillary surface forces, and with all macropores emptied, the wetted soil contains its maximum content of available water. This condition is called field capacity. Water movement in micropores is governed by tension forces of adhesion of water molecules to soil capillary surfaces and by cohesion of water molecules to each other. Movement of water in micropores is slow and in the direction of greatest tension forces that coincides generally with the driest soil. Tension differentials are generated by water evaporating from the soil surface and also by growing transpiring plants. Water moves from larger micropores into plant roots, through conducting tissues in roots, stem, and leaves, finally passing into the atmosphere in much the same way that water evaporates into the atmosphere from any wet porous surface. When rate of water loss from the plant exceeds the general rate of water absorption into plant roots, the plant wilts, temporarily slowing or even stopping plant growth processes. If additional water is added to the soil by precipitation or irrigation, soil moisture tensions decrease, the plant is again able to absorb water from the soil, and growth resumes. Absorption of water by the plant may also resume after sufficient time has elapsed for capillary tensions in the soil to equilibrate as long as equilibrium tension forces in the root system zone of the soil are lower than those exerted by the plant.

The fertile silt loam soil referred to above provides optimum water availability for plant growth because surface water infiltrates the soil quickly through macropores and, in addition, there is good capacity to retain added water in the micropore system. This soil responds quickly and effectively each time water is added, which maximizes the benefits of added water for plant growth. A sandy soil could absorb more water more quickly, but because retention capacity is low, most of the added water would be lost to plants growing in this soil. Conversely, a clay soil may have much more retention capacity in its greater proportion of micropore volume than the silt loam soil, but because water moves so slowly in the smaller capillaries of micropore space, most of the added water runs off the surface. This decreases the water available for plant growth and increases potential for soil erosion.

With few exceptions, water is almost always a limiting factor in crop growth. Even where total precipitation is adequate for crops being grown, the seasonal distribution may not coincide with stages of plant development that are critical for attaining maximum crop yields. As a result, irrigation use has increased in the United States and, where water and technology are available, throughout the world. To the extent that yields of crops are increased by additions of water, irrigation has had the effect of adding extra hectares to the world's agricultural land base. Depending on the geographical area involved, water is distributed to a specific location by ditches, flumes, or pipelines where it is applied to the crop by flooding, in furrows, by sprinklers, or by drip systems. Flooding and furrow applications are less efficient than sprinklers or drip systems. Optimum soil moisture range for maximum crop growth is from 80 to 100 percent of field capacity. This level is easiest to maintain with sprinklers or drip systems.

Water-use efficiency in crops is usually improved by keeping levels of available essential nutrient elements high. Current high-intensity agricultural practices in developed countries depend on large applications of chemical fertilizers to stimulate crop growth in moderately fertile soils or to correct specific nutrient deficiencies where this is necessary. Most frequently applied are nitrogen followed by phosphorus and potassium. Secondary attention is given to sulfur, calcium, and magnesium and to iron, manganese, zinc, copper, boron, molybdenum, cobalt, and chlorine when soil analyses indicate need for these elements. Fertilizer applications are necessary to replace nutrients removed in harvested crops or lost by soil leaching or erosion. In the United States, only about half of the nitrogen and a third of the phosphorus and potassium applied as fertilizers appear in harvested crops. Much of the remainder is found in sediments and water of streams, lakes, and ponds where biomass growth in aquatic ecosystems is increased, sometimes to detrimental levels. In the latter case, remedial measures are often required to restore water quality if consumption by municipalities is contemplated.

An alternative to the use of industrial forms of nitrogen fertilizer, which require very high expenditures of energy and fossil fuel as feed stock, is biological symbiotic and nonsymbiotic fixation of atmospheric nitrogen in soil. Legume crops are effective in fixing atmospheric nitrogen and plowing these crops into the soil may add several hundred kilograms of nitrogen per hectare to the soil (*see* SOILS, FORMATION OF). In tropical and subtropical countries, legume tree species are used in mixed cropping systems to supply nitrogen to adjoining crop plants and also to provide high-protein forage from the trees for farm animals. Legume seeding is used to supplement protein production in natural grassland pastures or on forestland newly cleared for pasture in tropical areas.

Soil acidity can influence availability of plant nutrient elements to crop plants. For most temperate-zone soils and conventional field crops a pH range from 6 to 7 is satisfactory. Acidophilus plants such as azalea, laurel, rhododendron, blueberry, and cranberry require acid soils in the range from 4 to 4.5 and may require applications of sulfur or sulfate fertilizer to prevent iron deficiency. For most other agricultural practices soil acidity is controlled by applying ground limestone or calcium oxide or hydroxide at rates that depend on the degree of acidity, other characteristics of the soil, and the crop being grown. In many areas, but especially in the topics, high levels of soluble aluminum are common when the pH value falls below 5.5 and the content of soil organic matter is low. Because roots of most crop plants cannot tolerate high levels of aluminum, the pH value must be raised by liming to precipitate the aluminum. Phosphorus availability is also a serious problem in many tropical soils in which the capacity to fix applied phosphorus is high because of soil aluminum. This capacity must be determined for each individual soil using chemical analyses.

Erosion

The most critical adverse environmental effect of agriculture is soil erosion. In the United States, about 40 percent of the sediment found in streams and reservoirs comes from agricultural land, and, if grazing land is included, the responsibility attributable to agriculture rises to over 50 percent. The situation in many areas of tropical and subtropical countries is especially grave where expanding populations of humans and grazing animals are increasing pressure on the land. The erosion problem emerges when land is cleared of its self-sustaining cover of native vegetation. Crop plants increase exposure of surface mineral soil to kinetic energy of falling raindrops during intense storms. All raindrops transfer some level of energy to soil but intense storms usually generate larger drops that break down soil structure and dislodge soil particles on impact. The combined effect of soil aggregate destruction and surface sealing by dislodged soil particles leads to reduced infiltration capacity and to increased overland flow.

Overland flow moves downslope as a sheet of water containing soil particles kept in suspension by the impact of falling drops. This is called sheet erosion. Movement of soil particles downslope depends on the degree of slope of the land, which in turn determines the velocity of the moving water. Because velocity of moving water determines particle carrying capacity, any irregularity in slope that increases flow rate also increases rate of soil removal. In time, a small channel is formed that tends to focus flow and energy. This is called rill erosion. With successive storms, rills increase in size as more soil is moved downslope and eventually rills can become gullies. In gullies, flow behavior becomes essentially fluvial in nature, undercutting banks that slump into the flowing stream, greatly increasing the dimensions of the degradation process. Moderate degrees of sheet and small rill erosion can usually be healed to some extent during cultivation, but once gullies have formed, extensive engineering techniques and structures are required to stop their advance. These are usually beyond the capabilities of an individual landowner.

Erosion control measures include cultivation across the slopes rather than up and downslopes, growing crops in narrow strips alternating row crops such as corn with cover crops such as timothy or clover, and using grass waterways to moderate runoff velocity. In more extreme cases, terraces are constructed along contours to move runoff downslope, gradually feeding it into a grass waterway. Another approach that is gaining favor is to reduce or eliminate tillage employed during growth of row crops such as corn. This practice keeps residue from the preceding crop on the soil surface instead of plowing it into the soil. The surface layer of residue absorbs drop energy, leading to dramatic reductions in soil erosion loss. Because crop residue also functions as a mulch, infiltration of precipitation increases and evaporation loss from surface soil decreases. In many cases increased crop yields result. In no-till or reduced-till systems crop seeding is done by drilling through the crop residue and weed growth is controlled by herbicide applications.

In areas of flat terrain such as coastal or outwash plains or in broad river valleys, soils are often poorly drained. These soils may be made agriculturally valuable by installing drainage systems to lower the water table, thus increasing rooting depth. This may be accomplished by open ditches or by burying perforated plastic or tile pipes in the soil to collect excess water and direct it to an outlet. Such installations are very expensive and generally are used only when high-value cash crops such as vegetables or grass sod for lawns are grown. With proper management poorly drained soils can be made to produce very high yields of specialty crops but careful attention must be paid to requirements of such crops. Often specific watering, fertilizer, liming, and pesticide regimes for different growth stages are involved.

Forestry

Compared with agriculture, forestry is a land use that is extensive rather than intensive because of the timescale involved in growing the crop. In naturally generated forests that may be several hundred years old, soils are likely to be close to the self-sustaining condition characteristic of the geographic location and the soil-forming factors found there. The eventual harvesting operation destroys soil stability, especially if all trees are removed over large areas. The scale of such a harvest exposes previously shaded soil to much higher surface temperatures that accelerate organic matter decomposition, leading to drastic release of nutrient elements downslope. Loss of organic matter and essential fauna causes quick breakdown of soil structure and, if the terrain is steep, each successive storm washes tons of sediment and logging debris into the drainage system. Damage is especially severe when extensive access road systems must be built into the landscape to remove timber. Regeneration of forests on severely eroded sites resulting from logging or fires is a long-time process requiring many decades before soil healing can begin. Current timber cutting policies are gradually changing to reduce the size of individual clear-cut patches and seeding grass to restore vegetative cover on logging roads after harvesting to minimize erosion damage.

In plantation forestry, cutting rotations are shorter than those used in natural stands. Nursery-grown seedlings are planted in rows to maximize land-use efficiency and harvested mechanically. In some cases genetically improved planting stock and fertilizer applications are used. Such practices increase pressure on soil productive capacity and growth decline in successive rotations of plantations of Monterey pine in Australia and New Zealand has been observed.

Many forested areas are managed for the water they yield. Where water is the primary product usually little or no timber cutting is done. This protection leads to the most reliable supply of high-quality water that can be expected from the precipitation that falls. In a protected forest of sufficient age, the soil will absorb nearly all of the rain that falls and the snow that melts. Subsurface water movement moderates extremes in streamflow from the area and the resulting health and vigor of the aquatic ecosystem maintain the high quality of the water flowing to the municipalities served by the forest. In forests where water is one of several management priorities, neither flow rate nor level of quality of water can be as consistent as the water supply from a protected watershed. Timber harvesting on a large scale, road building, large-scale recreational use, extensive livestock grazing, and mining operations all contribute to varying degrees to soil changes that reduce water yield and water quality.

Some types of land use are extreme examples of the degree to which the biological, chemical, and physical components of the soil ecosystem can be disrupted or destroyed. These include mining operations, deposition of industrial wastes, sewage sludge, dredge spoil or fly ash, operation of livestock feedlots, and other similar uses. Some of these situations involve presence or release of chemicals at toxic levels and some involve physical deficiencies such as slope stability, moisture-holding capacity, organic matter, or a living community of fauna and microflora. Contemporary environmental regulation may require some modest degree of soil restoration but in many cases these are short-lived. Reclamation techniques for restoration currently being used or studied are aimed at establishing native or introduced vegetation on the disturbed site. This may involve grading extreme slopes to match surrounding natural topography, erosion prevention measures, liming to reduce extreme acidity, application of sulfur or gypsum to reduce extreme alkalinity, adding organic matter such as plant residues (straw, leaves, sawdust, woodchips), animal manure, and municipal sewage sludge, replacing previously removed topsoil, prolonged use of fertilizers, and restoration of a wide variety of microbial and animal communities. Progress with this level of soil restoration has been slow with few examples of complete success. Major barriers that remain are primarily economic and political.

In the most extreme cases of soil degradation, more drastic measures must be employed such as the large-scale engineering techniques used on the U.S. Environmental Protection Agency Superfund sites. Such efforts are also very slow and extremely expensive. It must be said that attempts to restore soils on badly disrupted sites are not very encouraging considering the size and scope of the problem.

Bibliography

Brady, N. C. *The Nature and Properties of Soils*, 9th ed. New York, 1984.

Carlson, P. S. *The Biology of Crop Productivity*. New York, 1980.

Crosson, P. R. *The Cropland Crisis*. Baltimore, 1982.

Lal, R., and B. A. Stewart. *Soil Restoration. Vol. 17: Advances in Soil Science*. New York, 1992.

Lockeretz, W. *Agriculture and Energy*. New York, 1977.

Sanchez, P. A. *Properties and Management of Soils in the Tropics*. New York, 1976.

Spedding, C. R. W. *The Biology of Agricultural Systems*. New York, 1975.

GARTH K. VOIGT

SOILS, FORMATION OF

Formation of soil from regolith is a continuation and extension of the weathering process that forms regolith from bedrock. Bedrock is vulnerable to chemical and physical disintegration when it is exposed to the atmosphere at the surface of the earth and to water falling from the atmosphere. Water penetrates rock surfaces, dissolving constituent mineral elements, carrying them downslope. If water freezes, it may create or enlarge existing cracks by expansion, thus allowing deeper penetration of water and further fracturing, finally producing rock fragments. Salt crystallization and expansion may also contribute to fragmentation. All of these processes lead to increased reactive surface and continued fragmentation. Over centuries a mantle of unconsolidated rock debris accumulates that is called regolith. This mantle varies in thickness from a few centimeters to hundreds of meters depending on circumstances. Soil formation results from acceleration of this weathering process due to intrusion of living systems.

Soil, Regolith, and Bedrock

Soil is the uppermost layer of the regolith. Through time, this layer has reacted to a greater extent than underlying material to temperature, precipitation, and wind, all of which are manifestations of solar radiation and all of which are important in soil formation. The principal distinction between regolith and soil is the unique influence of living systems sustained by the process of photosynthesis that focuses additional extraterrestrial energy to carry the process of soil formation beyond what would be possible without life.

With few exceptions, life on Earth depends on photosynthesis. On land surfaces of our planet soil is the natural medium for photosynthesizing plants. Plant growth alters the soil and the regolith from which it developed. In situations that remain undisturbed for sufficiently long periods of time, vegetation and its biological associates, as well as the soil that supports this system, acquire distinctive characteristics that depend on climate and the nature and history of the regolith involved. Radiant energy of the Sun drives production of plant compounds that generally favor maintenance of the given climax ecosystem and its associated photosynthetic capacity. Consequently, the collective imprint of all of these factors and processes gradually becomes increasingly distinguishable from regolith which has, by that distinction, become soil. Boundaries between regolith and soil at the extremities of soil formation are usually indistinct compared with the separation of regolith from bedrock. Figure 1 is a diagrammatic summary of some relationships between soil, regolith, and bedrock in soil formation (Brady, 1984).

The history and composition of the upper few meters of regolith predetermine the nature of the soil that follows. This is especially true of the amount of sand, silt, and clay contained in a soil. In any given situation, regolith constitutes the major source of mineral elements required by wild plants for completion of their life cycles. These are potassium, calcium, magnesium, phosphorus, sulfur, iron, manganese, copper, zinc, molybdenum, boron, cobalt, and chlorine. Carbon, hydrogen, ox-

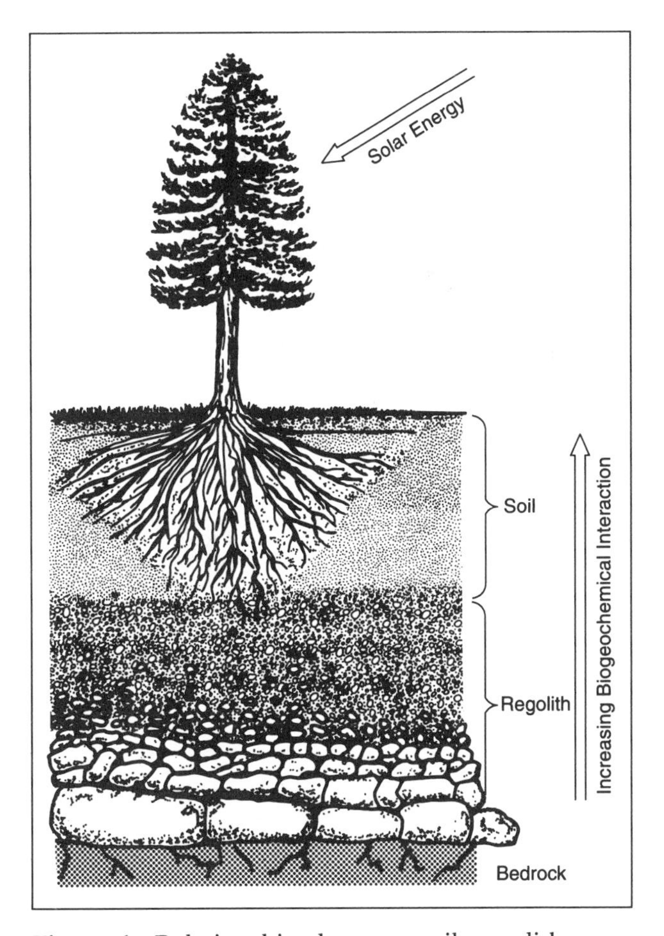

Figure 1. Relationships between soil, regolith, and bedrock in soil formation. Modified from Brady (1984).

ygen, and nitrogen are also essential and are generally accumulated from atmospheric sources as soil formation proceeds. Within a specific climatic zone, regolith containing relatively high concentrations of these mineral elements is more likely to form a productive soil supporting a diverse, more stable ecosystem than regolith with minute amounts of nutrient elements or where essential elements are lacking entirely. Regolith derived from igneous bedrocks with a variety of dark-colored base-rich minerals will generally form more productive soil than will result from regolith derived from rocks high in light-colored base-poor minerals or from sedimentary rocks that have experienced one or more weathering and erosion cycles and that now consist of reconsolidated grains of quartz and other very stable minerals. In the latter case, ecosystem progression may be retarded or arrested in a sub-climax stage that alters soil development.

Because regolith is generally incongruous in nature, its transformation to soil is also incongruous. As a result, early weathering stages are characterized by vegetation growth that may be sparse and scattered over the landscape, its distribution corresponding to the differential weathering rate of soil material being transformed. Bedrock in place may consist of minerals with a wide range of susceptibility and exposure to mechanical and chemical weathering. When bedrock has been subjected to physical or chemical disintegration and subsequently transported by gravity, wind, or water in liquid state or in the form of mountain or continental glaciers to a new location, regolith particles may range in size from boulders that may be a meter or more in diameter to clay which is 0.002 millimeters (mm) or less in diameter. Compared with intact bedrock, transported or residual material exposes much more reactive surface to weathering agents. As a result, chemical alteration and rock disintegration are usually far more rapid. Eventually, as more vegetation invades and dominates the site and biological inputs increase, weathering rates are amplified by the energy of photosynthesis and inorganic material becomes infused with life much more rapidly.

Soil and the Sun: Climate, Plant Succession, and Soil Formation

As vegetation in a wild ecosystem grows and develops above and below ground, it becomes more or less equilibrated with limits to growth imposed by general factors such as macro and micro climate, nutrient availability, and genetic composition of the living part of the system. On a worldwide scale the spectrum of plant formation types may range from tropical rain forest to temperate forest to grassland to desert to tundra depending primarily on the temperature-moisture factors of climate. In this sequence one would expect the soil weathering rate to be highest in the tropical rain forest and lowest in the desert and tundra. Biomass production, organic matter decomposition rates, leaching rates, clay formation, and soil acidity are generally higher in warm, wet forested areas. In desert soils, development is generally retarded by scarcity of water, whereas in tundra areas, low temperatures in the soil are usually the limiting factor. In all of these examples, plant growth adjusts to available resources required for growth and the ecosystems in each set of circumstances are essentially stable. Thus the impact of vegetation on soil formation is interrelated with climatic effects (*see* SOIL TYPES).

All plant formations alter the effects of macroclimate to some degree by shading the soil or reducing wind velocity. Because of their physical dimensions and the fact that, generally, trees grow where precipitation exceeds evaporation, mature forest ecosystems dominate the site they occupy to a greater extent than other lesser vegetation types. Soil formation reflects this. The time required to attain the climax stage of forest succession may vary from less than a century to two or three centuries or more, depending on the nature of the climate, vegetation, and presence or absence of growth-limiting conditions.

Plant Roots, Microbes, and Soil

During the time span indicated, the volume of the root system supporting the expanding vegetative cover increases in direct proportion to the increase in total above-ground biomass. As a result, more of the mineral substrate in the upper regolith is subjected to acidifying effects of root exudates and to organisms associated with root surfaces. Acidification results from saturation of soil moisture with metabolic carbon dioxide and with organic acids secreted by growing plant roots or produced during decay of plant detritus by heterotrophic decomposers. In such an environment, weathering of soil minerals is enhanced and mineral elements are more easily absorbed and translocated to metabolically active sites within the plant that produces more biomass. Increased mineralization by these processes concentrates elements for recycling and also replaces leaching losses of mineral elements from the ecosystem. Other aspects of soil formation such as clay synthesis and soil horizon development are also promoted. Conceptually then, soil formation extends to the depth and extent of root penetration.

Concentration of plant roots in soil generally decreases by several orders of magnitude with increasing soil depth, even in stable ecosystems. Conditions for root growth are usually best in the upper half-meter of any given soil. More than 75 percent of soil moisture and nutrient element absorption occurs here. It also includes the location of the most intense activity of detritus decomposers. In this zone of tremendous biological activity many fascinating relationships occur between plant roots and certain soil fungi, actinomycetes, and bacteria. Increases in numbers and activities of some of these organisms are found at or near the surface of young plant roots. Stimulation is related to presence in the root zone (rhizosphere) of metabolites produced by living plants that are subsequently released from root surfaces. Organisms in the rhizosphere use such metabolites as energy sources and by their own metabolism secrete additional substances that facilitate release of plant nutrient elements from primary mineral sources in soil and regolith.

In other cases, root surfaces or root hairs may actually be invaded and modified by specific microorganisms to form symbiotic relationships that accelerate nutrient element uptake by the host plant in exchange for energy sources for the invader. Other plant-microbe associations provide significant incorporation of atmospheric nitrogen into tissues of the host plant. Many leguminous plants combine with certain species and strains of bacteria to produce root nodules capable of fixing atmospheric nitrogen. Nonleguminous plants such as species of *Alnus* combine with some actinomycetes to form nodules with similar nitrogen-fixing capacity. Ultimately, all of these associations enrich the soil when plant tissues die and are decomposed. Such combinations are used in agricultural practice to maintain soil fertility (*see* SOIL TYPES AND LAND USE).

Roots of many tree species in forest ecosystems may eventually penetrate soil and regolith to depths of several meters, but such roots function primarily in providing physical anchorage for the forest stand. These roots interlace with roots of adjoining trees and, in some cases, roots of one tree may graft onto roots of an adjoining tree, thus promoting still greater stand stability.

In these and other ways, root systems and activities of early successional species facilitate development of more demanding subsequent stages of ecosystem progression, thus contributing to increased overall site productivity and the general process of soil development. Eventual death and decay of all or part of the root system lead to aggregate formation and increased water-holding capacity in lower portions of the substrate. Old root channels through the soil serve as excellent conduits for percolating water and for promoting gas exchange through soil and regolith. For normal temperate-zone rainfalls, root channels in forested areas are sufficiently effective in promoting percolation to hold overland flow to a minimum, thus increasing baseflow of groundwater and reducing soil erosion.

At regular and irregular intervals biomass from above-ground parts of the ecosystem fall to the surface of the ground, where they serve as energy and nutrient sources for the literally millions of heterotrophic consumer organisms living in the upper soil volume. Amounts and nature of these biomass additions depend on the particular plant formation type involved, which is broadly related to climatic zone.

Organic Matter Decomposers: Their Impact on Soil

In forest regions dimensions of fallen biomass may range from uprooted trees weighing many tons to a few milligrams of soluble substances leached from the living foliage of trees during an individual rainfall. Most common and predictable are seasonal additions of leaves, flowers, fruits, and small twigs. In the temperate zone of North America, amounts of such material dropped on the ground surface each year vary from a few kilograms to several tons per hectare. These tissues contain higher concentrations of essential nutrients, especially nitrogen, than large branches or stems. As soon as they are deposited, nutrient and energy-rich tissues are attacked by a host of heterotrophic consumers. Breakdown activities progress in sequence through larger animals such as earthworms of various sizes, ants and termites, sowbugs, beetles, millipedes and mites on down to microscopic organisms, including fungi, actinomycetes, and bacteria. Breakdown of this detritus is not necessarily a stepwise process because microbial predigestion usually precedes feeding by larger animals that also consume microbial cells and fungal hyphae. As detrital particles become reduced in size, they also are intimately mixed with mineral soil particles through ingestion and excretion by soil animals. Any remaining energy or resources are repeatedly attacked by smaller animals and, eventually, by microbial heterotrophs.

Excretions occur as aggregates that reflect the size of the animals that produced them. These aggregates are the fundamental units of soil structure. Because they are bound together by fungal hyphae or cemented by bacterial gums and gels, they impart exceptional stability to the general soil architecture. Thus, the decomposition process ultimately involves not only original detritus from the primary photosynthetic producer plant, but also includes remains of heterotrophic decomposers and their predators as by-products of simultaneous synthesis and breakdown resulting from all of the interactions among all of these organisms. With each overlapping wave of consumer organisms, remaining residues become more and more resistant to further breakdown under existing conditions of moisture and temperature until an equilibrium soil organic matter content is reached. This fraction, together with residual or newly synthesized clay minerals, constitutes the indispensable component of soil productivity. The whole process of organic matter breakdown, release of essential mineral elements recently obtained from primary mineral sources in weathering regolith, and, more recently from soil, reabsorption of these elements that further increase growth of the photosynthetic capacity, encapsulates the cyclical self-sustaining nature of soil formation. The organisms involved, the nature of these interactions, and the soil that is formed all depend on where these events take place and under what conditions (*see* SOIL TYPES).

How Old Are Soils?

The time required to attain any typical soil varies from fractions of a century to many centuries but it is unlikely that such estimates can be very precise. Most modern soil formation is generally considered to have taken place within the past 1 million years and, in areas of the most recent glaciation period of the late Pleistocene, within the last 15,000 years. In some tropical areas characterized by severe weathering conditions, soils may well be older than a million years. Certainly, all of the agents involved in modern soil formation as it is currently understood were in place during the Pleistocene period.

When a catastrophic disturbance occurs, as in a severe fire, windstorm, volcanic eruption, or landslide, any existing ecosystem equilibrium is partially or completely overwhelmed and progression must begin anew. Most existing soils have very likely experienced some degree of process disruption many times, which makes identification of a beginning point very difficult.

Bibliography

Brady, N. C. *The Nature and Properties of Soils*, 9th ed. New York, 1984.

Buol, S. W., F. D. Hole, and R. J. McCracken. *Soil Genesis and Classification*, 2nd ed. Ames, IA, 1980.

Hunt, C. B. *Geology of Soils*. San Francisco, 1972.

GARTH K. VOIGT

SOLAR SYSTEM

How did the solar system come about? When did it form? How common are planetary systems in the Galaxy and the universe? How diverse might they be? The first steps toward a scientific study of these questions became possible in the seventeenth century with Johannes Kepler's discovery of the laws of planetary motion and Isaac Newton's discovery of the universal law of gravitation, which established the basic parameters and regularities of the solar system. The current paradigm about the origin of the solar system can be traced to the eighteenth-century German philosopher Immanuel Kant (1724–1804) and the French mathematician Pierre-Simon Laplace (1749–1827). This paradigm, called the nebular hypothesis, posits a common origin of the Sun and the planets from an interstellar gaseous nebula. Other models have also been considered in the past. One in particular that received significant attention in the early half of the twentieth century is the "catastrophic model," wherein a chance close encounter of the Sun with a passing star drew out material from one or both stars that went into orbit around the Sun and eventually coalesced into the planets. However, a variety of evidence has been gathered in the last several decades that points strongly toward the nebular hypothesis.

Any model of our solar system seeks to explain the following gross features: (1) 99.8 percent of the solar system's mass resides in the Sun with less than 0.2 percent in the planets, whereas most of the angular momentum of the system resides in the planets; (2) the planets are found in widely spaced, nearly coplanar, circular orbits revolving around the Sun in the same (prograde) direction as the Sun's rotation; and (3) there are two broad categories of planets—the inner rocky ones and the outer gas giants. To this list can be added several other less obvious features that also hold important clues to the formation process: the 4.5 billion year (4.5 Ga) age for meteorites, the Moon, the Earth, and, by inference, the Sun (an age that is considerably less than the 15–20 Ga age of the universe); the evidence of heavy bombardment of all preserved solid surfaces of planets; the existence of dense rocky cores in the giant planets (inferred from the small deviations of their figures from perfect sphericity); the generally prograde rotation of the planets (but with some randomness in the orientation of their rotation axes); and the existence, orbital distribution, and composition of small bodies (planetary satellites, asteroids, comets, and interplanetary dust grains). These features support the "planetesimal hypothesis" for the formation of the planets: that is, the planets formed by the accretion of smaller bodies called planetesimals, rather than directly by the "gravitational fragmentation" of the nebula.

The detailed picture of planetary system formation is derived mainly from three sources: astronomical observations, data from meteoritics, and theoretical modeling. Astronomical evidence indicates that the formation of Sun-like stars generally occurs within large, slowly rotating interstellar molecular clouds where the temperature is low (about 10 K) and the density is high (greater than about 100 particles per cubic centimeter) relative to the background interstellar medium. These molecular clouds extend over large regions (several tens of parsecs in size), but contain smaller "cores" where the density is greater by a factor of 10–100. The cores slowly contract under their own gravity, and when a certain critical density is reached, a very rapid gravitational collapse occurs. The end of this phase of evolution yields a central condensation surrounded by a rotationally flattened distribution of matter. (Fragmentation of the cloud into binary or multiple stars may also occur if the initial cloud core has sufficient angular momentum.) As matter accretes into the center, its gravitational potential energy is converted to heat; the central region eventually becomes sufficiently hot and dense to prevent further collapse and to commence nuclear fusion of hydrogen into helium. This signals the birth of a star.

The solar system is believed to be the end product of this relatively common astrophysical process. In the last decade, observations of young stars have confirmed the presence of dusty disks with dimensions similar to that of the solar system. By compiling statistics over many young Sun-like stars of different ages, astronomers have found that the lifetime of these "proto-planetary" disks is 1–30 million years (Ma). The final fate of the dusty disks is not yet directly observable. Indeed, most of the stages relevant to the formation of planets occur over such small energy and spatial scales (by astronomical standards) that they are not easily observed. Therefore, many of these stages are known only by theoretical modeling.

Formation of the planets takes place within the disk of matter surrounding the proto-Sun, called the solar nebula. Modelers make the reasonable assumption that the initial cloud core is chemically well-mixed, that is, gravitational collapse does not induce significant chemical segregation. Therefore, the composition of the solar nebula is presumed to be very similar to that of the Sun itself, namely, 98 percent hydrogen and helium, with less than 2 percent elements heavier than helium. The solids in the solar nebula are in the form of dust grains, those surviving from the interstellar cloud as well as those condensing at later stages of the collapse of the cloud. Evidence from primitive meteorites and from interplanetary dust particles indicates grain diameters of 1/20th to 100 micrometers (μm). The total mass of solids in the present-day solar system is only about 0.2 percent of the mass of the Sun, the predominant gaseous fraction having been expelled from the system during its early history. The minimum mass of the primordial solar nebula can be estimated by augmenting the observed mass of solids with the mass of hydrogen and helium gas necessary to bring it up to solar composition; this exercise finds that the solar nebula had a mass at least 2 percent of that of the Sun.

Theoretical studies show that the solid grains in the solar nebula settle into the mid-plane of the disk in a thin layer whose density becomes high enough that the rate of their mutual collisions becomes significant. A large fraction of the dust grains coagulate, over a period of about ten thousand years, into solid bodies, called planetesimals, with diameters of 1–10 kilometers (km). Closer to the center, where temperatures remain high due to the proximity of the young Sun, the planetesimals are mostly rocky in composition; in the outer, cooler regions, volatile materials such as water ice and other ices are abundant and the composition of the planetesimals is rich in ices. Comets are believed to be the most primitive remnants of the icy planetesimals of the outer solar system that did not get incorporated into the planets.

The end product of the coagulation of dust grains is a large population of planetesimals distributed in a flat disk revolving around the Sun. The next stage is believed to involve the mutual gravitational interactions in this swarm of planetesimals that perturb the orbits of individual bodies, causing them to collide. Fragmentation and accumulation go on simultaneously. Smaller bodies are more easily fragmented due to their weaker gravity. Because the larger bodies have larger gravity and larger collision cross sections, they tend to grow rather than fragment, and they grow faster than smaller bodies. Thus, the largest body in the swarm at any particular distance from the center effectively sweeps up smaller bodies and carves out a clearing in its orbital path. The result of this phase of the evolution is a population of "planetary embryos" revolving around the Sun in relatively isolated orbits. The initial compositional dichotomy in the planetesimal disk is expected to carry over to this stage of the evolution: planetary embryos in the outer solar system would be richer in ice, while those in the inner solar system would be mostly rocky in composition. Computer modeling finds a few hundred Moon-sized planetary embryos at the end of the planetesimal accumulation process in the inner solar system; the situation in the outer solar system is similar, but the planetary embryos there are larger in size, perhaps as large as Earth.

The next stage in the inner solar system is relatively violent, involving high-velocity collisional accumulation of many large planetary embryos. The random tilts of the rotation axes of the planets are evidence of large impacts in the late stages of planetary accumulation. Computer simulations of this stage in planet formation show that a few bodies (four or five), similar to the present terrestrial planets, come to dominate the inner planet region in a timescale of about 100 Ma. The high impact energies lead to large-scale melting and differentiation of the terrestrial planets. Their thin atmospheres are a result of later outgassing from the solids rather than a remnant of the nebular gas. In the outer solar system, planetesimal accumulation continues to even larger masses. The ice-and-rock component of the cores of Jupiter, Saturn, Uranus, and Neptune is each approximately ten times the mass of Earth, the rest of each planet's mass being made up by a massive gaseous atmosphere. Calculations show that such large ice-rock cores would attract large amounts of nebular gas. The gaseous envelope would shrink owing to the large gravity of the total mass, allowing even more gas to be captured. This accumulation process would terminate when the giant planet exhausts the supply of nebular gas in its vicinity. The increasingly lower gas-fraction from Jupiter to Neptune is attributed to the longer timescale required at larger distances for accretion of planetesimals into the solid cores of these planets. The gaseous component of the solar nebula is depleted at later stages in the nebular lifetime owing to processes such as an energetic solar wind or evaporation due to radiation from the young Sun. Thus, by the time a large enough core is formed for Neptune, there is much less nebular gas available for this planet.

Each of the giant planets is accompanied by a satellite system akin to a miniature planetary system. The *Voyager* spacecraft reconnaissance of these systems showed the remarkable diversity of properties of the four satellite systems. Planetary satellites are generally classified into two types: regular and irregular. The regular satellites orbit in the same sense as the planet's rotation in well-spaced, nearly circular orbits in the planet's equatorial plane. The irregular satellites of the giant planets are found on highly elongated orbits far from the regular satellites and generally not in the planet's equatorial plane. Most of these bodies are quite small, except for Triton, the large satellite of Neptune. In the inner solar system, the Earth's Moon and the two satellites of Mars also fall in the irregular category. The giant planets also have small satellites and rings orbiting close to the planet, well within the Roche zone (where the planet's tidal forces would not allow the gravitational accumulation of large bodies). The clues to satellite formation are somewhat obscured by the fact that slow dissipation of tidal energy in these systems over several billion years has erased information about their initial rotation and their initial orbits. (Most satellites have their rotation synchronously locked to their orbital motion, just as the Moon, which always presents the same face to Earth, and many satellites are locked in orbital period commensurabilities with neighboring satellites.) The formation of regular satellites in a circum-planetary nebula is expected to be similar to the accumulation of planetary embryos described previously, but not simply a scaled-down version, for there are great differences of physical conditions between the circum-planetary environment and the solar nebula. For the irregular satellites, on the other hand, no single mechanism can be expected to explain them all. Their orbital characteristics suggest that most such satellites are captured planetesimals from the solar nebula. However, Earth's Moon is thought to be the result of a physical collision of the young Earth with a Mars-sized planetary embryo; a giant impact may be responsible for the formation of Pluto and its large satellite Charon at the edge of the planetary system.

The asteroids that orbit the Sun between Mars and Jupiter are the remnants of planetesimals and planetary embryos that failed to accumulate into a planet. The shapes and spins of these small bodies are evidence of the collisional evolution that occurred in this region. Meteorites are fragments of asteroids that fall to Earth's surface from time to time. Chemical and physical analyses indicate that these fragments were formed in bodies no larger than a few hundred kilometers in size. The exact reason for the absence of a full-sized planet in this region is not known, but it is speculated that perturbations from the giant planet Jupiter were either directly or indirectly responsible. These small bodies may hold important clues to the details of the formation of the largest planet in our solar system.

In seeking the origins of the solar system, it is tempting to focus on the regularities and deemphasize the anomalies, particularly because it is easy to be overwhelmed by the many details in this interdisciplinary subject. Our current picture is well founded in known physics and chemistry of astronomical phenomena and suggests that planetary systems like our own may be a common byproduct of the formation of Sun-like stars. However, it must be recognized that many aspects of the best current model are more accurately described as "hypotheses" rather than "theories"; they are driven by the particulars of our solar system alone. One need only look to the great diversity of satellite systems of the outer planets to acknowledge that perhaps nature has room for a great diversity of planetary systems as well. The recent discovery of planets around a pulsar (a rapidly spinning, very compact star) amply illustrates the point.

See also: EARTH, ORIGIN OF; METEORITES; MOON; OLDEST ROCKS IN THE SOLAR SYSTEM.

Bibliography

LISSAUER, J. J. "Planet Formation." *Annual Reviews of Astronomy and Astrophysics* 31 (1993):129–174.

LEVY, E. H., and J. I. LUNINE, eds. *Protostars and Planets III*. Tucson, AZ, 1993.

ROTHERY, D. "Origin of the Planets." In *The Planets*. Stony Stratford, Eng., 1994.

TAYLOR, S. R. *Solar System Evolution: A New Perspective*. New York, 1992.

WOLSZCZAN, A. "Confirmation of Earth-mass Planets Orbiting the Millisecond Pulsar PSR B1257+12." *Science* 264 (1994):538–542.

WOOLFSON, M. M. "The Solar System—Its Origin and Evolution." *Quarterly Journal of the Royal Astronomical Society* 34(1) (1993):1–20.

RENU MALHOTRA

SOLAR WIND

See Planetary Magnetic Fields; Sun

SOLID EARTH-HYDROSPHERE-ATMOSPHERE-BIOSPHERE INTERFACE

Since the early part of the twentieth century it has been traditional to treat the geochemistry of planet Earth in terms of its separate parts, the geospheres.

Most present models of Earth, from the center out, divide it into:

1. The inner core, solid and dense (13 g cm^{-3}), dominated by crystalline metal (nickel-iron).
2. The outer core, liquid and dense (11–12 g cm^{-3}), dominated by liquid metal.
3. The lower mantle, solid, density near 5 g cm^{-3}, dominated by magnesium-iron-silicon-oxygen (Mg-Fe-Si-O).
4. The upper mantle, solid, density 3.5–4.0 g cm^{-3}, dominated by Mg-Fe-Si-O.
5. The asthenosphere, essentially solid but perhaps with a small molten component, dominated by Mg-Fe-Si-O.
6. The lithosphere, a crystalline outer layer 100–300 kilometers (km) thick, solid, carrying the crust of Earth, with complex chemistry (the low melting fraction of the planet).
7. The hydrosphere, including all the systems of Earth dominated by liquid water.
8. The atmosphere, dominated by oxygen and nitrogen with many other gases in small quantities (e.g., ozone, methane, water, carbon dioxide, argon).
9. The biosphere—living matter dominated by carbon-oxygen-nitrogen-hydrogen (C-O-N-H) and including various crystalline and amorphous matter. More recently there is a tendency to use the word ecosphere—the total system of minerals, H_2O, and atmosphere—that contains and envelops all living matter.

The gross Earth model was built on observations of the accessible parts plus indirect observations of the deeper interior using data from studies of how seismic waves travel through the planet. The mean density, mass, and moment of inertia of the planet; the chemistry of meteorites; and studies of the states of matter over the pressure-temperature range of the planet also aided its development. Such models have been continuously refined with the increasing sophistication of seismic methods, global seismic networks, and the use of large computers to analyze data (Table 1). The new "global tomography" increasingly shows detail of the main boundary regions of Earth. Today this is being amplified by improved electrical methods that provide information on the resistivity of Earth materials to depths of several hundred kilometers. But while models improve in detail, and mantle structures grow more complicated, the gross pic-

Table 1. Mass of Earth and Components

Total mass	5.973×10^{24} kg
Mean density	5.515×10^{3} kg m^{-3}
Mass of inner core	9.7×10^{22} kg
Mass of outer core	1.85×10^{24} kg
Mass of mantle	4.0×10^{24} kg
Mass of crust	2.6×10^{22} kg
Mass of oceans	1.4×10^{21} kg
Mass of ice	4.3×10^{19} kg
Mass of freshwater	3.6×10^{17} kg
Mass of atmosphere	5.1×10^{18} kg
Mass of water underground	1.5×10^{19} kg
Mass of carbon in the biosphere*	6.0×10^{14} kg

* Note: This is an approximate figure—we now know that near-surface rocks with cracks and pores, and a temperature below 100°C, contain a wide variety of microorganisms. Their total mass is not known at this time. As many organisms are dominated by water, the total mass is larger.

ture established decades ago is still essentially valid. What changed dramatically in the last two decades of the twentieth century is our knowledge of Earth dynamics—how the planet changes with time and the interactions between the major geospheres. While recycling of materials in the near surface was widely recognized in the 1950s, the total picture has changed dramatically with greatly increased emphasis on significant recycling of materials at least to upper mantle depths. Also, the new observations on other planets in the solar system have sharpened our views on planetary evolution.

A Convecting Planet

Over the past decades new observations of Earth, particularly those from the ocean floors, have shown that Earth is an actively convecting body. Previously, many researchers concerned with the physical properties of the deeper parts had considered that some properties, particularly viscosity, would be too great to allow significant convection. Such considerations led to the great debate over the possibility of "continental drift."

Modern observations have resolved this debate. Data on other planets and in particular the Moon, Mars, and Venus indicated that the planets were rapidly heated during their accumulation from the dust cloud of the solar system. Planets were almost totally assembled in approximately 100 million years (Ma) after the birth of the solar system. Gravitational collapse would provide enough energy to almost totally melt a body the size of Earth. Energy would also be supplied by a host of short-lived radioactive isotopes in the early solar dust cloud. Separation of the Moon from Earth may have resulted from a major impact. Most of the larger planets could have been covered by magma (molten) oceans. With such a scenario, heavy materials, in particular liquid or solid metals, would rapidly sink to form the core, liberating gravitational energy. A heavy atmosphere rich in volatiles (H_2O-CO_2-Ar-N_2) would form in equilibrium with the surface molten zone of the planet.

We know that the planetary objects in the solar system (and most of the meteorites) formed about 4.6 billion years (Ga) ago. The oldest materials on the planet (some zircon crystals from Australia) are perhaps somewhat older than 4 Ga. The oldest rock masses on any significant scale are preserved from about 4.0 Ga ago. The record of the first 500 Ma is essentially missing, a tribute to the turbulence of the early planetary surface. But what is perhaps surprising is that the oldest rocks, with a few exceptions, are really quite similar in type to those of the modern Earth even if the quantitative relations and geologic structures are a little different.

The Earth system today, Earth dynamics, is powered by a number of heat sources. First, the surface environment is dominated by solar heating, which because of our planet's atmosphere and albedo, maintains surface water in the liquid state, and all present evidence shows that liquid water has been present for 4 Ga. This has a profound effect on the dynamics and chemical evolutions of Earth. Heat flows from the interior and this energy is supplied by residual energy from accumulation that perhaps mainly resides in the thermostat of the solid–liquid, essentially metallic core system from which the latent heat of crystallization will be slowly released; and by the radioactive decay of elements with long half-lives and in particular isotopes of potassium, uranium, and thorium in the mantle. Most Earth models consider that these sources can explain the heat flow observed at the modern surface.

For any regular object cooling from its surface, it is not difficult to determine if convection is occurring. For a uniform material cooling by conduction, heat flow should be uniform over the surface. If convection occurs, the coupled heat-mass transport will cause irregularity in the surface heat flow. Further, horizontal thermal irregularity in the body will cause an irregular topography of the surface. The most casual inspection of the topography of the solid surface of Earth shows gross irregularity, and further, the surface topography can be correlated with the expected irregularity in heat flow for a convecting body. Earth is cooling by conduction and convection and modern data show that the total energy flow is transmitted about equally by the two processes at the present time. Consideration of the thermal history of a cooling object also shows that in the past, convection must have been more important or the dominant process (for the present Moon and Mars, heat loss appears to be essentially by conduction only). We now are beginning to appreciate that the convection processes occurring on Earth drive the surface processes and are essential to the maintenance of the biosphere or ecosphere and the environment. Further, fluctuations in the convective system play a crucial role in changes of the surface environment. One of the remarkable features of our planet is that while there have been major fluctuations, to the best of our knowledge, Earth's hydrosphere has never totally frozen and never boiled. The system that supports life has been maintained more or less constant for 4.0 Ga or more.

Interactions: The Water-Cooled, Living Planet

Our planet is convecting—new hot materials reach the surface as we see in volcanic phenomena on land and even more at the great submarine ocean ridges. The study of global tomography by modern seismic techniques shows us that this complex mass-energy transport to the surface originates at great depths—hundreds of kilometers or even near the core mantle boundary.

When molten rocks (in the range of 1,000°C ± 200°C) cool near the surface, they cool, crystallize, and contract and crack. They become porous and permeable. As the surface of our planet is wet, fluids near the hot regions expand and convect, the classic cooling process. We can observe this in many regions of Earth, as in well-known volcanic hot spring regions (e.g., Iceland; Yellowstone National Park; Wairakei; New Zealand), but the greatest hot spring areas are the now famous "black smoker" regions near the ocean ridges.

We now know that the total mass of water heated by these processes (say, to 100°C) is 10^{15} kilograms (kg) (1,000 cubic kilometers, or km^3) per year. This water cooling flux compared to the mass of the oceans (1.4×10^{21} kg) shows that the entire ocean mass is processed through hot rocks in about 1 Ma. As Earth is about 4,500 Ma old, and as it was hotter in the past, it is obvious that this interaction is profound. Hot water dissolves minerals at much higher concentrations than cold water and this hot spring water carried into oceans or surface waters vast quantities of nutrients required by the living organisms (e.g., silica, SiO_2, and metals like Zn, Cu, Mn, Fe, and, at times, even gold and silver). Many of our great metal deposits were formed by such cooling processes. Such processes illustrate the vast extent of influences from the deepest parts of our planet on the living biosphere. And such processes add gases like CO_2 and H_2 to the atmosphere.

When hot rocks from the interior are water-cooled, they also absorb water and other surface materials. Consider the very familiar case of the volcanic rock basalt that comes from the deep mantle and makes up most of the floor of the oceans and islands such as the Hawaiian chain. In fact, basalt is the most common material of the crust of Earth as well as the inner planets.

Basalt is dominated by a few minerals: olivine, and iron-magnesium silicate; pyroxene; and feldspar. When these minerals are exposed to our atmosphere and hydrosphere, various reactions occur, resulting in carbonate, quartz, and clay minerals.

Such processes, which we call "weathering," are critical to the formation of the stuff we call soil but also play a major role in moderating our atmosphere and its content of carbon dioxide and even oxygen. It is interesting to note that the mass of oxygen in our atmosphere (20 percent of 5.1×10^{18} kg, about 1×10^{18} kg) reacting with basalt with 10 percent FeO, to form Fe_2O_3, would be totally removed by a volume of basalt of about 3×10^7 km^3, which is at present produced in about 2 Ma. On a "dead" planet we would have virtually no oxygen except a small amount produced by photochemical actions of solar ultraviolet on water vapor. Thus, many chemical species essential to life, such as CO_2, O_2, and even sulphate, are fixed by common minerals that come from the deep Earth.

We now know that at the sites of the great ocean trenches, old, cold, heavy, wet, volcanic, and sedimentary materials sink back into the deep Earth, at least to depths of about 700 km. This process is associated with the great earthquake events on our planet. The process recycles the surface rocks back into the mantle. Again, it is a profound process. Old rocks are removed and new ones appear. We now know that the entire crustal mass has been recycled many times in earth history (fortunately a few old pieces are preserved).

Life: The Biosphere

All the pieces of our planet interact but only in recent times have we begun to appreciate the significance to life of such interactions. The reactions of atmosphere and water on new rocks produce fertile soil and provide the host of macro (Ca, Mg, P, N, C, S, etc.) and micro (Cu, Zn, Mn, Fe, Co, etc.) nutrients required by the complex systems of the biosphere. All life requires water, often soil, with appropriate chemistry. In the key process of photosynthesis, plants and organisms use H_2O and CO_2 to produce complex organic molecules, our food supply. In return they provide oxygen and, as indicated above, without photosynthetic organisms we would not be able to survive. But on a dead, nonconvecting planet, where water and atmosphere would remove (leach) the nutrients, there would only be a very limited specialized biosphere.

Such complex interactions led to the Gaia hypothesis made popular by James Lovelock and others. Gaia suggests that we should consider the planet as a complex "living" interactive system. The concept started with the famous Earth scientist JAMES HUTTON (physician, farmer, naturalist) who in 1789 stated, "I consider the Earth to be a super organism and that its proper study should be by physiology." A recent statement by Lovelock defines the Gaia concept: "The whole system of life and its material environment is self-regulatory at a state comfortable for the organisms." Such a statement implies that "life" plays a crucial part in maintaining the environment itself.

There is no doubt that living organisms now with us play a critical role in the nature of our present environment. They provide the oxygen in our atmosphere that in turn provides the ozone in the stratosphere. They play a major role in the global budget of carbon, water, oxygen, nitrogen, sulfur, etc. Organisms form coral reefs, and synthesize many of the minerals in soils and sediments. And we now know that in some situations microorganisms live to depths of several kilometers. But some might say that the term "regulate" is too strong. When we have a great ice age, most would agree that the major cause (forcing) is the result of a low input of solar energy. If "organisms" were really smart, the moment cooling starts, they would immediately stop fixing CO_2 and produce an enhanced greenhouse. In fact what happens is the opposite. When solar energy is low, CO_2 also diminishes in the atmosphere. Organisms do respond to such changes in forcing, and without doubt our security lies in the vast biodiversity of the biosphere. Sun–Earth–Life have adapted to a changing system. The total dynamics of this geobiosystem is one of complex interactions that we must understand better.

It is interesting to reflect on the future. As our planet cools, more volatiles like water and carbon dioxide will be fixed in minerals in the cooler top few kilometers. There will be less volcanism and resupply of fresh rocks, and the biosphere may be reduced in mass. We could absorb the oceans in the minerals of the interior. Is this what happened on Mars? We must go and see.

Bibliography

FYFE, W. S. *Geosphere Interactions on a Convecting Planet: Mixing and Separation. Vol. 1: The Handbook of Environmental Chemistry*, ed. O. Hutzinger. Berlin, 1992.

JEANLOZ, R., and T. LAY. "The Core-mantle Boundary." *Scientific American* 268, no. 5 (1993):48–55.

LOVELOCK, J. *GAIA: The Practical Science of Planetary Medicine.* London, 1991.

NATIONAL RESEARCH COUNCIL. *Global Change in the Geosphere-biosphere.* Washington, DC, 1986.

SCHOPF, J. W. "Micro Fossils of the Early Archean Apex Chart: New Evidence of the Antiquity of Life." *Science* (1993):640–645.

WILLIAM S. FYFE

SORBY, HENRY C.

Henry Clifton Sorby (1826–1908) had a remarkable geologic career, ranking as a Titan of science. He was born on 10 May in Sheffield, England. A bachelor with independent income, he spent his life following wherever his curiosity happened to lead. Sorby never attended a university, never held any formal job, and lived isolated in a provincial backwater. He is the father of petrography—the microscopic study of rocks.

Sorby's interest in science was encouraged by his mother, who provided a private tutor. Sorby began publishing in agricultural chemistry, but in 1846 he obtained a copy of Playfair's book (*see* PLAYFAIR, JOHN) on JAMES HUTTON and was immediately converted to geology; he started work on alluvial deposits near his home, presenting his first paper in 1847. In 1848, at age twenty-two, Sorby met W. C. Williamson, a physician and lapidary, who made thin sections of petrified wood, teeth, and bones. Sorby learned the technique at Williamson's home and immediately started preparing thin sections of ordinary rocks; he was the first to realize the fundamental geologic importance of microscopy.

In 1849 he delivered a paper on his method at the Sheffield Philosophical Society, interpreting the local stratigraphic units, but was ridiculed for trying to "examine a mountain with a microscope." In 1851 he published the first article using polarized light to study rocks; in a cherty limestone, he used polarization to identify quartz, chalcedony, and calcite.

Sorby published the first petrographic investigation of a metamorphic rock (1856), measuring compression by the flattening of circular oolites and solving the origin of slaty cleavage. He determined mineral percentages for the first time by tracing projected images on paper, then cutting out the pieces and weighing them. In Sorby's 1858 paper he studied fluid inclusions and heated crystals to determine the temperature of original crystallization, and discriminated between crystals formed in magma and those from solution. In 1859 Sorby published a study of paleocurrents, having first noticed current structures in sandstones in 1847. At the time he had been engaged in observing sand deposition in a river and was struck with the application of his research to ancient rocks.

Sorby toured Europe in 1860, lecturing on his technique of making thin sections. In 1861 he met Ferdinand Zirkel, a young student, and described the technique to him; as a result, Zirkel immediately started making slides of igneous rocks. In 1863 Sorby published the first paper describing solution of grains at points of stress.

Passing from slides of igneous rock to meteorites, he proved that these meteorites he studied had been once molten by their crystal structure; his study of meteorites soon led him to an interest in man-made metals. Hence, in 1863 Sorby introduced the science of metallography when he recorded the first microscopic examination of polished surfaces of iron and steel. His classic paper in metallography came out in 1887; he defined techniques, components, and principles still valid today, and for this work he is considered the father of metallography. In 1865 Sorby invented the spectroscopic microscope, and gained criminological fame for detecting old bloodstains. Color and spectroscopy now absorbed his mind for over a decade.

Sorby's brilliant work on quartz varieties in sandstones and parent rocks appeared in 1877, and over eighty years passed before this work was much improved on. In 1879 Sorby founded the field of carbonate petrography in a paper describing the microscopic characters of fossils and recent shells, using his data to interpret the genesis of limestones. In 1880 he published another outstanding article on the petrography of sand grains and sandstones. He first discovered overgrowths on quartz grains, and in this paper he also discussed processes of metamorphism and recrystallization in schists.

From 1882 to 1897 he was a college president at the University of Sheffield, England, but he managed to study archaeology, painting, architecture, ancient languages, color phenomena, and so on. In 1872 he bought a yacht, and for many summers thereafter sampled marine sediments and fauna on the English coast. As a sideline to his coastal studies, Sorby walked 1930 km in four months of 1892 describing ancient buildings for his archaeological researches. He continued research on his floating laboratory until 1903 when he was paralyzed by a fall and became virtually an invalid.

Sorby's immobility channeled him back into geological microscopy, which he continued vigorously at home. In 1907 the Geological Society of London acknowledged him as the "Father of Petrography." While confined to his room, he reviewed his geological experiments of the previous sixty years and assembled them in one of his greatest papers (published posthumously in 1908), a landmark in quantitative geology. Sorby died in Sheffield in 1908.

Sorby championed free and unrestricted research, unsupported by financial backing for work in specific fields. He felt best results were obtained if the mind was allowed to wander freely into any subject. He was an initiator, not a consolidator, but Sorby's papers on limestones, slaty cleavage, crystal growth, and quantitative geology are still frequently cited.

Bibliography

Folk, R. L. "H. C. Sorby (1826–1908), the Founder of Petrography." *Journal of Geological Education* 18 (1965):43–47.

Higham, N. *A Very Scientific Gentleman: The Major Achievements of Henry Clifton Sorby.* New York, 1963.

Summerson, C. H., ed. *Sorby on Sedimentology.* Miami, FL, 1976.

———. *Sorby on Geology.* Miami, FL, 1978.

ROBERT L. FOLK

SOUTHERN LIGHTS

See Auroras

SPACE, FUTURE EXPLORATION OF

Space exploration is manifestly a child of the cold war's technology and intense political competitiveness. Following the end of the cold war in the early 1990s, space exploration appears to have lost much of its earlier priority, and, in the mid–1990s, future directions are unclear. The nature of the political order that emerges in the coming decades may set that direction and pace; developments in space exploration will also influence society but probably not as much as advocates might wish. On a longer timescale of fifty years or more, however, discoveries made in space and incremental technological progress are likely to cause space activity to play an increasingly significant role in the structures of technologically advanced nations—one that is both transforming and mundane—the same kind of role that civil aviation achieved by the end of the twentieth century.

Following the drama of the race between the United States and the former Soviet Union to send the first humans to the Moon in the 1960s, human explorers have had their horizons limited to orbital missions in space stations (*Skylab, Progress,* and *Mir*) and in the space shuttle. Although much has been learned about human performance in space (and human capabilities have been impressively demonstrated for carrying out space experiments and repairing robotic spacecraft), progress toward further human exploration of the Moon and the inner solar system has been modest. Following Project Apollo, solar system exploration has been exclusively the domain of robotic spacecraft—whose successes (and occasional failures) have been spectacular. The *Mariner 10* flybys of Mercury, the Viking orbiter and lander missions at Mars, the Venera, Pioneer, and Magellan missions to Venus, and the Voyager exploration of the outer solar system all the way to Neptune, have together transformed our understanding of the solar system. Meanwhile, telescopic observations, meteoritical analyses, and theoretical modeling have made us much more knowledgeable of the populations of comets and asteroids, and of their significance to the history and future of our own planet. Comet Halley returned to the inner solar system in 1986 to be greeted by an armada of robotic spacecraft from Japan, the former Soviet Union, and Europe flying in a coordinated program of exploration (which included important contributions by the United States) that may have established an important precedent for future space exploration.

The Soviet space program, prior to the dissolution of the Soviet Union, had the long-term goal of sending crewed flights to Mars. Cosmonaut flights with durations of up to a year in the *Mir* space station demonstrated that typical flight times to Mars could be tolerated by humans (albeit, not without a significant physical and psychic toll). The Soviets also established a heavy-lift launch capability—the Energya rocket—that spoke of serious intentions for future manned spaceflight. The present Russian space program has inherited a number of robotic Mars missions from that earlier program and the individuals and institutions involved are demonstrating remarkable progress toward actually carrying out those flights in spite of dramatic cutbacks in spending on space.

In the United States, before the end of the cold war, a rationale for resuming the human exploration of space had been developed by the presidential administration of George Bush and enunciated in terms of the goal of returning to the Moon and pushing on to Mars by the time of the fiftieth anniversary (2019) of the first Apollo landing. However, the rationale was never stated convincingly or forcefully enough for the U.S. Congress to appropriate funding to support this proposal, and it was eventually killed in the first year of the Clinton administration. A significant part of Congress's lack of enthusiasm was no doubt due to the rumored price tag of several hundred billion dollars. The U.S. space program's exploration effort, therefore, remains robotic. That program, too, is being scaled back as part of changing priorities within the National Aeronautics and Space Administration (NASA).

Elsewhere in the world, space exploration has enjoyed only relatively modest priority, probably because the United States and the former Soviet Union moved so far and so fast in their competition. Also, deep-space ventures call for the most powerful launch capability, for huge, globally distributed antennas, highly autonomous spacecraft, and there is high risk—placing space exploration beyond the resources of most nations. The European Space Agency (ESA) has accepted this cost and with Giotto's intercept of comet Halley and *Ulysses's* successful "solar polar" mission (the latter shuttle-launched and powered by a U.S.-provided radioisotope thermoelectric generator (RTG) in a cooperative program with NASA) has demonstrated a mature capability. ESA's plans to carry out the first comet rendezvous (the Rosetta mission) are, therefore, significant to space scientists throughout the world. Japan also has marshaled the resources to compete in certain exploration niches and to systematically acquire the necessary skills to do more. Missions to the Moon (penetrator landers), to Mars (a space physics orbiter), and to comets (dust sample return) figure in their plans. The French have shown notable enthusiasm for space exploration, which they have carried out imaginatively as part of the ESA program and, especially, as partners with the former Soviet Union in the exploration of Mars. Likewise, Germany has developed effective partnerships with both the United States (*Galileo*) and with the Russians (on their planned Mars missions).

Proponents of vigorous space exploration face increasing skepticism, in the United States especially, as they argue for the expenditure of the considerable resources needed to further explore space beyond Earth's bounds. It seems evident that there is no near-term monetary profit to be made by such exploration, or else private capital would suffice. Moreover, the role and purpose of humans (as opposed to robots) in space remain unclear to many. And fundamental scientific knowledge about the rest of the solar system appears irrelevant to the immediate needs of nations that have spent much of their treasure on fighting the cold war and now are engaged in close combat in the economic arena. Certainly the case for bold space exploration, by humans or robots, has not been made effectively since Apollo and, evidently, the first post-cold-war decade is not a good time to try. (Such advocacy might describe space exploration as an essential part of mankind's enlightenment through the search for knowledge, as a unique source of inspiration—especially for young people—as a natural realm for mutually beneficial international collaboration, and as a significant factor in the long-term economic and environmental well-being of our planet. But these are considered soft arguments in the absence of a specific threat to national or international security.) The attention of most governments is focused on immediate economic adjustment to a peacetime economy and to other long-postponed problems of society.

What then are the near-term trends and what enduring issues are likely to provide an upswing in the prospects for a renewal of interest in space exploration? As noted earlier, such exploration will continue to be dominated by robotic activity but somewhat less so by the U.S. and Russian programs. The U.S. program for the past two decades has been based on the launch capability of the Titan-Centaur and the Shuttle-Inertial Upper Stage (IUS) (including the one-stage IUS, or Transfer Orbit Stage) but is now being sized to the capability of the Delta launch vehicle and smaller rockets. For the foreseeable future, the Cassini mission (including ESA's Huygens Probe) to Saturn and Titan may be the last of the line of heroically scaled missions. This downscaling is taking place at the same time that the technological benefits (e.g., miniaturization of sensors and propulsion subsystems) from the recent Strategic Defense Initiative (SDI, or "Star Wars") are becoming available to civilian space programs; small-scale missions, therefore, need not lack substance. Some analogy may be drawn with the transition from mainframe computing to personal computing, though the analogy should not be pushed too far—space technology is certainly not demonstrating order of magnitude improvements every few years.

The U.S. planetary exploration program will have several thrusts. First is a program of small missions (the Discovery Program) targeted to planets and small bodies in the inner solar system—missions driven by technological innovation and public interest as much as by science. Mission scope will be constrained along the lines of the recent Clementine mission to the Moon (this mission, carried out by the U.S. Department of Defense, may be a precursor of others). A first Discovery mission will begin the exploration of Earth-approaching asteroids and following missions may demonstrate considerable diversity in terms of targets and technical implementation. The Mars Surveyor Program of small orbiters and landers is being planned, and eventually will become international in scope. In the outer solar system, a fast flyby mission to Pluto by a relatively miniature spacecraft will formally complete the initial reconnaissance of all the planets. Yet another exciting direction is being contemplated by a program of orbital telescopic missions to search for planets (including Earth-sized bodies) orbiting other stars in our galaxy.

The Mars Surveyor Program is new and just beginning to take shape, one that will be adjusted specifically to benefit from cooperative missions with partners from Russia, Europe, and Asia. Initially, the focus will be on the achievement of the science objectives previously planned for the *Mars Observer*, which was lost in September 1993. Assuming success for this endeavor and for the planned Russian Mars orbiter/lander missions scheduled for 1996, the data will then be available to plan subsequent lander missions, which are likely to include both mobile laboratories and sample returns.

The planet Mars has emerged as the focus of international scientific interest for several reasons: (1) it lies within what is believed to be our solar system's habitable zone (the range of radial distance from the Sun within which liquid water may have been stable at a planetary surface at some time or another in its history); (2) it has a complex geologic history with evidence of a past aqueous environment; (3) surface/atmosphere interactions are important on Mars and their study may provide clues about our own climate history; and (4) the planet is relatively easy to reach. Because the habitable zone of our solar system may have extended out to Mars's orbit, life may have evolved on Mars early in its history and there is a strong motivation to look for geochemical evidence of primitive early life, or even for fossils.

Presently we have little assurance that this particular scientific quest—to resolve the issue of whether life originated on Mars—will be straightforward. Certainly, on our own planet, even with all the means at our disposal, the paleontologist's task has been exceptionally difficult and progress has been slow. Evidence of ancient life is hard won and biotic and abiotic processes can sometimes leave similar records. No clear picture has yet emerged as to when, where, and how life evolved on Earth. At Mars, the likelihood that a relative handful of robotic spacecraft teleoperated from Earth—and even returning samples to Earth—will be able to meet the equivalent challenge does not seem great. Thus many scientists whose interests are focused on early Mars are hopeful that one day the unique talents and insights of in situ human explorers will be brought to bear on a problem that otherwise may not yield.

Given the current modest priority of space exploration throughout the world, one may suppose it unlikely that an impulse toward human exploration of Mars will emerge without powerful political motivations. What might they be? One such motivation is already in evidence in the proposed cooperation between the United States and Russia in building and operating an international space station—a desire to further the internal stability of Russia and to diminish the likelihood that Russian rocketry hardware and expertise will be exported to unfriendly nations. Another motivator could be a challenge from a relatively new space-faring nation such as China. The resources of near-Earth space—especially energy in a world where the population is set to increase dramatically and where environmental overload is widely feared—may in time entice the advanced nations to undertake serious exploitation; the mining of helium 3 (a potentially clean fuel for fission reactors absorbed as a component of the solar wind over 4.5 billion years, or Ga) from the lunar soil or the establishment of great "farms" of solar cells on the Moon are enterprises that will likely call for human skills. Taking a more positive point of view, the telescopic discovery of Earth-like planets in other solar systems could lead to an enthusiasm to build large telescopes on the Moon to characterize the atmospheres of such planets, looking for evidence of life. Other political and cultural motivators and other scenarios may be projected also—for just as each of the planets has proved more remarkable than scientists had dared predict prior to each first close flyby, so human history defies adequate prediction.

What might an advocate of both human and robotic exploration missions offer as an optimistic scenario for the future, independent of political speculation? First, coming technological advances will diminish the intimidating expense of travel to and beyond Earth orbit; recycled intercontinental ballistic missiles (ICBMs) may make such a contribution but a new generation of launchers (reusable single stage to orbit vehicles?) will be needed to make a fundamental change in costs. For interplanetary travel, solar electric propulsion (long promised and soon to be available in Europe) can achieve significant efficiencies, while for surface operations and Earth return missions, the use of in situ resources for chemical propellants can transform the economics of space flight.

Second, the advocate would note that the heavy cost of mission operations will increasingly yield to the power of the workstation and personal computer and to the widening availability of modern ("user-friendly") operations and control software—scientists and engineers will exert more direct control over their experiments and will do so from their offices rather than from specialized facilities at space centers. "Telepresence" operations of robots on the Moon, carried out by scientists and engineers on Earth with the immediacy of being actually present at the scene of activity, could dramatically increase productivity there. Together with technology to use in situ lunar oxygen, such operations could prepare the way for low-cost, low-risk human exploration even within the present budgets of the present space-faring nations.

The exploration advocate would also confidently expect that Mars would continue to surprise and delight us with discoveries germane to our own planet's evolution—discoveries that would lead us to escalate the pace of activity there. Sample return missions will likely begin the exploitation of Martian resources, specifically the creation of carbon-rich fuels and liquid oxygen for the return trip, and will demonstrate substantial economies. The use of in situ resources for fuels and, also, for life support (water, oxygen, and buffer gases for air) might allow human missions to Mars to be carried out within the current budgets of an increasingly international community of space explorers.

Our confidence in proceeding with long duration human missions will grow as we accumulate experience at an international space station; experiments with an artificial gravity centrifuge, with microgravity countermeasures, and with bioregenerative life support systems will set the scene for the first expedition to Mars, perhaps by the end of the first decade of the twenty-first century. The crew would rely on small nuclear reactors for primary power during their Martian surface exploration, which would not be a rushed thirty-day mission but could be, in principle, almost of indefinite duration. A Mars base would inevitably grow up around a well-chosen first landing site and the exploration of Mars would be carried out globally by in situ field geoscientists and by surrogate robots under telepresence control from the base. Within a decade or two of the base's establishment we would expect to have definitive evidence of whether life originated on more than one planet in our solar system.

Probably of more importance to most people's lives, low-cost access to space, high capability telepresence control of robots, and the accessibility of near-Earth resources would gradually make space operations routine, affordable, and safe, enabling many new applications that would rapidly be taken for granted. Our exploratory urges will not then disappear, however, but will likely turn to newly discovered Earth-sized planets many light years away—for at least one telescope capable of such discovery is already being developed and could be operating in high Earth orbit by the turn of the century.

Bibliography

GATLAND, K. *The Illustrated Encyclopedia of Space Technology.* Scottsdale, AZ, 1989.

LEWIS, J. S., and R. A. LEWIS. *Space Resources: Breaking the Bonds of Earth.* New York, 1987.

U.S. GOVERNMENT PRINTING OFFICE. *America at the Threshold: Report of the Synthesis Group on America's Space Exploration Initiative.* Washington, DC, 1991.

———. "Exploring the Moon and Mars—Choices for the Nation." In *Report of the Technology Assessment of the United States Congress.* Washington, DC, 1991.

GEOFFREY BRIGGS

SPECIAL METALS

Among the materials that we extract from Earth, the special or minor metals are the most difficult to characterize. Fuels, for example, are all valued for their energy production. The special metals, on the other hand, have no such common use. Their uses are far more diverse, sharing only the characteristic that market demand is small since each fills a specialized niche in our technologically based economy.

Properties, Production, and Uses

The important physical/chemical properties of each of these metals, their geological and geographical occurrence, uses, and present and future importance in the world economy are summarized. Tables 1 and 2 list the important physical properties of the special metals. Table 3 gives production statistics. Table 4 lists the major producing countries.

Data on special metals are limited because some companies keep production statistics confidential.

To accommodate the diversity of these commodities, each special metal is discussed individually:

- Antimony (Sb): In the United States, antimony is a by-product of silver mining, whereas in China and South Africa stibnite (Sb_2S_3) is mined in its own right. Production is dominated by China (Table 4) whose trade policy has resulted in a chronic oversupply. Large undeveloped resources of antimony exist; there is more than a sixty-five-year supply. Antimony is used as a flame retardant in plastics and textiles, an alloy in lead batteries, a corrosion-resistant additive to steel, and for low-friction bearings.
- Arsenic (As): Arsenic is a by-product of nonferrous metal smelting. It is a highly toxic metal with deleterious environmental impacts. The potential supply is enormous, with production dominated by the major refiners of copper, lead, and zinc. Arsenic is used as an additive in wood preservatives and as a toxic agent in pesticides, fungicides, and insecticides.
- Beryllium (Be): Traditionally beryllium comes from the mineral beryl ($Be_3Al_2Si_6O_{18}$), which is mined from coarse-grained, shallow intrusive igneous rocks called pegmatites—common in the eroded cores of many mountain belts. Major producers are Russia, in particular the transbaikal region, Kazakhstan, Brazil, Argentina, Zimbabwe, and Madagascar. The gem varieties of beryl, emerald and aquamarine, are mined in Brazil and Colombia. Beryl resources worldwide are extensive and are concentrated in Brazil, India, the former Soviet Union, and the United States. Beryl resources equal a one-hundred-year supply.

  Beryllium metal is also extracted from bertrandite, $Be_4Si_2O_7(OH)_2$. Commercial bertrandite at Spor Mountain, Utah, is found in volcanic tuffs that have been altered by hot aqueous fluids. This deposit, responsible for all of the U.S. beryllium production in 1994, has reserves equal to sixty years of production. Additional bertrandite resources are known in Utah, in volcanic-hosted gold mineralization, and at Thor Lake, Northwest Territories, Canada, in sodium- and potassium-rich granites similar to peg-

Table 1. Important Physical Properties of the Special Metals

	Symbol and Atomic No.	*Specific Gravity*	*Melting Point (°Celsius)*	*Resistance (μohm-cm)*	*Thermal Conductivity (W cm^{-1} K^{-1} at 0°C)*
Antimony	Sb_{51}	6.69	631	39.0	0.255
Arsenic	As_{33}	5.73	817	33.3	0.539
Beryllium	Be_4	1.85	1278	4.0	2.18
Bismuth	Bi_{83}	9.75	271	106.8	0.10
Cesium	Cs_{55}	1.87	28.4	20.0	0.36
Gallium	Ga_{31}	5.90	29.8	17.4	0.41
Germanium	Ge_{32}	5.32	937	46×10^6	0.67
Hafnium	Hf_{72}	13.31	2227	35.1	0.23
Niobium	Nb_{41}	8.57	2480	12.5	0.05
Selenium	Se_{34}	4.79	217	10^6	0.05
Tantalum	Ta_{73}	16.6	2996	12.45	0.57
Tellurium	Te_{52}	6.24	449	4.36×10^6	0.04
Zirconium	Zr_{40}	6.51	1852	40.0	0.23

Table 2. Important Physical Properties of the Rare Earths*

	Symbol and Atomic No.	*Specific Gravity*	*Melting Point (°Celsius)*	*Resistance (μohm-cm)*	*Thermal Conductivity (W cm^{-1} K^{-1} at 0°C)*
Scandium	Sc_{21}	2.99	1541	61.0	0.16
Yttrium	Y_{39}	4.47	1522	57.0	0.17
Lanthanum	La_{57}	6.15	921	5.70	0.13
Cerium	Ce_{58}	6.66	799	75.0	0.11
Praeseodymium	Pr_{59}	6.77	931	68	0.12
Neodymium	Nd_{60}	6.80	1021	64.0	0.16
Promethium	Pm_{61}	7.22	1080		0.18
Samarium	Sm_{62}	7.52	1077	88.0	0.13
Europium	Eu_{63}	5.24	1597	90.0	0.14
Gadolinium	Gd_{64}	7.90	1313	140.5	0.10
Terbium	Tb_{65}	8.23	1356		0.14
Dysprosium	Dy_{66}	8.55	1412	57.0	0.11
Holmium	Ho_{67}	8.80	1474	87.0	0.22
Erbium	Er_{68}	9.07	1529	107.0	0.19
Thulium	Tm_{69}	9.32	1545	79.0	0.14
Ytterbium	Yb_{70}	6.97	819	29.0	0.35
Lutetium	Lu_{71}	9.84	1663	79.0	0.24

* The rare earth elements are often referred to collectively as the REEs.

matites. Thor Lake also contains resources of niobium, tantalum, gallium, and rare earths.

Beryllium metal imparts hardness, strength, and corrosion resistance to copper for products such as aircraft bearings, computer parts, and even eyeglass frames. Its characteristic high melting point (Table 1) is critical to the aerospace industry for reentry vehicles and rocket motor parts. Beryllium mirrors are used in infrared detection devices. Beryllium metal, being transparent to X rays, is used for windows on X-ray tubes. Beryllium oxide makes a hard, chemically resistant ceramic used in insulators for high-power electronic circuitry, lasers, and microwave devices.

- Bismuth (Bi): Bismuth is a by-product of nonferrous metal smelting and refining. Production is small, limited to a few thousand tons per year. The majority of bismuth is consumed in the manufacture of medicines and cosmetics. The other major use is for low-melting alloys.
- Cesium (Cs): Cesium is found in two minerals, the lithium mica, lepidolite ($LiAl_3Si_3O_{10}(OH)_2$), and by atomic substitution in pollucite ($Cs_4Al_4Si_9O_{26} \cdot H_2O$). Most North American production is from pollucite mined at Bernic Lake, Manitoba. Several hundred thousand tons of pollucite reserves remain in this district, sufficient to meet demand for many years. As the most electropositive element in the periodic table, cesium is used as an oxygen-getter in vacuum tubes. It is also used in photocells and in cesium clocks for precision timing.
- Gallium (Ga): Gallium is a rare metal. Demand is small. No production data are available. It is a by-product of zinc production and is used in specialized metal alloys and in the electronics industry. Gallium arsenides are used in light-emitting and light-sensitive diodes. Great potential lies in the semiconductor and solar power industries. Gallium reserves are large.
- Germanium (Ge): Germanium is a by-product of zinc production. The United States imported 13.2 metric tons of germanium materials in 1992, predominantly from Germany, the People's Republic of China, Spain, United Kingdom, Russia,

Table 3. Production of Special Metals (in thousands of metric tons)

	U.S. (1992)	*World (1992)*
Antimony	19.7	75.7
Arsenic trioxide	0	47,600
Beryllium (as beryl)	7	4.8
Bismuth	n.a.	3.0
Cesium	n.a.	n.a.
Gallium	n.a.	n.a.
Germanium	n.a.	n.a.
Hafnium	n.a.	n.a.
Niobium (Columbium)	0	14.6
Rare earth elements	20.7	55.4
Selenium	0.26	2.01
Tantalum	0	0.35
Tellurium	0.04	0.19
Zirconium	108.2	807.0

n.a.: not available.

Note: Except for Se and Te, all data are from the most current *Annual Report of the U.S. Bureau of Mines.* Typical publication of these data is delayed 1–2 years for domestic and 2–3 years for foreign statistics. Se and Te data are from Hargreaves and others, 1994, and are for 1991.

Belgium, and Hong Kong; U.S. production is estimated at 13,000 kilograms (kg). Supply historically exceeds demand.

Germanium use expanded in the 1950s with the development of the transistor. Silicon-based integrated circuits replaced transistors in the 1970s with demand shifting to infrared detectors and fiber-optic communications. Single crystals of germanium are used in gamma-ray detectors.

- Hafnium (Hf) and Zirconium (Zr): Most special metals are intrinsically rare. Zirconium and hafnium are, however, relatively abundant in the earth's crust. They occur together as major constituents of the mineral zircon ($ZrSiO_4$) and in the less common mineral baddeleyite (ZrO_2). Zircon is common in silica-rich igneous rocks, especially granites and granite pegmatites, and in beach sands rich in heavy minerals. Hafnium replaces zirconium isomorphously in both minerals. The hafnium/zirconium ratio is about one-to-fifty in both minerals; thus, hafnium is a coproduct of zirconium production.

Two markets operate. The first is zircon as an industrial mineral. Zircon by-products of titanium oxide mining are concentrated for use as a foundry sand or as an additive to ceramics.

Metallic zirconium and hafnium are extracted from zircon and baddeleyite. Both metals are used in the nuclear power industry: zirconium as an anti-corrosion coating on fuel rods, hafnium for control rods. A small amount of zirconium oxide is used to synthesize cubic zirconia, whose high-light dispersion and hardness make it a "poor-man's diamond." Zirconia is highly refractory, raising the melting point and stabilizing the structure of many high-temperature ceramics.

- Niobium (Nb) (Columbium): This refractory, chemically stable metal was traditionally mined from rocks rich in columbite, $(Fe,Mn)Nb_2O_6$, which is always associated with the tantalum (Ta) mineral, tantalite ($(Fe,Mn)Ta_2O_6$), in silica-rich igneous rocks. Thus, niobium and tantalum are often discussed together. Today, most niobium production is from the mineral pyrochlore, $(Na,Ca)_2Nb_2O_6(O,OH,F)$, which contains little tantalum. Thus, niobium and tantalum production are no longer coupled.

Niobium reserves are vast, a thousand times annual consumption. They are located in Brazil, Canada, Nigeria, and Zaire. Eighty percent of demand is for additives to improve the strength and corrosion resistance of steel. It is used in high-temperature environments like jet engines, rocket nozzles, and exhaust manifolds. Niobium carbide is an effective abrasive.

- Rare Earth Elements (REEs): The rare earths include seventeen chemical elements: scandium (Sc), yttrium (Y), plus the fifteen elements of the lanthanide series (Table 2). The name is misleading. For example, scandium is more abundant than copper, although it is the least abundant rare earth; thulium (Tm) is more common than antimony. "Rare" refers to the practical difficulties of separating these metals from one another. Their chemical and physical behavior is so similar that obtaining pure rare earth metal or salts is difficult.

The rare earths are found together, most commonly in varying proportions in monazite, $(REE)PO_4$, xenotime, YPO_4, or bastnaesite, $(REE)CO_3F$. Monazite and xenotime are common in heavy mineral beach sands, which account for more than half the world's rare earth

Table 4. Major Producing Nations of Special Metals (in rank order with percent of world's total 1992 production given)

Antimony	China (59.2%), Russia (13.2%), Bolivia (8.6%), South Africa (5.3%), Kyrgyzstan (4.0%)
Arsenic	n.a.
Beryllium	U.S. (51.0%), former Soviet Union (22.6%), PRC (15.5%), Brazil (10.0%)
Bismuth	PRC (30.0%), Peru (15.3%), Mexico (18.0%), Australia (10.2%), Japan (12.0%), U.S. (2.6%), Canada (2.6%)
Cesium	n.a.
Gallium	n.a.
Germanium	n.a.
Hafnium	see zirconium
Niobium	Brazil (80.1%), Canada (15.7%), Zaire (3.2%)
REEs	U.S. (37.4%), China (30.7%), former Soviet Union (14.4%), Australia (7.0%), India (5.0%), Brazil (2.5%)
Selenium	Japan (26.7%), Canada (19.7%), U.S. (12.9%), Belgium (12.4%), Germany (5.5%), Philippines (3.0%), former Yugoslavia (2.7%)
Tantalum	Australia (52.4%), Brazil (17.0%), Canada (12.7%), Rwanda (9.3%), Zaire (4.0%)
Tellurium	Belgium (32.1%), Japan (32.1%), U.S. (18.7%), Canada (7.0%), Peru (4.8%)
Zirconium	Australia (37.2%), South Africa (28.5%), U.S. (13.4%), Ukraine (9.3%), Sri Lanka (3.7%), Brazil (2.5%), India (2.2%), PRC (1.9%)

n.a.: not available

Note: Except for Be, Bi, Se, and Te, all data are from the most current *Annual Report of the U.S. Bureau of Mines*. Be, Bi, Se, and Te data are from Hargreaves and others, 1994. Be data are for 1990; Bi, Se, and Te are for 1991.

production. Bastnaesite is found in rare magmatic, marble-like rocks called carbonatites. The major U.S. source of rare earths is Mountain Pass, California, and the huge deposit at Bayan Obo in Inner Mongolia consists of carbonatites. Mountain Pass can produce 27,000 tons/year of bastnaesite concentrate with reserves equal to about 150 years. Worldwide resources are vast with still indeterminate reserves at Bayan Obo and at the recently discovered Olympic Dam deposit in Australia.

The uses of rare earths are diverse and not

easily characterized in a short review. Annual consumption has increased steadily over the past decade and is likely to continue to do so as new uses of these metals develop. Rare earth oxides are catalysts in oil refineries and catalytic converters, colorizers in the glass industry, and important components in phosphors. Samarium (Sm), neodymium (Nd), praseodymium (Pr), and dysprosium (Dy) are mixed with cobalt and iron borides to make strong, permanent magnets. Scandium, neodymium, terbium (Tb), and yttrium are used in lasers. Lanthanum (La) is present in many semiconductors and low-temperature superconductors. Significant future uses of rare earths are expected in the communication/fiber-optic, superconductor, security (phosphors to create invisible bar codes), and battery industries.

- Selenium (Se): Selenium is a refinery by-product of nonferrous and precious metal mining. World reserves exceed a forty-year supply. Production likely amounts to several thousand tons per year. Selenium is used in electronics, in photocopier drums, and in pigments for ceramics, plastics, inks, and paints. Sunlight transmission in plate glass is reduced by additions of selenium.
- Tantalum (Ta): As noted (see niobium), tantalum is present in tantalite, $(Fe,Mn)Ta_2O_6$. Tantalum metal is refractory, resistant to corrosion, and a good conductor of heat and electricity. More than half the annual production of tantalum is for tantalum-based capacitors. Remaining demand is for superalloys in the aerospace industry and tantalum carbide for cutting tools and war-resistant parts. Reserves in Australia, Thailand, Nigeria, Canada, and Zaire are sufficient for at least one hundred years at current levels of consumption.
- Tellurium (Te): Tellurium is found in nonferrous and precious metal deposits and is a by-product of their production. Gold-silver telluride minerals are mined in many districts. Reserves are extensive. Annual consumption is limited to a few hundred tons; it is mostly used as a minor additive to improve the machinability of copper and stainless steel.

Uniqueness of Special Metals

Special metals are unique in several senses. First, annual production and consumption are quite limited, amounting only to tens, hundreds, or a few thousands of tons. Second, some are rare, meaning that they are limited in abundance in the earth's crust, whereas others are scarce, difficult, and expensive to acquire in the marketplace even though they may be relatively abundant. Special metal markets are uniquely unstable and unpredictable. Demand can vary significantly from metal to metal, from year to year. Limited applications often mean only a small handful of producers or purchasers are active in the market at any time. Finally, since many of the special metals are by-products or coproducts, production is strongly dependent on other, independent market factors. Many special metals are removed from the main product regardless of market demand to avoid price penalties that buyers would impose were the special metal left as an impurity in the main product. In other instances, the special metals are removed from the waste stream for environmental reasons. In both of these latter cases, the result is ready potential to meet increases in market demand from existing production.

Bibliography

Craig, J. R., D. J. Vaughan, and B. J. Skinner. *Resources of the Earth.* New York, 1988.

Guilbert, J. M., and C. F. Park. *The Geology of Ore Deposits.* New York, 1986.

Hargreaves, D., M. Eden-Greene, and J. Devaney. *World Index of Resources and Population.* Brookfield, VT, 1994.

U.S. Department of the Interior. *Annual Report of the Bureau of Mines.* Washington, DC.

P. GEOFFREY FEISS

SPONGES

See Colonial Invertebrate Fossils

STALACTITES AND STALAGMITES

See Caves and Karst Topography

STANDARD MODEL

See Cosmology, Big Bang, and the Standard Model

STARS

The subject of how stars are born from interstellar gas, how they evolve during their lifetime, and how they die is a fascinating puzzle that many astronomers have studied for most of the twentieth century. Our understanding of this subject has now reached a level that a fairly comprehensive picture (with some gaps) can be presented. From a galactic standpoint stars play an important role in the recycling and processing of interstellar gas. The next generation of stars are formed in part out of the ashes of older stars that have died. Stellar ashes are composed of materials like carbon, silicon, and iron that have been formed as a result of nuclear reactions inside stars and without which it would not be possible to form planets or allow even life itself to exist. This entry discusses the classification of stars, star formation, stellar evolution, and how stars end their lives (*see also* SUN).

Classification and General Characteristics

A star is a large gaseous body that is able to generate its own heat and light. Stars produce energy by both gravitational contraction (thereby releasing gravitational potential energy) and by nuclear fusion. During a star's life the majority of its energy production arises from nuclear fusion. In contrast, planets shine only as a result of reflected light from their parent star. (Planets do produce some internal heat as a result of radioactive decay of unstable elements, but they are incapable of sustaining nuclear fusion reactions.)

Stars are initially composed mostly of hydrogen (approximately 70 percent by mass) and helium (approximately 25 to 30 percent by mass). Elements heavier than hydrogen and helium (i.e., lithium to uranium and beyond) comprise at best only a few percent of the total stellar mass for the current generation of stars and even a lesser fraction for stars from earlier generations. For example, this fraction for the Sun has been deduced to be about 2 percent. (All the hydrogen and most of the helium in the universe are believed to have been produced in the Big Bang, while some helium and virtually all the other elements were later produced as a result of nuclear reactions in stars.) Except for some stars at the end of their life, such as white dwarfs and neutron stars, all the matter in a star exists in a gaseous state composed of molecules, neutral atoms, ions, and/or free electrons, depending on temperature.

Stars vary greatly in size. In terms of mass they range from about 0.08 to about 100 times the mass of the Sun. The Sun's mass is 1.989×10^{33} grams (g), which is 332,900 times the mass of Earth. Objects with a mass below the lower mass limit are unable to ignite the nuclear fusion reactions that convert hydrogen into helium. This inability prevents such an object from making the transition from being a protostar undergoing gravitational contraction to becoming a star where the forces of gravity are balanced by the internal pressures generated by nuclear fusion. An object that forms with a mass above the upper mass limit is vibrationally unstable; this condition causes it to lose enough mass to bring it within the upper limit. In terms of radius, normal stars show an even greater range of values compared to mass, ranging from about 0.1 to over 2,000 times the radius of the Sun. The solar radius is 6.960×10^{10} centimeters (cm), which is 109.1 times the radius of Earth. The largest star known, if placed in the center of our solar system, would extend almost to the orbit of Saturn.

Non-normal stars such as white dwarfs or neutron stars are even smaller than the lower limit for normal stars. The radii of white dwarfs are roughly about one-hundredth the size of the Sun (approximately the size of Earth) and the radii of neutron stars are believed to be around only 10 kilometers (km)!

In terms of brightness normal stars can shine with a luminosity ranging from 0.01 to 10^6 times that of the Sun. Stars that undergo explosions can temporarily become even brighter than the upper limit. The luminosity of the Sun is 3.85×10^{33} ergs per second, which means in one second the Sun releases the energy of 9.2×10^{10} one-megaton bombs. In general the more massive the star is, the more luminous it is. White dwarf stars and neutron stars can be less luminous than the lower limit of normal stars to the point of being essentially invisible. The surface temperatures of normal stars range from roughly 2,000 K to around 50,000 K with the Sun's surface temperature measured at 5,778 K. In comparison a room temperature of 27°C corresponds to only 300 K. White dwarfs and neutron stars can have surface temperatures in excess of and below the range encountered by normal stars. (For more information, see Böhm-Vitense, 1989.)

In the earliest studies, stars were classified in terms of their apparent brightness. One system devised almost two thousand years ago by the Alexandrian astronomer Claudius Ptolemy is still in use today, with some changes. It groups stars into steps of brightness called magnitudes in which the brightest stars were considered to be of the first magnitude and the faintest stars visible to the naked eye were considered to be of the sixth magnitude. Today, the system is more precisely defined so that a difference in five magnitudes corresponds to a brightness ratio of 100 and magnitude numbers range from -26.8 for the Sun to $+28$ for the faintest stars detectable with the space telescope.

While it is a useful classification system, the magnitude system does not shed much information regarding the intrinsic properties of stars. Much progress in the area was achieved, however, when stars were studied (beginning in the late 1800s) with a spectroscope. The spectra of stars showed different patterns of dark and sometimes bright lines that allowed stars to be classified into spectral classes of differing characteristics. Ultimately astronomers were able to deduce the elemental abundances, the temperature, and other physical conditions present on the surface of a star based on the line pattern and strengths of the lines in the stellar spectrum. The principal method of stellar classification used by astronomers today is the spectrographic system in which stars are grouped into one of seven main spectral classes or a few special categories. The main spectral classes form a temperature sequence, and in order of decreasing temperature the types are: O, B, A, F, G, K, and M (Figure 1). For example, O stars, being the hottest, show lines of singly ionized helium, while M stars, being the coolest, show the presence of molecules such as titanium oxide. Hydrogen lines are strongest for type A. Each spectral class is divided into potentially 10 subgroups designated by the digits 0 through 9, and there are additional descriptions to note whether, for example, a star is of normal, giant, or supergiant size. The Sun's spectral type is G2V. The V means it is a normal size star. (For additional details, see Kaler, 1986; Böhm-Vitense, 1989; and Kaler, 1989.)

Star Formation

Space between stars is not empty but actually contains gas and dust at very low densities, which is known as the interstellar medium. For reasons not currently well understood, concentrations of material from the interstellar medium occur which can become interstellar clouds that are able to maintain their identity by their self-gravity. These clouds range in size to small ones having a few solar masses of material and are about a light-year in size—a light-year is the distance light travels in one year, which is 3.09×10^{12} km—to the giant molecular clouds that can have a million solar masses of material and be over 300 light-years across. These clouds are the coldest natural objects in the universe, with a temperature as low as 10 K in the denser regions, and they are composed mostly of molecular hydrogen at a mean density of 100 to 300 molecules per cubic centimeter (cm^3). Most of the interstellar molecular mass in the Galaxy is contained in a few thousand giant molecular clouds. These clouds have lifetimes of 10^7 to 10^8 years. Interstellar clouds are able to resist gravitational collapse from a combination of magnetic pressure generated from the internal magnetic field present from the galactic background and turbulent pressure produced by the highly supersonic and turbulent motion of gas inside the cloud. How the turbulence originates and is maintained inside the cloud is one of the unsolved problems of star formation. The structure of the giant molecular clouds is not uniform and about 10 percent of cloud mass is contained in dense cores having a density of 10^4 hydrogen molecules per cubic centimeter and sizes ranging from 1/3 to 15 light-years across, which corresponds to a few to a thousand solar masses of gas.

Interstellar clouds as a whole or their dense cores can be induced to contract by different external and internal forces. External forces include the effect of shock waves from a nearby supernova and gravitational perturbations produced by interactions with nearby stars and other molecular clouds. If left alone, internal forces provide the ultimate limitation on the lifetime of the cloud. Current the-

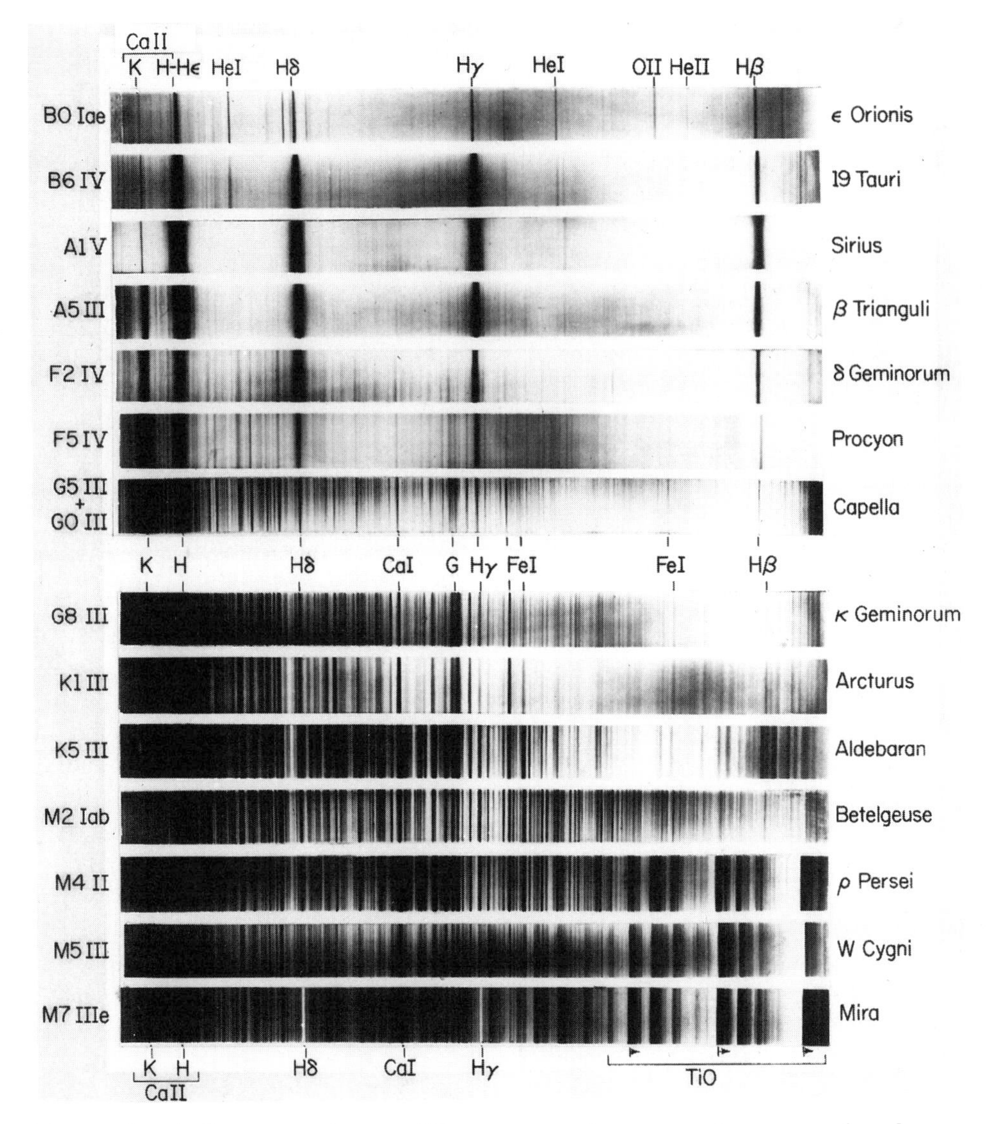

Figure 1. Stellar spectral types. The spectra of stars are shown ranging from hot B-type stars to cool M-type stars. The Roman numeral I indicates the star is a supergiant, a III indicates the star is a giant, a V indicates the star is of normal size, while two other numerals represent intermediate cases. Wavelength increases left to right going from the near ultra-violet to the green. Some of the lines due to various elements and molecules are identified (H: hydrogen, He: helium, O: oxygen, Ca: calcium, Fe: iron, and TiO: titanium oxide). Reproduced with permission from the University of Michigan.

ory predicts that the magnetic field can slowly diffuse out of the cloud or dense core, which means it eventually becomes unstable to gravitational collapse. Once begun, the process proceeds rapidly so that a star is formed on a timescale of 10^6 years. In the early phases of the collapse a stellar embryo forms in the center that is heated up due to shock collisions from the infalling gas. When the embryo has generated sufficient thermal pressure to begin to resist gravitational contraction, the embryo transforms into a protostar and continues to grow in size as it accretes more infalling mass from the cloud. However, the growth process is complicated by the fact that the infalling matter has angular momentum inherited from the cloud that ultimately prevents over 80 percent of the infalling matter from reaching the protostar. Instead of falling directly onto the protostar, much of the matter forms an accretion disk that orbits around the protostar. The disk flattens with time due to internal collisions in the gas, and eventually gas from the inner edge of the disk that lost enough angular momentum by frictional dissipation is able to fall onto the protostar. The infalling gas heats up the protostar's surface, which sets up a powerful stellar wind that blows away most of the infalling matter into a direction parallel to the protostar's rotation axis. This flow is bipolar and comes out of each pole of the rotating protostar. When most of the gas in the accretion disk is gone, a circumstellar disk of dust remains that has a diameter of 100 to 1,000 times the radius of Earth's orbit (1.496×10^8 km). This dust may coalesce to form planetesimals, and, over a timespan of 10^8 years, planets may form leading to the formation of another solar system. (For more information about cosmogony, see Sargent and Beckwith, 1993.)

The final mass of the protostar appears to depend on initial gas density and the strength of internal magnetic field in the cloud or core; the larger these quantities are, the more massive is the protostar. Massive stars like types O and B form almost exclusively in clusters, while low-mass stars like the Sun can form in either clusters or in isolation. Once stars are formed in a cloud, their luminous energy results in the cloud's dispersal so that only a few percent of the cloud's mass is converted into stars. The theory of star formation as presented here applies to the formation of single stars, however; over half of all stars are formed in binary or multiple star systems. The process of how binary or multiple star systems form is still under investigation (Stahler, 1991; Lada, 1993).

A protostar generates energy by gravitational contraction and accreting infalling matter. As the protostar grows in mass, the central temperature increases, eventually leading to the start of nuclear fusion reactions. Once the central temperature is around 10^7 K, hydrogen fusion reactions take place with such intensity that the gravitational contraction is slowed and eventually stopped. Once this happens, the protostar becomes a star.

Stellar Evolution

A star is a controlled thermonuclear reactor in which the force of gravity is perfectly balanced by the pressures generated by the nuclear reactions taking place in the interior. Energy is transported to the surface by either radiative diffusion or convection. Stars initially produce energy through various nuclear reactions that take four protons and convert them into a helium nucleus. This process is often referred to in astrophysics as hydrogen-burning. These nuclear reactions take place at temperatures of 10 to 50×10^6 K and densities of around 3 to 150 g/cm^3. (The density of water in comparison is 1 g/cm^3.) Despite these high densities matter is still in a gaseous state because it is nearly completely ionized. Because of the conversion of hydrogen into helium, stars are forced to evolve due to changes that take place in the interior as the hydrogen fuel is consumed. In this subsection only the evolutionary behavior of single stars will be described because the nature of binary or multiple star evolution is more complicated and not well understood. The core hydrogen-burning phase is the longest portion of a star's lifetime, occupying up to 90 percent of the total. Only around 10 percent of the total stellar mass undergoes the conversion of hydrogen into helium during this phase. Lifetimes range from a few million years for the most massive stars to almost a trillion years for the least massive stars. The Sun's core hydrogen-burning lifetime and current age are estimated to be, respectively, 10^{10} years and 4.5 billion years (Ga). It is one of the ironies of nature that the massive stars that have so much hydrogen fuel consume it so fast that these stars have the shorter lifetimes.

As the star is burning hydrogen, it slowly becomes brighter (Figure 2). Once all the hydrogen is exhausted in the stellar core, a hydrogen-burning shell forms around a helium core that undergoes gravitational contraction. Although the deep interior of the star is contracting, the outer layers of the star respond in the opposite manner and the star expands and cools to become a giant. The hydrogen-burning shell moves outward in mass as it consumes hydrogen and, correspondingly, the helium core continues to increase in mass while shrinking in size. During this and later evolutionary phases the star may pulsate, temporarily becoming a variable star. As the outer layers of the star become cooler during this and other later evo-

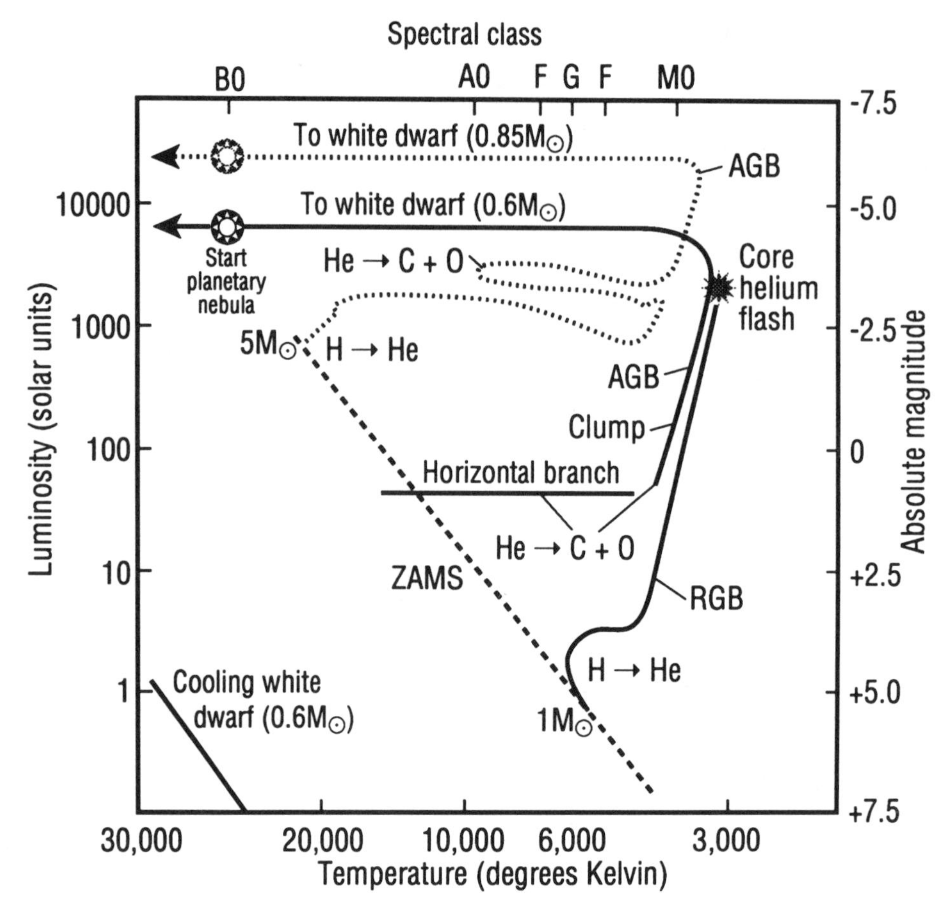

Figure 2. Evolutionary H-R diagram for 1 M⊙ and 5 M⊙ stars. A representation of the evolution of one and five solar mass (M⊙) stars in a luminosity vs. temperature diagram based on the theoretical calculations of Icko Iben, Jr. Such a graph is called an H-R diagram in honor of E. Hertzsprung and H. N. Russel who first proposed it. Stars begin their lives on the zero-age main sequence (ZAMS) producing energy by the fusion of hydrogen into helium. When hydrogen is exhausted in the core, a star moves from the ZAMS to the red-giant branch (RGB) until helium fusion begins. During core-helium fusion, the Sun moves onto the horizontal branch, while the 5M⊙ star loops to the left. Upon core-helium exhaustion these stars move onto the asymptotic-giant branch (AGB) until the outer layers of these stars are expelled into space; then they evolve into white dwarfs. From *Stars by Kaler.* Copyright © 1992 by Scientific American Library. Reprinted with permission of W. H. Freeman and Company.

lutionary phases, a convective layer may form that extends from the surface to so deep into the interior that material which has been processed by nuclear reactions is brought up to the surface. Once the helium core has grown to a sufficient size, it becomes possible for nuclear reactions involving helium to take place. For temperatures greater than 10^8 K, three helium nuclei can fuse to form carbon, and helium nuclei and a carbon nuclei can fuse to form oxygen. Stars with masses greater than about one-half of a solar mass will be able to burn helium in their cores. Stars less than this mass will consume all their hydrogen and then shrink in size, ending their lives as helium white dwarfs.

The core helium-burning phase is the second longest-lived phase of evolution, lasting on the order of one-tenth of the core hydrogen-burning phase. In addition to the helium-burning core the hydrogen-burning shell continues to operate. Once helium starts to burn in the core, the star stops expanding and it becomes hotter. Later when helium is exhausted in the core, a helium-burning shell forms around a contracting carbon-oxygen core and the star begins again to expand and cool while evolving into a supergiant. Stars with masses up to about nine solar masses will evolve with two active burning shells (i.e., hydrogen and helium) moving outward in mass around a carbon-oxygen core. As the star grows in size, it starts to lose mass due in part to the weaker gravitational force exerted on its outer layers. Mass loss becomes so great that stars less than about nine solar masses will be reduced to objects of mass less than 1.4 solar masses, which then evolve into and end their lives as carbon-oxygen white dwarfs (see Kaler, 1993, for the ultimate fate of the Sun).

Stars more massive than nine solar masses, while they do lose some mass, are able to attain successively hotter temperatures in their interior to ignite additional nuclear reactions. Their cores develop an onion-shell structure where the end products of one fusion process become the fuel for the next process. The next process after helium-burning is carbon-burning, followed by neon-burning, oxygen-burning, and finally silicon-burning, which leaves behind a central core of iron. It is not possible to fuse iron and produce additional energy. Instead, the iron in the core begins to disintegrate when the central temperature and density are around 10^{10} K and 10^{10} g/cm^3, which causes the central core to collapse and either leads to a supernova explosion and the formation of a neutron star or to the formation of a black hole (*see* COSMOLOGY, BIG BANG, AND THE STANDARD MODEL). During the supernova explosion, nuclear reactions are believed to take place that produce all the elements heavier than iron. The supernova explosion blows most of the mass of star (including material that has been processed by nuclear reactions) back into space. Examples of supernova are the event in 1054 that produced the Crab Nebula in Taurus, and the recent explosion in the Large Magellanic Cloud Galaxy known as supernova 1987A. (For a more detailed description, see Kaler, 1992, and Woosley and Weaver, 1989.)

The processes of star formation and stellar evolution are linked together because material that was in stars of a previous generation is recycled and used in a succeeding generation of stars. The existence of planets and life is due in part to the nuclear reactions that took place in stars to produce almost all elements heavier than helium. We as living creatures are made up in part of star dust.

Bibliography

BÖHM-VITENSE, E. *Introduction to Stellar Astrophysics. Vol. 1: Basic Stellar Observations and Data.* Cambridge, Eng., 1989.

KALER, J. B. "Origins of the Spectral Sequence." *Sky and Telescope* (February 1986):129–134.

———. *Stars and Their Spectra: An Introduction to the Spectral Sequence.* Cambridge, Eng., 1989.

———. *Stars.* New York, 1992.

———. "Giants in the Sky: The Fate of the Sun." *Mercury,* March/April 1993, pp. 34–41.

LADA, C. "Deciphering the Mysteries of Stellar Origins." *Sky and Telescope* (May 1993):18–24.

SARGENT, A. I., and S. V. BECKWITH. "The Search for Forming Planetary Systems." *Physics Today* (April 1993):22–29.

STAHLER, S. W. "The Early Life of Stars." *Scientific American* (July 1991):48–55.

WOOSLEY, S., and T. WEAVER. "The Great Supernova of 1987." *Scientific American* (August 1989):32–40.

STEPHEN ALLAN BECKER

STEWARDSHIP

See Research in the Earth Sciences

STILLE, HANS

Hans Stille is best known for his postulate that throughout geological history, orogenic deformation of Earth's crust occurred in discrete and relatively short pulses, each of which simultaneously affecting various regions of the globe. He gave a name, usually based on a locality, to each of these pulses that he called "phases." Moreover, he maintained that many of these phases created fold or thrust belts and fault arrays with characteristic orientations, which he also classified by name. Although Stille's postulates are no longer accepted and his classifications of orogenic events and orientations of tectonic features have become obsolete, they were attempts at dating phases of deformation, and in this respect he and his students stimulated the development of geochronology.

Stille was born in Hannover, Germany, on 8 October 1876. He attended primary and secondary schools in Hannover, collected fossils in local quarries, and worked in his own small chemical laboratory. He began studying chemistry at the Technische Hochschule Hannover, the local institute of technology, but soon switched to geology at nearby Göttingen University, where his studies were supervised by Adolf von Koenen. His dissertation, submitted in 1900, involved mapping and interpretation of stratigraphic and structural details of a portion of the Teutoburger Wald hills in northwestern Germany.

His first job (1900–1902) was at the Technische Hochschule in Hannover. As a member of the Prussian Geological Survey (1902–1908), he mapped in impeccable detail and described the geology of regions near his thesis area. Next, he occupied the chair of geology at the Technische Hochschule Hannover (1908–1913). Stille spent a few months in 1913 at Leipzig University and then taught at Göttingen University as his former mentor's successor (1913–1932). His scientific work was interrupted by World War I, when he served as an officer in the Prussian army (1914–1918). In the postwar years he began to attract a large number of students and became a popular supervisor of Ph.D. candidates after taking over the chair in geology at the Alexander von Humboldt University in Berlin (1932–1950). His popularity rose in part because his earlier students had established themselves in influential positions at universities, in the various geological surveys of Germany, and in industry, and had formed a network, the so-called Stille school, which tended to facilitate employment of its new members. However, students were attracted to him mainly by their admiration for his scientific work and their appreciation of his personality, a combination of formal, soft-spoken politeness, reticence, gruff kindliness, stern insistence on diligence and precision, and a lifelong concern for his former students' welfare and careers.

The Alexander von Humboldt University became part of the Russian occupation zone in 1945, but the new authorities allowed Stille to continue his work and to travel as extensively as he had done before the war. When Stille retired in1950, before the Berlin wall had been erected, the government allowed him to take all his possessions and papers to the West. He moved back to Hannover, where he enjoyed a long and productive retirement and continued to publish until shortly before his death on 25 December 1966.

Stille's thesis project and the areas he mapped as a survey geologist consisted of folded and unusually heavily faulted. Mesozoic sediments underlain by thick Permian salt deposits. Stille recognized that the structural style of that particular Mesozoic basin was strikingly different from that of Paleozoic basement blocks or that of the Alps. In 1917, he named this style "germanotype," ascribing its peculiarities to the effects of salt being buoyant and easily deformable; it rises toward the surface in salt ridges and domes, and between these features the overlying sediments subside to form local basins or fault-bounded grabens. Because of the spatial and temporal persistence of the principal unconformities throughout the mapping area, Stille concluded that deformation was restricted to definite, short time intervals (phases), and he began to name them, beginning with the designation as "Kimmerische" phase of tectonic activities approximately simultaneous with those of the Nevadan orogeny in North America.

Stille was also impressed with the prevalence of two distinct trends of fold axes and fault traces in his mapping area, one NNE-SSW, following the azimuth of the Rhine graben, the other NNW-SSE. Seeking clarity by nomenclature, he called these trends "Rhenish" and "Eggish" (from the Egge Hill range in northwest Germany). He attached significance to the fact that the strikes of basement-block boundary faults of the North German Mesozoic basin have these orientations. Later, place names were given to other azimuths of tectonic lineaments, either by Stille himself or by his pupils, until their angular ranges covered nearly all directions of the compass.

Stille concluded (1924) through literature research and personal observations that orogenic events fell worldwide into a limited number of short phases, and that they were separated by long intervals of quiescence. Each potential new phase was critically evaluated before it was either rejected as invalid or recognized and named, Stille himself always acting as the final arbiter. Despite his resistance to proliferation, evidence for more and more phases was discovered and had to be fitted into the phase list.

Eventually, Stille's system of classification covered the tectonics of whole continents; he interpreted North American geological history in his book *Einführung in den Bau Amerikas* (1940), but that book did not become widely known, at least partly because of the war. However, his views on the persistence of tectonic trends over geological time influenced many geologists in Europe and contributed to their resistance to the idea of continental drift.

Summaries of Stille's career were compiled on the occasions of his eightieth (Lotze, 1956), ninetieth (Bogdanov et al., 1966, with the only complete bibliography of Stille's 186 publications), and posthumous one-hundredth birthdays (Pilger, 1977); the present article draws heavily on these sources.

Bibliography

BOGDANOV, A. A., et al. "Hans Stille, k devianostoletiiu so dnia rozhdeniia" (to commemorate his ninetieth birthday). *Sovetskaia Geologiia* 10 (1966):111–120.

LOTZE, F. "Hans Stille, geb. zu Hannover am 8. Oktober 1876, zur Vollendung seines 80. Lebensjahres." In *Geotektonisches Symposium zu Ehren von Hans Stille*, ed. F. Lotze. Stuttgart, 1956.

PILGER, A. "Laudatio auf Hans Stille zur Wiederkehr seines 100. Geburtstages." *Zeitschrift der deutschen geologischen Gesellschaft* 128 (1977):1–9.

STILLE, H. *Grundfragen der vergleichenden Tektonik.* Berlin, 1924.

———. *Einführung in den Bau Amerikas.* Berlin, 1940.

GERHARD OERTEL

STOMMEL, HENRY

Henry Melson Stommel was born in Wilmington, Delaware, on 27 September 1920 and died on 17 January 1992. He was a prodigious oceanographer, and to a large extent he helped to create modern physical oceanography as a discipline in which theoretical ideas and models were developed and tested in the context of observation. Much of his professional life of forty-five years was spent at the Woods Hole Oceanographic Institution, in Woods Hole, Massachusetts. His characteristic approach was to take observations and his deep intuition as a guide to develop a mathematical model of phenomena that isolated and exposed simple, underlying physical processes and then communicate this understanding with enormous passion and warmth to his collaborators and colleagues. He brought a rare degree of harmony and collegiality to the field of oceanography.

Stommel graduated from Yale University in physics in 1942, and continued as a graduate student in astronomy, teaching celestial navigation to U.S. Navy V-12 students. As a conscientious objector to the war, he took a job at the Woods Hole Oceanographic Institution in 1944 as a research assistant. He gradually became involved in oceanographic research, eventually turning his attention to nearly all aspects of physical oceanography, participating actively in oceanographic research cruises. His interests extended from the nature of friction and dissipation in the ocean to the large-scale global circulation of water throughout the world's oceans. Most characteristic of Stommel's approach was his return to particular areas of oceanography repeatedly, making significant contributions several times. His published contributions range from reports of measurements and observations, and commentaries on particular fields, to papers of breathtaking originality and genius.

It is difficult to summarize briefly all of Stommel's contributions to oceanography. Perhaps the work that most clearly established his name in oceanography was his 1948 paper titled "The Westward Intensification of the Wind-driven Ocean Currents." The underlying scientific question is, Why is the Gulf Stream a swift narrow current along the western boundary of the North Atlantic when neither the wind nor the distribution of heating and cooling has such asymmetry? Stommel developed a theoretical model of an ocean driven by a symmetrical distribution of wind stress with westerlies at northerly latitudes and easterlies at southerly latitudes. He set about to compute the solution by hand—well before the advent of digital computers!—using a relaxation method that he had previously used for a tidal calculation. Starting from an initial guess of a symmetrical ocean circulation mimicking the wind stress, he found that the circulation began to intensify toward the west. A careful analysis showed that this westward intensification was due to the variation of the earth's rotation (Coriolis force) with latitude. Instead of continuing with these more general calculations, Stommel developed a much simpler model including the variation of the Coriolis term, which could be solved analytically, demonstrating the role of the variation of the earth's rotation in the westward intensification. This work marked the beginning of large-scale ocean circulation modeling. Stommel quickly realized the power of this approach, of isolating physical processes in mathematically simple models, and continued to develop this skill into a fine art, his hallmark.

Stommel often exhorted oceanographers to use theory and modeling to make predictions—to really *test* their ideas against nature. The deep abyssal circulation in the ocean was largely unobserved and assumed to be extremely sluggish. Stommel reasoned that turbulent mixing in the upper ocean would mix warmer water downward, and since the oceans were in more or less of a steady state, there must be a corresponding upward flow of cold water to balance this downward flux of heat. Since it was known that the sources of deep water were near the poles in the Atlantic, Stommel developed a model of the abyssal circulation driven by these sources of water. He found that, as with the wind-driven circulation, there must be an intense narrow current along the western boundary of the deep Atlantic, and he predicted a Deep Western Boundary Undercurrent. This prediction was borne out in the observations by John Swallow off Cape Hatteras in 1957 and was a triumph for theoretical oceanography. This fundamental contribution has led to an understanding of the global ocean circulation and the distribution of chemical tracers in the ocean, as well as the ocean's role in climate.

Together with his collaborators, Stommel developed a theory of oceanic thermocline, that region of transition between the warm upper ocean and cold abyssal ocean, recognizing the essential complexity of the problem in which the distribution of temperature and salinity both result from and constrain the velocity field in the ocean—the problem is essentially nonlinear. He would often entrain others in his scientific research by his engaging enthusiasm, and work with collaborators to bring new mathematical and observational techniques into the field of oceanography.

Stommel often returned to scientific problems many times during his career. His contributions to a fundamental understanding in a particular area would often draw many others who could build upon his simple model and ideas to develop more elaborate and complex models. He would invariably turn his attention to another question at this point, having little desire for elaboration. Often work on problems that Stommel addressed, such as the wind-driven circulation, stagnated after several years, as he became confronted by a deeper problem, such as the connection between the thermocline and the wind-driven circulation. After nearly thirty-five years, Stommel returned to this question and was able again with collaborators to make a fundamental contribution, showing how the conservation of potential vorticity allows one to explain both the vertical and horizontal distribution of density in a wind-driven circulation.

One question that puzzled Stommel was the fact that although variations in salinity and in temperature arise predominantly from atmospheric forcing (heating, cooling, evaporation, and precipitation), there is a strong correlation between temperature (T) and salinity (S) variations. These correlations persist over long times and large spatial scales, and can be used to identify characteristic water masses connected to source regions. One of Stommel's first papers in the 1940s used this correlation to infer salinity from temperature measurements in order to estimate density variations. Over the years, he often returned to the question of why T and S are correlated and sought unsuccessfully to develop a fundamental understanding of the underlying process. In his last paper, Stommel developed a simple stochastic model to demonstrate how localized rain squalls in the presence of a large-scale temperature gradient will produce exactly the kind of correlations observed in the ocean.

There are similar contributions in many areas of oceanography in which Stommel's deep physical insight and relentless urge to simplify mathematical and physical models have led to a fundamental understanding of processes. Stommel was a member of the National Academy of Sciences, and a foreign member of the Royal Society of London, the Soviet Academy of Sciences, and the French Academy of Sciences. He received numerous medals and prizes, including the National Medal of

Science and the Crafoord Prize of the Royal Swedish Academy of Sciences. Stommel maintained contact with many people, engaging them with his enthusiasm for understanding the phenomenal world and his passion to communicate it.

Bibliography

LUYTEN, J. R., J. PEDLOSKY, and H. STOMMEL. "The Ventilated Thermocline." *Journal of Physical Oceanography* 13, no. 2 (1983):292–309.

STOMMEL, H. "The Westward Intensification of Wind-driven Ocean Currents." *Transactions, American Geophysical Union* 29, no. 2 (1948):202–206.

———. "The Anatomy of the Atlantic Ocean." *Scientific American* 192, no. 1 (1955):30–35.

———. "The Abyssal Circulation of the Ocean." *Nature* 180, no. 4589 (1957):733–734.

———. "The Circulation of the Abyss." *Scientific American* 199, no. 1 (1958):85–90.

———. *The Gulf Stream: A Physical and Dynamical Description.* Berkeley, CA, 1958.

———. "Thermohaline Convection with Two Stable Regimes of Flow." *Tellus* 13, no. 2 (1961):224–230.

———. "Varieties of Oceanographic Experience." *Science* 139, no. 3555 (1963):572–576.

———. "Future Prospects for Physical Oceanography." *Science* 168, no. 3939 (1970):1531–1537.

———. "A Conjectural Regulating Mechanism for Determining the Thermohaline Structure of the Oceanic Mixed Layer." *Journal of Physical Oceanography* 23, no. 1 (1993):142–148.

STOMMEL, H., and E. STOMMEL. *Volcano Weather: The Story of 1816, the Year Without a Summer.* Newport, RI, 1983.

STOMMEL, H. M., and D. W. MOORE. *An Introduction to the Coriolis Force.* New York, 1989.

JAMES LUYTEN

STRATEGIC MINERAL RESOURCES AND STOCKPILES

Traditionally defined, strategic mineral resources are resources required for the successful conduct of war. In a broader sense, they are the resources needed to sustain a healthy, competitive economy through peace and international conflict. The aim of government mineral resource policies is to protect the country from political and economic pressures resulting from the interruption of supplies of essential raw materials. Strategic mineral stockpiles are national reserves of minerals or metals for immediate use in case of a shortfall in supply. The following analysis of strategic resources relates to metallic and nonmetallic minerals, but does not include fuels.

In the United States, concrete steps to deal with potential shortages of essential mineral resources were taken in the wake of World War I, in the Strategic and Critical Stock Piling Act of 1939, which mandated a National Strategic Stockpile of materials not found or produced in the country in sufficient quantities to meet demand in case of a national emergency. The term "strategic" relates to the relative national availability of a material, and the term "critical" denotes the importance of the material for military and industrial purposes (Kessel, 1990). In the following discussion, the term strategic will be used except when criticality is to be emphasized.

Of key concern in the debate of strategic mineral resources is the fact that the United States is heavily dependent on the import of a large number of critical metals and minerals, as, for example, chromium, manganese, tungsten, cobalt, and nickel for the manufacture of specialty steels, and platinum-group metals, titanium, aluminum, and a variety of other metals for the aerospace, electronic, and chemical industries. The yearly Mineral Commodity Summaries published by the Department of the Interior list over twenty essential mineral resources for which the United States is more than 50 percent import-dependent.

While import dependence does contribute to making a country vulnerable to disruptions of supply, there are a variety of factors that decrease this vulnerability. Among them are, for example, the density of the present-day worldwide trade network, the increasing economic interdependence of all countries, the multiple sources of supply for most mineral commodities, and the large number of possible means and routes of transport.

The Strategic Minerals and Metals

The major factors influencing the strategic importance of mineral resources are the criticality of the resource for industrial and military purposes, total national consumption, the proportion consumed for military purposes, and the sources of supply. Total consumption, or the need of the peace-time economy, is satisfied by free-market trade and by privately held industrial stockpiles. The availability of strategic mineral resources for military purposes, on the other hand, is addressed by strategic-resource policies and government stockpiles. These include resources for conventional military supplies and weapons, for example, manganese, chromium, nickel, cobalt, copper, and tungsten; for air transportation and high-technology weapons systems, as are titanium, aluminum, and magnesium; and for electronic, computer, and optical control or guidance systems, among them the platinum group metals, germanium, rare-earth elements, and others. Using a variety of criteria to assess the strategic and critical importance of mineral resources, the U.S. Army War College Strategic Studies Group in the mid–1970s developed a relative ranking of these resources in the "Vulnerability Index" (Szuprowicz, 1981) (Table 1).

An adequate supply of strategic minerals and metals, while essential, can be assured by an assortment of economic and political strategies. For a few metals, for example, copper and molybdenum, U.S. domestic production could satisfy demand in an emergency. For most of the other strategic mineral resources, however, the United States and most other industrialized nations rely heavily on imports (Table 1). For these, the vulnerability to disruptions of imports depends on several geographic and political factors, among them the number of producing countries, their political disposition toward the importing country, their political stability, and the safety of the transport routes. South Africa and the former Soviet Union (FSU) are key providers of several of the most critical mineral resources: together, they have a virtual monopoly on, and contain the bulk of global reserves of, chromium, the platinum-group metals, titanium, and manganese. In addition, they are essential suppliers of tungsten, nickel, and titanium. Most of the other important metals and minerals are, or could be, supplied by a variety of countries and are therefore less susceptible to disruption either directly or by a threat to transport routes (Table 1).

Strategic-Minerals Policies

In case of an emergency, the national demand for strategic mineral resources could be satisfied by domestic adjustments in demand or supply, by stockpiles of imported raw or processed materials, or by long-standing international arrangements. In view of the effectiveness of free international trade and the high costs of self-sufficiency (Kessel, 1990), it does not seem practical to assure the national supply by increasing domestic production of strategic minerals. The best national policies to minimize dependency will encourage efficiency of use, conservation, recycling, and substitution, all aided by the development of new technologies.

The traditional safeguard against shortages in the supply of strategic resources has been the National Defense Stockpile, established and revised in the Strategic and Critical Stock Piling Acts of 1939, 1946, 1979, 1984, and 1990. The stockpile, administered by the Department of Defense (DoD), holds reserves of over seventy materials, including twenty-five metallic and ten nonmetallic basic minerals, which could satisfy military and industrial demand in the short term. In addition, there is a large stockpile of surplus manufactured goods and military equipment.

Several developments since the establishment of the National Strategic Stockpile place the relationship between import dependence, vulnerability, and the need for stockpiles into perspective. A vastly increased international trade of resources and goods by multinational corporations has led to the development of interconnected global markets, and to an international economic interdependence which decreases the short-term and long-term vulnerability to trade disruptions for all members of the network. At present, any measures to secure strategic minerals that do not respond to market forces are not likely to succeed, and past efforts of producer countries to gain political and economic advantage through formation of cartels in the nonfuel industries have failed.

The effects of the broadening of the supply base are complemented by profound changes in the type of warfare the strategic stockpile was originally designed for. First nuclear deterrence and then the collapse of the east block have made large-scale, material-intensive, protracted conventional wars requiring full-scale industrial mobilization unlikely. The scenario has shifted toward smaller-scale, high-technology wars for which there is a large stockpile of weapons systems developed during the cold-war buildup and a reduced need for stockpiles of primary resources. Accordingly, the emphasis in the management of the national stock-

Table 1. Strategic Mineral Resources

Material	*Vulnerability Index**	*Use*	*Import Reliance† (%)*	*Major World Producers (%)*	*Major World Reserves*
Chromium	34	Stainless and heat-resisting steel	75	FSU‡ (51) S. Africa (43)	S. Africa (88) FSU (10)
Platinum group metals	32	Catalysts, electronics	93	S. Africa (51) FSU (44)	S. Africa (89) FSU (10)
Tungsten	27	Steel alloy, tool making, machinery	75	China (44) FSU (22) Korea (5)	China (44) Canada (13) FSU (10)
Manganese	23	Steel production (essential)	100	FSU (38) S. Africa (13) Brazil (11)	S. Africa (41) FSU (33) Gabon (11)
Aluminum/bauxite	22	Construction, transportation, electrical	97**	Australia (38) Guinea (17) Brazil (8)	Guinea (25) Australia (20) Brazil (13)
Titanium (metal)	20	Aerospace: jet engines, missiles	NA§	Australia§ Canada S. Africa	Brazil§ S. Africa Norway
Cobalt	20	Super alloys: industrial, aircraft turbines	84	Zaire (63) Zambia (12) FSU (11)	Zaire (42) Cuba (33) Zambia (11)
Tantalum	16	Electronics, transportation	89	Brazil (32) Australia (19) Thailand (17)	Thailand (33) Australia (21) Nigeria (15)
Nickel	14	Stainless steel, nonferrous alloys	75	Canada (27) FSU (26) Australia (10)	Cuba (37) Canada (16) FSU (14)
Mercury	11	Electrical, chemical, instruments	NA§	FSU (37) Spain (25) Algeria (11)	Spain (59) FSU (8) Mexico (3)
Tin	6	Containers, electrical, brass, bronze, solders	73	Brazil (18) Indonesia (16) Malaysia (16)	Malaysia (26) Indonesia (16) Brazil (15)

Sources: Szuprovicz, 1981; Kessel, 1990; Minerals Information Office, 1993.
* U.S. Army War College.
† Net import reliance as percentage of apparent consumption (calculated total demand).
‡ Former Soviet Union.
§ Given for bauxite.
** No reliable statistics available.

pile is shifting away from a "buy and hold" policy to the selling of excess holdings through the Defense National Stockpile Center. The sales are made at a slow rate, because of the complexity of the political decision-making process and to minimize their effect on the international mineral markets.

In the future, there is likely to be an increased reliance on international economic and political cooperation rather than on stockpiles to supply needed strategic and critical minerals, and on industrial stores and productive capacity to bridge times of international crisis.

Bibliography

KESSEL, K. A. *Strategic Minerals: U.S. Alternatives.* Washington, DC, 1990.

MINERALS INFORMATION OFFICE. *Mineral Commodity Summaries.* Washington, DC, 1993.

SZUPROWICZ, B. O. *How to Avoid Strategic Materials Shortages.* New York, 1981.

VAN RENSBURG, W. C. J. *Strategic Minerals. Vol. 1: Main Mineral-Exporting Regions of the World.* World Resources, Energy, and Minerals series. Englewood Cliffs, NJ, 1986.

———. *Strategic Minerals. Vol. 2: Major Mineral Consuming Regions of the World,* World Resources, Energy, and Minerals series. Englewood Cliffs, NJ, 1986.

HALF ZANTOP

STRATIGRAPHY

The stratigraphic record provides a database that is invaluable in the reconstructions of past climates, past environments, past oceans and atmospheres, and past life in the context of geologic time. The record that rock layers collectively comprises in the stratigraphic record may be read as clues in a grand mystery story. The clues must be assimilated and analyzed, and the story of the past constructed. Stratigraphy includes the observation and study of rock layers or strata and the interpretation of the features of those strata to develop a historical perception of Earth. Stratigraphy encompasses both the physical and biological attributes of strata and an understanding of the processes involved in forming strata.

Strata are formed from accumulations of sediment: dirt, volcanic ash and dust, volcanic rock, shells of organisms, and bits of rock ground and worn away from a parent sediment. Sediment commonly is swept by the action of wind and water to a site where it accumulates. Accumulations of sediment commonly are layered. Layers of sediment may accumulate on virtually any land surface or bottom of a lake, river, or the ocean.

Niels Stensen, a Danish physician who served in the court of the Duke of Florence in the seventeenth century, documented the basic principles of stratigraphy. In the hills of northern Italy, Stensen examined layered sedimentary rocks and the traces of ancient life contained in them—fossils. He published a series of four principles that utilize stratigraphy in reconstructing the history of the Earth. The first principle notes that the materials which comprise strata were initially sediment. Strata may contain the remains of organisms that were buried essentially as sediment. Second, sediments accumulated on solid surfaces. The bottom of any stratum thus conforms to the topography of the surface upon which it accumulates. The tops of most strata are horizontal. Stensen's third principle is perhaps the most widely cited. It states that in any sequence of strata, the oldest is at the bottom. This principle provided a sense of order and direction to determinations of relative age. Stensen also noted that rock layers seen on both sides of a river valley, for example, could be said to have been continuous originally. Thus, strata were continuous throughout the area in which they may be recognized at the time of accumulation.

A conspicuous characteristic of stratal sequences are the surfaces that separate essentially homogenous accumulations of sediment, the beds or stratum. Such surfaces are commonly called bedding planes. Spacing of bedding planes as well as the characteristics of the strata themselves reflect how sediment accumulates. Most bedding planes are the result of a pause in sediment accumulation. That pause may or may not be accompanied by removal (erosion) of some sediment. Certain bedding planes may bear features such as ripple marks, small-scale scours, grooves, tracks or trails of some organism, shrinkage cracks, or burrows. Many such features may be observed forming in modern environments. These features are evidence that, for a time, the surface of a bed or stratum was the scene of physical or biological activity, or both.

The presence of bedding planes reveals that sediment accumulation is an episodic process. Study of modern processes of sediment delivery to sites of accumulation shows the random nature of such episodes. In areas where rainfall and accompanying runoff from lands to streams and oceans is seasonal, for example, sediment removal and transfer to the site of accumulation may be primarily seasonal. In terrestrial accumulations, winds strong enough to carry sediment from its source to the accumulation site may be randomly distributed through the year or they may be primarily seasonal. Sediment accumulations in terrestrial environments may reflect wind direction and wind-strength patterns. Other terrestrial accumulations may reflect patterns in rain and related runoff from highlands to the site of accumulation. Such rains and related runoff patterns commonly reflect seasonality in climate. In areas that experience droughts, rain sufficient to transport sediment from its source to the accumulation site may be spaced at intervals of many years. Bedding surfaces in such accumulation areas may remain surfaces of nonaccumulation and even sediment removal for relatively long-time intervals before sediment accumulates upon them. Sediments that are the shells of organisms may fall to lake or ocean floors in patterns that reflect the bloom followed by life and death of large numbers of organisms. In areas located in such a way that organismal blooms are seasonal, then sediment derived from them will reflect that seasonality. In general, episodes of sediment accumulation may reflect random events, seasonal events, or cyclic events of durations different from seasons. Once sediment has accumulated, it is subject to removal by wind or water motion across the accumulation site. Removal is another aspect of the stratal record, for it alters the sediment accumulation record. Many bedding surfaces reflect incidences of sediment removal. The sediment record is thus comprised of what may be seen as small snapshots in the long continuum of geologic time.

If sediment accumulates in environments in which oxygen is available for organismal respiration and some form of nutrient for organisms is also available, then the sediment may be churned up by organismal activity. Such organismal activity is called bioturbation. In sediment rich in food for bioturbating organisms, the sediment may become virtually homogenized. When oxygen is depleted, sediment layers then will accumulate undisturbed. Many individual layers or strata are characterized by homogeneity imprinted by bioturbating organisms. With cessation of accumulation and consequent end of nutrient supply, the organisms leave.

Clearly, sediment layers and the bedding surfaces between them reveal much about environmental conditions at the site of accumulation. The nature of the sediment itself and the features left in it are also indicative of its origins and may be used to deduce something about climate and environment in the source area.

Geologic Time

The concept of geologic time is one of geology's most significant contributions to an understanding of the world around us. The development of a timescale proceeded from Stensen's basic principles. Those principles gave order and a sense of time to the study of strata and the traces of past life contained in them. The principle of lateral continuity of strata helped in recognizing stratal sequences in a broad area that had accumulated over relatively long intervals of time.

Fossils from local strata were collected by people in many parts of Europe for several centuries. Local stratal sequences were described and the positions of fossils within individual strata were noted by many collectors. These local successions of strata and their fossils were unlinked parts of the story of Earth and life history. An English surveyor, William Smith, provided the needed catalyst to link these clues and decipher a relative timescale. In his travels around England and Wales surveying canals, he noted that certain fossils always occurred in the same strata. Indeed, certain fossils characterized the strata in which they were found. Because the relative position of varying strata could always be determined by superposition and lateral tracing, fossils could be useful in predicting the occurrence of certain strata. Smith used these conclusions to develop a general succession of strata for England and Wales. He portrayed the stratal succession on a map and indicated those fossils he found most useful in tracing certain strata from place to place. Smith's map gave a sense of time direction to the succession of past life as seen in the stratigraphic succession. Nevertheless, Smith's work was scorned by geologists. In response, he commented that "fossils have long been studied as great curiosities, collected with great pains, treasured with great care, and at great expense, and shown and admired with as much pleasure as a child's rattle or hobby horse is shown and admired by himself and his playfellows, because it is pretty; and this has been done by thousands who have never paid the least regard to that wonderful order and regularity with which nature has disposed of these singular productions, and assigned to each class its particular stratum."

Farmers found Smith's map valuable because certain crops could be grown more effectively on soils developed over one rock type as opposed to another. Crops that grew well on somewhat acidic soils, for example, would not develop as well on soils over chalk.

A small group of geologists applied Smith's ideas in recognition of a relative timescale. Charles Lyell was a leader in this endeavor (*see* CHARLES LYELL). In the first volume of his *Principles of Geology* (1830), he wrote: "In order, therefore, to establish a chronological succession of fossiliferous groups, a geologist must begin with a single section, in which several sets of strata lie one upon the other. He must trace these formations, by attention to their mineral character and fossils, continuously, as far as possible, from the starting point. As often as he meets new groups, he must ascertain by superposition their age relative to those first examined, and thus learn how to intercalate them in a tabular arrangement of the whole." Lyell's directions are basic to recognition of time units based on fossils. Lyell's directions were applied widely to fossil-bearing strata. In time, sets of rock layers typified by major types of fossils came to be recognized. Superposition provided the time direction for the succession of such sets of layers. Each major set characterized by its unique association of fossils came to be the basis for a major unit in a relative sequence of such units. These became the primary divisions of a geological timescale based on the fossil content of strata. As the major sets of strata were clustered into broader units of geologic time, geologists turned their attention to more regional or local divisions of the major units. Lyell noted, for example, that among the youngest marine strata, the percentages of still-living mollusks decreased downward in any stratal succession commencing at the top. He used the percentage of still-living mollusk species as a mechanism for dividing a relatively long interval, the stratigraphically highest and thus youngest in the geological timescale. Patterns on occurrences of fossils were analyzed in many parts of the world and, using precise stratal positions as a guide, the broader or more general timescale units were divided into relatively small-scale units. A set of timescale units, once established for an area or broader region, became a sort of frame of reference. The fossils characteristic of each unit could serve as indicators of a time interval. They are used as a frame of reference for correlation tables. Such tables reflect time synchroneities of rocks. Fossils obtained from individual strata are matched with those considered characteristic of a time unit in the relative timescale. Time synchroneities or correlations among stratal units and even individual strata may be indicated on a correlation chart by matching fossil associations. Correlation charts may be developed for a broad area or region so that time equivalencies among rock units can then be used to depict regional or even global patterns at any time interval. Correlation charts are significant tools in recognizing changes in depositional environments through time.

Sea-Level Changes and Seismic Stratigraphy

The ancient Greeks found shells of marine creatures in strata on the tops of hills, far removed from the seashore, and ascertained that sea levels must have changed over time. Stratigraphers have continued to be fascinated by the comings and goings of seas across lands over the centuries. These major incursions of seas across land are called transgressions. Regressions mark retreats of the sea from the land. Transgressions and regressions are clear evidence of sea-level change. These events leave a record of significant environmental change in strata formed during them.

Mechanisms for sea-level change include seawater withdrawal by development of massive ice sheets, and ocean volume increase at times of ice sheet melting. Earth motions (tectonics) also play a role in sea-level changes, which are reflected in transgressions and regressions. Cooling or heating of plates by subcrustal sources causes plates to sink or rise that, in turn, influences sea levels. For example, marine environments shallowed significantly across most of North America in the early part of the Ordovician. Ultimately, much of mid-continent North America was exposed subaerially. The environmental changes resulted from gradual upwarping of the middle part of the North American (or Laurentian) plate by heating within the mantle beneath it. When mantle heating subsided, the middle part of the plate subsided as well and, gradually, marine environments returned. Stratigraphy, thus, may be used as an indicator of patterns in mantle heating and cooling.

Changes in sea position are studied most easily in strata that accumulate on plate or continental margins. Today, such areas include, for example, the Gulf of Mexico and the drowned continental margin or continental shelf of the Atlantic Ocean side of North America. The most complete stratal record of sea transgression and regression lies unseen under water and modern vegetation covering the stratal record. Because this remarkable record of sea-level motion is not available to visual inspection, tools were devised to study it. Under stimulus from the oil industry's interest in offshore exploration, seismic techniques were developed to explore the unseen strata. The methodology and terminology of seismic stratigraphy grew out of the oil industry's concern for linking stratal information gleaned from wells drilled into strata that are not visible. Indeed, an entire subdiscipline within stratigraphy (called seismic stratigraphy) has developed. Peter Vail and his colleagues at Exxon were pioneers in this development. By the 1980s, seismic stratigraphy dominated the entire science of stratigraphy. Seismic stratigraphy is applicable, nonetheless, to only a small fraction of the entire stratigraphic record.

Seismic stratigraphic studies make use of vibrations that travel through the strata beneath the surface of the earth (*see* GEOPHYSICAL TECHNIQUES). Vibrations may be generated naturally by earthquakes or by people creating sound waves. Seismic exploration using man-made vibrations was introduced as a tool in oil exploration in the 1920s. Oil exploration under the Gulf of Mexico provided the primary stimulus for use of this method.

Seismic surveying involves transmission of sound waves into the earth. As transmitted waves pass through strata of differing density, they bounce or are reflected back from various stratal surfaces, some of which may be bedding planes. Reflected waves are received by sensitive sound detectors known as geophones or hydrophones. The receivers may be clustered into groups or arrayed along a line. Geophone clusters are wired to a single line that transmits a signal to magnetic tape-recording equipment housed in a recorder vehicle. The data are analyzed by computer. Ultimately, vibrations received may be printed as sets of wiggly lines. The primary reflections are a response to density differences and thus indicate velocity contrasts along bedding and discontinuity surfaces. The episodic aspect of stratal accumulation results in development of certain bedding surfaces that may be strongly reflective. As a generated wave travels down through strata, many bedding surfaces and surfaces between strata of different density will reflect waves. Changes in lithologic aspects of strata observed in vertical stratigraphic sections are much finer in scale than are the wavelengths of induced seismic waves used to penetrate them or to reflect from stratal surfaces. Accordingly, much smaller-scale stratigraphic divisions may be recognized through inspection of rock layers at an outcrop or in a core taken from a well than may be recognized through analysis of seismic wave data.

The positions of bedding planes across which a significant density difference exists are revealed on seismic records. Although every bedding surface may reflect seismic waves, most are so closely spaced that they cannot be distinguished on a seismic record. The seismic record is a composite of reflections from several sets of bedding surfaces.

Seismic reflections may be analyzed and described by certain basic characteristics. Parallel layers suggest relatively uniform rates of deposition and subsidence. Change in the deposition rate or rate of subsidence is indicated by sets of lines that diverge. A relatively conformable succession of genetically related beds as determined by seismic reflection that are bounded by well-marked surfaces has been termed parasequences. Parasequence boundaries commonly may be interpreted as being the consequence of sea-level change. A parasequence may be a set of strata deposited during a single episode of submergence. If so, the parasequence commences with evidence of transgression. As the transgression attains maximum extent, land-derived sediment may have built out across the last stage of marine transgression. That change marks the top of the parasequence.

Parasequences may be grouped as sets (they are also called tracts, systems tracts, or generalized facies tracts). Three types of parasequence sets may be recognized (Figure 1): (1) a set formed when sediment delivery exceeds rate of subsidence and accumulation; (2) a set formed when rate of subsidence exceeds rate of sediment delivery and accumulation; and (3) a set formed when the pace of

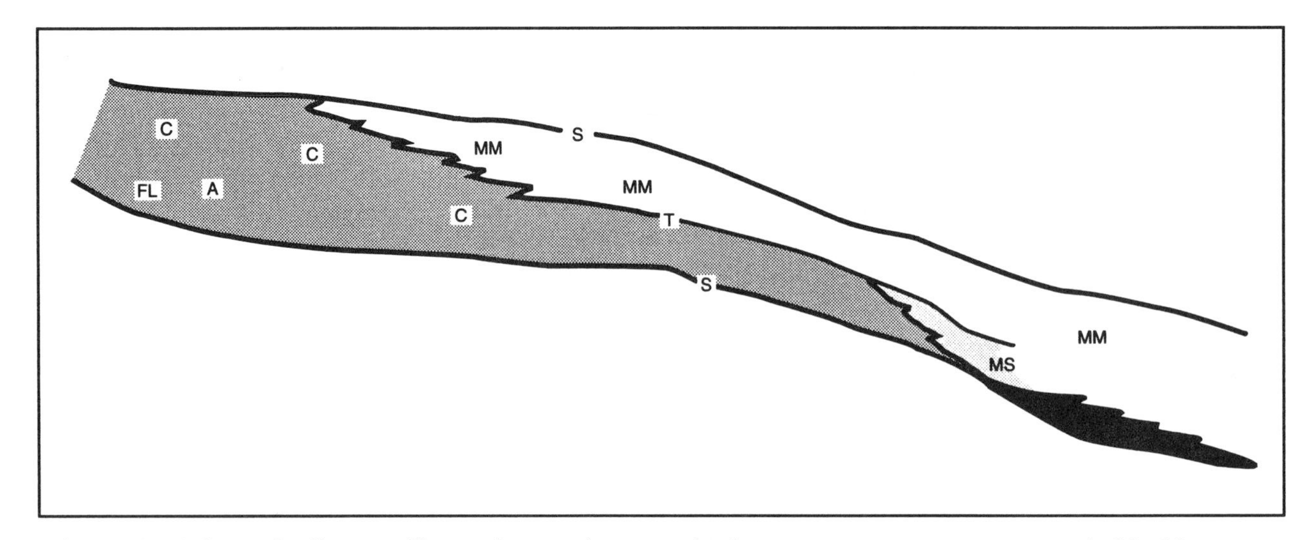

Figure 1. Schematic diagram illustrating terrigenous clastic parasequence sets across a shelf with shelf-slope break. Lowstand System Tract in which sediment supply is greater than rate of subsidence. Note position relative to Transgressive System Tract (T). Fluvial (FL), Alluvial Plain (A), Coastal Plain (C) [including beach, dune, deltaic, lagoonal, shallow subtidal sediments], slope (MS), and submarine fan (F) deposits are components of this set. Transgressive System Tract (T) includes thinly bedded mud rocks (commonly hemipelagites and pelagites) in condensed stratigraphic sections. This set formed when sediment supply and rate of subsidence about balanced. Highstand System Tract formed when rate of subsidence exceeded rate of sediment supply. Marine mudrocks which tend to be fossiliferous are common (MM). These rocks interfinger inshore with coastal plain materials. In the basin, these rocks interfinger with submarine fan and slope deposits. S is a surface or set of surfaces that may show evidence of subaerial exposure and erosion. Such surfaces tend to be good seismic wave reflectors.

sediment delivery and that of subsidence balance. Each such tract formed during a part of a transgression-regression cycle. Transgression commonly begins with set 2 and is followed by set 3. Regression begins with set 1. The tracts recognized as indicative of the major phases in a transgression-regression cycle comprise the largest stratal "packages" recognized in seismic stratigraphy. They are bounded by surfaces that are high-quality reflectors. Such surfaces may be traced from strata that accumulated in marginal marine and terrestrial environments into the marine basin environments. These aspects of seismic stratigraphy reveal continental shelf tectonism and, presumably, phases of cooling and heating of crust from mantle materials. Time synchroneity among seismically derived records of these events must be achieved through fossils obtained from well-core materials when a well is drilled into the sediments within the seismically determined stratal package. Fossil data from such wells may be used to confirm environmental changes postulated from the seismic record. Ultimately, traditional stratigraphic paleontology is used to confirm not only time-synchroneity among seismically defined stratal packages but also changes in depositional environments inferred from seismic data. Nevertheless, seismic stratigraphy has provided new insights into continental margin tectonism and the record of the pace of sea-level change.

The program of drilling and coring materials from the ocean floors has opened new vistas in evolutionary biology and in ocean current histories. Certain parts of deep ocean cores are essentially little else but shells of floating marine organisms. The millions of shells in such cores provide a wealth of data on change in form with time. The data from the shells are also valuable information in documenting positions of ancient ocean water masses. The deep-ocean sediment record may be studied only in the small glimpses obtained from core materials, yet it provides significant clues to ocean floor history and cooling of the crust beneath the oceans.

The record found in the strata is of vital significance to an understanding of past life and of the crustal history of Earth. It also provides important clues to heating and cooling patterns in subcrustal materials. Accordingly, stratigraphy is a fundamental cornerstone of geology.

Bibliography

BERRY, W. B. N. *Growth of a Prehistoric Time Scale; Based on Organic Evolution.* Palo Alto, CA, 1987.

FRIEDMANN, G. M., J. E. SANDERS, and D. C. KOPASKA-MERKEL. *Principles of Sedimentary Deposits.* New York, 1992.

LEMON, R. R. *Principles of Stratigraphy.* Columbus, OH, 1990.

LYELL, C. *Principles of Geology.* 3 vols. London, 1830–1833.

PHILLIPS, J. "Palaeozoic Series." In *The Penny Cyclopedia of the Society for the Diffusion of Useful Knowledge,* ed. G. Long. London, 1840.

STENO, N. *De solido intra solidum naturaliter contendo dissertationis prodromus.* Florence, 1669.

THACKRAY, J. *The Age of the Earth.* London, 1980.

VAIL, P. R., R. G. TODD, and J. R. SANGREE. "Stratigraphic Interpretation of Seismic Reflection Patterns in Depositional Sequences." In *Seismic Stratigraphy—Applications in Hydrocarbon Exploration,* ed. C. E. Payton. Tulsa, OK, 1977.

VISHER, G. S. *Exploration Stratigraphy.* Tulsa, OK, 1984.

WILLIAM B. N. BERRY

STRATIGRAPHY, HISTORY OF

Layers of rock composed of sediment particles and, in certain sites, fossils or the remains of once-living organisms reveal a record of Earth's biological and physical processes (*see* STRATIGRAPHY). This connection has been grapsed by observers of nature for centuries. More than five centuries before the birth of Christ, the Greeks recorded finding seashells similar to extant forms, but far inland from then-existing shores. They concluded that in an earlier period seas had spread much farther inland but had since retreated. The Greek historian Herodotus suggested that the Nile delta was created by sediments supplied to it by the Nile's waters. Aristotle taught that relative land and sea positions were not fixed but changing constantly. In succeeding centuries, Arab and Roman students of nature concurred with Aristotle's conclusions. Leonardo de Vinci, writing during the Rennaissance, pointed out that mud from alpine lands was transported to the sea where it entombed shells of marine animals in seafloor sediment. De Vinci also noted that, through time, soft sediment hardened to rock and eventually became land.

Basic principles of stratigraphy developed from these and similar observations. These principles were given definitive form by the Danish anatomist/geologist Niels Stensen, who moved to Italy in 1665 and adopted the name Nicolaus Steno. Based on his observations of layered rocks in the hills of northern Italy, Steno concluded in a 1669 monograph that (1) layered rocks in the hills had formed from sediments and that many contain remains of past life (fossils); (2) the bottoms of such rock layers conform to the shape of the surface upon which they accumulated; (3) in any sequence of such layers, the oldest is at the base and the youngest at the top; and (4) each layer originally was continuous throughout the locale in which it had accumulated. Steno's stratigraphic principles provided geologists (and evolutionary biologists) with a means of determining relative age through stratal sequence.

Steno's observations led to an understanding that strata had both a physical and biological aspect. Prominent features in many strata include either traces of organismal activity or the actual shells or bones of creatures.

Inquiry into the biological component of strata revealed both the remains of past life through time and the direction of its development. In *Origin of the Species,* CHARLES DARWIN explained that the stratigraphic sequence of fossils is the evidence for organic evolution, even though that record is somewhat tattered.

For centuries, the physical aspects of rock layers have afforded great wealth to many nations. Certain strata bear minerals of great value. Other layers may be quarried for building stones. Yet others may be burned or contain oil and gas that may be burned. Such layers have provided the energy upon which the growth of the industrialized world has depended. Still other strata have yielded the raw materials needed to produce glass or fine china, or the water that sustains human life and supports agriculture. As commercial uses of stratified rocks expanded during the late eighteenth and early nineteenth centuries, people recognized the need to transport one rock type (coal, for example) closer to other rock types (metal ores, for instance). Using barges to float raw materials through canals proved cost-effective. Building canals entailed unearthing rocks that lay beneath soils and forests and even beneath other rocks. Costs dictated that canal excavations be made through relatively soft, easily broken rock layers. Predicting where such layers could be found came from the observations of William Smith, a canal builder and engineer. In the late 1790s, Smith recognized that individual strata could be identified by their fossil content. Smith concluded that one could predict the succession of rock layers in a particular stratum by knowing the succession of fossil faunas. Smith's principle of faunal succession developed into a scale of relative time units based upon fossil strata.

Smith used the idea of identifying rock layers and their position in an overall succession of layers by their fossil content in the surveying work he carried out in different regions of England and Wales. In each area he surveyed, he could use the succession of fossils obtained from the individual layers he examined to establish a correlation of layers with the sequences in other areas studied. By plotting rock sequences on maps for each area surveyed, and by documenting correlations among strata seen in many regions, Smith constructed a geologic map of England and Wales. The map, published in 1815, showed the usefulness and validity of his principle of faunal succession. In the half century after the publication of that document, a geological timescale was developed. Of a hierarchical nature, it included units of relatively short duration as divisions of relatively long duration units; it also used the succession of fossils seen in strata that Smith had worked out initially. Each unit in the timescale was recognized by a unique association of fossils that could be found in many areas.

Timescale units allowed for a more complete understanding of the history of the earth and how life on it had been documented. Use of the timescale units to predict stratigraphic sequences led to improved geologic maps, which ultimately enhanced man's recovery of nature's wealth.

When stratal sequences that had been documented in many regions were graphically displayed in time correlation charts and diagrams, the general features of the record of past geographies, environments, and climates could be established. Timescale units based on fossils provide an established frame of reference. Newly discovered rock sequences that contain fossils (for example, in cores from the deep ocean) can be compared with known sequences. To do so, fossils from each stratum in the new area are compared with those characterizing each known timescale unit. Matches of similar fossils suggest time synchroneity and a time correlation is established.

The concept of facies developed from documentation of time correlations among strata over broad areas. Facies are strata of different physical attributes that accumulate in contiguous, time-synchronous environments. The passage of beach sands and marshes laterally seaward into deeper marine environments is an example of facies forming in modern conditions.

Thinking critically about the facies changes that occur when moving laterally from land to freshwater bodies or to the ocean leads to an understanding of the concept of base level. Base level in water is the position at which sediment accumulation is no longer possible because of depth. On land, base level is that position below which erosive processes are no longer effective. Base level changes occur when rise or fall of the sea influences shoreline and nearshore environments. Sea level rise and fall result in exposure or flooding events that create uniquely identifiable surfaces in stratigraphic sequences. Such surfaces bound sediment packages or sequences with distinctive physical attributes. These sequences are the basic building blocks of sequence stratigraphy. Because the bounding surfaces may possess uniquely identifiable physical properties, they appear as discrete lines on profiles generated by continuous seismic reflection profiling. These lines, which are considered surfaces in three dimensions, are the fundamental components of seismic stratigraphy, a tool found to be of great value in the quest for oil and gas along the continental margins of the world (*see* STRATIGRAPHY). Seismic profiling and seismic stratigraphy were developed by Exxon in the 1970s and 1980s under the leadership of Peter Vail. Study of numerous sets of seismic profiles revealed sets of stratal sequences that had formed during ancient rises and falls of sea level. Scrutiny of the sequences indicated that five different scales or orders could be distinguished. The broadest scale or first order sequences were described by L. L. Sloss in 1963 in analyses of mid-continent North American stratigraphy. Sloss's documentation of major stratal sequences bounded by unconformities that could be traced over great distances is a cornerstone of sequence stratigraphy.

Recognition of smaller-scale sequence sets or orders followed from both analyses of seismic profiles and outcrop successions in the 1970s, 1980s, and 1990s. The more precise scrutiny of strata by sequence stratigraphers has led to recognition of evidence for the record of rare or unique events in strata. Event stratigraphy encompasses the stratigraphic records of massive tsunamis, bolide impacts, widespread volcanic ash falls and their consequences, and changes from oxic to anoxic conditions over wide areas of the oceans.

Rhythmic repetition of clusters of strata reveals the relationship between certain unique environmental cycles and stratigraphy. Cyclic stratigraphy or cyclostratigraphy examines the stratigraphic record of changes in the Earth-Sun distance and changes in tilt of Earth's axis with respect to the planet's orbit around the Sun. Orbital forcing of climate and related environmental changes elucidated by the mathematician Milutin Milankovitch have left a record in the cyclic repetition of certain sets of strata.

Stratigraphy bears the record of many events in the geological and life history of Earth. It also ensures that nature's wealth continues to be recovered from stratified rocks. Accordingly, stratigraphy continues to be a field of inquiry that has an impact on the daily lives of much of the world's population.

Bibliography

BERRY, W. B. N. *Growth of a Prehistoric Time Scale,* rev. ed. Palo Alto, CA, 1987.

FRIEDMAN, G. M., J. E. SANDERS, and D. C. KOPASKA-MERKEL. *Principles of Sedimentary Deposits.* New York, 1992.

PAYTON, C. E., ed. *Seismic Stratigraphy—Applications to Hydrocarbon Exploration.* Tulsa, OK, 1977.

SCHWARZACHER, W. *Cyclostratigraphy and the Milankovitch Theory.* Developments in Sedimentology 52. Amsterdam, 1993.

SLOSS, L. L. "Sequences in the Cratonic Interior of North America." *Geological Society of America Bulletin* 74 (1963):93–114.

WILLIAM B. N. BERRY

STRUCTURAL GEOLOGY

Displacements of rock masses are the subject of structural geology. They may occupy domains of any size from microscopic, through mesoscopic (what can be seen in an individual outcrop), to megascopic (shapes and configurations accessible only by mapping), or regional (compiled from a collection of geological maps). Only when a domain approaches the size of a continent is the study of its displacements considered a topic apart, called tectonics (*see* TECTONIC BLOCKS; TECTONISM, ACTIVE; TECTONISM, PLANETARY). Displacement fields (with respect to an arbitrarily chosen reference frame) can be continuous (neighboring points remain neighbors) or not (across surfaces of discontinuity, points originally adjacent to each other become separated), and displacements in a given domain can be homogeneous (independent of position, hence the same throughout) or not. Homogeneous displacements of a whole domain result in its simple translation, like that of a rigid body. Inhomogeneous displacements generally cause points at some distance from each other either to move closer together or to become more widely separated; hence, an inhomogeneous displacement field in a domain generally involves deformation of the rock masses filling it. (A special case of an inhomogeneous displacement field is the rotation of a body about an axis without deformation.)

Faults

If one rock domain is displaced, or slips, with respect to another on a surface of discontinuity, the resulting surface is termed a fault; faults are classified by their orientation and the direction of slip. A fault has a strike (azimuth of a horizontal line on its surface) and a dip (angle between the steepest line on its surface and its horizontal projection). As long as the displacements on a fault are homogeneous in both adjacent domains, their relative movement is a simple translation of rigid blocks, and the slip is uniform. A fault with horizontal slip usually forms a vertical plane and is called a strike-slip fault (Figure 1a), and on inclined fault planes the slip of the upper hanging wall (the underside of the upper block) over the foot wall (the top of the lower block) is commonly either down the dip line in a normal fault (Figure 1b), or up the dip line of the fault in a reverse or thrust fault (Figure 1c);

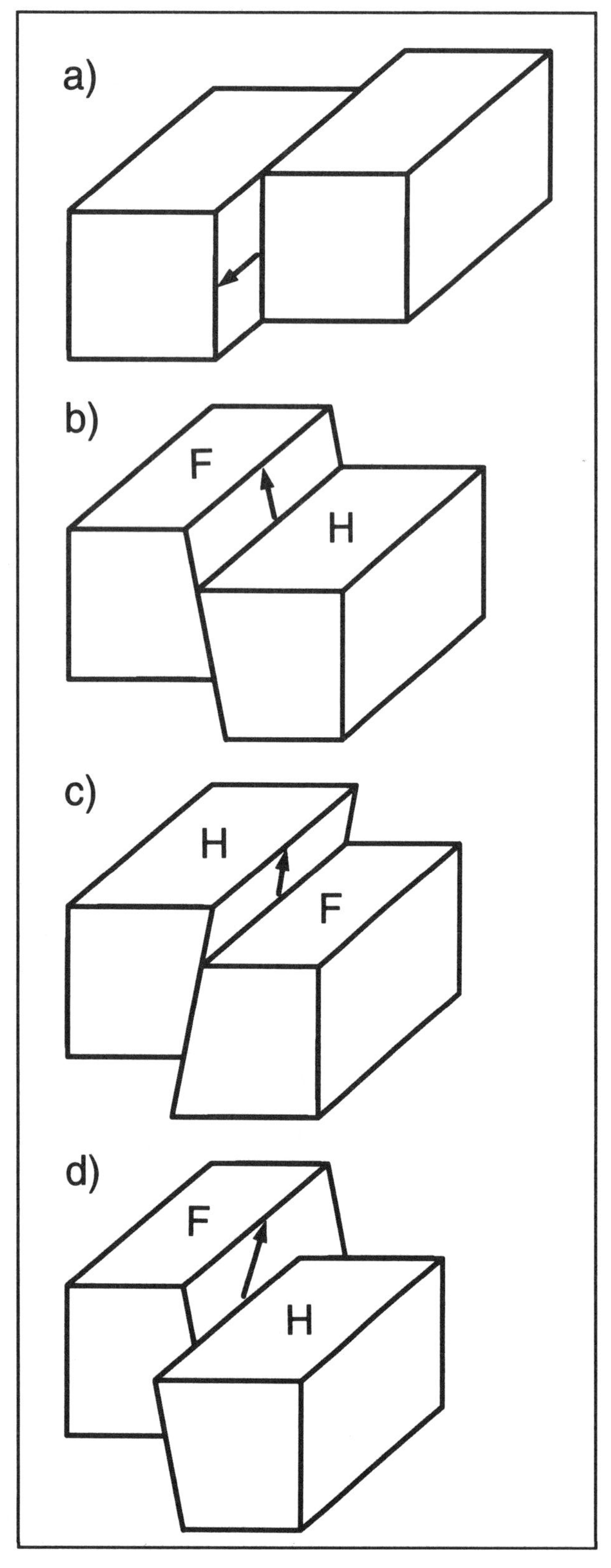

Figure 1. Types of faults. a. Strike-slip fault. Arrow–slip of left fault block relative to the right fault block. b. Normal fault. Arrow–slip of the footwall–F relative to the hanging wall–H. c. Reverse fault. Arrow–slip of the hanging wall–H relative to the footwall. d. Oblique fault. Arrow–slip of the footwall–F relative to the hanging wall–H, a combination of strike slip and normal slip.

oblique slip, however, is also common on faults of all orientations. Slip on a fault can be determined either if the offset of a discrete plane (say, a distinctive bed or a dike, demarcating a discrete line on each of the walls of the fault) is known and the mutual displacement has left scratch marks, or slickensides, on the two walls, or else if the offset of a discrete line is known (demarcating a discrete point on each wall); discrete lines usually consist of the intersection of two discrete planes (of real planes, like a bed with a dike; or of imaginary planes, like the extensions of two plane limbs of a fold; see below). Where the offset of discrete planes (for example, bedding planes) is the only indication of faulting, faults must be characterized by the separation, as seen on a particular, commonly horizontal, surface of the traces of a distinctive marker plane. Such a characterization provides limits for possible slip directions. Fault planes may be curved. This may be due to deformation of the fault together with its walls after the end of slipping or, in some instances, while the slip is occurring. On occasion fault planes are originally curved, but this puts severe restrictions on what slips are possible without the fault walls moving apart.

Deformation

If inside a continuous domain displacements change from place to place (i.e., if a displacement gradient exists), the material in that domain has been strained, rotated, or most commonly both. The strain and rotation can be calculated from a known displacement gradient. For the complete description of a strain (change of shape, volume, or both), a symmetric tensor quantity of the second rank (see the discussion of the stress tensor below), its six independent components must be known. Simple rotation does not affect shape or volume of a domain and requires only three independent components for its characterization. The assumption of constancy of volume (or in ignorance of actual volume change, the normalization of the strain to a change of shape at constant volume) reduces the number of independent components of the strain tensor to five. These five components are measurable where fossils or other features with a known original shape have been deformed, or where known original orientation distributions of flat or needle-shaped marker grains have been modified by a deformation. Usually, neither the original sizes of the fossils nor the original number of marker grains per unit volume are known, and thus the sixth strain component indicating volume change cannot be recovered from the evidence.

Stress

Strain is the response of a material, according to its properties, to stress. Of the forces exerted from the outside on the complete boundary of a small and compact material domain, only the portion that is in equilibrium needs to be taken into account in calculating the stress. Although the domain can have any compact shape, it is most convenient to consider one shaped like a small cube. The force acting on each of its faces by the surrounding material, divided by the area of the face, is called a traction (a vector), and each traction in turn is conventionally decomposed into a normal component perpendicular to the face and the two tangential (or shear) components parallel to its edges. Normal components are either compressive or tensile (tending to elongate the material across the plane). Collectively, the components for all faces of the domain constitute the stress tensor (of the second rank).

Mechanical properties determine how a material responds to stress. An elastic material assumes a strained state only as long as the stress is maintained and returns to the unstrained state when the stress is removed; elastic strain is not permanent. Stress causes a ductile material to be strained at a particular rate. If that rate is proportional to the stress, the material is said to possess a linear (or Newtonian) viscosity, but more commonly the relationship between stress and strain rate in rocks is nonlinear and is called a flow law. After cessation of the causative stress, a ductile material ceases to deform further, but the strain acquired during an episode of stress is permanent. Many rock materials have flow laws that, at stresses usual in Earth's crust, allow them to deform only slowly; superposed onto that gradual flow is, however, an elastic strain which does not change as long as the stress is constant. Sudden stress changes, such as those occurring during an earthquake, elicit sudden strains (*see* EARTHQUAKES).

Materials have strength to withstand a traction without breaking. The critical shear traction at which fracture occurs, the shear strength, is usually greater than the critical, tensile and normal traction, the tensile strength. The walls of a fracture subject to a traction with both compressive normal and tangential components resist slip by friction as long as the shear traction does not exceed a critical value calculated by multiplying the normal traction component with the coefficient of friction of the wall materials. Because incipient shear fractures cannot grow without mutual slip of their walls, frictional resistance effectively increases the shear strength on planes under compression.

Isotropic materials have the same mechanical properties in all directions. Therefore, strain and instantaneous strain rate in isotropic materials have the principal directions of greatest and least elongation (or rate of elongation) in the same directions in which the stress is least and most compressive. A third principal direction, both for stress and for strain, lies at a right angle to the plane defined by the first two. Few geological materials, however, are isotropic.

Anisotropy

Many materials, and all individual crystal grains, differ in their mechanical response with direction. Just as wood is stiffer in the direction of the grain than across it, so almost all rocks or rock assemblages differ in elastic stiffness and flow rate for a given stress, depending on the relative orientation of the principal stress directions in the deforming body. As a consequence, the principal strain or strain rate directions of anisotropic rock bodies do not, in general, parallel the principal directions of the causative stress.

Inhomogeneity

Most materials are not homogeneous throughout but are compound materials consisting of constituent subunits (grains, beds, and so on). Inhomogeneity of a material may give anisotropic properties to the compound body as a whole. A bedded sequence, with beds of different elastic properties and obeying different flow laws, would, taken as a whole, respond to stress like an anisotropic body, even if the rocks in each bed were isotropic (they rarely are). Crystalline rocks are commonly anisotropic because the crystals of each constituent phase are to a certain extent crystallographically aligned (they have a crystallographic fabric) or because grains with contrasting properties and elongated or tabular shapes are aligned according to their shape (they possess a shape fabric).

Deformation Instabilities

Inhomogeneous materials can react to stress in an unstable way. A relatively stiff layer embedded in a more easily flowing matrix, rather than thickening uniformly when a compressive stress is applied parallel to the layer, tends to buckle at a point that, by chance, is slightly thinner or slightly less stiff than the rest of the layer. Once initiated, the bending moment at the incipient fold hinge becomes catastrophically greater, and both the angle of deflection of the limbs to either side of the hinge and the length of the hinge, measured along the hinge line, tend to grow with increasing strain. At the same time, the stiffness of the stack of beds considered as a compound body, its resistance to layer-parallel shortening, decreases more and more. Once a first buckling fold exists, other incipient folds grow most rapidly at certain distances from the first and from each other. The favored distance depends on the thickness of a stiff layer, its stiffness contrast with the surrounding layers, and on the presence, properties, and spacing of other stiff layers in the vicinity. This selective fold growth causes the remarkable but never perfect rhythm of fold spacing observed in most fold belts, for example, the Alps or the Appalachian Mountains.

Analogous instabilities arise when a stiff layer within a less stiff medium is extended. Where the stiff layer is locally thinner, the tensile stress is concentrated because the same force acts on a smaller cross-sectional area, and thus this portion becomes even more stretched and thinned. The same effect is produced by a local stiffness defect. Although in this case the stress is initially not increased, the strain rate and, cumulatively, the strain are locally intensified; the result is a reduction of thickness and increase of stress, as before. In either case, a local pinch (double-sided groove) develops in the layer and grows deeper. Growth occurs also in the length of the pinch along a line perpendicular to the direction of extension. As with folds, the growth rate of additional pinches is most rapid at a certain distance from existing ones, the distance again depending on the layer thickness, its stiffness contrast with the medium, and the presence of additional layers. The resulting approximately rhythmical undulation of layer thickness is usually designated as pinch-and-swell structure although the thicker portions of the layer do not really swell. Initiation of pinch-and-swell generally requires greater stiffness contrasts than that of buckle folding. Related to pinch-and-swell structures is boudinage (from the French word *boudin*, for sausage) in which gaps in the stiff layer are filled with material from the less stiff medium. It forms where in response to an excessive tensile stress a stiff layer breaks into segments, or where the thin segments of a pinch-and-swell structure are further thinned until they vanish. This feature is displayed excellently in some metamorphic rocks.

Diapirs

An intrusion of magma (or of salt deposits) into overlying rock units is called a diapir and is usually due to the gravitational instability caused by a less dense rock material underlying a denser one. The density contrast may be original, or it may be the consequence of thermal expansion. In both cases displacements are caused by the buoyancy of the less dense rock body (comparable to that of a cork submerged in water), which produces the driving upward force. Diapiric intrusions come to a standstill when the stresses due to buoyancy are insufficient to continue the deformation both inside and out of the intrusive body, because either the materials have become less easily deformable (say by cooling of the magma) or the density contrast between the new surroundings and the diapir is less than it was initially. Salt diapirs have proven to be excellent structures for trapping oil and gas in adjacent, deformed marine sediments.

Fold Patterns

Depending on when the process of folding was arrested by a decay of stresses, folds can be gentle and open or so tightly compressed that their limbs are virtually parallel. The planes formed by the hinges (axial hinge planes) of a stack of beds can be upright (Figure 2) or inclined (Figure 3). In large portions of fold belts all hinge planes are inclined the same way (they have a constant vergence). The beds in folds may preserve their original thickness (this requires beds to be displaced with respect to one another by flexural slip, the outward bed for each hinge slipping toward it with respect to the inward bed, as in Figure 4), or the beds may swell at the hinges and thin in the limbs by flow of material toward the hinge regions (in flow folds, as in Figure 5). Folds may drape over the terminations of faults, or reverse and thrust faults may originate in the cores of folds too tightly compressed to fold further.

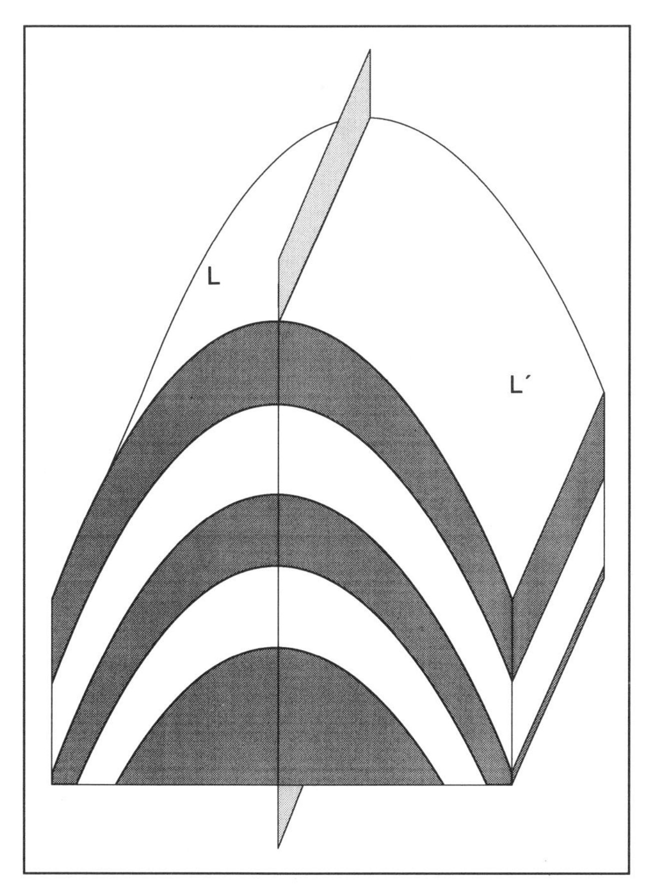

Figure 2. Upright fold with limbs L and L′. The vertical axial hinge plane (light gray) passes through the hinges (lines of greatest curvature) in successive beds (dark gray or white).

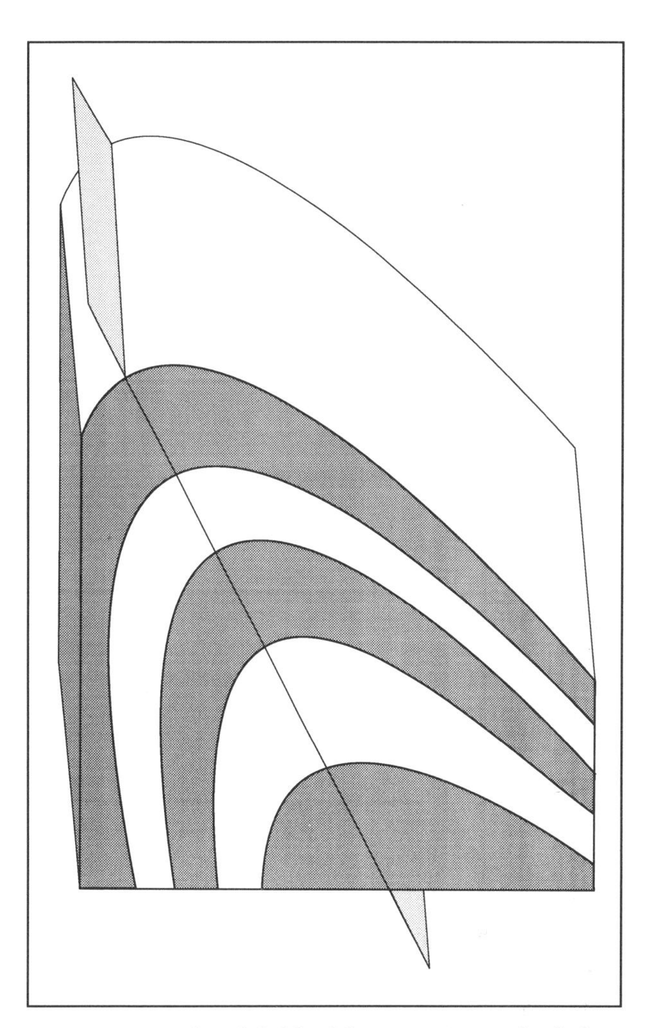

Figure 3. Inclined fold with vergence to the left. The axial hinge plane (light gray) passes through the hinges (lines of greatest curvature) in successive beds (dark gray or white) and dips to the right.

Paleostress

Faults, dikes, diapirs, folds, and other structural features result from ancient states of stress. Their orientations and distribution in space provide clues as to the magnitude and orientation of the former stresses. The age, as far as it is known, of rock units affected or unaffected by faulting, folding, and so on permits a bracketing of the time during which the stress existed. As a rule, these clues are insufficient for a complete analysis of the paleostress and its changes with time.

Fault planes indicate the orientation of planes having reached critical shear traction at the mo-

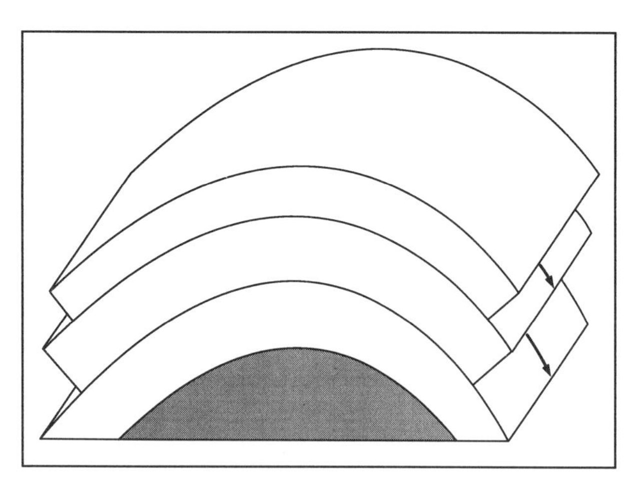

Figure 4. Flexural slip fold. The core of the fold (gray) consists of easily strained (soft) rock material. A series of overlying beds (white) have constant thickness; this requires that, in order to bend, they slip over each other so that the lower bed moves downward relative to the next higher bed; the thicker the bed, the greater the slip–arrows.

ment of fracture. For a given general state of stress and coefficient of friction, faults are most likely to break in one of two orientations; the potential fault planes are at a right angle to the plane containing the least and most compressive stress directions, and they are commonly inclined about 30° to the direction of greatest compression. Compression can be effectively relieved by a pore-filling fluid under pressure; the compressive force of the walls of a crack is partly supported by the fluid. If the fluid pressure equals the least compressive stress in the solid rock, the fluid can percolate into potential cracks, which then tend to form in the plan perpendicular to the direction of this stress. The stress orientations deduced from the orientations of faults and dikes, however, are valid only for the instant of breaking; they imply nothing about later

Figure 5. Flow folds, seen in cross section.

stresses and their orientations. Faults (and zoned dikes with multiple filling events) may have histories of later displacement that reflect stress regimes distinct from those that caused the initial fracture.

To estimate magnitudes of the stresses responsible for fault and dike formation, it is necessary to know the material properties under the conditions (temperature, pressure, fluid pore pressure, chemical reactions between solid and fluid) prevalent at the time of cracking. They usually are only incompletely known.

Clues as to the paleostress can only be derived from the observed (hence permanent) rock deformation. This, however, may be a cumulative deformation, caused by a former stress or a whole history of stresses changing over a prolonged time interval, causing various strain rates according to the flow laws appropriate for each rock type under each of the temporary conditions prevalent at various times during the deformation episode. Anisotropy and inhomogeneity of the deformed material may cause a lack of alignment between the principal strain rates and principal stresses, and the observed cumulative strain is the integral effect of the whole deformation episode consisting of time increments of various lengths with various strain rates. Furthermore, old rock units may have passed through more than a single compound episode. Conclusions from the observed strain as to the causative paleostresses and their durations are thus usually tentative and strongly depend on assumptions. Nevertheless, usually there are also numerous constraints limiting which assumptions can reasonably be made. The metamorphic grade of a rock (*see* METAMORPHIC PROCESSES; METAMORPHIC ROCKS) indicates possible pressures and temperatures in the rock's history; rock deformation experiments have furnished flow laws for many common rock types under a variety of conditions. Folds and faults that were formed simultaneously must be the responses to the same stress distributions, and postulated displacements must be kinetically possible (they must obey the compatibility conditions that a material domain may not at any time occupy the same space as another, nor may spaces have opened inside the crust that were not immediately filled).

Calculated Models

Enough computing power has recently become available to allow detailed, quantitative assumptions to be made about an initial distribution of rock bodies, their material properties, physical conditions, and stresses, and to calculate the outcome of the ensuing fracture, slip, and deformation processes in computer experiments. Such computer models are usually designed as finite element models (Lan and Hudleston, 1991) in which the effect of each new increment of strain on each member of a suitably constructed set of domains, consisting of simulated material, responds according to its assumed material properties to the slightly altered conditions caused by the previous increment. At present, certain simplifications are necessary to keep the volume of calculations within acceptable bounds; thus computer analyses commonly are made in only two dimensions, with the unrealistic implication that the investigated plane is typical for a body indefinitely extended perpendicular to that plane. This shortcoming may soon be overcome as computers become more powerful and easier to program.

Bibliography

HOBBS, B. E., MEANS, W. D., and WILLIAMS, P. F. *An Outline of Structural Geology*. New York. 1976.

LAN, L., and HUDLESTON, P. J. "Finite-Element Models of Buckle Folds in Non-Linear Materials." *Tectonophysics* 199 (1991):1–12.

MEANS, W. D. *Stress and Strain: Basic Concepts of Continuum Mechanics for Geologists*. New York, 1976.

SUPPE, J. *Principles of Structural Geology*. Englewood Cliffs, NJ, 1985.

GERHARD OERTEL

STRUCTURAL GEOLOGY, HISTORY OF

Structural geology is the study of the deformation of Earth as expressed by the geometry of folded and faulted rocks. Aside from deciphering the geometry of structures, the structural geologist is concerned with the path through which the rocks traveled during deformation (i.e., kinematics) and the stresses responsible for the deformation (dynamics). One of the earliest treatments of structural geology as a unified subject is found in Joseph LeConte's (1823–1901) *Elements of Geology*, written in 1877. Three components of structural geology, field mapping, experimental analysis, and petrofabrics, were first combined by Charles R. Van Hise's (1857–1918) famous treatise, *Principles of North American Pre-Cambrian Geology*, written in 1896 as part of the sixteenth annual report of the U.S. Geological Survey. One of the first texts devoted exclusively to structural geology, *Structural Geology*, was written by Charles K. Leith (1875–1956) in 1913. Each of these publications indicates how the three subdisciplines of structural geology (geometry, kinematics, and dynamics) evolved along parallel paths as a consequence of feedback among them.

The study of geometry dates from the discovery that sediments were deposited as horizontal layers, younger on older. Having deduced that horizontal layering was an intrinsic attribute of sedimentation, Nikolaus Steno (1638–1686) recognized that dipping layers were disturbed by convulsions from within the earth. For the next three hundred years the origin of earth stresses causing these convulsions was the central question of structural geology. An early idea was that the cause of tectonic deformation was outbursts of air or burning gas originating from Earth's fiery interior. By the latter half of the eighteenth century JAMES HUTTON (1726–1797) proposed that the earth's interior was molten, not fiery, and that igneous intrusions into the central portion of mountain chains pushed aside and tilted strata. Such diastrophic events were part of the so-called Plutonist theory that vertical forces caused mountain building. However, a second view, postulated by ABRAHAM GOTTLOB WERNER (1749–1817), was that dipping beds were attributed to sedimentation on the sides of mountains, to compaction of sediments, or to local slumping. This so-called Neptunist theory for the origin of rocks was a nondiastrophic explanation of structures.

Mapping of structures in outcrops led to the conclusion that folds were diastrophic (or tectonic) and many were caused by horizontal forces. In his exploration of the Alps, Horace-Bénédicte de Saussure (1740–1799) observed recumbent folds which required horizontal displacements of great magnitude. Using experiments on layered clay, Sir James Hall (1761–1832), the grandfather of experimental structural geology, showed that wrinkles in layered sediments were just as likely a consequence of horizontal stress on the ends of the beds. After the observations of de Saussure and the experiments of Hall, theories to explain large horizontal forces crept into the literature and the downfall of both the Plutonist and the Neptunist points of view was just a matter of time.

To explain the origin of horizontal forces, Léonce Élie de Beaumont (1798–1894) hypothesized that Earth's crust had cooled to an equilibrium temperature while the interior of the planet continued to lose heat. As the interior cooled, it contracted beneath a stable outer cover or crust. Eventually the crust became too large for its shrinking core and collapsed, a process involving horizontal shortening by folding as documented by both Saussure and Hall and by thrust faulting, a new concept arising from Arnold Escher von der Linth's (1807–1872) description of overpushing (*Überschiebung*) of overthrown (*überstrürzten*) Cretaceous on to Middle Tertiary Molasse in the Glarus Canton, Switzerland. The Élie de Beaumont theory of lateral folding by contraction was further used to explain the region behind a mountain chain. According to EDUARD SUESS (1831–1914), tangential compression (parallel to the earth's surface) of the crust generated lateral thrusts that override forelands to form mountains, whereas radial tensions (perpendicular to the surface) produced collapse in the hinterland.

Élie de Beaumont's contractional theory based on a rapidly cooling Earth was abandoned once geologists recognized that radioactivity in the crust was responsible for heat flow from the crust. However, horizontal forces still required an explanation and Felix Andries Vening Meinesz (1887–1966) looked to convection currents in the mantle as the source of horizontal forces necessary for crumpling the crust. About the same time ALFRED WEGENER (1880–1930) noted the remarkable similarity in shape between the east coast of South America and the west coast of Africa and proposed that the continents drifted like rafts on the seafloor. Most important, Wegener recognized that shifting continents could explain the late Paleozoic glaciation on several continents and the Carboniferous climatological zonation. ARTHUR HOLMES (1890–1965) was one of the first to propose that convection currents in the mantle furnished the motor needed to drive continental drift. Finally, developments in the early 1960s vindicated Wegener's ideas and allowed HARRY HESS (1906–1969) and a number of others to formulate a new paradigm, plate tectonics, which guides the modern structural geologist in the analysis of mountain ranges.

Mountain belts were first systematically studied in the 1830s when state governments in the United States supported detailed surveys throughout the Appalachian Mountains. William B. Rogers (1804–1882) and Henry D. Rogers (1808–1866) wrote an extraordinarily accurate account of asymmetrical and isoclinal folding plus steep thrusts in the Pennsylvanian Appalachians. James Hall's (1811–1898) study of New York State led to the conclusion that mountain belts have thicker and coarser sediments relative to the thin cover on adjacent cratons. JAMES DANA (1813–1895) amplified on Hall's mapping by suggesting that the boundary between ocean basins and continental crust was a weak zone where the horizontal forces imagined by Élie de Beaumont were relieved. The weak zone is a trough of thicker sediments called a geosyncline. Marcel Bertrand (1847–1907) demonstrated that certain unexpected contacts between rock units of different ages could be explained by large-scale overthrusts known as nappes, and developed a theory of continents accreting around older nuclei by juxtaposition of marginal bands. Albert Heim (1849–1937) was first to fully appreciate the extent and size of overthrust faulting in his analysis of the Glarus Thrust, which was described years before by Escher von der Linth. A major breakthrough came with Ernest M. Anderson's (1877–1960) correlation between stress orientation and fault mode: thrust, strike-slip, and normal and the attitude of associated dikes. Finally, a number of structural geologists devoted themselves to understanding the anatomy of entire mountain ranges. Of this group the best known include Émile Argand (1874–1940) for work in the Swiss Alps and East Asia, Hans Cloos (1885–1951) for work in Africa and Germany, Jean Goguel (1908–1987) for work on French Alpine tectonics, Vladimir Beloussov (1907–1990) for work in mountain ranges of the former Soviet Union, and John Rodgers for work in the Appalachian Mountains of North America.

Structural geology matured along a path leading from the definition of geometries, through analyses of kinematics, and to experiments defining tectonic stress. First, the geometry of more prominent structures such as large overthrusts was described by Escher von der Linth, among others. Next geologists recognized that small-scale and microscopic processes contributed to the formation of large-scale structures. ADAM SEDGWICK (1785–

1873) distinguished cleavage from bedding and proposed that the orientation of cleavage was a consequence of "crystalline forces." Daniel Sharpe (1806–1856) suggested that slaty cleavage was a consequence of mechanical flattening by compression at right angles to an applied force. HENRY C. SORBY (1826–1908), the grandfather of modern petrofabric studies, made extensive use of the petrographic microscope in proposing that pressure solution was responsible for the development of rock cleavage. Statistical means for the analysis of petrofabrics was extensively developed by Bruno Sander (1884–1979) as an aid to understanding the kinematics of deformation. Eleanor B. Knopf (1883–1974) is credited with introducing Sander's petrofabric techniques to the English literature. Francis J. Turner (1904–1985) recognized that some microscopic structures such as mechanical twins could be used to infer the orientation of the stress responsible for the deformation.

One of the most powerful aids in understanding kinematics is the strain marker. Strain in rocks is recorded by the distortion of such small-scale features as fossils, concretions, worm tubes, reduction spots, conglomerate pebbles, basalt pillows, and fibers about pyrite inclusions and in veins. John Ramsay led the way in developing many techniques for strain analysis using strain markers in rocks. When combined with the geometry of folds and faults, strain analysis provides a means for unraveling the kinematics of local structures as well as tectonic events on the scale of mountain ranges.

Part of structural geology concerns the discovery of rock properties that govern deformation within the lithosphere. When subject to stress, the lithosphere will deform or strain and it is the mechanical properties of rock that serve to link stress and strain or strain rate. Laboratory experiments proved a powerful tool for analysis of the kinematics of folding as first attempted by Sir James Hall in England and carried on by Baily Willis (1857–1949) in his famous studies of the folded Appalachian Mountains. Some of the initial work on dynamics (i.e., stress in rocks) was attempted with experiments by Auguste Daubrée (1804–1896), who produced shear planes (faults) in compression experiments, and by Frank D. Adams (1859–1942), who carefully defined the elastic properties of many rocks. In order to simulate pressure and temperature conditions within the earth, the development of pressure vessels by Theodore Von Karmon (1881–1963) and Percy Bridgman (1882–1961) was a major advance. David Griggs (1911–1974) was probably the most influential of a group of geologists working on specific geological problems using experimental rock deformation. This group, including John Handin (1919–1991) and William F. Brace, was able to define rock strength under both brittle conditions found in the upper crust and ductile conditions found throughout the lithosphere. Experiments led to a better understanding of earthquakes (e.g., the work of James Byerlee) and distribution of stress in the lithosphere through the construction of deformation mechanism maps (e.g., those of Ernest Rutter). Finally, one of the more noted disciples of the theory behind rock strength experiments, rock mechanics, was John C. Jaeger (1907–1979).

Geological modeling was first attempted through folding experiments in the early nineteenth century. The goal was to scale these experiments so that the artificial boundary conditions leading to stress and strain within the model were geologically realistic. M. KING HUBBERT (1903–1989) led the way in devising techniques for scale modeling to explain geological structures. Hubbert was also responsible for the theory of overthrust faulting with the aid of high pore pressure and, hence, low effective stress. The modeling of folds was refined by many people, including Arvid Johnson and Peter Cobbold. The modeling of mountain belts as wedges was pioneered by William Chapple (1932–1978) and refined by Dan Davis using a Coulomb criterion as the mechanical basis for the critical taper. Fracture mechanics, as championed by David Pollard, is another area where analytical modeling has allowed major advances.

With the advent of the high-speed digital computer, more and more analyses of geological structures involved modeling. The most recent major development in structural geology involves the restoration of fold and fault shapes to their original position with horizontal bedding. Such restoration, called balancing of cross sections, as pioneered by David Elliot (1931–1982), depends on the definition of some standard rules concerning the relationship between fold shapes and faults in the core of the folds. John Suppe is largely responsible for setting down a series of rules governing the development of folds above detachments in fold-thrust belts. These rules have been widely applied since about 1985 to better predict the position of subsurface structures by means of constructing realistic structures restored to their pre-deformation position.

In conclusion, structural geology has become a complex subject that ties physics and chemistry of the earth to the understanding of the geometry, kinematics, and dynamics of Earth's structures.

Bibliography

ADAMS, F. D. *The Birth and Development of the Geological Sciences.* New York, 1938.

FAILL, R. T. "Evolving Tectonic Concepts of the Central and Southern Appalachians." In *Geologists and Ideas: A History of North American Geology,* eds. E. T. Drake, and W. M. Jordan. Boulder, CO, 1985.

LECONTE, J. *Elements of Geology.* New York, 1877.

LEITH, C. K. *Structural Geology.* New York, 1913.

VAN HISE, C. R. "Principles of North American Precambrian Geology." In *16th Annual Report, U.S. Geological Survey (1894–1895).* Washington, DC, 1896.

TERRY ENGELDER

SUESS, EDUARD

Eduard Suess was born in London, England, on 20 August 1831. His father's family was from Vogtland, a region between Bohemia, Saxony, and Bavaria, and his mother was from Prague. Suess's father was in the wool business and the family spent several years in England. He thus learned English as a child and later also acquired fluency in French. These linguistic abilities were to prove a great asset for his later endeavors.

His interest in geology and paleontology developed after his parents had moved to Prague, where he became profoundly interested in fossils. Politics in the Austro-Hungarian Empire of the 1840s was dominated by the increasing activity of liberal movements in response to the sometimes repressive government of the Austrian statesman Metternich, who had risen from the diplomatic service to become Foreign Minister, and then Chancellor of Austria, in 1821. He is mainly known for his success in establishing a balance of the European powers after the Napoleonic wars, and for his continuous attempts to curb the rising tide of nationalism and liberalism. He was overthrown by the revolution of 1848. Young Suess sided with the liberals and joined the Legion Académique in Vienna, resulting in surveillance by the police and temporary imprisonment. Unable to pursue regular studies, he was appointed assistant at the Imperial Mineral Collection in Vienna in 1852. In 1855, he married Hermine Strauss, the niece of the director of the Museum of Natural History. They had three sons and two daughters; one of the sons, Franz Eduard, succeeded his father as chair of geology at the University of Vienna in 1911.

Fruitful years of research and publication followed. Suess covered various aspects of the paleontology of the Vienna Basin and adjoining areas, and devoted special interest to brachiopodes and cephalopodes. This work led to increasing interest in stratigraphy and sedimentology. In 1862, he published a small book titled *Der Boden der Stadt Wien* (The Subsoil of Vienna). This work attracted the attention of the city fathers who had been confronted with frequent epidemics of typhoid and cholera, as well as inundations by the Danube River. Suess was put in charge of the Water Supply Commission of Vienna and given the task to find safe sources of clean water for the city's growing population. Drinking water had been obtained from several wells in the young sediments of the Vienna Basin, but as a result of an insufficient sewer system, water was frequently polluted. Suess introduced the bold plan of bringing water from springs in the Alps of Lower Austria and Styria via an aqueduct 120 kilometers (km) long. With skill and tenacity, and against the resistance of bureaucrats and financiers, he succeeded in completing this project successfully in 1873, when the aqueduct was officially opened in the presence of the emperor, Francis Joseph. From this point on, the number of typhoid deaths fell to one-tenth of the previous figure, and Suess, a man of great integrity and modesty, considered this the most important project he ever conducted.

Above all, Suess sought to teach and conduct research, and in 1857 he was appointed associate professor of paleontology in the University of Vienna. In 1862, he moved to the chair of geology, the first chair of geology in Austria. Suess's success testifies to the flexibility of the university system during that period; he became professor without ever having completed formal training in the geosciences, or receiving a doctorate or "habilitation," the degree then (and still) considered a basic requirement for a professorship.

Like many leading earth scientists, Suess had started as a paleontologist but then moved toward stratigraphy and structural geology. It was in the latter field that he made his most important contribution, namely that orogenic zones (zones of mountain-building, e.g., the Alps, Rocky Mountains, Himalayas) are not the product of vertical movements induced by magmatic processes, but of horizontal thrusting over considerable distances.

Fieldwork in many parts of the Alps had led Suess to this concept. He was the first to recognize the northward thrust of the Mesozoic northern Calcareous Alps over their younger, tertiary Flysch-filled foreland. Another important step was the identification of the Glarner Doppelfalte (the Glarus Double Fold, Switzerland) as the product of northward thrust. Suess participated in the first major geological project in the eastern Alps, the establishment of a section from Passau in the north to Duino (near Trieste) in the south; he worked on the Brenner and Semmering railway sections, both major engineering feats of the day; and he studied stratigraphic problems of the southern foreland of the Alps, in the Vicenza area. In 1875, he published a small booklet titled *Die Entstehung der Alpen* (The Origin of the Alps), where for the first time the significance of horizontal movements of rock masses (nappes) was documented.

Suess's grandson, Hans Eduard (1909–1993), was trained as a chemist at the University of Vienna and moved to the United States in 1949. From 1955–1977 he was professor of chemistry at the University of California in San Diego. His paper with H. C. Urey on "Abundances of the Elements" was one of the most influential contributions to cosmochemistry.

Suess worked closely with Marcel Bertrand of Paris, who asked him to try and persuade Albert Heim—the doyen of Swiss geology and, until then, a defender of vertical tectonics—to accept this new concept. This was the birth of "Deckenlehre," or Nappe Tectonics, which revolutionized and stimulated geological thinking in the way that plate tectonics did almost one hundred years later.

In 1883, Suess commenced publication of his monumental, three-volume work *Das Antlitz der Erde* (The Face of the Earth, 1883–1909), which was eventually translated into English, French, Italian, and Spanish. In it, Suess applied the concept of horizontal tectonics (thrusts and nappes) on a global scale.

Parallel to these activities and a full teaching program, Suess started another major civil engineering project, the excavation of the bed of the Danube, with the aim of protecting Vienna from recurrent floods and to facilitate shipping. In 1869, he was a member of the delegation accompanying the Austrian Emperor Francis Joseph to the opening of the Suez Canal. In Egypt he met the head of dredging operations and invited him to come to Vienna in 1870 with his crew and equipment. The Danube Canal was opened in 1875. Since then, there have been no major inundations in Vienna.

In addition to his scientific and public works endeavors, Suess was active in politics. In 1863, he was elected a member of the Vienna City Council, where he was in charge of the Water Supply Commission. In 1869, he became a member of the Diet of Lower Austria and, in 1873, a member of Parliament. He always refused, however, invitations to join the upper house, or Herrenhaus. In these political capacities, he worked particularly for enlightened legislation affecting education, with emphasis on the concept of the popular university and on primary education. He contributed significantly to the preparation and promulgation of the Reichsvolksschul-Gesetz, the imperial law regarding primary education. All biographers praise Suess as an outstanding and always fair speaker, who set standards for the high quality of parliamentary debate in what we now recognize as the golden age of the Austro-Hungarian Parliament.

Suess had been elected a member of the Austrian Academy of Sciences in 1860 when he was only twenty-nine years old; he was its secretary from 1890–1901 and its president from 1901–1911. He was instrumental not only in organizing research projects and in turning the academy into a modern research organization but also in founding the International Association of Scientific Academies. After his retirement from the chair of geology in 1901, he devoted all his energy to the academy. He died on 26 April 1914 in Marz, Austria, just prior to World War I.

Suess stands out for his contributions to the advancement of geology, by the width of his horizon, by his intuition, and by his tremendous energy. He was an outstanding scientist and inspiring teacher, an exceptionally skilled organizer, and a successful politician. A number of scientific societies worldwide had elected him an honorary member. Suess refused to accept various medals and honors during his life. The honorary citizenship of the city of Vienna was the only exception.

Bibliography

ÖSTERREISCHISCHE AKADEMIE DER WISSENSCHAFTEN. "Eduard Suess, Geologe, Organisator und Politiker." In *Österreichische Naturforscher und Techniker*. Vienna, 1950.

WEGMANN, E. "Suess, Eduard." In *Dictionary of Scientific Biographies*, Vol. 13. New York, 1981.

EUGEN F. STUMPFE

SUN

The Sun is a star "close up." As seen from Earth, the Sun is the overwhelmingly dominant object in the sky in terms of the light and heat that irradiate Earth (Plate 41). The Sun is also dominant by other measures. For example, the Sun contains almost 99.9 percent of the total mass in the solar system, comprising the Sun and the planets—the mass of Earth represents only about three millionths of the total.

From the perspective of our galaxy, the Milky Way, the picture is just the opposite. The Sun is quite undistinguished among the billions of stars in the Milky Way. In fact, the Sun is a rather ordinary cool star, classified as a dwarf. It is located in the Galaxy between the Sagittarius and Perseus spiral arms (on the inner edge of a poorly defined structure called the Cygnus arm), at a distance of about 30,000 light-years from the center. This places the Sun a little more than half the distance from the center to the outer reaches of the galactic disk. The whole galaxy rotates (like other astrophysical objects) about its center, which gives the Sun an equivalent linear velocity probably somewhat greater than 200 kilometers per second (km/s).

Beginnings and End of the Sun

Observations and computer simulations indicate that stars form in an interstellar cloud where there is a denser-than-average accumulation of gas (mostly hydrogen) and dust grains, the process taking several million years. Mutual gravitational attraction of the cloud particles gradually increases the density, the region inexorably contracting by self-gravitation, and at an accelerating pace. As this collapse continues, the concentrated material forms an embryo star that heats up through the conversion of gravitational energy into thermal kinetic energy, the temperature sufficient to melt and vaporize the dust grains in the central (core) region. With a greater rate of collapse in this region (where the gravity is greatest) as compared with the outer regions, the core temperature rises still further, to more than 10^7 K (depending on the object's mass). At this tremendous temperature, nuclear fusion of hydrogen to helium begins, releasing an enormous quantity of radiant energy. This establishes a pressure gradient that stops further contraction, and the infant star is then in a state of quasi-equilibrium in which the outward pressure is just balanced by the weight of the overlying layers. While the radius, brightness, and rotation change gradually with time, the above equilibrium condition persists typically for at least several billion years (for stars of approximately the Sun's mass). Based on such theoretical models, and measurements such as the age of meteorites (which can be determined rather accurately), the age of the Sun is estimated to be 4.6×10^9 years, which, according to its stellar type, corresponds to a middle-aged star. The models predict an eventual expansion of the Sun, after the available hydrogen fuel is all spent, to a much greater size (the "red giant" phase) for a relatively brief period, followed by final collapse to a tiny white dwarf star (*see* STARS).

Current State of the Sun

The Sun rotates in the same sense of rotation as the orbital motion of the planets about the Sun. The rotation axis is inclined at an angle of 82° 52′ to the ecliptic, the plane of Earth's orbit about the Sun. The Sun's surface rotation rate varies with solar latitude. It is greatest at the solar equator (zero latitude), with a (sidereal) period of approximately twenty-five days, and about thirty-two days close to the poles. These values depend somewhat on what solar surface features are used to track the rotation. The average temperature at the solar surface (the photosphere; see below) is approximately

6,000 K, increasing inwardly to an estimated 1.5×10^7 K in the core. The mean density is 1.41 g cm^{-3}, with a range of approximately 2.8×10^{-7} g cm^{-3} at the surface to 150 g cm^{-3} in the center. The density and temperature in the center yield a central pressure of approximately 3×10^{11} standard Earth atmospheres. Because of the high temperatures involved, the Sun is gaseous throughout its volume, despite the high density in the core region.

The intensity of the radiation emitted by the surface of the Sun varies as a function of wavelength similar to a black body at the same temperature, as described by the Planck function. This radiation spans ultraviolet to far-infrared wavelengths and peaks in the green at a wavelength of about 0.5 microns (μm). The Sun also emits at radio wavelengths, and at much shorter wavelengths, including X rays and γ rays. The total radiation (luminosity) produced by the Sun is 3.845×10^{33} erg s^{-1}, which converts to an irradiance at the distance of Earth of 1.367×10^6 erg cm^{-2} s^{-1}, or 1,367 W m^{-2}. This value is known as the solar constant, a fundamental quantity since it represents the amount of energy impacting the terrestrial atmosphere, and constitutes the principal determinator of Earth's climate. Precise radiometric measurements from space reveal that the above value is not a true constant. Short-term variations at a level of 0.2 percent can occur over a period of days, associated with changes in activity features (discussed below) on the solar surface. Cyclical changes in activity over periods of several years produce corresponding irradiance variations at a level of about 0.1 percent. Variations over much longer intervals (centuries) appear possible, since it is thought that the characteristic activity level of the Sun can change on these timescales as well. In addition to such variations, there is a much larger long-term trend: according to the standard solar model, the Sun's luminosity has gradually increased by about 40 percent during its lifetime!

The spectrum of solar radiation is marked by thousands of dark lines (the Fraunhofer spectrum) that reveal the presence of most chemical elements, exceptions being primarily the highest atomic number, radioactive elements. Because of the temperatures involved, almost all of the element species have lost one or more electrons and are thus in the form of ions. Neutral atoms are also present in relatively small numbers, and simple molecules also exist in cooler regions of the solar atmosphere. Hydrogen is the most abundant element, with 92.1 percent of the total number of atoms (mostly in the form of ions), and helium with 7.8 percent. Hence, the sum of the abundances of all other elements amounts to only about 0.1 percent of the total. At the temperatures that prevail in stars similar to the Sun, and in less massive stars, the conversion of hydrogen to helium by the proton-proton chain reaction (*see* NUCLEOSYNTHESIS AND THE ORIGIN OF THE ELEMENTS) is thought to be the dominant source of nuclear energy production in the core. The models indicate that other low atomic weight elements are also produced, but with extremely small abundances. The higher atomic weight elements in the Sun suggest that they must have been formed in previous stars and released in prior cataclysmic events. These elements were then present in the material that later accumulated to form the Sun.

General Structure

Theoretical models suggest that the core region of the Sun, the site of nuclear energy generation, extends out to about 25 percent of the solar radius. In the high temperature and density in the core of the Sun, a pair of hydrogen nuclei, or protons, sometimes fuse together. Two of these new nuclei can in turn combine via an intermediate step to form a helium nucleus. But the sum of the masses of two protons and two neutrons that constitute a helium nucleus is slightly less than the masses of these separate particles by about 7 percent. By the equivalence of mass and energy, this small difference in mass is liberated as energy. Based on the current solar luminosity, the Sun is converting about five million tons of matter into energy every second. Above the core, in the radiative zone where energy is transported outward primarily by diffusion of radiation, the temperature, pressure, and density decrease uniformly out to about 70 percent of the radius. In the remaining outer envelope, the convection zone, energy is transported primarily by convective processes. The decreasing temperature in this zone allows the existence of complex ions and atoms, resulting in a high rate of atomic absorption and emission of the radiation as it interacts with the gas particles. This tendency to block the outward flow of radiation, due to the opacity of the gas, promotes large-scale convection that penetrates up to the apparent "surface" of the

Sun, the thin photosphere, as seen in ordinary (broad-spectral-band) visible light. The temperature decreases with height through the photosphere and the atmosphere above, from about 8,500 K ("lower photosphere") down to roughly 4,400 K at the temperature minimum in the "upper photosphere," about 500 km above the visible photospheric surface. High-resolution images of the photosphere, recorded over a broad spectral band, display a pattern of convective cells called granulation. Individual granules have average dimensions of about 1,100 km, with characteristic lifetimes of five to thirty minutes. Other observations reveal larger-scale cellular patterns, called mesogranulation and supergranulation. The latter structures have dimensions of approximately 30,000 km.

The most obvious structures in the photosphere are localized dark regions, or sunspots. Sunspots appear dark because the gas in this region is cooler, and hence it emits less light than the surrounding region, due to the presence of strong magnetic fields that inhibit the normal flow of convective heat. Sunspots have a central core, or umbra, usually surrounded by a penumbra, consisting of more or less radial filaments. Typical field strengths in the umbra are 2,000 to 3,000 Gauss (G). Figure 1 shows a large group of sunspots with surrounding granulation. The long dimension in this image covers almost one-tenth of the visible solar disk.

A few hundred kilometers above the temperature minimum, the atmosphere increases in temperature from about 6,000 to 8,000 K over a height range of 1,000 km in the layer called the chromosphere, while the density drops by two to three orders of magnitude. The chromosphere is too rarified to emit white light, and spectral line emission from several elements, including hydrogen, calcium, and magnesium, dominates. Figure 2 is an image of the Sun recorded with a special narrow-spectral-band filter in the light of the red hydrogen line (Hα). This shows the typical complex structure evident in the chromosphere. Areas of

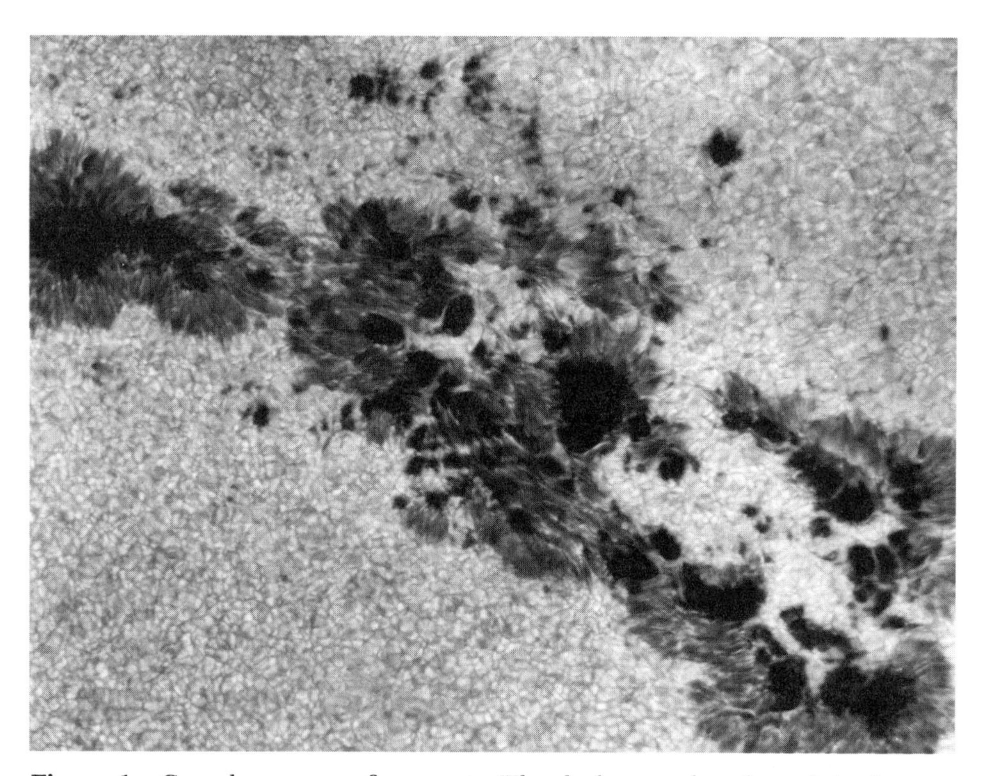

Figure 1. Complex group of sunspots. The dark central umbra of the larger spots is surrounded by the characteristic filamentary structure of the penumbra. The background field shows the cellular structure of granulation.

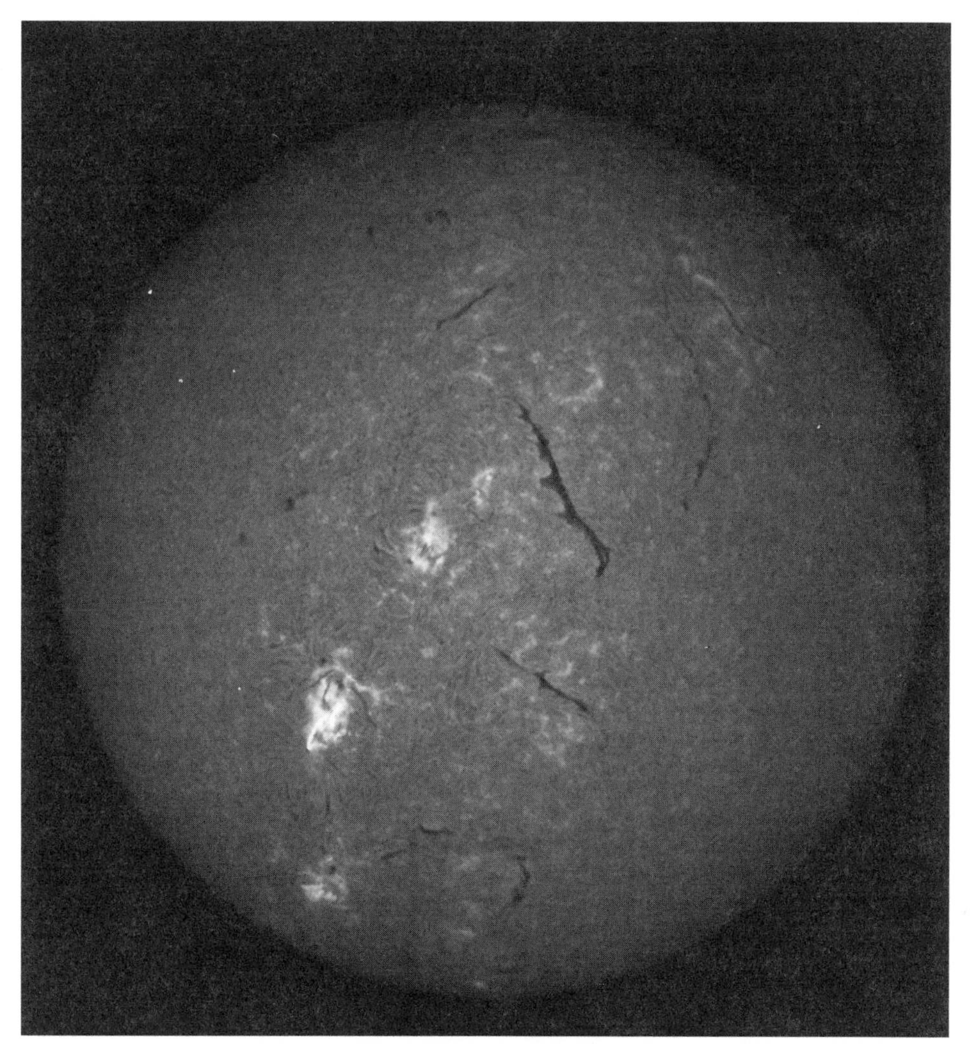

Figure 2. Image of the Sun recorded in the light of the red line of hydrogen (Hα). This shows the complex chromospheric structure, including the bright "active regions," and dark "filaments" that consist of gas suspended above the solar surface.

concentrated magnetic field, called active regions, appear bright. These occur primarily in active latitude zones on either side of the solar equator, the larger regions typically associated with sunspots as seen in the photosphere. On average, the number of sunspots and the occurrence of active regions reach a maximum about every eleven years. Half way between maxima the Sun usually shows far less activity and sometimes none. This sunspot cycle, illustrated in Figure 3, is also evident in other measurements of solar activity. Certain aspects of the magnetic polarities (fields directed into, or out of, the Sun) associated with sunspots reverse each activity cycle. Thus the Sun also has a twenty-two-year magnetic cycle, and there is evidence of longer cycles as well. Such cyclical changes represent a basic characteristic of the Sun (and of many stars) that arises ultimately from the interaction of convection and differential rotation in the interior that generates and then winds up and redistributes the internal magnetic fields.

Also evident in Figure 2 are dark filaments consisting of gas that has accumulated above the solar surface due to particular localized magnetic field configurations. Filaments have a temperature of about 10,000 K and appear dark because more

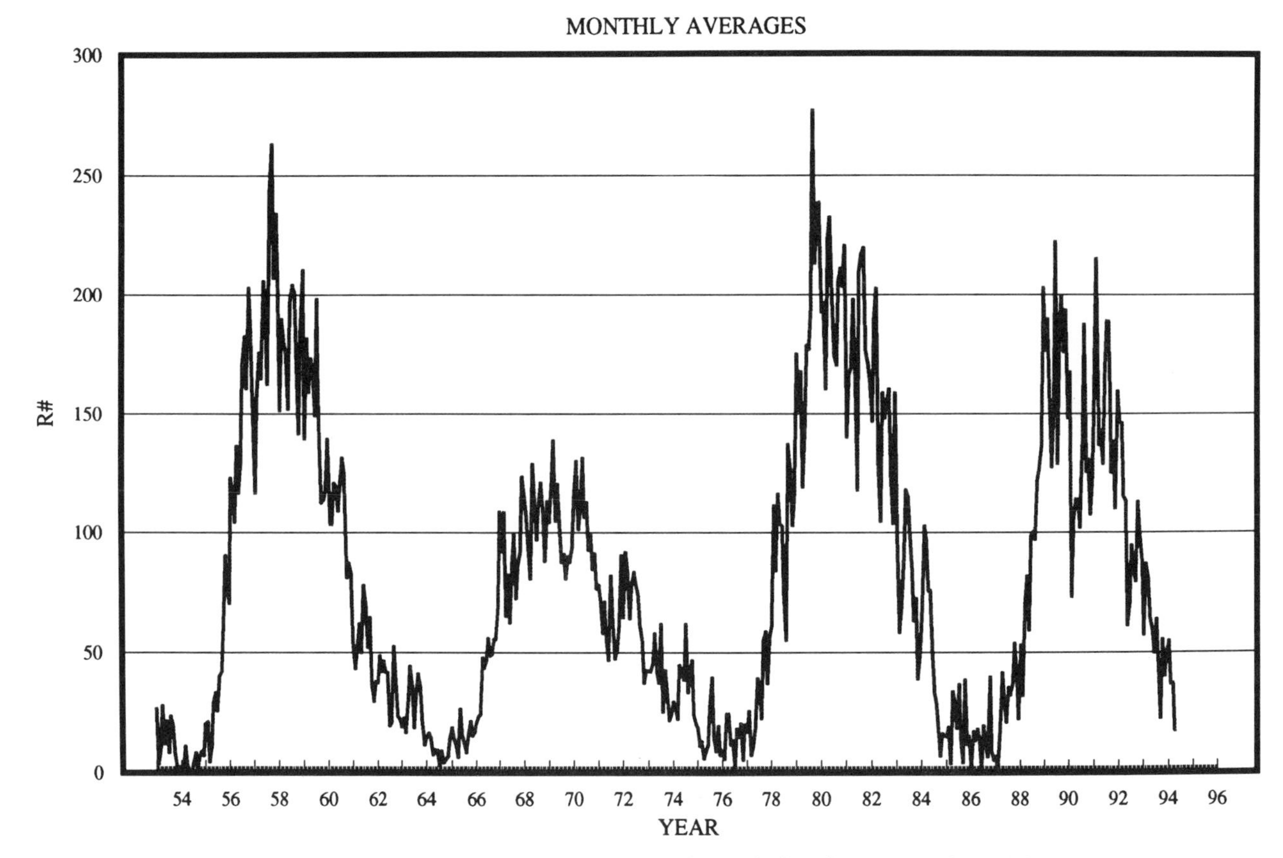

Figure 3. The sunspot cycle, which shows the systematic variation in magnetic activity on the Sun over a period of about eleven years, as measured by the number of sunspots visible on the Sun day by day.

light from below is absorbed by the filament gas than it emits. Seen at the limb, these same features are called prominences, which have typical arch shapes with several footpoints. During a characteristic lifetime of many days, large quiescent prominences often grow slowly in height to 100,000 km or more, and then become unstable and erupt, sometimes with sufficient velocity to allow the prominence material to escape into interplanetary space. On a smaller scale, the low chromosphere commonly shows small, spike-like protuberances, or spicules, which shoot up from the surface and then quickly fade, with lifetimes of about ten minutes.

Figure 4 shows a large, extended flare in a field of light and dark fibrils shaped by the chromospheric magnetic field distribution. Flares occur when the magnetic field in an active region becomes twisted and stressed due to convective motions, with the abrupt conversion of the magnetic potential energy to kinetic and thermal energy, resulting in a rapid increase in temperature, enhanced radiative output, and the acceleration of particles to high energies. Flares can result in sudden disturbances in Earth's ionosphere due to the effect of the high-energy radiation, while the stream of high-energy particles can be harmful to space operations.

Above the chromosphere, at a height a little more than 2,000 km above the photosphere, the temperature increases abruptly in the transition region, a distance of less than 100 km, to a million or more degrees in the outer atmosphere called the corona. The reason for such high coronal temperatures is not well understood, but the principal energy source is probably magnetic, in the form of magnetohydrodynamic (MHD) waves on the solar surface that deposit energy in the corona. A consequence of the lower gas density in the layers above the photosphere is that the pressure due to magnetic fields dominates that of the gas pressure. Due to the complex distributions of magnetic fields at these heights, images recorded at wavelengths of the chromospheric spectral line emissions display complex structures. Because of the high temperatures and low densities (about 1.7×10^{-16} g cm^{-3} at a height of 70,000 km) in the corona, the coronal atmosphere consists primarily of ions and free electrons. The white-light corona, as observed at a total solar eclipse, arises from photospheric light scattered by free electrons in the hot coronal plasma. Its spectral brightness is only a few millionths that of the photosphere. Another component of the observed corona is emission associated with spectral transitions of the coronal ions. Light scattered by interplanetary dust is a further component.

Images taken at the time of a total solar eclipse, when the disk of the Sun is hidden by the Moon, show complex loop structures in the inner corona; further out one sees primarily radial streamers. Sensitive measurements show these streamers extending well beyond ten solar radii. Such structures trace magnetic field lines originating at the solar surface. In addition to bright structures, X-ray images of the solar corona reveal coronal holes, regions where coronal emission is drastically reduced as compared with that from the average corona. These coronal holes correspond to open magnetic field regions, in contrast to the closed regions evident by low-lying coronal loops. Spacecraft have detected a continuous stream of coronal particles from the Sun called the solar wind, with both fast (approximately 700 km s^{-1}) and slow (approximately 300 km s^{-1}) components. The fast component originates principally in coronal holes where the particles are accelerated by a much greater amount than in surrounding regions. The slow component may represent the continuous expansion in streamers of coronal material away

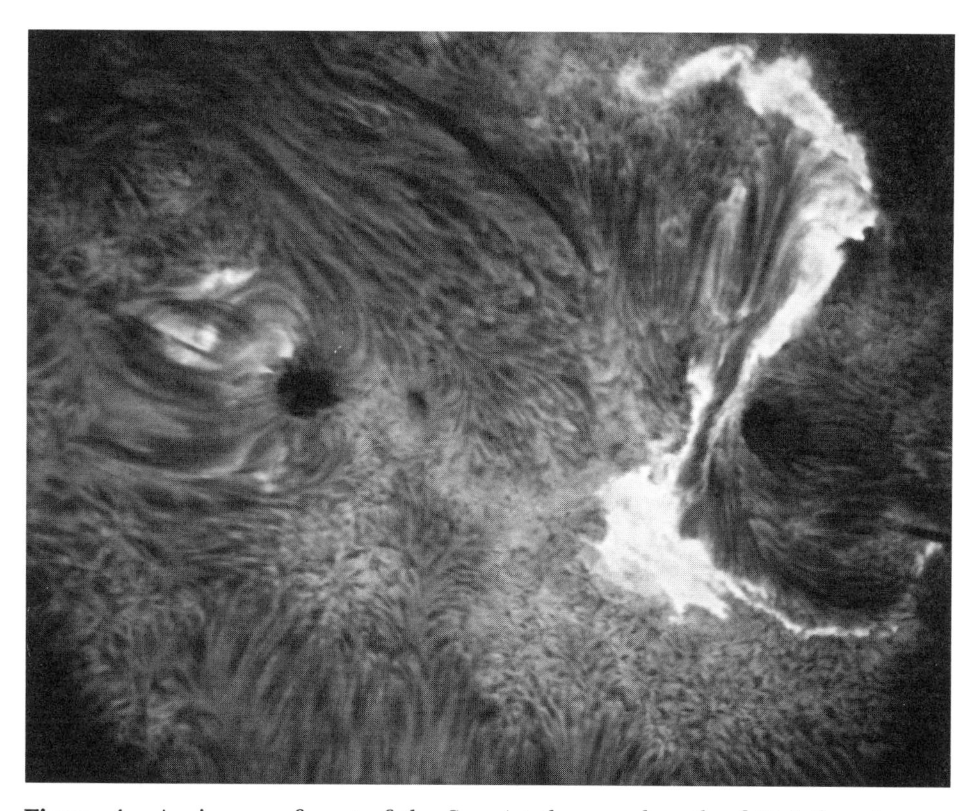

Figure 4. An image of part of the Sun (at the wavelength of Hα) that shows the occurrence of a large flare, as indicated by the extended bright regions. Two large sunspots, on either side of the figure, are just evident in this image that otherwise shows the complex structures that typify the chromosphere.

from the Sun. Following certain changes below, such as a prominence eruption, a much larger amount of material can suddenly leave the Sun in a coronal mass ejection (CME). Such events can cause magnetic substorms in Earth's outer atmosphere.

Indirect Observations of the Solar Interior

Knowledge about the internal structure and evolution of the Sun, described earlier, is based principally on advanced theoretical models. But some experimental checks are now possible. Neutrino particles are produced as a by-product in the nuclear fusion reactions in the solar core, and elaborate experiments on Earth have been set up to measure the flux of neutrino particles from the Sun. These measurements show only about one-third the number of neutrinos predicted theoretically. The cause of the measured deficit is uncertain, but could be explained by transmutations of neutrinos between different states, some of which would not be detected. Without such an explanation, the results would raise questions about the validity of stellar evolution theory.

Acoustic, or pressure, waves are generated and propagated within the Sun. These waves tend to be channeled around the Sun as a consequence of reflection at the lower part of the photospheric surface, and refraction in the solar interior that results in downward propagating waves being redirected back toward the surface. Waves are, therefore, trapped as in an acoustic cavity. Such waves cause subtle oscillatory vertical motions of the photospheric surface. Analysis of these motions and their horizontal scales reveals a large number of wave modes, with a dominant period of approximately five minutes. The field of helioseismology combines extremely sensitive measurements of these oscillations with theoretical models of the solar interior to explore physical parameters of the solar interior, such as the depth of the convection zone, structures within it, and the solar rotation rate with depth and latitude, issues of fundamental importance in refining solar and stellar models.

Table 1. Key Solar Parameters

Age	4.6×10^9 years
Diameter	1.392×10^{11} cm
Mass	1.991×10^{33} g
Mean density	1.41 g/cm^3
Spectral class	G2V
Effective surface temperature	5800 K
Mean distance from earth	1.496×10^{13} cm
Mean angular diameter	32.0 arc minutes

Bibliography

FOUKAL, P. *Solar Astrophysics.* New York, 1989.
NOYES, R. W. *The Sun, Our Star.* Cambridge, MA, 1982.
STIX, M. *The Sun.* Berlin, 1989.
ZIRIN, H. *Astrophysics of the Sun.* Cambridge, Eng., 1988.

RAYMOND N. SMARTT

SUPERGENE ENRICHMENT

Enrichment of metals in certain ore deposits occurs near the surface as a result of chemical processes involving the atmosphere, the biosphere, and the flow of surface and shallow meteoric waters. Such processes are called "supergene" to distinguish them from hypogene processes, those that concentrated ore minerals in the crust originally from solutions of deep origins in or below the crust. The term "secondary enrichment" is also sometimes applied to supergene enrichment. A conventional view held in considering supergene processes is that they preclude the further effects of erosion, which result in placer concentrations of the chemically inert and heavy minerals. Among important supergene enriched ores are those of aluminum (Al), manganese (Mn), iron (Fe), nickel (Ni), cobalt (Co), copper (Cu), lead (Pb), silver (Ag), and gold (Au). Supergene processes act in two different fundamental ways to affect ore concentration. One is by residual concentration of insoluble components of minerals and rocks, and the other by chemical solution, transport, and precipitation of enriching elements.

Residual supergene enrichment is responsible for concentration of rich ores of Fe, Al, Mn, and sometimes Au. These elements are concentrated in certain geological and climatic environments through solution and removal of more soluble chemical components of the rocks and minerals in which the elements were originally deposited. The process leaves behind concentrations of relatively insoluble ore minerals.

Enriched iron ores are derived by weathering and alteration of iron silicate and oxide minerals, leaching silica and enriching iron as magnetite, hematite, and limonite. A cross section of iron ore in the Mesabi Range (United States) is shown in Figure 1. The primary ironstones (iron silicate and iron chert strata) contain about 25 percent Fe, the enriched ores usually more than 55 percent Fe with specific gravity of the ore rocks increased by compaction of pore space. The results of surficial weathering produce a profile of different iron minerals in which the deep anhydrous (oxide) minerals are overlain by hydrated (hydroxide) minerals resulting from continued surficial weathering.

Transported enrichment results from the solution of minerals and movement of solutions of dissolved elements to a site of contrasting chemical conditions where they are precipitated as new minerals at greater concentrations. Numerous metals dissolved from hypogene ore deposits are so enriched and include Pb, zinc (Zn), and Ag. Deposits of Ni are produced in such manner from weathering of small amounts of Ni from obducted ocean

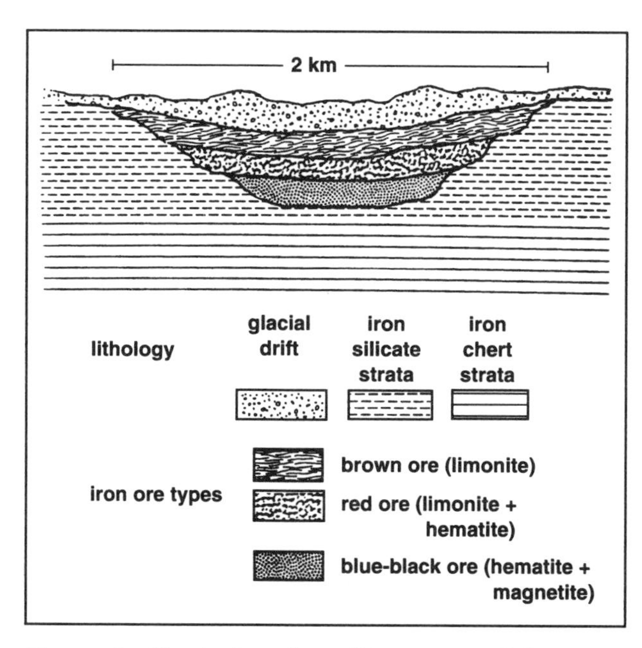

Figure 1. Vertical zoning of iron ores of the Mesabi Range (United States), adapted from Leith (1913). Horizontal scale is only approximate and the vertical scale exaggerated by 2x.

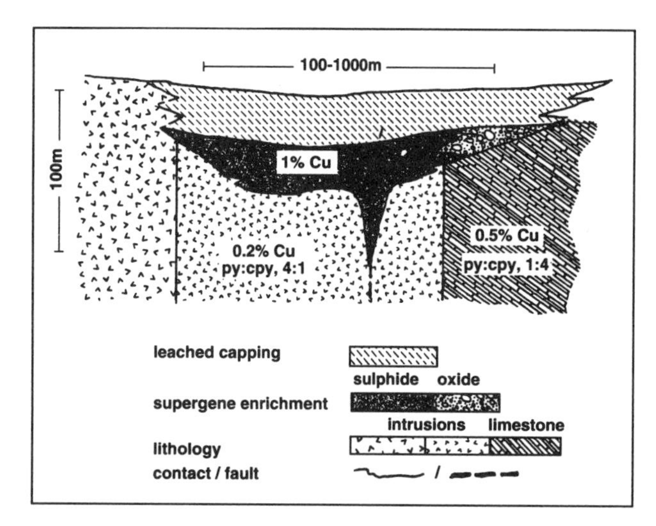

Figure 2. Idealized cross section through a supergene enriched intrusion-porphyry copper deposit with carbonate wall rocks showing the lensoidal cross section of such enriched bodies, which are equant in plan, in rocks of potassium-silicate alteration; common hypogene grades and pyrite : chalcopyrite ratios in such ores are shown for the respective rocks. The section shows a common habit of deep penetration of supergene fluids along faults or joints. The horizontal scale shows a common range of size for these orebodies; the vertical scale is general. Adapted and modified from Jerome (1966).

crust. The most important and significant deposits of supergene enriched metals, however, are those of Cu. A generalized cross section through a supergene enriched porphyry copper deposit is shown in Figure 2.

Under the effects of chemical weathering and biogenic activity, the pyrite and chalcopyrite of the hypogene copper deposit are dissolved to form acid, dissolving in turn other copper minerals. The copper moves downward to precipitate as chalcocite either directly or by replacement of other sulphides. Copper is ordinarily enriched by factors of 2x or greater. Where copper-bearing limestones altered to calcium-silicate minerals at intrusion contacts are involved, acid neutralization takes place over short distances and the metal precipitates as various copper-bearing sulphate, carbonate, and oxide minerals. Original rocks stripped of their metals are left behind and constitute a leached capping whose specific characteristics may indicate the former presence of hypogene copper.

The process of supergene enrichment of Cu is enhanced by the presence of sufficient pyrite to produce acid in supergene solutions, and by the existence of permeable rocks, which allow deep penetration reaction of the surface-related waters.

Bibliography

JEROME, S. E. "Some Features Pertinent in Exploration of Porphyry Copper Deposits." In *Geology of the Porphyry Copper Deposits, Southwestern North America*, eds. S. R. Titley and C. L. Hicks. Tucson, AZ, 1966.

LEITH, C. K. "A Summary of Lake Superior Geology with Special Reference to Recent Studies of the Iron-Bearing Series." In *Ore Deposits*, ed. S. F. Emmons. New York, 1913.

SPENCER R. TITLEY

SUPERNOVA

See Stars

SURFACE WATER, STORAGE AND DISTRIBUTION

For water to be available to all people at all times, it is necessary to devise systems for storage and delivery of that water. These systems must be able to provide water that is safe to drink and available at all times of the year. Water must not only be available for home use, but also for other uses such as commercial (car washes, beverage industries, etc.), irrigation, and firefighting needs. This entry examines the various methods of storing and delivering water to consumers.

Water Storage

A "water system" might be the water department of a large city with thousands of customers, a small subdivision with fifty homes, a campground, or a single cabin in the woods. In each case, water must be stored so that it is available to cover periods of high use (such as morning or evening bathing, toilet flushing, or dishwashing), periods of sustained use (such as irrigation or car washing), or emergencies (firefighting). The methods of water storage vary, depending on the type of water system that is involved.

Methods of Water Storage. Water has been stored, or, more accurately, has stored itself naturally since the beginning of time. Civilization, meanwhile, has devised a number of additional, man-made ways of storing water to meet its needs.

1. Natural storage. The obvious methods of natural water storage include oceans and icebergs, although this water is often inaccessible or inappropriate for domestic use. Rivers and lakes usually serve as above-ground natural storage areas for many human needs. Natural water storage also occurs underground, in water-bearing layers of sands and gravels called aquifers (*see* GROUNDWATER). Some aquifers are quite large and are recharged frequently by rainfall, snowmelt, or river water seepage. Many municipal water systems depend on wells that are drilled into productive aquifers. Water that is stored openly above ground is vulnerable to many sources of pollution, including fuel and chemical spills or surface water run-off from adjacent lands that may contain pesticides, animal manure, or other contaminants. Water stored in underground aquifers, on the other hand, is protected by the layers of soil above, and is generally considered safer from pollution by surface contaminants. One disadvantage in depending on underground stored water for a water supply is that you cannot see the water, so it is difficult to measure how much is present. In an above-ground lake or river water supply, the volume of available water can be determined at any time, and the effects of rainfall, snowmelt, and water use can be easily measured. Calculating the amount of available water in an underground aquifer, on the other hand, requires that water level measurements be taken periodically through observation wells, and even then the exact shape, depth, and porosity and permeability of the aquifer must be estimated.

2. Man-made storage. Dams and levees have for years represented our best efforts to control and store surface water flows. The large body of stored water behind a dam, for example, can provide a steady supply of water for city water supplies, irrigation systems, and electrical power generating plants. Canals and aqueducts have facilitated efforts to deliver this stored water to the areas where it is needed. One drawback to open surface water storage, however, is the large amount of space required. Most large cities, therefore, have constructed closed storage reservoirs to maintain the reserve of water that is necessary to cover fluctuations in water demand. A common recommendation is that the volume of storage be equal to two or three days of the water system's average use. It is also recommended that storage reservoirs be located at a high enough elevation to allow the water to flow by gravity to the distribution system. If applicable, storage for fire protection should be provided as well.

Wood reservoirs have historically been associated with smaller water systems such as campgrounds, parks, and small communities. They are less expensive than concrete or steel reservoirs. They are typically found in sizes of 454,000 liters (l) or less. Redwood is the most common type of wood employed for this purpose. Circular steel hoops provide perimeter support, and the natural swelling of the wet wood provides a near-watertight seal. Some leakage, however, can be expected. Also, if a wood reservoir is allowed to draw down to a near-empty condition during warm weather, the subsequent drying and shrinking of the boards could present leakage problems when the reservoir is again refilled.

Concrete reservoirs are the most durable and permanent type of reservoir construction. Concrete is relatively expensive, but is nearly maintenance-free (other than routine cleaning). Concrete reservoirs can be constructed to almost any size and to satisfy any condition. Since they are built in place and are usually rather large, a site with good access for construction vehicles is required. The expense of concrete reservoirs is due to the high cost of this type of construction.

Steel reservoirs generally have lower construction and installation costs than concrete, but they do require more maintenance. To protect against rust and corrosion, the exterior should be kept clean and painted. Interiors of steel reservoirs are commonly coated with an epoxy or enamel-type finish. Some coal-tar linings used in the past have apparently degraded over time, and are implicated in the release of small amounts of solvents into the stored water. Steel reservoirs are usually welded or bolted together and are manufactured in a variety of sizes. Small steel reservoirs can be manufactured off-site and then trucked and lifted into place.

Construction Features. There are a number of items that should be considered when designing or building a closed water storage reservoir. The design goals are to optimize operation and facilitate maintenance. An access hatch is usually located on top of the reservoir, and, in the case of large reservoirs, perhaps an additional hatch in the side that will permit access when the reservoir is drained for cleaning. A vent should be provided to allow air exchange, and it should be screened to keep out insects, birds, and windblown debris. Ladders usually provide exterior and, in some cases, interior access.

Water Distribution Systems

The energy required to distribute stored water comes from either gravity or pumps. The means for distributing this water comes from canals, aqueducts, ditches, and pipes.

Canals and Pipes. Canals and aqueducts are used to carry water long distances, usually for irrigation purposes. Canals can be lined with earth, clay, or concrete to retard seepage and increase flow. Some aqueducts have carried water across canyons, or, as in the Roman days, carried water above ground from the point of supply to the consumer. There are several types of piping material that can be used in water distribution systems. Each has its own advantages and disadvantages. The pipe material must have adequate strength to withstand external loads from trench backfill, traffic, and earth movement; high burst strength to withstand high water pressure; and smooth interior surfaces, corrosion-resistant exteriors, and tight joints.

Wood stave pipe represents some of the earliest pipe material used (other than hollowed-out logs). Wood staves were banded together with wire or steel and often coated on the outside with tar. Because of its inability to withstand high pressures and because of the newer types of pipe available, wood stave pipe is of mostly historical value, although some is still in use today.

Steel, galvanized iron, and ductile iron are the common metals used for water pipe. All are very strong and able to withstand high internal and external pressures. They are vulnerable to exterior corrosion, however, unless protective coatings or wrappings are applied prior to installation.

Asbestos cement pipe is manufactured from portland cement, long fibrous asbestos, and silica. Its main advantages are its ability to withstand corrosion and its good hydraulic flow characteristics. A major disadvantage is that it is heavy, brittle (and therefore difficult to tap), and is easily broken during construction or by impact loading. It is often used for the conveyance of sewer or storm water, but is seldom used for new waterline construction.

Polyvinyl chloride pipe (PVC) is currently the most common type of pipe used in distribution systems. It is available in diameters of 1.25 cm and larger, and in lengths of 3.05, 6.10, or 12.16 m. A main advantage is its light weight and flexibility, allowing for easy installation. A disadvantage is its inability to withstand shock loads. It is also a good idea to install a wire in the trench with any nonmetallic pipe, so that it may be located in the future with a metal detector.

Water is finally delivered to the consumer's home plumbing system, which usually consists of galvanized iron or copper pipe. Galvanized iron usually has threaded joints and fittings, whereas copper pipe is soldered together. Copper pipe is also quite malleable and can be easily bent to almost any configuration.

Bibliography

AMEEN, J. *Community Water Systems*, 5th ed. High Point, NC, 1980.

U.S. ENVIRONMENTAL PROTECTION AGENCY. *Manual of Individual Water Supply Systems*. Washington, DC, 1982.

VIESSMAN, W., and M. HAMMER. *Water Supply and Pollution Control*, 4th ed. New York, 1985.

SCOTT G. CURRY

SVERDRUP, HARALD U.

Harald Ulrik Sverdrup, a major figure in the development of modern oceanography, was born in Sogndal, Norway, on 15 November 1888. After serving as scientific chief on Roald Amundsen's *Maud* expedition to the Arctic (1918–1925), Sverdrup helped solidify Norway's already prominent position in geophysical science. Under Sverdrup's direction (1936–1948), the Scripps Institution of Oceanography at La Jolla, California, emerged as an internationally leading research center. Here, together with two colleagues, he wrote the bible of oceanography, *The Oceans: Their Physics, Chemistry and General Biology*. He left California in 1948 to direct the newly created Norwegian Polar Institute.

As a student at University of Oslo in 1908, he was enthralled with Vilhelm Bjerknes's visionary plan for establishing an exact physics of atmospheric and oceanic motions (*see* BJERKNES, J. A. B. AND V. F. K.). Sverdrup followed Bjerknes to Leipzig University as an assistant and doctoral student from 1913 to 1917; he produced or coproduced over twenty articles. His impressive early work provided research interests and methods that he sought to develop further: analyses of the general large-scale atmospheric and oceanic circulation as well as of the micro-scale factors mediating the energy and heat transfer that drive circulation.

When approached by Amundsen to join his much-delayed voyage to drift with the ice over the polar sea, Sverdrup responded affirmatively. With virtually no job prospects, he joined this patriotic expedition that could bring honor to Norway and help establish a foundation for a career. He also hoped to obtain experience with studying geophysical phenomena in the field. While drifting for several years in the ice, Amundsen's ship, the *Maud*, was to be a floating laboratory for studying terrestrial magnetism, northern lights, and conditions in the Arctic atmosphere and ocean.

Expeditions, Sverdrup soon learned, are often frustrating experiences. After he set out in 1918, numerous mishaps delayed for three years the actual drift across the Arctic. Further unfortunate wind and ice conditions kept them from reaching the central Arctic. The expedition was considered largely a fiasco. One opportunity did exist to save its honor: to convert the enormous amount of observation material into scientific reports. Bjerknes and Bjørn Helland-Hansen worked to bring Sverdrup to the new Geophysical Institute in Bergen, Norway.

Sverdrup recognized, however, the continued need to explore the polar sea to understand better its significance for hemispheric ocean currents. Little prospect existed for comparable expensive expeditions; he pursued alternatives. First, he engaged with an international project for using a Zeppelin airship for systematic polar exploration. When these costly plans stalled, he then joined forces in 1931 with explorer Hubert Wilkins to use a submarine to cross the Arctic under the ice.

Again, Sverdrup experienced a frustrating expedition. After many delays, the voyage came to an end when, getting ready to dive under the ice north of Spitsbergen, they discovered that the U-boat's diving rudder had disappeared. Still, Sverdrup managed to amass valuable observations.

Back in Bergen, Sverdrup began editing the last volume of *Maud* findings. Again he grew restless; he needed new data. Work from the *Maud* focused his attention on problems of heat and energy transfer between the atmosphere and the ocean. Laboratory and theoretical studies provided some clues, but Sverdrup wanted direct measurement. He could begin this research program by first studying in detail the heat budget above and below a layer of smooth snow. He and the Swedish glaciologist Hans Wilhelmson Ahlmann obtained funds to spend a summer camped on a glacier high in the Spitsbergen mountains. Having amassed a huge amount of data, Sverdrup extended the theory of geophysical turbulence and began to plan further studies on transport of heat and water vapor at the ocean surface. Then, in 1936 he was asked to become director of the Scripps Institution of Oceanography.

Sverdrup had previously received offers to come to the United States, where attempts to develop physical oceanography required foreign expertise. Sverdrup's incessant work and brilliant ability to find physical meaning in oceanographic observations had brought him an international reputation. To ensure that he remain in Bergen, Helland-Hansen had arranged that Sverdrup receive in 1930 the first professorship at the new Christian Michelsen Institute, which entailed full-time research opportunities. But now Helland-Hansen himself urged Sverdrup to embark on a three-year expedition to California.

Sverdrup's first impressions at Scripps were decidedly negative. The institution was oceanographic in name only. No possibilities existed for systematic work at sea. Without a clear oceanographic mission the institution merely developed as an umbrella for a number of independent specialties and laboratories. Personnel were largely demoralized. By 1937 Sverdrup had begun improving conditions significantly, including acquiring an oceangoing research vessel. In particular he recognized the need to find new patrons for supporting oceanographic studies. As had been the case with pioneering Swedish and Norwegian ocean science, the fisheries proved at the time to be the best avenue, and in particular, the local sardine economy. With assistance from the California Fish and Game Commission, he organized in 1937 the first systematic ocean study designed to understand conditions during the sardine spawning season. This preliminary investigation afforded clear insight into the nature of local currents and, equally important, provided Sverdrup with a vision for invigorating Scripps. He devised a comprehensive research program that integrated scientific and social concerns by linking the biological, chemical, geological, and physical oceanography of the eastern Pacific.

Sverdrup's overhaul of Scripps had begun, but fearing that his accomplishments might be undone if he left prematurely, he agreed to stay on until the end of 1941.

After the Germans invaded and then occupied Norway, Sverdrup accepted that he would have to stay longer. American entry into the war led to mobilizing Scripps for training new oceanographers and embarking, with massive government support, on innovative research programs that had both great scientific and military significance. After the war, Sverdrup was the chief oceanographer in a country about to invest heavily in oceanography, a field that had assumed new significance for national security. Nevertheless, within three years Sverdrup returned to Norway.

Immediately following victory in Europe, Sverdrup's friend Ahlmann became alarmed over the extraordinary measures being taken by the Soviet Union to step up polar research and colonialization. Americans also had become active in the Arctic. Ahlmann feared for the Nordic nations and especially for Norway's vital interests in the Arctic and Antarctic. Territorial claims were still disputed, and only vigorous activity could legitimate these claims. In dicussions with cabinet ministers, Ahlmann proposed a major institution to oversee polar research, surveying, and commerce. And such an institution required as a director an internationally respected scientist such as Sverdrup.

In 1948 Sverdrup came home to lead the Norwegian Polar Institute, to take over the planning of the Norwegian-Swedish-British Antarctic Expedition, and to assume an adjunct professorship in geophysics in Oslo. His activities in Norway until the time of his sudden death, in Oslo on 21 August 1957, were many and varied. He found it difficult to say no to tasks for which he felt qualified to help, and administration increasingly replaced research. These activities included reorganizing natural science training at the University of Oslo as well as leading a Norwegian aid program to India.

Sverdrup's most important published contributions brought acknowledged physical theory and great personal insight to the interpretation of observations while explicating geophysical phenomena such as trade winds; polar tidal currents; relationships between wind, sea, and swell; and the Pacific equatorial counter current. His influence lives on through the many editions of *The Oceans* as well as through his students, disciples, and colleagues at SIO, who helped shape postwar American oceanography both in style and substance.

Bibliography

DEVIK, O. "Minnetale over professor Harald U. Sverdrup." In *Det Norske Videnskops-Akademi i Oslo Årbok 1958.* Oslo, 1959.

FRIEDMAN, R. M. "The Expeditions of Harald Ulrik Sverdrup: Contexts for Shaping an Ocean Science." William E. and Mary B. Ritter Memorial Lecture, October 1992. Scripps Institute of Oceanography, 1994.

REVELLE, R., and N. MUNK. "Harald Ulrik Sverdrup—An Appreciation." *Journal of Marine Research* 7 (1958):127–138.

SVERDRUP, H. U., M. W. JOHNSON, AND R. H. FLEMING. *The Oceans.* New York, 1942.

SPJELDNAES, N. "Sverdrup, Harold Ulrik." In Vol. 13, *Dictionary of Scientific Biography,* ed. C. C. Gillispie. New York, 1970–1980.

ROBERT MARC FRIEDMAN

T

TAPHONOMY

The term "taphonomy" (derived from the ancient Greek *taphos*, or grave) was originally defined in 1940 by the Russian paleontologist I. A. Efremov as the study of fossilization processes. Some consideration of the processes affecting organisms' remains, from the time of their death to their final entombment in sediments, must precede any consideration of the fossil record, whether for systematic, evolutionary, or paleoecological study. Physical, chemical, and biotic factors of depositional environments serve as a series of "filters" through which paleonotological information must pass. These taphonomic processes bias the fossil record to varying but considerable degrees. As such, taphonomic study through most of the later nineteenth and twentieth centuries has been involved primarily with negative aspects, that is, the extent of information loss from the fossil record, and the often poor resemblance of modern communities to their preservable counterparts in sediments.

Live–Dead Studies and Time-Averaging

Considerable literature now exists on the biasing influences on the fossil record, based on comparative study of modern communities and their expression as potentially observable hard parts within the accumulating sediments. These many studies have led to a number of intriguing, and sometimes counterintuitive, conclusions. First, the readily preservable component of communities is highly varied, but ranges from a high of over 90 percent to lows of only a few percent of the total number of taxa within a community. The average values are generally about 30 percent preservable species in the marine communities. Virtually all fossil samples are biased, but at least for marine invertebrates most potentially preservable organisms are recorded as skeletal accumulations. Also, in most offshore marine environments only very local transport of skeletons takes place. For terrestrial vertebrates, plants, and the considerably more species-rich insects, chances for preservation are far lower (*see* FOSSILIZATION AND THE FOSSIL RECORD). Despite representation of most skeletonized invertebrates, the marine fossil record is highly biased with regard to ecological parameters such as the original rank abundance, dominance-diversity, and trophic group proportions. Results seem a bit more encouraging if biomass, as opposed to numbers of individuals or species, is used as a proxy for "importance" of a particular species in its original environment or community.

Perhaps the most intriguing and important result of live–dead studies in marine environments is that the skeletonized assemblage buried within the upper few centimeters of sediments is commonly richer in species than the predicted "easily preservable" (skeletonized) portion of the sampled live community. The addition to the death assemblage of species not present but alive in the sample 2 area is a key result of the phenomenon of time-averaging (Figure 1A). The potential fossil sample carries remains of many generations of organisms that represent distinctly different but patchy subcommunities. The patches migrate into and out of the local environment over the course of centuries or millenia.

While nearly all studies point to the critical role of rapid burial (Figure 1), a number of studies involving radiocarbon dating of shells prove that skeletons differing in age by as much as 3,000 years in tidal flats, or perhaps 5,000 years in offshore environments, may be intimately commingled in a single sample. Presumably, storm-induced physical reworking as well as bioturbation (or intense burrowing) serve to mix shells that were originally buried at different times. Provided that shells do not remain in the more destructive seafloor environment for any length of time, they may be preserved and moved through sediment or they may be exposed briefly and mixed with new shells and sediment.

Comparative Taphonomy

Taphonomic processes are largely, though not entirely, mediated by physical and chemical processes, and thus are more subject to uniformitarian study than are most biotic phenomena. Skeletal remains of organisms may be thought of as sedimentary particles. In many offshore marine environments they may be the only sizeable particles; hence, their mode of orientation, sorting, articulation, and other taphonomic properties may provide important insights into depositional processes otherwise undecipherable from the sediment. Moreover, burial of bodies and skeletons of organisms in particular sedimentary regimes may induce unique types of biogeochemical processes. Hence, aspects of early diagenesis can also yield important information about sedimentary and burial environments, particularly on the geochemical milieu of the sediments. From these sorts of considerations, a new comparative taphonomic approach emerged during the 1970s and 1980s. The field of comparative taphonomy focuses on what might be termed positive aspects of taphonomy, that is, the use of fossil preservational information in geology.

Taphonomic study is commonly divided into two subdisciplines: biostratinomy and fossil diagenesis. The former deals primarily with physical sedimentary processes affecting organism remains between the time of death and final burial. The latter deals with physical and chemical processes affecting the potential fossils after their burial. Generally excluded from taphonomy are very late-acting processes such as structural deformation, metamorphism, or weathering.

Biostratinomy. Early contributions to the field of biostratinomy involved actualistic studies of the decay of organisms in recent environments, especially the detailed work of Weigelt (1989) on vertebrate carcasses and of Schäfer (1972) on the "Aktuopaleontology," or modern taphonomy, of invertebrates and aquatic vertebrates in shallow marine environments of the North Sea. In the 1980s and 1990s, several researchers undertook field and laboratory experimental studies to document, more quantitatively, the rates and processes affecting disintegration of organisms with various combinations of environmental conditions. Critical studies include a long-term survey of vertebrate skeletal remains in subaerial environments of Kenya; research on transport and decay and burial of plant remains, and numerous experimental studies on the rates of disintegration of invertebrate skeletons. Biostratinomic studies begin with the understanding of mortality processes. In addition to normal mortality, via disease, aging, and so on, organisms may experience mass mortalities; paleontologists commonly observe the effects of such catastrophic mortality (Figure 1D). Floods, mudslides, storms, and volcanic eruptions may extinguish and rapidly bury the remains of terrestrial animals and plants. In marine environments, storm-generated effects are probably the major agents of preservable mass mortalities; excess turbidity, changes in water temperature and geochemistry, and rapid sedimentation during these events are particularly troublesome for benthic (bottom dwelling) invertebrates. Overturning of anoxic and/or toxic water masses may produce mass die-offs of fish (squids) or other nektonic organisms. This is apparently the case with the faunas of the famous Cretaceous fish beds and the extraordinary assemblages of aquatic and terrestrial organisms in the Eocene Messel oil shales of Germany.

Once the bodies of organisms accumulate on the substrate, whether by normal or mass mortality, the rate of burial becomes extremely critical to

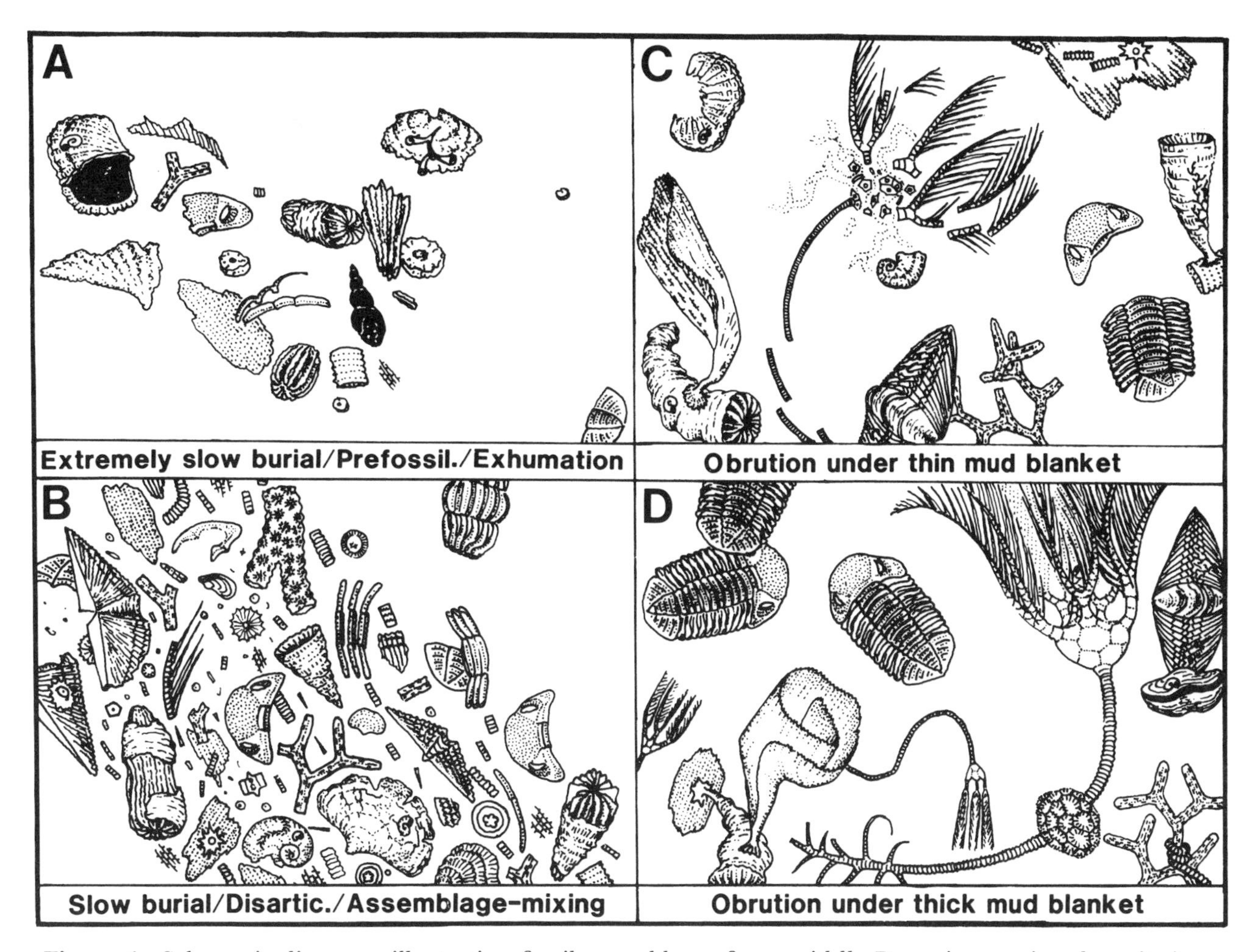

Figure 1. Schematic diagrams illustrating fossil assemblages from middle Devonian marine deposits in eastern North America. A through D illustrate hypothetical fossil assemblages accumulated under differing conditions of sedimentation. Note highly distinctive taphonomic attributes of each defining distinct taphofacies.

A. Lag deposit of skeletal debris accumulated during interval of long-term sediment starvation. Virtually all fossils are disarticulated and several display fragmentation or wear. Black material filling one articulated brachiopod and elsewhere represents phosphatic material deposited interstitially in sediment during times of sediment starvation. Fossils have been reworked on the seafloor.

B. Time-averaged assemblage accumulated during normal periods of low background sedimentation. Note fossils are in stringer-like accumulation reflecting minor current activity and that elongate skeletal elements are somewhat aligned parallel to these windows. Most shells are oriented convex upward in a hydrodynamically stable configuration. Few trilobite segments remain articulated, but most fossils are disarticulated, although they show little breakage or abrasion.

C. Assemblage reflecting mortality of organisms in rapid but not instantaneous burial; several hours to a few days have passed between the time of the depth of the organisms and burial. Note disarticulation of fragile arms of crinoids and partial disarticulation of trilobites.

D. Disarticulate of trilobites—Obrution deposits (rapidly buried assemblage) showing articulated crinoids and trilobites and closed brachiopod shells buried in life position.

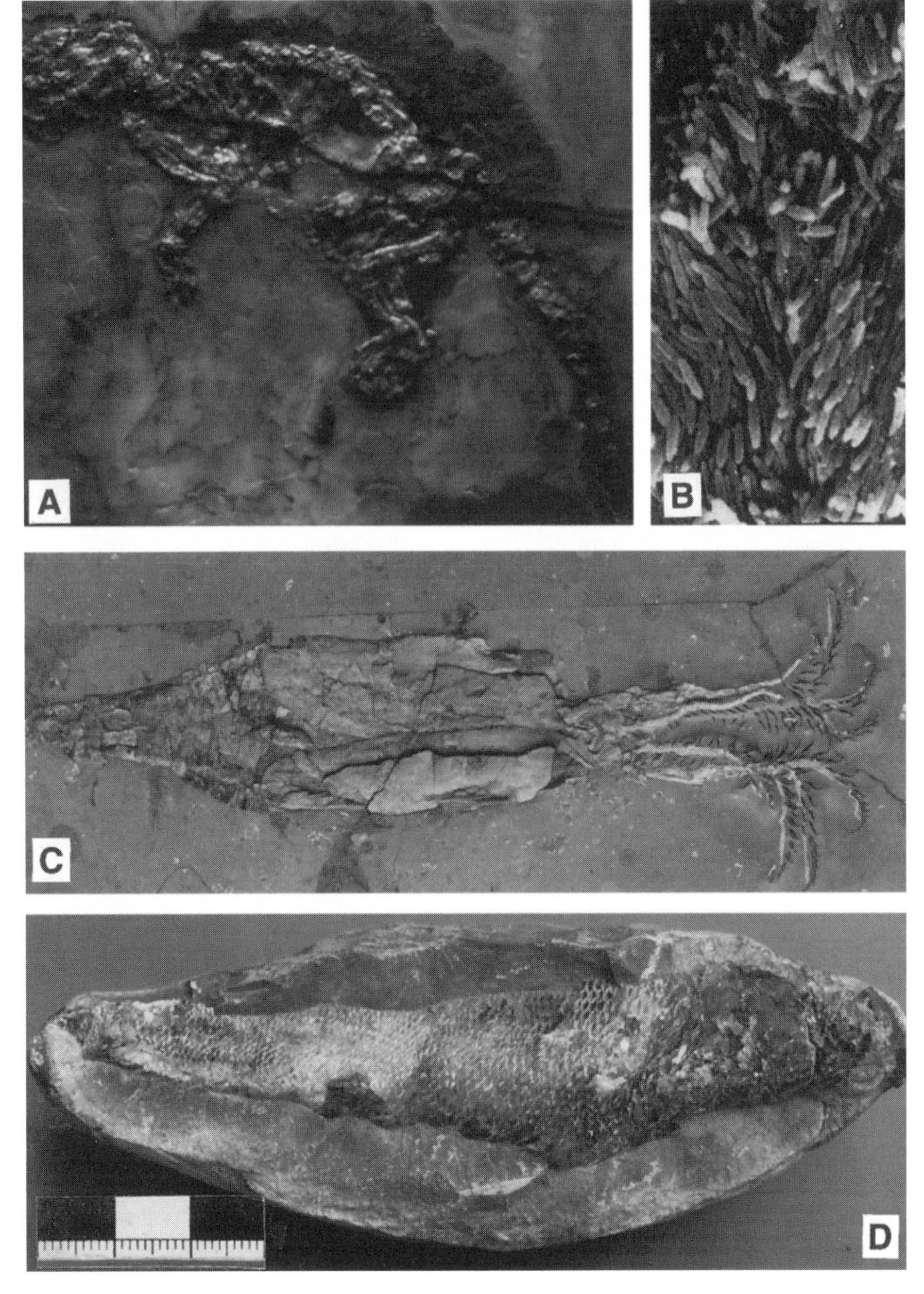

Figure 2. Examples of extraordinary fossil preservation illustrating unusual conditions of burial and early diagenesis.

A. Fossil rodent with completely articulated skeleton and traces of hair and skin; preserved in oil shale from an Eocene (50 million years, or Ma, old) lake deposit; Messel, Germany; specimen about 40 centimeters (cm) long.

B. Scanning electron micrograph of fossil skin from the Messel oil shale showing replacement of tissue by iron-impregnated fossil bacteria; each rod-like bacterium is about 1 to 2 microns (μm) long.

C. Complete body of a fossil squid from the Upper Jurassic Oxford Clay at Wiltshire, England; note preservation of tentacles, fins, and other tissues by replacement with phosphatic minerals; specimen about 15 cm long.

their taphonomic fate (Figure 1). Entombment in sediments is an absolute prerequisite for preservation. In the case of extraordinary fossil assemblages (conservation Lagerstätten), bodies of organisms are buried more or less instantaneously and with little or no later disturbance. Even in such cases, anaerobic bacterial decay and the scavenging activities of infaunal, sediment-feeding organisms (scavengers) rapidly degrade soft tissues, except under the most unusual circumstances, such as engulfment in amber or in ice. Normally, only mineralized skeletons or highly refractory organic compounds, such as lignin, stand much chance of representation in the permanent fossil record.

For most potential fossils, a number of processes intervene between death and burial of the hard parts. Once labile organic compounds are destroyed, the decay of slightly more refractory ligaments and connective tissues, together with scavaging effects, tends to disarticulate multielement skeletons, such as the bivalve shells of clams and, particularly, the intricate skeletons of echinoderms, arthropods, or vertebrates (Figure 2). The rate of disarticulation varies substantially, both with the type of skeleton and with the specific environmental conditions. For example, experiments show that the calices of some crinoids and the tests of echinoids may resist disarticulation for months, while the pinnulate arms of many crinoids commence disarticulation within a period of a few hours after the death of the organism (Figure 1C). Cool seawater temperatures can retard disarticulation rates by weeks to months; the level of oxygen in the environment has a significant but lesser effect.

Brachiopods with interlocking hinge teeth resist disarticulation, whereas bivalves with ligamental hinges have a tendency to splay open at the hinge upon decay of internal musculature; this splaying undoubtedly aids in the disarticulation process, but the elastic hinge ligaments are resistant to decay for periods of up to several weeks. Thus, the occurrence of "butterflied" bivalves, that is, those splayed open at the hinge but still articulated, is a good indicator of rapid, but not abrupt pulses of deposition of sediments.

Disarticulated parts and whole skeletons of organisms may be affected selectively by hydrodynamic processes. For example, different parts of vertebrate skeletons are transported differentially by streams. Light-weight bones, such as vertebrae, have a tendency to pass through the stream system as part of the bed load, whereas heavier long bones and teeth tend to accumulate as lag deposits on point bars. Marine shells may be sorted by size or shape through the action of currents and waves. Left-right sorting of the mirror image symmetrical valves of clams, for example, occurs in a wave swash zone and is a good indicator of a very shallow subtidal environment.

Even minimal, short-distance transport may produce selective orientation of fossils. Concavo-convex shells, such as the valves of clams and brachiopods (*see* BRACHIOPODS; MOLLUSKS), will settle selectively in a concave upward position if allowed to fall freely as when briefly lifted above the substrate by storm waves (Figure 1B). Pavements of shells in hydrodynamically stable, preferred convex-up orientations are more commonly observed and are indicators of current action. Elongate skeletal remains, such as nautiloid cephalopod shells, vertebrate bones, and graptolites, are commonly aligned on bedding planes with a more pointed or tapered end pointing upstream and may provide critical clues as to current directions (Figure 1B).

In highly turbulent environments, where shells

Figure 2. Continued

D. Uncrushed preservation of a marine fish within a concretion; Cretaceous (100 Ma old) Santana Formation, Brazil; fish carcass was buried very rapidly; subsequent geochemical changes from incipient decay of tissues within anoxic mud caused precipitation of calcium phosphate within the tissues and calcium carbonate (lime) cements in the mud, forming a hardened concretion that jacketed and protected the fish remains. Note scale.

Courtesy of Peter A. Allison, Postgraduate Institute for Sedimentology, University at Reading, England.

are frequently reworked, fragmented, abraded, and/or faceted, skeletons may be prevalent. Conversely, bioerosion and chemical corrosion become more important in offshore, low-energy environments. Clearly the conditions of skeletons, whether they are perfectly articulated or highly corroded, fragmented or abraded, provide clues as to the residence time of skeletal parts on the sedimentary interface and depositional environment.

Processes by which empty skeletal parts become infilled with sediments and finally entombed within sediments require much more study. Heavy shells may become entombed literally by sinking under their own weight, whereas lightweight shells more frequently reside on the sea bottom for longer periods of time. Complexly chambered skeletons, such as the phragmocones of cephalopod shells, may serve as sediment baffles, but in some cases even these phragmocones may become infilled with sediment via processes of draft filling, in which a current is intrained inward through the siphuncle by external currents flowing over punctures or holes in back parts of the shell.

Finally, very important clues as to sediment dynamics may be provided by prefossilized remains, that is, organism skeletons that have been buried and undergone early phases of diagenesis prior to exhumation, reworking, and finally reburial in new sediments (Figure 1A). These fossils may provide critical evidence for hiatuses in the geological record that would otherwise go unnoticed.

Fossil Diagenesis

Once entombed in sediment, or rarely, even while resting in part on the sediment water interface, the remains of organisms undergo chemical alterations that may aid or detract from their ultimate preservation. Under oxygenated conditions, bacteria rapidly degrade most soft proteinaceous and plant tissues. Only mineralized hard parts and refractory organic materials are apt to make it into the sediment. Chemical dissolution of skeletal components may also occur at or slightly below the sediment/water interface in many marine environments. The degree and timing of dissolution depend not only on the ambient pH and other geochemical factors of the environment, but also on the skeletal mineralogy (e.g., aragonite vs. calcite) and the degree of undersaturation in the water. In most environments, aragonite and high-magnesium calcite are much less stable than low-magnesium calcite. An intriguing example of selective dissolution is provided by the Jurassic Aptychus limestones of the Alps, in which only the calcitic aptychi (opercula or trap doors that cover the body chamber) of ammonites are abundant; the aragonitic shells themselves are almost completely missing due to very early dissolution. If the organic hydroxyapatite of vertebrate bones, conodonts, or other phosphatic skeletons is recrystallized to calcium fluoroapatite, as is common in many marine environments, then these become among the most geochemically resistant particles, and may form distinct solution lags when all carbonates have been dissolved by prolonged exposure to undersaturated seawater.

Sedimentary compaction provides a convenient dividing line between early and later diagenetic processes. Syndiagenesis or early diagenesis occurs before significant compaction of sediment and fossils has taken place. Dewatering and compaction of sediments, particularly in muds, may strongly deform potential fossil remains and may commence in the upper few meters of the sediment column. On the one hand, when dissolution of skeletal materials occurs prior to significant compaction in unconsolidated sediments, plastically deformed external molds or composite internal-external molds of the shells may develop. On the other hand, the appearance of brittle fracture patterns on internal molds of some fossil shells indicates that these shells remained intact up to the earlier stages of compaction.

Rapid burial of organic matter may lead to the development of geochemical micro environments; in some cases, where organism bodies are entombed extremely rapidly, evidence for decay gas buildup and explosion may appear in the fossils. There are even cases of shattered concretions that formed very early around decaying organic matter and were exploded by the buildup of decay gases. Many early mineralization phenomena are mediated by anaerobic bacteria. While anaerobic decay precedes somewhat more slowly than aerobic decay of organic matter, it is still relatively rapid and will degrade organic tissues within a matter of a few years or sooner. However, in some cases, the bacterial decay will lead to geochemical changes that aid in early mineralization. The products of anerobic decay may preserve vestiges of soft tissues by very early diagenetic mineralized films. For example, the extraordinary Messel fossils (Eocene of Germany), which include skin, and hair in mammals and the wings of butterflies, are actually preserved as pseudomorphs built of the siderite-replaced cells of iron-reducing bacteria. The famed Cambrian Burgess Shale soft-bodied fossils appear to represent very thin organic films complexed with coatings of authigenic illite clay.

Entombment of organic nuclei in otherwise organically poor but anoxic sediments may be particularly critical in promoting early mineralization. For example, where proteinaceous material is suddenly entombed in a sediment of low organic content, the organism bodies may serve as nucleating sites for pyrite or its precursors. Sulfate-reducing bacteria metabolize the protein of the organic matter and produce sulfides and bicarbonates as byproducts. The concentration of these substances near the decaying organic matter may mobilize cations such as calcium and ferrous iron, which then couple with the sulfide and bicarbonate, respectively, to form pyritic nodules or coatings and/or carbonate concretions. Such early mineralization may buttress the fossil molds against compactional deformation. Thus, the occurrence of noncompacted pyritic or concretionary steinkerns is a good indication of very early diagenetic mineral formation.

Formation of early diagenetic concretions may be triggered by the elevation of pH and/or the buildup of bicarbonate as a result of protein decay. Siderite nodules of the Mazon Creek (Pennsylvania, Illinois) and those from the Santana Formation (Cretaceous, Brazil) provide an excellent example of the early encapsulation of organism remains in relatively impervious jackets of concretionary carbonate. Phosphatic and glauconitic linings or infillings may also develop, and organic remains provide a nucleus in micro-oxidizing environments that overlie stable, reducing sediment interfaces. These are particularly diagnostic of very slow sediment deposition or even slight erosion. Once fossilization has taken place, the fossils may be reworked as rather resistant prefossilized clasts.

Obviously, diagenesis does not always end with early phases of burial. Later diagenetic alteration of fossils may substantially distort their original appearance or may accentuate features, particularly in trace fossils. Mineral replacements, dissolution, and perminearlization of fossils are common processes that must be carefully factored from early diagenetic phenomenon.

Taphofacies. The realization that fossil preservational modes provide a wealth of information on physical and chemical processes in environments has led to the notion of taphofacies, or taphonomic facies. These are bodies of sediment or rock that are characterized not by taxonomic composition but by the taphonomic condition of fossils (Figure 1). Skeletons of very similar type are preserved in distinctly different ways in different physical, chemical, or biological settings. By combining biostratinomic and early fossil diagenetic attributes of preservation, it is possible to characterize and distinguish taphofacies. These may occur in superficially similar lithofacies or sediments and may provide important clues to environmental differentiation that would not otherwise be recognizable. For example, in illustrating the concept of taphofacies, Speyer and Brett (1986) demonstrated that two common Devonian trilobites, which occur in a wide array of depositional environments, show distinctly different patterns of preservation that can be keyed to their position in sedimentary cycles, and thus are probably related to differential processes acting at different depth zones, as well as under varying sedimentation rates. From these sorts of observations Speyer and Brett modeled hypothetical taphofacies in relation to physical environmental energy and sedimentation rate. Obviously, a key feature that controls the taphonomic aspect of a given fossil assemblage is exposure time. Those environments that experience heavy rates of sedimentation typically show higher taponomic grades (i.e., most fossils are articulated, not fragmented or abraded), whereas environments characterized by lower rates of sedimentation show varying degrees of alteration of the skeletal materials. Depending upon whether these lower sedimentation environments occur in shallow, turbulent settings or in deeper waters, the fossils may be either heavily physically broken or abraded or chemically corroded or bored. The distinctive modes of preservation provide a fingerprint of ancient depositional environments. The relatively few studies of taphofacies in modern environments confirm that distinctive aspects of fossil preservation are characteristic of certain depositional settings, corroborating the validity of a taphofacies approach for ancient paleoenvironmental reconstructions.

Bibliography

ALLISON, P. A., and D. E. G. BRIGGS, eds. *Taphonomy: Releasing the Data Locked in the Fossil Record.* New York, 1991.

DONOVAN, S. K. *The Processes of Fossilization.* New York, 1991.

SEILACHER, A. "Biostratinomy: The Sedimentology of Biologically Standardized Particles." In *Evolving Concepts in Sedimentology,* ed. R. N. Ginsburg. Baltimore, MD, 1973, pp. 159–177.

SPEYER, S. E., and C. E. BRETT. "Trilobite Taphonomy and Middle Devonian Taphofacies." *Palaios* 1(1986): 312–327.

WEIGELT, J. *Recent Vertebrate Carcasses and Their Paleobiological Implications.* Trans. by J. Schaefer. Chicago, 1989.

CARLTON E. BRETT

TARS, TAR SANDS, AND EXTRA HEAVY OILS

Tar is a viscous organic liquid produced by the destructive distillation of wood, coal, or oil. The term is commonly, albeit incorrectly, applied to naturally occurring bitumen as well. A rock that is rich in organic matter, commonly a dark-colored and fine-grained mudstone, is termed a petroleum source rock. Bitumen is the organic matter in such a rock that is extractable with ordinary organic solvents. Bitumen constitutes only a small proportion of the organic matter in a petroleum source rock. The inextractable organic matter is termed kerogen. The thermal maturation of the kerogen accompanies increased burial and, hence, increased temperature to which the source bed is subjected. This leads to the conversion of the kerogen to the bulk of the bitumen, crude oil, natural gas, and ultimately, carbon and graphite. The products of this evolution—the bitumen and oil—may be altered at any stage of their evolution in various ways. Inorganic constituents of the rock, especially clays, presumably act as catalysts, enhancing all steps of the maturation process. Geologic time also affects the process to a degree that cannot be duplicated in the laboratory but may be inferred. The bitumen evolved from the organic matter is here referred to as tar.

Because tar and crude oil comprise a hydrocarbon continuum, their classification is arbitrary but is based upon certain physical and chemical characteristics. The boundary between heavy oil and conventional oil is commonly set at or near 20° API (sp g 0.934; density in degrees API converts to grams per cubic centimeter [g/cm^3] in the metric scale: $g/cm^3 = 141.5/131.5 + °API$), which is close to the lower API gravity limit for unenhanced recovery of oil from the ground. Tar is distinguished from oil by having a viscosity of 10,000 centipoise or cP (millipascal second or mPa · s; cP is equivalent to .001 pascal second [Pa · s] in the metric scale), which is the approximate viscosity limit for the recovery of bitumen in the subsurface without supplemental application of heat.

The kerogen in the source rock converts to large hydrocarbon molecules, which generally incorporate trace metals, such as vanadium and nickel, and nonmetals, such as nitrogen, oxygen, and sulfur as substituents. Thus, large, heteroatomic molecules, those which include one or more atoms other than carbon in the ring structure, form which, with increasing temperature, break up into smaller, lighter (lower molecular weight) molecules. The fragments of the heteroatomic molecules left behind have in them most of the metals and nonmetals, constituting the asphaltenes and resins that form most of the tar and significant proportions of the heavy oil.

Conventional crude oil is made up mostly of paraffin, naphthene, and aromatic hydrocarbons with few of the metallic and nonmetallic impurities. It therefore is comprised mainly of low molecular weight components and is of low viscosity. On the other hand, the heavy oil and tar, being constituted of heavy, heteroatomic molecules, are viscous and contain most of the undesirable constituents.

Some of the tar present in seepages and large deposits may be primary. The tar is considered to be primary when it has not been altered from the bitumen that initially evolved in the source rock.

Table 1. World Tar Resources by Country (million barrels)*

Country	*Demonstrated*	*Inferred*
North America		
Canada	1,697,359	831,100
United States		
Alaska		11,000
Alabama		7,970
California	6,210	2,540
Kentucky	1,720	1,690
Missouri-Kansas-Oklahoma	220	2,730
New Mexico	91	
Oklahoma	42	22
Texas	2,670	300
Utah	16,856	6,429–12,362
Wyoming	120	70
U.S. total	27,929	32,751–38,684
South America		
Trinidad	60	
Venezuela	62	
Europe		
Albania	371	
Azerbaijan	82	
Georgia	636	
Germany	315	
Italy	1,260	
Romania	25	
Russia	94,516	
Ukraine	599	
Asia		
Kazakhstan	2,112	
Kyrgyzstan, Tajikistan, Uzbekistan	48	
People's Republic of China	10,050	
Russia	91,779	
Turkmenistan	3	
Africa		
Madagascar	1,600	3,400
Nigeria	42,740	42,000
Zaire	30	
Middle East		
Syria	13	
Southeast Asia		
Indonesia	8	
Philippines	3	
World Totals	1,971,600	909,251–915,184

* Sources: Energy Resources Conservation Board [Alberta]. *Alberta's Reserves of Crude Oil, Oil Sands, Gas, Natural Gas Liquids and Sulphur*, ERCB ST 92-18, pp. 3-1–3-6. 1992;
Goldberg, I.S., ed., *Prirodnye bitumy SSSR, Zakonomernosti formirovaniia i razmeshcheniia* (Natural bitumens of the USSR, Regularities of formation and distribution), Leningrad, 1981;
Meyer, R. F., and J. M. Duford. "Resources of Heavy Oil and Natural Bitumen Worldwide." In *Proceedings Fourth UNITAR/UNDP International Conference on Heavy Crude and Tar Sands*, eds. R. F. Meyer and E. J. Wiggins. Calgary, Alberta, Canada, 1989.

Similarly, some of the heavy oil present in reservoirs today may be primary, or immature in terms of thermal evolution and still containing a large proportion of large, heteroatomic molecules rich in sulfur and trace metals. But at least 90 percent of the tar and heavy oil worldwide are the remnants of conventional oil that has been altered through a number of mechanisms: (1) biodegradation, whereby bacteria consume the lighter components, particularly the paraffinic hydrocarbons; (2) water washing, in which the most water-soluble components are removed; (3) inspissation, the simple evaporation of the lighter components when oil is exposed at or close to Earth's surface; and (4) deasphaltening, whereby the very large asphaltene molecules are precipitated from the oil by natural gas or natural gas liquids, commonly collecting at the base of the reservoir as a tar mat.

Thus, through any or all of these processes, the heavy molecules—in which most of the heteroatoms reside—are concentrated. The tar and the heavy oil are characterized by sulfur up to 5 weight percent and trace metals, notably vanadium and nickel, 500 parts per million (ppm) or more. Also contained in the heavy oil and tar are as many as forty other trace elements and a number of the major elements, including oxygen and nitrogen. In addition, the high viscosity of the substances makes their recovery and separation from the rock containing them both difficult and expensive. Processing of the product in the refinery is difficult and expensive because of the need to separate the sulfur and other atmospherically toxic elements. Also, the vanadium collects on the catalysts and eventually renders them ineffective.

Tar occurs at or near Earth's surface in a number of forms, ranging from highly viscous liquids to semisolids and solids. Quantitatively, however, at least 99 percent is in the form of natural asphalt, a very viscous liquid with a specific gravity approaching or exceeding that of water. Natural asphalt may be found as a nearly pure asphalt (tar) essentially free of mineral matter, as in the La Brea Tar Pits near Los Angeles or the Pitch Lake of Trinidad. More commonly, the tar is found in seepages or as large deposits in sandstones or limestones, as in Utah or Alberta, Canada. Such tar-impregnated rocks are called tar sands or, in the case of Alberta, oil sands.

Estimated tar resources for the world are given in Table 1. About 86 percent of the world's demonstrated resources are found in the province of Alberta. The next largest resources, 9.5 percent, are located in Russia, where they are about equally divided between European Russia and Asian Russia.

The inferred tar resources, those less accurately known than the demonstrated resources because of fewer drill holes and outcrop measurements, amount to about 32 percent of the total resource of 2,881,600 barrels. Alberta possesses about 91 percent of the inferred resources. The inferred resources of Canada, the United States, Nigeria, and Madagascar are estimated from yet incomplete geological surveys. In many other countries, additional studies will likely add inferred resources to the inventory and lead to some resources being moved from the inferred to the demonstrated category.

In most cases, country totals represent several individual occurrences of tar. Generally, each of these occurrences is sufficiently large to be exploitable with existing technology but seldom under extant economic conditions.

Resources known to be present in heavy oil fields are listed by area in Table 2. Information for fields in the former Soviet Union is sparse and therefore the data are understated. Of the total demonstrated resource, 55 percent is found in South America, 22 percent in the Middle East, and 13 percent in North America. The Eastern Venezuela Basin alone holds 44 percent.

Table 2. World Heavy Oil Resources by Area (million barrels)*

Area	*Demonstrated*	*Inferred*
North America	34,970	14,293
Central America-Caribbean	8	0
South America	143,457	172,500
Europe	5,005	23,074
Africa	2,620	0
Middle East	58,421	22,078
Asia	14,517	70,800
Southeastern Asia	2,466	3,589
World Totals	261,464	306,334

* Source: Meyer, R. F., and J. M. Duford. "Resources of Heavy Oil and Natural Bitumen Worldwide." In *Proceedings Fourth UNITAR/UNDP International Conference on Heavy Crude and Tar Sands*, eds. R. F. Meyer and E. J. Wiggins. Calgary, Alberta, Canada, 1989.

The resource levels of both tar and heavy oil are very large. However, the production of tar, notably from Alberta, and heavy oil, particularly in Venezuela and the United States, is small relative to the worldwide production of conventional oil. This is because production requires expensive technology in the form of thermally enhanced recovery from wells or recovery from surface or underground mining of the reservoir rock and subsequent processing. Thus, tar and heavy oil have enormous production potential but will require much higher prices per barrel in the future in order to be economically viable.

Bibliography

Chilingarian, G. V., and T. F., Yen, eds. *Bitumens, Asphalts and Tar Sands.* Amsterdam, 1978.

Hills, L. V., ed. *Oil Sands, Fuel of the Future.* Calgary, Alberta, Canada, 1974.

Rühl, W. *Tar (Extra Heavy Oil) Sands and Oil Shales.* Stuttgart, 1982.

RICHARD F. MEYER

TECTONIC BLOCKS

Tectonics refers to broad changes in the architecture of the outer part of the earth. Tectonic blocks (also called terranes) are individual packages of rocks that have moved along faults with respect to their neighbors. Tectonic blocks are major parts of tectonic plates. The continent-sized plates move at the surface of the earth and cause most earthquakes, volcanic eruptions, and mountain building (see separate entries on these subjects).

Tectonic blocks are distinguished from adjacent blocks by their contained rocks, which represent a particular geologic setting, such as an island arc, a submarine fan, the ocean floor, or a displaced continental fragment. The rocks of adjacent blocks may be of the same geologic age, so we cannot easily separate them on ordinary geologic maps, where a different color represents each age. The study of tectonic blocks requires most of the tools available to geologists. We use several methods to determine the age and environment of a block, along with its original geologic setting.

Four processes can lead to fault-bounded packages of rocks (Howell, 1995). They are rifting (tearing of either the continent or the seafloor), amalgamation (joining another tectonic block, usually in an oceanic setting), accretion (moving into the continental framework), and dispersion (movement along a major vertical fault).

Geologists use several methods to distinguish tectonic blocks and to interpret their history of movement. Sedimentary rock layers that lap across two blocks help to define the time when they came together, as does granite that intrudes the fault between the blocks and "stitches" them together. Metamorphic rock caused by heat and pressure that crosses a boundary provides a minimum age for the amalgamation of two blocks. And we can identify the time in the past when two blocks came together by the time when sand and gravel that were eroded off one of them were first deposited on the other. Finally, we can identify the former environment of a block by the fossils it contains, by climatic indicators such as desert or glacial materials, by preserved former shorelines, and by rock materials marking volcanism or strong mountain uplift.

Paleomagnetism can reveal the movement of tectonic blocks. When fresh lava chills after it has solidified, or when sediment settles out of water to form a sedimentary rock, the magnetic grains in these materials behave as compasses that point parallel with the earth's magnetic field (see GEOPHYSICAL TECHNIQUES). The resulting magnetization persists over long periods of geologic time. The earth's magnetic field resembles that of a bar magnet near the rotation axis of the earth. The magnetic grains in rocks record the general direction in map view toward a magnetic pole. After a tectonic block has turned as it moves, a magnetometer can detect the rotation of the block's magnetic field. The magnetic grains also record the inclination of the magnetic field as seen at the earth's surface. This inclination ranges from horizontal at the equator to vertical at a magnetic pole. We can attribute changes in the magnetic inclination recorded by a tectonic block to northward or southward movement of the block.

Tectonic blocks are ordinarily given unique names. Modern study of tectonic blocks began in 1977, when David L. Jones and collaborators named a western North American terrane "Wrangellia." Analysis of the terranes in western North America illustrates the general principles by which tectonic blocks have been assembled throughout the world.

About 85 million years ago (85 Ma) during the Cretaceous Period, North America captured Wrangellia at a subduction zone where one plate carrying the buoyant block descended beneath another. Wrangellia is believed to be an island-arc block formed in the tropics. After Wrangellia became plastered onto the continent, it moved toward the north along faults parallel with the coast and was broken into three pieces that now lie in Oregon, British Columbia, and Alaska.

Today's Baja California, in northwestern Mexico, may serve as a model for the northward movement of large tracts of land along western North America. Baja California is moving northwestward along California's San Andreas Fault, leaving behind new oceanic crust in the Gulf of California.

About 700 Ma, during Precambrian time, North America along with Australia and Antarctica was part of a much larger continent (Figure 1). A continental break divided this supercontinent along roughly the present west coast of North America (Hoffman, 1991). The separation of the supercontinent took an estimated 20 to 40 Ma. Then seafloor volcanic material moved in between the two halves to form oceanic crust. During the period of continental stretching, the crust of North America got thinner, forming a margin several hundred kilometers wide, which tapered toward the west. The tapered former edge of the continent is well preserved in Nevada, where richly fossiliferous rock layers of younger geologic periods blanket the former continental edge.

Several major island-arc blocks later joined western North America. The first was Sonomia, now centered in Nevada, which arrived about 250 Ma during the Triassic Period (Figure 2). Stikinia, now centered in British Columbia, arrived about 200 Ma during the Jurassic. Stikinia later moved toward the north and took with it part of Sonomia

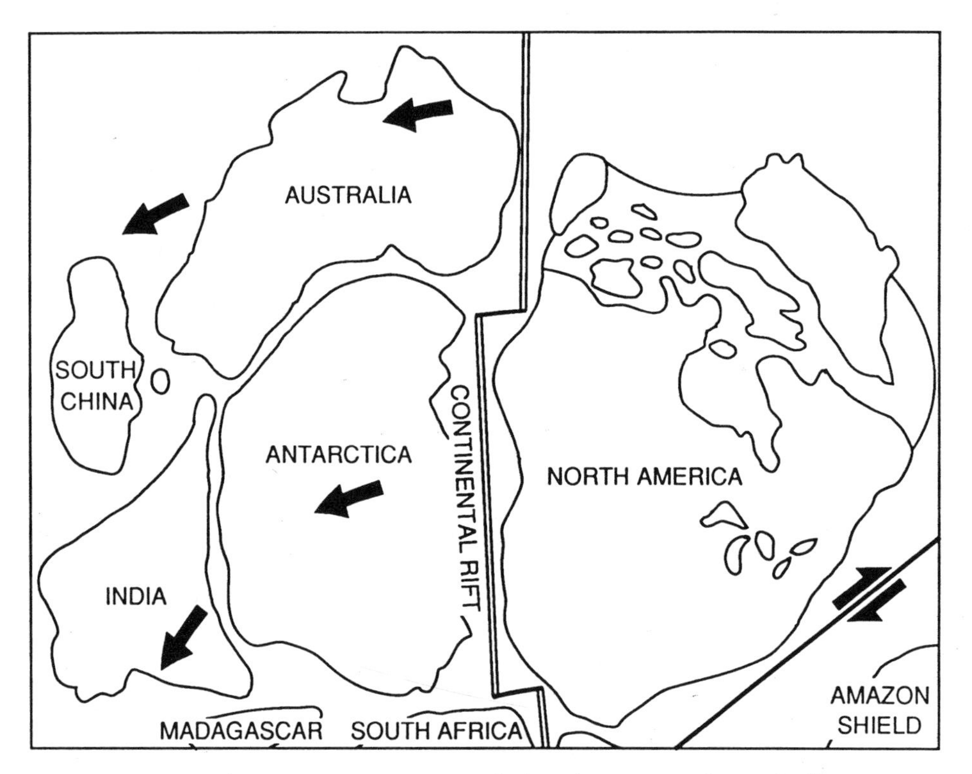

Figure 1. About 700 Ma during Precambrian time, a continental rift separated North America from a much larger continent. The tapered edge of North America that formed then began later to capture tectonic blocks, most of which were offshore volcanic arcs similar to Alaska's Aleutian Island chain of today.

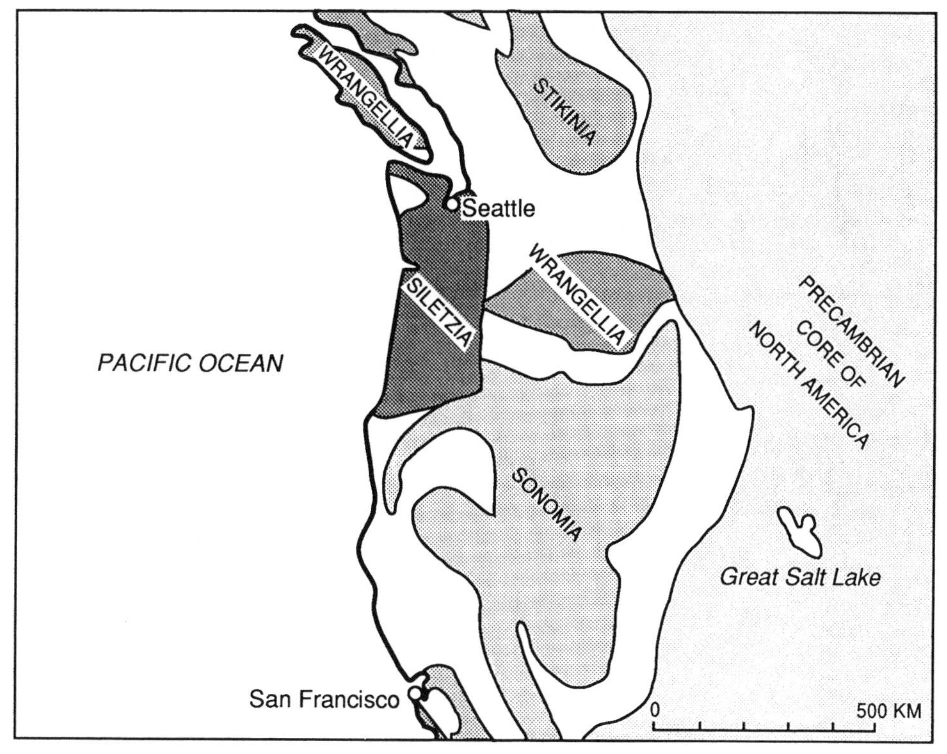

Figure 2. Several large tectonic blocks became attached to the Precambrian core of North America. First to arrive was Sonomia, then Stikinia. (Intervening onshore unpatterned areas are seafloor materials squeezed between the blocks.) After Stikinia joined North America, it moved northward, reexposing the Precambrian core of the continent to the ocean. Into this gap came Wrangellia, the final great island-arc block. Then followed a series of displacements toward the north that separated Wrangellia into several fragments, and left behind Siletzia, a tract of ocean floor. A later subduction zone established the Cascade Volcanic Arc along the east edge of Siletzia and understuffed and lifted it up to form the present Coast Range of Oregon and Washington.

and much of the former tapered continental margin of North America. This left a gap at Idaho where oceanic crust of the Pacific Basin lay side by side against the Precambrian core of North America.

The gap at Idaho was filled when Wrangellia arrived and lodged directly against the core of North America. A final great bite was taken from the continental edge about 60 Ma ago, in early Tertiary time, when nearly all of what then was coastal Oregon and Washington was carried northward toward Alaska (Moore, 1984). This left behind Siletzia, an ocean-floor block formed in place, the youngest piece of new ground in this part of North America.

Siletzia is made of thin crust of volcanic origin, and like the Gulf of California, it too initially was a marine gulf underlain by new oceanic crust on the seafloor. Soon, however, a new subduction zone originated along the continental margin, and the volcanoes of the Cascade Mountains came into being. When the new subduction began, a wedge of buoyant sediment and rock forced itself under the edge of Siletzia's new ground. The wedge lifted the edge of Siletzia above sea level, and today it forms the Coast Range of the Pacific Northwest.

The relatively recent understanding of the movement of tectonic blocks has changed the way geologists look at large areas of the earth's surface, and it has contributed to our comprehension of the way that continents form.

Bibliography

HOFFMAN, P. J. "Did the Breakout of Laurentia Turn Gondwanaland Inside Out?" *Science* 252 (1991):1409–1412.

HOWELL, D. G. *Principles of Terrane Analysis.* London, 1995.

JONES, D. L., N. J. SILBERLING, and J. W. HILLHOUSE. "Wrangellia—A Displaced Terrane in Northwestern North America." *Canadian Journal of Earth Sciences* 14 (1977):2565–2577.

MOORE, G. W. "Tertiary Dismemberment of Western North America." *Circum-Pacific Energy and Mineral Resources Conference Transactions* 3 (1984):607–612.

GEORGE W. MOORE

TECTONISM, ACTIVE

Active tectonics refers to tectonic movements that are expected to occur within a future time span of concern to society. It describes tectonic processes now active, taken over the geologic time span during which they have been acting in the presently observed sense, and the resulting geological structures. When we say "now active," we do not mean that you could go outside today and see it happen. Rather, the geologic evidence indicates that the process has been recently active and there is no evidence that the forces driving the process have ceased.

Because active tectonics involves dynamic processes operating today, it can contribute a data set that involves a considerably thicker slice of the earth than the study of tectonic processes that are no longer active. A surface fault may be mapped to depths of 10–20 kilometers (km) based on the planar distribution of earthquakes on the fault and on the distribution of crustal strain on and adjacent to the fault. Accordingly, active tectonics combines the fields of geology, seismology, and geodesy. Active faults and folds produce a variety of landforms, and their study provides information on the tectonic processes that formed them as well as the surficial processes that degrade them. Rates of deformation may be determined if Quaternary stratigraphic units involved in deformation are dated; thus, active tectonics includes a consideration of past Quaternary depositional processes under differing climatic regimes, and also of dating techniques limited to the Quaternary.

Present-day slip rates on active faults near plate boundaries, including those that do not reach the surface (blind thrusts), may be compared with the rates of motion between adjacent plates (time frames of 10^5–10^6 years) and motions based on tectonic geodesy (time frames of years or decades). For example, the rate of motion between the Pacific and North American plates has been determined based on plate tectonics, space geodesy, and the study of slip rates on individual geological structures; these rates are all consistent with one another.

The distribution of heat in the crust affects the way crust deforms, and this may be measured directly or on the basis of contemporary volcanic rocks. The presence of fluids in rock has two weakening effects: (1) pore pressure that acts in the opposite direction than load stress, thereby weakening the rock; and (2) chemical action of pore fluids with rock, which also weakens the rock. The distribution of Earth's gravity field produces additional constraints on the nature of active tectonic processes.

Active tectonics is in part observational: (1) the careful description and mapping of active faults and folds that have moved in late Quaternary time, including descriptions of displaced Quaternary stratigraphic units in subsurface excavations; (2) observations of the distribution of earthquakes in the crust, their mechanisms of displacement, and the information seismograms provide about strain release at the earthquake source; and (3) observation of the strain field as based on ground-based trilateration and leveling and space-based geodesy, principally the Global Positioning System (GPS), Very Long Baseline Interferometry (VLBI), and radar interferometry. It also includes laboratory-based studies of the behavior of rock at varying crustal temperatures, pressures, and strain rates, searching for analogs to natural systems. Old fault zones exhumed by erosion may be studied directly as analogs to earthquake sources that are buried too deeply for direct access. In addition, deformation may be studied by theoretical modeling. For example, theoretical models predict that an earthquake should be preceded by a short phase of increasing strain near the future earthquake source. If this change in strain could be identified prior to an earthquake, a short-term prediction could be issued.

The direct observation of earthquakes may be divided into three stages. During the twentieth century, seismic waves have been measured directly with seismographs, which, in the last few decades, have become highly sophisticated and much more widely distributed around the world. The seismographic stage was preceded by the historical stage of record keeping, up to three thousand years long in certain parts of the world but only two centuries long in many others. Recorded earthquakes in the pre-seismograph era could determine the time and general area of occurrence but commonly not the geologic structure that produced the earthquake. The historical stage was preceded by a stage in which individual earthquakes were recorded by geological evidence. The field of paleoseismology identifies earthquakes over time frames of 10^2–10^4 years based on geological data.

The historical and paleoseismological observations of earthquakes on a given fault show that earthquakes are not periodic but tend to occur in temporal clusters separated by long periods of quiescence. On some faults, the larger earthquakes tend to be of a particular size and displacement; such earthquakes are said to be characteristic. The study of repeated earthquakes on a given fault has as goals the long-range probabilistic forecasting of earthquakes and the determination of the maximum size of an earthquake expected on a given structure or in a given region. Although much progress has been made on the San Andreas fault and several other faults in California and the Wasatch fault in Utah, these goals have not yet been achieved even on those faults.

Bibliography

KELLER, E. A., and N. PRINTER. *Active Tectonics: Earthquakes, Uplift, and Landscape.* Upper Saddle River, NJ, 1995.

WALLACE, R. E., ed. *Active Tectonics.* Washington, DC, 1986.

YEATS, R. S., K. E. SIEH, and C. R. ALLEN. *Geology of Earthquakes:* New York, 1996.

ROBERT S. YEATS

TECTONISM, PLANETARY

Tectonics is the area of study within geology that investigates the origin and development of structural or deformational features found on planet and satellite surfaces. Tectonic or structural features such as faults and folds form when sufficient stress (force per unit area) causes fracture and deformation of all or part of the lithosphere, which is the rigid outermost part of a planet or satellite. Study of tectonic features commonly involves investigating the geometry, kinematics or deformation through time, and the dynamics or stresses that led to failure. Finally, given that structural features form in the lithosphere, whose thickness and mechanical properties are a function of the thermal gradient, strain rate, thickness, and composition of the crust (the outermost chemical layer of a planet or satellite), their characteristics can be used to understand the thermo-mechanical evolution of a planet or satellite (*see* TECTONISM, ACTIVE; TECTONIC BLOCKS).

In the most general sense, relative tectonic activity among the solid planets and satellites correlates with size. For the five terrestrial (or silicate) planets, Earth, the largest planet, is the most active, followed by Venus, Mars, Mercury, and the Moon (the smallest body). For the outer planet satellites, larger bodies such as Io, Europa, Ganymede (all satellites of Jupiter), and Triton (satellite of Neptune) tend to be more tectonically and geologically active than smaller bodies such as Iapetus, Tethys, Dione, Mimas (all three Saturnian satellites), Oberon, Umbriel, and Titania (all Uranian satellites). Significant anomalies do exist, however, such as the small, but active Miranda, Ariel (both Uranian satellites), and Enceladus (satellite of Saturn), and the large, but inactive Callisto (satellite of Jupiter). The general relation between size and tectonic activity is understandable considering that most geologic and tectonic processes are driven by the internal heat engine of a planet or satellite, and large bodies take more time to cool and probably have had greater contributions to heating from accretion, differentiation, and radioactivity than small bodies.

Tectonic features on a solid surface can be used to determine the stresses that led to their formation. In general, three types of stress states give rise to three classes of tectonic features: extensional, compressional, and strike slip (Figure 1). Extensional tectonic features form when the lithosphere is pulled apart or extended. Compressional tectonic features form when the lithosphere is compressed or shortened. Strike-slip faults form when material is moved side by side or parallel to the fault strike (the direction of the fault plane intersection with the horizontal). As a result, if the slip of a fault observable in spacecraft images can be determined, then the orientation of the stresses that led to its development can be deduced.

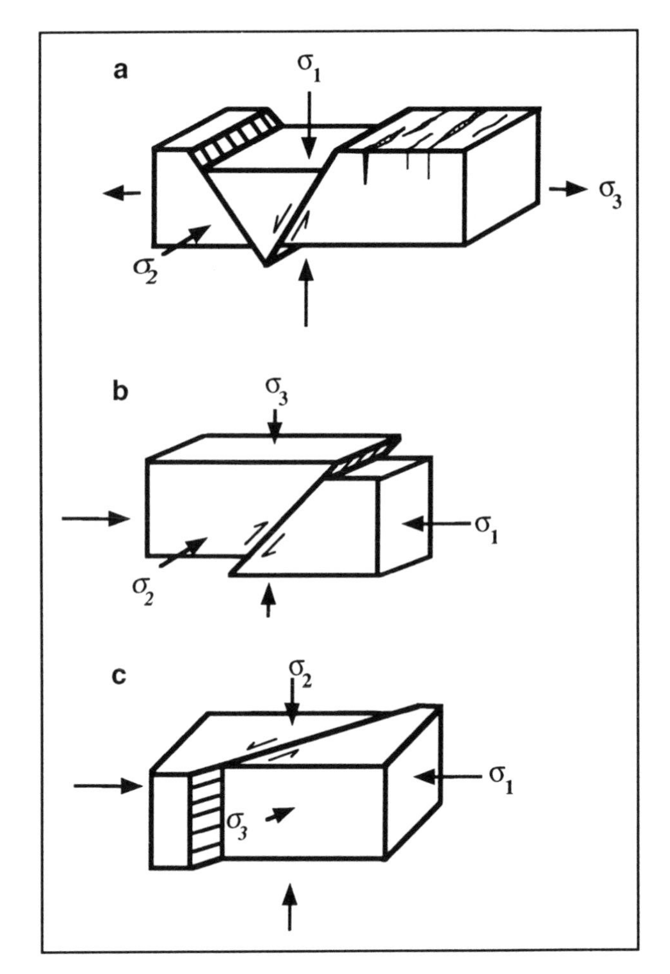

Figure 1. Block diagram showing the stress states responsible for the three classes of tectonic features. a. Extensional tectonic features (graben, shown on the left, and tension cracks, shown on the right), which typically result in linear troughs, form when the maximum compressive stress, σ_1, is vertical; tension cracks require that the minimum stress, σ_3, be tensile (as shown). b. Thrust faults (pictured) and other compressional tectonic features, which typically result in linear topographic highs, form when the minimum principal stress is vertical; the maximum compressive stress is horizontal and perpendicular to the strike of the fault. c. Strike-slip faults, in which motion occurs parallel to the surface strike, form when the intermediate stress, σ_2, is vertical.

Extensional Tectonic Features

Extensional tectonic features are the most common structures found on the surfaces of the terrestrial planets and outer planet satellites. Extensional tectonic features are characterized by the presence of normal faults and tension cracks. Normal faults form when the area above a dipping fault plane (the hanging wall) moves down relative to the area below the fault (the footwall); dip is the maximum vertical angle between the fault plane and the horizontal (*see* STRUCTURAL GEOLOGY). Tension fractures are narrow vertical cracks in which both sides separate without a floor. Extensional tectonic features form when the maximum compressive stress is vertical (due to gravity) and the smallest stress is horizontal and perpendicular to the surface strike of the structure (Figure 1). Slip or motion on extensional faults generally produces a topographic low, such as a trough, valley, or basin.

Extensional tectonic features can be grouped into a number of subsets based on their width and complexity. The least complicated is a simple graben (Figure 2). A graben is a normal fault-bounded trough; a simple graben is bounded by two inward dipping normal faults whose downdropped floor is flat and untilted (tens to a few hundred meters deep). Simple grabens are found on all the terrestrial planets and many of the outer planet satellites. They are characteristically a few kilometers wide and are bounded by steeply

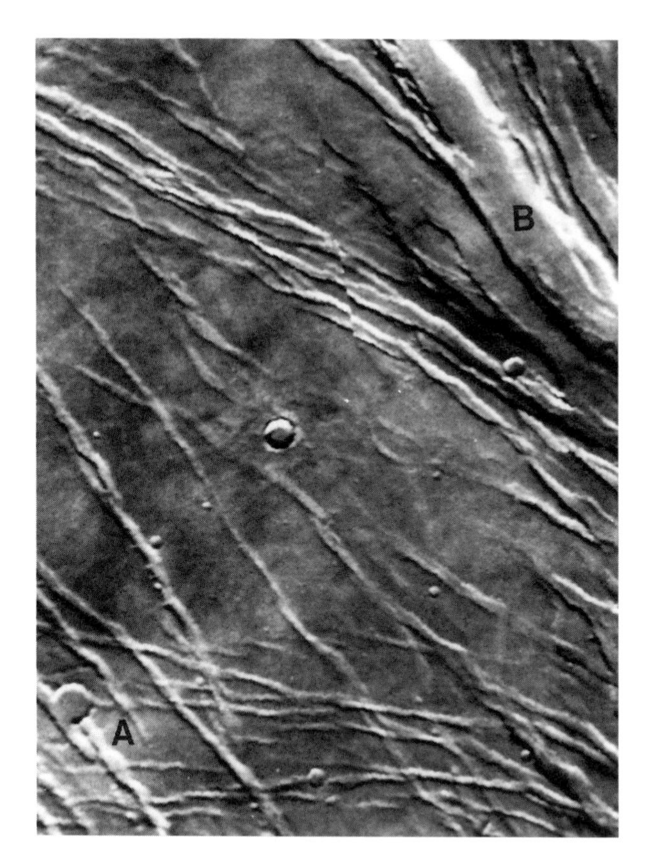

Figure 2. Simple and complex grabens on Mars. Simple grabens (A) are bounded by two inward dipping normal faults (which dip in the direction of the surface scarp). They are a few kilometers wide and a few tens of meters deep. Complex grabens (B) are about 30 km wide and bounded by multiple border and interior faults.

dipping normal faults (approximately 60°) that extend a few kilometers deep within the crust. Tension cracks, particularly evident on Mars and Europa, are associated with many grabens and are often opened in response to the pressurized injection of volcanic magma or water. Complex grabens (Figure 2) are wider, deeper, more complex structures, whose faults extend deeper into the lithosphere, roughly comparable to their width (tens of kilometers wide). Rifts are larger structures where the entire lithosphere has ruptured under extension. On the terrestrial planets these features are over 100 kilometers (km) wide and up to 10 km deep. Rifts are found on Earth (e.g., African rift system, western United States Basin and Range, Rio Grande rift in New Mexico, Rhinegraben in Germany, and Baikal rift in Siberia), Venus, and Mars and have complexly faulted interiors, which commonly contain deposits of sediments and/or volcanics. Although rifts are substantially less numerous than simple grabens, they can be large, impressive structures. Perhaps the most impressive rift in the solar system is Valles Marineris, Mars, which is 2,000 km long, up to 100 km wide, with floors downdropped to depths of 10 km below their rims (Figure 3). The extension or increase in width measured across extensional structures is up to 10 percent.

Compressional Tectonic Features

When the minimum compressive stress is vertical, reverse faults form whose trend is perpendicular

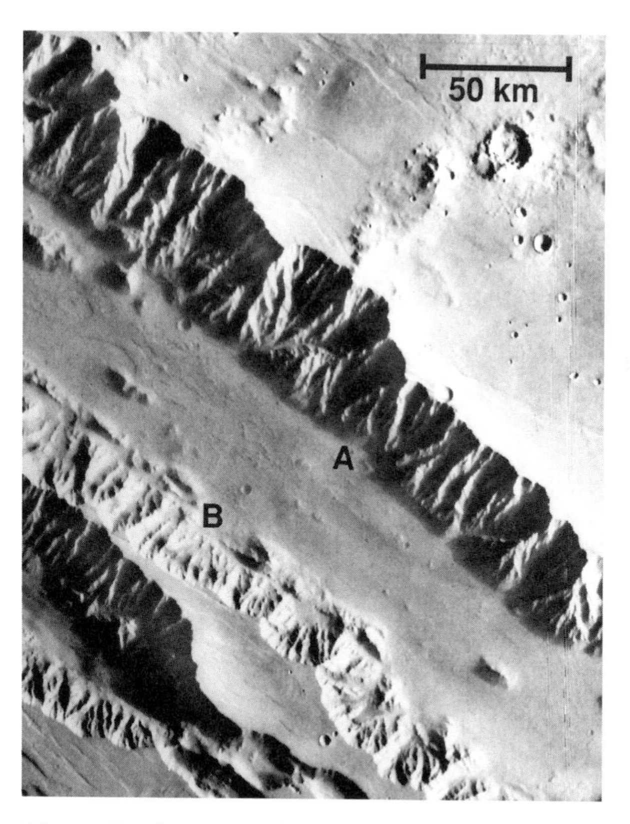

Figure 3. Canyons of Valles Marineris, Mars, which are interpreted as rifts (where the lithosphere has ruptured in extension). The canyon shown is bounded by two inward dipping normal faults (note fault scarps at base of cliffs, A, B). It is 60 km wide and 10 km deep.

to the maximum compressive stress (which is horizontal). Reverse faults dip beneath the raised block, or the hanging wall of the fault moves up relative to the footwall (Figure 1). Thrust faults are shallowly dipping reverse faults. Slip or motion on faults of this type generally produces a topographic high, such as a ridge or mountain belt. A variety of compressional tectonic features are commonly found on the planets and satellites, which include, in order of increasing size and shortening: wrinkle ridges, ridges, lobate scarps, and compressional mountain belts.

Wrinkle ridges are the only compressional tectonic feature found on all of the terrestrial planets (including the Moon). They range from linear to arcuate asymmetric topographic highs that show considerable complexity (Figure 4). Wrinkle ridges are typically composed of a few basic elements: a broad, gentle topographic rise (with widths of tens of kilometers and slopes of only a few degrees), a superposed hill or ridge—a narrower hill, commonly less than 10 km wide, comprises the "ridge" of wrinkle ridges, and a crenulation—a small wrinkle in the surface that makes up the "wrinkle" of wrinkle ridges. The remarkable resemblance of wrinkle ridges to a variety of analogous structures on Earth suggests that they form by compressional folding and faulting of surface materials. Shortening across wrinkle ridges is estimated to be up to 10 percent. Wrinkle ridges tend to form in groups that accommodate small amounts of regional shortening (up to 1 percent). Larger ridge belts on the plains of Venus are also interpreted as compressional folds and faults. Lobate scarps on Mercury are large-scale thrust faults that accommodate lithospheric shortening due to planetary cooling and contraction. Cycloid-pattern ridges on Europa (the second nearest Galilean satellite of Jupiter) may be smaller thrust or reverse faults.

Compressional mountain belts (found on Earth and Venus) occur where significant horizontal motions of the crust result in complexly folded and faulted lithospheric terrains (Figure 5). On Earth, these mountain belts (e.g., Cordillera and Rocky Mountains in the western United States and Canada, Appalachians in the eastern United States, Himalayas in India and Asia, Andes in South America) represent zones of concentrated lithospheric shortening (typically many hundreds of percent). In the Himalayas, the crust has been roughly doubled in thickness. On Earth, compressional mountain belts result from the collision of

Figure 4. Wrinkle ridges in southern Serenitatis on the Moon. Ridges deform lava flows (A) into broad arches (up to a few hundred meters high) and hills with crenulations. Wrinkle ridges are interpreted as resulting from thrust faulting and folding of surface materials. The width of the lava flow at A is approximately 30 km.

continents or the strong horizontal coupling between the downgoing oceanic and overriding plates. On Venus, which appears to lack plate tectonics, the compression and shortening are likely imposed by tractions from subcrustal mantle flow.

Strike-Slip Faults

Strike-slip faults occur when the intermediate stress is vertical (Figure 1). In this case, movement along the fault is parallel to the surface strike of the fault, which is oblique to the maximum and minimum stress directions (both horizontal).

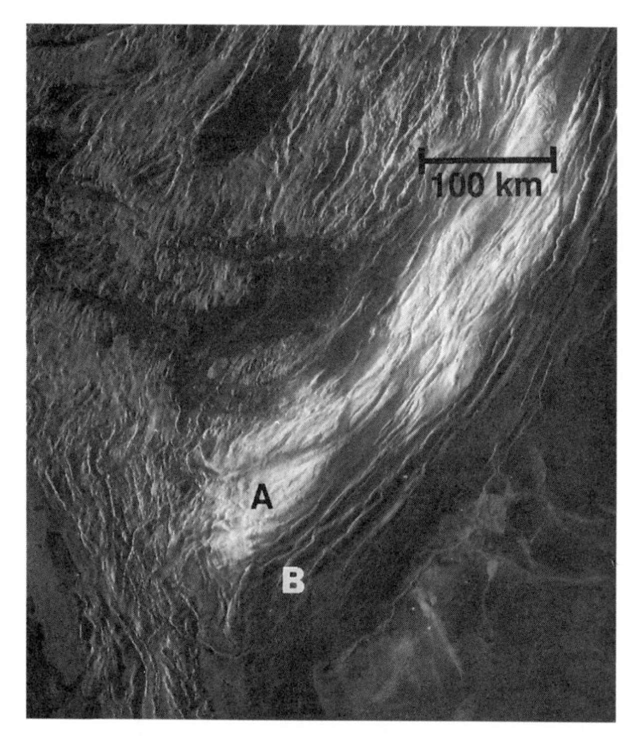

Figure 5. Radar image of compressional mountain belt (southern Akna Montes) on Venus. The highest part of the mountains is the radar bright area (A), which is a few kilometers above the plains. Note curvilinear fold and thrust fault forms (resembling wrinkle ridges) that extend into the adjacent plains (B).

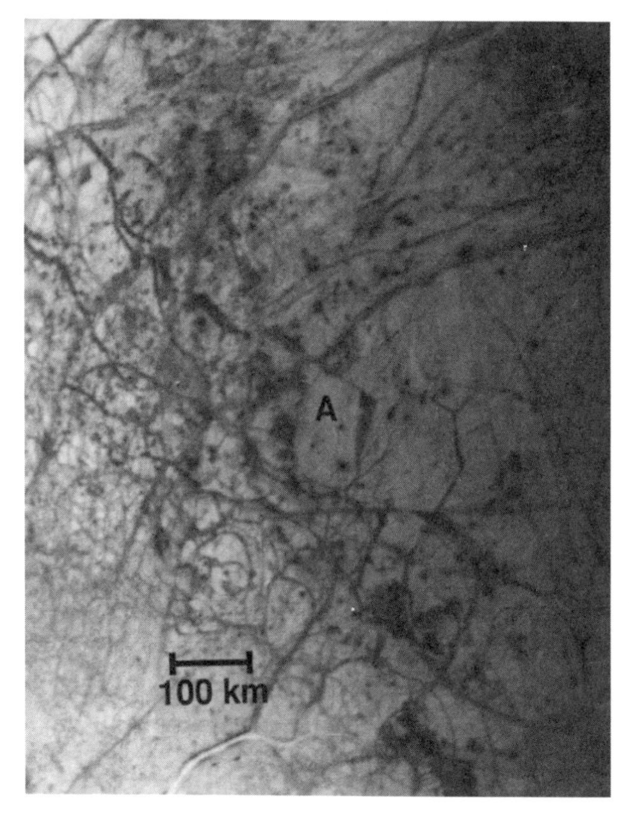

Figure 6. Zone of counterclockwise rotated lithospheric blocks (50–100 km wide) on Europa (e.g., A), bounded by dark wedge-shaped bands (tension cracks; right side of A) and transform faults (strike-slip faults connecting wedge-shaped bands; top of A). The fact that the blocks can be rotated back without gap or overlap indicates that the wedge-shaped bands are tension cracks that have opened up an amount equal to their width (10–30 km).

Strike-slip faults are remarkably uncommon tectonic features on the planets and satellites; no other solid planet or satellite has nearly the number or variety of strike-slip faults found on Earth (which may be due to the laterally confined nature of planets and satellites, which are comprised of but one plate, and in contrast to the free boundaries of mobile plates on Earth). On the planets and satellites, strike-slip faults can be divided into large-scale transform faults (defined as a special class of strike-slip faults along which the displacement suddenly stops or changes form) that separate thin lithospheric blocks (such as the San Andreas fault in California, bounding the Pacific and North American plates) and smaller strike-slip, or tear faults, that accommodate smaller displacements within deformed regions.

On Europa, rotated lithospheric blocks 50–100 km in size are separated by wedge-shaped bands that appear to be tension cracks and narrow linear dark bands that appear to be transform faults with 10–30 km of displacement (Figure 6). In many areas on Ganymede (the third closest Galilean satellite of Jupiter), grooves, interpreted to be grabens, terminate in a "T" against an older groove that bounds the heavily cratered terrain. This relation requires that extension of the younger grooves was accommodated by strike slip (order of 10 km) on the older bounding groove between the undeformed heavily cratered terrain and the extending grooved terrain, analogous to intraplate transform faults on Earth.

On the ridged plains of Mars a number of stepping strike-slip faults (each with less than 1 km slip), analogous to transfer faults on Earth, link wrinkle ridges, which appear to be compressional steps or push-ups. A number of escarpments on Mars, such as Gordii Dorsum, have an association of anticlinal folds, push-up structures, and shears that have been used to suggest lithospheric strike-slip faults with a few tens of kilometers of slip, suggestive of ancient transform faulting. No clear examples of strike-slip faults have been found on the Moon or Mercury; distributed shear may be accommodated across zones of compression or extension on Venus. Tentative identification of strike-slip faulting on Ariel and Triton has been suggested, as well as large-scale strike slip on Ganymede.

Tectonic Activity and History

The tectonic features described above have been a manifestation of the varied tectonic and geologic histories of the terrestrial planets. The Moon, Mercury, and Mars are bodies with single thick rigid lithospheric plates that date back to early in the history of the solar system. On the Moon, grabens and wrinkle ridges formed within and around basins that were depressed under the load of volcanic material that filled the basins and flexed the lithosphere. On Mercury, global cooling and contraction have produced a series of compressional lobate scarps. On Mars, a huge bulge called Tharsis has produced a hemisphere-wide system of radiating grabens and rifts, and concentric wrinkle ridges. Earth is the most tectonically active planet with a rich diversity of tectonic features ultimately produced by the rapid lateral motion of large, thin lithospheric plates (plate tectonics). Venus is intermediate between the single-plate planets and Earth, with many Earth-like tectonic features, such as folded and faulted mountain belts, yet it probably lacks plate tectonics. The outer planet satellites have a remarkable range of tectonic activity. The larger satellites tend to have had more protracted tectonic histories than the smaller satellites; most have had tectonic histories characterized by single-plate lithospheres that appear to have been dominated by extensional stresses.

Bibliography

BERGSTRALH, J. T., E. D. MINER, and M. S. MATTHEWS, eds. *Uranus*. Tucson, AZ, 1991.

BURNS, J. A., and M. S. MATTHEWS, eds. *Satellites*. Tucson, AZ, 1986.

KIEFFER, H. H., B. M. JAKOSKY, C. W. SNYDER, and M. S. MATTHEWS, eds. *Mars*. Tucson, AZ, 1992.

VILAS, F., C. R. CHAPMAN, and M. S. MATTHEWS, eds. *Mercury*. Tucson, AZ, 1988.

MATTHEW P. GOLOMBEK

TEKTITES

Tektites are natural glasses that have been known to mankind for many centuries. Australian Aborigines and Central European Paleolithic cultures used them for tools and probably as jewelry; in Indochina they were found in temples; and they appear in Chinese records of the tenth century. Josef Mayer, in 1787, and CHARLES DARWIN, in 1844, were the first to introduce tektites to modern science. Tektites are chemically relatively homogeneous, often spherically symmetric objects that are in general several centimeters in size, and occur in four strewn fields on the surface of the earth (Table 1): the North American, Central European (moldavite), Ivory Coast, and Australasian strewn fields. Tektites found within such strewn fields are related to each other with respect to age and their petrological, physical, and chemical properties. Over the last two centuries, several main theories of tektite origin were developed: glassy meteorites; eruption of lunar or terrestrial volcanoes; ejecta from meteorite impacts on the Moon or on Earth. Only the latter theory, formation during impact on Earth has emerged as being consistent with the data.

Tektites form three morphological groups: (1) normal or splash-form tektites; (2) aerodynamically shaped tektites; and (3) Muong-Nong-type (or layered) tektites. Groups (1) and (2) differ only in their appearance and some of their physical characteristics. The aerodynamic ablation results from partial remelting of glass during atmospheric reentry after it was ejected outside the terrestrial atmosphere and solidified through quenching. The shapes of splash-form tektites (spheres, drop-

Table 1. Summary of Information on the Four Tektite Strewn Fields

Strewn Field	*North American*	*Central European*	*Ivory Coast*	*Australasian*
Tektites in strewn field	Bediasites Georgiaites	Moldavites	Ivory-Coast tektites	Australites, Indochinites, Thailandites, Philippinites, and others
Geographical localities of tektite finds	Texas, Georgia, Barbados, DSDP Site 612	Czech Republic, Slovakia, Austria, Germany	Ivory Coast	Australia, Thailand, Laos, Cambodia, Vietnam, Indonesia, China, Philippines, etc.
Microtektites present	yes	no	yes	yes
Age (Ma)	35.4	15	1.09	0.77
Area (10^6 km^2)	10	0.3	4	50
Total mass (10^6 t)	300–42,000	5?	20	2000?
Age of source terrain (from Sm-Nd, in Ga)	0.7	0.9	1.90	1.11
Sedimentation age (from Rb-Sr, in Ma)	400	15	0.95	175
Possible source crater (country)	underwater? east of North America	Ries Germany	Bosumtwi Ghana	Cambodia? Laos?
Source crater diameter (km)	30?	24	10.5	50–100?

lets, teardrops, dumbbells, etc., or fragments thereof) result from the solidification of rotating liquids, and not ablation. Muong-Nong-type tektites, named after the type locality in Laos, are commonly considerably larger than normal tektites (up to 24 kilograms, or kg) and are of chunky, blocky appearance. Muong-Nong-type tektites show a layered structure with abundant vesicles and are less depleted in volatile elements (e.g., the halogens, Zn, Cu) than splash-form tektites.

Since the mid–1960s microtektites have been found in deep-sea cores of three of the four strewn fields (Glass, 1990). They are generally less than 1 millimeter (mm) in diameter and show a wider variation in chemical composition than tektites on land; however, their average composition is very close to that of "normal" tektites. Microtektites are important for defining the extent of the strewn fields, for constraining the stratigraphic age of tektites, and to provide evidence regarding the location of possible source craters.

The chemical composition of tektites is almost identical to the composition of the terrestrial upper crust (Taylor, 1973; Koeberl, 1986). Of particular use in establishing such a relationship are trace elements: the ratios of, for example, Ba/Rb, K/U, Th/Sm, Sm/Sc, Th/Sc, K versus K/U, in tektites are practically the same as those in upper crustal rocks. The absolute abundances of the rare earth elements (REE) in tektites as well as their chondrite-normalized abundance patterns are very similar to those of shales or loess, and have the characteristic shape and total abundances of REE distributions in the post-Archean upper crust. None of the trace element ratios or REE patterns are close to those of lunar or other extraterrestrial materials, for which an abundance of comparative data exists. The REE patterns of upper crustal rocks such as shale, sandstone, greywacke, granites, and related rocks are the result of geological processes on or close to the surface of the earth, through mixing, weathering, erosion, and transport. These processes do not operate, or operate much less efficiently, on water- and atmosphereless bodies. The REE patterns of tektites are therefore a strong argument against an origin from lunar volcanoes (e.g., O'Keefe, 1976).

The occurrence of relict minerals in some tektites provides another link to sedimentary source rocks (Glass, 1990). Lechatelierite—the amorphous remainder of partly digested quartz grains—indicates quartz-rich precursor rocks. Muong-Nong-type tektites contain various unmelted relict inclusions: zircon, chromite, quartz, corundum (plus SiO_2), rutile, and monazite, all showing evidence of various degrees of shock metamorphism. The type of mineral inclusions present, as well as their size and shape, suggests that a fine-grained, well-sorted sediment was the tektite parent material. Independent evidence for shock comes from the presence of coesite in Muong-Nong tektites. Coesite, shocked minerals, and vesicular impact glass were also found in several microtektite-bearing layers in cores from the Australasian and North American strewn field.

The study of the abundances and ratios of various isotopes, especially the Rb-Sr and Sm-Nd isotopic systems (*see* ISOTOPE TRACERS, RADIOGENIC), is important in the study of tektites and their source rocks. All tektites have distinct negative ε_{Nd} and large positive ε_{Sr} values that are uniquely characteristic of old terrestrial continental crust. The Nd model ages of tektites reflect the age of the source terrain of the target rocks, that is, the age of partial mantle melting to form new crust. All tektites have Precambrian crustal source terrains. Sedimentation or weathering can disturb the Rb-Sr, but not the Sm-Nd isotopic system. The Rb-Sr ages range from very young sediments for the moldavites (recent at the time of the impact; 15 million years or Ma ago) to rocks of about 0.95 billion years (Ga) for the Ivory Coast tektites. For the Australasian tektites, the Rb-Sr system indicates that the sediments that were later melted to form tektites were weathered and deposited about 167 Ma ago and probably comprised Jurassic sediments, which are common throughout Indochina.

The water content of tektites is very low at about 0.002–0.02 wt. percent. The low water content is typical for impact glasses and can be used as convincing evidence for an origin by impact. However, even the low water content in tektites is several orders of magnitude higher than that of lunar rocks. Contrary to earlier belief, it is possible to drive water out of the parent sediments in the short time available for the tektite production. Atomic bomb glass—which originated in a short-time, high-temperature event from local sediments—is very dry (0.007 wt. percent H_2O). Bubbles in tektites contain residual terrestrial atmosphere at pressures less than or equal to a third of atmospheric pressure. The N_2/Ar ratio, as well as the isotopic ratios of $^{40}Ar/^{36}Ar$, $^{36}Ar/^{38}Ar$, $^{82}Kr/^{84}Kr$, $^{129}Xe/^{132}Xe$, $^{84}Kr/^{132}Xe$, and others, agree very well with the respective atmospheric ratios. To explain the contents of dissolved heavy noble gases in tektites, the glass must have solidified at low ambient pressure, equivalent to a height of about 40 km in the atmosphere.

Cosmogenic radionuclides provide further proof of a terrestrial origin of tektites and direct evidence on the nature of the sedimentary precursor and the target stratigraphy. ^{10}Be in tektites cannot have originated from direct cosmic ray irradiation in space or on Earth, but can only have been introduced from sediments that have absorbed ^{10}Be that was produced in the terrestrial atmosphere. The concentration of ^{26}Al is of crucial importance: if tektites were ever exposed to cosmic radiation in space to produce the observed ^{10}Be concentration, then the $^{26}Al/^{10}Be$ ratio must be between 2.7 and 5.4, which is significantly different from the observed value of less than 0.07. The concentration of ^{10}Be in the environment is a strict function of the depth from the surface. Mixing of a 200-meter (m) column of bedrock into the ^{10}Be-containing surficial cover explains the ^{10}Be concentrations observed in tektites. Tektites are thus made exclusively from the top few hundred meters of the target stratigraphy (Koeberl, 1992).

Numerous suggestions and educated guesses have been made regarding the location of the possible source craters for the tektite strewn fields. Reliable links have been established between the Bosumtwi (Ghana) and the Ries (Germany) craters and the Ivory Coast and the Central European fields, respectively. For the Australasian field, many proposals for possible source craters were made and later discounted. However, the occurrence of the large Muong-Nong-type tektites and the observation that a quantity of both impacts debris and microtektites in cores all over the Australasian strewn field increase toward Indochina, supporting a location in Indochina. The North American tektite source crater is not known, but is likely to be located at or near the eastern coast of the North American continent, perhaps underwater, on the continental shelf; the southern end of Chesapeake Bay recently has been suggested.

The production of tektites must require special impact conditions because otherwise more than just four tektite strewn fields would be known on

Earth. It is possible that low-angle impact is important because of the asymmetric distribution of tektites within a strewn field. Tektite production has to occur before the main excavation phase of the crater formation has started. The resulting melts are superheated and probably distributed with the expanding vapor plume of the impact over distances of up to 6,000 km. Although some open questions will keep tektite research vigorous, it has become clear that the terrestrial impact theory is in best agreement with the data.

Libyan Desert Glass

Libyan Desert Glass (LDG) is an enigmatic natural glass found in an area of about 6,500 km^2 between sand dunes of the southwestern corner of the Great Sand Sea in western Egypt, near the border with Libya. A report by F. Fresnel dates back to 1850, but the first scientific descriptions were made by P. Clayton and L. Spencer in 1934. The glass is irregularly shaped with signs of erosion. Its age, determined by the fission track method, is around 29 Ma. Chemically, LDG is remarkably homogeneous with about 96.5–99 wt. percent SiO_2. Major and trace element abundances fall into a narrow range. Many trace elements, such as the rare earth elements, occur at levels typical for upper crustal rocks. The majority of workers favor an origin of LDG by impact, although no associated impact crater has yet been identified. It is clear, however, that LDG originated at high temperatures because it contains baddeleyite, a high-temperature breakdown product of zircon. In addition, dark streaks in LDG contain enrichments in iridium and other siderophile elements, which is in agreement with a meteoritic component from an impact origin.

Bibliography

GLASS, B. P. "Tektites and Microtektites: Key Facts and Inferences." *Tectonophysics* 171 (1990):393–404.

KOEBERL, C. "Geochemistry of Tektites and Impact Glasses." *Annual Reviews of Earth and Planetary Science* 14 (1986):323–350.

———. "Geochemistry and Origin of Muong Nong–type Tektites." *Geochimica et Cosmochimica Acta* 56 (1992):1033–1064.

O'KEEFE, J. A. *Tektites and Their Origin.* New York and Amsterdam, 1976.

TAYLOR, S. R. "Tektites: A Post-Apollo View." *Earth Science Reviews* 9 (1973):101–123.

CHRISTIAN KOEBERL

TELESCOPES

The first telescope used for astronomical purposes was created by Galileo Galilei in 1610, when he took the design for a spyglass produced by a Dutch lensmaker and turned the contraption toward the nighttime sky. Galileo's small telescope was powerful enough to reveal the phases of Venus, the four largest moons of Jupiter, the rings of Saturn, sunspots on the Sun, craters and mountains on the Moon, and the stellar composition of the Milky Way, thus revolutionizing astronomy as it was known in the early seventeenth century. Astronomers today still rely on telescopes to unravel most of the universe's secrets.

Any telescope has three major purposes. The first is to gather more light than an astronomer can do with only the naked eye. The amount of light that a telescope can gather depends on the square of its diameter. For example, the 500-cm telescope on Mount Palomar near San Diego gathers four times as much light as the 250-cm telescope on Mount Wilson near Los Angeles. Astronomers therefore are always interested in building bigger telescopes because they can then gather more light and see further into the universe (and since light travels at a finite speed, looking further into the universe means looking further back into time). The second purpose of a telescope is to resolve fine detail, thus increasing the clarity of the image viewed. The amount of resolution depends on the diameter of the telescope, so again astronomers prefer larger telescopes so they can see more detail. The third major purpose of a telescope is to magnify the image over what one can see with the naked eye. Without optical aid, the planets just look like bright stars in the night sky, but even with a small telescope one can easily see the rings of Saturn and the polar caps of Mars. The magnification depends both on the telescope itself as well as the eyepiece used during observation.

Two major types of optical telescopes exist: refractors and reflectors. Refractors, such as the one Galileo used, utilize glass lenses to converge light to a point. Reflecting telescopes, which bounce light off of mirrors, were first designed by Issac Newton in 1668. Many subtypes of telescopes currently exist, but all are permutations of the basic refractor or reflector design (Figure 1).

The two largest refracting telescopes in the world are the 102-cm refractor at the Yerkes Ob-

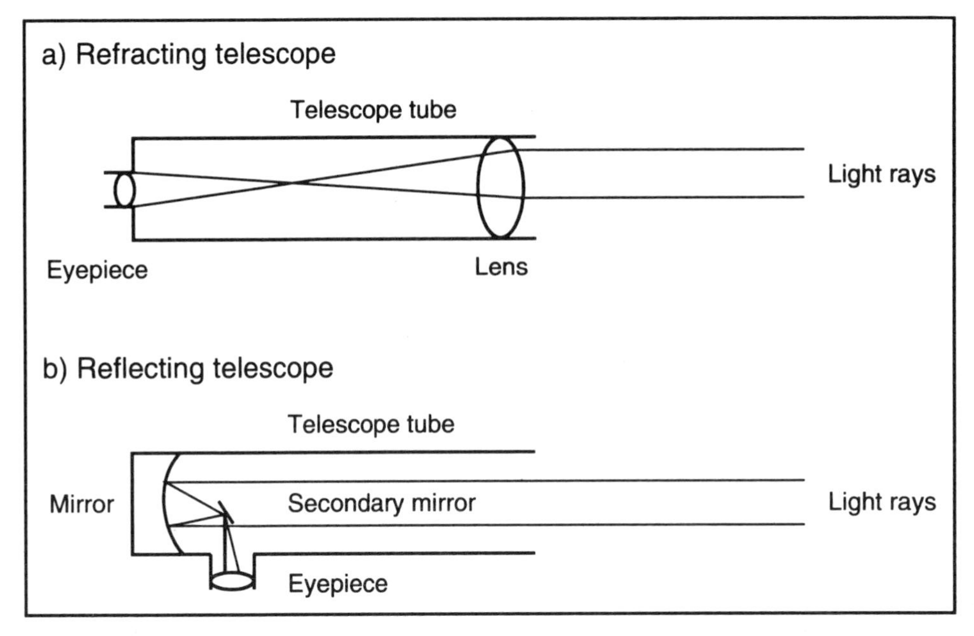

Figure 1. a. A simple refracting telescope that uses lenses to bend light to the eyepiece. b. A Newtonian-style reflecting telescope that bounces light off a primary and a secondary mirror to the eyepiece. Magnification of the image occurs at the eyepiece.

servatory in Wisconsin, and the 91-cm refractor at Lick Observatory in California. Refracting telescopes with lenses larger than about 100 cm are difficult to construct and maintain for two reasons. First, the lenses used in refracting telescopes must be optically pure so that light can pass through the lens without distortion. The difficulty of producing an optically pure lens increases as the lens size increases, so the Yerkes Observatory refractor is currently the largest clear lens that has been produced. A second problem with large refracting telescopes comes from the fact that the lens is supported by its edges within the telescope tube. With no back support possible for the lens, gravity can cause large lenses to deform under their own weight.

The largest telescopes in use today are the reflecting telescopes. Since light is reflected off the surface of a reflecting telescope, the internal clarity and the existence of support behind the mirror are of no concern for the resulting image. The major constraint in producing ever-larger reflecting telescopes is generally cost, which increases dramatically as the size of the mirror increases. Among the largest reflecting telescopes in the world are the 5-m Hale telescope on Mount Palomar in California, the 6-m reflector on Mount Pastukhov in the Caucasus Mountains of Russia, and the 10-m Keck telescope on Mauna Kea in Hawaii. Another 10-m telescope is being constructed next to the Keck telescope on Mauna Kea and is expected to be operational by late 1996. The two 10-m telescopes can be used separately or in tandem, acting like one big reflector with a mirror diameter of at least 15 m.

Hawaii and Chile are prime telescope locations because of the excellent atmospheric conditions created by the high mountain ranges in these places. Several large reflecting telescopes currently are being constructed in these locations and elsewhere around the world. Four 8-mm reflectors (light-gathering equivalent to one 16-m mirror) will make up the Very Large Telescope atop Cerro Paranal, about 130 km south of Antofagasta, Chile. Farther south in Chile, a 6.5-m mirror will be housed in the Magellan telescope planned for operation of Cerro Manqui in 1997. A clone of the 6.5-m Magellan telescope will be built on San Pedro Martir in Baja California, Mexico, and the 8.3-m Subaru mirror at the Japan National Large Telescope will be in place and operational on Mauna Kea, Hawaii, by 2000. Two 8.1-m reflecting telescopes are being planned for the peaks of Cerro

Pachon in Chile and Mauna Kea in Hawaii, with operations beginning in 1998. In addition, two 8.4-m reflectors (called the Columbus Project) will begin operation in 1997 atop Mount Graham in Arizona and an 11-m segmented mirror called the Spectroscopic Survey Telescope is being built at the McDonald Observatory on Mount Locke, Texas (completion expected in 1996). Updated articles on telescopes under construction and being planned can be found in monthly magazines such as *Sky and Telescope* and *Astronomy*. In particular, the reader is directed to the "Decade of Discovery" series in several 1991 issues of *Sky and Telescope* and the "Temples in the Sky" series by Serge Brunier in the February, June, and December 1993 issues of *Sky and Telescope*.

Several technological advances in the past twenty-five years have led to new designs for reflecting telescopes. The first occurred in 1979 when the Multiple Mirror Telescope (MMT) began operation on Mount Hopkins in southern Arizona. The MMT was the first telescope to use several smaller mirrors in concert to give the equivalent light-gathering power of a larger mirror. The smaller mirrors each weighed less than a single large mirror, and each was made even lighter by the use of a honeycomb structure, as opposed to one of solid glass, in the mirror itself. Computers aligned each of the mirrors to allow them to act in tandem as one larger mirror. The design worked well but is now being upgraded to a 6.5-m single mirror. More recently, the production of large telescope mirrors has been facilitated by two major advances: (1) construction of the honeycomb support structure, thus reducing the weight associated with a solid glass mirror; and (2) development of the "spin-cast" technique whereby the correct shape of the mirror is produced by spinning molten glass until it solidifies. This spinning produces a concave shape to the mirror, resulting in less effort expended in grinding the mirror to the correct parbolic shape. These two advances allow large mirrors to be produced relatively cheaply and more quickly.

Many telescopes also have been designed to observe in regions of the electromagnetic spectrum other than visible wavelengths. Radio wavelengths can pass through Earth's atmosphere, so radio telescopes have been used on Earth's surface since 1936. Radio telescopes are generally parabolic in shape and work on the same principle as reflecting telescopes. However, since radio waves have much longer wavelengths than visible light, a solid receiving dish (such as a mirror) is not necessary—the dish can be composed entirely of wires, making radio telescopes much lighter than their optical counterparts. Radio telescopes can also be built much larger—the largest single-dish radio telescope is the 305-m dish of the Arecibo Observatory, actually built in a natural depression in the Puerto Rican mountains. The disk is not movable because of its size and construction. The largest movable radio telescope is a 100-m dish in Bonn, Germany. However, many radio telescopes do not use separate parabolic dishes—sometimes the telescope may be nothing more than wires stretched across an area. In other cases, such as with the Very Large Array (VLA) near Socorro, New Mexico, the telescope is composed of a series of small parabolic dishes that can be used in tandem, a technique known as interferometry.

Optical reflecting telescopes located on high mountain peaks above much of the water vapor in Earth's atmosphere also can be used to study infrared wavelengths. In particular, many of the telescopes atop Mauna Kea in Hawaii are used in infrared studies of the universe.

Other regions of the electromagnetic spectrum can only be observed from above the protective shield of Earth's atmosphere. Since the 1970s, orbiting telescopes have been used to study X-ray, ultraviolet, and gamma-ray wavelengths, in addition to providing unhampered views in the optical and infrared regions of the electromagnetic spectrum. The 2.4-m Hubble Space Telescope (HST), launched in 1990, is the best known of these orbiting observatories. A flaw in the mirror shape prevented detailed observations with the telescope until corrective lenses were installed during a NASA space shuttle mission in 1993. The HST can make observations in optical, ultraviolet, and near-infrared wavelengths. A second orbiting telescope is the Compton Gamma-Ray Observatory, which was launched in 1991 and conducts observations in the gamma-ray region of the electromagnetic spectrum. Two more telescopes are planned to complete the four-telescope Earth-orbiting "Great Observatories" program: the Advanced X-Ray Astrophysics Facility (AXAF), scheduled for a 1998 launch, and the Space Infrared Telescope Facility (SIRF), currently facing congressional approval for a 2001 launch date. Other orbiting observatories include Rosat (a German-American-British X-ray mission launched in 1990), the

Extreme Ultraviolet Explorer (launched in 1992 by the United States), Japan's Advanced Satellite for Cosmology and Astrophysics (an X-ray satellite launched in 1993), the Russian Granat X-ray Observatory (launched in 1989), the joint Japanese-American Infrared Telescope in Space (launched in 1995), the Infrared Space Observatory (to be launched in late 1995 by the European Space Agency), and the U.S. Submillimeter Wave Astronomy Satellite (planned for a late 1995 launch).

In addition to ground-based and Earth-orbiting spacecraft, plans are now being made to establish telescopes on other bodies in the solar system. The Moon in particular is a prime telescope site because of its low gravity and lack of an atmosphere. In addition, radio telescopes established on the lunar farside would not be affected by terrestrial radio signals since the Moon itself will block out such emissions. Such telescopes will be a future generation of the automated telescopes just now beginning to be operational here on Earth.

Bibliography

BURBIDGE, G., and A. HEWITT, eds. *Telescopes for the 1980's.* Palo Alto, CA, 1981.

FLORENCE, R. *The Perfect Machine: Building the Palomar Telescope.* New York, 1994.

MUIRDEN, J. *Astronomy Handbook.* Englewood Cliffs, NJ, 1985.

PARKER, B. *Stairway to the Stars.* New York, 1994.

SMITH, R. W. *The Space Telescope.* New York, 1993.

NADINE G. BARLOW

TERRESTRIALIZATION

Organisms live in three major environments: marine, freshwater, and terrestrial. The last two comprise the continental or nonmarine habitat. Despite absence of evidence that the continental environment is any less old than the marine, it has always been assumed that life arose in the sea, and that freshwater and terrestrial habitats were secondarily colonized with entry onto the land from freshwater populations. The larger number of phyla and classes of marine life compared with freshwater and terrestrial supports this assumption, indicating that many marine organisms never adapted to any other way of life, even though the greatest number of species reside on land. Thus many common marine organisms, including echinoderms, brachiopods, bryozoans, and corals (*see* ECHINODERMS; COLONIAL INVERTEBRATE FOSSILS; BRACHIOPODS), have few or no freshwater or terrestrial representatives. Marine algae follow the same evolutionary pattern. Moreover, marine fossils for predominantly marine groups with freshwater and terrestrial representatives are always oldest, sometimes preceding by millions, even hundreds of millions, of years the first nonmarine fossils. While it is commonly believed that no major group of organisms evolved in freshwater, a number show significant adaptive radiations there. Likewise, major terrestrial groups, such as vertebrates, and arthropods, such as insects and spiders, have had significant adaptive radiations on land, although both evolved in the sea. Mosses, their relatives, and tracheophytes similarly have diversified on land into many families, genera, and species, although their phylogeny traces back to freshwater green algae.

Fossil evidence supports aspects of this scenario. Fossils in the marine ecosystem go back about 3.5 billion years (Ga) into the Precambrian (*see* HIGHER LIFE FORMS, EARLIEST EVIDENCE OF). At the beginning of the Cambrian (570–590 million years, or Ma) nearly all major skeletonized marine invertebrate phyla appeared and marine vertebrates are known by the end of the Cambrian. During the early Paleozoic (Cambrian, Ordovician, Silurian) the seas teamed with diverse, complex multicellular animal and algal life. During an equivalent interval of time following the Cambrian, evidence for continental life is meager: hundreds of millions of years were to pass before evidence exists for similar diversity and complexity on land or in freshwaters. Little is known of early stages of terrestrialization including the nature of earliest organisms and the time when they breached the terrestrial habitat. Moreover, with few exceptions burial sites occur in water. Unless it is possible to determine that fossil remains belonged to an aquatic or terrestrial organism, not always easy during early stages of continentalization, it may not be possible to know where they lived.

Only now are a few pieces of the puzzle beginning to fall into place. Fossils have extended the record of land plant life downward 60–70 Ma into the early Middle Ordovician from the previous Silurian/Devonian benchmark, narrowing the gap that separates life's beginnings in the sea from colonization of the nonmarine habitat. But newly discovered fossils that seem to represent photosynthetic, chlorophyll-bearing bacteria (cyanobacteria) suggest that they may have lived on land as early as the Late Precambrian, well before plant life. Before the Middle Ordovician, however, only a scattered handful of records exist for nonmarine life, few expressly tied to the land surface. Beginning in the Middle Ordovician, microfossils provide continuous evidences for terrestrial plant life: not until the Late Silurian are there continuous evidences of terrestrial animals. This new information tells us that terrestrialization seems to have taken place slowly for a lengthy time. It also leads us to believe that the sequence of early terrestrialization may not reflect the organization of modern terrestrial ecosystems: with tracheophytes first at the base of the food chain, followed by herbivores that feed on tracheophytes, and last by predators that feed on the herbivores. Significantly, fossils largely responsible for change in view of the antiquity and character of early continental ecosystems are microscopic remains, commonly the original substances of the organisms themselves, freed from rocks by chemical digestion.

Precambrian

Fossil evidence confirms speculation that cyanobacteria were likely first organisms in continental habitats: among the earliest are stromatolites constructed by cyanobacteria in Archean lakes about 2.7 Ga ago. Minute fossils that resemble fecal pellets made by freshwater, planktonic microcrustaceans such as copepods have been found in Precambrian lake deposits as old as 1.3 Ga. Fossil evidence suggests that cyanobacteria may also have made it onto the land surface by 1.2 Ga ago in the late Precambrian. Cyanobacteria have the capacity to withstand ultraviolet radiation and survive today in environments that may mimic dry ground habitats of early Earth. Modern cyanobacteria form extensive surface mats in some desert areas and would have been important ground surface stabilizers in the absence of rooted plants. Because of microscopic size they could have been spread widely by wind to colonize suitable sites. Although there is no fossil support, the photosynthetic cyanobacteria may have been accompanied by other unicellular bacteria, green algae, and fungi and by animal-like unicellular organisms (protists, protozoa), some of which feed on cyanobacteria, fungi, and green algae. Microbial and algal mats may have supported communities of small annelids, nematodes, and arthropods that fed on the mats, their decomposition products, and each other.

Cambrian

No time equivalent evidence for nonmarine, multicellular plants, or animals exists during the Cambrian explosion of shelly marine invertebrates and vertebrates, but trails and resting marks in rocks—apparently subaerial surfaces of mudflats and sandy flats adjacent to shoreline—indicate that some animals, most likely arthropods, were able to venture briefly out of the water possibly in search of food, or for mating purposes like modern horseshoe crabs.

Ordovician

Earliest evidence for potentially fully terrestrial organisms begins with tough ,alled Mid-Ordovician spores that resemble spores produced by some modern land plants: the oldest, from Arabia, appeared about 470 Ma ago. The spore-producers are believed to have been more like bryophytes (mosses, hornworts, hepatics) than tracheophytes, today's dominant land plants. Spores are reproductive devices borne by wind that spread the spore-producer widely over the earth's surface by Late Ordovician-Early Silurian times, circa 440 Ma. They provide the only direct evidence for land plants for 60–70 Ma, until sometime in the late Early Silurian. Oldest evidence for possible terrestrial animals are burrows in Late Ordovician fossil soils. The burrow-makers like the spore-producers are otherwise unknown, but were possibly similar to millepeds that commonly tunnel through soils in search of food.

Silurian

In the earliest Silurian, but a little later in time, scraps of actual animals (bristles, spines, leg fragments, and possible body-covering cuticles) are found associated with the spores in rocks made from sediments apparently deposited in a river. The fragments are similar to body parts of arthropods, the dominant terrestrial invertebrates (for example, insects, spiders, mites), that also include marine crabs and lobsters and crustaceans common in both freshwaters and the sea (*see* ARTHROPODS). By late Early Silurian time, spores provide the earliest possible evidence for tracheophytes, today's dominant land plants (*see* PTERIDOPHYTES; ANGIOSPERMS; GYMNOSPERMS). Some 10–15 Ma later, in the early Late Silurian, large fossils of tracheophyte-like plants and possible tracheophytes are found. These early "experiments" gradually disappeared as tracheophytes gained ascendency with a major adaptive radiation some 10–15 Ma later in the late Early Devonian (Gedinnian). Earliest possible evidence for Ascomycetes (yeasts and other common terrestrial and aquatic fungus, many parasitic) as well as large remains tentatively identified as hepatics are found in the Late Silurian. Increasing evidence of terrestrial animal life is associated with the Late Silurian spores and large plant remains. Some seem to represent excrement like that produced by arthropods that feed exclusively on fungal hyphae (tubular, thread-like fungal filaments). Others are actual body remains, all arthropods. Thus members of this important terrestrial phylum have left the oldest fossil animal remains on land although some, such as the oldest millepeds, may still have lived in freshwaters. Late Silurian arthropods include myriapods, possible millepeds and centipeds, and spiders.

Devonian

We still know relatively little about terrestrial life in the Devonian (400–360 Ma ago), especially the Early Devonian. Major discoveries seem likely as we continue to dissolve rocks in acids and bases to recover microscopic plant and animal remains. But diverse plant and animal communities are beginning to be recognizable in terms of modern organisms. Devonian arthropods include mites, spiders, and spider-like creatures, centipeds, millepeds, pseudoscorpions and true scorpions, some of which may still have been aquatic, and the earliest wingless insects. Mollusks had also made their way into the continental habitat, possibly still aquatic. Plants include "true" tracheophytes, some allied to still living lineages, tracheophyte-like plants of uncertain relationship, and fossil hepatics similar to modern *Pallavicinia*. By the Late Devonian, amphibians had evolved from wholly aquatic vertebrates and terrestrialization was well underway with the appearance of reptiles in the early Carboniferous.

This highlight picture indicates that during a time when the seas were exploding with life no equivalent evidence exists for continental life. Can such seeming anomalies be explained? Were land surfaces largely barren as some have speculated? Did it take a long time for marine organisms to develop the physiological and morphological adaptations that made life on Earth's surface possible? Many modern organisms easily move between sea and freshwater environments, but physiological barriers have kept entire major groups of organisms in the sea and off the land surface. Many organisms never developed adaptations to overcome the difficult marine to nonmarine transition; others developed them only slowly. Another possibility is that early land inhabitants were soft-bodied and incapable of fossilization. Still another is that nonmarine rocks suitable for the preservation of fossils are now so comparatively scarce that limited opportunity exists to find evidence of early nonmarine life.

Good reasons exist to believe that shallow continental waters and land surfaces were physically inhospitable to most multicellular organisms with limited food supply, limited shelter, raw rocky soils without organic matter, and so on. Some disagreement also exists about the level of atmospheric oxygen and the shielding effects of the ozone layer for ultraviolet radiation on surface-dwelling organisms in the absence of shelter. Cyanobacteria, single-celled algae, and protozoa as well as small animals may well have been able to survive in environments unameliorated by higher plants and animals but would be unlikely to leave decipherable fossil records. In these circumstances the soil ecosystem may have been among the earliest to develop, harboring small microbiota.

Speculation abounds with regard to the routes that marine organisms may have taken into the nonmarine habitat. Theoretical expectations favor estuaries and rivers that drained into them as a principal route ashore. Some organisms may have come ashore via spaces between beach sands and shoreline swamps and other environments at the interface of the marine and nonmarine habitats. Such environments can be seen as favoring the gradual physiological adaptations necessary for organisms accustomed to the high salinities of seawater to make the transition to low salinity freshwater and hence ultimately to the land surface. "Oases" formed by cyanobacteria and algal mats on the land surface may have been colonized by some microscopic and single-celled organisms blown in from the sea.

Only an extended fossil record can provide the answers to these and other puzzling aspects connected with terrestrialization that continue to plague those interested in such questions.

Bibliography

GRAY, J. "Evolution of the Freshwater Ecosystem: The Fossil Record." In *Paleolimnology: Aspects of Freshwater Palaoecology and Biogeography*, ed. J. Gray. Amsterdam, 1988.

GRAY, J., and A. J. BOUCET. "Early Silurian Nonmarine Animal Remains and the Nature of the Early Continental Ecosystem." *Acta Palaeontologica Polonica* 38 (1994).

GRAY, J., and W. SHEAR. "Early Life on Land." *American Scientist* 80 (1992):444–456.

HORODYSKI, R. J., and L. P. KNAUTH. "Life on Land in the Precambrian." *Science* 263 (1994).

LITTLE, C. *The Terrestrial Invasion.* Cambridge, Eng., 1990.

JANE GRAY

THERMODYNAMICS AND KINETICS

The science of thermodynamics began when nineteenth-century scientists discovered that heat and work could be transformed into each other and that there were rules that applied to such transformations. The broadest definition of thermodynamics would be the study of how matter responds to changes in temperature. Chemical reactions consume or create heat when chemical bonds are broken or made. Because of this, chemical changes can be described using the laws of thermodynamics and, in particular, the concept of equilibrium. When a system (i.e., a collection of things of interest to us) is in its equilibrium state, there is no tendency for the system as a whole to change its state. This means that there is no driving force for chemical reactions to occur that changes the nature of the members of the system.

The basic tenets of thermodynamics are contained in three laws that sound deceptively simple. The first law states that the energy of the universe is a constant. The second law states that the entropy of the universe will increase in any spontaneous process. The third law is that the entropy of a perfect crystal at absolute zero temperature is zero.

The third law gives the best intuitive explanation of entropy, S, as a measure of disorder. At absolute zero temperature (−273°C), no motion other than the groundstate vibrational motion is possible, so we can locate the atoms in the crystal with minimal uncertainty. In addition, if the crystal is perfect, there is no geometric disorder. Once the temperature goes up, disorder increases because the atoms in the crystal can vibrate about their lattice positions.

The energy of a system, E, can be increased by adding heat to the system or by doing work on the system. In chemical systems the concept of enthalpy, $H = E + PV$ (P = pressure, V = volume), is more useful because the heat given off or taken up during a chemical reaction is measured by the change in enthalpy of the reaction. The condition for equilibrium in the chemical system is then that the Gibbs free energy, G, for the system is at a minimum. The Gibbs free energy is defined as $G = H - TS$ (T = absolute temperature). For a reaction that occurs at constant temperature, $\Delta G = \Delta H - T\Delta S$. The condition for a spontaneous process is then that the change in G must be negative so that the free energy of the final state is less than that of the starting condition.

While the study of thermodynamics can help us decide what the stable final condition of a collection of objects will be, it cannot tell us how fast a system that is not at equilibrium will react in response to its conditions. To understand the rate of change of chemical systems, we need to understand the kinetics of reactions. Many natural systems in the earth sciences consist of collections of chemical compounds that are far from equilibrium and have been frozen into their disequilibrium state because of a rapid change in temperature. Glassy volcanic rocks, which will be discussed in more detail below, are a good example of this type. In many cases, it is the failure to attain equilibrium quickly that allows the history of a geologic process to be preserved. In other cases, we are relying on establishment of equilibrium quickly at high temperature to ensure validity of models that help us understand processes taking place deep within Earth.

In the earth sciences, thermodynamics provides the essential key to understanding dynamic processes that involve chemical reactions. A good example of this is the interaction of carbon dioxide between the oceans and the atmosphere as well as the fate of carbon dioxide used by marine organisms to make carbonate skeletons. In the earth sciences the carbon dioxide cycle is very complex because it involves photosynthesis by plants, which removes CO_2 from the atmosphere, and the burning of fossil fuels, which adds CO_2 to the atmosphere, as well as many other interactions. For our example, we will only consider the local equilibrium between air and water and assume the concentration of atmospheric CO_2 remains constant. In the water layers near the top of the ocean, life is abundant so long as there are sufficient nutrients. Microorganisms remove calcium carbonate from the seawater to form their protective skeletons. This removal lowers the concentration of carbon dioxide in the seawater and moves the seawater-air system away from equilibrium. The system adjusts by spontaneously exchanging some CO_2 across the air–water interface from the air into the water. The laws of thermodynamics tell us that the final amount of CO_2 gas dissolved in the seawater will depend on its partial pressure in the atmosphere above the water, or more correctly on its thermodynamic activity or fugacity, which is a measure of its effective concentration. As long as the rate of exchange of CO_2 across the air–water interface is equivalent to or faster than the rate of removal of CO_2 from the water by marine organisms or other removal processes, the seawater will have the equilibrium concentration of CO_2.

When the marine organisms die, they fall to the bottom of the ocean and accumulate in the sediments. Their skeletons may be preserved as calcium carbonate in the sediments, depending on the degree of saturation of the water above the sediments. If the effective concentrations (activities) of Ca^{++} and $CO_3^{=}$ ions in the water above the sediments are high enough for the calcite or aragonite solubility product to be exceeded, the skeletons will be preserved as fossils. If, on the other hand, the water is undersaturated with respect to the mineral from which the skeleton is made, the calcium carbonate will dissolve. Kinetics of dissolution will determine how long remnants of the skeletons will remain.

In the earth sciences, thermodynamics can often be used to draw inferences about processes that occurred in the past and about parts of Earth and the solar system that we cannot directly sample. In the late 1960s Gast (1968) showed that the contents of trace elements in volcanic rocks could be used to estimate the chemical composition of the source region of the magmas from which the rocks were formed (*see* TRACE ELEMENTS). When a rock begins to melt, the trace elements in the rock partition themselves between the solid and liquid phases. For an ideal system, the behavior of trace elements obeys Henry's law—that is, the activity of the trace element in any phase is directly proportional to its molar concentration (concentration in moles of substance per liter of fluid or kilogram of solid) in that phase—and the partition coefficients of trace elements between phases are independent of the absolute concentration of the trace element. One can measure partition coefficients in laboratory experiments and then use these data and the assumption of Henry's law to convert the trace element data in volcanic rocks into an estimate of the composition of their source region. In practice, this produces only a first approximation because nonideal solution behavior must be accounted for, but the ideal system model produced a very good idea of the relationship between magma type and degree of partial melting, as well as a picture of the distribution of trace elements in the mantle.

Volcanic rocks are formed from magmas that reach the surface of Earth. The rate of ascent of the magma from the point in the mantle where it separates from its parent mantle rock and the temperature of the overlying mantle through which it rises affect the crystallization of the melt into mineral phases. An extremely rapid ascent with explosive eruption will result in the production of ash fragments that are frozen droplets of magma. Ash eruptions are associated with magmas that have high silica concentration, which attain high viscosity when they lose their water content during eruption. Magmas with lower silica concentration usually have less water originally and do not erupt explosively. They rise more slowly through the mantle and partially crystallize as they rise. These rocks can range from mostly glass with a few crystals, to mostly crystalline with a little glass between the mineral grains. In any case, the glassy portion of all of these rocks is not in thermodynamic equilibrium because glass is not a stable phase in the strict thermodynamic sense. The glass will eventually devitrify and form crystalline phases. The rate of devitrification—that is, the kinetics of the process—will be greater if the temperature is high and if water is present. Both of these parameters help in the movement of chemical species through the glass and aid in formation of the new crystalline phases. Because the glass devitrifies at temperatures much lower than those at which the original crystals formed during magma ascent, it is possible to have a final mineral assemblage that is not in thermodynamic equilibrium. The early-formed crystals will be much less reactive than the glassy phase and will not be able to equilibrate with the newly formed crystals. Much later in time, when a geologist examines the resulting rock, the texture of the mineral grains and the existence of the disequilibrium assemblage will provide the evidence of the origin and history of the rock. (See also Lasaga and Kirkpatrick, 1981.)

Thermodynamics is fundamentally a description of how objects react to changes in their environment. All igneous rocks are formed by partial melting of preexisting rocks, followed by crystallization, of which part or all usually takes place during initial cooling of the rock. For intrusive rocks, the crystallization process takes place at elevated pressures because the magma does not reach the surface of the earth, but crystallizes within the crust. Slow crystallization allows large mineral grains to form, and as the temperature cools, the diffusion rate of chemical species through the crystallized grains becomes low enough that the core of the mineral can no longer equilibrate with the residual magma. This produces a mineral in the final rock that shows concentric bands of composition. Each of the bands reflects the composition of the magma at the time the band formed, with suitable adjustment for changes in composition caused by diffusion during the crystallization process. From studies of these zoned crystals, the detailed process of crystallization of the rock can be deduced.

Minerals that formed at high temperatures and/or pressures will not necessarily represent the phases that are stable at 1 atmosphere pressure and 20°C. The process of weathering of rocks represents the slow attempts of igneous and metamorphic rocks to adjust to temperatures and pressures that are lower than those at which they formed. Because reaction rates decrease as temperatures decrease, the processes of weathering are dominated by kinetics of reactions. In recent years, the phenomenon of dissolution of minerals has received extensive study, both in laboratory experiments and in theoretical studies. (See Lasaga and Kirkpatrick, 1981; Brady and Walther, 1989.)

In iron meteorites, which are thought to represent the centers of a parent body that was later broken apart by impact, the response of the iron-nickel (Fe-Ni) alloy to changes in temperature has been used to determine the cooling rates of the parent bodies and, thus, to obtain an estimate of the parent body size. The class of iron meteorites called octahedrites has a chemical composition for which a single alloy phase is not stable. The metal partitions itself into fine plates or lamellae of two different alloys—kamacite, which is low in Ni, and taenite, which is high in Ni. As temperature decreases, the equilibrium favors higher nickel in the taenite phase. Nickel diffuses into the taenite grains from the adjacent kamacite regions, but the diffusion distance is limited because of slow diffusion within the solid phases. If the diffusion coefficients for Ni in taenite and kamacite are known, the shape of the Ni diffusion profile in the taenite can be used to calculate the cooling history of the alloy and, thus, of the meteorite's parent body. This is another good example of the increase in information that we can obtain when kinetics are slow and equilibrium is not achieved. If diffusion were fast in the alloys, no cooling history would have been preserved.

Bibliography

Brady, P. V., and J. V. Walther. "Controls on Silicate Dissolution Rates in Neutral and Basic pH Solutions at 25°C." *Geochimica et Cosmochimica Acta* 53(1989): 2823–2830.

Broecker, W. S., and V. M. Oversby. *Chemical Equilibria in the Earth.* New York, 1971.

Gast, P. W. *Trace Element Fractionation and the Origin of Tholeiitic and Alkaline Magma Types. Geochimica et Cosmochimica Acta* 32 (1968):1057–1086.

Lasaga, A. C., and R. J. Kirkpatrick, eds. *Kinetics of Geochemical Processes.* Vol. 8, *Reviews and Mineralogy.* Mineralogical Society of America. Washington, DC, 1981.

VIRGINIA M. OVERSBY

THORIUM

See Uranium and Thorium

TITAN

The unmanned spacecraft, *Voyager 1*, made a close flyby of Titan during its pass through the Saturnian system in November 1980. The Voyager missions and subsequent investigations revealed many new discoveries about Titan and its atmosphere. Results from Voyager and current research both from groundbased observations and orbiting spacecraft [International Ultraviolet Explorer (IUE), Hubble Space Telescope (HST)] on Titan's atmosphere and surface are summarized below. The joint NASA-ESA (European Space Agency) Cassini mission, scheduled for launch in the mid to late 1900s, is the next spacecraft mission designed to investigate Saturn and Titan. A brief overview of this mission with emphasis on prospects for answering current questions about Titan concludes this entry.

Titan was discovered in March 1665 by Dutch astronomer, Christiaan Huygens. Titan is the largest of Saturn's moons, but in many ways it is more like a planet than a moon. It orbits Saturn at a distance of 5.0 × 10^8 meters (m) (twenty Saturn radii or 20 R_S) every 15.95 days. Its surface is masked by an extensive atmosphere that was first detected by GERARD KUIPER (1944, 1952) but most information about its structure and chemistry has developed since 1971. One influential study of John Lewis (1971) showed that Titan's low mean density required an interior model much richer in ices than that of the terrestrial planets, despite other similarities between these bodies and Titan.

Titan's size (radius = 2.575 × 10^6 m or 40% of Earth's radius) and average density (1,881·kg m^{-3} are nicely bracketed by those for Jupiter's Galilean satellites, Ganymede and Callisto. Whereas these Jovian moons are believed to consist of rock (silicates and iron compounds) and water ice (25–50 percent by mass), Titan's interior is more likely to be equal parts (approximately 52:48 by mass) rock and either ammonia or methane ice ($NH^3 \cdot H_2O$ and $CH_4 \cdot nH_2O$).

Titan's Atmosphere: Chemistry and Structure

Titan is the only moon with a significant atmosphere (surface pressure = 149.6 ± 0.2 × 10^3 pascals or Pa). Its atmosphere is denser than that of all the terrestrial planets, with the exception of Venus. *Voyager 1* images of Titan show thick, dark-orange or brown, stratospheric haze layers that totally obscure its surface (Figure 1). The hazes are composed of complex organic polymers that are probably the end products of methane (CH_4) photochemistry. Titan's photochemistry is a function of two processes: photodissociation of methane and dissociation of nitrogen (into N^+) by energetic particles from Saturn's magnetosphere. The N^+ ion then dissociates the methane into nitriles and higher chain HCN polymers. Titan's highly reduc-

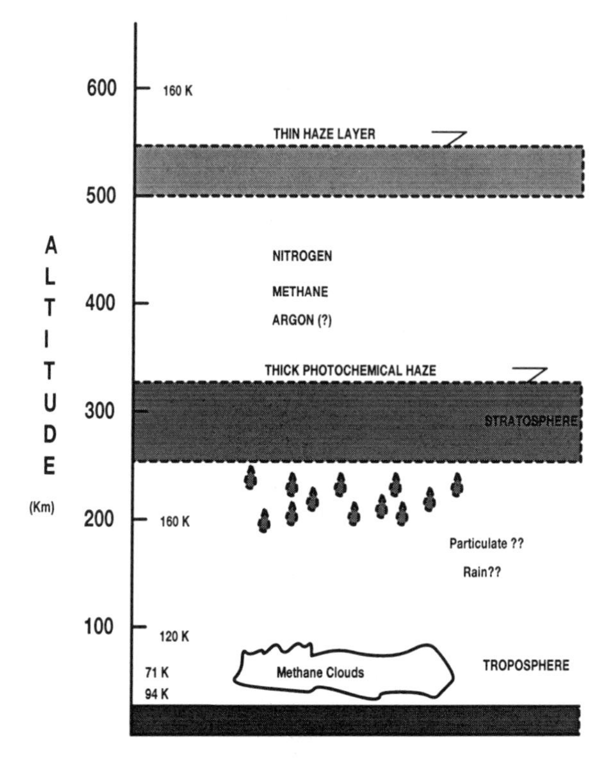

Figure 1. Schematic diagram of the Titan atmospheric cloud structure.

ing atmosphere suggests a striking analogy with what might have been the early Earth atmosphere. The discovery of complex organics reinforced this concept, suggesting that Titan's atmosphere is a natural laboratory for studying prebiotic evolution.

The *Voyager* Infrared Interferometer Spectrometer (IRIS) and radio occultation measurements inferred a mean molecular weight of nearly 28 atomic mass units (amu, where 1 amu is defined as 1/12 the mass of an atom of the ^{12}C nuclide). The most likely candidates for the dominant component were therefore nitrogen (N_2) and carbon monoxide (CO). The UVS (ultraviolet spectrometer) experiment on the *Voyager* spacecraft measured strong UV emissions indicating nitrogen as the atmosphere's principal component (Strobel and Shemansky, 1982). Two possible sources for the nitrogen are primordial or photodissociation of ammonia (NH_3). The latter origin would require sufficiently warm, early surface temperatures to ensure an adequate NH_3 vapor pressure (Owen, 1987).

Initial estimates of the mean molecular weight ranged from 27.8 to 29.4. Since the mean value (28.6) is higher than that for pure nitrogen, investigators have suggested other gases to account for the difference. The additional gas needs to be significantly heavier than pure nitrogen and spectroscopically inactive. A favored candidate is primordial (nonradiogenic) argon (Ar). While estimates of the necessary argon mixing ratio are reasonable, the question of whether or not argon exists is still unanswered.

The estimate of Titan's surface temperature, believed to be approximately 94 K, is somewhat controversial and could be as high as 101 K (Lellouch et al., 1989). Titan has a minor greenhouse warming caused primarily by pressure-induced opacity of N_2, CH_4, and H_2 (McKay et al., 1991). The atmospheric temperature declines to a minimum of 71 K at the tropopause (pressure = 128 mbar; height = 42 kilometers or km). It then rises rapidly to between 160 and 170 K until levels of about 200 km are reached, at which point it levels off to a value of about 186 ± 20 K at the exobase (1,600 km). The exobase refers to the base of the exosphere, which is defined as the uppermost region of a neutral planetary atmosphere. Titan's effective temperature, the temperature a perfect absorber would emit were its flux integrated over wavelength and surface area, is 84 K.

The tropospheric temperatures are at least 5 K from the condensation temperature of N_2 so it is unlikely that nitrogen clouds can form. The condensation of CH_4 clouds, however, is possible if the CH_4 mixing ratio below them exceeds 1.5 percent by volume. Accurate determination of the methane abundance is therefore very important for understanding the thermal structure of the atmosphere and possible surface states. The relevant factors which constrain the methane mole fraction and methane condensation near the surface are discussed in Lindal et al. (1983), Eshleman et al. (1983), and Flasar (1983) in detail. An upper limit for the methane abundance near the surface of 3 percent is derived by Eshleman et al. (1983) for the scenario in which methane condensation occurs at 15 km. Flasar's analysis leads to a surface pressure for methane of 0.11 bars and methane condensation commencing at about 3 km above the surface. IRIS data analyzed by Lindal et al. (1983) place limits on the methane mixing ratio between 1 and 2 percent at the tropopause. The above results provide constraints on atmospheric thermal structure and surface models (Lunine et al., 1989).

An interesting characteristic of Titan's atmosphere is its permanent, noncyclic evolution. Methane is permanently dissociated as a result of three basic processes which cause the heavy molecule dissociation products to: (1) form in the stratosphere; (2) fall down to the temperature minimum that acts as a cold trap where they condense and form aerosols; and then (3) fall to the surface. There must, therefore, be a source or reservoir of methane at or below the surface that resupplies it to the atmosphere.

Other important constituents, in order of abundance, include molecular hydrogen (H_2) and hydrocarbons such as ethane (C_2H_6), propane (C_3H_8), and acetylene (C_2H_2). Carbon monoxide (CO) and carbon dioxide (CO_2) have been detected from Earth at levels of 60–150 parts per million (ppm) and 1.5 parts per billion (ppb), respectively. An upper limit of less than 1 percent neon measured in the Titan atmosphere implies that the atmosphere cannot be captured (Owen, 1987).

Titan's upper atmosphere was studied in great detail by the *Voyager 1* UVS instrument. The temperature of 186 K is sufficient to assure the escape of hydrogen and helium. This process is rate-limited by diffusion from below. The observed mixing ratio for H_2 is compatible with this rate and with its proposed source, namely photolysis of methane.

The escaping gas is likely to be a mix of H and H_2 and it goes into orbit about Saturn to form a torus. The size and extent of this torus were estimated to range from 8 to 25 R_S (Saturn radii) with a vertical extent of 14 R_S. It is believed to envelope Titan's orbit at 20 R_S.

Titan's Surface: Are There Oceans?

Given the presumed conditions of temperature and pressure on Titan's surface, methane should be liquid. The temperature profile derived from the radio-occultation experiment corresponds to a dry adiabatic gradient, making the existence of a pure methane ocean unlikely. Lunine et al. (1983) suggested an ethane ocean with significant amounts of methane and other trace constituents as an alternative. The composition of any liquid ocean on Titan's surface is highly model- (hypothesis-) dependent. For example, the uncertainty in surface temperature can lead to extremely different ocean depths (Lellouch et al., 1989), and the amount of the Titan surface assumed to be covered can significantly affect the extent and composition of a Titan ocean.

Photochemical models (Yung et al., 1984) are constrained to produce stratospheric column abundances of hydrocarbons that are consistent with *Voyager*'s IRIS measurements. There is, however, uncertainty in our understanding of both aerosols and simpler hydrocarbons. A model in which most of the products of methane photolysis reach the surface as aerosols, rather than as liquids (Lunine et al., 1989), is also a possibility.

The only direct probes to the surface are by microwave radio emission and radar. While the results from these measurements are uncertain (Muhleman et al., 1990; Wagener et al., 1988), Muhleman et al. (1990) interpret their recent radar measurements as ruling out the possibility of a deep methane or methane-ethane ocean on the surface. Clearly, however, the amount of methane present in the atmosphere or at the surface is dependent on the amount of uncertainty in the thermal profile and surface temperature.

Titan's Origin

There are three basic scenarios for Titan's origin: cold accretion, a scenario in which there is little or no temperature rise during the formation process; hot accretion in the absence of a dense gas; and hot accretion in the presence of a thick, gravitationally bound primordial atmosphere. In the latter scenario the energy of accretion has to be eliminated by convection through this atmosphere.

The first scenario, cold accretion, requires an absence of gas (atmosphere) and it must occur over a long timescale in order to satisfy the criterion of little or no temperature rise during formation. Since the accretion process is generally expected to occur quite rapidly, and since the impact processes involved generally result in significant heat being retained, the cold accretion scenario is unlikely.

The second and third scenarios both have difficulty explaining the significant atmosphere of the present day Titan. In the gas-rich environment, however, the massive primordial atmosphere can provide an environment which is warm enough for the ammonia (NH_3) to convert to N_2. In the gas-free case no atmosphere is likely to form unless highly volatile ices of CH_4, N–2, or CO are included in the accretion material.

The Cassini Mission

Titan's atmosphere and surface will be studied in more detail by the unmanned Cassini mission scheduled to arrive on Titan in the year 2003. This mission will go to Saturn (distance from Earth = 9.546 astronomical unit or AU, where 1 AU is equal to the distance between Earth and the Sun), using gravity assists from Earth and Jupiter. A probe named Huygens, named after Titan's discoverer, will be supplied by ESA and dropped into Titan's atmosphere. It will take the probe approximately three hours to parachute to Titan's surface. If the probe survives the landing, then it will continue to transmit valuable data for about another thirty minutes. Cassini will also carry a Titan Radar Mapper which it will use to map the surface of Titan during its numerous flybys of the moon. A summary of a preliminary payload for the probe is given by Lunine (1990).

Bibliography

BARBATO, J. P., and E. A. AYER. *Atmospheres. A View of the Gaseous Envelopes Surrounding Members of Our Solar System.* New York, 1981.

BAUER, S. J. "Titan's Atmosphere and Atmospheric Evolution." *Advances in Space Research* 7 (1987):565–569.

ESHLEMAN, V. R., G. F. LINDAL, and G. L. TYLER. "Is Titan Wet or Dry?" *Science* 221 (1983):53–55.

Flasar, F. M. "Oceans on Titan?" *Science* 221 (1983):55–57.

Hunten, D. M. "Titan's Atmosphere and Surface." In *Planetary Satellites*, ed. J. A. Burns. Tucson, AZ, 1977.

Hunten, D. M., M. G. Tomasko, F. M. Flasar, R. E. Samuelson, D. F. Strobel, and D. J. Stevenson. "Titan." In *Saturn*, eds. T. Gehrels and M. S. Matthews. Tucson, AZ, 1984.

Kuiper, G. P. "Titan: A Satellite with an Atmosphere." *Astrophysical Journal* 62 (1944):245.

———. "Planetary Atmospheres and Their Origin." In *The Atmospheres of the Earth and Planets*, ed. G. P. Kuiper. Chicago, 1952.

Lellouch, E., A. Coustenis, F. Raulin, N. Dubulos, and C. Frere. "Titan Atmosphere Temperature Profile: A Reanalysis of *Voyager 1* Radio-Occultation and IRIS 7.7 Micron Data." *Icarus* 79 (1989):328–349.

Lewis, J. S. "Satellites of the Outer Planets: Their Physical and Chemical Nature." *Icarus* 15 (1971):174–185.

Lindal, G. F., G. E. Wood, H. B. Hotz, D. N. Sweetman, V. R. Eshleman, and G. L. Tyler. "The Atmosphere of Titan: An Analysis of the *Voyager 1* Radio Occultation Measurements." *Icarus* 53 (1983):348–363.

Lunine, J. J. "Titan." *Advances in Space Research* 10 (1990):1137–1144.

Lunine, J. J., D. J. Stevenson, and Y. L. Yung. "Ethane Ocean on Titan." *Science* 222 (1983):L73.

Lunine, J. J., S. K. Atreya, and J. B. Pollack. "Present State and Chemical Evolution of the Atmospheres of Titan, Triton and Pluto." In *Origin and Evolution of Planetary Satellite Atmospheres*, eds. S. K. Atreya, J. B. Pollack, and M. S. Matthews. Tucson, AZ, 1989.

McKay, C. P., J. B. Pollack, and R. Courtin. "The Greenhouse and Anti-Greenhouse Effects on Titan." *Science* 253 (1991):1118–1121.

Owen, T. "How Primitive Are the Gases in Titan's Atmosphere?" *Advances in Space Research* 7 (1987):551–554.

———. "Titan." In *The New Solar System*, eds. J. K. Beatty and A. Chaikan. Cambridge, MA, 1990.

Strobel, D. F., and D. E. Shemansky. "EUV Emission from Titan's Upper Atmosphere: *Voyager 1* Encounter." *Journal Geophysical Research* 87 (1982):1361–1368.

Wagener, R., T. Owen, W. Jaffe, and J. Caldwell. "The Surface Emissivity of Titan at 2 cm." *Bulletin of the American Astronomical Society* 20 (1989):843.

Yung, Y. L. "An Update of Nitrile Photochemistry on Titan." *Icarus* 72 (1987):468–472.

Yung, Y. L., M. Allen, and J. P. Pinto. "Photochemistry of the Atmosphere of Titan: Comparison Between Model and Observations." *Astrophysical Journal Supplement* 55 (1984):465–506.

CINDY CUNNINGHAM

TITANIUM DEPOSITS

Economic deposits of titanium (Ti) contain the minerals rutile (TiO_2), ilmenite ($FeTiO_3$), and/or altered ilmenite. The deposit types vary widely; the two most important are placer deposits formed along former shorelines and ilmenite-rich portions of igneous masses called anorthosite-ferrodiorite massifs.

Uses and Refining

The commercial uses of titanium are diverse. The highest-value use, as titanium metal, consumes only a small fraction of mineral production. The more voluminous use is as pigment.

Titanium metal is used in applications that take advantage of its high strength-to-weight ratio (for example, in aircraft) or its resistance to corrosion and heat (as in desalinization equipment). However, titanium metal is difficult to purify and to work and is therefore expensive.

Titanium-based pigments take advantage of the high index of refraction (2.6 to 2.9) and the resulting extreme whiteness of micro-crystalline rutile, which impart great opacity to titanium-based paints.

Refining of mineral concentrates can be by chlorination (to form $TiCl_4$) or by digestion in sulfuric acid. Two auxiliary processes are smelting to form a high-TiO_2 slag and leaching of ilmenite to form synthetic rutile.

Deposit Types

The economic geology of titanium differs from that of most metals in that the deposit value is less a function of elemental enrichment than of ore mineralogy. An enrichment of titanium in the minerals sphene or titaniferous magnetite currently has no economic value. In contrast, an occurrence of rutile can be an economic deposit at less than average crustal abundance of titanium if all other factors are optimal (as they are in some Australian shoreline placer deposits). Accordingly, economic deposits of titanium minerals are those that provide some mineral enrichment of ilmenite, altered ilmenite, and/or rutile. Production from anatase ore is still on an experimental basis. Only a few deposit types, described in greater detail by Force (1991), are economic.

Shoreline Placers. More than half the world's titanium mineral supply comes from shoreline placer deposits. Production is from geologically young deposits (Miocene through Holocene) having little overburden because such deposits are less cemented and easier to find. However, few active beaches are mined because of competing land uses. Modern mining methods require large deposits (about 1 million tons of saleable concentrate). Economic mineral grades for little-altered ilmenite are typically about 4 percent (by weight), whereas those for rutile are 1 percent or less. Other minerals recovered from the same deposits include zircon, monazite, and the aluminosilicate minerals—kyanite, andalusite, or sillimanite. The countries where such deposits are currently mined are Australia, the United States, South Africa, India, and Ukraine (Table 1).

Concentration in these shoreline placers is a result of hydraulic processes in the wave swash zone that sort mineral grains by their specific gravities (and those of rutile and unaltered ilmenite are 4.6 and 4.9, respectively). (These concentration processes are described in PLACER DEPOSITS.) The individual enrichments of dense minerals are small; the requisite volumes that permit mining commonly represent closely nested storm deposits, or eolian removal of enriched material for storage in dunes.

Factors other than placer enrichment that determine the potential of a shoreline sand as a placer ore of titanium minerals include: (1) source area (rutile and ilmenite tend to come from high-grade metamorphic terranes); (2) weathering history (rutile and ilmenite are residually enriched by intense weathering); and (3) transport history (little-diluted supply of altered ilmenite and/or rutile).

Table 1. World Production of Titanium Minerals in 1993 in Thousand Metric Tons on a Contained-TiO_2 Basis

	Rutile	*Ilmenite, Altered Ilmenite, and Derived Ti-Slag*
Australia	200	850
Brazil		40
Canada		600
China		75
India	10	160
Malaysia		190
Norway		300
Sierra Leone	140	
South Africa	70	800
Sri Lanka		35
United States	145	350
Former USSR		150

Modified from Slatnick, 1994.

Igneous Ilmenite Deposits. Second among deposit types in titanium mineral supply are igneous ilmenite deposits. These occur in close association with the ferrodioritic portions of andesine anorthosite-ferrodiorite massifs (i.e., nonlayered masses of plagioclase-rock and Fe-, Ti-rich diorite), 1,700 to 900 million years (Ma) in age. Modern mining methods require large deposits, typically containing 30 percent by volume or more of separable ilmenite, or larger percentages of oxide-mineral mixtures that can be smelted. Magnetite and/or apatite are recovered or recoverable from some deposits. The countries where such deposits are mined are Canada and Norway (Table 1); the United States has large resources in this deposit type also.

Ore tends to occur in disseminated and/or cumulate-like layered deposits in ferrodiorite, and as dikes in anorthosite. Ore textures have suggested to some authors that titanium oxide and some other chemical components became immiscible during the cooling of ferrodioritic melts, and accumulated mechanically at the bases of ferrodiorite bodies in the host rocks. This opinion is not unanimous.

Other Types. Some current or recent production of titanium minerals comes from fluvial placer deposits (Sierra Leone) and deeply weathered alkalic igneous rocks rich in titanium (Brazil). Other resource possibilities include rutile in eclogites (i.e., ultra-high-pressure metamorphic rock) and submerged placer deposits now on continental shelves.

Bibliography

FORCE, E. R. *Geology of Titanium-mineral Deposits.* Geological Society of America Special Paper 259. Boulder, CO, 1991.

SLATNICK, J. A. *The Availability of Titanium in Market Economy Countries.* U.S. Bureau of Mines Information Circular 9413. Washington, DC, 1994.

ERIC R. FORCE

TOPOGRAPHY

See Caves and Karst Topography

TRACE ELEMENTS

The abundance of an element in a mineral, rock, ore deposit, meteorite, river, hot spring, and so on is expressed in one of the units of concentration (e.g., weight percent, moles per litre, etc.), sometimes normalized to a standard. The major elements are in the range approximately 1 to 100 wt. percent, the minor elements 0.1 to 1 percent, and the trace elements at lower concentrations, often expressed as parts per million (ppm) or grams per ton (g/t).

In the early days of the twentieth century trace elements were usually referred to as the "rare elements," as opposed to the "common" ones. But rarity should not be construed as the opposite of familiarity, for some of the rarest elements are familiar to all (gold, platinum, mercury), while some of the least rare are unknown to the person in the street (rubidium, argon, lanthanum). The most familiar are those which form natural simple chemical compounds that can be readily identified, whereas the least familiar occur in a so-called dispersed or camouflaged state as dilute components in compounds or mixtures of common elements.

Because of their rarity, many trace elements were discovered late and knowledge of their geochemical behavior and distribution developed slowly. By the mid-twentieth century, however, few remained undocumented: the major research contributions came from the work of VICTOR MORITZ GOLDSCHMIDT and his collaborators. At the present time, every element can be measured at concentration levels down to a few ppm under favorable circumstances, and many to a few parts per billion (10^{-3} ppm or ppb): measurement with high precision, however, often is still a problem.

Trace elements are interesting to scientists for several reasons. The first is perhaps commercial value, as exemplified by most of the familiar metals such as lead (Pb), copper (Cu), zinc (Zn), tungsten (W), nickel (Ni), gold (Au), silver (Ag)—but not iron (Fe) or aluminum (Al), which are major elements in most rocks. In common rocks these metals occur at parts-per-million levels or less, but they can be exploited from ore deposits where natural processes have concentrated them. Another interest arises from the biochemical role played by some elements. For example, small concentrations of cobalt (Co) in pasture are essential to the health of grazing animals such as sheep, and if the concentration in the soil is less than about 1 ppm Co, the animals will become sick; however, as with most heavy metals, too much Co will induce a toxic response. Another trace element of biochemical importance is iodine, small amounts of which are essential to mankind; deficiency leads to goiter.

Another reason why trace elements are studied intensively is for their use in helping solve geological problems, that is, interpreting a past process from present evidence or post-diction. In processes such as magma crystallization, evaporite formation, sediment accumulation, and rock melting (anatexis), the behavior of major elements is dictated by availability and by external controls such as temperature (T), pressure (P), water supply, and so forth, but trace elements are extraneous to the process and their distribution among minerals may provide useful clues to what took place. For example, certain rhyolite lavas may be products either of partial melting of continental crust or of fractional crystallization of a basic magma from the mantle. The major element composition will not help much in choosing between these alternatives, but if the abundances of the rare earth elements (REE) are known, it may be possible to choose by comparing the relative abundances of samarium (Sm) and europium (Eu); even more useful might be the abundance of ^{87}Sr, relative to its sister isotope ^{86}Sr (*see* ISOTOPE TRACERS, RADIOGENIC).

Trace Elements in Crystal-Liquid Equilibria

To understand the significance of trace elements it is necessary to understand their distribution in the various earth systems, that is, the lithosphere, hydrosphere, biosphere, and atmosphere. The last of these may be disregarded because its only components are gases and particulates, which are not important for most trace elements; the participation of trace elements in the biosphere is important, as already mentioned, but most of the processes are short-term effects, related to life-cycles of organisms, and will not be discussed here.

The distribution of trace elements in the lithosphere and hydrosphere is dominated largely by solution and adsorption chemistry. Beginning with the igneous and metamorphic rocks, the most important process is crystallization of minerals, forming from a liquid phase that might be a magma or a solution. The behavior of an element in these circumstances is governed by its partition coefficient, D^{i-l}:

$$D^{i-l} = \frac{C^i}{C^l} \qquad (1)$$

where i and l refer to a mineral and the medium (liquid) from which it is growing, and C indicates the activity of the element or its compound; in practice, the activity is usually not known and the concentration is used in its place. This definition assumes that the two phases i and l are in chemical equilibrium. If they remain in equilibrium throughout, the process is known as equilibrium or batch crystallization; if the two phases are in equilibrium only momentarily, and each crop of newly formed crystals is effectively unable to react with the remaining liquid, the process is described as fractional or Rayleigh crystallization.

Partition coefficients may vary with T and P; they may also vary as the concentrations of other elements change but, if they do not, the process is said to be ideal or Henry's law crystallization. Such behavior is more common with trace than with major elements.

If several minerals are crystallizing simultaneously, in constant relative proportions $P_1, P_2, \ldots, P_i, \ldots, P_n$, then the behavior of the element is determined by the bulk partition or whole-rock coefficient, D^{WR}, where

$$D^{WR} = P_1D^{1-l} + P_2D^{2-l} + \ldots P_nD^{n-l} \qquad (2)$$

The value thus obtained may be used to calculate the concentration of the element in the liquid and solids as crystallization proceeds. This is most easily expressed using F, the fraction of liquid remaining. For batch crystallization the expression is

$$C^L = \frac{C_0}{F + D(1 - F)} \qquad (3)$$

and for fractional crystallization the expression is

$$C^l = C_0F^{(D-1)} \qquad (4)$$

where C_0 is the initial concentration of the element.

The behavior of the element thus depends on the magnitude of D^{WR}. For example, consider a basic magma crystallizing to form a basalt in the mineral proportions given in Table 1 (partition coefficients from Allègre et al.).

The value of D^{WR} may be used in equation 3 or 4 to calculate the changing liquid composition from an initially completely liquid magma (F = 1.0) as crystallization proceeds toward complete solidification (F = 0.0).

The discussion has so far been concerned with the freezing and crystallization of liquids to form minerals. At depth within Earth's crust and mantle, solid rocks may undergo melting to generate magma. The behavior of a trace element depends on the principles just discussed and, although it is not true that melting is the exact opposite of freezing, the concentration changes are expressed in equations symmetrical with equations 3 and 4, although somewhat more complex, because the melting solid is not a homogeneous phase like a crystallizing liquid (details may be found in Henderson, 1982).

The changing course of trace element concentration in the magma or liquid during fractional crystallization is shown in Figure 1 for different values of D. Three kinds of behavior are possible for both equilibrium and fractional crystallization. When $D \approx 1$, the element concentration does not change much as the magna evolves. When $D > 1$, the element accumulates in early mineral fractions and later liquids are depleted; the element is said to be compatible. When $D < 1$, the opposite takes

Table 1. D^{WR} for Ni in a Basalt Magma

Mineral	*Pi*	D^{i-l}	PiD^{i-l}
Olivine	0.25	13	3.25
Orthopyroxene	0.15	6.6	0.99
Clinopyroxene	0.25	4	1.00
Plagioclase	0.30	0.26	0.08
Magnetite	0.04	12	0.48
Apatite	0.01	0	0
D^{WR}			5.80

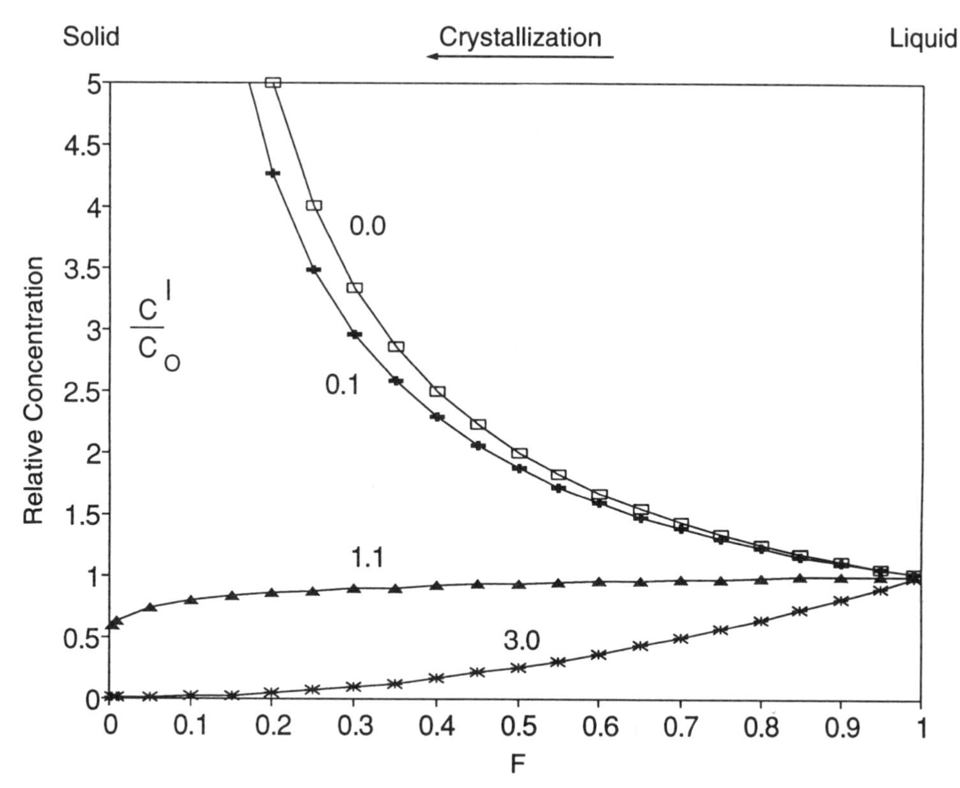

Figure 1. Rayleigh fractionation curves showing how relative concentration of a trace element changes with F, for four values of D.

place and the element is incompatible: Figure 1 makes it clear that if D is of the order of 0.1, the behavior of the element is essentially the same as if it all stays in the liquid and none goes into the minerals (D = 0).

It should be noted, however, that compatible and incompatible behavior depends on what minerals are crystallizing: in Table 1 it is seen that Ni has a partition coefficient >1 for olivine but <1 for plagioclase, and the mineral proportions are such that the whole-rock coefficient determines that Ni will behave compatibly. But if the plagioclase proportion were 0.90, as might be the case for an anorthosite magma, the Ni would be incompatible and its concentration would increase in residual liquids.

Although incompatible behavior is thus not an invariable property, the trace elements behave in common igneous rocks with some predictability and are frequently classed into two main groups:

LILE, or large ion lithophile elements, e.g., Cs^{2+}
HFSE, or high field strength elements, e.g., Ta^{5+}

Measurement of Partition Coefficients

It has long been believed that knowledge of partition coefficients would help solve many petrological problems, making use of the theory just described, with measured D-values; these are obtained in two principal ways.

The first is to use rocks with adjacent minerals that have crystallized in equilibrium, for example, a garnet-biotite gneiss, or where a mineral has crystallized in equilibrium with a magma which then froze to a glass, for example, plagioclase crystals in an obsidian. The minerals and glass are usually separated mechanically and analyzed individually for the element of interest, although separation is unnecessary if an electron microprobe (*see* GEOCHEMICAL TECHNIQUES) can be used; the ratio of the concentrations measures the D-value.

The second method is experimental, carried out by making a synthetic mineral react to equilibrium, usually at high temperature and high pressure, with the appropriate liquid or silicate melt containing a known amount of the trace element (*see* PET-

ROLOGIC TECHNIQUES). After freezing, the components must of course be analyzed separately.

Although simple in theory, there are many practical difficulties with both these methods, the principal one being the difficulty of achieving equilibrium. Also, other elements present cause major variations in D-values, for example, $D^{olivine\text{-}glass}$ for Ni was found by one research group to vary between 4 and 24 as the proportions of SiO_2 and MgO in the glass varied between the common values in basalt and andesite. During the 1980s many careful studies have confirmed such striking variations and have provided explanations for them.

Trace Elements in Minerals

The magnitude of a partition coefficient is determined by the bonding behavior of the element. The value of D^{i-l} for olivine in Table 1 is 13, which means that Ni has a much stronger preference for olivine rather than silicate melt.

Most samples of olivine accordingly contain some Ni, though feldspar never does; by contrast feldspar usually contains Ba, which is not the case for olivine. The reason why one trace element accumulates in mineral A but not in mineral B was only discovered in the early twentieth century as a consequence of the elucidation of chemical structures using X-ray diffraction, and is really quite simple.

Most common minerals, such as silicates and carbonates, have structures consisting of different arrangements of the oxygen anion O^{2-}; this has a large radius (1.40 Å) and occupies over 90 percent of the mineral volume. Regularly spaced cavities or "holes" occur in the framework and are the sites of the metallic elements. These sites are of different sizes, depending on the configuration of the O^{2-} ions; a site of a particular size will accept any metallic ion of that size. This is the basis of trace element geochemistry. For example, there are two metal sites in olivine (called M1 and M2) and both of these accept ions of radius 0.5–0.8 Å; metallic ions in this range include Mg^{2+}, Fe^{2+}, and Ni^{2+}, but does not include Ba^{2+}, which is three times larger. Orthoclase feldspar, by contrast, has a metal site that only accepts the large alkali and alkaline earth ions.

The same principle applies to other families of minerals, such as the sulphides, the framework in that case being built of the anion S^{2-}.

This description of the bonding of metallic elements in minerals is somewhat oversimplified but explains many of the major phenomena of geochemistry.

Trace Elements in Sedimentary Rocks

In addition to fragments of preexisting rocks, most sediments contain new minerals formed under ambient P-T conditions in the presence of water, often during diagenesis, as the soft sediment is transformed into rock. Trace elements may be trapped by crystallization or by adsorption. The first of these proceeds by the mechanisms discussed above, and the liquid is a watery solution or brine; thus, for example, rock-salt (NaCl) forms by evaporation of brines and always contains trace quantities of Rb^{1+} and Br^{1-}, substituting for the Na^{1+} and Cl^{1-}, respectively. Similarly calcite ($CaCO_3$), formed by many invertebrate organisms such as corals, contains small amounts of Pb, Zn, Sr, and other elements of about the same size as Ca.

Fine-grained particles, floating in estuaries and in the sea, have unsatisfied surface charges and will readily absorb dissolved metals, even at very low concentrations; the kinds of elements scavenged in this way depend on the oxidizing/reducing conditions. The adsorbed metals are not strongly held but commonly such particles recrystallize to form clays and incorporate the trace elements; many rare elements thus occur in shales, some sufficiently concentrated to constitute ore deposits (e.g., uranium [U] and phorphorus [P]) (*see* MINERALOGY, HISTORY OF; DIAGENESIS).

Bibliography

ALLÈGRE, C. J., et al. "Systematic Use of Trace Element in Igneous Process. Part 1: Fractional Crystallization Processes in Volcanic Suites." *Contributions to Mineralogy and Petrology* 60 (1977):57–75.

HENDERSON P. *Inorganic Geochemistry*, 1st ed. New York, 1982.

MASON, B. *Victor Moritz Goldschmidt: Father of Modern Geochemistry*. The Geochemical Society, Special Publication No. 4. San Antonio, TX, 1992.

DENIS M. SHAW

TRACE FOSSILS

There is a sense of morbidity concerning fossils. They represent corpses, fragments of skeletons, death. Not so, however, with trace fossils: these are structures produced by the life activities of animals and plants. Central within the definition of trace fossils (or ichnofossils) are the structures produced by animals burrowing within or walking over sedimentary substrates: that is, biogenic sediment structures. As such, the study of trace fossils (ichnology) bridges between paleontology and sedimentology. Also included are borings and scrapings of organisms in hard substrates (*see* BIOEROSION), and fossil excrement or coprolites. Some geologists include rooting structures of plants and stromatolites produced by microbes.

The process of churning sediment by burrowing animals is known as bioturbation. Animals live within sediment for a number of reasons. Important among these considerations are (1) protection from predation or physical disturbance; (2) the ability to lay in ambush for passing prey; (3) the advantage of burrowing two openings to the sea floor, offering efficient canalization of respiration water; (4) utilizing water current for feeding on organic matter and organisms in suspension; (5) ingesting the substrate sediment for animals that are deposit feeders; and (6) culturing microbes for food from the organic content of the substrate.

These different activities demand different designs of burrow and, consequently, the morphology of the trace fossil offers evidence for the type of behavior that created it. Indeed, Seilacher (1967) described trace fossils as "fossil behavior" and organized them in an ethological classification. Seilacher's classification has been extended, and there are now nine ethological "classes" of trace fossils.

Domichnia are the simple structures that result from stationary burrows constructed for living purposes. Corresponding behavior may be suspension feeding or ambush predation. Examples are the vertical tubular burrows known as *Skolithos linearis* and the U-tube *Arenicolites* ichnospecies.

Cubichnia are resting traces, made by animals digging shallowly into the substrate and remaining stationary for a time. Five-rayed *cubichnia* called *Asteriacites* ichnospecies are produced by starfish and brittle-stars, and almond-shaped depressions (*Lockeia amygdaloides*) are created by bivalves.

Repichnia are locomotory trace fossils simply indicating travel from one place to another. *Repichnia* range from structures caused by burrowing sea urchins, producing *Scolicia prisca* (Figure 1), to trackways trampled by dinosaurs.

Fodinichnia include a combination of two activities: more or less stationary habitation, as well as deposit feeding. Such trace fossils are quite varied; they may be branched, open tunnel systems (*Thalassinoides suevicus*), complicated spiral structures (*Zoophycos* ichnospecies), or groups of fingerlike probes (*Dactyloidites ottoi*; Figure 2).

Pascichnia represent locomotion together with deposit feeding. The animals that create these trace fossils are "grazers" that produce from sim-

Figure 1. *Scolicia prisca*, a trace fossil produced by a burrowing echinoid, viewed from below. Eocene deep marine turbidites, Spain. Natural size.

Figure 2. *Dactyloidites ottoi*, produced by an unknown animal as it exploited the sediment for food by a series of probes. Viewed from above. Cretaceous shallow marine sands, West Greenland. Natural size.

ple, winding traces (*Planolites montanus*) to complicated, regularly meandering structures (*Nereites* ichnospecies).

Praedichnia are rare, the product of acts of predation. The most common examples are the drillholes in mollusk prey shells by carnivorous mollusks (*Oichnus simplex*). These belong to the category of borings (*see* BIOEROSION).

Agrichnia are systems of burrows that are more complicated than *pascichnia*; they are considered open, meandering, or network structures that function as traps for wandering prey or as farms for culturing microbes for food. *Paleodictyon minimum* is an example of a network farm (Figure 3).

Fugichnia are disturbances within sediment produced by sudden escape behavior. The most common form is the reaction of an animal to sudden burial under a blanket of sediment. The buried animal seeks upward to the new seafloor, not always successfully.

Equilibrichnia are produced by animals that inch upward during periods of deposition, or downward during erosion, thereby keeping pace with the movements of the seafloor. A characteristic U-tube having such vertical adjustment structures of this kind is called *Diplocraterion parallellum* (Figure 4).

While trace fossils can commonly be identified with a behavior pattern, it is much less easy to identify the nature of the architect. Closely similar structures can be produced by radically different animals showing the same behavior. In particular,

Figure 3. *Paleodictyon* ichnospecies, a deep marine network burrow that functions as a microbe farm. It is a shallow tier structure and has been cut through subsequently by a burrow related to *Thalassinoides*, produced at a deeper tier. Viewed from below. Cretaceous turbidites, Poland. Natural size.

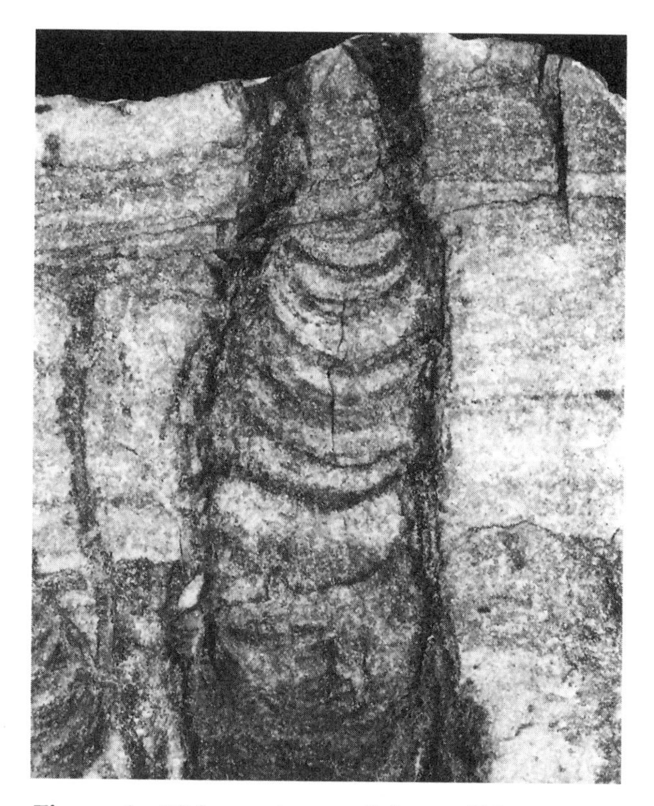

Figure 4. *Diplocraterion parallelum,* a U-burrow that has descended into the seafloor by stages with the growth of the burrower. The two funnel-shaped openings at the seafloor are visible at the top of the figure. Lateral view. Cambrian shallow marine, Norway. Natural size.

subtle variation in the consistency of the substrate (its grainsize, organic matter content, degree of dewatering) plays an important role in determining the types of behavior available to its inhabitants.

Trace fossils are classified with a binomial nomenclature which resembles that of the organisms that produce them. The genus-rank term is called ichnogenus (abbreviated ichnogen. or igen.) and the species-rank term ichnospecies (ichnosp. or isp.). The names of trace fossils (ichnotaxa) and those of tracemakers (biological taxa) are conceptually different and should be segregated carefully. Ichnogeneric names commonly carry the suffix "-ichnus" as a distinguishing feature.

Behavior patterns have evolved more slowly than the species of the trace-making animal. This means that recurring trace fossil assemblages are more stable and may be compared over longer intervals of time. This allows us to see common features in, say, Paleozoic and Cenozoic environments, on the basis of animal behavior, despite the fundamental biological differences in the faunas of these eras.

Trace fossils have the advantage over body fossils (skeletons of actual organisms) in being essentially in situ structures. A reworked trace fossil is a rarity and is easily identified as such. The bodies of echinoids can easily be transported from their living environment, whereas the burrowing traces of echinoids remain in place.

In contrast to body fossil taphonomy, however, is the mutual destructiveness of trace fossils. More deeply digging animals in a community burrow through and obliterate the structures produced by more shallowly digging species. This phenomenon is known as tiering: the endobenthic community subdivides the habitat in a series of horizontal niches. Different behavior patterns are characteristic of these tiers according to the rigorous vertical zonation of the substrate. Oxygen tension, content of organic matter, microbial zonation, and distance from the seafloor are all factors that affect descent into the sediment. And because the activity in the deeper tiers destroys structures produced in shallower tiers, the deep traces have a better preservation potential than the shallow. In this way, the preserved assemblage of trace fossils emphasizes deeper structures at the expense of shallow. The structure imposed on sedimentary rocks by bioturbation is known as an ichnofabric. There are several features of the ichnofabric that are of value to the geologist as paleoenvironmental indicators. These include: (1) the quantity of bioturbation, that is, the amount of disturbance of the primary structure of the sediment that has taken place. This gives us information about the relative rates of the processes of bioturbation and deposition. (2) The diversity of trace fossils in an assemblage or ichnofabric, which reveals the maturity of the burrowing community in terms of behavior. From this we learn something of the biological aspect of the depositional environment. In a stable, predictable environment, a highly diverse, mature community can develop. In a stressful, unpredictable setting, however, low-diversity pioneer communities prevail.

The large majority of trace fossil studies deal with marine sediments. Marine endobenthic communities have become far more sophisticated than brackish or nonmarine ones. The reason for this is the generally greater continuity of the marine environment and its relative predictability, in contrast to restricted lacustrine, fluviatile, or estuarine environments. Nevertheless, characteristic trace fossil assemblages are now recognized in eolian settings, as well as in lakes and streams. Terrestrial trace fossils are dominated by footprints and trackways of tetrapod vertebrates, attention focusing particularly on dinosaur tracks.

Bibliography

Bromley, R. G. *Trace Fossils: Biology and Taphonomy.* London, 1990.

Ekdale, A. A., R. G. Bromley, and S. G. Pemberton. "Ichnology. The Use of Trace Fossils in Sedimentology and Stratigraphy." *SEPM, Society for Sedimentary Geology, Short Course Notes 15.* Tulsa, OK, 1984.

Maples, C. G., and R. R. West, eds. "Trace Fossils." *Short Courses in Paleontology 5.* Knoxville, TN, 1992.

Pemberton, S. G., J. A. MacEachern, and R. W. Frey. "Trace Fossil Facies Models: Environmental and Allostratigraphic Significance." In *Facies Models: Response to Sea-level Changes,* eds. R. G. Walker and N. P. James. St. John's, Newfoundland, 1992.

Seilacher, A. "Fossil Behavior." *Scientific American* 217 (1967):72–80.

RICHARD BROMLEY

TRITON

Triton, the largest of Neptune's moons, was discovered in 1846 by W. Lassell, shortly after the discovery of Neptune. Prior to the 1989 *Voyager 2* spacecraft encounter with Neptune, very little was known about Triton beyond a few significant observations from Earth of its orbit and spectrum. The *Voyager* encounter advanced our picture of Triton from a mere point of light in a telescope to an exotic world with unusual geologic processes, active geysers, an atmosphere, and seasons. Figure 1 shows a mosaic of high-resolution *Voyager* images. The surface geology is clearly unique, unlike that of any other satellite in the solar system.

Triton Observations Before *Voyager*

Triton's orbit is one of the well-established observations from Earth-based telescopes. Triton travels in a retrograde (clockwise) circular orbit around Neptune. This retrograde orbit is uncommon for a large satellite and led observers to speculate that Triton may not have formed originally with Neptune, but may have been captured gravitationally by Neptune at some time after the planet's formation. If indeed Triton was captured, its initial orbit would have been elliptical. Tidal energy would have worked to circularize Triton's orbit over a time span on the order of 500 million years (Ma).

Triton's current orbit gives its seasons unusual attributes. The period of Neptune's orbit around the Sun is 165 years and its obliquity (the angle between the spin axis of a body and the perpendicular to its orbital plane) is 28.8°. Triton's orbital plane is inclined 20° to Neptune's equatorial plane and precesses about Neptune's pole approximately every 688 (+/−40) years. As Figure 2 shows, the combination of these two effects causes the subsolar point on Triton to trace a latitudinal path that repeats approximately every 165 × 688 years. The subsolar latitude at solstice varies from +/−10° to +/−52°. (As a comparison, on Earth the Sun's latitudinal path moves sedately between −23.5° and +23.5° every year.) At the time of *Voyager*'s encounter the subsolar latitude was −45°, the furthest south the Sun has been in 470 years.

The detection of methane by Cruikshank and Sylvaggio (1979), and the detection of nitrogen in Triton's near-infrared spectrum (Cruikshank et al., 1984), lent particular significance to Triton's strange seasons. It was postulated that polar caps would form from these volatiles, which would freeze out at Triton's cold temperatures.

Higher spectral resolution observations in the near infrared made in 1989 showed that the CH_4 spectral features were clearly due to both the gaseous and the condensed states of methane, and that the N_2 absorption feature at 2.15 microns (μm) could only be caused by a several-meter pathlength through solid N_2. There was no spectral signature of water ice, the dominant constituent of most outer planet satellites. Water ice was nonetheless expected to be a significant fraction of Triton's bulk composition; thus, the spectral data indicated that most of the surface had to be covered with at least a veneer of methane and nitro-

Figure 1. This is the highest resolution mosaic of Triton images obtained by *Voyager 2,* acquired shortly before the spacecraft's closest approach. The southern hemisphere is sunlit. The latitude of the subsolar point on Triton at the time of the *Voyager* flyby was −45°; thus, the north polar region was in darkness, making it impossible to detect a potentially forming bright, north polar cap. Visible in this mosaic of images of Triton is the latitude of the boundary thought to mark the edge of the bright, south polar cap. The equatorial region is also bright. North of the equator is a uniform, relatively dark region. Surface wind streaks are scattered across the southern hemisphere, and the two plumes are visible near the bottom of the mosaic. The "cantaloupe terrain" is seen on the west side of Triton. Evidence of cryovolcanism is apparent on the east side. This is the oldest terrain, as evidenced by the number of craters. Global lineaments are believed to be graben that are the source of much of the extruded material.

gen ices. These surface frosts and ices could mask or confuse the true composition of the underlying crust.

Viewed from an Earth-based telescope, Triton is an unresolved point of light. It could have been large and dark or small and bright, and it would have the same appearance through a telescope. Therefore, prior to *Voyager,* estimates of Triton's radius ranged from 1,100 to 3,200 kilometers (km). This uncertainty led to a corresponding uncertainty in its density—would *Voyager* find a body composed of mostly water ice, like other satellites

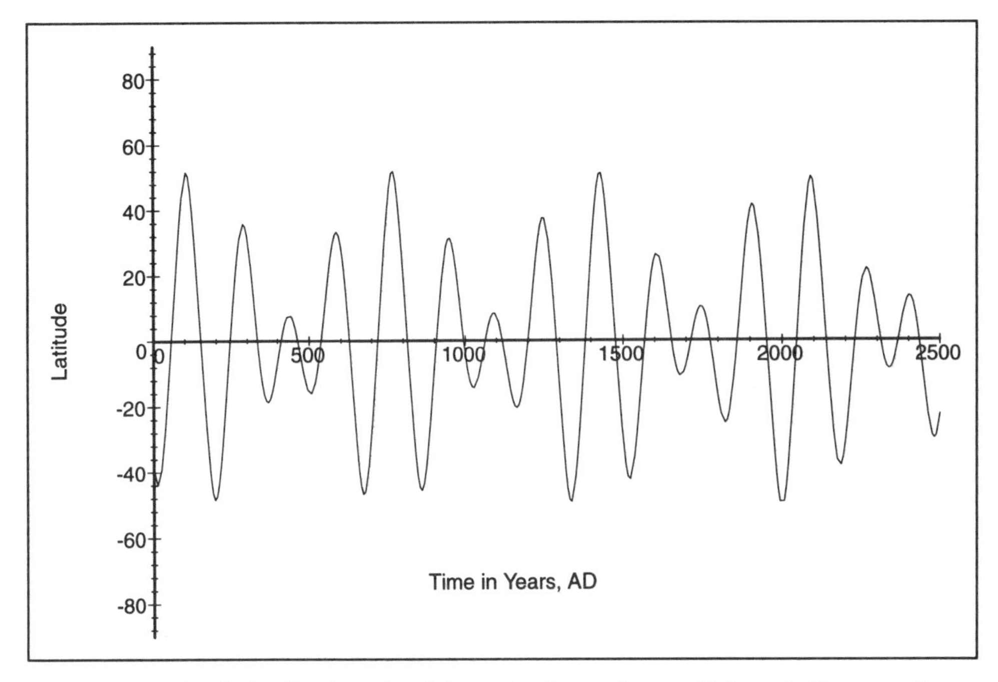

Figure 2. The latitudinal path of the subsolar point on Triton is illustrated. The path appears to repeat approximately every 660 years because 660 is an integral multiple of 165.

in the outer solar system, or one containing a substantial amount of rocky material?

The *Voyager 2* Encounter

As *Voyager 2* approached Neptune and radioed distant pictures of Triton to Earth, estimates of Triton's radius decreased as estimates of its albedo increased. Ultimately Triton's radius was determined to be 1,350 km, with a spherical albedo of 0.9. This was the cold bright extreme of the many pre-*Voyager* models.

The *Voyager* flyby of Triton yielded a wealth of new data. High-resolution images were shuttered by *Voyager*'s cameras, and the spacecraft's ultraviolet spectrometer determined that Triton's dominant atmospheric constituent is N_2, with a CH_4 mole fraction of only 2.5×10^{-6} (Broadfoot et al., 1989). The atmospheric pressure was determined to be 1.6 +/−0.3 pascals (Pa) (Tyler et al., 1989). Triton's very cold surface temperature was just barely detected by the infrared detector at 38 K +3/−4 K (Conrath et al., 1989), for a surface emissivity of 1.0. (If Triton's surface emissivity is 0.5, the temperature estimated goes up to 41 K +3/ −5 K.)

The images revealed that the northern and southern hemispheres of Triton had distinctly different appearances (Figure 1). The sunlit southern hemisphere was very bright, with complex, mottled albedo features. The visible portion of the northern hemisphere was somewhat darker, with more subdued, small-scale albedo variations. The dominant north–south contrast was attributed to the emplacement of a bright south polar cap (Smith et al., 1989). *Voyager*'s images thus seemed to confirm the predicted existence of polar caps on Triton.

Unlike most outer planet satellites, Triton has few craters, indicative of a geologically "youthful" surface. The material strength required to support the local topographic relief seen in the images is consistent with the strength of water at Triton's very cold temperature. With Triton's radius known, its density could finally be calculated. Triton's density is 2.054 grams per cubic centimeter (g/cm^3); thus, in addition to water Triton contains a substantial amount (about 70 percent mass fraction) of silicate rock.

Limb hazes and terminator clouds were immediately obvious in the images of Triton. Also apparent were dark wind streaks on the surface that were preferentially oriented to the northeast. Careful processing of images taken of the same territory with different viewing angles led to the astonishing discovery of two dark eruption plumes or "geysers" rising vertically to 8 km. At this altitude, the geyser plumes stream off to the west over 100 km.

Triton's Geology and Evolution

Lacking heavily cratered terrains, Triton's surface has clearly been resurfaced since the period of heavy bombardment that took place early in the history of the solar system. Only geologically active bodies such as Earth and Mars retain little or no trace of this period—most outer solar system satellites are covered with impact craters. Altogether only about 150 craters were unambiguously identified on Triton's surface. The number and size of the craters identified on the oldest terrain unit on Triton's surface are consistent with an age of about 2 billion years (Ga), approximately equal to that of the mare on Earth's moon. Other units are younger, probably less than 1 Ga.

Impact crater morphologies give us insight into the rheology (deformation and flow of material) of Triton's crust. On Triton, impact craters up to 12 km in diameter have a simple bowl-shaped morphology. Craters larger than 12 km in diameter, up to the largest crater observed (27 km diameter), have a complex morphology with flat floors and central peaks. The transition diameter from simple to complex craters and depth/diameter ratio are consistent with a water ice or ammonia–water ice crustal composition. Triton's craters are pristine with sharp rims, and no examples of relaxed impact craters were observed. This is also indicative of a relatively rigid surface. Other possible ices (such as the CH_4 and N_2 identified in Triton's spectrum) would have relaxed over a 1–2 Ga timescale.

The most unusual terrain on Triton has never been observed anywhere in the solar system—other than on the skin of a cantaloupe! This terrain was dubbed "the cantaloupe terrain" by the *Voyager* imaging team, and a new term has entered the geologic vocabulary. A large fraction of the portion of Triton imaged at high resolution is covered with this organized cellular pattern, which consists of a dense concentration of ridges and depressions, seen on the left side of Triton in Figure 1. Cells are preferentially 5 km and 26 km in diameter; thus, it is unlikely that this is highly degraded cratered terrain. Ridges between the cells have rough pitted crests. The stratigraphy is consistent in that adjacent cells interfere with their neighbor cells. This is in contrast to impact craters or igneous calderas, which crosscut each other and lack regular spacing or sizing.

The best earthly analogue may be salt domes known as diapirs (Schenk and Jackson, 1993). A diapir is produced by gravity-induced overturn of sedimentary beds of salt, driven by density inversions caused by either compositional or temperature contrasts. Comparative structural geology suggests that Triton's cells in the cantaloupe terrain may be diapirs which consist of at least two ices that have risen through a third ice. CO_2, H_2O, and ammonia-water are known or suspected constituents of Triton's crust, and theoretically have sufficient density contrast for this process to happen on Triton.

It is clear from Figure 1 that Triton has global-scale lineaments of trenches or "graben," and ridges where viscous material has welled up in graben floors. Graben are tectonic features indicative of tensional stresses on a surface. In many places highly viscous flows as much as a few kilometers thick have extruded onto the surface. Morphologically these resemble features associated with basaltic rift volcanism and are also suggestive of flows on Uranus' satellite Ariel.

The western edge of Triton's eastern plains (visible in the upper center of Figure 1) contains lake-like features: low-lying walled plains with flat floors. The appearance of terracing on the flat floors may be indicative of fluid having been extruded several times onto the floor. A single impact crater in one of the frozen "lakes" is 15 km in diameter. Its morphology shows that the floor materials are rigid on geological timescales.

The plains further east experienced extensive eruptions originating from graben, which flowed out onto adjacent plains and formed hummocky (rolling deposits). This region has the greatest number of impact craters. Enigmatic features to the far right in Figure 1 with dark cores and very bright "haloes" have no solar system analogue. Their origin is unknown.

Triton's geology supports the hypothesis that global melting occurred and erased the record of the period of early heavy bombardment. This internal melting was driven by the dissipation of tidal energy, as its orbit was circularized, and internal radiogenic heating. The sequence of events believed now to have occurred in Triton's history is as follows:

1. Formation in the outer solar system in a heliocentric orbit
2. A close interaction with Neptune that resulted in gravitational capture into an elliptical retrograde orbit around the planet
3. Circularization of Triton's orbit by dissipation of tidal energy, which also resulted in global melting of the interior of Triton
4. Simultaneous differentiation, enabled by the global melting, to a rocky silicate core approximately 1,000 km in diameter, a water ice mantle/crust with some amount of CO_2 and ammonia-water, and a methane/nitrogen surface ice veneer
5. Subsequent re-solidification and the initiation of the record of geological processes such as impact cratering and formation of diapirs visible today

Cryogenic Geysers

The discovery of Triton's plumes, active geyser-like eruptions, was astounding. The term "geyser" has been generalized to Triton's plumes, although strict use of the term implies a knowledge of the source mechanism (a liquid to vapor transition). On Triton the source mechanism is not actually known.

The geysers were discovered with careful stereo processing of *Voyager* images taken of the same territory on Triton from different angles (Soderblom et al., 1990). This stereo analysis showed two plumes rising above the surface. (The same analysis also showed that the wind streaks ubiquitous in the southern hemisphere lie on or very near the surface.) There is a less certain identification of approximately a dozen other possible geysers.

Figure 3 shows one of the two best-observed plumes. A column of dark material can be seen rising vertically upward to 8 km. At that altitude the plume material becomes entrained in the prevailing wind and drifts off to the west over 100 km. This stream of material casts a shadow, from which

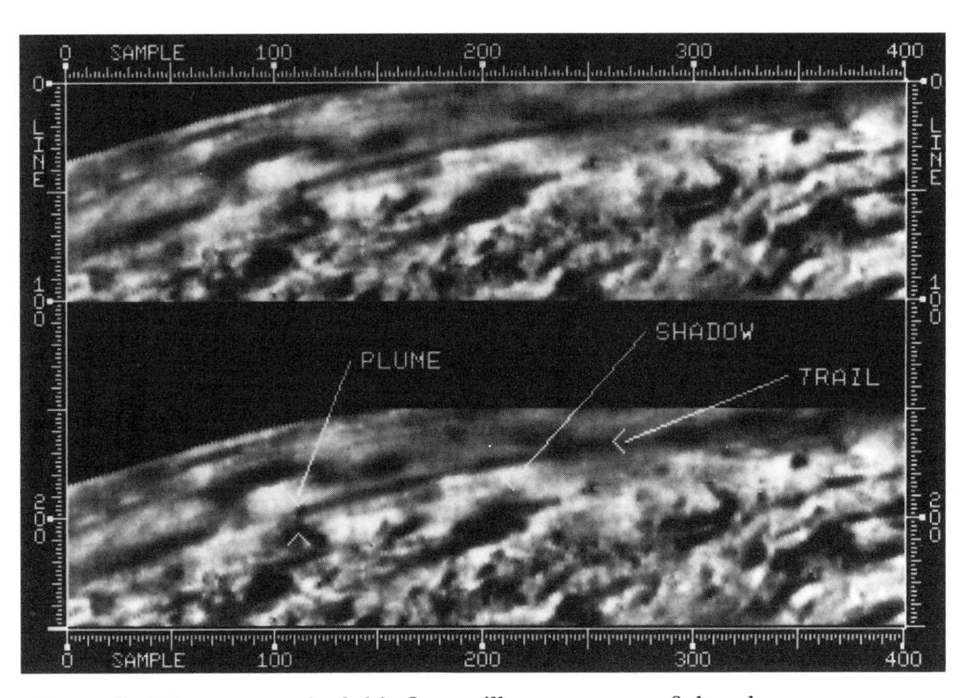

Figure 3. The top panel of this figure illustrates one of the plumes discovered by *Voyager 2*. The arrows in the bottom panel point out the source of the plume, the top of the plume, its tail, and the shadow it casts on the surface.

it is possible to estimate the optical depth of the wind-carried plume material. By tracking intermittent clots of material, the wind speed at this altitude was calculated to be 10–20 meters per second (m/s). The particle size can be estimated by noting that there is no evidence of settling over this distance. Given 10 to 20 m/s winds, the size of a particle that can be held aloft over this distance turns out to be less than 5 μm. Knowing the optical depth and the particle size, the mass flux in the trailing clouds is estimated to consist of up to 10 kilograms (kg) of fine dark particles per second, plus possibly up to 400 kg/s vapor.

The surface wind streaks are almost certainly the remnants of plumes that at the end of their eruptive lifetime did not exceed the 1 km altitude of the atmosphere's boundary layer, and thus were carried northeast by the Coriolis force. All of the surface streaks probably formed in this Triton season because they would not be expected to survive the seasonal pole-to-pole volatile transport. The number of windstreaks and dark splotches in the southern hemisphere can be counted; dividing this number by one Triton season gives a geyser lifetime on the order of one Earth year.

The fact that geysers and wind streaks were only seen in the southern hemisphere led to the hypothesis that the geysers are driven by, or at least triggered by, solar insolation. The most popular model is solar-driven venting of nitrogen gas, bottled up under an icy layer of solid N_2. The vapor pressure of N_2 ice increases rapidly with increasing temperature, building up enough pressure to cause an eruptive plume. As the gas explosively decompresses, it could entrain dark particles from a darker subsurface layer. Sufficient solar energy collection to sustain such a plume for a year with the estimated mass flux is problematic, however, so other possibilities have been suggested as alternatives.

Other proposed mechanisms include geothermal sources, buoyant methane, and dust devils. Geothermal sources have the theoretical advantage of not requiring concentration of solar energy, but this mechanism requires nonisotropic internal heat flow. Plausible scenarios have been developed and require a very thick N_2 ice layer in the southern hemisphere. Buoyant methane plumes could certainly rise in Triton's nitrogen atmosphere, but would be unlikely to entrain dark particulates. Dust devils require high localized temperature gradients on Triton's surface, difficult to sustain for the required lifetimes. A definitive understanding of the source of Triton's fascinating plumes awaits a future Neptune mission.

Triton's Climate

Triton has a chilly surface temperature that ranges between approximately 34 and 41 K. Triton's atmospheric composition and pressure, and surface temperature measurements, are consistent with an N_2 atmosphere in vapor pressure equilibrium with N_2 ice on the surface. This tenuous atmosphere is substantial enough to support a variety of atmospheric phenomena including clouds and winds.

Ingersoll (1990) argues that many of the observable features on Triton's surface and atmosphere could be explained if Triton has a Mars-like atmosphere (*see* ATMOSPHERES, PLANETARY). On Mars, carbon dioxide condenses in the winter hemisphere and sublimes in the summer hemisphere seasonally, and atmospheric pressure is determined by the amount of CO_2 sublimed into the atmosphere versus the amount condensed on the polar caps at any given time. The atmospheric pressure thus varies seasonally. On Triton the dominant volatile is nitrogen. Triton, like Mars, should experience seasonal volatile transport as N_2 moves between the North and South Pole.

Winds are driven by seasonal transport of nitrogen, as it sublimes in warm regions and condenses in cold. The orientation of the wind streaks left on Triton's surface can be attributed to the northward mass flux of the subliming south polar frosts, deflected eastward by the Coriolis force. The high-altitude westward wind may be caused by geostrophic winds driven by the temperature difference between the bright cold polar regions and the darker, warmer equatorial region.

Triton's Place in the Solar System

Stern (1991) postulates that Triton and Pluto are the remnants of a population of 1,000 to 10,000 similar bodies that originally formed in the outer solar system in proximity to Uranus and Neptune. These small rock/ice planets got ejected one by one to the outer realms of the solar system by the gravitational scattering of Uranus and Neptune. Only Pluto, Charon, and Triton remain near today precisely because of their unique orbits. Pluto is in a 3:2 orbital resonance with Neptune, which ensures Pluto and Charon will never get any closer than 18 astronomical units (AU) to Neptune, much too far away for gravitational scattering to be a threat. Now a satellite deep in Neptune's gravitational well, Triton is also protected from future perturbations that could have sent similar small planets to exile in the Kuiper Disk.

The attributes of this original population may be reflected in the two bodies for which we have good data: Triton and Pluto. Triton and Pluto are similar in a number of respects: they are about the same size; their densities are comparable and indicative of a substantial fraction of rock; both have surface ices composed predominantly of N_2, CH_4, and CO; both have tenuous atmospheres; and the length of a day is comparable. Triton's surface geology, influenced strongly by its early period of tidal heating, is probably unique.

The same tidal forces that are causing the Earth's moon, in a prograde orbit, to spiral slowly away from Earth, act in an analogous manner but opposite direction on retrograde Triton. Now a satellite of Neptune, Triton will remain in orbit until it slowly spirals into Neptune, predicted to occur 5 to 10 Ga from now.

Bibliography

BROADFOOT, L., et al. "Ultraviolet Spectrometer Observations of Neptune and Triton." *Science* 246 (1989):1459–1465.

CONRATH, B., et al. "Infrared Observations of the Neptunian System." *Science* 246 (1989):1454–1459.

CRUIKSHANK, D., and P. M. SYLVAGGIO. "Triton: A Satellite with an Atmosphere." *Astrophysical Journal* 233 (1979):1016–1020.

CRUIKSHANK, D., R. H. BROWN, and R. CLARK. "Nitrogen on Triton." *Icarus* 58 (1984):293–305.

INGERSOLL, A. "Dynamics of Triton's Atmosphere." *Nature* 344 (1990):315–317.

SCHENK P., and M. P. A. JACKSON. "Diapirism on Triton: A Record of Crustal Layering and Instability." *Geology* 21 (1993):299–302.

SODERBLOM, L. A., et al. "Triton's Geyserlike Plumes: Discovery and Basic Characterization." *Science* 250 (1990):410–415.

SMITH, B., et al. "Voyager 2 at Neptune: Imaging Science Results." *Science* 246 (1989):1422–1450.

STERN, S. A. "On the Number of Planets in the Outer Solar System: Evidence of a Substantial Population of 1000 km Bodies." *Icarus* 90 (1991):271–281.

TYLER, G., et al. "Voyager Radio Science Observations of Neptune and Triton." *Science* 246 (1989):1466–1473.

CANDICE HANSEN

U

UNIFORMITARIANISM

See Famous Controversies in Geology

UNIVERSE

See Cosmology, Big Bang, and the Standard Model

URANIUM AND THORIUM

Uranium occurs in the earth in three isotopes: ^{238}U (99.28 percent), ^{235}U (0.715 percent), and ^{234}U (0.005 percent). Each is naturally radioactive and each can be fissioned under suitable conditions. The tremendous energy of nuclear fission accounts for the two main uses of uranium, electric power and military weapons. Nuclear power generates 20 percent of the world's electricity. History suggests that fission is one of the safest forms of energy production (Cohen, 1990), but negative public perception has slowed its use in industry.

Uranium is widespread in small quantities in nearly all of Earth's crust and water. A carcinogen in large doses, uranium and its fission products constantly produce low levels of natural radiation to which humans are exposed without significant injury.

Uranium is concentrated in a wide variety of ore deposits, hosted by nearly every major rock type, and found in scores of countries on five continents. The major ore deposit types are Precambrian quartz-pebble conglomerate deposits, Proterozoic unconformity deposits, Proterozoic breccia-complex deposits, and Phanerozoic sandstone deposits. The main geochemical feature controlling uranium source, transport, and deposition is its solubility in the oxidized (U^{6+}) form and its relative insolubility in the reduced (U^{4+}) state.

Geochemistry

Uranium in nature always forms compounds with oxygen, such as oxides, hydroxides, silicates, and phosphates; there are about 160 individual minerals containing uranium (Steacy and Kaiman, 1978; Fleischer, 1983). The most common uranium mineral is uraninite (including its varietal form, pitchblende), which is black, cubic, and with compositions ranging from UO_2 to U_3O_8. High-temperature uraninite is typically coarsely crystalline, whereas pitchblende, the low-temperature variety, is fine-grained and massive or sooty. Secondary minerals, formed by weathering of uraninite, are commonly vividly colored in hues of yellow, orange, and green.

Uranium concentrates in molten rock (magma) in preference to crystalline solids. Consequently, during Earth's 4.6 billion year (Ga) evolution, uranium has migrated upward with magma into the upper crust. Today, the average concentration of uranium in upper crustal rocks is about 2 parts per million (ppm) (Turekian and Wedepohl, 1961).

U^{6+} is soluble and typically has concentrations of 0.3 to 10 parts per billion (ppb) in seawater, 0.06 to 10 ppb in fresh surface water, and 1 to 120 ppb in groundwater, and local concentrations can be much higher. Dissolved uranium typically forms uranyl $(UO_2)^{2+}$ complexes with carbonates, dissolved organic matter, hydroxides, and phosphates (Langmuir, 1978). Uranium typically dissolves by the oxidation and dissolution of primary uranium minerals during weathering of uranium-rich granite and volcanic ash. Dissolved uranium precipitates in ore deposits typically by reduction of U^{6+} to U^{4+} and dissociation of uranyl complexes. Reduction of uranium is always coupled to oxidation of other material, such as organic matter, hydrogen sulfide (H_2S), ferrous iron (Fe^{2+}), or methane (CH_4), commonly aided by bacteria.

Ore Deposits

Uranium ore deposits are extremely varied but are grouped by the rocks in which they occur: Precambrian rocks of several types, igneous rocks and veins of various ages, and Phanerozoic sedimentary rocks (Nash et al., 1981). Deposit types have characteristic ages that reflect stages in the geologic and biologic evolution of Earth. Uranium deposits are unknown in Archean rocks older than 2800 million years (Ma), presumably because the crust had not differentiated sufficiently to concentrate uranium. Quartz-pebble conglomerate deposits, formed by the accumulation of uraninite grains in streams, are found only in sedimentary rocks deposited before Earth's atmosphere became rich in oxygen, which would have caused the uraninite to dissolve. Abundant atmospheric oxygen, beginning in the Early Proterozoic about 2200 Ma, promoted uranium oxidation, dissolution, and transport in solution, and this led to uranium concentration in sediments by chemical precipitation. The spread of land plants in the Devonian, about 400 Ma, provided organic matter essential to formation of Carboniferous to Tertiary sandstone uranium deposits. Deposits formed at Earth's surface tend not to be preserved, and therefore surficial deposits are less than a few million years old.

Deposits in Precambrian rocks are found in shield areas of Canada, South Africa, and Australia. Quartz-pebble conglomerate deposits are composed of uraninite grains that accumulated in Late Archean to Early Proterozoic conglomeratic stream sediments between 2800 and 2160 Ma. Unconformity deposits occur at unconformities between metamorphosed Early Proterozoic rocks and unmetamorphosed Middle Proterozoic sandstone and originally formed between 1740 and 1330 Ma (Cumming and Krstic, 1992). Graphite in the metasediments suggests that methane reduced the uranium. The one representative of the breccia-complex type, Olympic Dam in Australia, is the world's largest uranium deposit (the deposit also produces substantial amounts of copper, gold, and silver). It formed at 1590 Ma probably by boiling and mixing of fluids accompanied by intense fracturing (brecciation) of the rocks (Oreskes and Einaudi, 1992). Ultrametamorphic deposits consist of disseminated uranium in granitic rocks. Uranium concentration was related to melting during very high-grade metamorphism.

Igneous rock and vein deposits have a wide range of ages (from Precambrian to Tertiary), locations, and associations. Vein deposits occupy faults and fractures. The largest vein uranium deposit in the United States, in Colorado, formed at 69 Ma probably by degassing of CO_2 and dissociation of uranyl carbonate complexes (Wallace and Whelan, 1986). Deposits associated with igneous rocks are in alkali syenite intrusions, contact metamorphic aureoles, and volcanic rocks.

Uranium deposits in Phanerozoic sedimentary rocks are in sandstone, black shale phosphate rock, and recent alluvium. Tabular and roll sandstone uranium deposits produced 95 percent of U.S. uranium through 1993. Worldwide, sandstone uranium deposits are in Permian-Tertiary fluvial, arkosic, locally bleached, redbed sandstones. Roll uranium deposits, mostly Tertiary in age in Wyoming and Texas, are elongate with characteristic C-shapes in cross section. They formed from uranium-bearing, oxidized groundwater that was reduced by pyrite, organic matter, or H_2S at oxidation-reduction fronts. Tabular uranium deposits, typically in Triassic and Jurassic rocks of the southwestern United States, form nearly concordant layers in locally reduced host sandstone. Uranium was probably reduced by amorphous organic matter that originated at the surface as humic acid and precipitated at an interface between fresh and saline water (Sanford, 1992). Phosphate mining in Florida and Morocco yields by-product uranium. Low-grade disseminated uranium in black shale from Sweden and the eastern United States is a potential resource although not currently economic. Surficial uranium deposits form from surface water and shallow groundwater. Arid-lands surficial deposits form in regions such as western Australia and Namibia where uranium precipitated by evaporation and dissociation of uranyl carbonate complexes. Wetlands surficial deposits form in humid temperate regions, such as Washington state, where uranium is adsorbed by peat.

Production

World production for uranium averages 50,000 metric tons annually. The former Soviet Union, Canada, Australia, and the United States produce 60 percent of that total, and twenty-four countries account for the remainder. The United States is the largest consumer, but Belgium and France consume the most per capita, for nuclear generation of electricity. World resources in 1991 were 2 million metric tons. With the end of the cold war, the dismantling of nuclear warheads may eventually provide additional uranium for power generation and could contribute 20 percent of world supply through 2015.

Thorium

Like uranium, thorium has naturally occurring radioactive isotopes and was concentrated by upward movement of magma; unlike uranium, thorium compounds are relatively insoluble. Thorium is recovered from the mineral monazite, a rare-earth phosphate, (Ce, La, Nd, Y, Th)PO_4. Monazite is an accessory mineral in granitic igneous and gneissic metamorphic rocks. Concentrations of monazite in beach sands are the principal commercial source of thorium (Barthel, 1991). Such deposits are in India, Brazil, and the southeast United States. Thorium is also found in veins, pegmatites, and intrusions of syenite and carbonatite.

The high melting point of thorium oxide (3,300°C) accounts for many of its uses. Thorium is used in molds for high-temperature casting, elements in microwave magnetron tubes, mantles for incandescent lanterns, aerospace alloys, welding electrodes, ceramics, and nuclear power. Australia, India, Brazil, the former Soviet Union, and China are major producers. The only U.S. production is from mining of titanium- and zirconium-rich sand in Florida.

Bibliography

Barthel, F. H. "Thorium Deposits." In *Gmelin Handbook of Inorganic and Organometallic Chemistry*, eds. F. H. Barthel, F. J. Dahlkamp, R. Ditz, B. Sarbas, P. Shubert, and W. Töpper, 8th ed. Berlin, 1991.

Cohen, B. L. *The Nuclear Energy Option—An Alternative for the 90s.* New York, 1990.

Cumming, G. L., and D. Krstic. "The Age of Unconformity-related Mineralization in the Athabasca Basin, Northern Saskatchewan." *Canadian Journal of Earth Science* 29 (1992):1623–1639.

Fleischer, M. *Glossary of Mineral Species*, 4th ed. Tucson, AZ, 1983.

Langmuir, D. "Uranium Solution-Mineral Equilibria at Low Temperatures with Applications to Sedimentary Ore Deposits." *Geochimica et Cosmochimica Acta* 42 (1978):547–569.

Nash, J. T., H. C. Granger, and S. S. Adams. "Geology and Concepts of Important Types of Uranium Deposits." *Economic Geology*, 75th Anniversary Volume (1981):63–116.

Oreskes, N., and M. C. Einaudi. "Origin of Hydrothermal Fluids at Olympic Dam: Preliminary Results from Fluid Inclusions and Stable Isotopes." *Economic Geology* 87 (1992):61–90.

Sanford, R. F. "A New Model for Tabular-Type Uranium Deposits." *Economic Geology* 87 (1992):2041–2055.

Steacy, H. R., and S. Kaiman. "Uranium Minerals in Canada: Their Description, Identification and Field Guides." In *Short Course in Uranium Deposits: Their Mineralogy and Origin*, ed. M. M. Kimberley. Mineralogical Association of Canada, Short Course Handbook 5, 1978, pp. 107–140.

Turekian, K. K., and K. H. Wedepohl. "Distribution of the Elements in Some Major Units of the Earth's Crust." *Geological Society of America Bulletin* 72 (1961):175–192.

Wallace, A. R., and J. F. Whelan. "The Swartzwalder Uranium Deposit, III: Alteration Vein Mineralization, Light Stable Isotopes, and Genesis of the Deposit." *Economic Geology* 81 (1986):872–888.

RICHARD F. SANFORD
GARY R. WRINKLER

URANUS AND NEPTUNE

Uranus and Neptune are the seventh and eighth farthest planets, respectively, from the Sun and are the fourth and third most massive planets in the solar system. Their large masses, sizes, and hydrogen-rich compositions place them in the class of Jovian, or Jupiter-like, planets. This entry will address their atmospheres, weather, interiors, magnetic fields, and origins.

Both Uranus and Neptune orbit the Sun in roughly circular orbits, inclined less than 2° to the plane of the ecliptic. Their physical characteristics are summarized in Table 1. Uranus's rotation axis lies almost in the plane of its orbit. Since the rotation axis of a planet maintains a fixed orientation in space as a planet orbits the Sun, Uranus's polar regions alternate between almost continuous sunlight and darkness for half of each eighty-four-year orbit.

Uranus was the first planet discovered in modern times. William Herschel (*see* HERSCHEL FAMILY) swept the planet up in his telescope while conducting a search for double stars from his backyard in Bath, England, in 1781. Although at first he thought he had found a comet or a "nebulous star," others soon realized that the fuzzy disk in his telescope was a new planet. Although not "discovered" until 1781, Uranus is faintly visible to the human eye. It was even included on a star map in 1690—as a star. Such observations helped to refine knowledge of Uranus's orbit and pointed to small differences between the planet's actual orbital velocity and mathematical predictions.

To explain these deviations, it was soon postulated that a new eighth planet was disturbing Uranus's orbit. Neptune was subsequently discovered in 1846 by German astronomer Johann Galle, based upon positions calculated by John Couch Adams and Urbain Leverrier. As with Uranus, Neptune had been observed prior to its official discovery. Galileo included the planet as a moving star in his sketches of Jupiter during 1613 when Neptune was near Jupiter in the sky. Galileo's failure to recognize the object as a new planet is somewhat puzzling, but without star charts, the task was not simple.

Even with modern telescopes Uranus and Neptune both appear as small, faint, bluish-green disks with no discernible atmospheric features. Until the exploration of Uranus and Neptune in 1986 (see Plate 42) and 1989 by the *Voyager 2* spacecraft, very little was known about the two planets. During its brief encounters, *Voyager* imaged the planets' atmospheres, probed their atmospheric composition and temperature, and measured the planets' magnetic fields. In doing so, *Voyager* multiplied our knowledge of these planets tremendously (see Plate 43).

Atmospheres

The visible atmospheres of Uranus and Neptune consist of a mixture of about 84 percent hydrogen, 14 percent helium, and 2 percent methane gas.

Table 1. Properties of Uranus and Neptune

Quantity	*Uranus*	*Neptune*
Mass (g)	8.679×10^{25}	1.024×10^{26}
Equatorial radius (km)	25559	24764
Rotation period (hours)	17.24	16.11
Mean distance from Sun (10^6 km)	2,871	4,497
Orbital period (years)	84.0	164.8

Since these planets have no solid surfaces, locations in the atmosphere are identified by atmospheric pressure. At a pressure of 1 bar, equal to the air pressure at sea level here on Earth, the temperature of Uranus's and Neptune's atmosphere is about 75 K (−325°F). Just below the 1 bar pressure level on both planets are thin clouds of condensed methane. Below these clouds, at pressures of about 3 bar, are thicker cloud decks. These clouds may consist of frozen hydrogen sulfide (H_2S).

The ratio of carbon to hydrogen in both planets' atmospheres is higher than found in the Sun by about a factor of 30. This enhancement of carbon likely resulted from the impact of many carbon-rich comets into both atmospheres during the formation of the planets. In these cold atmospheres other cometary constituents like ammonia and water would quickly form ices after the initial heat of impact dissipated. These ices would sink to deeper warmer regions below the levels visible to *Voyager*. Since these constituents cannot be detected by spacecraft or ground-based telescopes, their relative abundance is unknown. Trace amounts of carbon monoxide and nitrogen have been observed in Neptune's atmosphere. Radio telescopes can provide compositional information about the deep atmospheres (up to several tens of bars) of Uranus and Neptune, but uncertainties in key parameters used to interpret these data limit the technique's usefulness.

Uranus and Neptune appear blue-green for several reasons. One factor is that blue light is more easily scattered by atmospheric gases than red light. This process, Rayleigh scattering, is responsible for the blue sky on Earth. Because it is not so strongly scattered, red light can penetrate deeper into the atmospheres of both planets where it is absorbed by methane. Finally the H_2S cloud particles absorb red light preferentially. Thus the sunlight that is scattered by the atmosphere and cloud deck has had much of its red wavelengths removed by the methane and cloud particles. The light returned to observers is thus bluish-green.

Below the visible atmosphere, the gas density, pressure, and temperature increase with depth. The rate of temperature increase is steep enough to allow mixing or convection. Thus energy is transported from the warm, deep interiors of both planets to the atmosphere. Above the 1 bar level the temperature on both planets falls to about 55 K at several tenths of bars and remains approximately constant to pressure of about 0.01 bar. Above this pressure, the temperature rises with altitude on both planets. Even though it is farther from the sun, Neptune's upper atmosphere is warmer at these pressures than that of Uranus. Also above 0.01 bar a haze of very small particles is seen on both planets. The haze results from a chain of chemical reactions involving the destruction of methane by solar ultraviolet light, ultimately producing more complex hydrocarbons, such as ethane (C_2H_6) and acetylene (C_2H_2). These compounds condense in the cold upper atmospheres of both planets, forming the haze layer.

Weather

Uranus is an almost featureless planet. Before enhancement, even images returned by the *Voyager* spacecraft appear quite bland. Only computer processing of the images reveals small plumes and clouds rotating around the planet. These clouds indicate that, even though the Sun does not rise at one pole or set at the other for decades, strong winds blow from east to west at equatorial and middle latitudes. Unlike Jupiter and Saturn, Uranus has no internal heat source warming the atmosphere from below. Since Uranus receives only 0.27 percent of the sunlight that falls on Earth, the observed lack of a dynamic atmosphere is not surprising.

In contrast to Uranus, the atmosphere of Neptune is less hazy and displays complex, rapidly changing cloud systems. This observation was one of the great surprises of the *Voyager* mission since Neptune receives two and half times less sunlight than Uranus. Dominating Neptune's face is a Great Dark Spot, reminiscent of Jupiter's Great Red Spot. Neptune's spot lies at 22°S latitude, is the size of Earth, and rotates counterclockwise. Nearby westward winds blow at 2,200 kilometers (km) per hour with respect to the interior of the planet. Numerous other, smaller spots and clouds, some brilliantly white, are apparent against Neptune's faintly banded bluish disk. As on Jupiter and Saturn, winds on both Uranus and Neptune are zonal. This means they blow along lines of latitude.

Measurements of temperature across the face of both Uranus and Neptune show very little temperature contrast. Unlike Earth where the poles are much colder than equatorial regions, the atmospheric temperature varies by only a few degrees over the similar range in latitudes. This implies that even on Uranus, atmospheric motions are efficient at redistributing heat around the planet.

Interior Structure

The interior structure and composition of a planet give clues to its formation and evolution in time. The internal structure of Earth is deduced from the study of seismic waves created by earthquakes. Since no such information is available for Uranus and Neptune, their interiors must be studied theoretically.

Interior models indicate that the atmosphere of each planet continues downward with roughly the same composition to about 80–85 percent of each planet's radius. The compression of each point in the atmosphere by the weight of overlying atmosphere causes the temperature, pressure, and density to increase with depth. By 85 percent of the radius the pressure has reached 300,000 bars and 3,000 K (5,400°F). Under these conditions hydrogen and helium behave more as hot, dense liquids than as gases. Below about 85 percent of the radius the composition of the planets appears to change. The density rises to 1 g/cc (the density of water), indicating a change from a mixture of hydrogen, helium, and methane to a mixture of water, methane, and ammonia. This envelope continues with roughly the same constituents to very deep in the planets.

The bulk density of Uranus and Neptune, 1.3 and 1.6 g/cc, reflects the water, methane, and ammonia composition of their interiors. Most of the mass of both planets consists of these materials. Very deep in their centers, where pressures reach 6 million bars and 7,000 K, both planets may have a "rocky" core about the mass of Earth. The core, if present, is likely composed of silicates, iron, magnesium, and other elements typically found in terrestrial rocks and the inner planets. It is also possible that Uranus's and Neptune's inventory of rocky material is distributed through their water, methane, and ammonia envelopes and does not form a distinct core. Both types of interior models are theoretically acceptable.

Although the interior structures of Uranus and Neptune are similar, they are not identical. Neptune is more massive than Uranus. Its greater mass compresses the constituents of its interior to higher pressures. Thus the density of Neptune is everywhere higher than the density of Uranus. But it appears that the interior compositions of the two planets are quite similar.

Heat Flow

Jupiter, Saturn, and Neptune all radiate to space more energy than they absorb from the Sun. Jupiter radiates 1.7 times as much energy, Neptune 2.5 times. Since Neptune is 6 times farther from the Sun than Jupiter, and thus receives 36 times less solar energy, the total energy radiated by Neptune is much less than that radiated by Jupiter. Uranus radiates no more than 1.1 times the energy it receives from the Sun. This energy comes from the gradual contraction and cooling of the planets. After their formation, the giant planets were much warmer than they are today. As they radiate their energy away into space, they cool and contract. Energy from their interiors is transported to the cooling layers of the atmosphere by convection. The rate of cooling at depth is controlled by the rate the atmospheres radiate energy into space.

Both Uranus and Neptune, however, radiate less energy into space than theoretical models would predict. It is possible that slight changes in composition with radius in their deep interiors inhibit the action of convection. This would slow transport of energy from their interiors to atmospheres and account for the almost nonexistent heat flow of Uranus and the low heat flow from Neptune.

Magnetic Fields

Uranus and Neptune also share peculiar magnetic fields. The magnetic fields of Jupiter and Saturn, like the magnetic field of Earth, are closely aligned with the spin axis of the planet. In other words, the fields appear as if a giant bar magnet were centered within the planets with its long axis closely aligned with the planetary rotation axis. Instead, Uranus and Neptune's magnetic fields are both strongly offset and tilted dipoles. This means that the imaginary bar magnet in their interiors points far from the rotation axis (59° at Uranus and 47° at Neptune) and is also displaced from the center of the planet.

The magnetic fields of planets are believed to be generated by a combination of planetary rotation and convective motions in the conductive regions of their interiors. Earth's molten iron core, for example, is believed responsible for its magnetic field (*see* EARTH'S MAGNETIC FIELD). By subjecting samples of an "artificial Uranus" mixture to high temperatures and pressures, their conductivity can be measured. These experiments show that constituents in the "icy" mantle of Uranus and Neptune are quite electrically conductive in these planetary interiors. One possible explanation for the strange magnetic fields of these planets is that the electrically conducting region is quite close to the surface of the planets, compared to the other Jovian planets and Earth.

Origin

Uranus and Neptune both formed in the protosolar nebula, about 4.55 billion years (Ga) ago. Observations of the planets, computer modeling, and evidence from comets and the other planets point to the following series of events.

The formation of Uranus and Neptune began with the accumulation of a solid core of icy and rocky materials from the abundant solids present in the low temperatures of the outer solar nebula. Closer to the newly formed Sun, volatile compounds such as ammonia and methane likely existed in a vapor phase and did not accumulate into solid bodies. As the planets grew, they began to accumulate hydrogen and helium gas from the solar nebula. The greater masses and larger atmospheres of the outer planets allowed them to capture more and more of the comet-like planetesimals that orbited far from the Sun. The accumulation of these objects was responsible for the formation of the "icy" mantles of the two planets, which now comprise the majority of their masses.

Since Uranus and Neptune are farther from the Sun than Jupiter and Saturn, it is likely that their process of accumulation proceeded more slowly. This is because orbital periods increase with distance from the Sun and planetary growth depends upon encounters between orbiting bodies. By the time the solar nebula began to dissipate, perhaps due to a strong solar wind from the Sun, Jupiter and Saturn had already accumulated a great deal of hydrogen and helium gas from the nebula. Uranus and Neptune, however, had not had time to grow sufficiently large to capture a great deal of gas. These two planets may thus represent an intermediate phase of giant planet evolution, frozen in place over the history of the solar system.

Uranus and Neptune have had a similar story, except that late in Uranus's formative process it may have been struck by an especially large protoplanet. Calculations show that the collision of an object with one to two times the mass of Earth with Uranus could tip the planet's rotation axis upon its side, accounting for Uranus's peculiar obliquity. Such a giant collision may also have created a disk of material around the planet, from which Uranus's satellites formed.

Bibliography

BEATTY, J. K. "Getting to Know Neptune." *Sky and Telescope* 79 (1990):146.

BEATTY, J. K., and A. CHAIKIN, eds. *The New Solar System.* Cambridge, Eng. 1990.

HARTMANN, W. K. *Moons and Planets.* Belmont, CA, 1993.

STONE, E. C., E. D. MINER, et al. "The Voyager 2 Encounter with the Uranian System." *Science* 233 (1986):39.

———. "The Voyager 2 Encounter with the Neptunian System." *Science* 246 (1989):1417.

MARK S. MARLEY

URBAN PLANNING AND LAND USE

Urban planning and land use is an early twentieth-century concept in which the land is evaluated for its ability to sustain and benefit the activities of its human occupants with minimal damage to the environment. Urban planners are those responsible for designing the interface between people and nature. That design must make full use of geologic descriptors and quantification of all elements related to the land, its characteristics and properties, and those of water, surface and subterranean. Engineering geologists and a variety of other earth scientists provide the geologic data, commensurate with their experience and capabilities.

Urban planning grew from many of the concepts and philosophies of the nineteenth and early twentieth-century naturalists, as tempered by the human-use values of the Arts and Crafts movement, which reflected simplicity of design to represent nature and to serve the basic needs of people. Urban planning recognizes the need for people to have access to the land for various uses. Land use is categorized by the human activities that can be accommodated with maximum interfacing consistent with minimal damage to the environment. Unlike the naturalist concept, which promotes minimal use of the land, urban planning strives to place human need just ahead of the movement to restrict human use of the land.

Urban planning came to flower on the west coast of the United States, mainly in California, where abundant land and water were found to combine with the temperate climate to produce a nearly ideal habitat for people. By 1900 those interested in an orderly growth of land use had embraced what was termed city planning. The first National Conference on City Planning was convened in 1909. California soon became the showplace for orchestrated urban development. In 1915 the state legislature enacted two landmark laws: the Municipal Planning Act, providing for the formation of municipal planning commissions, and the Los Angeles County Flood Control Act, providing for the creation of the first county flood-control agency, and specifying a major land use that depended on urban planning.

The Los Angeles City Planning Commission was formed in 1922, and a county commission was in place by January 1923. As a result of the California Conservation and Planning Act of 1947, counties were required to adopt master plans containing elements that sought to provide for orderly land utilization, including location of transportation and utility routes, standards for subdivisions and housing plans, for recreation, and for conservation of natural resources.

By 1924, the 16th Annual Meeting of the National Conference on City Planning reported that 250 cities and towns in the United States were operating with zoning provisions and that complete-town subdivision were in place at Forest Hills Gardens, Long Island (N.Y.) and at the Country Club District of Kansas City (Mo.) Whole new towns were identified as Palos Verdes (Calif.), Clewiston (Fla.), Mariemont (Ohio), Westminster (Pa.), Alco (Tenn.), Three Rivers (Tex.), and Longview (Wash.). Land was being reclaimed mainly from low-lying areas at Chicago, La Crosse and Madison (Wis.), Memphis, Minneapolis, St. Louis, and Worcester (Mass.).

Engineering geologists were able to imagine myriad ways of using geologic studies in these planning efforts. Generally, reliance on such studies was not the case, however, as much of the site characterization depended on the land developer and only minimal planning requirements were met, mainly as nonengineered activities. As is often the case in employing science for the public good, geology came to be applied in a significant way only after passage of legislation, in this case the Federal Housing Act of 1954 and its subsequent amendments. Under section 701 funds were released to entities of state and municipal government, mainly in the late 1960s, continuing until about 1980. State funds tended to be utilized to support state geological surveys for a variety of assessments based on county to regional areas. State geological employees performed most of this work, and many of the reports survive today as formal publications or as recoverable open-file reports. Block grants went to city, county, and regional boards, as well as to districts and authorities, utilizing the services of local geological and planning consultants.

There is no master catalog of urban planning reports containing geological input or of geologic reports for the purposes of supporting urban planning. The two main sources are the publications lists of the various state geological surveys (reasonably complete) and the National Technical Information Service (NTIS) of the U.S. Department of Commerce, Springfield, Virginia. The latter contains random reports and this repository contains only those that were received at the behest of the federal agencies. Coverage is known to be spotty on federal documents in general before 1965 and imperfect until about 1975. Most urban planning reports contain large drawings, often printed in color, and therefore NTIS xerographic facsimile copies commonly are difficult to interpret and use.

Geologic reports or geologic content for urban planning uses come in a variety of categories (Table 1).

Geologic content of urban planning reports is found in the form of geologic baseline data for county or regional master plans and as supplemental reports developed to meet special needs or threats. Again, as is the case with locating the individual urban planning documents with geologic content, there is no standard outline, format, or content for geologic input. The closest organiz-

Table 1. Geologic Content of and for Urban Planning Reports

Content	*Purpose*	*Elements*
Character of land	Optimize land use	Soil, weak-rock, rock properties Depth to ground water Terrain characteristics
Natural resources	Preservation of economic resources	Industrial mineral deposits Construction material deposits Energy-related deposits Sole-source aquifers
	Preservation of natural scenic areas	Unique landforms Fossils
	Historic areas	Unique mining spoil Unique quarries
Geologic constraints	Reduce threat to public	Swelling soil Shrinking soil Slope stability Naturally unstable ground Active tectonic faults Nontectonic faults River flooding Flash flooding Debris flows River erosion Seacoast erosion Seacoast inundation Potential volcanism
Man-made hazards	Reduce threat to public	Abandoned mined lands/ground Acid mine drainage Former toxic waste disposal sites Contaminated surface/ground water Dam safety
Facility siting	Enhance safety of facility performance	Energy facilities Water-retention facilities Waste management facilities Underground space

ing principle is the fact that geologic information applicable to urban planning is most similar within individual physiographic provinces and subprovinces.

There is no doubt that the U.S. Geological Survey (USGS) has performed the lead role in establishing the potential breadth and depth of studies to support urban planning. From its Denver and Menlo Park, California, regional centers, the USGS has contributed extensive professional time to develop generic studies of the Denver Urban Corridor and the San Francisco Bay. Dozens of formal reports are available for the Denver corridor and literally a few hundred have been published or placed in the Open-File category for the San Francisco Bay area. The Bay area in particular benefited from a massive U.S. Department of Housing and Urban Development study known as the *San Francisco Bay Region Environment and Resources Planning Study*, consisting mainly of maps in the USGS "Miscellaneous Field" maps series.

Of the state geological surveys, perhaps those of

California, Kansas, Illinois, Iowa, Oregon, Texas, and Utah are the leaders in the general area of geology applied to urban planning. Key citations are included in the references cited below. Several other sources of specific state or city information are especially valuable, although not funded directly for use in urban planning. Galster's 1989 edited two-volume summary of the engineering geology of the state of Washington is an outstanding reference source for that state, and the Cities of the World Series of standard-format journal papers in the *Bulletin* of the Association of Engineering Geologists provides the most readily available source of urban geologic information for the eighteen cities in print (as of early 1993).

In a worldwide sense, the United Nations Educational, Scientific, and Cultural Organization (UNESCO) has held numerous conferences dealing with geology as a part of urban planning, mainly in and for the Asian countries and with little involvement by North American geologists. In Europe, there is a strong tradition of development of urban geologic databases and studies, not so much from the standpoint of urban planning as it is known in North America, but for primary use by civil engineers and, in a secondary sense, for the more limited urban planning found in that region.

Book and formal report references shown below are by no means exhaustive; they represent a fair sampling of the types of engineering geologic topics and contributions to urban planning and indicate the general distribution by sponsor. A typical example from each source is listed. Most of the domestic works, however, were at least partially funded by federal dollars.

Most available engineering geologic studies for urban planning were completed in a fifteen-year period from the 1960s to 1980. Deterioration of the American urban infrastructure has begun to capture the imagination and attention of the U.S. Congress, yet the budgetary restrictions required to attempt to balance the national budget will likely defer future federal spending in this area and user states, counties, and municipalities will need to rely on practitioners to locate available sources of geologic and engineering geologic references, to borrow or compile the new data that will always be required for urban planning. Urban planning is such a positively entrenched concept that it will survive and will require additional engineering geologic data, albeit at a lower rate in the deficit-plagued governments of the 1990s and beyond.

Bibliography

ALLEN, P. M. "Urban Geology of the Interstate Highway 35 Growth Corridor from Hillsboro to Dallas County, Texas." *Baylor Geological Studies* 28. Waco, TX, 1975.

AMERICAN GEOPHYSICAL UNION. *Environmental, Engineering, and Urban Geology in the United States.* Guidebooks to the 28th International Geological Congress, 2 vols. Washington, DC, 1989.

CUSHING, E. M., and R. M. BARKER. "Summary of Geology and Hydrologic Information Pertinent to Tunneling in Selected Urban Areas." U.S. Department of Transportation, Rept. DOT-TST-75-49. Washington, DC, 1974.

LEOPOLD, L. B., B. B. HANSHAW, and J. R. BALSLEY. "A Procedure for Evaluating Environmental Impact." *U.S. Geological Survey Circular* 645. Washington, DC, 1971.

MCGILL, J. T. "Growing Importance of Urban Geology." In *Focus on Environmental Geology,* ed. R. Tank. New York, 1973.

NICHOLS, D. R. "Earth Sciences and the Urban Environment." U.S. Geological Survey Pamphlet. Washington, DC, 1975.

SLOSSON, J. E. "Engineering Geology—Its Importance in Land Development." Urban Land Institute, Technical Bulletin 53. Washington, DC, 1968.

UTGARD, R. O., G. D. MCKENZIE, and D. FOLEY, eds. *Geology in the Urban Environment.* Minneapolis, MN, 1978.

ALLEN W. HATHEWAY

UREY, HAROLD

Harold Urey's scientific interests and accomplishments transcended any single discipline, encompassing chemistry, geophysics, and planetary science. He carried out undergraduate research on Missoula River protozoa at the University of Montana; in 1924, as a graduate student at the University of California (Berkeley), he published an important paper in the *Astrophysical Journal* on the stability of the hydrogen atom; during this period he published the first of a long series of papers on the application of statistical mechanics to chemical equilibria; as a post-doctoral fellow at the Niels Bohr Institute for Theoretical Physics, he published a paper on the theory of the atomic structure of the heavy elements. Urey was one of a number of young scientists who attempted to explain the anomalous Zeeman effect by the hypothesis of the electron spin. With Arthur E. Ruark he published *Atoms, Molecules and Quanta* (1930), one of the early textbooks on quantum mechanics in the English language.

Urey was born in Walkerton, Indiana, on 29 April 1893 to Samuel Clayton and Cora (Reinoehl) Urey. His father, a minister and schoolteacher of modest economic means, died when Harold was six. He was raised on a farm by his mother and grandmother. After graduation from high school, he taught at a small country school for a year. His mother remarried and moved to Montana to ranch near Big Timber. Harold joined her in 1912 and taught school there for two years. At the age of twenty-one, Urey entered the University of Montana, where he came under the influence of Professor A. W. Bray in the biology department. He graduated in three years and went to work for the Barrett Chemical Company preparing toluene, a precursor of TNT, during the First World War. After the war he returned to the University of Montana as an instructor, but soon realized that it would be necessary for him to attend graduate school. Urey was accepted to the outstanding program in physical chemistry at the University of California, Berkeley, in 1921. At Berkeley he conducted research under the guidance of the noted physical chemist Gilbert N. Lewis, and received his degree in 1923. Most of Urey's doctoral thesis involved independent work, which he published as the sole author.

Urey is principally recognized for his discovery of deuterium, a work carried out in the fall of 1931 with the assistance of F. G. Brickwedde and G. M. Murphy. For this discovery Urey received the 1934 Nobel Prize in Chemistry. The discovery of deuterium was not accidental; it was the result of a brilliantly planned and carefully executed confirmation of the hypothesis by Urey and others of the existence of a stable heavy isotope of hydrogen in nature. Following the discovery of deuterium, Urey engaged in sophisticated chemical engineering, the separation of the isotopes of the light elements (principally carbon and nitrogen) by the multiplication of small isotope effects in chemical exchange equilibria. In this work he enjoyed the collaboration of Karl P. Cohen, John R. Huffman, Clyde Hutchison, Jr., and Harry G. Thode, among others. Here again he was guided by predictions using quantum statistical mechanics.

Urey was diverted from scholarly pursuits by active participation in the U.S. effort to build an atomic bomb during World War II. He was a major protagonist of the program, leading research programs that resulted in industrial plants for the production of heavy water, ^{10}B, and ^{235}U, the latter by gaseous diffusion. All three of these enriched isotopes were produced for the atomic bomb program: heavy water in ton quantities as a rector moderator, ^{10}B as a neutron absorber, and ^{235}U as the fuel for the uranium bomb. Although he was raised as a pacifist, he supported the Allied military effort against Nazi fascism and Japanese militarism.

At the end of the war Urey resolved to return to scholarly activities and have no further connections with the military. He resigned from Columbia University, where he had been appointed Associate Professor of Chemistry in 1929, and joined the newly established Institute for Nuclear Studies at the University of Chicago. His first effort was to update and extend calculations he made a decade earlier on isotope effects in chemical exchange equilibria. His work was simplified and carried to much higher accuracy by advances in the theory made by Jacob Bigeleisen and Maria Goeppert-Mayer during World War II. Urey made the interesting observation that the enrichment of ^{18}O over ^{16}O in carbonate ion in exchange equilibrium with liquid water decreased by 0.016 percent/°C near 25°C. He was led to predict that "accurate determinations of the ^{18}O content of carbonate rocks could be used to determine the temperature at which they were formed" ["The Thermodynamics of Isotopic Substances," *Journal of the Chemical Society* (1947):562–581]. All that was required was the reliable development of assaying $^{18}O/^{16}O$ ratios, normally 2×10^{-3}, with a precision of 0.02 percent of the ratio and the demonstration that the record was preserved both in geological time and in the analysis. Urey accomplished all of these with the assistance of H. A. Allen, S. Epstein, H. Lowenstam, C. R. McKinney and J. M. McCrea. They created the new science of paleotemperature measurement (*see* ISOTOPE TRACERS, STABLE). For this achievement Urey received the Arthur L. Day Medal of the Geological Society of America. Recent advances have been made by Cesare Emiliani, Samuel Epstein, and Harmon Craig, all alumni of the Urey school. Epstein and his students have applied these methods to anthropology and paleontology.

Another major shift in Urey's scientific interests came early in his Chicago days when he read Ralph Baldwin's book *The Face of the Moon* (1949). Urey applied his mastery of chemical thermodynamics to the question of the origin and development of the Moon, and subsequently to other planets and to meteorites. From here Urey was led to ponder the origin of life on Earth. He hypothesized that lightning could lead to the synthesis of amino acids, the building blocks of life, in a reducing atmosphere containing ammonia, methane, and water. A graduate student, Stanley L. Miller, was successful in identifying amino acids when electrical discharges were passed through such a gas mixture in 1954. Its relevance to the origin of life is questionable, since a reducing Earth atmosphere at the time life began is no longer widely accepted. At this time Urey also recognized the importance of developing and improving geochronological timescales, apart from those recorded by the radioactive decay of the heavy elements. With Urey as his advisor, and with the assistance of M. G. Ingraham and R. J. Hayden, Gerry Wasserburg worked on $^{40}K/^{40}Ar$ dating. Subsequently Wasserburg made fundamental contributions to geochronology using high-precision isotopic analysis of microgram samples (*see* ISOTOPE TRACERS, RADIOGENIC). His paper with Hans E. Suess on "The Abundance of the Elements" provided crucial data for the subsequent work of W. A. Fowler and others on the origin of the elements.

Urey had the vision in 1957 to see the potential for cosmological research that could result from the newly created American space program. He was loud in support of lunar exploration by manned and unmanned vehicles. He was at the Johnson Space Center in Houston when the astronauts returned with lunar samples. At age seventy-six he was an active participant in the analysis of rocks brought back from the Moon; his principal interest was evidence for life on the Moon, an interest that has not been substantiated (*see* MOON).

Urey was ever cognizant of the privations of his childhood. He was concerned about the welfare of his students and was generous to his associates. During the 1930s he championed democratic causes and was outspoken in his opposition to fascism. He was active in relocating refugee scientists to the United States. In all of this he was joined by his wife Frieda Daum Urey, a woman of great stature. Harold Urey died in La Jolla, California on 5 January 1981.

Bibliography

ARNOLD, J. R., J. BIGELEISEN, and C. A. HUTCHISON, JR. "Memoir Harold Clayton Urey: April 29, 1893–January 5, 1981." In *Biographical Memoirs of the National Academy of Sciences* 68. Washington, DC, 1995.

BRICKWEDDE, F. G. "Harold Urey and the Discovery of Deuterium." *Physics Today* 35 (Sept. 1982):34–39.

UREY, H. C., and A. E. RUARK. *Atoms, Molecules, and Quanta.* New York, 1930.

UREY, H. C., and H. E. SUESS. "Abundances of the Elements in Planets and Meteorites." *Handbuch der Physik* 52 (1958):296–323.

JACOB BIGELEISEN

USEFUL MINERAL SUBSTANCES

A mineral is defined as a naturally occurring, solid, inorganic substance having a definite or narrow range of chemical composition and a characteristic crystal structure. Of the more than three thousand known minerals, only about two hundred or so are of economic interest, because either they contain metal(s) that can be extracted or they can be utilized for their bulk properties or energy-producing reactions. The term mineral substances (or mineral commodities) is used in a broader sense to include all earth materials of established or potential economic value, whether they are inorganic or organic in origin and whether they are solids, liquids, or gases. Some of these are minerals in the strict sense, some are rocks (aggregates of different minerals in various proportions), and others, such as crude oil and natural gas, are mixtures of natural organic compounds, but all were formed by natural geologic processes in the past and are mined from the earth (*see* RESOURCE USE, HISTORY OF).

Classification

Our dependency on mineral substances is not readily appreciated because we invariably use them in the form of processed and manufactured products rather than as the raw materials in which form they were mined from the earth. Mineral substances recovered from the earth may be classified in various ways. A simple scheme of use-based classification, with examples, is presented in Table 1. The three main divisions of this classification are: (a) metallic mineral substances; (b) nonmetallic mineral substances; and (c) energy-related mineral substances (*see* INDUSTRIAL MINERALS).

Metallic mineral substances include various metals that are extracted from specific minerals of appropriate chemical composition called ore minerals. The extraction of a metal from its ore mineral(s)—for example, iron (Fe) from its two most important ore minerals, hematite (70.0 percent Fe) and magnetite (72.4 percent Fe)—usually involves a complex series of steps ranging from simple crushing and grinding of the material to finer sizes to separation of the desired metal(s) from remaining material by sophisticated metallurgical processes. Further subdivision of the group is based on usage as governed by the properties of the metals. Precious metals such as gold and silver are expensive because of their high demand for jewelry and investment relative to availability. The group of ferrous metals includes, besides iron, metals primarily used for alloying with pig iron, recovered from a blast furnace, to make steels possessing distinctive properties, such as high-temperature hardness (tungsten-steel, molybdenum-steel, cobalt-steel), corrosion resistance (chromium-steel, nickel-steel), or wear resistance (manganese-steel). Nonferrous metals, such as copper (Cu), lead (Pb), and zinc (Zn), are so called because they are not used to make steel alloys (*see* NONFERROUS METALS). They are also referred to as base metals, a term that arose in the Middle Ages because these metals were considered inferior to the precious metals and, therefore, less desirable. In reality, metals of this group are in great demand for a wide variety of products such as chemical alloys (brass, bronze, pewter), storage batteries, transmission lines, utensils, paint pigments, plumbing fixtures, roofings and sidings, plating, and so on. Special metals have unusual properties that make them very suitable for specialized applications—for example, germanium (Ge) for semiconductors in electronic equipment, beryllium (Be) for light alloys used in high-speed aircraft, mercury (Hg) for vapor lamps. Although these metals are needed only in small quantities, their continued availability is of vital importance to our high-technology society.

Nonmetallic (or industrial) mineral substances

Table 1. Classification of Mineral Substances

Classes of Mineral Substances	*Examples*
A. Metallic Mineral Substances	
(a) Precious metals	Gold, silver, platinum-group elements (especially platinum)
(b) Ferrous (ferro-alloy) metals	Iron, manganese, nickel, cobalt, chromium, titanium, tungsten, molybdenum, niobium
(c) Nonferrous (base) metals	Copper, zinc, lead, tin, aluminum, mercury, cadmium
(d) Special metals	Antimony, arsenic, bismuth, zirconium, germanium
B. Nonmetallic (or Industrial) Mineral Substances	
(a) Precious stones and gemstones	Ruby, sapphire, emerald, diamond, amethyst, topaz, jade, pearl
(b) Building materials	Sand and gravels, stones, clays, gypsum
(c) Fertilizer and chemical materials	Saltpeter, phosphate rock, evaporites, caliche, borax, limestone
(d) Abrasive materials	Industrial diamond, garnet, corrundum
(e) Other industrial materials	Asbestos, mica, talc, barite, bentonite, talc, barite
C. Energy-related Mineral Substances	
(a) Fossil fuels	Coal, crude oil, natural gas, tar sand, oil shale
(b) Nuclear Fuels	Uranium, thorium

comprise minerals and rocks that are used for their bulk properties (physical, chemical, thermal, and electrical) directly or after some processing. Unlike metallic mineral substances, they are not processed for extracting metals. The most eye-catching of the nonmetallic mineral substances are the precious stones and gemstones (*see* GEMS AND GEM MINERALS), the gem varieties of specific minerals, but the most important in terms of the dollar value of annual production are the untreated rock products (e.g., crushed rock, sand, and gravel) and treated rock products (e.g., cement, plaster, bricks) that are classified as building materials. Fertilizer and chemical materials are the source of a large number of elements that are either vital to plant growth (e.g., nitrogen, phosphorous, potassium) or to the chemical industry (e.g., fluorine, chlorine, sodium), or both. Abrasive materials, of which diamond is the hardest, are used to cut, shape, grind, and polish all the modern alloys and ceramics. The last group, other industrial materials, includes all nonmetallic mineral substances that do not fit into the rest of the categories. Some examples are asbestos, used for its fire- and heat-resistant properties; barite for adding weight to drilling mud; fillers (such as talc, clays), which are added to a variety of manufactured products to give them special characteristics; refractory materials (such as magnesite, silica), used for furnace lining; and fluxes (such as fluorite, limestone), used in smelting operations to separate waste products (slag) from the metal concentrate (matte). Some earth scientists also include water and soil under nonmetallic mineral substances.

The three most familiar ingredients for energy production—coal, crude oil, and natural gas—are examples of fossil fuels, which represent past accumulations of plant and animal debris in sediments and their subsequent "cooking" (distillation) due to burial under younger sediments. Coal (carbon with varying degrees of impurities) and natural gas (essentially methane gas) are generally used directly. Crude oil, composed of a mixture of hydrocarbons (compounds of carbon and hydrogen) is refined to give various fuels (petroleum, aviation fuel, kerosene, heating oil, etc.) and nonfuel products (asphalt, wax, etc.). Tar sands are sedimentary rocks containing a very thick, semisolid, tarlike petroleum that is believed to represent immature petroleum deposits formed from the same organic material as lighter oils. Oil shales are fine-grained sedimentary rocks (not necessarily shales) that carry significant amounts of the waxy, insoluble hydrocarbons known as kerogen, which originated from the decomposition of plants, algae, and bacteria. Deposits of nuclear fuel metals, such as uranium and thorium, were not generated from organic debris; they are the raw material for reactions that can produce nuclear energy. The commercial nuclear reactors in operation today use uranium as the fuel and rely on the splitting of fissionable uranium atoms for the release of energy.

Bibliography

CAMERON, E. N. *At the Crossroads.* New York, 1986.
CRAIG, J. R., D. J. VAUGHAN, and B. J. SKINNER. *Resources of the Earth.* Englewood Cliffs, NJ, 1988.
MISRA, K. C. *Mineral and Energy Resources: Current Status and Future Trends.* University of Tennessee Studies in Geology 14. Knoxville, TN, 1986.

KULA C. MISRA

V

VENUS

Venus is the second planet from the Sun, and when it is at its nearest to Earth, it is the second closest object to Earth after the Moon. Despite its proximity to Earth, Venus is the last terrestrial planet to be explored, and prior to the 1960s, little was known about its geology, geophysics, atmosphere, and evolution. Yet it is particularly interesting to geologists familiar with Earth because it is the planet most similar to Earth in the solar system in terms of size (only 651 kilometers or km smaller in diameter than Earth's 12,755-km diameter), density and surface gravity (on the order of 0.9 and 0.8 of that on Earth, respectively), and relative position in the solar system. These characteristics mean that it is potentially the most Earth-like of all the planets. Many conditions on the surface are now known to be decidedly unlike those of Earth. Understanding how a planet that is so similar in many basic ways to Earth evolved to be so different geologically is a significant goal in the study of Venus with important lessons for understanding global environmental change on Earth.

Atmosphere

Unlike Mars where the atmosphere (with the exception of periodic global dust storms) is relatively clear, or the Moon (where there is no obscuring atmosphere), clouds perpetually shroud the surface of Venus. As a result, the surface of Venus cannot be seen by optical telescopes or conventional flyby spacecraft images. Very little was known about the atmosphere until spacecraft with special instruments were sent, and almost nothing was known about the actual physical surface until the development of radar techniques. Ground-based radar was the first able to establish that Venus rotates once every 243 Earth days, and that the rotation is reversed, or retrograde (east to west) from the sense on Earth and the other planets.

Passive observations of the microwave thermal emission from Venus, using radio telescopes and U.S. (*Mariner 2*, in 1962) and Soviet flyby spacecraft, suggested that the surface temperatures were high. Since then there have been numerous spacecraft missions to Venus, including landers as well as orbiters and flyby spacecraft (Table 1). Atmospheric entry probes, and Soviet Venera landers, directly measured the temperature and pressure within the atmosphere and proved that the surface is indeed hot (450°C), or about as hot as the surface of a catalytic converter on an automobile. It is believed that this non-Earth-like characteristic of the surface reflects the early development of an atmosphere that consists mostly of carbon dioxide (about 97 percent) and a "runaway greenhouse effect." The greenhouse effect refers to a condition on Venus in which solar heating of the upper atmosphere at the short wavelengths of visible light

Table 1. Chronological Summary of Spacecraft Missions to Venus

Spacecraft	*Country*	*Arrival*	*Type*	*Observations Made*
Mariner 2	USA	14 Dec. 1962	Flyby	Measured high sfc temperatures, magnetometer
Venera 4	USSR	18 Oct. 1967	Entry Probe	Atm measurements to 25-km altitude
Mariner 5	USA	19 Oct. 1967	Flyby	Measured high sfc temperatures, magnetometer
Venera 5	USSR	16 May 1969	Entry Probe	Atm measurements to 12-km altitude
Venera 6	USSR	17 May 1969	Entry Probe	Atm measurements to 16-km altitude
Venera 7	USSR	15 Dec. 1970	Lander	Atm measurements to surface (475°C 90 bars)
Venera 8	USSR	22 July 1972	Lander	Atm measurements/surface/light levels/composition
Mariner 10	USA	5 Feb. 1974	Flyby	Images of cloud tops, magnetometer
Venera 9	USSR	20 Oct. 1975	Lander	Atm measurements/images/composition surface
Venera 10	USSR	25 Oct. 1975	Lander	Atm measurements/images from surface
Pioneer-Venus	USA	4 Dec. 1978	Orbiter	Global topography/100-km resolution
Pioneer-Venus	USA	9 Dec. 1978	Entry Probe(4)	Atmosphere measurements to surface
Venera 11	USSR	21 Dec. 1978	Lander	Atm composition
Venera 12	USSR	25 Dec. 1978	Lander	Atm composition
Venera 13	USSR	1 March 1982	Lander	Surface color images/composition
Venera 14	USSR	5 March 1982	Lander	Surface color images/composition
Venera 15	USSR	Oct. 1983	Orbiter	Radar images northern 25 percent/2-km resolution
Venera 16	USSR	Oct. 1983	Orbiter	Radar images northern 25 percent/2-km resolution
Vega 1	USSR	15 Dec. 1984	Lander/Balloon	Surface composition/atm balloon probe
Vega 2	USSR	21 Dec. 1984	Lander/Balloon	Surface composition/atm balloon probe
Galileo	USA	10 Feb. 1990	Flyby	Images of cloud tops
Magellan	USA	10 Aug. 1990	Orbiter	Global radar images/120-m resolution

radiation results in warming of the atmosphere, but the longer wavelengths of thermal radiation in the lower atomosphere cannot penetrate the atmosphere and re-radiate to space; the temperature of the atmosphere thus continually increases. First recognized from the study of Venus, the greenhouse effect is now discussed for Earth where it is recognized that industrial additions of carbon dioxide to the atmosphere pose potential environmental problems of similar global magnitude.

The surface is characterized by extreme surface pressures, which in combination with the high surface temperature required the technological development of atmosphere entry probes and surface landers that could survive the extreme pressure-cooker conditions of the surface long enough to take meaningful measurements and radio the results back to Earth. Entry probes and landers show that the surface pressure on Venus is 92 times that of Earth (92 bars or 9 MPa), or equivalent to pressures at about 1-km depth in the sea, and the atmosphere is water-free, consisting dominantly of carbon dioxide. These probes also determined that the clouds perpetually obscuring Venus's surface from Earth optical observation are at an altitude of about 45 to 70 km above the surface and consist of sulfuric acid droplets rather than water droplets, as is the case for Earth. The altitude of the base of the cloud deck is where the atmospheric pressure and temperature are similar to those at the surface of Earth (Figure 1). But on Earth an altitude of 60 km is well above the atmosphere and is nearly at altitudes occupied by orbital spacecraft. Most of the atmosphere and opaque atmospheric phenomena (e.g., clouds) on Earth occur at less than 20-km altitude. The high surface pressure on Venus results from both the greater thickness of the atmosphere and because carbon dioxide is a heavier gas than the primary gases of Earth's atmosphere (nitrogen and oxygen).

Systematic telescopic observation of the atmosphere in the late 1970s detected short-term global change in the sulfur content and related compositions of the atmosphere. Because of the scale and short-lived change in the composition of the upper atmosphere, the most likely interpretation of abrupt global changes of the observed magnitude is that of a large volcanic eruption that injected

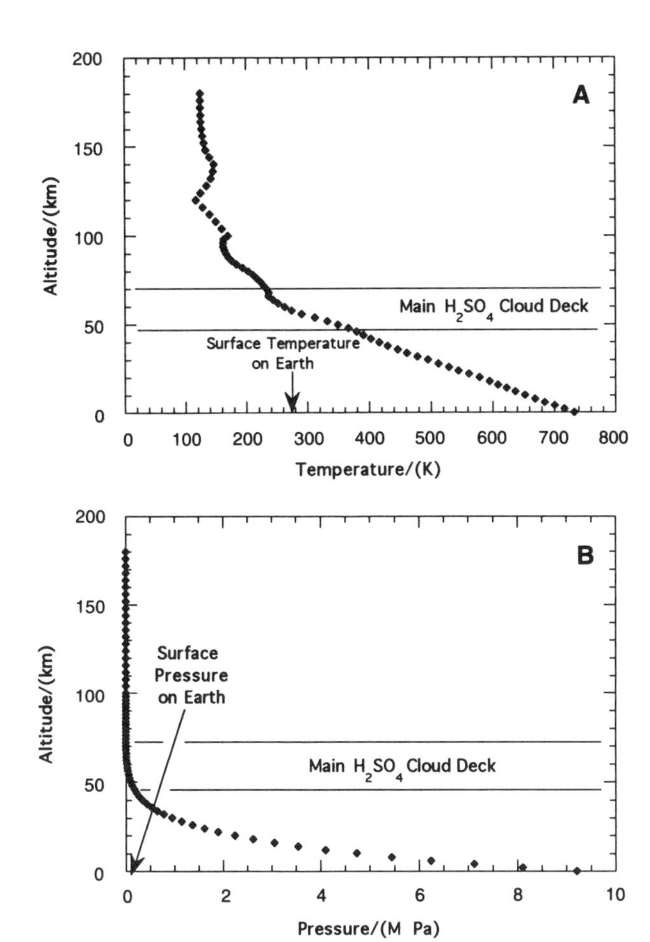

Figure 1. Plot of the variation in temperature and pressure with altitude in the atmosphere of Venus. These measurements were made by the Pioneer-Venus probes as they descended through the atmosphere. Note that the clouds that obscure the surface from view are between 45 to 70 km above the surface.

volcanic dust and gases into the upper atmosphere. Another transient and localized variation in atmospheric composition was detected by one of the entry probes of the Pioneer-Venus missions in the late 1970s. These early observations of possible atmosphere-related eruptions suggested that active volcanic processes are taking place on Venus.

Surface

Synthetic aperture radar images from the Russian *Venera 15/16* (covering the northern 25 percent of the surface of Venus) and the U.S. *Magellan* spacecraft, and altimetry obtained by the Pioneer-Venus orbiter and later by the *Magellan* spacecraft, have been obtained for 98 percent of the surface. Ironically, the surface of Venus is now more completely mapped, and with uniformly higher resolution, than the seafloor of Earth. The absence of running water on Venus means that it is a museum of volcanic and structural features perfectly preserved in a manner not possible on Earth. These data have shown that the surface is geologically young and complex and is characterized by more volcanic and structural features than previously seen on any planet other than Earth.

Global maps of altimetry show that the absolute range of elevations is much less than that of Earth (Figure 2) and that Venus is therefore relatively flat compared with Earth. More than 80 percent of the surface area lies within a kilometer of the mean planetary radius (6051.84 km). A relatively few highland regions rise from one to several kilometers above the mean planetary radius. However, these cover less than one-half the fraction (less than 15 percent) of the surface of Venus compared with that covered by continents (29 percent) on Earth. This means that, also unlike Earth, the global frequency distribution of elevations, or hypsometry, is unimodal, whereas Earth is characterized by a bimodal hypsometry (Figure 2).

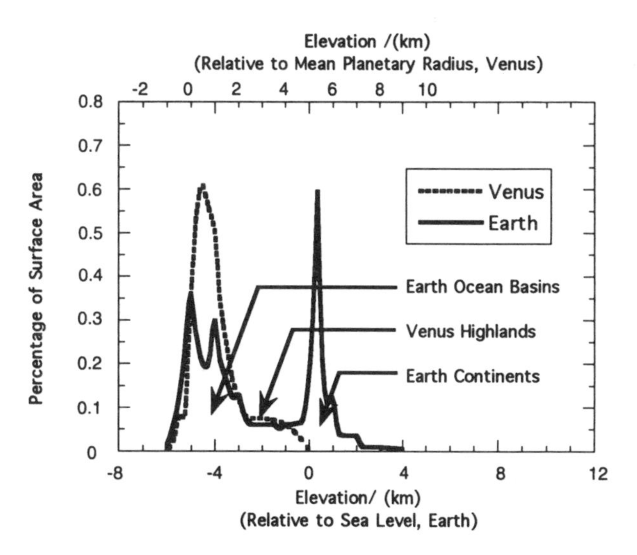

Figure 2. Plot of the frequency of different elevations (hypsometry) on Venus compared with that of Earth. A large fraction of the surface of Earth consists of high-standing continents and is the primary difference from Venus.

The highland regions are among the most complex surfaces and generally consist of a terrain that is faulted in orthogonal patterns. These regions are known as tesserae (Figure 3) after the Greek word for mosaics of tiles. There is some evidence that tesserae also represent the oldest preserved surfaces on Venus. The largest highland, Aphrodite Terra, is one such area. (Once actual surface characteristics began to be observed, the International Astronomical Union, which formally names features on other planets, established the convention of naming most features on Venus after mythical women. Impact craters and large calderas are named after historic women.) Another of the continentlike highlands, Ishtar Terra, is surrounded by mountain belts that rise from 6 to 11 km above the mean planetary radius and appear to have formed from compression and buckling of the surface, similar to orogenic belts on Earth. Akna Montes, which rises 6 km above the surroundings, is an example. Because the high surface temperature would cause the rocks forming these mountains to flow and flatten over a few million years, their very existence implies that they must either still be active and forming or that the lithosphere is unusually strong and that it can support them over geologically long periods of time. Low ridges of possibly similar origin in a range of sizes occur singly or in "belts" throughout the plains areas (Figure 4).

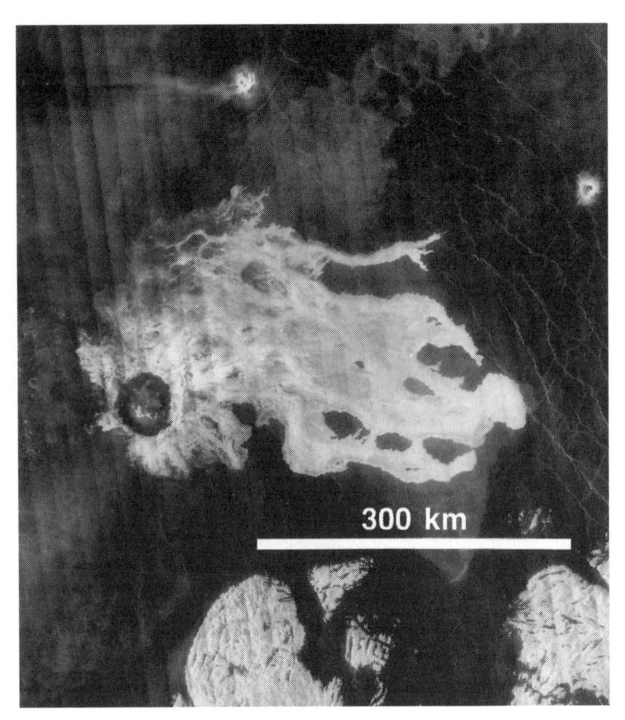

Figure 4. *Magellan* radar image showing characteristic arcuate ridges common throughout low elevation plains (on the right) and a large impact crater. The well-preserved morphology and unusual fluidal ejecta flows are typical of impact craters on Venus and may result from interaction of the impacting body and crater ejecta with the atmosphere.

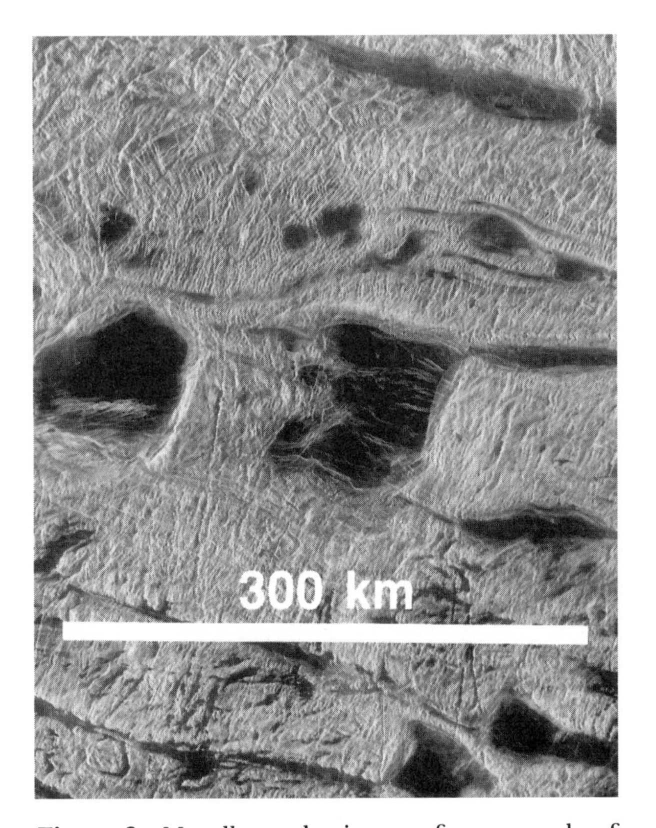

Figure 3. *Magellan* radar image of an example of complexly faulted and deformed terrain known as tessera. This radar image (and all others in the following figures) is illuminated from the left.

Faults, fractures, and immense rift valleys are present in abundance. One rift valley, Diana Chasma, is similar in size to the great East African Rift Valley and, likewise, probably formed due to stretching and pulling apart of the surface. On Earth, erosion and sedimentation quickly obscure all but the latest structures associated with such rifts, but on Venus the absence of erosion means that all of the structural details are perfectly preserved (Figure 5).

Volcanoes are ubiquitous on Venus, but are particularly concentrated in the hemisphere centered near longitude 250°E, much like hot spots are on Earth that are abundant in the Africa-centered hemisphere, and the many volcanoes that occur in the Tharsis regions of Mars. Large volcanoes characterized by patterns of flows radiating for tens to hundreds of kilometers are common, as are patterns of lava flow fields, extensive low-lying regions of lava plains, and lava channels. One lava channel is longer than the largest rivers on Earth. Low-relief domical edifices, many less than several kilometers in diameter, are globally abundant and number in the hundreds of thousands. Some vol-

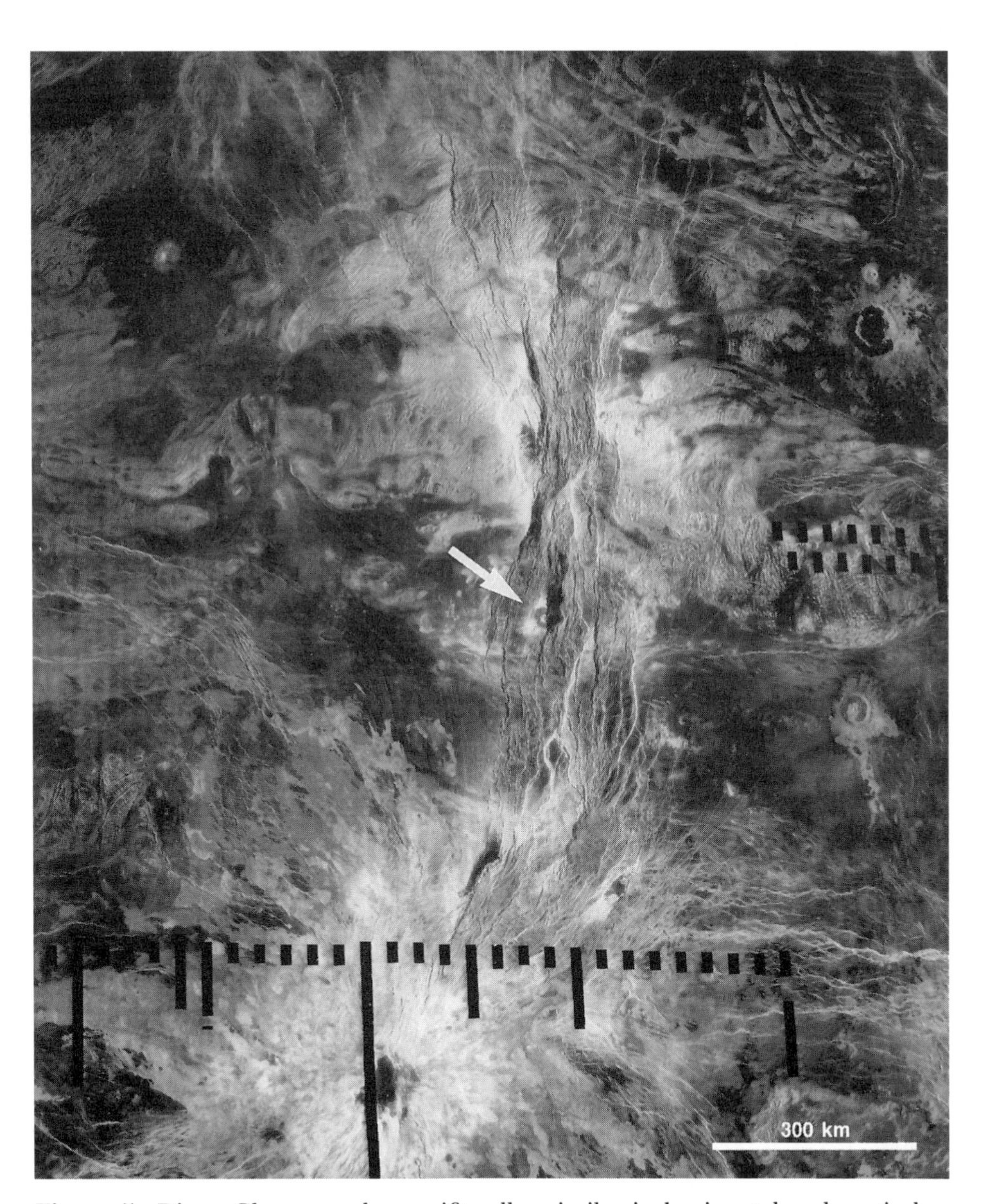

Figure 5. Diana Chasma, a large rift valley similar in horizontal and vertical dimension to the African Rift Valleys of Earth. The rift is delineated by the numerous parallel bright and dark lines running from top to bottom through the image. The arrow indicates an impact crater on the west margin of the rift that is cut by rift faults.

canoes appear similar to those formed from eruption of thick, viscous lavas on Earth. Additional volcanic features include calderas similar to those on Earth, although generally much larger; complex topographic annular features known as arachnoids; and circular structural patterns up to several hundred kilometers across with associated volcanism known as coronae. Many of these are generally thought to represent local formation of large and deep magma reservoirs. Stellate fracture patterns associated with volcanoes are common (Figure 6) and may represent the surface deformation associated with radial dike-like magma intrusions.

Impact craters are about as numerous on Venus as they are on continental areas of Earth, and are thus not as common as they are on most other planets. Only about 900 have been identified on Venus. Meteors smaller than a certain size disintegrate on entering the atmosphere. As a result, impact craters smaller than 2 km are infrequent. Morphologically, craters on Venus resemble those on other planets with several exceptions related to the interaction of the crater ejecta with the dense atmosphere. These include extensive parabola-shaped "halos" much like fallout from plumes associated with volcanic eruptions on Earth. These open to the west and possibly record the interaction of the upward expanding cloud of crater ejecta with the strong global easterly winds. Many craters are characterized by large lava-flow-like features that may represent molten ejecta flowing outward away from the crater after the impact (Figure 4). Impact craters also appear nearly uniformly distributed, unlike most planets where large areas of different crater abundance indicate variations in age of large areas of their surfaces. Based on estimates of their rates of formation on surfaces in the inner solar system, impact crater statistics indicate an average surface age on Venus

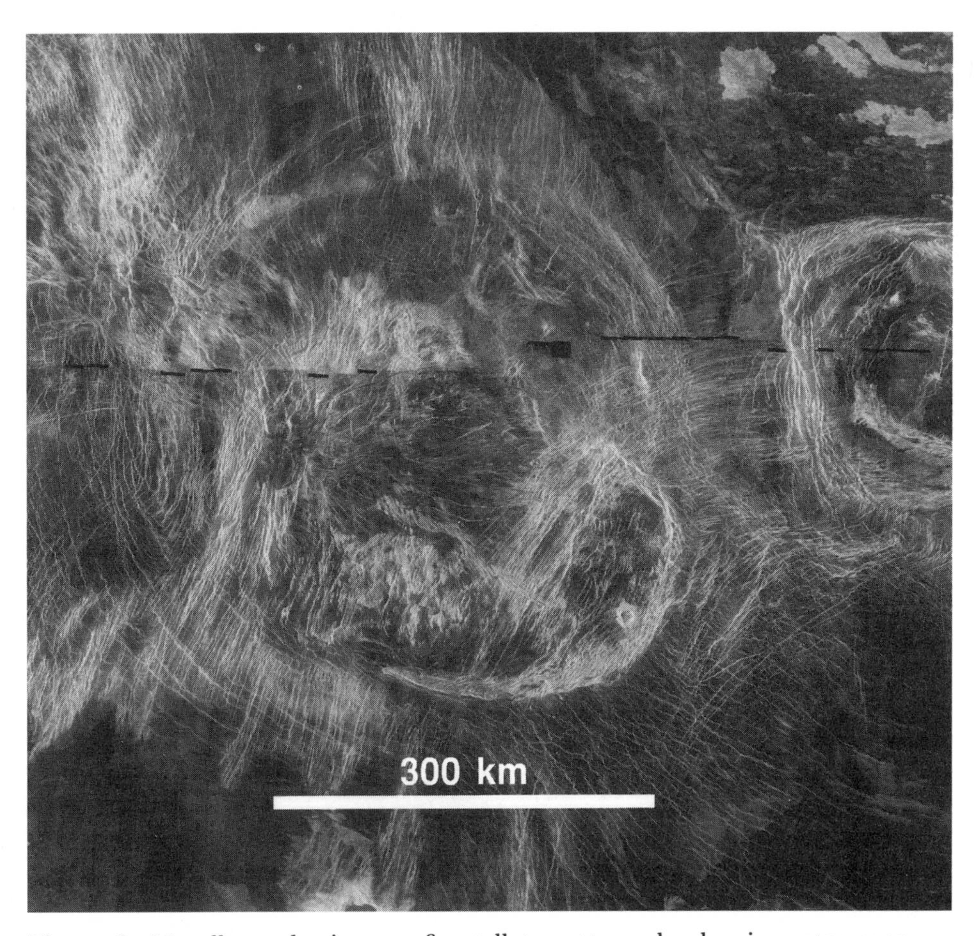

Figure 6. *Magellan* radar image of a stellate-patterned volcanic center over 200 km in diameter.

of about 500 million years (Ma). Either most of the surface was formed over 500 Ma ago in a catastrophic resurfacing event and volcanism has been much reduced since that time, or continual, widespread, and evenly spaced volcanism and tectonism remove craters with a rate that yields an average lifetime of the surface of 500 Ma.

By carefully tracking spacecraft orbits, variations in gravitational acceleration associated with differences in mass on and beneath the surface can be detected. On Venus, this technique reveals that the strength of the gravity is directly proportional to the surface topography, in contrast to Earth, where mass associated with topography is generally compensated underneath by lower density "roots." In general this means that many large-scale topographic features are attributable to either direct support of topographic masses without a low-density root, or by topography originating from bulging up of the surface through convective processes in the deep interior of the planet. If the first type is assumed, it may indicate that the lithosphere has considerable strength such that a low-strength layer at the base of the lithosphere (or the asthenosphere, as occurs on Earth) is not present. Such a result could originate from either a lower mantle temperature or the lower water content of Venus, both of which would reduce the partial melting that is responsible for an asthenospheric low strength layer on Earth. Small amounts of water in mantle minerals play a role in lowering silicate melting temperatures and are responsible for the asthenosphere's low strength on Earth, resulting in the detachment of plates from the mantle. An absence of an asthenosphere at the base of the lithosphere on Venus may be one reason why plate tectonics, as we understand it on Earth, does not appear to be present on Venus, although there is evidence for large-scale lateral movement of parts of the surface.

The surface of Venus was examined by the Soviet space program Venera landers that returned both images and chemical information about the rocks at several sites. Despite the dense cloud cover, surface images showed that the illumination is equivalent to a cloudy day on Earth. The relatively rocky surroundings show little erosion, but are typically characterized by small amounts of soil filling low areas between rocks. Most of the areas observed appear to be volcanic lava flow surfaces. The measured chemical compositions of surface materials are indistinguishable for the most part from tholeiitic and alkali basalts typical of ocean basins and hot spots on Earth. One site, however, suggested an even more evolved, alkali-rich composition similar to trachytes or syenites. The presence of alkali basalts and evolved alkalic rocks implies that the geological and geochemical evolution of Venus may have proceeded much further than it has on Mars or the Moon.

Because of the relative youth of the surface, only the last 20 percent of the history of Venus appears to be preserved and little is known about most of the surface geologic history. The geological complexity and young surface ages of both Venus and Earth, relative to smaller terrestrial planets, can be attributed to their more dynamic thermal evolution, their warmer and more mobile interiors, extensive surface deformation (tectonism), and copious production of mantle melts (volcanism) over extended periods of geological time.

Bibliography

BURNHAM, R. "Venus: Planet of Fire." *Astronomy* 19, no. 9 (1991):32–41.

CATTERMOLE, P. *Venus: The Geological Story.* Baltimore, MD, 1994.

HUNTEN, D. M., L. COLIN, T. M. DONAHUE, and V. I. MOROZ, eds. *Venus.* Tucson, AZ, 1983.

"Magellan at Venus." *Journal of Geophysical Research.* Special Volume on Results of Magellan Mission to Venus 97, nos. E8 and E10 (1992):13063–13675 and 15921–16382.

PLAUT, J. J. "Venus in 3-D." *Sky and Telescope* 86, no. 2 (1993):32–37.

LARRY S. CRUMPLER

VERNADSKY, VLADIMIR I.

Vladimir Ivanovich Vernadsky was born in Saint Petersburg, Russia, on 12 March 1863 and died in Moscow on 6 January 1945. His parents were Ukrainian, his father a political economist who held several academic positions. His early schooling was in Kharkov, where his father was appointed manager of the State Bank in 1868. The family returned to Saint Petersburg in 1876, and he entered the university there in 1881, specializing in chemistry, geology, and mineralogy. He spent the years 1888–1890 in graduate study in France and Germany. He returned to Russia in July 1890 and was appointed to a professorship in crystallography and mineralogy at Moscow University.

In the fall of 1886 Vernadsky married Natasha Staritsky, the daughter of a high official in the czarist government. Their son, George, was born 1 September 1887; he later became a distinguished historian of Russia at Yale University.

During his professorship at Moscow University, Vernadsky developed excellent laboratory facilities and taught hundreds of students who became outstanding workers in mineralogy, geochemistry, and allied sciences. He made many field trips throughout the Russian Empire and also traveled extensively abroad, examining mineral collections in Europe and in North America.

In 1911 Vernadsky resigned from Moscow University in protest against the reactionary policies of the Minister of Education. He returned to Saint Petersburg to a position in the Academy of Sciences, to which he had been elected in 1906. His later work and life were largely based in that institution.

During World War I Vernadsky devoted his time and energies to the study of radioactivity and to the organization of expeditions to Lake Baikal and eastern Siberia in search of radioactive minerals and rocks. In 1915 he realized that insufficient information was available on strategic minerals and he organized the Committee for the Study of Natural Productive Forces, commonly known as KEPS, to which he recruited not only the academy staff but also geologists from the universities and government departments.

In 1917 Vernadsky was diagnosed with tuberculosis and moved first to Poltava and then to Yalta in the Crimea. In 1918, during a brief period of Ukrainian independence, he helped to organize and was the first president of the Ukrainian Academy of Sciences. In 1921, when the White armies were defeated, he left Russia and lived in France for five years. During this time he gave a seminar course in geochemistry at the Sorbonne and wrote two pioneer books: *La géochimie* (1924), his synthesis of this new science based on his Sorbonne seminar, and *La biosphère* (1924), a pathbreaking book defining the concept of the biosphere and outlining what was known and what needed to be known in this field.

The Soviet government had a keen appreciation both for Vernadsky's enormous scientific prestige and his moral authority. In 1925 it encouraged him to return to Russia by awarding him a newly created chair in the academy, with the freedom to pursue his own lines of research. He accepted this chair and he and his wife returned to Leningrad in March 1926. He worked with renewed interest and was instrumental in creating many vital committees in the academy. He organized the Committee for the Study of Meteorites, of which he was the first chairman, and the Committee on the Study of Frozen Ground, which was of immense importance in dealing with the problems of permafrost.

His interest in radioactivity continued throughout his life. He was particularly concerned with the role of radioactive elements in geology and their distribution in the earth's crust. He realized the significance of the discovery of atomic fission, and in June 1940 he urged the academy to undertake the study of uranium minerals in connection with the production of atomic energy.

Vernadsky's last years were saddened by the horrors of the German invasion during World War II and the loss of his wife, his constant and helpful companion. His name lives on in the Vernadsky Institute of Geochemistry in Moscow, the premier institute for geochemical research in the former Soviet Union.

Bibliography

Bailes, K. E. *Science and Russian Culture in an Age of Revolutions: V. I. Vernadsky and His Scientific School, 1863–1945*. Bloomington, IN, 1990.

BRIAN MASON

VIEWS OF THE EARTH

When the Apollo astronauts transmitted pictures of planet Earth viewed from space, our home planet took on a whole new perspective (see Plate 44). Earth appeared as a beautiful and fragile ecosystem in the black ocean of space. Those photographs marked a turning point. Since 1960 astronauts and Earth-orbiting satellites have been documenting the evolution of planet Earth both photographically and with an increasingly sophisticated array of imaging techniques.

Modern remote sensing—collecting information by observing a planet from space—began with the Ranger and Surveyor missions, which were designed to provide more detailed views of the lunar surface than anything that could be obtained with a telescope, in preparation for the Apollo lunar landings.

The informational value of the view from above is not new. Early civilizations understood the advantage of the high ground. The fortified hill towns of medieval Italy provide a prime example of strongholds from which defenders could keep the surrounding countryside under surveillance. Aerial photography, the forerunner of remote sensing, commenced when Gaspard Felix took the first photograph from a balloon, 80 meters (m) above Paris, France, in 1858. Military applications have driven the development of both aerial photography and remote sensing. The first serious application of aerial photography was the use of photographs taken from balloons to track Confederate troop movements during the American Civil War. Aerial photography from airplanes, which was introduced in 1909, was used extensively during World War I.

Between the two world wars, cartography, geological mapping and prospecting, and forestry began to rely on aerial photographs. Impetus for further development came during World War II. In 1940 the planned German invasion of Britain was forestalled when many ships were observed from the air along the English Channel. Late in the war, night bombers were equipped with an imaging radar for night "vision." This was the forerunner of the radar-imaging systems flown on aircraft, satellites, and spacecraft today. Radar has been used to map the geology of the ocean floors and the *Magellan* spacecraft used radar to penetrate the dense clouds surrounding Venus and provide a global topographic map of our nearest planetary neighbor (*see* VENUS).

During the 1950s false-color infrared imagery, which uses the transmission of spectral reflectance values, was applied to vegetation mapping. From there, it was a short step to crop, land-use, and environmental monitoring. The first weather satellite was launched in 1960, joining a number of unpublicized "eye in the sky" reconnaissance satellites.

Astronauts have carried cameras into space since the early days of manned spaceflight. The *Gemini* spacecraft traveled sufficiently far from Earth to provide sweeping planetary perspectives, revealing large-scale geological features and weather phenomena. Similarly amazing pictures were returned by the Apollo and Skylab crews, and thousands of photographs of our home planet are brought back at the conclusion of each space shuttle mission. These photographs are used for geological and environmental research, for teaching geography, geology, oceanography, and meteorology, and are widely used in publicity aimed at raising public awareness concerning the delicate environmental balance of Earth's ecosystem.

Applications

Looking at Earth from space makes meaningful global monitoring possible. Data can be obtained very rapidly for large areas of the planet, and comparatively inaccessible regions can be investigated quickly and efficiently. Remote sensing usually applies to both the surface and atmosphere of a planet. Meteorology is a major application, profiling atmospheric temperature, measuring wind velocity, and tracking storm systems. Much progress has been made in understanding the manner in which the oceans, which cover 70 percent of the surface of Earth, contribute to the balance of our finely tuned biosphere (*see* EARTH OBSERVING SATELLITES).

Satellites chart ocean current movements, map sea surface temperatures, and continue to contribute to developing a better understanding of the El Niño effect, for example, by providing a global view of oceanic and atmospheric systems. Astronauts have contributed photographs of wave packets, current margins, and plankton blooms that can only be seen from the distance of space. Satellites can image and track the movement of plankton, assisting fishing fleets in tracking shoals of fish to their feeding grounds. Similarly, endangered species such as blue whales can be better protected if wildlife services are alerted to the feeding grounds on migration routes. Marine navigation is assisted by monitoring the distribution and motion of sea ice.

There are many geological applications of remote sensing. Geologists use satellite imagery to identify rock types, to prospect for petroleum and mineral deposits, to understand complex, major-scale geological features, and to observe crustal motions over time. At a more local scale, images from space are used to monitor volcanic events, for example, observing activity at the vent, mapping the movement of a debris flow, and tracking an ash cloud as it spreads kilometers from the volcano and climbs into the upper atmosphere. Astronauts flying over the Earth's surface at 490 kilometers (km) a minute see snapshots of geologic evolution. They can see how volcanoes grew, eroded, and sank back into the ocean as atolls in the French Society Islands, or how tectonic plate movement under the Red Sea is moving the Arabian peninsula away from Africa (see Plate 45).

Satellite imagery is used routinely in updating maps. "Perspective" images, produced by computer techniques that combine data sets from a satellite, provide elevation data. Remote sensing is used extensively in agriculture, forestry, and botany, monitoring the extent, health, and type of vegetation cover. Imagery can reveal a wealth of data including soil type and water content for the prediction of crop yield, and can provide early warning of pest infestation.

Imagery from space provides accurate and up-to-date information on water resources and is used to predict reservoir capacities. This capability has monitored the continuing drought crisis in the Sudan. Winter snowfall data are analyzed to determine meltwater run-off from snow in spring. This is how the multi-year California drought was declared at an end in the depths of the winter of 1993. Imagery also assists in determining where to lay pipelines for essential resources, reducing cost, surveying, and construction time.

Environmental monitoring is a major contribution of space-based observation techniques. Shuttle astronauts regularly photograph the Betsiboka Estuary in Madagascar, which runs red with the rich soil carried down from the interior. Forest clearing has resulted in the rapid erosion of the topsoil and land cleared for agriculture will be useless in a matter of years. Disaster control has been greatly enhanced by imagery from space, providing warnings of hurricanes, blizzards, dust, and sandstorms, and monitoring floodwaters and pollution. Satellite images contributed to the environmental impact assessment of the Chernobyl nuclear power plant disaster. Shuttle astronauts photographed extensive smoke plumes from the many oil-well fires set during hostilities in Kuwait in 1991.

In 1993, NASA's Earth Radiation Budget Experiment (ERBE) satellite provided conclusive evidence that the June 1991 eruption of Mount Pinatubo in the Philippines had resulted in a temporary 1°F (a half of 1°C) cooling in global surface temperature. ERBE measurements showed that stratospheric aerosols produced by volcanic gases and ash from the Pinatubo eruption were responsible for increased reflection of solar radiation and the resultant cooling effect (see Plate 46).

Remote-Sensing Platforms

Remote-sensing systems can be classified as active or passive. Further categorization relates to the wavelength of the radiation to which sensors respond. Radar is an active system, illuminating the subject of study with its own "radiation." Passive sensors include photography and infrared scanning systems (*see* EARTH OBSERVING SATELLITES).

Shuttle astronauts provide the only natural color photographs of Earth from space. The shuttle flies at an altitude of between 260 and 610 km above Earth's surface and surface features as small as 20 m across can be seen if a 250 millimeter (mm) lens is used. Earth-observing satellites have two basic operating orbits. Sun-synchronous satellites orbit at approximately 550–950 km altitude, and geosynchronous satellites are located at 41,000 km altitude. Sun-synchronous satellites pass close to the poles in orbits that move progressively around Earth. They are inclined relative to the Equator to ensure correct timing for global coverage. For example, *Landsat 5,* which was launched by the United States in March 1984, passes over the same location at 9:30 A.M. local solar time every 16 days. The Thematic Mapper (TM) on board *Landsat 5* can provide 30 m resolution. (The size of the smallest element that can be identified is 30 m on a side.) The French SPOT satellite, launched in 1985, passes overhead at 10:30 A.M. every 26 days. SPOT also achieves a resolution of 30 m.

The motion of geosynchronous satellites, which orbit 41,000 km above the surface, is parallel to the rotation of Earth. The satellite remains above a fixed point on Earth's surface. Such satellites can monitor an entire hemisphere. The resolution is coarse but these systems are ideal for weather satellites monitoring major atmospheric circulation patterns. The GOES weather satellites operated by the

National Oceanic and Atmospheric Administration (NOAA) fall in this category.

Remote-sensing capabilities have expanded enormously due to technological innovation and the development of sophisticated image processing techniques over the thirty-year history of this space-based science. Launch of NASA's Earth Observing System will mark a major development in remote sensing, with the establishment of a dedicated, integrated, Earth-monitoring system. Many applications of remote sensing have been perfected to a point where they can be extrapolated to obtain information about the surfaces and atmospheres of other planets. For example, data from the *Clementine* spacecraft lunar mapping mission in 1994 provided new information on specific rock types within lunar craters, establishing the existence of strata of different rock types on the Moon.

Bibliography

DRURY, S. A. *A Guide to Remote Sensing.* Oxford, Eng., 1990.

FRANCIS, P., and P. JONES. *Images of Earth.* Englewood Cliffs, NJ, 1984.

Space-Based Remote Sensing of the Earth. Report to Congress prepared by NOAA and NASA. Washington, DC, 1987.

STRAIN, P., and F. ENGLE. *Looking at Earth.* Atlanta, GA, 1992.

PATRICIA DASCH

VOLCANIC ERUPTIONS

Why do volcanoes erupt? Are there ways to forecast when they will erupt? Can we do anything to prevent eruptions? How many volcanoes are there around the world? Which might erupt? In the earth sciences, a favorite subject is volcanic eruptions. Few people in the United States have ever witnessed the active processes of an erupting volcano, and photographs of volcanic eruptions have generated perhaps more questions than they have answered.

Volcanology has made tremendous progress in the past thirty years because new instrumentation has combined with the theory of plate tectonics to make volcanism much more quantitative. Important eruptions—such as Mount Saint Helens (United States 1980), Nevado del Ruiz (Colombia, 1985), and Mount Pinatubo (Philippines, 1991)—have provided good examples of the different natures of volcanoes and captured the attention of the public and of the officials responsible for hazard mitigation.

The different types of eruptions are well known from famous examples. Vesuvius (Italy) erupted in 79 B.C.E. and buried the cities of Pompeii and Herculaneum because of its Plinian (explosive) eruption style. Pliny the Elder, a military leader who helped to evacuate some of the citizens, died from the effects of the ash and gas. His nephew, Pliny the Younger, who wrote one of the first volcanological reports, describes the sustained, highly explosive column that towered about 30 kilometers (km) over the volcano for almost a day. Mount Saint Helens and Pinatubo are modern examples of this eruption style, which usually ejects magmas of silicic composition. For two thousand years Stromboli (Italy), known as "the Lighthouse of the Mediteranean," had many sustained nonexplosive eruptions (i.e., magma fountains to 300 meters or m above the crater) and serves as the namesake for similar volcanic eruptions around the world. These often occur at cinder cones, small basaltic volcanoes that erupt for short periods (lasting about one decade) and are hence known as monogenetic. Peléan eruptions, named after the terrible 1902 eruption of Mt. Pelée (Martinique) that killed 28,000 people, are violently explosive and produce pyroclastic flows. These hot clouds of ash and gas move great distances downslope, at velocities as great as 100 kilometers/hour (km/h), because of low internal friction. These eruptions are also associated with volcanoes that erupt silicic magmas. Hawaiian-style eruptions, best known to the people of the United States, are characterized by fluid basaltic magmas that flow rapidly down valleys as lava, causing fires and burying property.

Most of the 1,500 active volcanoes of the world (Simkin and Siebert, 1994) occur in the circum-Pacific "Ring of Fire" that surrounds the Pacific Ocean basin. They owe their existence to the subduction of the ocean floor (plates) beneath the margin of the continents or island arcs that surround the basin. Once subducted to depths of approximately 150 km, the oceanic plates become unstable and partially melt to produce basaltic magmas. These magmas are buoyant and rise to produce chains of volcanoes parallel to the coastline (e.g., the Andes of South America and the Aleutians of Alaska). A few volcanoes exist in the middle of oceanic plates (such as Hawaii) because isolated hot spots in the mantle produce "plumes" of molten magma that rise through the overriding plate. Other volcanoes are found where plates have broken and are moving apart (mid-ocean or continental rifts) and the mantle upwells and produces magma. Most of these volcanoes are underwater and their eruptions are poorly known. The causes of eruptions are not well constrained for individual events but are ultimately related to the rise of magma. Earth experiences fifty to sixty subaerial eruptions somewhere each month.

Eruptions of volcanoes cause large death tolls for a variety of reasons. Krakatau (Indonesia) erupted in 1883 and killed more than 36,000 people despite the fact that almost no one lived on the island of the volcano. The collapse of the mountain, however, triggered tsunamis (tidal waves) that traveled many kilometers and destroyed coastal cities. Virtually the entire population of St. Pierre, Martinique, was killed by the eruption of Pelée because the hot pyroclastic flows destroyed everything in their path. The relatively small Plinian eruption of glacier capped Nevado del Ruiz killed about 25,000 people in distant towns because melting ice caused lahars (volcanic mudflows) that rapidly descended from the high summit. By far, lava flows are the least dangerous eruptive style, a fact little known by people accustomed to images of burning property. Lava flows usually represent the last stage of an eruption, in which the gases have escaped and little energy is left in the magma. The velocity of lava is usually low enough and the areas covered quite small that people can walk away to safety.

Forecasting is the term used by volcanologists to describe their efforts to warn people about potential eruptions. Volcanologists are not able to predict an eruption in the same way that meteorologists can measure the probability of temperature, rainfall, or the severity of hurricanes. Seismic events, deformation, and gas geochemistry are three independent areas that are used to recognize patterns of activity associated with approaching volcanic eruptions. Earthquakes that occur beneath volcanoes are special in physical character because they are caused by the movement of the magma and the release of gases. Harmonic tremor is one example of seismicity known to be associated with magma and is often used in forecasting an eruption. Deformation is the change in the shape of a volcano, which can be recognized by electronic tiltmeters and laser distance measurements that are caused as magma moves up into the edifice of a volcano before eruptions. Gases are the driving force behind explosive eruptions because they expand powerfully as the pressure confining a magma is released. Direct sampling or remote sensing of gases has been used to recognize reawakening of volcanoes, because these gases are premonitory of eruptions. A very small number of volcanoes around the world have been so carefully monitored, for a long enough time, so that scientists can forecast eruptions with considerable confidence.

Mount Saint Helens was recognized—because of careful geological mapping of deposits and dating of older eruptions—as the U.S. volcano most likely to erupt in the lower forty-eight states before the end of the twentieth century. The first sign of activity was an earthquake and a small explosion in the crater on 27 March 1980. For almost two months, explosive eruptions and powerful earthquakes were studied. One special quality was the near absence of volcanic gases released by the eruptions. This told scientists that the eruptions were "phreatic" in nature, indicating that the magma was still sealed from the surface and only boiling groundwater was being released despite the notable deformation of the volcano, which produced a bulge on the upper slope of the cone. After almost no change in activity, a major earthquake signaled by seconds the beginning of an extremely explosive eruption, which began as a section of the northern flank of the volcano collapsed. A lateral blast sent old rock debris and magma at very high velocity to distances of 30 km, knocking down huge trees, stripping the soil away, and rolling large bulldozers. The sector failure became a debris avalanche, which traveled for about 20 km down a large valley. A lahar formed because melting glacier ice and stream water combined with the debris, flowing another 30 km. After a few minutes, a Plinian eruption column grew to about 24 km and lasted for nine hours. Ash fell over the eastern part of Washington and then in Idaho and Montana as well.

In 1991, Mount Pinatubo produced a powerful Plinian eruption of approximately 5 to 7 cubic kilometers (km^3) in volume. The recognition by scientists from the Philippine and the U.S. government geological agencies to the hazards of future eruptions was immediate and their rapid response mitigated the threat to people. The largest group at risk, those living at nearby Clark Air Force Base, were evacuated just three days before the major eruption of 15 June. One important aspect of the volcanic eruptions, however, happened anyway. The gases that drive explosive eruptions can be carried to high elevations and enter the stratosphere, the upper atmosphere located beyond normal weather activity, and these gases cannot be removed quickly. The Pinatubo eruption injected approximately twenty million metric tons of sulfur dioxide into the atmosphere (other gases, such as water and carbon dioxide, are even more abundant in such eruptions), and such events are known to cause temperature cooling in the stratosphere. Sulfur dioxide absorbs some of the incoming ultraviolet sunlight and can cause changes in the climate patterns, which can alter the overall world climate. Some major historic eruptions have almost certainly produced unusually severe winter weather, such as the 1815 eruption of Tambora in Indonesia.

The most difficult task faced by volcanologists is to make their scientific data useful for mitigating volcanic hazards. The lack of universal rules applicable to all volcanoes makes forecasting highly uncertain and reduces the confidence of public officials in scientific ideas. Despite the knowledge that Mount Saint Helens was active and that certain areas were seriously at risk, government officials weighed the economic impact of preventing the logging industry from maintaining normal production. After a few weeks of small eruptions, the warnings of volcanologists were dismissed and several hundred men were back at work. It was extremely lucky that the major eruption occurred on a Sunday morning, or many loggers may have been killed. When Nevado del Ruiz increased gas emissions and caused much more seismic activity for about one year, the Colombian scientists and some international experts worked quickly to assess the hazards of future eruptions. They recognized that the town of Armero was located in a position that had allowed it to be destroyed by lahars in historic eruptions. Public officials focused attention on the high probability that a serious eruption might devastate the town. However, in the ninety years since the previous eruption memory of its consequences had faded. The perceived negative economic impact of educating the public about hazards was seen as much more important than the possibility of an eruption, so authorities avoided decisions concerning what action should be taken. About two hours after the eruption caused the lahar, a wall of mud, rocks, and logs measuring 45 m in height left the canyon at 40 km/h and within minutes destroyed almost all of the town buildings and killed more than 85 percent of Armero's residents.

As scientists, volcanologists often find themselves in the difficult position of trying to explain realistically to civil authorities, who lack scientific training, what a pattern of activity on a nearby volcano might mean in terms of hazards. It is possible to trigger large-scale panic or a false sense of confidence, if there are misunderstandings. Even more likely, the officials are forced to make very important decisions based on the concept of probability for certain types or scales of eruptions that might be happening in different time frames. There is a serious need for the discussions and decisions of officials to be made in collaboration with experts, so that the overall impact of natural hazards can be balanced appropriately with the social considerations. Volcanology is developing rapidly but it is still unable to provide accurate predictions of future eruptions with unqualified success.

Bibliography

DECKER, R., and B. DECKER. *Mountains of Fire, the Nature of Volcanoes.* Cambridge, Eng., 1991.

SIMKIN, T., and L. SIEBERT. *Volcanoes of the World,* 2nd ed. Tucson, AZ, 1994.

SMITHSONIAN INSTITUTION. *Bulletin of the Global Volcanism Network.* Washington, DC.

STANLEY N. WILLIAMS

VOLCANISM

On Earth, volcanism, or volcanic activity, takes many more forms than it does on the other terrestrial planets because a wider range of magma types (*see* IGNEOUS ROCKS) reach the surface, each with different physical and chemical properties. Another reason for the diversity of volcanism is that Earth's hydrosphere allows rising magma and water to interact explosively in many environments. If the water is deep enough, as in the ocean basins, hydrostatic pressure stifles explosive activity, and pillow and sheet lavas develop; pillows are a mass of bloblike and tube-shaped pods of lava. About 70 percent of all magma reaching Earth's surface does so under the oceans. Therefore activity occurring in the submarine environment and the volcanic rocks formed there are the most common on our planet, yet they remain the least well known for obvious reasons. The restricted range of basalt magma types erupted on the ocean floors (*see* PLATE TECTONICS) suggests submarine volcanism and its products are quite uniform worldwide.

In the subaerial environment, on land and around its shallow-water fringes, volcanism is either effusive—that is, lava-producing—or explosive, activity that produces volcanic ash (more correctly called pyroclastic material or tephra). The explosivity of eruptions depends largely on magma properties, including viscosity, which is dependent on chemical composition and temperature (both somewhat related) and gas content, and upon magma-water interaction. On our wet planet, magma meets ground or surface water almost everywhere that it rises surfaceward, leading in many cases to the explosive intermixing of a hot (600–1,200°C) viscous liquid and a cold one. The deeper under the surface that this interaction takes place, the less likelihood of surface explosions.

It is difficult to divorce explosive from effusive activity as they occur simultaneously at most active volcanoes, so they are described together while discussing the major types of volcanic activity. The types take their names from representative volcanoes or volcanic eruptions, and they grade into each other to some extent. In approximate order of increasing explosive power, the types are as follows.

1. Hawaiian activity forms widespread, thin, black flows of basalt lava. Liquid basalt lava is runny, or of low viscosity, at one end of the spectrum of physical properties of magma types. These flows behave somewhat like water, flowing down volcanoes from the vent area and filling in depressions and valleys. The surfaces of the lava are either smooth and ropy (termed *pahoehoe*) or blocky (called *aa*; both words are Hawaiian), depending on viscosity and shearing relationships in the flow. Stacks of basaltic lava flows build up shield volcanoes, like Mauna Loa and Mauna Kea (both over 3,900 kilometers or km high) on Hawaii. Because the bases of the volcanoes are 7,300 meters (m) below sea level, these monsters are the highest single edifices on Earth's surface. Hawaiian activity often occurs along fissures, lines of vents that are semicontinuous sites of magma emission during eruptions. Some fissures in Iceland are over 100 km long and represent a similar kind of activity to that occurring along mid-ocean rifts (*see* PLATE TECTONICS), as Iceland can be considered to be formed of oceanic crust exposed above sea level. A more extreme variety of Hawaiian activity is the flood basalt volcanism that has occurred spasmodically throughout Earth history, perhaps related to the rise of mantle plumes. In this type of volcanism, vast amounts of lava (from 100,000 to more than 1 million cubic kilometers or km^3) erupt over about 1 million years (Ma) in a relatively small area. These create a pile of flows so heavy that they depress the crust (*see* ISOSTASY), leading to large areas of flat topography (e.g., the Columbia River plains of Washington and Oregon). To our knowledge, this type of activity is not occurring on land today. Explosive activity accompanying Hawaiian volcanism is generally quite mild due to the gas-poor character of basalt magma. It forms fire fountains that exceptionally reach about 1,000 m high. These build small cones and raised areas of spatter (molten lava bombs that flatten and adhere together when they land); in most cases finer ash deposits are not formed in abundance. Intra-plate hot-spot volcanoes like Reunion in the Indian Ocean feature Hawaiian activity. Under these volcanoes magma rises from beneath through dike and sill networks (*see* PLUTONISM). It is sometimes held in shallow reservoirs or chambers en route; the latter form small gabbro (mafic or iron- and magnesium-rich) plutons when they solidify. When occurring along plate spreading centers such as mid-ocean ridges, sheet-dike complexes will form under fissure zones as well as plutons like the famous Skaergaard intrusion in Greenland.

2. Strombolian activity is akin to Hawaiian as it forms both lava flows and pyroclastic deposits, and is also characterized by basaltic to andesitic (intermediate) magma, but there are subtle yet important differences. Magmas forming strombolian eruptions are slightly more viscous (sticky) and gas-rich, causing a higher proportion of material to fragment into pyroclasts. Bigger cones, called scoria cones, therefore form around strombolian vents. Scoria is frothy (vesicular) dark-colored pyroclastic material, as opposed to the better-known pumice, which is light-colored, frothy, and of lower density. Strombolian activity also displays pulsating fire fountains from a single vent, higher eruption columns (the convecting cloud above the vent), and a wider dispersal of scoria and ash fallout. Lava flows issuing from strombolian vents are usually shorter, thicker, and have more blocky *aa*-type surfaces than their Hawaiian counterparts; magma viscosity is again the controlling factor.

3. Vulcanian volcanism is a catch-all term covering quite a wide range of explosive behavior. Often there is accompanying effusive activity. Closely related is so-called Pelean activity, the name for the notorious Mount Pelée that caused the instantaneous death of thirty thousand people in the town of Saint Pierre on Martinique (Lesser Antilles) on 8 May 1902. Vulcanian eruptions involve moderately viscous and gas-rich magma that may shatter with great force, fragmenting itself into tiny pieces of hot ash. This, and the high rate at which the ash and pent-up gases are emitted from the vent, lead to towering eruption columns that commonly reach from 12 to 20 km in altitude. Some of the coarser pyroclastic material falls around the vent area, building a low cone, but most rises high in the column, promoting wide dispersal of fallout as it blows downwind as an ash plume. Lavas produced during vulcanian activity are viscous and blocky. They tend to clot around the vent, forming a lava dome, or dribble down the sides of the cone as thick tongues. Lava domes often cause vent blockage, which helps to build up gas pressure in the magma below, leading to a similar situation to shaking a bottle of carbonated beverage with one's finger over the top. Push out or remove the lava, or your finger, and vigorous emission occurs! Sometimes the gas escapes sideways, due to a plug of lava over the vent, forming a directed blast. This phenomenon, and the ground hugging, hot flows of pyroclastic material that it spawned, caused the utter devastation at Saint Pierre. A similar blast, although triggered by collapse of the volcano, produced the equally devastating blast flow at Mount St. Helens at 08:32 A.M. Pacific Standard Time on 18 May 1980. An unpopulated region lay in the path of the blast flow and the death toll was limited to thirty to fifty unfortunate people, but 5,500 square kilometers (km^2) of planted, commercial forest were flattened. Not all vulcanian eruptions are so violent, and mild, semicontinuous explosions of andesitic magma from composite (or strato-) volcanoes are probably the world's most common type of volcanic activity. Because intermediate (andesitic) magmas dominate volcanism around Earth's subduction zones—e.g., the Pacific Ring, or Rim, of Fire; *see* PLATE TECTONICS—composite cones such as Mount Fuji (Japan) are Earth's most abundant and most active large volcanoes. At any one time about ten to twelve composite volcanoes are usually producing mild vulcanian explosions, venting water, carbon dioxide, and sulfur and chlorine gases into the atmosphere. Lava domes formed of andesite or dacite magma also tend to undergo small vulcanian explosions almost continuously while they are in a growing stage. Composite volcanoes can grow to large volumes and have long life spans (up to 1 Ma) involving many individual eruptions. Inside and below, dikes and stocks of diorite or granodiorite are emplaced. Exposed examples of these intrusions host some of Earth's richest ore bodies (*see* MINERAL DEPOSITS, IGNEOUS) of precious metals (e.g., gold, silver, copper, lead, and zinc).

4. Plinian activity and related types occur when more silica-rich, highly viscous, gas-rich magma such as dacite or rhyolite magma erupts, causing huge columns of ash and gas that shoot up to heights of 40 km into the atmosphere. Great eruptions such as Mount Pinatubo (1991), Krakatau (1883), and Tambora (1815) are most often of this type. Widespread fall deposits of pumice and fine ash form downwind of the vent. The lowermost parts of the eruption columns may collapse to form pyroclastic flows along the ground, depositing the volcanic rock ignimbrite or ash-flow tuff. Vast sheets of ignimbrite are among the largest volcanic deposits, rivaling flood basalt provinces in volume and area, and are testimony to eruptions that yield, in extreme cases, in excess of 2,000 km^3 of magma. Often before and after these great events, gas-poor parts of the underlying magma leak to the surface to form silicic (or felsic) lava domes. These are usually steep-sided, oval blobs of crumbling, blocky lava up to 600 m in elevation. Plinian eruptions often occur from, and cause, caldera volcanoes because the volume of magma withdrawn is large enough that significant evacuation of the subjacent magma chamber occurs, temporarily leaving a void into which the superstructure of the volcano collapses, forming deep, oversize craters known as calderas. Large silicic magma bodies reside under calderas, and if these freeze at the end of the volcano's life cycle, they become stocks or batholiths of granitic rock.

As noted above, surface or ground water can modify all of these eruptive styles. Those discussed represent magmatic activity, because the characteristics are dominated by magma properties, but in many instances phreato- (or hydro-) magmatic activity occurs (*phreatos* is Greek for ground water). Compared to their magmatic counterparts, the columns in these eruptions are steam-laden, the magma is more highly fragmented, and the resulting pyroclastic deposits are more fine-grained. Very often the water in the deposited ash causes extensive chemical alteration, forming a hardened (indurated) deposit, one type of tuff.

Over forty thousand volcanoes are recognized on Earth. A volcano is simply a place on the earth's surface from which magma and/or magmatic gases issue. Most have a vent or vents, the actual orifice from where the magma is released, surrounded by a cone or ring of pyroclastic deposits or lava. Volcanoes can be classified in several ways but there are always many exceptions to every rule. The most abundant individual volcanoes are monogenetic cones and related features. These are small volcanoes that erupt only once, during the event that creates them, and they may form by both phreatomagmatic and magmatic activity. Scoria cones, tuff rings and cones, and spatter cones are common types; maars are bowl-shaped craters with low rims of tuff. Many monogenetic volcanoes form on bigger polygenetic volcanoes, when they are sometimes called adventive or parasitic cones. Some grow on the flanks of larger shield volcanoes, but in other cases, a whole region of individual scoria cones, sometimes numbering in the tens of thousands, develop over time. The reason such volcanic cone fields develop rather than a single volcano, or a few larger ones, is poorly understood. Most examples are situated on thick lithosphere, implying that depth to the magma source region and lithospheric structure are controlling factors. A dike complex must feed magma to such scoria cone fields from the magma source beneath. Polygenetic, or complex, volcanoes grow by the addition of material from many eruptions. The major types are shields, composite (or strato-volcanoes), and lava-dome complexes, all lava-dominated volcanoes. There is a fallacy that composite volcanoes have approximately equal parts of lava and pyroclastic material; ash deposits very rapidly erode from steep volcano sides and most composite cones are composed of lava flows and debris eroded from the lavas. Large conical volcanoes composed mainly of pyroclastic deposits are very rare. If a polygenetic volcano has gone through a caldera-forming stage, then it may possess, or be called, a caldera. Most volcanologists consider that a caldera must have a component of collapse due to eruption or magma withdrawal to warrant the name, but large craters formed by erosion were recognized as calderas in the past. Calderas can be as small as 2 km in diameter, and the largest on Earth are about 70 × 40 km in size. Many calderas have no positive topographic relief (i.e., there is no volcanic edifice), rather they are basins, or holes in the ground. As these are often the largest examples, and as collapse is usually along regional structural weaknesses (e.g., faults, graben), they are sometimes called volcano-tectonic depressions. An example is the 35 × 40 km caldera occupied by Lake Tanpo, North Island, New Zealand, a volcano that erupts rhyolite magma.

Volcanoes can be either active or inactive, the latter class traditionally having been divided into dormant or extinct. However, because during the past few decades at least five volcanoes considered to be extinct have produced major eruptions, this classification has obvious problems! The main one is that some volcanoes have very long periods between eruptions, longer than human history or memory. A better, simpler classification is to consider volcanoes to be live (will erupt again) or dead; but proving that a volcano is dead or alive is a costly, time-consuming process involving much volcanological detective work.

Volcano types discussed above are those occurring on land. Submarine volcanoes probably outnumber the subaerial total by several tens of thousands but we know little about them. Most are called seamounts, but this term is not specific to volcanoes because some undersea mountains of nonvolcanic origin also share this name. The term guyots has also been used. Those few that have been explored by drilling, dredging, remote imaging, or submersible dives are shield-like. This is not surprising considering that they are dominantly basalt and that the deep water environment favors effusive activity.

Volcanoes can also be classified according to their tectonic setting (*see* PLATE TECTONICS).

Bibliography

DECKER, R., and B. DECKER. *Mountains of Fire*. Cambridge, MA, 1991.

FRANCIS, P. *Volcanoes: A Planetary Perspective.* Oxford, Eng., 1993.

STEPHEN SELF

VOLCANISM, PLANETARY

Volcanism has been one of the major processes shaping the surfaces of the terrestrial planets. Volcanic features have been identified on the Moon, Venus, Mars, Jupiter's moon Io, Neptune's moon Triton, and, possibly, Mercury. Most of the information we have on extraterrestrial volcanism has been obtained by remote-sensing spacecraft. Since sample-collecting missions are expensive and complex, the study of extraterrestrial volcanism relies mainly on the interpretation of imaging, spectroscopic, and topographic data. This interpretation takes into account what we know about volcanism on Earth and also theoretical considerations about how different conditions on each planet, such as gravity and lithospheric thickness, affect the occurrence and the type of volcanic activity (*see* VOLCANISM).

One important difference between terrestrial and extraterrestrial volcanism is plate tectonics. Plate tectonics does not appear to have operated on the other planets, except perhaps in the case of Venus. The absence of plate tectonics is thought to have led to significant differences between volcanoes on Earth and those on other planets. For example, Mars has much larger volcanoes than the Earth does, and this may be attributed at least in part to the lack of plate movement. Unlike the Earth, eruptions on Mars could have occurred repeatedly in the same location for a long period of time, resulting in much larger structures being built.

At present, most of the available data on extraterrestrial volcanic features are morphological in nature, having been obtained in the form of images. It is, therefore, particularly important to be able to relate the morphology of a volcanic landform to eruption dynamics and magma properties. The standard approach is to use terrestrial data to develop physical models that predict how a particular factor (e.g., eruption rate) will affect a particular morphological characteristic (e.g., length of a lava flow). The model can then be applied to other planets to work out eruption and magma parameters from morphological characteristics, taking local conditions into account. Major factors that are known to determine the morphology of volcanic landforms are summarized in Table 1.

Magma compositions are best obtained by analysis of samples returned to Earth; so far, this has only been done for the Moon (*see* METEORITES FROM THE MOON AND MARS). Very few, but valuable, in situ measurements by lander craft have been done on Venus and Mars. Besides those, limited data on surface compositions have been obtained by spectral analysis using both ground-based telescopic observations and instruments aboard spacecraft.

Table 1. Some Important Factors Determining the Morphology of Volcanic Landforms

Planetary Variables	*Magma Properties*	*Eruption Properties*
Gravity	Viscosity	Eruption rate
Lithostatic pressure variation with depth	Temperature	Density contrast between magma and crust
Atmospheric drag	Density	Volume erupted
Atmospheric pressure	Magma composition	Duration of eruptions
Renewal rate of crust	Volume and composition of volatiles (e.g., H_2O)	Slope of terrain
Planetary composition	Crystal content and size	Ejection velocity
Presence of surface liquids	Surface tension	Number, size, and shape of vents
Erosion and burial rates	Yield strength	
Atmospheric cooling		

Modified from Whitford-Stark, 1982.

Such data have been obtained mainly at low resolutions, though this is changing as a result of rapid improvements in instrumentation.

The following sections will review what is known about volcanism on the Moon, Venus, Mars, Io, and Triton, and about the possible occurrence of volcanism on Mercury. It is important to note that apart from Earth (*see* VOLCANISM), active volcanism has only been identified on Io and Triton, although there have been suggestions that activity may be taking place on Venus and on Jupiter's moon Europa.

Moon

Volcanism has been a major process on the Moon (*see* MOON). The dark, smooth areas of the Moon's surface, called maria, are the result of extensive effusions of lava that occurred during a long period in the Moon's history, peaking between 4.0×10^9 years (billion years or Ga) and 3.2 Ga ago. The maria are mainly found on the nearside of the Moon, where the lavas filled up preexisting basins formed by huge meteorite impacts, such as the Imbrium basin. Mare Imbrium (Figure 1) is one of the Moon's youngest maria and shows clear outlines of lava flows. The eruptions that formed the maria are thought to have been quiet (i.e., nonexplosive) effusions of long, low-viscosity lava flows. The equivalent style of volcanism on Earth is called flood basalt.

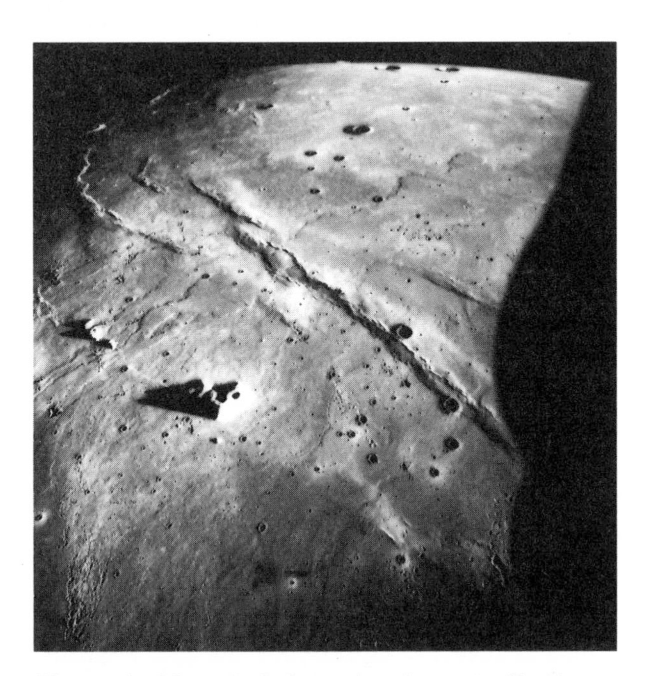

Figure 1. Mare Imbrium view from *Apollo 15*. Note the lava flow running from top right to bottom left and the mare ridges from the top left to bottom right.

The lunar surface lacks large volcanic structures, such as the Hawaiian shield volcanoes, or steep-sided composite volcanoes, such as Fuji. The absence of these features may be due to the high effusion rates of most lunar lavas and to the rarity of explosive activity. Individual shields can only build up when the mean distance flowed by the lavas from a given source area is substantially less than the mean spacing between various source areas. This was generally not the case on the Moon, so "seas" of lavas were formed instead. A few shieldlike structures have been recognized on the lunar maria, but they are small (a few kilometers in diameter) in comparison with the lava flows (which can be hundreds of kilometers long). Also seen on the maria are a few small structures, such as domes and cones, which are thought to have been formed by pyroclastic (explosive) activity. Some small craters thought to be volcanic in origin have also been recognized, their morphology being significantly different from that of the majority of lunar craters which were formed by impact.

The most distinctive features seen on the lunar maria, apart from lava flows, are sinuous rills and mare ridges. Sinuous rills (Figure 2) consist of winding channels that may have a rimless pit at one end. They are interpreted as collapsed lava tubes or drained lava channels. In terms of size, they are considerably larger than their terrestrial lava tube counterparts, which are at most a few kilometers long. In comparison, Hadley Rille on Mare Imbrium is over 130 kilometers (km) long and 5 km wide in places. It has been proposed that the size difference between sinuous rills and terrestrial lava tubes and channels is due to the higher discharge rates of the lunar lavas, and also to the lower gravity on the Moon.

Mare ridges (Figure 1), also called wrinkle ridges, are prominent features on the lunar maria, which can be tens of kilometers long. They are thought to have been formed by a combination of volcanic and tectonic processes. There are several types of ridges; some are thought to be compressional features, while others may have resulted from the piling up of lavas erupted along a fissure.

The collection of lunar samples by the Apollo missions was a great step forward in our under-

Figure 2. View of Hadley Rille, one of the Moon's sinuous rills, from *Apollo 15*.

standing of the Moon. Laboratory analysis of these samples has vastly improved our understanding of lunar volcanism in terms of chemistry and mineralogy. Analysis of mare lavas showed that they are basaltic, but there are important differences between these and terrestrial basalts. Most notable are the absence of detectable H_2O in the lunar samples and the higher abundances of iron, magnesium, and titanium. The lunar lavas also lack the alterations found in terrestrial basalts due to chemical weathering and hydrothermal activity. Dating of samples from both the maria and the highlands also provided invaluable calibration points for the crater counts that are the standard way of dating planetary surfaces using remote-sensing data.

Mercury

Mercury has only been visited by one spacecraft, *Mariner 10*, in 1974. Images from *Mariner 10* revealed that Mercury's surface is, to a first order, similar to that of the Moon. Unlike the Moon, however, there are no vast expanses of maria on Mercury, and no definite evidence that active volcanism has ever taken place (*see* MERCURY). The large number of craters on Mercury's surface, plus the existence of the solar system's largest known impact basin, Caloris, indicates that, like the Moon, Mercury has had a complex history of meteorite impacts.

Mercury does have areas of smooth plains (Figure 3) that may have been formed by volcanic activity, or may have been the result of melting caused by impacts. No telltale volcanic features such as lava flows, shields, or domes were identified on the *Mariner 10* images. It is possible, perhaps likely, that such features exist, but they are beyond the resolution limit of the *Mariner 10* camera (1 km at best). A new mission to Mercury that can acquire higher-resolution images will probably be needed before the volcanic origin of the smooth plains can be established.

Figure 3. Mercury's smooth plains viewed from *Mariner 10*.

Venus

Volcanism on Venus is widespread and extremely varied (*see* VENUS). Over 80 percent of the Venusian surface is composed of volcanic plains and edifices and the remaining regions are probably deformed volcanic deposits. Great progress in the study of the surface of Venus was made by means of high-resolution radar images obtained by the *Magellan* spacecraft. Since Venus has a permanent, thick cloud cover, it is necessary to use radar to map the surface.

Venusian volcanic edifices have a wide range of sizes and shapes. The most common features are small shield volcanoes that are generally less than 200 meters (m) high and have diameters between 2 and 8 km. Less common are the larger shield volcanoes that tend to have extensive lava flows associated with them. One such shield is Sif Mons, which is 300 km in diameter and 1.7 km high. Shields such as Sif Mons appear to be basaltic in composition, but the presence of some relatively steep-sided lava domes (Figure 4) suggests that more evolved lavas, similar to rhyolites and dacites on Earth, may have erupted on Venus. The only data available on the surface composition of Venus have been obtained by five Soviet Venera landers. Geochemical data from all but one landing site indicated the surface is similar in major-element composition to that of tholeiitic and alkali basalt lavas on Earth; however, data from the remaining Venera site are more consistent with an intermediate to silicic composition. It is likely, therefore, that Venus has lavas of varied compositions.

There have been suggestions that pyroclastic activity has also taken place on Venus, and that some radar-dark areas are ash deposits mantling radar-bright plains. Since the high atmospheric pressure on Venus is expected to inhibit pyroclastic volcanism unless the volatile content of the magma is high, these proposed pyroclastic deposits may indicate that volatile-rich magmas have erupted on Venus.

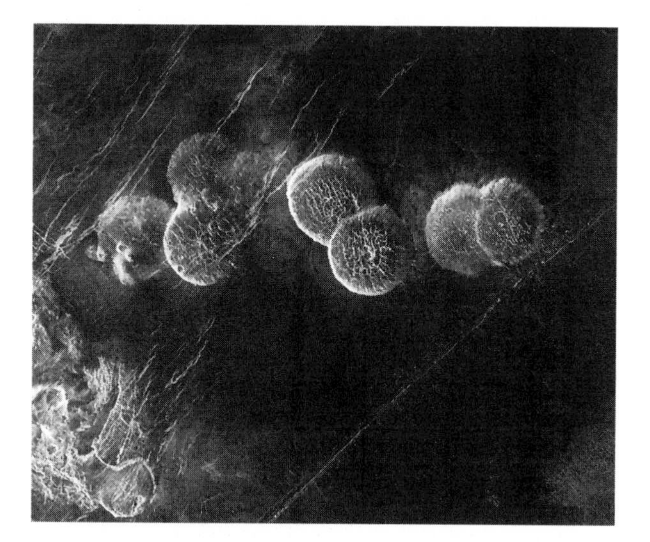

Figure 4. *Magellan* image of lava domes on Venus. Each dome is about 25 km in diameter.

Some of the most enigmatic volcanic features on Venus are termed coronae. These are circular to elongate structures with diameters 200 to 1,000 km, characterized by annuli of concentric ridges surrounding an elevated center. They are thought to be the result of local plumelike mantle upwelling. The Venusian surface also has very long sinuous channels, some of which resemble the lunar sinuous rills. One of the channels, Hildr Fossa, is 6,800 km long and is the longest channel so far found in the solar system.

Mars

Mars has a wide variety of volcanic landforms, which include the solar system's largest known volcano, Olympus Mons (Figure 5). This enormous shield volcano is about 600 km in diameter and 25 km high. Most of the volcanoes on Mars are located on the Northern Hemisphere, which is very different from the Southern Hemisphere. The Northern Hemisphere is formed by relatively young lava plains dotted with volcanic structures, some of which, like Olympus Mons, are spectacularly large. The Southern Hemisphere appears to be much older, as shown by the heavily cratered terrain, and the volcanic structures located there also appear to be more degraded than those in the Northern Hemisphere. The differences between the two hemispheres are accentuated by the fact that the southern plains are 1 to 3 km higher than the Mars datum (average level), while most of the Northern Hemisphere stands below the datum. There are, however, two prominent bulges on the northern plains, which are the volcanic regions of Tharsis and Elysium. The Tharsis bulge is about 8,000 km across, with a summit elevation of 10 km above the datum, while the Elysium region is a smaller but still significant bulge. The origin of these regions is still uncertain. Tharsis is surrounded by a network of fractures and it is clear that it has played a major role in the tectonic evolution of Mars. It has been suggested that its origin is linked to mantle convection associated with the separation of the planet's core (*see* MARS).

Figure 5. *Viking* image of the giant shield volcano Olympus Mons on Mars.

Martian volcanoes are concentrated in the Tharsis and Elysium regions in the Northern Hemisphere. Apart from Olympus Mons, the most conspicuous volcanoes are the giant shields Ascreus, Pavonis, and Arsia Montes that sit atop the Tharsis bulge. Like Olympus Mons, these volcanoes have shallow flank slopes, numerous visible lava flows, and are topped by complex-looking calderas. Also located in the Northern Hemisphere is Alba Patera, a peculiar, shallow volcano 1,600 km across and 6 km high, topped by two nested calderas surrounded by graben. "Patera" is a collective term for a variety of unusual "saucer-shaped," shallow features that often have a central caldera. Other volcanoes on Mars are the domes, usually referred to as tholii (singular-tholus). These relatively small, steeper-sided volcanoes may have experienced explosive activity, or may have been built by more viscous lavas.

The oldest Martian volcanoes are located on the southern highlands. Some of these volcanoes, such as Tyrrhena Patera, have numerous radial channels on their flanks which appear to be the result of extensive erosion. These volcanoes may be similar to ash shields on Earth and may have been formed by eruptions through regolith containing groundwater. Later lavas could have carved the radial channels on the flanks.

The volcanic plains on Mars are similar to the lunar maria, having features such as wrinkle ridges, sinuous rills, and overlapping lava flows. Many of these flows seem to originate several hundred kilometers from the volcanoes, suggesting the presence of vents, fissures, and feeder dikes on the plains. Source areas for specific flows, however, have not been identified in the presently available images. The unfortunate loss of *Mars Observer* precludes more detailed knowledge at this time.

Io

Io, the innermost of Jupiter's four GALILEAN SATELLITES, is of great interest to the study of extraterrestrial volcanism because it is one of the two extraterrestrial bodies where active volcanism is known to occur. Io's volcanic activity was first detected on images returned by the *Voyager 1* spacecraft in 1979, which showed a spectacular plume rising some 200 km above the surface. Several other "hot spots" were observed by the two *Voyager* spacecraft. Since Io is a small body, similar in size and density to Earth's moon, it was apparent that a heat source other than radioactive decay was needed to drive such intensive volcanism. It is thought that tidal heating generated by the gravitational interactions of Jupiter and the other Galilean satellites cause melting of Io's interior and volcanic activity.

The second *Voyager* spacecraft arrived at Io eighteen months after *Voyager 1*. Differences in hot spot (plume) activity between the two encounters and monitoring of hot spots from Earth since that time have revealed that there are two major types: persistent hot spots, such as the major hot spot known as Loki, and short-lived activity, such as Pele, a major plume observed by *Voyager 1* that had ceased to be active by the time *Voyager 2* arrived. Ground-based monitoring of Io since the *Voyager* encounters has considerably extended our knowledge of Io's activity. Using infrared astronomical observations, temperatures of individual volcanic spots have been measured and new spots have been detected.

One of the most interesting issues about Io's volcanism is the relative importance of silicates and sulfur on Io's upper crust and volcanic melts. Io's similar size and density to Earth's moon suggests a predominantly silicate composition. However, the presence of at least a thin layer of sulfur on Io's surface is indicated by spectral measurements obtained by Voyager and also from ground-based observations. *Voyager* detected SO_2 gas over the erupting Loki volcano and ionized sulfur was detected in the Io torus, the trail of neutral and ionized particles that Io leaves behind in its orbit. Io's surface colors as revealed by *Voyager*, predominantly red, orange, and yellow, were considered by some to be evidence of the presence of sulfur, since sulfur cooled from different temperatures commonly exhibit such a range of colors. More precise calibration of the Voyager data, however, indicated that Io's colors are in fact mostly yellowish-green, rather than orange-red. These new colors do not rule out the presence of sulfur; however, interpretations of composition based on color have been contested on the basis that the exact colors of sulfur can be altered by several factors that are unknown for Io, including the presence in the sulfur of even small amounts of other materials.

Probably the strongest argument supporting a predominantly silicate composition comes from the presence on Io of high mountains, steep scarps, and caldera walls. Because sulfur has a low thermal conductivity and a low melting point, it would become ductile at relatively shallow depths on Io and steep structures would not be present, as they would not be self-supporting. This does not preclude the possibility that sulfur exists as a thin layer on Io's surface. In fact, eruptions of silicate magmas could cause sulfur deposits on the surface to melt and produce flows.

Many planetologists are now tending to conclude that both silicate and sulfur volcanism are taking place on Io. However, the relative roles of these materials on Io's crust and volcanic activity are not likely to be fully understood until future spacecraft missions make further observations of the surface and the volcanic activity. Results from the *Galileo* spacecraft, scheduled to arrive at the Jovian system in late 1995, are eagerly awaited.

Triton

Triton, one of Neptune's satellites, has a young, geologically active surface and thin atmosphere (*see* SATELLITES, MIDSIZED). Active volcanism, in the form of plumes, was first revealed by *Voyager 2* during its flyby in 1989. Although not as spectacular as the plumes on Io, the two dark plumes observed by *Voyager* erupting over Triton's south polar cap are quite remarkable. The plumes rise vertically to an altitude of about 8 km, at which point they bend sharply to become dark horizontal clouds extending some 100 km over the surface. It appears that, like on Earth, plumes on Triton cease to rise when they reach an altitude at which they are no longer buoyant. The 8-km limit on Triton may mark the tropopause. The wind at this altitude probably causes the plumes to streak out horizontally to the surface. Nitrogen is thought to be the driving gas for the plumes, although methane is also considered a possibility. The energy source of the plumes is thought to be supplied from within Triton, either from the background heat flow or from the intrusion of relatively warm icy lavas beneath the colder polar cap. Either mechanism could vaporize the ice to produce nitrogen gas. Triton's plumes can be considered analogous to geysers on Earth, where volcanic heat causes rapid vaporization of water to steam.

Ice Volcanism

It is thought that a type of volcanism termed "ice volcanism" may have occurred on some satellites of the outer solar system, such as Europa and Ganymede (*see* GALILEAN SATELLITES). Ice is the major component of these bodies and phase changes in the ice as the satellites evolved could have caused fracturing and faulting of the crust, releasing subsurface liquids. This flooding can be thought of as a type of volcanism in which water and ice slush are released instead of magma.

In conclusion, our solar system has a variety of bodies that are or have been extremely active volcanically. While many of the volcanic landforms seen on other planets can be interpreted as analogues of features on Earth, the specific and diverse conditions on each body will exert a major influence in the style of volcanism and the resulting landforms.

Bibliography

CARR, M. H., H. MASURSKY, R. G. STROM, and D. E. WILHELMS. *The Geology of the Terrestrial Planets.* NASA Special Publication 469. Washington, DC, 1984.

GREELEY, R. *Planetary Landscapes.* Boston, 1987.

ROTHERY, D. A. *Satellites of the Outer Solar System.* Oxford, Eng., 1992.

WHITFORD-STARK, J. L. "Factors Influencing the Morphology of Volcanic Landforms: An Earth-Moon Comparison." *Earth Science Reviews* 18 (1982):109–168.

ROSALY LOPES-GAUTIER

W

WALCOTT, CHARLES

Charles Doolittle Walcott was born in central New York State on 31 March 1850, and became interested in the local fossils at an early age. He spent summers at Trenton Falls, northeast of Utica, and developed a detailed knowledge of the Middle Ordovician fossils and geology. He left the Utica Free Academy at age eighteen, without graduating, and then worked as a hardware clerk. At twenty, he moved to the farm of William Rust in Trenton Falls. Walcott assisted on the farm, and collected fossils jointly with Rust for commercial sale. In 1873 they sold a collection to JEAN LOUIS RODOLPHE AGASSIZ at the Museum of Comparative Zoology in Cambridge, Massachusetts. Walcott spent a week at the museum unpacking the collection and Agassiz encouraged him to study trilobites, a major group of extinct arthropods. Seven years later, Walcott made another sale of fossils to Alexander Agassiz.

Walcott made the transition from collector to scientist. In 1876, he published the first definitive study of trilobite limbs. He discovered the legs of trilobites in unique material from the Rust farm. A few specimens preserved in a thin limestone layer had contracted and rolled up while dying; later their limbs were replaced by calcite. Walcott studied these legs by cutting the rock and making thin sections, both difficult hand operations.

James Hall, state paleontologist of New York, employed Walcott late in 1876 as a special assistant because of his general knowledge of fossils. It was Walcott's first professional position, and it lasted little more than a year, but he remained in Albany, publishing on the correlation of the rocks around Utica with those at Cincinnati and on the growth series of a trilobite, only the second study of its kind.

In 1879, Walcott joined the newly formed U. S. Geological Survey (USGS)—*see* GEOLOGICAL SURVEYS—as a temporary geological assistant and was immediately sent to southern Utah. He measured a stratigraphic section approximately two miles thick, from the Pink Cliffs southward into the Grand Canyon. During this work, he located the Permian-Triassic boundary on the Kanab Plateau. As a result of his field investigations, he received a permanent position. He was also able to complete a fundamental work on trilobite limbs, published in 1881 by the Museum of Comparative Zoology.

In 1880, he was with a field party led by Arnold Hague, studying the Eureka, Nevada, mining district. He returned to Eureka in 1882 for more material and his results were published in 1887 as a USGS Monograph. The fall and winter of 1882–1883 were spent in the Grand Canyon, where he studied Cambrian and Precambrian rocks and measured another stratigraphic section of more than two miles thickness. Combined with his 1879 study, this was the longest section measured by one person.

Walcott next moved into eastern New York and western New England to investigate the "Taconic problem," the question of whether or not the strata exposed in the Taconic Mountains constituted a distinct geologic system. This was a major controversy in American geology for decades. By 1887, Walcott had collected convincing evidence from fossils that these strata were younger rocks which had been misinterpreted, and he removed the Taconic concept from consideration.

During the 1880s, Scandanavian geologists had noted that the sequence of trilobites zones they used in the Cambrian was reversed in America. In 1888, Walcott determined that the American zonation was incorrect. In 1890, Walcott made another dramatic discovery when he found the oldest known vertebrate remains. It would be more than half a century before older fish were found.

In 1894, Walcott succeeded JOHN WESLEY POWELL as the third director of the USGS and headed the agency for thirteen years. Throughout this time, he continued to write on Cambrian rocks and fossils, but extended his investigations downward into the older, enigmatic Precambrian. Meanwhile, the Geological Survey flourished, particularly increasing its activities in water investigations, topographic mapping, and geology. In addition, for nearly a decade and beginning in 1897, the USGS studied and mapped the national forest reserves. In 1902, the Reclamation Service was placed under the Geological Survey with Walcott also supervising its activities. The foundations for the future Bureau of Mines were laid by Walcott, and he served as an advisor on conservation to President Theodore Roosevelt.

Late in 1901, Walcott played a key role in the founding of the Carnegie Institution of Washington, significant for its fundamental and continuing contributions in earth sciences. He served as secretary of the new organization for four years and continued on the executive committee for two decades. Walcott is directly responsible for the establishment of the Geophysical Laboratory, world-renowned for studies of igneous rocks (*see* BOWEN, NORMAN L.).

In 1907, Walcott became the fourth Secretary of the Smithsonian Institution. He immediately instituted a research program in western Canada, where Cambrian deposits were known but little studied. For nearly two decades, Walcott collected and measured stratigraphic sections in this area (*see* HIGHER LIFE FORMS, EARLIEST EVIDENCE OF).

His most notable scientific achievement was discovery of the Burgess Shale fauna in western Canada, the single most important fossil locality in the world. Walcott made a massive collection through several field seasons and described a host of bizarre animals, for this fauna contains many soft-bodied organisms not normally preserved in the geologic record. Walcott was able to demonstrate how incomplete the fossil record is when it is based only on organisms which developed hard parts. He was also able to demonstrate that such diverse groups as holothurians and annelid worms have a vastly longer geologic record than had ever been suspected.

Another significant investigation was that of Precambrian algae. In connection with these studies, Walcott found the first authentic evidence of ancient bacteria. His publications after 1907 fill five volumes of *Smithsonian Miscellaneous Collections*, most of which are devoted to rocks and fossils of western Canada.

In 1912, Walcott and F. C. Cottrell founded the Research Corporation, dedicated to providing funds for science. In 1915, recognizing that the United States needed more research in aviation, he founded the National Advisory Committee for Aeronautics, and served as its chairman from 1919 until his death. From 1917 to 1923, Walcott was president of the National Academy of Sciences and helped found the National Research Council, partly to aid the war effort. Walcott is also credited with the building of the Freer Gallery of Art before Freer's death. At the time of his death on 9 February 1927, he was engaged in a campaign to increase the endowment of the Smithsonian Institution.

Bibliography

DARTON, N. H. "Memorial of Charles Doolittle Walcott. *Bulletin, Geological Society of America* 39 (1928):80–116.

YOCHELSON, E. L. "Charles Doolittle Walcott, 1850–1927." *Biographical Memoirs, National Academy of Sciences* 39 (1967):471–540.

ELLIS L. YOCHELSON

WASTE DISPOSAL, MUNICIPAL

One environmental challenge tied to human existence has transcended the ages. The need for people to deal with the debris of their daily lives has dogged humanity in general, and civilization in particular, from its inception. Prehistoric and nomadic peoples could afford to engage in the practice of discard and move. Settled populations, out of necessity, developed more sophisticated means of managing trash. As early as 500 B.C.E., Greek villages established local dumping sites for refuse. Those dumping grounds were primitive prototypes of, and are direct ancestral links to, the highly engineered landfills of modern times. While waste management technology has advanced over the millennia, it has recently been overwhelmed by the relentless surge in the capacity of humans to generate trash. The result is that, today, people everywhere are facing a garbage crisis.

Municipal solid waste (MSW)—commonly called trash or garbage—is the predominant component of the larger waste stream known simply as solid waste. MSW, the focus of this entry, includes durable goods, nondurable goods, containers, packaging, food wastes, and yard trimmings derived from residential, commercial, institutional, and industrial sources. Examples of MSW items include newspapers, clothing, diapers, cans, bottles, food scraps, boxes, office paper, cafeteria trash, and shipping pallets. Despite the name, MSW invariably contains liquid components, usually as residues in containers.

Tables 1 and 2 provide an overview of the MSW stream.

Annually, Americans generate over three-quarters of a ton of MSW per person, with ubiquitous packaging (paper, plastic, glass, or metal) constituting the single greatest component of the waste stream. It has been estimated that about 25 percent of MSW is inorganic and thus resistant to incineration or composting. About 60 percent of MSW arises from residential sources, with the balance from commercial.

In the United States, MSW is regulated by the Environmental Protection Agency (EPA) under Subtitle D of the Resource Conservation and Recovery Act. Subtitle D establishes stringent technical standards (see 40 Code of Federal Regulations Parts 240–259) for the environmentally safe operation of solid waste disposal facilities (e.g., landfills and incinerators). Subtitle D allows EPA to delegate MSW management authority to states that have developed MSW management plans acceptable to the agency. Many states have done so and are therefore major players in MSW regulation.

Other types of solid waste (e.g., sewage sludge, agricultural waste, used oil, mining waste, construction and demolition debris) and hazardous or radioactive waste are not covered here. The management of MSW discussed here will focus on waste minimization and the environmental impacts of the two major disposal options. Review of the complex issues of MSW collection, transport, storage and processing can be found in the monographs cited at the end of this entry.

The primary objective of solid waste management is the protection of public health. Improperly managed trash is a serious public health threat that can lead to outbreaks of disease. Garbage is rife with human pathogens as people annually dispose of tons of human (e.g., diapers) and animal (e.g., cat litter) wastes. MSW is an excellent growth medium for pathogens and a breeding ground for the vectors (e.g., mosquitoes, flies, and rodents) that transmit them. This disease threat gave rise to the concept of a "sanitary landfill." MSW is also managed for aesthetic reasons—it is both unsightly and foul-smelling. MSW represents a logistical problem

Table 1. Municipal Solid Waste Generation (1990) Product Type

	Tons (Millions)	*Percent (by Weight)*
Containers and packaging	64	33
Nondurable goods	52	27
Yard trimmings	35	18
Durable goods	28	14
Food waste	13	7
Other	3	1

Table 2. Municipal Solid Waste Generation (1990) Material Type

	Tons (Millions)	*Percent (by Weight)*
Paper products	73	38
Yard trimmings	35	18
Metals	16	8
Plastics	16	8
Glass	13	7
Food	13	7
Wood	12	6
Textiles	6	3
Rubber and leather	5	2
Other	5	2

as both its mass and volume are enormous (annually on the order of hundreds of millions of tons and cubic yards, respectively).

Until the mid–1980s, MSW was managed with a simplistic, "end-of-the-pipe" approach. Final disposal of MSW (e.g., landfilling, ocean dumping, uncontrolled surface dumping, or incineration) was the primary focus of its management. Increasing environmental consciousness led to the recognition that such practices were not benign. People began to ask, "What is ultimately happening to this stuff upon disposal?" The answers that came back were not reassuring. In the case of incineration and landfilling, the answers about associated risks were forthcoming and manageable. For ocean dumping and uncontrolled surface dumping the risks were either unknown or known to be too great, respectively, and both practices were outlawed in the United States.

The recognition that our MSW problem is overwhelming "end-of-the-pipe" control has brought about the development of a philosophy of integrated waste management that stresses the virtue of waste minimization. If materials traditionally destined for disposal could be diverted from the waste stream then the magnitude and severity of the remaining materials management problem would be diminished. This principle is the driving rationale behind incorporation of waste minimization into the MSW management scheme. Disposal will continue to be needed for some portion of the waste stream. The objective of integrating waste minimization with disposal practices is to lessen the dependency on these practices, all of which currently entail some undesirable societal, political, and environmental risk. Today, the preferred strategy of waste management is to first maximize efforts to avoid waste, then practice optimized disposal options on that portion of the waste stream that cannot be avoided.

The two primary waste minimization techniques are source reduction and recycling. In addition to reducing the amount of waste disposed, both techniques result in resource and energy conservation by reducing the need to generate goods from raw materials. Since both energy and resource consumption create waste, waste minimization has a compounded positive effect on long-term waste reduction. Waste minimization is beneficial when it reduces either the amount or the toxicity of the waste stream.

Source reduction is the process of avoiding unwanted or unneeded material production. Quite simply, if a product (or by-product) is not created at its source, then there is no need for its disposal. Source reduction is the best form of waste management. Source reduction can be practiced by manufacturers and consumers. Manufacturers can optimize production processes to minimize waste or scrap formation, reduce packaging, and substitute less toxic components in products (e.g., substitution of soy-based inks for inks containing the toxic metal cadmium). Consumers can seek out products that avoid unnecessary packaging (e.g., tomatoes can be purchased unpackaged), purchase in bulk to minimize packaging, reuse containers, and substitute less toxic products (e.g., cedar chips are an effective substitute for synthetic chemical mothballs). Reuse lessens the demand for new products and is perhaps the simplest form of source reduction.

Communities have encouraged source reduction by supporting dissemination of information about such opportunities, establishing economic incentives for source reduction such as "pay-as-you-throw" charges based upon the amount of trash hauled away, and instituting bans on packaging considered burdensome (e.g., fast-food "clamshells").

Recycling is predicated on recognizing residual value in a used item. An item becomes waste once its value has been dissipated. It is important that we be sure that all useful value has been dissipated before disposal. By withholding an item from the waste stream as long as possible, the stream's growth is managed. Both manufacturers and consumers can recycle. Recycling in the manufacturing process (known as pre-consumer recycling) can be practiced when, for example, scrap material is returned to the feed stock. This is commonly done with thermoset plastic trimmed from final products. Post-consumer recycling involves reprocessing a used material into a new product. This reprocessing consumes less energy and natural resources than manufacturing from raw materials. For example: each ton of recycled glass saves 41 liters (l) of fuel oil; each ton of recycled newsprint saves seventeen trees. About 17 percent (by weight) of the entire MSW stream in the United States is recycled. The corresponding percentages for individual components are: paper (29 percent), metals (23 percent), glass (20 percent), yard wastes (12 percent), and plastics (2 percent). Yard wastes are recycled through composting.

Recycling rates are affected by factors including government actions, market forces, and disposal options. Many local and state governments are setting mandatory recycling objectives requiring up to 50 percent reduction in MSW streams. Governments also practice preferential purchasing of recycled goods, impose surcharges on products made entirely from raw materials, and provide tax breaks for the use of recycled goods.

The market forces that affect recycling are the traditional ones of supply and demand. It is essential that a recycling program's ability to supply materials be matched to a market demand for them. Numerous recycling efforts have foundered due to poor planning in this regard. To this day, developing reliable markets for recycled goods in a dynamic and rapidly fluctuating economy remains one of the greatest challenges to recycling. A dilemma that illustrates the complexity of our MSW problem has developed since the 1980s. Recycling increasingly removes material from the waste stream, but new disposal facilities (either incinerators or landfills) need a steady supply of waste to recover their substantial construction and operating costs. These facilities usually charge fees based on the tonnage of MSW disposed. The vast majority of MSW escapes source reduction or recycling and must be disposed. Landfilling and incineration are the primary disposal techniques in the United States. Disposal invariably entails releasing waste, or some waste residue, into the environment. These releases raise the following questions: What risks to human health or the environment do they pose? What is the nature of the released material? How much is released, how toxic is it, what components of the environment are impacted as a result, and can these impacts be mitigated?

Landfilling is the dominant disposal method, accounting for over two-thirds (more than 130 million tons) of the waste disposed in the United States. Traditionally the technique was practiced as nothing more than "crush and bury." Such practice did little to protect the environment from the waste or its degradation products. All components of the environment (i.e., groundwater, surface water, soil, and atmosphere) have experienced pollution as a result of uncontrolled landfills. Rainwater percolating through MSW creates a toxic leachate containing heavy metals, organic acids, and pathogens. This leachate can contaminate ground and surface waters. Decomposing MSW produces methane gas that can accumulate and create an explosion hazard. There have been instances of buildings constructed on top of abandoned landfills exploding as a result of methane infiltration. Also, landfills represent a topological stability hazard because they undergo subsidence as their contents settle. MSW in uncontrolled landfills emits foul odors and noxious gases.

Modern landfill design attempts to manage these problems. Regulations require new landfills to be double-lined and have leachate collection systems. The liners are thick (30–50 mil), chemically resistant, flexible plastic membranes set over layers of low permeability clay. Monitoring systems are installed to detect leaks. The leachate is treated to remove contaminants before release. Landfills are also capped with impervious materials to significantly reduce the amount of infiltration. Landfill gas is collected and treated, and is sometimes sold as a fuel supplement.

The siting of a landfill involves many social, political, and engineering issues. The engineering concerns include topography (e.g., slope), soil properties (e.g., permeability and plasticity), depth to bedrock and groundwater, and distance to surface water, wells, floodplains, and wetlands. Modern landfills are heavily engineered, extremely expensive to construct and operate, and designed to segregate waste from the environment. They are effectively high-tech tombs for MSW. Materials do not degrade to any significant extent in a landfill. With both oxygen and water denied to the waste, virtually no chemical or biological degradation occurs. Well-engineered landfills that are over twenty years old have been examined and their contents often appear to be as fresh as the day they were emplaced. Newspapers are completely readable and foodstuffs look like they just came from the kitchen.

Incineration consumes about 16 percent (32 million tons) of MSW disposed. The primary advantages associated with incineration are volume reduction and energy recovery. The combustion of MSW converts its organic content primarily to carbon dioxide (CO_2), leaving an ash whose volume is only 10 percent of the original waste volume. The amount of solid material to be managed (usually by landfilling) is thus significantly reduced. Virtually all incinerators use the heat they release to produce steam that generates electricity. One ton of combusted MSW produces 400–500 kilowatt hours (kWh) of electricity (typically, 1 barrel of oil produces 600 kilowatt hours). Public utilities are required to purchase this electricity.

The two most common types of incinerators are mass-burn and refuse-derived fuel (RDF). Mass burn incinerators combust MSW as received without processing. RDF incinerators burn MSW that has had noncombustibles removed and the combustibles shredded for enhanced incineration. The removed noncombustibles must be managed either through recycling or landfilling. RDF incineration produces more energy per unit weight of waste and less ash then mass burn.

Incineration creates secondary waste streams, both solid and gaseous, that must be managed. The ash collected at the top (fly ash) and bottom of the incinerator are composed of oxides of silicon and aluminum, as well as potentially toxic heavy metals and products of incomplete combustion (PICs). The latter two constituents are the basis for concern about the risks of incinerator ash. The debate over what is acceptable management of incinerator ash persists.

Emissions from incinerators include gases and particulate matter. Particulate matter is fly ash that escapes capture by pollution control devices (e.g., fabric filters or electrostatic precipitators) and is discharged to the atmosphere. The gaseous emissions include carbon dioxide, water, nitrogen oxides, sulfur dioxide, hydrochloric acid, and PICs. Alkaline scrubbers remove the acidic components from emissions. The PIC emissions remain a persistent concern with incinerators. PICs are a complex mixture of organic compounds that are derived from incomplete oxidation during combustion or formed as reaction products during combustion. The hazardous class of chemicals known as dioxins are one of the PICs of most concern.

Several approaches are employed to reduce the toxicity and amount of the secondary waste from incinerators. Removal of metals from MSW prior to incineration reduces heavy metals in the ash and particulate emissions. Optimum control of the combustion conditions can reduce PIC formation. The completeness of combustion is regulated by the following combustion zone parameters: (1) temperature (typically 1,600–1,800°F); (2) residence time (typically 1.5–2.0 seconds); (3) turbulence/thoroughness of mixing; and (4) amount of oxygen. Finally, new technologies for capturing emissions and treating ash are reducing these threats.

This entry serves to clearly indicate that the problem of MSW management is extremely complex, not amenable to a single technological solution, requires coordination of a number of management practices, involves societal and lifestyle issues, and will be with us for some time to come.

Bibliography

DENISON, R. A., and J. RUSTON, eds. *Recycling and Incineration—Evaluating the Choices.* Washington, DC, 1990.

LUND, F., ed. *The McGraw-Hill Recycling Handbook.* New York, 1993.

NEAL, A., and J. R. SCHUBEL. *Solid Waste Management and the Environment: The Mounting Garbage and Trash Crisis.* Englewood Cliffs, NJ, 1987.

OWEIS, I. S., and R. P. KHERA. *Geotechnology of Waste Management.* London, UK, 1990.

PFEFFER, J. T. *Solid Waste Management Engineering.* Englewood Cliffs, NJ, 1992.
U.S. ENVIRONMENTAL PROTECTION AGENCY. *Decision-Maker's Guide to Solid Waste Management.* EPA/530-SW-89-072. Nov. 1989.
———. *Characterization of Municipal Solid Waste in the United States: 1992 Update.* EPA/530-R-92-019. July 1992.
———. *The Consumer's Handbook for Reducing Solid Waste.* EPA/530-K-92-003. Aug. 1992.

CHRISTOPHER VAN CANTFORT
CLYDE FRANK

WATER QUALITY

In addition to being essential to life, water is important in many human activities—energy production, agriculture, industry, and recreation. Rapid growth in these activities has placed high demands on available water and has often threatened its quality (Global Tomorrow Coalition, 1990). Water quality refers to standards that must be met for water to be used for a specific purpose. These standards differ for different uses. Water has to be of a certain quality, that is, it must meet certain standards, for it to be used as drinking water. Different water quality standards apply, however, to water used for irrigation, for industrial applications, or, importantly, for maintaining a certain aquatic habitat.

Water quality is described by physical, dissolved chemical, and microbiological characteristics. Physical characteristics of water quality include temperature, turbidity (presence of particulates in the water) and pH (acidity of the water). Physical aspects may not be harmful in themselves, but may interfere with the water's intended use. Elevated temperatures decrease the dissolved oxygen in water, perhaps affecting fish populations. Particulates such as silt or iron oxides interfere with water treatment and may make the water unfit for certain industrial uses. Acidic (low pH) water is corrosive and may damage plumbing or cause elevated levels of lead and copper in the water.

Dissolved chemicals occur in water at the atomic or molecular scale and are not visible. All natural waters contain dissolved chemicals (see below); however, some chemicals may result from surface activities and may cause water quality problems (*see* POLLUTION OF LAKES AND STREAMS). Dissolved chemicals may include metals, pesticides, solvents, and other potentially harmful constituents. Long-term exposure to these chemicals may be harmful to internal organs such as the liver or kidneys, may damage the nervous or circulatory system, or in some cases, may lead to higher risk of cancer.

Microbiological criteria used for determining water quality include the presence of pathogenic organisms associated with waterborne disease: bacteria, viruses, and protozoa. These organisms may cause illnesses from gastrointestinal distress to more serious diseases such as typhoid, cholera, and hepatitis. Illness can occur following a single exposure to the organisms, such as from drinking contaminated water.

Drinking Water Standards

Quality standards for drinking water supplied by public water systems have been established. In the United States, the Safe Drinking Water Act requires the U.S. Environmental Protection Agency (EPA) to establish maximum contaminant levels (MCLs) in drinking water for contaminants that pose a risk to human health. Table 1 lists the drinking water standards established by the EPA (Pontius, 1992). Concentrations are in milligrams per liter (mg/l), a unit equivalent to parts per million (ppm). With the exception of nitrate and the microorganisms, where a single exposure at the MCL may cause health problems, a lifelong exposure at the MCL is required to produce health concerns.

The contaminants in Table 1 are listed in the following categories:

- Microbiological: Bacteria (e.g., *Legionella, Salmonelia typhi,* and *Vibrio cholerae*), viruses (e.g., hepatitis A and Norwalk virus), and protozoa (e.g, *Giardia lamblia* and *Cryptosporidium*).
- Inorganic Chemicals: The heavy metals, such as cadmium (Cd), chromium (Cr), cobalt (Co), mercury (Hg), nickel (Ni), and so on; the nonmetals, such as arsenic (As) and selenium (Se); and nutrients, such as nitrate (NO_3^-). Although these constituents occur naturally, concentrations approaching the MCL generally indicate pollution.

Table 1. Drinking Water Quality Standards as Established by the U.S. Environmental Protection Agency

Contaminant	*MCL*	*BAT*
Microbiological		
Total/fecal coliforms	>0 organisms	Disinfection
Giardia lamblia	>0 organisms	Filtration/disinfection
Cryptosporidium	>0 organisms	Filtration/disinfection
Inorganic Contaminants		
Barium	2	a, b, f
Cadmium	0.005	a, b, e, f
Chromium	0.1	a, b, e, f
Mercury	0.002	a, b, e, f
Nitrate (as Nitrogen)	10	a, b
Nitrite (as Nitrogen)	1	a, b
Selenium	0.05	a, b, e, f
Antimony	0.006	b, e
Beryllium	0.004	a, b, e, f
Cyanide	0.2	a, b, g
Nickel	0.1	a, b, f
Thallium	0.002	a
Radionuclides		
Radium-226	20 picocuries/l	a, b, f
Radium-228	20 picocuries/l	a, b, f
Uranium	0.02	a, e, f
Radon	300 picocuries/l	h
Alpha particles	15 picocuries/l	a, b, e
Photons/beta particles	4 millirems/yr	a, b, e
Volatile Organic Chemicals (VOCs)		
Benzene	0.005	c, d
Vinyl chloride	0.002	c, d
Carbon tetrachloride	0.005	c, d
1,2-Dichloroethane	0.005	c, d
Trichloroethylene (TCE)	0.005	c, d
1,1-Dichloroethylene	0.007	c, d
1,1,1-Trichloroethane (TCA)	0.2	c, d
para-Dichlorobenzene	0.075	c, d
cis-1,2-Dichloroethylene	0.07	c, d
1,2-Dichloropropane	0.005	c, d
Ethylbenzene	0.7	c, d
Monochlorobenzene	0.1	c, d
trans-1,2-Dichloroethylene	0.1	c, d
Styrene	0.1	c, d
Tetrachloroethylene (PCE)	0.005	c, d
Toluene	1	c, d
Xylenes	10	c, d
o-Dichlorobenzene	0.6	c, d
Dichloromethane	0.005	c, d
Hexachlorocyclopentadiene	0.05	c, d
1,2,4-Trichlorobenzene	0.07	c, d
1,12-Trichloroethane	0.005	c, d

Synthetic Organic Chemicals (SOCs)		
Pesticides:		
2,4-D	0.07	d
Ethylene dibromide (EDB)	0.00005	c, d
Heptachlor	0.0004	d
Heptachlor epoxide	0.0002	c, d
Lindane	0.0002	d
Methoxychlor	0.04	d
Pentachlorophenol	0.001	d
Toxaphene	0.003	d
2,4,5-TP (Silvex)	0.05	d
Alachlor	0.002	d
Aldicarb	0.003*	d
Adicarb sulfoxide	0.004*	d
Aldicarb sulfone	0.002*	d
Atrazine	0.003	d
Carbofuran	0.04	d
Chlordane	0.002	d
Dibromochloropropane	0.0002	c, d
Dalapon	0.2	d
Dinoseb	0.007	d
Diquat	0.02	d
Endothall	0.1	d
Endrin	0.002	d
Glyphosate	0.7	i
Oxamyl	0.2	d
Picloram	0.5	d
Simazine	0.004	d
Other SOCs:		
Benzo(a)pyrene	0.0002	d
Di(2-ethylhexyl)adiapate	0.5	d
Di(2-ethylhexyl)phthalate	0.006	d
Hexachlorobenzene	0.001	d
Polychlorinated biphenyls (PCBs)	0.0005	d
2,3,7,8-TCDD (Dioxin)	5E-08	d
Disinfection By-Products		
Total trihalomethanes	0.1*	l
Total haloacetic acids	*	l
Asbestos	7 million fibers/l	e, j, k
Lead	0.015**	k
Copper	1.3**	k

Note: Concentrations (MCL = maximum contaminant level) are given in milligrams/liter (mg/l = ppm) except as noted. Also given in the table are the best available technologies (BAT) for treatment.

* Final values had not been set as of July 1993.

** Action values: 90% of residences served by public water system must be less than this value.

Treatment Techniques: (a) ion exchange; (b) reverse osmosis; (c) packed tower aeration; (d) granular activated carbon; (e) coagulation/filtration; (f) lime softening; (g) chlorine oxidation; (h) aeration; (i) oxidation; (j) direct and diatomite filtration; (k) corrosion control; (l) reduction of total organic carbon.

- Radionuclides: Uranium, decay products of ^{238}U and ^{232}Th (radon and isotopes of radium), and decay particles emitted during radioactive decay (photons and alpha and beta particles).
- Volatile Organic Chemicals: The VOCs are the most common organic contaminants in drinking water. These include solvents [e.g., trichloroethylene (TCE) and perchloroethylene (PCE)] and components of gasoline, commonly referred to as the BTEX compounds: *b*enzene, *t*oluene, *e*thylbenzene, and *x*ylene (Domenico and Schwartz, 1990).
- Synthetic Organic Chemicals: The SOCs include pesticides, PCBs, and dioxin. Although many of the pesticides on the list have been withdrawn from use, others continue to be used.
- Disinfection By-Products: When chlorine, the most common disinfectant used by public water systems, interacts with dissolved organic matter in water, disinfection by-products are produced (Table 1). These chemicals are suspected of being harmful at levels above the MCLs. Disinfection is needed, however, to control waterborne disease in public water systems, particularly those using surface water sources.
- Asbestos: This contaminant occurs as a particulate in water. Although natural sources of asbestiform minerals do occur, the common source of asbestos in drinking water is asbestos cement water pipes, used for distribution lines prior to the mid–1980s.
- Lead and Copper: Although lead (Pb) and copper (Cu) may be related to the natural source of the water, the most common source of these constituents in drinking water is lead and copper pipes, lead-based solder (banned in the mid–1980s), or brass fixtures. This is particularly so where the water is slightly aggressive or corrosive (i.e., has a low pH).

In addition to health standards, secondary standards have been developed that deal with how palatable the water is, for example, the water's taste, odor, or appearance. Table 2 lists secondary standards for a number of components found in drinking water.

Agricultural Use

Water is used for livestock watering and for irrigation. Table 3 gives recommended limits for certain constituents in water used for livestock. For most constituents, levels in Table 3 are higher than those used to assess human potability in Tables 1 and 2. Livestock can, for example, tolerate significantly higher total dissolved solids than can humans. Exceptions are nitrate and fluoride, which are the same for livestock and humans.

Table 2. Secondary Standards for Drinking Water, Based on Aesthetic Reasons

Although the constituents below may not pose any significant risk to health, they may cause the water to be less palatable as a result of taste, odor, or appearance. The concentrations given below are the levels at which most individuals would be able to recognize the presence of these constituents. More sensitive individuals may detect the presence of these constituents at even lower concentrations. In other cases, concentrations exceeding the levels below may interfere with some desired use of the water. Concentrations are given in milligrams per liter (mg/l) unless otherwise noted.

Constituent	*Recommended Maximum Concentration*
Aluminum	0.05–0.2
Chloride	250
Color	15 color units
Fluoride	2
Foaming agents	0.5
Hardness	250
Iron	0.3
Manganese	0.05
Odor	3 threshold odor numbers
pH	6.5–8.5
Silver	1
Sulfate	250
Total dissolved solids (TDS)	500
Zinc	5

The chemical quality of water is important in irrigation, for example, total dissolved solids (TDS) (Hem, 1985). If the TDS in irrigation water is high, salt buildup in the soil is possible because only water and nutrients are used by the plant's roots. If the concentration of salts in the soil becomes too high, it interferes with the plant's uptake of water and nutrients. Removal of salts from soil is problematical. A common method of removal, leaching the salts through precipitation and/or irrigation, may lead to contamination of groundwater. Ele-

Table 3. Recommended Water Quality Standards for Agricultural Uses

Taken from Freeze and Cherry (1979) and Hem (1985). Values represent maximum recommended concentrations in ppm.

Constituent	*Livestock*	*Irrigation Water*
TDS		700
Poultry	2,860	
Pigs	4,290	
Sheep	12,900	
Horses	6,435	
Dairy cattle	7,150	
Beef cattle	10,100	
Nitrate as nitrogen	10	NA*
Arsenic	0.2	0.1
Boron	5	0.75
Cadmium	0.05	0.01
Chromium	1	0.1
Fluoride	2	1
Lead	0.1	5
Selenium	0.05	0.02

* Not applicable.

vated TDS and nitrate concentrations are common in groundwater from areas of extensive irrigation.

Industrial Applications

Quality requirements for industrial water vary widely. In some cases, such as in ore processing or for cooling, quality is not critical. In other cases, as in the manufacture of high-grade paper or pharmaceuticals, water quality is extremely important (Hem, 1985). The U.S. Federal Water Pollution Control Administration (1968, as cited in Hem, 1985) has published water quality requirements for various industrial processes. These "standards" exhibit a wide range, for example, total dissolved solids from less than 1 mg/l for feedwater in high-pressure boilers to 1,000 mg/l for use in wood chemicals. It is possible to treat water to achieve industrial quality standards (see below). Treatment is expensive, however, and because large quantities of water are used, treatment may be economically unfeasible.

Water Ecosystems

Water quality in lakes and streams is important because dissolved constituents may affect individual organisms, and because contaminants may be concentrated up the food chain. This may affect the capability of higher organisms to function adequately, including the ability to reproduce. It is difficult to establish water quality standards for aquatic ecosystems because the effects of water composition on these systems are not well understood and may often affect one organism differently than another. For example, algae are more susceptible to copper contamination than some other species (Stumm and Morgan, 1981).

Other factors that influence the toxicity of trace metals to aquatic organisms include: (1) pH, which controls the chemical form of the metal in solution; (2) the number of organisms present, which controls the metal available to any one organism; (3) the amount of particulates in the water, which tend to concentrate the metals on their surfaces; and (4) factors that control the solubility of the metal (Stumm and Morgan, 1981).

Two characteristics used to evaluate the "health" of a stream or lake are dissolved oxygen and dissolved nutrients concentration. Oxygen is necessary to maintain a viable aquatic ecosystem. Pollution may be indicated by a high biochemical oxygen demand (BOD), indicating that available oxygen may be consumed in reactions involving organic matter in the water.

Nutrients of concern are nitrogen and phosphorous, which can produce high populations (e.g., algal blooms) in streams and lakes. Blooms are followed by massive organism die-off as a result of competition, and the decaying organic matter consumes available oxygen (Drever, 1982). A common source of BOD and nutrients is sewage. Normal sewage treatment processes are effective at reducing BOD to low levels, but are relatively ineffective at nutrient removal (Drever, 1982).

Water Treatment

Table 1 lists treatment options for those chemicals in which drinking water standards have been determined. The more common treatment options are briefly described below as a function of contaminant groups (Consumer Reports, January 1990; Water Treatment Handbook, 1985).

Microorganisms. Disinfection is the primary tool for inactivating microorganisms. The most common disinfectant is chlorine, although ozone and ultraviolet light are also used. If the chlorine is provided adequate contact time with the water (e.g., in a storage tank), the majority of pathogens will be unable to inflict disease. Particulate and dissolved constituents in water can significantly interfere with the disinfection process.

Successful treatment for these organisms usually follows several steps. Coagulants such as alum are added to the raw water and the pH is adjusted to promote flocculation, the aggregation of the particulate matter in the water. A sedimentation stage allows the larger of the flocculated particles to settle. The water is then passed through a filter that removes the remaining particulates as well as many of the larger protozoans. The filtered water is then subjected to disinfection prior to entering the water system's distribution lines. There are many variations on this process, as dictated by the individual water chemistry and requirements of the system (Viessman and Welty, 1985).

Synthetic Organic Chemicals (SOCs). Organic compounds are those in which carbon is a major constituent and is important in determining the structure of the molecule. Dissolved organic molecules tend to attach themselves (i.e., adsorb) to organic carbon particles. SOCs can therefore be removed as water flows through granular activated carbon (GAC). The charcoal is granulated to increase surface area and the sorption process.

Volatile Organic Chemicals (VOCs). These are organic chemicals that are volatile (i.e., tend to escape to the atmosphere). Aeration is used to treat water containing VOCs. Contaminated water is pumped into the top of a tower packed with material designed to decrease drop size. As the water cascades down through this material, air is blown through the tower transporting the volatile constituents into the atmosphere. The packed-tower aeration method is supplemented by passage of the water through a GAC filter followed by disinfection to inactivate organisms introduced during aeration.

Inorganic Chemicals. Inorganic chemicals are removed by ion exchange and reverse osmosis. In ion exchange, contaminated water flows through a resin that is in the form of small spheres. As water moves in and around these spheres, contaminants are removed by exchanging with harmless constituents on the resin. This process is similar to a water softener in which calcium and magnesium, which cause the water to be "hard," replace sodium on the exchange resin.

In reverse osmosis, contaminated water is held at an elevated pressure against a membrane through which water molecules can diffuse. Water molecules are attracted to the membrane forming a layer several molecules thick. Dissolved positive and negative ions, and large organic molecules, are unable to pass through the water molecule layers. As a result, most of the contaminants are left behind as the water diffuses through the membrane.

Surface Water Versus Groundwater

Surface water (e.g., lakes, streams, etc.) often differs in its chemical composition from groundwater (Freeze and Cherry, 1979; Drever, 1982; Hem, 1985). To understand why this is so, we need to explore the relationship of these two freshwater reservoirs on Earth. Surface water and groundwater are part of the hydrologic cycle, the process by which water evaporated from the ocean's surface is carried over land where it falls as precipitation. This precipitation may evaporate, run off as surface water, be used by plants, or sink into the ground to recharge groundwater.

Rainwater is very dilute (Hem, 1985), but it does have some dissolved species (Table 4). The sources of these constituents include anthropogenic activities (e.g., gases such as SO_4, NO_2, CO, and HCl from the burning of fuels, metallurgical process, etc., and particulates from industrial and vehicular emissions) and natural sources (e.g., terrestrial dust, sodium chloride, and other salts from the sea surface). An important characteristic of natural precipitation results from the interaction of water and carbon dioxide in the atmosphere. These two constituents combine to form carbonic acid (H_2CO_3), making natural precipitation slightly acidic (pH = 5.7 [Krauskopf, 1980]). (Note that this is not the same as the "acid rain" that results from the interaction of rain and emissions and produces even more acidic conditions harmful to the environment—*see* POLLUTION OF ATMOSPHERE.)

This slightly acidic character of precipitation and the bipolarity of the water molecule make water an excellent solvent, a characteristic that causes weathering (i.e., decomposition) of rocks. Rainwater will dissolve rock material as it flows over it and as it percolates through it. The dissolving of rock material occurs more readily below the surface because the water is in contact with the rock material for longer times.

Typical compositions of surface water are given in Table 4 (analyses 1–6). Note the significant variation in waters from different areas. The actual composition of surface water will depend on the relative contributions from precipitation, runoff, and groundwater. Other contributing factors include biological processes, evaporation, and irrigation return flow (*see* POLLUTION OF LAKES AND STREAMS).

Groundwater obtains its composition from reactions with rock material in the aquifer. These reactions include mineral dissolution and precipitation,

Table 4. Typical Compositions of Precipitation, Surface Water, and Groundwater

All analyses are reported in mg/l. See references listed below for additional analyses.

Constituent	*1*	*2*	*3*	*4*	*5*	*6*
Ca	0.8	nd[b]	3.51	40.7–83.6	38	394
Mg	1.2	0.2	0.54	7.2–10.8	10	93
Fe	nd[b]	nd[b]	nd[b]	nr[c]	0.02	nr[c]
Mn	nd[b]	nd[b]	nd[b]	nr[c]	nr[c]	nr[c]
Na	9.4	0.6	0.41	1.4–98.7	20	333[d]
K	nd[d]	0.6	0.79	1.2–7.4	29	[d]
SiO_2	0.3	nd[b]	2.36	3.7–5.5	7.9	17
HCO_3	4	3	11.7	13.5–152	113	157
Cl	17	0.2	0.25	1.1–178.2	24	538
SO_4	7.6	1.6	2.5	36.0–77.8	51	1150
TDS[a]	38	4.8	nr[c]	nr[c]	232	2610
Constituent	*7*	*8*	*9*	*10*	*11*	*12*
Ca	56	0.78	32	58	88	17
Mg	12	0.29	12	13	7.3	1.7
Fe	nr[c]	nr[c]	0.01	0.04	0.02	0.33
Mn	nr[c]	nr[c]	nd[b]	1.3	nd[b]	nd[b]
Na	7.9	1.34	30	23	19	7.4[d]
K	1	0.28	5.2	2.8	2.8	[d]
SiO_2	nr[c]	2.73	49	10	24	29
HCO_3	160	328	220	101	320	69
Cl	12	0.14	7.9	39	13	1.1
SO_4	53	0.1	11	116	6.7	6.9
TDS[a]	nr[c]	nr[c]	257	338	322	98

[a] Total Dissolved Solids; [b]nd = not detected during analysis; [c]nr = not reported; [d] equals Na plus K.

1. Rain, Menlo Part, CA, January, 1958 (Hem, 1985). 2. Snow, east of Lake Tahoe (Hem, 1985). 3. South Cascade Lake, Washington (Drever, 1982). 4. Rhine River: Swiss Alps to Germany-Holland border (Drever, 1982). 5. Mississippi River near New Orleans (Hem, 1985). 6. Pecos River near Artesia, New Mexico, discharge-weighted average, 1949 (Hem, 1985). 7. Average groundwater from Florida limestone aquifer (Freeze and Cherry, 1979). 8. Springs, Sierra Nevada, Calif. (Freeze and Cherry, 1979). 9. Groundwater, basalt aquifer, Umatilla County, Oregon (Hem, 1985). 10. Groundwater from sand and gravel aquifer, Kanawha County, West Virginia (Hem, 1985). 11. Groundwater from sandstone aquifer, Rice County, Kansas (Hem, 1985). 12. Groundwater from metamorphic rock, Burke County, North Carolina (Hem, 1985).

reactions between constituents in the water and those on mineral surfaces, oxidation–reduction reactions, and mixing with other groundwater or surface water. Groundwater from different aquifers can often be recognized based on their distinct chemical composition. Groundwater samples from wells in aquifers along the path that groundwater flows often record increasing amounts of dissolved solids. Subtle differences in proportions of these compounds provide clues as to what chemical reactions have taken place in the subsurface and the kind of aquifer–for example, limestone versus volcanic rock–in which the water has been in contact (Domenico and Schwartz, 1990). Because of the geologic complexity of aquifers and the variation in the physical and chemical environment in the subsurface, groundwater shows a wide variation in composition (Table 4).

Summary

Water is a complex chemical solution. The composition of any given water will reflect physical, chemical, and biological processes that have affected the water since it fell as precipitation. The impact of human activities on a water's composition is important in many areas. Water quality standards are designed to evaluate whether or not a given water has an appropriate composition to be used for a particular purpose. Because uses differ markedly (e.g., drinking water versus industrial uses) there is no one set of water quality standards. The composition of water can be altered through a number of treatment technologies, thereby improving its quality.

Bibliography

Consumer Reports. "Fit to Drink?" (Jan. 1990): 27–42.

Domenico, P. A., and F. W. Schwartz. *Physical and Chemical Hydrogeology.* New York, 1990.

Drever, J. I. *The Geochemistry of Natural Waters.* Englewood Cliffs, NJ, 1982.

Faure, G. *Principles of Isotope Geology,* 2nd ed. New York, 1986.

Freeze, R. A., and J. A. Cherry. *Groundwater.* Englewood Cliffs, NJ, 1979.

Global Tomorrow Coalition. "Fresh Water." In *The Global Ecology Handbook,* ed. W. H. Corson. Boston, 1990.

Heath, R. C. *Basic Ground-Water Hydrology.* U.S. Geological Survey Water-Supply Paper 2220. Washington, DC, 1989.

Hem, J. D. *Study and Interpretation of the Chemical Characteristics of Natural Water.* U.S. Geological Survey Water-Supply Paper 2254. Washington, DC, 1985.

Moyer, R. A., ed. *Water Treatment Handbook. A Homeowners Guide to Safer Drinking Water.* Emmaus, PA, 1985.

Pontius, F. W. "A Current Look at the Federal Drinking Water Regulations." *Journal of the American Water Works Association* 84 (1992):36–50.

Shelton, T. B. *Interpreting Drinking Water Analysis. What Do the Numbers Mean?* Publication Number E156, Rutgers Cooperative Extension, Rutgers University. New Brunswick, NJ, 1991.

Stumm, W., and J. J. Morgan. *Aquatic Chemistry: An Introduction Emphasizing Chemical Equilibria in Natural Waters.* New York, 1981.

Viessman, W., Jr., and C. Welty. *Water Management: Technology and Institutions.* New York, 1985.

DENNIS O. NELSON

WATER SUPPLY AND MANAGEMENT

It hardly seems possible that on a planet with 75 percent of its surface covered by water, that one of the toughest challenges facing humankind would be the management of its water resources. Population growth, pollution, and intense usage, however, have put a strain on the reserve of clean freshwater. International and local efforts in water management are required to assure future generations a clean and abundant supply of water.

The main focus of this entry is to discuss the global distribution of usable surface and groundwater, and the management issues needed to balance the requirements of humans with the needs of the environment. All too often in the past the needs of development have overlooked the environmental impact.

Distribution of Usable Water

Only 3 percent of water on the planet is considered freshwater—that is, water that can be used for human consumption or agricultural needs. Of this 3 percent, over two-thirds is trapped in the polar icecaps and snow cover of Antarctic and Greenland. Only about 0.25 percent of all water is contained in the rivers and lakes from which most human water consumption derives.

The global distribution of freshwater supply is highly concentrated. In fact twenty-eight of the largest lakes in the world contain over 85 percent of the total lake volume on earth. Most of these lakes are located in North America and Asia. Lake Baikal, in the former Soviet Union, contains almost 25 percent of all the world's lake water by volume. Continental river runoff quantities show a similar distribution. Asia and South America combine to account for almost 56 percent of the annual river runoff worldwide, while Europe accounts for 7 percent, and Australia only about 1 percent. The Amazon River basin alone produces almost 20 percent of the annual global runoff. Probably a more accurate measure of a nation's surface water reserve is the quantity of water per capita. Again, there is a great discrepancy between the nations with a large reserve per capita and those with little reserve. Nations with large reserves tend to find economic development easier than those with lower reserves. Norway and Canada lead all nations each with approximately 1.18 billion liters (l) per capita. The United States has about 9 million liters per capita, while India and China each have just over 2 billion liters per capita.

Groundwater resources are more difficult to quantify (*see* GROUNDWATER). The quantity of water that can be drawn from a well can vary significantly within short distances. It is known, however, that in many parts of the world the water table is dropping due to an overuse of the supply. It is estimated that one-fifth of the irrigated cropland of the United States is watered by groundwater that is pumped faster than it can be recharged. Other areas of the world are experiencing the same dilemma. Saudi Arabia, for instance, has depleted its aquifers to a point where pumping costs now exceed the cost of desalinization of saltwater.

Management Issues

Management of the earth's freshwater resources, to assure both quantity and quality for humans while maintaining or improving the environment, is an extremely complex issue. An increasing population base, overuse of resources, pollution in rivers, lakes, and underground aquifers, agricultural needs, industrial needs, and environmental needs must all be considered when designing a management plan for freshwater resources. The needs of one area can have a profound impact on another. To better comprehend the cause and effect of each water-related use, each issue needs to be addressed separately.

Population Pressure. It is estimated that by the year 2000, 50 percent of the earth's population will live in urban areas. Supplying this population with clean water for drinking and sanitation purposes is a major problem for water managers. Most cities are now near the maximum capabilities of their present systems, and will require large capital investments to meet future demands. Mexico City, for example, is experiencing extreme population growth that has created a water management crisis. In 1920 Mexico City had a population of 1 million people. By 1980 the population had grown to 15 million and by the turn of the century it is expected to exceed 25 million. The expanding population base has forced the city to extract water from a source more than 200 kilometers (km) away and at an elevation 2,000 meters (m) lower. Since most of Mexico City's water rates are subsidized, the cost of delivering water to the city has put a strain on future economic growth of the country.

Environmental Impact. In the past the environmental impact has often been overlooked. In the past, the environmental impact of supplying water to the population base has often been overlooked. The interrelation of water quantity and quality has taken a back seat to the need to deliver water for domestic and agricultural uses. For instance, the need to supply water for cotton production in the former Soviet Union has all but dried up the Aral Sea. An almost 100 percent diversion of the Amu Darya and the Syr Darya, the rivers that feed into the Aral Sea, has lowered the surface area of the sea by almost 65 percent while the salinity has tripled in the last sixty years resulting in the death of all twenty-four species of fish that had resided in the lake. The mismanagement of the Aral Sea has led to the destruction of the local fisheries and created a health hazard for the residents of the area.

Water Quality and Health. Effective management of the earth's freshwater resources requires focusing on both the quantity and quality of the water. If a source of freshwater is contaminated, it serves little purpose. The connection between industrial and agricultural practices and the quality of the water we consume was rarely made in the past. Fertilizer heavily dosed with nitrate was and still is applied to croplands draining to a watershed or shallow aquifers. Many tributaries to the Amazon River have excessively high levels of mercury, a legacy of the gold mining process. Many of the fish species in the Amazon basin now have mercury levels that exceed what is considered safe for human consumption. Although the issue of chemical contamination lately has been brought to the forefront, the more imminent threat, particularly in developing countries, is the threat of bacteriological contamination. More than 250 million new cases of water-related diseases are reported worldwide each year resulting in over 10 million deaths. Water treatment can remove most of the contaminants, but for developing countries the cost can be staggering.

Economic Development. Water is essential for a nation to sustain economic growth. Industry needs water for cooling and heating processes, transporting goods, as well as for air conditioning and cleaning. Unfortunately, many nations are now experiencing water shortages that can cripple any hopes of economic growth. It is estimated that some eighty countries, representing 40 percent of the world's population, already have serious water shortages at some point in the year. Nations will need to become more flexible with the management of their resources. Alternate methods such as conservation and water reuse need to be implemented to reduce the dependency on the supply side. Israel's water management policy may become a model for other nations to copy. Israel has two major supplies to the country. The first is a pipeline known as the National Water Carrier from the Sea of Galilee to the north, and the other is a well field that catches water flowing underground from the West Bank to the Mediterranean. Overpumping of the latter source has reduced the aquifer to a point below sea level, causing an intrusion of saltwater into the wells that has rendered most wells useless. Today water managers are looking at two alternate means of supply. The first group stresses the need for conservation and remaining flexible. This could mean shifting water resources away from agribusiness and, consequently, the need to retrain displaced workers in this field. The second group thinks that treated wastewater could be used to irrigate farmlands, thereby reducing the freshwater needs of agriculture. It is possible, and probably beneficial to the nation, that both options be implemented. If successful, the Israeli alternative to seeking another water supply may become a model for other nations to adopt.

Agriculture. As the population base increases, so does the demand for agricultural products. Today, over two-thirds of the freshwater needs are used for irrigation. Reservoirs must be constructed to retain water runoff during times when rain is heaviest. Because of the growing concern for the environmental impact such reservoirs generate, new reservoir projects are few and far between. Water tables have dropped drastically in areas were recharge of the aquifers was unable to meet the demands of the irrigated croplands. The pumping cost to retrieve water from the lower water table has driven many farms out of business. It is estimated that one-fifth of all irrigated land in the United States is supplied by aquifers that are being pumped faster than they can be recharged. The cost of retrieving water from the declining water levels in the Ogallala aquifer in the Texas High Plains, for example, reduced the area of irrigated lands by 34 percent between 1974 and 1989. In some areas farmers can make more money selling water to nearby cities than by raising crops.

Politics of Water Management

As the need for water to meet the demands of an expanding population base and to fuel economic growth increases, so will tensions between nations sharing water sources. It is probable that as sources become scarce, conflicts between nations over the use of water supply will increase. In the Middle East competition for scarce water supplies has raised tensions to the point of war. In the mid–1960s Arab states announced plans to divert the headworks of the Jordan River. At the time Israel was looking to build its own water carrier system, part of the National Water Carrier system. Israel considered the Arab states' action to be in violation of its sovereign rights and launched military strikes against the Arab headworks diversion project.

Many conflicts over the water rights of individual nations have occurred in the past and will undoubtedly continue into the future. Water managers will need to look at all factors, particularly how a new proposed water supply may affect their neighbors, before constructing diversion channels or dams.

Meeting Future Water Needs

As the world population grows to perhaps 10 to 15 billion sometime in the the twenty-first century, new management approaches must be implemented to meet the demands of the population while maintaining or improving the condition of the environment. Some possible alternatives to developing new sources need to be implemented. For instance, expanded conservation measures, particularly in the area of agriculture, should be a national goal. Water reuse may potentially supply both domestic and agricultural needs. Experiments in land application of treated sewage water for agricultural use have shown excellent results. Theoretically, there is no reason why treated wastewater could not be retreated to meet health standards for human consumption, although the public opposition may prove to be an insurmountable obstacle. Domestic reuse of "gray" or "dirty" water (water used in washing dishes or bathing), is now in effect in water-restricted areas as a means of watering home lawns and shrubbery. Innovations in water storage such as aquifer recharge (pumping water into a well during times of high surface runoff, and pumping the water out during high-demand periods), are another alternative to the traditional reservoir storage.

Future management must also look at the quality of the water supply and the effects on the environment. Watershed management and wellhead protection need to be developed at national and international levels. Cooperation between nations in the management of their shared resource must be initiated to avoid future international conflicts.

Bibliography

AKKAD, A. A. "Conservation in the Arab Gulf Countries." *American Water Works Association Journal* 82 (May 1990):40–50.

GLEICK, P. "An Introduction to Global Fresh Water Issues." In *Water in Crisis,* ed. P. Gleick. New York, 1993.

SALVATO, J. A. *Environmental Engineering and Sanitation,* 3rd ed. New York, 1982.

MICHAEL WHITELEY

WATER TREATMENT

See Water Supply and Management

WATER USE

Next to oxygen, there is no compound found on Earth that is more essential to life than water. Without water there would be no life as we know it. Unfortunately this valuable resource is not distributed evenly nor is it available in sufficient quantity and quality in all regions of the planet. Three percent of Earth's total water is fresh and most of that is stored in the ice caps of Antarctica and Greenland or in deep underground aquifers that are either technologically inaccessible or accessible only at great cost. Only 0.3 percent of the total available freshwater is practically and economically available for human use. This water is found primarily in surface water bodies (lakes and rivers) and shallow underground aquifers. The history of human activity is well documented by the struggle to locate, develop, and distribute usable water for a myriad of essential and nonessential human activities.

This entry focuses on the different ways freshwater is used and how those uses vary around the world. The reasonable use of freshwater as a finite natural resource requires an appreciation of the environmental consequences associated with its development and extraction, and its relation to the global and regional hydrologic cycle. Such environmental concerns are highlighted and appropriate protection and conservation strategies are emphasized.

Domestic and Industrial Uses

Water is essential for human and animal health and sanitation. About 63 percent of an average adult human's body weight is water, and approximately 1.3 liters (l) of water must be consumed daily to maintain normal health and vitality. Living without water for five to ten days can be fatal. It is, therefore, essential that water be made available in sufficient quantity and quality to provide for these basic needs. See Table 1 for a comparison of water

Table 1. Example of Average Daily U.S. Water Use

Location	*Liters per Day*
Single family home (per resident)	173–259
Restaurant with toilet facilities (per patron)	24–35
Schools	
With cafeteria, gymnasium, showers (per pupil)	86
Without cafeteria, gymnasium, showers (per pupil)	52
Hospital (per bed)	864–1,382
Self-service laundry (per wash)	173
Swimming pool (per swimmer)	35
Highway rest area (per person)	17

needs for some typical domestic uses of water in the United States.

Drinking, bathing, cleaning, and waste disposal comprise the major domestic uses of water in most parts of the world. Water that is withdrawn from rivers, wells, and springs, is commonly treated before it is distributed to individual homes and communities where it is made available for domestic use. In many parts of the world human and household wastes are removed separately from water supplies via sewerage systems, septic tanks, or other waste collection systems. Unfortunately, in many developing countries the local river or lake may be used for all these activities, a condition that may have grave consequences for the life and vitality of the affected community. Water is used for many forms of sport and recreation. Municipal swimming, pools and water amusement parks require large amounts of water to operate and maintain a sanitary environment for swimmers. In addition to artificial structures, natural bodies of water are used for recreational boating, swimming, and other water sport activities, as well as transporting people and goods. Recreational fishing is a global activity that requires the cooperative management of limited multiuse water resources. Often the interests of energy production, agriculture, and the fishing industry are in conflict over use of limited water resources and best management practices.

In some parts of the world water is an integral part of religious rites and ceremonies. Many religions use water in ritual baptisms that are considered rites of passage. Some religions consider certain bodies of water (lakes or rivers) sacred.

Water has numerous industrial uses as well. In fact, water can be as critical to the survival of a nation's economy as it is to the survival of the human body (Table 2). Many industries use water to convey raw materials to production facilities and transport finished products for sale via navigable waterways. The lumber industry, for example, uses water in rivers, streams, artificial flumes, and ponds to float and store raw logs for processing in lumber mills. Other industries use water to convey products through a series of production stages or in the actual production of the product itself. Paper mills require very large amounts of water to make paper and in the treatment and disposal of generated waste products. Food and pharmaceutical processors require water of high quality in the manufacture of products designed for public consumption. Mining of minerals and ores often require large amounts of water in the extraction process as well as in transportation and reclamation. Common to all these industrial operations is the need to provide a healthy and sanitary environment for the employees who work there, and this can only be accomplished by providing an adequate and safe water supply for drinking and cleaning.

Agricultural Uses

The discovery of oil and new pumping technologies in the twentieth century created an environment favorable for the use and potential overuse

Table 2. 1990 Estimated Total Freshwater Withdrawals in the United States

Water Use	*Million Liters per Day*
Public supply	133
Domestic	12
Commercial	8
Irrigation	473
Livestock	16
Industrial	67
Mining	11.5
Thermoelectric	453

Source: *Estimated Use of Water in the United States in 1990*. USGS Circular 1081, 1993.

of surface and groundwater sources for the irrigation of food crops. As a result, every year approximately 2,700 cubic kilometers or km^3 (1 km^3 = 913 billion liters) of water are pumped from the world's rivers and underground aquifers to irrigate crops. To provide enough water for large-scale food production, large amounts of water are pumped and often transported great distances. Two major food producing regions of the world, California's Central Valley and the Aral Sea basin of the former Soviet Union, would be barely arable without a major commitment in cost to use and transport the limited water resources in those regions. In 1990 commercial irrigation accounted for approximately two-thirds of global freshwater use.

Intensive large-scale crop irrigation has taken its toll on global water sources as well as on the irrigated land itself. Poor management practices and inefficient use of water have resulted in the depletion of once abundant aquifers by overpumping, causing extraction costs to skyrocket. In many areas cropland has become waterlogged due to overwatering. Overwatering may cause local water tables to rise into productive root zones thus depleting young plants of necessary oxygen needed for maturation. In arid regions with high evaporative water loss, soils may become saline enough to inhibit growth and may further concentrate toxic elements like selenium, which may kill birds and fish in the affected areas.

The breeding and raising of livestock also requires adequate amounts of potable water. A typical dairy cow requires 121 liters of water per day for drinking and servicing, while a horse requires approximately 42 liters. Water is also extensively used for cleaning and sanitation and in storing and disposing of animal wastes either in the form of sewage lagoons or water-carried sewerage systems.

Energy Production Uses

Water and energy are closely related. On the one hand, considerable amounts of energy must be used to extract and transport water, sometimes over great distances, from regions of abundance to regions of need. The twentieth century has seen extensive water transfer projects designed to sustain growth in semiarid and arid regions of the planet where freshwater sources are limited. On the other hand, water is directly and indirectly connected with the development and generation of energy. The amount of water needed to generate energy depends entirely on the type of fuel used in the process.

Energy produced from fossil fuels (oil, coal, natural gas, and synthetic sources) as well as nuclear sources require enormous amounts of water for fuel processing and especially power plant cooling. Common to all these production processes is the conversion of water into steam to drive electric turbines. Water used in these facilities must be condensed in cooling systems designed to recycle the process water through the turbines. Such water needs required siting production facilities near large bodies of water such as rivers, lakes, or even oceans. In 1990 the United States withdrew 270 km^3 of freshwater and saltwater to cool such power plants, an amount equal to 40 percent of all freshwater used globally.

Water is used in coal extraction, combustion, waste disposal refining, and transportation. Coal slurry pipelines have been constructed that use large amounts of water to carry coal in suspension great distances from where it is mined to power plants. The water used in the arid southwestern United States comes from underground aquifers where excessive withdrawals may exceed the rate of recharge of these limited supplies. Reclamation of mined areas also requires water for revegetation of the scarred land.

Traditionally, oil and gas production has used moderate amounts of water, mostly in the exploration and drilling processes. As oil supplies have diminished, however, more water-intensive methods have been developed to extract the oil. These methods include the use of water flooding of oil reservoirs to increase oil flow to wells and the injection of steam into oil fields in an attempt to heat up the oil and thus reduce viscosity, leading to a quicker recovery. Water is also required for oil refining and of course for cooling at power generating facilities.

In 1990 approximately 12 percent of global energy demands was supplied by nuclear power. In several major industrial countries including France and South Korea this percentage of energy produced locally can be as high as 50 percent. In addition to cooling reactor water at power generating plants, uranium mining, reclamation, milling, refining, and enriching all require the use of ample amounts of freshwater. The environmental concerns related to the use of nuclear power are well publicized and focus primarily on health and safety, thermal pollution of streams used in reactor cooling systems, and withdrawal of large amounts of water from stream flows.

In regions with abundant surface water, hydroelectric power can be generated by the energy from falling water. Water directly turns turbines to generate electricity as it falls through spillways in hydroelectric dams. To produce sufficient amounts of energy this way many streams and rivers must be dammed to control water movement, at great financial and environmental cost. The construction of such dams can alter the local hydrologic balance by creating standing water reservoirs in place of the normal free flowing water ecosystems. Such dramatic changes can displace wildlife populations and increase the consumptive use of water by increasing evaporation of water from the surface area of reservoirs and by increasing seepage from underlying dam foundations. Dams can also kill fish by poisoning them with high levels of nitrogen produced in the turbulent falling waters, by entrainment into turbines, and by disturbing their normal migration patterns by creating physical barriers to their normal movement.

In 1990 hydroelectric dams generated approximately 20 percent of worldwide electricity. Hydropower constitutes 70 percent of the electricity produced in Central and South America and 20 percent for the United States and Canada combined. Use of hydroelectric power appears to be on the decline in industrial countries due to the environmental constraints mentioned above and on the increase in many developing countries.

Wastewater Reuse

As available water resources are depleted and/or polluted, recycling wastewater has become a widespread conservation and economic strategy. Wastewater produced by residential, industrial, and agricultural activities can be treated and reused for the same or another purpose depending on the water quality desired.

Municipal wastewater (sewage treatment and storm sewer effluents) is by far the most widely available and reliable source for reuse in most developed countries. Its end use depends on how much it can be treated. Some uses of reclaimed municipal wastewater in the United States include the production of portable drinking water, irrigation of seed and some food crops, orchards, vineyards, lawns, golf courses, and recreation areas.

Aquifer storage and recovery projects have been developed where reclaimed water is injected directly into groundwater aquifers or spread over the surface of natural groundwater recharge areas for later use. In some coastal areas, wastewater is injected between saline and freshwater aquifers to prevent or restrict saltwater intrusion due to overpumping. Reclaimed wastewater is also used in fire protection, refuse and soil compaction and in the development of recreational impoundments. Industrial wastewater is reused for cooling water, boiler feed water, and wash water.

In 1990 the U.S. Geological Survey estimated that 3.2 billion liters per day of reclaimed wastewater were used in the United States.

Although it is not possible to describe all global uses of freshwater here, it is easy to understand and appreciate how this valuable natural resource is essential to human existence. As a limited natural resource, all reasonable and prudent safeguards should be used to protect and conserve it. As worldwide population increases it will be even more difficult to provide an adequate supply of high-quality water. Strategies for chemical and biological pollution control, improved land use practices, and efficient use and management of water will be the vanguard of the future.

Bibliography

Environmental Protection Agency. *Manual of Individual Water Supply Systems.* Washington, DC, 1985.

Gleick, P. *Water in Crisis.* New York, 1993.

Keller, E. *Environmental Geology.* Columbus, OH, 1976.

Middlebrooks, E. J., ed. *Water Reuse.* Ann Arbor, MI, 1982.

U.S. Geological Survey. *Estimated Use of Water in the United States in 1990.* Washington, DC, 1993.

KURT D. PUTNAM

WEATHERING AND EROSION

Over time, Earth's internal geologic forces determine the composition and distribution of different rock types (*see* EARTH AS A DYNAMIC SYSTEM, META-

MORPHIC PROCESSES; METAMORPHIC ROCKS; IGNEOUS PROCESSES; IGNEOUS ROCKS; SEDIMENTS AND SEDIMENTARY ROCKS, CHEMICAL AND ORGANIC; SEDIMENTS AND SEDIMENTARY ROCKS, TERRIGENOUS). Rocks are made up of crystalline compounds called minerals (*see* MINERALS AND THEIR STUDY); silicate minerals dominate Earth's crust (*see* MINERALS, SILICATES). Most of these rocks and minerals were formed, or at least turned into their present lithified form, at depths of several kilometers or more beneath Earth's surface (*see* IGNEOUS ROCKS; METAMORPHIC ROCKS; SEDIMENTOLOGY; DIAGENESIS). Internal geologic (tectonic; *see* PLATE TECTONICS) processes eventually cause some of these materials to be force upward, where they are ultimately exposed to Earth-surface conditions. Conditions at Earth's surface are strongly influenced by the distribution of temperature and moisture, which are in turn governed by the response of the atmosphere and hydrosphere to the Earth's major external source of energy, the Sun (*see* EARTH'S ATMOSPHERE). When rock of any kind is exposed to Earth surface conditions, it is subjected to weathering and erosion. The nature of weathering processes and products depends both on the chemical and physical properties of the starting material (rock), and on the conditions of temperature, moisture, and so on, to which the material is exposed.

Weathering is the physical (mechanical) and chemical breakdown of rock in place. Where present, organisms may be active agents of both physical and chemical weathering. The material formed when rock is broken down exclusively by abiotic processes is called regolith. Material that has been physically and chemically altered by weathering, and which supports vegetation, is called soil (*see* SOILS, FORMATION OF). Vegetation and the source of water both are at the land surface, and the material at the surface has been exposed to these agents the longest. Consequently, the intensity of weathering (the degree to which the material has been weathered) is greatest at the surface, and diminishes with depth. This vertical variation produces a weathering profile. At different depths, as organic compounds decompose, or as soil solutions encounter minerals that were depleted from shallower levels, different chemical reactions form different weathering products. This creates mineralogical, chemical, and textural layering (soil horizons) within the soil profile. Erosion is the removal of the (usually weathered) material.

Physical Weathering

Physical weathering involves the mechanical breakdown of the rock, usually without chemical modification. The major processes of physical weathering are outlined below.

Crystal (Frost) Wedging. A given mass of ice (at a temperature just below the freezing point of water) occupies a slightly larger volume than the same amount of water (at a temperature just above the freezing point). Thus, when water freezes, it expands slightly. If this water is confined in cracks in the rock, the expansion of the water makes the fractures grow, breaking down the rock into smaller fragments. This process is most effective if it is repeated many times. This freeze-thaw action is most common in polar climates, where temperatures at and near the rock surface repeatedly fluctuate back and forth through the freezing point of water.

Thermal Expansion. Like most solids, rock expands slightly as it is heated. Rocks can be affected by insolation (solar) heating, and by heat derived from combustion of naturally occurring organic fuels (e.g., forest fires). Insolation heating is most effective in hot deserts, where temperatures fluctuate widely (hot days, cold nights) and frequently (daily or seasonally). Because rock does not conduct heat well, only the outermost part is affected; if this expands significantly relative to the cooler interior, fractures separate the outer shell from the interior. Separation of a thin outer shell from the remainder of the rock surface is called exfoliation.

Hydration–Dehydration. Some minerals can absorb and desorb water when the relative humidity of their environment changes. Some of these expand slightly when they absorb water, and shrink when they lose water. The associated volume changes create stresses in the surrounding material that can cause exfoliation. This kind of weathering is especially important in temperate and subtropical climates with both wet and dry seasons.

Stress Release. When rocks are formed in Earth's interior, they are confined and compressed by the weight of the overlying rock. When the overlying material is removed by erosion, the confining stresses are released, and the rock expands slightly in a direction perpendicular to the newly exposed surface. The resulting fractures usually occur as multiple sets, parallel to the land surface, but at different depths in the rock. This is referred to as sheeting or exfoliation.

Salt Weathering. When water containing dissolved solids evaporates, the dissolved materials are precipitated as mineral crystals in fractures or between the individual grains or crystals of the rock. The growth of these crystals exerts stresses on the rock, expanding fractures and separating grains from one another. Salt weathering is most common in marine coastal regions and deserts with occasional rainfall, settings where water containing abundant dissolved material can occur on, and be evaporated from, the surface of the rock.

Root Wedging. Plant roots often penetrate fractures in rock. As the roots grow, they pry the rock apart.

Chemical Weathering

Chemical weathering involves transformations of the primary minerals of rocks by a variety of chemical reactions. Major types of reactions involved in weathering are given below.

Solution. The chemical species (elements or ions) that make up the minerals are removed in dissolved form. No solid residue is left behind.

Hydration. As noted above, some minerals can take up water from their surroundings into their structure. The mineral's composition is changed (water is added to the structure, and to the formula, of the mineral); minerals typically expand upon hydration.

Hydrolysis. When a chemical reaction can be written so that a hydrogen ion (H^+) is consumed (or a hydroxyl ion is produced), the reaction is a hydrolysis reaction. Another way of looking at hydrolysis is to think of it as the consumption of acid (the hydrogen ion is the "active ingredient" common to acids). There are multiple sources of hydrogen ion to hydrolysis reactions in geological systems, including the dissociation of water, and of sulfuric, nitric, carbonic, and humic (organic) acids. Sulfuric and nitric acids are important in the phenomenon of acid deposition (more commonly known as acid rain); carbonic acid is important in the greenhouse effect.

Oxidation. When an element or ion loses an electron, it is oxidized (reduction occurs when an electron is gained). Of the major rock-forming elements, only iron is oxidized or reduced.

Chelation. Organic complexes originate from biological activity in soils. Formation of these organic complexes greatly increases the solubility of elements that would otherwise be very immobile, such as iron and aluminum. Thus, chelation is closely related to biological weathering (in fact, it is the major mechanism of podzolization; *see* SOIL TYPES).

Carbonation. When carbonic acid is consumed, enough bicarbonate can be produced to allow the precipitation of carbonate minerals. This results in the addition to the weathered material of carbonate that was not originally present in the parent material.

Mineral Stability and Relative Rates of Weathering

Chemical weathering reactions have primary rock-forming minerals and a dilute solution of ions and acids in water (usually rainwater, or water that has interacted with vegetation) as reactants. Products can include residual (unreacted) primary minerals (usually those that are more resistant to weathering), secondary minerals (except in the case where primary minerals are completely dissolved), and dissolved products, which are exported from the weathering profile as water percolates through and reacts with it (*see* GROUNDWATER). Weathering reactions of most silicate materials involve: (1) consumption of acid; (2) destruction of the primary mineral; (3) formation of a secondary mineral; and (4) removal (leaching) of dissolved products in solution. Chemical reactions proceed more rapidly at higher temperatures, and both the supply of acid and the ability to remove dissolved products are proportional to the amount of water percolating through the weathering profile. Thus, climate factors, temperature, and moisture, exert a significant influence on the rate of chemical weathering reactions, the extent to which elements are leached away, and the composition and nature of the resulting soil horizons and soil profiles (*see* SOIL TYPES; SOIL TYPES AND LAND USE).

The vulnerability of individual minerals to weathering depends on their chemical composition. Ions of different chemical elements are bonded into crystal structures with different bond strengths, and exhibit differential mobility in the weathering environment. The sequence of relative elemental mobility in weathering for major elements is:

$$\text{Ca, Mg} > \text{Na, K} > \text{Si} > \text{Fe, Mn} > \text{Ti, Al}$$

(The position of iron and manganese is for their oxidized forms. Reduced forms are highly mobile.) In general, nonsilicate (e.g., carbonate) minerals weather more readily than silicate minerals (*see* MINERALS, SILICATES; MINERALS, NONSILICATES). Primary minerals rich in calcium, magnesium, and iron are very weatherable. They generally weather first and most quickly, and therefore tend to be easily destroyed during weathering. Primary minerals containing aluminum, sodium, and potassium are moderately weatherable; they weather more slowly and commonly survive weathering. Secondary minerals are enriched in the least mobile elements (e.g., iron, aluminum), and are the least vulnerable to further weathering.

Erosion

Erosion is the removal of material from one place to another. Physical erosion is the movement of solid (particulate) material; chemical erosion is the removal of material from the landscape in dissolved form. Most of the material available to be physically moved at Earth's surface consists of soil and regolith.

Erosion is accomplished by the movement of water and air. If rain falls sufficiently fast and/or abundantly that water does not percolate into the soil or regolith, runoff results (*see* EARTH'S HYDROSPHERE; GROUNDWATER). Surface runoff capable of eroding loose, granular material includes sheet wash and rill erosion; localized erosion produces gullying. Wind can also erode loose, unprotected material. Erosion rates are controlled by the amount of speed of water or wind, the slope of the land being eroded, and by the presence and nature of vegetation. Erosion products can be transported as solids (particles of various sizes, e.g., sand, gravel, or clay, which when deposited, become sediment), or in dissolved form (*see* EARTH'S GLACIERS AND FROZEN WATER; GEOLOGIC WORK BY STREAMS; GEOLOGIC WORK BY WIND; LANDSLIDES AND ROCKFALLS; SEDIMENTOLOGY; SEDIMENTS AND SEDIMENTARY ROCKS, CHEMICAL AND ORGANIC; SEDIMENTS AND SEDIMENTARY ROCKS, TERRIGENOUS).

Weathering and Economic Resources

Soil, which is the basis of agriculture, owes its physical properties and chemical composition to weathering. Under extreme conditions of intense or prolonged leaching, the weathering profile consists exclusively of oxides of aluminum and/or iron. This kind of weathering produces a soil type that can be used as an ore of iron (laterite), and produces the only ore of aluminum (bauxite). Some economic clay deposits (for structural products, ceramics, and specialty applications) are produced by less extreme leaching.

Weathering also has adverse effects. Pollution of urban atmospheres makes airborne moisture more chemically aggressive toward the minerals in building and monument stone. This is especially important in densely populated urban areas with culturally and historically important structures built of stone. Although damage may be minor, volumetric, or purely economic, the loss of cultural material can be devastating and irreversible.

Weathering, Erosion, and the Sedimentary Cycle

Weathering processes in sediment source areas modify the texture and mineralogical composition of the exposed geological materials, altering or destroying primary rock-forming minerals prior to the erosion, transport, and deposition of the granular material. Highly weatherable primary minerals (those rich in calcium, magnesium, and iron) are easily destroyed during weathering, and usually enter the "residuum," or coarse detrital fraction, only if mechanical erosion is very fast relative to weathering and soil formation. Primary minerals containing aluminum, sodium, and potassium weather more slowly. They commonly survive weathering to be eroded, forming sand- and silt-size fraction of soil and terrigenous (land-derived) clastic (physically transported particulate) sediments (*see* SEDIMENTOLOGY; SEDIMENTS AND SEDIMENTARY ROCKS, TERRIGENOUS). Secondary minerals are typically very fine-grained, and occur in the clay size-fraction of soil and sediment (*see* SEDIMENTOLOGY). Dissolved products may be transported into lakes or oceans, where they may be chemically and/or biochemically precipitated to form chemical sediments (*see* SEDIMENTS AND SEDIMENTARY ROCKS, CHEMICAL AND ORGANIC). Thus, weathering strongly influences the composition of sediments and sedimentary rocks (*see* SEDIMENTOLOGY).

Weathering and Geochemical Cycles

Greenhouse Effect. Silicate-mineral weathering reactions are fundamental processes that determine the contributions of continental crustal weathering to global geochemical cycles. Silicate minerals weather by hydrolysis, thereby consuming naturally occurring acids, including carbonic acid. Because carbonic acid forms by the reaction of carbon dioxide with water, reactions that consume carbonic acid remove carbon dioxide from the atmosphere. Silicate weathering is responsible for over half of the carbonic acid consumption by continental weathering. Weathering is therefore an important component of the global long-term carbon cycle and associated environmental consequences such as global ("greenhouse") warming (*see* GLOBAL CLIMATIC CHANGES, HUMAN INTERVENTION, and GLOBAL ENVIRONMENTAL CHANGES, NATURAL).

Acid Rain. Mineral weathering reactions are also fundamental processes in the response of landscapes, soils, ecosystems, and surface waters to acid precipitation. The rates of the reactions are especially important. More reactive minerals react more rapidly with environmental acids, neutralizing acidity; less reactive minerals are less effective at neutralizing acid. Because different minerals weather at different rates, the mineralogic composition of bedrock and soil minerals is a fundamental control on environmental acidification. Landscapes developed on carbonate bedrock (limestone, marble), calcareous sandstones, shales, or carbonate-rich glacial materials contain sufficient readily reactive mineral matter to mitigate the effects of acid deposition. Landscapes underlain by crystalline silicate bedrock (especially granite and gneiss) are more sensitive to acification than are carbonate-dominated landscapes, because silicate minerals react slowly with the acid in throughgoing solutions. The precise identity of the silicate minerals present controls on how much or how little acid the weathering reactions can consume.

Weathering and Erosion Through Geologic Time

Weathering and erosion have operated as important processes in the rock cycle for as long as Earth has had an atmosphere and a hydrosphere. The oldest known terrestrial rocks, which date back to when Earth was barely one-tenth its present age, are metamorphic rocks believed to have been derived from sedimentary precursors (*see* OLDEST ROCKS IN THE SOLAR SYSTEM). Thus, weathering and erosion have been creating the granular materials that make up sediments and sedimentary rocks for nearly the entire history of Earth.

The precise nature of that weathering has probably evolved with time. Some evidence for this comes from the study of ancient preserved soil profiles (paleosols; *see* FOSSIL SOILS); other evidence comes from the character of the sediments and sedimentary rocks derived and eroded from ancient weathering profiles (*see* SEDIMENTOLOGY). Oxidized iron minerals become abundant in paleosols and sedimentary rocks about 2 billion years (Ga) old and younger. This suggests that Earth's atmosphere lacked sufficient oxygen to create these minerals prior to 2 Ga. As very little oxygen is needed to oxidize iron, oxygen levels in the atmosphere before 2 Ga ago must have been substantially lower than present levels (*see* LIFE, ORIGIN OF).

Changes in the texture and composition of sediments and sedimentary rocks may have coincided with the appearance of vascular land plants (*see* GYMNOSPERMS; PTERIDOPHYTES). Land plants may have accelerated mineral weathering by producing soil organic acids and chelating agents. Vegetation probably also changed the character of erosion; the root networks of forests and grasslands protect the underlying soil from erosion, allowing chemical weathering to proceed to a greater extent before material is eroded than would be possible in the absence of land plants.

Weathering and Erosion on Other Planets

The temperature of Earth's surface is near the freezing/melting point of water. The formation of sedimentary rocks for essentially the entire history of Earth indicates that temperature conditions remained within the range favorable for liquid water throughout Earth's history. Surface conditions on other planets permit interactions between exposed rocks and planetary atmospheres, but under conditions considerably different from those on Earth (*see* WEATHERING AND EROSION, PLANETARY). "Weathering" under the dense atmosphere of Venus takes place at temperatures approaching the melting point of lead (*see* VENUS), whereas weathering on the present-day surface of Mars takes place under conditions of atmospheric pressures much lower than those at Earth's highest mountaintops, and temperatures below those of Antarctica. There is evidence that Mars may have experienced erosion by running water, but it has not been firmly established whether this took place under an earlier, much denser Martian atmosphere (liquid water cannot exist at present Martian atmospheric pressures), or as a result of a brief episode of sudden heating of volcanic or meteorite-impact origin that melted permafrost-like underground ice (*see* MARS). Minerals that may represent the products of weathering occur in some meteorites believed to have originated from Mars (*see* MARS; METEORITES; METEORITES FROM THE MOON AND MARS). Similar products of low-temperature alteration also occur in carbonaceous chondrite meteorites, indicating the presence of liquid water on the parent bodies of those meteorites (*see* METEORITES).

Bibliography

BERNER, E. K., and R. A. BERNER. *The Global Water Cycle: Geochemistry and Environment.* Englewood Cliffs, NJ, 1987.

BERNER, R. A., and A. C. LASAGA. "Modelling the Geochemical Carbon Cycle." *Scientific American* (March 1989): 74–81.

BRICKER, O. P., and K. C. RICE. "Acidic Deposition to Streams." *Environmental Science and Technology* 23 (1989):379–385.

DIXON, J. B., and S. B. WEED, eds. *Minerals in Soil Environments,* 2nd ed. Madison, WI, 1989.

MOHNEN, V. A. "The Challenge of Acid Rain." *Scientific American* (August 1988): 30–38.

NAHON, D. B. *Introduction to the Petrology of Soils and Chemical Weathering.* New York, 1991.

OLLIER, C. D. *Weathering,* 2nd ed. London, 1984.

MICHAEL A. VELBEL

WEATHERING AND EROSION, PLANETARY

The natural alteration of the physical or chemical state of rock materials on a planetary surface is termed weathering (*see* WEATHERING AND EROSION). Weathering is concerned with alteration only; the detachment and transport of any rock are separate processes. When water, wind, or another agent of transport removes weathered or unweathered surficial materials, the process is called erosion. Processes that move weathered debris downslope under the influence of gravity are collectively called mass wasting; examples include creep, slump, slide, and rock avalanche. Mass wasting will not be covered in this entry. Weathering and erosional processes, operating over the spectrum of planetary environments encountered in our solar system, produce a wide range of distinctive products and landforms.

Weathering

Weathering mechanisms fall into two broad categories: chemical and physical. Chemical weathering is a set of decompositional processes in which the chemical composition of the minerals comprising the original surface material is altered. Disintegration of surface materials is caused by weathering-induced volume changes and the attendant distortion and weakening of the minerals adjacent to the weathering site. Chemical weathering can facilitate the disintegration of polycrystalline rocks by destroying one or more constituent minerals, leaving the unweathered minerals unsupported. An example is the disintegration of granite by the chemical weathering of its feldspars into clays. The remaining quartz and other minerals resistant to chemical weathering then disaggregate because the matrix containing them has been destroyed, forming a debris called grus.

Chemical weathering processes include oxidation, hydration, carbonation, hydrolysis, base-exchange, and chelation. All require the presence of at least some liquid water and all have their reaction rates controlled by water abundance. The most common example of oxidation is the change of iron minerals from the ferrous (Fe^{2+}) to the ferric state (Fe^{3+}), resulting in the formation of ferric oxides that are both stable and insoluble. Hydration is the incorporation of water molecules into the surface minerals. Carbonation and hydrolysis are the breakdown of surface minerals facilitated by the addition of carbon dioxide and the hydroxyl radical, respectively. Base-exchange involves the mutual transfer of positively charged ions (cations) between surface minerals and aqueous solution. Chelation is an organic process in which metallic cations are removed from surface minerals and tightly bound into hydrocarbon molecules.

Physical weathering is a set of disintegration processes that fragment surface materials but do not alter their composition. One type of mechanical fracturing, called sheeting, commonly occurs due to differential expansion when pressure is released from a rock mass. Another cause of mechanical failure is repeated thermal expansion and contraction. Neither type of mechanical weathering requires the presence of water or other agent. A common type of mechanical weathering occurs where foreign crystals form and grow in microfractures on the surface materials, wedging them apart. Water is particularly efficient at breaking rocks; enormous forces are generated by freezing water in a confined space and the attendant forcing of water into microfractures by capillary action, a process termed hydrofracturing. Also important is salt weathering, a general term for similar disruption caused by a variety of water-soluble salts, where the expansion force that fractures the rock arises either from the expansion of the growing crystal or the expansion caused by hydration of the salt. Biological activity can cause both chemical and physical weathering. An example is a growing root. The wedging action of growing roots is a type of mechanical weathering, and chemical reactions are facilitated by chemical changes caused by nutrient uptake. A type of chemical/mechanical weathering important on most solid planetary surfaces but not often considered because it is insignificant in the terrestrial environment is alteration caused by micrometeoroid impact and the interaction of the surface with high-energy particles and radiation.

Erosion

"Erosion" is a broad term for a set of processes in which weathered and unweathered materials are detached from their original location and transported away by some agent. The most common agents of erosion are running water, flowing ice, the wind, lava, and impact (*see* GEOLOGIC WORK BY STREAMS; GEOLOGIC WORK BY WIND). The capability of running water to erode and transport large quantities of material is familiar to most everyone. Overland movement of water, or sheet flow, tends to produce a network of closely-spaced channels called rills. Channel flow produces a wide variety of diagnostic erosional and depositional features, including braided channels, meanders, oxbow lakes, and point bars. Subsurface flow of water can also contribute to erosion; the undermining of the rocks above caused by erosion where migrating water emerges from the ground is called sapping. Sapping channels have few tributaries and amphitheater-shaped heads. Glacial erosion produces very distinctive features by plucking, scouring, and grinding, including U-shaped valleys, hanging valleys, roches moutonnées, cirques, cols, arêtes, and horns. Examples of wind erosion include the formation of ventifacts, surface rocks abraded by the natural sandblast action of wind-transported sand, and the formation of streamlined ridges called yardangs, where wind erosion is aided by weathering and running water. Yardangs can be as large as a few tens of meters high and kilometers in length. Erosion by flowing lava is less common, but the thermal and mechanical erosive capability of moving lava is considerable. Impact cratering, the alteration of a planetary surface by the impact of interplanetary debris, can redistribute large quantities of surface materials over a very wide area.

Planetary Environment

The relative importance of each of the processes mentioned above and the exact nature of the resulting landforms depend on the surface environment of the body in question. All of the processes operate on Earth, but in most locations the dominant weathering and erosion processes are those dependent on surface water. For example, the rate of impact cratering on Earth is so low compared to the rate of erosion by other means that the original form of all but the youngest impact craters has been obliterated. Only in the most arctic or arid terrestrial locations is some other process more important. On other planets, water-related weathering and erosion do not overwhelm other processes.

Airless Case. The simplest case to consider is that of a surface with essentially no atmosphere present, including Mercury, the Moon, and most of the outer planet satellites. Weathering and erosion processes that rely on water or air cannot operate. The only kind of chemical weathering possible under airless conditions are the surface changes that occur due to the interaction between the surface and high-energy particles and radiation. Mechanical weathering is limited to thermally induced volume changes and impact cratering. Impacts not only produce craters but they also create ejected debris with a wide range in sizes. Over time, a thick layer of impact-generated debris called a regolith will cover the surface and protect it from all but the largest impacts. If the surface dates from the early solar system when large impacts were much more common than today, and no other erosional process has operated to cause resurfacing, then the regolith layer might be up to a kilometer or more thick, in which case it is termed a megaregolith.

There are two primary sources of differences between impact processes and products on different bodies in the solar system: the kinetic energy of the impactor and the composition of the surface being impacted. In general, a given target will produce the following features if subjected to a series of increasingly energetic impacts. Craters formed at relatively low impact energies are bowl-shaped, while those formed at somewhat higher energies are shallower compared to their diameters. As impact energy increases, central peaks then central rings of peaks are formed. The highest-energy impacts produce multiringed basins with diameters of hundreds of kilometers (Figure 1). The impact energies where the morphologic transitions occur are primarily a function of the size (gravity) of the target body. For high-energy impacts, impactor and target physical properties are less important than for impacts with lower energy; however, target properties are very important for post-impact modification. Surface composition will also affect initial crater formation, especially for intermediate and smaller impacts.

Figure 1. Mare Orientale on our Moon is a relatively young example of a large (900 km), multiringed impact basin. Similar features in various states of erosion can be found throughout the solar system. This is a mosaic of *Lunar Orbiter* images.

Venus. Venus represents the other extreme of the set of planetary surface conditions from airless bodies. Although the bulk properties of both planets are similar, the venusian surface environment is very unlike that of Earth. Atmospheric pressures and temperatures at the surface are very hostile, ranging from about 40 bars (690 psi) and about 650 K (710°F) at the summit of Maxwell Montes (11 kilometers or km) to about 107 bars (1,570 psi) and about 757 K (903°F) at the lowest elevations (about −2.5 km), with most of the venusian surface having conditions near 92 bar (1,350 psi) and approximately 733 K (869°F). Numerous chemical weathering reactions are possible under these conditions and the determination of the details of chemical reactions between the surface and the atmosphere is a topic of considerable study. Obviously, weathering and erosion processes that require the presence of liquid water cannot operate on Venus.

Radar imaging systems aboard orbiting spacecraft, particularly the *Magellan* mission, have revealed a wide variety of volcanic, tectonic, and impact features. Also found were many erosion features, among them four different types of erosional channels carved by some sort of liquid, including one old channel that can be traced for 6,800 km; several types of wind streaks (Figure 2); sand dune fields; and possible yardangs. Surface wind speeds are low, but low wind speeds are sufficient in the dense venusian atmosphere to induce particle motion and present winds can probably account for the observed aeolian features. It is not known if the channels on Venus required running water and, hence, a more clement climate in the past. The wind can apparently sculpt venusian landforms, but on the smaller scale, it is not clear how well sand-laden venusian winds abrade surface materials. It is possible that rocks impacted by moving sand actually gain mass as material broken off the impactor adheres to it.

The presence of a large number of relatively uneroded impact basins and craters on Venus indicates that the rate of weathering and erosion on Venus must be lower than on Earth. Crater populations indicate that the surface of Venus is on the order of half a billion years old; in most terrestrial environments, erosion would obliterate an impact crater in only a few tens of thousands of years.

Mars. Much of the martian surface is very old, retaining the imprint of the high rate of cratering by large impactors in the final stages of the formation of the solar system. Some areas are young, and others have been (repeatedly) resurfaced relatively recently. The martian atmosphere, less than 1 percent as dense as the terrestrial atmosphere, is at present so rarefied that liquid water cannot exist as a liquid on the surface; if present it would evaporate. Yet the martian surface was not always so dry; there is abundant geologic evidence of surface water in the martian past, including river channels

Figure 2. These venusian wind streaks are located near the crater Mead and indicate a wind direction from the upper left to the lower right. Individual streaks exceed 60 km in length. This is a radar image returned by the *Magellan* spacecraft.

Figure 3. Terrestrial versions of the martian runoff channels shown here almost always indicate the overland flow of (rain) water. However, the present martian climate does not permit liquid water to exist on the surface, therefore, the climate must have been different when these channels were cut. This is a *Viking Orbiter* image with a width of 250 km.

(Figure 3), lake and shoreline deposits, and other features analogous to their terrestrial counterparts. We have also seen from martian orbit numerous landforms indicating aeolian erosion, transport, and deposition. These features may or may not be relicts of a previous environment. Even present-day conditions allow a much greater degree of weathering and erosion than on an airless body, although reactions requiring bodies of liquid water or water as an agent of transport do not presently operate.

The details of the chemistry of martian surface materials are imperfectly understood. Chemical reactions do occur; dust observed in the martian atmosphere has been found to be chemically weathered basalt fragments and the addition of water to the soil in the *Viking* caused the liberation of large quantities of oxygen and produced ambiguous results on one of the *Viking* biological experiments. Low temperatures and the lack of liquid water at the surface will tend to retard, but not prevent, many chemical reactions. The mechanical weathering processes that do not require running water will operate to some degree on Mars, although typically at a much lower rate than on Earth. One exception may be wind erosion. The rarefied martian atmosphere must have much higher winds (approximately ten times greater) than those on Earth to impart enough momentum to initiate and sustain particle motion. Once set in motion, however, the particle is accelerated to much higher speeds than on Earth and is, hence, much more capable of causing erosion.

Bibliography

BLOOM, A. L. *Geomorphology: A Systematic Analysis of Late Cenozoic Landforms*, 2nd ed. Englewood Cliffs, NJ, 1991.

CARR, M. H. *The Surface of Mars*. New Haven, CT, 1981.

GREELEY, R. *Planetary Landscapes*, rev. ed. Boston, 1987.

HUNTEN, D. M., L. COLIN, T. M. DONAHUE, and V. I. MOROZ, eds. *Venus*. Tucson, AZ, 1983.

KIEFFER, H. H., B. M. JAKOSKY, C. W. SNYDER, and M. S. MATTHEWS, eds. *Mars*. Tucson, AZ, 1992.

STEVEN H. WILLIAMS

WEGENER, ALFRED

Alfred Lothar Wegener was born in Berlin, Germany, on 1 November 1880. He died on 10 November 1930, at approximately 71°N, 44°W on the Greenland Ice Cap.

Wegener is known to earth scientists today as the author of the theory of continental displacements ("continental drift"), which he proposed in 1912 and developed in successive editions of his book *Die Entstehung der Kontinente und Ozeane* (The Origin of Continents and Oceans). Trained as an astronomer, Wegener received his Ph.D. in 1905. Most of his published research was in meteorology and atmospheric physics. He authored a standard text on atmospheric thermodynamics and the physics of clouds (*Thermodynamik der Atmosphäre* in 1911) and was an expert on atmospheric layering and turbulence. His numerous publications, well over 150 in number, included a dozen books, among them one on the origin of lunar craters as impact structures, *Die Entstehung der Mondkrater.* (The Origin of Lunar Craters in 1921) and a pioneering text on paleoclimatology, *Die Klimate der geologischen Vorzeit* (Climates of the Geological Past in 1924) co-authored with his father-in-law, the climatologist Wladimir Köppen (*see* CLIMATOLOGY).

Wegener played a significant role in the exploration of Greenland and made three expeditions there: 1906–1908, 1912–1913, and 1929–1930. During all of these expeditions, he pursued research in glaciology as well as meteorology and climatology; during the final expedition, he lost his life while trying to resupply a beleaguered base in the middle of the ice cap. Wegener made contributions to atmospheric optics, atmospheric acoustics, the physics of the ionosphere, and was a pioneer aeronaut—having set a world record for time aloft in a free balloon (fifty-two hours) with his brother, the meteorologist Kurt Wegener (1878–1964), in 1906. He held academic positions at Marburg (1910–1918), Dorpat (1919), Hamburg (1919–1923), and finally a professorship in meteorology and geophysics at Graz, Austria (1924–1930). In World War I, he served as a captain of infantry on the Western Front (1914–1915) and in the Army Weather Service (1915–1918) in the East after having been (twice) seriously wounded. In 1913, Wegener married Else Köppen, who supported his scientific work with translations and published after his death a number of works about his travels and researches. Together they had three children: Hilde, Käthe, and Charlotte.

The theory of continental displacements occurred to Wegener in 1910–1911. His office mate at Marburg had received a world atlas for Christmas, and Wegener was struck by the map of the Atlantic Ocean. More astonishing than the matching coastlines of Africa and South America was the match at the submerged continental shelves: this indicated to him that the match was not an accident of sea level, but the result of some physical process. In pursuit of an answer, he read texts in geology, paleontology, and geophysics, and eventually became convinced that the solution to many problems in these sciences could be found in the splitting and drifting apart of great continental fragments—creating new oceans as they split and destroying old ones as they collided (*see* CONTINENTS, EVOLUTION OF). The result, published in 1912, was a hypothesis that changed little in its fundamental outlines over the next seventeen years.

The theory of continental drift, and the long controversy surrounding it, are more easily understood when one realizes that it was not a geological theory at all, but a geophysical and astronomical theory of the earth. Wegener was indifferent to the phenomena of geology and paleontology, except insofar as they provided supporting evidence for the physical theory, based in classical thermodynamics. Wegener's view of the earth was like his view of the atmosphere and the ocean: a series of spherical shells of increasing density downward, marked by sharp discontinuities in temperature and in physical properties at the boundary surfaces. The outer crust of the earth, once continuous and now fragmented into sections, floated buoyantly on the next interior layer of the earth, represented by the ocean floors, because of the slightly lower density of the crustal material. The principle supporting this flotation, called ISOSTASY, was confirmed by gravity measurements that showed ocean floors to be of material heavier than the continents. The boundary between the crust and mantle was identified by Wegener as the Mohorovičić Discontinuity—a surface from 16–80 km below the surface discovered from the refraction of seismic waves measured in a 1909 earthquake. Working from experimental data in solid mechanics, Wegener proposed that at a depth where the mantle would be devoid of strength, the floating bergs of crust would still be well above their melting points and could plow through the mantle without disintegrating. The implication of this for geological history was that continents were constantly splitting and drifting apart, with folded mountain ranges such as the Alps, Himalayas, and Andes forming by compression at the leading edge of a drifting continent, or through continental collisions. The picture was complicated by the movement of the pole of rotation, but Wegner's 1924 book on ancient climates brings together the motion of the poles and the displacement of continents to develop a history of continental arrangements in the past, explaining much anomalous paleontological data—such as coal deposits from tropical plants in very high latitudes.

The theory of continental drift was hotly debated in the 1910s and especially in the 1920s. Wegener never settled on a mechanism to drive the splitting continents and always maintained that the theory would be proven or disproved solely by measuring changes in the longitude of fixed stations on continental platforms. Though it brought Wegener much celebrity and notoriety, continental displacement theory was, for him, always a sideline to his own fundamental research. The plans for his final expedition in Greenland (1929–1930) were devoted to meteorology and glaciology (including the seismic determination of ice cap thickness) and do not even mention continental drift.

Wegener was a man of great physical strength, noted for his patience, his fearlessness, his inspirational style of leadership, his good humor, and his love of tobacco. Colleagues remembered him for his rapid and intuitive grasp of physical problems and for his tireless energy in the collection and reduction of huge sets of quantitative data in his meteorological and glaciological research. Germany's center for oceanographic and polar research, in Bremerhaven, is fittingly dedicated to his memory.

Bibliography

KÖPPEN, W., and A. WEGENER. *Die Klimate der geologischen Vorzeit.* Berlin, 1924.

SCHWARZBACH, M. *Alfred Wegener, the Father of Continental Drift.* Madison, WI, 1986.

WEGENER, A. *Thermodynamik der Atmosphäre.* Leipzig, 1911.

———. *Die Entstehung der Mondkrater.* Braunschweig, 1921.

———. *The Origin of Continents and Oceans,* 1929 4th ed., trans. by J. Biram. New York, 1964.

MOTT T. GREENE

WERNER, ABRAHAM GOTTLOB

Abraham Gottlob Werner was born on 25 September, 1749 in Wehrau, Silesia (present-day Germany). In 1769 he began his academic career as a freshman at the Bergakademie (Mining Academy) in Freiberg, Saxony, and by 1770 he had been made an honorary member of the Leipziger Ökonomische Gesellschaft (Economic Society of Leipzig). In 1771 he transferred to the University of Leipzig where he began his study of the natural sciences, languages, and the law. He completed his studies in 1774 and in the same year published the book *Von den äußerlichen Kennzeichen der Fossilien* (On the External Characteristics of Fossils). In 1775 Werner became a researcher and teacher at the University of Freiberg. During his time in Freiberg he was often visited by well-known scientists, such as G. Forster, who took part in James Cook's voyage around the world.

In 1791 Werner published the book *Neue Theorie von der Entstehung der Gänge mit Anwendung auf der Bergbau, besonders den freibergischen* (New Theory of the Origin of Veins with Application to Mining, Especially at Freiberg), which quickly made him well known. It is interesting to note that, at this time, one of Werner's pupils was Alexander von Humboldt, who was interested in Werner's theory on mining and would go on to change geoscience significantly.

Werner's following years were filled with a long-standing, forcefully presented argument over the topic of "neptunism." This theory, developed by Werner, attempted to explain that all minerals and rocks, with the exception of volcanic rocks, are crystallized from water-based solutions and later deposited as sediment. The opposite viewpoint, "plutonism," held that all the minerals and rocks, as well as the emergence of earth formations, were created by volcanic action.

Werner's belief in neptunism led him to the incorrect conclusion that basalt, too, must have formed from a water-based solution as a sedimentary rock. Werner saw plutonism as secondary, caused by burning and smoldering coal-fields. The discussion of the genesis of basalt, which after Werner's death was determined with the help of his students, Johann Voigt, A. von Humboldt, and L. von Buch, also argued against Werner's viewpoint. Neptunism was ultimately discarded as a viable theory.

In addition to these theoretical discussions, Werner also introduced the term "mineral" and, in a general sense, the discipline of mineralogy. In 1790 he defined the following disciplines: Mineralogie (mineralogy); Mineralogische Chemie (mineralogical chemistry); Geognosie (geology); Mineralogische Geografie (mineralogical geography); and Ökonomische Mineralogie (economic mineralogy). With these early classifications Werner contributed to the study of mineralogy as an independent scientific discipline.

Werner played an important role in the development of petrography and stratigraphy in Germany. With the increasing number of new minerals being discovered in the eighteenth century, it became necessary to develop a new and more precise classification system. In 1816 Werner presented a new system of mineral classification. His classification was used frequently, and it certainly opened the way for mineralogical studies as an independent science. Werner's idea was to observe and explain the minerals, not just to describe them. It was clear to Werner that a more theoretical and practical method of scientific explanation would be useful. In this respect, he was a pioneer in the more exact observation and explanation of nature and a founder of modern mineralogy.

Using his empirical methods, Werner discovered the minerals pyrrhotite (1789), vesuvianite (1795), aragonite (1796), and Karinthin (1817; a type of hornblende, the name is no longer used). He also used his system of classification to describe the minerals zircon (1783), apatite (1786), boracite (1789), graphite (1789), olivine (1790), augite (1792), anhydrite (1804), zoisite (1805), grossular (1811), and helvite (1817).

Werner made many interesting contributions to the practical and theoretical explanation of the geological world. His method of observation and classification is an example of the modern scientific method and a basis for modern geology. His part in the controversy between neptunism and plutonism, although his theory did not survive, led to a more exact explanation of the natural world. His books, articles, and training at the mining university of Freiberg influenced both his and later generations. When Werner died on 30 June 1817 in Dresden, he was well known to the international scientific community as one of the pioneers of the modern understanding of the nature of the earth.

Bibliography

GEIKIE, A. *The Founders of Geology*. London, 1897, 1905.

WERNER, G. A. *On the External Characters of Minerals*. Urbana, IL, 1962.

WILFRIED SCHRÖDER

WHITE DWARFS

See Stars

WILSON, J. TUZO

John Tuzo Wilson was born in 1908 in Ottawa, Ontario, the national capital of Canada, to a family of Scottish descent (on his father's side) and Huguenot-American descent (on his mother's side). Growing up in the Rockliffe Park area, he attended Elmwood School, Ashbury Collegiate, and Lisgan Collegiate in succession. His university undergraduate education began in the Honors Mathematics and Physics Program at the University of Toronto in 1926. Wilson became fascinated with geology through his involvement in various fieldwork beginning in the summer of 1924, and in the summer of 1927 he enjoyed looking for gold north of Lake Superior with Neil Odell (a geologist who had been lionized for his efforts in the 1924 Mount Everest Expedition). In the fall of 1927, Wilson requested a transfer into geology. His mentor in the Department of Physics, Lachlan Gilchrist, managed to dissuade him of this, but he was thereafter allowed to follow an undergraduate degree that consisted of an "ill assorted mixture of geology and classical physics courses," as he once described it. He obtained his B.A. in 1930, for which he was awarded the Governor General's Medal for the best degree at the university in his graduating year. The degree was conferred in honors physics and geology, the first such degree ever awarded by the University of Toronto, but one which has remained a possible course of study since that time. His graduate work was completed at Cambridge University and Princeton University; he received a Ph.D. from the latter in 1936.

Following his military service in 1946, Wilson resumed his academic career when he accepted appointment as professor of geophysics at the University of Toronto, a position in which he succeeded his mentor Gilchrist. He was appointed full professor with tenure in the Department of Physics, where the geophysics group continues to flourish, in no small part due to the base of excellent faculty that Wilson personally recruited, including Sir Edward Bullard, who lived briefly in Toronto and served as chair of the department. At this time, Wilson's personal science had yet to develop truly significant creative energy, but he had begun to travel extensively and amass the incredible personal catalog of earth structures, the synthesis of which was to make his name. He was elected to fellowship in the Royal Society of Canada in 1948. When he met Sam Carey, a passionate proponent of Wegener's idea of continental drift, at the University of Tasmania in 1950, Wilson found the idea outlandish. Yet he traveled everywhere—to India, Malaysia, Singapore, South Africa, and New Guinea—always adding to his first-hand knowledge of global geology. At this time, he developed simultaneously with A. E. Gill of McGill University the idea that Precambrian shields could be divided into a discrete set of "provinces," an idea that had in fact been first suggested by Arthur Holmes, whom Wilson had met in 1948. With two students, R. M. Farquhar and R. D. Russell, Wilson measured isotopic ages for these provinces to show how significantly different in age they were from the rocks of the Proterozoic. Together they published the results of this work in an important paper in the *Handbuch der Physik.*

Through the first decade after his return to the University of Toronto, Wilson had become more and more deeply involved in international science through the International Union of Geodesy and Geophysics (IUGG), for which he served as chairman of the finance committee (1948–1954) and as vice president (1954–1957). By the time of the International Geophysical Year of 1957–1958, he had become president of the IUGG (a post that he held until 1960), and in this capacity he presided over the General Assembly that was held in Toronto in 1958. At the time of this assembly, Wilson was fifty years old, but his personal scientific blossoming was only just beginning.

Wilson's intellectual horizon was reached with the stimulation provided by Harry Hess, whose epochal 1962 paper, "History of Ocean Basins" (a chapter in a book published by the Geological Society of America), expounded the idea of seafloor spreading. It was this paper that caused Wilson to revise his earlier view of mobilism as "outlandish." The suddenness of this change of mind is made even clearer by the fact that just the previous year he had published a paper in *Nature* entitled "Some Consequences of the Expansion of the Earth." Thereafter began the continuous flow of papers from Wilson that are now seen as constituting much of the intellectual fabric of plate tectonics—not the analytical fabric, all of which would be supplied by others, but rather the conceptual fabric, which took Hess's idea of seafloor spreading and put it to work in the context of a model, constrained by the constancy of Earth surface area, that incorporated all that was known about the large-scale structural geological features of the planet. A series of papers followed: in 1963, in *Nature,* Wilson published "Evidence from Islands of the Spreading of Ocean Floors"; in 1963, in the *Canadian Journal of Physics,* "A Possible Origin of the Hawaiian Islands"; in 1965, in *Nature,* "A New Class of Faults and Their Bearing on Continental Drift"; in 1966, in *Nature,* "Did the Atlantic Close and Then Reopen?"; and, in 1968, in the *Transactions of the Royal Society of Canada,* "A Revolution in the Earth Sciences," which was the Bancroft Award Lecture. These papers contained three ideas that have proven seminal to our understanding of the surface kinematic aspects of the mantle convection process, the body of ideas that has come to be referred to as plate tectonics: the idea of the mantle plume, the idea of the transform fault, and the idea of the cycle of ocean basin opening and closing (now usually referred to as the Wilson cycle). Quoting from the excellent textbook entitled *Plate Tectonics* (1986), by Alan Cox and Robert Brian Hart:

> The intellectual breakthrough that established plate tectonics was based upon a simple new geometrical insight. In a five-page article published in *Nature* in 1965, J. Tuzo Wilson noted that movements of the Earth's crust are concentrated in narrow mobile belts. Some mobile belts are mountain ranges. Some are deep sea trenches. Some are ocean ridges. Others are major faults. Earlier geological maps of these long linear features showed many of them to come to dead ends. Wilson postulated that the dead ends are an illusion: the mobile belts are not isolated lineations but rather are all interconnected in a global network. This network of faults, ridges and trenches outlines about a dozen large plates and numerous smaller ones, each comprising a rigid segment of the lithosphere (that is, the top layer of the Earth).

Wilson is justifiably seen as the master synthesizer who assembled all of the fragmentary pieces of knowledge that were being collected in the early and mid–1960s and brought them together into a conceptual whole.

For his creative output in the years 1963–1968, Wilson received a very large number of awards of great distinction, beginning in 1968 with the Logan Medal of the Geological Society of America, the Bucher Medal of the American Geophysical Union, and the Penrose Medal, also of the Geological Society of America. In 1974, he received the J. J. Carty Medal, the highest award of the U.S. National Academy of Sciences, of which he had been elected a Fellow in 1968, the same year in which he was elected to fellowship in the Royal Society of London. In 1984, as the importance of his intellectual contribution came to be even more deeply understood, he was awarded the Gold Medal of the Royal Canadian Geographical Society, the Wollaston Medal of the Royal Society of London, the Vetlesan Prize of Columbia University (long considered the equivalent of the Nobel Prize in the earth sciences), and the first award of the J. T. Wilson Medal, now the highest award of the Canadian Geophysical Union. In 1980, he received both the Ewing Medal of the American Geophysical Union and the M. Ewing Medal of the Society of Exploration Geophysicists of the United States. In 1989, he was awarded the A. Wegener Medal of the European Union of the Geosciences and in the same year the prestigious Killam Award of the Canada Council. He is justly regarded as having been a giant of the plate tectonics revolution.

Bibliography

Cox, A., and R. B. Hart. *Plate Tectonics: How It Works.* Cambridge, MA, 1986.

Peltier, W. R. "J. Tuzo Wilson: 1908–1933." *E.O.S.* 75, no. 52 (December 27, 1994): 609–612.

W. RICHARD PELTIER

WOMEN IN THE EARTH SCIENCES

Jonathan Cole (1979) observed that until 1971 the largest compendium of biographical information about living American scientists was titled *American Men of Science,* although women were included among the listings. This fact underscores the very recent redressing of the former status of women in all scientific fields, and is particularly appropriate to the discussion of women in the earth sciences.

The first flowering of American earth science came in the form of exploration fever, when the riches of the land for agriculture, including the river systems, and potential mineral resources, were documented by the fledgling eastern seaboard states. These efforts were coordinated through state surveys, with Massachusetts being the first in 1841 (Socolow, 1988). In the 1870s, the U.S. Geological Survey (USGS) extended the mapping efforts to the Pacific (Rabbitt, 1979, 1980, 1986). The mining industry as well as agriculture and transportation required accurate information about the soils, rocks, mountains, and streams in order to capitalize on these natural endowments. In turn institutions, such as Sheffield Scientific School at Yale, were established to educate and train individuals and to research improved methods for exploration, extraction, and utilization of the hydrological and geological largess. The only women active at these earliest stages were in roles that reflected their status in society: they were "clerks" or recorders of the accomplishments of others, usually husbands or other male relatives and friends. Today, state surveys and the USGS coexist and integrate their efforts. There is topographic, hydrologic, and geologic coverage for the entire nation, including Hawaii and Alaska, at a scale of 1:250,000, a remarkable compilation of data. The contribution of women to earth sciences is now obvious and diverse, encompassing all occupational niches—government, industry, and academe (Table 1). Many former barriers to women have disappeared; those who wish can work underground, in space, at sea, or with a "field party" in remote areas. As in the past, women are more likely to be found in the laboratory or office than in the field, although there is an increase in female consultants who can use their observational geological training and expertise in exploration, at mines or at well heads. Consultants can choose their occupational activities and times of employment, options that grant personal freedom but may also reflect the demands of a family schedule as well as the difficult job market worldwide in the 1990s.

Table 1. Geoscience Occupational Statistics

	Total Employed	*Male*	*Female*
1978	54,000	47,900	6,100
1988	94,200	83,000	11,100

National Science Foundation, January 1992.

The initial openings for women as practicing earth scientists came for those who, with advanced education and training, demonstrated their commitment and capabilities. In the United States, Florence Bascom (B.S., University of Wisconsin; Ph.D., Johns Hopkins University, 1893) became the first woman employed by the USGS when she joined Roland D. Irving, lecturer at the University of Wisconsin and researcher with the USGS, in the 1880s to study outcrops in the Lake Superior region. When M. Carey Thomas opened Bryn Mawr, a women's college near Philadelphia, she solicited Bascom, then at Ohio State University, to teach the popular study of "natural history." This marked not only the foundation of geological sciences at Bryn Mawr (1895) but undoubtedly served to legitimize the admission of women in geology at other universities, where only a few had pursued graduate study (Rossiter, 1981). These few women with advanced degrees found employment mostly at female undergraduate institutions (Ida Olgivie, B.A., Bryn Mawr, Ph.D., Columbia University, at Barnard College Columbia; Mignon Talbot, B.A., Ohio State, Ph.D., Yale University, at Mount Holyoke College), and as curators of paleontological and mineral collections at museums (Winifred Goldring, B.A., M.A., Wellesley College, at the New York State Museum, site of the New York State Geological Survey).

The first half of the twentieth century saw little change in the general attitudes toward women geopractitioners although in retrospect, as described in their obituaries and mentioned briefly below, several individuals were clearly independent scholars. The range of their contributions heightens awareness of some basic changes in the culture of the United States, and documents the second flowering of American earth science. Laboratory investigations became the modus operandi to explicate geoscientific observations. Notable contributions were made to petrology by Eleanora Bliss Knopf (B.A., Ph.D., Bryn Mawr, 1912) as an adjunct in research in the geology department at Yale University. Margaret Foster (B.A., University of Illinois, 1918; Ph.D., American University, Washington, D.C., 1936), a chemist in the USGS, who in addition to her seminal contributions on the composition of groundwater in the southern Atlantic coastal plain, became a member of the Manhattan project (1942) aiding the efforts to devise methods to analyze uranium and thorium. In 1962, Foster published on clay mineral exchange in soils. The

USGS hired women geoscientists as researchers in its hydrological and geological divisions during World War II. Tiasia Stadnichenko, a Russian by birth and educated at Vassar College (B.A., 1952), became the USGS expert on the geochemistry of coal.

Although deeply committed and advancing knowledge in the geosciences, women were few in number as members of scientific societies, as presenters at professional society meetings, or as authors in journals (such as the *American Journal of Science*, *American Mineralogist*, and *Bulletin of the Geological Society of America*). There are a few famous individuals such as Inge Lehman (1888–1993), a Dane, who proposed the existence of the earth's inner core from seismological observations and calculations in 1936. She helped to found the Danish Geophysical Society and was awarded an honorary doctorate at Columbia University in 1964. They were isolated, some would say marginalized (Cole, 1979), in second-tier positions, hardly noticed in the dynamic, expanding, broadly based earth sciences (e.g., petrology, paleontology, glaciology, sedimentology, petroleum and structural geology). Women scientists attracted few acolytes, and those that entered earth sciences were mostly from the women's colleges. The fewest number of doctorates in geoscience were granted to women in the 1950s and 1960s ("Climbing the Academic Ladder: Doctoral Women Scientists in Academe," Committee on the Education and Employment of Women in Science and Engineering NAS/NRC, 1979) when opportunities for women during the war years had ended and men returned to the universities and industry. Yale, for example, awarded the Ph.D. to Mignon Talbot in 1904, but by 1968 had awarded only five more to women. Female faculty members in departments of geology and related fields at the larger, coeducational universities were few until the late 1970s.

Undoubtedly as a direct response to the women's movement and to several important pieces of equal access legislation in the late 1960s and early 1970s directed at institutions seeking federal funds for education, along with formal avenues through which women could seek redress of gender discrimination, there was a four-fold increase of women students in earth sciences. In 1972–1973, women made up 14 percent, and by 1978–1979, 22.6 percent, of all students enrolled for advanced degrees in two hundred geoscience departments in U.S. colleges and universities. A 1970 survey showed that more than 60 percent of male geologists and geophysicists were employed in petroleum and petroleum-related industries whereas about half of all women geoscientists were in academe, although roughly an equal percentage of men and women (20 percent) held Ph.D.'s. These women were employed in the lower ranks, and therefore had lower salaries than men. In the 1970s there were few highly visible women in the field of earth sciences, none at the prestigious academic institutions, nor in the upper ranks of the USGS or state surveys.

The growing number of female scholars hoping for careers in geology benefited the few professional women geoscientists who were now offered opportunities to be associate editors on journals and to participate in the councils and committees of the professional societies. Such recognition and visibility was promulgated and sustained by the diligence of groups such as the Women Geoscientists Committee of the American Geological Institute (AGI), later the Association for Women Geoscientists (Moody and Marvin, 1976). The committee actively provided support for both newcomers and experienced professionals as it circulated lists of women as potential speakers and conferred awards at the annual national meeting of the Geological Society of America (GSA). By the late 1970s, the status of women in the geosciences was changing (Schwarzer, 1979).

By 1981 the first female head of a state survey was appointed: Genevieve Atwood (B.A., Bryn Mawr, 1968; M.A., Wesleyan, 1973) as director of the Utah State Survey. In 1986, Pricilla Grew (B.A., Bryn Mawr, 1962; Ph.D., Berkeley, 1967) became director of the Minnesota State Survey. Susan M. Landon (B.A., Knox College, 1968; M.A. State University of New York, Binghamton, 1975) became the first female president of the American Institute of Professional Geologists in 1990. Doris Curtis (B.A., Brooklyn College, 1933; Ph.D., Columbia, 1949), after a long career in sedimentology—predominantly with the petroleum industry, first with Shell Oil Co. in Houston, Texas, and then as an independent consulting geologist—became president of the Society of Economic Paleontologists and Mineralogists in 1978–1979, and AGI president in 1980–1981. She was also the first woman elected president of the GSA (1989).

The National Research Council created a committee to increase the participation of women in science and engineering, and thus utilize their ex-

Table 2. Gender and Occupational Category (%)

	Male	*Female*
Geologists, geochemists	95.6	4.4
Geophysicists	93.9	6.1
Petroleum engineers	95.5	4.5
Mining engineers	97.8	2.2
Engineering geologists	95.2	4.8
Hydrologists/hydrogeologists	89.7	10.3

North American Survey of Geoscientists, AGI, 1992.

pertise. It has already documented that the proportion of doctorates granted to women in all fields of science has risen dramatically in the last decade. National Science Foundation (NSF) graduate fellowship awards in earth sciences to women exceeded those for men for the first time in 1991. The marked increase of women geoscientists, perhaps due to a heightened interest in the environment in the 1980s, is reflected in the AGI 1986 North American Survey of Geoscientists. Those data showed that female respondents had, in aggregate for all fields designated (Table 2), less than fourteen years experience and, not surprisingly, were found at the lower income levels both for industry and academe.

Encouraging women to study earth science and the support that is now available to women have had the desired effect. Thirty percent of the student enrollment and of the degrees granted in earth sciences, both at the undergraduate and graduate levels, are awarded to females (Claudy, 1992). In addition Wallace (1992) shows that approximately one-third (3,889) of the USGS workforce is comprised of women, 36 percent of whom are in the geosciences either as professional scientists or technicians. The upper grades remain mostly male (6:1) but these figures represent a steady growth in numbers of women as scientists and as administrators over the past twenty years. The status of women in the geosciences is indeed evolving.

Bibliography

CLAUDY, N. *North American Survey of Geosciences.* American Geological Institute. Washington, DC, 1992.

COLE, J. R. *Fair Science: Women in the Scientific Community.* New York, 1979.

HENDERSON, B. C. "As You Might Guess, Men Are Paid More." *Geotimes* 20, no. 3 (1975):30.

MOODY, J. B., and B. M. URSULA. "Professionalism among Women and Men in the Geosciences." *Journal of Geological Education* 24 (1976):166–171.

RABBITT, M. C. *Minerals, Lands and Geology for the Common Defence and General Welfare.* Vol. 1, *Before 1879* (1979); Vol. 2, *1879–1904* (1980); Vol. 3, *1904–1939* (1986). Washington, DC (1979–1986).

ROSSITER, M. W. "Geology in Nineteenth-century Women's Education in the United States." *Journal of Geological Education* 29 (1981):228–232.

SCHWARZER, T. F. "The Changing Status of Women in the Geosciences." *Annals of the New York Academy of Sciences* 323 (1979):48–64.

SOCOLOW, A. A., ed. *State Geological Surveys: A History.* Columbus, OH, 1988.

WALLACE, J. H. *Status of Women in the U.S. Geological Survey.* Washington, DC, 1992.

H. CATHERINE W. SKINNER

INDEX

E

J

Q

R